A Treatise
on
Limnology

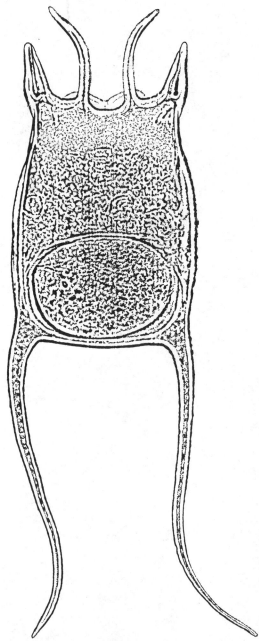

Keratella quadrata from the plankton of Lago Maggiore, showing the immense relative size of the amictic egg (from a photograph by Livia Tonolli Pirocchi).

A Treatise

on

Limnology

VOLUME II
INTRODUCTION TO LAKE BIOLOGY
AND THE LIMNOPLANKTON

G. Evelyn Hutchinson

DEPARTMENT OF BIOLOGY
YALE UNIVERSITY

JOHN WILEY & SONS

New York • Chichester • Brisbane • Toronto

By G. Evelyn Hutchinson

A TREATISE ON LIMNOLOGY, Volume I, *Geography, Physics, and Chemistry*

20   19   18   17   16   15   14   13

Library of Congress Catalog Card Number: 57-8888
Printed in the United States of America
ISBN 0 471 42572 9

But if, retaining sense and sight, we could shrink into living atoms and plunge under the water, of what a world of wonders should we then form part! We should find this fairy kingdom peopled with the strangest creatures—creatures that swim with their hair, that have ruby eyes blazing deep in their necks, with telescopic limbs that are now withdrawn wholly within their bodies and now stretched out to many times their own length. Here are some riding at anchor, moored by delicate threads spun out from their toes; and there are others flashing by in glass armour, bristling with sharp spikes or ornamented with bosses and flowing curves; while, fastened to a green stem, is an animal convolvulus that by some invisible power draws a never-ceasing stream of victims into its gaping cup, and tears them to death with hooked jaws deep down within its body.

Close by it, on the same stem, is something that looks like a filmy heart's-ease. A curious wheelwork runs round its four outspread petals; and a chain of minute things, living and dead, is winding in and out of their curves into a gulf at the back of the flower. What happens to them there we cannot see, for round the stem is raised a tube of golden-brown balls, all regularly piled on each other. Some creature dashes by, and like a flash the flower vanishes within its tube.

We sink still lower, and now see on the bottom slow-gliding lumps of jelly that thrust a shapeless arm out where they will, and, grasping their prey with these chance limbs, wrap themselves around their food to get a meal; for they creep without feet, seize without hands, eat without mouths, and digest without stomachs.

Time and space, however, would fail me to tell of all the marvels of the world beneath the waters. They would sound like the wild fancies of a child's fairy tale, and yet they are all literally true. . . .

C. H. Hudson in *The Rotifera; or Wheel-Animalcules, both British and Foreign*. By C. T. Hudson, LL.D. Cantab., F.R.S., assisted by P. H. Gosse, F.R.S., vol. I, pp. 3–4. 1889.

Preface

The enormous mass of information about the inhabitants of lakes has necessitated treating the biological aspects of limnology in two parts, the division being mainly dictated by convenience.

Volume II begins with a long introductory chapter intended to relate the fresh-water biota to the rest of the living world. This is followed by a short formal discussion of the division of the communities of lakes into appropriate parts, with definition of that minimum of ecological terms which appears to be essential to biological limnology. The rest of the book is devoted to the plankton, all aspects of which are treated, except productivity in the strict sense of that word; this matter will be discussed along with the equivalent aspects of other parts of the lake community after they have been considered in Volume III.

It seems reasonable to suppose that the third volume will also be able to contain the typological, stratigraphic, and developmental aspects of limnology. These subjects inevitably involve the investigator with the lake as a repository, if not a mirror, of human history, and such archaeological considerations will make a natural ending to the whole work.

The more the manuscript has grown, the more impossible has it appeared to consider even all the really significant works. Many excellent investigations, which constitute the lowest floor, if not the foundations of the structure, are present only by implication in treatment of later works. Among the latter, the most that can be hoped is that no significant kind of phenomenon has been omitted, and that good papers not specifically mentioned may be fairly easily found by reference to the works starred in the bibliography.

Parts of the more general sections of the book were the subject of the J. A. Nieuwland Lectures at the University of Notre Dame.

I am indebted as previously to a host of friends who have helped me in many ways. I would particularly remember the sympathetic encouragement of those who have held the Directorship of the Laboratory in which I have worked, particularly, since Volume I was published, Dr. Edgar J. Boell and Dr. Donald F. Poulson. In the company of so many other scholars I am again indebted to the John Simon Guggenheim Foundation for a fellowship which has enabled me to write and study in Europe. Part of the present volume was written in Palermo in the Istituto Zoologico of the University at that ancient center of scientific learning as the guest of the Reverend Professor Giuseppe Reverberi, whose kindness and that of his whole laboratory I am particularly happy to acknowledge. I am also deeply indebted to the American Academy in Rome for receiving within their courts an investigator in a strange and alien field which at first could have appeared to have no relation to Classical Studies, though the relationship will appear at the end of the third volume. To my friends Dr. Vittorio and Dr. Livia Tonolli, in again expressing my thanks, I can say, not only on my own behalf, that no two people have done so much for limnologists as well as for limnology.

To Dr. D. J. Hall, Dr. Susan C. Hantschmann, Dr. R. G. Stross, Dr. D. W. Tappa and to Dr. H. Werntz, I am grateful for the opportunity to use material prior to publication.

Dr. Ruth Patrick, Dr. W. T. Edmondson, Dr. Gordon Riley, Dr. E. S. Deevey, Jr., Dr. Ramon Margalef, Dr. Vittorio and Dr. Livia Tonolli, and Dr. Clyde Goulden have read various chapters and have greatly improved them. My indebtedness to Dr. Edmondson and to Dr. J. L. Brooks will be apparent to the critical reader. I would also acknowledge the help of Dr. Olga Sebestyén, the *commère* of Hungarian limnology. I am indebted to Dr. Ursula M. Cowgill for help and criticism of many kinds. Dr. A. J. Brook has given me most useful information on desmids and Dr. J. W. Lund on other phytoplanktonic algae. Many of the illustrations of the present work have been prepared by Miss Martha M. Dimock, Mrs. Nancy Kimball, Miss Lorraine Larison, Mrs. Eleanor Wangersky, and Mr. William Vars. I am indebted to Dr. G. E. Fogg for Figure 84 and am proud to present Mrs. Jane Marshall's beautiful platypus in Figure 56. I am much indebted to the Cornell University Press for permission to reproduce part of the

material in Figure 4 and to the E. Schweizenbart'sch Verlagsbuchhand-
lung of Stuttgart for their kindness in allowing a reproduction of the
material in Figures 96, 97, 98, 101, 102, and 103 from Professor
Huber-Pestalozzi's volumes in *Die Binnengewässer*.

The work of preparing the manuscript for the press has been enormously facilitated by Miss Susan Cornwell, to whom I owe particular
gratitude. Much of the indexing has been the work of Mr. Joel
Meyerson, Mr. James H. Unterspan and Mr. Frederick H. Hyde. I
am indebted to Mrs. Patricia Tappa for the care she has expended on
the typing of the manuscript.

To all these people as well as to those mentioned in the preface to
the earlier volume, including particularly my wife, I extend my thanks.

New Haven, Connecticut   G. Evelyn Hutchinson
March 1966

Contents

CHAPTER *18*

The Nature and Origin
of the Fresh-water Biota

It is appropriate to introduce the biological aspects of limnology with an account of the composition of the fresh-water flora and fauna and of their probable modes of origin. A considerable but concise literature on the physiology of adaptation to a fresh-water environment exists. Much of this literature is easily accessible and well known to most biologists. The taxonomic and biogeographical aspects of the problem in contrast are discussed in a very diffuse body of writing, which obviously cannot be fully digested by any one investigator. The present account is therefore likely to be incomplete and overselective in some respects, and in others it runs the risk of repeating, though in a highly condensed form, material that is easily available. Its documentation, moreover, may be uneven; when good reviews exist, they have been extensively employed and reference to them alone is given.

STATEMENT OF CLASSIFICATION ADOPTED

In order to present the material to be considered in an intelligible way, it is necessary to adopt some sort of classification of organisms which embraces the whole living world. Such a procedure would have seemed natural in the eighteenth century and even up to the time of Ernst Haeckel (e.g., Haeckel 1866, 1894–1896). In the present century attempts to produce an all-embracing classification have been few and for the most part unsatisfactory. The latest and most comprehensive, that of Copeland (1956), has great value, but it is too idiosyncratic to be adopted unchanged. Dougherty's (1955) schematic presentation is clearly the best of the briefer treatments to appear so far and is the basis of the system used here.

The difficulties that must be faced are due not only to the truly biological problems encountered in the natural delimitation of the higher categories of taxonomy, but also to the lack of coordination between botanical and zoological nomenclature and to the fact that it is still not agreed how far the principles of priority and usage should determine the names of the highest taxa employed.

The following attempt at a comprehensive classification is likely to be unsatisfactory. It is drawn up partly because it is necessary for the purposes of this book and partly in the hope of stimulating botanists, microbiologists, and zoologists to achieve some kind of agreement in the nomenclature, if not the nature, of the various organisms that cross the boundaries of the three fields that such investigators study.

It is necessary throughout to distinguish clearly the existence of great evolutionary branches, or *clades* in Huxley's (1958) terminology, which may transgress several *grades* of evolutionary complexity. Ordinarily, taxonomic recognition is given primarily to clades, but whenever these branches converge on a series of forms of simple grade, related to a presumed common ancestor, it is often tempting and indeed useful to give this primitive assemblage taxonomic status. Thus we divide a major group into minor groups, of which all but one are parts of clades of high grade, whereas the remaining minor group is composed of an assemblage of the lower-grade parts of each clade. Most of the difficulty in classification is inherent in this practice, particularly when the point of origin of some clades in the lower-grade group is clear and that of others is debatable. This is apparent in the discussion of the Protista.

The classification given includes only the independent true organisms, though it is realized that the successful cultivation of a virus in a cell-free medium would formally alter the status of that group. In nature it may perhaps be assumed that the difference between an organism and a virus is, ecologically at least, fairly sharp.

Kingdom and grade Monera. The first major dichotomy, and the only really satisfactory one, is between those organisms that lack true plastids (mitochondria and chromoplastids), and in which no clear mitotic figure with an achromatic spindle and chromosomes appear at nuclear division, and those organisms that have these features. The first group, which includes the blue-green algae, or Myxophyceae, and all the organisms somewhat indiscriminately regarded as bacteria, is usually called in contemporary literature the kingdom Monera, as now defined:

> No true plastids; flagellum, if present, simple and not formed of eleven strands; nuclear structures still somewhat problematical but usually regarded as procaryote, not exhibiting a complete mitotic figure at division; single-celled[1] or colonial organisms, cells ordinarily minute (usually 1–10 μ in diameter), if colonial with very little differentiation between cells except in some cases in the formation of fruiting bodies. Kingdom (and grade) MONERA.

The kingdom consists of at least four well-defined classes, which probably can be grouped (Pringsheim 1949; Dougherty 1955; Stanier 1959) in pairs as *motionless or gliding organisms,* which never possess flagella, and as *swimming organisms* in which simple flagella or modifications thereof occur in at least some stages of a great proportion of species. The two groups ultimately may well be given the formal status of phyla, but since no certain designations for these supposed phyla exist no names are used in the present work.[2]

Phylum 1 (gliding monerans)	Class Myxophyceae
	Class Myxobacteriae
Phylum 2 (swimming monerans)	Class Eubacteriae
	Class Spirochaetae

Van Niel and Stainer (1959) suggest that the peculiar colorless gliding *Vitreoscilla* is a link between the two classes of gliding monerans.

A few groups of Monera must still be left *incertae sedis.* Among them, the causative agent of bovine pleuropneumonia and certain allied parasitic and free-living forms are of particular biological interest, for the young stages of the nonparasitic *Mycoplasma laidlawii* (Fig. 1), of diameter about 0.1 μ, are the smallest free-living organisms yet discovered.

---

[1] See footnote 3 on page 5.

[2] There is some biochemical evidence (Scher and Vogel 1957) that the Myxophyceae are more closely related to the Gram-negative bacteria than to the gram-positive families Micrococcaceae and Bacillaceae. The latter group possess, in common with the Protists and multicellular organisms, ornithine S-transaminase, which is absent from the blue-green algae and from Gram-negative bacteria. It is evident from this work that the old distinction between Gram-positive and Gram-negative microorganisms may involve much more fundamental metabolic characters than would be apparent from most modern classifications of the Monera.

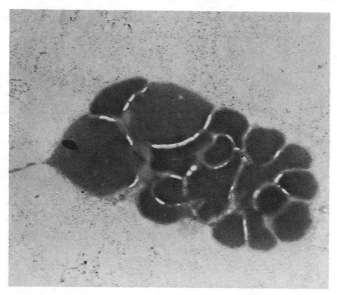

FIGURE 1. *Mycoplasma laidlawii;* reproducing cell that has
fragmented into a number of small individuals, the smallest
under 0.5 μ across, growing on formvar over agar. Individuals not growing on a film would be much less flat and of
smaller diameter at this stage ($\times$ 13,500, photograph by cour-
tesy of Dr. Harold J. Morowitz).

Although Tyler and Barghoorn (1954) believed that the remarkable
biota of the Gunflint formation of lower Middle Huronian age, which
constitute by far the best preserved Precambrian fossils so far described,
could be referred to aquatic fungi (i.e., Phycomycetes) as well as
to the blue-green algae, the minute size of all the cells and filaments
discovered perhaps suggest that if we knew the details of the internal
structure, all of these organisms would appear as monerans. The
histograms published (Barghoorn and Tyler 1965) after this book
went to press seem to confirm such a conclusion.

Kingdom and grade Protista. The organisms with true mitochondria
and, in the case of photosynthetic plants, chromoplastids, in which
the nuclei undergo some sort of recognizable mitotic division, are far
harder to classify than are the Monera, mainly because they consist
not merely of an extraordinarily diversified assemblage of simple and
for the most part unicellular forms but also of several clades of multicellular organisms, which can be traced from relatively simple be-

ginnings up to the highest plants and animals. The degree to which
the various classes reach the grade of functional multicellularity, and
the extend of the gaps within any lineage between successive grades,
differ greatly from clade to clade. The only really satisfactory definition of what is currently regarded as the grade, or perhaps kingdom,
Protista is that it consists of those organisms with true plastids, and
with nuclei dividing mitotically, that are not higher plants (Bryophyta
and Tracheophyta), sponges, or multicellular animals. Bearing this
in mind, we may make a more formal definition along the following
lines:

> Mitochondria and, in the photosynthetic forms, chromoplastids
> present; flagella (or cilia), if present, never simple, normally
> composed of two axial and nine (perhaps fundamentally double)
> peripheral strands, single-celled[3] or colonial organisms, each cell
> ordinarily at least 5 μ in diameter (but there are many excep-
> tions)[4]; if colonial, usually with little cell differentiation; nuclear
> membrane often retained during mitosis. Kingdom (and grade)
> PROTISTA.

As has been indicated, the criterion of true functional multicellularity
is to some extent arbitrary, since it is both conventional and convenient
to place the lowest land plants and their fresh-water allies, such as
the phylum Bryophyta, in the kingdom of the Metaphyta or higher
plants of multicellular grade, whereas all the brown algae are included

---

[3] Since in possession of nucleus, mitochondria, often flagella or cilia and other
cell organelles, there is a perfectly definite topological similarity between the
non-colonial protist body and the cells of multicellular organisms, the argument
that the protist body is to be regarded as acellular appears to me to be artificial, however multicellularity may have come about. The acellularity of the
protist body, if accepted, implies that in constructing the language of bioligy
there is an axiom that the bodies of organisms may be said to consist of n cells,
where n is either apparently zero or some positive integer greater than one, but
that it cannot have the value of unity. This seems unnecessarily complicated,
if taken at face value, and to imply certain definite evolutionary hypotheses, if we
ask why the axiom is adopted. However plausible such hypotheses may be,
they should not be incorporated into biology by means of an axiomatic
linguistic convention.

[4] The smallest is perhaps *Micromonas pusilla* (Manton and Parke 1960). I.
Manton (1959) has given a fascinating account of this organism, which is 1.0 to
1.5 μ long and has a single chloroplast and single mitochondrion. It was
originally referred to *Chromulina*, then put in the Chlorophyta. As this book
goes to press, it has been associated with some other small forms as Prasino-
phyceae, a class that in the classification here adopted may require the status
of a phylum.

in the Protista. This arrangement would never have been adopted if the land plants had not advanced beyond the grade of organization represented by the mosses and liverworts. A similar difficulty is raised by the higher fungi or Carpomycetes and in a rather less acute form by the higher red algae.

Among the animal groups the divisions are somewhat more satisfactory, mainly on account of our ignorance. Good evidence points to the sponges being derived from the choanoflagellates, but in the embryology of some of the calcareous species there is more than a hint of an affinity with the lower Chlorophyta. The origin of the true metazoa is moreover still vigorously debated. Such disagreements permit us at least to make clear taxonomic distinctions in the region of uncertainty.

Classification of the Protista. The classification finally adopted was arrived at in the following way. The Rhodophyta, or red algae, were first separated as a phylum characterized by the lack of flagella and possession of phycoerythrin and, in some cases, phycocyanin. In both characters[5] the red algae show some similarity with the Myxophyceae, and though Fritsch (1945) is inclined to doubt the significance of such resemblance some recent authors, notably Dougherty and Allen (1959) have emphasized the isolation of the group and the possibly very primitive nature of its simplest members.

The Chlorophyta are then separated, though on less satisfactory grounds. In possessing chlorophylls *a* and *b,* certain characteristic xanthophylls, simple flagella (Fig. 2), cellulose cell-walls, and starch as a storage product, they clearly form a coherent group, though some of these characters, but never all together, are shared by other groups. In their biochemical characters the green algae in general resemble the higher plants, both phyla of which are usually regarded as derived from the Chlorophyta. Following Smith (1950), two classes are recognized, though the presence of certain special xanthophylls in the Siphonales perhaps suggests that they should be given more than ordinal rank. The Euglenineae, which share with the Chlorophyta the possession of chlorophyll *b* but differ radically from the latter phylum in their lack of cellulose and starch, in their peculiar xanthophylls, and in their flagellar structure, are tentatively given the rank of a phylum Euglenophyta. Smith (1950) recognizes such a group of the same rank as the Chlorophyta. This conclusion is reinforced

---

[5] The phycoerythrin and phycocyanin of the Myxophyceae differ somewhat from those of red algae. The red algae may have chlorophyll *d*, not known elsewhere.

by Vogel's (1959a,b) discovery that *Euglena gracilis* synthesizes lysine by a path involving aminoadipic acid, as in the higher fungi, whereas in all bacteria so far studied, in the green algae, and in the higher green plants, the synthesis proceeds by a different path involving α-ϵ diaminopimelic acid.

The algal and flagellate forms now remaining constitute a perplexing problem. If we start with an organism such as *Nereocystis* or any other large brown seaweed, we can progress down the scale to the lowest filamentous brown alga, such as *Ectocarpus*. These forms seem, from the nature of their gametes, to be clearly related to the Chrysophyceae (cf. Manton 1952) and other unicellular brown and yellow forms. From them we can progress by easy stages through the animal flagellates to such a flagellate amoeboid form as *Mastigamoeba* and so to typical amoebae, testaceous rhizopods, and ultimately to the foraminifera, or, if we prefer another path from the simpler animal flagellates, we can reach the fantastic polymastigine forms living symbiotically in the alimentary tracts of termites. To place *Nereocystis* or *Laminaria* in the same phylum as these complex flagellates and the foraminifera will probably satisfy neither botanist nor zoologist, though at no point in our imaginary journey have we crossed a boundary of a fundamental kind. Common sense seems to demand some sort of division, and on the whole the gap which separates the simple filamentous brown seaweeds, which bear definite sporangia and may exhibit a true differentiation into somewhat heteromorphic sporophyte and gametophyte, from the allied flagellate forms, seems the logical one to emphasize.[6] The Phaeophyta, with a single class, are therefore separated as a third phylum of Protista. Though placed before the Sarcomastigophora in the tabular classification which follows, they are, of course, as clearly derived from that phylum as the higher plants are from the Chlorophyta.

We may now conveniently consider the fungi. There is no reason to suppose that the group is monophyletic. A connection between the higher fungi and the Rhodophyta has long been suspected on the grounds of certain similarities in their fruiting bodies combined with a lack of flagellate stages in both groups. Both characters may well be due to convergence, but they are accepted by several modern

---

[6] Against this, the most practical arrangement, may be set the fact that the Phaeophyceae share or are supposed to share (Seward Brown raises some doubt in a personal communication) chlorophyll *c* with the diatoms and dinoflagellates, and fucoxanthin with the diatoms. More work on the pigments of the flagellate yellow-brown forms is urgently needed (see, however, Goodwin 1964).

workers, including Bessey (1950) in one of the best modern works on the morphology and classification of the group. It seems certain at any rate that the higher fungi should be placed by themselves as a phylum, but the exact position of this phylum in the Protista is no doubt debatable. Following Bessey, the name Carpomycetes is used. The lower fungi, ordinarily grouped as Phycomycetes, clearly have a different origin. Copeland regards the group as polyphyletic and places the forms with a single posterior whiplash or acroneme flagellum (Chytridiales) in a phylum, the Opisthokonta, to themselves, whereas those with an anteriorly directed tinsel or pantoneme flagellum, with (Oomycetes) or without (Hyphochytriales) a posteriorly directed whiplash, are put near the brown algae. Bessey treats the Phycomycetes as a class, in which the Oomycetes occupy a central position, one or other flagellum being lost in the other two orders. As this book goes to press, Vogel (1964) has given biochemical evidence for the Opisthokonta being allied to the Euglenophyta and Carpomycetes; Copeland's arrangement, insofar as it relates to the Opisthokonta, is accepted on the basis of Vogel's work.

Having separated the Rhodophyta, Carpomycetes, Euglenophyta, Opisthokonta, Chlorophyta, and Phaeophyta as distinct phyla, we are left with three very coherent groups formerly treated as belonging to the Protozoa, namely the Ciliophora, the Sporozoa, and the Cnidosporidia, and an enormous number of amoeboid and flagellate organisms, along with groups such as the Phycomycetes and Bacillariophyceae, which are obviously related to such flagellate organisms. The Ciliophora, Sporozoa, and Cnidosporidia, for reasons given in a later paragraph, may certainly be awarded the rank of phyla. In the remaining group, in spite of its enormous diversity, it seems impossible, as already indicated, to make any clear fundamental divisions. This group has therefore been treated as a phylum, here termed the Sarcomastigophora, following Honigberg and Balamuth (1963), though these authors include within it a number of groups here separated as distinct phyla but exclude certain taxa commonly regarded as typical plants. The division of the Sarcomastigophora also differs from that of Honigberg, Balamuth, Bovei, Corliss, Gojdics, Hall, Kudo, Levine, Loeblich, Weiser, and Wenrich (1964), mainly because of the greater emphasis placed here on some biochemical characters and in the adoption of a simpler scheme which attempts not to beg the question of polyphyletic origins. No distinction, moreover, is made between flagellate and purely rhizopod classes. All the classes accepted by Grassé are adopted, except the Euglenineae, here treated as a phylum; it is almost certain that further work will show that some of the smaller

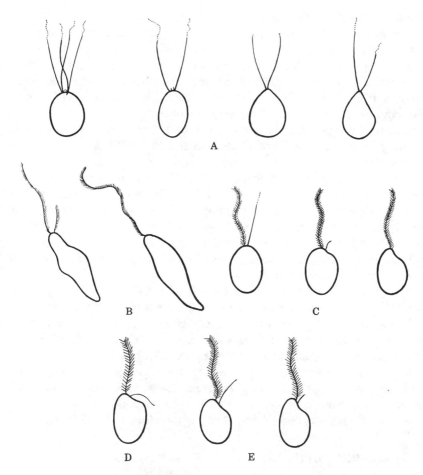

FIGURE 2. Flagellar types in the Protista. *A* (*Carteria, Polytoma, Ulothrix, Dictyococcus*) acroneme and simple flagella in the Chlorophyta; *B* (*Distigma, Euglena*) stichoneme flagella in the Euglenophyta; *C* (*Synura, Uroglena, Mallomonas*) pantoneme and acroneme flagella, the latter undergoing reduction in some lines, in Chrysophyceae; *D* (*Ochromonas*) comparable pantoneme and simple flagella also in Chrysophyceae; and *E* (*Tribonemia, Botrycium*) the same in Xanthophyceae (mainly from Vlk 1938).

groups of flagellates should be fused.   Omitting the inadequately known Chloromonadineae, the phylum as here conceived contains a basic group of organisms with a pantoneme and an acroneme flagellum, (Fig. 2) primitively probably only with chlorophyll a[7] and with a very varied assemblage of xanthophylls.   A connection with the Rhodophyta is possibly indicated by the occurrence of phycoerythrin and phycocyanin-like pigments in the Cryptomonadineae.

One line which develops chlorophyll c, if that substance is a single entity, gave rise to the Phaeophyta and Bacillariophyceae, which have fucoxanthin, and to the Dinophyceae, which have their own xanthophylls.   The phototrophic forms in the Sarcomastigophora therefore more or less correspond to Smith's (1950) Chrysophyta.   The phylum Sarcomastigophora, as here understood, differs from the Chrysophyta mainly in that it contains a vast series of animal and fungal forms, the more primitive members of which assemblage seem to be merely apochlorotic derivatives from yellow-brown flagellates that botanists would regard as chrysophytes.

The justification for the separation of the Ciliophora as a phylum is that in this group we really have, in the nuclear structure, something new and peculiar.[8]   The Sporozoa are given a like rank because we have no real knowledge of their affinities.   The Cnidosporidia are certainly an independent group (Grassé 1952), possibly allied to the Dinophyceae through cnidocyst-bearing members of the latter group such as *Polykrikos;* they may even represent a degenerate parasitic branch of the group that gave rise to the coelenterates.

The final classification of the Protista therefore is as follows:

Phylum RHODOPHYTA	Class Bangioideae
	Class Florideae
Phylum CARPOMYCETES	Class Zygomycetes
	Class Ascomycetes (+ Hypho-
	mycetes *pro magna parte*)
	Class Basidiomycetes
Phylum EUGLENOPHYTA	Class Euglenineae
Phylum OPISTHOKONTA	Class Archimycetes
Phylum CHLOROPHYTA	Class Chlorophyceae
	Class Charophyceae

---

[7] Strain (1958) in his most recent work seems rather doubtful about chlorophyll e from certain Xanthophyceae (see, moreover, Goodwin 1964).

[8] The formally comparable nuclear organization in some foraminifera can hardly have any phylogenetic implications.

Phylum PHAEOPHYTA   Class Phaeophyceae

Phylum SARCOMASTIGOPHORA[9]   Class Cryptomonadineae
                                 Class Chloromonadineae
                                 Class Xanthophyceae
                                 Class Ebriideae
                                 Class Silicoflagellata   flagellate
                                 Class Coccolithophorida   and
                                 Class Chrysophyceae   allied
                                 Class Bacillariophyceae   classes
                                 Class Phycomycetes
                                     (*s. str.*)
                                 Class Dinophyceae
                                 Class Zooflagellata

                                 Class Acantharia
                                 Class Heliozoa   actinopod classes
                                 Class Radiolaria

                                 Class Lobosa
                                 Class Filosa   rhizopod
                                 Class Granuloreticulosa   classes
                                 Class Mycetozoa

                                 ?Class Acrasiae   *incertae*
                                 ?Class Plasmodiophorales *sedis*

Phylum CILIOPHORA   Class Ciliata

Phylum SPOROZOA   Class Gregarinomorpha
                                 Class Coccidiomorpha
                                 Class Sarcosporidia
                                 ?Class Haplosporidia (and some
                                     other parasites *incertae sedis*)

Phylum CNIDOSPORIDIA   Class Neosporidia

The Protista presumably had a long Precambrian history. If Tyler and Barghoorn (1954) are correct in their belief that some of the Gunflint fossils are aquatic fungi, the kingdom is known from Huronian times. A better early Precambrian protist is perhaps provided by *Corycium enigmaticum* of slightly younger age from Finland, which may have been an organism (Rankama 1948) and may have consisted of large single cells, each like those of the green alga *Hydrodictyon*. The later Precambrian radiolarians are very dubious organisms.

---

[9] Bovee and Jahn (1965) give a vastly improved classification of the rhizopod and actinopod forms which should be substituted.

Multicellular grade: kingdoms Metaphyta and Metazoa. The truly multicellular plants and animals with definite differentiation of adjacent cells to form functionally specialized tissues, constitute at least two and probably four or five independent groups. All the organisms included in these groups have nuclei which exhibit typical mitosis, usually with disappearance of the nuclear membrane, all have true mitochondria, and, when flagellate or ciliate, the flagella or cilia have a basic eleven-strand structure. That they arose from organisms which would be placed in the Protista is certain; beyond this there is no universal agreement. It is usual and convenient to recognize two kingdoms, one for the plant and one for the animal forms, but it must be remembered that in all probability neither are monophyletic.

> Holophytic except in a few cases of saprophytes and parasites of obvious general affinity; chlorophylls *a* and *b,* and the xanthophylls usually found in the Chlorophyta, present in the chloroplasts; starch the usual storage product; no major cavity, except as a secondary structure in some flowering plants, in the body; cellulose cell walls present; motility, when present, not due to elongate cells containing contractile fibrous protein structures; haploid and diploid alternating generations always recognizable, though the haploid generation is much reduced in the Tracheophyta; male gamete in Bryophyta with two anteriorly set flagella, in the lower Tracheophyta with many flagella (or cilia). Kingdom META-PHYTA.[10]

Two clades, regarded as phyla by Tippo (1942), whose classification is here followed, are recognizable. They are now usually regarded as having separate origins in the Chlorophyta.

Phylum	BRYOPHYTA	Class Musci
		Class Hepaticae
		Class Anthocerotae
Phylum	TRACHEOPHYTA	
Subphylum	PSILOPSIDA	Class Psilophytineae
Subphylum	LYCOPSIDA	Class Lycopodineae
Subphylum	SPHENOPSIDA	Class Equisetineae
Subphylum	PTEROPSIDA	Class Filicineae
		Class Gymnospermae
		Class Angiospermae

> Holozoic; cell wall absent; cells arranged around at least one epithelially lined cavity, absent only in a few parasites of obvious eumetazoan affinities; motility in part or wholly due to elongate

---

[10] Many botanists prefer Embryophyta but the name Metaphyta is used here as coordinate with Metazoa.

cells containing contractile fibrous protein structures; adult typically diploid; if metagenetic, not with alternating haploid and diploid generations; sperm usually with a single posterior flagellum. Kingdom METAZOA.

Classification of the Metazoa. Two subkingdoms are ordinarily recognized; these almost certainly have quite different origins in the Protista and are associated only because the sponges are a relatively small group, functionally comparable to the true animals rather than to the plants. It is by no means certain that the true animals are monophyletic.

Central cavity (spongiocoel), or derivative outpocketings of this cavity, lined with an epithelium of flagellate collar cells; principal apertures (oscula) exclusively exhalent, inhalant apertures minute, scattered and very numerous; contractile cells only in oscular sphincter and around other apertures, not universally present. Subkingdom PARAZOA.

The embryology of some Calcarea suggests derivation from an apochlorotic member of the order Volvocales (Chlorophyta); the collar cells equally strongly suggest an origin among the order Choanoflagellata (Zooflagellata). Only one living phylum is recognized; the extinct early Palaeozoic Archaeocyatha possibly constitute a second phylum, though they were clearly very different from sponges. The sponges themselves are well represented from the Cambrian to the present and may possibly have left in *Eospicula cayeuxi* (see de Laubenfels 1955) fossil traces of Precambrian age.

Phylum   PORIFERA   Class Calcarea
                                      Class Demospongiae
                                      Class Hexactinellida
Phylum   †ARCHAEOCYATHA[11]
                incertae sedis

Central cavity (*gastrocoel*) present except in a few parasites and other organisms of obvious metazoan affinity; no collar cells; principal aperture of gastrocoel, either both ingestive and egestive, or two principal apertures present. Subkingdom EUMETAZOA.

---

[11] In the present chapter no class names are given for extinct animal phyla of no direct limnology interest or not always when an animal phylum consists of a single class. In the latter too many purely nomenclatorial problems are involved to justify an attempt to produce a list at a time when the whole matter of the naming of higher animal taxa is under discussion. No use has here been made of taxa intermediate between subphylum and class, though these may in some cases be illuminating.

Two views of the origin of the Eumetazoa are widely held. According to the first, they are derived from a spherical flagellate colony which invaginated to produce a gastrula-like protocoelenterate. According to the second theory, a process of internal differentiation produced a multicellular acoel-like flatworm from a multimacronucleate ciliate. The relative virtues and defects of the two theories have been admirably reviewed by Hanson (1958), who strongly supports the second view. If it is adopted for the bulk of the Eumetazoa, the problem of the origin of the coelenterates remains. Hadzi (1953), who has been the main proponent of the second view in recent years, derives the Anthozoa from the flatworms; the simplicity of the Hydrozoa is thus believed to be secondary. Alternatively, the coelenterates may have had a separate origin from the Protista, as Huxley (1958) appears to suspect. The present writer is very sceptical of Hadzi's deviation of the coelenterates and feels that an origin from cnidocyst-bearing dinoflagellates, such as *Polykrikos,* with the cnidosporidia as a degenerate parasitic representative of the protocoelenterate line (cf. Grassé 1952) is conceivable.

Another possibility, namely Cain's (1959) suggestion that the ciliates are degenerate flatworms, would, of course, permit recognition of a real affinity between the two groups without disturbing the conventional Haeckelian view of the origin of the Metazoa.

According to the ciliate theory of the origin of the flatworms, the well-known members of the Mesozoa (Orthonectida and Dicyemida) are most reasonably placed, as has often been done, near the Platyhelminthes. On the alternative protocoelenterate theory they may be persistent or paedogenetic planulae. Of the other organisms that have been placed in the Mesozoa, all seem to be protistan with the single exception of *Salinella,* if it ever existed. This animal, which as an adult (Frenzel 1892) is said to have consisted of a single layer of cells which formed a barrel-shaped structure with an axial gut *provided with mouth and anus,* was described from cultures made from the deposits of Argentinian salt lakes. It obviously does not fit into any scheme of phylogeny now accepted, and to the writer it has a distinctly Piltdown quality. Except for placing the Mesozoa among the bilateral forms, the classification adopted follows Hyman (1940, 1951a,b, 1959) in broad outlines, though a few minor changes in the classes accepted have been made.

The Eumetazoa were probably represented in the Cambrian seas by all the principal phyla except the Chordata. Indications of worm burrows are not uncommon in the later Precambrian. There is a single record of problematic protomedusan form, which may be the cast of a bursting bubble (Cloud 1960), from the Algonkian of the Grand Canyon, and it seems that rocks yielding several genera of coelenterates that may well be Pennatulaceae (Anthozoa, Alcyonaria) in Australia, South Africa, and England are also perhaps of late Precambrian (Glaessner 1959) age. The south Australian locality, Ediacara, has also yielded other fossils, which include the beautiful annelid *Spriggina* (Glaessner 1958) conceivably near the ancestor of the trilobites as well as the supposed coelenterate *Dickinsonia* placed in the problematic class Dipleurozoa.

Radiate Phylum
     Phylum COELENTERATA
        Subphylum CNIDARIA   Class Hydrozoa
                                         Class Scyphozoa
                                         Class Anthozoa
                                         †Class Dipleurozoa
                                             incertae sedis
                                         †Class Protomedusae
                                           incertae sedis

        Subphylum CTENOPHORA[12]
    Bilateral Phyla
      Acoelomate
        Phylum PLATYHELMINTHES   Class Turbellaria
                                           Class Trematoda
                                           Class Cestoda
       Phylum RHYNCHOCOELA (NEMERTINA)
       Phylum MESOZOA *incertae sedis*
    Pseudo-coelomate
       Phylum ACANTHOCEPHALA
       Phylum ASCHELMINTHES   Class Rotifera
                                           Class Gastrotricha
                                         Class Kinorhyncha
                                         Class Priapulida[13]
                                         Class Nematoda
                                         Class Nematomorpha
      Phylum ENTOPROCTA (KAMPTOZOA)   *incertae sedis*
    Eucoelomate, protostomatous, nonlophophorate
       Phylum MOLLUSCA   Class Monoplacophora
                                         Class Amphineura
                                         Class Gastropoda
                                         Class Bivalvia
                                           (Lamellibranchiata
                                           or Pelecypoda)
                                         Class Scaphopoda
                                         Class Cephalopoda

---

[12] Komai (1963) finally regards the ctenophores as coelenterates but constituting a separate subphylum.

[13] See, however, Shapeero (1961), who concludes that the body cavity is a true coelome.

Phylum SIPUNCULOIDEA
Phylum ANNELIDA[14]
    Class Echiuroidea
    Class Archiannelida
    Class Polychaeta
    Class Clitellata

Phylum ARTHROPODA
    Subphylum PARARTHROPODA
    Class Onychophora
    Class Tardigrada
    Class Pentastomida
      incertae sedis

    Subphylum INSECTA[15]
    Class Myriapoda
    Class Hexapoda

    Subphylum †TRILOBITOMORPHA
    †Class Trilobita
    †Class Merostomatoidea
    †Class Marellomorpha
    †Class Pseudocrustacea
    †Class Arthropleura
      incertae sedis

    Subphylum CHELICERATA
    †Class Eurypterida
    Class Xiphosura
    Class Arachnida
    Class Pycnogonida[16]
      incertae sedis

    Subphylum CARIDA[17]
    Class Crustacea

Eucoelomate, protostomatous, lophophorate
Phylum POLYZOA (BRYOZOA)
    Class Phylactolaemata
    Class Gymnolaemata

Phylum PHORONIDA
Phylum BRACHIOPODA
    Class Ecardines
    Class Testicardines

Eucoelomate, deuterostomatous
Phylum CHAETOGNATHA

---

[14] Hyman regards the Echiuroidea as a separate phylum. The Archiannelida as here understood include the Nerillidae, Polygordiidae, Protodrilidae, Saccocirridae, and Dinophilidae. The group may be, but it is not necessarily, polyphyletic; the Histriobdellidae and the Myzostomida are placed in the Polychaeta. In regard to the propriety of uniting the leeches and oligochaets as Clitellata (Michaelsen 1928) there can be no doubt whatever. The apparently very primitive fresh-water psammobiont worm *Rheomorpha neiswestnovae* may be an oligochaet (Ruttner-Kolisko 1955) though it is possible that it will need a new group.

[15] This is the usage of Remington (1955); it has good historic precedents, even if it is a little disconcerting to those of us brought up to regard an insect as an animal with an exoskeleton and six legs. The further, seemingly old-fashioned, division follows the most recent discussion of Manton (1964), some compromise between Remington's and Manton's schemes will doubtless ultimately be reached.

[16] Tiegs and Manton (1958) tend to favor assigning the pycnogonids to the Pararthropoda.

[17] Carides of Haeckel, but without the Trilobitomorpha or lower Chelicerata.

Phylum ECHINODERMATA
   Subphylum PELMATOZOA

        †Class Cystoidea
        †Class Eocrinoidea
        †Class Paracrinoidea
        Class Crinoidea
        †Class Carpoideae
        †Class Machaeridea
         incertae sedis
        †Class Cyamoidea
         incertae sedis

   Subphylum ELEUTHEROZOA

        Class Asteroidea
        †Class Auleroidea
        †Class Somasteroidea
        Class Holothuroidea
        Class Ophiuroidea
        Class Echinoidea

Phylum HEMICHORDATA

        Class Enteropneusta
        Class Pterobranchia
        †Class Graptolita[18]

Phylum BRACHIATA[19]
        Class Pogonophora
Phylum CHORDATA
   Subphylum UROCHORDATA

        Class Larvacea
        Class Ascidiacea
        Class Thaliacea

   Subphylum VERTEBRATA[20]

        Class Cephalochordata
        †Class Cephalaspides
        Class Petromyzones
        †Class Pteraspides
        Class Myxini
        †Class Pterichthyes
        †Class Coccostei
        †Class Acanthodii
        Class Elasmobranchii
        Class Holocephali
        Class Dipnoi
        Class Teleostomi
        Class Amphibia
        Class Reptilia
        Class Aves
        Class Mammalia

---

[18] Hyman rejects this affinity; the subject is clearly in need of more study.

[19] See Ivanov (1955); erected for the single recently discovered class Pogonophora by which name these extraordinary animals are best known.

[20] Modern opinion seems to incline to a relatively close connection between the tunicates and the other Chordates, and to a belief that *Amphioxus* is a specialized derivative of a very primitive vertebrate. Division of the Vertebrata into superclasses Acrania for the Cephalochordata, Agnatha for the next four classes and Gnathostomata for the rest is reasonable. Within the last named group Berg ([1940], 1947) retains the Pisces as a series, including the seven classes Pterichthyes to Teleostomi. The reptiles may well be polyphyletic, and the monotremes may well be separable as a class from the true mammals.

THE EVOLUTIONARY EURYHALINITY OF THE VARIOUS GROUPS OF MONERA AND PROTISTA

Salinity relationships of the Monera and Protista. Although a vast period of time has elapsed since the main divisions of the Monera and Protista were established, it is desirable to scrutinize the various groups carefully to see if they throw any light on the original habitat of the ancestors of the animals and plants. In this scrutiny particular attention will be paid to what is here called evolutionary adaptability or, specifically, *evolutionary euryhalinity* (Hutchinson 1960). By high evolutionary adaptability is meant the condition in which relatively small genetic changes, permit great changes in the type of environment that can be occupied. A group of individuals that can migrate from one environment to the other without any genetic change obviously has a maximum adaptability; one that can make the change by a single mutation would ideally provide the next category. Such a refined scale, of course, is impractical, but the kind of data that is available does permit some sort of judgment in many cases. When a group is composed of genera, some of which contain both marine and fresh-water species, it will be obvious that an amount of genetic change equivalent to that involved in forming a species is sufficient to permit migration from fresh to salt water or vice-versa; such a group will be regarded as having high evolutionary adaptability or, specifically, euryhalinity. When a large group exists exclusively in fresh or in salt water, it may be presumed that relative to the difference in medium its evolutionary adaptability is very low.

In order to use data on evolutionary euryhalinity as evidence of the habitats in which major groups arose, it is necessary to suppose that organisms that appear to be primitive have an appreciable probability actually of being so. The evidence for this, of course, is inadequate. In any given case it is always simpler, in default of other evidence, to assume that primitive characters represent lack of modification rather than specialization followed by secondary reduction. It is very doubtful if we may use the simplicity postulate or Occam's razor as a method of assessing a nonanalytic probability even of the vaguest kind. There is, however, a little empirical evidence, such as the occurrence of what are practically the same species of *Pediastrum* (Fig. 3) in lower Tertiary, probably Eocene, sediments in Sumatra (Wilson and Hoffmeister 1953) and in the Cretaceous of both Asia and North America (Evitt 1963) as are known today, that some unicellular forms have an extra-ordinary persistence in time and thus must have reached an evolutionary dead end as fully adapted to some persistent type of environment

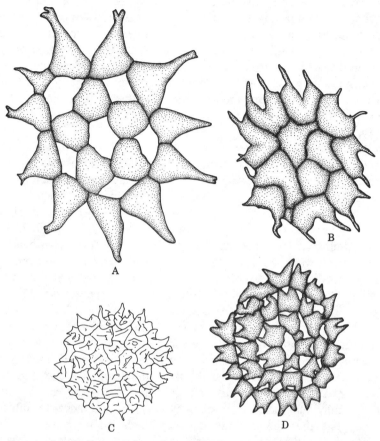

FIGURE 3. *Pediastrum* from Lower Tertiary and Cretaceous sediments. (cf. fig. 91). *A. P. clathratum* type (described as *P. kajaites*) Lower Tertiary, Sumatra. *B. P. boryanum* type (described as *P. bifidites*) Lower Tertiary, Sumatra. *C. P. boryanum* type, Upper Cretaceous, California. *D. P. duplex* type (described as *P. delicatites*) Lower Tertiary, Sumatra. (*A, B, D* Wilson and Hoffmeister; *C,* Evitt.)

as their nature permits. A comparable situation may perhaps be indicated by the occurrence of a few genera of specialized flagellates in both termites and the xylophagous cockroach *Cryptocercus,* which must have diverged in the late Paleozoic (see Grassé 1952 for discussion). If, as the fossil *Pediastra* seem to indicate, several species of the very small assemblage of fresh-water unicellular Chlorophyta known from the Cretaceous and early Tertiary have persisted virtually unchanged for several tens of millions of years, there is nothing too improbable in supposing that in the over-all array of modern Protista a fair proportion of the relatively small number of apparently primitive forms do actually indicate something about the nature of the earliest representatives of the group several hundred million years ago.

Free-living Monera. At least among the four well defined orders of the Eubacteriae and among the Myxophyceae there is abundant evidence of the possibility of very wide evolutionary adaptation to variations in salinity.

In the Eubacteriales species or strains are known that live in a complete range of habitats from ordinary dilute fresh waters up to saturated salt solutions. Although most of the bacteria actually inhabiting the open ocean or its sediments are distinct and, when isolated, relatively stenohaline species, at least sometimes these species are clearly representatives of close allies in fresh water. ZoBell and Upham (1944) point out the close affinities of *Flavobacterium marinovirosum* with the fresh-water *F. devorans* and those of *Serratia marinorubrum* with the fresh-water *S. indica.* Though they consider the floras on the whole to be distinct, they note that of sixty species isolated from the ocean fifty-six became, after prolonged laboratory culture, markedly euryhaline and could be grown on fresh-water media that they would not tolerate on initial isolation. Some systematic variation was noted in the ease with which this process took place, species of *Bacillus* and *Micrococcus* adapting more easily than those of *Pseudomonas* and *Vibrio.* ZoBell (1946) notes that in old stock cultures of marine bacteria individual cells may differ greatly in tolerance to dilution. This suggests that the adaptation occurring is mainly genetic. Pratt and Waddell (1959) have studied the process in somewhat greater detail and conclude that a reduction in NaCl concentration may be tolerated by a proportion of the individuals in a population if certain other ions are maintained at full strength, but not if the medium is greatly diluted with distilled water. At least in the initial stages of adaptation maintenance of the original marine concentration of magnesium may be important. Doudoroff (1940) studying the adaptation of *Escherichia coli* to salt water found that under some conditions a proportion of the individuals could adapt reversibly but that some selection for resistant genotypes also seems to have occurred.

Korinek (1926) noted, and ZoBell (1946) has confirmed the observation, that fresh-water bacteria tend to be much more euryhaline than freshly isolated marine species. Nevertheless, typical fresh-water species are rarely found in the sea. The initial stenohalinity of unadapted marine bacteria is moreover equally evident when tested in dilute or in concentrated media. The obligate polyhaline organisms, such as *Pseudomonas salinaria* and *P. cutirubra,* which occur whenever organic matter exists in the presence of much salt, as on salted hides, salted fish, or in concentrated salterns or brine pools, are clearly specially adapted to such concentrated habitats. These observations suggest, though they certainly do not prove, that the marine bacteria have cell contents more or less isotonic with sea water and that they are poikilosmotic, with little osmoregulatory capacity, though the ionic composition must be regulated. Fresh-water bacteria are presumably much more homoiosmotic, but the passage from one condition to another, though it must often involve genetic change, does not imply any fundamental reorganization of the structure or physiology of the organisms undergoing adaptation.

Detailed physiological studies are curiously scarce. Many Gram-negative bacteria can be plasmolyzed, and bacteria grown in fairly concentrated media often burst when transferred too rapidly to dilute media. During the process of plasmolysis the cell membrane apparently pulls away from the sheath. A difference in the relation of the sheath to the cytoplasm may underlie the fact that Gram-positive bacteria do not plasmolyze. Mitchell (1949) concluded that the osmotically active layer is about 0.01 μ thick at or just below the cell membrane. Knaysi (1951) estimates that the cells in a mature culture of *Escherichia coli* had an internal osmotic pressure of not more than 7.5 atm. when the medium had an osmotic pressure of 6 atm.; such an excess turgor pressure of about 1.5 atm. is doubtless typical of nonmarine forms.

The Rhodobacteriales are found in both fresh and salt water. At least the photosynthetic sulfur bacteria appear more commonly under somewhat saline conditions, since in a reducing environment more H_2S is likely to be formed from saline waters containing sulfate than from most fresh waters.

The Actinomycetales are mostly either parasitic or soil organisms. Among the free-living genera several are represented in both fresh and salt water. *Micromonospora* is likely to be a particularly important member of the group in the hydrosphere owing to the diversity of carbon sources that it can use. Laboratory culture of marine forms suggest that many are markedly evolutionarily euryhaline (Roach and Silvey 1959).

Among the Chlamydobacteriales the only certain genus, *Sphaerotilus,* is listed by ZoBell (1946) as occurring in the sea, presumably on the basis of species earlier referred to other genera now regarded as synonyms. The genus is better known from fresh water.

The free-living Spirochaetes are known from both fresh waters and the ocean.

Among the gliding organisms most of the Myxobacterieae appear to be saprobionts adapted to subaerial life, growing on the surface of cow dung and the like and producing aerially distributed spores in cysts or fruiting bodies. Two genera, known in soils, have also been recorded from the sea, where they are probably associated with decaying seaweeds. Of these *Sporocytophaga* produces isolated spores rather than discrete fruiting bodies, but *Cytophaga* is not known to produce spores of any kind. The lack of records of such organisms from fresh waters is probably accidental and due only to lack of investigation.

The Myxophyceae exhibit a diverse type of adaptive radiation. This group is predominantly one of fresh waters, but in all five orders recognized by Fritsch (1945) some marine species are known. Many genera (e.g., *Anacystis, Oscillatoria, Phormidium*) contain both marine and fresh-water members. In a few cases the same species appears to occur in both fresh waters and the sea; *Lyngbya maiuscula* and several species of *Dermocarpa* which are widespread marine forms appear to enter fresh waters mainly in the tropics (Geitler and Ruttner 1936), but in *Phormidium tenue* (Hof and Frémy 1933) a great tolerance within inland waters is found in temperate regions. The strictly marine members of the group belong mostly to the Pleurocapsales, which are largely epiphytes referrable to the genera *Pleurocapsa* and *Radaisia*. There are, however, a number of marine forms in the Chroococcales, which is certainly the most primitive order, and in the Chamaesiphonales; the filamentous Nostocales contribute, among other genera, *Trichodesmium* and *Katagnymene* to the marine plankton and *Spirulina* and *Nodularia* to that of saline inland localities. Marine forms appear to be least well developed in the Stigonematales, which constitute the most specialized members of the class and in which Fritsch (1945) cites *Mastigocoleus testarum* as the only marine species.

Any over-all interpretation of the conditions in the Monera is difficult because of the complete lack of indications to suggest which group is the more primitive. All that can be said at present is that as a whole the Monera exhibit very high evolutionary adaptability, not merely in regards to salinity, but also in making transitions from aquatic environments to soil. As will appear later in discussion of the bacteriology of lakes in Volume III, most aquatic bacteria are associated with surfaces, and the Monera typically are probably organisms living at liquid-solid interfaces. This perhaps suggests that some of the gliding forms may really be primitive. Fritsch (1945) suspects that the great development of terrestrial forms among the blue-green algae and particularly among the lowest order, the Chroococcales, indicates a more or less terrestrial habitat for the ancestral members of the group. It is tempting to speculate that the oldest complete organism may have lived at the bottom of very shallow bodies of water, or on rock surfaces kept moist by trickling water or spray, under conditions of moderate and variable salinity.

Protista. The various groups of Protista are evidently more diversified than are the Monera.

Rhodophyta. As a whole, the group is characteristically marine, and all the more specialized orders, often grouped as Euflorideae, are almost exclusively so. The more primitive class, the Bangioideae, contain in the Bangiaceae a few fresh-water members such as *Kyliniella latvica* and some very euryhaline species such as *Bangia atropurpurea,* known from estuaries and from Lake Balaton, whereas the Porphyridiaceae are mainly terrestrial, euryhaline, or fresh-water (e.g., *Porphyridium aerugineum*). The Porphyridiaceae are probably reduced and less primitive than the Bangiaceae, but the class Bangioideae is evidently fairly adaptable.

In the Florideae, the most primitive order, the Nemalionales, contains a number of fresh-water genera, including the widely distributed *Batrachospermum, Lemanea,* and *Thorea,* the North American *Tuomeya,* the Philippine *Nemalionopsis,* and the Australian *Nothocladus,* as well as odd fresh-water species in such normally marine genera as *Acrochaetium* (= *Chantransia*). Among the higher red algae there are great numbers of marine genera and species but almost no fresh-water forms. *Hildenbrandia rivularis,* a member of a marine genus in the Cryptonemiales, which forms red or brown crusts on stones in places of low light intensity in both streams and lakes, appears to be the principal exception.

Carpomycetes. Terrestrial, but there are a very few freshwater lichens, *Hydrothyria* being an American genus found in streams; *Dermatocarpon* and *Staurothele* are slightly less exclusively aquatic forms often found on stones in the littoral of lakes. The fungal components of these plants belong to the phylum.

Chlorophyta. This group as a whole seems to be principally fresh-water. One of the higher orders, the Siphonales, which is probably a member of a special class, has become marine, but fresh-water forms are found among its more primitive members. The other terminal group, the Charophyceae, is almost exclusively fresh-water with a few brackish forms such as *Chara baltica.* The fairly specialized Oedogoniales and Conjugales are also fresh-water with a small proportion of brackish-water species. The less specialized orders contain many marine species and genera, and one family, the Ulvaceae of the Ulotrichales, is mainly marine but has retained considerable evolutionary adaptability, for both fresh-water and marine species of *Enteromorpha,* as well as some that are very euryhaline, are known. The most primitive order, the Volvocales, which contains the motile unicellular forms, exhibits clear evidence of such adaptability. Species that have been referred to *Chlamydomonas*[21] occur in the sea, though the genus is primarily fresh-water, but *Dunaniella* is notorious for its capacity to live in saline waters.

Euglenophyta. Primarily a fresh-water phylum and in general probably heteroauxotrophic and so limited to waters relatively rich in organic matter or in contact with sediments containing abundant bacteria. A few little-known marine forms (Schiller 1925), such as *Chlorachne* and *Ottonia,* which seem to be more primitive than the fresh-water members of the group, require further study.

Opisthokonta. The species of the single order, the Chytriales, are largely parasites of fresh-water algae and of terrestrial plants.

Phaeophyta. A few fresh-water genera (*Pleurocladia, Heribaudiella, Bodanella*) belong to the most primitive order, the Ectocarpales. *Bodanella* plays a limited role in the phytobenthos of some large European lakes (Zimmermann 1928).

Sarcomastigophora. The Cryptomonadaceae consist of relatively few genera, but at least in *Rhodomonas* both fresh-water and marine races or species are known. Among the motile species of Xanthophyceae, referrable to the order Heterochloridales, there appear to be fresh-water, brackish, and marine genera and perhaps species of the same genus living in fresh and in salt water. The taxonomy of the group is not well known and some of the included species may really belong elsewhere. It is clear, however, that the order as a whole exhibits considerable evolutionary adaptability. Of the higher, more strictly algal members of the class most are fresh-water, but *Halosphaera* and *Meringosphaera* are marine genera with fresh-water allies.

The Chloromonadaceae consist of a few, mostly monotypic, genera occurring in both fresh and brackish waters.

The Chrysophyceae form a large and extremely diversified group of organisms. In such a genus as *Ochromonas* autotrophic, heterotrophic, and phagotrophic nutrition is possible in the same species. In

---

[21] Some of the marine forms reported are apparently members of *Dunaniella* (Ryther, 1956).

some allied forms such as *Monas* photosynthetic pigments have been lost, whereas in other flagellate members of the class such as *Dinobryon* nutrition appears to be primarily autotrophic. Likewise in locomotor structure there is immense diversity; the primitive condition is probably biflagellate, one flagellum being pantoneme or tinsel, the other acroneme or whiplash, but they may also be equal or unequal in length, and one may be lost or additional flagella added. Amoeboid forms are known, as are nonmotile and purely algal species. In a single genus *Hydrurus*, which lives in cold streams, a macroscopic plant body consisting of feathery branched tufts is developed. It is reasonable to regard the Chrysophyceae as a central group near to the points of origin of the other classes of the phylum. As a whole the group inhabits fresh waters but in the larger and better known motile genera marine as well as fresh-water species occur.

The Bacillariophyceae presumably represent a specialized development of the Chrysophyceae, in which group siliceous cyst walls are commonly found. The very peculiar *Phaeodactylum tricornutum*, which seems to link the diatoms with the Chrysophyceae (Lewin 1958), is primarily an inhabitant of tide pools and may be somewhat euryhaline. The diatoms presumably arose in the sea; throughout the group many genera are characteristic of marine and others of fresh water, often of particular kinds of fresh water, but often both marine and freshwater congeneric species occur (e.g., *Asterionella, Nitzschia, Melosira*), and in a few cases, such as that of the curious *Bacillaria paradoxa*, normal inhabitants of the sea can invade fresh waters.

The Coccolithophorida are predominantly marine, but fresh-water species occur in *Acanthoica* and *Anacanthoica*, genera also containing marine species, and in *Hymenomonas*, apparently defined solely to contain certain fresh-water species. These genera all belong to the supposedly most primitive family, the Syracosphaeridae. The other two classes of flagellates with elaborate mineral skeletons, the Ebriaceae and the Silicoflagellata, are entirely marine and largely known from fossils.

The most primitive Dinophyceae consist of a few genera of biflagellate monads, referrable to the family Desmomonadinaceae. A fairly complete series of morphological transitions can be traced from these organisms through slightly less primitive families as the Prorocentraceae to the typical unarmored dinoflagellates such as *Gymnodinium*. Of the three genera accepted by Chatton (1952) in the Desmomonadinaceae, *Desmomastix* is fresh-water, *Haplodinium* brackish, and *Pleromonas* marine. The great bulk of the Dinophyceae are marine, but species of the larger genera of both unarmored, such as *Gym-*

nodinium, and armored, such as *Peridinium* and *Ceratium,* dinoflagellates contribute significantly to the fresh-water biota. There are, in addition, a few peculiar fresh-water genera.

The Phycomycetes, which are doubtless of Chrysophycean or comparable origin, are primarily a fresh-water group, but the fungal flora of the ocean is quite inadequately known.

The Zooflagellata, when apochlorotic organisms clearly related to one or other classes of phytoflagellate (e.g., *Polytoma* near *Chlamydomonas, Monas* near *Ochromonas, Astasia,* an apochlorotic *Euglena*) are excluded, consist of a vast assemblage of parasites and a relatively small number of free-living forms. Most of the latter are found in the more primitive orders of the subclass Protomonadina, namely the Choanoflagellata, the Bicoecidea, and the Bodonidea, though a few heterotrophs appear in the order Trichomonadina of the subclass Metamonadina. Among the free-living forms most have been described from fresh water, but within the larger families there is a fair representation of marine species. This is most noticeable in the Choanoflagellata, which also show the clearest Chrysophycean affinities, though they certainly are not ancestral to the other orders.

We now turn to the rhizopod members of the phylum; the classification of the naked amoebae of the class Lobosa is unsatisfactory, but it is quite clear that both fresh-water and salt-water forms exist. The thecate members of the Lobosa are almost all fresh-water, though *Antarcella atava* is a brackish species. The class Filosa consists of a single naked genus *Penardia,* which occurs in fresh water, and a great number of testaceous forms, most of which are fresh-water, a few (*Euglypha laevis, Trinema lineare*) that are brackish, and one of the two species of *Gromia* (*G. oviformis*) marine.

The class Granuloreticulosa contains a few naked genera, sometimes placed as pseudoheliozoans, *Arachnula* being both fresh-water and brackish, *Gymnophrys* apparently fresh-water and marine, *Biomyxa* fresh-water and *Pontomyxa* marine. The less specialized testaceous genera, often termed unilocular foraminifera and mainly grouped in the Allogromiidae and Microgromiidae, are largely fresh-water but have a number of marine species; the highly specialized multilocular foraminifera are exclusively marine.[22]

The true Heliozoa are regarded by Tregouboff (1953) as comprising two orders which are probably unrelated. The Actinophrydia contain the fresh-water *Actinosphaerium,* the marine *Camptonema,* and *Actinophrys,* of which two species are fresh-water, one marine and one

---

[22] *Nonion* occurs in the saline L. Niris in Iran (Löffler 1953).

(*A. sol*) found in both habitats. This group is probably related through the aberrant genus *Vampyrellidium* to *Arachnula* and other primitive granuloreticulose forms. The other order, the Centrohelidia, also contains both fresh-water and marine members and is probably derived directly from the Chrysophyceae through genera such as *Actinomonas* and *Dimorpha* which are often placed as pseudoheliozoans in the order Protomyxidea.

The other actinopod classes, namely the Acantharia, probably descended like the Centrohelidia directly from Chrysophyceae, and the Radiolaria, are exclusively marine.

Ciliata. The Ciliata are a very large group, all major subdivisions of which appear to possess a considerable degree of evolutionary adaptability. It is not possible to detect any trend in preference for fresh or salt water in most major subdivisions of the group, and even the Tintinnida which seem to be the most completely specialized for marine life, contain a number of fresh-water species. The most primitive ciliates, the Gymnostomida, and the most advanced orders, placed in the Spirotricha, all contain both fresh-water and marine genera and species. If there are more fresh-water forms known, it is probably merely because they have been more carefully studied.

The two remaining phyla, which are exclusively parasitic, occur in marine, fresh-water, and terrestrial hosts and give no hints to the possible medium inhabited by free-living ancestors.

Physiological considerations. In view of the fact that the most elementary way to counteract the increase in volume which occurs when a body of some solution enclosed in a semipermeable membrane is immersed in a hypotonic solution is to provide the body with a rigid covering that can exert a counterpressure, it is a matter of some importance to note that the phenomenon of evolutionary euryhalinity is exhibited just as clearly by naked animal forms, such as the amoebae, as by phytomonads with cell walls, such as the Volvocales. A simple mechanical control is important in the more complex plant forms, but it cannot be the primitive method of maintaining a given volume in the face of an osmotic gradient.

In nearly all the motile fresh-water protists the main method of maintaining a constant volume is almost certainly the contractile vacuole, now universally regarded as an osmoregulatory organelle (Kitching 1952 and references therein). A particularly neat and relevant demonstration of the activity of such structures is provided by Guillard's (1960) study of a strain of *Chlamydomonas moewusii* lacking these organelles, the parent wild-type form having two. The mutant strain behaved as an obligate stenohaline brackish-water organism, unable to divide in a medium of osmotic pressure less than about 1 atm., though the wild type could live in ordinary much more dilute fresh water. Neither the mutant nor the wild type could exist in media of osmotic pressure above about 5 atm., though marine species lacking contractile vacuoles can do so. It is evident that the fresh-water species of *Chlamydomonas* differ from their marine allies in at least two ways. First they require a lower internal concentration though one that is hypertonic to their normal medium; second, by the contractile vacuole they can maintain the implied difference in osmotic pressure. Loss of the contractile vacuole merely limits the organism to a range of external concentrations which we may roughly suppose correspond to workable internal concentrations, but since this range is evidently lower for fresh-water than for marine species loss of the vacuole does not convert a fresh-water into a marine species. For this to happen it would be necessary also for the permissible range of internal pressures to be raised. The occurrence of closely allied species in fresh and salt water indicates that little evolutionary adjustment would be needed to produce the change, which clearly cannot correspond to any fundamental alteration of cell physiology.

With regard to the contractile vacuole, reversion to a type possessing such a structure occurred in about one in 10^5 cells in Guillard's cultures of his mutant. This may have been due to back-mutation, but not necessarily to the wild-type allele, as the reverted form had larger vacuoles which contracted less regularly than those of the wild type. According to Kitching (1938), the structure is absent from most marine rhizopod protists but present in most flagellates and ciliates in the sea. It tends to be lost in fresh-water algal forms with heavy cell walls. The simpler types of contractile vacuole are probably not greatly different from other noncontractile vacuoles not involved in osmoregulation; there is some evidence, summarized, for instance, by Lloyd (1928), that they may arise *de novo* in cells not possessing them under ordinary conditions. Both the possibility of forming such vacuoles and the possibility of cell metabolism proceeding normally at different osmotic pressures must be involved in the high evolutionary euryhalinity of the lower protists.

It is important to realize that although the cell sap of the marine algae may be more or less isotonic with sea water its composition is normally very different. Even when a marine plant appears to have little capacity to regulate its water content, it has a very great capacity to regulate the internal concentration of the various ions present (Table 1).

A study by Droop (1958) of *Skeletonema costatum* and of various littoral euryhaline unicellular algae has indicated the rather surprising

TABLE 1. *Concentration in cell sap in millemols of various inorganic ions compared with that of medium (Osterhout 1933)*

	Sea Water	*Valonia macrophysa*	*Halicystis* sp	Pond Water	*Nitella* sp
Cl⁻	580	597	603	0.9	90.8
SO₄⁻	36	? tr	tr	0.3	8.3
H₂PO₄⁻	—	—	—	0.0002	3.6
Na	498	90	557	0.2	10.0
K	12	500	6.4	0.05	54.3
Ca	12	1.7	8	0.78	10.2
Mg	57	? tr	16.7	1.69	177.7

fact that even *S. costatus,* a pelagic marine diatom, which has a fairly restricted salinity range, is living in its normal habitat in water containing an amount of sodium above the optimal concentration. All the euryhaline species studied had lower sodium requirements than *S. costatum.*

It may be noted that in passing from the sea into fresh water the total quantity of CO_2 available for photosynthesis is reduced, but since the pH of the sea is higher than that of most inland waters the CO_2 in the sea is present mainly as bicarbonate. A greater proportion of the smaller amount present in inland waters therefore is much more available to those photosynthetic organisms that cannot use the bicarbonate ion. Unfortunately, the distribution of the capacity to use this ion is quite inadequately known (cf. page 309).

It is also probably of importance that the phosphate content of sea water ordinarily remains greater than that of most inland waters when there is considerable algal growth, whereas available iron and accessory organic compounds are likely to be more plentiful in inland waters in which the ratio of the drainage basin to the free water is normally much higher than even in enclosed arms of the sea. These variations in nutrient supply may well prove to be as important as salinity in regulating the occurrence of some groups of lower organisms (cf. page 340).

General remarks on the distribution of salinity requirements in the Protista. Excepting the fungal phyla and the Euglenophyta, in which the few marine forms are poorly understood, every major branch of the Protista has in its more primitive divisions both fresh-water and marine species and so may be regarded as primitively possessing high evolutionary euryhalinity. This is true of the Bangioideae and Nemalionales among the Rhodophyta, the Ectocarpales among the Phaeophyta, the Volvocales among the Chlorophyta, the phytomonads, except those with highly specialized siliceous skeletons, the most primitive Dinophyceae, and the amoeboid groups with or without unspecialized tests. Some clades, notably the Bacillariophyceae, the Heliozoa, and to a lesser extent the Dinophyceae and the Ciliata, have retained a high degree of evolutionary adaptability throughout their evolutionary history. Other clades, notably the Rhodophyta, the Siphonales among the Chlorophyta, the Phaeophyta, the Foraminifera, the Acantharia, and the Radiolaria, have become highly specialized to a marine existence. Yet other clades, notably the Conjugales and Charophyceae among the Chlorophyta and to a much more limited degree the higher algal members of the Chrysophyceae, have become specialized almost entirely in fresh water. The general picture is emphatically not one of primitive adaptation to the ocean, followed by invasion of fresh waters, but rather a primitive evolutionary lability, followed in some cases by profound specialization, mainly to a marine environment. This specialization, which has produced the typical assemblages of organisms that we regard as characteristic of the ocean on the one hand and of fresh water on the other may well have been initially related to biotic and physical factors rather than merely to variations in osmotic pressure. If we remember that the same sort of evolutionary adaptability appears characteristic of the less specialized Monera, the picture becomes even more impressive. There is nothing in the empirical evidence to support the oft-repeated dictum that life began in the sea.

THE COLONIZATION OF FRESH WATERS BY THE METAPHYTA

If, as is now commonly believed, the simpler liverworts, which include certain aquatic forms, are secondarily simplified, it is probably safe to conclude that the entire aquatic metaphyte flora is of terrestrial origin.

Bryophyta. Among the mosses the hygrophytic species of *Sphagnum* are often more or less submerged in the littoral zones of small lakes and bog pools. The Fontinalaceae (*Fontinalis* and *Dichelyma* in the Holarctic, *Hydropogon* and *Hydropogonella* in tropical America, *Wardia* in South Africa) are characteristic water mosses; *Fontinalis antipyretica* occurs in many situations from the margins and banks of lakes to great depths and in the brackish waters of the Baltic as well as in very soft fresh water. Many species occur only in running water and some are characteristic of acid streams from volcanic and hydrothermal springs. Some mosses are free floating, notably in the genera *Amblystegium* and *Drepanocladus* (*A* of Fig. 4).

FIGURE 4. Fresh-water bryophyta and lower tracheophyta. *A. Drepanocladus,* a free-floating moss (Fassett 1940). *B. Ricciomorpha,* a pleustonic liverwort (Fassett 1940). *C. Lepidotis inundata,* widespread in very wet bogs and probably the most aquatic of the club mosses (W. Hooker 1861). *D. Equisetum fluviatile,* a widespread emergent aquatic horsetail (Muenscher 1944). *E. Azolla caroliniana,* a pleustonic heterosporous fern, North America (Muenscher 1944). *F. Salivina rotundifolia,* another pleustonic heterosporous fern, native to tropical America (Muenscher 1944). *G. Marsilea quadrifolia,* a rooted heterosporous fern with floating leaves, widespread in Old World and introduced to eastern North America (Fassett 1940). *H. Ceratopteris pteridoides,* a pleustonic homosporous fern from tropical and subtropical America. (All $\times \frac{9}{16}$ except *G*, which is about $\times \frac{1}{3}$ and *H* which is $\times \frac{3}{16}$.)

Several pleustonic (*B* of Fig. 4) liverworts (*Ricciella, Ricciocarpus*), which may be associated with *Lemna* and the Salviniaceae, are known.

Lower Tracheophyta. It is ecologically convenient, if morphologically out of date, to consider together the three lower subphyla with the class Filicineae of the Pteropsida, for in all these groups, as in the mosses, a flagellate sperm, implying at least a momentary aquatic life for the male gamete, is retained. In spite of this, all groups are now basically terrestrial, though in this one sense not fully adapted to dry land. The aquatic species are in fact almost certainly secondary inhabitants of fresh water. All are evidently of ultimate chlorophycean and so presumably of fresh-water origin.

In the Lycopsida a few species, notably *Lepidotis inundata,* (*C* of Fig. 4), are almost entirely aquatic, and the Isoetaceae or quill worts contain species of *Isoetes* that grow completely submerged and play some role in the ecology of lakes. In the Sphenopsida there are a few aquatic species of *Equisetum,* such as *E. fluviatile* (*D* of Fig. 4).

In the Filicineae (*E–H,* Fig. 4) there are three unrelated families of aquatic ferns, namely the homosporous Ceratopteridaceae, and the heterosporous Marsileaceae (water clover), and Salvinaceae. The last named family includes the two important neustonic genera *Salvinia* and *Azolla*.

Angiospermae. No gymnosperms are truly aquatic, though many grow in swamps; the angiosperms, however, are very well represented. Some authorities have supposed that the whole group of monocotyledons may have originated in fresh water.[23]

The main genera of monocotyledonous plants growing in fresh water are found in the Butomaceae and Hydrocharitaceae of the order Butomales, in the Potamogetonaceae of the Potamogetonales, and in the Zannichelliaceae of the order Najadales. Of all these plants perhaps *Potamogeton,* by nature of its ubiquity and diversity, is the most important.

There are also numerous isolated aquatic genera in other families, such as *Pistia* in the Araceae. To the nonbotanical reader it may come as a shock to learn that the duckweeds, or Lemnaceae, a family which in *Wolffia* and *Wolffiella* contains the smallest known (Fig. 5) flowering plants and which is of considerable ecological importance, is also generally believed to be derived from the Araceae, better known as the family containing the arum lilies and jack-in-the-pulpits.

---

[23] For a critical discussion see Arber (1920), one of the classics of hydrobiology; however, see J. Hutchinson (1959) whose classification is followed here.

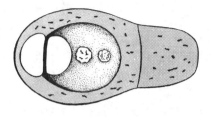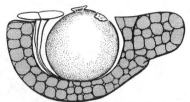

FIGURE 5. *Wolffia brasiliensis* (× 70), the smallest flowering plant; the specimen illustrated both from above and in section carries two anthers, one partly covering the ripening fruit. The plant was first described from Matto Grosso, *Victoria amazonica* also occurring in the locality; *singulière bizarrerie de la nature d'avoir semé ensemble ces deux végétaux.* (Weddell 1849.)

The aquatic dicotyledons are found mostly in the lower orders, notably the Ranales, which includes, apart from many aquatic species of eurytopic genera such as *Ranunculus,* two aquatic families. One of these, the Ceratophyllaceae contains a single genus with three species but is important because of the wide distribution and commoness of *Ceratophyllum.* The other family, the Nymphaceae, includes the familiar white and yellow water lilies *Nymphaea* and *Nuphar,* the gigantic South American *Victoria* (Fig. 6), and the lotus *Nelumbo.* Many isolated aquatic genera are found in other orders, particularly in families containing many hygrophytes, for a complete series of transitions from terrestrial through swamp-inhabiting to aquatic plants can be detected over and over again. Examples of such isolated genera are the curious floating *Aldrovandra* in the insectivorous sundew family, the Droseraceae, and the trap-bearing *Utricularia* and its allies, a group of great biological interest, in the Lentibulariaceae. A few completely aquatic families of isolated position are known, the most remarkable being the Podostemataceae (Fig. 7), which consists of creeping plants that spread over the surfaces of stones in streams, particularly in the tropics. All of these plants are clearly of ultimate terrestrial ancestry. In the production of *turions,* or specially protected and detachable buds that survive unfavorable circumstances, many water plants (Fig. 8) parallel fresh-water sponges and bryozoans.

In inland waters the flowering plants are to a large extent the ecological representatives of the larger littoral and sublittoral algae of the sea. The fresh-water flowering plants, however, are usually limited to a zone of greater light intensity than that which can be colonized by certain large attached algae, which often show chromatic adaptation in both lakes and the ocean. The fresh-water angiosperms are, in

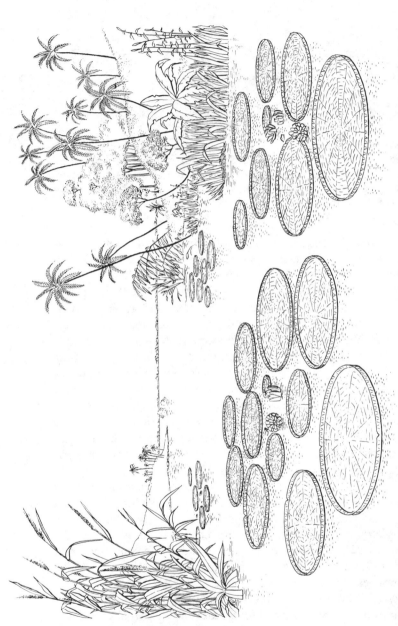

FIGURE 6. *Victoria amazonica* (foreground leaves c. $\times \frac{1}{10}$), the giant water lilly of tropical South America, the largest fresh-water plant, formally described as a member of a new genus by Lindley in 1837, the year of Queen Victoria's accession, though known to some earlier naturalists. Haenke, in 1803, fell to his knees when he first saw it; to D'Orbigny it was the most beautiful known plant, and to Spruce the great leaves resembled "a number of green tea trays floating, with here and there a bouquet between them." The creamy white flower with a crimson center is said to rise 10° C. above the ambient temperature through metabolic heating, dispersing, no doubt, the heavy perfume. After fertilization the flower sinks and the fruits develop below water; the seeds when ground yield an edible flour. (Caspary 1891; for history see Allen 1854, for biology Arber 1920.)

34

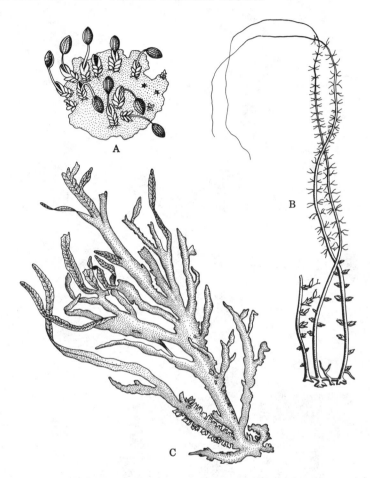

FIGURE 7. Representatives of the Podostemaceae, perhaps the most specialized of the aquatic phanerogams and often exhibiting a striking convergence of form toward that of some of the higher marine algae; found mainly in tropical regions in running water, with the thallus-like root attached to or spread over rocks and stones. *A. Zeylanidium olivaceum* ($\times$ 2), flat thallus with endogenous flowering shoot, Oriental region. *B. Dicraea elongata* ($\times$ $\frac{2}{9}$), Ceylon, creeping and floating roots, the latter bearing flowers. *C. D. Stylosa* ($\times$ $\frac{1}{2}$), Ganges region, branched, band-like thallus. (Warming 1891.)

FIGURE 8. Asexual reproduction by protected buds in fresh-water benthic organisms. *A*. The flowering plant *Myriophyllum verticillatum* with turions (*t*). *B*. Sponge *Spongilla lacustris* with a gemmule (*g*). *C*. Polyzoan *Plumatella repens* with statoblasts. (*s*) both *in situ* and isolated. (Arber, Bowerbank, Potts, Allman.)

fact, the ecological equivalents of the Ulvaceae, the Siphonales, and some of the Phaeophyceae rather than of the Rhodophyceae.

A few flowering plants have invaded the domain of the algae in the sea. The marine genera of seed plants belong to *Halophila, Enhalus,* and *Thalassia* in the Hydrocharitaceae (Butomales), *Zostera* and *Phyllospadix* in the Zosteraceae (Apogetonales), *Posidonia* in the Posidoniaceae, a family of the peculiar Juncaginales, and several genera, notably *Cymodocea,* in the Zannichelliaceae (Najadales). All are clearly of fresh-water origin.

THE METAZOAN FAUNA OF FRESH WATERS

In striking contrast to what was noted in the cases of the Monera and Protista, in which all the major groups appear to be derived from highly adaptable but not specifically marine ancestors, most of the fresh-water Metazoa seem to be either clearly marine or, like the higher aquatic plants, clearly terrestrial in origin. Only perhaps in the flatworms, as indicated in a later section (page 46), is there a suggestion of the evolutionary euryhalinity which is so conspicuous among the more primitive Protista, and even then the available evidence is ambiguous. The problems presented to an animal by life in fresh water are very different for a marine invader and one from the land; it is primarily a matter of water and salt regulation for the first and of respiration for the second. In the following account[24] these origins are noted briefly; the osmotic and the respiratory problems encountered by invaders from the two primary sources of the fresh-water fauna are discussed in greater detail on subsequent pages.

Porifera. Sponges are common in fresh waters but are limited to two families, the Spongillidae of widespread distribution and the

---

[24] Much of the information given below in condensed form can be extracted, often, however, with considerable difficulty, from the great standard treatises on the animal kingdom. At the time of writing the whole field is almost entirely covered by the monumental surveys of Hyman (1940 et seq.), Kükenthal and Krumbach (editors, 1923 et seq.) and Grassé (editor, 1948, et seq.); all are incomplete, but between them and the various more modern volumes of Bronn's *Tierreich* almost the whole of the field is available. References are given only when material is discussed that is not in these works and not referred to elsewhere in the present work. The table of marine and fresh-water representation of the main groups of the animal kingdom given by Carpenter (1928) contains many errors. Figures are given in the present chapter mainly of organisms of great interest in relation to the mode of origin of the fresh-water biota but of little further significance in later chapters. This inevitably leads to an overemphasis of curiosities and rarities; many of the common work-a-day species appear later.

Baikalian Lubomerskiidae, with which *Ochridaspongia* from Lake Ohrid and *Metschnikovia* from the Caspian may perhaps be associated. Both families are silicious and belong to the order Haplosclerina of the class Demospongiae. Both families appear to be allied to the marine Haliclonidae (= Renieridae), but it is not impossible that they originated independently; the relevant literature is summarized by Brooks (1950a).

The osmotic pressure of the cell fluid of *Spongilla lacustris* is equivalent to about that of 27 mM. NaCl, or nearly four times that of the fresh-water medium in which the specimens studied were living (Zeuthen 1939). During gemmulation the osmotic pressure rises greatly, apparently because of the production of small organic molecules. Contractile vacuoles are known in *Spongilla* and *Ephydatia* but not unequivocally in any marine sponge (Jepps 1947), though clear vacuoles which do not contract can be seen in the choanocytes of at least some marine species.

Some littoral marine sponges are moderately euryhaline; Hartman (1958) has demonstrated experimentally species differences in this respect in *Microciona*.

Coelenterata. This group is primarily marine, and truly fresh-water forms only are found in one of the classes, namely the Hydrozoa.

Within the Hydrozoa there is unfortunately no unanimity in regard to the most primitive group. If we omit the Siphonophora, which are irrelevant to the present discussion, we may recognize a series of medusoid forms passing from the Trachymedusae with direct development through the group named the Limnomedusae by Kramp, in which there is a small sessile polyp, to the Corynidae and Tubulariidae of the Anthomedusae or Gymnoblastea, but the direction of evolution involved is by no means clear. Kramp and Bouillon regard the Trachymedusae as derived from the Gymnoblastea through the Limnomedusae. The series, however, can easily be read in the opposite direction, as implied, for instance, by Hyman; the present author feels that this is the more reasonable interpretation, in spite of the immense authority of the proponents of the opposite view. The Narcomedusae are presumably an offshoot of the early trachyline stem, whereas the Calyptoblastea (Leptomedusae) are reasonably derived from the Gymnoblastea.

The fresh-water coelenterates are found in the rather aberrant family of Gymnoblastea, the Hydridae, as one or two species of *Cordylophora* in the otherwise marine gymnoblast family Clavidae, in the Limnomedusae, many of which are marine or brackish, and as a single aberrant member of the Narcomedusae. The Limnomedusae are of particular interest because they introduce the phenomenon of a relatively coherent group, clearly of marine origin, which appears to be able rather easily, though for no known physiological reason, to make the evolutionary transition to fresh water. Cases of such apparent pre-adaptation to invasion of fresh waters will be met with in other groups.

The family Hydridae in the Gymnoblastea contains a few genera of wide distribution, the genus *Hydra* containing a number of species of fairly localized range. A medusoid stage is lacking and the life history is of a characteristic fresh-water type, asexual budding occurring throughout most of the year, with seasonal production of resistant eggs.

The only other gymnoblasts that may fairly be considered inhabitants of fresh water are one or possibly more species of *Cordylophora*, a genus of the otherwise marine family Clavidae. A single widespread species *C. caspia* (= *lacustris*) has been recognized by most authorities in brackish waters from many parts of the world; in some localities, as in the Norfolk Broads, it enters almost or, in certain North American fluviatile localities, apparently quite fresh waters (Davis 1957). There, are, however, few chemical data from such habitats of *Cordylophora*, and it would not prove surprising if in all of them the chloride concentration were a little higher than is usual in ordinary inland waters. At least some populations of the hydroid appear to be unable to live in quite fresh no less than in full strength marine water. Among the other names given to brackish or fresh-water members of the genus it is possible that *C. japonica* (Itô 1951) refers to a distinct species.

The order Limnomedusae (Fig. 9) contains the families Moerisiidae, Olindiidae, and perhaps the marine Proboscidactylidae. The first family contains three genera (Picard 1951): *Halmomises* (= *Moerisia*), *Odessia*, and *Ostroumovia*. *Halmomises* occurs as *H. lacustris* in a lagoon in Trinidad (von Kennel 1891), as *H. pallasi* in the northwest Caspian (Derzhavin 1912), and as *H. lyonsi* in the Birket el Quarun (Boulenger 1908); it is also apparently known from the Ganges delta (Picard 1951). *Odessia maeotica* occurs typically in the Sea of Azov, as f. *ostroumovi* in some Black Sea limans, as f. *gallica* in littoral lagoons on the Mediterranean Coast, and as f. *marina* in the Mediterranean (Picard 1951) and at Casablanca. The third genus, with a single species *Ostroumovia inkermanica*, appears in estuaries and limans on the northern coast of the Black Sea. Though capable in one instance of living in sea water, the whole group seems to consist primarily of brackish-water forms with a Tethyan type of distribution.

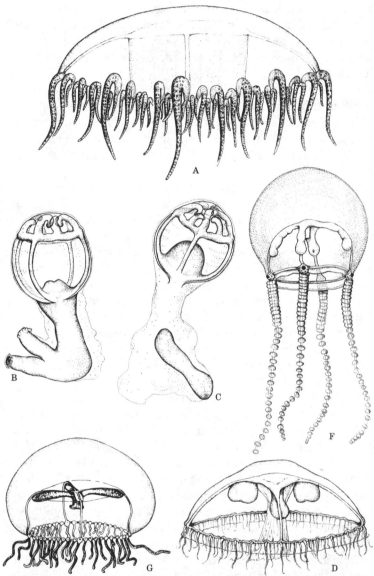

FIGURE 9. Limnomedusae; representatives of two families, the Olindiidae with marine and fresh-water species and the Moerisiidae with mainly brackish species, some from inland but somewhat saline waters. The whole group appears preadapted to water of low salinity, for no other medusae have been able to invade such dilute habitats. *A. Limnocnida victoriae*, widespread in Central Africa, adult medusa. *B. C.* Hydroids of

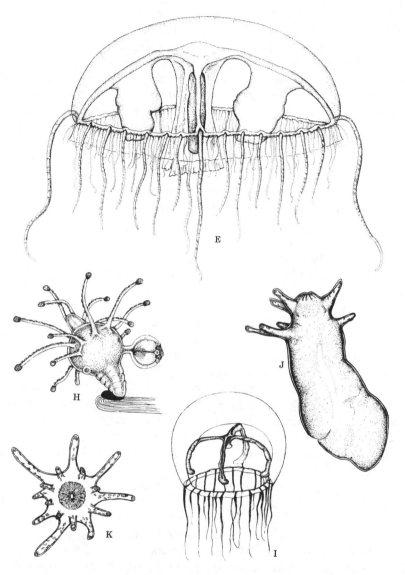

L. tanganjicae, Lake Tanganyika, *C* with frustule. *D. Craspedacusta sinensis,* headwater valleys of Yang-tze-kiang. *E. C. sowerbyi,* presumably originally Chinese, now widespread in Europe and America. *F. Halmomyses* (= *Moerisia*) *lyonsi* Birket-el Quarun. *G. H. Odessia maeotica* f. *gallica,* coastal lagoons in Mediterranean, medusa and hydroid. *I. Ostroumovia inkemanica,* Black Sea limans. *J. K. Calpasoma dactyloptera,* a problematic fresh-water limnomedusan hydroid. (Boulenger, Bouillon, Kramp, Picard, Fuhrmann.)

The Olindiidae exhibit considerable evolutionary euryhalinity, *Olindias* and *Gonionemus* are marine; the two closely allied genera *Craspedacusta* and *Limnocnida* are capable of life in purely fresh-water environments and do not tolerate appreciable salinities.   At the present time *Craspedacusta sowerbyi* is widely spread throughout the temperate regions of the world; in the New World at least most occurrences are in artificial waters.   Its original appearance in the *Victoria amazonica* tank at Kew, and the later discovery that *Microhydra ryderi* from North America is the polyp of *Craspedacusta,* has led to a general belief that the animal is of New World origin, but it seems now far more likely to be derived from China (Sowerby 1941; Kramp 1950). In China the species (in part perhaps represented by a subspecies *kiatingi* of rather doubtful validity) is common in the Yang-tze-kiang valley, and in the tributaries of the same river (Gaw and Kung 1939a,b; Kramp 1950) a second valid species *C. sinensis,* is found.   There is possibly a third species (*C. iseana*) in Japan (Uchida 1955).

Wherever found, the polyps of *C. sowerbyi* appear to live mainly in running water, the medusae in lakes (Dejdar 1934; Lytle 1960), where they swim to the surface and then sink slowly, usually oral side up, with extended tentacles that may come in contact with and capture zooplankton. From the various accounts summarized by Kramp it seems likely that optimum natural conditions for the animal may well have been provided by the fluviatile lake system (Vol. I, Fig. 41) of the Yang-tze-kiang, in which water levels are so variable that a basin may contain a lake at flood stages but be transversed by a river at low water; alternatively, a river may be a torrent at flood stage but a series of quiet pools in the dry season.

The genus *Limnocnida* was first made known from Lake Tanganyika (Günther 1893) where its existence provided a zoogeographic sensation and gave support to the now abandoned marine relict hypothesis regarding the fauna of that lake.   There appear (Bouillon 1957a) to be three valid species in Africa: *L. tanganjicae* from Lake Tanganyika, *L. victoriae* from a large part of Central Africa, westward to the Niger, northward to Lake Victoria, southward to the Limpopo, and, more recently, possibly artificially spread to the vicinity of Johannesburg and Pretoria, and *L. congoensis* from the Congo basin (Bouillon 1957b).   A fourth species, *L. indica,* occurs in India (Annandale 1912; Gravely and Agharker 1912; Rao 1931, 1933).

The hydroid stages of both genera bud to form a colonial organism of several morphological individuals.   Such colonial organisms bud off frustules which can form new polyps; they can also form resistant internal buds which are capable of withstanding rather unspecified unfavorable conditions. A third type of budding produces the sexual medusae. A single introduction of the polyp will lead to the production of unisexual populations of the medusae. This is doubtless the explanation of the occurrence of but one sex of *Craspedacusta* medusae in some American localities. In *Limnocnida tanganjicae,* but in no other species, medusae are budded from the manubrium of the adult.

The little known and peculiar hydroid described by Fuhrmann (1939) as *Calpasoma dactyloptera,* from a fresh-water aquarium at Neufchatel, Switzerland, has, according to its describer, nematocysts like *Craspedacusta,* that is, microbasic heterotrichous euryteles, and so may be related to *Craspedacusta* and *Limnocnida,* or, as Lytle suggests (personal communication), be a polymorphic form of the former.

Polypodium hydriforme, recognized as a narcomedusan solely by its possession of atrichous isorhizae and no other type of nematocyst, is as an adult a small actinula-like hydroid that walks oral surface down on its tentacles. This polyp bears gonads. At some stage in its early development the animal infects the sterlet *Acipenser ruthenus,* reaching the ovary and forming a stolon in an ovarian egg, which escapes when the egg is laid. The polyps are budded off from the stolon. The animal is found in the Volga and the rivers entering the Black Sea and is presumably of Sarmatian origin.

No osmoregulatory organelles appear to have been described in the cells of the fresh-water coelenterates. *Cordylophora caspia,* which must have cell contents hypertonic to the medium in which it lives, appears not to be able to invade absolutely fresh water; this suggests that it is limited by the inefficiency of a salt uptake mechanism. Certain fairly euryhaline brackish water Scyphozoa and Anthozoa are known. *Aurelia aurita* lives in water of salinity between 5 and 6 per mille in the Gulf of Finland, but the scyphistoma stage cannot develop below just over 6 per mille (Segerstråle 1951). Miyawaki (1951) found among the Anthozoa that *Diadumene luciae* can survive exposure to 7.5 per cent sea water, that is, a salinity of around 2.5 per mille, mainly by secretion of mucous over its surface. Such a case throws no light on the normal mechanisms of adaptation to fresh waters by coelenterates.

Platyhelminthes. The modern classification of the flatworms departs considerably from the older view retained in most textbooks and even in Hyman's (1951a) comprehensive survey. According to Westblad (1948; cf. Karling 1940), the Turbellaria, with which we are primarily concerned, should be divided into two orders, the Archoophora and the Neoophora. These taxa are accepted, essentially as grades, by Ax (1963). The former group consists of those Turbellaria in which the egg is endolecithal and not supplied by yolk from yolk glands; the latter has at least some separation of vitellarian and germarian cells, if not a quite separate yolk gland and ovary, and consequently an ectolecithal egg.

The Archoophora (Ax 1963) consist of five suborders, the Acoela (in an extended sense including the Hofsteniida and the Nemerto-dermatida), the Proplicastomata, the Polycladida, the Macrostomida, and the Catenulida. The Acoela are small,[25] often minute animals, and all live in the sea. At first sight it is reasonable to suppose that the primitive turbellarian, whatever its exact character may have been, was a small marine animal, lacking osmoregulatory organs.

The other three suborders of the Archoophora are distinctly more elaborate. The Polycladida are normally moderate or large worms, some of which, at least, have a poorly developed protonephridial system. All are marine except *Limnostylochus borneensis* from fresh water in Borneo (Stummer-Traunfels 1902), evidently a rather spectacular crimson red animal, up to 3 cm. long, that can swim or glide at the surface of the pools it inhabits. In view of the lack of analyses of the waters of these habitats the status of *Limnostylochus* as a fully adapted fresh-water animal is not unequivocal. The Macrostomida and Catenulida, formerly placed in the Rhabdocoela, are small worms with well-developed protonephridia. The Macrostomida include *Pro-macrostomum*, a primitive form, supposedly with acoele affinities, from Lake Ohrid (An-der-Lan 1939; Stankovič 1960), *Macrostomum*, a large genus mainly but not exclusively of fresh-water species, and two marine genera, whereas the Catenulida are entirely fresh-water.

The Neoophora comprise two main groups, the Alloiocoela and the Neorhabdocoela. All of these animals have more or less elaborate generative organs and protonephridia, though the latter are best developed in the fresh-water species. The Alloiocoela, not accepted by Ax (1963) as a useful taxon, contain a number of families arranged in three main series. The most primitive, the Lecithoepitheliata, contain a few marine members and the important fresh-water family Prorhynchiidae. In the Cumulata the family Protomonotresidae contains two fresh-water and one marine genus; several strictly marine genera are arranged in related families, whereas in the Plagiostomidae, *Plagiostomum*, a large genus mainly living in the sea has *P. lemani* in European lakes, *P. lacustre* in Tanganyika, *P. evelinae* in Brazilian fresh waters, and the generically barely separable *Hydrolimax grisea*

---

[25] The largest is the Caspian *Anaperus sulcatus*, which is 12 mm. long. This animal may have had a fresh-water history during Pleistocene freshening of the Caspian.

in springs and streams in a limited area of the eastern United States (Hyman 1938; Westblad 1955). A third family of Cumulata, the Baicalarctiidae, contains only the type genus and species *Baicalarctia gulo* in Lake Baikal, an animal clearly not related to the other fresh-water forms. The Seriata contain among their less specialized representatives *Otomesostomum* and *Bothrioplana,* fresh-water genera placed in separate monotypic families; the higher Seriata contain the Paludicola formerly placed as triclads and consisting of the peculiar North American cavernicolous Kenkiidae, the widespread epigean Planariidae and Dendrocoelidae, and an extraordinary assemblage of large forms from Lake Baikal, apparently referrable to the last-named family (Kozhov 1963).

The Neorhabdocoela are arranged in four divisions. The Dalyelloida contain a number of fresh-water species in the Dalyellidae and the isolated genus *Bresslauilla* in the marine family Graffillidae which occurs in fresh and brackish water in Europe. The second division, the Typloplanoida, consists of a few small marine families and the very large fresh-water Typhloplanidae, which includes *Typhloplana, Castrada, Praenocora,* and *Mesostoma,* to name the more important genera. There are also a few terrestrial forms. The third division of the Neorhabdocoela, the Kalyptorhyncha, contains mostly marine sand-living animals, but *Gyratrix hermaphroditus* is an extraordinarily euryhaline organism which lives both in fresh waters and the sea, and there are fresh-water species in the families Polycystidae and Koinocystidae. The fourth series, the Temnocephalida, are ectocommensal, mainly on fresh-water animals. At least the case of *Gyratrix hermaphroditus* (Meixner 1938; Kromhout 1943) is more reasonably interpreted as an invasion of the sea from fresh water than the reverse (Fig. 10).

Taken at its face value and assuming only one invasion when quite closely allied animals occur in fresh water, even if they

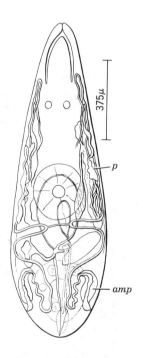

FIGURE 10. *Gyratrix hermaphroditus,* the fresh-water form, shows a well-developed excretory system, almost or entirely lost in the marine form, with ampullae (*amp.*) and paranephrocytes (*p*) which are absent in brackish-water populations. (Kromhout 1943.)

have many marine relatives, at least thirteen independent invasions are needed to give the fresh-water flatworm fauna from a supposedly primitive marine Archoophoran.

Two alternative possibilities may be considered. If the simplest Turbellaria are really related to the ciliates, they may have retained much of the osmoregulatory capacity and evolutionary euryhalinity of the Protista. It must, however, be admitted that it is the morphologically apparently least specialized Acoela that are most clearly restricted to marine habitats.

If the Turbellaria are the descendants of a primitive coelomate ancestor which also gave rise to the worms and mollusks and which in producing the flatworm lost its coelom much as arthropods, leeches, and some mollusks have done, as Remane (1963) and Ax (1963) believe, it is conceivable that the Turbellaria are of fresh-water origin, a hypothesis that would explain the retention of the protonephridial system and internal fertilization. Because of the great similarity of the spermatozoa of all marine invertebrates with external fertilization and the specialized diversity of the sperm of the Turbellaria, some special circumstances in the evolution of the reproductive processes of the group are almost certainly indicated. A fresh-water history may well have provided such circumstances.

Rhynchocoela. The nemerteans are clearly a marine group. They are sometimes derived from the Platyhelminthes and because they possess both anus and circulatory system may have originated from the flatworms at about the point at which the higher schizocoel metazoa had their origin. The presence of a protonephridial system presumably indicates an original habitat, perhaps of varying salinity, like that of the lower flatworms. The simple reproductive system is, however, of an obviously marine type. *Geonemertes* is terrestrial and *Prostoma* with at least half a dozen species is a widespread fresh-water form (Stiasny-Wijnhoff 1938) in the Northern Hemisphere and in South America. The North American species (Fig. 11), usually termed *P. rubrum*, is regarded by Stiasny-Wijnhoff as invalid and based on *P. lumbricoides* and *P. graecense*, both also European. Whatever

FIGURE 11. *Prostoma rubrum,* the common fresh-water nemertean of North America. (Coe.)

its taxonomic status, this worm is functionally hermaphrodite. It secretes a mucous sheath into which both eggs and sperm are discharged and from which in a few days elongate ciliated larvae emerge; these larvae grow without metamorphosis into the adult worms. There is a tendency to protandry in some specimens, and two worms may lie side by side in the same mucous cyst (Child 1901; Coe 1943). Other species are dioecious or always protandrous. The life history of *P. rubrum* is interesting as a hint of the kind of process that must also have occurred in the early history of the Annelida Clitella. The genus *Prostoma* forms, with the hardly separable marine *Tetrastemma,* the family Tetrastemmidae. Hermaphroditism and even viviparity occur in the marine genus, which may in a sense be preadapted in this way to fresh-water life.

In addition to *Prostoma,* five other species of marine affinity are known in fresh water. *Planolineus exsul* from fresh waters in Java (De Beauchamp 1929) is a heteronemertean, as is *Siolineus turbidus* from near Santarém in the Amazon basin (du Bois-Reymond Marcus 1948) and no doubt also the problematic *"Nemertes" polyhopla* of the Lake of Nicaragua (Schmarda 1859). *Otonemertes denisi,* a metanemertean has been described from Tonle Sap, Cambodia (Dawydoff 1937)., *Malacobdella aurita,* a bdellonemertean is parasitic on a fresh-water mollusk in Chile (Blanchard 1847).

Pseudocoelomate phyla. Hyman (1951b) treats as pseudocoelomate three phyla, the Acanthocephala, the Aschelminthes, and the Entoprocta. The first of these are parasitic, occurring as adults in marine, fresh water and terrestrial vertebrates. Their affinities are admittedly problematic. The Entoprocta are a small phylum formerly associated with the Polyzoa, to which some authorities continue to ally them. Their status as a phylum really depends on the uncertainty of their affinities. Here, as often elsewhere, phyla are delimited by ignorance, classes by knowledge.

The Aschelminthes, which provide the interesting problems from the standpoint of this work, are divided into five or six classes, the Rotifera (= Rotatoria), Gastrotricha, Nematoda, Nematomorpha, Kinorhyncha (= Echinodera), and perhaps the Priapuloida. Many zoologists are doubtful about the affinities of any of these groups. It is possible that the Acanthocephala along with the last two of these classes should be united as Rhynchohelminthes, separated from the Aschelminthes which would combine the rotifers, gastrotrichs, and nematodes. Some doubt about the position of the Nematomorpha would then appear (cf. Hyman 1959). The rotifers stand somewhat apart in their specialized trophic organs and certain other anatomical features. They have retained to a varying degree external ciliation

and possess a well-developed protonephridial system. The Gastrotricha also retain external ciliation but in most structural respects are
nearer nematodes than rotifers; a protonephridial system is developed
only in the order Chaetonotoidea. The other groups have lost all
external ciliation and in the nematodes internal cilia as well; a striking
thickening of the cuticle has taken place and this cuticle is molted as
the animal develops.

Salinity relationships of the Rotifera. The Rotifera are a fairly
large class divided into three orders and containing about 1800 species,
of which Myers (1936) believes some 10 per cent are reasonably
euryhaline or halophil. Bērziņš (1951) concludes that only about
a quarter of these species are normally restricted to salt water. With
few exceptions the marine species are confined to the Monogononta.
The most important of these exceptions is provided by the order
Seisonacea, which contains a single genus *Seison* with two or three
species, all epizoic on the marine crustacean *Nebalia*. *Seison* is the
only rotifer in which an unreduced male is known, and the reproductive
system is in general more primitive than in the other two orders.
It is therefore probably a fundamentally primitive rotifer, even though
its external characters are rather peculiar, owing no doubt to its epizoic
mode of life.

The order Bdelloida are mostly littoral benthic fresh-water animals,
some of which have become virtually terrestrial, living in mosses and
lichens. The semiterrestrial species are active only when wet but can
withstand desiccation and in the desiccated or anabiotic condition are
apparently absolutely cold-resistant, recovering after exposure to liquid
helium. All are parthenogenetic, the male being unknown in the
group. A single marine species, *Zelinkiella synaptae,* in the family
Philodinidae, is epizoic on holothurians.

The majority of the rotifers belong to the Monogononta, and it
is in this order that most of the fully marine species are found. Nearly
all of these strictly marine forms are members of relatively specialized
genera such as *Synchaeta, Trichocerca, Keratella,* and *Notholca.* It
is not unlikely that the main limitation imposed on these forms is
trophic rather than chemical; at least some of the marine rotifers appear
to be commonest when much organic matter, such as sewage, and
so many bacteria are present. Individuals of some species are apparently capable of tolerating very great and sudden salinity changes
(P. Gray personal communication). There can be little doubt from
the typical life history of the Monogononta, in which a series of
parthenogenetic generations, hatched from subitaneous eggs, is succeeded
by a sexual generation whose reduced males and mictic females produce

resting eggs, that the group became established early in their history in fresh water and that the species now found in the sea are secondarily marine.

Salinity relationships of the Gastrotricha. The Gastrotricha are divided into two orders; the Macrodasyoidea, which lack protonephridia, are exclusively marine, and the Chaetonotoidea, which possess protonephridia and contain more than three quarters of the two hundred known species of this class, are almost entirely confined to fresh water. The Macrodasyoidae appear to be the more primitive group, since they are hermaphrodite, while except for two hermaphrodite genera, *Neodasys* and *Xenotrichula,* both marine, the Chaetonotoidea are all parthenogenetic, with no males or male organs. A few brackish and marine species are included in the larger genera of the Chaetonotoidea.

Salinity relationships of the Nematoda. The Nematoda are an enormous group; about ten thousand species have been described, but it has been estimated, largely on the basis of host specificity of the parasitic forms and of the number of possible hosts, that this figure is at least an order of magnitude too low.[26] The fresh-water species, though numerous, are still quite inadequately known.

The class is conveniently divided into the lower and largely free-living Adenophorea or Aphasmida and the more specialized, largely parasitic Secernentea or Phasmida. The arrangement here adopted follows Goodey (1951). More elaborate classificatory schemes (e.g., Maggenti 1963) are useful in detailed phylogenetic studies but throw little further light on environmental relationships. Among the aphasmid nematodes the suborder Enoplina of the order Enoplida is primarily marine, though with fresh-water species in the Oncholaimidae, Ironidae, Tripylidae, and Mononchidae, which must represent several invasions. The larger genera, such as *Ironus,* which contain fresh-water species, also contain soil forms. The other enoplid suborder, the Dorylaimina, are primarily fresh-water and soil nematodes, though a few marine species are known. *Dorylaimus* itself contains about two hundred species, of which about two dozen are strictly inhabitants of fresh water with almost as many in very wet soil, three are marine, a few live in moss, and nearly all the rest are typical soil nematodes. *Actinolaimus* is a less numerous but important genus, found exclusively in fresh water. The Mermithida live, as adults, in soil or fresh water

---

[26] Estimates as high as half a million species have been given, for instance, by Hyman (1951b), but they imply an excessive degree of host specificity and are almost certainly too great.

but have a parasitic larva, usually in insects. The other orders of Aphasmida are mainly marine. The Chromadorida contain in the Chromadorina a few odd fresh-water species, mostly in genera also containing marine species, of which Goodey (1951) lists five; *Puncto-dora ratzeburgensis,* a member of a monotypic genus, is a characteristic fresh-water form in Europe, which probably feeds on diatoms. The Monhysterina again are mainly marine but in the Plectidae and Axonolaimidae contain a number of fresh-water genera; there are also a few in related families. *Monhystera* contains mainly fresh-water and soil species. *Prismatolaimus* is another important fresh-water genus. The Desmoscolecida are a peculiar marine group with a single fresh-water species, *Desmoscolex aquaedulcis* (Fig. 12), from drip water in a cavern near Laibach in Yugoslavia. This animal is clearly a member of the peculiar hypogean fauna (pages 191–193) of marine affinities, which appears to reach its maximum development in the countries around the Mediterranean.

The orders placed in the subclass Phasmida are mostly parasitic; *Macrolaimus aculeatus* (Cephalobiidae) in the Rhabditida is recorded in fresh water from Paraguay, whereas *Hemicycliophora aquaticum* in the Tylenchida is known from the Lunzer See, and several species of *Criconemoides* are found in fresh-water mud, probably feeding on roots. All are related to soil species; *Dolichodorus heterocephalus* found in Douglas Lake and other American fresh waters is probably a plant parasite.

In general, the limited information available suggests that invasion of fresh waters by nematodes has proceeded in part directly from the sea and in part indirectly by species of ultimate marine origin which have become members of the fauna of soil or damp decaying vegetation.

A special case is provided by the few species of *Turbatrix* and *Panagrellus* which live in fermenting liquid media often in artificial conditions; *T. aceti,* the vinegar eel, is the best-known species, which, according to Goodey, tolerates a pH range of 1.6 to 11. A closely allied form, regarded as a variety *dryophila,* lives in white slime-flux on oaks. Species of *Panagrellus* may occur in the same type of environment; one, *P. nepenthicola,* lives in the pitchers of *Nepenthes* in Indonesia.

The Nematomorpha are a very small and not certainly monophyletic group in which the young stages are parasitic in Arthropods. Of the two orders, the Gordioidea are fresh-water as adults, the Nectonematoidea, marine. At least the larval stage of the Gordioidea is reminiscent of the two remaining classes of Aschelminthes, which are en-

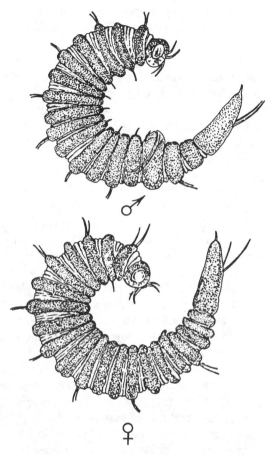

FIGURE 12. *Desmoscolex aquaedulcis,* a fresh-
water member of a peculiar marine genus of
Nematoda, from a dripwated pool in Krška-jama,
a cave in Laibach, Slovenia. (Stammer.)

tirely marine, though two of the six species of Priapuloida can live
in the dilute sea water of the western Baltic.

If we consider the Aschelminthes as a whole, there can be little
doubt that the basic ancestral form was marine, since in all three
classes in which fresh-water forms occur there is a strong suggestion
that the marine habitat is characteristic of the more primitive forms;
moreover two classes live only in the sea. The hypothetical ancestor no doubt possessed protonephridia. It is possible that their
loss in the Macrodasyoidea is concomitant with the small size and

partly uncuticularized epidermis of these animals. The Rotifera which must have entered fresh waters early in their evolution have become one of the most successful groups of fresh-water animals, some of which could return to the sea. The Nematoda, having acquired a thick cuticle, no doubt obtained some degree of protection against any kind of unfavorable chemical environment, even though their respiratory needs still had to be met. It would be interesting to know exactly how the vinegar eel, for instance, deals with the immense variations in the fluid that it must almost inevitably take into its alimentary tract. The invasion of nematodes into fresh waters was doubtless a dual process, in part directly from the sea and in part from terrestrial species that had become adapted to damp soil. It is interesting that the Kinorhyncha, in spite of their cuticle and protonephridial system which would seem to be, as in nematodes, possible preadaptations, have not been able, as far as is known, to enter fresh water.

Salinity relationships of the Entoprocta. The phylum Entoprocta (= Kamptozoa) contains only three families. Most of the species belong to two families that live in the sea. The third family, the Urnatellidae, contains only two fresh-water species of *Urnatella* (*A* of Fig. 13): *U. gracilis* is found mainly in running water in eastern North America, from Pennsylvania north to the Great Lakes, although there have been several recent records from Europe;[27] the other is *U. indica* from India. The disjunct distribution suggests that the family was once more important in inland waters. No agreement has been reached on the position of the phylum in the animal kingdom, but it is reasonable to suppose that it is of marine origin.

Mollusca. There can be no doubt that this group is fundamentally marine. The Cambrian Monoplacophora are evidently marine fossils, and the discovery of a living genus, *Neopalina,* surely the noblest survivor from the past to have been made known during this century, confirms the generalized nature of the class. Among the other classes the Amphineura, Cephalopoda, and Scaphopoda are exclusively marine, as are most of the families of Bivalvia and the lower streptoneurous Gastropoda.

The Bivalvia of inland waters. The classification of the bivalves is still in a somewhat labile state. In the following discussion the arrangement of Thiele (1925–1926), as given by Haas (1955), has been used as a basis, but certain recent modifications have been adopted in the appropriate places.

---

[27] Damas (1939) Pennak (1963); in spite of Thienemann's (1950) uncertainty, it is reasonable to suspect artificial introduction.

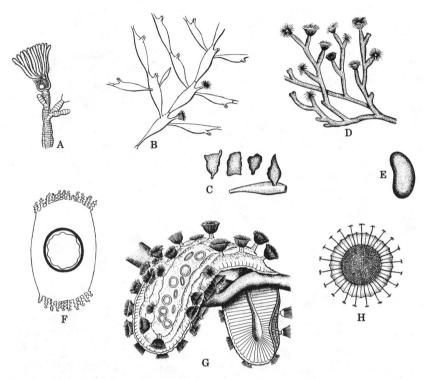

FIGURE 13. *A*. Entoprocta: *Urnatella gracilis* (× c. 20), usually in rivers. *B*. Polyzoa Gymnolaemata: *Paludicella articulata* (× c. 8), the commonest fresh-water member of the class Gymnolaemata. *C*. Hibernacula of same. *D*. Polyzoa Phylactolaemata: *Fredericella sultana* (× c. 8), a rather simple and widespread member of the class Phylactolaemata. *E*. statoblast of same (× c. 40). *F. Lophopodella carteri* statoblast (× c. 28). *G. Cristatella mucedo* (× c. 3), the motile colony of the most specialized fresh-water polyzoan. *H*. statoblast of same (× c. 20). (Rogick, after various sources; cf. also Figure 8.)

Haas considers that we can regard the fresh-water lamellibranchs as falling into two categories—the old well-adapted groups which have produced a great diversity of species and the small and supposedly recent invasions which have contributed little to the fresh-water fauna of the earth as a whole. The two categories, however, are not sharply distinguishable. In the first we have the Unionaceae and most of the Sphaeriaceae, which make up the bulk of the fresh-water bivalve fauna of the world, and two invasions of brackish water in the Ponto-Caspian area which contributed little to the true fresh-water fauna.

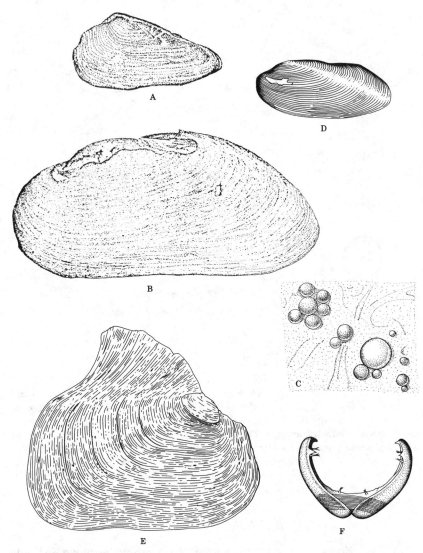

FIGURE 14. Fresh-water Bivalvia. *A. Carbonicola* (= *Anthracosia*) *acuta*
(× 1), a supposedly fresh-water species from the Coal Measures of England,
a member of a group perhaps related to the precursors of the Unionaceae;
B. Margaritifera margaritifera (× c. ⅔), (Margaritiferidae), the holarctic
fresh-water pearl mussel; *C.* pearls of same, from Scotland, *in situ* (× 9/4); *D.*

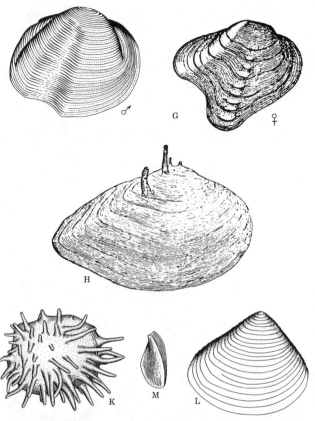

Unio pictorum (× c. ⅔) (Unionidae), palaearctic, the shells formerly used as receptacles for watercolor paints; *E. Proptera alata*, (× ½) widespread in St. Lawrence and Mississippi drainage; *F.* Glochidium of same; *G. Epioblasma (E.) bilobum*, (× ½) Ohio River, ♂ and ♀ to show extreme sexual dimorphism in external form; *H. Elliptio (Canthyria) spinosa*, (× ⅔) Altamaha River, Georgia, shell with remarkable projections, supposed to act as anchors; *I. Mutela bourguignati* (× 1), Lake Victoria; J. Fryer's larva of same (× 96); *K. Aetheria heteromorpha* (× ⅔), an African member of a family of oyster-like species found in very rapid water in isolated tropical areas; *L. Corbicula fluminea* (× ⅔) China and introduced to North America, a member of the Corbiculidae, of the superfamily Sphaeriaceae; *M. Scaphula pinna* (× 1), Tenasserim River, India, a member of the Arcidae and the only fresh-water Taxodont. (*A*, Hind; *B, D*, Wesenberg-Lund; *C*, Home; *E, G, H, L*, Clench from various sources; *F*, Lefevre and Curtis; *I*, Smith; *J*, Fryer; *K*, Simroth; *M*, Cooke.)

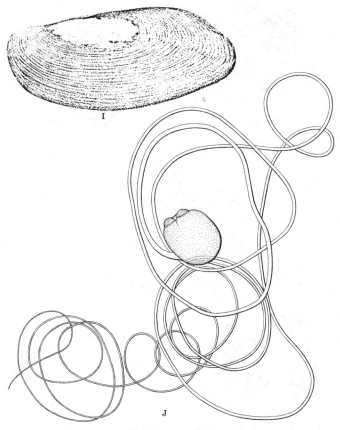

FIGURE 14.   (*Continued*)

In the second category we have mainly odd fresh-water genera, often with brackish allies, in otherwise marine families.

The Unionaceae (Fig. 14) are an immense group belonging to the suborder Schizodonta of the order Eulamellibranchiata.   The Schizo- donta also include the Trigonaceae, a marine superfamily, largely fossil, which originated in the Devonian but has some modern representatives. The classification of the Unionaceae has proved extremely difficult. The best system appears to be that of Modell (1942, 1949), modified so far as the Australasian species are concerned by McMichael and Hiscock (1958).   Modell recognizes thirty-nine subfamilies, arranged in four families, the Mutelidae, Elliptionidae, Margaritiferidae (= Margaritanidae), and Unionidae.   The Mutelidae are Ethiopian,

Neotropical, and Austral and, according to McMichael and Hiscock, include probably all but one of the Australian Unionaceae. The family was apparently widely distributed at the close of the Paleozoic but disappeared from the Holarctic in the Mesozoic. At least one species appeared in Australia during the Triassic and is apparently referrable to the Australasian subfamily Velesiunioninae. The Elliptionidae, not recognized as a family by earlier workers, are today exclusively North American, but the fossil subfamily Trigonodinae is known from the Trias of both Europe and North America. The Unionidae also appear in the Triassic and are widely distributed but have only one doubtfully placed genus, *Haasodonta,* in Australia. The restricted Margaritiferidae appear to occur mainly in the Northern Hemisphere. The Holarctic *Margaritifera* was an important source of pearls in Northwestern Europe during early historic times (*B,C,* Fig. 14). The four families have produced a bewildering number of genera and species in many parts of the world, reaching a climax of specialization in the Mississippi Valley and in the river drainages to its East in the unglaciated areas of North America (*E,G,H,* Fig. 14). There is, however, no reason to suppose that the group is of American origin. The great fresh-water clam fauna of southeastern North America may rather be compared to the specialized faunas of ancient lake basins and must be due largely to the long period of relative stability that the group has been able to enjoy in a region that has suffered neither glaciation, desiccation, nor submergence beneath the sea.

The Unionaceae as a whole are presumably related to the fossil Anthracosidae (*A* of Fig. 14) (Devonian-Trias), the dominant fresh-water bivalves of the late Paleozoic, but their place of origin at that remote time is, of course, unknown.

The sexes are normally separate in the Unionaceae, and some of the more specialized genera exhibit striking sexual dimorphism in shell characters (*G* of Fig. 14). The eggs are held in brood pouches variously formed from the gills of the female. The exact arrangement of these structures provides important taxonomic characters when it can be ascertained, but it is perhaps subject to more rapid parallel evolutionary changes than has often been supposed.

The characteristic larva in all families, with the exception of the Mutelidae, appears to be the *glochidium* (*F* of Fig. 14), an ectoparasite for a time on fishes or in one case the amphibian *Necturus*. In a single South American subfamily of Mutelidae, the Glabarinae, an apparently more primitive and probably nonparasitic *lasidium* larva is found, and a new type (*I* of Fig. 14) of parasitic larva has recently been described by Fryer (1959) for *Mutela bourguignati* in Central

Africa. Further study of the larvae of South American and Australian species may produce considerable modification of the classification.

The superfamily Sphaeriaceae is perhaps allied to the Schizodonta, but it is now usually placed in the much more extensive suborder Heterodonta. It consists of three families, the Corbiculidae, the brackish and largely tropical Cyrenoididae (= Cyrenellidae), and the Sphaeriidae. The Corbiculidae are (*L* of Fig. 14) mainly tropical and appear to exhibit some evolutionary euryhaline adaptability; *Corbicula* (= *Cyrena*), *Polymesoda* (*s. str.*) and some species of *Batissa* are fresh water, whereas *Polymesoda* (*Geloima*) and other species of *Batissa* and *Villorita* are inhabitants of brackish water, largely in southeast Asia. The Sphaeriidae are a very widely distributed family of small and taxonomically perplexing fresh-water bivalves, with their present center of distribution in the Northern Hemisphere, though extending to New Zealand. Some species are important in the limnobenthos, but many inhabit ponds or marginal wet moss and a few are known under wet leaves and have become almost terrestrial. The Sphaeriidae are hermaphrodite and ovoviviparous, the eggs developing in a marsupium; the brood is normally very small. At least some species appear to be annuals, but with two breeding seasons.

The rather isolated family Dreissenidae in its own superfamily of the Heterodonta consists mainly of brackish-water species, among which *Congeria* and the closely allied *Mytilopsis* occur chiefly along the coasts of the warmer seas, though *Mytilopsis* extends north to New Jersey on the Atlantic coast of North America. *Congeria* was an important member of the fauna of the Sarmatian Sea and gave rise in the Ponto-Caspian basins to *Dreissena*[28] during the Pliocene, whereas the allied *Dreisseniomya* developed in fresh-water basins in the Balkans. *Dreissena* (Fig. 15) has continued to be an important element in the Caspian and more dilute water fringing the Black Sea, and one species *D. polymorpha* certainly was established as a fresh-water species throughout the Volga, lower Danube, and intermediate drainage systems before 1800. It may have been present earlier throughout much of western Europe, but if so it apparently had disappeared before the end of the eighteenth century. Since 1800 the species has spread throughout most of the major river systems of western Europe, ap-

---

[28] This appears to be the correct name. *Dreissena* P. Beneden 1835 = *Dreissenia* H. Bronn 1848 = *Dreissensa* A. Moquin-Tandon 1855 = *Dreissensia* H. Bronn 1862 = *Dreissina* Sowerby 1839 = *Dreyssensia* Oppenheim 1891 = *Driessena* P. Beneden 1835 = *Driessenia* anon 1835 = *Driessensia* Dollfus and Dautzenberg 1886 = *Driessina* Scudder 1882 = ?*Dreistena* Boué 1840. As Ronald Firbank said, in an entirely different context, this should be a warning to us all.

parently not from the Volga, but from foci of introduction in the West. It appeared (Haas 1955) in London in 1820, in Rotterdam in 1826, and in Hamburg in 1830. The spread westward and southward from the Danube may represent an independent artificial extension of range. *D. polymorpha* like the rest of the family is a mussel anchored by byssus threads and though mainly an inhabitant of rivers retains a veliger larva.

A somewhat similar history is exhibited by the Cardiidae of the same region. The late Miocene fauna of the Sarmatian Sea included several species groups of *Cardium* which gave rise during the Pliocene to a number of characteristic genera (Fig. 15). Among them *Limnocardium* from Pliocene deposits in Hungary and *Budmania* from Croatia apparently were fresh-water forms. Among the brackish-water genera *Monodacna, Didacna,* and *Adacna* are still an important element in the fauna of the Caspian, the Sea of Azov, and the more dilute Black Sea limans (summary in Haas 1955; see Kelecnikov 1950 for recent palaeontology; Borcea 1924). No existing member of the Cardiidae has been able to spread into fresh water in the manner of *Dreissena polymorpha,* though, apart from the endemic Ponto-Caspian genera, species of *Cardium* are often fairly euryhaline and *C. edule* has once been reported in fresh water.[29]

Many of the less important invasions which have not added greatly to the fresh-water fauna have, however, taken place at a sufficiently remote period to produce genera characteristic of fresh water in groups that are otherwise marine.

Scaphula (*M* of Fig. 14), a member of the Arcidae and so the only fresh-water taxodont lamellibranch, is widespread in the Ganges, Jumna, and Tenasserim Rivers in India, occurring 2500 km. upstream from the ocean, and is certainly a fully adapted member of the fresh-water fauna.

Several genera of Mytilidae, in the order Anisomyaria, have apparently become adapted to fresh water. *Sinomytilus,* occurring in southeast Asia and reaching at least 1000 km. from the sea, is a subgenus of *Mytilus.* From the same region *Brachidontes, Limnoperna,* and the doubtfully distinct Australian *Fluviolanatus* are close to *Modiolus;* these forms perhaps are not all fully adapted to fresh water.

---

[29] Bateson (1889) found this species in Lake Ramleh No. 2, near Alexandria, Egypt, in water that he believed to be fresh. No analysis was given and the nature of the locality suggests the possibility of the presence of saline water under a superficial, much fresher layer.

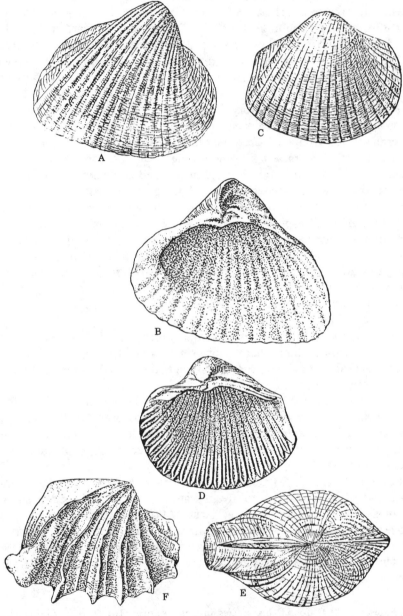

FIGURE 15.   Pontocaspian bivalves and their derivatives in inland waters.   *A, B.*
Didacna trigonoides (Haas after Ostroumoff); *C, D, E. Limnocardium banati-*
cum (Haas after Brusina), early Pliocene, Hungary, shell suggesting strong
development of syphons; *F. Budmania meisi* (Haas after Brusina), late

The family Glaucomyidae with its single (Owen 1959) genus, *Glauconome* (= *Glaucomya*), is apparently a member of the heterodont superfamily Veneraceae. It is widely distributed in brackish waters from Southeast Asia to Australia and is frequently regarded (Haas 1955) as entering the fresh-water fauna of that area, though most of the detailed records appear to refer to coastal localities.

The genus *Rangia* which occurs in the lower parts of various rivers entering the Gulf of Mexico is sometimes referred to a special family, though Thiele places it in the Mactridae. Apparently it is not fully adapted to fresh waters. *Tanysiphon*, another mactrid (Owen 1959), formerly placed with *Glauconome*, seems to occupy comparable habitats in India.

Profischeria, a subgenus of *Iphigenia* in the Donacidae (Tellinaceae), is apparently a true fresh-water form living in West Africa. The typical *Iphigenia* (*s. str.*) has brackish-water species in tropical areas.

The remarkable *Novaculina gangetica*, certainly a true fresh-water animal, once placed with the razor-shells, forms with the apparently

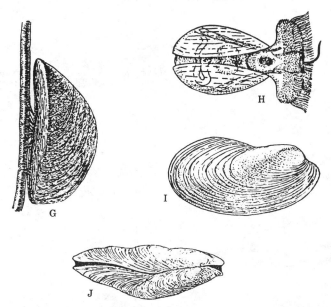

Pliocene, Croatia, a still more specialized form from inland waters; *G. Dreissena polymorphus; H.* Late larva of same (Wesenberg-Lund after Meisenheimer); *I. J. Dreisseniomya žujoviči* (Haas after Brusina), a fresh-water derivative, from the Pliocene of Serbia (all × $\frac{2}{3}$).

marine *Sinovaculina* from eastern China a subfamily, the Novaculininae, which Yonge (1949) places in the Asaphidae (= Psammobiidae).

Anticorbula and *Erodona,* members of the Aloididae (Myaceae), are not unimportant in the fauna of the Amazon, *Erodona* also occurring in other South American rivers draining to the Atlantic and (*E. afra*) at the mouth of the Congo.

A species of boring mollusk, *Martesia rivicola,* referred to a well-known marine genus of the Pholadidae, apparently occurs in wood in the fresh waters of Borneo. *Nausitoria dunlopei* (Fig. 16), originally from trunks of trees in potable and allegedly soft water of a distributary of the lower Ganges (Wright 1864) is apparently widespread in southeast Asia and is probably the same as a species cultivated for food in supposedly fresh waters near Bangkok (Moll 1936). The normally marine *Teredo minima* is also recorded by Seurat (1933) from inland waters.

It is difficult to estimate the number of invasions involved in the existing lamellibranchs, but, if we allow one for the Unionaceae, two for the Sphaeriaceae, one for *Dreissena polymorpha,* one for *Scaphula,* two for the Mytilidae, one for *Profischeria,* one for *Novaculina,* one for the Aloididae, one for *Nausitoria*, and one for *Martesia*

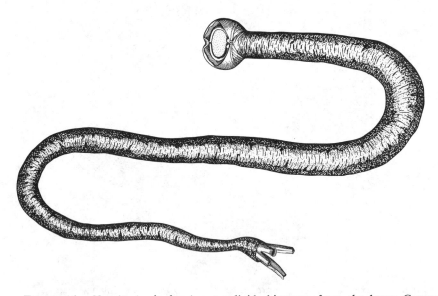

FIGURE 16. *Nausitoria dunlopei,* a teredinid shipworm from the lower Ganges region, Thailand, and elsewhere in southeast Asia, living in wood immersed in supposedly fresh water. (Wright 1864.)

rivicola, we have twelve.   This excludes *Rangia, Glauconome, Tany-siphon,* and *Teredo minima* as not truly adapted fresh-water animals. It is certain that the numbers of invasions of the same order of magnitude exhibited by the Turbellaria, though the number of species of marine lamellibranchs that potentially might take part in invasions would appear to be much greater.

The Gastropoda of inland waters. Among the Gastropoda there are two very different types of invasion.   The fresh-water prosobranchs seem to have originated from ordinary sea snails of various kinds, whereas the equally important fresh-water pulmonates are probably derived from a marine littoral group that became amphibious inhabitants of the shore line by transforming the mantle cavity into a lung as the gill disappeared.   The terrestrial stylommatophorous pulmonate snails had a like origin but perhaps not from the same littoral species that gave rise to the fresh-water families.

The only family of Archaeogastropoda to contain fresh-water species is the Neritidae (*A* of Fig. 17).   In this family some members are marine (*Nerita*), others are widespread in brackish or fresh-waters (*Theodoxus*), and *Neritodryas* from the Phillipines is terrestrial. Fresh-water species appear to have existed at least through most of the Tertiary.   An endemic genus *Lepyrium* (*B* of Fig. 17) has developed in the Cahawba and perhaps the Coosa rivers of Alabama.

In the Mesogastropoda the superfamily Cyclophoraceae contains only terrestrial, amphibious, and fresh-water species.   The Viviparidae are an important and widely distributed family of fresh-water snails (*D* of Fig. 17); the allied Lavigeriidae contain a single genus *Lavigeria* endemic to Lake Tanganyika, and the Pilidae (= Ampullariidae) are large but usually not fully aquatic tropical snails (*C* of Fig. 17).   The Valvataceae, with a single family and genus *Valvata* (*E–C,* Fig. 17) and no obvious existing marine allies, are, like the Cyclophoraceae, somewhat isolated.   Members of both the Viviparidae and Valvatidae appear in the Jurassic.   The Rissoaceae contain the very important family Bulimidae (= Hydrobiidae, = Amnicolidae), apparently closely allied to the marine Rissoidae and of Mesozoic origin.   Many species still inhabit brackish water, and the family was of great importance in the Sarmatian Sea.

The classification of the Bulimidae (*H* of Fig. 17) is difficult, and no agreement among authorities appears to have been reached on the number of subfamilies that should be recognized.   A number of widespread genera (*Hydrobia, Amnicola,* and *Potamopyrgus*) of small species have allied to them certain genera (*I* of Fig. 17) of restricted distribution, some of which are of limnological interest.   Lake Ohrid

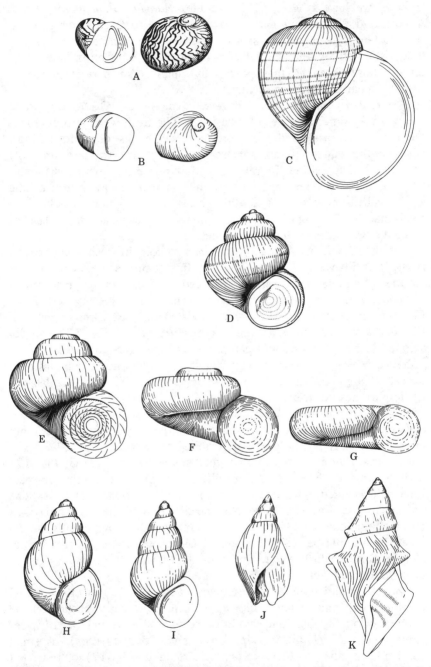

FIGURE 17. Fresh-water prosobranch Gastropoda. *A. Theodoxus danubialis,* (× 2), a member of a fresh-water but often euryhaline genus of Neritidae widely

contains endemic species of *Pyrgula,* some of which have their closest relatives among pliocene fossils. The important Baikal endemics *Benedictia,* which includes large snails, and *Kobeltocochlea* are often separated as a subfamily Benedictiinae but Thiele (1925) places them in his Hydrobiinae. The Ponto-Caspian Micromelaniinae include *Micromelania, Caspiella,* and *Clessiniola;* Ohrid endemic snails have been referred to the first-named genus. Thiele places *Baicalia* in this subfamily, but with *Liobaicalia* this genus is often placed in a separate subfamily, the Baicaliinae. Here, also, the mesozoic fossil *Paleobaicalia* from fresh water deposits in Siberia presumably belongs. Finally the Buliminae (*H* of Fig. 17) include the widespread *Bulimus* (= *Bythinia*).

A third superfamily of Mesogastropoda, the Cerithaceae, contains several marine families, the Potamiidae living in brackish water, the Planaxidae in which *Planaxis* is a littoral marine, *Quadrasia* (*J* of Fig. 17), a fresh-water genus, and the very important fresh-water family Tiaridae (= Melaniidae). The Tiaridae are mainly tropical and are divided into eight subfamilies by Thiele; among them the Pleurocerinae, which are important in North America (*K* of Fig. 17) are usually given family rank by American malacologists. The Paramelaniinae include the numerous and remarkable Tanganyika endemic snails with the exception of *Neothauma* in the Viviparidae and *Lavigeria* in the Lavigeriidae.

Ancylodoris from Lake Baikal, described by Dybowski (1900), is a mythical animal apparently based on the mislabeling (Lindholm 1927) of a marine onchidorid.

Genuine fresh-water opisthobranchs do, however, exist in the genus *Acochlidium,* which, with a few marine and brackish-water genera, forms the superfamily Acochlidiacea, related to the Aeolidiacea. *A. weberi* occurs in fresh water in Sumba and perhaps in brackish water in Flores; *A. amboinense* (Fig. 18), originally from Amboina, also

---

distributed in the southwestern Palaearctic; *B. Lepyrium showalteri* (× 3½), sometimes put in its own family, endemic to certain rivers in Alabama; *C. Pomacea paludosa* (× 1), a large amphibious snail of the family Pilidae from tropical America; *D. Viviparus viviparus* (× 1), Europe; *E. Valvata piscinalis* (× 5½), palaearctic; *F. V. pulchella* (× 8), northern and central Europe; *G. V. cristata* (× 11), palaearctic, three species showing progressive flattening of the spire; *H. Bulimus tentaculatus,* (× 3½), Holarctic, a widespread member of the Buliminae; *I. Lartetia quenstedti* (× 8½), Swabian Alps, a small hypogean member of the Hydrobiinae; *J. Quadrasia hidalgoi* (× 2), a fluviatile member of the otherwise marine Planaxiidae from the Philippine Islands; *K. Io fluvialis,* (× 1), a striking member of the Pleuroceritinae, endemic to the Tennessee River system. (Ehrmann, Clench, Crosse.)

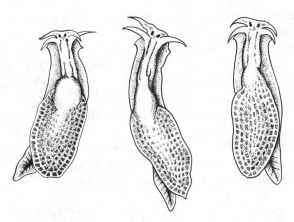

FIGURE 18. *Acochlidium amboinse,* a green fresh-water opisthobranch from Ambonia eastward to the Palau Islands. (Bayer and Fehlmann 1960.)

occurs in potholes in a rocky stream 50 to 60 m. above sea level in the Palau Islands (Bayer and Fehlmann 1960).

The true fresh-water Pulmonata belong to the suborder Basommatophora which also includes a number of families (Ellobiidae, Otinidae, Amphibolidae, Gadiniidae, and Siphonariidae) characteristic of the marine littoral. These largely littoral families vary in their degree of specialization. *Siphonaria* has a gill, but the gill-less Amphibolidae live in salt or brackish water. Most, however, are more or less amphibious; *Carychium* in the Ellobiidae is a very widespread terrestrial snail. There can be little doubt that the fundamental pulmonate character of the group was developed by amphibious species on the marine shore, but very early in the evolution of the Basommatophora an invasion of fresh waters by gill-less lunged snails must have occurred because the Chilean fresh-water genus (*A* of Fig. 19) *Chilina* (Chilinidae) alone among the pulmonates has a relatively streptoneurous nervous system (*B* of Fig. 19). From such a stock the New Zealand *Latia* in the Latiidae and the four cosmopolitan families Physidae, Lymnaeidae, Planorbidae, and Ancylidae (*C–F,* Fig. 19) must have been derived. The western American limpet-like *Lanx* (*G* of Fig. 19) is sometimes separated from the Ancylidae as a distinct family. The pulmonates are poorly known until the Mesozoic, but the important aquatic families appear in the Jurassic.

The other group of pulmonates, the Stylommatophora, have one primitive but peculiar and shell-less marine littoral family, the On-

cidiidae, but are otherwise nearly all terrestrial, though the Suc-
cineidae, well-known snails living on emergent fresh-water vegetation,
may be regarded as semi-aquatic.

It would seem probable that the existing fresh-water gastropod fauna
represents six or seven invasions, though in the Bulimidae an initial
adaptation to brackish water may have been followed by movements
into freshwater by several groups, possibly in some cases followed
by a return to brackish conditions.

The main invasions possibly occurred a little later than in the bivalves
but a fauna of relatively modern appearance and affinities was clearly
established in both groups by the middle of the Mesozoic. The additions that have occurred since that period are, both in the Gastropoda
and the Bivalvia, apparently to some extent connected with the history
of the Sarmatian Sea.

Annelida. The Echiuroida are exclusively marine. The other
classes fall into two groups. The Polychaeta and Archiannelida which
are primarily marine have a few fresh-water species; the Clitellata, consisting of the oligochaets and leeches, are mainly fresh-water or terrestrial with a limited number of obviously secondary marine species.
The seemingly very primitive *Rheomorpha neiswestnovae* (Fig. 20)
from the sandy beaches of lakes in eastern and central Europe requires
further study.

Among the Archiannelida *Protodrilus spongioides,* a member of a
large marine genus, was described by Pierantoni (1908) from a fresh-
water aquarium at Naples and probably came from the River Sarno.
In the absence of salinity data its status as a quite fully adapted fresh-
water organism is uncertain.

The most remarkable fresh-water archiannelid is certainly *Troglochaetus beranecki* (Fig. 21), which occurs in hypogean waters in the
Alps and is clearly allied to *Thalassochaetus* in the family Nerillidae
(Ax 1954). *Troglochaetus* may be regarded as a member of the
assemblage of hypogean organisms of marine origin that have entered
underground waters of the lands from the interstitial water of marine
sand (see page 193).

The number of euryhaline polychaets is considerable but their distribution among the families of the class is very uneven. According
to Hartman (1938), about fifty species occur in the Nereidae. The
subfamily Fabricinae of the Sabellidae also exhibits euryhaline adaptability. Among the twelve species of free-living polychaets, either errant
or sedentary, which can be fairly regarded as completely adapted to
fresh water, ten are members of these two groups.

Among the Polychaeta Sedentaria there are certainly three and prob-

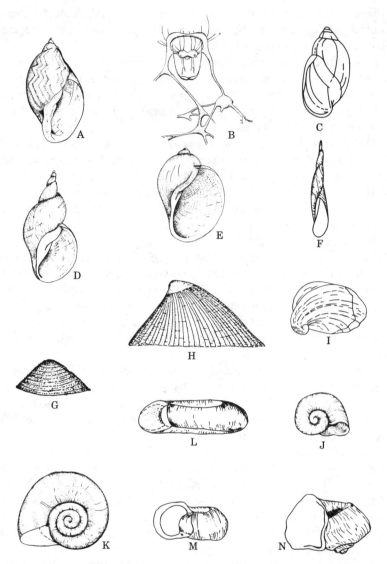

FIGURE 19. Fresh-water pulmonate Gastropoda. *A, Chilina strebeli* (× $\frac{6}{5}$) Chile, a member of the most primitive genus of fresh-water pulmonate snails, retaining *B,* a streptoneurous nervous system; *C, Physa gyrina* (× 1½), a widespread North American species of the sinistral family Physidae; *D, Lymnaea stagnalis* (× $\frac{5}{7}$), Holoarctic; *E, Radix auricularia* (× 1), Palaearctic, introduced to North America; *F, Acella haldemani* (× 1), upper Mississippi and St. Lawrence drainages, three representatives of the normally dextral Lymnaeidae showing possible variation in form in family; *G, Lanx newberryi,* (× 1), north-

ably five quite definitely fresh-water species; four belong to the sub-family Fabricinae. This subfamily (Hartman 1951) consists mainly of very small worms living in the surface layers of mud in tidal basins, enclosed seas, and estuaries.[30] Nine genera, including sixty to seventy species, are found exclusively in salt water. The genus *Manayunkia* includes two species that appear to be fully marine and five that are found in brackish waters, in estuaries in western Europe, in the Baltic in the Caspian, and in Chilka Lake in India. At least in the last-named locality the animal may at times be exposed to very dilute water. There are, in addition, two species (Fig. 22), *M. speciosa*, from rivers in Pennsylvania and New Jersey and from the benthos of Lake Erie and Lake Superior (Pettibone 1953), and *M. baicalensis*, from Lake Baikal and other adjacent Siberian waters, that are exclusively fresh-water animals. The allied *Monroika africana*, from 150 km. up the Congo, beyond tidal effects, and *Caobangia billeti*, from fresh waters in Tonkin, are also possibly fully adapted fresh-water animals. Observations by J. P. Moore (Johnson 1903) indicate that adult individuals of *Fabricia sabella* can be acclimated to fresh water, whereas specimens of *Manayunkia speciosa* can be somewhat less perfectly acclimated to sea water.

The fifth fresh-water sedentary polychaet is *Marifugia cavatica* (Fig. 23), a member of the Serpulidae which inhabit calcareous tubes in underground waters of the karstic areas of Yugoslavia. In some localities frequented by the worm there is an underground connection to the sea, though the habitat of *Marifugia* appears to be hard but

---

[30] The unwary limnologist should note that *Fabricia limnicola* described in this paper is not a lake species but comes from estuarine intertidal mud flats. It is not clear whether a somewhat Homeric usage is intended or whether a *lapsus calami* has occurred.

---

west United States, a limpet-like member of a monotypic family, allied to the Lymnaeidae by its radula; *H. Rhodacmaea (Rhodocephala) filosa* ($\times$ 16), Mississippi drainage and adjacent areas, a minute fluviatile fresh-water limpet, with a pink apex to shell, a member of the Ancylidae, a family containing nearly all of the fresh-water limpet-shaped pulmonates; *I, Amphigyra alabamensis* ($\times$ 10); *J, Neoplanorbis tantillus* ($\times$ 10), members of two genera of Ancylidae endemic to the Coosa River, Alabama, in which the shell has a spiral form; *K–L, Tropicorbis havanensis* ($\times$ 1), a large ramshorn snail, a member of the fundamentally sinistral Planorbidae ranging from Louisiana into South America; *M, Helisoma anceps* ($\times$ 1½), a member of a widespread genus of Planorbidae; *N, Carinifex newberryi* ($\times$ 1), western United States, a planorbid in which the ultra-dextral shell has a well-marked spire, set to indicate its morphological relationship with the flat species of the family. (Various sources after Thiele and Clench.)

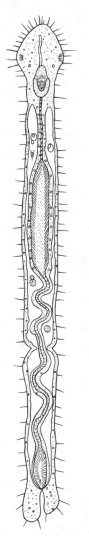

FIGURE 20. *Rheomorpha neiswestnovae,* a primitive worm of uncertain position, possibly an extremely primitive oligochaet, occurring in sand on the shores of Lago Maggiore and also the River Ora in Russia. Ruttner-Kolisko 1955.)

quite fresh water. In other localities a connection with the sea is less obvious, and presumably we may consider the species as another member of the peculiar cavernicolous fauna, living in fresh water but of marine origin, which is characteristically developed in the lands bordering the Mediterranean.

Among the free-living Polychaeta Errantia it seems certain that at least three and probably seven occur in fresh water. The three certain species belong to the subfamily Namanereinae of the Nereidae. Hartman (1959a, b) in her latest treatment synonymizes a number of previously separated species, admitting the following only. *Namanereis quadraticeps* (Fig. 24) is a cosmopolitan brackish species known inland in Sumatra in waterfalls (Feuerborn 1931 as *Lycastopsis catarractarum*) and even in wet leaves; another fresh-water population from the Dutch West Indies is admitted by Hartman (1959b) as a separate species *N. hummelincki. Namalycastis abiuma* has a wide distribution in brackish-water localities and has been recorded under a number of names from fresh waters in Hawaii (as *Lycastis hawaiiensis*), Sumatra (as *L. ranauensis*), and in the lower Amazon basin (as *L. siolii*). In at least the second (Feuerborn 1931) and third (Corrêa 1948, Sioli 1951) cases there is no doubt that the animal is fully adapted to fresh water, though in Sumatra it appears, like *N. quadraticeps,* to tend to a semiterrestrial habitat. The third species is *Lycastoides alticola* from the Sierra Laguna, California, at an altitude of 2000 m. (Johnson 1903). A fourth species *"Lycastis" geayi,* of uncertain generic position, may be a fresh-water species, known from the Ouanary River, French Guinea, (Gravier 1901). Further

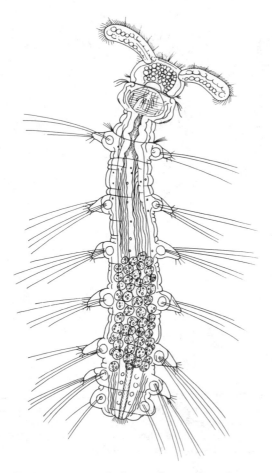

FIGURE 21. *Troglochaetus beranecki,* a hypogean fresh-water archiannelid of the family Nerillidae from the Swiss Alps. (Delachaux.)

work perhaps will indicate that some of the species synonymized by Hartman are worthy of some sort of taxonomic recognition. In *Namalycastis* there is a nectochaeta larva, with a few postprostomial segments, described by Feuerborn from fresh water in Sumatra. In *Namaneresis* a few very large eggs are produced, hatching late. Both genera appear to be hermaphrodite. Nothing seems to be known of the biology of *Lycastoides alticola* which as the sole representative of its genus, living in a montane habitat, is perhaps the most adapted fresh-water polychaet.

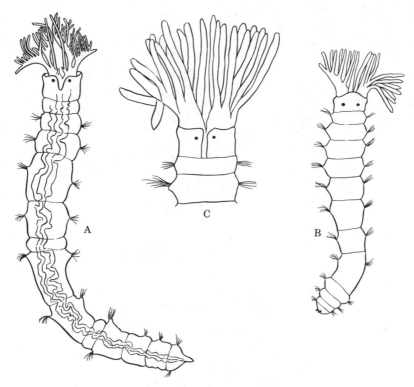

FIGURE 22.   Fresh-water species of *Manayunkia,* removed from their tubes.
A. M. baicalensis from Lake Baikal and other Siberian localities ($\times$ c. 10),
(Kozhov 1947), dorsal view; *B. M. speciosa,* a much smaller species from
eastern North America and the Great Lakes ($\times$ 33), ventral view; *C.* the
same ($\times$ c. 50), dorsal view (Pettibone 1953).

The remarkable Californian *Nereis (Hediste) limnicola* (= *lighti*)
was first described from Lake Merced near San Francisco, but it is
now known from a number of salt, brackish, and apparently fresh
localities in the same region (Hartman 1938; Smith 1950, 1953, 1958).
The water of Lake Merced contains 90 mg. Cl per liter, and though
it is regarded as fresh by Smith (the lake in fact is used as a reservoir)
this quantity of chloride is much greater than would be expected in
a nonpolluted inland water not draining from saline sediments (see
Vol. 1, page 561).   Though in its osmotic relationships *N. limnicola*
is almost a fresh-water animal, it is quite likely that it and some of the
other coastal fresh-water polychaets are limited to areas in which
the chloride content is enhanced by airborne sea salt.   *N. limni-*

cola is hermaphrodite, self-fertilizing, and viviparous, the minute but fully developed worms being liberated by localized rupture of body segments of the parent; the latter probably recovers from the process. Smith's important work on the temperature relations of osmoregulation in this species is discussed later (pages 164–166). There is a little evidence that racial differences in salt tolerance may occur in this worm.

Three species of *Nephthys* have been reported in fresh water; the evidence is best for *N. oligobrachia.* This species, first recorded from the chemically variable Chilka Lake, is known from a number of brackish waters and apparently in fresh-water localities in Central China (Okuda 1943); unfortunately, no analyses have been given of the fresh waters inhabited by the worm. An allied species, *N. polybrachia,* also known from Chilka Lake, is recorded by Fauvel (1932) from fresh water at Whaupoo, 13 km. below Shanghai. The third species,

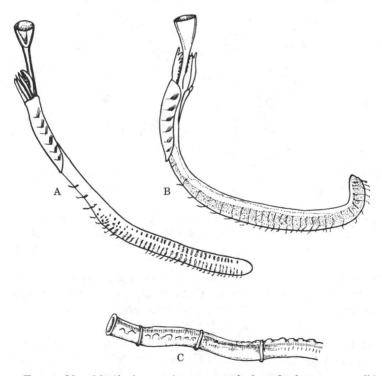

FIGURE 23. *Marifugia cavatica,* a cavernicolous fresh-water serpulid from the karst region of Yugoslavia. *A.* ♀ removed from tube (× 7); *B.* ♂ (× 8); *C.* apex of tube (× 6). (Absolon and Hrabe 1930.)

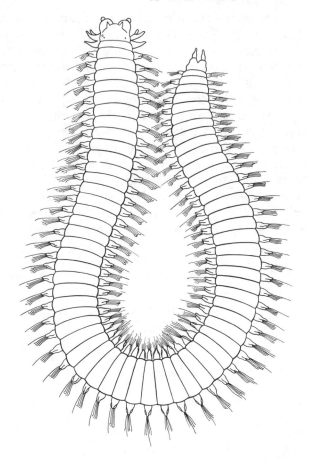

FIGURE 24. *Naṃanereis quadraticeps* (× 10). A
polychaet found in waterfalls and in wet leaves near
lake margins up to 500 km. from the coast in Java and
Sumatra (Feuerborn), but elsewhere in brackish water.

N. fluviatilis, was described by Monro (1937) from Arroyo de Pando,
which drains into the Río de la Plata in Uruguay. With it occurred
Leptonereis pandoensis and the capitellid *Heteromastus similis,* the
latter a Chilka Lake species, evidently of immense distribution. Monro
compares his locality, in which the water is said merely to be potable,
with the Ouanary River in French Guinea from which Gravier described
not only the *Namanereis* already discussed but the capitellid *Eisigella
ouanaryensis.* It is probable that both the African and South American
localities are at least as salt as Lake Merced.

Nineteen polychaets are recorded from Chilka Lake, of which eight occurred at times in water of specific gravity recorded as 1.000 Among the nineteen only *Heteromastus similis* and the two *Nephthys* occur in supposedly fresh water elsewhere. It is possible that the actual sites occupied by the worms were more saline than the water for which the density is given, and although some species (e.g., *Manayunkia spongicola* by Hartman 1951) are recorded as fresh-water polychaets by other authors, this is doubtful. Comparable situations are probably implied by the records of *Nereis seurati* from Gambir Island and the Aru Island and *N. nouhuysi* from the Celebes. We may also conclude that *Mercierella enigmatica,* recorded from supposedly almost fresh water in California and in Tunis (Seurat 1927), is merely very euryhaline but not a facultative fresh-water animal in the strict sense. It occurs in many brackish localities throughout the world, and in the hyperhaline Lake of Tunis (55.5 per mille of NaCl) the tubes of this worm form reefs dangerous to navigation. It may well be one of the most euryhaline of all animals of marine origin but apparently does not breed in fresh water (Hartman 1959c).

To the free-living polychaets occurring in fresh waters must be added the four species of *Stratiodrilus* (Fig. 25), epizoic or perhaps ectoparasitic on fresh-water malacostraca in the Southern Hemisphere. *Stratiodrilus tasmanicus* occurs in Tasmania, and *S. novae-hollandiae* in New South Wales, both on *Euastacus; S. haswelli* is found on *Astacoides* in Madagascar and *S. platensis* on *Aegla* in temperate South America (Harrison 1928). The allied *Histriobdella* lives on lobsters in northern seas.

The complete list of polychaets, which can be fairly certainly referred to the fresh-water fauna, thus appears to be

Nereiidae:	*Namanereis quadraticeps
	N. hummelincki
	*Namalycastis abiuma
	*Lycastoides alticola
	"Lycastis" geayi
	Nereis (Hediste) limnicola
Nephthyidae:	Nephthys oligobranchia
Histriobdellidae:	*Stratiodrilus tasmanicus
	*S. novae-hollandiae
	*S. haswelli
	*S. platensis
Sabellidae:	*Manayunkia speciosa
	*M. baicalensis
	Monroika africana
	Caobangia billeti
Serpulidae:	*Marifugia cavatica

Even in this list the species marked with an asterisk seem to have a superior claim to inclusion than the others.

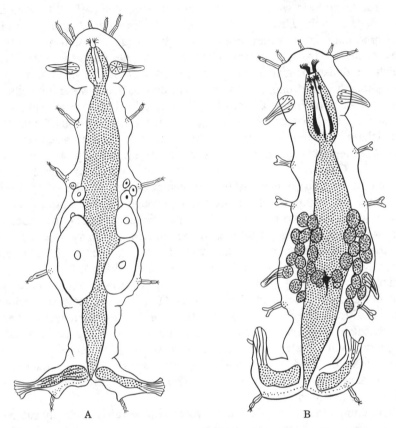

FIGURE 25.   A. *Stratiodrilus novae-hollandiae* from *Euastacus serratus,*
New South Wales; *B, C.* ♂ and ♀ *S. haswelli* from *Astacoides
madagascarensis,* Madagascar; *D. S. platensis* from *Aegla laevis,*
Uruguay. [Harrison 1928, *D* after Cordero *A, B, C,* (× c. 75); *D*
(× c. 50).]

The Clitellata are clearly a group with a long history in inland
waters.   The apparently very primitive Aeolosomatidae and the more
obviously oligochaetan Naididae appear to be exclusively fresh-water
animals.   A number of marine species of Enchytraeidae are known,
though the family is mostly terrestrial.   The families Tubificidae,
Lumbriculidae, Phreodrilidae, and Haplotaxidae are mainly fresh-water
with a few marine members.   The larger oligochaets or earthworms
are probably derived from a *Haplotaxis*-like fresh-water ancestor
(Stephenson 1930).   Several genera of the predominantly terrestrial
Glossoscolecidae, namely *Alma, Drilocrius,* and *Criodrilus* are fresh-
water animals living in the mud of swamps and lake bottoms.   All

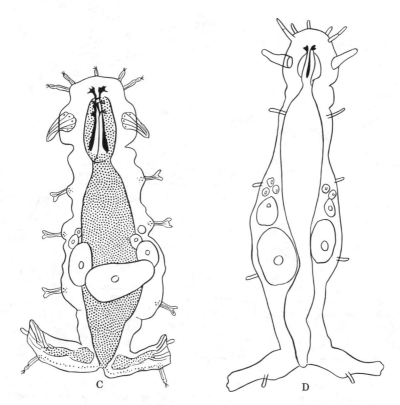

C   D

FIGURE 25. (*Continued*)

are regarded by Stephenson as secondarily aquatic, as is *Eiseniella* in the Lumbricidae. If the latter family were really derived from these aquatic forms, as Stephenson believes, it is secondarily terrestrial, and *Allolobophora lacustris* from the littoral benthos of Lake Ohrid is thus a tertiary aquatic animal. *Eiseniella,* which ordinarily occurs in swamps, is found quite deep in the benthos of this lake. There is at least one genus of truly terrestrial earthworms (*Plutellus*) which has invaded the sea.

The family Lumbriculidae is a particularly interesting group of mainly fresh-water oligochaets. It contains a number of species endemic to Lake Baikal, some with affinities to the fauna of Lake Ohrid, and some in endemic genera, among which *Agriodrilus vermivorus* (Fig. 26) is of great importance, for its structure gives a clearer idea of the origin of the leeches than does that of any other living form. The allied families Branchiobdellidae and Acanthobdellidae, parasites on fresh-water crayfish and on salmon, respectively, have been referred

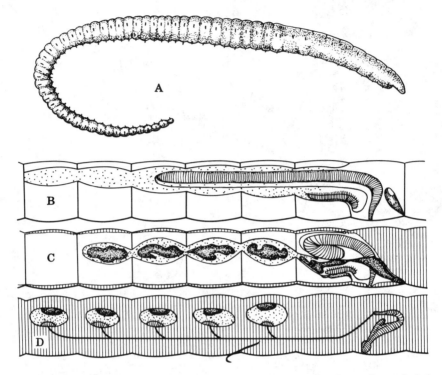

FIGURE 26. *Agriodrilus vermivorus,* a lumbriculid oligochaet from Lake Baikal, shows the probable mode of evolution of the leechs. *A.* whole worm natural size; *B.* diagram of the male genitalia of the related *Rhynchelmis* with testis anteriorly in segment XI and a large sperm sac produced backward, formed from the posterior wall of the segment; *C.* the same in *Agriodrilus* with the testicular tissue invading the sperm sac which is moniliform and the coelom of the first ten segments largely filled with connective tissue; *D.* the process completed in a leech with secondary small testes developed from the sperm sac and the entire coelom occluded. (Michaelsen.)

to the leeches by some authorities, to the oligochaets by others. Most of the leeches have remained fresh-water animals, some have become terrestrial, and a few genera of Ichthyobdellidae have entered the sea as fish parasites, having more continual relations with their hosts than is usual among the fresh-water forms. The osmotic relations of these animals should be investigated.

The environmental relations of the Arthropoda. The oldest known and only fossil onychophoran, *Aysheaia,* from the Middle Cambrian of British Columbia, seems certainly to have been marine, a number of specimens having been found in admirable condition in what is otherwise ordinarily regarded as a marine fauna. All modern species of Onychophora are terrestrial but require very low saturation deficiency of water vapor. The tardigrades are minute, and therefore presumably degenerate, but may be members of the same stock. They are best known as fresh-water animals, though they often frequent the water retained on wet moss and other minute aquatic habitats. A few aberrant and, in some respects, primitive forms occur in the sea, and it would not be unreasonable to look for an ancestry among the marine Protonychophora, as typified by *Aysheaia.*

The subphylum Insecta, which includes the classes loosely spoken of as myriopods and insects, appears from the embryology and morphology of its lower members to have close affinities with the Onychophora. The main history of the Insecta has clearly taken place on land, and all the aquatic species have an obvious terrestrial ancestry.

The Myriopoda except for a few marine littoral centipedes (Cloudesley-Thompson 1948), are all terrestrial. Among the Hexapoda a number of species of Collembola, a group that is on the whole very sensitive to low humidity, are associated with the surface films of both fresh and salt water and one or two are true marine animals descending below the tide marks. The other primitive wingless groups of Hexapoda appear to lack aquatic species.

Among the lower Pterygota the nymphal stages of three of the most primitive orders, the Ephemeroptera or mayflies, the Odonata or dragonflies and damselflies, and the Plecoptera or stone-flies, are characteristically and superficially not dissimilar fresh-water animals (*A–D,* Fig. 27). Among the endopterygote orders aquatic larvae occur in the Megaloptera, Trichoptera (*F, G,* Fig. 27) in the Sisyridae of the Neuroptera, in a few Lepidoptera, and in a great variety of forms in a number of families (*I–K,* Fig. 27) mainly nematocerous, of the Diptera. Among such Diptera the Culicidae and the Tendipedidae (= Chironomidae) are of peculiar importance.

In several families of both Hemiptera (*A–E,* Fig. 28) and Coleoptera there are aquatic adults as well as juvenile stages, though in the holometabolous Coleoptera the pupal stage is often passed in a chamber in the soil of the bank of the pond, lake, or stream inhabited by the larva and adult. On the whole, the aquatic Hemiptera (*F–L,* Fig. 28) both submerged and neustonic, are the most elaborately adapted fresh-water adult insects. Sporadic, more or less aquatic adults, at least of one sex, occur in some other orders, the most interesting being those parasitic Hymenoptera (*M* of Fig. 28) that enter the water to oviposit on aquatic hosts. Thorpe (1950) gives a fascinating account of some of the more peculiar adaptations involved.

The vast assemblage of fresh-water insects, of which the larger families contain hundreds of species, contrasts strikingly with the feeble

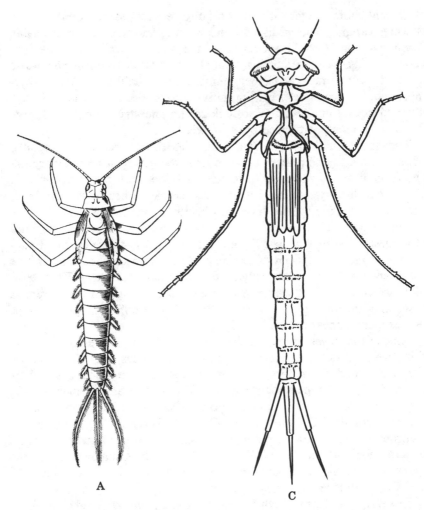

FIGURE 27. Aquatic nymphs and larvae, showing a certain similarity of form in
the lower orders, including the endopterygote Neuroptera, but an extraordinary
diversity of form, rivaling the adult Heteroptera, in the larvae of the Diptera
and to some extent the Coleoptera. *A. Phthartus rossicus*, Permian of Russia,
a fossil mayfly (ephemeropterous) nymph, a very primitive type with eight
pairs of simple gills; *B. Somatochlora metallica* ($\times$ 4), nymph of a boreal
European dragonfly (Odonata Anisoptera); *C. Enallagma cyathigerum* ($\times$ 5),
nymph of a common Palaeartic zygopteran, a damsel fly of American writers,
though the adult of this species is, as its trivial name suggests, certainly Rosetti's
"dragonfly . . . like a blue thread loosened from the sky"; *D. Chloroperla cydippe*
($\times$ 12), nymph of a North American stonefly (Plecoptera); *E. Sialis* sp. ($\times$ 3½),
larva of a widespread genus of the Neuroptera Sialodea; *F. Leucotrichia* sp.,
considerably enlarged, curiously physogastric larva of a caddisfly (Trichoptera)

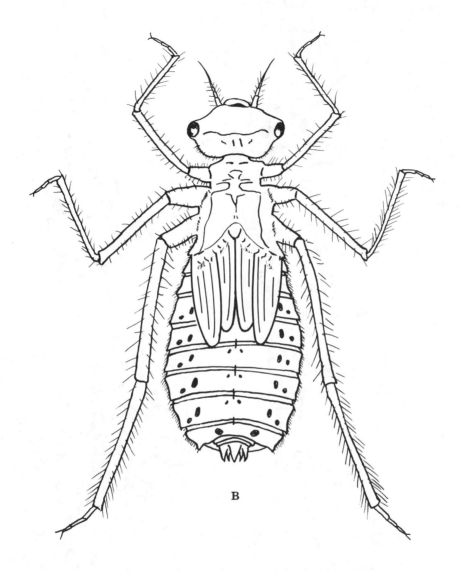

B

of the family Hydroptilidae; *G*. case of the same; *H. Ectopria* sp. larva of a beetle of the family Psephenidae, showing extraordinary convergence in form with trilobites; *I. Deuterophlebia* sp., larva (Diptera) found locally under stones in torrential mountain streams in the Holarctic; generally fairly pale with two rows of suckers on prolegs; *J. Neocurupira hudsoni*, New Zealand, with a single median set of suckers, a member of the widespread Blepharoceridae (Diptera), found, like the Deuterophlebiidae, in mountain torrents, but ordinarily on the surface of stones, and, deeply pigmented, this specimen bears the commensal and typically vermiform tendepedid larva of *Dactylocladius commensalis; K. Eristalis tenax* larva or rat-tailed maggot (Diptera), with posterior respiratory tube cf. Nepidae in Fig. 28. (Handlirsch, Lucas, Claasen, Ross Gurney and Parfin, Pulikovsky, Tonnoir, Miall.)

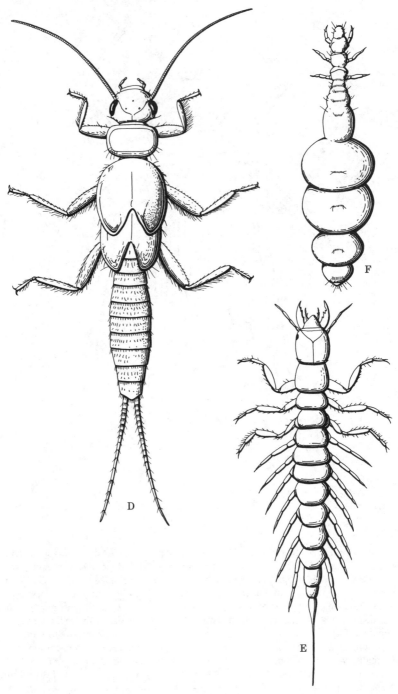

FIGURE 27. (*Continued*)

82

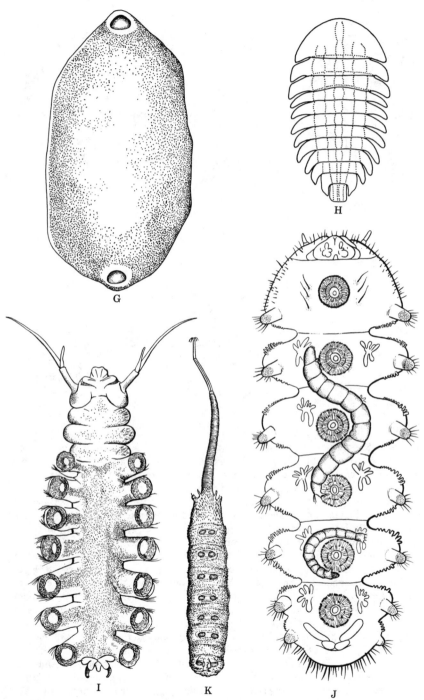

FIGURE 27. (Continued)

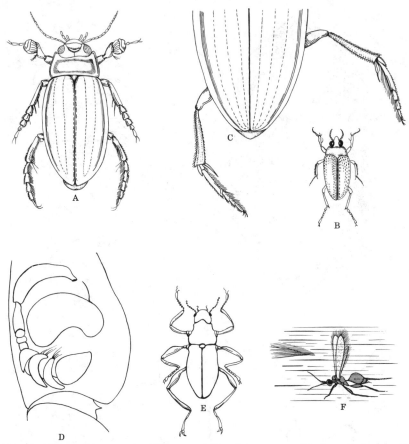

FIGURE 28. Adult aquatic insects. *A–E*. Coleoptera, showing moderate diversity of form with considerable variation in other aquatic specializations and great number of species (about 760 fully aquatic in North America). *G–M*. Hemiptera showing greater diversity of form in a group of fewer species (about 350 fully aquatic in North America); the same generalization prevails in the terrestrial species. *F*. Hymenopteran representing the very few adult aquatic insects outside the beetles and true bugs. *A. Dytiscus marginalis* ♂ Europe, posterior legs moved together; *B. Hydrophilus piceus* (× ¾), Europe and Western Asia, posterior part showing legs used alternately as in terrestrial walking (diagrammatic, from memory); *C*. antenna of same, used to break surface film and put air reservoir in connection with atmosphere; *D. Cnemidotus caesus* (× 2.5) of the family Haliplidae, small, bottom-living and feeble swimmers; *E. Ancyronyx variegatus* a member of the Elmidae, a family of small walking species living under waterlogged wood, stones, and the like, with plastron respiration; *F. Notonecta lactitans* (× ³⁄₂), vicinity of Cape Town, South Africa, breeding in winter in lowlands and migrating to mountains in summer, an outlying member of a widespread genus; *G. Benacus griseus* (× 0.35), North America, one of the largest aquatic hemiptera; *H. Abedus* sp., North America, ♂ carrying eggs glued to clytra, a habit found in several genera of Belostomidae; *I. Ranatra fusca* (× 0.35), a North

84

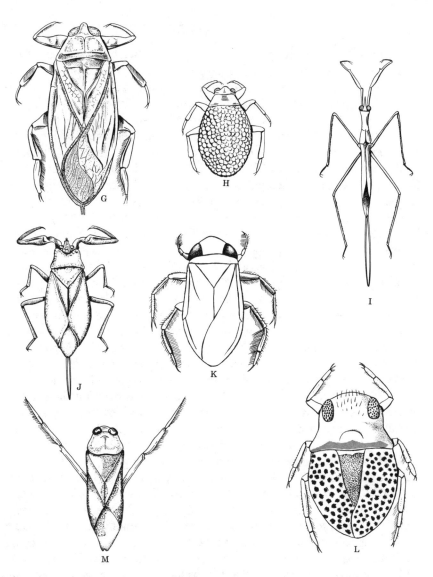

American member of an elongate genus, procryptically resembling a piece of stick, in fact a water stick-insect; *J. Nepa apiculata* ($\times$ ¾), eastern North America, a contrasting member of the same family Nepidae, commonly known as a water scorpion, with raptorial front legs and posterior breathing tube; *K. Diapreporis baryphala* ($\times$ 3), South Australia; insects of this genus are a favorite food of *Ornithorhynchus* and are very primitive members of the cosmopolitan, specialized, and very numerous family Corixidae; *L. Paskia minutissima* ($\times$ 20), littoral of Lake Tanganyika, a member of the endemic subfamily Idiocorinae of the family Helotrephidae; apneustic and with the head fully fused to the thorax dorsally; *M. Cataphractus cinctus* ($\times$ c. 7), supposedly from both Europe and North America, one of the few acquatic Hymenoptera, swimming with its wings and as a larva parasitizing the eggs of *Notonecta*. (Miall, Leach and Sanderson, Hungerford, Esahi and China, and original.)

development of the group in the sea. A number of Coleoptera (von Lengerken 1929) occur in the intertidal zone. The European *Aepus* and *Aepopsis* and the South American *Thalassotrechus,* all of the sub-family Trechinae (family Carabidae), and the staphylinid *Micralymma* may be mentioned. Most of these beetles are related to terrestrial rather than aquatic forms, though hydrophilids occur in saline coast pools and some accidental occurrences at sea are known (summary in Gunter and Christmas 1959). The marine hemipteran *Aepophilus* found on the coast of western Europe intertidally and subtidally under stones certainly belongs to the family Saldidae and so is allied to animals living mainly on very wet mud rather than to the truly aquatic bugs (Leston 1956). There is a possibility (Sailer 1948) that *Tricho-corixa reticulata* (Hemiptera, Corixidae), which lives in saline waters in western North America, and *T. verticalis,* which has been found alive in the sea (Hutchinson 1931; Gunter and Christmas 1959), have been able to colonize distant islands in part by drifting across the ocean. *T. verticalis* may occur in tide pools, and the lack of success in colonizing the sea is presumably caused by biotic rather than chemical factors. Numerous surface bugs (*Halovelia, Trochopus,* and *Hermatobates*) are found in coastal waters; *Halobates* is a tropico-politan pelagic form that lives on the ocean surface quite remote from land. All are doubtless derived from species that lived on the surfaces of fresh waters. Whether they have any sensory equipment, such as tarsal chemoreceptors, which indicate to the insect that it has or has not left salt or fresh water, is not known.[31]

A considerable number of dipterous larvae inhabit salt water, and, particularly among the Tendipedidae, species passing their early stages in the littoral zone of the sea have been discovered. (See Edwards 1926). One extraordinary case of a completely marine member of this family is known. *Pontomyia natans* lives among *Halophila* plants in the lagoon within the fringing reef at Apia, Samoa. The adults of both sexes are entirely aquatic. The male is motile and swims with its elongate anterior and quite well-developed posterior legs; the wings are reduced in a curious way, reminiscent of some *Drosophila* mutants, and the intermediate legs are short. The female is vermiform. The larvae clearly exhibit affinities with *Calopsectra.* Buxton (1926), who indicates, not quite accurately, that the Diptera alone among insects

---

[31] As far as experiments on the surface of tap water in a tooth glass, performed in a hotel at Panjim, former Portuguese India, are significant, it would seem that *Halobates* is not discommoded by a fresh-water substratum. This matter needs reinvestigation; the converse experiment may prove interesting.

can stand salinity as great as that of the sea, supposes that the group has had little success in invading the ocean because the food available there is so different from that of terrestrial or fresh-water habitats, dominated by spermatophytes. He considers that it is probably significant that the only fully submarine insect should be associated with one of the few marine flowering plants.

The Chelicerata, which include the living Xiphosura and Arachnida as well as a great diversity of fossil forms, are almost certainly derived from the Trilobitomorpha, a group also highly diversified in the early Palaeozoic and, as far as is known, entirely marine.[32]

The Xiphosura have remained marine to the present time, though *Carcinoscorpius* is said to enter fresh water in the Ganges delta (Annandale 1922) and must be very euryhaline, presumably with considerable osmoregulatory capacity. The Eurypterida are now usually regarded as having come from the sea. The Ordovician and some Silurian species were marine, but from the Devonian until the extinction of the group in the Permian, the Eurypterida were almost entirely confined to fresh water (Kjellesvig-Waering 1958). The most primitive Arachnida are clearly related to the Eurypterida, and at least some of the Paleozoic scorpions seem to have been marine; the evidence is well discussed by Størmer (1963). All existing aquatic Arachnida are, however, of obviously terrestrial origin. Apart from the mites (Fig. 29), represented mainly but not exclusively by the Hydrachnellae in fresh and by the Halacaridae in salt water, the number of such aquatic arachnids is limited. The most perfectly adapted is the famous palaearctic spider *Argyroneta aquatica* (Fig. 30), which lives permanently below the surface of fresh-water ponds in the palaearctic, storing air in a web beneath the water. A number of ground-living species of spiders run out onto the surface of quiet water, but none seems to be exclusively neustonic as are many Hemiptera. Several intertidal and partly neustonic marine spiders and one false scorpion are known (Bristowe 1930, 1931).

The fresh-water Crustacea. The final subphylum of living Arthropoda, containing the single class Crustacea, has remained primarily aquatic, and we may suppose that all or almost all of the numerous fresh-water representatives of the group have had marine ancestors without intervening terrestrial stages. Because of the importance of the group to the limnologist, a rather detailed account of the evolution is desirable.

---

[32] The problematic Arthropleurida of the late Palaeozoic were presumably fresh-water or terrestrial, but their affinities are obscure.

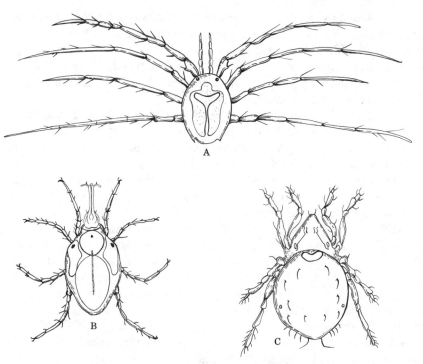

FIGURE 29.   Aquatic mites.   *A. Unionicola crassipes,* Europe, one of the more
limnetic species of the aquatic superfamilies (Hydrachnellae) of Parasitengona,
(× 32); *B. Lohmannella falcata,* Europe, a fresh-water member of the Halacaridae (× 50); *C. Hydrozetes lacustris,* Holarctic, a fresh-water member of
the Oribatei, (× 70).   (Soar, Newell, Michael.)

Origin and classification of Crustacea.   The origin of the Crustacea
is still obscure.   The segmentation and superficial structure of the appendages of the anterior end seem easily homologized with those of
the Tracheata, but even here the resemblance of the crustacean and
insect mandibles is that of analogy rather than homology (Manton
1964) and in the postcephalic structures little close relationship can
be made out.   Tiegs and Manton (1958) in their recent and very
learned review tend to suggest a closer relationship between the
Crustacea and Trilobitomorpha than between the Crustacea and the
Insecta, and it is by no means impossible that some of the fossils
placed as Pseudocrustacea and supposed to have a trilobitan limb are
nearer to the Crustacea than has often been believed.
    Much of the difficulty inherent in the problem of the origin and
early phylogeny of the Crustacea is due to the scarcity of well-studied

FIGURE 30. *Argyroneta aquatica,* the Palaearctic water spider in its air-filled web.

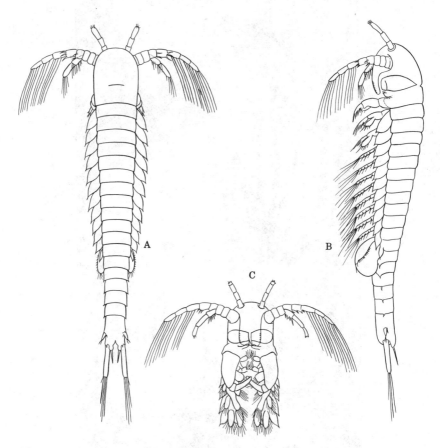

FIGURE 31. *Lepidocaris rhyniensis. A.* female from dorsal surface; *B.* the same lateral; *C.* male ventral ($\times$ c. 30). (Scourfield.)

and clearly interpretable fossils. At the present time, on purely morphological grounds, the two most primitive known Crustacea appear
to be the allegedly Devonian *Lepidocaris rhyniensis* (Fig. 31), which
presumably inhabited a peculiar chert-depositing water, fed perhaps by
a mineral spring, and the four living members of the Cephalocarida
from mud bottoms in shallow water on both coasts of North America,
the Caribbean, and Japan. Even *Lepidocaris* is far too young to be
ancestral; paradoxically, some of the features by which the Cephalocarida differ from *Lepidocaris* appear to suggest that these recent forms
are morphologically far more primitive than the Devonian fossil.

It is reasonable to suppose, as Tiegs and Manton have emphasized, that the ancestral crustacean was a small animal, though perhaps larger than the primitive genera just mentioned. No animal that was not moderately well cephalized would be admitted to the class. Sanders (1963) believes that the Cephalocarida have antennules, antennae, mandibles, and first maxilla which have undergone modification in the adult, whereas the second maxillae are represented by unmodified trunk limbs. In this respect the group differs markedly from all other living Crustacea and from *Lepidocaris*.

In both *Lepidocaris* and the Cephalocarida there is a postgenital region which lacks appendages. The evidence from the Malacostraca and perhaps from the Notostraca suggests that initially the abdomen bore appendages. There is a hint in *Lepidocaris* that the caudal furca may represent such an appendage, with a rudimentary appendage on the penultimate segment. It is usual, however, to regard the furca as developing from the postsegmental telson. The marine Devonian *Vachonia* (Lehmann 1955) seems to have a complete series of appendages to the end of the trunk, but since the trunk does not extend beyond the carapace it may be reduced from a notostracan condition.

The trunk presumably was originally a narrow elongate structure, probably with well-developed pleura on each pedigerous segment. If the head was wide, as in the Cephalocarida and the whole trunk both pedigerous, as in the Malacostraca, and with pleura throughout, the primitive crustacean was doubtless quite trilobite-like in appearance.

The great divergence in opinion has related to the appendages. It is reasonably certain that the free-swimming filter-feeding forms with typical phyllopodia are not primitive. The discovery of *Lepidocaris*, in which the posterior appendages are simple biramous structures comparable to those of copepods and the anterior appendages are progressively more phyllopodan, at first suggested the correctness of the view that the primitive form had a simple biramous appendage, with both rami set at the apex of a protopodite of more than one podomere. A new possibility is opened up by the Cephalocarida. Here the endopodite is a well-developed structure on the limbs of the first seven trunk segments. It consists of a ramus of six well-defined podomeres, the most distal of which bears three large spine-like structures. The middle one is freely flexible and a muscle is said to enter it; it is therefore possibly a seventh podomere. The endopodite is attached to an undivided protopodite which carries several endal lobes to form a gnathobase and which may well be equivalent to several podomeres. On the outer side of this protopodite is inserted an exopodite apparently of four podomeres, of which the proximal bears a pseudepipodite. The general structure of this limb is nearest that of the Syncarida, which are the most generalized Malacostraca. Sanders thinks that the so-called biramous appendages of the posterior pedigerous segments of *Lepidocaris* are derived from this type of limb by a suppression of the true endopodite, as indeed occurs on the eighth trunk segment of *Hutchinsoniella* in the Cephalocarida. The phyllopodan limb is then developed from this reduced type. The same process was supposed, perhaps rather less probably, to have occurred in the Copepoda, the original exopodite becoming the "copepodan endopodite" and the pseudepipodite, in some cases after jointing, becoming the "copepodan exopodite." In the malacostracan biramous appendage the original endopodite and exopodite are retained. Functionally, in *Hutchinsoniella* the role of the spinous endopodite is to stir the superficial layer of sediment, edible particles from which are pushed toward the mouth by the gnathobases. In a free-swimming filter-feeder this function disappears and the endopodite becomes unnecessary. In benthic forms, notably the Ostracoda and Malacostraca, it is retained as a walking leg.

TABLE 2. *Classification of the Crustacea*
(Fossil forms entirely *incertarum sedium* omitted)

Subclass CEPHALOCARIDA
Subclass MALACOSTRACA
Division Phyllocarida
    Order Leptostraca
    Order †Nahecarida
Division Syncarida
    Order †Palaeocaridacea
    Order Anaspidacea
    Order Bathynellacea
Division Pancarida
    Order Thermosbaenacea
Division Peracarida
    Order Spelaeogriphacea
    Order Mysidacea
    Order Cumacea
    Order Tanaidacea
    Order Isopoda
    Order Amphipoda
Division Eucarida
    Order Euphausiacea
    Order Decapoda
        Suborder Natantia
        Suborder Reptantia
            Section Palinura
            Section Astacura
            Section Anomura
            Section Brachyura
Division Hoplocarida

TABLE 2. *Classification of the Crustacea (Continued)*

Subclass BRANCHIOPODA	Order †Lipostraca
	Order Anostraca
	Order Phyllopoda
	Suborder Notostraca
	Suborder †Acercostra ca
	Suborder †Kazacharth ra
	Suborder Conchostraca
	Suborder Cladocera
Subclass MAXILLOPODA	?Order †Archicopepoda
	Order Copepoda
	Suborder Calanoida
	Suborder Misophrioida
	Suborder Monstrilloida
	Suborder Harpacticoida
	Suborder Cyclopoida
	Suborder Notodelphyoida
	Suborder Caligoida
	Order Branchiura
	Order Mystacocarida
	Order Cirripedia
	Suborder Thoracica
	Suborder Ascothoracica
	?Suborder Apoda
	Suborder Rhizocephala
Subclass OSTRACODA	Order †Leperditacea
	Order †Beyrichiacea
	Order Myodocopa
	Order Podocopa

Whatever the exact details of the earliest crustacean, it is reasonable to suppose, with Tiegs and Manton (1958), that it was marine, small, benthic, and not a filter-feeder on plankton. Dahl (1956, 1962) considers that four lines (Table 2) diverged from such an animal. The first line gave rise to the Malacostraca which form a coherent subclass. Another line led through *Lepidocaris* (Lipostraca) to the Anostraca or fairy shrimps and presumably also to the three groups which Dahl admits to the Phyllopoda (cf. Preuss 1951), namely the Notostraca, Conchostraca, and Cladocera; all of these animals in which there is a tendency for the originally numerous trunk limbs, primitively with gnathobases, to form filter chambers, are placed in a subclass Branchiopoda. A third branch, which constitutes Dahl's subclass Maxillopoda, without gnathobases on the thoracic limbs and with the maxillules and maxillae forming the filter in filter feeders, evolved perhaps through the extinct Archicopepoda and gave rise to the Copepoda, with the

Branchiura, the peculiar interstitial marine Mystacocarida, and possibly the Cirripedia as side branches. The fourth subclass of Dahl's classification contains only the Ostracoda, the origin of which group is entirely obscure. Dahl's classification, slightly emended as the result of Sanders' most recent work, appears to be much the most natural one yet proposed.

The subclass Malacostraca is usually divided into two series, the Leptostraca and the Eumalacostraca. The propriety of this division, which is based formally on the Leptostraca having seven abdominal somites, the Eumalacostraca six, is dubious, since the most primitive mysids have seven embryonic somites. It is doubtless best, following Siewing (1959), to put the Leptostraca in a division Phyllocarida coordinate with the other divisions of the Malacostraca. The few living species are marine. The fossils referred to the order Nahecarida, which have allegedly five to nine abdominal segments, lived in the Palaeozoic and early Mesozoic seas and so need not be considered here.

The rest of the Malacostraca are now arranged in (Siewing 1958a, 1959) five divisions, the Syncarida, Pancarida, Peracarida, Hoplocarida, and Eucarida. Not all modern workers (e.g., Barker 1959) accept Siewing's Pancarida for the Thermosbaenacea, and recent discoveries certainly have tended to blur the distinctions that might be made between the first three divisions.

The Syncarida. The Syncarida, which in their appendages appear to be the most primitive Malacostraca, consist of three orders, the Palaeocaridacea, the Anaspidacea, and the Bathynellacea (Brooks 1962). The Palaeocaridacea (*A* of Fig. 32) are exclusively fossil, with eight free thoracic and six abdominal segments. They appeared in the Upper Mississippian and extend to the Permian; all but perhaps the latest members of the group are believed to have been marine. The Anaspidacea appear in the Permian of Brazil where *Clarkecaris brasilicus* is marine; the only other fossil member of the order, *Anaspidites antiquus,* is from a fresh-water Triassic formation in Australia. There are three living families, the Anaspididae of Tasmania, the Koonungidae (*D* of Fig. 32) of both Tasmania and the mainland of Australia, and the Stygocaridae (*E* of Fig. 32), hypogean animals from the southern Andes and their foothills (Noodt 1963). This exclusively austral order differs from the northern fossil Palaeocaridacea in that it has the first thoracic segment fused to the head. *Anaspides* (*B* of Fig. 32) and *Paranaspides* (*C* of Fig. 32) are large shrimp-like forms, *Paranaspides* being limnetic; the other two families contain small and cryptozoic if not hypogean animals. Although they are clearly somewhat reduced, the Stygocaridae, in the presence of a minute rostrum, a cleft telson with indications of a furca, and a lacinia mobilis or comparable process on the mandible, are reminiscent of Malacostraca outside the Syncarida. The third order, the Bathynellacea, are minute hypogean (*F* of Fig. 32) or interstitial animals found sporadically in Eurasia, Africa, and South, but so far not in North, America. They clearly represent an early divergence from the other Syncarida, since they have retained a furca and eight free thoracic segments, though in other ways they are much reduced. Six genera have been described, all from fresh water except *Thermobathynella amyxi* which comes from a brackish interstitial habitat at the mouth of the Amazon.

The Pancarida. The Pancarida are a group consisting only of two genera (*A, B,* Fig. 33) the thermobiont *Thermosbaena* from the hot-springs of Tunis, and *Monodella* from Italy, Yugoslavia, Palestine, and Texas. All are minute, hypogean, blind, animals, some species of which are tolerant of saline water.

The Peracarida. The Peracarida, a large and very important group of crustacea with a great number of fresh-water species, are divided at present into six orders, the exact affinities of which are not easily determined. Apart from the curious Spelaeogriphacea, erected for *Spelaeogriphus lepidops* (*A* of Fig. 34), a peculiar hypogean fresh-water species from a cave on Table Moutain, Cape Town, Union of South Africa (Gordon 1957), all the orders contain far more marine than fresh-water forms; in the Cumaceae and Tanaidaceae it may in fact be doubted whether any cases of complete adaptation to fresh waters are known.

The fresh-water Mysidacea. The order Mysidacea, which presumably represents a relatively primitive stock[33] in the Peracarida, includes nearly four hundred living species, of which there appear to be more than fifty[34] known from brackish and about twenty-five from fresh water, or water of salinity less than 0.5 per cent.

In spite of these rather impressive statistics, given by Stammer and roughly corrected for later discoveries, it may be doubted whether most of the fresh-water species are really fully adapted. Many are quite euryhaline and some appear to be restricted to fresh-water localities near the coast, where the problem of sodium chloride intake is likely to be easier than further inland.

---

[33] Judgment of the primitiveness of the Mysidacea depends somewhat on whether the carapace is regarded as a primitive feature of the Malacostraca.

[34] The brackish- and fresh-water species known up to 1936 are enumerated by Stammer (1936) and later American records are given by Banner (1953). See also Tattersall (1951).

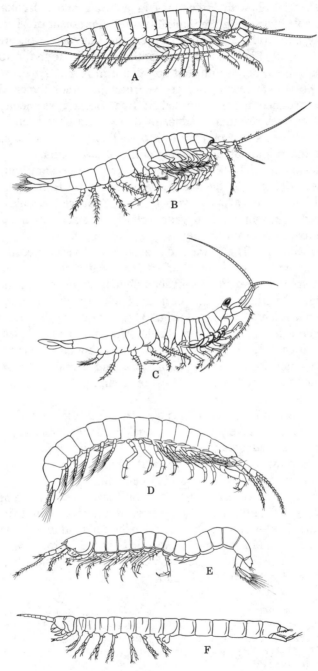

FIGURE 32.   For descriptive legend see opposite page.

The most widely distributed fresh-water mysid is certainly *Mysis relicta,* now (Holmquist 1959) regarded as specifically distinct from the brackish and marine arctic *M. oculata (A–E,* Fig. 67). *M. relicta* has entered lakes in many parts of the Holarctic region formerly connected with or covered by proglacial lakes. Its distribution and ecology are discussed on a later page.

Eleven species, mostly of *Paramysis,* extend varying distances up river from the Ponto-Caspian basins; some are clearly fully adapted fresh-water animals, as indicated in the discussion of the Ponto-Caspian crustacean fauna (page 217).

Most of the other species recorded from fresh waters have coastal distributions. The best studied case is that of *Neomysis integer* (= *vulgaris*) which appears to occur in Europe in water of salinity 0.1 per cent up to normal sea water, though it is not common or successful in fully marine environments (Tattersall and Tattersall 1951). It is known in coastal lakes that otherwise contain an exclusively fresh-water fauna. It is reasonable, however, to suppose that these localities are somewhat enriched in salt by windborne sea spray (Volume I, pages 544–546). Other species such as North America, *N. intermedia* from eastern Siberia to Japan, and *Acanthomysis awatchensis* from Japan, China and the Pacific Coast of North America behave similarly. The same is probably true of the European *Mesopodopsis slabberi,* of *M. orientalis* from the regions bordering the Bay of Bengal, and of *Diamysis bahirensis* from the eastern Mediterranean, in Lake Scutari. *Diamysis americana* occurs in ditches in Dutch Guinea (Tattersall 1951). *Teganomysis novae-zealandiae* in brackish waters on the New Zealand coasts has been recorded from Lake Waikare. Several species of monotypic coastal genera, *Gangemysis assimilis* in the Ganges, as much as 960 km. from the ocean but mainly in brackish water, *Nanomysis siamensis* in slightly brackish and fresh waters in Siam, and *Taphromysis louisianae* from a sup-

---

FIGURE 32. The Syncarida. *A. Acanthotelson stimpsoni* (× 3.3), restoration, a fossil syncarid of the order Palaeocaridacea, Pennsylvania of Illinois, apparently a marine animal; *B. Anaspides tasmaniae* (× 2), pools and streams of the central plateau of Tasmania; *C. Paranaspides lacustris* (× 2), shallow benthic region of the Great Lake of Tasmania. *D. Koonunga cursor* (× 7.3), from small pools near Melbourne, Australia, a member of the Koonungidae of the order Anaspidacea; *E. Parastygocaris andina* (× 10), interstitial water in banks of streams, Uspallata, Argentina, a member of the Stygocaridae of the order Anaspidacea. *F. Parabathynella fagei* (× 24), a southern European member of the hypogean order Bathynellaceae. (Brooks, Manton modified, Calman, Noordt, Delamare-Debouteville.)

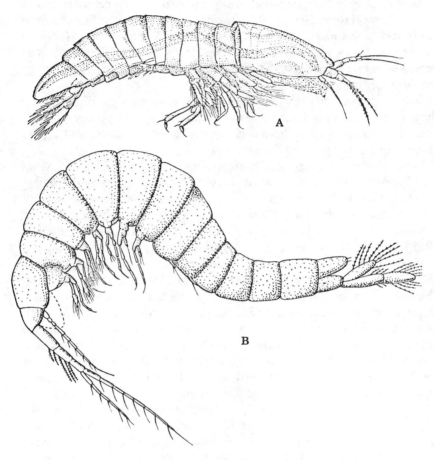

FIGURE 33. Members of the two genera of Thermosbaenaceae. *A. Thermosbaena* (× 14), El Hamma, Gabes, Tunis; *B. Monodella stygicola* (× 33), L'Abisso, Castromarina, Italy. (Delamare-Deboutteville after Absolon and Bruun.)

posedly fresh-water ditch in Louisiana (Banner 1953), are known; these species seem to be well adapted, but it may be doubted if all can tolerate the low sodium chloride content of ordinary inland fresh waters.

Finally, a few Mysidacea are known in caves. *Heteromysis cotti,* with reduced eyes, is recorded in a tidally fluctuating pool in a cave 6 km. from the coast of Lanzarote, Canary Islands. The blind *Lepidops servatus* occurs in brackish water in a collapsed cave in

Zanzibar; the related *Spelaeomysis bottazzii,* also a blind form, was found in slightly salt water in a cave near Otranto, Italy. In the same region the specialized *Stygiomysis hydruntina,* made the type of an independent family by Caroli (1937), is found; a second species (*B* of Fig. 34) is known in the West Indies (Gordon 1960). An interesting pair of species (Tattersall 1951) occurs in Central America. *Antromysis anophelinae* occurs in the holes of a shore crab above high tide mark in Costa Rica, probably in brackish water; it apparently lives in the dark but has pigmented eyes. *A. cenotensis*

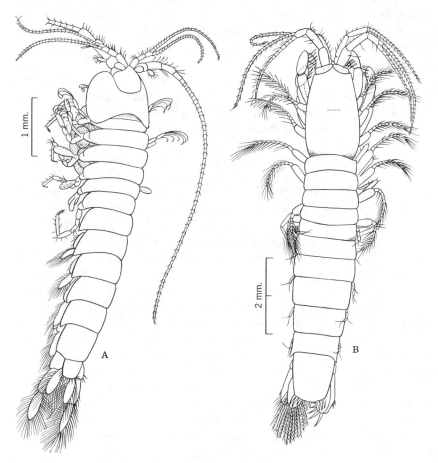

FIGURE 34. *A. Spelaeogriphus lepidops,* the sole representative of the Spelaeo-griphaceae; *B. Stygiomysis holthuisi,* an aberrant cavernicolous mysid from the West Indies, congeneric with a species in southern Italy. (Gordon.)

is a fully hypogean fresh-water species from Yucatan without eye pigment. *Troglomysis vjetrenicensis* (Fig. 64) from quite fresh water in a cave near Zarala, Yugoslavia (Stammer 1936) is certainly a fully adapted fresh-water animal. The occurrences of these cavernicolous mysids are of considerable interest in that they suggest progressive stages in the evolution of fresh-water cave animals from marine ancestors. *Lepidops servatus* and *Spelaeomysis bottazzii* belong to the Lepidopidae; all other existing fresh- and brackish-water mysids except *Stygiomysis* belong to the Mysidae.

It is quite possible that early in the history of the Mysidacea the more primitive suborder Lophogastridea, now exclusively marine, had fresh-water representatives. A number of late Palaeozoic fossils appear to belong to the Lophogastridae; at least *Notocaris* from the upper Dwyka shale near Kimberley, South Africa (Broom 1931), was probably a fresh-water genus.

The general picture presented by the Mysidacea is one of apparent easy penetration into brackish or *almost* fresh water but striking lack of success as fully adapted fresh-water animals.

Supposed fresh-water Cumacea and Tanaidacea. The Cumacea are well represented by species (Fig. 70) of the family Pseudocumidae in the Ponto-Caspian basins and occur in the almost entirely fresh waters of the Volga delta; they are, however, apparently unable to move upstream (page 216). A single species *Lamprops korroensis* which occurs along the coasts of Kamchatka, Southern Sakhalin, and the northern Kurile Islands is said to live in fresh water (Derjavin 1930; Uéno 1936) in some of its stations; a situation similar to that of some of the euryhaline brackish species of *Neomysis* is probably indicated by the data.

The Tanaidacea are represented in fresh water by three or four species. *Tanais fluviatilis* known from the Argentine, builds tubes of sand on the shells of Mutelid mussels (Giambiage 1923). *T. stanfordi* occurs in fresh-water lakes on Kunasiri in the Kurile Islands (Stephensen 1936; Uéno 1936), the same species also living in the brackish lagoon of Clipperton Island. *Nototanais beebei* is recorded from the stomach of the catfish, *Pimelodus clarias,* in British Guinea. There is perhaps a hypogean species in Madagascar (Paulian and Delamare-Deboutteville 1956). The environmental tolerances of all these animals require further study; most or all are probably dependent on slightly more saline water than is usual far from the coast.

The fresh-water Isopoda. The Isopoda are usually divided into eight suborders: the Gnathida, Anthurida, Asellota, Phreatoicidea, Flabellifera, Valvifera, Epicaridea, and Oniscoidea. The first named is a small exclusively marine group, the last an immense group of terrestrial animals. The other six suborders have numerous fresh-water representatives, which are clearly derived from many invasions.

The Asellota are partly marine (Stenetriidae and most Jaeridae) but include the important fresh-water Holarctic family Asellidae. This family, widely distributed in both North America and Palaearctic Eurasia, gives rise to a great number of local forms in southern Europe and a few in Lake Baikal and has a considerable number of hypogean species. Among the Jaeridae, *Jaera italica* is recorded from supposedly fresh water on the Dalmatian coast (Karaman 1953), and there is a fresh-water species from the Azores. Two minute blind genera, *Protojanira* from South Africa and *Heterias* from Victoria, Australia, may represent an old Austral invasion by this family. The subfamily Microparasellinae is a peculiar group of minute animals living in interstitial water. Some (*Microcharon, Microparasellus*) (*D, E,* Fig. 35), occur in ground water inland in southern Europe, but *Microcharon* is also found in marine interstitial waters both in the littoral zone and at moderate depths (Spooner 1959a). One genus, *Protocharon,* is recorded from a fresh-water spring, covered by the sea at high tide, on the island of Réunion, and some other marine littoral genera are known. The whole group (Chappuis, Delamare-Deboutteville, and Paulian 1956; Spooner, 1959a) appears to be characteristic of interstitial waters of a wide range of salinity with little differentiation between the species living in fresh and salt waters.

The Anthurida are represented in the subfamily Anthurinae by the hypogean *Crurigens fontanus* in New Zealand, by *Cyathura milloti,* probably not fully adapted, in the spring on Reunion just mentioned, and in the subfamily Microcerberinae by several minute species of *Microcerberus* (*C* of Fig. 35). This genus, which has fresh-water interstitial representatives in the Balkans, Italy, and North Africa, also has marine species on the coasts of the Mediterranean, West Africa, Madagascar, India, and the Bahamas. Chappuis, Delamare-Deboutteville, and Paulian (1956) call attention to the strong positive thigmotaxis of a small anthurid such as *Cyathura* and suggest that invasion of terrestrial groundwaters from the interstitial groundwater of the ocean would be encouraged by thigmotactic behavior.

The fresh-water members of the Flabellifera appear to consist of three ecologically different groups. There are first a certain number of species which are found, like *Neomysis integer* and other animals already discussed, in supposedly fresh water in the vicinity of coastlines. *Cirolana browni* from Cuba (Van Name 1936) may belong in this category, but most of the known cases refer to the Sphaeromidae.

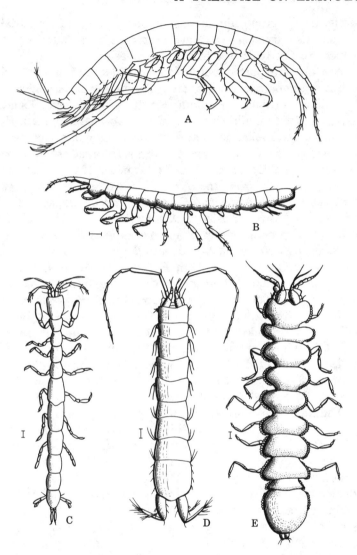

FIGURE 35. Minute fresh-water species of peracaridan Crustacea belonging to genera occurring in both marine and continental interstitial waters. *A. Bogidiella a. albertimagni* (Amphipoda: Bogidiellidae), Rhine Valley, Alsace (Hertzog 1933); *B. Ingolfiella acherontis* (Amphipoda: Ingolfiellidae), Skoplje, Jugoslavia (Karaman (1933b); *C. Microcerberus stygius* (Isopoda, Anthuridae), Skoplje, Yugoslavia (Karaman 1933a); *D. Microcharon latus* (Isopoda, Microparasellinae), Skoplje, Yugoslavia (Karaman 1940); *E. Microparasellus puteanus* (Isopoda, Microparasellinae) Skoplje, Jugoslavia (Karaman 1940). Natural lengths indicated by lines.

Unfortunately, the taxonomy of several of the most interesting species is somewhat confused. Menzies (1954) has examined the distribution and ecology of certain species formerly placed in *Exosphaeroma* but separated by him as *Gnorimosphaeroma,* among which *G. o. oregonensis* is a marine to brackish water animal with a salinity range of 9.1 to 30.9 per mille found along the Pacific coast of North America. *G. o. lutea* (Fig. 36) is described as a dilute water subspecies occurring from Alaska to Oregon in water of 0.48 to 2.20 per mille salinity. Menzies thinks it is conceivable that the two so-called subspecies are actually environmental modifications, but this seems rather unlikely. Riegel (1959b) believes that they are distinct species; since they are more or less sympatric, subspecific status is the least likely kind of difference between them. A marine intertidal species, *G. noblei,* is known on the coast of California, and a supposedly fresh-water species on San Nicolas Island, associated with *Physa virgata,* has been described as *insulare.* Other members of the genus are known on the Asiatic coasts

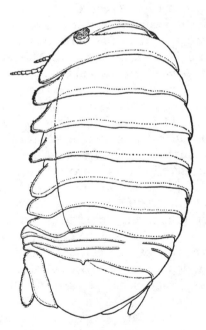

FIGURE 36. *Gnorimosphaeroma oregonensis lutea,* a fresh-water sphaeromid from the northwest coast of North America (× 30). (Menzies.)

of the Pacific, *G. ovata* being marine from the Japanese Sea and *G. chinensis* fresh-water from Shanghai. A species referred to *oregonensis* but probably actually new (Menzies 1954) was associated with *Tanais stanfordi* on Kunasiri in the South Kurile Islands (Uéno 1936) and presumably has ecological preferences comparable to these of *lutea*.

Riegel (1959a) finds that *G. o. lutea* from both fresh and estuarine waters is much more euryhaline than *G. o. oregonensis,* which cannot live in fresh water. When *G. o. lutea* is kept in supposedly fresh water at 16° C., the body fluids lose salt and become isosmotic with about half-strength sea water, whereas *G. o. oregonensis* is strikingly more poikilosmotic. It is not certain, however, that the normal habitat of *lutea* near the coast would not contain rather more chloride than most fresh waters.

Other Sphaeromids are found in coastal brackish water, and *Sphaeroma terebrans,* a well-known tropicopolitan species boring in wood, is recorded from fresh water in Florida (Richardson 1897).

The second group of fresh-water Flabellifera consists of cavernicolous species. Several species of hypogean *Cirolana,* referrable to a special subgenus *Speocirolana,* occur in Mexico; *C. cubensis* is found in Cuba, and a species referred to the Old World marine genus *Conilera* as *C. stygia* is recorded in wells at Monterey, Mexico. Two species in endemic genera, *Cirolanides texensis* and *Creaseriella anops,* are present in Texas and Mexico, respectively. In the Old World *Typhlocirolana* is found in underground waters in the Balearics, North Africa, and the region of the Dead Sea and the Gulf of Haifa (Por, 1962). *Sphaeromides* occurs in caves in France and Istria and *Faucheria* is present in ground waters. In the Sphaeromidae a peculiar group of genera, *Monolistra, Microlistra,* and *Caecosphaeroma,* with several subgenera, found in central France, northern Italy and east into Yugoslavia, is apparently close to the marine *Paravireia typicus* from the Chatham Islands; the various genera are placed in a tribe Monolistrini. The fresh-water species of this tribe (Fig. 63) appear now to live in regions close to the northern shore of the Tethyan Sea (Hubault 1938).

A third ecological group consists of various species of Excorallaridae and particularly of Cymothoidae, parasitic on the external surface, buccal membranes, and gills of fresh-water fishes (*A, B,* Fig. 37). *Ichthyoxenus* and *Livoneca* are known as parasites of this sort in the Old World, *I. sinensis* occurring on *Carassius* and *L. parasilura* on *Parasilurus* in China. In the New World it seems probable that *Excorallana berbicensis* from the rivers of British Guiana and several Cymothoidae of the genera *Telotha* and *Livoneca* behave in this way; the unsatisfactory information is summarized by Van Name (1936).

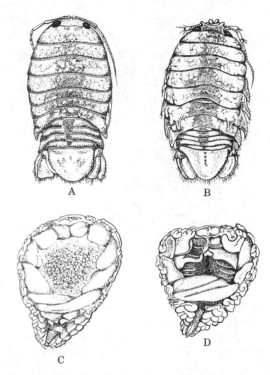

FIGURE 37. Fresh-water parasitic Isopoda. *A, B,* ♀, ♂ *Tachaea lacustris* (× c. 5). Sumatra and Java in lakes, a member of the cymothoid subfamily Corallaninae; *C. Probopyrus giardi* (× 3.36), ♀ from the fresh-water prawn *Palaemon placidus,* Sumatra, bearing on its posterior end the small hyperparasitic ♂; *D, Probopyrus bonnieri* (× 2.25), widely distributed in the East Indies on *Palaemon lar* and *P. javanicus;* apparently euryhaline.

The Phreatoicidea are regarded by Dahl (1954) as a very ancient offshoot of the Flabellifera. They are exclusively fresh-water animals, save for a few species that have adopted a burrowing, almost terrestrial habit. The order contains three families, the Amphisopidae and Phreatoicidae, almost or quite confined to Australia and New Zealand and the Nichollsiidae in India. Twenty-one living genera are recognized by Nicholls (1943, 1944) in Australasia. Outside that area one amphisopid, *Mesamphisopus* (*A* of Fig. 38) occurs mainly in ancient

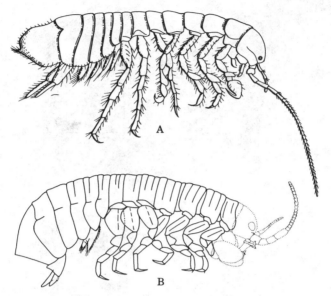

FIGURE 38.   Representatives of the Isopoda Phreatoicidea.
A. Mesamphisopus capensis, Table Mountain, Union of
South Africa.   (Barnard   1914.)   *B.   Protamphisopus
wianamattensis,* a Triassic fossil from Australia.   (Nicholls
1943.)

swampy valleys of an uplifted mesozoic peneplain in the southernmost
part of Africa (Barnard 1927), *Nicollsia* is a ground-water form in
the Gangetic plain of India (Chopra and Tiwari 1950, Tiwari 1958).
A Triassic fossil *Protamphisopus* from Australia probably belongs to the
Amphisopidae.   The group is evidently an old one, possibly originally
distributed throughout Gondwanaland, whatever the boundaries of that
rather mysterious ancient continent may have been.

The Valvifera are represented by *Pentidotea lacustris* from fresh
waters in New Zealand (Thomson 1879), possibly by *Synidotea fluviatilis* from Travancore, India (Pillai 1954), and most significantly by
the glaciomarine *Saduria entomon* (Fig. 66).   The last-named species
is discussed later in more detail (pages 181, 210); at least some of
its races appear to be fully adapted.

A few of the Epicaridea parasitic on other Crustacea have invaded
fresh water.   *Probopyrus* (*C, D,* Fig. 38) is widespread on the fresh-
water shrimps of the genus *Macrobrachium* in the warmer part of
the New World.

The fresh-water Amphipoda. The remaining group of the Peracarida, the Amphipoda, which probably had their origin from the early
ancestors of the division at a very remote time, possibly from a form
resembling the Syncarida in their lack of a true carapace, contains many
hundreds of fresh-water species, though the marine members of the
order are still more numerous. Four suborders are usually recognized.
The Hyperidea and Caprellidea are exclusively marine. The Gammaridea, an immense group, contains the majority of the order and all
but three of the fresh-water species. The Ingolfiellidea includes
the single genus *Ingolfiella* with six or seven species, one from deep
water in the Davis Strait, one from the coast of Siam, one from interstitial littoral water on the coast of Peru, two, *I. acherontis* (*B* of
Fig. 35) and *I. petrovskii,* from ground water in Yugoslavia, and
one (*I. leleupi*) from cave waters in the Belgian Congo. The last-
named species is very large, more than a centimeter in length, possibly
secondarily (Siewing, 1958b); the other species, one or two millimeters
long, belong to the same kind of interstitial fauna as the Microparasellinae and Microcerberinae; a form almost identical with *I. acherontis*
has in fact been discovered in the English Channel (Spooner 1959b).
The disjunct distribution suggests considerable antiquity (Ruffo 1951)
in spite of the rather specialized and reduced nature of these animals.

The Gammaridea,[35] according to Barnard (1959), contain 3146
species, of which 232 occur in Lake Baikal, 400 in other epigean
fresh waters, and 50 in hypogean waters; 88 are terrestrial. Most
of the fresh-water species belong to the Gammaridae. The contributions of the other families to the fresh-water fauna are as follows.
The Hyalidae contain two species of the otherwise marine genus *Hyale*
in wells near Zanzibar which are quite likely not fully adapted. The
Hyalellidae contains the important fresh-water genus *Hyalella,* widely
distributed in the New World with a number of endemics in Lake
Titicaca; the related *Parhyalella* is marine. Many authorities refer
both the Hyalidae and Hyalellidae to the Talitridae, a widespread terrestrial family. In the Pontogeniidae Bousfield (1958) has described
two species of *Paramoera* from spray pools and estuarine parts of
streams on the Pacific coast of Canada. The associated fauna can
be typically fresh-water, but it may be doubted if these animals are
fully adapted. In the Haustoriidae the marine genus *Pontoporeia* has
given rise to one species, *P. affinis,* found in brackish waters in the
Arctic with fully adapted and probably subspecifically distinct lacustrine
populations both in Europe and North America. The distribution of

---

[35] The divisions into families follow J. L. Barnard (1958).

this animal is discussed later (page 206). The Calliopiidae have one epigean and one hypogean fresh-water species of the endemic genus *Paraleptamphopus* in New Zealand but are otherwise marine. The Corophiidae contain *Corophium curvispinum,* which is making its way across northern central Europe from the rivers draining into the Ponto-Caspian basins, from which it originated. *Corophium spinicorne* occurs in tidal fresh waters on the Pacific coast of North America but is presumably not fully adapted. In the same family *Kamaka* likewise has a brackish and coastal species along the margins of the Okhotsk Sea and a fully adapted species *K. biwae* in Lake Biwa, Japan (Uéno 1943). Two peculiar families of hypogean species are known. The Hadziidae are confined to southern Europe, but the very minute Bogidiellidae, known from various European localities (*A* of Fig. 35), with a coastal species of *Bogidiella* from Brazil and an undescribed genus from the English Channel (Spooner 1959a, b), is an amphipodan ecological equivalent of the Isopodan Microparasellinae and Microcerberinae.

The Gammaridae are the only large family of Amphipods containing more fresh-water than marine species. There appear[36] to be about thirty genera and rather under 200 species known from the sea, excluding the dilute waters of the Ponto-Caspian basins, fourteen genera and fifty-six species from these Ponto-Caspian waters and their influents, thirty-five genera with 232 species in Lake Baikal and associated rivers, and about forty genera with some 300 species from the inland waters of the rest of the world. Many of the last-named group of species are hypogean. Recent physiological and ecological work, to be discussed in a later section, makes the history of the Gammaridae peculiarly instructive.

The enormous Baikalian fauna, to be discussed and illustrated in greater detail when lacustrine endemism is considered in Vol. III, is certainly derived from a group of species rather than a single ancestor, and some members of the group appear to have been related to the ancestors of the Ponto-Caspian forms (Fig. 68). According to Bazikalova (1945), *Crypturopus* and *Homocerisca,* two of the more primitive littoral Baikalian genera, are probably allied to the Ponto-Caspian *Niphargoides* (*H* of Fig. 68); the Baikalian *Brandtia* and

---

[36] These figures are approximations based on J. L. Barnard (1958). Probably the splitting of genera, particularly outside the Ponto-Caspian and Baikalian faunas has been an excessive. At the same time it is possible that recognition of the specific status of what has been supposed to be races or subspecies has been inadequate.

Hyalellopsis may be related to the Ponto-Caspian *Axelboeckia* (*A* of Fig. 68), and the relatively unspecialized *Eulimnogammarus,* which has many species in Lake Baikal, is probably allied to *Dikerogammarus* (*E* of Fig. 68) of the Ponto-Caspian region, to *Ostiogammarus* of southeastern Europe, and less closely to the widespread marine littoral genus *Marinogammarus.* At least a considerable part of this fauna is believed by Martinson (1958) to have developed in a series of basins, originally connected with the ocean, which ran across Asia from Korea westward as far as Fergana. A considerable evolution of a fresh-water fauna occurred in such basins from the late Cretaceous throughout the early Tertiary, and this fauna contains mollusks apparently allied to both Baikalian and Caspian species. The affinities of the Baikalian and Ponto-Caspian Gammaridae are probably due either to a parallel development from identical or very closely allied late Cretaceous marine species, living throughout the Tethyan Sea, or to derivation from brackish- or fresh-water ancestors that had begun their evolution in the freshening central Asiatic basins.

Pallasea quadrispinosa (*1* of Fig. 67) a species of Baikalian origin, has managed to penetrate westward across northern Siberia and Russia to the lakes of the Baltic region, probably by way of late glacial ice lakes. Some species of the Ponto-Caspian fauna have colonized the influents of the Black Sea and Caspian and a few have reached considerable distances along these rivers (see page 216).

Bousfield (1958) thinks that *Crangonyx,* mainly in North America, *Synurella* and, by implication, the hypogean *Niphargus* and its allies in the Old World are old Tertiary or Mesozoic invaders. The same is doubtless true of the hypogean forms outside the Holarctic.

Sarothrogammarus asiaticus, from fresh waters at an altitude of about 2700 m. in Turkestan, is a member of a genus known in brackish waters and sometimes in streams near the sea, in the Mediterranean region, Madeira, and the Azores (Dahl 1958). The closely allied *Neogammarus festae* is a marine littoral Mediterranean form. It can hardly be doubted that *S. asiaticus* represents an ancient independent invasion from a *Neogammarus*-like stock.

The various South African species of *Paramelita,* with possibly one Australian representative, are supposed by Schellenberg (1937) to be derived from an invasion by the marine *Melita,* though Barnard (1940) is not quite convinced of this.

A series of invasions, at least into brackish water, by *Anisogammarus* appears to be taking place on the west coast of North America (Bousfield 1958).

The very widespread subgenus *Rivulogammarus* (Fig. 39) of the

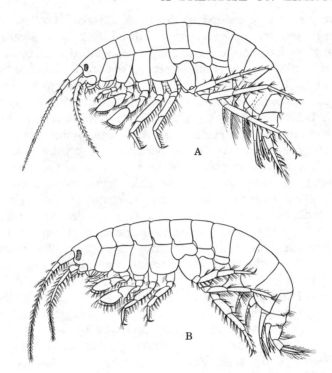

FIGURE 39. Two species of *Gammarus* belonging to the
predominantly fresh-water subgenus *Rivulogammarus*. *A,*
G. (*R.*) *pulex,* Europe westward into central Asia, not
naturally occurring in Ireland; *B, G.* (*R.*) *deubeni,*
fresh-water in Ireland, Isle of Man, and parts of the
extreme southwest of England; otherwise brackish, and
apparently displaced in fresh-water by *G.* (*R.*) *pulex;*
G. (*R.*) *deubeni* is a hairier species with wider tolerances
but lower reproductive rate than *pulex.*   (Sars.)

Holarctic is presumably derived from the predominantly brackish *Gammarus* (s. str.).   The invasion of these forms is doubtless the most, but not necessarily very, recent to have added significantly to the fresh-water gammarid fauna.   The details of the salinity requirements of the different species involved are discussed in greater detail from a physiological point of view on a later page (see pages 200–203).

The freshwater epigean Gammaridae are apparently confined to temperate climates, though outside Lake Baikal the most complicated pattern of speciation appears to be in *Rivulogammarus* in the southern part of its range in the Mediterranean area.   The appearance of fresh-water Gammaridae of presumably separate marine origins in the damp temperate parts of both hemispheres suggests that the absence of these animals in the tropics is due to physiological or possibly competitive limitations rather than to strictly historical zoogeographic causes. *Austroniphargus* from 2600 m. in Madagascar is probably a form of hypogean origin, as it lacks eyes. The hypogean species are more widespread than the epigean occurring in India and the Philippine Islands. Some belong to otherwise epigean genera such as *Gammarus* or *Crangonyx* and others to widespread typically cavernicolous genera such as *Niphargus* in Europe and *Stygobromus* in North America. A few hypogean genera have very restricted distributions, among which the extraordinary *Stygodytes balcanicus* from Popocopolje in Yugoslavia, an animal 5 cm. long, is the largest freshwater amphipod known outside Lake Baikal; it is clearly a specialized derivative of *Niphargus* (Fig. 40). Several species of the latter genus have entered the cold dark regions at the bottom of lakes. *Gammarus* may be nektoplanktonic in some lakes lacking predatory fish and littoral benthic in others in which fish occur. There are often puzzling discontinuities in the distribution of these animals (Uéno 1934b; Hynes 1954).

The fresh-water Eucarida. The more primitive of the two primary divisions of the Eucarida, namely the Euphausiacea, are exclusively marine, but the large and diversified Decapoda[37] have contributed greatly to the fresh-water fauna.

In the Decapoda Natantia a few Penaeidae and Sergestidae apparently occur in quite dilute brackish or even fresh water; in the former family *Penaeus duorarum* is reported from Lake Ahémé in Dahomey, (Lefèbre 1908), a supposedly fresh body of water, whereas more certainly Aldrich (1962) has found that the sergestid *Acetes paraguayensis* extends far up into the Amazon basin in Peru. Most of the fresh-water prawns belong to the Caridinea. Here in the Alphaeidae two West African species of *Alpheopsis* are known. *A. haugi* in Gaboon and *A. monodi* in the coastal region of the Cameroons. Of far greater importance are the almost exclusively fresh-water Atyidae (Fig. 41) and the Palaemonidae which show great evolutionary euryhalinity.

---

[37] The classification adopted is that of Borradaile (1907), accepted by most modern workers. Recently an entirely different arrangement, due to Beurlen and Glaessner, has found favor among some investigators, mainly palaeontologists. The relative merits of the two schemes are considered by Balss (1957) who concludes that the older classification does more justice to internal anatomy. It is probable that a compromise will ultimately be evolved.

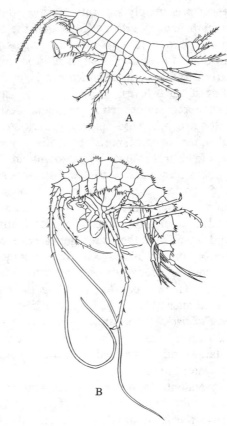

FIGURE 40.   Two related subterranean
isopod. *A. Niphargus k. kochianus*
(× 6.7), known from many wells and
caves in the lower Rhineland, south into
the Jura Mountains; *B. Stygodytes
balcanicus,* Popovopolje, Yugoslavia, the
largest fresh-water amphipod outside
Lake Baikal, natural size. (Schellenberg, Absolon.)

The Atyidae are distributed throughout the warmer regions of the
earth and include a number of quite heavily built forms which, in
the tropics, probably occupy the types of niche occupied by crayfish
in temperate regions. The family is closely allied to the deep-sea
Oplophoridae and Nematocarinidae. The most primitive genus
Xiphocaris is an epigean fresh-water form living in the West Indies.
The other relatively primitive genera retaining exopodites on all the

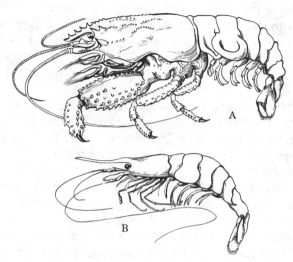

FIGURE 41. Fresh-water decapods of the family
Atyidae. *A. Atya crassa* (× ½), Central America,
a large crayfish-like species; *B. Xiphocaris elongata*
(× $\frac{9}{10}$), widespread in the West Indies, a primitive
prawn-like member of the family. (Bouvier.)

pereiopods are *Paratya* distributed from Japan southward to New
Zealand and present on a number of isolated islands, *Antecaridina*
in brackish cave waters and saline lakes on Lau Island Fiji, and
Palaemonias from Mammoth cave in Kentucky. Several other cavernicolous genera are known. The family obviously
is an old inhabitant of fresh water, and the
habitat of *Antecaridina* is probably irrelevant to
the marine origin of the group. Three endemic
genera with eleven species occur in Lake Tanganyika. There is considerable metamorphosis in
the ontogeny of the family; the youngest stage
of *Atyaephyra* has but three pereiopods and no
abdominal appendages.

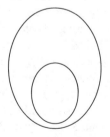

The other fresh-water prawns occur in the
family Palaemonidae. Here most species of the
subfamily Palaemoninae occur in dilute waters and
many species in fresh localities. *Palaemonetes
varians* is a brackish species of northern Europe;
the very closely allied *P. antennarius,* which lays
far fewer, though larger, eggs (Fig. 42) is a fresh-
water form of southern Europe. It is possible

FIGURE 42. Inner
line, profile of egg
of the saltwater
*Palaemonetes var-
ians;* outer line,
the same of the
fresh-water *P. an-
tennarius* (both
× 20). (Boas.)

that comparable relationships exist among other species, but in *Macro-brachium*, with numerous species in the Americas, the egg size, which is variable from species to species (cf. Holthuis 1952), is apparently not sharply correlated with salinity. Some species are supposedly somewhat catadromous. A number of cavernicolous genera are known (Holthuis 1956).

The Decapoda Reptantia include the three families of crayfish in the Astacura, a single fresh-water genus and family of Anomura, and a large number of fresh-water Brachyura. Only in the Astacura are the marine species less numerous than the fresh-water.

The fresh-water Astacura belong to three families, the Astacidae (= Potamobiidae) of the Northern Hemisphere, the Parastacidae (Fig. 43) of wide but disjunct distribution in the Southern Hemisphere, and the Austroastacidae only from Australia.

The distribution of both the Astacidae and the Parastacidae is rather peculiar. The Astacidae contains *Astacus* and *Austropotamobius* in the Palaearctic eastward to Turkestan, *Pacifastacus* on the Pacific slope of North America, *Cambaroides* in eastern Asia from the Amur basin into Korea and Japan, and several genera often placed in a special subfamily, the Cambarinae, naturally living in North America east of the Rocky Mountains. The American crayfish have undergone an extraordinary amount of speciation in the Mississippi basin and in the areas adjacent, more than two hundred species and subspecies in five genera now being recognized (Chase, Machin, Hubricht, Banner, and Hobbs 1959a). The family appears in the Neocomian lake deposits of China and Mongolia (van Straelen 1942).

The Parastacidae are found in South America (*Parastacus*) mainly in temperate regions, in Madagascar (*Astacoides*), in New Zealand (*Paranephrops*), and most prominently in Australia and Tasmania, in which regions there are thirty-one species in seven genera. One species of *Cherax* is found in New Guinea. The Murray lobster, *Euastacus serratus*, may reach a length of half a meter and is doubtless the largest known fresh-water crustacean. The rather aberrant family Austroastacidae contains but a single genus and occurs only in Victoria and South Australia (Clark 1936).

The Anomura are nearly all marine. Two species of *Callianassa*, *C. turnerana* from the Cameroons and *C. grandidieri* from Madagascar, are recorded, however, from fresh water; neither are likely to be fully adapted and *C. turnerana* seems to be catadromous (Monod 1927). The abundance of this species in the Río dos Camaroès (Bay of Cameroon) was responsible for the name of the part of Africa it inhabits. In the Galatheidea the family Aeglidae, with a single genus

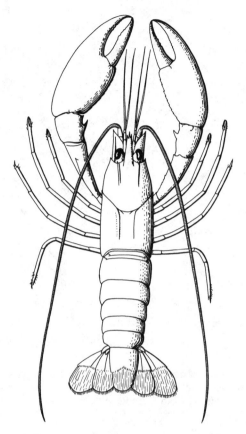

FIGURE 43. *Cherax dispar* (× 0.9), a member of the Parastaciidae or southern Crayfishes, Queensland, Australia. (Rich 1954.)

Aegla (Fig. 44), has no less than eighteen species in the temperate fresh waters of South America (Schmitt 1942).

The Brachyura contain a large number of fresh-water crabs, found mostly in the warmer parts of the world and so generally but by no means exclusively (van Straelen 1942) occupying regions in which crayfish are absent. The majority of fresh-water species belong to the Potamonidae (Fig. 45), which are characteristically fresh-water animals, and to the Grapsidae, which are largely littoral or brackish animals with a number of fresh-water and terrestrial forms, notably in southeastern Asia. The various genera of the Potamonidae have speci-

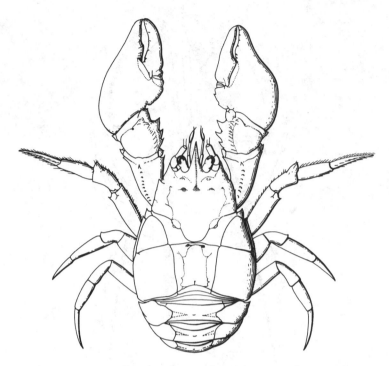

FIGURE 44. *Aegla platensis,* one of the eighteen species of a genus of characteristic South American fresh-water anomuran decapod Crustacea. (Adapted from Schmitt 1942.)

ated extensively; *Potamon* (s. lat.) has more than a hundred species in the Old World tropics and *Pseudothelphusa* about seventy in the Neotropical region and Central America.

In the Grapsidae at least *Eriocheir* is catadromous, a condition usual in land crabs but rare in fresh-water species, and several species such as *Pseudograpsus crassus* are recorded from both fresh waters and the ocean.

Among the crabs in other families *Potamocypoda pugil* in the Ocypodidae is a fresh-water form in Malaya; its nearest allies inhabit the shores of the same region. Several species of the family Hymenosomidae, which, unlike the previously mentioned crabs, belong to the Brachygnatha Oxyrhyncha rather than the Brachyrhyncha, have invaded fresh waters; *Halicarcinus lacustris* occurs in New Zealand and South Australia, *H. wolterecki* in the Philippines, *Rhynchoplax introversus* in China, and *R. kempi* in Irak. The family contains just

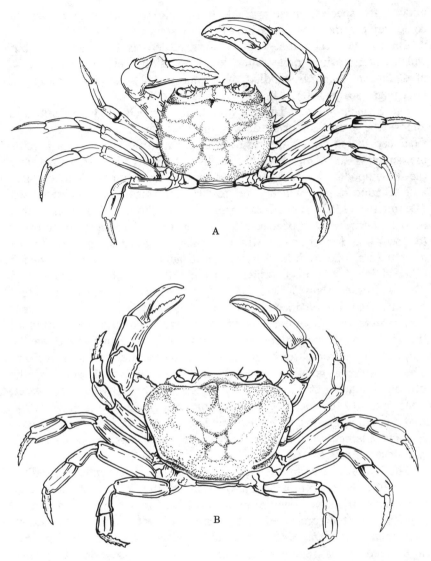

FIGURE 45. Fresh-water crabs. *A. Platytelphusa armata* (× ¾), open water Lake Tanganyika; *B. Potamon lirrangensis* (× ⅗), central Africa, including margins of Lake Tanganyika. (Capart 1952.)

under fifty species, mostly brackish or marine, but is obviously in some way preadapted to invasion of fresh waters.

Among the crabs that cannot be regarded as fully adapted, but which penetrate into quite dilute localities, is the very interesting case of *Callinectes sapidus,* studied by Odum (1953) and discussed on a later page (see p. 190).

The Branchiopoda. Turning now to the Branchiopoda, we find that almost the entire subclass occurs in inland waters, the few marine Cladocera being clearly the result of a return to the sea. Dahl regards the subclass as composed of two living orders, the Anostraca and the Phyllopoda, the latter comprising the Notostraca, Conchostraca, and Cladocera to which the fossil notostracan-like Acercostraca (Lehmann 1955) and Kazacharthra (Novojilov 1957) presumably must be added. It is probably convenient to give *Lepidocaris* an order to itself, the Lipostraca. Although such a classification may be the most natural one morphologically, the Cladocera stand apart from the other living Phyllopoda, which resemble the Anostraca ecologically. The Anostraca, Notostraca, and Conchostraca are almost all larger animals than any cladoceran and are in general inhabitants of temporary waters, which either dry up or freeze solid at some season of the year. A number of species occur in somewhat mineralized waters, particularly in temporarily filled closed basins in semi-arid regions. The anostracan genus *Artemia* is polyhalophil, but in spite of the obvious capacity of members of the Anostraca to live in salt water, none occur in the sea. It is reasonable to suppose with Bond (1934) that the three groups of branchiopodans containing large species have become adapted to waters in which large predators, particularly fish, are most unlikely to exist.

The Notostraca certainly lived in the Permian, apparently in localities not unlike those occupied today. The superficially notostracan genus *Protocaris* from the lower Cambrian of Vermont cannot certainly be ascribed to the crustacea, let alone the Notostraca. The Lower Devonian *Vachonia,* differing from the Notostraca in their lack of a postpedigerous region and with the carapace covering the whole body, is referred to a new suborder, the Acercostraca, by Lehmann (1955). It appears to have been marine.

The Conchostraca have a long geological history, certainly appearing in the Devonian. Whether the lower Cambrian *Fordilla* or the group of genera of the same age (*Bradoria, Walcottella,* etc.) placed in a separate order Bradorina by Raymond (1946) are Conchostraca, Ostracoda, or belong to some quite different major groups cannot at present be decided.

The palaeontological history of the Anostraca is fragmentary, fascinating, and perplexing. Apart from certain Tertiary fossils clearly related to living forms, four Palaeozoic genera appear possibly to be involved. One, *Branchipusites,* is based on a specimen so fragmentary as to warrant no consideration. *Gilsonicaris* from the lower Devonian of Germany is believed by van Straelen (1943) to represent a marine anostracan, but the fossil adds little to our understanding of the ancestry of the group. The other two fossils are *Rochdalia* from the middle Carboniferous of Britain, possibly a fresh-water animal, and *Opabinia* from the Burgess Shale (Middle Cambrian) of British Columbia, presumably a marine form. The two genera appear to be related. *Opabinia* is the better known of the two, but it is by no means adequately understood. In both genera there appears to be a complete set of pleura,[38] the last pair forming a fan with the telson. Under each pleuron an appendage apparently bears a number of narrow lamelliform projections. This needs further study. Raymond (1935) and Størmer (1944) interpret the structure as comparable to that of the pre-epipodite of the trilobites. If this is correct and the main ramus of the limb is supposed to have disappeared, *Opabinia* belongs in the Pseudocrustacea and has nothing to do with the Anostraca. The animal, however, is extraordinarily anostracan-looking, with pedunculate eyes and probably with a copulatory modification of some head appendage in the male; its exclusion from the crustacea does not appeal to all modern workers (Linder 1945; see also Dahl 1956). All we can safely conclude is that the larger Branchiopoda have had a long history in inland waters but *may* have relatives among the marine fauna of the first half of the Palaeozoic.

The Cladocera are among the most characteristic of fresh-water animals. A few species have invaded the sea secondarily: *Bosmina coregoni maritima* in the Baltic, a characteristic brackish form (Purasjoki 1958), and *Penilia* in the family Sididae, an obvious secondary invader of the ocean.

Of greater significance in the marine fauna are the genera *Podon* and *Evadne* in the family Polyphemidae of the superfamily Polyphemoidea, several species of which are widespread in the marine neritic plankton. In the Ponto-Caspian region an extraordinary development of the Polyphemidae (Fig. 46) has taken place, the Caspian supporting not only single endemic species (Sars 1902; Mordukhai-Boltovskoi 1962) of *Polyphemus, Podon,* and *Evadne,* but a number of endemic

---

[38] These pleura are the outer parts of the trunk appendages of Hutchinson (1930).

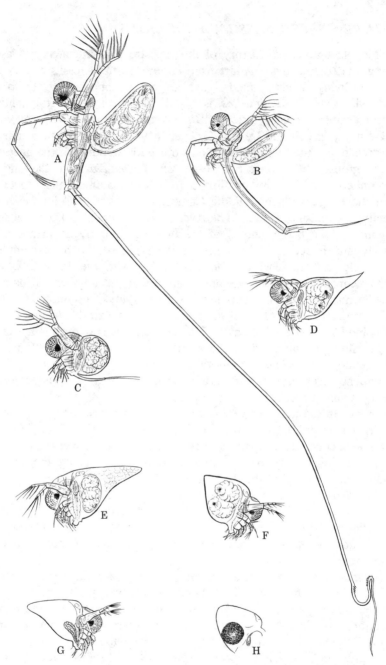

FIGURE 46. Caspian Polyphemoidea. *A. Cercopagis socialis* ♀ (× 11.25); *B. Apagis cylindrata* ♀ (× 13); *C. Polyphemus exiguus* ♀ (× 36.25); *D. Evadne anonyx* ♀ (× 18.75); *Podonevadne camptonyx* ♀ (× 27.5); *F. P. trigona*, parthenogenetic ♀ (× 26.5); *G.* the same, ♂; *H,* the same, brood pouch with sexual eggs (Sars).

genera, namely *Cercopagis* with ten species, *Apagis* with four, *Podone-vadne* and *Corniger* each with three species, and also the monotypic *Caspievadne*. Several of these animals also occur (Mordukhai-Bol-tovskoi 1964) in the more dilute waters fringing the Black Sea, notably the Sea of Azov; *Corniger lacustris*, according to Mordukhai-Boltovskoi (1964), perhaps identical with *C. bicornis* from both Caspian and Pontic basins, is known from Lake Gödjik in Armenia (Spandl 1923). Recently *Podon* has been recorded from Tso Nyak on the western boundaries of Tibet (Goulden in press) under circumstances not likely to involve confusion of specimens or labels. The wide accidental trans-port of the more halophil Polyphemoidea is therefore quite probable, and the occurrence of a Ponto-Caspian species in Lake Gödjik does not necessarily imply an early Caspian or Sarmatian transgression for which there is no geological evidence.

In addition to the genera mentioned in the preceding paragraph, the Polyphemidae include *Bythotrephes* which has three species in many of the deeper lakes of the Palaearctic region. Because they are Clado-cera and have a typical fresh-water life history, the Polyphemoidea doubtless originated in fresh water, but early adaptation to varying salinity in the Sarmatian or Ponto-Caspian basins may have facilitated the evolution of the modern truly marine species.

The Maxillopoda. The third line, constituting the subclass Maxillopoda, contains the Copepoda, the Branchiura, the Mystacocar-ida, and possibly the Cirripedia. The Mystacocarida consist only of the genus *Derocheilocaris* with a few species, which are peculiar elon-gate inhabitants of the interstitial water of marine sand. The group is little known, but it would not be surprising if it turned up in the interstitial fauna of hypogean continental waters in the same manner as the Microparasellinae, Microcerberinae, Ingolfiellidae, and Bogidiellidae.

The Copepoda, an immense group, contain the most important plank-tonic crustacea, both in the sea and in fresh waters, and a vast number of benthic and parasitic forms. Their classification has been the subject of much discussion. For the purposes of the present work the major divisions of Gurney (1931), largely following Sars, have been em-ployed. At a generic and specific level, however, Gurney generally tended to be a "lumper," and the finer divisions employed by Kiefer and other European students of the group appear to be more illu-minating. It seems likely that no student has yet dared to go as far as nature herself in this regard (Price 1958).

Gurney recognizes seven major groups of Copepoda, here given sub-ordinal rank, namely, the Calanoida, Misophrioida, Monstrilloida,

Harpacticoida, Cyclopoida, Notodelphyoida, and Caligoida, all of which are more widely represented in the sea than in fresh waters. The Misophorioida, Monstrilloida, and Notodelphyoida, each containing but one or two families, are exclusively marine. This leaves two exclusively free-living suborders, the Calanoida and the Harpacticoida, the Cyclopoida with both free-living and parasitic, but mainly marine, members, and the exclusively parasitic Caligoida, with fresh-water representatives. There can be no doubt that the group is primarily of marine origin, but a large number of fresh-water invasions have undoubtedly occurred.

The Calanoida contain about twenty-five families, of which six have fresh-water species (Fig. 47). According to Gurney's (1931) classification, these six families can be arranged in four superfamilies: (1) the Centropagoidea, which contains numerous fresh-water species in a single one of its families, the Centropagidae, (2) the Calanoidea, which contains a single fresh-water subspecies of rather doubtful validity referrable to one of the most widespread marine genera of the single component family, the Calanidae, (3) the Paracalanoidea, in which the family Pseudocalanidae contains, according to Gurney (1931) and Wilson (1959), the North American and Siberian fresh-water planktonic genus *Senecella,* though Marsh (1933) refers this animal to a special family, the Senecellidae, and (4) the Temoroidea, which contains three familes that include a number of fresh-water species, the Temoridae, Pseudodiaptomidae, and Diaptomidae. It is evident that the various fresh-water genera are not all closely related and that the two most important families containing such genera, namely the Centropagidae and the Diaptomidae, are particularly far apart. Moreover, the discussion of the Centropagidae to be given later suggests that members of the family have invaded fresh waters on at least four and possibly more occasions at widely separated times and places. Some unknown but widespread character of the Calanoida evidently preadapts the group to a greater or lesser extent to life in inland waters.

Senecella, the only member of the Paracalanoidea involved in the invasion of fresh waters, is a North American and Siberian genus with a distribution reminiscent of *Coregonus* in these areas; it is probably comparable to the glaciomarine elements discussed later (page 205), but it differs from them in not having migrated westward into Europe.

The only member of the Calanidae recorded (Stålberg 1931) from an inland locality is *Calanus finmarchicus telezkensis* from Lake Telezker in the Altai, a locality lying about as far from the ocean as any place on earth. Stålberg claims that there is no possibility that his

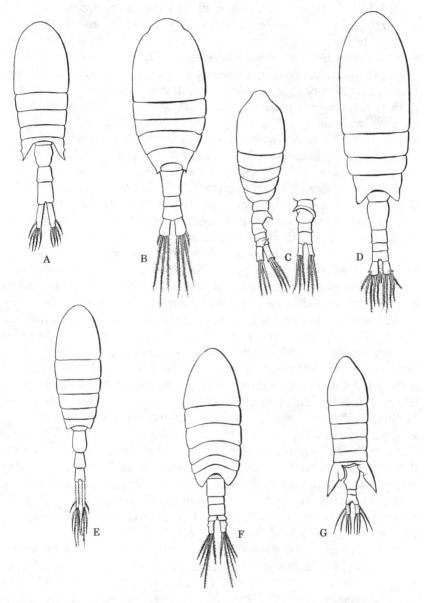

FIGURE 47. Calanoid copepods. *A. Eurytemora velox* (Temoridae); *B, Heterocope borealis* (Temoridae); *C. Epischura nordenskiöldi* (Temoridae), with urosome of ♂; *D. Senecella calanoides* (Pseudodiaptomidae); *E. Limnocalanus m. macrurus* (Centropagidae); *F. Osphranticum labronectum* (Centropagidae); *G. Boeckella orientalis* (Centropagidae); all ♀ (× 25) unless otherwise stated (Sars, Wilson).

record is the result of confusion of material or of labels. *Calanus finmarchicus* is, of course, a marine species, very abundant in the North Atlantic and adjacent seas. The supposed central Asiatic sub-species differs in quite minor characters. The occurrence of such an organism in a fresh-water oligotrophic lake in the Altai Mountains is so extraordinary that, in default of further observations, a certain degree of scepticism regarding the validity of the record may perhaps be pardoned, though it is comparable to the finding of *Podon* in a plankton sample from Tso Nyak, considered in the preceding section.

Freshwater invasions by copepods. Having dismissed this puzzling problem unexplained, we proceed to the illuminating cases of the Temoroidea and Centropagoidea. It is convenient ecologically, if not entirely proper taxonomically, to begin with the Temoroidea, and specifically with the predominantly marine family, the Temoridae, for in this family the various stages in the invasion of fresh waters are most clearly demonstrated. *Temora, Temoropsis,* and *Temorites* are marine genera. In the genus *Eurytemora,*[39] which appears to be actively invading inland waters at the present time, there are purely marine species such as *E. herdmanni* and *E. americana.* There are brackish-water species such as *E. thompsoni* from Nova Scotia, the Sea of Okhotsk, and a brackish pool in Sussex (Lowndes 1931) and the widespread *E. affinis* which occasionally appears in fresh waters and which is distributed along the coasts of the North Atlantic, Baltic, and Caspian as well as on the Pacific coast of North America. There are euryhaline species in brackish water but also penetrating far into fresh waters, such as *E. velox* of the rivers and lakes draining into the Caspian, the North Sea and the Baltic. In the Baltic region *E. lacustris* is a purely fresh-water form, though Wilson (1932) records it, apparently correctly, from brackish ponds near Woods Hole. An-other fresh-water species, *E. wolterecki,* is known from Turkey. In closed inland basins such as the Caspian, *E. grimmi,* as well as *E. affinis* and *E. velox,* is found, and in central Asia *E. composita,* apparently allied to the Arctic *E. raboti,* inhabits Issyk-Kul.

In Europe *E. affinis* lives in salt or brackish water along the coasts of the Baltic, North, and Irish seas, and the English Channel at salinities varying from 33 to 0.1 per mille; it may also, perhaps rarely, occur in quite fresh water. The species is known, too, on both the Pacific and Atlantic coasts of North America, in which regions it has occa-

---

[39] Three revisions of the genus appeared independently and almost simultane-ously (Gurney 1931; Wilson, 1932; and Marsh 1933); since then no one has attempted to clear up the discrepancies between these three accounts.

sionally been found in truly fresh water. Willey (1923) thinks that in Lake St. John, Quebec, *E. affinis* is a relict form derived from the latest marine transcursion in that area; the species, however, is present in the St. Lawrence into which the outflow of Lake St. John discharges. The more marine form of *E. affinis* is often separated as *E. hirundinoides;* Gurney, however, concludes that this supposed species is merely an inconstant, small, slender variety of *E. affinis. E. velox,* in Europe, penetrates farther into fresh waters than *E. affinis* and, at least in the east of England, is not found in the more estuarine regions of rivers inhabited by the latter species. Since in some localities *E. velox* is obviously euryhaline, Gurney thinks that there may be several physiological races involved, but it is also possible that the outcome of competition between *E. affinis* and *E. velox* is dependent on salinity. The observations of Lowndes (1935) suggest that *E. velox* occurs primarily among weeds, to which it can attach itself while feeding. Elton (1927, 1929) has commented on the frequent appearance of the species in artificial waters of recent origin. He thinks that *E. velox,* which is supposed to produce resting eggs,[40] at least in Scandinavia (Ekman 1907), may be able to colonize new localities more rapidly than can *Eudiaptomus gracilis,* which does not produce resting eggs. Elton supposes that once *E. gracilis* does get into a locality previously invaded by *E. velox* the latter species is doomed. Lowndes (1929, 1930) disputes these conclusions and Gurney (1931) notes that the two can indeed co-occur; he thinks, however, that the frequent occurrence of *E. velox* in artificial ponds does demand some explanation. Such an explanation could be only along the lines suggested by Elton, namely that *E. velox* is to some extent what may be called a fugitive species (Hutchinson 1951). It is not impossible, moreover, that the presence of *E. affinis* in quite fresh water and *E. lacustris* in brackish water in eastern North America may be due to the absence of competition with *E. velox.*

Although *Eurytemora* seems to show the invasion of fresh waters by species initially adapted to brackish conditions actually in the process of occurring, and, in the case of *E. velox* in Britain, being facilitated by new environments created by man, the other fresh-water genera of the Temoridae are by now far more completely adapted to inland waters. *Heterocope* has indeed one species, *H. caspia* in the Caspian, but it is also found up the Volga as far as Saratov and in the relict Lake Abrau in the Crimea (Rylov 1935). The other

---

[40] There is no evidence that such eggs can hatch after drying, though by analogy with the Diaptomidae it is admittedly possible.

five species are all fresh-water, *H. soldatovi* in Eastern Siberia, *H. septentrionalis* in Alaska, and the other three in Europe. *Epischura* is entirely fresh-water and has three or four species in North America and three in Siberia, including *E. baicalensis,* the dominant member of the zooplankton of Lake Baikal. *Lamellipodia fluvatilis,* originally regarded as an *Epischura,* is limited to North America immediately north of the Gulf of Mexico.

Although *Eurytemora* is mainly a genus of Arctic and North Temperate brackish water and the various Centropagidae of the Southern Hemisphere, to be discussed later, have brackish-water representatives in that region, the genus *Pseudodiaptomus* is the most important brackish water copepod in the tropics, though the distribution is rather irregular; moreover, several species are found in China and Japan and along the American coasts in temperate latitudes. The genus constitutes, with *Calanipedia,* the family Pseudodiaptomidae; *Calanipedia aquae-dulcis* (commonly, but incorrectly, known as *Poppella guernei*) is a euryhaline form from the brackish waters of the Mediterranean coast and the Ponto-Caspian region, sometimes occurring in entirely fresh waters and known in small artificial waters as far inland as Kiev, which implies remarkable powers of dispersal (Rylov 1935).

Kiefer (1938b) in the most recent catalogue of the species of *Pseudodiaptomus* lists thirty-four, some of which were formerly included in a separate genus *Schmackeria.* To these must be added *P. (Pseudodiaptallous) euryhalinus* from the coast of California near La Jolla (Johnson 1939). This species, though it is found only in the sea when the waters of coastal lagoons have been carried out by heavy floods, can tolerate salt concentrations far greater as well as much smaller than those of the ocean, having been known to breed at salinities of 1.8 to 68.4 per mille. It is presumably limited to coastal lagoons and estuaries and normally excluded from the sea by factors other than the chemistry of the water. Among the other species eight are exclusively marine, three can occur both in the sea and in coastal brackish water, thirteen are found only in brackish water, one in both fresh and brackish, and nine entirely in fresh water. In spite of the immense range of salinity tolerated by *P. euryhalinus,* it appears that when several species occur in the lower reaches of a river they may exhibit marked salinity preferences. In the Amazon (Dahl 1894; Wright 1936) *P. gracilis,* and *P. richardi* ascend furthest up into fresh water and are followed by *P. marshi* and finally by *P. acutus,* which, though definitely associated with a lower more saline (11.8 per mille salinity according to Dahl) region of the river than the other species, is nevertheless not a member of the marine fauna. It is unfortunate that there are so few chemical data relating to this distribution, nor is there any way of assessing whether the observations are to be explained in terms of simple salinity tolerance or of competitive phenomena. An even greater assemblage of species is known from the Bay of Bengal than from the Amazon, and more than one third of the known members of *Pseudodiaptomus* are recorded from southeastern Asia. Most of the purely fresh-water species occur in this area, and few are found far from the coast. The distribution, as Tollinger (1911) and Burckhart (1913) emphasize, clearly indicates penetration of fresh waters from the sea rather than a movement of a fresh-water genus seaward. It is quite likely that *Pseudodiaptomus* is an old inhabitant of brackish coastal waters which enters fresh waters sporadically as occasion offers, though it has contributed little to the fresh-water fauna of the earth as a whole.

The third family of the Temoroidea, namely the Diaptomidae, which inhabit inland waters exclusively, have numerous species, some of which are mentioned many times in this book and a few of which occur characteristically in saline and alkaline lakes. Two subfamilies, the Paradiaptominae, exclusively Old World and largely African, and the Diaptominae of very wide distribution are distinguished by Kiefer.

A remarkable set of invasions in calanoid copepods has occurred in the family Centropagidae. Here two genera, *Isias* and *Centropages*, are entirely marine, save that one species, *C. hamatus*, is estuarine and enters brackish water. *Limnocalanus* occurs as *grimaldi* in brackish water in Greenland, Spitzbergen, and Nova Zemlja, the Barents, Kara, and Nordenskjöld seas, and on the coast of Alaska; it also occurs in the Baltic and Caspian. It has given rise to a purely fresh-water eulimnoplanktic form, *macrurus*. The two forms clearly deserved only subspecific rank as *L. m. macrurus* in fresh and *L. m. grimaldi* in brackish water. The nominotypical *L. m. macrurus* is found mainly in the Baltic countries within the boundaries of the late Glacial Yoldia Sea but is also known from Ennerdale Water in the English lakes and in a number of deep lakes in eastern North America. The presence of the species in some of the North American localities cannot be explained in terms of marine transgression, and its occurrence in Ennerdale Water also implies active migration, presumably from an ice-lake in the Irish Sea. Jaschnova (1929) has observed the penetration of *Limnocalanus* 15 km. up from the mouth of the Lower Dwina. The most interesting feature of the adaptation of *Limnocalanus* to fresh water is the fact that it is accompanied by morphological changes which Ekman insists are most pronounced in those lakes longest cut off from the sea and least in those most recently. The changes appear to have been essentially the same in Europe and in North America, though there is no certainty that they were simultaneous. A second species *L. johanseni* occurs in tundra ponds in Arctic America and in apparently a fully adapted fresh-water animal.

Two other genera of Centropagidae, namely *Osphranticum,* with the single species *O. labronectum* in the Mississippi and in western North America (Marsh 1933), and *Sinocalanus,* with half a dozen named forms in need of taxonomic revision from the lower Hwang-ho, the Yang-tse, and interconnecting water courses (Burckhardt 1913), have established themselves in the fresh waters of parts of the Northern Hemisphere. Although *Sinocalanus* is not too distantly related to *Limnocalanus,* the two invasions are almost certainly independent of that of *Limnocalanus* and of each other.

The most striking penetration of the Centropagidae into fresh waters has, however, led to the development of a considerable series of fresh-water species in the Southern Hemisphere (see especially Brehm 1939). A number of genera of the family occur in the brackish and fresh waters of southern South America, Australia, New Zealand, and some of the sub-antarctic and Antarctic islands; unlike some other groups of similar distribution, these genera are not found in South Africa. *Boeckella,* the most widespread of the southern genera of Centropagidae, is found both in South America and in Australia and New Zealand but also has a well-characterized (Kiefer 1937) species *B. orientalis* in Mongolia. The genus has in *B. pooponensis,* and the doubtfully distinct *B. rahmi,* species adapted, no doubt secondarily, to saline lakes. An analysis of the South American assemblage of species by Loeffler (1955) indicates a definite distribution pattern in the affinities of the species, the most northern one, *B. occidentalis,* being closest to *B. orientalis.* Loeffler, however, is not willing to commit himself as to whether invasion of South America was from the north. The other genera have more restricted distribution. *Parabroteas* is South American. *Pseudoboeckella,* with a number of species in southern South America, has one species, *P. enzii* which occurs at the southern end of the continent, in the Falkland Islands, and South Georgia and extends from these regions to the adjacent Ludwig-Philipp Land on the coast of the Antarctic continent; a second species, *P. brevicaudata,* is found in Patagonia, the Falkland Islands, New Amsterdam, and Kerguelenland. *Gladioferens* from Australia and New Zealand occurs in brackish estuaries as well as in quite fresh waters; *G. brevicornis,* if not the other species, is markedly euryhaline. *Metaboeckella* and *Hemiboeckella* from Australia are monotypic fresh-water genera. *Calamoecia* (= *Brunella*) from New Guinea, Australia,

Tasmania, and New Zealand (Jolly 1955; Bayly 1961, 1962b), is a fresh-water genus with one species, *C. salina* (Nicholls 1944), adapted, no doubt secondarily, to life in a salt lake. It appears probable also (Bayly 1961) that *C. tasmanica* is found mainly in water relatively close to the coast, like some of the imperfectly adapted species of other groups of crustacea, but again this is probably secondary.

It is difficult to estimate how many separate invasions of Centropagidae may have taken place in the Southern Hemisphere. It is reasonable to regard the family as particularly prone to give rise to brackish and fresh-water species as occasion arises. Part of the occasion is doubtless the general rarity of Diaptomidae in the regions in which the southern fresh-water Centropagidae have been most successful.

The Harpacticoida constitute about twenty-five families of benthic copepods, almost all of which are predominantly marine. Brackish-water species occur in about half the families and in some cases may be found associated in coastal localities with members of the fresh-water fauna. The truly fresh-water members of the group (Gurney 1932; Borutskii 1952) are confined to six families, among which the Phyllognathopodidae (Viguierellidae) contain a single genus *Phyllognathopus* with several species: the best known *P. viguieri* is an excessively eurytopic animal recorded from the leaf axils of bromeliaceous plants, including the pineapple, mainly in hothouses, in lakes and springs, in moss, in decaying banana leaves, and once in a coal mine in the Saar Valley. The ecological significance of these records is obscure. The possibly related family Chappuisiidae in Europe contains but a single genus, with two hypogean species. The interesting genus *Schizopera* of the family Diosaccidae ordinarily appears in brackish water but has a remarkable species flock in Lake Tanganyika.[41] The Ameiridae contain several exclusively marine genera and two in which fresh-water species occur. In *Nitocra* there is a complete set of transitions from entirely marine species through brackish-water (e.g., *N. hibernica*) to exclusively fresh-water forms; *N. subterranea* and *Nitocrella* are hypogean. The closely allied Canthocamptidae are difficult to classify; Borutskii, following earlier European workers, recognized four subfamilies in inland waters and Gurney, five groups of genera (excluding the Ameiridae which he includes in the Canthocamptidae). For the purpose of this book it is probably sufficient to indicate that

---

[41] There may be some impropriety in regarding this very slightly saline lake as fresh water, but at times when the outlet has been running for a long time it has presumably been quite fresh.

the genus *Mesochra,* which is represented in the delta of the Amu-Daria by *M. aestuarii aralensis,* is marine and brackish and that the others form a coherent group of a dozen fresh-water genera, many with wide distributions and numerous species. The subgenus *Baikalocamptus* of *Canthocamptus and Baikalomoraria* of *Moraria* are striking endemic groups in Lake Baikal with two and fifteen species, respectively (Borutskii 1952). The peculiar genus *Parastenocaris,* the only member of its family, contains a great number of species, mainly hypogean.

It is reasonable to suppose that three to six invasions of fresh waters are represented by this assemblage. All the animals involved are benthic in a broad sense. Gurney points out that, in view of the great number of littoral, benthic, marine Harpacticoida and the frequency of brackish-water forms, the group has not been a very successful invader of fresh waters and emphasizes the difficulties of the process.

The free-living Cyclopoida, or Cyclopoida Gnathostoma, contain the marine Oithonidae, the marine and brackish-water Cyclopinidae, and the Cyclopidae. In the last-named family *Euryte* is marine and *Halicyclops,* brackish with some species apparently in purely fresh water. The remaining genera, with a great array of species, are primarily members of the fresh-water fauna; a few species of fresh-water origin occur in moderately saline lakes.

The Cirripedia. Often placed near the Ostracoda because of the bivalve shell of their so-called Cypris-larva, such an affinity is probably ruled out by their strikingly biramous appendages, and Calman's suggestion, accepted as possible by Dahl (see page 94) of a connection with the Copepoda is perhaps more likely. Although several supposed Cirripedes (other than the Machaeridia, now regarded as echinoderms) have been described from the Palaeozoic, the group does not appear unequivocally until the Triassic. All the free-living forms are marine, but two genera of Rhizocephala, namely *Sesarmoxenos* on *Sesarma* in the Andaman Islands (Annandale 1911) and Java (Feuerborn 1931) and *Ptychascus* (Fig. 48) on *Sesarma* and *Aratus* in the Amazon Basin (Boschma, 1933, 1934) have evolved as parasites of fresh-water crabs of marine affinities.

The Ostracoda. Their origin is completely obscure, though it is reasonable to suppose that the group has evolved as an independent benthic assemblage of small animals. The valves are known in bewildering profusion throughout almost the entire geological record, but in the absence of other structures the status of the earlier forms is uncertain. The belief held by certain palaeontologists that the char-

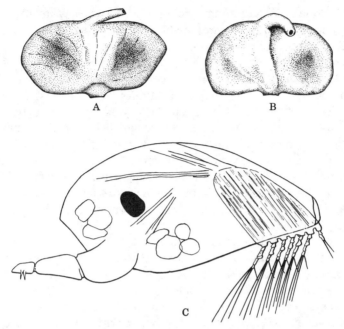

FIGURE 48. *Ptychascus glaber,* River Amazon. *A* and *B* adults from the crabs *Sesarma benedicti* and *Aratus pisonii* respectively. *C.* Cypris larva from brood pouch of adult. (Boschma 1933.)

acteristic Palaeozoic forms referred to the orders Leperditacea, and Beyrichiacea, which appear in the Ordovician and Silurian, are derived from Conchostracan-like Cambrian ancestors is extremely unlikely if the former are really Ostracods and the latter really phyllopodan. All of the earlier orders of Ostracoda were marine. The existing Ostracoda are usually separated into two orders, the Myodocopa and Podocopa, all the fresh-water species belonging to the Podocopa. The majority of such species are in the family Cypridae, of which some subfamilies are marine; the Candoninae occur in both salt and fresh waters, and the Iliocyprinae and the large subfamily Cyprinae are found only in inland waters. The small family Darwinulidae, with a single genus *Darwinula,* is fresh-water and has been recorded from Mississippian times to the present. The Cytheridae, which contain the great majority of marine Podocopa, include also the fresh-water *Limnicythere* as well as three fresh-water genera of peculiar habitat: *Elpidium* in bromeliaceous phytotelmata in Brazil, *Entocythere* in the branchial cavity of *Cambarus,* and *Sphaeromicola* on the hypogean *Caecospharoma* in the caves of France. These four genera are not closely related. The entire group of Podocopa presumably has invaded fresh waters on at least five and probably more occasions.

Fresh-water Polyzoa. The Polyzoa or Bryozoa consist of two classes. One, the Phylactolaemata, contains but few genera (*D–H,* Fig. 13), often widely distributed, of animals occurring only in inland waters. They exhibit a type of life history comparable (Fig. 8) to that of fresh-water sponges, the resting stages being internal buds provided with resistant walls and known as statoblasts. In some cases the resistant walls contain air-filled cavities facilitating flotation or are armed with hooks suggesting dispersal caught on the plumage of birds. Sexual reproduction, resulting in one of the rare types of free-swimming fresh-water larvae also occurs. On the whole this class is more primitive than the other class of Polyzoa, the Gymnolaemata. Its members do not produce mineralized skeletons and so exhibit no fossil record, but is probably safe to assume that they represent a very ancient assemblage of fresh-water animals. The Gymnolaemata contains a vast number of marine species, but in one of its orders, the Ctenostomata, there is a group of genera characteristic of brackish- and fresh-waters: *Hislopia,* ordinarily in brackish water, has a species in Lake Baikal; *Victoriella* which occurs in both marine and inland saline waters has a species in Tanganyika. In the same lake the endemic *Arachnoidea raylanksteri* belongs to a genus found in the sea in Indonesia. *Paludicella* (*B* of Fig. 13) is widespread in fresh waters, whereas *Pottsiella* is apparently endemic to southeastern North America. These fresh-water Gymnolaemata probably represent more than one invasion by a stock that is especially adaptable to low salinity.

The remaining proterostomatous phyla, the Sipunculoida, Phoronida, and Brachiopoda, of which the second and third are probably allied to the Polyzoa, are exclusively marine. In the Brachiopoda, a group containing an immense number of well-known fossils, we are probably justified in assuming that no truly fresh-water species have existed, though some authorities (cf. Robertson 1957) have believed that *Lingula* lived in quite dilute water in the middle Palaeozoic.

Echinodermata, Pogonophora, Hemichordata, and lower Chordata. All of these lower deuterostomatous animals are marine, and in the Echinodermata for which the fossil record is admirable, it is probable that no fresh-water members of the phylum have existed. It is possible, though by no means certain, that the Phoronida, Brachiopoda, and Polyzoa are very distantly related to the deuterostomatous phyla. If this can ever be established, it would seem possible that the triploblastic metazoa are divided into two groups, one of which, containing the nonlophophorous protostomes, adapts somewhat more easily to fresh waters than do nearly all the members of the other group of primitively lophophorous proterostomes and deuterostomes. The major adaptation that has taken place in this second group is, however, of paramount importance.

The Vertebrata. The stage set for the origin of the vertebrates, the dominant group of animals in the sea, in fresh waters, and on land at the present time, has naturally been a matter of great interest. It is, however, paradoxical to find that in the one subphylum that has probably evolved within Phanerozoic time no unanimity has been reached in regard to a marine as opposed to a fresh-water origin.

The oldest fossil vertebrate appears to be represented by part of a jaw with teeth, named *Archeognathus primus* by Cullison (1938), from the Lower Ordovician Dutchtown formation of Missouri (see also Miller, Cullison, and Youngquist 1947). The Dutchtown is an unequivocal marine deposit. Conodonts, some of which may be vertebrate teeth, are known in somewhat older Ordovician rocks.

The only fossil that throws any light on the origin of the vertebrates is *Ainiktozoon,* from the Ludlovian (Silurian) of Lesmahagow in Scotland, which looks like a tunicate to which is attached what seems to be a primitive vertebrate tail (Scourfield 1937). The *Ceratocaris* bed which yields *Ainiktozoon* also produces a rich but puzzling association of fossils, both marine and terrestrial. The occurrence of *Ainiktozoon* in the Ludlovian must be due primarily to exceptional conditions of fossilization. It is probably significant that on a previous occasion during the formation of the Middle Cambrian Burgess Shale of British Columbia, in which extraordinary preservation of soft-bodied animals occurred, a great variety of both benthic and pelagic animals was fossilized but nothing remotely like a vertebrate seems to have been present.[42] The reasonable conclusion is that the radiation of deuterostomatous animals that produced the true vertebrates, with the ancestors of *Ainiktozoon* and perhaps the tunicates, took place in the later part of the Cambrian.

Since this radiation has left no known contemporary fossils, it is impossible to deny categorically that it took place in fresh water, as many investigators have thought. The arguments against a marine

---

[42] This statement is based not only on the published evidence but also on a cursory examination of the collection in the U.S. National Museum in 1929; other investigators who have studied the material have apparently come to the same conclusion.

origin are apparently threefold. Chamberlin (1900; Barrell 1916; Berrill 1955) believed that the typical fusiform vertebrate form with a true tail implied a direct response to a fluviatile environment; stripped of its original Lamarkian elements, this argument implies that the ancestors of the tunicates and *Amphioxus* were fresh-water. It also seems to imply that metameric swimming animals could not develop in quiet water (for the opposite view see Berry 1925; G. M. Robertson 1950; J. D. Robertson 1957). *Ainiktozoon* might represent a persistent tunicate-vertebrate intermediate, but, whatever its nature, its form was about as unfluviatile as can be imagined for any animal with a tail.[48] The second argument is that all early vertebrates occur in atypical marine, estuarine, or truly fresh-water deposits. In the case of the atypical marine deposits it is believed that fresh-water animals were carried into shallow estuarine water, coastal seas, or lagoons. Denison (1956) and Robertson (1957) have reviewed the stratigraphic and palaeontological arguments and find them specious. Finally, it has been claimed that the glomerular kidney (Smith 1932, 1953; Marshall and Smith 1930) implies a fresh-water origin for those chordates that possess it. In its original form this theory was partly grounded in a belief in the empirical evidence supposedly provided by actual fossil associations. It has become widely accepted, partly because of the elegance of the physiological researches of Smith and partly because of the skillful literary style and unbounded enthusiasm for the kidney that characterize his more general writings. In the later writings of Romer (1955) the physiological argument from the glomerular kidney now appears conclusive enough to permit rejection of marine occurrences of the earliest vertebrates, so that a degree of circularity has entered the argument. Full details may be found in Robertson's review. The uncertainty of any single line of evidence, taken by itself, is neatly demonstrated by the myxinoids, the only vertebrates in which the blood is both approximately isotonic with the medium and has a sodium chloride concentration that accounts for most (about 90 per cent) of this osmotic pressure, varying linearly with the concentration in the medium (McFarland and Munz 1958, who review all previous data). In this respect the myxinoids resemble marine invertebrates. All the other marine vertebrates, which have probably passed through a fresh-water stage in evolution, have different osmotic relationships. We may therefore suspect the myxinoids of

---

[48] This extraordinary organism has been hardly considered since it was described more than twenty years ago. Berrill (1955) does not mention it, though it goes far to support his phylogenetic thesis but not his views on the habitats in which the phylogenesis occurred.

having an exclusively marine history at least as metazoa. Yet they possess a glomerular kidney.

The available evidence on the medium in which the vertebrates originated would appear to consist solely of the following propositions, strongly set forth by Robertson.

The known deuterostomatous phyla other than the chordates, and one of the two subphyla of the latter, are exclusively marine.

All the unequivocal occurrences of Ordovician vertebrates are in sediments which are either marine or at least show no indications of not being marine. This is apparently true of nearly all the Silurian occurrences also.

The physiological evidence supplied by the myxinoids is equivocal; it is at least as well interpreted in favor of a marine as of a fresh-water origin.

The reasonable conclusion that the vertebrates arose in the sea does not contradict the hypothesis that all the known living vertebrates other than the myxinoids have probably gone through a fresh-water stage in their evolution.

Apart from the myxinoids, all the existing marine vertebrates give evidence of ultimate fresh-water origin. This is shown in the marine Petromyzones and Teleostomi by the marked hypotonicity of the blood to sea water and in the elasmobranchs by the isotonicity of the blood being mainly due to a strikingly high concentration of urea and trimethylamine oxide rather than to sodium chloride as in the myxinoids and marine invertebrates. Without going into the rather complicated palaeontological history of the early fishes, on which much paleoecological work remains to be done, it is reasonable to derive all existing forms from fresh-water ancestors living in the late Palaeozoic. During the Devonian the evidence of a considerable fresh-water vertebrate fauna is unequivocal. In marine bony fishes the existence of a swim-bladder almost certainly bears out the physiological evidence, for such a structure is clearly a derivative of a lung. The vast majority of fish exhibiting aerial respiration are fresh-water forms living in tropical swamps or less commonly in rivers which are reduced in dry seasons to a collection of independent stagnant pools. It is from some such fresh-water fish that the vast teleost population of the sea presumably was derived; the lung became a hydrostatic organ and the blood retained to a great extent a low sodium chloride concentration and osmotic pressure, though now hypotonic rather than hypertonic to the medium.

The elasmobranchs solved the migration back into sea water by their peculiar secondary isotonicity. Smith (1936) lists more than fifty species of sharks and rays that have been recorded in fresh waters, mostly in tropical regions; many of the records merely imply temporary sojourn of marine species in rivers, but a number of fully adapted rays as well as a landlocked shark are known. The latter animal, *Carcharhinus nicaraguensis,*[44] is an abundant fish in the Lake of Nicaragua and in some of the rivers of the drainage basin. It is clearly derived from the marine but very euryhaline *C. leucas* which often ascends rivers (Bigelow and Schroder 1948). *Pristis perotteti,* the well-known southern sawfish of the Atlantic, occurs in a morphologically unmodified form in the same lake. Many typical rays live in rivers, some casually, some permanently. Occurrences of this sort are particularly common in the rivers draining into the Bay of Bengal and eastward throughout southeast Asia and in South America where the fresh-water sting-rays (Fig. 49) of the family Potamotrygonidae are important. Such a distribution is reminiscent of that of the Pseudo-diaptomidae and to a lesser degree that of the river dolphins.

The oldest teleostomes, the osteolepid Crossopterygia, appear in the lower Devonian and could have been close to the ancestors of the Dipnoi; they may also have given rise to the other teleostomes and the Amphibia. Such animals presumably had, like their descendants, a lung or open swim bladder, as Westoll (1943) believes. The living Dipnoi occur either in tropical swamps (*Protopterus, Lepidosiren*) or in seasonally variable rivers (*Neoceratodus*). Studies by Carter and Beadle (1930, 1931), in particular have emphasized the high probability of the origin of air-breathing mechanisms (*C* of Fig. 50) in tropical swamps in which oxygen is a limiting factor. Most investigators, impressed by the redbeds of the Devonian, have supposed that air-breathing fish developed in response to seasonal aridity. The hypothesis that they appeared in response to what Carter and Beadle call an aerochrotistic habitat in which dissolved oxygen in severely limiting is, by analogy with the present, more probable and demands more consideration from palaeontologists than it has so far received.

Whatever the exact type of habitat, the origin of the lung or physostomatous swim bladder is far more likely to have taken place in a fresh-water than in a marine environment. Later various Crossopterygia took to the sea. The only existing genus, *Latimeria,* has a large, lung-like, open swim-bladder which has become filled with fat (Millot and Anthony 1958).

---

[44] Dr. Daniel Livingstone has expressed verbally some doubt of the distinctness of *C. nicaraguensis,* as he suspects that the supposed characters are based partly on sex differences.

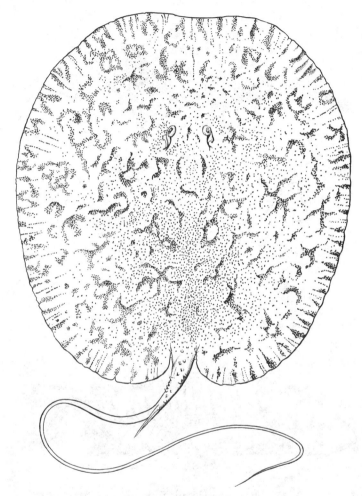

FIGURE 49. *Disceus thayeri,* headwaters of the Amazon above Teffé, Brazil, a sting-ray of the neotropical fresh-water family Potamotrygonidae (Garman 1913).

The Teleostomi Actinopterygii are divided by Berg (1940, [1947]) into sixty orders. None of the older supraordinal divisions is retained, for when the fossil forms are included there seems to be no real break in the series. The most primitive order, the Polypteriformes, which contains only the African *Polypterus* and *Calamoichthys,* is now fresh-water but may have had marine Eocene representatives; the large late Palaeozoic and Mesozoic assemblage referred to the Palaeonisci-

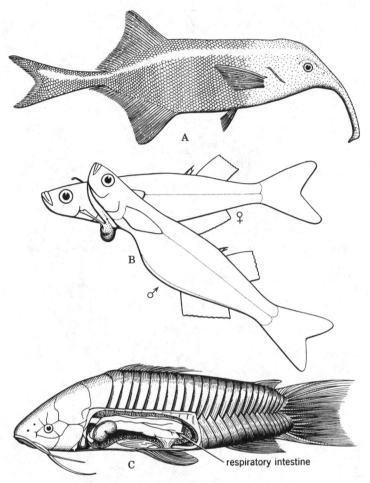

FIGURE 50. Extreme specialization in fresh-water fishes. *A.
Gnathonemus ibis,* Nile, a member of the Mormyridae, a family
notable for its peculiar facies, enormous cerebellum, and electrical
location mechanism.   *B. Gulaphallus mirabilis,* Philippine Islands, a
copulating pair of a member of the Phallostethidae, a family in
which the pelvic apparatus has moved far forward under the head
and in the male has become modified as an asymmetrical copulatory apparatus.   *C. Hoplosternum litorale,* an inhabitant of the
swamps of the Gran Chaco, Paraguay, in which the intestine has
become modified as a lung; such modifications make possible a
diverse assemblage of fishes in aerocratistic tropical swamps,
(Boulenger, Villadolid and Manacop, Carter and Beadle.)

formes was largely fresh-water but a number of marine fishes of a similar grade of organization occur.[45]   The existing Acipenseridae and Polyodontidae are fresh-water or anadromous; the allied Chondrosteidae from the middle Mesozoic are found in marine sediments.   These fish are a secondary largely cartilaginous offshoot of some group near the Palaeoniscids.   The existing fresh-water bowfin (*Amia*) and the predominantly inland gar pikes of the family Lepisosteidae have some fossil fresh-water relatives but other allied forms seem to have been marine.   In the orders ordinarily placed in the teleosts the same lack of pattern in the over-all direction of evolution to and from fresh-water forms is equally evident.   The Clupeiformes (Malacopterygii), the lowest teleost order, are largely marine, but sporadic genera such as *Alosa* are anadromous or fresh-water; the same is true of the whole suborder Salmonoidei, whereas the Esocoidei, the Notopteroidei, and the odd isolated monotypic African families Phractolaemidae and Cromeriidae are purely fresh-water.   A comparable situation exists in the Perciformes (Acanthopterygii), which are mainly marine but contain the important families Percidae, Centrarchidae, and Cichlidae which are primarily fresh-water.   The Cypriniformes by contrast are a predominantly fresh-water order but with some marine forms.

Although most of the fresh-water fish fauna is composed of species belonging to the three teleost orders just mentioned, there are in freshwaters a certain number of peculiar and high specialized groups, some of which are given ordinal rank and whose evolutionary history is very uncertain.

The peculiar order Mormyriformes, consisting of the Gymnarchidae and Mormyridae (*A* of Fig. 50) found in the fresh waters of Africa, are electric fish navigating by sensory appreciation of distortions in their own electric fields and exhibiting many anatomical specializations, notably an enormous development of the cerebellum, particularly in the Mormyridae, in which it is proportionally larger than in any other vertebrates.

The Cyprinodontiformes are an interesting order containing, in the Amblyopsidae, a series of increasingly hypogean genera from North

---

[45] The writer cannot help commenting on the extreme difficulty of obtaining information about the probable habitats of fossil fishes without referring to original stratigraphic accounts, an impossible task in preparing a work such as this.   Admittedly, there is often doubt, but in many cases there is not; only the fact that some Mesozoic formations are well known to him, has allowed him to determine in some cases whether a fish is likely to have been marine or fresh water.   Yet in most cases this is almost all one can learn of the fossil species biologically.

American and, in the family Cyrinodontidae, a number of coastal, brackish-, and fresh-water species, including *Lamprichthys* (Lamprichthinae), endemic to Tanganyika, and *Orestias* (Orestiinae), endemic to Lake Titicaca. The related Andrianichthyidae occur only in lakes in the Celebes. Also contained in the order are several allied families of small viviparous fish which occur in tropical America. The allied Phallostethiformes (*B* of Fig. 50) has two families in southeast Asia which possess a fantastic copulatory organ below the head in the males. These examples, along with the extraordinary evolutionary phenomena presented by the Cichlidae in the large lakes of central Africa (to be considered in Volume III), may be mentioned simply to demonstrate the remarkable evolutionary developments that can occur among the teleosts in fresh waters.

The amphibia contain a small number of perennially aquatic forms and a vast assemblage of terrestrial species which return to the waters to breed and which have aquatic larvae. A few members of the group are entirely terrestrial. *Rana cancrivora* in southeastern Asia and *Bufo viridis* in southern Europe can tolerate high salinities, the former species achieving osmotic equilibrium by means of urea, as in the elasmobranchs (Gordon, Schmidt-Nielsen, and Kelley 1961; Gordon 1962). The group is certainly of fresh-water origin and is derived from lung-breathing early Devonian crossopterygian fishes.[46]

The oldest reptiles must have been derived from amphibia with fresh-water larvae; the status of the Cotylosauria is in fact uncertain in this respect. Though most members of the class have been terrestrial, there has been a tendency to reinvade both fresh waters and the sea. Usually the marine lines have undergone greater specialization than the fresh-water, probably because of the greater stability of the marine environment which has permitted more time for the undisturbed evolution of any given line.

The Mesosauria were probably fresh-water reptiles derived from the Cotylosauria and living in South America and South Africa during the Permian.

The Chelonia probably originally evolved as swamp-living animals, and more existing genera are aquatic than terrestrial. They are mainly fresh-water, but the marine Chelonioidea and Dermatochelyoidea are, as is true of other amniotic groups in the sea, more specialized for aquatic life than are the fresh-water forms. None has become independent of the land in reproduction.

In the subclass Lepidosauria to which the modern lizards and snakes belong, there was in the Eosuchia an early aquatic specialization, most

---

[46] Westoll (1943) gives a good review.

strikingly developed in the Thalattosauria of the Trias. The extinct Mosasauridae appear to have been marine lizards which become specialized to a number of types of habitat in the Cretaceous seas. The only modern lizard that provides any ecological parallel is *Amblyrhynchus* from the shores of the Galapagos Islands, a member of the much more primitive Iguanidae. A few lizards habitually living partly in fresh water appear to be known; the Cuban *Deiroptyx vermiculatas* and some other iguanids dive into streams when disturbed and in some cases are able to run over water surfaces (Barbour and Ramsden 1919).

The snakes have sometimes been regarded as derived from aquatic lizards, and at least some of the early Tertiary fossil species seem to have been marine. It is, however, much more likely that they were originally terrestrial and perhaps fossorial (Bellairs and Underwood 1951). Many more or less fresh-water snakes are known in the Colubrinae, notably in the genus *Natrix*. The subfamily Acrochordinae of the Colubridae contains the oriental river snake *Acrochordus* and there are other fresh-water genera in the allied Xenoderminae. The Homalopsinae in southeastern contain numerous fluviatile and estuarine viviparous fish-eating snakes.

In the Elapidae there is a water cobra, *Boulengerina,* from central Africa (Fig. 51). The sea snakes, or Hydrophiidae, are closely allied to the Elapidae and are the most specialized aquatic serpents. Except for *Laticauda,* the most primitive genus of the family, which lays eggs on small islands, these snakes are viviparous. Though most species are neritic bottom feeders, *Pelamis* is a wide-ranging pelagic snake. *Laticauda crockeri* is known only from Lake Tungano, a brackish lake (Cl 3.40 per mille) in a cryptodepression 3 km. from the sea on Rennell Island in the Solomon group. The breeding habits of

FIGURE 51. *Boulengerina annulata stormsi,* an aquatic snake allied to the cobras, from Lake Tanganyika, with the nominate race in other parts of central Africa. (Boulenger.)

the genus would easily lead to invasion of such a locality; the species appears to be well defined (Slevin 1934). *Hydrophis semperi* from Lake Bombon, Luzon, Philippine Islands, is another lacustrine hydrophiid; though Smith (1926) gives it specific status, it appears to differ from the widespread marine *H. cyanocinctus* only in coloration and small size.

The Crocodilia represent the main aquatic line of the subclass Archesauria. Existing forms are included in the Crocodylidae, in three subfamilies, and are mainly fresh-water. *Crocodylus porosus,* however, is an estuarine animal, often known at sea in southeastern Asia; as is pointed out on a later page, it may be better adapted to a saline medium than are the other living species of the family. The long and complicated history of the crocodiles is not easy to interpret in environmental terms, but at least the Thalattosuchia of the Jurassic and Cretaceous were marine. The other orders of Archosauria were mainly terrestrial and often bipedal, though the sauropod dinosaurs are supposed to have been amphibious.

The wholly extinct subclass Ichthyopterygia, including only the ichthyosaurs, was apparently entirely marine and, as far as is known, viviparous. The ichthyosaurs were the most specialized of all aquatic reptiles.

In the extinct subclass Euryapsida the order Sauropterygia, consisting of the more primitive notosaurs and the more specialized plesiosaurs, was certainly marine; the rather distantly related Placodontia seem to have specialized as mollusk feeders in the middle and upper Triassic of Europe. The final subclass of reptiles, the Synapsida, was terrestrial.

No bird has become independent of a terrestrial substratum for breeding, so that even the most aquatic or aerial species must return to land. A vast number of birds are more or less aquatic, and among the groups specialized for aquatic existence there is considerable differentiation into marine and fresh-water birds, the physiological significance of which is considered later (page 174). All penguins (Sphenisciformes), auks, murres and guillemots (Alciformes), and albatrosses, shearwaters, and petrels (Procellariiformes) are exclusively marine, as are the Sulidae, Phaetonidae, and Fregatidae among the Pelecaniformes. In the last named order, however, the cormorants (Phalacrocoracidae), though mainly marine, have species that invade fresh waters, and the pelicans (Pelecanidae) likewise frequent fresh as well as salt water. The terns and gulls (Lariformes), like the cormorants, were evidently originally littoral marine birds which have here and there invaded, or are invading, fresh waters. To a naturalist unprepared for the event (Fig. 52) nothing can be more surprising

FIGURE 52. *Sterna hirundo tibetana* in an idealized western Tibetan landscape. (Martha M. Dimock.)

than an encounter with the Tibetan tern *Sterna hirundo tibetana* in south central Asia. Several other species and subspecies, however, are primarily inland water birds, notably *Sterna aurantia* on large rivers in northern India and *S. albifrons pusilla* in the same general area. The grebes (Podicipidae) are almost exclusively, and the related loons and divers (Colymbidae) mainly, fresh-water birds, as are the majority of swans, geese, and ducks (Anatidae). The last-named group, however, has some purely marine genera such as the eider *Somateria*. Most of the large forms that wade rather than swim, such as the herons, are fresh-water birds, but the lovely *Ajaia* often haunts marine lagoons, and most flamingos (Fig. 53), though lacustrine, are limited to somewhat saline and sometimes excessively alkaline lakes, either littoral-marine or inland. The aquatic rails, water hens or gallinules and coots (Rallidae) are fresh-water birds, as are the Heliornidae, the extraordinary jacanas (*Jacana, Hydrophasanius*), which walk on floating vegetation (Fig. 54), and the dippers (*Cinclus*), the only truly aquatic passerines. The last named, along with some of the ducks and several groups of sea birds use their wings (Fig. 55) for swimming underwater; Goodge (1959) gives a good account, summarizing earlier work.

Among the mammals, more or less specialized fresh-water species occur in the monotreme *Ornithorhynchus,* the insectivores (*Neomys* in the Palaearctic, *Potamogale* in central Africa, *Limnogale* in Madagascar), artiodactyls (Hippopotamidae), carnivores (Lutrinae), and rodents (some species of *Arvicola, Neofiber, Ondatra,* and particularly the Castoridae). The primarily fresh-water species are less morphologically specialized than the three great groups of marine mammals; *Ornithorhynchus* (Fig. 56) and the beavers are doubtless the most modified for aquatic life.

The marine mammals, with the exception of the remarkable sea-otter *Enhydra* of the northern Pacific, belong to the Pinnipedia, the Cetacea, and the Sirenia. All three groups have fresh-water representatives. The Sirenians are strictly littoral and the manatee, *Trichechus,* is just as much at home in the rivers of western Africa as in salt water.

The seals may possibly be diphylectic (McLaren 1960); at least the Phocidae seem to be related to the otters by such Tertiary fossils as *Potamotherium* and *Semantor,* and so may have acquired their initial aquatic adaptation in fresh waters, perhaps, as McLaren supposes, in the large inland basins of Asia postulated by Martinson (1958). Several lacustrine populations of Phocidae are known and are discussed in relation to evolutionary problems in Volume III. If McLaren's

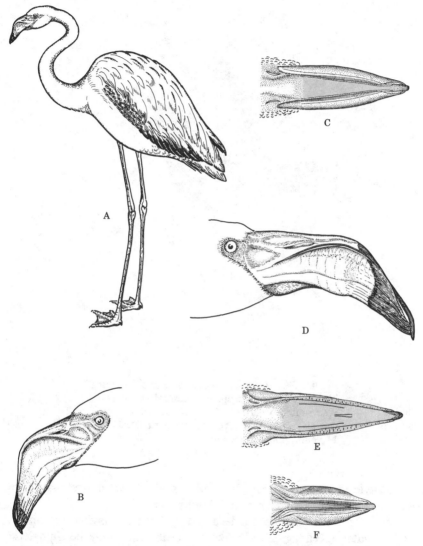

FIGURE 53. Flamingos (Phoenicopterygidae), a family of microphagous birds frequent in inland basins. *A. Phoeniconaias minor,* the lesser flamingo of Central Africa, feeding primarily on Myxophycean phyloplankton, *B.* lateral view of head; *C.* dorsal view of bill showing lamellae; *D.* lateral view of head of *Phoenicopteryx roseus,* Europe, Africa, and parts of Asia, feeding mainly on crustacea inhabiting littoral mud; *E.* dorsal view of same, the coarser lamellae not exposed below the wide maxilla; *F. Phoenicoparrus jamesi,* Andes, primarily a diatom feeder, dorsal view of bill with a very narrow maxilla and fine widely exposed series of lamellae. (*A, B, D* redrawn from various sources, *C, E, F,* Jenkin.)

FIGURE 54. *Jacana spinosa,* a more or less neustonic bird, widespread in central and northern South America. (M. M. Dimock from various sources including skins in Peabody Museum, Yale University.)

phylogenetic ideas are correct, the Baikal seal *Phoca sibirica* may have passed its entire history as a seal in inland basins!

A case can perhaps be made for the Cetacea having originated in fresh water (Kellogg 1928); like the seals, they may be diphyletic (Kleinenberg 1958). Only the toothed Cetacea occur in inland waters. A number of dolphins are known to enter river mouths, particularly in the tropics: *Sousa sinensis* and *Orcaella fluminalis* in the oriental region are particularly characteristic (*E, F,* Fig. 57); *S. teuszii,* from the Cameroon River and Senegal was long supposed, but incorrectly, to have been herbivorous. Of even greater interest is the extraordinary family of fresh-water dolphins: the Platanistidae, consisting of *Inia* in Amazon (*A* of Fig. 57), *Stenodelphis* in the Amazon and

La Plata rivers, *Lipotes* (*D* of Fig. 57) in the Tung Ting Lake, 950 km. up the Yangztse River in China, and *Platanista,* (*B* of Fig. 57) in the Ganges and Indus. These animals, particularly the last two, are very peculiar; *Lipotes* is asymmetrical even in the caudal region and has very minute eyes (Hinton 1936); *Plantanista* has developed extraordinary maxillary crests (*C* of Fig. 57) and does not even possess crystalline lenses. Allied Tertiary genera are known that indicate a former wide distribution for the family, which seems at one time to have been partly marine, though since it is basically rather primitive

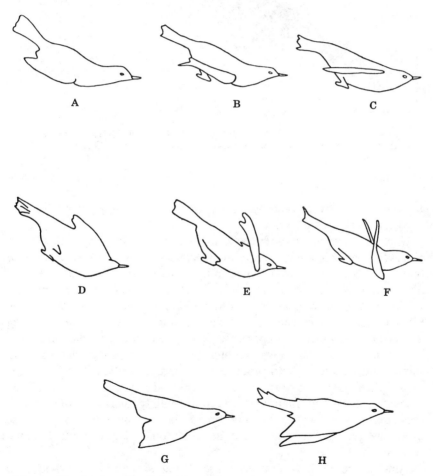

FIGURE 55. Underwater flight of the dipper *Cinclus mexicanus. A-D,* recovery phase; *E-H,* propulsive phase. (Goodge.)

FIGURE 56. *Ornithorhynchus anatinus,* the duckbill or platypus, an aquatic member of the reptile-like mammals or Monotremata. (Original drawing by Jane Marshall.)

a continuous fresh-water ancestry for the living genera is not inconceivable.

General features of amniote adaptation to aquatic life. It will be observed that throughout the amniote classes of vertebrates only those that have been able to become fully aquatic are those that are or have been viviparous. This is true of the Hydrophiidae and the Ichthyosauria, as well as more obviously the mammals, of which the Sirenia and the Cetacea are wholly aquatic. Once the aquatic egg has been abandoned it apparently is not reinvented in the amniotes, and it is evident that even among the viviparous mammals a solid substratum for mating and other aspects of reproduction is relinquished with difficulty. All groups that are really specialized for aquatic life among reptiles and mammals are now mainly marine, even though they may have had fresh-water ancestors and may, as in the case of a few sea snakes, seals, and at least some Cetacea, invade fresh waters from the sea.

Among the birds a few very specialized structural adaptations to a partial aquatic life have been developed, notably in the straining devices (Fig. 55) for plankton feeding in the flamingos (Jenkin 1957)

FIGURE 57. Fresh-water Cetacea. *A. Inia geoffrensis,* Amazon; *B. Platanista gangetica,* Ganges; *C.* Skull of same; *D. Lipotes vexillifer,* Tung-Ting Lake, China; *E. Sousa sinensis,* Amoy R.; *F. Orcaella fluminalis,* Irrawaddy R. *A–D,* belongs to the peculiar fluviatile Platanistidae, *E–F,* fluviatile members of the predominantly marine Delphinidae. (Compiled from all available sources by Martha M. Dimock.)

or in the semineustonic feet of the jacanas. It would seem likely that the long period needed for their development would be possible only in an organism that can move easily by land or air from one aquatic locality to another as the environment changes.

In general, when a completely aquatic life is achieved, extreme specialization would perhaps be possible only in the more stable environment of the ocean, though occasionally very large persistent lakes such as Baikal and Tanganyika, or some of the largest river systems, might provide environments stable enough to permit a degree of specialization comparable to those of the marine fauna in purely aquatic animals, but apparently only in invertebrates.

PHYSIOLOGY OF ADAPTATION TO FRESH WATER

Although it now appears most unlikely that the Monera and Protista were ever limited to the ocean, there seem to be no groups of multicellular fresh-water organisms, with the possible exception of some lineages in the Turbellaria, that have had an exclusively fresh-water history as Metazoa. It is evident that when we can trace the origin of a fresh-water multicellular group we may find that it is of marine origin (sponges, coelenterates, aschelminths, annelids, mollusks, Crustacea, and lower vertebrates) or of terrestrial origin (flowering plants, insects, arachnids, and amniote vertebrates). The physiological problems of invasion from the sea and from the land (page 183) are quite different.

Considering first only the marine invasions, it is evident that they are distributed irregularly and apparently in a superdispersed manner throughout the animal kingdom. Some groups, such as the echinoderms, cephalopods, and probably the brachiopods, have been totally unable to add to the fresh-water fauna; other groups do so with considerable ease. Moreover, when a group such as the polychaets is considered, it seems certain that special families such as the Nereidae, and within that family a few genera allied to *Namanereis* (= *Lycastis*), are prone to adapt particularly easily, even though they have added little to the fresh-water fauna as a whole. In more general terms, it is tempting to suppose that certain groups are more or less preadapted in the sense of fulfilling some but not all the necessary conditions for fresh-water life before they encounter fresh water, whereas other groups are not. Unfortunately, at the present time we can only guess at the probable nature of most of such preadaptations. It would be reasonable to suppose from the study of the following sections that they might include in many cases (a) a less than average over-all permeability to water and electrolytes, (b) an ability to perform cellular work at an osmotic pressure lower than that of sea water and to maintain the correct ratio of ions in the cells at various concentrations of the circulating fluid, (c) a well-developed ability to excrete water, so maintaining a constant internal osmotic pressure, and (d) an ability to develop some sort of absorbing mechanism which can take up salts against a diffusion gradient. It will appear from the discussion that the first of these characteristics may well be the least important but that in the course of evolution a fifth requirement, or perhaps a specialization of the third, developed only late in the process of adaptation, namely a capacity to secrete urine hypotonic to the blood, becomes an important condition when the permeability of the body wall to water remains high.

In addition to these primary problems, there is evidence that the extreme variability in temperature of inland waters may prove a barrier to migration from sea to fresh water. There is also good reason for believing that life histories with pelagic larvae, so usual in the sea, are inappropriate in fresh waters, though some differences of opinion have existed as to why this should be so.

The osmotic problem. All marine invertebrates, except perhaps the few insects and other arthropods of terrestrial or fresh-water origin, possess body fluids that have approximately, though often not exactly, the same osmotic pressure as that of the medium in which they live. When a small change in the concentration of the external medium takes place, there is an equivalent change in the concentration of the internal fluids. It is obvious that in passing from sea water to the very dilute water of rivers and lakes the internal fluids of such a strictly poikilosmotic marine organism would be diluted until the internal salt concentration fell below that necessary to ensure the integrity of the biochemical mechanisms of the organism. To enter fresh waters a considerable degree of homoiosmotism must be evolved. The details of the physiology of this evolution have been set out so ably in a number of recent works by some of the most eminent of comparative physiologists and biochemists that a brief undocumented summary is all that is needed in this book. The interested reader may consult Baldwin's (1948) admirable little book, the rather more technical summary of Florkin (1949), the basic monograph of the late August Krogh (1939), the theoretical studies of Potts (1954b), the reviews of Beadle (1943, 1957) and of Robertson (1960), the physiological part of the excellent book on the brackish biota by Remane and Schlieper (1958), and the first chapter of Nicol's (1960) work on the physiology of marine animals. A new book by Potts and Parry (1964) has appeared in time to emphasize the amateur nature of what follows in this chapter, but too late to have been of any use in preparing the present account.

Ionoregulation as a universal process. It cannot be too strongly emphasized that as long as we remain on a cellular level, studying either a single protist immersed in any natural medium or the cells and tissues of a metazoan in contact either with the outer medium or with the internal environment in Claude Bernard's sense, we shall always find marked differences in ionic composition even if the total osmotically active concentrations are the same inside and outside the cells. In general, in marine organisms, potassium will be relatively higher within the cells than outside, sulfate and perhaps magnesium relatively lower. The higher intracellular K:Na ratio is, of course, involved in the functional activities of contractile and conducting cells if not all other types of function. In the cell nucleus, however, sodium rather than potassium may be differentially absorbed. Some degree of regulation is therefore universal.

When we consider the internal environment of those metazoa in which there is an intercellular fluid of any kind, we find a great degree of variation in the independence of the ionic ratios even when the internal and external media are isotonic, but the most accurate work suggests that nearly always some regulation in ionic proportions occurs (Robertson 1949). In the fluid of the gelatinous mesogloea of *Aurelia aurita* there is a slight concentration of potassium and a marked reduction in sulfate, the other major ions adjusting to maintain an isosmotic steady state. In the perivisceral fluid of echinoderms a slight relative enrichment of potassium is usual. In the coelomic fluid of marine polychaets a comparable enrichment of potassium and in some cases a reduction in sulfate have been observed. In the higher invertebrate phyla the effects are usually more striking. In marine mollusks there is often a relative increase in potassium and calcium and sometimes a marked decrease in sulfate. In the higher crustacea there is a tendency toward reduction of the relative concentrations of potassium, magnesium, and sulfate, with an increase in calcium and a most noticeable relative increase in sodium. There is some evidence, reviewed by Robertson (1960), that decreasing blood magnesium is correlated with increasing activity. *Eledone* evidently resorbs potassium, calcium, and magnesium from its renal organ, whereas in *Homarus* there is resorption of potassium and calcium and active excretion of magnesium and sulfate. The details, for which Robertson's fundamental paper may be consulted, vary greatly from genus to genus. The important point is that some degree of isosmotic ionic regulation appears to be almost if not quite universal.

As Krogh (1939) and others have intimated, ionoregulation between isotonic solutions inside and outside the organism may well be the phylogenetic basis on which the later mechanism of osmoregulation is built, though in the Protista and Monera, if the comparative argument of the earlier sections of this chapter are acceptable, it may be difficult or impossible to give primacy to either aspect of the regulatory process.

Types of euryhaline adaptation. It is evident that some completely marine animals are more euryhaline than others. This probably often means that the dilution of the body fluids by a given amount is fatal to some forms but tolerated by their relatives, which in this sense are preadapted to dilute water. *Arenicola,* which inhabits quite dilute parts of the Baltic, is poikilosmotic but euryhaline (Schlieper 1929); so to a considerable extent are *Macoma baltica* and *Mytilus edulis.* Some such tolerance of internal dilution is presumably a prerequisite for the evolution of a fresh-water organism, but since the abundant biota of coastal seas, varying from say 25 to 40 per mille salinity, is typically marine, there is obviously little difficulty in achieving it. A factor of almost two in the concentration range of internal fluids within a genus if not a species is clearly often possible in poikilosmotic animals. This is certainly true of *Asterias rubens,* even though local races may perhaps be involved (Schlieper 1929).

The typical euryhaline organisms of estuaries and coastal brackish waters represent a more definite adaptation. In them there are mechanisms by which dilution is prevented as the external concentration decreases. Such animals have thus become at least to some slight extent osmotically independent of their environment; they are at least partly homoiosmotic. *Marinogammarus obtusatus* provides an excellent example (Fig. 58), *Nereis diversicolor* another.

All modern work indicates that this process does not involve, at least in ordinary cases, the development of an absolutely impermeable cuticle.[47] This indeed could hardly be expected, since some surface must be exposed for respiration, and membranes freely permeable to O_2 are likely to be somewhat permeable to H_2O. It is probable that two rather different types of adaptation are involved in different groups.

In the lamellibranch Mollusca, in which there is inevitably a great area of ctenidium exposed to the external medium, the adaptation consists fundamentally in an extraordinary dilution of the internal

---

[47] Certain insect larvae (see page 183) are impermeable, and impermeable stages are known in some eggs, such as those of the trout.

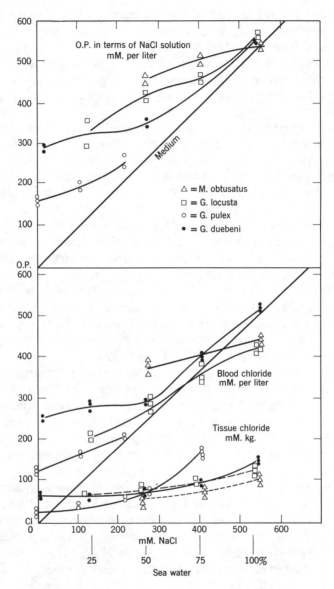

FIGURE 58. Upper panel, variation in osmotic pressure of blood with concentration of medium in *Marinogammarus obtusatus, Gammarus (G.) locusta,* G. *(Rivulogammarus) deubeni* and G. *(R.) pulex.* Lower panel, blood chloride and tissue chloride of same (Beadle and Cragg.)

medium (20–46 mM. per liter) and a dilution of the cell fluids almost, but not quite, as extreme. In *Anodonta,* which apparently has the lowest blood concentration known in any animal, the concentration of the muscles was found by Potts (1958) to be somewhat more than half again as great as that of the blood. The difference was in part due to greater concentration of potassium, as would be expected, and in part also of amino acids. In spite of this greater concentration, the animal's cells must be able to work at an unusually low internal concentration of ionic and other osmotically active substances. *Anodonta* secretes an excessively dilute urine, hypotonic to the blood. This particular kind of adaptation corresponds to the theoretical possibility of conserving salts not only by renal resorption but also by maintaining a low permeability to salts, but not to water, over the surface of the body, which approaches an ideal semipermeable membrane. Potts (1954b) shows theoretically that this is an efficient situation in fresh-water animals if it can be evolved.

The alternative scheme of adaptation is shown by the fresh-water crab *Potamon* and by various brackish forms in which a relatively low permeability to water as well as to salt, coupled with strong capacity to absorb ions actively from the environment, permits the retention of a high blood concentration with an almost or quite isotonic urine.

Potamon niloticus has been extensively studied by Shaw (1959b,c). The animal can survive for at least three weeks in 50 per cent sea water, for one to more than three weeks in 75 per cent sea water, but for less than four days in 100 per cent sea water. There is probably some variation among the species of the genus in this respect, Duval (1925) having found the south European *P. edulis* able to live in full-strength sea water. The blood concentration in *P. niloticus* remained at about 280 mM. up to 50 per cent sea water and then rose rapidly; animals near death in full-strength sea water were approximately isotonic with the medium. Death is possibly due to dehydration of some of the tissues, for the concentration of water in the muscles fell at least 25 per cent. There is, under these conditions, a slight rise in amino acids and a more definite rise in sodium, but these increases are insufficient to keep the muscles in osmotic equilibrium with the blood. *Carcinus maenas* transferred from less to more concentrated sea water liberates a much greater quantity of amino acids in its muscles and so controls their water content more effectively. Urine production is very low, less than one per cent of the body weight per day, whereas for *Anodonta* (Picken 1937; Potts 1954a) the rate is of the order of a quarter of the body weight per day at ordinary temperatures (15–18°). In what is presumably a later stage in the evolution of fresh-water decapod Crustacea, the Astacidae and *Gammarus fasciatus* have developed an hypotonic urine and so appear to be moving toward a condition comparable to that of *Anodonta*. It is, however, extremely doubtful, as Potts (1954b) points out, that *Anodonta,* with a large and permeable gill surface, could have ever passed through a stage comparable to *Potamon* or to the katadromous *Eriocheir;* the cost of maintaining a steep osmotic gradient across the gills would have been prohibitive.

Except in the river crabs, Gammaridae, Astacidae, and the fresh-water mussels, we do not really know much about the systematic distribution of these two extreme modes of adaptation nor of the intermediate situations that presumably occur. The matter is interesting from the standpoint of historical or evolutionary ecology, for we might suspect that the type of adaptation shown by *Anodonta* in which external medium, internal medium, and tissues would all have to undergo concomitant though not equivalent dilution would most easily take place in slowly freshening basins, leading through a condition like that of *Arenicola* in the Baltic to the passive production of a fresh-water animal, whereas the *Potamon* type of adaptation, which puts a small initial demand on the tissues, would be possible in any kind of brackish environment of variable salinity. A detailed study of some of the mollusks, such as *Scaphula* or *Novaculina,* that appear to have invaded tropical rivers would be interesting; the critical environmental factor may be the very gentle salinity gradients that must occur at the mouths of some of these rivers. It is at least obvious that the initial stages will tend to be harder for an animal with an enormous epithelial surface than one with an exoskeletal surface, even though the latter is far from impermeable all over. This may explain the much greater number of invasions, probably about forty, that appear to be recorded in the Crustacea than in the Mollusca, where less than twenty more or less successful attempts have been made by a group more numerous in marine species than the Crustacea.

Whatever the exact course of evolution, the internal medium in truly fresh-water animals is apparently never so dilute as the external and the body wall never completely impermeable to water, so that an osmotic gradient must always exist with an incoming stream of water. Moreover, however impermeable to salts the body wall may become and however efficient the excretory system may be in resorbing ions, salts must be taken from the environment since growth mus⁺ take place. The physiological problems to be faced, which are at least twofold, involve the excretion of the incoming water and the retention or uptake of salts from a very dilute medium.

The existence of the incoming stream of water has been critically demonstrated in *Procerodes* (=*Gunda*), in *Nereis diversicolor,* in *Carcinus maenas,* and in various purely fresh-water animals. At least in the first two organisms osmoregulation is dependent on the presence in the medium of adequate calcium without which water enters the animal so fast that the regulating mechanisms cannot cope with the incoming stream. Pantin (1931) pointed out that invasion of hard waters may be possible to animals that would be unable to enter soft waters from the sea. In *Procerodes* the excretion of water takes place from the gastrodermis; in *Nereis diversicolor* the site is unknown. In *Carcinus maenas* the rate of excretion by the antennary gland is increased as the concentration of the medium falls. The blood becomes hypertonic to the medium, but the urine is isotonic with the blood, so that the excretion of the incoming water inevitably produces a loss of salts. The manner in which salt is taken up is discussed in greater detail in a later section.

Energetics. The theory of the energetics of osmoregulation has been considered by Potts (1954b) for an animal with an ideal semipermeable boundary, an excretory mechanism of varying capacity to resorb salts, and a salt absorbing mechanism that permits a steady state to be maintained even if the urine is isotonic with the blood.

The minimum work required to maintain a steady state is given by

$$(1) \qquad W_0 = R\theta_K p_a a(C_b - C_m)\left(C_u \ln \frac{C_u}{C_m} + C_b - C_u\right)$$

where W_0 = work in calories per hour
$\quad$ R = gas constant
$\quad$ θ_K = absolute temperature
$\quad$ p_a = mean permeability in moles per square centimeter per hour
$\quad$ a = area of organism
$\quad$ C_b = concentration of blood
$\quad$ C_m = concentration of medium
$\quad$ C_u = concentration of urine

The most efficient practical condition is for the urine to be isotonic with the medium ($C_u = C_m$); the least efficient is for it to be isotonic with the blood ($C_u = C_b$). The minimum work required to maintain a steady state therefore varies between

$$(2) \quad W_0 = R\theta_K p_a a(C_b - C_m)^2 \quad \text{and}$$

$$W_0 = R\theta_K p_a a(C_b - C_m)\left(C_b \ln \frac{C_b}{C_m}\right)$$

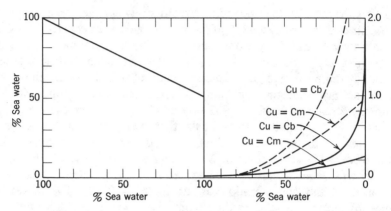

FIGURE 59. *Left panel:* Relation of blood concentration to concentration of medium in a moderately homoiosmotic marine animal undergoing dilution of the medium, half the blood concentration being retained in fresh water. *Right panel:* Osmotic work as a function of concentration implied by blood concentration, $(C_u = C_b)$ when the urine is isotonic with the blood and $(C_u = C_m)$ when the urine is isotonic with the medium. At low concentrations, as in fresh water, the latter is a far more efficient condition. Broken lines give the same situation for the hypothetical condition in which all the salt is retained, rather than half, when the animal is in fresh water, generally increasing the work. (Potts, modified.)

If C_b is supposed to fall, but less rapidly than C_m as dilution takes place, according to the line in the left-hand panel of Fig. 59, the limits of the minimal work will fall between the two lines of the right-hand panel of the same figure. It will be observed in this realistic example considerable dilution is needed to produce any significant difference between regulation with urine isotonic or hypotonic to the blood, but as the great dilution of quite fresh water is approached the production of hypotonic urine becomes increasingly economical.

In fresh water in general C_m will be small compared to C_b and we may write the limits as between

$$(3) \qquad W_0 = R\theta_K p_a a C_b{}^2 \quad \text{and} \quad W_0 = R\theta_K p_a a C_b{}^2 \ln \frac{C_b}{C_m}$$

A similar value for the required work can thus be obtained by reducing C_b or by reducing p_a. The former corresponds to the situation in *Anodonta* and doubtless other soft-bodied invertebrates, the latter to the condition in *Eriocheir* and *Potamon* with isotonic urine and to *Astacus* with hypotonic urine. Potts computes for specimens weighing 60 grams wet (excluding shell in *Anodonta*) the values given in Table 3. In assessing the last row of figures it must be remembered that though the minimal required work represents a small increment of the total metabolism of the whole animal it may represent a large increment of the metabolism of the regulatory organs; the process, moreover, is certain not to be completely efficient. It is evident, however, that the mechanisms employed by *Anodonta* and by *Eriocheir* can be of approximately equal physiological efficiency, though, as pointed out, the *Anodonta* mechanism is evolutionarily much less probable.

The increased metabolic rate often observed on transfer of a euryhaline animal from salt to less salt water has frequently been taken as evidence of the magnitude of the osmoregulatory work, but this is not necessarily the case. In *Nereis diversicolor* statistically significant changes in metabolism have not been found (Schlieper 1929; Beadle 1931, 1937; Krogh 1939). In this case, however, cyanide abolishes the capacity to perform the regulation, which therefore seems to depend on oxidative metabolism. Presumably the increased work required for osmoregulation is small relative to the experimental error and to random variation of the metabolic rate of the animal. Similarly, in *Eriocheir* no changes in metabolic rate dependent on the concentration of the medium have been noted, but a marked increase in respiratory rate has been found in *Carcinus maenas* when transferred from normal to diluted sea water. Since this animal secretes urine isotonic with the blood, such an increase has been attributed to the work of taking up salt against a concentration gradient. A decrease in oxygen uptake, amounting to 40 to 50 per cent, has been recorded in the fresh-water crayfish *Astacus* on transfer from fresh water to sea water diluted to a salinity of 15 per mille (Schwabe 1933; Peters 1935). Such a figure is much greater than the minimum calculated by Potts. The haemolymph, initially isotonic with sea water of salinity 14 per mille, rises in concentration slightly, becoming isotonic with

TABLE 3 *Work done in maintaining osmotic gradients across the body wall in a fresh-water bivalve and in two crustaceans*

	Anodonta	Eriocheir[a]	Astacus
p_a	14.2	0.16	0.23 ml. hr^{-1}
W_{os} done at body surface	0.0131	0.176	0.0288 cal. hr^{-1}
W_{oe} done in excretory organ	0.0014	0	0.0079 cal. hr^{-1}
W_o total $(W_{os} + W_{oe})$	0.0145	0.176	0.0367 cal. hr^{-1}
W_o relative to total metabolism	1.2	1.3	0.3%

[a] Potts corrected by Werntz (1957).

water of 20 per mille salinity. Excess osmotic pressure is thus decreased by the transfer and the need for water regulation is indeed reduced. Peters, however, found that not only the renal tissue, which in *Astacus* certainly recovers salts, but also muscle and hepatopancreas participated in the metabolic decrease, so that it is quite unjustifiable to attribute the whole of the difference in metabolic rate in fresh and salt water to the changed amount of work required to keep the blood hypertonic to the medium. Life in fresh waters may be more expensive than in salt, but not entirely, possible even not mainly, because of the over-all osmotic work that must be done across the body wall in a dilute medium.

Distribution and role of excretory organs. It has long been known that in many groups of fresh-water animals the structures that excrete water are better developed than those of allied marine forms. The presence of contractile vacuoles in fresh-water protozoa and their rather general absence in most marine rhizopods, some groups of flagellates, and a few of ciliates provides a familiar, if often exaggerated, example (Kitching 1938). There can be little doubt that the vacuole is primarily an osmoregulator in fresh water. This is neatly demonstrated by Guillard's (1960) finding that the fairly euryhaline fresh-water *Chlamydomonas moewusii* is converted into a stenohaline brackish organism by the mutational loss of the vacuole; a further adjustment of the cell physiology to operate at a higher ionic concentration would be needed to produce a marine form. The slow pulsation of the vacuole in the marine ciliates that possess it probably merely keeps pace with sea water ingested with the food. The generally better development of a protonephridial system in fresh water as contrasted with marine Turbellaria provides a comparable example. Grobben (1880) pointed out the same sort of phenomenon in the better development of the antennary gland in the nauplius of the fresh-water *Cyclops* than in the marine *Cetochilus,* and many other investigators have concluded that such differences are rather widespread in the Crustacea.

Whether the development of a protonephridial system at some level above the acoel Turbellaria indicates that the vast array of animals constituting a large part of the animal kingdom, which are or seem to be descended from forms with such a water-excreting mechanism, had an ultimate fresh-water origin may well be debated, but the hypothesis obviously cannot be categorically denied and even has a certain plausibility, though as indicated earlier (pages 134–135) this type of argument may be specious. When we find in closely related organisms excretory organs that are more elaborately developed in fresh than in salt water species, we are probably justified in suspecting that the elaboration is primarily concerned with resorbtion of salt rather than the ultra-filtration of water. This is clearly indicated in the Astacidae, when compared with the poikilosmotic *Homarus,* though it must not be forgotten that in such a poikilosmotic form the excretory organ does play a part in regulating the ionic proportions in the body fluid. We evidently find a similar situation in the fresh-water Gammaridae, in which the anatomical observations that the marine forms have a less elaborate antennal gland than the fresh-water (Schwabe 1933; Hynes 1954) are confirmed by Werntz (1957), who, in addition, obtained direct evidence of almost isotonic urine in *G. oceanicus* at all dilutions but every hypotonic urine in *G. fasciatus* in fresh and dilute sea water. Many crustacea apparently dispense successfully with this adaptation. No anatomical evidence of elaboration of the antennary gland is given by the fresh-water *Mysis relicta* as contrasted with the salt-water Mysidacea or by fresh- and salt-water species of *Palaemonetes* (Schwabe 1933). In *Potamon,* moreover, it is known that the urine is almost isotonic with the blood. As Potts (1954b) has indicated theoretically, a resorption mechanism will always be energetically advantageous, if not essential, when a semipermeable animal with a large exposed area, such as a ctenidium, becomes adapted to quite fresh water. In the case of the Crustacea, which in general are much less permeable than mollusks, the need for such an adaptation is less pressing.

The uptake of chloride and other ions. If a water-regulating mechanism, which permits excretion of the incoming stream that must pass from a fresh-water environment across any semipermeable part of the body wall of the organism, is an essential feature of fresh-water animals, some method of retaining or utilizing the limited quantity of inorganic ions within or without such organisms in a very dilute medium is no less important, and the degree of development of these methods is likely to determine rigorously the limits of colonization of lakes and rivers.

In the adult animal a completely efficient resorption of ions by the excretory system is in theory all that is required; in practice this is doubtless infinitely hard to achieve. In a growing animal an external source of ions is obviously necessary, though perhaps the food can sometimes act as such a source for growth, as it has been supposed to do for maintenance in the eel in fresh water and in the Anostraca and Notostraca studied by Krogh (1939). In all other cases, and it must be admitted the crustacean exceptions may be more apparent than real, some mechanism for active salt absorption independent of the food appears to exist.

The first hint of such a mechanism seems to have come from Fritsche's (1916) study of the osmotic pressure of the body fluids of *Daphnia magna*. He found that animals in medium of concentration 16 mM. osmolal[48] had a mean blood concentration of 156 mM.; animals kept in water of concentration 210 mM. had a mean blood concentration of 242 mM. An excess pressure over the medium is here clearly maintained, though with increasing difficulty as the medium becomes concentrated, as indeed is now known to be true in general in fresh-water animals. It is therefore probable, as Frische pointed out, that salt is taken up by the growing animals. Critical demonstration of the uptake of salt from the medium by *Carcinus maenas* was given by Nagel (1934). Krogh (1939) and his co-workers, Wikgren (1953), Koch and Evans (1956), Shaw (1959a,c), and some other investigators have now established the process as all but universal. The interesting features concern the sites of the uptake, the rate of uptake and, from an ecological point of view, the minimum concentration that permits equilibrium.

Krogh (1939) has shown that mechanisms for the independent uptake of Na^+ and Cl^- exist in the leech *Haemopsis sanguisuga*. Wernstedt (in Krogh 1939) demonstrated the uptake of NaCl by all the fresh-water mollusks she studied, including representatives of the lamellibranchs and of prosobranch and pulmonate gastropods. *Viviparus viviparus* could reduce the concentration to 0.105 mM. or 3.6 mg. Cl per liter. *Dreissena* and *Unio pictorum* were almost as effective, but *Lymnaea stagnalis* seems to have required a minimum concentration of eight or nine times this amount. Such observations suggest that variations in chloride in inland waters might be ecologically of considerable importance.

In the catadromous mitten-crab *Eriocheir sinensis,* when in fresh water, there is again evidence that cation and anion uptakes are independent and quite selective (Krogh 1939). The site of the uptake is almost certainly the gill. *Astacus* appears to absorb salt less rapidly than *Eriocheir,* which, in view of its renal absorbtive mechanism, is not unreasonable. The crayfishes in which, at least in *Cambarus* (Maluf 1940), all gills are involved in salt uptake can, however, take up sodium chloride from a much lower concentration than can the mitten crab, for although the crab stops absorbing at 0.2 to 0.4 mM. Cl, or 0.2 to 0.5 mM. Na (Koch and Evans 1956), *Astacus* which loses

---

[48] These figures are given on the basis of a depression of freezing point of 1.86° C. for an ideal nonelectrolyte of concentration 1 molal. Krogh uses 1.0° C. for 0.293 mole Cl per liter, which gives rather under half the figures used above.

about 0.15 mM. g.$^{-1}$ hr.$^{-1}$ of sodium, can be in equilibrium with a medium containing 0.02 to 0.09 mM. Na. The latter concentration, corresponding to 0.5 to 2.0 mg. Na. per liter, is likely to be available in inland waters, whereas the low but definite salt requirements of *Eriocheir* are doubtless met by most large Chinese and European rivers, but not by the least contaminated fresh waters.

In the Potamonidae Shaw (1959c) found that *Potamon niloticus* loses about 0.8 mM. Na. g.$^{-1}$ hr.$^{-1}$ and about one-sixteenth this amount of potassium. It can maintain its salt concentration in a medium containing at least 0.05 mM. per liter Na and 0.07 mM. per liter K. Another species *P.* cf. *johnstoni* was found to have a sodium requirement that permitted equilibrium at 0.02 mM. per liter Na and apparently occurred in more dilute waters than *P. niloticus.* At least in *Potamon,* sodium and presumably chloride concentration appear therefore to have ecological significance. The uptake of chloride if not of sodium occurs through the first three gills; in the fresh-water prawns it appears to take place through the gills and the inner surface of the branchiostegite (Koch 1934; Ewer and Hattingh 1952).

Among the lower Crustacea it is reasonable to suppose that the branchiae of *Daphnia,* which contain silver-reducing cells, are, by analogy with the anal papillae and anal gills of Diptera, the salt-absorbing organs. A few experiments by Krogh appeared to indicate that the anostracan *Branchipus* and notostracan *Lepidurus* have no mechanism for taking up salt except from the food. Panikkar (1941a), however, has demonstrated that in the anostracan *Chirocephalus diaphanus* although survival is impossible for more than two or three days in tap water containing but a trace of chloride, in the absence of food, there is clear evidence that a chloride-absorbing mechanism is present on the bracts of the appendages. It is merely less efficient than that of other fresh-water crustacea. The same condition no doubt obtains in other anostracans and phyllopods, including those studied by Krogh.

The dipterous larvae, in general, have organs that appear to be salt-absorbing and which have been termed anal papillae or anal gills. They are found in both Nematocera and Brachycera and are not limited to aquatic forms. They reduce silver from dilute silver nitrate solutions. Wheeler (1950) has shown that in *Drosophila* the region in question accumulates iodine strongly. It is probable that chloride is also concentrated and that for this reason AgCl is precipitated and reduced.[49]

In the aquatic nematoceran larvae (*Culex, Aedes, Tendipes*) salt

---

[49] Krogh (1939), however, thinks Ag$^+$ is taken up for Na$^+$.

uptake (Koch and Krogh 1936; Koch 1938; Wigglesworth 1933a,b,c, 1938) occurs through the so-called anal gills. The organs are also permeable to water. Wigglesworth has demonstrated experimentally that the degree of development of the anal gills in *Culex* and *Aedes* is controlled by the salinity of the medium. Many observations indicating reduction of the anal gills of *Tendipes* in saline water have been made, though factors other than NaCl content may be involved. The so-called ventral gills or ventral tubes of *Tendipes* are not involved in chloride uptake.

Histological study by silver staining suggests that in the Anisopteran dragonflies patches of epithelium on the rectal gills are involved in chloride absorption from the medium, whereas three plaques in the prerectal ampoule probably remove chloride from the urine.

Among the fresh-water fishes the ability to take up salt from the medium varies somewhat. The goldfish *Carassius auratus,* particularly well studied by Krogh, gave evidence of independent and highly selective uptake of cations and anions, apparently through the gills. Wikgren (1953) concludes that a carp losing about 5 per cent of its salt content per day could probably maintain salt balance if it constantly fed on healthy *Daphnia magna,* but that in *Salmo* and *Lampetra* in fresh water, from which the loss is more rapid, the source in the food would be inadequate. Any fish during relatively short periods of starvation is likely to have to use active uptake from sources other than the food.

The minimum equilibrium values recorded in the laboratory for various animals may be arranged in the categories of Table 4.

How far these figures can be considered ecologically significant is doubtful. Wikgren suspects that both long period adaptation and possibly the existence of local races may be involved. Only in the case of Shaw's (1959c) study of *Potamon* do field data suggest that distributions may be limited by the efficiency of salt uptake at great dilution. It is, however, reasonable to suppose that the limits of dilution that brackish-water species, such as the flounder in Table 4, can tolerate are set by efficiencies within the range of the table. The common phenomenon, noted again and again in the systematic presentation earlier in this chapter, of sporadic species of brackish or marine animals apparently overlapping the typical fresh-water fauna in distribution but not invading fresh waters far from the coast receives a rational explanation if we assume genetic differences in the efficiencies when different species with different evolutionary histories are compared.

The effect of temperature. Smith (1957), studying *Nereis (Hediste) limnicola,* found that above about 5 per mille Cl, the chloride

TABLE 4. *Minimum concentrations of sodium or chloride required in the medium to maintain ionic equilibrium in various fresh- and brackish-water animals*

0.00–0.05 mM. Cl or Na	*Astacus astacus*	(K. Cl)
	Austropotamobius pallipes	(S. Na)
(= 0.00–0.09 mM. osmolar ideal	*Potamon* cf. *johnstoni*	(S. Na)
nonelectrolyte)	*Lampetra fluviatilis*	(W. Cl)
	Rutilus rutilus	(K. Cl)
	Carassius carassius	(W. Cl)
	Carassius auratus	(M. Cl)[a]
	Rana esculenta	(K. Cl)
0.05–0.10 mM. Cl or Na	*Anodonta cygnea*	(K. Cl)
	Viviparus viviparus	(K. Cl)
	Potamon niloticus	(S. Na)
0.10–0.20 mM. Cl	*Unio pictorum*	(K. Cl)
	Dreissena polymorpha	(K. Cl)
	Perca fluviatilis	(W. Cl)[a]
0.2–0.4 mM. Cl	*Eriocheir sinensis*	(K. Cl)
	Gasterosteus aculeatus	(Danish fresh-water race; K. Cl)
	Pleuronectes flesus	(W. Cl)
0.40–0.80 mM. Cl	*Lymnaea stagnalis*	(K. Cl)
	Ameiurus sp.	(K. Cl)
	Ambystoma sp. larva	(K. Cl)
0.80–1.60 mM. Cl	*Haemopsis sanguisuga*	(K. Cl)
	Libellula sp. nymph	(K. Cl)
	Acerina cernua	(K. Cl)

K = Krogh (1939); M = Meyer (1948); W = Wikgren (1953); S = Shaw (1959a,c). Na = sodium uptake; Cl = chloride uptake.

---

[a] Krogh puts *Carassius auratus* in the 0.10–0.20 mM. class and gives an impossibly high figure for *Perca fluviatilis*.

content of the coelomic fluid is approximately that of the medium but that at lower external concentrations the animal exhibits marked hyperosmotic regulation. In water containing only 0.006 g. Cl per liter, which is within the normal range of fresh waters in many regions, the mean chloride content of the coelomic fluid is as high as 3.6 g. per liter at 18 to 20° C., and the animal behaves essentially as an inhabitant of fresh water. If the temperature is reduced to 0.5–1.5° C., regulation is less perfect, the internal chloride content falls to 3.0– 3.4 g. per liter, and the animals appeared weak and sluggish. It is probable that if the internal chloride falls below 3.0 g. per liter death will occur. The animals in water near freezing are therefore at their limit with regard to hyperosmotic regulation. Since the worms live better in Lake Merced water than in sea water diluted to the same chlorinity, ions other than sodium and chloride presumably are involved, as would be expected. It is reasonable to suppose that *N. limnicola* is just adapted osmotically to fresh waters at the temperatures it encounters in its natural range, but that if the winter climate were more severe it would have to make further physiological adjustments to survive ·prolonged periods of nearly freezing temperatures. Smith thinks that this final adjustment may not have been possible in *Nereis diversicolor* in Europe and that the problematic aspects of the distribution of this worm in the Baltic may be explained by a failure to regulate in very dilute water during the winter.

Among the crustacea there is a considerable body of information. Otto (1937) found in *Eriocheir sinensis* that the haemolymph of animals brought into fresh water between 0 and 6° C. froze at a mean temperature of −1.262° C., corresponding to 0.678 osmolal or 0.370 molal Cl, whereas the haemolymph of animals at 24 to 25° C. froze at a mean temperature of −1.097° C., corresponding to 0.589 osmolal or 0.322 molal Cl. Other investigators have found a similar dilution of the haemolymph at high temperatures in prawns. These results suggest that a rise in temperature increases the stream of water entering and leaving the animal more than it increases salt uptake, a conclusion that is perhaps unexpected. In *Gammarus oceanicus* in dilute water and in *G. fasciatus,* Werntz (1957) finds the urine production doubled by raising the temperature from 15 to 25° but gives no information on concentration changes. The increase in the water stream, however, is obvious in this case. Otto (1934) suggests that the dilution of the haemolymph sometimes limits the animal at the warmer edge of its range and concluded from rather imperfect studies that distribution of the Zuiderzee crab *Rhithropanopeus harrisii tridentata,* which is absent from brackish water in south Holland, is a possible example of such limitation. Later physiological studies by Kinne and Ratthauwe (1952) confirm the dilution of the haemolymph at high temperatures and give additional evidence of poor viability in warm dilute water. Panikkar (1940) believes that the lowered osmotic pressure of the haemolymph at high temperatures observed when *Palaemonetes varians* and *Leander serratus* are brought into fresh water implies that at such temperatures a dilute haemolymph is desirable; the data do not rule out the possibility that it is merely inevitable. Nevertheless, the experimental work on the salinity tolerances of prawns does suggest that low concentrations are better tolerated at high temperatures even though the changes in rate of excretion of water relative to uptake appears to result in a steady state with a more dilute haemolymph at high than at low temperatures.

Broekema (1942), however, working with *Crangon crangon,* found the haemolymph to be more concentrated at 20 than at 4° C. at low salinities, though less at high salinities. The animal normally migrates into warmer, more dilute water in summer.

At quite low temperatures it is possible that the whole regulatory mechanism may become inoperative. In fresh-water crabs Shaw (1959c) found the sodium uptake mechanism to be inhibited when *Potamon niloticus* was brought from 24 to 14° C. or when the ambient oxygen concentration was reduced to about 1 mg. per liter. He believed that these changes had ecological significance.

Riegel (1959a) found evidence at low temperatures of considerable movement of water into and out of *Gnorimosphaeroma,* whereas at 16° C. only salt appears to move in osmotic response to salinity changes in the medium. This suggests a more perfect type of regulatory response at the higher temperature.

Wikgren, who gives a good review of earlier work not considered here, studied the salt balance of the crayfish *Astacus astacus,* the lamprey *Lampetra fluviatilis,* and the carp *Carassius carassius.* Although the crayfish is much less permeable to water than the two vertebrates and excretes a less dilute urine, the temperature relations of all three are comparable. A greater loss of chloride occurs near 0 than at 10° C., and a temporary increase in chloride loss can be induced by a sudden rise and fall in temperature. This effect of temperature shock at least indicates that the mechanisms involved are not under the control of the simplest physicochemical processes.

In *Lampetra,* which Wikgren investigated most thoroughly, the cold-induced chloride loss is a transitory phenomenon, and a slow adaptation, probably involving a lowering of permeability to ions, reduces the chloride loss to the original very low rate. Witgren calls attention to earlier work on winter and summer frogs which indicates a comparable adaptation.

Ecologically, we may suspect from this work that when the ionic uptake mechanism is not perfectly developed invasion will be easier at high than at low temperatures. In the higher forms that are most completely adapted various compensations permit the animal to regulate near freezing temperatures. These findings are concordant with the general observations that marine invasion is most likely to occur in the tropics. In certain special cases in which dilution of the internal medium is observed at high temperatures it is possible that this dilution may occasionally set an upper limit to the temperature tolerance of the animal exhibiting it. On the whole, when the haemolymph can be diluted without fatal results at a high temperature, this dilution will be energetically favorable to the animal. In *Eriocheir* the minimum work at a given concentration (cf. Potts 1954b) would be increased by a factor of 298/276, or 1.08, in going from 3 to 25° C.

whereas dilution would at least reduce the work by the square of the ratio of the concentrations of the haemolymph at the two temperatures or, as in Otto's 1937 experiments, by $(1.097/1.262)^2$, or 0.756. It is evident that for such an animal the dilution of the haemolymph is likely to be a real economy insofar as Potts' theory (equation 3) is applicable.

Special physiological aspects of osmoregulation in the lower vertebrates. The osmoregulatory mechanisms of cyclostomes and fishes have been discussed in detail by Black (1957), by Beadle (1957), and, in relation to endocrine factors, with great learning by Pickford and Atz (1957). A relatively undocumented summary is therefore all that is needed in the present work.

The blood of the myxinoid cyclostomes, or hagfishes, as has already been indicated, is almost isotonic with the medium; survival is possible only over a rather restricted salinity range. The blood contains approximately the same amount of chloride as does sea water, but with rather more sodium and phosphate and less magnesium, calcium, and sulfate (Robertson 1954). The kidney is a segmental glomerular opisthonephros. Though various authors, including Black in her recent review, believe that this structure implies a fresh-water origin for the group, Robertson gives cogent arguments against such a view, and there is really no more reason to believe that the myxinoids were originally fresh-water organisms than that the crustacea with their maxillary or antennary glands, which can act as ultrafilters, were of lacustrine or fluviatile origin.

The lampreys may be anadromous or purely fresh-water. The rather meager evidence for *Petromyzon marinus* indicates that, unlike the hagfishes but like the teleosts, the blood is markedly hypotonic to the medium when the animal is living in the sea. In fresh water the blood may be a little less concentrated than in the ocean (Sawyer in Black 1957), but it is, of course, very hypertonic to the medium. The urine in fresh water is markedly hypotonic to the blood. Wikgren found that the body surface of *Lampetra fluviatilis* in fresh water is rather more permeable than that of the carp and that the rate of chloride loss is considerably greater than in the carp or in *Astacus*.

The capacity to regulate hyposmotically in the partly anadromous *Lampetra fluviatilis* is apparently lost when the animal migrates from estuarine into fresh waters. (Morris 1956). Such animals become poikilosmotic in concentrations higher than 200 mM. osmolar and survive poorly above 300 mM. osmolar. The purely fresh-water *L. planeri* has a slightly lower blood concentration, becomes isosmotic at 150 mM. osmolar, and dies at 160 mM. osmolar in the adult stage but is somewhat more resistant as an ammocoetes larva (Hardisty 1956). The loss of hyposmotic regulatory power is apparently correlated with a degeneration of chloride-secreting cells. Though no certain palaeontological evidence is available, it would be not unreasonable to regard the lampreys as an old fresh-water group derived possibly from fresh-water ostracoderms, as has already been indicated.

The marine elasmobranchs are usually supposed to possess blood slightly hypertonic to the medium; Robertson (1954), however, thinks that this conclusion is due to technical errors and that the blood is nearly isotonic with sea water. The high osmotic pressure is due largely to urea and to smaller amounts of trimethylamine oxide. Chloride is excreted by the rectal gland (Burger and Hess 1960; Burger 1962; Fänge and Fugelli 1963).

The modern fresh-water elasmobranchs have less concentrated blood than the marine but more concentrated than freshwater teleosts. The chloride content is about the same in the two groups, the difference being due mainly to urea. The only reasonable explanation of the peculiar osmotic relationships of the elasmobranchs is that it is secondary, the result of readaptation of fresh-water ancestors to the sea.

A great deal of work has been done on teleostomes, but it is confined almost entirely to the teleosts. The anadromous sturgeons, as far as is known, resemble the anadromous members of the teleosts; the sterlet *Acipenser ruthenus* is a completely fresh-water fish.

In the sea the blood of teleosts has a salinity of about 9 to 12 per mille, or one third of that of sea water. The blood of strictly fresh-water teleosts has about half the concentration and chloride content of that of marine species. Many of the less sensitive fresh-water species can live in water of any concentration hypotonic to the blood, and similarly a number of marine species survive in diluted sea water so long as it is hypertonic to the blood. Some fresh-water species such as the carp can be acclimated to dilute sea water more concentrated than the blood. Such fish, like the lampreys that have lost their capacity for hypotonic regulation, become completely poikilosmotic but cannot survive full-strength sea water. In the carp, Duval (1925) observed survival in up to about 50 per cent sea water.

The euryhaline teleosts, many of which are anadromous or catadromous, present a variety of situations, the commonest of which may be described as *regulatorily poikilosmotic*.[50] Movement from fresh

_____

[50] Pickford and Atz (1957) speak of these fishes as poikilosmotic, but the range of blood concentration is much less than would be observed in an animal such as *Arenicola*, and the blood is regulated in both media.

to salt water or vice versa leads to a change in blood concentration from that characteristic of one medium to that characteristic of the other. This is found, with varying details, in the salmonids, in immature (yellow) eels in experiments, and in sturgeons. In a few cases there is strict homoiosmosis, even when an animal moves or is moved from completely fresh to completely salt water. The anadromous race of *Gasterosteus aculeatus* retains a *fresh-water* type of blood in salt water so long as it is immature but cannot live, when in breeding condition, in salinities that are tolerated earlier in the life history. The silver, or immediately prereproductive eel, in sea water appears to retain a fresh-water blood concentration. In contrast, Bergeron (1956; see also Burden 1956) finds that *Fundulus heteroclitus* is strictly homoiosmotic over a wide range of salinities and retains a characteristically *marine* blood concentration, though in this case, at least in some fresh waters, the chloride content is lowered (Black 1948) and the chloride lost must be replaced by other anions.

The ordinary mechanism of hyposmotic adaptation in the sea is for fish to drink sea water. Sodium and chloride are actively excreted from the gills and the divalent ions by the kidney. The urine flow is very small, and in some groups the kidney has become aglomerular. Water must be removed osmotically from any permeable surface, the rates of drinking, salt excretion, and water loss being such that the concentration of the blood is kept at a steady state hypotonic to the medium. The site of chloride excretion is usually believed to be certain cells (Keys and Willmer 1932) in the gills. The chloride-excreting function of these cells has not been unequivocally established; they are rich in carbonic anhydrase which may play a part in the active excretion of chloride. The mechanism of hyperosmotic adaptation is fundamentally the excretion of a very dilute urine, hypotonic to the blood, as a result of efficient renal salt excretion. There is also ordinarily a non-enteric mechanism for salt uptake from the environment, the over-all efficacy of which has already been discussed. It is possible, at least in some cases, that the salt-absorbing mechanism is identical with the salt-excreting mechanism, but operating in the reverse direction. Though most fresh-water fishes have a glomerular kidney, at least in the case of the pipe fish *Microphis boaja* in Siam, migration into fresh water has been possible for a species with an aglomerular kidney (Grafflin 1937).

The true euryhaline fishes, whether homoiosmotic or regulatory poikilosmotic, must be able, according to their environment, to utilize to some extent both mechanisms just described. A really full analysis has been made in relatively few cases. In eels in sea water the typical marine type of regulation occurs, but in fresh water salt uptake is apparently only enteric. The body wall of intact fresh-water eels is impermeable to both water and salts. Immature fresh-water eels are somewhat poikilosmotic but regulatory. Silver or sexually mature migrating eels apparently maintain a fresh-water blood concentration on entering the sea.

The best studied case is that of *Fundulus heteroclitus* in which Bergeron (1956) found complete homoiosmosis, with the blood osmotic pressure corresponding to $\Delta = -0.75°$ C. over a wide range of salinities ($\Delta = -0.01$ to $1.77°$ C.). In water below $\Delta = -0.04°$ C. the fish is a typical fresh-water animal, save for its high blood concentration; in water above $\Delta = -0.75°$ C. it is a typical marine teleost. In the intermediate range the fish drinks like a marine teleost, and its chloride cells paradoxically appear to be organized to excrete rather than absorb chloride, having the cytology of those of a marine fish. Bergeron concludes that in this wide range of intermediate salinities a dilute urine must be produced; the chloride content of the blood is maintained in the face of salt excretion by drinking the medium.

The functional interrelation of the processes involved in osmotic regulation in fish is complicated and probably varies from group to group. The integrity of the mucus layer on the surface of the animal has often been regarded as important, but Krogh (1939) failed to confirm earlier work which demonstrated its great significance in the eel.

Some experimental work suggests nervous control of osmoregulation in fishes. Endocrine factors have also been suspected; curiously enough, the kind of water and salt regulation that we might deduce from the known roles of the pituitary and adrenal cortex in higher vertebrates is not established in fishes. In some cases the thyroid appears to be involved in osmoregulatory physiology. Thyroid feeding reduces the salinity tolerance of nonbreeding anadromous *Gasterosteus,* though this cannot be general, for the increase in regulatory capacity in the silver eel is coincident with an increase in thyroid function. In *Fundulus,* moreover, there is evidence that the thyroid is not involved in osmoregulation (Harris in Pickford and Phillips 1959). The only really clear case of endocrine control of osmoregulation is in *Fundulus heteroclitus.* In this fish hypophysectomy destroys the capacity to regulate in fresh water, though it has no effect on hyposmotic regulation in sea water. Chloride loss in fresh water is greatly increased, and the fish become asthenic and die in a few days or weeks (Burden 1956). The effect can be abolished by injection of the brei of *Fundulus* pituitary and to some extent by that of *Perca flavescens,* but not by that of the marine *Pollachius*. Pickford and Phillips (1959) have shown, most interestingly, that at least in part prolactin is involved, a finding that is of obvious evolutionary significance. Hypophysecto-mized fish in fresh water were found by Burden to have fewer mucous cells on the gills than have intact individuals.

It is evident from the scattered observations on other fish that the rather peculiar situation in *Fundulus* cannot be generalized to all euryhaline teleosts. When the whole mechanism of osmoregulation is understood it will presumably be found to consist of a number of superimposed homoiostatic systems, different adjustments in varying parts of the whole mechanism in different species giving comparable over-all end results. In the present state of our knowledge it looks as though the whole mechanism will turn out to be more complex than that of the fresh-water Crustacea or Mollusca. This apparently greater complexity, however, may be due merely to an over-all greater understanding of vertebrate than invertebrate physiology, which would permit relevant questions to be asked in greater detail about the killifish than about the crayfish.

Hypotonic blood in marine crustacea, the problem of migration from fresh water to the sea. The general rule that truly marine invertebrates have body fluids isotonic with their marine environment has some quan-titatively minor but qualitatively important exceptions. In certain grapsid and ocypocid crabs the blood is strikingly hypotonic to ordinary sea water, at least during a large part of the molting cycle (Baumberger and Olmstead 1928). In at least some cases in which the animal lives on high localities on rocky coasts (*Pachygrapsus crassipes*), on mud flats (*Uca* spp.), or in mangrove swamps (*Heloecius*) this hypo-tonicity seems associated with a partly terrestrial habitat. In such examples (Edmonds 1935, Jones 1941) striking hypotonicity is retained in concentrated sea water, and Jones supposes that the adaptation is largely to the concentrated salt water that might be present in the gill chambers after evaporation in the air. The development of this regulation, however, is not a necessary condition for some degree of euryhaline hypertonic regulation in dilute water. All the genera just mentioned can do this, but so can species of *Hemigrapsus* that are isotonic with the medium in normal or concentrated sea water. Many species of the various subfamilies of the Grapsidae, notably the marsh crabs of the subfamily Sesarminae, live in dilute water.

In the genus *Leander* the blood concentration is about 82 per cent of sea water in *L. serratus* and 75 per cent in *L. squilla* according to the interesting studies of Panikkar (1941b). This author believes it to be probable that these prawns, which are closely allied to the fresh and brackish water genera of the Palaemonidae, have been derived from fresh or brackish ancestors that reinvaded the sea. However, Broekema (1942) found *Crangon crangon,* which is a member of a marine family but which enters dilute waters in summer, to show some osmotic regulation, having not merely hypertonic blood in dilute water but hypotonic blood in sea water over 21.5 to 23 per cent, the exact value depending on temperature. This suggests that development of a relatively small degree of regulation, appropriate to this animal's life history, can confer on it a blood hypotonic to ordinary sea water when it returns to that medium. Odum (1953), moreover, has found the same phenomenon in the locally catadromous crab *Callinectes sapidus.*

It is most unfortunate that nothing is known about other invertebrates, such as the marine earthworm *Plutellus* (= *Pontodrilus*), which may have had a history of migration from fresh waters to the sea, or the various species of marine rotifers and Cladocera. Though Beadle and others have believed that an early fresh-water history should be reflected in later osmoregulatory behavior in the sea, it is by no means obvious that regulatory capacity would always be retained over a long stretch of evolutionary time if it had ceased to be of any value.

Regulation in hyperhaline environments. The haemolymph of *Artemia,* ordinarily living in hyperhaline environments, is hypertonic to the medium in 10 to 25 per cent sea water, in which it can survive, but it has hypotonic haemolymph at higher environmental concentrations up to saturated NaCl. Regulation is much as in teleosts; the animal drinks the medium and excretes salt from the branchiae of the first ten thoracic legs (Croghan, 1958a,b,c,d).

Salinity relationships in aquatic reptiles, birds, and mammals. It has been pointed out by Zeuthen (in Krogh 1939) that all large flying birds must drink in order to stabilize their body temperatures, which otherwise would become impossibly high. In oceanic birds this implies drinking sea water. The forms feeding on teleosts (Sulidae, *Pelecanus, Phalacrocorax*) have some advantage in that their food is hypotonic to the ocean, but many of the largest species, notably among the albatrosses, feed on squids, which have isotonic body fluids. Lockley (in Krogh 1939) has observed *Puffinus, Hydrobates, Fratercula,* and *Alca,* members of exclusively marine families, drinking sea water, though not extensively. It is obvious that if any appreciable water intake by oceanic birds does occur a very chloride-rich urine would be expected. This, however, is clearly not the case, and Hutchinson (1950) has pointed out that the low chloride content of Peruvian guano indicates that the birds, mainly *Phalacrocorax bougainvillei* and *Sula variegata,* which produce such excreta, are losing no more chloride from the kidney than they would obtain from the hypotonic fishes they eat.

This paradoxical situation has been resolved by the recent remarkable work of Schmidt-Nielsen and his associates which demonstrates clearly that marine reptiles and birds can dispose of excess salt by excreting it from glands associated with the nasal region.

The marine iguana, *Amblyrhynchus cristatus,* of the Galapagos Islands blows a cloud of mist from its nostrils which Schmidt-Nielsen and Fange (1958) find can be hypertonic to sea water after salt administration.   The very curious habit of weeping exhibited by female marine turtles after they land to lay their eggs, which is the basis of a well-known character in *Alice in Wonderland,* is also explained by Schmidt-Nielsen and Fange who find that in *Caretta caretta* a large salt gland, opening in the orbit, secretes a salt solution hypertonic to sea water after injection of NaCl. A comparable gland is known in *Chelone midas* and another rather less well developed in the brackish *Malacolemys terrapin.*   Histologically, these glands resemble the nasal salt glands that the same group of workers has studied in birds. Among the sea snakes at least *Hydrophis ornata* has a histologically similar gland above the eye, and nasal glands occur in *Enhydris* and some other sea snakes, though curiously not in the pelagic *Pelamis.* The same authors suspect that the nasal gland in the Crocodilia, described as large and rather acinous in the sea-going *Crocodylus porosus,* will prove to be a salt gland.[51]

In the birds nasal salt glands which excrete a solution hypertonic to sea water have been demonstrated in *Phalacrocorax auritus* (Schmidt-Nielsen, Jorgensen, and Osaki 1957, 1958), in *Spheniscus humboldti* (Schmidt-Nielsen and Sladen 1958), and in *Larus argentatus* (Fange, Schmidt-Nielsen, and Osaki 1958).   These birds are as well adapted to life at the sea surface as marine teleosts are to life in ocean water. In the plovers (Charandriidae) Bock (1958) has concluded that cranial differences in the nasal region are also correlated with life on the margins of fresh or salt water; the large nasal gland of sea birds was in fact well known before Schmidt-Nielsen's studies indicated its very interesting physiological significance. The rather

---

[51] The position of such a structure could be studied in the skulls of mesozoic reptiles. At least in the ichthyosaurs, as Dr. J. T. Gregory has shown me on a partly disarticulated skull, there must have been a fair amount of room for an orbital gland.

sharp divisions of various orders of aquatic birds into marine and fresh-water groups thus receives a reasonable physiological explanation.

Among the marine mammals Irving, Fisher, and McIntosh (1935) have shown that the ichthyophagous marine seals have no great difficulty in maintaining their water balance from the water, actual and metabolic, of their food, though, as Krogh points out, during lactation a severe strain must be placed on the water balance of seals and whales in salt water, accounting no doubt for the high concentration of the milk of these animals. Species feeding on poikilosmotic marine invertebrates may have some special salt-regulating mechanism (Fetcher 1939; Fetcher and Fetcher 1942), but unlike marine birds they are active out of the water only when breeding, so that ordinarily they have no thermoregulatory problems and do not drink. McLaren (*personal communication*) suggests that lactating females of some species of seals breeding on snow-covered ice may eat snow; at least under such circumstances difficulty in thermoregulation would not be important.

The problem of physiological races. In various places in the preceding discussion there have been indications that populations may exist which are morphologically inseparable or closely allied but which are physiologically different in their salinity tolerances. Thus some of the data on the tolerances of *Nereis limnicola* and of *Nereis diversicolor* might be explicable on the basis of physiological races (page 73). Gordon (1962) found evidence of the phenomenon in *Bufo viridis*. Beliaev and Birstein (1944) found that individuals of *Dikerogammarus haemobathes* from the Volga regulated better in very dilute water than did those from the Caspian. The same is evidently true of *Saduria e. vetterensis* in contrast to *S. e. entomon*. In view of Sexton's (1928) claim that with sufficiently careful acclimation the various British littoral brackish and fresh-water species of *Gammarus* can be brought to live in either fresh water or sea water, the role of individual history in determining physiological differences is likely to be great. In the various cases that have been discussed genetic differences may well occur; in the evolutionary processes that have taken place indeed they must have occurred, but at the present time specific evidence is incomplete and indirect.

Parallel evolution in the descent of lacustrine animals. A series of cases, which in one genus, namely *Saduria,* certainly involves the production of populations better adapted to fresh water than were their marine ancestors, also involves the evolution of characteristic fresh-water subspecies distinguishable by morphological characters. A particularly puzzling situation is presented by some of these cases in which there seems to be an immense amount of evolutionary parallelism, the same fresh-water race being developed independently over long periods of time in separate isolates from a brackish species.

The most striking example concerns the so-called glaciomarine relict species of the Baltic area and other glaciated parts of the Northern Hemisphere. The zoogeography of these animals is discussed later in this chapter (see pages 205–213).

Limnocalanus is an arctic brackish water copepod of wide distribution, which occurs also in the Baltic (*L. macrurus grimaldii*). In a number of lakes that have been isolated from the Baltic by postglacial rebound completely adapted fresh-water races (*macrurus*) of characteristic form have developed. This has also happened in the Great Laurentian Lakes and many of the larger Canadian lakes. Ekman (1913, 1914), who has studied the matter with great care in Scandinavia, finds that the degree of differentiation quite clearly depends on the age of the lakes, being least in Mälar, which was separated from the Baltic about 1200 A.D., and greatest in Siljan, which was separated late in the Yoldia-sea stage of the Baltic basin. In Vetter, Vener, Fryken, and other lakes separated from the Ancylus lake, forms intermediate between the extreme *macrurus* of Siljan and the transitional *grimaldii-macrurus* of Mälar are found (Fig. 60).

The degree of development in the direction of *macrurus* apparently depends not on the present salinity but on the length of time that the populations have been exposed to fresh water. Ekman concluded that the genotype over a long period of time must be directly influenced by the environment; this conclusion will hardly appeal to modern workers. As Segerstråle (1957a) implies, the direction of evolution is not necessarily constant. If *Limnocalanus* reached the Baltic as the fresh-water *macrurus,* the salt-water *grimaldii* developed in the Littorina sea but has achieved a form comparable to that in other more or less saline waters such as the Caspian. It is just possible that part of the phenomenon discovered by Ekman is due to an early invasion by *macrurus* with subsequent invasion and partial introgression by the brackish water *grimaldii*.

Outside Sweden it is possible that the rate of development of *macrurus* from *grimaldii,* if this process is involved, is not quite so constant as might be supposed from Ekman's work, for the population of *Limnocalanus* in Ennerdale Water must be as old as that of Siljan, if not older, yet it appears to be a less developed form of *macrurus* than exists in the Baltic lake. Even in Sweden the correlation of form and age probably admits some exceptions.

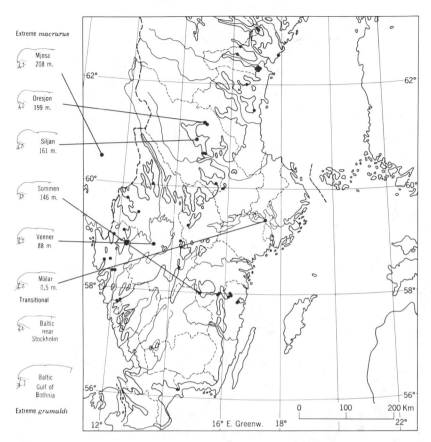

FIGURE 60. Distribution of *Limnocalanus* in Sweden, with a selection of profiles of the head showing parallel evolution of the short elevated head of the extreme *macrurus* populations, for example, in Mjosa and Siljan, two of the oldest lakes, and the very little change in Mälar, a fresh water lake but one cut off only since the early Middle Ages. Elevations of lake surfaces are given below figures of the heads.

The cottid fish *Myoxocephalus quadricornis* L. provides a very interesting set of examples of subspeciation (Fig. 61), comparable to, but more complicated than that of *Limnocalanus*. Like *L.m. grimaldii*, the species has a circumarctic, marine, brackish distribution at high latitudes. In the arctic two subspecies occur, *M.q. hexacornis* from the northern part of the Bering Sea and *M.q. labradoricus* (Girard) from most of the rest of the coasts of the Arctic. In the Baltic typical *M.q. quadricornis* is found. This subspecies is distinguished from most of the others by its four large horn-like excrescences on the head;

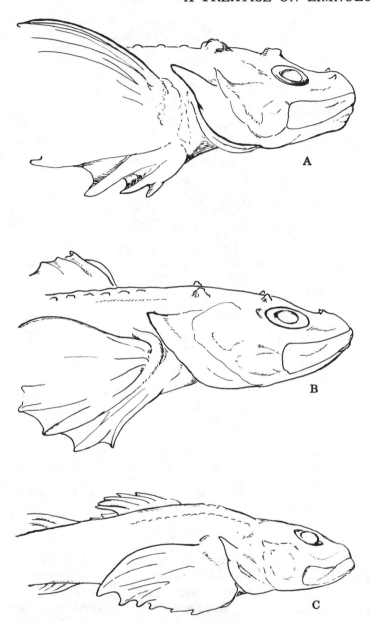

FIGURE 61.   Lacustrine races of *Myoxocephalus quadricornis*.   *A.*
M. q. lönnbergi, Lake Ladoga, isolated about 2000–2500 B.C.   *B.*
M. q. relictus, Lake Vetter, isolated somewhat before 6500 B.C.   *C.*
M. q. asundensis isolated before 7500 B.C.   (Lönnberg, Berg.)

M.q. labradoricus has reduced horns, particularly in the female, but *M.q. hexacornis* resembles the typical form in the head armature (Berg and Popov 1932). Segerstråle (1957a) finds that the Baltic population overlaps *labradoricus* considerably, though he admits that well-developed horns are commoner in the Baltic than in the Arctic populations, possibly as a direct effect of higher temperature. Isolated populations are known in sixteen lakes in Sweden (summary in Thienemann 1950), in a number of Finnish lakes, including Ladoga, and in Lake Onega in Russia. Nearly all of these populations are more or less differentiated. The differentiations mainly involve the presence or absence of the horns, which may be environmentally determined, the presence or absence of spinous scales along the sides of the body, the size of the eye, and the relative length of the caudal peduncle.

The least modified populations are those of Lake Mälar in Sweden (Lönnberg 1932) and Lake Ladoga in Finland and Russia (Berg and Popov 1933). Mälar, as we have noted, was isolated from the Baltic about 1200 A.D., Ladoga from the Littornia Sea about 2000–2500 B.C. Both populations exhibit some reduction of the cephalic horns and the spinous scales when compared with the Baltic fish; the reduction is somewhat greater in *M.q. lönnbergi* from Lake Ladoga than in the Mälar form. The longitudinal diameter of the eye in these populations is about 16˙ to 18 per cent of the length of the head, as in adult Baltic specimens.

In the other lakes there is always a considerable reduction of the cephalic horns. In *M.q. vaenerensis* from Lake Vener, isolated early in Ancylus times, about 6500 B.C., small but definite cephalic horns are found. In *M.q. frykenensis,* which must have been derived from *vaenerensis* rather than directly from the Baltic stock, these structures are absent, at least in the female. In *M.q. relictus* from Lake Vetter, which may have been separated rather earlier than Lake Vener, the horns are better developed than in any land-locked populations except those of Ladoga and Mälar. In general, the spinous scales show a degree of development in these races comparable to that of the cephalic horns.

Among populations apparently isolated from the Yoldia Sea or Baltic Ice Lake Stages not later than 7500 B.C., *M.q. oernensis* in Lake Örn has slight horns, whereas *M.q. asundensis* in Lake Åsunden lacks these structures. Rudimentary horns may occur in the small *M.q. pygmaeus* of Lake Puruvesi. *M.q. onegensis,* which was probably derived from a part of the population in the Arctic Ocean before the Yoldia Sea had developed (cf. Sauramo 1939) and so may be the oldest isolated population, also lacks horns. The eye is very large, at least 24% of the length of the head, in *kallavesensis, borkensis, pygmaeus,* and *onegensis* and is almost as well developed in *asundensis* (22.3%) and *relictus* (20.1 to 22.8 per cent). This condition, along with the lack of horns and the reduction of the spiny scales, is regarded by Lönnberg as the retention of a juvenile character.

There can be little doubt that, in general, isolation in fresh waters has led to the development of smooth, unarmored, large-eyed, often small, juvenile-like forms. The rate of development of such forms in the different basins, however, seems very irregular. Lönnberg thinks that low calcium content and poor food as well as the freshening of the water are implicated.

Insofar as the changes have taken place in response to the lacustrine environment, the Mälar population indicates, as in *Limnocalanus,* that they are not entirely environmentally determined modifications but must involve some true evolution on a genetic basis. It has been pointed out by Lönnberg that the Baltic population must have gone through a time when the fresh water of the Ancylus lake was its only habitat. Poorly preserved fossil material (Nathorst in Lönnberg 1932) from the Ancylus lake has indeed been supposed to be *relictus.* It is worth noting, however, that *M.q. labradoricus,* which has smaller horns than *M.q. quadricornis,* must frequently live at higher salinities, apparently up to 24 per mille, than prevail in the eastern Baltic, in which *M.q. quadricornis* occurs ordinarily in water under 5 per mille salinity. The development of horns or their disappearance can therefore hardly be looked on as an inevitable result of changes in salinity. It is possible, in fact, that the hornless condition in lakes is partly determined by low hypolimnetic temperatures. Insofar as the lacustrine populations are true genetic subspecies, it is conceivable that they reflect a process of gene fixation in small populations when some form of selection has ceased to operate. The change, in general paedomorphotic, might of course rescue *Myoxocephalus quadricornis* from sculpindom and permit it to start in a new direction; it may therefore be invidious to speak of it as degeneration, as Lönnberg (1933) does.

European ichthyologists (Berg and Popov 1933; Lönnberg 1932) consider the deepwater sculpin *Triglopsis thompsoni* of the Great Lakes and part of Canada indistinguishable generically from *Myoxocephalus* and probably another landlocked form of *M. quadricornis.* The supposed characters of *Triglopsis* are the same as the characters separating the most modified from the least modified subspecies of *M. quadricornis.*

A third case is provided by the group of animals included in or allied to *Saduria* (= *Mesidotea*) *entomon,* most recently studied by Gurjanova (1946).

TABLE 5. *Proportions of body and of telson in various
races of* Saduria entomon

			Length/Breadth	Base of Telson/ Length of Telson
orientalis	(Shantar Is.)			
	♂ immature	25–40 mm.	3.10 ± 0.17	0.669 ± 0.027
	♂ mature	40–60 mm.	3.36 ± 0.14	0.612 ± 0.032
entomon	(Baltic)			
	♂ immature	25–35 mm.	3.14 ± 0.13	0.551 ± 0.026
	♂ intermediate	40–70 mm.	3.48 ± 0.11	0.494 ± 0.021
	♂ large mature	70–80 mm.	3.52 ± 0.10	0.487 ± 0.015
caspia	(Caspian)			
	♂ immature	25–40 mm.	3.27 ± 0.21	0.530 ± 0.029
	♂ mature	40–54 mm.	3.50 ± 0.21	0.500 ± 0.007

Gurjanova considers the most primitive member of the genus to
be *S. sibirica,* a littoral but strictly marine arctic form. From this
species or its immediate precursor in the Tertiary arctic basin two lines
are believed to have developed. One invaded deep water, giving rise
to *S. megalura* and *S. sabini;* the other entered brackish water.
Among the brackish forms *S. (e.) orientalis*[52] from the Bering Sea
and the Sea of Okhotsk lives in almost full-strength sea water and
is the widest form, with a wide telson, rough exoskeleton, and only
moderate development of hairs on the epimera and legs. From it
a morphological series can be traced through *entomon* of the Baltic,
glacialis of the estuaries of the rivers of Siberia draining into the Arctic
Ocean, and *caspia* in the Caspian Sea to the strictly fresh-water *vetterensis* of Lake Vetter. The Caspian and Lake Vetter forms are much
smaller as adults (max. length ♂ 54 and 52 mm., respectively) than
the other races which grow to more than 80 mm. The animals of
more dilute waters tend to be less sculptured, hairier, and narrower,
with a narrow telson. Since it appears from Table 5 that the adults
have a generally narrower body form at all salinities, the very narrow
vetterensis (breadth base telson/length telson = 0.40) cannot, in this
case, be interpreted as paedomorphic.

In view of the wide separation of the different races of the animal
and of the general tendency for elongate forms to develop in the
more dilute localities, it seems probable that *Saduria* is similar in some

---

[52] It is not entirely clear if Gurjanova regards *orientalis* as specifically or only
subspecifically separable from *entomon.*

ways to *Limnocalanus* and *Myoxocephalus*. There is also a suggestion
of comparable phenomena in *Gammaracanthus* and *Pontoporeia affinis*
when the populations of Lake Ladoga are compared with brackish
marine populations on the one hand and those of Lake Onega and
other lakes older than Ladoga on the other (Lomakina 1952).

Although it seems reasonably certain that this group of glaciomarine
animals has a considerable capacity for evolutionary response to fresh
water in a strictly limited set of directions, the phenomena certainly
require further study.

Significance of small size. Apart from the highly complicated
adaptations of euryhaline fishes, there seems to be a tendency for
migration from marine to fresh-water environments or vice versa to
be easier for small animals than for large. Such a tendency could
underlie the prevalence of evolutionary euryhalinity in the Monera
and Protista, though it cannot explain its loss in various lines of the
latter kingdom. In the Metazoa the tendency is exhibited by the lower
flatworms, and apparently by some rotifers, as well as by the small
amount of change apparently needed to convert a member of the marine
interstitial fauna (see pages 101, 107) into one of the fresh-water interstitial fauna. This apparent correlation may prove fortuitous, but, if it
is not, it is most curious, since it would be supposed that a high
surface:volume ratio would be less rather than more favorable to an
organism in the process of changing its chemical environment (Hutchinson 1960). Further study is obviously needed.

A possible dynamic pattern in the metazoan body plan. Finally,
because of their possible ecological significance, it is desirable to call
attention to the brilliant speculations of Willmer (1956, 1960) on
the fundamental plan of the metazoan embryo. Willmer supposes that
the basic pattern of such an embryo is an animal pole of ciliated
or flagellated cells which tend to take up cations and pass them into
the blastocoele and a vegetal pole of amoeboid cells which take up
water. The animal cells are believed to excrete water and the vegetal
cells are believed to excrete cations into the medium. The liquid
content of the blastocoel is therefore an internal environment regulated
by the organism in a simple way. The two types of cell are compared
with the two phases of the protistan *Naegleria,* which in soil or in
more concentrated culture media is an amoeboid organism but which
develops flagellae and becomes free-swimming when the medium is
diluted. Willmer believes that such a change is a model for differentiation in embryogenesis.

Willmer's hypothesis, though based mainly on indirect evidence,
unifies a large number of embryological, histological, and physiological

facts.   Although it does not necessarily imply a fresh-water origin for the metazoa, it would seem that the postulated mechanisms would be more likely to develop in small bodies of low but variable salinity than in a medium as constant in composition as modern sea water.

PHYSIOLOGICAL PROBLEMS OF PASSAGE
FROM LAND TO FRESH WATERS

The problems to be faced by a terrestrial animal entering fresh waters are in general entirely different from those encountered by a marine invader.   Until it enters freshwater, the marine animal has had no occasion to be disturbed by the passage of water across its boundary; when such a problem is solved in the ways discussed in the preceding sections, the organism is normally left with respiratory surfaces that are as suitable in fresh water as in salt.   Terrestrial animals, however, have had to develop a high degree of impermeability to prevent evaporation at their surfaces, and in doing so they have frequently hidden their respiratory epithelia deep in recesses such as lungs or tracheae which can easily be kept saturated with water vapor.

The high degree of impermeability of the cuticle of even fairly soft aquatic insect larvae is indicated by Beadle and Shaw (1950) in their investigation of the larva of *Sialis lutaria*.   The main osmotically active material in the haemolymph of this animal is nitrogenous, apparently consisting of amino acids.   About 0.15 to 0.35% NaCl is present; five weeks of starvation in distilled or tap water were needed to reduce this amount below 0.10 per cent.   Unlike the condition in aquatic fly larvae, no chloride absorbing organs exist, and the diffusible material in the haemolymph is retained solely by the low permeability of the body wall.

The greater mass of oxygen in unit volume of air than in saturated water, 0.29 g. per liter in air at $0°$ C. and 1 atm. pressure against 0.014 g. per liter in water in equilibrium with 1 atm. at the same temperature, allows a smaller respiratory surface to meet the needs of the terrestrial animal.   The lower density and viscosity of air permit any respiratory cavity to be ventilated easily.   When a terrestrial animal enters fresh water, the problem of its water balance may become simpler; it certainly does not become more difficult, but its respiratory needs are harder to satisfy.

The aquatic amniote vertebrates, primarily mammals and reptiles, which have retained aerial respiration, come to the surface to breathe. Many adaptations of interest have developed which ensure an adequate store of air, obtained by rather infrequent excusions to the surface.

Nearly all of the organisms exhibiting them are marine and outside the scope of this book.

The pulmonate mollusks usually remain air breathers. Some have developed haemoglobin which permits respiratory exchange through the skin at very low oxygen tensions in the medium. The facultative use of the mantle cavity, normally a lung, as a gill chamber is known in some deepwater Lymnaeidae; these are mostly cold-water forms, presumably with a low metabolic rate.

The most important examples of the various ways in which the respiratory problem has been solved are found in the insects, to which group the present section is primarily devoted. The account given here is relatively condensed and undocumented, for excellent extended summaries are already in existence in the well-known book on insect physiology by Wigglesworth (1950) and in Thorpe's (1950) fascinating summary of the problems of plastron respiration.

Very small arthropods encounter no particular problem in aquatic respiration, since diffusion through the skin without the development of specialized respiratory surfaces is sufficient to meet their needs. In the simplest cases, the simplicity certainly always being secondary, tracheae, if present, are filled with liquid, and the organism in its respiratory physiology is roughly equivalent to a nematode or small oligochaet. Examples are provided by the water mites, the young larvae of the dipterous families Tendepedidae and Simuliidae and of the aquatic moth *Acentropus*.[53] The first instar nymphs of the Corixidae provide comparable cases in which numerous capillary blood sinuses beneath the thinly cuticularized body wall presumably enhance the respiration exchange. Some of these animals, notably the tendepedid larvae, develop projections of the body wall that have been called blood gills. The anal blood gills of the Tendepedidae, however, are certainly concerned with chloride uptake, as has been indicated. Under ordinary conditions the ventral blood gills play no part in respiration (Fox 1920), though it has been found by Harnisch (1930, 1937) that they come into action as respiratory surfaces during liquidation of an oxygen debt. Many genera of the family contain a haemoglobin with a very low loading tension and, as discussed in Volume III, are among the most important animals living in the oxygen-deficient bottoms of lakes.

---

[53] The peculiar general *Paskia* and *Idiocoris* endemic to Lake Tanganyika and constituting the subfamily Idiocorinae of the Helotrephidae (Esaki and China 1927) are apneustic and presumably completely aquatic in their over-all respiration, but it is not known whether they have gas-filled tracheae.

The development of the gas-filled tracheal system into a relatively elaborate mechanism for aquatic respiration is the commonest respiratory adaptation in the aquatic nymphs of the hemimetabolous insects and also occurs in many holometabolous larvae. In some cases loops of the tracheal system are found in special projecting tracheal gills; in others a rather elaborate system of anastomosing branches lies beneath the whole of the body wall. It is essential in such cases that the walls of the tracheal system be sufficiently rigid to withstand any hydrostatic pressure met with by the insect without a significant decrease in volume. In a rigid air-filled cavity within the body of a respiring insect the utilization of oxygen will lower the pressure and at the same time increase the proportion of nitrogen. Oxygen will diffuse in from the water to restore the equilibrium pressure. Once the gas is inside the tracheal system gaseous diffusion provides a relatively efficient mechanism for distribution in a small animal; it is this fact that underlies the retention of the gas-filled tracheal system in aquatic insects. If, however, the system were not rigid, it would decrease in volume as the oxygen was used up, no pressure difference would be set up, and there would be a continual tendency for the partial pressure of nitrogen to rise above that in equilibrium with water. Nitrogen would then tend to be lost to the medium as oxygen was lost to the respiring tissues, and the system would collapse completely.

Precisely the same principle applies to the rather large number of insects that have hydrofuge surfaces and swim partly enclosed in a bubble of air, as first pointed out by Ege (1915).

In such animals the bubble confers buoyancy on its possessor, which usually rises passively to the surface unless active swimming or holding fast to a submerged object permits the insect to remain under water. The smaller Notonectidae of the subfamily Anisopinae are exceptional in that they can remain suspended in free water by means of a mechanism of buoyancy control which is dependent on the presence of an organ composed of haemoglobin-containing cells richly supplied with tracheae (see page 256). Ege found that in cold water in winter, or in small species of Corixidae in summer, respiration through the bubble is effective over very long periods of time without ascent of the insect to the surface. It is worth noting that some of these insects seem able to live in quite deep water; *Sigara* (*Lasiosigara*) *lineata* is recorded by Hungerford (1948) in depths up to 11 m. in Lake Erie. Taking, as did Ege, the invasion coefficient of O_2 as three times that of N_2, the four volumes of nitrogen that correspond to the one volume of oxygen present may be replaced gradually by no less than twelve volumes of oxygen. The whole bubble, therefore, is equivalent in

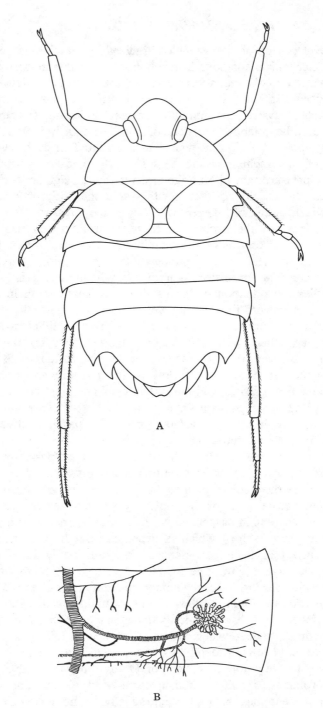

A

B

FIGURE 62. For descriptive legend see opposite page.

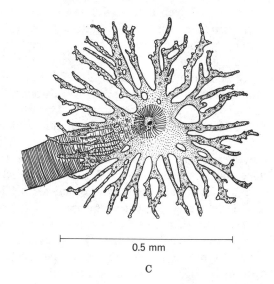

0.5 mm

C

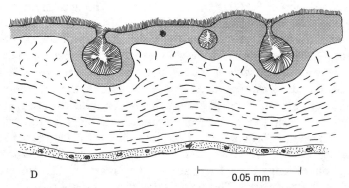

D
0.05 mm

FIGURE 62. *A*. Dorsal view of *Aphelocheirus aestivalis* ♂, from a specimen from the River Teme, coll. E. J. Pearce (P. M. Yale Univ.), × 10; *B*. Spiracular rosette on third abdominal sternite with part of right tracheal trunk dissected from above; *C*. Spiracular rosette, × 90; *D*. Section through same showing ventral hair pile and openings of rosette channels, × 550. (*A* original, others from Thorpe and Crisp slightly modified.)

respiration to thirteen times its own volume of air. That the nitrogen does act in this way is shown not only in theory, which because of the virtual nature of invasion coefficients may require some correction, but also by Ege's experiments, confirmed by Thorpe (1950), which indicate that *Notonecta* without access to the surface can survive in water saturated with air for seven hours, with oxygen for only thirty-five minutes, and with nitrogen for five minutes. The journey of a bubble-bearing insect to the surface is thus as much, if not more, a journey to obtain nitrogen as to obtain oxygen. The mechanism in *Notonecta* works well only below 15 to 20° C. (Popham 1961).

The ultimate refinement in the development of a tracheal respiratory system for use under water would consist in making the bubble rigid while maintaining an air-water interface which would provide maximum permeability. Such a device could eliminate journeys to the surface indefinitely and by permitting the volume of the bubble to be small, provided its surface was not reduced, would allow the insect to become benthic without having to hold to its substrate actively. This specialization, which at first sight might appear improbable, has actually been developed by those insects that possess a true gaseous plastron. The most extraordinary is the rheophil (Pearce 1945) water bug *Aphelocheirus* (Fig. 62). In the adult of this insect the body is covered with a fine pile of hydrofuge hairs bent over at the tips. Each hair has a length of 5 to 6 μ and the hairs are set 0.6 μ apart, giving a density of about 2,500,000 hairs mm^{-2}. The gas layer held by these hairs is stable up to a hydrostatic pressure of 4 to 5 atm., which represents a depth far greater than the insect is every likely to penetrate in nature. The oxygen that diffuses into the plastron enters the tracheal system by very specialized stellate spiracles. Some comparable cases, as in the elmids *Stenelmis* and *Cylloepus* and in the aquatic weevil *Phytobius,* are known in other groups of insects; numerous. intermediate conditions between the replaceable bubble and the plastron are also found. The whole problem has been considered by Thorpe and Crisp (1947a,b,c, 1949), whose work constitutes a comparative physiological investigation of unparalleled elegance. The main results are also accessible in the review by Thorpe (1950) already mentioned.

OTHER PHYSIOLOGICAL PROBLEMS PRESENTED BY LIFE IN FRESH WATERS

Apart from the adaptation to low salinity and the respiratory problems that have been discussed in the preceding sections, other difficulties must be overcome by an organism entering fresh waters. Some of these difficulties have in fact appeared to be fundamental to previous students of the subject.

The problem of larval forms. Sollas (1884), who believed that initial adaptation to low salinity usually occurred passively in gradually freshened basins cut off from the ocean, considered the critical event in the evolution of a fresh-water fauna to be the suppression of delicate free-swimming larvae that would be swept downstream as soon as the newly adapted fauna started moving upstream. Since this suppression would imply that the embryo must hatch at a later stage , it would necessitate a greater supply of yolk to compensate for food that the free-swimming larva might have obtained for itself and the reduction in number of eggs as the size increased. There can be no doubt that such a process has been widespread in the evolution of the fresh-water fauna, the mollusks and decapod crustacea providing good examples. The well-known case of *Palaemonetes,* in which the northern marine *P. varians* lays small eggs and the closely allied southern inland-water *P. antennarius,* lays large eggs which hatch at a later stage, perhaps is the most striking. More study of other fresh-water prawns, however, may show that this example is not so significant as it once was supposed (see page 114). The differences in life history between the lobsters (Nephropsidae) and the fresh-water crayfishes (Astacidae and Parastacidae) or between the marine crabs and fresh-water crabs (Potamobiidae) are of a similar kind. Both genera of fresh-water Rhizocephala, the only known nonmarine cirripedes, hatch as cypris-larvae (Fig. 48) rather than nauplii, but this is also true of the marine genera *Clistosaccus* and *Sylon* (Boschma 1933, 1934) from the Arctic, where pelagic larvae are also generally suppressed, as well as in the very peculiar genus *Thompsonia.*

There is in some cases a suggestion of preadaptation. *Namanereis* has very large eggs and presumably direct development in both fresh and salt water, though *Namalycastis* is hatched from smaller eggs as a nectochaeta larva with but three parapodial segments, even in the fresh waters of Sumatra.

The general absence of motile larvae probably prevents the development of a rich animal lasion in fresh waters (Hutchinson 1964b).

There are some striking exceptions to the general rule of the suppression of larval forms. The nauplius is retained in the Copepoda, which are among the most successful invaders of fresh waters, in striking contrast to the suppression of larvae in fresh-water decapod Crustacea. The highly successful invader *Dreissena polymorpha* has a typical lamellibranch larva, which is particularly paradoxical, as Gurney (1913) points out, in that this species is largely an inhabitant of rivers. Many brackish estuarine animals retain their larvae even though they have to maintain, as species, a precarious position in a rather limited region at the mouth of a river. In these cases the tidal cycle may be of considerable assistance in returning some of the larvae just leaving the estuary to the upper reaches inhabitable by the species, but it can hardly be as effective as the abolition of the larval stage.

There are, moreover, important invertebrate groups, notably the Cephalopoda, in which there is no small pelagic larva and at the same time no fresh-water species. Because of these divergences from what would be expected from Sollas' theory, Needham (1930) has suggested another and far more plausible explanation of the general reduction of larval stages and the concomitant decrease in numbers and increase in size of the eggs of fresh-water organisms. Basing his argument on the considerable uptake of various inorganic ions from the sea water by the eggs of cephalopods and the larvae of echinoderms and other marine invertebrates, he supposes that in the early stages of fresh-water animals there would be a real difficulty in acquiring mineral nutrients and that its most practical solution would be for the egg to contain an adequate supply of inorganic constituents and for the young individual to hatch at a late stage when it could, at once, use whatever mechanisms for uptake and retention of salts are available to the adult (cf. also Krogh 1939). This hypothesis has the great advantage over that of Sollas in providing a reasonable explanation of the catadromous invertebrates such as *Eriocheir sinensis* or *Callinectes sapidus* which ascends rivers in Florida to a salinity of 25 mg.[54] Cl. per liter but must descend to breed (Odum 1953).

The Caspian fauna, to be discussed at greater length later, is of considerable interest in connection with the problem of the invasion of rivers. In the Volga delta a large number of endemic peracaridan Crustacea inhabit water that is nearly fresh. The geological evidence strongly suggests that during some glacial ages the Caspian was a lake with an outlet, so that the entire endemic fauna must have been adapted to even fresher water. Therefore no chemical barrier exists to the ascent of the Volga. Moreover, the Peracarida carry their eggs and have no larval stages, so that the barrier postulated by Sollas does not operate. Yet among the forty-one Mysidacea, Cumacea, and Amphipoda known in the delta only five have penetrated naturally into the greater part of the Volga. It is evident that other barriers besides those of decreasing salinity and increasing current must be crossed before the successful colonization of fresh water can be begun by a marine invader.

The temperature barrier. More than a century ago von Martens (1857, 1858), in a work on the fresh-water fauna of Italy, concluded that invasion of fresh waters by marine genera is today largely a tropical phenomenon. The data available when von Martens wrote strongly suggested that in the mollusks, fishes, and perhaps the Crustacea the

---

[54] Or 12 mg. Cl. per liter where other electrolytes are high.

ratio of the number of families common to fresh water and the sea to the number that are purely marine increases steadily with decreasing latitude. Von Martens concluded that the stability of temperature rather than the high temperature of tropical fresh waters was mainly involved. There can be little doubt that there is a considerable element of truth in von Martens' contention, though it has been criticized by some subsequent writers. Pelseneer (1905) has objected that equally stable conditions occur in the Arctic at very low temperatures. Actually, this is probably not a real objection to the validity of the theory, since a considerable number of arctic Crustacea, the so-called glaciomarine relict species of the Baltic, are cold stenotherm, brackish animals of ultimate marine origin which easily invade fresh waters. Pelseneer has also pointed out that a disproportionate number of the cases contributing to von Martens' statistics came from the Ganges, the Malayan region, Indonesia, and southern China, regions in which very gentle salinity gradients occur because of the discharge of large rivers draining the tropical rain belts. Such a salinity distribution may be more significant than constant temperature as a condition of migration.

Pelseneer (1905), who attributed to Issel[55] (1901) the hypothesis that adaptation to fresh water is physiologically facilitated by high temperatures, disputed the idea on the ground that permeability increases with temperature. Modern work, already discussed (see pages 164–8), though it has not fully solved the problem, does at least suggest that since osmoregulation is an active process it can be more efficient at high than at low temperatures, in spite of increased permeability.

It must not be forgotten, however, that in the most thermally variable parts of the temperate zone, the littoral areas of shallow protected seas are likely to show almost as great a temperature range as the surface waters of small lakes (Fig. 63) and that a eurythermal marine fauna, somewhat limited in composition but no doubt often richer than the adjacent fresh-water fauna, does exist. This fauna might, if other barriers did not exist, well give rise to fresh-water species.

Hypogean fauna of marine affinities. One particularly interesting set of examples, which contributed a little to von Martens' argument, is provided by the rather numerous animals of marine origin found in the cool waters of caves. At the time von Martens wrote only the atyid *Troglocaris* from the Adelsberg Caves and the sphaeromid *Monolistra,* first described from the same region, were known. Subsequently a large and most remarkable fauna has been discovered.

---

[55] The paper referred to apparently does not contain any reference to the matter.

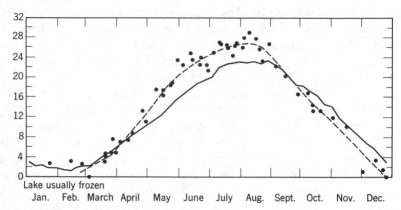

Lake usually frozen

Jan. Feb. March April May June July Aug. Sept. Oct. Nov. Dec.

FIGURE 63. Solid line means weekly temperature of the sea, Baxter's Beach, Pine Orchard, Conn., on Long Island Sound (E. S. and G. B. Deevey, 1946–1956). Points, individual surface temperatures, 1935–1939, Linsley Pond, North Branford, Conn., inland about 6 km. from Pine Orchard; broken line fitted by eye to the points to give an approximation to the mean surface lake temperature. In the summer the surface of the lake tends to be about 4°C. warmer than the sea; the latter has once reached 27.1°C. at the station studied, or within 2°C. of the maximum lake temperature of 29.0°C. during the period under consideration. The comparison seems reasonable for a small lake and a protected arm of the sea lying close together. The temperature range is clearly greater in the lake, but not very much so.

As far as is known at the present time, there appear to be two different assemblages of hypogean aquatic animals of marine affinities (Chappuis and Delamare Deboutteville 1954; Chappuis, Delamare-Deboutteville, and Paulian 1956; Delamare-Deboutteville 1960).

One type consists of relatively large animals, usually eyeless and depigmented and somtimes with elongate sensory appendages but otherwise of ordinary morphology. These animals appear to have entered the hypogean fauna by being isolated in gradually freshened waters in marine coastal caves. Several are known, such as *Heteromysis cotti* from a salt-water cave on Lanzarote, Canary Islands, *Lepidops servatus* from brackish hypogean water in Zanzibar, *Spelaeomysis bottazzii* from slightly salt water in a cave near Otranto, and the peculiar hippolytid shrimp *Barbouria poeyi* from similar localities in Cuba, in which this process actually appears to have produced a localized brackish-water endemic form. If such animals became tolerant of quite fresh water in an area formed of limestone, with many subterranean channels, that was undergoing uplift, a fairly well distributed fresh-water hypo-gean species might result.   There is little doubt that this was the mode of origin of the prawn Typhlocaris, presumably derived from an extinct marine animal likely to enter marine caverns, which lost its eyes several times in producing the four modern species (Fig. 64).   A similar history may be assumed for *Troglocaris,* for the monolistrine Sphaero-midae, for the cavernicolous Cirolanidae, for *Troglomysis,* and perhaps for the polychaet *Marifugia.*   These occurrences in lands bordering the Mediterranean are reasonably certain to be related to Tertiary shorelines as the southwestern part of the Tethys became the modern Mediterranean Sea (Hubault 1938).   Certain species in Central America and in Madagascar presumably indicate comparable histories (taxonomy in Holthuis 1956).

It is interesting that Derouet (1952) finds *Caecosphaeroma virei* to be extremely euryhaline, tolerating the whole range of salinities from fresh to ordinary sea water.   The blood is usually slightly less concentrated than that of *Gammarus locusta* in the more saline part of the range.   There is no striking increase in tissue chloride in such media, as there would be in *G. pulex.*   Since the animal has presumably been an inhabitant of fresh, if very calcareous, waters for some tens of millions of years, this euryhalinity is most remarkable.

The members of the second type of hypogean animal of marine affinities (Fig. 35) are interstitial forms, minute, elongate, and with reduced appendages.   These animals probably are all very thigmotactic. Here belong the Microparasellinae, the Microcerberinae, the Bogidiel-lidae, the Ingolfiellidae and, less typically, *Troglochaetus* (Fig. 21). There seems to be little morphological distinction between fresh- and salt-water species; in most cases members of the same genus occur in both the fresh- and salt-water habitats.   The interstitial environment, though it provides a chemical gradient at the coast, is otherwise proba-bly continuous from the ground waters of river valleys through estuarine deposits and beaches to the waters of permanently covered ocean sand. The fauna is apparently widely distributed, with little geographic separa-tion of genera, but it is still so poorly known that nothing significant can yet be said about it.   The peculiar nematode *Desmoscolex aquae-dulcis* (Fig. 12) may be a somewhat atypical member of the interstitial fauna or at least may have entered fresh water by this route.   The same is probably true of the Thermosbaenacea, though little is known of their history.   The interstitial Bathynellidae, though morphologically comparable, are presumably of fresh-water origin but perhaps show a slight tendency to seaward movement.   It would not be surprising if ultimately an interstitial fresh-water mystacocarid were found.

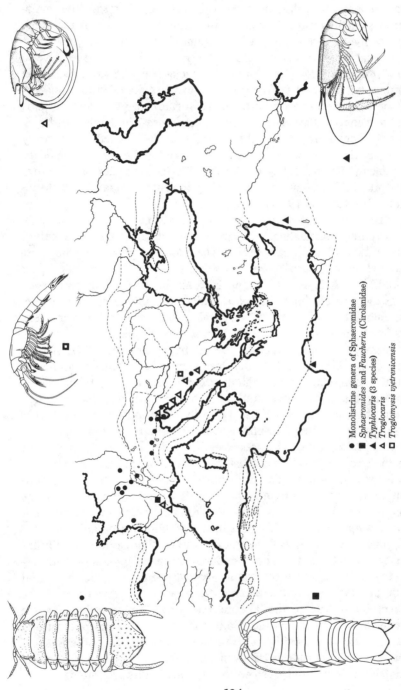

FIGURE 64. Distribution of various hypogean crustacea in Europe, relative to the supposed Tethyan shorelines (broken line). The species illustrated are *Monolistra* (M.) *absoloni*, in the Sphaeromidae, *Sphaeromides raymondi* in the Cirolanidae; *T. galilaea* in *Typhlocaris*, *T. schmidti* in *Troglocaris*, and the mysid *Troglomysis vjetrenicensis*.

• Monolistrine genera of Sphaeromidae
■ *Sphaeromides* and *Faucheria* (Cirolanidae)
▲ *Typhlocaris* (3 species)
△ *Troglocaris*
□ *Troglomysis vjetrenicensis*

Other physical barriers. The decrease in density as sea water is diluted leads to an inevitable increase in excess density between fresh-water homoiosmotic animals and their medium. Certain devices which permit some marine organisms to reduce their excess density are unavailable in fresh water. How far such differences in density relations are of importance to nonplanktonic animals is doubtful. Robertson (1957) thinks that a loss of armor and a general reduction in ossification probably occurred to offset density differences when ostracoderms and fishes became adapted to fresh water in the late Silurian or early Devonian.

The turbidity of estuarine waters may be a significant factor limiting the ascent of estuaries, as Gurney (1913) has suggested. Pennak (1963) has pointed out that since sponges are intolerant of suspended sediment they are not likely to have invaded fresh waters by this route. In the turbid waters of some estuaries, notably of the Ganges (Kemp 1917; Annandale 1923) both fish and Crustacea may show coloration and even some structural characteristic reminiscent of deep-sea animals, presumably because of the darkness of very turbid water.

There are certain biochemical differences between fresh- and salt-water animals which may be of some general significance. The absence of trimethylamine in fresh-water fish may be connected with osmotic and salt-content regulation. The differences in fat which is especially rich in 16- and 18-carbon unsaturated acids in fresh-water fishes, the change in visual purple from the porphyropsin (vitamin A_2, retinene$_2$) to the rhodopsin (vitamin A_1, retinene$_1$) system when a vertebrate leaves fresh water, the almost complete absence of luminous fresh-water animals, and perhaps the far less brilliant coloration of the mollusks of fresh water, even when their shells are as heavy and ornamented as is usual in the sea, should, when interpreted, throw light on the history of the invasion. At present such facts are isolated and not understood (see Florkin 1949; Harvey 1952).

Biological barriers. By far the most important difficulty that any invader has to meet is that almost always other species have by now got ahead of it. Initially the fresh-water metazoan fauna must have arisen by invasion into a virgin territory with no suitable niches filled. This process may have been easy. Today all niches are almost always already occupied, and the invader will seldom be able to make any progress unless some disturbance exterminating old inhabitants or creating new niches has occurred. This will be apparent at various points in the examples that are given of the invasion process as observed in nature. It will again be apparent in a much later chapter (Volume

III) when Olson's (1952) beautiful work on faunal succession in the lakes, streams and dry areas of a lower Permian delta is considered.

ACTIVE AND PASSIVE EVOLUTION OF THE FRESH-WATER BIOTA

Apart from occasional direct transference of extremely euryhaline organisms from one environment to the other, it would appear that two general patterns of adaptation are possible.

In the first the adaptation involves an *active* invasion of fresh water from the ocean or sometimes in the opposite direction. In the second pattern adaptation takes place *passively* in a slowly freshened body of water which has been cut off from the sea. Though most students of the matter have considered active invasion through rivers to have been the more important process, a few investigators, notably Sollas (1884) and Derjavin (1924), have felt that the fresh-water biota originated almost entirely by the passive process of adaptation.

Active invasions. Wherever we find a series of relatively closely allied species, some living in the sea, some in estuaries or other kinds of littoral brackish waters, and some in contiguous fresh waters, we may suspect that a process of active invasion has at some time taken place. Except in rapidly changing environments, it is not likely that we can catch the process taking place, but the final pattern may be expected to give some information about the processes that have occurred. Examples could be obtained from a number of groups already discussed, but only in one case, namely the Gammaridae of the coasts around the North Atlantic basin, is there an adequate array of species, well understood taxonomically, with plenty of good physiological and ecological work. Before discussing this case, the general theoretical expectations may be briefly examined.

The origin of the fresh-water fauna as a special case of ecological zonation. In both lakes and the ocean related species occupying different depth zones are frequently found, just as altitudinally limited species or subspecies are commonly encountered in terrestrial mountain systems. It is most unlikely, however, that the animals are reacting to depth or altitude as such; various ecological factors such as light intensity, nature of substrate, degree of turbulence of the fluid medium, or temperature may be correlated with depth or altitude. It is probable, moreover, that most species can occupy a more extended habitat, corresponding to a more capacious ecological niche, when they are present by themselves than when they compete with close relatives. Since the direction of competition is often dependent on environmental factors and since the only possible equilibrium in ordinary competition is based on a single species inhabiting each niche, competition occurring in an area in which the environmental factors are varying continuously results in a division of the area into discontinuous zones, occupied by different species (Gause and Witt 1935). The boundaries of these zones correspond to values of the environmental variables at which the outcome of competition changes.

Consider a species S_1 which occupies a certain range in the gradient when it is by itself. Suppose that in the middle of the range conditions are optimal but that for a certain distance x on either side of the optimum wandering individuals can still settle down and leave some offspring. Provided that the possible range of adaptation exhibited at any time by the species is not coincident with the entire range of the gradient and that adaptation depends on many genes, selection for adaptation to the low values of the gradient when a member of the population moves down will be balanced by selection for adaptation to high values when an individual moves up. If no barrier to movement exists, the position of the optimum will remain unchanged. Now consider a second species S_2 with comparable properties to S_1 but with a lower optimum. Suppose that the upper range of S_2 overlaps the lower range of S_1. Let both species coexist. Competition will occur and an interspecific boundary will be set up in the overlap zone. It is now apparent that at least for one species, and usually for both, selection is operating on a population distributed asymmetrically with respect to the optimum. We should therefore expect the optimal environmental requirements to diverge. Brooks (1950a) has concluded that zonally distributed but otherwise sympatric species arise first by isolation, so that two species with slightly different ecological tolerances are produced, then by reinvasion of each other's territory, and finally by the operation of selective divergence of the kind just described. If species S_1 is a slightly homoiosmotic animal in an estuary, it will become more definitely homoiosmotic as selective forces act on it asymmetrically and so may become capable of moving up the estuary. It will be apparent that, given the degree of moderate euryhalinity often encountered in those marine stocks that seem preadapted to brackish water life, such as the Nereididae, some of the Calanoida, or the Gammaridae, a process of the kind described is likely to result in the evolution of quite successful invaders of dilute waters. In this sense the fresh waters of the world are only the marginal zones of the ocean. Once a physiologically satisfactory solution to the problems of life in fresh waters has been achieved by any species it is still most unlikely that that species will become a permanent member of a fresh-water fauna, because in almost every environment today all the niches are occupied.

It has often been pointed out that even thalassohaline brackish waters have a far less diverse fauna than either the oceans or fresh water. There are clearly two adaptive peaks corresponding in a general way to a very low salinity and to an ordinary marine salinity. As Pennak (1963) points out, variability in the brackish habitat, as a sort of ecotone between the two major aquatic habitats is probably involved. It is important also to remember that part of the small brackish-water fauna that appears to be invading fresh water may have long been fixed as a euryhaline assemblage derived from the sea, living in brackish waters, and occasionally adventuring into fairly fresh environments.

A considerable number of cases already discussed probably provide examples of this type of evolution. The most suggestive are doubtless found in the Pseudodiaptomidae of tropical rivers and in the species of *Eurytemora* in the North Temperate region, though in neither one does much information exist beyond the mere records of ranges. *Eurytemora velox* may provide one of the rare examples of an animal caught in the last scene of the act of adaptation to fresh water; the final process may here be facilitated by the reduced competition provided by incomplete faunas in many new man-made basins. In the one case, that of the Gammaridae, for which we have a variety of kinds of information, it is doubtful if the process of invasion is at the moment actively occurring.

Active invasion in the Gammaridae. As indicated in an earlier section, there is evidence that the Gammaridae of fresh waters have arisen as the result of a number of invasions by species living in the littoral zones of the sea. In discussing these invasions, only the phenomena on the coasts of Europe and North America will be considered, since the other putative invasions involving *Sarothrogammarus,* the Baikalian and Ponto-Caspian species, and *Paramelita* in South Africa are not well enough understood to add materially to the discussion.

In western Europe a group of littoral species (Sexton and Spooner 1940) referred to *Marinogammarus* are found intertidally among stones or under pieces of seaweed. *M. obtusatus,* living in the zone between mid-water and low-water neap tide and penetrating only a short way into estuaries, appears to be the most strictly marine of the common members of the genus. In the same genus *M. marinus* lives normally somewhat higher up the intertidal region, and though widely distributed along the coast frequently has well-developed estuarine populations which occur mainly in the upper part of the tidal region. *M. stoerensis* and *M. pirloti* appear to be species that prefer diluted sea water but not estuarine conditions; they occur most frequently where small fresh-water streams or seepage flow out over a stony intertidal zone. We have, therefore, in *Marinogammarus* a hint of two rather different types of penetration. *M. stoerensis* and *M. pirloti* seem to be adapted to variable salinity but to maintain their preferences for an ordinary stony beach, whereas *M. marinus* alone has become estuarine. In general, these two approaches, the seep or small perennial stream and the large tidal river, present different problems. The salinity at the bottom of an estuary, particularly for burrowing forms, may be constantly high over a large part of the tidal reach, but silting may produce a physical environment very different from the normal marine littoral. The most littoral estuarine form, moreover, will be exposed at low tide. The small stream or littoral seep may provide extreme salinity variation but a perennial cover of water and little difference in other aspects of the physical environment from the neighboring tidal regions of the beach. Salt-marsh pools (Pearse 1950) may offer conditions that allow some passive adaptation perhaps as an interlude in the more active process.

The better studied case, namely of zonation in estuaries, mainly involves the nomotypical subgenus of the genus *Gammarus*. In the south of England Spooner (1947) regards the typical sequence in passing up an estuary as *Gammarus (G.) locusta* (Linn.), with the less strictly aquatic *M. marinus* nearer the high-tide marks, *G. (G.) salinus,* replaced further upstream in some cases by *G. (G.) zaddachi,* and finally giving place to *G. (Rivulogammarus) pulex* in quite fresh water. In Scotland *G. oceanicus* replaces *G. locusta* in the series. The typical habitat of *G. salinus* has a salinity varying greatly with the tide: the mean value is about 17 per mille; the maximum, not above 31 per mile. *G. zaddachi* is usually found where the low-water salinity is below 1 per mille, though at high water values up to 15 per mille may be encountered. In some localities, notably the Elbe, near Hamburg, *G. zaddachi* appears to occur in entirely fresh water.[56]

Segerstråle (1950) finds a series in the Gulf of Finland comparable to that described by Spooner, but the salinity gradient being very gentle and relatively constant apparently permits a great deal of overlap. *Gammarus locusta* occurs only in water of salinity in excess of 5 to 6 per mille. *G. oceanicus* and *G. salinus* occur together in water

---

[56] Spooner (1947, p. 33) says this is also true of lakes in Northern Ireland but later (p. 41) includes no such localities in his list of material examined. Segerstråle repeats this statement, but its factual basis seems questionable. Can there perhaps be some confusion with *G. deubeni?* Kinne (1954) records *G. zaddachi* from fresh water in Germany and Denmark.

down to 2.5 per mille, but *G. oceanicus* is commoner in the more saline and *G. salinus* in the less saline regions within the 2.5 per mille isohaline.   *G. zaddachi* is the only species of *Gammarus* to penetrate to the eastern end of the gulf where the salinity is less than 2 per mille.   Nothing significant can be said about the upper outer limits of these species, but all can co-occur in littoral water around 6 per mille.   In the same general region but in deeper water *Pontoporeia femorata* is common on mud bottoms in water of salinity of more than 6 per mille; it is replaced by *P. affinis* in less saline water.

In small streams running into the sea a different pattern is found largely because of the presence of *G. (R.) deubeni*.   In streams on South Rona in the Inner Hebrides, Beadle and Cragg (1940) found *Marinogammarus obtusatus* between tide marks remote from the streams, with *M. finmarchicus* at the lowest intertidal point examined.   *M. marinus* accompanied by *G. zaddachi* occurred almost to the neap high-water mark.   Above this the only *Gammarus* was *G. deubeni,* which overlapped the upper part of the *M. marinus-G. zaddachi* zone and reached into completely fresh water.   In the South Rona streams *G. deubeni* reached only a little less than 2 m. vertically above the high-tide mark, though it is known to occur in quite fresh water in Ireland, the Isle of Man, and the extreme western end of Cornwall, as the common fresh-water amphipod of these regions.   As Hynes (1954) has shown, there can be little doubt that *deubeni* has been replaced by *pulex* as a fresh-water species in the rest of Britain.

Beadle and Cragg found that *Marinogammarus obtusatus* has some power of osmoregulation but cannot survive rapid transfer to sea water diluted by more than 50 per cent.   *Gammarus (G.) locusta* can survive transfer to 25 per cent sea water and *G. (Rivulogammarus) deubeni,* from brackish water, can live in pure sea water or in fresh water. *Gammarus (R.) pulex* can stand rapid transfer only to about 40 per cent sea water.   The variations in osmotic pressure and chloride contents of the haemolymph and the tissues are indicated in Fig. 58. It will be observed that on dilution of the medium both *M. obtusatus* and *G. locusta* die when the chloride of the haemolymph is still much greater than in normal *G. pulex* or in *G. deubeni* in fresh water. The significant differences in the species must therefore lie in the relation of the tissues to the composition of the blood.   If the osmotic pressure in the tissues is assumed to be that of the haemolymph, the percentage of this supposed tissue osmotic pressure due to chloride can be calculated.   This percentage is plotted against blood chloride in Fig. 65.   *M. obtusatus* apparently loses chloride rapidly from its

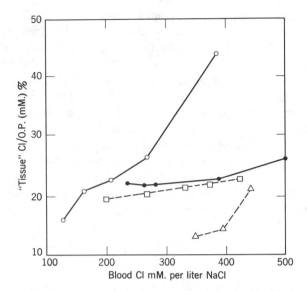

FIGURE 65. Osmotic pressure due to tissue chloride as percentage of total osmotic pressure of blood plotted against blood chloride for the species of Fig. 58. (Beadle and Cragg.) Triangles, *M. obtusatus,* squares, *G. locusta,* solid circles, *G. deubeni,* open circles, *G. pulex,* as before.

cells as slight dilution of its haemolymph takes place. *G. pulex* gains chloride in the cells as the concentration of the haemolymph increases and may conceivably die of chloride intoxication. The two more euryhaline species, *G. locusta* and *G. deubeni,* evidently have a cellular mechanism for regulating the composition of the cell fluids in spite of changes in the haemolymph. Such a mechanism must be an important part of the physiological equipment of euryhaline brackish animals. It is possible that the variation in oxidative metabolism with variation in the concentration in the medium may depend in some forms not only on changes in work needed for over-all osmoregulation but also in the work needed to maintain differences in cellular composition against gradients from cell to haemolymph.

Werntz (1957) has studied a comparable group of four species on the American coast, namely *Marinogammarus finmarchicus, Gammarus (G.) oceanicus, G. (G.) tigrinus,* and *G. (G.) fasciatus.* The first named was obtained from a rocky coast, the second from an intertidal fresh-water seep, the third from a fresh-water seep high between the tide marks, and the fourth from fresh water. Both *M. finmarchicus* and *G. oceanicus* regulated in an almost identical manner, comparable to Beadle and Cragg's *G. locusta*,[57] but both survived in greater dilutions than the last named. *G. fasciatus* behaved like *G. pulex* and *G. tigrinus* not unlike *G. deubeni*. Sodium ions were responsible for about 45 per cent of the total concentration of the blood of both *G. oceanicus* and *G. fasciatus* at all concentrations studied. *G. oceanicus* produces a urine isotonic with the blood at all blood concentrations; *G. fasciatus*, at least in dilute media, with blood normally less concentrated than that of *oceanicus* in the same medium, produces a urine hypotonic to the blood. Urine production was much greater in *G. fasciatus* than in *G. oceanicus* and in both species was faster at high than at low temperatures. As with *G. locusta, G. deubeni,* and *G. pulex,* so with the American species; the more marine *G. oceanicus* has a shorter nephridial canal than the fresh-water *G. fasciatus*.

Using Potts's (1954b) thermodynamic treatment, Werntz calculates that at a molal concentration of 0.61 (21 per mille salinity sea water) *G. oceanicus* does at least 16.3×10^{-3} cal.$^{-1}$ g.$^{-1}$ (wet) hr.$^{-1}$ in maintaining its high salt concentration against loss of salt in urine, whereas *G. fasciatus* does only 0.81×10^{-3} cal. g.$^{-1}$ hr.$^{-1}$. At 0.03 molal concentration the minimum values are respectively 37×10^3 and 4×10^{-3} cal. g.$^{-1}$ hr.$^{-1}$. The high internal concentration maintained and apparently needed by *G. oceanicus* in dilute sea water is thus 20 to 10 times as expensive to maintain as the lower concentration in *fasciatus*. Werntz supposes that in full-strength sea water the salt uptake mechanism is almost entirely inactive, even in those euryhaline forms that possess it. As dilution takes place it is activated at a fairly high salinity in the more marine forms but at a lower salinity in the fresh-water species. In the marine species the curve relating internal to external osmotic pressure usually declines suddenly as the lethal dilution is approached. The evolution of *Gammarus fasciatus* into a truly fresh-water animal probably provides as clear a case of active invasion as can be deduced from distributional data, since its immediate allies in *Gammarus* (s. str.) are brackish-water animals. We are probably not justified in assuming that *G. deubeni* has invaded fresh waters relatively recently; its external morphology, if not its physiology, relates it to *Rivulogammarus*, which must be an ancient component of the European fresh-water fauna. It is possible that *G. deubeni* is derived from fresh-water

---

[57] According to Wentz there may be unresolved taxonomic problems involving *G. locusta.*

ancestors and has acquired a secondary euryhalinity that permits it to survive in saline water in competition with the more rapidly reproducing *G. pulex.* Only when *G. pulex* is absent is *G. deubeni* not forced toward the sea. Speculation about the origin of the species, however, is rather uncertain, since *deubeni* has a haploid chromosome number of 27 rather than the 26 characteristic not only of *pulex* but of most other members of the genus.

Although there are morphological differences in the antennary glands of *G. pulex,* compared with *G. locusta,* and of *G. fasciatus,* compared with *G. oceanicus,* which differences are probably involved in chloride retention in fresh water, it is possible that given sufficient time most of the species of *Gammarus* can be acclimated to a much greater range of salinity than is ever occupied in nature. Such acclimation seems to have been achieved for six species by Sexton (1928), including as extremes *G. locusta* in fresh water and *G. pulex* in sea water. If this is really the case, much of the regular zonation of the species found in nature must be due to the kind of competitive phenomena that have been postulated to explain the evolutionary processes of active invasion.

Passive invasions. The best contemporary evidence is derived from the Baltic and the Ponto-Caspian basins, but in neither case does a detailed examination indicate the striking process of passive adaptation that might be expected on cursory study.

Passive adaptation in the Baltic. The history of the Baltic basin has been discussed briefly in Vol. 1, but the best general summaries are those of Segerstråle (1957a,b). At the present time almost all of the Baltic has a salinity of less than 15 per mille and all but the central deep area a salinity of less than 10 per mille (Fig. 66). The chemical composition is essentially that of dilute sea water. At the heads of the Gulfs of Finland and Bothnia dilute water of a salinity less than 3 per mille occurs. Compared with the North Sea, the fauna of the Baltic is extremely depauperated. The characteristically marine sea anemones (fifteen species in the North Sea), Sipunculoidea (fifteen species in the North Sea), Amphineura (at least thirteen species in the North Sea), and Echinodermata (seventy species in the North Sea) are absent in the central basin and to its north and east. In general, the groups present are represented by but a few per cent of the number present outside the Skagerrak. Thus the three hundred and thirty marine amphipods are reduced to twelve species. More than half the species of animals living in the central Baltic penetrate into the Finnish and Bothnian gulfs. A certain number of purely brackish-water species, particularly among the smaller Crustacea, occur,

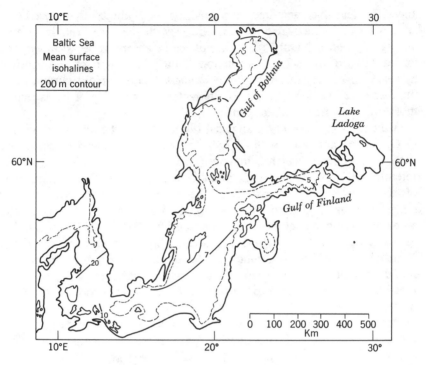

FIGURE 66. Map of Baltic Sea with isohalines. (Segerstråle.)

though none is endemic; and a limited number of fresh-water species such as *Radix ovata baltica* appear among the mollusks and *Abramis brama, Esox lucius, Lota lota,* and *Perca fluviatilis* among the fish. These brackish- and fresh-water elements, however, do little to compensate for the large number of marine species that drop out as the salinity falls to 15 or 10 per mille so that the total number of species present is much less than in the sea.

In addition to the ordinary marine, fresh-water, and widely distributed brackish species, a peculiar group of animals, commonly called glacial relict or glaciomarine relict species, lives in the Baltic basin, or in lakes that have been cut off from the basin or lie just to its south. These animals appear superficially to be of recent marine origin but to have become, in a number of localities, completely adapted to fresh water. The group was discovered by Lovén (1862) as an element in the fauna of the large Swedish lakes and since then has been the subject of a great many studies by Scandinavian and Russian zoologists. The whole literature has been admirably reviewed in a very important paper by Segerstråle (1957a), which inevitably is the basis of the present account. Some of the same animals occur in the glaciated parts of North America; the most recent study of these occurrences is that of Ricker (1959).

The interest of the Baltic fauna in providing supposed examples of passive immigration centers entirely on these so-called glaciomarine relicts. Because conditions for passive immigration in the Baltic area should have been particularly favorable, the mere fact that it seems to have been performed only by a rather peculiar assemblage of animals suggests that considerable caution should be exercised in taking the examples at face value.

The animals involved, some of which have or will appear in other contexts in this book, are the following.

Phoca hispida annelata, the Baltic seal, with lacustrine representatives in Lake Ladoga (*ladogensis*) and Lake Saimaa (*saimensis*), is a poorly defined subspecies of the arctic *P. h. hispida.* Allied species occur in the Caspian (*P. caspica*) and in Lake Baikal (*P. sibirica*) but are very distinct and probably old inhabitants of these basins. Seals of the genus *Phoca* often appear to enter fresh water without difficulty, and although the two lacustrine races in Lake Ladoga and Lake Saimaa raise micro-taxonomic problems of interest there is no reason to believe that these animals bear on the problem of passive invasion.

Osmerus eperlanus, the smelt, is an anadromous fish that easily forms landlocked populations. Although it may have entered the Baltic by the same route as the species discussed subsequently it has no clear bearing on the problem.

Coregonus (Leucichthys) albula in Europe is a fresh-water fish, but it has allopatric representatives in Siberia and Alaska which are anadromous and so again does not bear on the problem.

Coregonus (Coregonus) spp. Several species of white fish which occur in the Baltic and adjacent lakes have had a complicated history, brilliantly elucidated by Svärdson (1957), whose work will be discussed in Volume III in connection with introgressive hydridization and evolutionary problems in lakes. These fish appear to have reached the Baltic by the same routes as those used by the other so-called relict species, but, at least in the case of *C. nasus,* the invasion appears to have taken place too late for the proglacial lakes postulated to explain the migration routes, to have been of any value. Again these cases do not bear on passive immigration into fresh water.

Myoxocephalus quadricornis is an arctic fish occurring mainly, as indicated by the material collated by Segerstråle, in brackish or estuarine water at salinities less than 24 per mille. The fish is common in the Baltic, and many isolated populations are known in Swedish, Finnish, and west Russian lakes, but it does not occur south of the Baltic or in British lakes. An analogous lacustrine form is found in some of the deep lakes of glaciated North America. The nature of the fresh-water races has already been discussed (pages 177–180).

Limnocalanus macrurus occurs as *L. m. grimaldii* in brackish estuarine waters along the coasts of the Arctic Ocean, in the Caspian, and in the Baltic itself. The fresh-water form *L. m. macrurus* is found in various lakes, not only in those cut off from the Baltic but in Poland in basins beyond and above any Baltic transgression. It also occurs in Ennerdale Water in England and in many large lakes in glaciated North America.

Mysis relicta (*A, C, E,* Fig. 67, *A* of Fig. 69) has a distribution comparable to that of *Limnocalanus macrurus,* save that the Baltic population is not morphologically differentiated and that it occurs in far more localities south of the Baltic transgression in Denmark and North Germany and in several Irish localities. It is widespread in suitable habitats in glaciated North America. It was formerly supposed that *M. relicta* represented a fresh- and brackish-water form of *Mysis oculata* (*B, D,* Fig. 69). More recently another widespread northern species, *M. litoralis,* morphologically and to some extent ecologically intermediate, has been described. The meticulous researches of Holmquist (1959), however, have shown the three widespread and some rarer and more local forms to be quite separate species. *M. relicta* is largely a fresh-water animal except in the Baltic; the few arctic marine records are probably of animals washed out of fresh or dilute brackish waters. Four allied but well-differentiated species occur in the Caspian (*B, C,* Fig. 69), but there is no systematic variation of any of these animals with salinity as there is in *Limnocalanus* and *Saduria.*

Pontoporeia affinis (*F, G, H,* Fig. 67) in its typical form is found in the estuaries of rivers flowing into the Arctic Ocean and in the Baltic and is probably derived from the marine *P. femorata.* Lacustrine populations are distributed in Europe much as is *Mysis relicta,* but they are not found in Great Britain. The species is an important member of the benthos of the Laurentian Great Lakes and of the large lakes to the north and west; it also occurs in Lake Washington on the Pacific coast of North America. The north European lacustrine form is distinguished by Segerstråle as *lacustris,* but it is not regarded as having subspecific rank. A rather paedomorphotic male form occurs in the American lacustrine populations. A subspecies *P. a. microphthalma* inhabits the Caspian.

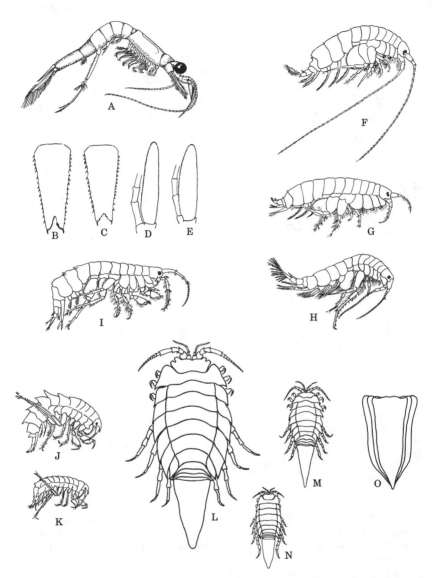

FIGURE 67. Baltic crustacea of the so-called glaciomarine type and some of their allies. *A. Mysis relicta. B. M. oculata* telson, Greenland. *C. M. relicta* telson, Sweden. *D. M. oculata,* exopodite and base of endopodite of second antenna. *E. M. relicta,* the same. *F. Pontoporeia affinis,* Gulf of Finland ♂. *G.* The same ♀ *H. P. affinis* f. *brevicornis,* Lake Nipigon. *I. Pallasea quadrispinosa. J. Gammaracanthus l. loricatus. K. G. lacustris. L. Saduria sibirica,* Karisch Sea. *M. S. entomon vetterensis,* Lake Vetter. *N. S. e. caspia* Caspian. *O.* Outer line, telson *S. sibirica* immature; middle, telson of *S. e. entomon;* inner, telson *S. e. vetterensis,* adult. (Tattersall, Holmquist, Segerstråle Sars, Ekman.)

207

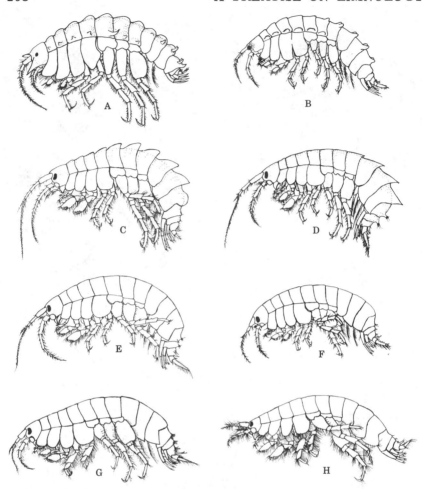

FIGURE 68. Caspian Gammaridae. *A. Axelboeckia spinosa* (× 2.5). *B. Gmelina kusnezowi* (× 2.9). *C. Amathillina cristata* (× 2.9). *D. Dikero-gammarus caspius*, (× 2.9). *E. D. haemobaphes* (× 2.5). *F. Chaeto-gammarus warpachowskyi* (× 6). *G. Pontogammarus robustoides* (× 2.5). *H. Niphargoides caspius* (× 3.3). (Sars.)

Gammaracanthus lacustris (*K* of Fig. 67) is a limnetic amphipod found in deep lakes cut off from the Baltic; in the main basin it has become extinct, presumably being fully adapted to the fresh water Ancylus lake and unable to return to a euryhaline condition when the sea broke in to form the Littorina sea. The animal is clearly derived from the arctic *G. loricatus* (*J* of Fig. 67) which has a brackish

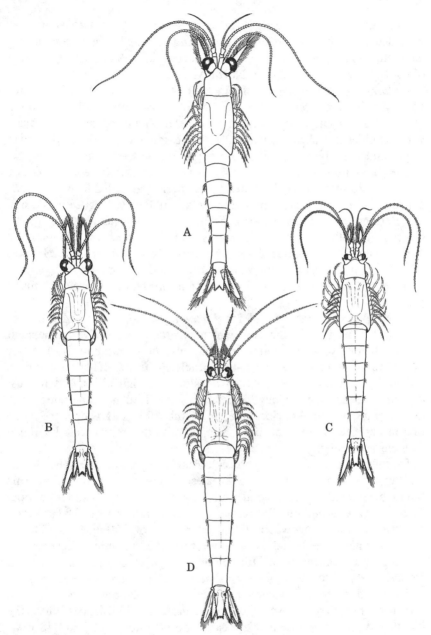

FIGURE 69. *A. Mysis relicta B. M. caspia C. M. microphthalma.* Both Caspian species are related to *M. relicta* and perhaps represent multiple invasions from the north. *D. Paramysis (Metamysis) ullskyi,* an endemic Caspian species adapted to fresh water that has penetrated far up the rivers entering the Ponto-Caspian basins. (Sars.)

subspecies *G. l. aestuariorum*. A small form *G. l. caspius,* somewhat intermediate between *loricatus* (s.st.) and *aestuariorum* occurs in the Caspian. *Gammaracanthus* does not occur in western Europe south of the Baltic nor in the lakes of North America.

Pallasea quadrispinosa (*I* of Fig. 67) is an amphipod with a European distribution comparable to that of *Pontoporeia affinis,* though in the Baltic itself it is confined to coastal waters of the inner basin with salinities not above 6 per mille. The genus is clearly of Baikalian origin and must be regarded as a member of the fresh-water fauna.

Saduria entomon (*L–O* Fig. 67) is represented in the Arctic by *S. e. glacialis* in the brackish waters of estuaries; an allied form *orientalis* appears in almost full-strength sea water in the Sea of Okhotsk and the Bering Sea. There is also, as has been indicated, a race in the Caspian. In the Baltic region the typical *S. e. entomon* occupies the main basin; the slenderer *S. e. vetterensis* is known in several lakes cut off from the Baltic but not in any locality south and above the maximum transgression. It is not found in North American inland basins.

It is apparent that, except for *Pallasea quadrispinosa,* which is of fresh-water origin and the seal which is probably largely independent of the chemistry of its habitat and may also be of ultimate fresh-water origin, these animals are euryhaline brackish forms. Their occurrence in the modern Baltic presents no physiological difficulties, but it must be remembered that all have presumably survived a fresh-water stage at the time of the Ancylus lake; if at this time salt water had been present at the bottom of the basin, it would almost certainly have become anaerobic.

In only one case has a physiological comparison between fresh- and brackish-water races of a species of the so-called glaciomarine fauna been made. Lockwood and Croghan (1957) found that both *S. e. entomon* from the Baltic and *S. e. vetterensis* from Lake Vetter can live in dilute sea water with a chlorinity of 600 mM., or 21 per cent, in which the blood is isotonic with the medium. Transferred to half this concentration, the blood is hypertonic to the medium. In nature the blood of *S. e. entomon* had a mean concentration of 335 mM. per liter in water of concentration 122 mM. per liter and 285 mM. per liter in water of concentration 75 mM. per liter; the blood of *S. e. vetterensis* had a mean concentration of 234 to 245 mM. per liter in fresh water of negligible chlorinity. As Krogh (1939) had earlier found, *S. e. entomon* dies in fresh water in which *S. e. vetterensis* can live. The varying history of the species in becoming passively adapted from fresh to salt water and vice versa clearly involves physiological adaptation as well as morphological change.

Four species, *Limnocalanus macrurus, Mysis relicta, Pontoporeia affinis,* and *Pallasea quadrispinosa,* designated by Segerstråle (1957a) as group I, are distributed in lakes south of the Baltic within the outer moraines of the last glaciation. Many of their habitats lie at considerably greater elevations than the maximum level of any body of water occupying the Baltic basin in late or postglacial times. This distribution is in accord with the idea that these four organisms moved along the retreating ice margin from the east in a series of inter-connecting proglacial lakes,[58] which because of the thickness of the ice cap lay at considerably greater elevations than the deglaciated Baltic at any stage in its history. This hypothesis, primarily due to Hogböm (1917), implies that the species in question entered the lakes south of the Baltic somewhat earlier than into the Baltic itself. It also provides a formal explanation, and was indeed put forward specifically for this purpose, of the presence of slightly modified or identical forms of *Pontoporeia* and *Limnocalanus* in the Caspian, since ice lakes held at a high level just north of the headwaters of the Volga might discharge into that river. It is, however, now believed more likely that the invasion of the Caspian took place from an ice lake formed across the present drainage of the Ob and Yenisei Rivers during the penultimate glaciation and so earlier than the invasion of the Baltic region. Actually the presence of four fairly modified relatives of *Mysis relicta* in the Caspian strongly suggests multiple invasion at different times during the Pleistocene. Invasion of certain lakes in southern Norway is explained by Segerstråle, following Holtedahl (1924) and Mathiesen (1953), as due to dilute water flowing out of the Yoldia Sea, which though in contact with the ocean must have contained much water of low salinity just as the Baltic does today. Outside the restricted Jaeren area in Norway no relict species are known in any lakes draining into the sea to the west of Scandinavia. The problem of the occurrence of *Limnocalanus* in Ennerdale and of *Mysis relicta* in this lake and in five lakes in Ireland is still unsolved. The presence of *C. (L.) albula* in some of these lakes suggests to Segerstråle that the route was a purely fresh-water system. At least in the case

---

[58] Cf., Volume I, pages 56, 88. Credner (1887) notes that Rink found seals in a proglacial lake in Greenland; this may well imply the presence of other animals and of plants, which constitutes a food chain. More observation of the faunas of proglacial lakes would be useful, though it is possible that no modern examples contain dammed up brackish-water Crustacea.

of Ennerdale, there is no possibility of postglacial marine transgression; if completely passive invasion is required, a high-level fresh-water ice-lake somewhere in the St. George's channel must be postulated. Segerstråle supposes that the two species, with C. (L.) albula, reached this area by a series of proglacial lakes against the margin of the last ice sheet to span the North Sea.

With regard to the other animals (group II) under consideration, namely Myoxocephalus quadricornis, Gammaracanthus lacustris, and Saduria entomon, to which Phoca hispida must perhaps be added, it has been supposed that either they arrived too late in the Baltic area to get into the high-level lakes to the south or that they were ecologically unable to survive somewhat warmer conditions in the areas in which they are not found. Segerstråle (1957a) strongly supports the latter argument of ecological restriction on the grounds that Gammaracanthus occurs in Lake Kenozero on the western edge of the Onega ice lake, which seems to be the most plausible proglacial lake from which the whole fauna can be derived. This point of view is perhaps strengthened by the restricted but perfectly definite distribution of Limnocalanus in Polish lakes, but not in those of Germany or Denmark, south of the Baltic. Before these localities were discovered Limnocalanus would have been associated with Gammaracanthus and Saduria rather than with Mysis.

With regard to the origin of the group as a whole, Segerstråle follows a suggestion of Pirozhnikov (1937) that all members arose in a proglacial lake in which water from the Arctic Ocean was impounded at higher and higher levels between the Ob and Yenisei rivers during the penultimate (Riss) glaciation. After adaptation to dilute but very cold water in this lake the various species are supposed to have passed the last interglacial as brackish-water arctic animals, which they still largely are, and then have been distributed eastward to America and westward to the Baltic by way of proglacial lakes during the last glaciation. Holmquist (1959), who accepts an Asiatic fresh-water origin for Mysis relicta not unlike that postulated for the origin of the hair seals, would put the development of this animal back farther in time than the Riss glacial, actually as far back as the Oligocene. There is nothing in the facts to suggest that she is wrong and that the same hypothesis may not be applied to the other species. There is also really nothing except tradition to postulate an ice lake as the site of the origin of the group, all members are evidently frequently estuarine, all may have become adapted to dilute waters in exactly the same way as the brackish Gammarus. The subsequent travels of the various species provide one of the most fascinating histories in the whole of

TABLE 6. *Analyses of ocean, Baltic, Black Sea, and Caspian (Dittmar, Schmidt, Kolotoff, and Lebedintzeff, as recomputed by Clarke, 1924)*

	Ocean	Baltic	Black Sea	Caspian Main Basin	Caspian Gulf of Karabogaz
Cl	55.29	55.01	55.12	42.04	50.26	53.32
Br	0.19	0.13	0.18	0.05	0.08	0.06
SO_4	7.69	8.00	7.47	23.99	15.57	17.39
CO_3	0.21	0.14	0.46	0.37	0.13	—
Na	30.59	30.47	30.46	24.70	25.51	11.51
K	1.11	0.96	1.16	0.54	0.81	1.83
Rb	—	0.04	—	0.02	—	0.06
Ca	1.20	1.67	1.41	2.29	0.57	—
Mg	3.73	3.53	3.74	5.97	7.07	15.83
Salinity $\%_0$	33.01–37.37	7.215	18.26–22.23	12.94	164.0	285.0

zoogeography, but there is clearly nothing in that history that implies the development of fresh-water animals passively by the gradual freshening of a basin. The only peculiarity of the group would seem to be that after the initial adaptation distribution has been mainly by way of lakes, active movement upstream being particularly difficult, though downstream movement, as of *Mysis relicta* in the River Shannon (Tattersall and Tattersall 1951), is evidently easy.

Passive invasion in the Ponto-Caspian region. The fauna of the Caspian[59] is in general largely a highly modified marine fauna, which is certainly derived from the Sarmatian Sea, of which the Caspian, after many vicissitudes, is the lineal descendant. This fauna is found living in an immense lake or inland sea, which exhibits extraordinary salinity gradients. In the extreme north, in the vicinity of the Volga delta, the water is usually almost fresh, though sometimes masses of brackish water may be blown inshore toward this region. In the Gulf of Karabogaz saturated brine is found, whereas in the greater part of the central and southern Caspian the water has a salinity of less than half that of ordinary sea water and a rather different percentage composition (Table 6).

Most of the major biological deficiencies exhibited by fresh waters in general are also characteristic of the fauna of the Caspian Sea. There are no echinoderms, no sipunculoids, and no cephalopods in

---

[59] By far the most accessible accounts of the Caspian, at least for the western reader, are those of Knipowitsch (1922) and of Zenkevitch (1957, 1963).

the Caspian fauna, and the coelenterates, polychaets, and bryozoa are very poorly represented. A few sponges, of which *Metschnikovia* is an endemic genus presumably of Sarmatian origin, may be allied to the Lubomirskiidae of Lake Baikal, *Ochridospongia* of Lake Ohrid, and the marine Haliclonidae (= Renieridae) (Martinson 1940). The limited turbellarian fauna includes a couple of endemic acoels, certainly of marine and presumably of Sarmatian origin; one of these, *Anaperus sulcatus* Beklemischew (1914), a member of a genus also known elsewhere, has the distinction of being 12 mm. long and so the largest known acoel.

The mollusks, which are clearly primarily of Sarmatian ancestry, consist mainly of lamellibranchs of the genera *Dreissena,* the endemic *Adacna* and *Didacna,* and of twenty or so species of gastropods of the hydrobiid subfamily Micromelaniinae, though *Planorbis* and *Theodoxus* (= *Neritina*) also occur. Apart from *Dreissena polymorpha,* no Ponto-Caspian mollusks appear to have extended their range beyond the Volga delta and the equivalent localities around the Black Sea. There is a large benthic crustacean fauna, consisting mainly of amphipods (Fig. 65), with some isopods, a number of Mysidacea, and all but two of the known members of the Cumacean family Pseudocumidae. These peracaridan Crustacea are of particular interest in the present context and are discussed at greater length later.

The planktonic crustacea of the Caspian consist largely of the copepod *Calanipedia aquae-dulcis* and of endemic polyphemid cladocera already discussed. The majority of existing marine cladocera are related to these animals, which may have colonized the sea from the Ponto-Caspian or earlier Sarmatian basin; this hypothetical addition to the marine fauna does not bear very much on the origin of fresh-water organisms.

The fishes of the Caspian include (Berg 1916, 1933, 1948 [1962]) a certain number that are obviously of fresh-water origin, namely in the Cyprinidae: *Rutilus rutilus caspicus* a subspecies of a widespread palaearctic species, *R. (Pararutilus) frisii kutum* a subspecies of a species that extends west into the upper Danube, *Vimba vimba persa* of like zoogeographic relations, *Barbus brachycephalus caspius* with the nominate subspecies in the Sea of Aral, and *Chalcalburnus c. chalcoides,* a member of a monotypic genus, with subspecies west to the upper Danube.

The sturgeons are well represented by *Huso huso, Acipenser nudiventris, A. güldenstädti, A. stellatus,* and in the influent rivers *A. ruthenus.* The zoogeographic status of these fishes is not entirely clear.

Among forms of obvious marine ancestry the Caspian pipefish

Syngnathus nigrolineatus caspius is perhaps of Mediterranean origin; it is now well adapted to fresh water, entering the lower Volga just as the nominate subspecies enters rivers flowing into the Black Sea.

Apart from these fish and several species of *Gobius,* the ichthyofauna consists of a number of members of endemic genera or species groups of obvious marine and presumably Sarmatian origin. The most important belong to the Clupeidae (*Alosa; Clupeonella*) and to the Gobiidae (notably species of *Mesogobius* and *Benthophilus*). The Caspian lamprey *Caspiomyzon wagneri* may be associated with this faunal element. The two clupeid genera occur in the Sea of Azov and in the limans and river mouths of the Black Sea. The variation in gillraker number in the Caspian species of *Alosa* suggests that a species flock of about half a dozen members (Berg 1913, 1948 [1962]), varying in feeding habits, is present in the Caspian.

There is a remarkable Arctic element in the Caspian fauna; the whitefish *Stenodus l. leucichthys,* and the Crustacea, *Pseudalibrotus* spp., *Saduria e. caspia, Gammaracanthus loricatus caspius, Pontoporeia affinis, microphthalma, Limnocalanus m. grimaldii,* and four species of *Mysis,* namely *M. caspia* (*B* of Fig. 69), *M. microphthalma* (*C* of Fig. 69), *M. amblyops,* and *M. macrolepis.* The four species of *Mysis* are allied to *M. relicta* and *M. oculata.* Each has a rather different distribution. *M. microphthalma* is apparently bathypelagic in 300 m. to the bottom at 927 m., occurring in both central and southern parts of the basin. The other species seem to be more benthic, *M. caspia* in 65 to 400 m. and *M. amblyops* in 203 to 927 m. occurring in both the central and southern region, whereas *M. macrolepis* is known only from 200 to 300 m. in the southern part of the basin. All four species are evidently ecologically differentiated, as is to be expected.

The Caspian whitefish *Stenodus l. leucichthys* is barely separable from *S. l. nelma* of the Arctic. The *Saduria* and *Limnocalanus* are essentially indistinguishable from Baltic forms.

The seal *P. caspica* is apparently morphologically and serologically well differentiated; it is now believed to be of Sarmatian rather than of directly Arctic origin. The palaeontological data are summarized by McLaren (1960). Taliev and Bazikalova (1934) find the Baikal seal much closer serologically to the Arctic seal *Phoca* (*Pusa*) *hispida* than to the Caspian animal; the Baikal race, however, differed considerably from the seal of Lake Ladoga (see page 205). It is unfortunate that more combinations were not tried, since the serological titers are, of course, not commutative.

The slightly modified arctic elements in the Caspian fauna may be regarded as having reached their present habitat by way of ice lakes, as already indicated. There is no reason, however, to suppose that this process was limited to the end of the last glacial age. *Mysis* may have invaded the area several times after earlier glacial ages than that responsible for the presence of the other less differentiated crustacea. Several of these early invasions would account for the presence of four species all apparently derived from ancestors near *M. oculata, M. litoralis,* and *M. relicta.*

The part of the fauna of marine origin is evidently very euryhaline. The gobies are well-known invaders of fresh waters in various parts of the world. *Alosa* is anadromous and gives rise to landlocked populations in many regions. The same is true of the lampreys, sturgeons, and salmonids.

Behning (1924) gives a list of forty-four peracarid Crustacea occurring in the fresh waters of the Volga delta, among which twenty-one occur also in the southern part of the Caspian and four more reach the central part of the basin. Half the benthic Crustacea of the Volga delta can therefore live under a wide range of salinities. It is by no means unexpected that a considerable part of those of Sarmatian origin should be capable of living in fresh waters, for during the Baku, Chosar (Khararsk), and Khvalynsk ages, which presumably correspond to the Mindel, Riss, and Würm glacial ages, the Caspian stood at a high level and drained into the Euxine basin, now occupied by the Black Sea. At least during the second of these high stages, when the Dardanelles were cut as a river channel leading to a greatly lowered Mediterranean, the entire Caspian must have been much more dilute than at present. The true Caspian fauna has therefore had to survive at least one and probably more dilute stages since its isolation from the Sarmatian Sea. It is reasonable to suppose that the chemical composition of the medium is a relatively unimportant factor in determining modern distributions, though it is possible that adaptation to the water in the lower Volga, which contains about 10 mg. Cl per liter does not imply a salt-absorbing mechanism that is good enough to permit an animal free entry into fresh waters of lower chloride content but otherwise of comparable composition.

The most important study of the subsequent movement of animals from the fresh water of the northern Caspian coast inland is Behning's (1924) work on the peracaridan Crustacea of the Volga. The peculiar Caspian Pseudocumidae are clearly well adapted to the very dilute water of the delta, seventeen species being found therein and six being confined to the northern Caspian. None of these species penetrates more than 200 km. up the Volga. Ten Caspian species of Pseudo-

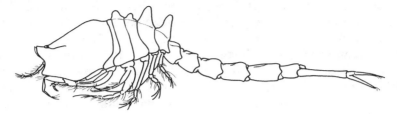

FIGURE 70. *Pterocuma pectinata*

cumidae are also known from the limans fringing the Sea of Azov and *Pterocuma pectinata* (Fig. 70) from the limans along the coast of the Black Sea (Derjavin 1925), but none occurs in the main basins of the Euxine. It is evident that the whole Caspian assemblage is adapted to fresh or brackish water but is not adapted to ascend rivers or colonize inland waters in general.

The Mysidacea are able to penetrate much farther up the Volga than are the Cumacea. Stammer (1936), whose taxonomic arrangement is largely followed, lists twenty-six species from the Caspian, of which seventeen belong to *Paramysis,* a genus also occurring in the Mediterranean and like *Pseudocuma* presumably of Sarmatian origin. The whole assemblage is obviously euryhaline. Among the twenty-six species ten are sufficiently euryhaline to occur in both the northern and middle Caspian, but they do not include the four species of *Mysis* near *M. relicta* which may well have had fresh-water ancestors. Six species occur in the Volga delta; only one, *Katamysis warpachowskyi,* has a range as limited as the Cumacea. Two species have an immensely wide distribution; *Paramysis intermedia,* which reaches to the lower part of the Kama River, is widespread in the Euxine limans and occurs in the mouths of the Don, Danube, Dneiper, and Kuban rivers. The other species (*D* of Fig. 69), *Paramysis (Metamysis) ullskyi (Paramysis strauchi),* is even more widely distributed, for it reaches more than 2800 km. from the mouth of the Volga. It is also found in the lower parts of the Ural and Don rivers. At least in the Volga, it would seem to be a fully adapted member of the fresh-water fauna, for it extends far north of any earlier Caspian transgression. It does not, however, appear to have entered lakes. Stammer indicates that in the laboratory the species requires about 6 mg. O_2 per liter.

Among the twenty-five Amphipoda known from the delta region, eight reach at least 1000 km. up the Volga, and *Dikerogammarus haemobaphes* (*E* of Fig. 68), *Gammarus macrurus,* and *Corophium curvispinum* (*sub devium*) penetrate more than 2000 km. The last-named species is apparently actively moving westward, for it appeared as far west as the Müggelsee near Berlin in 1912, though certainly was absent from this well-studied lake in earlier years (Wundsch, 1912). It has been confused with *C. sowinskyi,* so that the small differences between these two species have erroneously led to a belief that the western migrating population had also undergone some morphological modification.

Behning (1925) found that the number of eggs produced at one time by *Corophium curvispinum, Dikerogammarus haemobaphes,* and *Paramysis ullskyi* is greater in the more truly fluviatile populations than in the lower Volga. An increase in egg number might be expected in the first two species, which increase in size from the delta to Saratov. In the lower Danube, and the associated lakes and limans of the Black Sea, Cărăusu (1943) has found considerable variation in egg number in amphipods of the same species in different environments. *Ponto-gammarus obesus* of about 5 to 6 mm. length seems to have clutches of as many as eight eggs in the Danube and up to as many as eleven in a lake, in contrast to eighteen in the Volga. In *P. sarsii* there also seem to be larger clutches in the Volga than in the Danube. In all of these amphipods absolute size, which may increase with salinity, is clearly determinative of high clutch size.

In *P. ullskyi,* however, the increase in clutch size takes place in the upper part of the river in spite of a slight decrease in the size of the female. This interesting fact suggests that the familiar reduction of clutch size observed when fresh-water species are compared with their marine allies is probably an effect of adaptation to water chemistry and not to water movements. It may perhaps be suspected that in the lower reaches competition between individual young is of paramount

TABLE 7. *Mean length of reproductive female and mean clutch size of three Crustacea of Caspian affinities in the Volga* (*data from Behning 1925*)

	Corophium sowinskyi		Dikerogammarus haemobaphes		Paramysis ullskyi	
	l. ♀	clutch	l. ♀	clutch	l. ♀	clutch
Lower Volga	4.14 mm.	6	10.3 mm.	18	14.7 mm.	10.5
Saratov	4.68	10	10.8	28	14.2	18.5
Upper Volga and Oka	4.61	11	10.6	30	14.4	15

importance and that those females that produce the most vigorous young leave most descendents, whereas in the more fluviatile environment, in which loss of young swept down by the current is likely to be greater, the maximum number of descendents left *in situ* are those of females that produce the most eggs. It is unfortunate that no data on the geographical variation of egg size, which we would expect to be maximal in the small clutches of the lower Volga, are available.

There are a few cases of survival of Ponto-Caspian elements in relict lakes in the vicinity of the Caspian and Black Seas, but the process has played a far less important part in this region than in the area of the Baltic transgression, mainly because of the more arid climate and the tendency toward astatic and saline waters in such relict basins. The best examples of the occurrence of Ponto-Caspian animals in a lake isolated from the Caspian are in Lake Tscharchal. This lake lies just above mean sea level at a distance of 400 km. north of the Caspian and occasionally drains by a very intermittent outlet into the Ural River. The water, according to Behning (1928b) is about one tenth as salt as ocean water, which unlike the Caspian, it resembles in relative composition. The plankton consists mainly of fresh-water forms but includes as a common constituent the typically Ponto-Caspian Cladoceran *Podonevadne trigona. Clupeonella delicatula tscharchalensis* is also known from the middle Volga, whereas the nominate subspecies occurs in the mouth of the Volga, the Ural, the rivers draining into the Black Sea, and in the open water of both the Black and the Caspian.

It is evident from the history of the Caspian that certain groups of marine animals, notably the Mysidacea and other peracarid Crustacea and the various families of mollusks of Sarmatian origin, adapt far more easily to fresh waters than do other groups and constitute the only good case of passive adaptation to very dilute water on a large scale. However, once the adaptation has been made, great barriers, in part no doubt due to the physical environment of rivers, in part to competition with the existing fauna, prevent the adapted species from becoming normal components of the fresh-water fauna. Only *Dreissena polymorpha* and *Corophium curvispinum* have been able to cross, probably with human help, into rivers not draining into the Ponto-Caspian basin; in addition *Paramysis intermedia, P. ullskyi, Gammarus macrurus,* and *Dikerogammarus haemobaphes* seem to have become widespread in the Volga. This contribution is not a great one from a basin in which conditions were optimal for the passive adaptation of marine animals to fresh water.

Origin of the biota of hyperhaline waters. One other matter remains to be considered briefly. Many closed basins in semi-arid regions contain water as salt or saltier than the ocean. Insofar as such basins are inhabited at all, the organisms found in them are apparently nearly all of fresh-water origin (Beadle 1943). This may appear paradoxical at first sight, but it is evident that the dispersal mechanisms available to many fresh-water organisms ensure that inland saline lakes are more likely to receive immigrants from fresh waters than from the sea. Moreover, such saline lakes are frequently *athalassohaline,* to use Bond's (1935) useful term, with ionic proportions greatly different from sea water, so that mere marine origin may not give an invader much physiological assistance. Fresh-water animals with relatively impermeable surfaces may in fact be potentially better ancestors of the fauna of extremely salt lakes than any marine animal with a highly permeable body wall would be. The most characteristic animals of such saline lakes are the brine shrimp *Artemia;* in the New World waterbugs of the genus *Trichocorixa;* various mosquito larvae such as *Aedes detritus* and *A. natronius,* and a great number of larvae of salt flies (Ephydridae); in the Old World the copepod *Arctodiaptomus salinus;* the rotatorian *Brachionus plicatilis,* and in various local faunas other insects and Crustacea of limited distribution. The flora ordinarily consists of blue-green algae such as *Spirulina* and *Nodularia,* of the motile green alga *Dunabiella,* and, in localities in which sulfate is being reduced in the mud, of photosynthetic bacteria. All the lower organisms belong to evolutionarily euryhaline rather than to specifically fresh or salt water groups.

Artemia salina can maintain an almost constant blood concentration hypotonic to the medium, but it fails to live when the medium is diluted to the concentration of the blood (Medwedeva 1927). *Aedes detritus* can maintain an almost constant haemolymph concentration, whatever the concentration of the medium. A few forms, such as *Palaeomonetes,* of relatively recent marine origin are known from saline inland waters. This animal is able to regulate osmotically over an immense range of hypertonic and hypotonic media (Panikkar 1941b), but it is to be noted that when it occurs in inland salt water, as in North Africa, the populations are probably of immediate fresh-water origin. The evolution of fresh-water animals into inhabitants of waters hypertonic to the haemolymph is well shown in the Corixidae (Claus 1937). In this family *Sigara (Subsigara) distincta* cannot control a slow rise in internal concentration as the external salinity is raised and dies when the haemolymph is about isotonic with the medium at a salinity of about 19 per mille. *S. (Halisigara) lugubris,* which is normally found in salt marsh pools, can maintain a relatively constant internal concentration over a wide range of salinities and is able to live permanently in water somewhat more saline than that tolerated by *S.* (*S.*) *distincta.*

Beadle (1943), moreover, has pointed out that since many salt lakes are astatic, fresh-water animals, which have resting stages capable of withstanding desiccation, are at an advantage over marine forms not merely in having stages permitting transport into the salt lakes but also in surviving when the lakes dry up temporarily. Certain aspects of this matter will be treated at greater length when the limnology of semi-arid regions is considered in Vol. III.

SUMMARY

The aquatic members of the lowest groups of organisms, the Monera, without plastids or mitochondria, and the more primitive members of the Protista or single-celled organisms with such organelles, cannot be unequivocally regarded as either of marine or fresh-water origin. They easily cross the boundaries between marine and fresh-water environments by small evolutionary changes and are designated as evolutionarily euryhaline. Life surely began in water, but there is no evidence that it began in the sea.

In the Protista a number of lines leading from such evolutionarily euryhaline organisms became adapted to life in the ocean, notably the red algae, brown algae, the Siphonales among the green algae, the Foraminifera, and, of more doubtful ancestry, the Acantharia and Radiolaria. Fewer lines are specifically fresh-water, but the Conjugatae among the green algae and all but the most primitive Euglenophyta are important examples of such adaptation. Most of the animal protists other than those just mentioned and many of the smaller plant protists retain a great degree of evolutionary euryhalinity.

The higher plants probably represent several invasions of the land by green algae, doubtless from fresh water. All the aquatic forms are presumably secondary.

The origin of the Metazoa is still uncertain, but, except perhaps in the flatworms, all the phyla clearly have a marine ancestry. The fresh-water metazoan fauna is therefore all or almost all of marine origin, though the very important contribution made by the insects has come into fresh waters from the sea by way of the land.

The various phyla or classes have made very unequal contributions, and within any one of them it is not unusual to find evidence of several invasions by closely allied taxa, although the greater part of the group has contributed little or nothing. Thus in the Polychaeta nearly all of the fresh-water species belong either to the Nereidae or to the sabellid subfamily Fabricinae; nearly all of the Coelenterata in fresh water belong to the Hydridae, which have no living marine representatives, or to the Olindiidae. Some sort of preadaptation to fresh-water environments, distributed in a very superdispersed way, must exist.

Great contributions to the fresh-water fauna have been made by the various groups of Turbellaria among the Platyhelminthes, by the Rotifera, chaetonotoid Gastrotricha, and Nematoda among the Aschelminthes, by both Gastropoda and Bivalvia but by no other classes of Mollusca, by the Clitellata in the Annelida, by the Crustacea and Pterygota among the Arthropoda, and by the Vertebrata. No contribution has been made by the Echinodermata, Brachiopoda, or several smaller phyla. There is no good reason for supposing that the vertebrates arose in fresh water, but nearly all living vertebrates must have had a fresh-water stage in their phylogeny. The dominant group in the sea today, the teleostome fishes, thus appears to have been derived from fresh waters.

The fundamental physiological distinction between fresh-water animals and those that have passed their evolutionary history as Metazoa entirely in the ocean lies in the fact that the latter contain body fluids approximately isotonic with the medium whereas in the fresh-water animals the mineral content is well below that of the sea water, though much higher than that of the fresh waters in which they live. When a marine animal in this sense is brought into dilute water, it is internally diluted if it survives; if it is brought into concentrated water, it is internally concentrated. Such an animal is said to be *poikilosmotic*. A fresh-water animal brought into slightly more saline water maintains at first its initial osmotic concentration; it is *homoiosmotic*. As the medium becomes more concentrated there may be a slight increase in internal concentration; with still greater external concentration, if the animal does not die, as it usually does, it tends to become poikilosmotic. Brackish water and estuarine animals are also usually poikilosmotic at high concentrations and more or less homoiosmotic at low concentrations. The poikilosmotic animals usually contain body fluids and always contain cell fluids, which though isotonic with the medium contain different proportions of ions to those found in sea water. Ionic regulation thus is more primitive than osmotic regulation, and this more primitive process may underlie the more specialized process.

In almost all cases, since it is necessary to maintain some surface permeable to oxygen and so in general to water, homoiosmotism must involve an incoming stream of water which can be controlled only by doing work in pumping it out again. Life in fresh water therefore is more energetically expensive than in the sea; in actual practice the increase in energetic requirements often seems to be disproportionately great. There is clear evidence that cyanide-sensitive oxidative metabolism may be involved and that calcium deficiency in the environment may increase the permeability to water to an extent that prevents adaptation.

Two extreme possibilities appear to be open to an animal adapting from sea to fresh water. In one, adopted by the fresh-water lamellibranchs of the family Unionidae in which there is an immense naked gill surface, the internal fluids are reduced enormously in concentration so that the osmotic gradient becomes minimal. The water that enters the animal is then excreted as urine, hypotonic to the blood, produced in large quantities.

In the other, best known in fresh-water crabs in which the gill surface is small and covered by a thin exoskeleton, the incoming stream of water is reduced and a small amount of urine isotonic with the blood is excreted. In the crayfishes it is probable that the evolution of hypotonic urine is a refinement added to an evolutionary history originally comparable to that of the fresh-water crabs, but it is most unlikely that the same could be true of the Unionidae.

In all animals in which there is any salt loss in excreting water, and in all animals that grow, a mechanism for taking up salt against a concentration gradient is a necessary requirement for fresh-water life. In some cases the sodium chloride in the animal's food may be sufficient for maintenance but almost always special organs are developed for this function, often in association with respiratory surfaces. It is probable that a large number of animals, particularly Crustacea, which have solved the osmotic problem, are unable to migrate far inland from the coast because of their requirements for several milligrams of chloride per liter in the medium. Even though an increase in temperature increases the rate of water intake and excretion, it is probable that the over-all salt- and water-regulating mechanisms work best at high temperatures. It is possible that in some cases the entry of a species into almost fresh water is limited by low winter temperatures.

Among the lower vertebrates, all of which save the myxinoids have probably passed through a fresh-water stage in their ancestry, the species returning to the sea have retained an electrolyte content of the blood lower than that of the medium. In the elasmobranch fishes the deficit in osmotic concentration is made up with urea and tri-methylamine oxide. In the other marine fishes the blood is hypotonic to the medium; the fish keeps drinking water and excreting chloride through the gills. This mechanism ordinarily maintains the blood of marine teleosts at a concentration well below that of sea water but twice that of the blood of fresh-water teleosts. Some fish migrating from sea to fresh water or vice versa may change their blood concentration, but in other cases they do not. There is evidently a complicated control system involving the endocrine organs, notably prolactin production by the pituitary in *Fundulus*. Among the higher vertebrates, in reptiles and birds, species may become secondarily associated with either fresh or salt water. In salt water there is usually a considerable excess of salt, taken in with the food, that is excreted as tears by an orbital gland.

An appreciable part of the fresh water biota is of terrestrial origin. Here, at least among the animals, the main problem involves respiration rather than water and salt equilibrium. The diffusion of a gas through a gas is much easier than through a liquid, and the oxygen content of air per unit volume is much greater than is ever likely in the case of well-aerated water. Air breathing, once it has been evolved, is therefore much more efficient than the breathing of dissolved oxygen. The most perfect adaptation to aquatic life consists of mechanisms for the breathing of gaseous oxygen permanently underwater. The most remarkable cases are provided by the plastron mechanisms of some insects.

Apart from the obvious needs for chloride uptake, the control of incoming water, and for oxygen, it is probable that the fresh-water environment imposes other conditions on invading organisms. It has long been supposed that in temperate regions the variability of temperatures in fresh waters will be so much greater than in adjacent seas that a real barrier to invasion is present. This may have been exaggerated, but the large number of animals in hypogean waters that are clearly descended from marine ancestors suggests that the stable temperatures in caves and ground waters may have facilitated the process. The invasion of hypogean waters is probably of two kinds: passive adaptation in marine caves which are being isolated by uplift and receive water from the land and active invasion of interstitial waters. The first involves various larger Crustacea, the second the very small Microparasellinae, Microcerberinae, Bogidiellidae, and Ingolfiellidae.

It has frequently been supposed that the presence of small planktonic larvae limits invasion of fresh waters by some groups. This is probably true, but it is as likely to be due to the difficulty that such larvae, hatching very early in embryonic development, would have in taking up salt as in maintaining their position against a current.

A little evidence suggests that small organisms may adapt more easily than large animals. There is a little evidence of physiological races adapted genetically to different salinities but morphologically indistinguishable. When, as in several so-called glacial relict organisms, adaptation to fresh or salt water has apparently occurred in several different regions and involves both physiological and structural characters, the various forms often show striking morphological parallelism.

The most complete analysis of the processes of active invasion, which probably involve the competitive production of zonation in salinity gradients, can be made in the case of the Gammaridae, but even here the information is far from complete; there is no evidence, moreover, to suggest that the process is actually proceeding even in this case at the present time.

Passive invasion has been supposed to be of great importance. The cases that have been emphasized in the past include the so-called glaciomarine elements in the Baltic region. However, they seem to have been adapted to very dilute water before they reached the Baltic, and the evidence that they provide of passive addition to the fresh-water fauna is specious. Striking passive adaptation has probably occurred in the Caspian, but the subsequent invasion of ordinary fresh waters, which must be largely an active process, has added only four Crustacea to the upper Volga and only one crustacean and one mollusk to the fauna of areas north and west of the drainage of the rivers running into the Ponto-Caspian basins.

The biota of inland hyperhaline basins consists either of simple organisms of high evolutionary euryhalinity or of higher organisms that have been chiefly derived from fresh water.

CHAPTER *19*

The Structure
and Terminology of Lacustrine
Biological Communities

Before proceeding to any detailed discussion of the biology of lakes it is necessary to deal with the rather elaborate terminology that has grown up to designate the various parts of the biological community within a lake. In presenting this material, which may appear to many, with some justification, to belong in the more arid regions of philology rather than in natural science, the criteria adopted in the acceptance or rejection of a term are primarily the following.

Utility. Many terms have been proposed in ecology in case they might be needed or to complete classificatory schemes of uncertain applicability. As far as possible only terms that are actually useful in scientific discourse will be admitted. The discriminating logophile will find some extraordinary examples of ecological nonce-words still-born on the pages of Klugh (1923).

Translatability. Terms based on colloquial or dialect usage not clearly explained in ordinary bilingual dictionaries are to be avoided.[1]

---

[1] The reader may imagine the difficulties of a student who has learned English in his mature years coming across expressions such as "bang-bang theory" or "weak lumpability," both used in contemporary writings on highly respectable branches of knowledge. But "muck," so beloved by some down-to-earth limnologists and defined by Webster first as dung or manure (as in the familiar "muck-raker") and second as impure decomposing peat or black swamp soil, is almost as bad. In its dictionary meanings it certainly does not apply to the dung of tendipedid larvae.

When new words are needed, the convention of forming these from classical roots is still the best procedure, since it involves a minimum of local or national prejudice and produces words for which conventions of assimilation exist at least in all important Indo-European languages. On this matter Klugh was sound.

Euphony. This is obviously a matter of aesthetics; *de gustibus non est disputandum.* Yet most would probably agree that even at the cost of redundancy *epineuston* is a better word than *epimicropleuston.* However, words that are obviously excessively truncated, such as Shelford's *ece,* seem too small to carry their meaning, and few beside its author have preferred ece to biotope, which, when clearly defined, appears to have the same meaning.

DEFINITION OF CERTAIN GENERAL ECOLOGICAL TERMS

Population and assemblage. The term *population* is used exclusively to mean a collection of individuals, all of one species, within a defined area or volume. A collection of populations of different species, all living within a single defined area or volume, is called an *assemblage* in discussions of commonness and rarity in Chapter 21.

Biotope, biocoenosis, and ecosystem. The term *biocoenosis* was introduced by Möbius (1877) to designate the entire biological community of a defined area in which individuals of various species live side by side and persist by virtue of their reproductive activity. The term has been used in many different senses by subsequent workers (see particularly Dahl 1908; Enderlein 1908; Hesse 1924; Friedrichs 1927).

The term *biotope* was introduced by Dahl to designate the kinds of terrestrial or aquatic environments[2] in which organisms occurred.

It appears that most authors have regarded the biocoenosis as the community occupying a biotope, but this point of view is not quite universal; Hesse (1924) considers that a single biotope can contain more than one biocoenosis. It has also usually been assumed that the biotope and biocoenosis should have some sort of natural limits. Much of the difficulty that has developed with regard to both terms is in finding an operational definition of these limits. The difficulties have been enhanced by attempts to include in the definitions properties that might be empirically established after prolonged study but are useless in definition, for they do not enable us to determine rapidly whether we have a biocoenosis before us. Thus Friederichs emphasizes the self-regulatory properties of the biocoenosis. Certainly these prop-

---

[2] "Gelände-und Gewässerarten"; there is really no concise English equivalent.

erties are important, but if the biocoenosis is to be regarded as a convenient unit in biological studies it must be recognizable by properties other than those that have been demonstrated only after considerable investigation in a limited number of cases.

In order to achieve any sort of satisfactory definition, it is necessary to consider what may be called the mosaic nature of the environment. With the possible limiting exception of a tripton-free isopycnal surface in a highly stable stratified zone of a lake or the ocean, no part of the environment will ever be entirely uniform or even form part of a uniform gradient. This lack of uniformity will ordinarily be expressed as an irregular variation in structure, whether as small impermanent vortices or solid inanimate objects such as sand grains, stones, fragments of dead organisms such as leaves, or, relative to one species, the variously distributed individuals of other species. These *elements of the environmental mosaic* are ordinarily spatially repeated in a more or less irregular but not necessarily random way. Each element, moreover, will itself have certain structural and often irregular features so that in any given inhabitable space a hierarchical arrangement of *grain size* of detail will distinguish the environmental mosaic. If we consider, for instance, a notommatid rotifer living in one of several isolated weed beds in the littoral of a lake the immediate neighborhood of the rotifer at any moment might consist of the divided leaves of *Utricularia* or *Myriophyllum* plants, the stems to which these leaves are attached, epiphytic diatoms, specimens of *Hydra,* Cladocera, and other rotifers, various fish, drifting dead leaves, and temporary vortical structures in the water. To such a rotifer the existence of another weed bed, separated perhaps by the delta of a small stream from the one in which it is living, will be irrelevant. If we could consider an average specimen of the particular species of rotifer under discussion in the first weed bed, we would find that it had a certain mean range comparable to a mean free path, which would be larger than the dimensions of the elements of the environmental mosaic relevant to the species. If we now consider a territorial fish, a specimen coming into breeding condition might search along part of the littoral to find a suitable unoccupied breeding territory; in doing so it might pass several isolated weed beds of the kind just considered as habitats for littoral rotifers. The largest relevant elements in the environmental mosaic of the fish will obviously be much larger than the largest relevant to the rotifer.

Similarly, on land the crevices in the bark of a tree might be large relevant mosaic elements to a psocopteran, with the same relation to its mean range as are the whole trees and the spaces between them to that of a deer.

At the boundary of the lake or at the edge of the wood the conditions change. Even if a frog passes from lake to shore several times in the course of its life history, the disposition of lakes is most unlikely to be such that the distance between them is small compared with the mean range of the frog in the course of its life.

An area will be called (Hutchinson 1957, 1959b) *homogeneously diverse* relative to a motile species S if the relevant mosaic elements are small compared with the normal[3] ranges of the individuals of the species S, *heterogeneously diverse* if the mosaic elements are large relative to these ranges. The littoral of a lake will be heterogeneously diverse in relation to a notommatid rotifer, homogeneously diverse in relation to a sun fish.

At least in horizontal directions, any segment of the biosphere that can be usefully defined as an ecological unit should be homogeneously diverse in relation to the larger motile organisms present. Since the investigator is motile and of the same order of magnitude as the largest animals usually encountered, there is generally no difficulty in recognizing the major boundaries between units. In general, such boundaries will correspond to obvious changes in the substrate and to marked changes in vegetation. The main difficulties will come when the largest motile animal is wide ranging and eurytope. It is probable that a forest containing many small lakes is homogeneously diverse with respect to the moose, which is a relatively amphibious ungulate. The lakes, however, will not contain organisms of primarily aquatic adaptation that range in an equivalent way through large tracts of forest. Some discretion, which implies a certain degree of arbitrariness, is inevitably involved in the delimitation of units. This arbitrariness will depend on the interests of the investigator. Although there is an evident asymmetry in the example of a forest containing small lakes and moose, it is probably best to recognize the smaller units as primary divisions if the investigation is mainly directed toward them.

Although nothing is said about mutual interaction within any unit, the definition of homogeneous diversity implies contiguity or motility within the unit and so the *possibility* of interaction, though not that it occurs.

Vertical diversity cannot be treated in the same way as horizontal diversity because it is always determined by the vectorial nature of solar radiation and the gravitational field rather than by the irregular disposition of material objects. For the purposes of this book the vertical extent of aquatic biotopes and ecosystems is arbitrarily taken as the whole region below the surface down to the limit of the deepest

---

[3] This is intended to exclude directional migratory movements.

sediments actually laid down in any body of water under consideration. A comparable arbitrary vertical limitation will also be necessary on land.

The definitions to be adopted then are as follows.

A *biotope* is any segment of the biosphere with convenient arbitrary upper and lower boundaries, which is horizontally homogeneously diverse relative to the larger motile organisms present within it (in botanical work this ordinarily means relative to a nonmechanized botanist performing a random walk); the diversity will be partly biological and not a prior physical property of the space. An *ecosystem* (Tansley 1935) is the entire contents of a biotope. A *biocoenosis* is the totality of organisms living in a biotope, or the living part of an ecosystem.

All questions of interdependence, other than the possibility of its occurrence, self-regulation, and progressive maturation or succession are regarded as empirical, not deductively answerable from the definitions.

It is to be observed that if we are dealing with terrestrial communities the biocoenosis may occupy a biotope of enormous extent, of many thousand square kilometers, as in the northern coniferous climax forest; but within this climax forest any lake or any significant area passing through successional stages would constitute a different biotope and have a different biocoenosis, provided that the mosaic elements represented by such areas were large compared with the mean ranges of the motile organisms present. Any given biocoenosis occupies a biotope of definite area, so that two similar biocoenoses, say coniferous forest, may be separated by a dissimilar biocoenosis. In this the term differs from *formation,* which is a class of similar biocoenoses, at least on land.[4] The biocoenosis thus represents a more empirical unit than the *biome* (Clements and Shelford 1939), which is an ideal construction containing not only climax and successional stages but various arrested stages resulting from deflected succession.

All comparison between terrestrial and aquatic biocoenoses tend to be unsatisfactory because the changes over a few meters in a vertical direction in water are much greater than on land. Within a vertical range easily navigable by a fish within a few minutes, and habitually traversed by migrating copepods during a few hours of the evening,

---

[4] Dice (1952) regards the biocoenosis as equivalent to the association but he seems to use the latter word in a more abstract and extensive sense than is usual in plant ecology. His *stand* is almost the equivalent of association as used here in an empirical sense.

we may have as great a difference in physical conditions as that distinguishing the coastal plain and alpine region of mountains in the tropics, in which two terrestrial regions no species of plants or animals may be found in common. In lakes the very divergent associations of organisms at different depths have often been regarded by plant ecologists as successional stages, since the lake appears inevitably to be filling in. In the ocean the comparable zones are regarded as biomes by Clements and Shelford, even though they are usually not trophically independent of the organisms above them. The biome concept therefore does not seem very satisfactory in hydrobiological studies, and the term has been little used by investigators in the field.

Association, niche, and habitat. The only divisions of the biocoenosis that seem to have real significance in ecological limnology are what would be termed *associations* by plant ecologists, particularly in Europe. Such parts of the biocoenosis are those that can be characterized by dominant or at least characteristic species.[5] The concept of the association is definitely useful for both plant, animal, and mixed communities. It inevitably has an arbitrary size parameter which cannot be dealt with by a concept of homogeneous diversity but must remain relative to the visual field of the investigator. The association is recognized as a patch, which on examination proves to be caused by the presence of a number of specimens of a particular species close together or to the texture of an area defined by a more scattered but nevertheless spatially restricted distribution of organisms against a nonliving background. The fact that associations so often have definite limits, although the factors determining them often vary continuously, is doubtless due primarily to the external determination of the direction of competition, as Gause and Witt (1935) first pointed out in a formal manner. The term association ordinarily is used for an assemblage of species that recurs under comparable ecological conditions in different places. There is, however, no logical reason why some associations should not have a unique representation in nature. To deny that such an assemblage is an association is to introduce hidden theory into the meaning of the word.

Chodorowski's (1959) term *taxocene* for a group of species, all members of a supraspecific taxon and occurring together in the same association, is most convenient. We may thus speak of the benthic taxocene of tendipedid larvae, the epilimnetic planktonic taxocene of

---

[5] The question why a few species are common and many rare, a condition essential in the characterization of an association, is a fundamental one and is discussed at greater length in Chapter 21.

calanoid copepods, or the littoral taxocene of monocotyledonous flowering plants. The populations of the species of a taxocene have obviously been defined as an assemblage or part thereof.

Within any biotope we may also recognize a series of habitats which may be characterized in terms of the various species present. The *habitat* of a species, within the geographical range, may be regarded as operationally defined by specifying those parts of the ecosystem that must be present in a biotope in order for the species to occur. The habitat is regarded as having spatial extension.

The *niche* of a species is defined purely intensively. It is assumed (Hutchinson 1957, 1959b) that all the variation of the factors required to define a habitat can be ordered linearly on the axes of an n-dimensional coordinate system. If the species S_1 requires that the variable X' have values between X_1' and X_2' . . . we can define a hyperspace N_1; any point within N_1 corresponds to values of the variables X', X'' · · · which permit the species to occur. This hyperspace is called the *fundamental niche* of the species. The space of which the niche is a part is called the niche space or, symbolically, the N-space.

The habitats of two species, being two volumes of the physical space of the biotope (B-space) defined by the presence within them of certain parts of the ecosystem, can overlap or be coextensive. It is considered by many ecologists, including myself (see pages 356–357), that under ordinary circumstances at equilibrium two species cannot co-occur in the same niche (*principle of competitive exclusion*). For example, if we consider two species (S_1, S_2) of planktonic copepod or rotifer, one feeding on large and one on small algal cells, and both requiring an identical range of temperature and chemical composition, the habitat of one species would be a volume of water in ordinary B-space of the required physicochemical properties, with small algae suspended in it; the habitat of the other could be the same water with larger algae. Any volume of water might contain both algae and so be a habitat for both species of zooplankton. In the ideal niche diagram, however, the hyperspaces N_1 and N_2, defining the fundamental niches of the two species, would be separated by the different values of the food size on the axis along which food size is measured. More elaborate, if still two-dimensional cases, are easily envisioned (Fig. 71).

Strata and zones. In this book the lake is ordinarily considered as a single ecosystem, in a single biotope, supporting a single biocoenosis. The ecosystem provides many habitats and niches, and the biocoenosis is divisible into many associations of both plants and animals. The division of the biosphere directly into biotopes and the subdivision

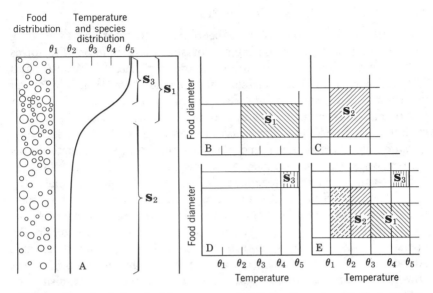

FIGURE 71. Analysis of niche space occupied by three hypothetical species differing in their temperature tolerances and the size of food taken. *A*. The vertical distribution of temperature, food organisms, and the three species S_1, S_2, and S_3. The tolerance of S_1 by itself, between θ_2 and θ_5°, eating small food. *C*. The same for the cold stenotherm but more efficient S_2 between θ_1 and θ_3°C., eating small and intermediate food. *D*. The same for the warm stenotherm S_3 between θ_4 and θ_5°C., eating large food. *E*. This shows only the realized niches; S_2, overlaps S_1 in the niche space S_1, excluding it from the hypolimnion, but is unable to realize its whole potential niche as there is no water cooler than θ_2 in the lake; S_1 and S_3 can coexist as they eat different food.

of the biotopes into habitats on a purely empirical basis may appear crude to those who, if they really exist, have been accustomed to building up biotopes into subbiochores, biochores, superbiochores, and biocycles and to subdividing the biotopes into facies, zones, and strata. The two last-named categories, however, for qualitatively recognized divisions along the quasi-horizontal and vertical axes, respectively, are useful. In limnology, as we leave the shore and approach the center of the lake along the bottom, we pass over several *zones* which represent intersection of *strata* with the bottom. Other convenient terms refer primarily to parts of the lake biocoenosis, and, since they depend on the fact that one boundary of a lake is a liquid-gas interface, the other a liquid-solid interface, they cannot be part of a general ecological scheme of classification.

THE LACUSTRINE BIOCOENOSIS

Several detailed attempts have been made to treat the biota of a lake as composed of an assemblage of species of different life forms within the biocoenosis. The most ambitious of these is due to Gams (1918), who provides a classification of life forms suitable for all organisms. The fundamental distinction in this classification is whether the organism is adnate or attached at a surface, rooted in a solid medium, or free. The three assemblages are termed, respectively, *ephaptomenon, rhizomenon,* and *planomenon.* As far as aquatic forms are concerned, the ephaptomenon consists of the plants which Warming (1895) called nereids,[6] together with equivalent sessile animals; the rhizomenon consists of the ordinary rooted higher vegetation, and the planomenon of the free forms which Gams distinguishes as *plankton,* drifting in water, *pleuston* at the surface, *edaphon* in the interstitial water of soils, and *tacheion,* actively moving organisms. All are divided into subclasses, the aquatic tacheion comprising the crawling organisms or *herpon* and the swimming organisms or *nekton.*

Warming (1923) has produced a modified scheme solely for plants, in which the aquatic forms are divided into free forms, or *planophyton,* which constitute the plankton and pleuston, and attached forms or *benthos.* The benthos is divided into *herpobenthos* to include blue-green algae and diatoms in the top layer of soft sediments, *rhizobenthos* for the rooted vegetation, *haptobenthos* for the *epilithon* and *epiphyton* attached to solid surfaces, respectively, of stones and other plants, and *endobenthos* for boring algae.

The schemes of both Gams and Warming have much to recommend them, but they have not been greatly used by other investigators perhaps because they are too ambitious and so not without some degree of artificiality.

Naumann (1917a) emphasizes what seems to me to be a sounder approach, namely to divide the lake biocoenosis into those forms associated solely with the liquid medium, those forms associated with the lower, solid-liquid interface (which really penetrates the mud to the bottom of the biotope), and those associated with the liquid-gas interface. In the liquid phase we have a complete set of transitions between organisms such as nonflagellate phytoplankton, which are completely at the mercy of the movements of the water and have become adapted

---

[6] To a zoologist a most unhappy term, fortunately later (Warming 1923) abandoned by the author in favor of *haptobenthos.*

mainly as passively suspended organisms, through flagellated or ciliated motile but small and weak unicellular forms, ciliated metazoa such as the rotifers, muscular forms such as the Crustacea which move great distances vertically but still must be influenced by turbulence, up to the fish for which turbulent movement, at least in a lake, can seldom be an impediment to directed locomotion. The nonmotile forms are *plankton* in the strictest sense of the word; the forms completely independent of turbulent movements are *nekton*. Most of the assemblage is to a greater or lesser degree really nektoplanktonic, but custom, technique of study, and intellectual convenience ordinarily lead to an extended use of the word plankton to include everything living in the free water except vertebrates and the larger insects and Crustacea. Any distinction is arbitrary. When in dealing with motile but still small forms partly controlled in their behavior by gentle water movements it is desired to emphasize this aspect of such organisms, the term *nektoplankton* will be used.

Plankton, seston, and tripton. Hensen's (1887) original definition of plankton ($\pi\lambda\alpha\gamma\kappa\tau\acute{o}s$, wandering) included all particulate organogenic material, living or dead, passively drifting in the water. Later workers have refined this concept and have inevitably added to the terms used in describing such as assemblage of particular matter. For both technical and theoretical reasons it is often desirable to consider together all of the particulate material present in the free water. This collectively is termed (Kolkwitz 1912) *seston.* The seston consists of *bioseston,* or plankton and nekton, which latter is ordinarily quantitatively negligible, and of *abioseston* or *tripton* (Wilhelmi 1917). The tripton may be of autochthonous or allochthonous origins, termed *eutripton* and *pseudotripton,* respectively.

A bewildering number of names for various kinds of seston, plankton, and tripton have been introduced. Prefixes denoting size (Schütt 1892; Lohmann 1911) are frequently employed and are defined, as by Naumann (1931), in Table 8.

This classification is arbitrary and not often completely followed. The term nannoplankton was originally introduced to include everything not retained by a townet and is usually so used. Most authors regard the total seston or plankton as consisting roughly of net plankton (macro-, meso-, and some microplankton) and nannoplankton (some micro-, nanno-, and ultraplankton of the scheme in Table 8). Centrifuge plankton is often considered as synonymous with nannoplankton, though with proper sampling it is essentially the total plankton. The tendency recently, in line with the current craze for alphabetical abbreviations, has been to refer to the algae of the ultraplankton, of

TABLE 8. *Size classes of seston*

Name	Linear Dimensions
Megaloseston (plankton or tripton)	n cm.
Macroseston (plankton or tripton)	1 mm.–1 cm.[a]
Mesoseston (plankton or tripton)	0.5–1.0 mm.
Microseston (plankton or tripton)	0.06–0.5 mm.
Nannoseston (plankton or tripton)	0.005–0.06 mm.
Ultraseston (plankton or tripton)	0.0005–0.005 mm.

[a] Naumann says "minimalgrösse mehrere Zentimeter" for megaloseston, "einign Zentimetern" as the upper limit for macroseston. For formal tabular purposes 1 cm. may be taken as a convenient rough dividing line. Lohmann originally defined megaloplankton as the assemblage of plankton visible on board the investigating vessel. The fresh-water medusae are probably the only limnetic animal megaloplankton, and they are small by marine standards; various plants in ponds such as *Utricularia* are truly megaloplanktonic.

the order of magnitude 0.001 mm., as μ-algae or μ-flagellates. In all cases the prefix *phyto-* signifying plant and *zoo-* signifying animal may be employed. Individual members of the plankton are correctly termed *plankters* (Burckhardt 1920).[7]

Among the other terms that have been employed *euplankton,* implying a permanent planktonic community, *meroplankton,* implying organisms with temporary planktonic phases or stages, and *pseudoplankton,* for accidental plankters, are often useful. The distinction between the *limnoplankton* of large lakes and the *heleoplankton*[8] of ponds is useful, and the term *potamoplankton* for the plankton of rivers is not infrequently employed; for other terms the philological reader may consult Wilhelmi (1917) and Naumann (1931). Multiple prefixes, as in *eulimnoplankton,* are occasionally useful but should be employed very discreetly.

Benthos. The organisms associated with the solid-liquid interface are ordinarily termed *benthos* (Haeckel 1891). There seems to be no reason to use the term *pedon* for limnobenthos, restricting *benthos* to the marine biocoenoses. Benthos obviously may be divided into *phytobenthos* and *zoobenthos.* Size categories have been little used; Moore (1939) and Welch (1952) distinguish only between macroscopic and microscopic benthos.

[7] Similarly, the adjective planktic would be etymologically sounder than planktonic, but usage unequivocally favors the latter, which is the only form accepted by OED.

[8] Tychoplankton is often incorrectly used; it is an absolute synonym of pseudoplankton.

Warming's (1923) terms, extended to cover animal forms when necessary, are useful, namely, *rhizobenthos* for the rooted vegetation, *herpobenthos* for the inhabitants of sediments, *haptobenthos* for forms adnate on solid surfaces, and *endobenthos* for the few forms boring in solid material—in fresh water mainly algae in calcareous rock. These terms, however, overlap other useful terminology. Hapto-benthos is synonymous with *periphyton* in its wide use. The term peripython was introduced by Behning in 1924 (see Behning 1928a) for the plant growth on buoys, ships, and moorings in the Volga; in the ocean this would be referred to as *fouling,* a term that in translation is likely to lead to confusion. Roll (1939) uses periphyton essentially in the sense of Gams' "Nereiden" and Warming's haptobenthos and divides it into *epiphyton* which forms scattered com-munities and *lasion* (Meuche 1939) which forms a thick matted com-munity. Periphyton and epiphyton presumably refer to assemblages of plants, though it has probably been uncertain whether an epiphyte is a plant growing on something or something growing on a plant; the term, which refers to a plant growing on a plant, was originally purely botanical.

Herpobenthos, which refers in the widest conceivable sense to the creeping of organisms through the substrate, overlaps the *psammon* (Sassuchin, Kabanov, and Neiswestnova 1927), or the community liv-ing in sand. As far as the relation to the substrate is concerned, the following usages seem reasonable and are employed.

Rhizobenthos: rooted in the substratum but well extended into the aqueous phase.

Haptobenthos: adnate to solid surface; the terms *epiphyton* ("Aufwuchs" of German authors), *epilithon,* and *lasion* ("Bewuchs") are employed as informal subdivisions when useful.

Herpobenthos: growing or moving through mud.

Psammon: growing or moving through sand; the category can be subdivided according to the relation of the sand to the lake margin (Vol. III).

Endobenthos: boring in a solid substrate.

When the term benthos is used in an unqualified sense, in relation to deep water, it ordinarily implies herpobenthos.

Zoobenthos is often classified as sessile or fixed, which category roughly corresponds to haptobenthos, vagile or wandering, which corre-sponds to the herpobenthos (s. str.) and nektic, or swimming from place to place. The nektic benthos is really an intermediate category between true benthos and nekton. It is quite likely that a considerable number of supposed benthic species of highly motile groups are benthic by day and nektoplanktonic by night. The term *merobenthos* may be used to designate such organisms. As will appear later, *Mysis relicta* and the larvae of *Chaoborus* belong to the meroplankton by night and the merobenthos by day. In dealing with the microscopic benthos, Bigelow (1928) distinguishes between an ooze film assemblage which is truly herpobenthos and an associated ooze film assemblage which swims immediately above the bottom and is really a localized part of the nektoplankton. The term *planktobenthos* is applied to such small forms in that they really belong to the part of the biocoenosis in the free water but are limited to the free water in the vicinity of the bottom. It emphasizes their nature intermediate between true herpobenthos and what would ordinarily be regarded as plankton.

Pleuston. For that part of the biological community that is associated with the air-water interface the term *pleuston* is used. The word was introduced by Schröter and Kirchner (1896) to designate the whole assemblage of floating plants, both submerged (e.g., *Utricularia*) and interfacial (e.g., *Lemna*). The submerged forms, however, are, really megaloplanktonic. Naumann (1917a) introduced the term *neuston* to designate the assemblage of microorganisms associated with the surface film. Gams (1918) used *pleuston* to cover the whole assemblage at the surface, excluding the megaloplanktonic submerged forms. He distinguished between the macropleuston and micropleuston, the macropleuston being essentially the pleuston of Schröter and Kirchner, the micropleuston, the neuston of Neumann. If one term is to be used, pleuston (πλεω, sail, go by sea, swim, float) would seem to be etymologically slightly preferable to neuston (νεω, swim, cognate of the Latin *nare*) as well as having priority. In general, the large organisms of the macropleuston appear to have an upper dry surface and a lower wet surface, whereas the micropleuston is associated with one side or the other of the surface film. It is therefore convenient to use neuston as a short equivalent of micropleuston to permit the use of prefixes, as suggested by Geitler (1942), the *hyponeuston* living on the undersurface, the much more important *epineuston* on the upper surface of the film.

Carpenter (1928; Welch 1952) uses superneuston to designate the assemblage of macroscopic animals walking on, but with most of the body raised above, the surface film. In the present system of terminology these animals, such as the Gerridae and Hydrometridae, should be regarded as part of the pleuston, the most correct designation for them being *epipleuston* (Rapoport and Sánchez 1963).

The various prefixes applied to plankton or seston can be used with pleuston or neuston. In particular, the term *meropleuston* is convenient, notably in its adjectival form *meropleustonic* for organisms such as *Scapholeberis* or *Gyrinus,* which are morphologically adapted to special kinds of interfacial life but do not occupy the interface continuously.

ZONATION

In delimiting the parts of the lacustrine biotope, the term *pelagial,* derived from oceanography, is usually employed adjectivally with reference to the free open water.

The areas of the bottom from the shore down to the deepest point exhibit a series of zones. Although there is fair agreement regarding the nature of these zones, there is almost no agreement regarding what they should be called. The first two zones, the *epilittoral* and the *supralittoral* (Sernander 1917; Neumann 1928) present no problems; they are really outside the lake. These zones in fact are paralimnetic in the sense of Hutchinson (1937b), to whom the *paralimnion* is the entire spaciotemporal collection of objects and events which constitute the environment of the lake and act on it. Unfortunately, this term is used by Allee and Schmidt (1951) as a synonym of the littoral zone as employed in this book. In the epilittoral the existence of the zone as a continuous part of the basin profile not occupied by water is no doubt often dependent on cutting at the effluent. The term *eulittoral,* which refers to a zone that is certainly part of the lake, is accepted by Lenz (1928) and Rüttner (1940, 1952b) and may be adopted. The main difficulty is based on the entirely different application of *sublittoral* by botanists and zoologists; possibly Rüttner's wide use of the term for everything between the eulittoral and the profundal symbolizes his great contributions to both biological aspects of limnology.

The present writer feels that Lenz's (1928) compromise scheme is the only possible basis for a solution of the nomenclatorial problems involved but I would add the term *infralittoral* used recently in marine studies (Pérès 1957; Hedgpeth 1957) for all that part of the littoral, defined widely as the zone of rooted or adnate (i.e., *Fontinalis*) higher vegetation, which lies below the eulittoral. In many cases the infralittoral will be divisible into an upper zone of emergent vegetation, a middle zone of floating-leaved vegetation, and a lower zone of submerged vegetation. These zones, however, are often not fully or even partly developed on exposed or steep coasts, and sometimes, as in the *Chara* zone in the Lünzer Untersee figured by Rüttner, a submerged zone is intercalated between the emergent and floating-leaved zones. In some deep, clear lakes in which *Fontinalis* occurs in many meters of water, to at least 100 m. in Crater Lake, Oregon, it may be convenient informally to refer to the deep infralittoral for that part of the lower infralittoral with a well-developed adnate vegetation.

The profundal is usually defined as the whole of the bottom below the limit of well-developed zones of vegetation, that is, of Charales, bryophytes, and vascular plants. Some photosynthetic Monera and unicellular algae may occur in its upper region, but in general they are herpobenthos and the bottom is recognizable as having an exposed layer of fine sediment. This transitional region with scattered algae, often adjacent to the metalimnion of a stratified lake and covered in many cases with littoral material is the microelittoral of Thomasson. It is probable, however, that the photosynthetic bacteria penetrate to

TABLE 9. *Zonation of the lake bottom*

Definition of Zone	Preferred Name	Synonyms
1. Entirely above water level and uninfluenced by spray	Epilittoral	None
2. Entirely above water level but subject to spraying by waves	Supralittoral	None
3. Between highest and lowest seasonal levels; often a zone of disturbance by breaking waves	Eulittoral	None
4. Emergent rooted vegetation present	Upper infralittoral	Littoral (Ekman 1915; Thienemann 1925; Lundbeck 1926; Lenz 1928; Eggleton 1931) Sublittoral (Sernander 1917; Naumann 1928) Sublittoral (Rüttner 1940, 1952b)
5. Floating-leaved rooted vegetation present	Middle infralittoral	"
6. Submerged rooted or adnate macroscopic vegetation present	Lower (and deep) infralittoral	Littoral (Lundbeck, Thienemann, Lenz, Eggleton) Macroelittoral (Thomasson 1925; Naumann etc.) Sublittoral (Ekman, Rüttner)
7. Transitional zone; photosynthetic forms if present usually scattered; ordinarily moneran and algal herpobenthos, occasional massive development of blue green algae	Littoriprofundal	Sublittoral (Thienemann, Lundbeck, Eggleton) Microelittoral (Thomasson, Naumann) Eprofundal, Lenz
8. Unvegetated fine bare mud	Profundal	Profundal of most authors

(Zones 4–6 are bracketed together as **Littoral**.)

considerably greater depths than has often been supposed and the final extinction of such organisms downward is doubtless very gradual. The sharpest boundary is thus probably at the bottom of the zone of macroscopic vegetation. The intermediate zone is here called the *littoriprofundal* because there seems to be no agreement whether it is best associated with the zones grouped as littoral or with the profundal (Table 9).

In the actual investigation of any lake some empirical modification of the definitions will normally be necessary. The term *abyssal* has been used by Ekman (1917) for depths below 600 m. The utility of the term may be doubted. There is little in common to the great depths of Lake Baikal, the Caspian, and the meromictic central African Tanganyika and Nyasa. As a special application, in discussing Lake Baikal, the term may have some utility, but to set the limits of the abyssal zone at 600 m. is arbitrary.

TERMINOLOGY OF SEASONAL CHANGE

A great deal of confusion has spread through the terminology of seasonal change. In general, three different, if often related, phenomena are involved. In the first place a population reproducing by a uniform method may arise to a maximum and then, as mortality exceeds natality, fall to a minimum. This process may be repeated several times in the course of an annual cycle of the seasons. For a species that exhibits one maximum per year, Ostenfeld's (1913) term *monacmic* is used; when there are two, three, or many maxima, the equivalent terms *diacmic, triacmic,* or *polyacmic* are appropriate.

When an organism exhibits a single generation per year, it is termed *univoltine;* when it has two or three generations, it is *bivoltine* or *trivoltine;* when there are many, it is *multivoltine.* Terms having these meanings are essential, particularly in the discussion of the zooplankton, but no unambiguous words appear to be available in the planktological literature. The recommended terms, originally applied to silkworms, are standard expressions for the annual number of generations in entomology and there appears to be no reason not to extend them to other organisms. In the rare cases in which an organism requires two or three years for its development it is spoken of as a *biennial* or *triennial;* when the life span lasts a number of years, it is a *perennial.*

When a population of any organism alters its mode of reproduction, ordinarily to produce resting stages, often by a sexual process after asexual reproduction, such as binary fission, parthenogenetic egg production, or budding, it is described as *monocyclic* if the process occurs once during the annual cycle, *dicyclic,* if twice, *tricyclic,* if three times, and *polycyclic,* if many times. This has been the accepted usage for the multivoltine rotifers and cladocera. Bivoltine copepods, in which both generations are sexual but in which one generation produces substaneous or rapidly hatching eggs, the other resting eggs, are clearly monocyclic, and not dicyclic, by the usage established for the groups in which two types of reproductive activity are more usual. Most dicyclic species are diacmic, but the production of a maximum, though often a necessary condition for a change in reproductive activity, is clearly not a sufficient condition; the species in question may in fact have only one sort of reproduction. The phenomenon of monacmy, diacmy, or polyacmy is clearly distinct from monocycly, dicycly, or polycycly and should not be confused nomenclatorially. Moreover, univoltine species, ordinarily monacmic, cannot be monocyclic in a strict sense, though it is quite likely that occasionally part of a population is univoltine and acyclic, another part bivoltine and monocyclic. Since the terms monocyclic and dicyclic have been widely used to mean monacmic and diacmic and sometimes univoltine and bivoltine and then by implication to mean acyclic and monocyclic in the sense of the foregoing definitions, the terminology here adopted may cause confusion to some, but it seems to be the only one that can be uniformly applied to all the groups of organisms with which the limnologist must deal.

SUMMARY

In order to provide a terminological framework for the ecological parts of this book some redefinition of terms commonly used has proved necessary. In the course of this redefinition it is important to realize that much terminology has been developed in relation to the size of particular organisms, often the investigator himself. To obviate some of this subjectivity, the term *homogeneously diverse* is used to express any area whose mosaic elements are small relative to the range of any particular organism under investigation.

A *population* is any unispecific collection of organisms in a discrete space or area.

An *assemblage* is a collection of co-occurring populations.

A *biotope* is defined as any segment of the biosphere with convenient arbitrary upper and lower boundaries, which is horizontally homogeneously diverse in relation to the larger motile organisms present within it.

An *ecosystem* is the entire contents of a biotope.

A *biocoenosis* is the totality of organisms in a biotope.

The *biocoenosis* on land is equivalent to the *formation* of botanical ecologists, save that it is empirically restricted to a particular area and not the class of all similar areas. Within a biocoenosis we may recognize *associations* of species. Within a biotope we may recognize *habitats* of species. The term *niche* is used in an abstract and purely intensive sense to designate the requirements of an organism abstracted from the spacially extended habitat. The habitats of two species may overlap completely; it is empirically probable that at equilibrium their niches never do.

Each lake is regarded as an ecosystem which can be divided horizontally into *strata* and along the bottom into *zones*.

The organisms of a lake are conveniently grouped ecologically into those associated with the free water, those associated with the solid-water interface, and those at the surface film.

The free water contains *plankton,* partly if not wholly controlled in its movements by turbulence, and *nekton,* which in a lake can undertake swimming movements in any direction in spite of turbulence. Most animal plankters are actually *nektoplanktonic.* The entire mass of suspended matter in a volume of free water is called *seston,* the nonliving part, *tripton.* All of these terms may be qualified by prefixes denoting size classes (Table 8). The *euplankton* constitutes the permanently planktonic species, the *meroplankton* those planktonic only at certain times in their life histories.

The organisms of the solid-water interface are *benthos.* The following terms appear to be useful in characterizing this assemblage.

Rhizobenthos, rooted in substratum
Haptobenthos, adnate to solid surfaces
Herpobenthos, growing or moving through mud
Psammon, growing or moving through sand
Endobenthos, penetrating a solid substratum

When the haptobenthos is well developed, particularly on objects projecting into the free water, it is called *lasion,* fouling, or *Bewuchs.* Many *merobenthic* animals may also be meroplanktonic.

The assemblage of organisms at the surface film is called *pleuston;* the *micropleuston* is often designated *neuston* and may be *hyponeuston* below the film or *epineuston* above it. Some large organisms living in air above the film on which they walk or run are best termed *epipleuston.* Many *meropleustonic* animals which reach the surface film from below are adapted to live at the surface for part of their lives.

The bottom area in contact with the water exhibits striking zonation. The following terms are recommended to describe the zones ordinarily present.

Epilittoral, completely above the influence of the water
Supralittoral, above water but receiving spray
Eulittoral, between high and low seasonal levels
Infralittoral, permanently covered but with rooted or adnate macroscopic vegetation, often divisible into upper (emergent vegetation), middle (floating vegetation), and lower (submerged vegetation)
Littoriprofundal, a transition zone with scattered adnate algae
Profundal, bare sediment below biogeochemical compensation point

In the terminology of seasonal change, if a population rises to a maximum and declines once in an annual cycle, it is *monacmic;* if twice, *diacmic,* etc.; if many times, *polyacmic.*

If a species has a single generation per year, it is *univoltine;* if two, *bivoltine,* etc.; if many, *multivoltine.*

If a generation takes two or more years, the ordinary terms *biennial* or *perennial* will be used.

If a *bivoltine* or *multivoltine* species changes its mode of reproduction, ordinarily to produce resting stages, once a year it is *monocyclic;* if twice, *dicyclic,* etc.   If no change occurs, it is *acyclic.*

CHAPTER 20

The Hydromechanics

of the Plankton

According to the principle of Archimedes, any body immersed in a liquid will be buoyed up by a force equal to the weight of the displaced liquid. The resultant force of gravity acting on the body will therefore be

(1) $$F = gkd^3(\rho' - \rho)$$

where kd^3 is the volume of the body, of which d is some appropriate linear measure, g the acceleration due to gravity, and ρ' and ρ, the densities of the body and of the liquid, respectively. The expression $(\rho' - \rho)$ is conveniently termed the *excess density*.

In general, a body can remain suspended in the liquid only if one of three conditions is met. In the first place, if the excess density is zero, there is no force acting on the body and therefore no tendency for it to depart from a position of rest in the liquid. In the second place, if an external force is applied to the body it may, even if the excess density is positive, be moved upward as rapidly as it tends to sink passively. This in effect is what happens to some parts of a population of suspended organisms which are distributed by the turbulent movement of the water. In the third place, by actively swimming the body may exert a force with a vertical component sufficient to balance the resultant force of gravity **F**.

Actually, any body suspended in water is acted on not only by gravity and by the external or physiological forces just mentioned but also by the pressure of the water, which, being uniform in all directions, is irrelevant to the present discussion, and by the resistance of the water which depends on the relative motion of the body through the water. At rest, in relation to the medium, the resistance is zero, but as the body begins to fall or rise through the water resistance increases with added velocity until it balances the forces tending to accelerate the body. In general, therefore, a falling body in a resistant medium reaches a terminal velocity at which the resistance of the medium exactly balances gravity and other forces acting on the body. It will therefore easily be seen that the greater the resistance developed as the body begins to sink the smaller the other forces required to maintain in position any body for which the excess density departs from zero. For a body of given volume the resistance will largely depend on its shape. Any change that tends to increase resistance to vertical falling will also tend to increase resistance to swimming upward. The problem of increased resistance as a method of remaining suspended in the water is therefore primarily of importance to those small and truly planktonic organisms for which the excess density is small and that, sinking slowly in still water, are now carried up and now down in small moving elements of water and are thus largely maintained afloat in nature by turbulent movements, ultimately caused by wind. It will therefore be convenient in the present chapter to consider mainly the problems of density and density difference, of resistance and sinking speed, and of turbulent diffusion.

THE DENSITY OF FRESH-WATER ORGANISMS

Experimental determinations. Because of the considerable technical difficulty of obtaining reliable data, the available information on the density of fresh-water organisms is extremely meager. It has long been known that certain species are provided with oil droplets or gas vacuoles which reduce their density below that of the medium so that they tend to float, whereas other organisms, indeed the great majority of the planktonic species in fresh waters, sink in undisturbed water. The organisms with gas vacuoles and gas-filled hydrostatic organs present certain special problems and are best considered separately. In a few cases it would seem that the cells of planktonic algae, even in the absence of visible gas vacuoles or oil drops, have almost the same density as the medium. Gross and Zeuthen (1948) believe indeed that this is the normal condition in marine diatoms living under optimal conditions. They conclude that because, in the formation of resting spores, the density of such diatoms increases above that of sea water, the cell sap that is expelled during spore formation must have a density below that of sea water. They consider that this low density is achieved in the same way in diatoms as in the cells of *Valonia,* in which the specific gravity is adjusted by the exclusion of divalent ions from the cell sap. The planktonic marine dinoflagellate *Noctiluca miliaris* is believed by Krogh (1939) to float because part of its sodium is replaced by NH_4^+, which is possible at the low *pH* of its cell sap, though Gross and Zeuthen think rather that the absence of sulfate is involved. Denton (1960) has summarized a number of cases of lowering of density by substances other than gases in marine organisms. It is difficult to see how any of these mechanisms, except occasionally oil inclusions, could operate in a fresh-water organism. There is, moreover, a considerable body of evidence that marine diatoms in nature sink at about the same rate as fresh-water species. Nevertheless, Ruttner (1930) found that the minute *Mallomonas akrokomos,* unlike the other green and yellow algae, including diatoms, that he observed, probably did not sink in fresh water. Similarly, Grim (1939) implies that the cells of *Cyclotella melosiroides* and other small species of the genus do not sink significantly when quite healthy, though the mean rate of sinking of living and dead specimens of small species of *Cyclotella* is apparently within the range of that of other diatoms, all of which he found to sink. Utermohl (1925) also found that all the diatoms he investigated sank, but the minute blue-green alga *Dactylococcopsis* passively descended at a rate of only 1 mm. hr.$^{-1}$, which implies almost the same density as the medium. It is possible that this form has gas vacuoles.

The results just presented suggest that nearly all fresh-water diatoms and probably most other phytoplanktonic organisms, except the blue-green algae, are denser than the medium and sink when undisturbed. The only important alga, other than the planktonic Myxophyceae that form water blooms, in which the density is certainly less than that of water is, as Klebahn (1895) pointed out long ago, the anomalous green alga *Botryococcus brauni.* In this organism the density is reduced by the presence of the inclusion of oil-droplets. It has frequently been supposed that oil droplets play a similar part in reducing the density of diatoms. This is likely in the few cases already noted in which fresh-water diatoms appear to remain suspended indefinitely in the water. In any consideration of the diatoms it must be remembered that variations in the uptake of SiO_2, dependent on the supply of soluble silicate, must influence the density markedly. Einsele and Grim (1938), from measurements of volume and SiO_2 content have computed that the density of *Fragilaria crotonensis* may vary between 1.1 and 1.45. The quantity of oil needed to compensate for the presence of a heavy siliceous shell is probably much greater than any healthy diatom could contain, but as indicated later the secretion of a sheath may greatly lower the over-all densities of these organisms.

It is evident that exact knowledge of the density of the fresh-water phytoplankton is greatly needed; the gaps in our knowledge of the zooplankton are also great but not so absolute as in that of the plant forms. In the study of the densities of aquatic animals two methods have proved to be fairly successful. Luntz (1928) used a procedure comparable to that often employed in mineralogy. A number of solutions of sodium chloride or of sugar of known density were prepared and observations were made whether the narcotized rotifers under examination floated or sank in them. Because of the obvious likelihood of osmotic disturbances, the consistency of the data obtained with sugar and salt solutions is rather surprising but suggests that considerable confidence can be placed in the results. Hamilton (1958) used a more modern method, involving a graded mixture of bromobenzene and liquid paraffin, in his work on *Holopedium*. Lowndes (1938, 1942) has introduced an elegant if somewhat exacting method which is theoretically quite free of objection for marine organisms, though in fresh-water organisms it does, like Luntz's method, involve immersion in a salt solution. The essential procedure is as follows. A specific gravity bottle of known volume is weighed empty, filled with salt solution, and weighed again. The contents are poured into silver nitrate solution and the AgCl precipitated is collected and weighed. The organism under investigation is now placed in the bottle, which is again filled with the salt solution and weighed. The solution is poured off through a filter into silver nitrate and the organism is washed on the filter with a mixed sodium and calcium nitrate solution, isotonic with the salt solution employed. The second AgCl precipitate is now weighed. By knowing the concentration of the chloride from the first two weighings of the bottle and the first weighing of AgCl it is possible to compute from the difference in the two AgCl precipitates the volume kd^3 of the organism. The difference in the weight of the bottle with solution, with and without the organism, gives $kd^3 (\rho' - \rho)$, and since ρ can be determined if the volume of the bottle is known ρ' can now also be determined. The method is obviously most suitable for marine organisms which can be kept in sea water throughout the experiment except during the brief washing with isotonic nitrate, which in practice proves quite innocuous.

Luntz and Hamilton express their data as density at 18° C. Lowndes expresses his as 1000 ρ'/ρ, which he calls the sinking factor. In Table

TABLE 10.   *Densities of fresh-water animals*

Rotatoria (Luntz 1928)	
Euchlanis triquetra, reared at 8° C.	1.025 at 18° C.
Euchlanis triquetra, reared at 25° C.	1.020 at 18° C.
Euchlanis triquetra, wild, taken in summer	1.025 at 18° C.
Brachionus quadridentatus, wild, taken in summer	1.025 at 18° C.
B. quadridentatus f. *rhenanus*, wild, taken in spring	1.025 at 18° C.
B. quadridentatus f. *rhenanus*, reared at 8° C.	1.025 at 18° C.
B. quadridentatus f. *rhenanus*, reared at 25° C.	1.020 at 18° C.
Crustacea (Lowndes 1938, 1942; Hamilton 1958)	
Chirocephalus diaphanus	1.011 at 11.3° C.
Daphnia pulex	1.017 at 7.0° C.
Holopedium gibberum, with jelly	1.0015 at 18° C.
Holopedium gibberum, without jelly	1.014 at 18° C.
Eudiaptomus gracilis	1.023 at 8.0° C.
Gammarus pulex	1.066–1.088 at 20° C.
Candona candida	1.025 at 7.4° C.
Chordata	
Gasterosteus aculeatus	1.003 at 10° C.
Gasterosteus aculeatus	1.002 at 20° C.
Rana temporaria tadpole	1.014 at 9.6° C.

10, Luntz's values are given exactly as he states them, whereas Lowndes's sinking factor has been divided by 1000 to give for a fresh-water organism almost the same value as ρ'.

The values of the excess density for fresh-water planktonic animals are evidently generally likely to lie between 0.015 and 0.025. The high values given by Jacobs (1943) for the density of protoplasm, which would imply much higher values of the excess density in fresh water, are apparently derived exclusively from marine and mainly from benthic organisms. They are presumably the source of the excessive estimates for the density of fresh-water plankton given in some recent European works. The high values given for *Gammarus pulex* suggest that the fresh-water benthos tends to be denser than the plankton, which is not the least surprising. Similar high values (1.031 to 1.095) can be computed from the data of Levanidov (1945), whose determinations of volumes and weights of various fresh-water insects and of *Asellus* were obtained by a method due to Gaievskaia, comparable to that employed by Lowndes. The low values for the single teleost fish studied are to be expected, for the presence of a swim bladder permits almost exact equilibration.

Considerably more information would be useful in the study of the fresh-water zooplankton, for Gross and Raymont (1942) have found interesting changes during the life history of *Calanus finmarchicus,* in which marine copepod there is a marked density minimum in the fifth preadult instar.

Eyden (1923) found that the sinking speed of *D. pulex* showed a diurnal variation, being greatest shortly after sunrise. In experiments done in August the minimum sinking speed was at midday and was followed by small irregular changes during the afternoon and night. In April the sinking speed declined slowly and regularly from just after sunrise to about 4 A.M. on the following day. In small specimens (1.5 to 2.0 mm.) in April the sinking speed of narcotized animals with the antennae closed at the time of maximum density was 0.61 cm./sec.$^{-1}$, at the time of minimum density 0.43 cm. sec.$^{-1}$ The corresponding figures for large specimens (2.3 to 3.0 mm.) at the same season were 0.98 cm. sec.$^{-1}$ and 0.80 cm. sec.$^{-1}$ Eyden believes that the high density in the early morning is due to the animals feeding in the surface layers of the pond in which she collected them at this time of day, the *Daphnia* supposedly coming up from the bottom of the pond at about sunrise. The full-fed specimens are supposed to be the densest; as digestion and egestion occur, the density is supposed to fall. If Allen's relationship (see pages 271–272) applies in this case, as Brooks and Hutchinson (1950) believe it does, the ratios of the sinking speed and of the excess densities will be related as

$$\frac{\vartheta_{\max}}{\vartheta_{\min}} = \left(\frac{\rho'_{\max} - \rho}{\rho'_{\min} - \rho}\right)^{\frac{2}{3}}$$

For the small animals the ratio of maximum to minimum excess density is therefore 1.69; for the large it is 1.35. Since a change in density from 1.015 to 1.025 will cause the excess density to increase by a factor of 1.67, the observed changes in sinking speed in the course of a day imply changes in the density of the *Daphnia* within the range of Table 10. Eyden also cites an example of the decrease in sinking speed from 0.74 cm. sec.$^{-1}$ to 0.50 cm. sec.$^{-1}$ when the animal gave birth to a brood of young. This corresponds to a decrease in excess density by a factor of 0.56. Fox and Mitchell (1953), however, were entirely unable to confirm these results, either on the effect of feeding or on that of eggs in the brood pouch. Hantschmann (1961) also found that the eggs did not affect sinking speed. Eyden's observations, on very few animals, may be incorrect, but it is possible that packing of the gut with partly digested diatoms might raise the density of a *Daphnia* in nature.

Coefficient of thermal expansion of aquatic organisms. It is most unfortunate that there is so little information on the effect of a change of temperature on the density of organisms. The pair of observations on the stickle-back *Gasterosteus aculeatus* made by Lowndes appear to constitute the entire body of data on intact animals or plants, though it is by no means self-evident that the coefficient of expansion of an organism without an equilibrating mechanism would be identical with that of the medium in which it is living. Du Bois-Reymond (1914), studying the legs of frogs, concluded that these organs have a cubical coefficient of thermal expansion between 20 and 30° C. of 0.00075 and that mammalian muscle must have about the same coefficient. If we accept the value of 0.00075 for the cubical coefficient of thermal expansion of plankton as well as of muscle, it will be evident that on passing from a lower temperature to one 10° C. higher the density of the organism will fall about 0.0075, whereas that of the medium will fall only about 0.0025. With an initial excess density at the lower temperature of 0.02, the excess density at the new higher temperature will be 0.015. Such a decrease in excess density is proportionately greater than the decrease in viscosity of 0.010 to 0.008 poise in going from water at 20° C. to water at 30° C. Such reasoning led du Bois-Reymond to consider that in Ostwald's (1902) discussion of the floating of plankton too much emphasis had been placed on the viscosity changes and not enough on the concomitant changes in excess density. Later work (Wilkie 1953), indicates, as might be expected, a smaller value for the cubical coefficient for the gastrocnemius of *Rana temporaria* between 0 and 20°, namely 0.00016; over this range the value for water would be comparable. In view of the excellent agreement with theory assuming that the temperature effect is due to viscosity alone, obtained by Brooks and Hutchinson (1950) in a consideration of the available data for the rate of passive sinking of small *Daphnia,* it would appear that by obtaining his numerical value, which is doubtless somewhat too high, from frog muscle in which the proportion of bound water is quite likely to be much higher than in intact *Daphnia,* du Bois-Reymond exaggerated the significance of the coefficient of cubical expansion of planktonic organisms. The subject nevertheless deserves much more careful study.

Gas vacuoles and gas-filled hydrostatic organs. Planktonic species of a number of genera of the blue-green algae (*Gomphosphaeria, Anacystis, Oscillatoria, Lyngbya, Nostoc, Anabaena, Anabaenopsis, Aphanizomenon, Gloeotrichia* and *Calothrix*) frequently contain minute and often irregular bodies termed pseudovacuoles (Fig. 72). Good accounts of these structures have been given by Fogg (1941) and Fritsch (1945). The subject has been little advanced during the last

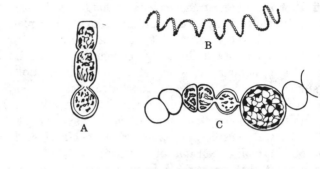

FIGURE 72. Gas vacuoles or pseudovacuoles in planktonic blue-green algae. *A.* Basal cells of a filament of *Gloeotrichia echinulata* (× 824). *B.* Filament of *Anabaena spiroides* (× 115), some but not all cells with gas vacuoles. *C.* Part of the same, highly magnified (× 824). *D. Aphanizomenon holsaticum* filament (× 824), with pseudovacuoles in several cells, including a heterocyst. (Kebahn.)

two decades. The nature of the pseudovacuoles has been a matter of controversy, but it now appears to be widely accepted that they are gas vacuoles, filled mainly with nitrogen. Destruction of the pseudovacuoles by pressure or other methods leads to an increase in the density of the alga which generally floats while the pseudovacuoles are present and sinks when they are absent. Klebahn (1922) found that *Gloeotrichia,* after destruction of the pseudovacuoles, had a density of about 1.007; if the vacuoles had occupied about 0.7 per cent of the volume of the alga, its density would therefore have been identical with that of the medium. Klebahn concluded that the actual volume occupied by pseudovacuoles was about 0.8 per cent of the total volume of the alga. It is quite certain that if the pseudovacuoles are responsible for the flotation of these organisms, as would seem to be the case, their density must be so low that only a filling of gas can account for the observations. Apart from occurring in those blue-green algae that form superficial water blooms, species of both the Myxophyceae and of the bacteria living on anaerobic mud or sapropel are known to produce pseudovacuoles (Lauterborn 1915); they may also be well developed in characteristically hypolimnetic plankters such as *Lyngbya compressa* (Utermohl 1925). It has been supposed by Canabaeus (1929) that when the environment becomes sufficiently unfavorable at the bottom of a lake or pond pseudovacuoles may appear in the blue-green algae inhabiting the deepest water; the appearance of the pseudovacuoles raises the algae out of the unfavorable region. This production of gas vacuoles to raise an organism from an anaerobic or toxic region is not without parallel, but the method of vacuole production suggested by Canabaeus, namely, the production of free molecular nitrogen by the fermentation of phycocyanin, though unique, is extremely improbable. Because of the possibility of the production of gas vacuoles in populations of blue-green algae in special circumstances, it is probable that some cases of the sudden appearance of such algae at the surface of lakes may be due to the induced vacuole formation in populations previously existing in the deeper water. By contributing to the flotation of the common blue-green algae, the pseudovacuoles certainly play a part in the development of water blooms, one of the most conspicuous phenomena of limnology.

A few of the rhizopod protozoa are also known to produce gas vacuoles that reduce their densities. Bles (1929), who summarized the older work of *Arcella* (Fig. 73), believed that in *A. discoides* the gas in the vacuole is oxygen, a conclusion recently confirmed critically by Cicak, McLaughlin, and Wittenberg (1963). It appears under any circumstances that are likely to reduce the supply of oxygen to the cytoplasm. In nature the effect of the production of the gas vacuole, as Bles suggests, would be to assist the animal to move upward

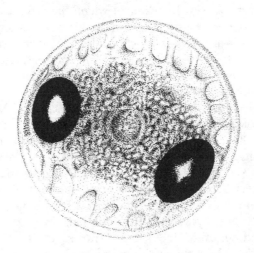

FIGURE 73. *Arcella* containing gas vacuoles,
(Cicak, McLaughlin and Wittenberg.)

when the water in contact with the bottom sediments became anaerobic. Large populations of various species of *Difflugia,* a genus also known to produce gas vacuoles, have been recorded in the plankton of lakes in summer. These occurrences are discussed later (see page 492). It is reasonable to suppose that the production of vacuoles in *Difflugia* by a mechanism comparable to that elucidated by Bles in *Arcella* is of some importance in concentrating these species in the epilimnetic plankton. It is to be noted that although reducing conditions may apparently produce gas vacuoles in both blue-green algae and in testaceous rhizopods and although the adaptive significance of rising in the water may be comparable in both groups the mechanism of the production of the vacuoles and the nature of the gaseous filling, if existing accounts are to be trusted, differs greatly in the two groups.

Among the Metazoa the most remarkable hydrostatic organs are the swim bladders of fishes and the tracheal organs of certain larval diptera. A consideration of the former structures is hardly in place in a discussion of plankton. The larvae of *Chaoborus,* which have the most highly developed tracheal hydrostatic organs, can, however, be regarded as meroplanktonic and are of sufficient limnological interest to justify a short discussion of the physiology of density change. It is also desirable, to stimulate further research in the subject, to mention one or two other arthropodan cases of some interest.

The hydrostatic organs (Fig. 74) of *Chaoborus* = (*Corethra* auctt.) consist of two pairs of kidney-shaped sacs, developed from the main tracheal trunks. They contain gas that is in equilibrium with the mixture of gases dissolved in the water in which the larva is living (Krogh 1911). The volume of the sacs can be altered considerably, and in this way the over-all density of the animal can be exactly adjusted to equal that of the medium and can be compensated after a meal of organisms denser than water. The entire organism constitutes a self-regulating Cartesian diver. Damant (1925) found that the gas could be reduced to 91 per cent and increased to 122 per cent of its original volume by altering the density of the medium. The mechanism of this volume change is the uptake of water by the walls of the hydrostatic organs (von Frankenberg 1915; see also later papers summarized in Wigglesworth 1939). Similar hydrostatic organs occur in the less specialized genus *Mochlonyx* and in the various genera closely related to *Chaoborus* but now regarded as separable from it.

A rather curious hydrostatic function has been described in the aquatic oribatid mites of the genus *Hydrozetes* (C of Fig. 29) by Newell (1945). These mites, when mechanically stimulated or when the illumination incident on them is suddenly reduced, produce a bubble

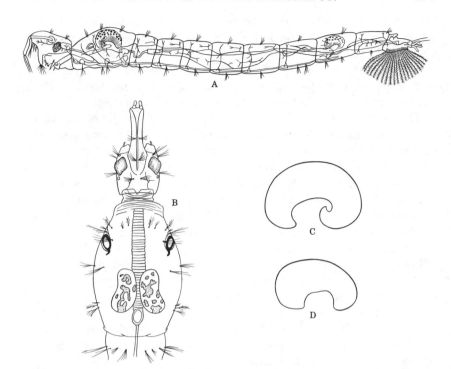

FIGURE 74.   *A. Chaoborus plumicornis,* full-grown larva ($\times$ 12).   *B.* Anterior
end of same from above ($\times$ c. 21) (von Frankenberg).   *C.* Anterior air-sac
of an individual equilibrated in a mixture of water and alcohol of specific
gravity 0.994 ($\times$ c. 29).   *D.* The same air sac after re-equilibration in a salt
solution of specific gravity 1.003, the change in volume amounting to about
25 per cent of the volume at the low density.   (Damant.)

of gas in the mid-gut which reduces their density sufficiently to permit
them to rise passively in the water.   It is therefore possible that any
mite detached from its normal substratum, which apparently is the
aquatic vegetation of ponds, would on falling to the bottom be
sufficiently stimulated by the falling light intensity to produce the bubble
and so be carried back to the level of its proper habitat.   Such an
example has certain formal analogies to the gas-vacuole formation by
blue-green algae and by *Arcella* under unfavorable conditions.

   Certain cases are also known in the arthropods in which the volume
of the external bubble of air carried for respiratory purposes can be
altered to adjust the density of the organism.   The aquatic beetles
of the family Elmidae are able to rise from the bottom or sink from
the surface by altering the pressure and so the volume of the air

that they carry. This is done partly by movements of the sixth, seventh, and eighth abdominal segments and partly by changes in the volume of the subelytral air spaces. Thorpe and Crisp (1949), who elucidated the physiology of this and of many other aspects of the respiratory physiology of these beetles, found that one species, *Riolus cupreus,* can rise and fall dozens of times in quick succession.

The small waterbugs *Anisops* and *Buenoa* of the family Notonectidae, which can remain poised at a constant depth in the water, certainly use their respiratory gas bubbles as hydrostatic organs. Careful observation suggests that after considerable activity on the part of the animal the system composed of animal and bubble is likely to become denser than water, doubtless because of the loss of oxygen from the bubble to the tissues and the rapid solution in the medium of the CO_2 produced. After a short period of slow passive sinking, the system regains its original low density, identical with that of the medium. It is reasonable to suppose that the haemoglobin organ or tracheal gland described in these insects by Hungerford (1922) and by Poisson (1926) is involved in storing a supply of oxygen that can be used to maintain the volume of the bubble and so the low density of the system between excursions by the insect to the surface. Experimental studies by Miller (1964), show that CO abolishes the buoyancy and so confirm the hydrostatic function of the haemoglobin organ.

Production of a bubble is recorded by the pedal region of *Hydra* (Kepner and Thomas 1928) and is under active investigation by Slobodkin (personal communication).

THE MECHANICS OF PASSIVE SINKING

Theory of a body falling in an undisturbed viscous liquid. The problem of the rate of sinking of a body in a viscous medium is one of unexpected complexity. The forces acting on such a body may be resolved into the force of gravity tending to accelerate the particle in a downward direction and the resistance (**R**) of the medium which experience shows increases with the velocity of the particle. When the resistance counterbalances the gravitational force, the body sinks with a constant velocity called its *terminal velocity* ($\hat{v}$). In theory, infinite time is needed to achieve the terminal velocity, but in practice the velocity is in nearly every case of interest approached so rapidly that the period during which the velocity is increasing need not be considered, and we may write,

$$R = kd^3(\rho' - \rho)g \tag{2}$$

where, as before, kd^3 is the volume of the body, ρ' its density, and ρ is the density of the liquid. The problem therefore, in the first instance, reduces to a determination of the relation between **R** and $\hat{v}$. The most convenient, if incomplete, elementary introduction to the subject is by way of dimensional analysis, as used, for example, by Allen[1] (1900).

Assume that **R** varies as unascertained powers of the density (ρ) and of the viscosity (μ) of the medium and the diameter (d) and the the velocity ($\hat{v}$) of the falling sphere

$$(3) \qquad \mathbf{R} = k'\, d^p \hat{v}^q \rho^r \mu^s$$

or, in dimensional form.

$$(4) \qquad [\mathrm{MLT^{-2}}] = [\mathrm{L^p}][\mathrm{L^q T^{-q}}][\mathrm{M^r L^{-3r}}][\mathrm{M^s L^{-s} T^{-s}}]$$

whence

$$r + s = 1$$

$$p + q - 3r - s = 1$$

$$-q - s = -2$$

and solving for p, r and s in terms of q

$$(5) \qquad \mathbf{R} = k'\, d^q \hat{v}^q \rho^{q-1} \mu^{2-q}$$

or, since we are considering terminal velocities, due to the action of the force of gravity on the particle

$$kd^3(\rho' - \rho)g = k'\, d^q \hat{v}^q \rho^{q-1} \mu^{2-q}$$

or

$$(6) \qquad \hat{v}^q = K d^{3-q}(\rho' - \rho)g\rho^{1-q}\mu^{q-2}$$

This equation may be expected to apply to bodies of any form, K varying with the shape of the body.

If $q = 2$, the term for viscosity disappears

$$(7) \qquad \mathbf{R} = k'\, d^2 \hat{v}^2 \rho$$

and

$$(8) \qquad \hat{v}^2 = K d(\rho' - \rho)\rho^{-1}$$

This is Newton's original formulation. He found it to be unsatisfactory empirically and realized that some unformulated term for what we now call viscosity would be necessary. Nevertheless, Newton's law

---

[1] In reading the older literature, it is important to bear in mind that what was formerly termed sinuous motion is now known as turbulent motion.

does apply quite well in certain cases in which a large body is falling rapidly in a fluid of low viscosity (Allen, 1900).

If $q = 1$, we have

(9) $$R = k' \, d\vartheta\mu$$

and

(10) $$\vartheta = Kd^2(\rho' - \rho)g\mu^{-1}$$

which is the generalized form of the expression obtained by Stokes for a very small sphere falling slowly through the viscous medium. Newton's law is probably of no interest to the planktologist whose objects of study either obey Stokes's law or the intermediate law corresponding to $q = \frac{3}{2}$ formulated on empirical grounds by Allen (1900):

The falling sphere. Stokes (1851), by classical hydrodynamic methods, obtained for a small sphere falling slowly through a viscous medium

(11) $$R = 6\pi\mu r\vartheta \, \frac{fr + 2\mu}{fr + 3\mu}$$

where f is the coefficient of sliding friction. When no slipping occurs, $f = \infty$.

(12) $$R = 6\pi\mu r\vartheta$$

(13) $$\vartheta = \frac{2}{9}gr^2(\rho' - \rho)\mu^{-1}$$

which are of the same form as (9) and (10) obtained by the looser method of dimensional analysis. If slipping of the liquid over the sphere were very great and f approached zero, the terminal velocity would approach one and a half times the value given by (13). The experiments of Arnold (1911), as well as those of many other investigators not specifically concerned with the problem of sliding friction, indicate that for solid bodies falling in ordinary liquids, no error is introduced by taking $f = \infty$.

The most important problem to be considered in relation to the sphere is that of the limits within which Stokes's law is valid. In general, the conditions for the validity of the law may be best expressed in terms of Reynolds' number R_e where

(14) $$R_e = \frac{dv\rho}{\mu}$$

d being, as before, an appropriate linear dimension which for the sphere may be taken as the diameter, or $2r$. The general conclusion of a large number of investigators (summaries in Falkenhagen 1931;

Wadell 1934; Dallavalle 1948; and numerous other older works) is that if

$$R_e < 0.5$$

deviations from Stokes's law may be neglected for all practical purposes when a particle is falling in a large, deep body of water. Correction formulas for bodies falling in vessels of restricted radius and depth have been developed by various investigators whose results are summarized by Schmiedel (1928) and by Falkenhagen (1931). More recently the matter has been reinvestigated by McNown and his associates (McNown and Malaika 1950). In a consideration of the sinking of planktonic organisms in the laboratory, the use of these correction formulas may be necessary, but they are not involved in the theory of plankton in nature.

It is also necessary to point out that more accurate treatment along the lines initiated by Stokes, but not neglecting terms of higher order, has been attempted by many authors. The older results are summarized by Falkenhagen and a more recent solution is given by Goldstein (1929). All the more accurate expressions for resistance involve powers of the velocity greater than the first and are therefore intractable in practice and of no particular value in this book.

The figures in Table 11 give the rate of fall of a sphere in millimeters per second, calculated from Stokes's law, for different values of the diameter and excess density, the viscosity being taken as 0.01 poise.

The data to be presented in a later paragraph show that in general biological objects less than 0.5 mm. in diameter will have passive sinking speeds in the nonturbulent waters of laboratory vessels that imply values of R_e within the range over which Stokes's law is valid. Bacteria, noncolonial planktonic algae, and many species of colonial algae, nearly all protozoa, and most rotifers sinking passively in quiet waters may therefore be expected to exhibit in their movement the propor-

TABLE 11. *Sinking velocities of spheres in water of viscosity 0.01 poise* (mm. sec.$^{-1}$)

$(\rho' - \rho) =$ 0.0001	0.001	0.002	0.005	0.01	0.02	0.05
d						
0.01 mm. 0.00000545	0.0000545	0.000109	0.000273	0.000545	0.00109	0.00273
0.10 mm. 0.000545	0.00545	0.0109	0.0273	0.0545	0.109	0.273
0.50 mm. 0.0136	0.136	0.273	0.681	1.36		
1.00 mm. 0.0545	0.545			$R_e > 0.6$		

tionalities implied by (10). For a diatom of diameter 100 μ, for example, sinking at a speed of 25×10^{-3} cm. sec.$^{-1}$ at 20° C.,

$$R_e \simeq \frac{0.01 \times 25 \times 10^{-3} \times 1}{0.01} = 0.025$$

A nonspherical form, however, may cause considerable deviation from the absolute values of the sinking velocities derived from (13) and presented in Table 11. In contrast to the phytoplankton, protozoa, and rotifers, the descent of the larger planktonic Crustacea is likely to take place at velocities implying a value of Reynolds' number outside the range within which Stokes's law is valid. The empirical evidence, however, suggests that the smaller Crustacea do sink according to the generalized form of Stokes's law implied by (10), as far as relation to viscosity is concerned.

The effect of departure from the spherical form: theoretical studies.
The simplest way (McNown and Malaika 1950) to consider the effects of changes in shape is to suppose that a sphere of radius r_s is deformed to give a new figure which encounters a resistance $\mathbf{R}_a$ given by

(15) $$\mathbf{R}_a = 6\pi\mu\phi_r r_s \hat{v}$$

where ϕ_r is termed, by a modification of Ostwald's usage, the *coefficient of form resistance.*[2] Since the volume and density of the body are unchanged by the deformation, at the terminal velocity

(16) $$\hat{v}_a = \tfrac{2}{9}gr_s^2(\rho' - \rho)\phi_r^{-1}\mu^{-1}$$

whence

(17) $$\phi_r = \frac{\hat{v}_s}{\hat{v}_a}$$

where $\hat{v}_s$ is the terminal velocity of the original sphere.

It is to be noted that if a sphere be constructed with radius r_a to fall with a terminal velocity $\hat{v}_a$, then

(18) $$\phi_r = \frac{r_s^2}{r_a^2}$$

---

[2] Ostwald's form resistance is actually $\phi_r(\tfrac{2}{9}gr_s^2)^{-1}$, but the use of a dimensionless number, conveniently referred to the sphere as unity, seems preferable. Comparing (10) and (13), $\phi_r^{-1} = 18K$. Dallavalle (1948) separates the linear dimension but includes g in the constant describing the effect of form,

$$\hat{v} = k_b d^2(\rho' - \rho)g^{-1}$$

where, for a sphere, k_b has the numerical value of 54.4 cm. sec.$^{-2}$

Some authors, notably Kunkel (1948), have given the results of experiments in terms of the ratio of r_s to r_a, r_a being termed the *nominal radius*. This is logical if one is interested in the use of sedimentation rates as a method of determining nominal radii. To the planktologist who usually has other more direct means of measuring linear dimensions but who is ignorant of, though interested in, velocities of passive sinking the use of the square of the ratio or coefficient of form resistance as already defined will prove the most practical procedure.

For certain simple solids, such as the ellipsoid and disk, a fairly exact theoretical treatment is possible. The ellipsoid was considered by Oberbeck (1876) whose theory has been developed more recently by McNown and Malaika (1950). Various approaches which attempt greater precision than that of Oberbeck's have also been elaborated; the mathematical reader will find them in standard works such as Lamb (1932) and Falkenhagen (1931), but they are of little importance from the standpoint of this book.

In an ellipsoid of semidiameters a, b, and c, falling in the direction of the a-axis, the coefficient of form resistance is given by

$$(19) \qquad \phi_r = \frac{\vartheta_s}{\vartheta_a} = \frac{8}{3}\,(abc)^{-\frac{1}{3}}(B_1 + B_2)^{-1} = \frac{8}{3}\,[r_s(B_1 + B_2)]^{-1}$$

where

$$B_1 = \int_0^\infty \frac{1}{\sqrt{(a^2 + \xi)(b^2 + \xi)(c^2 + \xi)}}\, d\xi$$

and

$$B_2 = a^2 \int_0^\infty \frac{1}{(a^2 + \xi)\,\sqrt{(a^2 + \xi)(b^2 + \xi)(c^2 + \xi)}}\, d\xi$$

ξ being an arbitrary variable which disappears when the definite integrals, between the limits 0 and ∞, are evaluated. The forms of the resulting expressions for B_1 and B_2 depend on the relative magnitudes of the three semidiameters and have the dimensions of the reciprocal of length, so that ϕ_r remains a dimensionless number. McNown and Malaika have evaluated the integrals for a number of special cases, and some of their results are given graphically in Fig. 75. In addition to the curves (solid lines) given by McNown and Malaika, a curve (broken line) has been drawn for a series of spheroids, falling with their axes of rotation horizontal, the lowest of the solid lines (b/c = 1) representing the equivalent cases when the axis of rotation is vertical.

Provided that the two semidiameters normal to the direction of motion are nearly equal, it will appear that a range of values for the semidiameters exists such that $\phi_r < 1$ and that the ellipsoid falls slightly

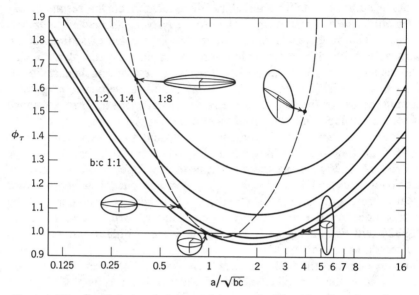

FIGURE 75. Coefficients of form resistance of ellipsoids of various shapes, with axes a vertical in plane of paper, b horizontal in plane of paper, c normal to plane of paper. Line for b:c = 1 gives values for spheroids, oblate on left, prolate on right, falling with axis of rotation vertical, while broken line gives values for the same spheroids, but with prolate on left and oblate on right, falling with axis of rotation horizontal. (Modified from McNown and Malaika.)

faster than the sphere. In the oblate spheroid or ellipsoid of rotation in which b = c, the lowest curve of Fig. 75 indicates that $\phi_r < 1$ when the ellipsoid falling along its major axis is 1 to 3.8 times as long as it is wide. The effect, however, is never great; the minimum value of ϕ_r is 0.955 and in general for values of a/b slightly in excess of unity the ellipsoid of rotation, moving parallel to its long axis, will meet almost the same resistance as the sphere of like volume. Ludwig (1928) noted this in his study of the resistance encountered by a *Paramecium* swimming, which may be treated approximately as an ellipsoid three times as long as it is wide. The coefficient of form resistance of such an ellipsoid would be 0.97, but Ludwig also attempted corrections for the rotation of the animal, which is probably not important in passive sinking, and for the surface properties, which are of questionable significance.

The question might be considered whether the elongate prolate form is more efficient for flotation than is the flat oblate form. Stated in so general a way the question is indeterminate, but the application of certain reasonable restrictions suggested by biological considerations may permit some answer to be obtained. First it is obviously only significant to compare bodies of the same volume. Second, it may be supposed that the diameter transverse to the center of the long axis cannot be less than, say, 2a, to accommodate structures such as the nucleus of the cell. Third, it may be supposed that the bodies under examination are always oriented to give a maximum coefficient of form resistance; it is pointed out later that this is preferred if such an orientation exists.

Comparison of the prolate and oblate spheroids may be made by the use of Fig. 75. If the oblate spheroid has a minor axis of length 2a and a radius of rotation of am, the volume will be $\frac{4}{3}a^3m^2$. If the comparable prolate spheroid has a radius of rotation a and the length of the axis of rotation is 2an, the volume will be $\frac{4}{3}\pi a^3n$, whence $n = m^2$. In the first figure the two axes normal to the direction of motion will be identical and the coefficient of form resistance can be read off the lowest solid curve, for b/c = 1, in Fig. 75, the value of $a/\sqrt{bc}$ being $1/m$. In the prolate spheroid b/c = m^2, but again $a/\sqrt{bc}$ = $1/m$. Since the broken line for the prolate spheroid set horizontally diverges rapidly from the lowest curve, which on the left of the diagram refers to the oblate spheroid set horizontally, it is evident that the prolate form is the more efficient in reducing velocity.

This conclusion, however, takes no explicit account of the changes in surface area, which may have great indirect significance in changing the excess density on passing from one figure to another.

The ratio of the surface of a prolate to that of an oblate spheroid of equal volume, using the above conventions for the lengths of the axes, is given by

$$(20) \quad \frac{2\left\{1 + \dfrac{m^4}{\sqrt{m^4-1}}\sin^{-1}\dfrac{\sqrt{m^4-1}}{m^2}\right\}}{2m^2 + \dfrac{m}{\sqrt{m^2-1}}\ln\dfrac{m+\sqrt{m^2-1}}{m-\sqrt{m^2-1}}}$$

This expression has the values set down in Table 12 and converges on $\pi/2$.

If we consider the transformation of an oblate into a prolate organism, the surface area will be increased by an amount equal to the ratio for the appropriate value of m; in an extreme it can be as much as 157 per cent of the original area. If the whole of the excess density is due to a surface layer, which is almost the case in diatoms, part of the

TABLE 12.   *Ratio of areas of prolate to that of*
oblate spheroids of identical volume

m	
2	1.17
3	1.31
4	1.41
5	1.45
7	1.49
10	1.53
∞	1.57

advantage in the increase in form resistance might be annulled by an increase in excess density.   This would, of course, happen only if the dense surface layer had a constant thickness.   Actually, comparison of the values of (20) given in Table 12 with the ratios of the form resistances indicated by the left-hand part of the broken line to those given by the lowest solid line of Fig. 75 indicates that at all values of m in excess of 2, or of $a/\sqrt{bc}$ less than 0.5, the transformation will raise the form resistance even if the excess density is increased in proportion to the increase in surface.   In any actual example it is possible that the cell with greater area per unit volume would have a somewhat thinner wall, so that the comparison just made probably represents an extreme limiting case.   Provided the organism is falling in the position of maximum resistance, it is probable that the elongate prolate form will always be the more efficient, though as the form becomes more cylindrical, the theory on which this conclusion is based breaks down.

A flat circular disk may be considered as the limiting case of the ellipsoid of rotation when the length of the axis of rotation approaches zero.   Gans (1911) has adapted Oberbeck's treatment, and a more elaborate approach not rejecting second-order terms also exists.   The approximate theory is reasonably satisfactory if $R_e < 0.5$, and Schmiedel (1928) has shown that if appropriate corrections for the effects of the wall of the vessel are made the more precise theory agrees well with experiment when $R_e \leq 1$.   When the thickness of the disk is a small fraction $1/m$ of its diameter, the approximate treatment of Gans for a disk falling in a direction normal to its surface gives

$$(21) \qquad\qquad \phi_a = 0.932 \sqrt[3]{m}$$

and for a disk falling in a direction parallel to its surface

$$(22) \qquad\qquad \phi_b = 0.622 \sqrt[3]{m} \left(\frac{3\pi}{3\pi - 8/m} \right)$$

The last expression, when the disk is very thin, may be written

$$(23) \qquad \phi_b = \tfrac{2}{3}\phi_a$$

The values derived from (21) and (23) are apparently reasonably satisfactory when $m > 30$, and those derived from (23) differ little from what would be obtained by extrapolating the right-hand part of the broken curve of Fig. 75. For positions between the vertical and horizontal, provided that the disk moves vertically, the values of ϕ_r will lie between ϕ_a and ϕ_b. In whatever position a small disk may be falling, it will therefore descend more slowly than will the equivalent sphere of the same volume and density. For a disk one hundred times as wide as it is thick the time taken to fall a given distance, if it is falling edge on, will be 2.9 times or, if falling broadside on, 4.3 times that taken by the sphere of the same volume and density.

A long cylinder falling with its long axis horizontal cannot be treated like the figures already considered for even at the lowest velocities the liquid carried along with the cylinder tends to increase and the inertial term does not disappear from the first-order approximation. The resistance encountered is therefore a function of R_e.

Lamb (1932) developed the theory from which, following White (1946; see also Munk and Riley 1952), we can obtain for the terminal velocity of a cylinder falling in a liquid of unit density

$$(24) \quad \vartheta + \left\{ \frac{141 d_c{}^2 (\rho' - \rho)}{\mu} \right\} \log \vartheta$$

$$= (0.869 - \log d_c + \log \mu) \left\{ \frac{141 d_c{}^2 (\rho' - \rho)}{\mu} \right\}$$

which, when the values of d_c, μ, and $(\rho' - \rho)$ are known, can be solved graphically. The results for two values of $(\rho' - \rho)$ and μ are plotted in Fig. 76 against d, the diameter of the cylinder. White found experimentally that when the length was well over five times the diameter it made little difference to the sinking speed, as is assumed in the mathematical treatment. The velocities given are dependent on the cylinder sinking in an indefinitely large volume of water. If the distance from the cylinder to the boundary is not more than 500 d_c, at some velocities a generalized Stokes's law, with a value of ϕ_r independent of density or viscosity, holds. This situation, however, is of no interest in lakes, though it may be significant in laboratory studies of sinking plankton organisms. However, in the size range of most cylindrical organisms, with diameters not greater than 100 μ or 0.01 cm., it can be seen from Fig. 76 that a horizontally placed cylinder falls very nearly according to a gen-

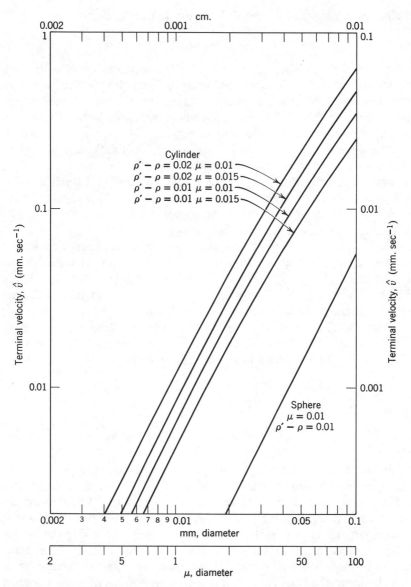

FIGURE 76. Rate of free fall of indefinitely long cylinder falling in a horizontal position, as function of diameter for two approximately limiting values of viscosity and excess density. Note scales of graph are given in millimeters and millimeters per second, as more convenient in limnology, though the quantities in equation (24) on which they are based, are in C.G.S. units.

eralized Stokes's law when R_e is less than 0.1, which will ordinarily be the case. In such circumstances the cylinder will fall roughly ten times as fast as the sphere of equal diameter. The rate of descent of a cylinder of diameter d_c will be roughly equal to that of a sphere of diameter $3.5d_c$ and such a cylinder will have a volume as great as that of the sphere if its length is just over $7d_c$. Clearly a cylinder can always be constructed to fall more slowly than the sphere without being inconveniently long, but it is always possible that increased excess density, if the cell walls are much denser than the interior, as in diatoms, will reduce the efficiency of the long thin forms.

The effect of departure from the spherical form; experimental studies on nonbiological objects. A number of studies of the falling of symmetrical metal bodies of various shapes in viscous lubricants have been made by McNown and Malaika (1950). They conclude that when the Reynolds' number is well below unity the sphere, the cube, the cylinder of length equal to the diameter, and the bicone of like proportions all fall at about the same velocity if their nominal diameters are equal. In the region $R_e \leq 0.5$ the differences due to form in this series are not more than 10 per cent of the mean sinking rate. An oblate spheroid whose axis of rotation has a length one fourth of the transverse diameter, a bicone of like proportions, a circular disk of diameter four times its thickness, and a square plate of side length four times the thickness are also found to fall at approximately the same rate when set with the major diameter horizontal. The form resistance of the bicone, however, is slightly greater and of the disk slightly less than that of the other figures, and the rectangular projection of the square plate apparently leads to a very slightly greater form resistance than that of the round disk. The ellipsoid is found to have a coefficient of form resistance of about 1.36 in these experiments which compares well with the value of 1.38 derived from Fig. 75. The observed values for the other forms apparently diverge from this expected value only by a few per cent.

A prolate spheroid with the axis of revolution four times the transverse diameter, a rectangular prism four times as long as it is thick or wide, a cylinder four times as long as it is wide, or a bicone of like proportions again fall at about the same speed when set with the long axis horizontal; when $R_e \leq 0.5$, the form makes almost no difference. The coefficient of form resistance is for all bodies 1.32, whereas for the ellipsoid a theoretical value of 1.30 is obtained from Fig. 75, so that the agreement of experiment and theory is excellent.

In all of these cases, after due correction for the diameter of the vessel has been made, the generalized form of Stokes's law is obeyed up to $R_e = 0.05$ and, within 10 per cent, up to $R_e = 0.5$. Though perhaps this agreement is not so good as might be hoped for in the region of greatest biological interest, within any one series of experiments, with a given ratio of length to breadth, the relative effects of form when the different shapes were compared appeared to be constant up to $R_e = 0.5$. It is certainly reasonable to suppose that only general proportions are important in determining form resistance.

Kunkel (1948) has made an interesting study of the velocities of passive sinking of systems composed of small glass beads immersed in a viscous oil. Though he was concerned primarily with the effect of aggregation on estimates of particle size made from determinations of sinking speeds, his artificial aggregates of beads bear marked resemblances to certain biological forms. In view of his particular interests, Kunkle gives his results in terms of the ratio of the nominal radius (r_s) of the sinking object to the radius of a sphere of the same density, sinking according to Stokes's law at the same velocity. These values have been converted into values of the coefficient of form resistance by means of (18).

For systems of glass beads cemented together to produce straight moniliform chains, set horizontally in the oil, the experimental results indicate that ϕ_r is essentially a linear function of the number of beads. In the region investigated (one to eight beads)

$$(25) \qquad\qquad \phi_r \simeq 0.837 + 0.163b$$

where b is the number of beads. The solid line in Fig. 77 represents this equation; below it the broken line, derived from Fig. 75, indicates the values of the form resistance for ellipsoids of rotation, falling with the major axis horizontal, plotted against the ratio of the major to the minor diameter. McNown and Malaika, as has just been indicated, have shown that cylinders and other elongate bodies fall at approximately the same velocity as ellipsoids of like length and breadth. It is therefore surprising to find that Kunkel's moniliform rods should fall so much less rapidly than the equivalent ellipsoids. In view of the prevalence of linear moniliform colonies of cells in the bacteria and filamentous blue-green algae, these observations are not without biological interest.

A flat triangle formed of three beads has a coefficient only a little less than that of three beads in a row, but a hexagon of six beads with a seventh in the middle, set horizontally in the fluid, is definitely a much less efficient arrangement for flotation than a horizontal chain of seven beads in a row. Six beads arranged as two superimposed triangles

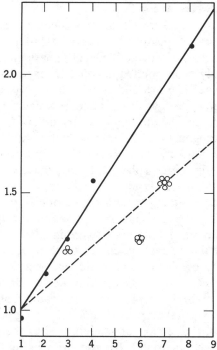

FIGURE 77. Coefficient of form resistance
determined experimentally for moniliform
chains of one to eight beads (solid line,
black circles; broken line ellipsoids from
Fig. 75) and for groups of beads indicated by clusters of open circles. (Data
of Kunkel.)

pointing in opposite directions provide, in their approach to a spherical
form, a case in which we might suppose that ϕ_r would approach unity;
the observed value of 1.31 seems extraordinarily high. It is evident
that further studies on figures formed of small spheres would be of interest.

Several determinations for the falling of small particles of irregular
shape such as crushed quartz or coke, collected from the literature by
Dallavalle (1948), indicate that within the region in which Stokes's law
is obeyed the sphere falls about 1.5 times as fast as the average random-
shaped body; we may therefore conclude that it has a coefficient of form
resistance of about 1.5. At high values of Reynolds' number it appears that the sphere may fall in a turbulent medium less rapidly than

the irregular body; it is unfortunate that no information exists for cases in which R_e lies between 10 and 20.

Little experimental work exists on the effect of microscopic roughness on the velocity of passive sinking of small particles in viscous media; such roughness, however, is usually considered to be unimportant. It has already been pointed out that all experimental work tends to imply that the coefficient of sliding friction in Stokes's original equation (8) is very large, so that the fraction in which this quantity occurs in both numerator and denominator hardly differs from unity. Mere alteration of the roughness of a body without any essential change in shape is therefore unlikely to increase the resistance. According to Arnold (1911), badly pitted spheres of Rose metal sink in Colza oil at the same velocity as perfectly smooth spheres, provided the diameters are identical. This result, if general, is obviously of considerable importance in limiting the possible interpretations of the adaptive significance of the surface structures of certain plankton organisms; the short spinous granulation on *Ceratium* or the pubescence of *Keratella hispida* immediately come to mind in this connection.[3]

Preferred orientation when falling through a viscous medium. A well-known theorem in hydromechanics indicates that a body moving in a liquid will tend to take up a position with its greatest area of projection normal to the direction of motion. Thus the disk will tend to fall with its flat sides horizontal, the rod, with its long axis horizontal. It has, however, been pointed out by Gans that if the falling body is small enough so that the inertial effects are negligible and the movement is purely translatory, the body will have no preferred position if it is symmetrical about three axes at right angles. Such a body when falling slowly through a viscous medium will maintain any arbitrary initial orientation. McNown and Malaika confirmed this conclusion experimentally, finding that the metal bodies they used in their experiments maintained any position, provided that the requirements of symmetry were met. Very small asymmetries, however, led the body to take up a preferred position with its greatest area of projection normal to the direction of fall. A marked displacement of the center of gravity from the approximate center of symmetry will cause orientation with the end nearer the center of gravity downward. Kunkel found that tear-shaped bodies and long plates of metal weighted at one end fell with the weighted end downward in the Stokes's law region.

---

[3] I am indebted to Sir G. I. Taylor for discussion and confirmation of the conclusion that surface roughness of this kind is of no importance in regulating sinking speed.

Because of the very considerable differences between the rate of passive sinking at low values of Reynolds' number that can be induced by changing the position of a thin disk or any of the more extreme ellipsoids, it is obvious that any mechanism leading to the assumption of a preferred position with a maximum projection across the direction of motion will materially decrease the sinking speeds. Lowndes (1937, 1938) has concluded also that the inverse situation may be true of certain planktonic organisms, the development of excrescences leading to a particular projection becoming maximal and so determining the posture of the organism; in the case of phototrophs the posture might be optimal in photosynthesis. It is highly probable that in very small organisms such as diatoms slight departures from perfect triaxial symmetry may be of great importance in forcing the cell to take up a preferred position of maximum form resistance. Elongate diatoms are often somewhat curved. Riley (personal communication) has noticed such diatoms changing position while falling, becoming more horizontal and even falling with the zigzag motion usual for larger bodies with a somewhat turbulent wake, so easily observed when a plate accidentally falls into the water. Kunkel noted that small steel disks dropped in viscous oil often become tilted and acquire a lateral component in their direction of fall. This type of movement may be important in flat platelike diatoms (see page 294).

Lund (1959a) has found that *Oscillatoria agardhii* var. *isothrix,* single cells of *Asterionella formosa,* and filaments of *Melosira italica subarctica* however, assume a vertical position in nonturbulent water and thus are not set for a minimum speed of sinking. He suggests that the main function of form resistance imparted by any special shape is to prevent the organism from immediately becoming vertical if it is displaced by turbulent movements so conferring on the cell a lateral component of movement.

Brooks and Hutchinson (1950) noted that a narcotized *Daphnia,* with its antennae extended and so constituting an axis of rotation, could apparently take a variety of positions while falling, ranging from the long axis of the animal being almost horizontal to this axis being almost vertical. Hantschmann (1961) found that small specimens of *D. schødleri* lie obliquely with the tail spine almost horizontal, whereas intermediate and large species adopted a more vertical position.

Allen's results for intermediate values of R_e. Before turning to the experimental studies that have been made on planktonic organisms, it is necessary to consider the most satisfactory elementary attempt to apply (4) to bodies of such size and such velocity that Reynolds' number is well in excess of 0.5, though still below the range at which inertia becomes dominant and viscosity, insignificant. This attempt is due to Allen (1900), who concluded from the study of the descent of small spheres falling in liquids that over a range of values of R_e in excess of those for which Stokes's law applies the value of q in (4) may be taken with a considerable degree of accuracy to be 3/2.

The general equation for the terminal velocity becomes

(26) $$\hat{v} = Kd_a(\rho' - \rho)^{\frac{2}{3}}g^{\frac{2}{3}}\rho^{-\frac{1}{3}}\mu^{-\frac{1}{3}}$$

The terminal velocity thus varies directly as a linear dimension d_a, as the cube root of the square of the density difference, and inversely as the cube root of the density of the medium and as the cube root of the viscosity. Over a moderate range of water temperature the cube root of the density will be essentially constant. The linear dimension d_a is given by

$$d_a = d - cd'$$

where the critical diameter d' is computed by Allen as

$$d' = \sqrt[3]{\frac{g\mu^2}{2g\rho(\rho' - \rho)}}$$

and c depends on the shape of the body, being 0.4 for a sphere and 0.28 for an irregular particle (cf. Dallavalle 1948). As pointed out by Brooks and Hutchinson, cd' is probably of the order of 0.1 mm. for the fresh-water plankton organisms to which Allen's results can be applied.

The theory of the gelatinous sheath. One special type of structure that does not necessarily involve a change in external form, or merely one of magnitude, must be considered. This is the gelatinous sheath often developed around planktonic organisms and widely believed (Wesenberg-Lund 1908; Huber-Pestalozzi 1938; Ruttner 1940, 1952b) to reduce the rate of passive sinking of the organisms that possess it.

The gelatinous sheaths of the phytoplankton (Fig. 78) have been considered by Naumann (1925), who notes their presence in most of the nonmotile planktonic algae. Such sheaths, for instance, are found in all the planktonic blue-green algae except some members of filamentous genera *Oscillatoria, Lyngbya,* and *Aphanizomenon.* Many of the nonflagellate green algae secrete these sheaths, which are most notably developed in the desmids, the elongate processes of the hemicells of such forms as *Staurastrum* and *Staurodesmus* often being completely embedded in jelly. Among the diatoms the sheath is well developed in *Cyclotella, Fragilaria,* and *Stephanodiscus,* which are regarded by Naumann as having a gelatinous capsule (Hüllgallerte). The star-

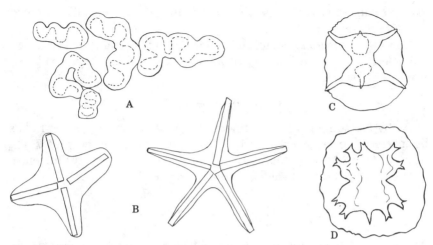

FIGURE 78. *A.* Gelatinous sheath of *Anabaena flos-aquae. B.* Cells of *Tabellaria* connected by jelly. *C. Staurastrum cuspidatum* var. *maximum.* D. S. *furcigerum.* (Naumann, Ruttner.)

shaped colonies of *Asterionella* and *Tabellaria* usually (Voigt 1901) have a gelatinous membrane (Schirmgallerte of Naumann) filling at least the inner parts of the sectors between the cells; in *Tabellaria* this membrane is formed by the fusion of the basal parts of the typical gelatinous capsules that enclose the cells. The star-shaped colony thus actually has a parachute-like form which has generally been regarded as an adaptation to a floating planktonic life. A few other forms, mainly colonial desmids and diatoms, have gelatinous material binding together the cells of the colony. Naumann regarded this secretion as constituting a third type of gelatinous structure (Verbindungsgallerte).

Some of the solitary flagellated planktonic algae may secrete temporary gelatinous capsules when dividing, and a few cases of members of the zooplankton forming such capsules also are known. The most remarkable is that of *Holopedium* (*C* of Fig. 149), a member of the Cladocera which resembles tapioca in a townet. Among the rotifers some races of *Brachionus angularis* secrete a gelatinous capsule during the summer (*E* of Fig. 243).

Although it is possible that gelatinous capsules may increase the difficulties of zooplankton attempting to eat the algae possessing them, as Naumann has suggested, the function of these capsules is usually supposed to be related to the problem of flotation and the reduction of sinking speed. In view of the fact that, other things being equal, an increase in diameter increases the velocity of passive sinking, it is necessary to investigate the theoretical validity of the generally accepted adaptive character of the sheath.

Consider a spherical organism of radius r and density ρ'. The terminal velocity of passive sinking, if Stokes's law is to be obeyed, is given from (13) by

$$(27) \qquad \qquad \hat{v}_1 = \tfrac{2}{9}gr^2(\rho' - \rho)\mu^{-1}$$

Now, by the addition of a gelatinous sheath of density ρ'', let the radius be increased to $\hat{a}r$, where $\hat{a} > 1$. The new volume will be $\tfrac{4}{3}\pi r^3 \hat{a}^3$, the new mass $\tfrac{4}{3}\pi r^3 [\rho' + (\hat{a}^3 - 1)\rho'']$, and the new density therefore $[\rho'' + (\rho' - \rho'')/\hat{a}^3]$. The new velocity is given by

$$(28) \qquad \qquad \hat{v}_2 = \frac{2}{9} gr^2 \hat{a}^2 \left[\rho'' - \rho + \frac{1}{\hat{a}^3} (\rho' - \rho'') \right] \mu^{-1}$$

The effect of increasing the radius will be most easily appreciated by considering a quantity ϕ_r, defined as before as $\hat{v}_1/\hat{v}_2$.

$$\phi_r = \frac{\hat{v}_1}{\hat{v}_2} = \frac{\hat{a}(\rho' - \rho)}{\hat{a}^3(\rho'' - \rho) + (\rho' - \rho'')}$$

If the gelatinous sheath decreases the rate of sinking, ϕ_r will exceed unity and

$$\hat{a}(\rho' - \rho) > \hat{a}^3(\rho'' - \rho) + (\rho' - \rho'')$$
$$\hat{a}(\rho' - \rho'') - (\rho' - \rho'') > \hat{a}^3(\rho'' - \rho) - \hat{a}(\rho'' - \rho)$$
$$(\rho' - \rho'')(\hat{a} - 1) > \hat{a}(\hat{a}^2 - 1)(\rho'' - \rho)$$

whence

$$(29) \qquad \qquad \frac{\rho' - \rho''}{\rho'' - \rho} > \hat{a}(\hat{a} + 1)$$

If the density difference between the organism and the jelly (internal density difference) is not greater than twice the density difference between the jelly and the medium (external density difference), the inequality cannot be satisfied by any value of $\hat{a}$, which is by definition greater than 1; for such values of the densities the sheath would be ineffective as a flotation mechanism. For all ratios of the internal density difference to the external density difference greater than two there exists a range of values of $\hat{a}$ over which addition of the jelly will reduce the sinking speed. If the jelly has the same density as the medium, it is possible to reduce the sinking speed to any desired degree merely by increasing the thickness of the sheath. The nonmathematical reader wearied by the analyses of the present chapter will probably find this result intuitively obvious. It is doubtful, moreover, if an organism in an indefinitely thick rigid gelatinous sheath would be as efficient meta-bolically as a naked cell. It would seem likely, however, that the gen-erally accepted idea of the function of the sheath is likely to be correct, particularly in the case of diatoms, in which, as we have seen, the den-sity (ρ') may be well over 1.1 and possibly as great as 1.45.

The only experimental work relates to *Holopedium gibberum,* studied by Hamilton (1958). Here the animal without the jelly had a density at 18° C. of 1.014, comparable to that of *Daphnia;* with its jelly its density was only 1.0015. The density of the jelly was 1.0002, so that the conditions for reduction of sinking speed can be satisfied. Ac-tually the narcotized complete animals sank at 18° C. at a mean rate of 0.22 cm. sec.$^{-1}$, but when the jelly was removed they sank at 0.31 cm. sec.$^{-1}$.

Normal viscosity and structural viscosity. Margalef (1957a) has questioned the completeness of the kind of theory already outlined on the ground that it is valid only for objects of zero electric charge, lack-ing any surface-active materials at their surfaces. Neither assumption is likely to be true in planktonic organisms. He supposes that surface-active groups and electric charges, by altering the degree of hydration of the exterior of the falling organism and so the structure of the water in its immediate vicinity, may alter the effective viscosity and so the sinking speed. Margalef gives evidence of a considerable spread in values of sinking speed from $\leq$ 0.2 mm. sec.$^{-1}$ to $\geq$ 1.0 mm. sec.$^{-1}$ with a mode at 0.4–0.6 mm. sec.$^{-1}$ for cells of *Scenedesmus obliquus* from the same culture. He also points out that *Rhizosolenia, Stichococ-cus,* and most cladocera are hydrofuge, whereas most green algae and copepods are hydrophil. Among crustacea, the hydrofuge forms tend to be freer of epibionts (Sebestyén 1951). Margalef also finds that *Scenedesmus obliquus* photosynthesising in the light at *p*H 7.2 are markedly anaphoretic, whereas individuals kept in the dark are mildly cataphoretic. The effect is inappreciable with *Stichococcus bacillaris* at *p*H 6.9; at 7.2 movement to the anode occurs but with much individ-ual variation. It is evident that great differences in the interaction be-tween the surfaces of organisms and the medium do occur, as do differ-ences in charge, dependent on the physiological condition. The only direct observations on change of sinking speed quoted by Margalef, namely a decrease in sedimentation rate of *Chlorella* when a few drops of butanol or isopropanol are added to suspensions of the cells of this alga, suggests to him that metabolites lost into the water may aid in the flotation of algae by altering the structural viscosity in the neighbor-hood of the cell. Margalef's most interesting ideas suggest an impor-tant line of investigation, but until such investigation is undertaken it is uncertain how far they will prove to be quantitatively significant. His further idea that the electrostatic properties of cells may directly alter sedimentation rates in the presence of natural electric fields in the water must also be tested empirically; he admits, however, that this effect is not likely to be great.

Empirical studies of fresh-water planktonic organisms.   Experiments have been conducted with living fresh-water planktonic algae by Fritz (1935), who used *Cyclotella bodanica* (diameter 32–62 μ, mean 49 μ), *Fragilaria crotonensis* (mean length of cell 68 μ), and *Asterionella formosa* (mean length of cell 68 μ).   The second and third species were used only in colonies.   For the first species, which appears to have sunk at a rate of 0.021 cm. sec.$^{-1}$ at 20° C. and 0.017 cm. sec.$^{-1}$ at 6° C., Reynolds' number must be of the order of 0.01, and it is not likely to be very different for the other longer but more slowly sinking forms. Fritz claims only relative significance for his results, which are presented mainly in graphical form but which appear to refer to the number of centimeters fallen by the average individual per hour.   The results of Table 13 were recomputed as well as may be, to centimeters per second, from measurements of his histograms.

Agreement with theory is not very good, but the divergence is not systematic.   It is reasonable to suppose that the low sinking speeds of these organisms give great opportunities, in rather long experiments, for disturbance by feeble accidental turbulent movements of the medium. Fritz notes that diatoms fixed with acetic-sublimate mixture sank three to four times faster than live diatoms.   This is presumably due to an increase in density as mercuric ions are absorbed and perhaps also to destruction of the sheaths.

TABLE 13.   *Observed and computed sinking speeds of diatoms in quiet water*

	Temperature (°C.)	Viscosity	Sinking Speed
Cyclotella bodanica	6	0.0147 poise	17×10^{-3} cm. sec.$^{-1}$
	20	0.0101 poise	21×10^{-3} cm. sec.$^{-1}$
calc.	20	by Stokes's law	25×10^{-3} cm. sec.$^{-1}$
Asterionella formosa	6	0.0147 poise	8.6×10^{-3} cm. sec.$^{-1}$
	20	0.0101 poise	11.1×10^{-3} cm. sec.$^{-1}$
calc.	20	by Stokes's law	12.6×10^{-3} cm. sec.$^{-1}$
Fragilaria crotonensis	6	0.0147	5.9×10^{-3} cm. sec.$^{-1}$
	20	0.0101	10.5×10^{-3} cm. sec.$^{-1}$
calc.	20	by Stokes's law	8.6×10^{-3} cm. sec.$^{-1}$

In addition to these experiments, Grim (1951) reports that Einsele and Grim[4] found that *Tabellaria* sank in capillary tubes at 17–18° C. at rates of 2 to 4.5×10^{-3} cm. sec.$^{-1}$; most specimens sank at 2.2 to 3.1×10^{-3} cm. sec.$^{-1}$. In nature without the restraint of the narrow tube employed in the experiments the sinking speed would have been a little greater. Lund (1959) found that the mean sinking rates of various populations of *Asterionella formosa* at 0° C. were 0.14 to 0.37 10^{-3} cm. sec.$^{-1}$, whereas the individual observations gave a range of 0.08 to 0.97 10^{-3} cm. sec.$^{-1}$ The temperature difference is far from enough to account for the difference from the results of Table 13. *Melosira italica subarctica* sank at a much faster rate in Lund's experiments; 0.60 to 2.43×10^{-3} cm. sec.$^{-1}$

In spite of the conclusions of Gross and Zeuthen (1948), about which Lund expresses some scepticism, that a variety of healthy marine diatoms are of exactly the same density as the sea water in which they live, Riley, Stommel, and Bumpus (1949, cf. page 284) concluded that a sinking rate between 2.9 and 4.2 10^{-3} cm. sec.$^{-1}$ at 0° C., or twice these rates at 25° C., would be needed to account for observed rates of organic production in the Atlantic, given any reasonable value of the eddy diffusivity. These values imply a range of sinking rates of 2.5 to 7.2 m. day^{-1}. Riley, Stommel, and Bumpus quote from the literature values for marine diatoms of 0.06 to 30 m. day^{-1}, whereas the various laboratory observations for fresh-water species just given imply rates of from 0.07 to 18 m. per day, a range that is roughly comparable with that derived from marine experience.

Grim (1939), who studied the actual movement of diatoms in Lake Constance, concluded that living specimens of the small species of *Cyclotella,* mainly *C. melosiroides,* do not sink appreciably when alive, at least under optimal conditions; there is, however, evidence of sinking of the dead and less healthy cells of this species at rates up to 7 to 8 m. day^{-1}. *Fragilaria crotonensis* sank at rates up to 7 to 9 m. day^{-1} and *Synedra acus delicatissima* on different occasions sank 3 to 20 m. day^{-1}. These maximal rates of sinking are achieved only in the relatively non-turbulent waters of the hypolimnion; in the epilimnion, because of the turbulence of the water, the apparent sinking rates are markedly less. In general, Grim concluded that the sinking velocity for live diatoms increased in the following order: *Cyclotella* small spp., *Asterionella formosa, Synedra acus angustissima, S. a. delicatissima, Cyclotella comta* and *socialis, Diatoma elongata, Fragilaria crotonensis,* and *Cy-*

---

[4] The paper referred to, apparently Einsele and Grim (1938), seems not to contain these data.

clotella bodanica. For dead diatoms the order was *Cyclotella* small sp., *Synedra acus delicatissima, S. a. angustissima, Asterionella formosa, Diatoma elongata,* and *Cyclotella comta* and *socialis.* Discrepancies in Grim's statements are probably due to differences in the condition of the organisms on different dates. The most obvious fact is that the very small *Cyclotellae,* which have a volume of less than 700^3 μ, remain floating much longer than the larger *C. socialis* and *C. comta,* which have volumes of 2000 to 4000 μ^3. *Asterionella,* when alive, sinks less rapidly than living *Synedra,* but dead cells of the former sink more rapidly than the latter.

The relationship of the rate of sinking of the rotifers *Euchlanis triquetra* and *Brachionus quadridentatus* f. *rhenanus* to the density difference $(\rho' - \rho)$ can be investigated by using the results of Luntz (1928, 1929) obtained by rearing at different temperatures and already presented for animals of different densities. According to Stokes's law (13), the velocity should be directly proportional to the density difference. Because the velocities involved are less than 0.1 cm. sec.$^{-1}$, the viscosity, at 18° C., the temperature of all the experiments, about 0.01 poise, and the dimensions of the animals less than 0.05 cm.,

$$\mathrm{R_e} \leq \frac{0.1 \times 0.05 \times 1}{0.01} = 0.5$$

and we may expect the law to hold approximately (Table 14). The agreement between theory and observation would appear to be quite satisfactory, particularly when it is remembered that different animals with different histories, are involved in the comparisons.

Luntz also found that wild vernal specimens of *B. q. rhenanus* without posterior spines and with a density of 1.025 sank at a rate of 0.050 cm. sec.$^{-1}$, whereas the typical form of *B. quadridentatus* with posterior spines, collected in the summer but also having a density of 1.025, fell

TABLE 14. *Observed and computed sinking speeds of rotifers*

	ρ'	$\hat{v}$
Euchlanis triquetra	1.025	0.0862 cm. sec.$^{-1}$
	1.020 obs.	0.0658 cm. sec.$^{-1}$
	1.020 calc. $= 0.0862 \times \frac{20}{25}$	0.0690 cm. sec.$^{-1}$
Brachionus quadridentatus	1.025	0.0495 cm. sec.$^{-1}$
f. *rhenanus*	1.020 obs.	0.0386 cm. sec.$^{-1}$
	1.020 calc. $= 0.0495 \times \frac{20}{25}$	0.0396 cm. sec.$^{-1}$

at 0.041 cm. sec.$^{-1}$ The presence of the spines in the typical forms
thus had almost the same effect on flotation as the decline of density at
high temperatures in cultures of the spineless variety *rhenanus*.

The falling of *Daphnia* has been investigated by Eyden (1923),
Bowkiewicz (1929), Brooks and Hutchinson (1950), Fox and Mitchell
(1953), and by Hantschmann (1961).   Brooks and Hutchinson re-
analyzed the data of their predecessors.   Narcotized specimens of small
limnetic *Daphnia galeata mendotae* with their antennae open fall at
velocities that clearly vary with the square of the length of the animal,
exclusive of the tail spine (*A* of Fig. 79) and inversely with the viscosity (*C* of Fig. 79).   Both relationships conform to Stokes's law,
even though it would seem likely that in this case R_e is greater than
unity, in the largest animal as great as 4.43.   For *Daphnia pulex* falling with antennae closed Eyden's data, though rather meager, indicate that the velocity is much more nearly proportional to the linear
dimensions than to their squares.   If Allen's relationship is considered
to apply, the velocity should be proportional to $(d - cd')$, where cd'
is probably about 0.08 mm., which will have little effect on the apparent
direct proportionality between velocity and length.   A slight correction
for the size of the tube in which Eyden's experiments were conducted
should, in theory, have been applied; this is impossible without knowing
the diameter of the tube, but if such a correction could have been made
it would probably have offset the change in slope due to allowing for
cd'.   Allen's relationship therefore appears to hold and moreover implies that the velocity varies inversely as the cube root of the product
of the density and viscosity.   Only the variation in viscosity will be
quantitatively significant.   The data of Bowkiewicz, at least for
his living narcotized *D. magna* falling with spread antennae (*C* of
Fig. 79) where $R_e \leq 16$ indicate as far as they go, such a situation.
Unfortunately, Bowkiewicz believed that his data implied an inverse
relationship with the square root rather than the cube root of the viscosity and that such an inverse relationship is given by Newton's law
of movement in a resistant medium.   As indicated earlier, Newton's
law contains no viscosity term.   Although small *Daphnia* thus appear
to sink according to Stokes's law and large *Daphnia* according to Allen's
relationship, Brooks and Hutchinson observed that specimens of
Daphnia dubia, slightly longer than the *D. g. mendotae* that they employed, were behaving in an intermediate way.   These animals sank
faster than *D. g. mendotae*.

Using a considerable range of sizes, Hantschmann (1961) also found
that *D. schødleri* longer than about 1.6 mm. fell at rates that were directly proportional to the length, essentially according to Allen's rela-

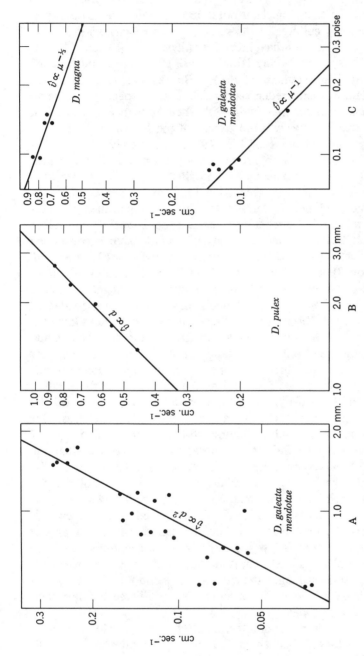

FIGURE 79. *A.* Sinking sped ϑ (cm. sec.$^{-1}$) of *Daphnia galeata mendotae* falling with open antenna, plotted against length *d*, without tail spine, falling approximately according to the square of *d* as in Stoke's law, from the data of Brooks and Hutchinson. *B.* The same for the larger *D. pulex* from the averaged data of Eyden, the fall being proportional to the length *d* rather than to *d*². *C.* The same plotted against viscosity as dependent on temperature for *D. g. mendotae* falling inversely as the viscosity and for the much larger *D. magna* falling inversely as the cube root of the viscosity, from the data of Brooks and Hutchinson and of Bowkiewicz.

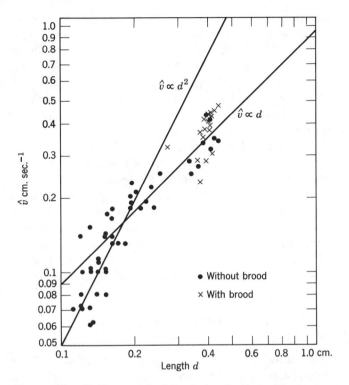

FIGURE 80. Sinking speed of *Daphnia schødleri* with antennae open, as a function of length (d); the line marked $\hat{v} \propto d^2$ indicates fall according to Stokes' law, that marked $v \propto d$ indicates fall approximately according to Allen's relationship. The absence of an effect of eggs or embryos in the brood pouch is noticeable. (Hantschmann.)

tionship, but the smaller specimens sank at speeds not greatly differing from (Fig. 80) what would be implied by Stokes's law.

Eyden has made a few observations on the effect of the spread of the antenna. Two experiments were conducted in which the rates of sinking of an animal with both antennae closed and with both open were compared. The individual values indicated a reduction in sinking speed when the antennae are extended from 73.1 to 66.3 per cent, with a mean of almost exactly 70 per cent, of the values with the antennae both folded. Seven experiments showed a reduction in sinking speed from 74.8 to 86.4 per cent with a mean of almost exactly 79 per cent, when animals with a single antenna extended are compared with those with both folded.

Hantschmann (1961) found that removal of all the lateral setae of the second antennae did not change the sinking position of a narcotized animal sinking with outspread antennae but increased the sinking speed by 22 per cent.   She also found that removing the apical setae of the dorsal ramus caused the sinking animal to tilt backward into an oblique position; removing the apical setae from both rami caused the animal to sink with its ventral side upward.   In these positions the animal actually sank more slowly, the rate being reduced to 88 and 68 per cent, respectively, of that of the unoperated animal.

These experiments, together with those of Luntz which compare spined and spineless (*rhenanus*) specimens of *Brachionus quadridentatus* taken in nature, are the only ones that permit any judgment of the effectiveness of projections of the body wall as adaptations facilitating planktonic life, and though it is obvious that spines and extended antennae do reduce the sinking speed it seems curious, at least in the case of the antennae, that the effect is so small.   It is unfortunate that no data on this matter are available for the Cladocera within the range of Stokes's law.

THE DISPERSION OF PLANKTON IN NATURAL TURBULENT WATERS

The role of turbulence in the dispersion and flotation of the plankton. Though it must have been obvious to many of the earlier workers who noted the sinking of phytoplankton in undisturbed vessels that water movements play a major part in keeping the phytoplankton suspended in the illuminated zone, the earlier statements on the matter are vague and sometimes not very intelligent.   The realization of the importance of turbulence developed slowly mainly because an understanding of the significance of turbulent movement in natural waters has itself grown rather slowly, almost entirely within the present century.   Zacharias (1895) found that in the laboratory dead *Melosira* filaments sank about one meter in fifty minutes, but he considered that in nature the time might be three or four times greater because of the disturbance of the water.   Whipple (1896) was perhaps the first to realize that wind-generated turbulence, though he did not call it that, is essential to the welfare of living diatoms.   He showed that if a long glass tube were filled with lake water and the tube suspended, with the upper end closed, in a lake, the diatoms would tend to fall out of the undisturbed water in the tube.   Whipple also believed that the feeble development of the diatom population in Lake Cochituate and other basins near Boston in 1895 was due to a spell of calm, very hot weather which stabilized the water in the absence of wind and allowed the diatoms to fall out of the shallow epilimnion into deep layers in which they had inadequate light. Lozeron (1902) also realized very clearly the need for external forces to keep the phytoplankton suspended but relied primarily on thermal convection currents. Quite recently Ruttner (1940, 1952b) could stress the great significance of vertical turbulent motion without an available quantitative theory of the interaction of such turbulence and of passive sinking. The beautiful work of Riley, Stommel, and Bumpus (1949), primarily developed for the ocean but certainly applicable in its broad outline to lakes, has now largely filled this lacuna.

Riley, Stommel, and Bumpus start with an equation of the general type which describes the rate of change of some nonconservative property in the hydrosphere.

$$(30) \qquad \frac{\partial \hat{p}}{\partial t} = A \frac{\partial^2 \hat{p}}{\partial z^2} - \hat{v} \frac{\partial \hat{p}}{\partial z} + \hat{w}\hat{p}$$

where $\hat{p}$ is the phytoplankton concentration, $\hat{v}$, the terminal sinking speed, in this case of the phytoplankton under nonturbulent conditions, and $\hat{w}$, the production rate coefficient, which is a constant supposed to indicate the sum of all biological processes per unit of phytoplankton; as before, t is time, z, depth, and A, the coefficient of turbulence, or specifically eddy diffusivity.

Experience shows that vertical patterns of distribution occurring in nature usually change much more slowly than they would if only vertical turbulence or passive sinking were involved. The observed natural situations may therefore be considered as approximating to steady states and may be treated by setting (30) equal to zero.

It is necessary to introduce certain simplifying assumptions before proceeding to a further examination of (25). A and $\hat{v}$ are considered constants independent of depth; the production rate coefficient, however, cannot be so regarded. The simplest form of its variation is to assume that from the surface to some depth z_l, defining the euphotic or trophogenic zone, $\hat{w}$ is equal to a positive constant $\hat{w}_1$ and that from depth z_l to an indefinite depth $\hat{w}$ is equal to a negative constant $\hat{w}_2$. Writing

$$a = \frac{\hat{v}}{2A}, \qquad b_1 = \sqrt{a^2 - \frac{\hat{w}_1}{A}}, \qquad b_2 = \sqrt{a^2 - \frac{\hat{w}_2}{A}}, \quad \text{and} \quad b_3 = -i b_1$$

Riley, Stommel, and Bumpus find that a nonvanishing steady-state distribution cannot exist if $\hat{v}^2 > 4A\hat{w}_1$. The condition for such a steady-state distribution is

$$(31) \qquad \frac{b_3(a + b_2)}{b_3{}^2 - a b_2} = \tan b_3 z_l$$

The complete solution for the phytoplankton distribution with depth is given for layer $z = 0$ to $z = z_l$ by

(32) $\hat{p}_1 = Ee^{az - b_2 z_l}(b_3 \cos b_3 z + a \sin b_3 z)$

and for layer $z = z_l$ to $z = \infty$ by

(33) $\hat{p}_2 = Ee^{az - b_2 z}(a \sin b_3 z_l + b_3 \cos b_3 z_l)$

Furthermore, it can be shown that in order that the phytoplankton concentration may never be negative at any depth

(34) $$0 < z_l < \frac{\pi}{b_3}$$

By taking arbitrary values of the production rate coefficients $\hat{w}_1$ and $\hat{w}_2$ and of the depth of the euphotic zone z_l it is possible by substituting in (31) to study the variation of A and $\hat{v}$ that permits the steady-state condition. In general, it is found that there is a value of A for which $\hat{v}$ is maximal, and that the eddy diffusion rate which permits this maximum sinking rate in a steady state population is optimal for production. It can be shown, moreover, that if production, consumption, and eddy diffusivity remain constant any increase in sinking rate must be compensated by an increase in the depth of the euphotic zone if a steady-state condition is to be maintained.

The population curve obtained by evaluation of (32) and (33) with arbitrary values of the constants show that when A is low there is a sparse population at the surface and a great increase near the bottom of the euphotic zone. When the eddy diffusion rate is high, the quantity of phytoplankton close to the surface increases greatly. These distributions, which are as reminiscent of the conditions in lakes (Figs. 81, 82) as they are of the ocean for which they were developed, are in no way dependent on variations of sinking speed with depth as is often supposed. Riley, Stommel, and Bumpus point out that if the temperature of a body of water rises the sinking speed rises; the immediate effect is to decrease the concentration of plankton at the surface. If a steady state is to be maintained, the depth of the trophogenic zone must be increased, but in general this is likely to happen, for the transparency of the water is to a large extent controlled by plankton, planktogenic detritus, and pigments of planktonic origin. They thus believe that the final effect of an increase in sinking speed is to produce a condition of low plankton concentration, high transparency, and a thick trophogenic zone. This condition is characteristic of tropical oceans and some temperate regions of the sea in summer. It is not typical of deep tropical lakes, which might be expected to show a like phenomenon,

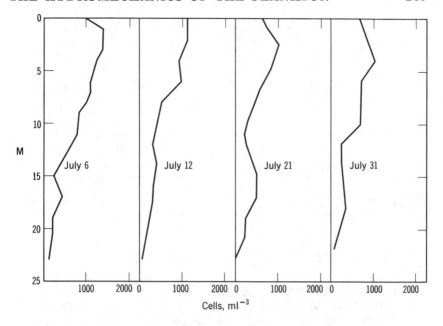

FIGURE 81. Vertical distribution of *Tabellaria* in the Zeller Basin of the Unter-see during July 1939; the data for July 21 are used in the analysis of Table 15. (Grim 1951, very slightly smoothed.)

because they are usually meromictic. It is quite likely, however, that the rather low plankton concentrations and high transparencies observed in some temperate lakes in early summer exemplify the process that Riley, Stommel, and Bumpus describe. It is uncertain whether their conclusion that the large size of the diatoms in spring blooms in the sea may permit the maximal sinking speeds, and so a steady state at the maximal rate of production, applies to lakes.

Application of the theory to Tabellaria *in the Untersee.* It should be possible to apply (30) directly to the analysis of the dynamics of the phytoplankton in the epilimnia of lakes, provided that A does not change with depth and that other variables are adequately known. Unfortunately, this is seldom, if ever, the case. As a purely illustrative example, we may study the distribution of *Tabellaria*[5] in the Untersee below Lake Constance, using the data collected by Grim in 1939. During July and early August (Fig. 81) the population changes little in

---

[5] Recorded as *T. fenestrata* but presumably actually a form of *T. flocculosa* (cf. Knudson 1952).

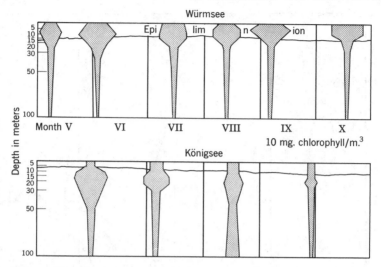

FIGURE 82. Top panel: distribution of chlorophyll in the Würmsee throughout the summer; planktonic maximum maintained in the thick (15–20 m.) turbulent epilimnion. Lower panel: distribution in much more sheltered and more transparent Königsee in which the optimal light intensity for photosynthesis lies below the shallow (5– 10 m.) epilimnion and the phytoplankton maximum cannot be maintained against a tendency to sink in the relatively nonturbulent hypolimnion. (Gessner.)

numbers or distribution, so that as a first approximation a steady state may be assumed. It is, however, continually dividing and sedimenting; from a study of the detailed changes in the deep water Grim concluded that the diatom was increasing by cell division, and at the same time decreasing by sedimentation, by 7 per cent a day. This corresponds to a division rate $\hat{w}$ of 0.81×10^{-6} cell cell^{-1} sec.$^{-1}$. The sinking rate $\hat{v}$ at the probable temperature of $18°$ C. has a mean value, as we have seen, of about 2.6×10^{-3} cm. sec.$^{-1}$. From the distribution of the diatom on July 21 values for the various terms of (30) can be very roughly obtained. It is useless to consider depths below 7.5 m., because (30) gives negative values of A. Actually, there is certainly much variation of turbulence with depth near the lower boundary of the epilimnion because of internal waves or seiches, and the equation is inapplicable. The necessary condition (31) for a steady state, namely $\hat{v}^2 \leq 4A\hat{w}_1$, can be applied, indicating that $A \geq 2.08$. The approximate values given in Table 15 may in part be obtained from Grim's curve or they may be computed from that curve with the aid of data for $\hat{v}$ and $\hat{w}_1$ already given.

The value for A at 4.5 m. is obviously too low; that at 6 m. is possible, though it may be argued that it is arbitrary to accept such a value merely because it lies between the impossible and the meaningless. The difficulty undoubtedly arises from the inaccuracy of the estimates of $\partial^2 \hat{p}/\partial z^2$. The example, however, is presented mainly to indicate how the theory can be applied, at least potentially, to an actual case.

It is to be noted that a diatom sinking at a rate of about 10×10^{-3} cm. sec.$^{-1}$, a value suggested as possible by Fritz's experiments (cf. also Margalef 1961), and reproducing at the same rate as the *Tabellaria* actually present could not form a stable population under conditions permitting the observed stable population of that species. For the hypothetical species, $\hat{v}^2 \simeq 10^{-4}$ and $4A\hat{w} = 8.4 \times 10^{-6}$. If populations of both *Tabellaria* and the hypothetical fast-sinking diatom had developed earlier in the season at a higher value of A of 30 cm.^{2} sec.$^{-1}$ or more, the hypothetical species would sink out of the water as the turbulence declined, leaving a steady-state population of *Tabellaria* in possession of the lake. It is quite possible that much of the seasonal succession of the phytoplankton is due to this interrelation between turbulence and sinking speed.

Apparent sinking speeds. In general, the eddy diffusivity in a lake will vary greatly with the depth and will decline rapidly at the lower boundary of the trophogenic zone. It is therefore evident that although the processes just described can account qualitatively for the kind of distribution of phytoplankton that is often observed, in practice they

TABLE 15. *Application of the theory of Riley, Stommel, and Bumpus to the* Tabellaria *population in the Zeller Basin of the Untersee below Lake Constance*

	300	450	600
$\hat{z}$ (cm.)	300	450	600
$\hat{p}$ (cells cm.$^{-3}$)	1000	900	700
$\dfrac{\partial \hat{p}}{\partial z}$ (cells cm.$^{-3}$ cm.$^{-1}$)	0	-1.0	-1.5
$\hat{w}\hat{p}$ (cells cm.$^{-3}$ sec.$^{-1}$)	0.81×10^{-3}	0.73×10^{-3}	0.57×10^{-3}
$\hat{v}\dfrac{\partial \hat{p}}{\partial z}$ (cells cm.$^{-3}$ sec.$^{-1}$)	0	-2.6×10^{-3}	-3.9×10^{-3}
$A\dfrac{\partial^2 \hat{p}}{\partial z^2} = v\dfrac{\partial \hat{p}}{\partial z} - \hat{w}\hat{p}$	-0.81×10^{-3}	-3.3×10^{-3}	-4.5×10^{-3}
$\dfrac{\partial^2 \hat{p}}{\partial z^2}$ (cells cm.$^{-3}$ cm.$^{-1}$ cm.$^{-1}$)	?	-5×10^{-3}	-1.7×10^{-3}
A (cm.^{2} sec.$^{-1}$)	—	0.66	2.6

must be greatly modified by other factors.   In this respect Riley, Stommel, and Bumpus' theoretical phytoplankton curves with maxima in the deeper part of the trophogenic zone bear the same sort of relation to actual curves of like form in lakes as Munk and Anderson's (1948) calculated temperature curves do to actual temperature curves in stratified lakes.

The most striking demonstration of the effect of the vertical variation of turbulent diffusion is provided by Grim's calculations of the apparent sinking speeds of certain diatoms in the various depth zones of Lake Constance (Table 16).

The sinking speeds in the hypolimnion presumably approach the values to be obtained in still water.   In the upper levels the sinking velocities decrease in spite of the decrease in viscosity in the epilimnion. This must be caused by the continual upward return of part of the sinking population by vertical turbulent mixing.

Tabellaria was found to sink in the Zeller Basin of the Untersee at a mean apparent rate of 1.5 m. day^{-1} in the top 20 m. or 1 m. day^{-1} in the top 7 m., rates of the same order of magnitude as those of the other species in Lake Constance (Grim 1951).

Gessner (1948) has pointed out that if the thermocline lies high in a lake so that a significant part of the trophogenic zone lies in the region of low turbulence below the epilimnion there is likely to be a progressive loss of plankton by passive sinking throughout the summer and the whole productivity of the lake may be reduced.   This condition is to some extent realized by the Königsee, compared with the Würmsee in Bavaria.   In the Würmsee, in which the thermocline lies at a depth of 15 to 20 m., the chlorophyll content below unit area declines little

TABLE 16.   *Apparent sinking speeds of diatoms in Lake Constance*

	Synedra acus delicatissima		Fragilaria crotonensis	Cyclotella spp. small dead cells
	main population	residual small cells		
10 m.	0.5 m. day^{-1}			
15 m.	—	} 0.4–0.5 m. day^{-1}	1.2–1.5 m. day^{-1}	2.5 m. day^{-1}
20 m.	1.0–1.5 m. day^{-1}			3.0 m. day^{-1}
25 m.}	4–5 m. day^{-1}			3.0 m. day^{-1}
30 m.}				3.5 m. day^{-1}
50 m.		} 0.7 m. day^{-1}	3 m. day^{-1}	5.0 m. day^{-1}
100 m.}	20 m. day^{-1}	3 m. day^{-1}	5 m. day^{-1}	7.0 m. day^{-1}
200 m.}			7 m. day^{-1}	8.0 m. day^{-1}

during the summer and the median point in the distribution of phytoplankton is always above 20 m. and relatively constant. In the
Königsee, in which the thermocline lies between 3 and 8 m. and the
median of the plankton lies between 25 and 40 m., the chlorophyll
content falls throughout the summer (Fig. 82). Further analysis of
these cases would have to follow the theoretical conception of Riley,
Stommel and Bumpus (1949).

Special relationships between water movements and sinking velocity.
An interesting special case has been investigated theoretically by Stommel
(1949), who considers the movement of plankton falling through the
wind-induced elongate rotating cells that Langmuir has shown are
formed by the action of the wind on water surfaces. Consider the
movement of a particle in such a rotating cell, the cross section of which
corresponds to the plane of the paper in Fig. 83. For the purpose of
the present discussion, the movement down wind perpendicular to the
plane of the paper need not be considered and the position of the
particle may be specified by the coordinates (x, z), x being measured
horizontally across the paper and z, vertically from the surface following the usual convention. The horizontal velocity of the particle is
therefore dx/dt the vertical dz/dt. Let the fluid motion of the water
in the cell be specified by a stream function of such a nature that the
horizontal velocity is given by $\partial \Psi / \partial z$ and the vertical by $-\partial \Psi / \partial z$. If
the velocity of passive sinking of the particle is $\hat{v}$, then

$$(35) \qquad \frac{dx}{dt} = \frac{\partial \Psi}{\partial z}; \qquad \frac{dz}{dt} = -\frac{\partial \Psi}{\partial x} - \hat{v}$$

The trajectory of any particle may now be specified by a function Ψ_1,
where

$$(36) \qquad \frac{dx}{dt} = \frac{\partial \Psi_1}{\partial z}; \qquad \frac{dz}{dt} = -\frac{\partial \Psi_1}{\partial x}$$

whence

$$(37) \qquad \Psi_1 = \Psi + \hat{v}x$$

the trajectory being a curve along which Ψ_1 is constant. The simplest
stream function to consider is, in the case of a convection cell,

$$(38) \qquad \Psi = \Psi_0 \sin x \sin z$$

whence

$$(39) \qquad \frac{\Psi_1}{\Psi_0} = \sin x \sin z + \frac{\hat{v}x}{\Psi_0}$$

Convergence   Divergence   Convergence

FIGURE 83. Behavior of three different types of plankton organism in a system of Langmuir spirals. The paths represent trajectories of plankton organisms. *A*. Sinking speed of organism just sufficient to permit descent everywhere in the system; organisms such as blue-green algae, passively ascending at the same speed, would follow the reversed trajectories but on reaching the surface would collect in the convergences as wind rows. *B*. Sinking speed half that of *A* so that part of the population is continually swirled round in a region of retention *centered on the intersection of the vertical plane of the divergence* and the horizontal plane of the axes of rotation of the spirals (Stommel). *C*. Organisms

when $\hat{v} >> \Psi_0$, the trajectories are nearly vertical lines, but as $\hat{v}$ approaches Ψ_0 they become more and more curved. When $\hat{v} = \Psi_0$, all the particles just settle out, taking the paths shown in *A* of Fig. 83. When $\Psi_0 > \hat{v} > 0$, the particles, though always falling relative to the circumambient water, may be rising in relation to the surface of the lake. In such a case the closed trajectories shown in *B* of Fig. 83, drawn for $\hat{v} = 0.5 \Psi_0$ are developed. Ideally, if particles with a sinking velocity less than Ψ_0 are distributed at random through the system, those that enter the parts of the system in which the trajectories are open will fall out of the cells but those that fall into the regions of retention will circulate indefinitely. In such regions of retention the concentration of the particles will be greatest in the middle in the vertical planes of the divergences, and will fall off as the boundary of the region of retention is approached. If particles of different falling speeds are involved, those with smaller sinking speeds will tend to be retained in a greater space than those with greater sinking speeds. In actual practice there is sure to be enough turbulent displacement for the accumulations of particles in the regions of retention to be slowly dissipated. When $\hat{v}$ is negative and the particles tend to float the whole process may be inverted, so that if $0 > \hat{v} > -\Psi_0$ accumulation in regions of retention are possible but not if $\hat{v} \leq -\Psi_0$. Usually such particles would collect in convergences as windrows (Fig. 84).

Stommel notes that Neess has observed that plankton samples taken in a lake by towing up- or downwind are often more variable than those taken by towing across wind. He concludes that up- or downwind the sampling of areas of retention and intervening areas from which the plankton has sunk is likely to be far less perfect than when the net is drawn across the cells indiscriminately.

Verber, Bryson, and Suomi (1949) have considered a second special case of plankton falling or rising in a column of water in an ascending or descending current, as in a divergence or convergence. The rate of passive descent $\hat{v}$ is not regarded as constant but as a function of the

---

actively seeking to maintain themselves at a certain position at, or in the figure, somewhat below the surface and just able to reach that position from any point below it, the lower parts of the trajectories starting as the reverse of those in *A*, but in the upper part progressively departing from this pattern according to the phototaxis or other regulatory behavior, producing Ragotskie and Bryson's *accumulation in the convergence*. All three species could be present simultaneously, though that of *A* would tend to disappear by sinking.

FIGURE 84.   Windrows of *Gloeotrichia* on Lake Eken.   (Fogg.)

viscosity, itself dependent on the temperature at depth z.   Considering only passive movement,

$$(40) \qquad \frac{d\hat{p}}{dt} = \frac{\partial(\hat{v}\hat{p})}{\partial z} = \hat{v}\frac{\partial \hat{p}}{\partial z} + \hat{p}\frac{\partial \hat{v}}{\partial z}$$

If the velocity of the vertical current is w, the quantity of plankton delivered into unit volume by the current will be $w\ (\partial \hat{p}/\partial z)$; as before, biological processes may be regarded as proportional to the population and be written $\hat{w}\hat{p}$.   Thus the rate of change of the concentration as influenced by all forces is

$$(41) \qquad \frac{d\hat{p}}{dt} = \hat{w}\hat{p} + \hat{p}\frac{\partial \hat{v}}{\partial z} + v\frac{\partial \hat{p}}{\partial z} + w\frac{\partial \hat{p}}{\partial z}$$

Neglecting the biological factors and considering the events of the top centimeter of a lake containing colonies of a bloom-forming algae, Verber, Bryson, and Suomi estimate the following values as reasonable. For a bloom-forming, colonial, blue-green alga such as *Anacystis*,

$\hat{p}$ 5 colonies cm.$^{-3}$

$\dfrac{d\hat{p}}{dz}$ 1 colony per cm.$^{-3}$ cm.$^{-1}$

$\hat{v}$ from Fritz (1935) very roughly of the order of 0.001 cm. sec.$^{-1}$

$\dfrac{d\hat{v}}{dz}$ from Fritz results on diatoms falling at different temperatures and

   from assumption of a temperature gradient of 0.2° C. cm.$^{-1}$, very roughly of the order of 10^{-5} cm. sec.$^{-1}$ cm.$^{-1}$

w from observations on Lake Mendota can be the order of 0.05 cm. sec.$^{-1}$

It is clear that the rate of increase of the bloom at the surface is determined almost entirely by the rate of upward movement of water and the rate of vertical movement of the algae and that the variation in the rate of movement due to viscosity changes is of little importance. It is evident under some circumstances that although a considerable quantity of algae might be brought to the surface by water movement there must be compensatory horizontal streaming to remove the water from the site of upwelling; such movements are likely to be far from negligible in removing the algae from the site of upwelling to other parts of the surface of the lake. If, however, as in the situation discussed in the next paragraph, the organisms in the upwelling water can maintain a particular depth, lasting local concentration is possible.

Ragotzkie and Bryson (1953) have considered such an organism, which by means of phototactic responses maintains a given small depth regardless of moderate vertical water movements (C of Fig. 83). It is evident that at a convergence the organisms will be concentrated, for they can be moved passively horizontally, but not vertically, so that they are not removed with the down-sinking water. Similarly, at a divergence the plankton at any given depth below the surface but still in the euphotic zone will be depleted. The distribution pattern achieved will, within a pair of rotating Langmuir spirals, tend to be the opposite of that predicted by Stommel's theory based only on sinking speed. Presumably different organisms may exhibit different responses, whether sinking or maintaining a fixed position. Moreover, it would be reasonable to expect different types of behavior by day and night. Other more empirical aspects of life in circulation cells are discussed later (see page 796–797); the entire subject is likely to develop considerably in the next few years.

SINKING SPEEDS AND NUTRIENT UPTAKE

The whole of the preceding discussion is based on the premise that the first requirement of a phytoplanktonic organism is to remain afloat. This is a gross oversimplification. If an autotrophic organism were to remain suspended at rest in an undisturbed body of water, it would rapidly utilize the nutrients dissolved in its immediate vicinity. The rate of division of such a cell would depend on the rate at which nutrients could diffuse from the main mass of water far from the cell into the impoverished shell of water in its immediate vicinity. If, however, the cell started to sink through the water, it would continually encounter regions from which nutrients would not have been removed. Within the euphotic zone, therefore, a nonzero sinking speed is advantageous to any organism not capable of spontaneous movement.[6] The advantage will be offset by losses into the region below the euphotic zone, and the upper limit for the existance of a steady-state population will, as we have seen, be set by the condition that $\hat{v}^2 = 4A\hat{w}_1$.

The problem has been examined by Munk and Riley (1952). The approach employed is essentially that developed for the study of the heat exchange between small objects and a streaming fluid medium in which they are immersed. The theory is unexpectedly difficult, but in some special cases experimental studies are available that extend, and to some extent confirm, the theoretical treatment. Results relating to heat transfer, of course, are easily applied to the problem of the diffusion of solutes.

Munk and Riley consider the absorption time τ_n to be the time taken for a cell containing n' g. g.$^{-1}$ of a limiting nutrient to absorb an equal quantity of this nutrient from a medium containing n g. g.$^{-1}$ in excess of the concentration at the external cell surface, here taken as zero. Since such absorption will be a rough condition for cell division, τ_n^{-1} will be a measure of the division rate.

Then, for cells of four different standard forms, we have the following expression:

Circular disk, diameter d, thickness md, set horizontally

$$(42a) \qquad \tau_n^{-1} = \frac{n}{n'}\left(\frac{\rho' - \rho}{\rho}\right)^{2/3} \tau_0^{-1} m^{-1} \, \mathbf{F}_d$$

Sphere, diameter d,

$$(42b) \qquad \tau_n^{-1} = \frac{n}{n'}\left(\frac{\rho' - \rho}{\rho}\right)^{2/3} \tau_0^{-1} \, \mathbf{F}_s$$

Cylinder, diameter d, length indefinite,

$$(42c) \qquad \tau_n^{-1} = \frac{n}{n'}\left(\frac{\rho' - \rho}{\rho}\right)^{2/3} \tau_0^{-1} \, \mathbf{F}_c$$

Rectangular plate, with d, thickness md, long axis inclined $\theta°$ to horizontal,[7]

$$(42d) \qquad \tau_n^{-1} = \frac{n}{n'}\left(\frac{\rho' - \rho}{\rho}\right)^{2/3} \tau_0^{-1} (m^{-2} \sin \theta)^{1/2} \, \mathbf{F}_\rho$$

---

[6] Negative sinking speeds, balanced by turbulent mixing, doubtless occur in the blue-green algae.

[7] This assumes no motion when the plate is horizontal. When vertical, the expression (43d) becomes isomorphous with the others, but the corresponding expression for the velocity does not. This case is not really comparable with the other three.

where

(43)
$$\tau_0 = \left(\frac{\mu^2}{g^2 K}\right)^{\frac{1}{3}}$$

which, assuming that the diffusivity of the nutrient K is 2.10^{-5} cm.^{2} sec.$^{-1}$ and the viscosity μ is 0.01 cm.^{2} sec.$^{-1}$, may be taken as 57.7.

The four functions F_d, F_s, F_c, and F_p, which give absorption rates relative to the other parameters, can be obtained from the theory of heat transfer and from experiments in terms of a relative length d/d_0, where

(44)
$$d_0 = \left(\frac{72\mu K}{g}\right)^{-\frac{1}{3}} \left(\frac{\rho' - \rho}{\rho}\right)^{-\frac{1}{3}}$$

or for water at 20°C., $0.24 \times 10^{-2}(\rho' - \rho)^{-\frac{1}{3}}$cm. The functions are plotted on a double logarithmic grid in Fig. 85. The upper part of the curve for a very small sphere is derived from theory, which, however, is in part defective; it appears to be continuous with the experimental results for a larger sphere. The curve for the disk is theoretical but cannot be safely extrapolated to the more interesting larger diameters. The curve for the cylinder is purely empirical, being based on heat loss from thin wires in flowing fluid. Only in the flat plate are theory and experiment concordant, and here the most interesting case, of fall in a horizontal position, is excluded.

The slope of the line for a plate is -1; the absorption rate is doubled as the width of the plate is halved. For a cylinder the slope is -1.2, for a large sphere, -1.1, and for a disk or a small sphere, -2. Thus halving the diameter more than doubles the absorption rate for a large sphere or cylinder and multiplies it by four for a small sphere or disk.

Munk and Riley now proceed to set up several possible models. In the first (A of Fig. 86) the relative excess density $(\rho' - \rho)/\rho$ is taken as 0.0001, corresponding to a nearly negligible sinking speed. This is believed to apply roughly to the conditions in actively dividing well-fertilized cultures of marine diatoms and so probably to natural populations in the sea at the time of the rise of a bloom. It is extremely doubtful if such a low relative excess density could characterize a fresh-water diatom in which only oil droplets could reduce the density to a value so close to that of the medium. A condition of this sort, however, may be implied by some of Grim's (1939) observations on *Cyclotella comensis melosiroides* as well as other small species. The limiting nutrient is taken to be phosphorus at a concentration of 1 μg-atom per liter, or 3.1×10^{-8} g. g.$^{-1}$ in the medium and 0.0001 g. g.$^{-1}$ in the cell. The absorption rates are plotted against d in A of Fig. 86,

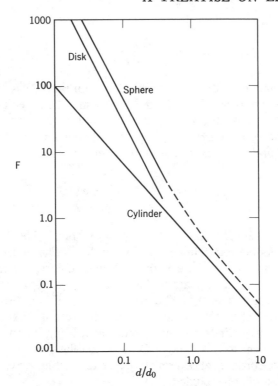

FIGURE 85. Relative absorption rate F, as a function of the relative diameter d/d_o, where, for water, d_0 is given by $0.0024(\rho' - \rho/\rho)^{-\frac{1}{3}}$. The disk and sphere are treated theoretically for small diameters and the sphere, empirically for large diameters. The cylinder is known empirically primarily from heat exchange studies of wires. (Munk and Riley.)

the plate being taken as five times as wide as deep and the disk as three times as wide as deep. The plate is supposed to fall at an angle of 10 degrees with the horizontal. In this example the upper limit of d for one division per day, assuming that doubling the limiting nutrient content causes division, varies from 76 μ for the width of the plate to 300 μ for the diameter of the disk. More than 90 per cent of marine diatoms have relevant dimensions within these limits; as far as can be determined from Huber-Pestalozzi and Hustedt (1942), the largest fresh-water diatom that can be regarded as planktonic is *Campylodiscus tanganicae,* a disk-shaped organism with a diameter as much as 256 μ.

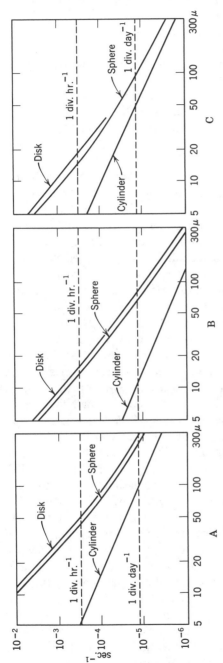

Figure 86. Rates of absorption for bodies of various shapes and sizes. The ordinate gives the reciprocal of the time needed to double the content of a critical material and so permit cell division. The points at which the solid curves cut the broken lines gives the upper limit of the diameter d, permitting one division per hour, or per day, under the conditions of the three graphs. These figures represent a somewhat more concrete expression of the theory than Figure 85. They require concentrations of a limiting factor, assumed for argument to be phosphorus, and a relative density difference to be set. *A.* $(\rho' - \rho)/\rho = 0.0001$, phosphorus limiting, at 1 mg. at m.$^{-3}$ or 3.1×10^{-8} g. cm.$^{-3}$; maximum diameter permitting 1 division per day, 105 μ for the cylinder, 260 μ for the sphere, and 300 μ for the disk three times as wide as thick. *B.* $(\rho' - \rho)/\rho = 0.0001$, phosphorus limiting, at 3.1×10^{-9} g. cm.$^{-3}$; maximum diameter permitting 1 division per day, 10 μ for the cylinder, 61 μ for the sphere, and 65 μ for the disk. *C.* $(\rho' - \rho)/\rho = 0.02$, phosphorus limiting, at 3.1×10^{-9} g. cm.$^{-3}$; maximum diameter permitting 1 division per day, 44 μ for the cylinder, and 95 μ for the sphere, with the disk near the latter figure. Munk and Riley believe that the ocean at times may approximate *A*, in freshwaters the larger density difference, and lower phosphorus concentrations suggest conditions between *B* and *C*.

In a second model the external nutrient concentration is reduced to one tenth of that in the first, the relative excess density being the same; this merely increases the absorption time by a factor of ten. For a plate or cylinder the width must not exceed 10 μ, and for a sphere or disk the diameter must be less than 61 and 65 μ, respectively, to permit one division per day (B of Fig. 86).

In a third model the relative excess density is 0.02, with the low nutrient concentration of the second model; the maximum dimensions permitting one division per day vary from about 50 to 100 μ. The disk cannot be treated in this case, but it is unlikely to differ much from the sphere (C of Fig. 86).

Munk and Riley consider that the conditions of the second model are unrealistic and that larger species than indicated by the model occur in the sea under conditions of severe nutrient depletion. In fresh waters, in which the relative excess density is certainly ordinarily greater than 0.0001, the limiting nutrient concentration is usually less than 3.1 mg. P m.$^{-3}$, as in this model. Some intermediate condition between the second and third models therefore is quite likely to approximate the usual situations in lakes. It is interesting to note that in the fresh-water diatoms that can be treated as disks, in the genera *Cyclotella, Stephanodiscus* (most species), and *Campylodiscus,* twenty-two of the thirty-two regarded as planktonic by Huber-Pestalozzi and Hustedt have maximum diameters not in excess of 50 μ; the few very large forms in *Campylodiscus* appear to be of rather uncertain ecology. Among the cylinders, mostly in *Melosira* and *Synedra,* the proportion with a maximum width under 10 μ is, of course, small though by no means negligible; they occur particularly in *Synedra,* whereas there is every reason to suspect that the relative excess density in *Melosira* is greater (pages 458–462) than in most other fresh-water plankton diatoms. It is therefore evident that the second of Munk and Riley's models, or some intermediate between the second and third model, would by no means be unrealistic for many fresh-water populations.

Munk and Riley have computed the sinking speed $\hat{v}$ for the various forms that they consider. Then, by using the condition defined by equation (31), it is possible to calculate the theoretical production coefficient ($\hat{w}'$) required to balance loss by sinking at any value of A. From the absorption time the upper limit of the rate of division under any specified conditions can also be computed. It is assumed, however, that no cell divides more than twice a day and that this division can take place only in the top fifth of the euphotic zone; beneath this level the rate falls off exponentially. The other conditions are those of the third model, with the relative excess density 0.02 and

TABLE 17. Production coefficient $\hat{w}$, calculated for a falling phytoplankton cell of relevant linear dimension d, of excess density 0.02, in medium containing 3.1 mg. m.$^{-3}$ of a limiting nutrient, compared with the production coefficient $\hat{w}'$ needed to maintain a stable population at two different values of A, the coefficient of turbulence. Below the horizontal bars no stable population can exist (Munk and Riley 1952).

	Sphere			Cylinder			Plate		
	$10^6\hat{w}$	$10^6\hat{w}'$ $A = 10$ cm.^{2} sec.$^{-1}$	$10^6\hat{w}'$ $A = 50$ cm.^{2} sec.$^{-1}$	$10^6\hat{w}$	$10^6\hat{w}'$ $A = 10$ cm.^{2} sec.$^{-1}$	$10^6\hat{w}'$ $A = 50$ cm.^{2} sec.$^{-1}$	$10^6\hat{w}$	$10^6\hat{w}'$ $A = 10$ cm.^{2} sec.$^{-1}$	$10^6\hat{w}'$ $A = 50$ cm.^{2} sec.$^{-1}$
d									
5	9.2	1.8×10^{-5}	3.6×10^{-6}	9.2	4.2×10^{-3}	8.4×10^{-4}	9.2	5.6×10^{-2}	1.1×10^{-2}
10	9.2	3.0×10^{-4}	6.0×10^{-5}	9.2	4.9×10^{-2}	9.8×10^{-3}	9.2	2.2×10^{-1}	4.4×10^{-2}
20	9.2	4.8×10^{-3}	9.6×10^{-4}	9.2	5.1×10^{-1}	1.0×10^{-1}	9.2	9.0×10^{-1}	1.8×10^{-1}
30	9.2	2.5×10^{-2}	5.0×10^{-3}	8.9	2.2	4.4×10^{-1}	7.8	2.0	4.0×10^{-1}
40	9.2	8.1×10^{-2}	1.6×10^{-2}	7.3	$\underline{5.6}$	1.2	6.6	3.6	7.2×10^{-1}
50	9.2	2.0×10^{-1}	4.0×10^{-2}	6.1	11	2.2	5.6	$\underline{5.6}$	1.1
60	9.2	4.2×10^{-1}	8.4×10^{-2}	5.3	20	$\underline{4.0}$	4.9	8.1	1.6
70	8.0	7.8×10^{-1}	1.6×10^{-1}	4.7	32	6.4	4.4	11	2.2
80	7.0	1.3	2.6×10^{-1}	4.2	48	9.6	4.1	14	2.8
90	6.3	2.2	4.6×10^{-1}				3.8	18	$\underline{3.6}$
100	5.6	$\underline{3.3}$	6.6×10^{-1}				3.5	22	4.4
150	3.7	17	$\underline{3.4}$				2.6	50	10
200	2.9	58	12						

299

the phosphate or limiting nutrient at 3.1 mg. m.$^{-3}$ The results are given in Table 17.

Table 17 shows the great advantage of a sphere over a cylinder of the same diameter when the interaction of nutrient uptake, sinking speed, and turbulence are taken into account. It is probable that a disk would behave essentially like a sphere in this respect. Munk and Riley assume that when $(\hat{w} - \hat{w}')$ is positive a stable population could be maintained by predation on this excess production. They further assume that a sphere of diameter $3d$ is as easily eaten as a long cylinder of diameter d or a plate of width d and thickness $0.2d$. Then, for any size categories, equivalent in this way so far as predation is concerned, we can obtain the relative predation rate. Thus, for $d = 90$ μ for a sphere or 30 μ for a cylinder or plate and $A = 10$ cm.^{2} sec.$^{-1}$, we find

$(\hat{w} - \hat{w}')$ sphere	4.1	cylinder	6.7	plate	5.8	mean 5.5
	5.5		5.5		5.5	
excess over mean	-1.4		1.2		0.3	

If we assume that the predation relation postulated is approximately correct, these excesses give a measure of the success of different shapes for different values of d in coping with nutrient uptake, sinking, turbulent redistribution, and predation. It will be noted that as a result of the assumptions about predation the sphere, and also probably the disk have lost their advantage over the long thin forms.

The resulting values are plotted against d for $A = 10$, a reasonably high value for a small body of fresh water, for the cylindrical shape in Fig. 87, and are compared with the relative distributions of size in marine centric diatoms of cylindrical form and in fresh-water centric diatoms of the same form.[8] It is evident that though the diameters of marine diatoms appear rather smaller than might be expected from Munk and Riley's model the fresh-water forms tend to have still smaller diameters, in line with the lower values of the turbulence coefficients expected in lakes. Munk and Riley also made similar curves for the spherical and plate-like diatoms. These curves are not reproduced, since a really comparable selection of fresh-water species proved impractical. In both cylinders and spheres they found that a few marine

---

[8] If the cylindrical pennate form were included, the curve would be displaced even more to the left by the various more or less cylindrical species of *Synedra,* but the shapes of pennate diatoms are so variable that too great an arbitrary element seemed to be involved in including a selection from the pennate genera. Munk and Riley prepared their frequency diagram of sizes somewhat differently; the same source (Cupp 1943) has been used for marine species.

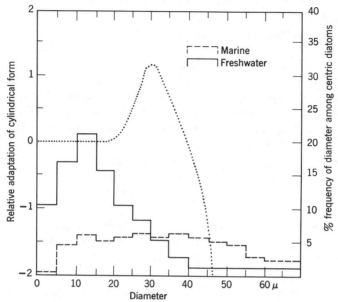

FIGURE 87. Dotted line, relative adaptive value of elongate cylindrical form in marine diatoms as a function of diameter, according to the theoretical calculations of Munk and Riley. Broken line, histogram of diameter frequency in marine genera, solid line, histogram of fresh-water genera. The modal diameter for marine species embraces the theoretical mode, and the distribution is slightly skewed with very few, very small, and relatively rather more very large species. The modal diameter for fresh-water species is clearly smaller and the distribution is skewed in the opposite sense to that in the marine taxocene.

species are impossibly large according to the theory. These are mainly members of *Rhizosolenia* and *Coscinodiscus* and are believed to have abnormally low specific gravities and nutrient contents because of the presence of very large vacuoles. If they contained large vacuoles of liquid of nutrient content and specific gravity identical to that of the medium, their existence would be possible. In fresh water it is very unlikely that the vacuolar contents could have a specific gravity as low as that of the medium, so that the apparent absence of this aberrant class of large species is reasonable. The existence of abnormally large diatoms in the sea is presumably due to selection favoring large size, if it is permissible on other grounds, thus making predation by small predators difficult. In fresh water this type of evolution is probably not able to go to such extremes as in the sea. In all other ways the smaller diatoms are likely to have an advantage over the large.

It must not be forgotten that vertical sinking, turbulent redistribution, and diffusion are not the only possible mechanisms that can bring a cell into contact with a nutrient molecule. Motility of the cell will have the same effect. In any case in which the vertical component of the path is not downward, which is quite possible in phototactic forms, motility is more advantageous than passive sinking. When passive sinking is uninterrupted by turbulent water movements, it is likely to be fatal. It will be pointed out later (page 349) that motility may be significant in precisely this way to auxotrophic flagellates swimming in unproductive waters deficient in accessory organic nutrient compounds.

SUMMARY

The density of most fresh-water planktonic organisms is 1.01 to 1.03 times that of the medium in which they float. Diatoms without gelatinous sheaths and benthic animals are denser. The coefficient of thermal expansion of organisms is almost entirely unstudied and urgently requires investigation. A limited number of organisms, namely, most planktonic blue-green algae, some rhizopod protozoa, and the mites of the genus *Hydrozetes,* can by producing bubbles reduce their density below that of the medium. Teleost fish can achieve quite accurate equilibration by means of the swim bladder, and some aquatic insects can achieve a comparable degree of adaptation by controlling the volume of internal gas-filled hydrostatic organs (*Chaoborus*), of gas bubbles held by pubescent plastra (Elmidae) or under the elytra (*Anisops, Buenoa*).

Most planktonic organisms, because of their slight positive excess density, sink slowly when placed in undisturbed water. Theoretical studies coupled with a consideration of purely physical experiments suggest that biological objects of mean diameter not more than 0.5 mm. will fall according to Stoke's law, the velocity of sinking rapidly reaching a value that varies inversely with the viscosity of the medium, directly with the excess density of the object over the medium, and directly as the square of the diameter or of some appropriate linear dimension. Bacteria, most planktonic algae, and nearly all rotifers would be expected to behave in this way, though the influence of Brownian movement cannot be neglected in dealing with the smallest bacteria. There is little experimental evidence bearing on the sinking of these groups; such as is available for rotifers and for the smallest crustacea is in good accord with theory, but the data on algal settling are in rather poorer agreement, probably because of the difficulty of maintaining the medium undisturbed during the long periods required for the experiment. It is possible also that the viscosity to be considered when very small cells are sinking should be a structural viscosity, modified by the chemical and electrostatic properties of the cell surface which could control the arrangement of water molecules in the vicinity of the cell.

Stokes's law in its complete form was derived for a sphere. Cubes, tetrahedra, and spheroids of moderate ellipticity do not diverge greatly from spheres of similar volume in their sinking speeds. The coefficient of form resistance or the reciprocal of the ratios of the sinking speeds of any object to that of a sphere of the same density and volume varies from 2.9 to 4.3 for a disk one hundred times as wide as it is thick, according to whether it is edge on or broadside on. In figures of moderate eccentricity a prolate spheroid falls more slowly than an oblate of the same volume, both set to encounter maximum resistance. A cylinder can always be constructed to fall more slowly than a disk or sphere of similar volume but it may have to be very long. These conclusions coincide with the frequency of long filamentous and acicular forms in the plankton, but in particular cases special factors, notably the uptake of nutrients, may often make other forms more efficient. Surface roughness or microsculpture on the cell walls or exoskeletons of planktonic organisms is unlikely to have a significant effect in reducing sinking speeds when the organisms are small enough to obey Stoke's law.

Bodies symmetrical about three axes at right angles and about the center of gravity, if small enough to obey Stokes's law, have no preferred orientation and fall in the position in which they are set. A very small asymmetry, however, will cause such a body to take up a preferred position which corresponds to the minimal velocity of descent unless the center of gravity is markedly eccentric. It is suggested that curvatures and other slight asymmetries in nonmotile phytoplankton cells may sometimes act as adaptations to reduce the sinking speed by ensuring fall in the position of minimal velocity.

The evidence that exists from experiments on organisms 1 or 2 mm. in diameter, such as moderate-sized specimens of *Daphnia,* suggests that instead of following Stokes's law the velocity is nearly directly proportional to the linear diameter, directly proportional to the cube root of the square of the excess density, and inversely proportional to the cube roots of the density and the viscosity of the medium. These relationships are known to occur over a narrow range of Reynolds' numbers greater than unity. Viscosity changes due to ordinary seasonal changes in temperature obviously will not affect such organisms very much.

Gelatinous capsules can reduce the sinking speed of an organism, provided that the difference in the density between the organism and capsule is at least twice the difference in density between the capsule and medium.   The production of such gelatinous capsules is apparently a common adaptation to flotation and is likely to be of particular significance to diatoms because of their high density.   The capsule is no doubt disadvantageous in limiting the uptake of dissolved nutrients.

Experimental studies have been made on the effectiveness of spines and comparable processes on the sinking speeds of certain planktonic animals.   In *Brachionus quadridentatus* the presence of the large posterior spines reduces the sinking speed about 18 per cent.   In *Daphnia pulex* the opening of one second antenna caused a reduction of about 20 per cent in the sinking speed and of both antennae about 30 per cent.   Removing the lateral hairs of both second antennae of *D. schødleri* increases the sinking speed by 22 per cent.

Experiments and observations in lakes confirm the conclusion that those members of the phytoplankton that are not less dense than water sink continuously, and the association as a whole depends on turbulent water movements to maintain its position.   It is possible to develop a deductive theory for a population of phytoplankton reproducing and sinking in a stratified body of water.   By making certain reasonable simplifying assumptions it can be shown that the interaction of turbulence, sinking, and reproduction produces a distribution in which there is a minimum in the population at the surface and a maximum in the lower part of the epilimnion.   This distribution agrees qualitatively with what is often observed in nature.   It can be shown that conditions under which a steady-state population of one species may exist do not necessarily imply the same possibility for a population of another faster sinking species.   When turbulence falls in the spring as the lake warms and becomes stratified, such a process would limit the specific composition of the phytoplankton that could form stable populations.

Although it is ordinarily assumed that the sinking of a nonmotile planktonic organism is disadvantageous, this is not necessarily true in turbulent water in which the sinking permits more rapid nutrient uptake and so faster division than would be possible for a stationary cell.   Spherical and flat discoidal forms apparently are at a considerable advantage over cylindrical forms when nutrient uptake by an organism sinking in turbulent water is considered.   This advantage, however, may be offset by a liability to predation.   Small size is metabolically advantageous unless some device for reducing excess density is present;

large size, however, may be an adaptation against predation by small predators. The largest marine diatoms, much larger than any in fresh waters, apparently can exist only by virtue of special mechanisms for reducing excess density; such mechanisms are not available in fresh waters. Fresh-water planktonic diatoms of a given shape tend to be smaller than the marine species, even though there is much overlap. This is to be expected from theory, in that the conditions for large size at any value of the excess density are high nutrient concentration and turbulence. In general, the sea is more turbulent and richer, at least in phosphorus, than the majority of lakes.

Motility by flagellate species in the phytoplankton may confer on its possessors the nutritional advantages of sinking through water without the risk of irreversible descent into the unilluminated and relatively nonturbulent hypolimnion.

CHAPTER 21

The Nature and Distribution
of the Phytoplankton

The detailed consideration of the biology of lakes reasonably begins with the phytoplankton, for this assemblage of organisms constitutes the greater part of the photosynthetic producing level in all but the shallowest lakes. The whole of the rest of the biological community therefore depends to a very large extent on the planktonic plants.

COMPOSITION OF THE PHYTOPLANKTON

The taxonomic enumeration of the phytoplankton of the inland waters of the world has been undertaken by Huber-Pestalozzi (1938, 1941, 1942, 1950, 1955, 1961), whose treatment, at the time of writing, covers all groups except the green algae. The work, which, when complete, will be one of the monuments of limnology, is of particular importance in that it summarizes in single volumes, with reproductions of the original figures, information about many small organisms that are hard to examine and impossible to preserve.

Among the earlier works the various volumes of the *Süsswasserflora Deutschlands* are useful; in particular, those on the Volvocales (Pascher 1927) and the Tetrasporales and Protococcales (Lemmermann, Brunnthaler, and Pascher 1915) are devoted to the green algae not yet treated by Huber-Pestalozzi, as are the two volumes of Smith (1920,

1924) on the phytoplankton of the Wisconsin lakes. The standard works of West and Fritsch (1927), Fritsch (1935, 1945), a superb general account of all algae, and of Smith (1950), who treats all American fresh-water genera, are doubtless properly familiar to the reader. Prescott's (1951) account of the algae of the Laurentian Great Lakes, though it omits the diatoms and, most unhappily, the desmids, is a fine regional flora. Patrick's forthcoming work on the diatoms of North America will be indispensible and is eagerly awaited. The recent second edition of Ward and Whipple's Fresh-Water Biology (Edmondson 1959) gives concise keys to the North American genera and species of blue-green algae (Drouet 1959), to the genera of diatoms (Patrick 1959), and to the genera of other fresh-water algae (Thompson 1959). For the species of desmids the incomplete work of Krieger (1937–1939) and West and West (1904, 1905, 1908, 1912; West, West, and Carter 1923) still hold their own, but the latter cannot safely be used outside western Europe. Smith (1924) gives an important treatment of Wisconsin species. The planktonic forms are still in need of some revision; the papers of Teiling (1948) and Brook (1959a,b) are of great significance. A detailed study by Prescott of the American desmids is anticipated.

OCCURRENCE AND NUTRITION OF FRESH-WATER PLANKTONIC ALGAE

In this chapter only the more significant planktonic algae are considered, chiefly in order that the zoological or geological reader may gain some idea of the taxonomic position of the main planktonic organisms to be mentioned later. As far as possible these forms have been illustrated so that the later discussion can be related to a visible entity and not merely to a named abstraction. As much physiological information as is directly relevant to the occurrence and distribution of the various groups considered is also given. Emphasis has been placed primarily on chemical factors, since the two seasonally variable physical factors, namely light and temperature, are discussed at length in Chapters 22 and 23. Much important information is given by Fogg (1964) in a work that appeared too late to use.

General nutritional requirements. It must be remembered that the cultivation of many species has proved difficult, and many others so far have not been brought into culture. This may mean that some of these species have peculiar chemical requirements, but it is fairly clear, in view of the work of Provasoli and his associates, that the provision of light and sources of C, N, P, S, K, Mg, Si, Na, Ca, Fe, Mn, Zn, Cu, B, Mo, Co, V, and the three vitamins thiamin, cyanocobalamin, and biotin should enable almost any alga to grow if it is possible to find the right concentrations and the right physical environment. With the techniques now available any further trace elements needed are presumably always present as impurities. Toxic effects from excess of most minor inorganic nutrients are clearly as significant as deficiencies, and very little is known of the effect of turbulence and the inevitable unnatural existence of the walls of the culture vessel in promoting or inhibiting algal growth. The great advance made by the use of artificial chelators such as EDTA has permitted the culture at very low ionic concentrations of many very sensitive forms which in nature are presumably dependent on slow diffusion of nutrients from the mud-water interface or on materials chelated by inadequately understood organic compounds in lake waters.

Before presenting the available details, the kinds of problems for which the investigator should be prepared may be briefly discussed.

Nature of carbon sources. For a strict phototroph, dependent only on carbon dioxide in the light for a supply of carbon, there are presumably five possible sources of the compound in lake waters, namely, free CO_2 in solution, H_2CO_3, HCO_3^-, $CO_3^=$ (Volume I, page 660–671) and carbon dioxide as carbamino carboxylic acid complexes. The last-named compounds can be produced from $CaCO_3$ and amino acids (Smith, Tatsumoto and Hood 1960) in alkaline waters containing bicarbonate and have the general form

$$
\begin{array}{c}
\text{H} \quad \text{H} \quad \text{O} \\
| \quad\; | \quad\; \| \\
\text{R—C—N—C} \\
| \qquad\qquad \backslash \\
| \qquad\qquad\quad \text{O} \\
| \qquad\qquad\; / \\
\text{C—O—Ca} \\
\| \\
\text{O}
\end{array}
$$

where R- is the remaining part of a protein, peptide, or amino acid. It is not impossible that at least part of the colloidal $CaCO_3$ recorded by Ohle (Volume I, pages 670–671) is actually complexed calcium carbonate of this kind.

At equilibrium, regardless of the pH, the quantity of free CO_2 is a function of the gaseous phase in contact with the water, its pressure and temperature, and of the concentration of salts in solution. Both from theory and as far as can be ascertained empirically the quantity of free CO_2 present will ordinarily lie between 0.4 and 1 mg. per liter, or roughly 1 and $2.5 \times 10^{-5}M$ per liter. The quantity of H_2CO_3 is likely to be about one hundredth of this amount and probably is not significant in the present context. In ordinary hard waters a quantity of HCO_3^- greatly in excess of the free CO_2 will also be present, and in the waters of the rare, excessively alkaline lakes of $pH > 10.5$, in which, however, some algae may flourish, most of the carbon dioxide will be present as $CO_3^=$. It is obvious that any plant that can take up bicarbonate ions and liberate from them, within its cells, CO_2 available for photosynthesis will have an enormous competitive advantage in hard waters over species that cannot. The phanerogams of hard water that have been studied can do this, whereas the aquatic moss *Fontinalis* cannot (Steemann Nielsen 1946, 1947; Ruttner 1947 1948). Even among plants that can use bicarbonate it is a less effective carbon source per molecule than is CO_2, but the possible vastly greater supply of bicarbonate far outweighs this in hard water at ordinary atmospheric CO_2 pressures. In general, hard water lakes contain a much greater standing crop of phytoplankton than do soft, and this must often imply in nature, as in fertilization experiments, that CO_2 in some form is limiting. There are less data on HCO_3^- utilization by phytoplankton algae than by higher plants. *Chlorella pyrenoidosa* cannot use bicarbonate to a significant degree (Steemann Nielsen and Jensen 1958; Felföldy 1960). The green algae *Scendesmus quadricauda,* at least in young, actively growing cultures (Österlind 1947, 1948), and *Kirchneriella contorta* (Felföldy 1960) can use bicarbonate, whereas *Coelastrum microporum* and *Chlorocloster terrestris* seem to photosynthesize most rapidly at pH 10.7, at which $CO_3^=$ is about three times as abundant as HCO_3^-. Presumably these last two species use carbonate ions (Felföldy 1960). The inability to use bicarbonate is apparently due to low permeability to HCO_3^- or the lack of an active transport mechanism to bring the ion into the cell; at least in the case of *Fontinalis* it is not due to the lack of the carbonic anhydrase system that liberates CO_2 from the bicarbonate ion.

The studies of Smith, Tatsumoto, and Hood (1960) on the significance of carbamine carboxylates call attention, as has just been indicated, to another possible source of CO_2 in alkaline waters. These compounds decompose spontaneously in acidic and also in very alkaline ($pH \geq 13$) solutions but are stable in the pH range ordinarily found in hard water. Though it is not likely that much more than $10^{-5}M$ per liter of amino groups able to form such compounds will be present in ordinary unpolluted lake water, it appears that CO_2 in the form of carbamine carboxylates is far more available for photosynthesis than is CO_2 as such. It is therefore conceivable that in both the sea and in ordinary hard water lakes, in which the pH will be close to optimal for the formation of these compounds, carboxylates may play a significant role. Since it is known that transfer of CO_2 in animal body fluids is mediated by the formation of such substances, it is quite possible that they also play a part in the initial transfer of CO_2 into plant cells. The immense preferential uptake of carbon from $CaCO_3$ complexed as alanine N-carboxylate, over that from free CO_2 in the experiments of Smith, Tatsumoto, and Hood on *Nitzschia closterium, Platymonas* sp, and a marine species of *Chlorella* would be explicable if such a process were occurring.

It is evident that in hard-water lakes free CO_2, bicarbonate, and possibly carbamine carboxylate may all be present as sources of CO_2. Not all plants can use bicarbonate; it is also conceivable that the relative ease of utilization of CO_2 and carboxylate may differ from species to species. The form of CO_2 source present, which will vary with pH and apparently also with the amino nitrogen concentration, may turn out to be of considerable qualitative importance in determining what species of phytoplankton are present in a given locality. This might happen directly or by regulation of the direction of competition.

The ecological importance of obligate and facultative phototrophy or heterotrophy also clearly requires more study, since such phenomena are likely to be involved in controlling the outcome of interspecific competition. Obligate phototrophy presumably involves impermeability or lack of transport mechanisms present in facultative heterotrophs (cf. Wetherell 1958).

Concentration of the medium. Myers (1951) and Provasoli (1958, 1960) have pointed out that the media most successful for the cultivation of fresh-water phytoplankters are generally quite dilute. Thus Chu (1942) found for *Pediastrum boryanum* that the optimal salinity or concentration of all required ions was about 100 mg. per liter, whereas *Chlorella* can grow in much more concentrated solutions. Some species apparently confined to acid water, low in calcium and other inorganic ions, might actually be limited more by total concentration than by hydrogen ion concentration or some particular cation. The ratios of monovalent to divalent cations and of potassium to sodium and magnesium to calcium, however, may prove to be significant. Much additional work is needed.

Limiting nutrients of major significance. It is commonly believed that the quantity of phytoplankton that can develop in any water is more likely to be determined by the concentration of combined nitrogen and of phosphorus than by any other factor. Little attention has been given to the study of the variation of the ratio of total phosphorus to total combined nitrogen; that of the soluble phosphorus to inorganic combined nitrogen in lake waters is, of course, implicit in many published data. Hutchinson (1941) concluded that in Linsley Pond the ratio of inorganic combined nitrogen to inorganic combined phosphorus was, on different occasions, 50–220:1 by mass or 11–49:1 by atomic proportions. In the seston of the lake the ratio was 9.4–25.5:1 by mass or 4.2–11.5:1 by atomic proportions, much as in marine plankton. This would suggest that phosphorus is limiting. However, the stationary concentration of both phosphorus and combined nitrogen in inorganic form in solution is so small that the addition of either potassium nitrate or potassium phosphate produced little increase in the chlorophyll content in a large bottle of lake water hung at the surface, whereas the addition of both substances produced spectacular increases. Hutchinson (1944) concluded that apart from combined nitrogen, phosphorus, and, for diatoms, silica there was little likelihood of other nutrients limiting the phytoplankton in Linsley Pond or other small, moderately hard-water, productive lakes. Thomas (1953), in an extensive study of thirty-two lakes in the Alps in which the uptake of nitrate and phosphate was determined from sterile lake waters inoculated with plankton with and without enrichment with 20 mg. NO^-_3 and 2 mg. PO^{---}_4 per liter, found that only the two elements in question were significantly involved. He concluded that phosphorus tends to be the limiting factor in winter but that nitrate may be in summer. This is in accord with what is known about the seasonal distribution of nitrate (Volume I, page 871). Later work by Thomas (1955), to be considered in relation to lake typology in Volume III, suggests that phosphorus may be the primary limiting factor in unproductive lakes and that in more productive lakes the two elements, varying more or less together, tend to play comparable roles in determining the quantity of phytoplankton. The role of silica in limiting diatom populations has been briefly indicated in Volume I (pages 797–798) and is discussed again in detail in Chapter 22.

A deficiency, or more probably the unavailability, of magnesium was observed by Goldman (1961) in C^{14} fixation experiments with water from Brooks Lake, Alaska.

Qualitative aspects of the forms of nitrogen present: the significance of molybdenum. As already pointed out (Volume I, pages 851–852), there is some evidence that certain green algae and diatoms prefer nitrate and others ammonia as a source of nitrogen. Since it is now known that the reduction of nitrate involves an enzyme system in which molybdenum is an essential constituent, it is possible that the apparent variations were due to variations in molybdenum concentration in the culture media employed. In the Euglenophyta, however, there is certainly considerable variation in the capacity to use various nitrogen sources. Dusi (1932) believed that the six species he studied obtained nitrogen from a complex peptone medium, *E. pisciniformis,* using no other source. *E. deses* could use single amino acids and *E. anabaena* the same organic sources and ammonia. The other species studied could use nitrate but in *E. gracilis* and *E. klebsii,* less efficiently than ammonia. Only *E. stellata* seems to utilize all sources equally well. More recently Provasoli (1958) has indicated that the Euglenophyta as a whole tend not to use oxidized nitrogen. The variation in ammonia content may play a part in determining the incidence of these and some other organisms, but it must be remembered that the species of *Euglena* nearly always need cyanocobalamin and that an appreciable amount of ammonia is likely to be present only when a variety of organic decomposition products, including vitamins, is also available.

The incidence of such blue-green algae as fix molecular nitrogen is likely to be determined in part by the supply of molybdenum, which is required in much greater quantities in nitrogen fixation than in nitrate reduction. Goldman (1960) has obtained clear evidence of a limitation of the phytoplankton production in Castle Lake in the Klamath Mountains by molybdenum deficiency. The chemical data suggest that interference with fixation rather than with nitrate reduction is more probable, though the organisms involved were not isolated. In this particular lake it is possible that the molybdenum deficiency is due to competition with *Alnus* bushes, which have nitrogen-fixing root-nodule bacteria, growing round the shore and taking up the element from superficial groundwater draining into the lake.

Minor nutrients. Iron is probably often taken up from ferric hydroxide (Volume I, page 714); a reserve may be associated with the yellow limnohumic acids. Available iron or other trace metals appear often to be limiting in the sea (e.g., Ryther and Guillard 1959), and by analogy this may also be expected in large deep lakes lacking organic matter in solution; in such lakes, however, the phosphorus content is usually likely to be less than in the open ocean. Clear experimental evidence of the role of iron deficiency in limiting the phytoplankton productivity of marl lakes in Michigan in which calcium carbonate is suspended, presumably in part due to photosynthetic precipitation, has been obtained both in experiments in bottles (Schelske 1962) and in fertilization of entire lakes (Schelske, Hooper, and Haertl 1962). When added as the iron sodium salt of N-hydroxyethylethylene diamine triacetic acid, iron caused a four- or fivefold increase in photosynthetic activity and led, in the whole lake, to a plankton bloom, after which the iron content was reduced below the original concentration in the water. It is probable therefore that the addition of the chelator improved the utilization of the iron already present in the lake. Comparable effects appear to occur in the sea. Copper is also associated with organic matter and in this form is unlikely to be deficient or present in excess. The possible production of specific metal complexing compounds by algae has already been discussed briefly (Volume I page 896). The zinc cycle urgently needs study, particularly in view of Provasoli and Pintner's (1953) observations on *Synura*. Rather surprisingly, Bachmann and Odum (1960) find in various marine algae that the rate of uptake of radioactive Zn^{26} is proportional to gross photosynthetic oxygen production. The element is not accumulated in the dark and the amount retained by the plant is proportional to net photosynthetic productivity. No progress appears to have been made in the study of the forms of manganese present in lake water since Harvey's work discussed in Volume I (page 805). The small amount of information available about the boron and vanadium requirements of algae is considered in greater detail in later paragraphs. The need for cobalt is intimately associated with the production and utilization of vitamin B_{12} and related substance; this is further discussed in relation to vitamin requirements after the detailed data have been presented.

Planktonic Myxophyceae. The fresh-water planktonic blue-green algae (Fig. 88) belong partly to the coccoid family Chroococcaceae of the order Chroococcales and partly to the filamentous families Nostocaceae, Rivulariaceae, and Oscillatoriaceae of the order Nostocales. *Aulosira planctonica* is often placed in the family Microchaetaceae of that order, and a few unimportant occurrences in the plankton of filamentous blue-green algae of other families have been reported.

The whole group is difficult taxonomically; Drouet's (1959) most recent presentation differs in important ways from that of Huber-Pestalozzi. In particular, Drouet and Daily (1956), in a revision of the coccoid families, have reduced more than a thousand specific and subspecific taxa to thirty that they deem worthy of nomenclatorial recognition. It is too early to decide how far this drastic treatment will prove acceptable. In general, in this book the nomenclature and taxonomic usage of recent standard monographs has been followed explicitly, but in the Chroococcales this has often not been possible, for several authors attribute special significance to supposed species that Drouet and Daily believe to be merely transitory ecologically determined forms.

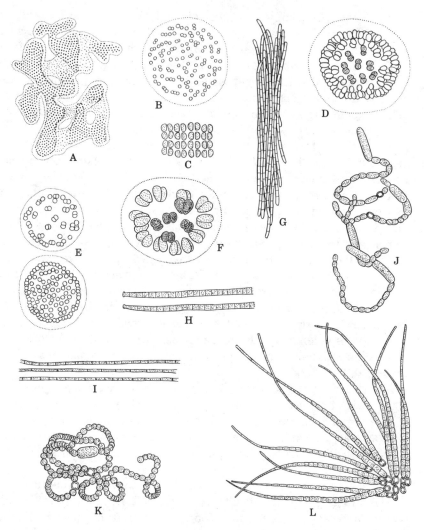

FIGURE 88. Common planktonic blue-green algae. *A. Anacystis cyanea*
(× 90), the abundant species, forming water blooms, formerly referred to
Microcystis aeruginosa. *B. Anacystis incerta* (× 500), usually referred to
Aphanocapsa elachista var *conferta,* one of the smallest-celled members of the
Myxophyceae. *C. Agmenellum quadruplicatum* (× 500), a species in which
the cells divide successively at right angles, but in a single plane, formerly
known as *Merismopedia glauca*. *D. Gomphosphaeria lacustris* (× 413), the
form usually referred to *Coelosphaerium naegelianum* (the latter specific name
has priority if its suspected identity with *lacustris* can be proved.) *E.* The
form usually referred to *Coelosphaerium kuetzingianum* but regarded by
Drouet and Daily as identical with the preceding. *F. Gomphosphaeria aponina*

In Chroococcaceae the important planktonic genera are *Anacystis* (= *Gloeocapsa, Microcystis, Chroococcus, Aphanocapsa,* etc.) *Gomphosphaeria* (= *Coelosphaerium*), and, to a lesser degree, *Coccochloris* (= *Aphanothece, Gloeothece,* etc.). In *Anacystis, A. cyanea* (better known in *Microcystis* or *Polycystis* as a *flos-aquae, aeruginosa, ichthyoblabe* or *viridis*) is a common constituent of massive summer water blooms. *Anacystis montana* (= *Aphanothece elachista*) with its form *minor* (= *pulverea*) is another planktonic species. The numerous planktonic records of *Chroococcus dispersus* and *C. limneticus* are referrable, according to Drouet and Daily, to *Anacystis thermalis* f. *major.* In related genera *Coccochloris elabens* (often referred to *Polycystis, Aphanothece,* or *Anacystis*) and *C. peniocystis* are often planktonic, as is *Agmenellum quadruplicatum,* which is usually referred to one of several species of *Merismopedia.*

The Nostocaceae include the plankter *Aphanizomenon holsaticum* (= *flos-aquae*), the predominantly halophil *Nodularia,* and, most important, the planktonic species of *Anabaena.* The Rivulariaceae include *Gloeotrichia, G. echinulata* often forming water blooms. In the Oscillatoriaceae there are planktonic species of *Oscillatoria* (notably *O. rubescens, O. prolifica, O. agardhii, and O. rileyi*), of *Lyngbya* (notably *L. limnetica*), and of *Schizothrix,* whereas *Arthrospira* may be prominent in inland saline lakes.

Pelogloea bacillifera, which has been regarded as a green bacterium related to or identical with *Chlorobium,* apparently belongs in the Myxophyceae (Pringsheim 1953; Van Niel and Stainer 1959). It has been recorded as the dominant phytoplankton in Scaffold Lake, Wisconsin, and Sodon Lake, Michigan (Newcombe and Slater 1950). The exact nature of the organism in these two lakes, however, is far from clear.

Though the blue-green algae as a whole give an impression of flourishing best in warm and nutrient-rich water, there are many exceptions to such generalization. The main phytoplankton of Crater Lake, Oregon, one of the most transparent lakes in the world, is a species of *Anabaena* (Utterback, Phifer, and Robinson 1942). The

---

($\times$ 413). *G. Aphanizomenon holsaticum* ($\times$ 200), better known as *A. flos-aquae,* a very common filamentous species producing water blooms. *H. Oscillatoria prolifica* ($\times$ 413), probably the commonest planktonic species of the genus. *I. Lyngbya limnetica* ($\times$ 500), a very thin filamentous species. *J. Anabaena flos-aquae* ($\times$ 200). *K. A. circinalis* ($\times$ 200), important planktonic species producing water blooms, at least *A. circinalis* appearing of great significance as a nitrogen fixer. *L. Gloeotrichia echinulata* ($\times$ 200), one quarter of a colony of radiating filaments. (Smith.)

algal flora of the few fresh-water lakes of Antarctica (Vinogradov 1957) appears to be mainly myxophycean; *Lyngbya, Phormidium,* and *Schizothrix,* as well as the green alga *Mougeoia,* are recorded. Among the planktonic species of the class several members of *Oscillatoria* are eurythermal or more often seemingly cold stenothermal (see page 441), mainly in lakes which are somewhat modified by drainage from human settlement. Massive water blooms of *Anacystis, Aphanizomenon,* and *Anabaena* tend to develop in temperate regions in productive lakes during the warmest months and such blooms are indicators of eutrophy in the strict sense of that much abused word (see pages 379–380). It is, however, paradoxical that such blooms usually develop when inorganically combined nitrogen in solution is low and inorganic phosphate, undetectible. *Gomphosphaeria lacustris* forms large populations in late summer or autumn in some of the large deep lakes of Central Europe which are otherwise not characterized by myxophycean plankton.

Culture studies throw some, but at present not very much, light on the characteristic occurrences of blue-green algae. As already indicated, a number of colorless organisms are probably apochlorotic heterotrophic members of the group, related to the Oscillatoriaceae. Among the photosynthetic members there is probably much variation in the nature of the utilizable carbon sources; whereas *Oscillatoria rubescens* seems to be a normal phototroph (Vollenweider 1950), Goryunova (1955), finding that *O. splendida* grew poorly in her pure cultures, concluded on further study that the organism is actually a phagotroph, needing bacteria at least as a supplementary source of food. *Cylindrospermum* sp. and the endophytic *Nostoc punctiforme* appear to be facultative heterotrophs (data assembled by Saunders 1957). In striking contrast to such findings, Kratz and Myers (1955) found *Anacystis marina* (sub *nidulans*), *Anabaena variabilis,* and *Nostoc muscorum* to be strict phototrophs using CO_2 reduced photosynthetically as their only carbon source.

The species that have been carefully studied all grow best in neutral or, more usually, alkaline media. In *Anacystis cyanea* Gerloff, Fitzgerald and Skoog (1952) found little growth below pH 8 and optimal growth as high as pH 10. *Coccochloris peniocystis* behaved similarly (Gerloff, Fitzgerald, and Skoog 1950b). Kratz and Myers (1955) found that *Anacystis marina* grow well between pH 7.4 and pH 9.0; *Anabaena variabilis* and *Nostoc muscorum* showed a slightly wider tolerance, from pH 6.9 to pH 9.0. It may be suspected that the species growing best in quite alkaline water use the bicarbonate ion as a carbon source in photosynthesis.

It is probable that all the blue-green algae need appreciable amounts of sodium, and Provasoli (1958) suggests that the presence of sodium in sewage and other artificial influents may in part underlie the association of blue-green algae with mild pollution, noticeable in most more or less urbanized lake districts. The quantities of magnesium and of calcium required are evidently variable. Arnon (1958) regards all four major cations as macronutrients for *Anabaena cylindrica,* though Gerloff, Fitzgerald, and Skoog (1950b) found *Coccochloris peniocystis* (= *Gloeothece linearis*) to grow almost maximally in a medium containing but 0.13 mg. Mg. per liter and only traces of calcium as impurities. The trace elements certainly required are Fe, Mn, Cu, Zn, Mo and Co. Eyster (1952) found *Nostoc muscorum* to require boron. The quantity of molybdenum needed by nitrogen-fixing forms is probably always high and this element is clearly important ecologically.

Cobalt is essential and can be used in inorganic form by all fresh-water species studied so far, though Holm-Hansen, Gerloff, and Skoog (1954) found that cyanocobalamin (vitamin B_{12}) was utilized vastly more efficiently than ionic cobalt. A number of species have been shown (Robbins, Hervey, and Stebbins 1951) to produce considerable amounts of cobalamines in culture; Benoit (1957) suspects the group to be an important source of such substances in the economy of lakes. The need for vanadium and for chloride has not yet been established in the Myxophyceae.

No critical evidence has been obtained that any of the score or more of fresh-water species of blue-green algae studied in culture require specific organic substances, though the marine *Phormidium persicinum* requires B_{12} (Pintner and Provasoli 1958). Important species that form waters blooms, *Anacystis cyanea, Aphanizomenon holsaticum,* and *Gloeotrichia echinulata,* and several species of *Anabaena,* have been cultivated for prolonged periods without any organic supplement (Gerloff, Fitzgerald, and Skoog 1950a,b, 1952; Fogg 1942; Kratz and Myers 1955), even though Rodhe (1948) had earlier concluded that *G. echinulata* required a thermolabile substance present in soil extract. The impression that the group is associated with waters rich in organic matter (Pearsall 1932) may be valid, but the association clearly does not imply that such compounds are needed in the nutrition of the Myxophyceae as a whole, and in some cases, as already indicated, sodium rather than organic matter derived from sewage may be involved.

Many, though by no means all, of the Nostocaceae and some of the other filamentous blue-green algae can fix molecular nitrogen (cf.

Volume I, page 847; Dugdale, Dugdale, Neess, and Goering 1959; Neess, Dugdale, Dugdale, and Goering 1962; Dugdale and Dugdale 1962). The species so far studied in cultures free from bacteria are not characteristic members of the plankton, but there is field evidence strongly suggesting that *Anabaena circinalis* can be an important planktonic nitrogen fixer (Hutchinson 1941; Dugdale and Dugdale 1962). Some of Guseva's (1937) observations on cultures that were not bacteria-free indicate the same for *A. lemmermanni*. Not all species of the genus *Anabaena* can fix molecular nitrogen, for Kratz and Myer found the faculty lacking in *A. variabilis*. The general incidence of blooms at times of high temperature, high light intensity, and low nutrient concentrations is apparently exhibited equally by species such as *Anabaena circinalis* which almost certainly fix nitrogen and by those such as *Anacystis cyanea* and *Aphanizomenon holsaticum* which do not.

Oscillatoria rubescens can use either nitrate or ammonia; Vollenweider (1956) also found good growth on asparagine and peptone as nitrogen sources and irregular growth with albumin; these experiments, however, are not said to have been in bacteria-free culture, and it is not really clear what sources of nitrogen were available.

Guseva (1937, 1939) obtained some evidence that *Aphanizomenon* is inhibited in nature by high concentrations of manganese. Gerloff and Skoog (1957a), in a study of *Anacystis cyanea,* have found antagonism in culture between manganese and calcium such that in the presence of 0.5 mg. Ca per liter growth was reduced to 43 per cent of that in the control by the addition of 1 mg. Mn per liter, whereas in the presence of 10 mg. Ca per liter the reduction was only to 89 per cent of the control. Similarly, addition of 4 mg. Mn per liter in the low calcium culture was completely inhibitory, but in the high calcium culture the growth was 69 per cent of that in the control. Gerloff and Skoog suspect that occasionally, under conditions of rapid mixing after summer stratification, inhibitory concentrations of manganese might be present in soft-water lakes; in such natural bodies of water containing a diversified competing flora the apparent effects of this inhibition might well be greater than in unialgal laboratory cultures.

The work of Gerloff and Skoog (1957b) indicates that under optimal conditions *Anabaena cyanea* uses about 4.5 mg. N and 0.12 mg. P to produce 100 mg. of dry alga. Under other conditions, with phosphorus limiting, nitrogen contents as high as 9.1 per cent are possible. In a series of lakes they found that the alga had nitrogen contents of 5.43 to 7.61 per cent. They believed, however, that these values are not really comparable to those derived from laboratory cultures because the sheath material disintegrates much more readily in nature than in culture. Applying an appropriate correction, they conclude that the natural populations are equivalent to laboratory-grown algae containing 3.88 to 5.02 per cent nitrogen and 0.16 to 0.42 per cent phosphorus. They believe therefore that in the six lakes in Wisconsin studied during August *Anacystis cyanea* was limited in its growth by nitrogen rather than by phosphorus and was in fact taking up an excess of the latter element. Vollenweider has also concluded from the distribution of *Oscillatoria rubescens* that nitrogen is usually limiting for this alga; he suggests that the limitation may occur at a higher threshold at high than at low light intensities, as is known to be true in *Lemna* (White 1937).

The impression that might be gained from this work, that blue-green algae, which are indeed normally rather rich in the element, would require quite impossible amounts of nitrogen to produce a bloom, is specious. Gerloff and Skoog found a roughly linear dependence of yield on nitrogen supply, other nutrients being adequate. If, as seems probable from their results (Fig. 89), this linear relationship can be extrapolated to the origin, we may expect that the continual passage

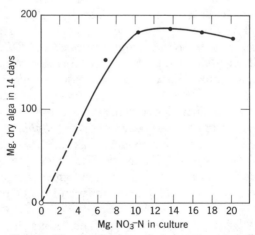

FIGURE 89. Production of *Anacystis cyanea* in culture containing initially different quantities of nitrate nitrogen, after 14 days growth, from the data of Gerloff and Skoog. The plateau in production is presumably due to limitation by some nutrient other than nitrogen or phosphorus.

of combined nitrogen from the shallow water-mud interface into the free water, in which the alga maintains a low concentration of the nutrient, could lead to the final development of a very considerable bloom.

Planktonic Chlorophyta. The green algae contain a great number of morphologically diverse organisms of varied ecology. The planktonic species for the most part belong either to one of the two lower orders, the Volvocales (including Tetrasporales of many authors) and the Chlorococcales (= Protococcales of many authors; cf. Fritsch 1935), in which at least the gametes are flagellate and motile, or to the Conjugatae, a much more specialized order in which the gametes are amoeboid and whose unicellular or loosely colonial species are commonly called desmids. The two major groups of planktonic green algae, the Volvocales (Figs. 90, 91) and Chlorococcales (Figs. 92, 93), on the one hand, and the desmids (Fig. 94), on the other, seem in general to have different physiological requirements and so different ecological preferences. Outside these two groups of planktonic Chlorophyta the anomalous genus *Botryococcus,* occasionally found in great quantity in the plankton of a highly diversified series of lakes, is the most important planktonic member of the group. It appears (Chu 1942) to be autoauxotrophic but to require fairly high concentrations of nutrients for optimal growth.

Planktonic Volvocales and Chlorococcales. The important genera, namely *Carteria, Chlamydomonas, Eudorina, Volvox, Elaktothrix, Gloeocystis,* and *Sphaerocystis,* among the Volvocales, and *Tetraedron Oocystis, Ankistrodesmus, Kirchneriella, Selenastrum, Dictyosphaerium, Pediastrum, Coelastrum, Crucigenia,* and *Scenedesmus,* among the Chlorococcales, are most likely to be found in the plankton of ponds or shallow, fertile lakes; there are, however, many exceptions, *Oocystis,* which may be the dominant phytoplankter in deep unproductive lakes, is perhaps the most important; *Gloeocystis* and *Sphaerocystis* may be important in the soft water of unproductive lakes.

Species of Volvocales and Chlorococcales in culture are frequently facultatively heterotrophic. A list of such green facultative heterotrophs compiled from the literature Saunders (1957) contains species of *Chlamydomonas, Chlorella, Chlorogonium, Chlorococcum, Coelastrum, Cystococcus, Nannochloris, Scenedesmus,* and *Stichococcus.* Both *Chlorella* and *Scenedesmus* have for many generations been grown in glucose in the dark. Apochlorotic genera, such as *Polytoma,* comparable to *Chlamydomonas,* and *Prototheca,* comparable to *Chlorella,* occur in both orders. It is probable (Provasoli 1958) that the facultative heterotrophs, when living autotrophically, require higher concentra-

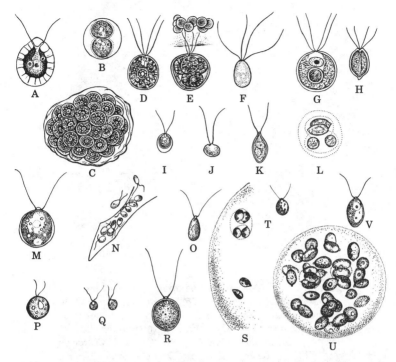

FIGURE 90.   Unicellular Volvocales of limnological interest.   *A. Haematococcus pluvialis* (× 350), motile cell.   *B.* The same, dividing.   *C.* Mass of daughter cells from division of aplanospore, enclosed in thick membrane.   *D. Carteria lohammari* (× 500), motile cell.   *E.* The same dividing, anchored to *Anacystis*.   *F. C. wisconsinensis* (× c. 500), apparently limnoplanktonic in Wisconsin.   *G. C. globosa* (× c. 500), widespread in small bodies of water, a winter heleoplankter in some North American localities.   *H. Scherffelia pelagica* (× 600), coldwater plankter in some Swedish lakes.   *I. Chlamydomonas globosa* (× 450), euplanktonic.   *J. C. epiphytica* (× 500), free-swimming plankter or in sheath of *Anacystis*.   *K. C. tremulans* (× 600), planktonic in Sweden.   *L.* The same, nonmotile form in gelatinous capsule.   *M. C. opisthopyren* (× c. 500), a large spring species in some Swedish lakes.   *N. C. dinobryonis* (× 500), small planktonic form aggregating in empty *Dinobryon* cell walls.   *O. C. gloeophila* (× 700), an inhabitant of the gelatinous matrix of *Conochilus* and of other similar animal and plant structures.   *P. C. polypyrenoidosa* (× 500), planktonic in northern Wisconsin.   *Q. C. grovei* (× c. 500), the smallest member of the genus, in the nannoplankton of ponds.   *R. C. hypolimnetica* (× 400), a rather large species among the generally small planktonic members of the genus, hypolimnetic in Swedish lakes.   *S. C. planctogloea* (× 335), portion of a large spherical colony.   *T.* Swarm spore of same (× 570).   *U. C. quiescens* (× 335), colony of nonmotile cells.   *V.* Swarm spore of same (× 570).   (Huber-Pestalozzi, after Prescott, Rodhe, Skuja, Snow, and West.)

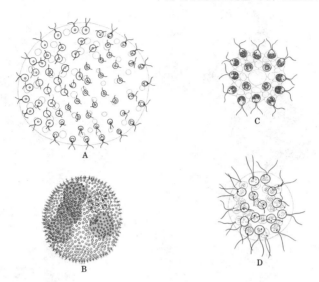

FIGURE 91. Colonial planktonic Volvocales. *A.*
Pleodorina californica (× 267), rare but remarkably
differentiated colonial form, facultative plankter. *B.*
Volvox tertius (× 177), limited distribution but at
least a facultative plankter. *C. Gonium pectorale*
(× 333), heleoplanktonic. *D. Eudorina elegans*
(× 333), widespread eurytopic facultative plankter.
(Smith.)

tions of inorganic nutrients than do the strictly phototrophic species.
The evidence, mainly derived from *Scenedesmus obliquus* and *Chlorella
pyrenoidosa,* indicates that Fe, Mn, Cu, Zn, Ca, Mo, and V are essential
as trace elements (Arnon 1958). Boron may also be required
(McIlrath and Skok 1957). The genus *Volvox* (Pintner and Provasoli
1959) has a rather high requirement for chelated iron.

If the optimum concentration of vanadium proves to be as high
as 10 to 20 mg. m.$^{-3}$ (Arnon and Wessel 1953; Arnon 1958) or even
one tenth of this concentration, the element may prove limiting.
Sugawara, Naito, and Yanada (1956) in six Japanese lakes, found
0.1 to 1.7 mg. m.$^{-3}$ (the mean value is 0.74 mg. V m.$^{-3}$), almost all
in solution. Molybdenum is required apparently only when nitrate
is the source of nitrogen; the quantity needed is small compared to
that used by nitrogen-fixing blue-green algae. For *Scenedesmus obliquus* a concentration of 0.1 mg. Mo m.$^{-3}$ appears optimal (Arnon
1958).

Both the Volvocales and the Chlorococcales seem to be predominantly autoauxotrophic, but a considerable minority of species (37.5 per cent of the 40 so far studied) are known that require either thiamin of B_{12} or very occasionally both (Provasoli 1958). The species studied up to 1958 are conveniently tabulated by Provasoli (1958). It seems that apochlorotic heterotrophs are more likely to need thiamin than are green forms, but no such generalization can be made about B_{12}. No green alga so far appears to require an external source of biotin. In their general auxoautotrophic nature the two lower orders of the Chlorophyta appear to differ from the Euglenophyta and from the flagellate pigmented groups included in the Rhizoflagellata (Chrysophyceae, Dinophyceae, and Cryptophyceae) in which nearly all species need B_{12} or thiamin and some biotin as well. At least *Chlorella,* and doubtless other auxoautotrophic green algae, synthesize cobalamines and are a source of such compounds in lakes. Among

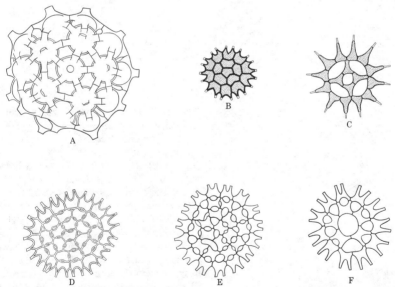

FIGURE 92. Planktonic colonial Chlorococcales. *A. Coelastrum cambricum* ($\times$ 413). *B. Pediastrum boryanum* ($\times$ 167). *C. P. simplex* var *duodenarium*. *D. P. duplex* ($\times$ 167). *E. P. duplex* var. *clathratum*. *F. P. duplex* var. *reticulatum* ($\times$ 167). The figures of *Pediastrum* may be compared with Fig. 3; all species of the genus are doubtless ecologically separable and in South Africa the more open forms of *P. duplex* appear to occur in the least mineralized localities, but the over-all picture is clearly complicated, though not understood. (Smith.)

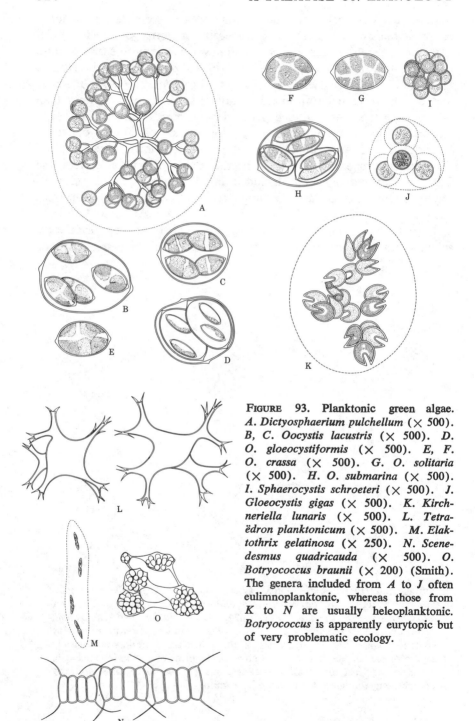

FIGURE 93. Planktonic green algae. *A. Dictyosphaerium pulchellum* (× 500). *B, C. Oocystis lacustris* (× 500). *D. O. gloeocystiformis* (× 500). *E, F. O. crassa* (× 500). *G. O. solitaria* (× 500). *H. O. submarina* (× 500). *I. Sphaerocystis schroeteri* (× 500). *J. Gloeocystis gigas* (× 500). *K. Kirchneriella lunaris* (× 500). *L. Tetraëdron planktonicum* (× 500). *M. Elaktothrix gelatinosa* (× 250). *N. Scenedesmus quadricauda* (× 500). *O. Botryococcus braunii* (× 200) (Smith). The genera included from *A* to *J* often eulimnoplanktonic, whereas those from *K* to *N* are usually heleoplanktonic. *Botryococcus* is apparently eurytopic but of very problematic ecology.

species requiring an external source of B_{12} *Volvox globator* and *Volvox tertius* may be mentioned (Pintner and Provasoli 1959).

Planktonic desmids. The term desmid is used in limnology to designate those members of the Conjugales that are either strictly unicellular or in which, if filamentous, the cells of the filaments are only loosely connected. The group contains the loosely filamentous Gonatozygaceae, the Mesotaeniaceae, or saccoderm desmids and the Desmidiaceae or placoderm desmids. The last-named family contains most of the species.

The Mesotaeniaceae are nearly all sphagnophil or occur on wet surfaces. Two only, namely *Spirotaenia lemanensis* (*A* of Fig. 94) and *Roya cambrica* var. *limnetica,* are known from lake plankton. Both genera of Gonatozygaceae, *Gonatozygon* and *Genicularia,* have planktonic members, but they are rare and unimportant.

The Desmidiaceae, or placoderm desmids, are unique among unicellular plants in that their cells are divided into two *semicells* united at an *isthmus* of varying width, almost undetectible in *Hyalotheca* and very narrow compared to the maximum width in *Staurastrum* and *Micrasterias.* The semicells are often of elaborate symmetrical shape, drawn out into processes. There are always three and often more well-defined symmetry planes, except in a few abnormal cases in which one semicell has a form characteristic of a certain number of planes, the other of another number; these are termed by Teiling *Janus* forms. In *I* of Fig. 94 a biradiate semicell, characteristic of a form of *Staurastrum chaetoceras* with three symmetry planes, is united to a triradiate semicell, characteristic of a form with four symmetry planes (Teiling 1950).

Most but not all the genera of desmids contain a few planktonic species; by far the most important assemblages occur in *Staurastrum* and *Staurodesmus* (= *Arthrodesmus* p.p. and *Staurastrum* p.p. auctt.). A more limited number of species of *Cosmarium, Closterium,* a few species of *Xanthidium,* and at least some forms of *Micrasterias mahabuleshwarensis* are eulimnoplanktonic (*L* of Fig. 94). Among the colonial and filamentous forms *Spondylosium planum, Cosmocladium saxonicum,* and several species of *Hyalotheca* are known from the plankton, though according to Brook only the first named is really euplanktonic. Species of *Desmidium, Sphaerozosma,* and *Onychonema* have also sometimes been recorded as living planktonically. In nearly every example the benthic-littoral and particularly the sphagnicolous members of these genera vastly outnumber the planktonic. Many of the genera in fact are very large; *Closterium* has more than 80 species, *Euastrum,* about 150, and *Cosmarium,* several hundred. The specific

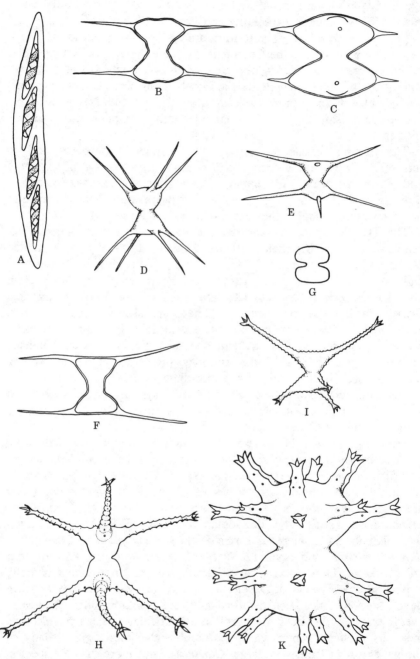

FIGURE 94. Desmids of limnological interest. *A. Spirotaenia lemanensis*

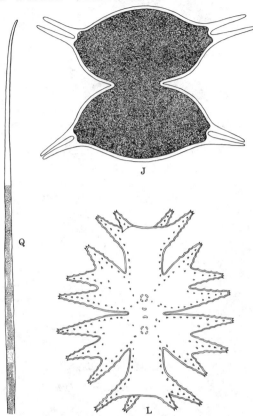

FIGURE 94.   (*Continued*)

($\times$ 333), a planktonic saccoderm desmid. *B. Staurodesmus subtriangularis* ($\times$ 347), characteristic of very soft waters. *C. S. megacanthus* ($\times$ 347), a soft-water species. *D. S. jaculiferus* ($\times$ 347), quadriradiate form. *E.* The same tri-radiate *excavatus form*. *F. S. jaculiferus* var. *excavatus* biradiate form ($\times$ 347). *G. Cosmarium bicculatum* ($\times$ 400), a minute simple planktonic species. *H. Staura-strum pingue,* a eurytopic planktonic desmid. *I. S. chaetoceras,* Janus form, one hemicell being biradiate, the other triradiate. *J. S. longispinum* ($\times$ 340), an acid-water species. *K. S. artiscon* ($\times$ 347), ordinarily heleoplanktonic in acid water. *L. Micrasterias mahabuleshwarensis* var. *wallichii* ($\times$ 286), *unusually large and elaborately dissected for a planktonic species. M. Xanthidium antilopaeum* ($\times$ 347), predominantly benthic littoral. *N.* Zygospore of same ($\times$ 267). *O. X. antilopaeum* var. *hebredarum* ($\times$ 347), a predominantly planktonic form. *P. X. antilopaeum* var. *depauperatum* ($\times$ 347), another planktonic form with not only fewer but shorter spines. *Q. Closterium aciculare* var. *subpronum* ($\times$ 267), one of the few facultititive limnoplanktonic members of the genus (most of one hemicell omitted). *R. Spondylosum planum* ($\times$ 334), a filamentous colonial

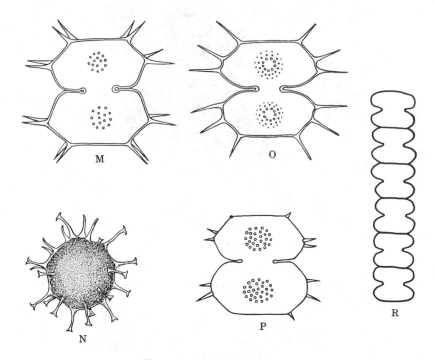

FIGURE 94. (*Continued*)

planktonic desmid. (Krieger after Reverden, West and West, G. M. Smith, Brook, Teiling.)

diversity of local, littoral, benthic, desmid floras can be quite extra-ordinary, as is indicated by a few examples in a later paragraph. The planktonic species constitute a small assemblage compared to the benthic, and all students of the desmid plankton have noted that in small and weedy lakes species better known as benthic-littoral organisms are likely to occur as facultative plankters. Moreover, among the euplanktonic species there are many that appear to be morphologically close to benthic taxa. Thus in *Staurastrum* the eulimnoplanktonic (Fig. 95) *S. cingulum* is apparently connected (Brook 1959a,b) by the varieties *obesum* and *affine* with the benthic *S. gracile*. Other examples in the genus are given by Brook. In *Xanthidium* nominate *X. antilopeum* (M of Fig. 94) is apparently benthoplanktonic in the littoral, whereas the varieties *hebridarum* and *depauperatum* are euplanktonic (O, P, Fig. 94). Examples, particularly from *Staurastrum*, could undoubtedly be multiplied, but most of them involve unresolved

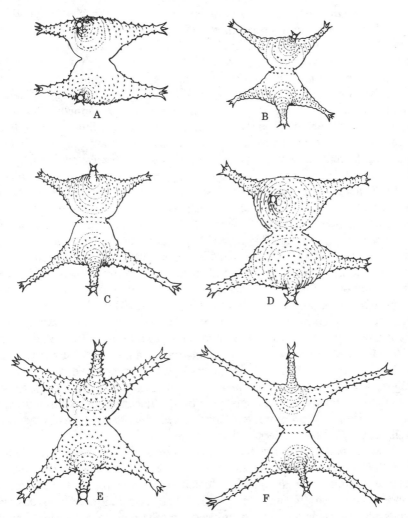

FIGURE 95. The mainly littoral benthic *Staurastrum gracile* and transitions to the eulimnoplanktonic *S. cingulum,* of which the less extreme var. *obesum* appears to be primarily characteristic of mesotrophic waters. *A. S. gracile* Littleton Reservoir, Renfrewshire, Scotland. *B. S. cingulum,* small form near *gracile,* Eye Brook Reservoir, Northamptonshire, England. *C. S. cingulum,* dichotypical, with the upper semicell near *gracile,* Loweswater, English Lake District. *D. S. cingulum* var. *affine,* Loch of Girlista, Shetland Islands. *E. S. cingulum* var. *obesum,* Derwentwater, English Lake District. *F. S. cingulum,* extreme limnoplanktonic form, Loch Lomond, Scotland (all × 600). (Brook.)

taxonomic complexities. The implications of the affinities of benthic-littoral and planktonic taxa have been variously interpreted. West and West (1909) regarded such affinities as due to ordinary evolutionary processes operating over long periods of time. Griffiths (1928), in a study of the flora of artificial lakes, which he assumed could be populated only by the original local flora, believes that desmid plankton evolved from that of bogs and other terraqueous habitats in the course of thirty or forty years. Brook (1959a) thinks that some at least of the characteristic planktonic desmids may be phenotypic ecological modifications of benthic-littoral species, which can be produced during a single annual cycle.

In general the euplanktonic taxa are distinguished from their benthic-littoral allies in having less elaborate ornamentation, smoother outlines, slenderer form (Teiling 1947), and in some cases, as in *Cosmarium* (*G* of Fig. 94), small size (Krieger 1933). Since there is a general tendency for the desmids to secrete a gelatinous covering, which is more strikingly developed in the planktonic species, and since the detail of the more elaborate forms is often too minute to be of much hydromechanical significance (see page 270) and tends to disappear in the planktonic taxa, it is very unlikely that the characteristic elaboration of the desmids is related to flotation. West and West (1904) have suggested that the more elaborate forms found, for instance, in *Micrasterias* and in the multiradiate species of *Staurastrum* might present difficulties to any small animal predator attempting to eat them. There is also the possibility that the dissected surfaces of such desmids promote easy exchange of material across the cell walls in very dilute nutrient-deficient water.

Geochemical determination of desmid floras.   There can be no doubt that the general occurrence of large numbers of species of desmids is usually correlated with very low calcium or calcium and magnesium concentrations in the ambient water.   The association of rich desmid floras with ancient igneous rocks was demonstrated in a classical paper by West and West (1909).   Pearsall (1921) concluded the determining factor to be the ratio $(Ca + Mg)/(Na + K)$, but there is no certain reason to believe (Hutchinson and Pickford 1932) that it is not the absolute concentration of calcium or of calcium and magnesium that is important.   Strøm (1926) agreed in general that the desmid plankton characterizes regions poor in lime except when fixation of cations by organic matter can lower the regionally characteristic alkaline earth content of the water.

It is possible that this type of limitation is merely an extreme case of the restriction of fresh-water algae to a certain low range of inorganic concentration or salinity. Provasoli (1960) believes that such limitation will be found to be very significant in explaining the distributions of fresh-water algae.

The incidence of rich desmid floras, either in the plankton or in the littoral benthos, is thus evidently dependent on the inorganic composition of the water. Although a great many species of desmids are sphagnophil, brown water rich in humic material is not essential to the development of a rich flora; as West and West point out, the richest desmid flora known to them, from the mountains of North Wales, comes from waters not rich in peat extractives, but though because they drain from old palaeozoic rocks such waters are certainly very soft. In other parts of the world the correlation with ancient rock formations breaks down. The desmid flora of Florida, for example, is exceedingly rich; though most of the state is underlain by limestone, the surface soil is much leached, sandy, and acid. An abundance of desmids occurs when lakes or ponds lie in shallow basins in leached sand above the limestone; when the lakes are deep or are fed by springs from the limestone, the water is hard and the group is much less conspicuous (Prescott and Scott 1952). Prescott and Magnotta (1935) also record the interesting case of Calhoun Lake, Albion, Calhoun County, Michigan, in which the higher vegetation, including *Chara* sp., is of the kind to be expected in a hard-water lake but in which the *p*H was 5.8. It is possible that the lake was passing from a hard to a soft phase and that the *Chara* plants were set in a thin layer of acid mud over a more alkaline deposit. A single slide made from washings of *Chara* showed at least thirty species of *Cosmarium*.

Pearsall (1921) attempted to group the planktonic desmids of the English lakes according to the calcium concentration, or total alkalinity, tolerated. His data, however, were limited, and subsequent work has not confirmed the details of his scheme. Brook (unpublished) found that among thirty-three eulimnoplanktonic species studied in Britain only *Staurastrum longispinum* (*J* of Fig. 94) and *Staurodesmus subtriangularis* (*B* of Fig. 94) appeared to be limited to waters containing not more than 15 mg. HCO_3^- per liter (i.e., not more than 12.3 mg. $CaCO_3$ alkalinity per liter). Pearsall had put at least a dozen species into his two categories that are included in such a range of alkalinity. Of these species it appears probable from his own work, and that of Brook and of Smith (1924), taken in conjunction with the analyses of Juday, Birge, and Meloche (1939), that at least *Staurodesmus jaculiferus* (*D* of Fig. 94), *S. megacanthus* (*C* of Fig. 94), the heleoplanktonic *S. glabrus* and *S. aversus, Staurastrum anatinum, S. arctiscon*

(*K* of Fig. 94), and the more or less facultative or heleoplanktonic *S. ophiura* and *S. brasiliense,* with the more planktonic species of *Xanthidium* and perhaps of *Micrasterias,* are essentially characteristic of soft waters.   In the strikingly developed desmid plankton of many soft-water lakes eurytopic desmids such as *Staurastrum pingue,* (*H* of Fig. 94), also known in association with blue-green algae and numerous diatoms in hard and very productive waters, are often present. The striking facies of the desmid plankton is perhaps increased by the general rarity of blue-green algae, the rather special nature of the associated green algae (*Sphaerocystis, Gloeocystis*), and certainly by the frequent presence of pseudoplanktonic desmids, which are easily derived from the littoral region or from bogs on the courses of influents. It must not be forgotten, however, as Järnefelt (1952) has insisted, that this characteristic facies is largely a property of the net plankton. In samples containing the whole phytoplankton a very large number of smaller forms may outnumber the desmids volumetrically.

The limitation of many species to waters poor in electrolytes, which are likely to occur in somewhat circumscribed areas, may explain the fact that the desmids appear to be rather less cosmopolitan than are most unicellular fresh-water algae.

In spite of the fair number of soft-water plankters included in the group and the enormous variety of sphagnophil and terraqueous forms occurring in water deficient in calcium and ordinarily of low productivity, the existence of eurytopic species and some requiring a high calcium concentration indicates that although something in the fundamental nature of the group must predispose the desmids to life in dilute water deficient in alkaline earths the requirement for the group as a whole is not an absolute one.   It is also noteworthy that almost the only significant work done on desmid nutrition indicates that calcium is essential to more than one species of *Micrasterias,* to *Mesotaenia caldariorum* and to *Cosmarium botrytis;* the effects of deficiency are much easier to demonstrate in these forms, the only ones that have been studied, than in species of *Chlorella, Stichococcus, Euglena,* and other genera (Pringsheim 1926; Waris 1933).   In *Micrasterias rotata,* growing in an acid medium, the omission of calcium (about 37 mg. per liter) prevented cell division; the element could be replaced to some extent by strontium and feebly by barium.   *M. denticulata* perhaps has a lower optimal concentration (about 19 mg. Ca per liter) than *M. rotata.* (Pringsheim 1926).

In the genus *Closterium* some of the commoner species found as heleoplankton or pseudoplankton, such as *C. venus* or *C. aciculare* var. *subpronum* (*Q* of Fig. 94), occur in hard water.   In his enumeration of the plankton desmids of Wisconsin Smith (1924) records these two from Lake Monona and Mendota, respectively. As indicated in Chapter 22 (page 480), six species of *Staurastrum* regularly occur in the Lunzer Untersee, the water of which contains about 37.5 mg. Ca per liter and so is approximately saturated with calcium bicarbonate with respect to the CO_2 of the air. Several species of *Cosmarium*, members of the *flora hygropetrica* or association of microscopic plants on wet rock surfaces, prefer calcareous substrates; West and Fritsch (1927) mention *Cosmarium dovrense* and *C. microsphinctum* as examples. The very peculiar *Oocardium striatum*, a rare form in streams, is even a calcium carbonate depositing plant (West, West, and Carter 1923).

Krieger (1933) finds that among the three hundred and seventy-seven taxa of desmids recorded from all sorts of habitats in Indonesia no less than seventy-five are known at about pH 8 and of these eight occurred above pH 9. Some species, notably of the largely nonplanktonic genus *Cosmarium*, are known to be extraordinarily eurytopic. *Cosmarium granatum* occurs in the cold dilute waters in the Torneträsk region of Swedish Lapland and[1] (Rich 1932) as var. *africanum* in Ronde Vlei near Cape Town with a salinity about half that of sea water. *Cosmarium salinum*, furthermore, is supposed to be a brackish form (Hansgirg 1886). It is interesting that six species of *Cosmarium* occur in Lake Kivu, including the eurytope *C. granatum* and the endemic *C. kivuense* which is rather widespread in the lake. The water is alkaline and rich in sodium and magnesium carbonate.

Though a number of species of desmids have been cultivated, usually in culture media enriched with soil extract, the study of their nutrient requirements other than the need for calcium appears to have been neglected.

Planktonic Euglenophyta. Though the phylum consists of more than 600 species of photosynthetic form, as well as many that are apochlorotic, few are of any importance as members of the lake plankton. Among the free-swimming species some members of *Trachelomonas* (*C, D,* of Fig. 96) and *Lepocinclis* and a few of *Euglena* are widely distributed in the open water of lakes; in the last-named genus *Euglena acus* (*A* of Fig. 96), which occurs in bodies of water as large and deep as the Sea of Galilee (Komarovsky 1959), is probably the most eulimnoplanktonic species. The majority of the

---

[1] The record at pH 9.8 for Lakeside Vlei (Hutchinson, Pickford, and Schuurmann 1932, *sub* var. *subgranatum*) is erroneous as the desmid was not recorded in this locality at the time of the very high pH reading.

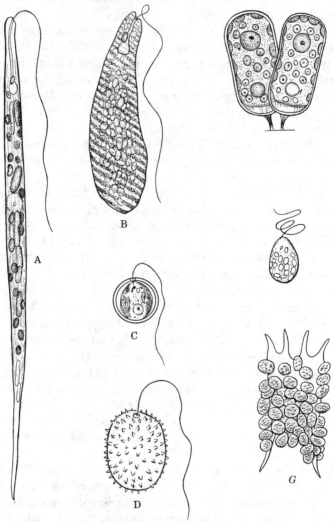

FIGURE 96. Planktonic Euglenophyta. *A. Euglena acus* (× 200),
the most truly planktonic. *B. E. sanguinea* (× 200), a red
species responsible for some blood lakes in the European Alps.
C. Trachelomonas volvocina (× c. 400), often heleoplanktonic.
D. T. hispida, widespread in plankton, usually in ponds and small
lakes. *E. Colacium calvum* (× 325), epiplanktonic on rotifers
and crustacea. *F. C. simplex* (× c. 500), free-swimming cell. *G.*
Nonmotile stage epiplanktonic on rotifers (× c. 125). (Modified
from Huber-Pestalozzi.)

species of *Euglena* and *Phacus* are found in small, somtimes very minute, bodies of water which often have a high organic content. *Colacium* (*E–G* of Fig. 96) is an exceptional epiplanktonic form, living, as a nonmotile adult, on various members of the zooplankton. Ammonia is for some, if not most, species the only inorganic nitrogen source; organic nitrogen sources can be utilized.

The group as a whole is facultatively heterotrophic; all species require either B_{12} or thiamin; most species require both vitamins.

Planktonic Cryptomonadineae. This small morphologically complex but biochemically possibly very primitive group of flagellates contains a few quite significant members of the nannoplankton of lakes. The genus *Rhodomonas* is particularly important. *R. lacustris* (*A* of Fig. 97) is often common in Lake Constance, in many of the Austrian Lakes, and Erken; it may be accompanied by the larger species *R. lens* (*B* of Fig. 97). *R. minuta* var. *nannoplanctonica* and *R. tenuis* (*C* of Fig. 96) are other planktonic members of the genus originally from lakes in Sweden, where *R. tenuis* appears to be a cold-water form. In *Cryptomonas, C. ovata* is a eurytopic species and *C. erosa*

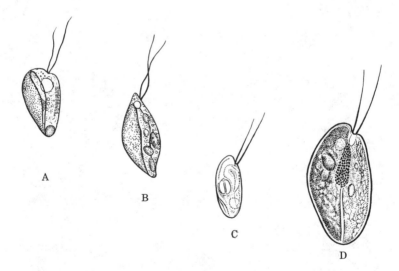

FIGURE 97. Planktonic Cryptomonads. *A. Rhodomonas lacustris* (× 1500), an important and relatively eurytopic member of the phytoplankton. *B. R. lens* (× 1500), often with *R. lacustris*. *C. R. tenuis* (× 1500), a small cold stenotherm species, Sweden. *D. Cryptomonas erosa* (× 825), large form. (Redrawn from Huber-Pestalozzi.)

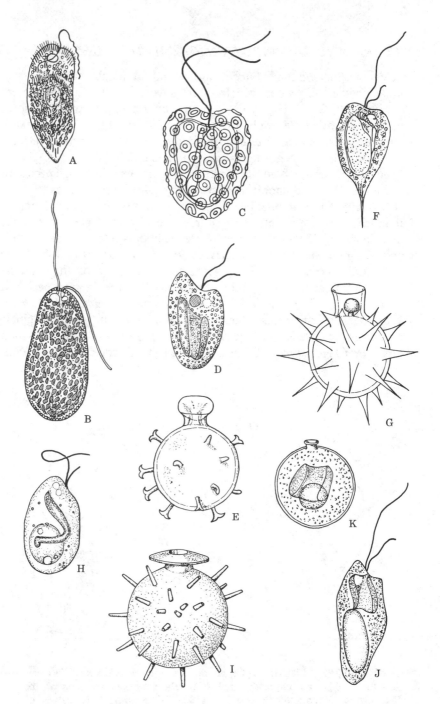

Figure 98. Two chloromonads (*A, B*); one of the fresh-water coccolitho-
phores (*C*); species of *Ochromonas* with cysts, all Western Europe
(*D–G*). *A. Gonyostomum semen* (× 400). *B. Vacuolaria virescens*
(× 600). *C. Hymenomonas roseola* (× c. 800). *D. E. Ochromonas
fragilis. F, G. O. stellaris. H. I. O. vallesiaca. J. K. O. granularis.* (Redrawn from Huber-Pestalozzi.)

(*D* of Fig. 96) is even more so; a number of the other thirty species in the genus are known in the phytoplankton.

The cryptomonads need B_{12} or thiamin, usually both, for optimal growth. None is certainly known to be autoauxotrophic. Chu (1942) cultivated *Cryptomonas ovata* with difficulty but obtained good growth in one of his inorganic media. It may be doubted, however, that all possible contamination by traces of cobalamin was excluded.

Planktonic Chloromonadineae. This small group is represented in the fresh-water plankton by the genus *Vacuolaria* (*B* of Fig. 98) and in some small bodies of water by *Gonyostomum semen* (*A* of Fig. 98). Very little seems to be known about the nutrition or biochemistry of the group.

Planktonic Xanthophyceae. The certain members of this group are rarely planktonic; the eulimnoplanktonic species are members of *Gloeochloris, Chlorobotrys, Pseudotetraedron,* and *Gloeobotrys. Tribonema* may be a facultative planker and *Ophiocytium* contains species, notably *O. parvulum,* which may occur in the plankton of small acid bodies of water, as well as *O. capitatum* f. *longispinum* (*J* of Fig. 99), found to be euplanktonic in Wisconsin (Smith 1920).

Planktonic Chrysophyceae. This group is complicated and by no means certainly monophyletic. The flagellate forms are usually arranged in three orders, the Chromulinales with a single flagellum, the Isochysidales with two equally long but by no means certainly identical flagella, and the Ochromonadales with unequal and different flagella. The order Rhizochrysidales includes nonflagellate forms presumably derived from organisms such as *Chrysamoeba* by the loss of the flagellum. The Chrysocapsales and Chrysosphaerales are palmelloid, and there are two orders of attached forms. The classification is obviously artificial and will probably have to be reconstructed as more flagella are examined with the electron microscope and more species are cultivated in sufficient quantity to permit a thorough study of their carotenoid pigments.

The Chromulinales contain a large number of flagellates which occur in small pools and a few are known from the plankton of lakes, among them *Chrysapsis agilis,* a minute member of a genus with a reticulate chloroplast, and several species of *Chromulina* are recorded from the plankton of the Lake of Lucerne and other Swiss and German lakes. *Chrysamoeba,* normally uniflagellate, but capable of becoming amoeboid, is known from both the Swiss lakes and some of the smaller lakes of Wisconsin. Several species of *Chrysococcus, Kephyrion,* and *Stenocalyx* produce large populations in ponds. It is quite likely that these genera, which belong to the Euchromulinaceae, contain species

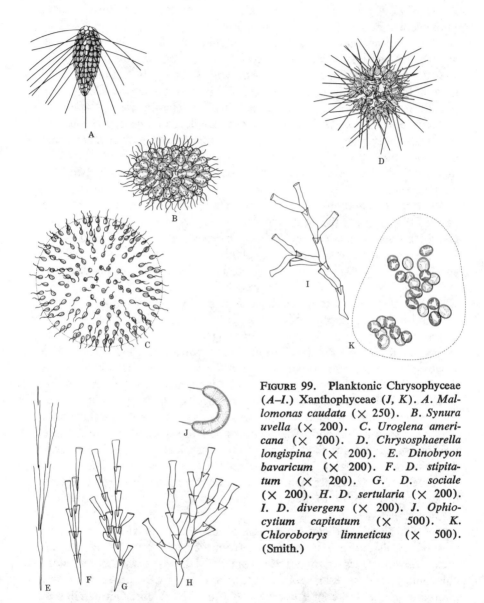

FIGURE 99. Planktonic Chrysophyceae (*A–I.*) Xanthophyceae (*J, K*). *A. Mallomonas caudata* (× 250). *B. Synura uvella* (× 200). *C. Uroglena americana* (× 200). *D. Chrysosphaerella longispina* (× 200). *E. Dinobryon bavaricum* (× 200). *F. D. stipitatum* (× 200). *G. D. sociale* (× 200). *H. D. sertularia* (× 200). *I. D. divergens* (× 200). *J. Ophiocytium capitatum* (× 500). *K. Chlorobotrys limneticus* (× 500). (Smith.)

that make a considerable contribution to the nannoplankton of many freshwaters. Several species of *Chromulina* are also important neustonic organisms. In the family Mallomonadaceae, the important genus *Mallomonas* contains more than fifty species, some of which appear to have a rather restricted ecology. Thus *M. akrokomos* seems to be a winter form in Europe, in both less and more productive lakes, though more abundant in the latter. *M. caudata* (*A* of Fig. 99) is apparently found mainly in less productive lakes, whereas *M. acaroides* is a summer form in the more productive localities. *Chrysosphaerella* is a colonial form, the cells of which bear silica rods (*D* of Fig. 99).

In the Isochrysales the most important planktonic genus is probably *Synura* (*B* of Fig. 99). The Coccolithophorida (*C* of Fig. 98), with a few primitive fresh-water genera (*Acanthoica* and *Hymenomonas*), are attached to this order by some investigators, including Huber-Pestalozzi (1941).

The Ochromonadales are the most important chrosomonads in the plankton of lakes. Most of the significant genera are colonial, *Uroglena* (*C* of Fig. 99) and particularly *Dinobryon* (*E–I*, Fig. 99) being the most widely distributed and best known. The unicellular forms such as *Ochromonas* (*D–G*, Fig. 98) and *Pseudokephyrion* are well known in the nannoplankton of ponds, and it is possible that when the sparse ultraplankton of unproductive lakes becomes better known various chrysomonads will prove to be important.

A few of the nonflagellate, amoebiform Rhyzochrysidales, notably *Rhizochrysis limnetica* and *Chrysidiastrum catenulatum,* are planktonic. Among the palmelloid genera of the Chrysosphaerales the genus *Stichogloea* is euplanktonic.

A large number of the Chrysomonadina produce siliceous cysts, the form and ornamentation of which are characteristic (*E, G, I, K* of Fig. 98). These cysts have great potential significance in palaeolimnology, but, in spite of some recent progress, at present more are known than can be assigned to the species that produce them. Nygaard (1956) has figured a great many of these cysts and has given details of their stratigraphic occurrence in the sediments of Lake Gribsø, although admitting that at the time when he wrote they throw little more light on the history of the lake than would be gained from a study of the ornithology of a region solely on the basis of eggs unassigned to definite species of birds. Leventhal (in press) has also found a great number of taxa of such cysts in the deeper sediments of Lago di Monterosi, deposited when the water was soft and probably slightly acid; they disappear almost entirely in the later eutrophic hard-water phase of this lake.

Nearly all the Chrysophyceae studied so far need one or more vitamins, the only known exception being *Stichochrysis immobilis*

(Provasoli 1958). The requirements are usually for thiamin or B_{12} or both, but a few species also need biotin.

Ecologically, the most striking nutritional character noted in the group is the obligate low-phosphorus requirement reported by Rodhe (1948) for *Dinobryon divergens* and for *Uroglena americana*. These findings are in complete accord with the field data to be reported later (pages 455, 456). It must be noted, however, that Rosenberg (1938) had concluded earlier that *Dinobryon* required a very high phosphorus concentration in culture. The whole problem of the cultivation of the genus in the laboratory is apparently a difficult one. Guseva's (1935) finding of a high phosphorus requirement for *Synura petersenii* is also to be noted.

Under laboratory conditions *Synura* appears to require much more zinc than do other algae cultivated in the same medium. The presence of a chelating agent makes comparison with the concentrations likely to occur in nature of no significance, but the comparative observations of Provasoli and Pintner (1953) on this matter are most striking and may well have ecological significance.

Planktonic Bacillariophyceae. The diatoms, certainly the most important members of the fresh-water phytoplankton, are nearly always present in significant numbers and in many, probably most, lakes are perennial dominants. It must be emphasized that most genera are benthic-littoral and that when planktonic species occur they are often in a minority in their genus. The group as a whole has commonly been divided into the centric diatoms exhibiting radial symmetry and the pennate diatoms which are bilateral in their fundamental structure. The most recent classifications (Hendey 1937; Patrick 1959) abandon this simple classification into two orders, but the descriptive terms centric and pennate are nevertheless convenient.

Among the centric forms (Fig. 100) the genera containing regularly euplanktonic species are *Cyclotella, Coscinodiscus, Stephanodiscus,* and *Melosira* (Coscinodiscaceae), *Chaetoceras* (euryhaline, Chaetoceraceae), and *Rhizosolenia* and *Attheya* (Rhizosolenaceae). Among the pennate forms (Fig. 101) the fresh-water planktonic species are included in *Tabellaria, Diatoma, Centronella* (one species euplanktonic), *Fragilaria, Asterionella* and *Synedra* (Fragilariaceae), *Cymbella* (mainly benthic, but *C. ruttneri* is euplanktonic in Indonesia, Cymbellaceae), *Denticula* (mainly benthic, *D. pelagica* in Indonesian plankton, Epithemiaceae), *Bacillaria* (euryhaline) and *Nitzschia* (largely benthic, Nitzschiaceae), *Surirella* (euplanktonic only in tropics), *Cymatopleura, Stenopterobia* (one planktonic species, Celebes), and *Campylodiscus* (mainly benthic, Surirellaceae).

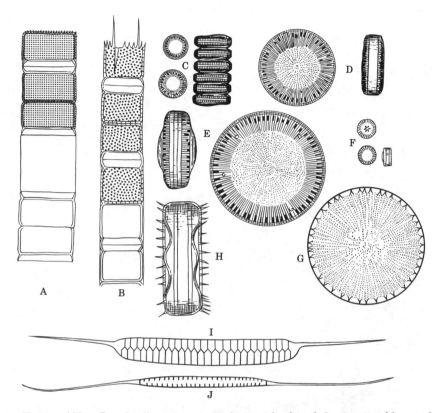

FIGURE 100. Centric diatoms. *A. Melosira islandica helvetica*, a widespread cold-water form, mainly in plankton in winter. *B. M. granulata*, a widespread species in the plankton of eutrophic lakes, mainly in summer. *C. Cyclotella comensis*, planktonic mainly in the larger subalpine and some small alpine lakes. *D. C. comta*, very widespread but by itself probably mainly in oligotrophic lakes. *E. C. bodanicola*, planktonic mainly in large subalpine lakes. *F. C. glomerata*, one of the smallest species in the plankton. *G, H. Stephanodiscus astraea*, widespread, mainly in eutrophic lakes. *I. Rhizosolenia eriensis*, mainly in large oligotrophic lakes. *J. R. longiseta*, apparently in more eutrophic localities than the last (all × 750 except *J*, × 375). (Huber-Pestalozzi and Hustedt.)

The most important genera in the temperate regions are those of the Coscinodiscaceae and the Fragilariaceae. There is obviously a tendency for some genera, notably *Surirella*, to some extent *Nitzschia*, and in a more limited way, *Cymbella*, which are strictly benthic in the temperate regions, to produce planktonic species in the tropics. The meaning of this is far from clear.

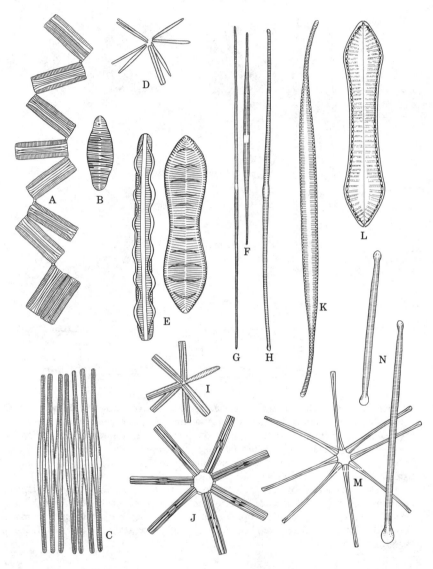

FIGURE 101. Pennate planktonic diatoms. *A. Diatoma vulgare* (× 333). *B.* The same (× 667). *C. Fragilaria crotonensis* (× 667). *D. Nitzschia holsatica.* *E. Cymatopleura solea* (× 500). *F. Synedra acus* (× 667). *G. S. acus angustissima* (× 667). *H. S. nunpens neogena.* *I. berolinensis* (× 667). *J. Tabellaria fenestrata* var. *asterionelloides* (× 356). *K. Stenopterobia belagica* (× 1650). *L. Surinella nyassae* (× 137). *M. Asterionella formosa* (× 310, 930). *N. A. gracillima* (× 930). (Huber-Pestalozzi and Hustedt.)

Many, if not all, the planktonic species of *Melosira* are really meroplankton and occur in the free water only when it is turbulent enough to prevent their settling. This very curious situation is discussed at length in a later section see (pages 458–462).

A small number of diatoms, most of which belong to the genus *Navicula,* appear to be facultative heterotrophs (Lewin 1953). Some members of *Nitzschia* also possess this faculty but evidently most do not; *Nitzschia putrida* is an apochlorotic obligate heterotroph. It is unlikely that heterotrophy will be found in the planktonic diatoms.

A large minority of the diatoms studied, in Provasoli's table (1958) seventeen of the thirty-seven species, require vitamins, usually B_{12}, sometimes thiamin, or both. Among those that do require B_{12} is the significant marine plankter *Skeletonema costatum;* of the fresh-water species that have been studied, the two most important eulimnoplanktonic forms, *Asterionella formosa* and *Tabellaria flocculosa,* are autoauxotrophic.

As indicated in Volume I (page 797), there is considerable variation in the optimal concentration of silicate required by different species of diatoms; the evidence that exists supports the reasonable hypothesis that the optimal concentration for planktonic species is lower, of the order of a few milligrams per liter or less, than that for benthic species which may grow best at a concentration of 30 mg. per liter.

It has already been pointed out that the fresh-water diatoms as a whole are smaller than the marine diatoms and that their size can be related to the environmental factors determining buoyancy, nutrient uptake, and predation. The largest planktonic species appear to be some in the great lakes of central Africa and in the larger lakes of Indonesia (taxonomic summary in Huber-Pestalozzi and Hustedt 1942; see also Thomasson 1955b).

Planktonic Dinophyceae. The dinoflagellates, though better developed in the sea than in fresh waters, contain a number of genera exhibiting marked evolutionary euryhalinity. Some of the species of these genera are important members of the plankton. Parenthetically, it may be remarked that Höll (1928) concludes that the separation of fresh-water and marine species is sharper than in the diatoms, with very few members of the group appearing in brackish waters; the great importance of certain brackish species, notably the red tide organisms, must not be forgotten.

In the Gymnodiniales (Fig. 102) species of the four genera *Amphidinium, Gymnodinium, Gyrodinium,* and *Massartia* are all known in lake plankton; the large genus *Gymnodinium* is the most important. In the Peridiniales (Fig. 103) the fresh-water planktonic genera are

Glenodiniopsis, Sphaerodinium, Hemidinium, Glenodinium, Gonyaulax, Woloszynskia, and, above all, *Peridinium* and *Ceratium,* but even in these genera marine species usually outnumber the fresh-water species. In a few quantitatively unimportant planktonic members of the Dinocapsales and Dinococcales the adult stages are nonmotile.

The nutrition of the dinoflagellates raises fairly complex problems which have been solved mainly for marine species, notably by Provasoli,

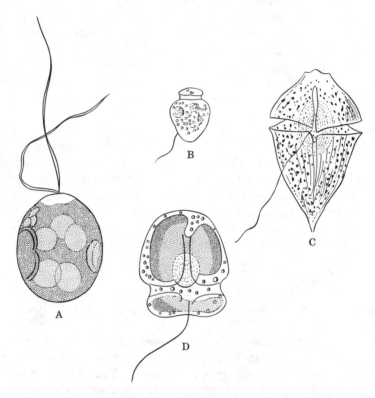

FIGURE 102. Dinophyceae, Adiniferae, and Gymnodiniales. *A. Desmomastix globosa* (× c. 1500), abandoned channels of Danube near Prague, a very primitive dinoflagellate of the subclass Adiniferae. *B. Amphidinium lacustre* (× c. 520), widespread in plankton of fresh and brackish waters of Europe, a species with numerous small brown chloroplasts. *C. Gymnodinium helveticum* (× 600), euplanktonic and holozoic, lacking chloroplasts and feeding on *Cyclotella,* small blue-green algal cells, and *Difflugia. D. Massartia plana* (× c. 1100), an autotrophic species from the plankton of Černé-jezero, a mountain lake in Šumava, Czechoslovakia. (Redrawn from Huber-Pestalozzi.)

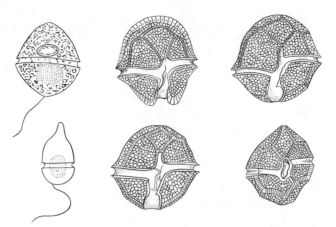

FIGURE 103. Dinophyceae: Peridiniales. *A. Glenodinium gymnodinium* (× c. 450), widespread in summer plankton of fairly hard waters, phototrophic. *B. G. apiculatum* (× c. 375), a phagotrophic species eating small diatoms, apparently mainly a winter form. *C. Peridinium willei* (× c. 410), eurytopic. *D. P. volzii* (× c. 440) hard-water species. *P. cinctum* (× c. 450), eurytopic. *P. gutwinskii* (× c. 420), an important plankter in certain eutrophic lakes in South East Asia. (Penard, Zacharias, Lefèvre, and Woloszynska.)

McLaughlin, and Droop (1957) and Provasoli (1958). Many of the Gymnodiniales are phagotrophic, and osmoheterotrophy is known for various species throughout the group. All but two of the eighteen species studied are alloauxotrophic; in nearly every case B_{12} is needed. In the solitary example of *Gyrodinium cohnii* Provasoli and Gold (1962) find independence of this vitamin but dependence on biotin and perhaps thiamin. *Gymnodinium brevis* and some species of *Amphidinium* need all three vitamins.

The ecological evidence assembled by Höll (1928) suggests that within the larger genera there are species adapted to acid, calcium-deficient and alkaline, calcium-rich waters as well as some that are eurytopic. Höll gives the following examples.

Water containing less than 10 mg. per liter Ca, pH below 6.0:

Peridinium raciborskii var. *palustre* apparently requiring water containing less than 5 mg. Ca per liter at $pH \leq 5.0$; variable organic (humic) content.

P. cinctum var. *tuberosum,* comparable to the preceding.

Gloeodinium montanum, not so oxyphil as the two preceding, but only in very humic water; phosphate optimum apparently very high.

Glenodinium lomnickii, in acidity and calcium and humic content of medium similar to the preceding, but without high phosphate optimum.

Peridinium lubieniense, similar to the preceding.

Glenodiniopsis uliginosum, transitional to the next group, in water up to *p*H 6.5 with Ca content up to 18 mg. per liter, but humic material high.

Water containing 10 to 18 mg. Ca per liter, *p*H usually 6.0 to 7.0:

Gymnodinium fuscum, humic material variable.

G. uberrimum, humic material variable; nutrient poor ponds.

Peridinium tabulatum, small humic ponds, slightly acid.

Gymnodinium palustre, apparently characteristic of clear unproductive slightly acid lakes.

Cystodinium cornifax, only in slightly acid humic ponds with high phosphate.

Water containing more than 20 mg. Ca per liter; *p*H more than 7.0; low or moderate humic content.

Diplopsalis acuta.
Kolkwitziella salebrosa.
Peridinium elpatiewskyi, quadridens, inconspicuum, munusculum, eximium, apiculatum, pygmaeum, volzi.
Glenodinium gymnodinium, penardii, cinctum, aciculiferum.
Gymnodinium helveticum, (?) *veris.*
Ceratium hirundinella (some forms).

Alkaline water, but often with fairly high humic content:

Glenodinium berolinense.
Gymnodinium neglectum, tenuissimum, aeruginosum.

Eurytopic:

Peridinium bipes, willei, cinctum.
Ceratium cornutum and some forms of *C. hirundinella.*

It is probable that some of the species in the first category require a low total mineral concentration in the medium.

There is some regulation by temperature. Among the species listed, the following appear to be cold-water forms:

Gymnodinium tenuissimum, veris, helveticum.
Glenodinium aciculiferum.
Peridinium quadridens, apiculatum.

The following species are warm water forms:

Kolkwitziella salebrosa.
Diplosalis acuta.
Glenodinium gymnodinium.
Peridinium elpatiewskyi.
Ceratium hirundinella.

It is possible that some of the correlations observed by Höll will not stand up to subsequent study, but on the whole his scheme is a convincing one, which, if combined with experimental work on the vitamin and trace-element requirements, might go far to define the specific fundamental niches of a group of phytoplanktonic organisms. Meanwhile, the general range of requirements of *Glenodiniopsis uliginosum, Gymnodinium fuscum, Peridinium cinctum,* and *P. willei* have been confirmed by Klotter (1955).

Interpretation of nutritional requirements of the phytoplankton. If we exclude apochlorotic species that are obligate osmoheterotrophs or phagotrophs, the phytoplankton can probably be divided into five categories on the basis of nutrition.

(1) Strict autoauxotrophic phototrophs, needing nothing that they cannot form from the products of photosynthesis and dissolved inorganic compounds and unable to use, as major sources of energy or matter, the organic compounds that may happen to be available. At least in some cases, such as that of B_{12} utilization by some blue-green algae, facultative uptake of vitamins may be possible. As far as is known, the great majority of planktonic fresh-water blue-green algae and most diatoms fall into this category.

(2) Autoauxotrophic but facultative heterotrophic phototrophs, able to live indefinitely, at least in some cases, in the dark on glucose, acetate etc. Many green algae among the Volvocales and Chlorococcales probably belong here. Provasoli (1958) concludes the facultative heterotrophs usually have higher optima for both phosphate and inorganic combined nitrogen. This may well explain the tendency for the lower green algae to be particularly abundant in highly productive waters, even though it seems unlikely that the concentration of easily assimilable organic compounds is likely to be high enough to be significant in the nutrition of such algae. A number of the very small μ-algae of various groups seem to be heterotrophic under the ice (Rodhe 1955).

(3) Alloauxotrophic but obligate phototrophs unable to use organic compounds as major sources of carbon or energy but dependent on an external source of some vitamins, generally a cobalamin. Some

dinoflagellates certainly belong here (Hutner and Provasoli 1955), including the fresh-water *Woloszynskia limnetica* cultivated by Provasoli and Pintner (1953) and Provasoli (1958). It is not unlikely that most Chrysophyceae and some fresh-water planktonic diatoms will also prove to belong in the category.

(4) Alloauxotrophic and facultatively heterotrophic. Such organisms, which require an external source of at least one vitamin and are able to use, though in the light are not dependent on, an external organic carbon source, include a few of the marine chrysomonads investigated, a few diatoms, some green algae, and nearly all the pigmented Euglenophyta studied so far.

(5) Alloauxotrophic and partly obligate heterotrophic, being unable to meet all major nutrient requirements by photosynthesis, though they can perform the latter function. This condition may involve phagotrophy, as apparently in *Oscillatoria splendida*. At least some strains of *Ochromonas malhamensis* (Hutner, Provasoli, and Filfus 1953) require an external source of organic matter for good growth in the laboratory, though other members of the genus such as *O. danica* (Aaronson and Baker 1959) can do much better as phototrophs under the culture conditions employed. Both species may need an organic hydrogen donator in photosynthesis. It is not unlikely that in such organisms the phagotrophic method of heterotrophy is a particularly efficient method not only of obtaining carbon compounds but specifically of rare vitamins such as biotin. This is clearly indicated by Aaronson and Baker for both species of *Ochromonas* fed on *Thiobacillus*.

Significance of alloauxotrophy. In the vast majority of cases the vitamins needed are a cobalamine, usually supplied as B_{12} in experiments, thiamin, and biotin, in that order of importance. This need is demonstrated by a diverse assemblage of flagellate forms, though so far no chlorophycean species requiring biotin is known. Provasoli speculates whether these vitamin requirements may not have a phylogenetic significance. In this connection it is important to note that the requirements are to some extent under environmental control, being greater in *Ochromonas* at high temperatures (Hutner, Baker, Aaronson, Nathan, Rodriguez, Lockwood, Sanders, and Petersen 1957). It is quite possible therefore that a species that appears to require a vitamin may actually be capable of limited but inadequate synthesis. This obviates the supposition that complete loss and subsequent regain of the capacity has occurred many times in the evolution of the phytoflagellates. Whatever may underlie the rather widespread need for the three vitamins cobalamine, thiamin, and biotin, rather than some others, it contrasts strikingly with the requirements of nonchlorophyllous tissues of flowering plants in tissue culture, which in general need thiamin, pyridoxine, and niacin (White 1954).

The known ranges of concentration of the three significant vitamins in solution in the open water of Linsley Pond (Volume I, pages 897–898; Benoit 1957) are

cyanocobalamin (*Euglena gracilis* assay) $0.06–0.075$ mg. m.$^{-3}$
thiamin $0.008–0.077$ mg. m.$^{-3}$
biotin $0.0001–0.004$ mg. m.$^{-3}$

The first two of these vitamins therefore can occur in quantities of the order of magnitude of one tenth of the lowest recorded soluble phosphate phosphorus concentrations in the lake water. Since in small ponds the cobalamin content can be as great as 2 mg. m.$^{-3}$ in late autumn (Robbins, Hervey, and Stebbins 1950), it is evident that the over-all range in inland waters is great and that we might expect large infertile lakes to contain much less cobalamine than Linsley Pond, just as small organic-rich ponds can contain much more. The quantity of cyanocobalamine in Linsley Pond, however, is below the optimum for the test organisms usually employed, so that there is a real possibility that in unproductive lakes the low concentration of vitamins might be limiting. Although the groups most likely to require vitamins, namely, the euglenophytes, dinoflagellates, chrysomonads, and cryptomonads, are probably best developed in fertile waters, all except the first-named groups have species clearly characteristic of dilute unproductive waters. Except for *Synura,* none of these rather definite oligotrophic organisms has yet been studied in culture, but in view of the all but universal need for cobalamines in the groups in question we may expect even such oligotrophic species to have a need for an external source of vitamins. In this connection it is interesting that at least the abundant cryptomonads and chrysomonads are mostly small and, more importantly, all are motile. Motility per se does not seem to be a prerequisite for life in the phytoplankton; the diatoms, which must all be denser as well as usually larger than the cryptomonads manage well without flagella. In view of Munk and Riley's work (see pages 294–301) on the relation of sinking speed to nutrient requirements, it is evident that motility, obviously a more efficient method of meeting rare molecules, even if the movement of the organism is at random, than downward sinking, which if continued inevitably leads to death, will have particular value to alloauxotrophic species in infertile waters (Hutchinson 1961).

SUMMARY

The photosynthetic phytoplankton consists of organisms of varying degree of autotrophy, though most of the species are largely or exclusively phototrophic. In general, we may expect that in the presence of suitable light and sources of C, N, P, S, K, Mg, Si, Na, Ca, Fe, Mn, Zn, Cu, B, Mo, Co, and V and the three vitamins thiamin, cyanocobalamin (B_{12}), and biotin nearly all algae can grow if the concentrations are correct and certain physical conditions are satisfied. At least a large minority of the planktonic algae clearly do not need any (autoauxotrophy) of the vitamins; most of the alloauxotrophic species require but one or two. It is probable that the maintenance of a very low but rather constant ionic concentration, which is often difficult in the laboratory, is an essential condition for the production of large populaons of many species of fresh-water phytoplankters.

The carbon source may be CO_2 in solution, or for some species (e.g., *Scenedesmus quadricauda, Kirchneriella contorta,* and probably many others) HCO_3 or perhaps $CO_3^=$; carbamine carboxylates, which may form by the complexing of calcium or other carbonates and amino compounds in alkaline waters are probably efficient sources of CO_2. A few planktonic algae are facultative heterotrophs; they are probably mainly pond forms and perhaps require higher concentrations of inorganic nutrients when phototrophic than do the obligate phototrophs. Among the inorganic nutrients, phosphorus, combined nitrogen, and, for diatoms, silicon appear to be the ordinary limiting substances; at least in the Alps and probably in other temperate regions phosphate is more likely to be limiting in winter, nitrogen in summer, except when nitrogen-fixing blue-green algae are present. Phosphorus may be more important relative to nitrogen in large unproductive lakes. There is some evidence that qualitative differences in nitrogen source may be important, but these differences have been reported largely without considering molybdenum, which is needed if nitrate is to be reduced and still more if N_2 is to be fixed. Nearly all, if not all, the Euglenophyta studied require reduced nitrogen. At least one case of limitation by low molybdenum concentration is known. Low or unavailable iron is certainly limiting in some marl lakes.

The Myxophyceae or blue-green algae, particularly the genera *Anacystis, Gomphosphaeria,* and *Coccochloris* in the Chroococcaceae, *Aphanizomenon* and *Anabaena* in the Nostocaceae, *Gloeotrichia* in the Rivulariaceae, and *Oscillatoria* and *Lyngbya* in the Oscillatoriaceae are of great importance in lake plankton. They are usually found in greatest abundance in productive lakes in summer, but when the nutrient concentration is low. Species of *Anabaena* are important nitrogen-fixing organisms. There is some evidence that high sodium may be conducive to large myxophycean populations. All the freshwater planktonic species seem to be phototrophs not requiring vitamins. They frequently form massive blooms and by development of pseudovacuoles can become less dense than water, floating at the surface. There is a suggestion that high manganese concentrations in the absence of calcium can limit blue-green algae in nature, but the detailed determinants of their occurrence are far from clear.

The planktonic Chlorophyta or green algae consist of three major groups of organisms of very different nature. The first group consists of the planktonic members of the two lower orders, the Volvocales and Chlorococcales, in which at least the gametes are flagellate. These seem to occur most abundantly in ponds or small productive lakes, *Pediastrum* and *Scenedesmus* being perhaps the most important, though species of *Oocystis* apparently are exceptional and can be dominants or subordinants in the plankton of deep unproductive bodies of water. Many species are known to be facultative heterotrophs, though in most lakes this property is unlikely to be significant. A large minority require thiamin or B_{12} or very rarely both.

The second group consists of the anomalous *Botryococcus* which has often been placed in other classes, though its pigments now show it to be a green alga. It can be extremely abundant, but under conditions that are so varied that nothing can be said of its ecological determination. It appears to be a phototroph not needing vitamins.

The third group of green algae consists of the planktonic desmids of the order Conjugales. A spectacular desmid flora develops in regions of dilute acid waters, low in calcium and magnesium. Many species are sphagnicolous, and the diversity of the littoral benthos and the flora of terraqueous habitats in such acid regions is quite extraordinary. In the lakes of these areas a desmid plankton consisting largely of species of *Staurastrum* and *Staurodesmus* is well developed. These genera are often accompanied by rarer and more spectacular species of *Micrasterias, Xanthidium,* and the chain-forming *Spondylosium, Cosmocladium,* and *Hyalotheca.* The rarer desmids in the plankton are usually adventitious, and there is evidence of derivation of planktonic populations, if not races, from littoral benthic forms. In addition to the typical desmids of acid waters, a few species of *Staurastrum* are regular members of the hard-water plankton of quite productive lakes and may at times be dominants. Various adventitious species of other genera occur under like conditions. The nannoplankton of acid lakes rich in desmids, consisting largely of flagellates and very small diatoms, may greatly outnumber the desmids, so that the common idea of a desmid plankton in soft-water regions is partly due to an artifact of the collecting gear employed.

The Euglenophyta, often abundant in small pools rich in organic matter, are unimportant in most lakes, though *Trachelomonas, Lepocinclis,* and a few species of *Euglena* are widely recorded, and *Colacium* is a widespread epiplankter on various members of the zooplankton. Ammonia may be the only significant inorganic source of nitrogen for most species; the group as a whole is facultatively heterotrophic and all species studied need thiamin, B_{12}, or both.

The Cryptomonadineae are a small group but contain in *Rhodomonas* and *Cryptomonas* several species of considerable importance in the nannoplankton of lakes. The species studied so far need thiamin, B_{12}, or both.

The Chloromonadineae are a small group, little known biochemically; *Vacuolaria* is the only important genus in the open water of lakes.

The Xanthophyceae are a diversified group with several eulimnoplanktonic genera. They are little known biochemically and in general are quantitatively unimportant.

The flagellate Chrysophyceae include in the Chromulinales a few genera with planktonic species; the most important is *Mallomonas,* which has a large number of species, many of rather restricted ecology. The genus *Synura* in the Isochrysales is also of some importance. The Ochromonadales include the very important genera *Uroglena* and *Dinobryon.* The group as a whole produces remarkable siliceous cysts which are sometimes found in great numbers in lake sediments; their specific characters are striking but many cannot yet be assigned to an organism known in the vegetative state. Almost all species studied need vitamins, usually thiamin, B_{12}, or both, but some also need biotin. In some genera such as *Ochromonas* a considerable degree of phagotrophy is developed; this may be the main way of obtaining vitamins. *Uroglena* and *Dinobryon* appear to be sensitive to an excess of phosphate, the quantity tolerated being less than the optimum for most other phytoplankters. This is important in determining their distribution and seasonal incidence.

The Bacillariophyceae, or diatoms, are probably the most important group in the fresh-water phytoplankton, the more significant genera being the centric *Cyclotella, Stephanodiscus,* and *Melosira* and the pennate *Tabellaria, Fragilaria, Asterionella,* and *Synedra.* There is a curious tendency for genera (*Surirella, Nitzschia,* and *Cymbella*), which are mainly or entirely benthic at high or moderate lattitudes, to have species in the plankton of tropical lakes. *Melosira,* which is really meroplanktonic, requires considerable turbulence to remain suspended but is reproductive only when in the plankton. A few species, mostly of *Navicula* and *Nitzschia,* are facultative heterotrophs; these are all benthic littoral organisms. A large minority of diatoms require B_{12} or thiamin and sometimes both. Fresh-water planktonic species, however, appear to be autoauxotrophic. The silicon requirement of planktonic species is probably lower than that of the benthic diatoms.

The Dinophyceae contain about a dozen genera, species of which contribute to the fresh-water plankton; *Gymnodinium, Peridinium, Cystodinium,* and *Ceratium* are the most important. The group is usually alloauxotrophic, nearly all species needing B_{12} and some thiamin and biotin as well. There is evidence of considerable chemical determination of the occurrence of certain species which may be confined to dilute acid or to more concentrated hard waters.

Excluding apochlorotic species which are probably not important in the eulimnoplankton, we may recognize five trophic groups of phytoplanktonic organisms:

(1) Strictly autoauxotrophic phototrophs, needing nothing that they cannot form from the products of photosynthesis and dissolved inorganic compounds and unable to use organic compounds as major sources of matter or energy. The majority of planktonic fresh-water blue-green algae and diatoms fall in this group.

(2) Autoauxotrophic but facultative heterotrophic phototrophs, which differ from the first category in their use of glucose, acetate, etc., in the dark. A number of Volvocales and Chlorococcales belong here. They are usually pond forms, and although it is doubtful that in most of their habitats there is a concentration of assimilable organic matter high enough for heterotrophic nutrition to be significant it is not unlikely that these plants also need rather high concentrations of inorganic nutrients when living as phototrophs.

(3) Alloauxotrophic but obligate phototrophs, unable to use organic compounds as major sources of matter or energy; cobalamine is usually required but thiamin or biotin may be needed. Some dinoflagellates certainly belong in this group, and it is probable that at least some chrysophyceaen planktonic genera, including *Synura,* and some diatoms will be found to belong to the category.

(4) Alloauxotrophic and facultatively heterotrophic, using simple organic compounds when present in the dark. A few diatoms, some chrysomonads and green algae, and almost all of the Euglenophyta belong here.

(5) Alloauxotrophic, with chlorophyll, but not fully autotrophic. Some species of *Oscillatoria* and some chrysomonads such as

Ochromonas malhamensis belong in this category, which is transitional to the obligate apochlorotic heterotrophs.

It is to be noted that even though the green and yellow algae are not closely related, and the green have clearer affinities with the higher plants than the yellow, a cobalamine, thiamine, and biotin are the vitamins needed by the alloauxotrophic unicellular forms, whereas in the higher plants tissue culture experiments emphasize the importance of thiamin, pyridoxin, and niacin, of which the last two seem not to be required by the alloauxotrophic algae.

The majority of the fresh-water plankters of groups seemingly requiring vitamins are small and flagellate; many such organisms are found in quite unproductive waters. It is suggested that motility is of value in bringing organisms into renewed supplies of organic or inorganic molecules existing in very low concentration. The other main method of dealing with nutrition at considerable dilution is sinking through turbulent water; even random movement is less dangerous than movement directed downward away from the light.

CHAPTER 22

Phytoplankton

Associations

T he most important feature of phytoplankton associations, though it is seldom appreciated and perhaps has never been fully explained, is that in nearly every case, even if only a few cubic centimeters of the environment are taken as a sample, a number of different species are present. Often one species, in greater abundance than any other, constitutes a dominant; sometimes there are one or more obvious subdominants, but, except in water characterized by very extreme ecological conditions, a little study will always reveal several rarer species in the presence of the dominant and subdominants. Since a similar experience is provided by the majority of areas covered with herbaceous vegetation, and often by forests as well, the remarkable nature of observations of this sort has seldom received the attention it deserves.

CONDITIONS FOR MULTISPECIFIC EQUILIBRIUM

If we imagine a uniform bounded biotope, in which a number of species are competing with slightly varying efficiency for the same resources, which they require in the same proportion, it is easy to show mathematically, provided no commensalism or symbiosis is occurring, that no equilibrium exists when more than one species is present. This result is general, in the sense of being independent of the form of the competition function $f(N_{i \neq m})$ in the equation,

$$(1) \qquad \frac{dN_m}{dt} = \frac{b_m N_m}{K_m} [K_m - N_m - f(N_{i \neq m})]$$

where N_i = the populations of the species present,

$\qquad N_m$ = the population of the mth species,

$\qquad b_m$ = the Malthusian parameter or unrestricted rate of reproduction of the mth species,

$\qquad K_m$ = the saturation population of the mth species living alone.

The ordinary elementary form of the equation for two competing species is

$$(2a) \qquad \frac{dN_1}{dt} = \frac{b_1 N_1}{K_1} \{K_1 - N_1 - \hat{\alpha} N_2\}$$

$$(2b) \qquad \frac{dN_2}{dt} = \frac{b_2 N_2}{K_2} \{K_2 - N_2 - \hat{\beta} N_1\}$$

where $\hat{\alpha}$ and $\hat{\beta}$ are linear competition coefficients. It is easy to show that in a deterministic system the only condition for coexistance of the two species at equilibrium is when

$$(3) \qquad \frac{K_1}{K_2} > \hat{\alpha}, \qquad \frac{K_2}{K_1} > \hat{\beta}$$

This implies that the effect of any individual of S_1 on an individual of S_2 is less than its effect on another individual of S_1 and vice versa. Although partial spacial separation is an obvious way of bringing this situation about, it can be achieved more generally by having some part of the niche of S_1 defined in the way previously given and not included in the niche of S_2 and *vice versa:*

The case in which

$$(4) \qquad \frac{K_1}{K_2} < \hat{\alpha}, \qquad \frac{K_2}{K_1} < \hat{\beta}$$

corresponds to mutual antibiosis; the species surviving at equilibrium depends on the initial concentrations.

The two cases

$$(5a) \qquad \frac{K_1}{K_2} > \hat{\alpha}, \qquad \frac{K_2}{K_1} < \hat{\beta}$$

$$(5b) \qquad \frac{K_1}{K_2} < \hat{\alpha}, \qquad \frac{K_2}{K_1} > \hat{\beta}$$

correspond to displacement of S_2 by S_1 and of S_1 by S_2 respectively. Cases in which the inequality signs are replaced by equality signs are excluded by the axiom of inequality, discussed in a later paragraph.

These formulations are independent of the form of the competition function $f(N_{i \neq m})$, provided they are always positive.

A stochastic theory (Feller 1939; Leslie, and Gower 1958), which indicates that in small populations there is a high probability of this result, also exists. In the very large populations ordinarily encountered in the plankton the determinate theory is certainly adequate.

The principle involved in this theoretical treatment is commonly referred to as Gause's principle, because experimental confirmation of its validity was first given by G. F. Gause (1934, 1935), following the mathematical investigations of Volterra (1926). It would appear, however, that the mathematical treatment for a pair of species was implicit in earlier work of Haldane (1924) and that the principle had been grasped intuitively from field studies by Steere (1894; cf. Rand 1952) and by Grinnell (1904; cf. Udvardy 1959). Hardin has shown that isomorphous theory was familiar to nineteenth century economists. In view of this gradual development of our understanding of the phenomenon, Hardin's (1960) neutral term, the principle of *competitive exclusion,* is perhaps a better name than one based on the several investigators who have contributed to the matter. It should be realized, however, that historically the great insight of Grinnell was unappreciated by his generation and that it was the work of Volterra and of Gause that first brought competitive exclusion squarely before ecologists.

Application to the phytoplankton. When we find, as in most phytoplankton associations, that the principle does not appear to hold, we may reasonably ask which of the postulates it is based on is erroneous. The following possibilities present themselves.

Firstly it might be argued that the competing species had exactly equal efficiencies in competition. This would be contrary to what Hardin (1960) calls the *axiom of inequality,* namely, that no two material systems are ever exactly alike. Riley (1963) feels that natural selection might operate to increase the efficiency of two species living in essentially identical environments so that these efficiencies approached asymptotically an upper limiting value set by the general biology of the organisms. If the organisms then entered into competition, the differences in efficiency might be so small that competitive exclusion would proceed at an undetectably slow rate. This could happen only when the demands of both species on the environment were initially identical and could be met in only one way. It is conceivable, however, in the phytoplankton. Since the rates of competitive displacement certainly vary, Riley's position seems theoretically possible in large populations in which the probability of random extinction is very small (see page 373, footnote 3).

Secondly, it may be supposed that in spite of appearances to the contrary the biotope is not really homogeneous and provides more than one niche. This question is discussed in the next section.

Thirdly, it is possible that the requirement that all competition coefficients be positive may not be satisfied, so that commensalism and symbiosis occur. This is probable in a rather limited array of examples.

Fourthly, it might be supposed that the assemblage of species in the phytoplankton never really approaches competitive equilibrium so that the prediction of theory implying equilibrium can never be relevant.

Nothing more can be said about the first possibility; the second and third possibilities are discussed in the next few paragraphs. The problem of nonequilibrium associations is fairly complicated and involves a long discussion of the kinds of distribution of numbers of individuals in the whole assemblage of species, given in the next section of this chapter.

Possible modes of physical niche diversification. It must be borne in mind that the problem as we have envisioned it might prove to be specious. It is supposed that apart from its inevitable light gradient the water of the euphotic zone is homogeneous, turbulent, and of zero stability. To a rough approximation this is certainly true of all lakes that develop a well-marked epilimnion. It is possible, however (cf. Lund 1964), that the approximation is too rough for the purposes of the present discussion. There is bound to be a detectable temporary thermal stratification near the surface on bright days in any but the most disturbed lake; chemical differences may be associated with the water surface (cf. Volume I, page 855), and the horizontal inhomogeneities associated with the Langmuir spirals in the wind drift certainly can effect motile and nonmotile organisms differently (see page 293), leading to a partial separation of populations, but it is difficult to believe that any of these effects could be persistent enough to have much significance in the environmental regulation of competition.

Predation as a possible means of niche diversification. If a pair of species S_1 and S_2 is competing in an environment in which S_1 would, in the absence of predation, completely displace S_2, it seems probable that the introduction of a predator S_3 feeding solely on S_1 might produce some niche diversification. If it is assumed that the populations of S_1 and S_3 come to equilibrium or oscillate with small amplitudes around a singular point at which equilibrium would be possible and the reduction of S_1 by the predator is great enough, its effect will be to liberate permanently some biotope space that S_1 would have occupied but is now available for S_2. The elementary and somewhat simplified mathematical statement of the rates of change of the populations of S_1, S_2, and S_3 is given by

$$(6a,b,c) \quad \begin{cases} \dfrac{dN_1}{dt} = \dfrac{b_1 N_1 \{K_1 - N_1 - \hat{\alpha} N_2 - \hat{\gamma} N_3\}}{K_1} \\[2mm] \dfrac{dN_2}{dt} = \dfrac{b_2 N_2 \{K_2 - N_2 - \hat{\beta} N_1\}}{K_2} \\[2mm] \dfrac{dN_3}{dt} = \hat{\delta} N_1 N_3 - d N_3 \end{cases}$$

where $\hat{\alpha}$ and $\hat{\beta}$ are competition coefficients (in this case, $\hat{\alpha} < 1 < \hat{\beta}$), $\hat{\gamma}$ is the predation coefficient, $\hat{\delta}$ is a suitable fraction of γ allowing for the efficiency of converting the prey into predator, and d the death rate of the predator in the absence of prey. The three derivatives are zero when

$$N_1 = \frac{d}{\hat{\delta}}$$

$$N_2 = K_2 - \frac{\hat{\beta} d}{\hat{\delta}}$$

$$N_3 = \frac{1}{\hat{\gamma}} \left(K_1 - \frac{d}{\hat{\delta}} - \hat{\alpha} K_2 + \frac{\hat{\alpha} \hat{\beta} d}{\hat{\delta}} \right)$$

Since both d and $\hat{\delta}$ must be positive, the conditions for this point being in the positive quadrant are

$$(7a) \qquad K_2 > \frac{\hat{\beta} d}{\hat{\delta}}$$

and

$$(7b) \qquad K_1 - \hat{\alpha} K_2 > \frac{d}{\hat{\delta}} (1 - \hat{\alpha} \hat{\beta})$$

Moreover d must be greater than $\hat{\delta}$; the considerable inefficiency of some zooplankters which pass many plant cells through their guts undigested (see page 554) may be involved here.

Commensalism and symbiosis. The possibility that some of the competition coefficients may be negative must be taken seriously. We have seen that some species of the phytoplankton require vitamins and that they are in general small motile forms. The smallness and motility would promote the uptake of mineral nutrients and accessory organic substances. If, as seems likely, at least in the open water of large lakes, the accessory organic substances are derived from the phytoplankton, a commensal situation between smaller motile alloauxotrophic and larger nonmotile, purely phototrophic species could be set up. This situation is an inherently probable one that may contribute to the phytoplankton diversity. A few examples of the stimulation of one species of phytoplankton by water in which another has been reproducing are given in the next chapter; it must be admitted, however, that inhibition is far more likely to occur than stimulation.

In spite of inherent possibilities that predation and commensalism, or perhaps symbiosis, may be producing diversity, it seems rather unlikely that the great variety usually observed could be wholly the result of these causes. It is therefore desirable to determine whether the phytoplankton association is really never in equilibrium. Before this can be done it is necessary to present certain methods of measuring the diversity of a multispecific assemblage.

The measurement of diversity. Before considering a theoretical approach which will throw light on the problem of equilibrium and nonequilibrium assemblages, it is convenient to examine three proposed empirical approaches to the problem of the distribution of specific diversity.

The index of diversity. The first widely used procedure is that put forward by Fisher, Corbet, and Williams (1943), who concluded that the distribution of species of insects in a collection supposedly made at random followed a limiting case of the negative binomial distribution, so that the numbers of species (S_n) represented by $N_n = 1, 2, 3, \ldots$ individuals is given by the series

$$(8) \qquad \alpha' x, \; \frac{\alpha'}{2} x^2, \; \frac{\alpha'}{3} x^3 \cdots$$

where x is a number very little less than unity and dependent on the size of the sample. For a large sample α is a constant called the *index of diversity*. For the rare species the distribution is very nearly the harmonic series

$$(9) \qquad \alpha', \; \frac{\alpha'}{2}, \; \frac{\alpha'}{3} \cdots$$

but as higher powers of x fall appreciably short of unity the numbers for the commoner species will be appreciably smaller than those given by the harmonic series. The biological assumptions behind this theory are very general, namely, that the number of individuals and species in any universe is finite, that the species are not all equally common, and that the distribution of commonness or rarity is of the simplest possible (i.e., Eulerian) form. The chief value of the theory is that it supplies, from collections in which the less common species have been taken whenever they were encountered, as is usual in qualitative floristic or faunistic work, a measure of the diversity of the fauna or flora under investigation.

Estimation as a lognormal universe. If the number of species with 1, 2, 3 · · · individuals followed the harmonic distribution and such numbers were grouped geometrically, the first group containing say all species having between 1 and 2 individuals,[2] the next group between 2 and 4 individuals, etc., the number of specimens for the first group would be 0.75 α' and the number in each successive groups or octaves would rapidly decline, converging on α' ln 2 or 0.693 α'. If the Fisher, Corbet, and Williams distribution rather than a strict harmonic distribution were followed, the number in each successive group or octave would slowly decline. Preston (1948) made the interesting discovery that when data for bird populations and catches of insects at moth traps were plotted in this way the curve descended not merely on the right but on the left also, a distinct mode being apparent whenever extensive data were available. This behavior appears to be quite regular and systematic, though it can be demonstrated only in the relatively few cases for which there are very large samples. As the sample increases, the curve moves across the paper from left to right, the mode becoming more and more apparent as more on the left-hand side of the curve is unveiled. This has been beautifully shown by Patrick, Hahn, and Wallace (1954) in counts of populations of diatoms which settled on slides in a Catherwood diatometer set in a stream. These workers counted an enormous number of specimens and were able to show in one example that after 8595 specimens, representing 157 species, had been counted, the mode lay at the right-hand end of the third octave, whereas after counting 35,092 specimens, representing 195 species, the mode had moved an octave to the right. Another comparable un-published case from the same laboratory is given in Fig. 104.

Preston concludes that the curves obtained in this way are log normal

---

[2] In practice, half those with one and half those with two species for the first group, half those with two, all those with three, and half those with four for the second group, etc. This method of plotting is, of course, equivalent to constructing a histogram of the number of species when the species are grouped by logarithms of specimens to the base 2. This choice of a base is dictated solely by convenience; in most cases, if a larger base is used, say 10 or even e, there are insufficient data to bring out the significant features of the resulting diagram. For the full procedure of fitting see Patrick, Hahn, and Wallace (1954).

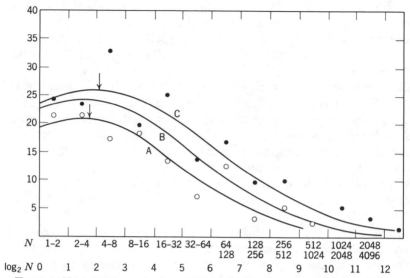

Figure 104. Preston or log-normal distributions of number of species having 1–2, 2–4, 4–8 individuals, thus constituting a scale of octaves or logarithms to the base 2, for diatoms attached to a diatometer slide immersed in Ridley Creek and collected October 5, 1956. *A.* Curve based on 5896 specimens, mode at 3.40, estimate for total assemblage, 188 species, individual points as open circles. *B.* Curve based on 15,732 specimens, mode at 3.81, estimate for total assemblage, 240 species. *C.* Curve based on 28,831 specimens, mode at 4.10, estimate for total assemblage, 263 species. (Patrick, unpublished.) Note movement mode to right as in Preston's treatment but also movement upward, which causes the increase in the estimate of the total number of species.

curves. If this is so the area under the whole curve from $-\infty$ to $+\infty$ is a measure of the total number of the species in the universe being sampled. MacArthur (personal communication), however, has noted that the Preston curves, when sufficiently unveiled, generally have a somewhat asymmetrical form; if any value of the ordinate above which the entire curve is in the wholly positive quadrant is taken, the area to the left of the mode will be a little greater than that to the right. This will imply a slight increase in the total theoretical population if the curve is fitted as log normal as more specimens are counted. Patrick, Hahn, and Wallace found this to be the case.

Information-theoretic approach to the diversity of the plankton. Margalef (1956, 1957b) has considered the problem of the diversity of the phytoplankton as a problem in information theory, and it is probable that for some purposes this approach will prove by far the most useful, since it is free from theoretical assumptions regarding the nature of the distributions involved.

If we imagine a monospecific assemblage, the identification of any single specimen will give as much information about the specific composition of the whole as does the identification of any other individual. If, however, every specimen in a sample belongs to a different species, each must be identified before the information provided by the sample is exhausted. In various intermediate situations the amount of information carried by a single individual will vary with the number of species and their relative frequency.

Ideally, the appropriate measure of information per individual is given by

$$(16) \qquad D_N = \frac{1}{N_s} \log_2 \frac{N_s!}{N_1!N_2! \cdots N_m!}$$

where there are m species and the total number of individuals of all species is N_s. If we have estimates of the number of individual species in a water sample of volume V_s, the information per unit volume is

$$(17) \qquad D_V = \frac{1}{V_s} \log_2 \frac{N_s!}{N_1!N_2! \cdots N_m!}$$

The results are given in *bits* (or *binits,* binary units of information), the use of the base 2 being dependent on the fact that in the theory information appears as a series of yes-no choices.

These expressions are clumsy to evaluate, though they can be made tractible by Sterling's approximation and by a table of ln $N!$ ($\log_2 N = 1.443$ ln N). Expression 16 has been evaluated by Margalef (1957b) for estuarine plankton in the mouth of the River Vigo and shows a maximum in the region in which fresh- and salt-water species are mixed.

A much easier approximate expression is

$$(18) \qquad D = - \sum_{i=1}^{m} p_i \log_2 p_i$$

where p_i is the probability of occurrence of the ith species. In an ordinary sample of specimens we may take $p_i = N_i/N_s$. This expression, commonly referred to as Shannon-Wiener information, has the same form as that for entropy in thermodynamics.

When the assemblage consists of a single species, $D = 0$; there is no uncertainty and no more information is gained when a random individual is identified. For a large number of species D is maximal when

all species are equally abundant, and the p_i therefore are equal.   In
such a case **D** increases as the number of species increases.

This particular form of information-theoretic diversity has been effectively employed by MacArthur (1955; MacArthur and MacArthur
1961) in studies of the stability of biological communities and in various
aspects of terrestrial ecology.   It is a convenient estimate of diversity
for comparison with environmental variables, which in diversified environments can often be expressed in the same forms (Crowell 1961;
MacArthur and MacArthur 1961).   It is less sensitive to the presence
of a varying number of very rare species than is the index of diversity
of Fisher, Corbet and Williams or the total log normal universe of
Preston.   This results in very little change in the diversity when, for
example (Table 18), the four separate seasonal lists of diatoms in
Braendegard Sø are compared with the whole planktonic list for all
seasons added together.   In all cases six perennial or almost perennial
species (*Cyclotella comta, Fragilaria construens, F. pinnata, Melosira
ambigua, Navicula gracilis,* and *Synedra acus* var. *angustissima*) contribute together about half the value of **D**.   The higher value for the
littoral flora, however, is noticeable.

Opportunistic and equilibrium species.   MacArthur (1957, 1958)
has approached the problem of number of species and number of individuals from a deeper biological standpoint than that adopted by previous investigators.   He first considers the abundance of a species S_m
at time t, namely $N_m(t)$.   Writing for the relative rate of increase or
decrease per unit of S_m.

$$(10) \qquad\qquad r_m(t) - \frac{1}{N_m(t)} \frac{dN_m(t)}{dt}$$

and integrating, we obtain

$$(11) \qquad\qquad \log N_m(t) = \log N_m(0) + \int_0^t r_m(t)\, dt$$

If we assume that the species to be derived from a community in
perennial approximate *equilibrium,* $N_m(0)$ may be taken as the mean
population of S_m at equilibrium and $r_m(t)\, dt$ as a small corrective term,
depending on the changes that have occurred as N_m has been displaced
a little from $N_m(0)$ and as it is drifting back to its equilibrium value.
Since in the equilibrium case $r_m(t)$ will obviously be density dependent
in magnitude and sign, $\int_o^t r_m(t)\, dt$ will vary around zero.

As an extreme alternative, we may consider species whose populations
are *opportunistic,* which rise rapidly to great numbers when conditions

are favorable and decline when they are not, the changes being due primarily to variable external factors such as the weather. In such a case the value of $N_m(0)$ will be a negligibly small initial population of no relevance to the later values of the accumulated integral, the magnitude of which will depend on the recent values of $r_m(t)$ and so on the factors on which these values depend.

Interpretations of the lognormal curve. It is known that in an opportunistic case the values of the accumulated integrals $r_i(t) \, dt$ become normally distributed if all species vary independently, that is, if density independence, both intraspecific and interspecific, is complete. This means, if the log $N_i(0)$ can be neglected, that the log $N_i(t)$ are normally distributed or the $N_i(t)$ lognormally distributed. This is one way, though perhaps not the only way, in which the Preston type of distribution might be established in nature.

Preston gave the equation of his lognormal curve as

$$(12) \qquad\qquad S_x = S_M e^{-[\delta(x_M - x)]^2}$$

and further noted that

$$(13) \qquad\qquad S_t = S_M \frac{\sqrt{\pi}}{\delta} = 1.77 \frac{S_M}{\delta}$$

where S_x = the number of species in an octave,

$\quad\ S_M$ = the number of species in the modal octave,

$\quad\ S_t$ = the total number of species in the universe under investigation,

$\quad\ x$ = the distance of the octave along the abscissa,

$\quad x_M$ = the value of x for the modal octave,

$\quad\ \delta$ = a number, which is given by $\delta = 1/\sqrt{2\sigma^2}$.

Preston pointed out that the value of δ is usually very close to 0.2. Actually, the values given in his paper, derived from populations of birds and of moths taken at traps, exhibit a range of 0.160 to 0.227. A comparable range of 0.193 to 0.275 is given by various studies on diatoms. These figures correspond to an over-all range in the value of σ of 2.57 to 4.42. Larger values are apparently given when very large populations, such as all the birds of North America, are considered and when it may be assumed that a great environmental diversity contributes to the dispersion of the assemblage.

It is reasonably certain that the approximately lognormal form obtained in such instances cannot be due to all the species being opportunistic, for in the case of birds there is adequate evidence that most are not. As MacArthur (1960) has pointed out, there is reason to believe that the same sort of lognormal distribution may arise by a process of evolutionary opportunism among competing species. We are therefore not restricted to the opportunistic case in examining the lognormal type of distribution.

It would seem likely that if the dispersion were very great so many species would have to be so common that all of their various habitats could not be represented adequately in the biotope space inhabited. At the same time, an equally large number of species would be so rare that their chance of being represented by even a single specimen in the biotope space would be small. There would, in fact, have to be excessive overpopulation by the commoner species to permit the existence of the rare species. This is in itself an unstable and paradoxical situation that we should not expect. If $\gamma = 0.05$ and we had $S_t = 100$, two million specimens would be counted in the average case to reach the mode when only half the species would have been discovered.

It is less clear why, with a total number of 100 species and a lognormal distribution with $\gamma = 0.4$, the situation is equally unrealistic. This distribution, when unveiled to the mode, would give twenty-one species in the first, sixteen in the second, eight in the third, four in the fourth, and one in the fifth octave, with about 190 specimens to be counted in the average case. It is evident that this represents too constant a number of species per individual. Unless we are dealing with complete opportunism, this uniformity would imply an excessively uniform division of the biotope into volumes corresponding to niches.

It is reasonable to suppose that what is called in the next section a type I distribution in ordinary diverse biocoenoses represents the limit of the stable distribution of specific habitat volumes under competition and that ordinarily this limit implies more small specific habitats than would be present when γ is well in excess of 0.3.

MacArthur's distributions for equilibrium species. If we consider an equilibrium population of the most extreme sort, we will not merely have an intraspecific density dependence but may expect interspecific interaction as well. In such a case it is necessary to make some assumption about the way that a limited number of species can be distributed among the niches provided by the habitat. The simplest hypothesis in accord with the principle of competitive exclusion is that each species has its own niche, but that the specific biotopes (in **B** of page 232) represent a random fractionation of the entire habitat, any kind of delimitation of biotope boundary being equally probable. On this assumption MacArthur (1957, 1958) has shown that the numbers of individuals will be distributed according to the formula that the expected abundance of the rth rarest species in a population of S_s species and N_s individuals is

$$(14) \qquad \frac{N_s}{S_s} \sum_{i=1}^{r} \frac{1}{S_s - i + 1}$$

The only arbitrary feature of this distribution, which we call the Mac-Arthur type I distribution, is the assumption that numbers of individuals are proportional to the sizes of the habitats occupied.

Vandermeer and MacArthur (in press) developed the theory for purely intraspecific competition. The comparable distribution for the rth rarest species is given by

$$(15) \quad N_s \left(\sqrt{S_s + 1} - \sqrt{S_s - i + 1} \right) \Big/ \sum_{i=1}^{r} \left(\sqrt{S_s + 1} - \sqrt{S_s - i + 1} \right)$$

In this case, the MacArthur type II distribution, the commonest species and the very rare would be rarer than in type I, whereas the numerical abundance of the species of moderate rarity in the rank order would be greater than in type I.

As far as is known, many vertebrates (King 1964) notably birds in a homogeneous diverse habitat (MacArthur 1957, 1960) and such large invertebrates as snails of the genus *Conus* on the reefs of Hawaii and Ceylon (Kohn 1959, 1960), follow the type I distribution. Other organisms studied in this way, namely birds in heterogeneously diverse habitats (MacArthur 1960), the microarthropods of soils (Hairston 1959), and diatoms on slides exposed in streams (Patrick and Mac-Arthur personal communication), follow neither type I nor type II, having greater populations of the commonest species than the type I distribution demands but not showing the greater abundances of the rarer species implied in type II. In view of the fact that MacArthur also investigated theoretically a situation in which each species received randomly delivered quanta of energy with equal probability, which gives an irregular distribution tending to equal abundance for all species after infinite time (type III distribution), it is convenient to call the observed distribution with excessive populations, relative to type I, of common, and deficient populations of all other species, type IV (Hutchinson 1961). The type I distribution presumably represents a limiting case as the excessive abundance of the commoner species of the type IV distribution is reduced by competition.

It is obviously a matter of the greatest interest that actual assemblages are distributed either according to type I or type IV. The explanations that have been given of type IV, however, are various. MacArthur has shown that one possible source of type IV distribution is the summation of type I distributions when the ratio of numbers of species to numbers of individuals varies from area to area studied.   This is in fact the explanation of type IV in assemblages of birds from a heterogeneously diverse area.   Hairston (1959) considers that the divergence of type IV from type I is a measure of the structure of the community sampled.   Insofar as this structure is an expression of clumping of individuals, so that the encounter with one specimen raises the probability that other specimens of the same species will be encountered, it could be the result of heterogeneity in the biotope, which would then not be homogeneously diverse, of aggregations due to reproductive activity in slowly moving organisms, which may often be an expression of a local transitory departure from equilibrium, or of complex adaptive interactions, either intraspecific or interspecific.   At least in the first and second of these alternatives the postulates implied in the type I distribution do not hold.

When the MacArthur type I distribution is plotted logarithmically by octaves following Preston, a curve is produced that is asymmetrical, with more species on the rarer side of the mode than on the commoner. If we have a symmetrical lognormal curve, we should therefore find common species in greater numerical abundance than on the MacArthur type I distribution, whereas the species of low rank order will be rarer than on the type I distribution.   Type IV distribution can therefore correspond to nonequilibrium or opportunistic assemblages, though this is not necessarily the only explanation of their occurrence.

Application to phytoplanktonic diatoms.   The application of the ideas of the preceding sections to the phytoplankton requires a large amount of data, consisting in general of counts of individuals of a great many properly separated taxa.   The only adequate data available appear to relate not to the phytoplankton as a whole but to the diatoms, since it is possible for an experienced worker in this group to identify every complete specimen accurately, which is by no means always true with the less easily preserved and less diagnostically diverse species of other groups.   Restriction of the treatment to a single class may well introduce errors, but since the diatoms share, in their silica requirements, a physiological property hardly developed in the rest of the phytoplankton the restriction is at least to a biologically natural entity.

The best data relating to natural populations of diatoms seem to be those of Foged (1954) for a series of lakes on the Danish island of Funen.   Among these lakes the flora of Braendegård Sø is the most extensive.   The data from this lake have therefore been used for a detailed analysis, though it is evident that the general results are confirmed by the diatom floras of Foged's other localities. Braendegård Sø is a shallow lake of area 1.18 km²; its greatest depth appears to be little more than 3 m., so that it is presumably never stratified except perhaps transitorily at the height of summer. At all seasons it is reasonable to suppose that the samples of the plankton association are derived from homogeneous, well-mixed water.

In Figs. 105 to 107 the results of the various methods of analysis are applied to four plankton samples from Braendegård Sø studied by Foged, collected on December 12, 1949, April 30, June 5, and September 13, 1950. In Fig. 105 the data are plotted according to the method of Fisher, Corbet, and Williams, with the theoretical distributions to be expected from the number of individuals counted and species determined. There is perhaps a tendency toward the occurrence of more rare species, represented by one or two specimens, than

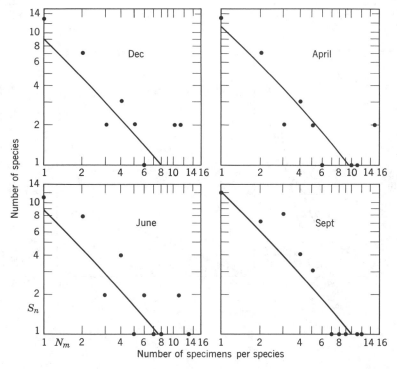

FIGURE 105. Number of species S_n exemplified by 1, 2, 3, . . . , (N_m) specimens in diatom counts of plankton of Braendegård Sø, Funen, Denmark, in December, April, June, and September 1949–1950, plotted against N_m; the lines give the expected distributions following Fisher, Corbet and Williams.

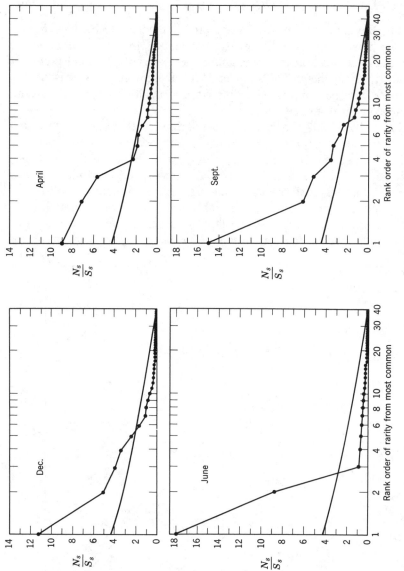

FIGURE 106. Relative number of individuals (N_s/S_s) arranged in rank order of increasing rarity, for the diatoms of Braendegård Sø, Funen, Denmark, in December, April, June, and September 1949–1950. The unmarked line corresponds to a MacArthur type I distribution; all the observed distributions are type IV.

would be expected, but the scatter of the points is too great to permit any positive conclusion.

In Fig. 106 numbers of specimens are plotted against the rank order of commonness, and the theoretical line for the MacArthur type I distribution is given. It is quite clear that on no occasion do the observed populations follow this distribution; the divergence appears to be greater during the warmer than the cooler months of the year. As already indicated, a lack of agreement with the type I distribution has been noted by Patrick and her associates in earlier work on diatoms settling on glass slides.

In Fig. 107 the numbers of species are plotted against populations arranged logarithmically in octaves, following Preston. A tolerable

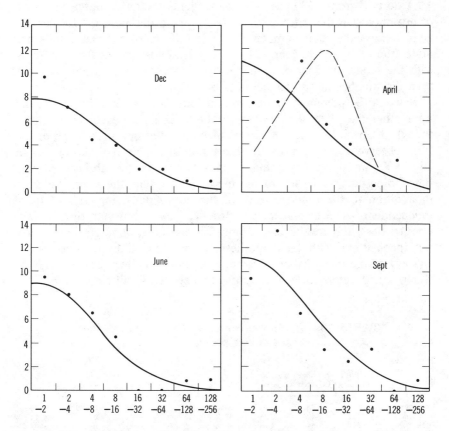

FIGURE 107. Distribution of number of species with populations in the intervals 1–2, 2–4, 4–8 for the same diatom populations in Figs. 105 and 106; the lines give the log-normal Preston distribution. The broken line in the April panel indicates the distribution for a rank order MacArthur type I distribution.

fit to the right-hand side of a lognormal curve is obtained in all cases, but the mode always lies almost at the origin, so that none of the left-hand side of the curve is unveiled. This also proves true when the records for all four dates are taken together and when the much larger littoral flora of the lake is considered.

The various parameters that can be derived from these modes of treatment are given in Table 18. In addition to giving the numbers of specimens of the different taxa actually counted, Foged lists a number of rare species noted occasionally in scanning much larger samples than the 500 specimens used for the counts. These are added to the number of species of which specimens were counted to give the column for total number of species observed. On the whole, the theoretical total (S_t) derived from (13) seems to be a slightly better empirical measure of diversity than the index of diversity, as the ordering of the former exactly follows the total number of species observed, whereas this is not quite true of the coefficient. The latter is almost identical numerically with the modal number S_m. In no case is the theoretical universe less than the number of species actually recorded.

It would be obviously wrong to conclude that the data just discussed imply that the phytoplankton represents a nonequilibrium assemblage, though it certainly does not represent an equilibrium assemblage of the kind observed among birds or some mollusks. Even if the curves of Fig. 107 are really lognormal curves, which in the absence of any descending part to the left of the mode is uncertain, there may be ways in addition to the random walk of the accumulated integrals of quite independent species that could produce the lognormal distribution. Nevertheless, the results of the analysis, as far as they go, are clearly not inconsistent with the hypothesis of a nonequilibrium assemblage. In view of the difficulty of understanding how a uniform mixed body of water can develop enough structure to permit the differentiation of as

TABLE 18. *Various estimates of diversity of assemblages of diatom populations*

	Index of Diversity	Total Number Counted	Total Number Observed	S_t	S_m	Position of Mode (octaves from origin)	$\delta = \dfrac{1}{\sqrt{2\sigma^2}}$	D (bits per individual)
December	9.2	38	56	58.5	7.8	0.25	0.238	3.29
April	11.7	44	81	88.7	11.0	−0.45	0.219	4.11
June	8.9	36	48	58.0	9.0	0.21	0.275	2.55
September	12.4	46	60	78.1	11.0	0.10	0.250	3.98
All plankton samples	15.5	74	110	144.2	15.7	−0.36	0.193	3.91
All littoral samples	26.9	140	189	254.1	25.8	0.08	0.180	4.79

many niches as there are species of autotrophic photosynthetic organisms living therein, the hypothesis of a nonequilibrium situation would appear to be reasonable.

The most likely sort of nonequilibrium system that can be postulated for the lake plankton is one in which a single species would, if given adequate time under constant conditions (see page 389), displace all the others but in which the conditions in fact never stay constant long enough for this to happen. The assemblage could be regarded as composed of species which have in general under the range of conditions experienced in the lake rather low competition coefficients, so that under constant conditions replacement is not too rapid. Admittedly, there is always a finite probability at any moment of the rarer species[3] becoming extinct (*cf.* Klomp 1961), but it must not be forgotten that even the rarest species actually found in plankton samples are likely to be represented in the lake by a quite fantastic number of individuals. A species present at a density of one individual per liter would ordinarily be quite undetectible in examination of the phytoplankton by routine methods unless it were a very large species living in a very sparse association. Yet, if such a species were confined solely to a layer 1 m. thick in a small lake of area 1 km.^{2}, the total population would be 10^9 individuals. If, because of the presence of other species in the plankton, competition were to become extremely severe, so that the numbers of specimens of the species were halved every day, it would take a month to remove the whole population. If the species were a thousand times commoner, one specimen occurring in 1 ml., which is still not common relative to what we might expect in a eutrophic lake, and the rate of decline involved halving the population in ten days, which is probably not unreasonable, it would take just a little more than a year for the species to disappear. Even in competitive exclusion, with only one species surviving at equilibrium, this rate of decline would probably be fairly high if Riley is right in supposing that a number of species have approached asymptotically an upper limit of adaptation as planktonic phototrophs.

---

[3] Or, for that matter, any species. For a species reproducing by simple fission, in which each specimen has at equilibrium an equal chance of dividing or dying, the probability of random extinction (Skellam 1955) is given by $p_e = [t/(1 + t)]^{N_0}$; time t is measured by number of generations and N_0 is the initial size of the population. For large populations and long times we may approximate by $t = -N_0/\ln p_e$. Thus for $N_0 = 10^9$, as in the example above, p_e would reach a value of 0.01 in 2.2×10^8 generations, which is likely to represent a period of the order of a million years, or much more than the ordinary lifetime of a lake. The process of random extinction of such a population is obviously very slow compared to the kinds of elimination that are likely to be observed actually occurring in nature.

Provided that a sufficient number of annually repeated possible combinations of light intensity, temperature, and chemical composition were present, each uniquely most favorable to a single species. which during the incidence of each set of conditions can multiply rapidly, a fairly diverse flora in a permanently nonequilibrium condition might result. We may define as the competition time that time taken to reduce by competition a maximal population of any species to some arbitrarily small fraction of the population. If the competition time were long compared to the period of changing environmental variables, the permanent diversity would be favored. The possible over-all composition of such a flora might be rather rigidly determined by the average properties of the lake and so over a long time would consist more or less of the same species. If random extinctions occurred it would be reasonable to suppose that, for a given lake of fairly constant physicochemical properties, reintroduction of a species that had once been moderately successful would more probably be successful than introductions of species that had not and, in the limiting case, could not live successfully in the habitats provided by the lake. Wherever there are resting stages, which in effect take the species out of competition and permit it to re-enter the arena at a later time, the danger of random extinction is greatly reduced. At least in the Chysophyceae, palaeolimnological studies (Nygaard 1956; Leventhal in press) suggest the possible existence of an extraordinary number of the order of one hundred sympatric species with such stages in a single basin. The vertical variation provided by even a small amount of stratification in the summer would greatly increase the number of niches available. Special cases of synergistic species might, as we have seen, also increase the diversity; the possibility that some species are limited by major nutrients and others at a given time by special trace elements or by vitamins, though it could not entirely abolish competition, for all algae need CO_2, phosphate, combined nitrogen, and light, might reduce the competition coefficients. It is just conceivable that the existence of separate predators for different species would be limiting and constitute a mechanism of niche diversification. Superdispersion of some, but not other, species by the turbulent structure (page 293) of the water might also produce some diversification. In spite of all the possibilities so far discussed, the investigator is left with an uneasy feeling that the problem of the diversity of the phytoplankton is still not fully solved.

Is the phytoplankton largely meroplanktonic? In view of this uneasiness, it seems legitimate to enquire (Hutchinson 1961) whether the phytoplankton, for instance, as it is exemplified by Foged's diatom populations, is really a properly delimited and genetically autonomous part of the biological community. If the continuous populations of many of the species were actually of limited extent and confined to fairly specific benthic littoral niches, but could under certain circumstances proliferate into immense planktonic populations fated to leave no descendants except from individuals accidentally falling into the already occupied benthic littoral biotopes, much of the paradox of the phytoplankton would disappear.

Admittedly, some species are almost certainly euplanktonic. It is reasonable, for instance, to suppose this of *Asterionella formosa*. For most diatoms, however, the evidence would appear to be far from conclusive; in *Melosira* the well-studied species (Lund 1954, 1955) are certainly meroplanktonic. This may well be true also of some, though perhaps not all, races of *Tabellaria flocculosa*. Dr. Ruth Patrick (*personal communication*) suspects that many littoral diatoms are brought into the open water by ascending when bearing bubbles of photosynthetically produced oxygen.

There is also a strong suggestion from the occasional existence of facultative planktonic populations of species such as *Staurastrum ophiura,* ordinarily a benthic species, that recruitment of the phytoplankton from the littoral benthos occurs also in the desmids. Brook (1959a) thinks that some of the supposed planktonic races of desmids may actually be environmental modifications produced in the course of a single season.

If we attempt to examine the relation of the area of the lake to the number of species present in the phytoplankton, we find in the best available data either an inverse or no apparent correlation. The modal number of species in the phytoplankton decreases with increasing area, as the present hypothesis might suggest, in Järnefelt's (1956) vast series of Finnish lakes. In the phytoplankton of fifteen lakes, covering an immense range of areas from 0.02 to 1129.7 km.^{2} in Indonesia, studied by Ruttner (1952a), the correlation coefficient of the logarithm[4] of the number of phytoplanktonic species to the logarithm of the area is —.019, obviously not significantly different from zero. It is quite obvious that the plankton of lakes does not increase in diversity in the same way as the bird, mammal, and ant faunas of small islands (Lack 1942; Hutchinson 1961; and, most notably, E. O. Wilson 1961) multiply with the increasing area of the islands.

If we suppose that the effect of the littoral on the composition of the plankton depends on the ratio of the length of the littoral to the area

---

[4] This is the appropriate elementary procedure, since negative values are impossible for both variables.

(in lakes of constant shape this will vary as $A^{-\frac{1}{2}}$) and also that the diversity of the littoral fauna depends on the diversity of the habitats it has to occupy, which will increase with the absolute length of the littoral (which varies as $A^{\frac{1}{2}}$), we will then have two antagonistic effects in increasing the size of a lake, which, assuming the simplest kind of relationship to obtain, will cancel out. It is unlikely that a simple linear proportionality would obtain with either $A^{-\frac{1}{2}}$ or $A^{\frac{1}{2}}$; the diversity of the shoreline would not increase indefinitely with length. It is clear, however, that we do not always have to assume an increase in species number with decreasing area if there is a large meroplanktonic component in the phytoplankton; a lack of any clear relationship is theoretically equally plausible.

This final suggestion, that the phytoplankton is largely meroplanktonic, is really tantamount to the most extreme opportunism or lack of equilibrium. It also implies that in a certain limiting sense the biotopes of the planktonic organisms are heterogeneously diverse, the species in question spreading out into an ultimately inappropriate region just as in plagues of rodents or Orthoptera more than the potentially permanent biotope is occupied at times of outbreak. The littoral benthic parts of the population, though metabolically active, would thus perform the same function as encysted resting stages in relation to the seasonal planktonic population (Hutchinson 1964b).

ECOLOGICAL CLASSIFICATION OF PHYTOPLANKTONIC ASSOCIATIONS

The annual variation in the plankton, to be discussed in detail in a later chapter, looks superficially like what terrestrial ecologists would term *aspection* and have in fact been so described by Panknin (1947). When we deal with small rapidly dividing organisms we may be observing a series of changes involving perhaps twenty to several hundred generations in the course of an annual cycle. The annual cycle is thus, in terms of generation time, the equivalent of perhaps up to ten thousand years in the successional history of organisms which are replaced as some forest trees every fifty or one hundred years. Moreover, the climatic changes estimated as mean temperature throughout the life of a single individual are for the small forms far greater between winter and summer than they are for the large forms during the different phases of postglacial time which have left their records as replacements of one species by another in a pollen diagram. The changes in significant lake chemistry during the months from the beginning of the spring phytoplankton bloom to the time when in early summer almost the whole of the stationary concentration of nutrients is exhausted is probably much greater than the changes in soil chemistry that are likely to have occurred in untouched habitats during the last several millenia. This point of view justifies at least the recognition of communities as we find them at any given moment as entities corresponding to associations rather than to the seasonal aspects of associations on land, even though they seem to us, whose lifespan is comparable to that of the shorter-lived forest trees, to be so transitory.

Dominant species and characteristic species. In attempting to classify phytoplankton associations, two rather different approaches are possible, though, as Thunmark's (1945) investigations clearly show, the two approaches are complementary rather than antagonistic.

If we consider the dominant species, which is responsible for the obvious characteristics of a phytoplankton assemblage, we are clearly dealing with a species on whose population small random variations have had little effect during the period of growth. In general, the niches occupied by dominant species in the abstract sense of the definition of a niche already given (page 232) are likely to be larger than are the niches of the species that are seldom or never abundant. Characterization of an assemblage by its dominant may give a clear idea of the general nature of the assemblage and some idea of the biotope in which it occurs, and it may indicate a good deal about the possible coactions occurring in the plankton. For instance, if we are told that a certain lake contains an *Anabaena circinalis* plankton, we suspect that the water has a pH greater than 7, that nitrogen is being fixed, that there may well be an obvious water bloom, and that the ratio of plant to animal plankton may be quite large. All of these conclusions are probability judgments and even so are rather crude. If we are told that there is an *Anacystis cyanea* plankton, we should come to much the same conclusions, save that regarding fixation of nitrogen.

If, however, we considered not the dominants but a certain number of rarer *indicator* species, we might learn a good deal more about the nature of the lake, though probably some information that might be derived from a knowledge of dominants would not be available. Thus in Thunmark's (1945) study of South Swedish localities we find that in a particular pair of cases, Växjösjön and Tynn, *Anacystis cyanea* was dominant in each but that in Växjösjön there were on different dates twenty-six to thirty-nine species of Chlorococcales and five to fifteen species of desmids, whereas in the plankton of Tynn both groups were represented by fifteen species. Study of a whole series of South Swedish lakes (page 390) indicated that Chlorococcales outnumbered desmids when the Secchi disk transparency is low ($\wedge\wr$ 1 m.), the pH high

($\geq$ 7.8), and the conductivity high (κ 10⁻⁶ $\geq$ 150), whereas the number of species of desmids is great relative to the number of species of Chlorococcales when the transparency is high ($\geq$ 4.5 m.), the pH low ($\leq$ 7.0), and the conductivity low (κ 10⁻⁶ $\leq$ 65). This relationship is more sensitive to external conditions than is the disposition of dominants, which can transgress a great range of the environmental variables but which also is by no means constant in the different parts of these ranges. In the regional example of South Sweden, in the less transparent more alkaline group of lakes, though there are always more than twice as many species of Chlorococcales as of desmids the dominant alga in June can be either *Anacystis cyanea* or *Melosira ambigua,* and none of the more desmid-rich softer water lakes ever has desmid dominants but rather species of diatoms, blue-green algae, or *Botryococcus.*

If the purpose of the designation of an association is to give in the simplest way an indication of the sort of phytoplankton present, the designation of the dominant, eurytopic though it may be, is doubtless the most concise procedure. If the designation of the association is to throw as much light as possible on the ecological variables involved in its determination, some designation based on a statistical enumeration of the less eurytopic but more characteristic species will obviously be desirable.

In the following discussion any phytoplankton assemblage present at a given depth on a given date will be regarded as an association according to the definition already given (page 231). Beyond such a formal definition the usage is clearly justified by the fact that certain associations of species are empirically far more probable than others; for example, *Melosira granulata* with *Anacystis cyanea* than either with a planktonic species of *Staurodesmus.*

The name of the dominant species or sometimes the dominant and subdominant will be used as designations for the association. If *Fragilaria crotonensis* were dominant and *Asterionella gracillima* subdominant, we speak simply of a *Fragilaria crotonensis-Asterionella gracillima* association. Messikommer (1927) used in such cases the association terminology of European students of plant synecology, writing of a *Fragilarieto crotonensis-Asterionelletum gracillimae,* but such usage has not been widely followed and seems to the writer to imply a priori too many analogies with terrestrial synecology, analogies which, if they exist, must be demonstrated empirically. In any given study it must not be forgotten that the dominant species numerically, volumetrically, and in physiological activity may be a member of the nannoplankton not collected by the techniques used in the investigation.

It may be convenient, with Thunmark and, less formally, with many workers before him, to characterize the plankton as a whole by both the dominant plant and animal form, as in the *Anabaena flos-aquae-Holopedium gibberum* association recorded in Allgunnen by Thunmark.

Finally, again following Thunmark, if we wish to indicate roughly the proportions of characteristic species as well as of dominants, we can speak of the extremely chlorophycean, moderately desmidean *Melosira ambigua-Daphnia cucullata* association of Häckebergasjön or the moderately chlorophycean extremely desmidean *Anabaena flos-aquae-Holopedium gibberum* association of Allgunnen.

When the dominants are taxonomically related in a way that seems significant ecologically, it is often useful to speak of a desmid plankton type or a myxophycean plankton type and the like. Such classes of associations are considered later at greater length (page 384–388). At the present time, at least, all such classes are regarded primarily as heuristic conveniences rather than as hierarchically defined formations.

Plankton associations and regional water types. The rather marked regional differences between the lakes and lake biota of the low-lying area around the Baltic on the one hand and of the mountains of northern and central Europe on the other impressed the earlier investigators of the limnology of Europe. Naumann (1917b) in particular realized that the differences expressed themselves largely in terms of a greater mass of phytoplankton, giving a distinct planktogenic color to the water and in quiet summer weather producing the appearance of dense superficial accumulations of phytoplankton, usually spoken of in contemporary English as water blooms, in the lakes of the cultivated lowlands. Such phenomena did not occur in the montane areas. Since the waters draining the ancient paleozoic rocks of the Scandinavian mountains were believed to be deficient in combined nitrogen and phosphorus, whereas those entering the lakes of the cultivated lowlands were presumed to be much richer in nutrient elements, Naumann (1919) used the term *oligotrophic* to designate what he regarded as the plankton formation of the montane lakes and *eutrophic* that of the lowland lakes. The terms were employed by Naumann not merely to designate two different kinds of lake plankton but also to express the nutrient differences between the waters in which that plankton occurred. Naumann's oligotrophic and eutrophic water types in fact were applied initially to pond and river as well as to lake water. In addition to these two terms, Naumann used *heterotrophic* for waters containing large amounts of available organic matter, usually derived by pollution and supporting a heterotrophic plankton in the ordinary sense of the word. Though the etymology and original usage of oligotrophic and eutrophic imply variation in nutrient content, at the time that Naumann introduced the terms there was almost no useful quantitative information about the actual concentrations of nutrients present. Naumann (1923), however, did realize that the experimental fertilization of an oligotrophic water could raise its phytoplankton crop.

Later, because the brown peaty waters of many of the lakes of the mountains of northern Europe seemed to support a different plankton association from that of the deep blue or blue-green oligotrophic lakes of central Europe, Thienemenn (1925) added a third category, the *dystrophic,* which was originally a lake rather than a water type. Most of Naumann's original oligotrophic waters were more or less rich in brown coloring matter and would have been regarded as dystrophic by Thienemann. In later work the system proliferated, and the distinction was inadequately made between water types defined at least ideally by their chemical composition and synthetic lake types in which morphometric and other nonchemical factors interact with the biologically significant chemical characteristics of the water and the sediments.

Among recent writers, Järnefelt (1952, 1953, 1956, 1958) has made the most determined effort to maintain the original meanings of the words eutrophic and oligotrophic, and his example is followed in this book. The problem of synthetic lake types will be discussed in Volume III.

For the purposes of the present discussion, a continuous series of waters from the extreme oligotrophic to the extreme eutrophic, with a more or less independent variation in organic color, may be assumed. It should be noted, however, that any very eutrophic water is likely to become somewhat colored as the result of the decomposition of plankton algae and other plants. The water color involved in the separation of Thienemann's dystrophic type is ordinarily regarded as due to extractives from bog soils and peat either at the lake margin or on the courses of its influents. There is still no agreement on the biological significance of such allochthonous color except in its action in reducing the penetration of light. The term dystrophic which is therefore rather unfortunate, suggests a more pathological condition than perhaps exists. The alternative *tyrfotrophic* (Naumann 1932), used by a few authors, is criticized by Järnefelt on etymological grounds. Järnefelt's term *chthoniotrophic* appears to be the soundest and will be used when needed.

Naumann (1919) spoke of *oligotrophic* and *eutrophic plankton formations,* each constituting a number of associations. As finally defined (Naumann 1931), the component members of the two formations are presented, not in terms of dominant species but rather by genera or larger taxa. These are loosely termed phytoplankton types. Naumann's classification, which was not given in a completely systematic form, was essentially as follows:

Oligotrophic formation
>Desmid or Caledonian (Teiling 1916) type
>Chlorophycean (*Botryococcus* and *Sphaerocystis*) type
>Chrysophycean (*Dinobryon* and *Mallomonas*) type
>Oligotrophic Peridinean (*Ceratium* and *Peridinium willei*) type
>Oligotrophic Diatom (*Tabellaria* and *Cyclotella*) type

Eutrophic formation
>Eutrophic Peridinean (*Ceratium* and *Peridinium* spp.) type
>Eutrophic Diatom (*Melosira* and *Stephanodiscus*) type
>*Pediastrum* type
>Myxophycean (including Chroococcacean) type

Though this classification may now appear incomplete and a little naïve, there is no doubt that it embodies a good deal of truth, for when lakes from regions other than those of southern Sweden are considered the same general pattern, though often with considerable geographical variation in species and genera, may be observed.

English Lake District. Pearsall (1921, 1932) summarizing his own work and that of his predecessors, found that in two (Wastwater, Ennerdale) of the three lakes that may clearly be regarded as oligotrophic the dominant algae of the net plankton were desmids with the colonial green alga *Sphaerocystis schroeteri* and *Peridinium willei*. In Crummock Water, the third of the lakes to be regarded as oligotrophic, the plankton association is somewhat intermediate between the desmid and diatom associations found in the region. In the more productive lakes *Melosira* is abundant in winter (see page 459), and a spring pulse of *Asterionella* is usually followed by *Tabellaria* and *Dinobryon* as the nutrients are exhausted in the early summer. In the most productive lakes (Windermere, Esthwaite Water) there is a striking development of Myxophyceae in late summer.

Comparing these results with Naumann's scheme, we clearly can have in the English lakes a typical desmid type of plankton; *Melosira* and Myxophycean types appear at different times of year in the more eutrophic lakes. *Melosira,* however, is a winter form, and the most important eutrophic diatom is *Asterionella,* which does not figure in Naumann's scheme. In the more eutrophic lakes several species of a group of green algae (*Tetraspora lacustris, T. limnetica, Dictyosphaerium pulchellum, Volvox aureus,* and *Eudorina elegans*) may be important.

The lakes of Finland. The monumental study of Järnefelt (1956) is particularly important, not only because of the great number of localities investigated but also because it takes into account the nannoplankton, omitted by many earlier investigators. The lakes considered are divided into two groups according to their productivity and again into two groups according to humic water color. To obtain clear-cut indications of the effect of productivity on the qualitative aspects of the phytoplankton the two categories based on phytoplankton productivity are selected arbitrarily to avoid a group of intermediate values. The four resulting classes are as follows:

(1) Eutrophic: water bloom formed; mean summer phytoplankton crop never less than 0.8 ml. m.$^{-2}$; low allochthonous water color.[5]

(2) Chthonio-eutrophic; as in (1), but water color allochthonous, moderate, or high, never less than 40 Pt units.

(3) Oligotrophic or oligochthonio-oligotrophic; water bloom never formed; mean summer phytoplankton crop never more than 0.15 ml. m.$^{-2}$; water color less than 15 or between 15 and 40 Pt units, respectively.

(4) Mesochthonio-oligotrophic or polychthonio-oligotrophic; as in (3) but water color 40 to 80 or more than 80 Pt units, respectively.

The modal number of species and the range in number of species are greatest in (1) and decline throughout the series. The modal number is greater in August than in the preceding or succeeding months and is greater in small than in large lakes.

In general, diatoms dominate the associations in the sense that they constitute more than 50 per cent of the phytoplankton volume in just over half the lakes studied. Only in the eutrophic lakes of class (1) do such diatom associations occur in a minority of lakes. There seems to be a tendency for diatoms to become increasingly dominant with increasing chthoniotrophy. In the more oligotrophic lakes *Cyclotella* is the main genus; in the more eutrophic, *Melosira* and often *Stephanodiscus.*

---

[5] Some productive lakes with a water color in excess of 40 Pt units are admitted to this category; presumably in all cases there is evidence from the littoral and influents that the color is autochthonous and planktogenic. In the darkest waters, that of Kouvalanjärvi, the color is said to be derived from diatoms. Järnefelt uses in this paper *eutrophid, oligotrophid,* etc., in place of the more familiar forms in *-ic.* There seems no pressing philological reason for abandoning the latter forms, nor scientific basis for a distinction between two series of adjectives. OED gives trophic in a physiological sense as dating in English from 1873; the Greek adjective is τρόφιμος, implying neither suffix.

The anomalous green alga *Botryococcus brauni,* placed at the time of Järnefelt's paper as a heterokont, was dominant only in oligotrophic waters.

Plankton dominated by Chlorococcales occurs primarily in the eutrophic categories, but one lake (Suonteenjärvi, No. 108) in the oligotrophic series had *Oocystis* sp. as dominant.

Chrysophycean plankton occurred in all kinds of lakes, but apparently more commonly in the more humic waters.

Two types of dinoflagellate association were recognized, one eutrophic with *Peridinium bipes* and *P. cinctum,* the other oligotrophic with *P. inconspicuum.* *Ceratium hirudinella* and sometimes *Glenodinium* were present in both associations.

The myxophycean and euglenophyte associations occurred largely in the two eutrophic categories.

Plankton dominated by desmids was seldom encountered and then as often in eutrophic as in oligotrophic waters. This unexpected result is in part because the whole of the phytoplankton rather than the net plankton was considered. In one case (Heinijärvi, No. 22) *Cosmarium bioculatum* was dominant, with blue-green algae as subdominants, in a eutrophic lake.

Mixed and intermediate associations occurred commonly, and the most frequent plankton types appear to have been the following:

Oligotrophic
    Oligotrophic diatom (mainly *Cyclotella*) type
    Oligotrophic diatom type with chrysomonads
    Botryoccus brauni type
    Desmid and Chlorococcales type
    Mixed type, diatoms, Chlorococcales, often with Myxophyceae
Eutrophic
    Mixed type, diatoms (mainly *Melosira* and *Stephanodiscus*)
        Chlorococcales, Myxophyceae, and Euglenophyta
    Chlorococcal type, sometimes with desmids
    Myxophycean type

Indonesia. It is interesting to note how the various types recognized in Europe can be compared with those of a well-studied tropical area. Ruttner's (1952a) summary for Indonesia gives the best data for such a comparison.

Danau Bratan in Bali is a characteristic desmid lake with dilute and slightly acid (*p*H 6.8) water. The dominant phytoplankter is *Staurastrum excavatum* var. *planktonicum,* the subdominant, *Melosira granu-*

lata, which suggests a more eutrophic environment than is usual in European desmid lakes. Many rarer species of desmids occur, as is usual in such lakes.

In the large deep lakes of Sumatra what may be regarded as oligotrophic diatom associations are found. *Denticula pelagica* is the dominant in the two main basins of Lake Toba and with *Synedra rumpens* var. *neogena* and *Oocystis crassa* in Lake Singkarak also. In the slightly shallower Lake Ranau *Synedra rumpens* var. *neogena* and *S. ulna* are dominant, with *Oocystis* sp.

In the smaller lakes of Java various combinations of the diatoms, *S. rumpens* var. *neogena* and var. *scotica, Nitzschia acicularis,* and *Cymbella turgida* with the blue-green algae *Dactylococcopsis fascicularis, Anabaenopsis raciborskii,* and *Lyngbya limnetica,* are characteristic dominants and subdominants. There also appears to be a rather characteristic dinoflagellate association with *Peridinium gutwinskii,* with or without *Ceratium hirundinella,* either being dominant. This association occurs in certain dams in west Java and is presumably comparable to the eutrophic dinoflagellate associations recorded in Europe. The rather peculiar association of the desmid *Cosmarium bioculatum* var. *minutum* as dominant with the blue-green alga *Dactylococcopsis fascicularis* as subdominant in Telaga Pasir, in the mountains of central Java, provides an interesting analogy with Heinijärvi in Finland.

A provisional classification of phytoplankton types. Taking the data already considered, along with those to be presented in Chapter 23, we may distinguish the following plankton types as reasonably common in the euphotic zones of lakes; an indefinite number of intergrading or mixed types may be expected. Each plankton type includes a variety of associations brought together primarily on the basis of taxonomic affinity. In some cases it may prove that there are no good reasons other than this affinity for including the various associations in a given type. Therefore the plankton types are not here considered as necessarily equivalent to formations in the phytosociological sense, though it is possible that some of them ultimately may be so regarded by those who find the term formation a useful one.

1. *Oligotrophic desmid plankton.* Dominants are usually species of *Staurodesmus* and *Staurastrum,* ordinarily associated with other desmids, including facultative plankters of benthic-littoral origin. Among the other green algae *Sphaerocystis schroeteri* and *Gloeocystis* may be present. Diatoms may include *Rhizosolenia morsa.* A number of associations of different species of *Staurodesmus* and *Staurastrum* must exist. A very oligotrophic example is provided by the sparse plankton of Lago Rocca near the Beagle Channel in Tierra del Fuego, where

Staurastrum corpulentum was the only common open-water phytoplankter (Thomasson 1955a). These two genera, and more rarely *Cosmarium,* are not the only possible dominants; *Hyalotheca dissiliens,* a colonial desmid, can be the dominant in oligotrophic mountain lakes, as in Amethyst Lake, Jasper, Alberta, Canada (Rawson 1953), in which it occurred with *Tabellaria.* In general, desmid plankton occurs in very dilute, unproductive waters, as in northwestern Europe, Devil's Lake, Wisconsin, and Mountain Lake, Virginia. A tropical montane example is provided by Danau Bratan in Bali, but it may be more eutrophic than the typical North Temperate localities; a rather characteristic desmid-diatom plankton, with *Melosira granulata, Melosira hustedti,* and *Staurastrum valdiviense* as the more abundant organisms, occurs in Lago Riñihue and Lago Panguipulli, Valdivia, Chile (Thomasson 1955a). The first-named species again suggests a relatively eutrophic locality.

2. *Oligotrophic diatom plankton.* At least in central Europe the most characteristic dominants appear to be species of *Cyclotella;* for example, *C. melosiroides* in the Millstättersee, *C. glomerata* in the Ossiachersee (Findenegg 1943b), and both species with *C. bodanica* in Lake Constance (Grim 1939). The associated species vary considerably. *Fragilaria crotonensis* can be very abundant; *Synedra minuscula* and *Rhizosolenia eriensis* may be the dominants in Lago Maggiore, with *Cyclotella comensis* subdominant (Ruttner 1959). The large subalpine lakes, however, are not to be regarded as oligotrophic in the full legitimate sense of the word. In Finland, in the clear, less chthoniotrophic unproductive lakes, an association of *Melosira distans,* with *Dinobryon divergens, D. bavaricum,* and often *Cyclotella stelligera* and *C. kützingiana,* is characteristic (Järnefelt 1956). In the large lakes of North America Rawson (1956), who has recently considered all the available data, concludes that *Asterionella, Tabellaria,* and *Melosira islandica,* often associated with *Dinobryon,* are the most usual phytoplankters in oligotrophic waters. A typical *Cyclotella* plankton, though sometimes found in the Laurentian Great Lakes, does not appear to be so characteristic as in Europe or to develop in the large, more northern lakes of Canada. In Indonesia *Denticula* appears to be the usual dominant in large and supposedly oligotrophic lakes. As indicated in the discussion of the eutrophic diatom plankton type, the occurrence of particular species of diatoms as dominants or subdominants appears, with a few important exceptions, to be relatively independent of productivity.

3. Botryococcus *plankton.* *Botryococcus brauni* is the dominant in some north European oligotrophic lakes, in which, according to Nau-

mann, it may be associated with *Sphaerocystis*. Järnefelt mentions a few lakes, for example Kyynäröjärvi (Järnefelt 1956, No. 141), in which *Botryococcus* was associated with *Dinobryon divergens* and *Peridinium inconspicuum*. In other parts of the world phytoplankton dominated by *B. brauni* is known, but it is not possible to form any idea of the ecological determinants involved. In temperate South America the alga appears as a dominant associated with oligotrophic organisms such as *Dinobryon* and *Staurodesmus* (Thomasson 1955a). Ruttner (1952a) found *Botryococcus* to be the dominant in one of the two shallower basins of Lake Toba in Sumatra, and Hutchinson Pickford and Schuurmann (1932) noted it as the commonest planktonic alga in Barberspan, Transvaal, a shallow and slightly mineralized alkaline lake.

4. *Chrysophycean plankton.* One of several species of *Dinobryon* may in extreme cases (Strøm 1926) be the only alga recorded in the plankton of north European mountain lakes. It may be the dominant also in large montane lakes such as Fagnano in Tierra del Fuego (Thomasson 1955a). The genus is frequently associated with *Tabellaria* and other diatoms and appears even in productive lakes when the nutrients are largely exhausted after the spring phytoplankton maximum. The evidence that other Chrysophyceae (*Synura, Uroglena,* and *Mallomonas*) are particularly likely to occur under comparable conditions is inadequate. Palaeolimnological evidence (Nygaard 1956; see also Leventhal in press), however, suggests that lakes may pass through stages in which a great number of different Chrysophyceae, known mainly from cysts, co-occurred for limited periods of time, suggesting characteristic and highly diversified chrysophycean associations. *Uroglena americana* seems to have the same low phosphorus requirement as *Dinobryon* (Rodhe 1948), and some, but by no means all, the species of *Mallomonas* have been regarded as characteristic of oligotrophic waters.

5. Oocystis *or oligotrophic chlorococcal plankton.* Certain large, unproductive lakes appear to have species of *Oocystis* as the principal members of their phytoplankton. This is true of Tso Moriri and Panggong Tso, the two largest, deepest, clearest, and bluest lakes in Indian Tibet studied by Hutchinson (1937a). In general, it seems from the data of Järnefelt that a number of species of *Oocystis* are distributed indifferently in eutrophic and oligotrophic waters in Finland, though in that country the same is true of species of *Crucigenia* and *Ankistrodesmus,* which would probably not be expected as dominants in large, deep, unproductive lakes. As already indicated, Järnefelt has noted one oligotrophic Finnish lake with *Oocystis* sp. as the dominant, and mem-

bers of this genus may be subdominants in the predominantly diatomaceous plankton of deep lakes in Indonesia.

6. *Oligotrophic dinoflagellate plankton.* According to Naumann (1917b), *Peridinium willei,* in some cases accompanied by *Ceratium hirundinella,* may be the dominant in oligotrophic waters, though Järnefelt considers *P. inconspicuum* as the characteristic oligotrophic species of *Peridinium* in Finland. Höll (1928), however, regards *P. willei* as one of the most eurytopic species of the genus and finds *P. inconspicuum* to be characteristic of the large lakes of northern Germany which he terms anorganotrophic, a category probably overlapping Naumann's eutrophic type.

7. *Mesotrophic or eutrophic dinoflagellate plankton.* Naumann considered plankton dominated by species of *Peridinium* other than *P. willei* to occur in more productive waters, whereas to Järnefelt the type is characterized by the dominance of *P. bipes* and *P. cinctum. Ceratium* and *Glenodinium* may co-occur with these two species. Lundh (1951) finds that *Ceratium* in Sweden is never abundant in lakes with a rich myxophyceae flora; it is probably characteristic of mesotrophic waters. Höll thinks that *P. bipes* and *P. cinctum* are eurytopic species like *P. willei.* The two categories of dinoflagellate phytoplankton are kept separate here mainly because Järnefelt, with his great experience in Finland, believes that such a separation should be maintained. The Indonesian waters dominated by *Peridinium gutwinskii* presumably belong here.

8. *Eutrophic diatom plankton.* Highly productive lakes, with plankton dominated, at least at certain seasons, by *Asterionella* spp., *Fragilaria crotonensis, Synedra* spp., *Stephanodiscus* spp., and *Melosira* spp., are well known throughout the temperate regions of the world. Most of these genera and many of their included planktonic species are common to oligotrophic and eutrophic waters. The case of *Asterionella* is particularly puzzling because to Rawson it is a characteristically oligotrophic organism in the large lakes of Canada, whereas in other regions, both in Europe and North America, stratigraphic studies suggest that (Volume III) its incidence is due to human disturbance of the basin. The somewhat uncertain distinctions between the two alleged species *A. formosa* and *A. gracillima* do not seem to be involved. Hustedt (1945) believes *Melosira granulata* to be the planktonic diatom most characteristic of eutrophic waters in Europe.

9. *Mesotrophic or eutrophic desmid plankton.* A few species of desmids, notably the so-called eutrophic species of *Staurastrum* (*S. chaetoceras, S. gracile, S. pingue,* and *S. planctonicum*) and one or two

species of *Cosmarium* (*C. regnesi* and *C. bioculatum*), may be dominant in lakes containing fairly high concentrations of calcium and giving evidence of greater productivity than is exhibited by the typical Cale-
donian desmid lakes. The Lunzer Untersee apparently belongs here,
at least during part of the summer (see page 417), and associations of
Cosmarium with blue-green algae are known as far apart as Finland
and Indonesia.

10. *Eutrophic chlorococcal plankton.* A number of genera, notably
Pediastrum and *Scenedesmus,* but also *Actinastrum, Ankistrodesmus,
Crucigenia, Dictyosphaerum,* and *Tetraedron,* may be abundant in
eutrophic waters, often in rather small lakes. The most usual dominants appear to be *Pediastrum* and *Scenedesmus.* The spinose projections on some of the species of the genera just listed give to the
chlorococcal plankton of ponds a distinctive facies (Zacharias 1899c).

11. *Myxophycean plankton.* Plankton dominated by *Anacystis,
Aphanizomenon,* or *Anabaena,* forming water blooms, is usual in the
more productive lakes of temperate regions in the summer and in many
shallow tropical localities. *Lyngbya* and *Gloeotrichia* occur under comparable conditions. Some species of *Oscillatoria,* notably *O. rubescens*
(see page 441), are cold-water forms which live in the upper hypolimnion in great abundance in summer and produce water blooms at the
surface during the winter; *O. prolifica* is another winter or perhaps
perennial form which is often a dominant, behaving in summer like
O. rubescens; it may be associated with *O. rileyi,* as in Linsley Pond,
but the last-named species seems to be less stenothermal (Hutchinson
and Setlow 1946).

In saline or alkaline lakes of high productivity there may be massive
blooms of *Nodularia spumigena,* as in Banagher Pan 3 in the Lake
Chrissie region of the Transvaal (Hutchinson, Pickford, and Schuurmann 1932), and in still more alkaline conditions *Arthrospira platensis*
(Rich 1931), which forms the food of the lesser flamingo *Phoeniconaias
minor* in Lake Nakuru. Several investigators in India (Ganapati 1940,
1960; Philipose 1960; George 1962) have described permanent water
blooms of *Anacystis cyanea* (sub. *Microcystis aeruginosa*) in artificial
bodies of water, notably temple tanks in South India. In the locality
studied by George there was a distinct seasonal variation in numbers of
Anacystis, which reached a maximum in July, followed by a minimum
in August. Almost no change in temperature or phosphate was involved in the reduction of the bloom, and George felt that dilution of
organic matter by rain may have been responsible. Since there is no
evidence that *Anacystis* is anything but a pure phototroph, this seems

unlikely. In the Madras temple tank described by Ganapati the seasonal variations, under more intensely tropical conditions, seem to have been less. The bloom apparently is almost monospecific. Although it is perhaps reasonable to suppose that excessive quantities of nutrients, derived from human or animal excreta, produce extreme conditions in which few species can flourish (Thienemann 1939, Patrick 1949), though in vast numbers, it is also perhaps legitimate to regard the association as an approach to a monospecific equilibrium, developed under conditions of constant high temperatures and a large nutrient supply, at least potentially in the sediments, in very slightly stratified bodies of water 1 to 3 m. deep.

12. *Euglenophyte plankton.* Dense blooms of species of *Euglena* usually occur in very small and organically polluted bodies of water rich in nonhomic organic matter; Klausener (1908) records cases of small alpine "blood lakes" which owe their color to massive blooms of *Euglena sanguinea.* In more ordinary small lakes both *Trachelomonas volvocina* and *Lepocinclis fusiformis* may occur as dominants; Järnefelt (1956) gives examples from Finland.

13. *Bacterial plankton.* The only cases of bacterial plankton in the euphotic zones of unpolluted lakes are provided by the occurrences of pigmented sulfur bacteria, usually in highly mineralized lakes in which a considerable supply of H_2S is being generated by the bacterial reduction of sulfate in the deeper water. One such lake, Son Sakasar Kahar, in the Punjab, in which the plankton consisted almost entirely of *Lamprocystis rosea,* has been described in Volume I (p. 774). The same organism has been reported as occurring abundantly below the ice of the Schliersee in Bavaria (Reindl 1912), presumably as the result of the production of H_2S at deeper levels in the lake. Smaller species of purple bacteria are known in saline lakes in Egypt. Two cases (see page 315) of supposed green bacterial plankton have been recorded but the organism in question is undoubtedly a blue-green alga.

The phytoplankton indices. Thunmark (1945) has proposed to use, in a quite general way, the ratio of the number of species of Chlorococcales to the number of species of desmids as a measure of the position of any plankton association in a series running from those characteristic of extremely unproductive soft transparent waters to extremely productive hard waters turbid with plankton. A comparable treatment of percentage composition, using instead numbers of individuals, had in fact been suggested by Pearsall (1921).

For fifteen South Swedish lakes, grouped in three geographical categories, the relevant data are set out in Table 19.

TABLE 19. *Properties of South Swedish lakes,*
and their chlorophycean indices

Lake Districts	Växjö and Lund	Västervik	Aneboda
Dominant	*Anacystis* (2 lakes)	*Anacystis* (2 lakes)	*Anacystis* (1 lake)
Phytoplankton	*Anacystis* or	*Anabaena* or	*Anabaena* or
	Scenedesmus	*Melosira ambigua*	*Tabellaria*
	(1 lake)	(1 lake)	(1 lake)
	Melosira ambigua	*M. ambigua*	*Botryococcus* or
	(1 lake)	(1 lake)	*Cyclotella*
		Fragilaria croton-	(1 lake)
		ensis (1 lake)	*Botryococcus*
		Pediastrum	(1 lake)
		(1 lake)	
Secchi Disk Transparency (meters)	0.82–0.38 m.	3.05–1.41 m.	6.04–4.48 m.
Water Color (Ohle methyl orange units × 2.8)	53–120	17–39	14–31
pH	7.8–8.4	7.2–8.2	6.8–7.0
Conductivity	155–296	66–178	39–60
Number of species of Chlorococcales	26–42	14–28	4–14
Number of species of Desmidiae	3–15	6–17	20–35
Chlorophycean Index	2.6–14	1.0–3.0	0.2–0.5

On the basis of the dominant phytoplankters there is little difference between the first and the second lake districts and not a very striking difference between the Aneboda district and the other two; the use of the Chlorophycean index, however, appears to separate the plankton communities as sharply as the transparency, taken as a rough measure of plankton density, or the conductivity, taken as a measure of total concentration. Conversely, these results show that the Chlorophycean index is not really a parameter describing the community, for in the Västervik area a value of the index between 2.0 and 3.0 can be given by communities dominated by *Pediastrum clathratum, Melosira ambigua, Anabaena,* or *Anacystis.* Apparently what the index does is to permit us to determine from the phytoplankton something about those properties of the lake as a whole which are ordinarily expressed in terms of oligotrophy or eutrophy and therefore supposedly in terms of primary productivity.

Nygaard (1949) has gone further into the matter and has proposed four other ratios or quotients, so that we now have

the myxophycean index = number of species of Myxophyceae/number of species of Desmideae

the chlorophycean index = number of species of Chlorococcales/number of species of Desmideae

the diatom index = number of species of centric diatoms/number of species of pennate diatoms

the euglenophyte index = number of species of Euglenophyta/number of species of Myxophyceae and Chlorophyceae

the compound index = number of species of Myxophyceae, Chlorococcales, centric diatoms, and Euglenophyta/number of species of Desmideae.

Since there is a tendency for green and blue-green algae to be summer forms, although the diatoms may flourish at any time of year, the indices other than the diatom quotient refer only to summer collections, preferably made in June, July, and August. The diatom quotient is supposedly applicable at any time of year. The indices, moreover, are to some extent relative to the mode of collection of the plankton; since the Chlorococcales include many species smaller than nearly all desmids, the Chlorophycean index and to some extent the compound index derived from counts made on total plankton with the inverting microscope or other sedimentation technique will be higher than if derived from net phytoplankton from the same localities.

Nygaard's data for a number of Danish Lakes, which can be separated into two categories, are given in Table 20.

TABLE 20.   *Phytoplankton indices for less productive and more productive groups of lakes*

	Myxo-phycean	Chloro-phycean	Diatom	Euglo-phyte	Com-pound
Less productive, more transparent (pH usually less than 7.0, Ca less than 10 mg. per liter.)	0.0–0.4	0.0–0.7	0.0–0.3	0.0–0.2	0–1
More productive, less transparent (pH usually over 7.0, Ca more than 10 mg. liter)	0.1–3.0	0.2–9.0	0.0–1.75	0.0–1.0	1.2–25

These figures indicate that although in the more heterogeneous series of lakes studied by Nygaard the Chlorophycean index, as used by Thunmark, and indeed the other simpler quotients are imperfectly diagnostic, the compound index does appear to give a rather clean separation into the less and more productive lakes as judged by other criteria, largely transparency. In general, Nygaard regarded lakes containing associations giving a compound index of less than 1.0 as unproductive and those giving an index of more than 3.0 as definitely eutrophic; the intermediate values implied mesotrophy or weak eutrophy.

Nygaard (1955) later attempted a direct comparison of the compound index with data on photosynthetic productivity (see Volume III), as determined by light and dark bottle experiments, in five very diverse Danish lakes (Table 21).

The first lake, Grane Langsø, which Nygaard says would be regarded as oligotrophic by most limnologists, clearly not only has a low productivity but also a low compound index. Similarly, the two highly productive lakes, legitimately regarded as eutrophic, have high compound indices. The other two localities, both acid and humic, do not fit into the linear pattern.

Apart from the rather unexpectedly high productivity per unit volume of the thin euphotic zone of the Kattehale Mose, the anomaly is largely due to the low diversity of the plankton in these chthoniotrophic locali-

TABLE 21. *Properties and phytoplankton indices of Danish lakes*

Lake	Grane Langsø	Kattehale Mose	Store Gribsø	Furesø	Frederiksborg Slotssø
Type	Oligotrophic	Chthonio-oligotrophic	Chthonio-oligotrophic	Eutrophic	Eutrophic
z_m	11 m.	1.5 m.	~11 m.	36 m.	5 m.
Maximum thickness euphotic zone	11 m.	1 m.	5 m.	12 m.	4 m.
Conductivity %20.10[6]	49–50	76–83	128–131	290–310	360–410
Mean gross productivity	m.$^{-3}$ 0.07 m.$^{-2}$ 0.62	0.24 0.20	0.175 0.58	0.17 1.88	1.21 2.58
Mean number of species in the numerator of compound index	3.75	3.75	3.25	16.0	30.8
Mean number of species of desmids	9.25	4.0	0.25	5.0	3.0
Compound index	0.3–0.6 %	0.7–0.85, ∞	4, ∞	2.3–4.0	9.0–11.7

ties. This permits random variation in the number of desmids to have an extreme effect. In Store Gribsø the actual ratios are 3:0, 4:1, 2:0, 2:0, 2:0 and in the Kattehale Mose 3:0, 0:0, 6:9, 6:7; only the last two ratios for the latter locality are reasonably informative. Even when the flora is not unduly restricted, the fact that the numerator is based on numbers of species of several groups, whereas the denominator is solely determined by the diversity of desmids, is a little unfortunate. This is the more so because the prevalence of a large number of desmids is probably directly determined by low calcium rather than by deficiency of nitrogen and phosphorus. Localities may exist in which a marked nutrient deficiency can occur in the presence of abundant calcium. When, as in the Scottish localities studied by Brook (1959a), there is a rich desmid flora in the less productive lakes, the compound index works well. Values calculated from Thomasson's (1955a) data for four lakes in Tierra del Fuego (chlorophycean 0.09 to 1.2, compound 0.27 to 2.8) doubtless reflect general oligotrophy in the region. However, in other areas notably northwestern North America (Rawson 1956), Lake Ohrid (Stankovič 1960), and Finland (Järnefelt 1956), the indices appear to be inapplicable.

Järnefelt has developed a rather different method, though it is implied to some extent by the theoretical remarks of the last paragraph. He found, in general, that the indices of Nygaard do not correlate with his classification of Finnish lakes into eutrophic, chthonioeutrophic, oligotrophic, and chthoniooligotrophic. He also found that on the whole the more eutrophic localities had a greater number of species. In a list of all the species of the plankton (Järnefelt 1952) in categories according to their occurrence in the various types of lakes a small number of species occurred in 0.7 times as many eutrophic as oligotrophic localities. The phytoplankters so listed are given in Table 22.

It is noteworthy that only one species (or rather closely allied species group) of desmid is included, whereas several species of *Staurastrum, Closterium,* and other genera occur in the much longer eutrophic list.

The species in Table 22 are taken as oligotrophic indicators. A list of at least ninety[6] species which occur four times as often in eutrophic as in oligotrophic lakes was also made by Järnefelt. The ratio of the

---

[6] Some entries refer to a genus followed by species. There seem to be some inconsistencies between Järnefelt's (1952) and (1956) papers. In the first *Anabaena flos-aquae* and *Aphanizomenon holsaticum* (*-flos-aquae*) are recorded as present in about 1½ times as many eutrophic as oligotrophic lakes; in the second they appear as eutrophic indicator species. Järnefelt also computes his quotient for a list of species three rather than four times as common in eutrophic than oligotrophic localities.

TABLE 22. *Oligotrophic indicator species*

Agmenellum quadruplicatum (= *Merismopedia glauca*)*
Crucigenia irregularis
Mallomonas allorgei
Dinobryon cylindricum
Staurodesmus spp. (= *Arthrodesmus incus*)
Dactylococcopsis smithii
Nitzschia gracilis
Kephyrion spp. (other than *spirale*)
Diceras spp.
Ochromonas sp.
Cyclotella operculata
Anacystis dimidiata (= *Chroococcus turgidus*)*
A. quadruplicatum (= *Merismopedia punctata*)*
Dinobryon bavaricum
D. divergens
Stichogloea olivacea
Cyclotella kuetzingiana

---

* Cf. Drouet and Daily (1956).

number of species (I_e) in the eutrophic list to the number (I_o) in the oligotrophic list is then used as one index of eutrophy; the volume of all specimens of species in the eutrophic list compared to that of all the specimens of species in the oligotrophic list gives another index.

Insofar as there is a definite tendency for the more eutrophic lakes to have more species, even if all the species known from oligotrophic lakes also occurred throughout the whole range of eutrophic lakes, it is obvious that the ratio of the number of species not on the oligotrophic list to the number on that list would tend to be correlated with the degree of eutrophy. The indices used by Järnefelt, however, go further than this, since some of his oligotrophic species such as *Dinobryon* spp. occur in eutrophic lakes only under conditions of temporary nutrient exhaustions (see page 455), and a few species on the list are probably really confined to oligotrophic waters. If in a given region a sufficiently large number of lakes have been studied, ratios of the kind computed by Järnefelt no doubt give, as he has claimed, a sensitive indicator of weak eutrophy, and, within the major categories recognized by the presence or absence of water blooms, his indices probably permit a significant linear ordering.

SUMMARY

The associations of phytoplankton species appear remarkably diversified, in view of the fact that the majority are competing for the same materials in a relatively uniform environment, in which conditions we would expect an approach to a unispecific equilibrium. Statistical analyses of planktonic diatom assemblages indicate that there may be a lognormal distribution of numbers of specimens per species and that the rank order of specimens per species follows a type of distribution known in a number of small organisms, particularly in heterogeneously diverse environments. This could be the result of failure to achieve equilibrium in the phytoplankton, though it is not necessarily implied by the distributions observed, it is at least a plausible explanation of the observed diversity, particularly when competitive exclusion is a slow process. It is possible that some synergistic effects between small efficient but vitamin-requiring species and larger less efficient species that can make an excess of vitamins could permit more equilibrium diversity than would at first be expected. The effect of grazing by herbivores able to eat only certain species in the phytoplankton may also be involved. On the whole, it seems more likely that the phytoplankton at least of small lakes is not entirely an autonomous part of the community, many species being recruited continually from littoral benthic habitats which might provide a great diversity of specific niches; benthic resting stages are doubtless also of importance in minimizing competition.

It is probably a mistake to regard the annual changes in phytoplankton as comparable to the aspection of vegetation on land. In some respects the changes are more like the responses to large scale climatic change in the postglacial of the terrestrial flora.

Phytoplankton associations may be designated by the name of the *dominant* or the dominant and one or more subdominant species. This designation may give a good general idea of the phytoplankton and of its principal possible coaction with zooplankters. Since most domiants will have developed large populations in spite of small random fluctuations in the environment, they may be expected to have larger niches than the rarer species. Enumeration of the rarer *characteristic* species can in some cases give more information about the lake, though in some ways less about its phytoplankton than mere reference to the dominant.

It has long been realized that there is some correlation between the quantitative abundance, and so ultimately the nutrient supply available, and the qualitative nature of the plankton. It seems possible, going from unproductive or *oligotrophic* to very productive *eutrophic* waters, oligotrophy and eutrophy ideally being defined in terms of available nutrient supply, to recognize a series of plankton types composed of one or more associations which may be loosely arranged as follows:

1. *Oligotrophic desmid plankton.* Dominants usually *Staurodesmus* spp. and acidophil *Staurastrum* spp., normally with many other desmids, often partly of littoral benthic derivation. *Sphaerocystis, Gloeocystis, Rhizosolenia, Tabellaria* may be associated. In dilute waters poor in alkaline earths and slightly acid.

2. *Oligotrophic diatom plankton.* *Cyclotella* spp. *Tabellaria* spp. or several other genera may be dominant; *Melosira granulata* absent, though other species of the genus may occur. The usual association in nutrient-poor lakes with neutral or slightly alkaline water.

3. Botryococcus *plankton.* The dominant is *Botryococcus brauni* which is found partly in oligotrophic lakes in temperate regions but also under a variety of conditions in tropical and subtropical regions. The incidence of the development of this organism as a dominant is not understood.

4. *Chrysophycean plankton.* One of several species of *Dinobryon* dominant; characteristic of nutrient-poor waters and often combined with type 2; frequently developed in otherwise productive lakes at seasons of nutrient depletion.

5. *Oligotrophic chlorococcal plankton.* Apart from *Botryococcus,* the only characteristic green algae dominant under conditions of low nutrient supply seems to be *Oocystis* spp. Sporadic in various parts of the world.

6. *Oligotrophic dinoflagellate plankton.* Dominant species of one or more species of *Peridinium;* in northern Europe apparently usually *P. inconspicuum.* Not clearly distinguished from the next.

7. *Eutrophic dinoflagellate plankton.* Dominants apparently *Peridinium bipes* or *P. cinctum,* with *Ceratium* and *Glenodinium.* In tropical regions (Indonesia) *Peridinium gutwinskii.* Not certainly distinguishable from 6.

8. *Eutrophic diatom plankton.* Dominants usually *Asterionella, Fragilaria crotonensis, Stephanodiscus astraea,* or *Melosira granulata,* but any of these except the last may occur in oligotrophic localities. The ordinary association of eutrophic lakes except at the warmest time of year.

9. *Eutrophic chlorococcal plankton.* *Pediastrum* or *Scenedesmus* the most usual dominants. Ordinarily in rather small lakes or ponds.

10. *Mesotrophic or eutrophic desmid plankton.* Characterized by eurytopic species of *Staurastrum* (*S. chaetoceras, S. gracile, S. pinque, S. planctonicum*) or more rarely by *Cosmarium* spp.

11. *Myxophycean plankton.* Dominants *Anacystis, Aphanizomenon, Anabaena,* and other blue-green algae. Usually found in the

warmest time of year in eutrophic localities, perennial and often essentially monospecific in highly fertilized tropical waters.

12. *Euglenophyte plankton.* Usually only in polluted water rich in nitrogenous organic compounds, though *Trachelomonas volvocina* or *Lepocinclis fusiformis* may occur as dominants in more ordinary small lakes.

13. *Bacterial plankton.* Photosynthetic bacterial plankton, composed of organisms such as *Lamprocystis* occasionally occurs in saline lakes into which much H_2S produced by sulfate reduction in the sediments is diffusing.

Many mixed types are possible, the commonest being combinations of diatom plankton with the others.

The rarer indicator species have been treated in several ways. A chlorophycean index, namely the ratio of number of species of Chlorococcales to the number of species of desmids and the compound index, or number of species of Myxophyceae, Chlorococcales, centric diatoms, and Euglenophyta to the number of species of desmids, both appear in some lake districts to be correlated with transparency, and so the total mass of seston, and to some extent with productivity. The use of these indices, however, requires discretion; they are certainly inapplicable in some regions. It is also possible to make for any region a list of organisms of more frequent occurrence in unproductive waters and another list of species of more frequent occurrence in more productive waters. The ratio, in the plankton of any lake, of the number of species on one list to the number on the other can be used to assess the position of the lake in a scale of productivities.

CHAPTER 23

The Seasonal Succession
of the Phytoplankton

To the casual observer the seasons in a lake appear to be mainly a matter of reflection. In the spring white, gray, and black give place to green; later, green gives place to orange and gold, which again darken in the early winter. At one time the reflected sky is mainly gray, at another mainly blue. The more perspicaceous perhaps may note the higher vegetation in the shallow water near the shore, growing and decaying in unison with its terrestrial counterpart. The fisherman will know the movements of his quarry, spawning in the shallows or ascending streams. In most cases these observers miss the most significant and deeply hidden changes. Only in very productive lakes, which produce at certain seasons, generally after midsummer, so dense a population of blue-green algae, that the water becomes turbid and vast masses of the plant may collect downwind against the shore, does a hint of the series of complex seasonal changes in the water become apparent to the unaided eye of the ordinary shoredweller or boatsman. This great, sometimes almost cataclysmic, productivity has long been called the flowering, breaking, or working of the waters. Griffiths, (1939) notes some examples from medieval writings, some typical, others of red organisms that produced "blood lakes."

Children have been told that they cannot bathe in a lake until the water has roiled, or turned over, an unfortunate expression likely to be misunderstood by the partly informed limnologist.

Though the production of a water bloom, or flowering, is observed only in highly productive lakes, it has been known since the investigations of the limnologists of the second half of the nineteenth century that all lakes show a remarkable series of changes in both the quantity and the kind of phytoplankton present in their waters. Moreover, the main hypotheses regarding the nature of the seasonal succession had already been advanced fifty or more years ago. It is therefore perhaps a little disconcerting to find that there is still a good deal of doubt about the interpretation of the seasonal changes observed in a particular lake or exhibited by any particular species. This does not imply inadequate industry and intelligence on the part of limnologists, as will be realized on reflecting how recent is our knowledge of the operation of seasonal events on land. Even urban biologists, surrounded by a few trees, pigeons, sparrows, starlings, and their own species, are confronted in spring by the effects of the increment of both radiation and temperature and may sometimes be at a loss to know how to disentangle them. In full natural communities on land the interactions are complex, and only within the present century has the role of light been adequately appreciated. Even on land it is easily recognized that the time of maximum isolation is not the time of maximum temperature. According to the aspect under consideration, the summer solstice can be either Midsummer's Day or the first day of summer. In the surface waters of a lake in the temperate regions this discrepancy is usually exaggerated by their high thermal capacity and the downward turbulent conduction of heat that is stored for a long time after the daily income of solar energy has begun to decline. In the Northern Hemisphere the maximum epilimnetic temperatures may therefore be expected at the end of July or even well into August, during which month the incident radiation may be about as great as in May but the temperature $10°$ C or more higher. This separation of the optical and thermal seasons is of considerable importance in interpreting the phenomena of phytoplankton succession. In the water, moreover, there is a third factor over and above temperature and illuminatoin, namely, the concentration of various substances required for plant nutrition or sometimes disadvantageous to particular species. The variation in such substances finds an analogy on land in the variation in rainfall, for there water is a substance of great nutritive importance often not present in adequate amounts. The number of seasonally variable nutrients in the hydrosphere, however, is much larger than on land. This chemical complexity, together with the difficulty of separating even under the rather favorable conditions provided by a lake the direct effect of solar radiation at the moment from its indirect and additive thermal effect, has made the progress of the understanding of the seasons in the water slow and fitful.

The phytoplankton consists almost exclusively of unicellular plants which in nature probably divide from two to ten times a month. At the higher division rates populations of 10^{15} to 10^{20} individuals can be produced from a single cell in the course of a season. Even if the asexual process of reproduction allows little opportunity for the appearance of new characters, the fact that we are dealing with increases and decreases of populations rather than with the growth and decay of a single generation of annuals or of a single year's production of leaves and flowers of perennials provides an opportunity for competition and natural selection to occur; in fact the replacement of one form by another throughout the season is essentially natural selection, the direction of which is controlled by the varying environment. What we witness annually in a lake is in some ways, as has already been suggested (page 376), comparable to the changes in forest cover during a period of postglacial climatic change rather than to the phenomenon of spring and fall in the life of the individual trees.

QUANTITATIVE VARIATION OF THE PHYTOPLANKTON AS A WHOLE

It is convenient first to consider the variations of the phytoplankton as a whole without regard to the individual species. It has long been known from the work of Calkins (1892), Whipple (1894), Apstein (1896), and other investigators that in many bodies of fresh water there is in the spring a great increase in the number of diatoms. Some of these diatoms, after disappearing or at least becoming very rare in the early summer, may reappear later in the year to give two-peaked distributions in time. In some lakes therefore a spring and an autumn maximum in diatoms is observable. Lake Cochituate and other localities studied by Whipple provide startling examples of this type of distribution. In other lakes the plankton rises again after an early summer minimum, not to an autumnal diatom maximum but to a late summer abundance of Myxophyceae which may be present in extraordinary quantities; floating to the surface, they produce the phenomenon of the water bloom. Much of the bloom may drift downwind and be deposited on the leeward shore. The double diatom maximum is well known in the sea, and its existence in lakes therefore raised no special surprise; it may well often have been sought for and sometimes found in lakes in which in truth it did not occur. The flowering is a typically lacustrine phenomenon but one confined to fertile landscapes and often indirectly of human origin.

With Lohmann's (1911) demonstration that in the sea the net plankton, on which the early ideas of periodicity were based, is only a small part of the total plankton, it became necessary to investigate this matter in fresh waters also. As in the sea, the plankton formerly known proved to be only a part of the whole, and the simple types of periodicity postulated as the result of studies of net plankton have proved to represent the true situation imperfectly. It is therefore essential to present the problem mainly in terms of the relatively few lakes in which the total phytoplankton volume, estimated from counts of all the important organisms, the organic seston, or the chlorophyll has been determined over at least an annual cycle. Many of the available data are presented in Fig. 108 which certainly exemplifies all the important types of seasonal variation except that dominated by immense winter maxima under ice. Small lakes and ponds showing this certainly exists, (see page 473), but no data have been published that permit an unequivocal graph of the whole seasonal cycle of the total phytoplankton to be drawn.

In the simplest cases, as exemplified by some of the Carinthian lakes studied by Findenegg (1943b), the variation consists merely of an increase in phytoplankton volume through the spring and early summer to a maximum in July and a decrease during the subsequent months to a minimum in winter. Such a progress is admirably demonstrated by the data for the trophogenic layers of the Millstättersee in 1935, though in the subsequent years there was also a well-marked secondary peak in October. The Faaker See in 1937 and, insofar as it was studied, the Klopeiner See in 1935 and 1936 exhibited the same pattern of a single summer maximum. The Ossiacher See in 1935 and the Weissensee diverge a little from the lakes already mentioned in having a prolonged maximum, with a hint of two peaks in the last-named lake. In 1936 a quite definite double peak was exhibited by the Ossiacher See, with the main maximum at the end of May and the beginning of June and a well-marked subsidiary maximum in September. Grim (1939) found in Lake Constance an almost identical pattern, with a winter minimum and two hardly separable summer maxima in June and August. It is not unlikely that the less productive of the lakes of the English Lake District, such as Wastwater and Ennerdale, which are deficient in plankton in winter (West and West 1912b, Pearsall 1932, and the Freshwater Biological Association 1949a), may prove to have a monaemic cycle with a single summer peak.

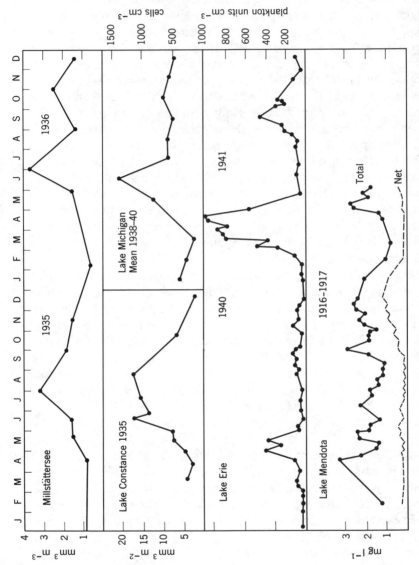

FIGURE 108. Seasonal variation of total phytoplankton in a variety of lakes. (Data from Findenegg, Grim, Daily, Chandler, Birge and Juday, Hutchinson, Riley, Vetter, Pennak.) For Linsley Pond, OS organic seston, CHL, chlorophyll.

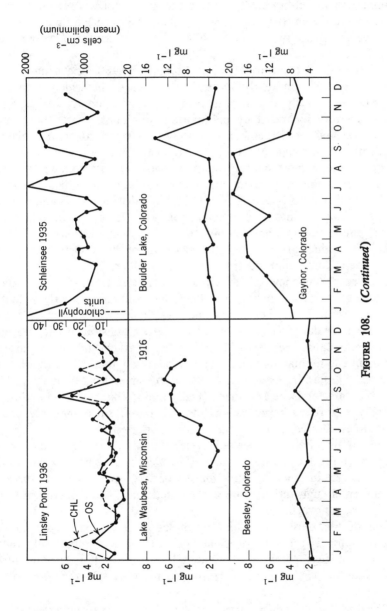

FIGURE 108. (*Continued*)

In Lago Maggiore (Berardi and Tonolli 1953) a single striking maximum in the chlorophyll of the trophogenic zone is recorded (Fig. 110); unlike the maxima in the large lakes of Switzerland and Austria, the greatest quantity of phytoplankton is present in October and November.

In Lake Ohrid (Kozarov 1957 in Stankovič 1960) the nannoplankton is minimal from early December to April, in which month it increases enormously to a maximum in late May and early June. Subsidiary maxima separated by unimportant minima occurred in August and at the end of September, but the main trend in the population is downward throughout the late summer and autumn. The net phytoplankton, almost entirely diatoms, show a maximum at the time of the spring maximum in the nannoplankton and then declines. The over-all pattern is one of a single extended irregular summer maximum breaking up into several smaller peaks. It is possible that some of the summer minima are determined by the grazing activity of rotifers.

It is tempting to suppose that a single pulse is characteristic of large, deep, cold, unproductive lakes; it is clear, however, that there is a marked tendency for the main maximum to split into earlier, late spring or early summer, and later, late summer or early autumn, maxima, as in Lake Constance and the Ossiacher See, or to break up more irregularly as in Lake Ohrid. An extension of the process, whatever its cause may be, will lead to two marked and well-separated vernal and autumnal maxima of the kind often found in the sea and perhaps still more frequently encountered in elementary textbooks. The available data for Lake Michigan (Daily 1938; Damann 1941), though probably not really representing total plankton, since they are based on samples filtered through silk disks supported on sand, appear to indicate a definitely two-peaked cycle, the first, or major, maximum being in June, the second, or minor, in October. The absolute minimum occurs in March, and the secondary minimum in July or September, though rather more accentuated than that observed in Lake Constance, is nevertheless not striking.

One of the most spectacular examples of a diaconic cycle in the total plankton is provided by Chandler's (1940, 1942a,b, 1944) data[1] for the centrifuge plankton of western Lake Erie. It is curious, however, that this cycle, which corresponds more closely than any other

---

[1] Chandler's results are presented in phytoplankton units, evaluated as one cell of *Navicula, Stephanodiscus,* or *Synedra,* one colony of *Coelastrum, Gomphosphaeria, Oocystis,* and *Pediastrum,* five cells of *Dinobryon,* eight cells of *Asterionella, Crucigenia, Diatoma* and of *Tabellaria,* and 300 μ of *Melosira* or filamentous blue-green algae or 100 μ of *Fragilaria.*

based on total plankton to the conventional conception of vernal and autumnal maxima, occurs in a rather atypical lake in which the quantity of plankton is to a considerable extent controlled by sudden changes in turbidity. Marked differences in the degree of development of the maxima occur from year to year and are certainly correlated in part with the turbidity changes, but in every year there seems to be some trace of the vernal and autumnal pulses. In extreme cases the differentiation of these pulses is marked and the two successive years of Fig. 108 demonstrate admirably the least and most developed seasonal succession that apparently can occur in the lake. Because turbidity acts on the phytoplankton by controlling illumination, it will be necessary to return to Chandler's work. In a later section (page 477), in which the effect of grazing on the phytoplankton is discussed, another remarkable example of a two-peaked cycle in Lenore Lake, Washington, is presented. It is indeed quite possible that the breaking up of a prolonged maximum into a series of pulses is as much determined by grazing as by more recondite chemical factors.

The cycles in those of the shallower and more productive lakes that have been studied present a more complex and diversified set of phenomena than the simple one- or two-peaked progressions, though it is to be noted that in some of the artificial lakes of Colorado, in which the maximum depth is often so small that thermal stratification is little developed or absent, Pennak (1946, 1949) has recorded changes in the total seston not unlike the simple progressions already considered. Some of these are illustrated in Fig. 108. More usually the productive and stratified lakes of temperate regions show seasonal cycles of greater complexity.

In Lake Mendota the mean total organic seston was determined by Birge and Juday (1922) from April 1915 to June 1917. In each year there was a marked spring maximum in April or early May. This, in the two summers completely studied, namely, 1915 and 1916, was followed by two subsidiary summer pulses, which fell in June and August 1915 and in May and July 1916. In each year a marked minimum in the organic seston at the end of August and beginning of September was followed by an autumnal pulse almost as striking as the vernal in late September and early October. In 1916, but not in earlier years, there was also a well-marked late autumnal or early winter maximum in late November and early December. The February and March minimum was in both years as well marked as the late summer minimum in August. It is of considerable critical interest to note that if the net plankton alone had been considered Lake Mendota would during 1916 and 1917 have been regarded as exhibiting only a low late spring maximum and a higher late autumnal and early winter maximum. Certain individual species of diatoms, notably *Asterionella,* would have been observed conforming to this two-peaked pattern.

Another well-studied example is provided by the Schleinsee, in which the total volume of phytoplankton present in the trophogenic zone during 1935 was computed by Vetter (1937). In this lake there is a striking late autumnal maximum in phytoplankton, coming just before the lake freezes. While the lake is frozen the total plankton slowly decreases. In the spring, just after the ice breaks, there is a moderate increase of an irregular kind, which is followed by an early summer minimum in June and a marked late summer maximum, though a true water-bloom does not develop. After a minimum at the end of August an early autumnal maximum almost as striking as the late summer maximum is developed. This first autumnal maximum is followed by a minimum in November, after which the plankton content rises to a new maximum just before the lake freezes, so completing the cycle. The four minima, in March, June, August, and November represent reduction of the population to about the same low level.

Linsley Pond has been studied by Riley (1939a, 1940) and by Hutchinson (1944); on the basis of these contributions, and of some unpublished data, a series of chlorophyll and organic seston determination is available from September 1935 to June 1938, except for a period from January to early April 1937. It is evident from these data, and from the plankton volumes computed for 1937–1938, that the most characteristic major maximum in this lake is in the late summer in August and September, when a water bloom was produced during the years of the investigation. Every spring, in April, there is evidence of some increase in phytoplankton, or at least of a fairly high level which descends to a minimum, most notable in the chlorophyll content, in June. In both 1936 and 1938 there was a remarkable maximum under the ice; in 1938 the evidence indicates that this was a superficial occurrence, limited to the top meter or two, but nevertheless it is an interesting and striking phenomenon.

Birge and Juday (1922) at the same time as they pursued their study of Lake Mendota made less extensive investigations in the other lakes of the Yahara drainage basin. An examination of Lake Monona and Lake Waubesa between May and November indicated an enormous increase in the plankton of these lakes from a minimum in June to a maximum in September or October. It is not clear if the values for the May organic seston, preceding the minimum, represent the end of a spring pulse with a large maximum in April. A striking example of a pure autumnal pulse occurring without much other seasonal change is provided by Boulder Lake, Colorado, studied by Pennak (1946, 1949).

A number of quite small ponds have been studied by Nygaard (1938) in Denmark. In some of these the seasonal variation of the total organic seston was very slight and quite irregular. Wherever marked variation occurred there was a tendency toward a two-peaked annual distribution with the major peaks in the late summer or autumn. Nygaard believed that lakes, in general, have a major maximum in spring and ponds, in late summer or autumn. This generalization probably has some statistical validity; it would seem likely that in passing from large unproductive to smaller or, at any rate, shallower and increasingly productive waters the tendency is for an initial early summer maximum to split into a main vernal and subsidiary autumnal maximum and that as the process continues the cycle may be modified in various ways, one being the accentuation of the summer or autumn maximum often with the development of a water bloom.

QUALITATIVE VARIATIONS AND THE CYCLES OF PARTICULAR SPECIES

From the large body of information that is available it has seemed desirable again mainly to consider those lakes in which both the net plankton and nannoplankton have been studied, though much of the available information on net plankton alone is not without importance and is discussed in later sections when the causes of the periodicity are considered.

The subalpine lakes of Carinthia. In the Millstättersee Findenegg (1943b) found the single main maximum due to the increase in *Cyclotella* (apparently *C. comensis melosiroides*) during June and July. In the later part of the season every year *Gomphosphaeria lacustris* increases considerably and with great regularity, maintaining a fairly high total phytoplankton volume or even giving rise to a secondary maximum at the end of October when the population of *Cyclotella* is declining. Dinoflagellates (*Peridinium willei* and *Ceratium hirundinella*) are of little quantitative importance but are present in greatest numbers at the time of the main phytoplankton maximum.

The Klopeinersee also shows a single aestival phytoplankton maximum, again due in part to *Cyclotella,* but in this lake dinoflagellates and other algae constitute a greater part of the population at the time of the maximum than in the Millstättersee.

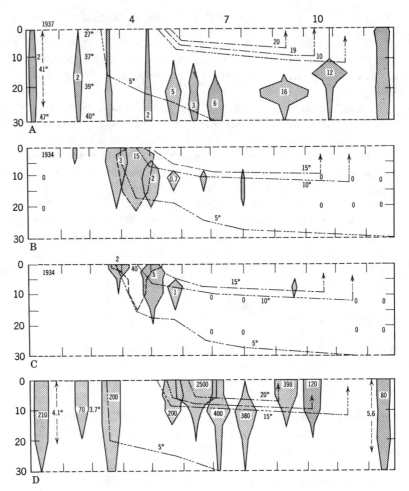

FIGURE 109. Seasonal variation of the main components of the phyto-plankton of the Wörthersee, for 1937 unless otherwise stated, illustrating Findenegg's observations and conclusions. Isotherms for 5, 10, 15, and 20° C. are indicated. (Figures indicate individuals per cubic centimeter; scale linear for *Oscillatoria,* proportional to cube root for other species.) *A. Oscillatoria rubescens,* a cold-water oligophotic species. *B. Dinobryon divergens* (1934), supposedly a cold-water polyphotic species, though doubtless also involving chemical determination. *C. Uroglena volvox* (1934), another chrysophycean of like ecology. *D. Cyclotella comensis melosiroides,* a perennial diatom with a diacmic cycle at the surface, mesothermal and polyphotic if not also involving other determining factors. *E. Ceratium hirundinella,* perennial but with a marked summer maximum suggesting a warm-water polyphotic organism. *F. Gomphosphaeria lacustris* (1936) perennial but with a strong autumnal maximum. *G. Dactylococcopsis smithii,* a late summer autumnal species. *H. Lyngbya limnetica,* autumnal and supposedly a rather warm-water oligophotic species. (Findenegg, modified and rearranged.)

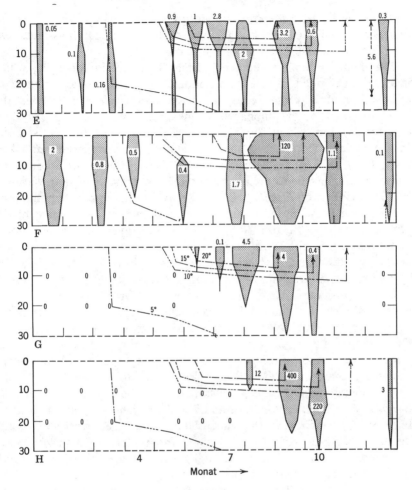

FIGURE 109. (*Continued*)

In the Ossiachersee the first phase of the prolonged maximum in April and May is due mainly to an increase in *Cyclotella,* the second, which may give a secondary maximum in September, to an increase in dinoflagellates and *Gomphosphaeria.* The main species of *Cyclotella* (in this lake *C. glomerata*) shows in a long series of years a marked tendency to a diacmic distribution, with maxima in the spring, evidently in May, and in the late summer or autumn, usually in September. Both maxima may be accompanied by a maximum in *Fragilaria crotonensis.* This species does not necessarily occur abundantly at both maxima of *C. glomerata* but it can do so, as in 1935. *Melosira granulata* usually occurs at the time of the autumnal maximum but can produce a small secondary maximum in May.

The data for the Faaker See indicate a not dissimilar cycle, which in 1937 exhibited a clear maximum in total phytoplankton at midsummer, due primarily to an increase in *Cyclotella* early in the year, followed by a dinoflagellate maximum in June and July. A less regular but essentially similar cycle occurs in the Weissensee. In the Wörthersee, one of the best studied of the Carinthian lakes, (Fig. 109), a comparable pattern of a rise in *Cyclotella* followed by a rise in dinoflagellates is superimposed on a rather irregular variation in the quantity of *Oscillatoria rubescens* in the upper hypolimnion, a species that usually accounts for more than half the volume of the phytoplankton in the top 30 m. *Dinobryon divergens*, though not a important alga in the Wörthersee, is extremely regular in its occurrence in the top few meters of water every spring. It is usually followed by a similar small transitory population of *Uroglena volvox*.

Lake Constance. The general picture (Grim 1939) is of a diatom plankton dominated by *Cyclotella melosiroides, C. glomerata, C. bodanica,* and *Synedra acus delicatissima.* Other species of *Cyclotella,* notably *C. comta* and *C. socialis,* are also of some quantitative importance. The only species not a diatom contributing significantly to the volume of the plankton is the Cryptomonad *Rhodomonas lacustris.* In some years *Asterionella formosa, Fragilaria crotonensis,* and *Tabellaria fenestrata* are important. In 1935, in which year the bulk of Grim's detailed work was done, the total plankton rose from a minimum in April to a maximum in the middle of June. There was a slight fall at the end of that month and thereafter a steady rise to a second maximum in August. The population then declined regularly to a low value at the end of the year. At the beginning of the period of investigation in March and April *Rhodomonas* accounted for half the rather small volume of plankton present. The rise to the first maximum was due mainly to the increase of *Synedra acus delicatissima,* which then declined as regularly as it had risen. It was almost immediately replaced by the dominant species of *Cyclotella,* which are responsible for the second maximum in August. The curve for the minute species of *Cyclotella (C. melosiroides* and *C. glomerata),* which are quantitatively the more significant rose more steeply than that for *C. bodanica* but both reached their maximum abundance at the same time. *Cyclotella comta* and *socialis,* which could not be separated in the quantitative estimations but which are much less important than the other species, continued to in-

SEASONAL SUCCESSION OF THE PHYTOPLANKTON   411

crease after the latter had started to decline. Among the other diatoms in 1935 both *Asterionella* and *Fragilaria* reached their maxima in May and would have appeared in quantity in net samples. *Dinobryon divergens* played a minor part in the lake and reached its maximum at the end of May at the same time as *Asterionella* and *Fragilaria*. *Ceratium hirundinella*, also never abundant, reached its peak in October.

Lago Maggiore. The great autumnal maximum (Berardi and Tonolli 1953) is due to the blue-green alga *Gomphosphaeria lacustris* (Fig. 110). The presence of this species as the dominant in the autumn is in line with what is known in other subalpine lakes, but in the examples already given it at most gives rise to a small secondary maximum. Diatoms, mainly *Fragilaria crotonensis*, show a series of pulses in the summer and during the early part of the *Gomphosphaeria* bloom. There is a small secondary myxophycean peak in March and April due to *Anacystis*.

Lake Ohrid. The spring maximum (Kozarov in Stankovič 1960) is characterized by the diatoms *Cyclotella ocellata* and *Stephanodiscus hantzschii;* in May these are replaced by the dinoflagellate *Cystodinium dominii* with its maximum at the end of the month and *Dinobryon divergens* maximal in late June. Green algae, notably *Oocystis rhomboidea* and *Didymogenes dukia* with *Staurastrum cingulum*, characterize the late summer plankton and Myxophyceae, including *Gomphosphaeria*, the autumnal assemblage. *Dinobryon* is present throughout the summer and autumn, a remarkable replacement of one species by another occurring (page 457). Among the less abundant species *Ceratium hirundinella* is a summer form confined to the upper epilimnion. Several cold-water diatoms are found, including *Asterionella formosa*, presumably represented by a cold-water race (cf. page 433), two possibly endemic species of *Cyclotella*, *C. fotti*, and *C. hustedtii*, *Stephanodiscus astrea*, and the eurytopic desmid *Closterium aciculare* all of which tend to behave much as *Oscillatoria rubescens* does in some of the lakes of Austria and Switzerland, surviving and in some cases flourishing in the hypolimnion in summer and also found at all depths in the colder seasons of the year. *C. aciculare* is known, as Stankovič emphasizes, not only in cold alpine lakes but also in Lake Toba in Sumatra (Ruttner 1952a).

Lunzer Untersee. The net and nannoplankton of the Lunzer Untersee were studied by Ruttner (1930) over a number of years, from 1908 to 1913. No clear indication of the total phytoplankton volume was given in this paper. The information relating to a particular species, however, is very rich, but, possibly because the data

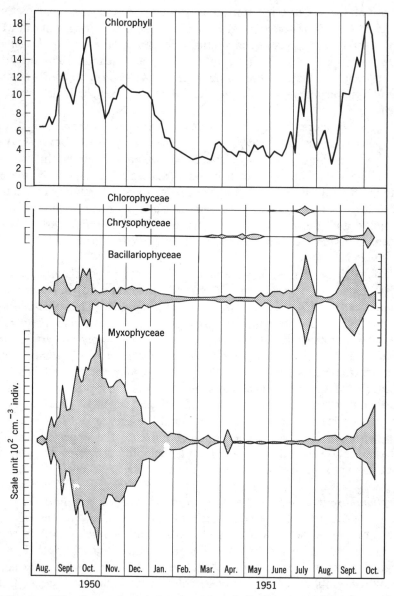

FIGURE 110. Seasonal variation to chlorophyll (mg · m.⁻³) and in main
components of the phytoplankton of Lago Maggiore (0 to 10 m.). Diatom
maxima are due primarily to *Fragilaria crotonensis,* myxophycean maxima,
to *Gomphosphaeria lacustris* with some *Anacystis* in September and April,
and chrysophycean maxima, to *Dinobryon sertularia.* (Berardi and Tonolli,
modified.)

for once are adequate in that samples from many years and a number of different depths are available, it is extremely difficult to give any concise account of Ruttner's conclusions.   It would seem likely, however, that there are two major categories of phytoplankton in the lake: those that are perennial and that oscillate in numbers without clear correlation with the seasons and those that are strictly seasonal, nearly all with maxima after midsummer (Fig. 111).   It is convenient to consider the perennial species first.

Rhodomonas lacustris (including *R. lens*) was absent from the lake on only one occasion, namely in February 1909.   Maxima are poorly marked and can occur at any time between June and October.   Ruttner (1937b) gives the mean temperature of the maxima of the species in the Lunzer Untersee as 10.2° C; in the other lakes of the Austrian Alps he finds similar temperature relationships.   Findenegg (1943b) gives an almost identical temperature, namely, 10.6° C, for the mean temperature of the maximum and indicates that the optimal range for the development of the species appears to be between 8 and 12°; in the lakes he studied the largest populations occurred in April and May.

Peridinium willei, in 1909 had its maximum in early September, followed by a minimum in October; in other years the minima occurred in August and the maxima, irregularly at various times of year.   The numbers present, however, were small.   Ruttner (1937b) gives the mean temperature at the times of plankton maxima in the Lunzer Untersee as 12.3° C.; for other Austrian lakes he gives 10.8° C. Findenegg found a slightly lower mean temperature for the maximum, namely, 10.5° C., but agrees that the species develops most strikingly between 10 and 15° C.

Cyclotella comta tends to have two maxima in the Lunzer Untersee. One is in the early winter at the circulation period, when the temperature is about 4° C., the other in the late spring or early summer at 8 to 12° C.   Both maxima are not necessarily well developed in any given year.   In January 1909 201 cells per cm.[3] were present; the species entirely disappeared in April, but in July 105 cells per cm.[3] were present.   In the next two years the winter maximum was negligible and a summer maximum occurred in June.   Findenegg indicates that the maximum is usually in July or August at a mean temperature of 13.5° C.   It is clear, however, that unless separate cold and warm water races are involved temperature cannot be directly employed to determine the time of maximum of this species.

Asterionella formosa in the Lunzer Untersee is conspicuously absent from the epilimnion in summer, though it appears to be perennial in the deep water.   Ruttner (1937a) has concluded that the race in-

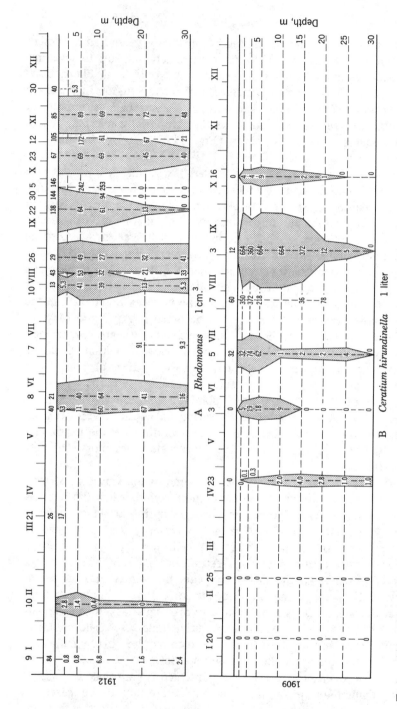

FIGURE 111. Seasonal variation in the Lunzer Untersee of *A*, the perennial *Rhodomonas lacustris* and *R. lens*, and of two strikingly seasonal species: *B. Ceratium hirundinella*, typically a summer plankter. C. *Mallomonas alpina* (Ruttner), which appears as an immense short-lived autumnal pulse. (Ruttner.)

414

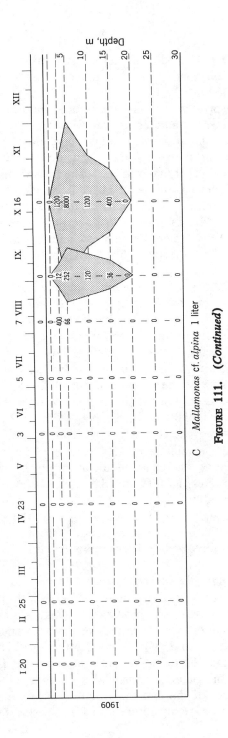

C *Mallamonas* cf. *alpina* 1 liter

FIGURE 111. (*Continued*)

415

habiting the lake is a cold stenotherm variety of the species but that in other lakes a eurytherm variety may have a quite different seasonal cycle (see page 433).

The most conspicuous seasonal forms are found in the late summer but some persist into the autumn or even winter.

Mallomonas cf. *alpina* develops large populations mainly at a depth of 5 to 10 m. during the period from July to October; cyst production terminates the period of abundance. The mean temperature of the maxima in the Lunzer Untersee is 11.3° C.; in the other lakes studied by Ruttner almost identical thermal relations were found. Findenegg, however, indicates that a form also doubtfully referred to *M. alpina* has its optimum development between 2.5 and 7° C., the mean temperature of the maximum being 6.5° C. There is probably some taxonomic confusion here.

Limited populations of *Dinobryon acuminatum* Ruttner developed in April and May, but not in every year. *D. divergens* occurred later in great numbers, with maxima at some time between May and October. In 1911 two maxima occurred: one in early July and one somewhat smaller in mid-October. In 1912 a subsidiary maximum in May was followed by the main maximum in July. In each year cyst formation occurred at the time of the maximum, but, if there were two maxima, cysts were produced only at the time of the major maximum. Ruttner concluded that the initial rise in the population in May usually takes place at temperatures between 7.9 and 10.8°C. and the most rapid increase before the maximum at 15.4 to 16.2° C. In his later paper he gave 13.4° C. as the mean temperature of the maximum in the Lunzer Untersee and 12.1°C. in the other lakes that he studied. Findenegg, on the other hand, gives 7.8° C. as the mean temperature of the maximum and 5 to 8° C. as the optimal range of temperature for the species. It is reasonably certain (see page 455–457) that temperature is not involved in determining the periodicity of *Dinobryon divergens*.

Ceratium hirundinella in the Lunzer Untersee is, as elsewhere in the North Temperate zone, a summer form, with its maximum in July or August. Cysts are produced at and after the maximum. The most rapid rate of increase in the population normally occurs when the temperature is 15 to 20°C. Later Ruttner (1937b) gave 12.3°C. for the mean temperature of the maximum in the Lunzer Untersee and 12.5° C. in the other Austrian lakes. Findenegg gives a slightly higher temperature for the maximum, namely 14.4° C., estimating that the optimal range of temperatures for the species falls between 12 and 19° C.

Cyclotella bodanica usually is an autumnal form which Ruttner believes to flourish at a somewhat higher temperature than *C. comta.* In some years, however, an increase in the population can occur at temperatures falling from 12 to 4° C. in the late autumn and early winter. Findenegg, who regards this species as occurring mainly between June and September, gives a mean temperature for its maximum of 11.8° C. and an optimal range of 10 to 15° C., which are lower than his corresponding values for *C. comta.*

Synedra acus delicatissima was not important in the earlier years of Ruttner's study but was a significant spring form from April to early June in 1912 and 1913. It disappeared from the epilimnion and later from the whole lake during the summer. It reappeared below 10 m. in October 1913. Later Ruttner (1937b) regarded *S. acus delicatissima* as having its maximum at a mean temperature of 6.3° C. in the Lunzer Untersee and at 5.2° C. in the other lakes that he studied. Findenegg gives the slightly higher temperature of 7.2°C. as the mean of the maxima in the lakes that he studied, regarding the species as developing best between April and July at temperatures between 5 and 10° C.

Staurastrum spp. dominated the late summer plankton of the Lunzer Untersee although, in keeping with the fairly high calcium content of the lake, few desmids of other genera are important. *S. cingulum,* (sub *paradoxum*) though a perennial in the lake, shows marked regular maxima in the late summer and autumn. Ruttner's data indicate that increases in the population can occur at 8 to 12° C. and at depths suggesting moderate to low light intensities. Ruttner later added that in the Lunzer Untersee and other Austrian Lakes he studied the mean temperature of the maximum was at 11.0° C. Findenegg, who regards the species as characteristic of August and September, gives 14.6° C. as the mean temperature of the maxima and 10 to 15° C. as the optimal range.

S. lütkemülleri[2] has a seasonal cycle similar to that of *S. cingulum,* but the rise in the population to the maximum is sharper with rapid increase only from 10 to 12° C. and with very little increase occurring at lower temperatures. Both species, but most notably *S. lütkemülleri,* reached their greatest density in the very hot year of 1911 and both showed their lowest maxima in 1912 when a fairly warm July was fol-

---

[2] For figures and an account of the unhappy career of this species see Teiling (1947). Brook (*in litt.*) thinks that the *"paradoxum"* of Ruttner's early work should be referred to *S. pingue* and that *lütkemülleri* may be a benthic derivative of that species. Ruttner, however, later used *cingulum* for the species earlier designated *paradoxum,* and for this book that name will be employed.

lowed by a cool late summer and early autumn. As later emphasized, it is evident that although the two species have almost identical requirements *S. lütkemülleri* does slightly better than *S. cingulum* at high temperatures.

S. manfeldtii var. *planctonicum* behaves in general like the preceding but can be absent in spring. Several rarer species, which are absent in spring but had autumnal maxima, also occurred in the lake, but none was numerous enough to permit any detailed conclusions comparable to those reached relative to the slight differences in thermal requirements between *S. cingulum* and *S. lütkemülleri*.

The English Lake District. The waters of the English Lake District are of particular importance in that they provided the basic data for the most detailed theory of plankton periodicity yet put forward, namely, that of Pearsall (1932). Unfortunately, the older work is based on subjective judgments of rarity or commonness or on estimations of the percentage of a given species in the net phytoplankton, without any certain knowledge that the associations as a whole were increasing or decreasing.

In considering the English lakes, the distinction between the more primitive rocky basins containing waters low in alkaline earths, such as Ennerdale and Wastwater, and the more evolved silted basins containing waters richer in the alkaline earths, such as Windermere and a number of the smaller lakes (West and West 1912b; Pearsall 1921, 1930), must be kept in mind (see Volume I *Frontispiece*). In the first type of lake a desmid plankton is immensely developed, in the second, a diatom—myxophycaean plankton—as already noted.

In Wastwater in 1908 and 1909 W. and G. S. West (1912b) found *Staurodesmus jaculiferus* f. *trigonus* which was abundant from June to November, to be the only species of desmid that ever became really common, though six other species of *Staurodesmus* and *Staurastrum* are recorded. There were two other important members of the phytoplankton. One was the diatom *Rhizosolenia morsa*, which rose in abundance from September to a great maximum at a temperature of 8.9° C. in December and then suddenly declined. The other species that provided a large maximum was the green alga *Sphaerocystis schroeteri*, which, becoming common in spring, reached its peak in October and November. The maximum concentration of phytoplankton is probably in June when *Staurodesmus jaculiferus* has already increased to a large population and when several other green algae, notably *Elaktothrix falcatus* var. *acicularis* G. S. West, *Oocystis lacustris, Gloeocystis gigas* var. *maxima,* as well as *S. schroederi,* are common. There may well be a secondary autumnal maximum, however, due to *Sphaerocystis* and *Rhizosolenia*. The winter plankton is evidently sparse indeed.

Pearsall (1932) found *Staurodesmus jaculiferus* to be the dominant desmid in 1928, the species being present in fair quantity much as in 1908–1909 from June to December. *Sphaerocystis schroeteri* perhaps appeared earlier than during the Wests' investigation, and *Mougeotia* filaments were commoner. The most striking difference, however, was the presence of *Gonatozygon monotaenium,* not recorded from Wastwater by the Wests, and the absence of *Rhizosolenia morsa,* which the Wests found to be an important member of the plankton.

A somewhat more elaborate pattern of succession, but probably essentially the same sort of variation in the total quantity of net plankton, is exemplified by Ennerdalewater, in which lake more species of desmids commonly occur than in Wastwater. In Ennerdale water West and West found a sparse winter phytoplankton, consisting mainly of a few species of desmids, none particularly common. Copepods were abundant in February, and it is possible that the phytoplankton minimum at this time is more apparent, relative to the zooplankton, than if considered per unit volume of water. In March two desmids, *Staurodesmus jaculiferus* f. *bigonus* and *Spondylosium planum,* start to increase. *Dictyosphaerium pulchellum* is the only other alga recorded as common at this season. In April at a temperature of 5.5° C. *S. jaculiferus* f. *bigonus* is the only common alga, but as the water warms up in May (surface temperature 10° C.) large populations of *Staurodesmus "incus"* and particularly of *Sphaerocystis schroeteri* appear. In June *Cosmarium subarctoum* first becomes common and increases to a great maximum in August, as do *S. "incus"* and *S. subtriangularis,* which, with *S. jaculiferus* f. *bigonus,* are responsible for the desmid maximum in August. The only diatoms ever to become common are *Cyclotella comta* in June, *Surirella robusta* in July, and most notably *Rhizosolenia morsa* in late summer and autumn, becoming the dominant alga in November. Dinoflagellates are found to be common only in summer and autumn, *Peridinium willei* having its maximum in July, and the rather less abundant *P. inconspicuum* in August. *Ceratium hirundinella* is present from May to November when the temperature exceeds 8.3° C. and has its maximum in September and October at 10 to 15° C. The only common bluegreen alga is referred to *Chroococcus minutus* in August, but *A. minima* and *Lyngbya limnetica* are evidently frequent in September and October, respectively.

It is not improbable from these observations that the general cycle of phytoplankton as a whole in both Wastwater and Ennerdale is not unlike that in some of the Carinthian lakes even though there desmids are not important. It is reasonable to suppose that there is a winter minimum and a main summer maximum in Ennerdale due to *Cosmarium,* and *Staurodesmus* and in Wastwater to *Staurodesmus* and various other green algae. It is quite likely, however, that a secondary maximum due to *Rhizosolenia* and, when present, to the blue-greens and *Ceratium* is developed later in the season.

Pearsall has given additional data for both these lakes and some others for 1921, 1922, and 1928, but he has recorded the percentage of the total phytoplankton belonging to particular species rather than the absolute abundance. Several interesting changes occurred between the period of the Wests' study and that of Pearsall's investigation. *Rhizosolenia* disappeared from Wastwater and appeared in Ennerdale only sporadically in 1928 and not at all in the other years. *Gonatozygon monotaenium* was of some importance in Wastwater in the late summer and autumn of 1928, and the dominant desmid in Ennerdale was *Staurastrum longispinum* at the time of Pearsall's study. Pearsall examined seven other lakes, finding in general a spring diatom plankton composed of *Asterionella* and *Tabellaria fenestrata,* the maximum of the *Asterionella* usually preceding that of the Tabellaria. Some lakes, notably Crummock Water, Loweswater, and, to a lesser degree, Windermere and Esthwaite, had also large *Melosira granulata* populations in spring; these lakes usually had blue-green algae, notably *Gomphosphaena lacustris* (sub *Coelosphaeum kützingianum*), as an important element in the late summer phytoplankton. In many lakes *Dinobryon divergens* occurred after the spring diatom maximum. Blue-green algae, as elsewhere, were most prominent in the late summer and autumn. The details of some of the successional processes observed by Pearsall, together with the more recent work of Lund are discussed in a later section (pages 443–458).

Lake Mendota, Wisconsin. The classical studies of Birge and Juday (1922) provide considerable information about the variation in the qualitative composition of the plankton (Fig. 112) between April 1915 and May 1917. The over-all variation of the total plankton has already been presented. The spring maximum is, in general, due to the development of an enormous population of *Stephanodiscus astraea,* which appeared in March or April and increased to an immense maximum a month later, rapidly declining and entirely disappearing before the beginning of July. *S. astraea* was in fact the most strikingly seasonal form in the lake. *Anacystis* (sub *Aphanocapsa delicatissima* West) occurs throughout the year at all depths, but in the spring the

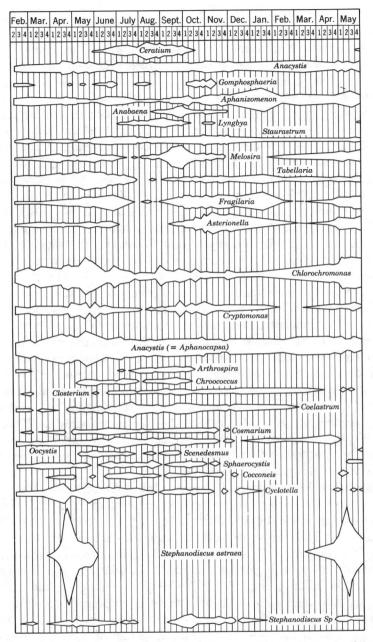

FIGURE 112. Seasonal variation of phytoplankton in Lake Mendota 1916–1917, relative abundances plotted in proportion to the cube roots of concentration. Note summer incidence of *Ceratium* as elsewhere, *Lyngbya* an autumnal form as in Austria, strong diacmic incidence of the three main diatoms in the net plankton (*Tabellaria, Fragilaria,* and *Asterionella*), and the extreme seasonal restriction of *Stephanodiscus astraea,* which forms immense spring blooms. (Birge and Juday.)

population rises characteristically before the *S. astraea* maximum falls a little as the diatom reaches its peak and then rises again as *S. astraea* declines. *Chlorochromonas minuta* Lewis and *Cryptomonas* spp. appear to reach their greatest development a little after *S. astraea*, so prolonging the spring maximum. *Cyclotella* sp. contributed to the spring maximum in 1915 and 1916 and in 1916 but not 1915 showed a double peak like that of *Aphanocapsa*. A second, undetermined species of *Stephanodiscus*, somewhat larger than *S. astraea*, rose and fell with the latter in the spring but unlike *S. astraea* reappeared in the autumn. *Asterionella*, which first appeared in the plankton in 1915, exhibited regular spring and autumn maxima and was absent in July 1915, minimal in February 1916, absent in July and August 1916, and minimal again in March 1917. *Fragilaria* and *Tabellaria*, for which records are available from 1911 onward until 1917, show a similar type of seasonal variation to that exhibited by *Asterionella*, save that they tend to persist in submaximal numbers throughout the summer, though truly absent for a short time in August 1916. The summer pulses in Lake Mendota were due in general to monads, *Gomphosphaeria Aphanocapsa*, and *Oocystis*, whereas at the autumnal maximum the phytoplankton contained monads, *Aphanocapsa*, *Gomphosphaeria* and those diatoms already mentioned as having a double-peaked type of seasonal distribution. A great winter increase in 1916 was mainly due to *Aphanizomenon*. Although a water bloom due to *Anabaena* was recorded as early as 1882 in the Madison Lakes by Trelease (1889), the phenomenon has become economically important only in this century; at present the greatest problem involves massive production of *Anacystis cyanea*.

Mansfelder See, Southwest Saxony. One of the earliest and still one of the most interesting studies on the seasonal cycle of the nannoplankton was made by Colditz (1914) on this curious, somewhat saline lake. The lake is about 5 km. long and just under 1 km. wide; the maximum depth is 7 m. Because of its great area and small depth, thermal stratification is not well marked. In winter a definite inverse stratification is apparent, with full spring circulation at the end of February and beginning of March; there is evidence throughout the summer of transport of heat into the deep water, but a feeble thermocline develops in June and July. Full circulation is probably re-established in August and continues until December. The water contains about 1 g. per liter of dissolved salts; the cations are mainly sodium and calcium, the anions mainly chloride, bicarbonate and sulphate.

Colditz was interested primarily in the nannoplankton as food for rotifers and crustacea; he therefore considered the nannoplankton obtained with the centrifuge and the animals in the net plankton and hardly discussed the net phytoplankton. All he says of it is that in July and August, when the nannoplankton content was minimal, there was a *Anacystis* bloom. As far as can be determined from his data, the year opens with *Cyclotella hyalina* as the dominant form. After decreasing slightly and then increasing, this species rapidly declines at the time of the vernal circulation. During April a large population of small myxophycaean colonies develops; these are presumed to be young *Anacystis*. In June at an epilmnetic temperature of 18° C., a very minute green alga, designated *Pleurococcus* (? = *Nannochloris*) *punctiformis* rises to a marked maximum and falls. The lake then produces an *Anacystis bloom*. In September a maximum in *Cyclotella meneghiniana* took place at a temperature between 9.3 and 11.3° C. This species was then replaced by *Cyclotella hyalina*, which appeared at the end of October, produced a maximum in November, than declined, and finally rose again at the end of the year to produce the condition observed at the beginning of the investigation.

Schleinsee. Vetter (1937) has given an excellent account of the variation in the plankton of this small and rather productive lake throughout the whole of 1935. The initial maximum, which was observed just before the lake froze in the middle of January, was due to *Cyclotella melosiroides,* by far the most important species present in January 1935. The population declined steadily until it had reached a negligible size in April. In late January and February a rather considerable population of *Asterionella formosa* developed under the ice mainly in the layers between 2 and 6 m. The rather sparse spring plankton consisted mainly of undetermined colonial blue-green algae, which were maximal at the end of March, *Dinobryon divergens* and *Cryptomonas* sp., maximal in April, and *Scenedesmus* cf. *spicatus* in early May. In late May and June *Cyclotella comta* became abundant and was associated with *Synedra acus angustissima* and other rarer diatoms. As these species disappeared, the total phytoplankton fell to a minimum level and then rose again as blue-green algae, mainly *Anacystis cyanea*, developed. The volume of the phytoplankton at the blue-green maximum in July was about the same as that in the winter *Cyclotella* maximum in January. After the blue-green maximum the total phytoplankton again declined to a minimum which was succeeded by a large autumnal maximum of *Cyclotella melosiroides* with *Cryptomonas* in the deeper water. These forms declined to a minimum in November, but *C. melosiroides* immediately started to rise again to the mid-winter maximum.

One curious feature of the Schleinsee is that although, as elsewhere,

Ceratium hirundinella is a summer form, not appearing at all in the lake until the middle of June and disappearing in October, it was here almost exclusively a species of the upper hypolimnion, which reached its maximum concentration at 9 m. on August 17 when 700 cells per cm.[3] were present. The temperature during the periods of the major increase in the population at this level was about 10° C. (Einsele and Vetter 1938). *Ceratium hirundinella* was here associated with the hypolimnetic blue-green algae *Lyngbya pseudovacuolata* and *Oscillatoria rosea,* which both occurred commonly in the 9-m. layer at the time when the *Ceratium* population was reaching its greatest development. The *Ceratium* was also associated with the upper part of the population of *Thiopedia rosea,* which, however, was better developed in general in slightly deeper, less oxygenated water containing H_2S. This occurrence provided a most curious contrast with the rest of our knowledge of the distribution of *C. hirundinella.*

Linsley Pond. The history of the total phytoplankton in the surface layers is known in detail from April 1937 to June 1938 and in the deeper water over a considerable part of this period; some information is available for the spring and early summer of subsequent years (Riley 1940; Hutchinson 1944). During the period in which intensive studies were made the main phytoplankton maxima, as indicated by the chlorophyll content, were in April 1937, August-September 1937, January 1938, and April 1938. There were several minor summer pulses which are also recorded in Fig. 108. The April phytoplankton at the time of the first maximum was dominated by *Synedra,* mainly *S. acus angustissima.* At the height of the maximum no less than 2880 cells per cm.[3] were present. *Asterionella* was relatively abundant, but the only other significant forms were minute Chroococcaceae, small centric diatoms, mainly *Stephanodiscus,* and great numbers of minute round cells, probably partly *Cyclotella,* partly green *Nannochloris*-like forms and badly fixed monads, and partly isolated cells from blue-green colonies. These small cells of undetermined nature were about ten times as abundant as the *Synedra,* but it is unlikely that their volume was as great as that of the diatom. *Dinobryon* was present on April 17, but disappeared two weeks later.

During May the *Synedra* declined in numbers; *Asterionella* reached a maximum of 212 cells/cm.[3] on May 17 at a temperature of 17.5°C. and then declined; the other forms oscillated irregularly. One June 1 at a temperature of 23.4° C. a marked maximum in *Dinobryon* corresponded to a minor maximum in the chlorophyll content; all other forms were scarce (Fig. 114). Later in the month only minute unidentified cells increased in numbers as *Dinobryon* declined.

Anabaena appeared early in July and *Fragilaria crotonensis* began to increase shortly afterward. The August maximum was dominated by these two algae in enormous quantities, the *Anabaena* forming a bloom. In September *Oscillatoria* appeared, and this alga became dominant during October. An increase in *Oscillatoria* under the ice in January was responsible for the great maximum in the surface waters in January, and the April maximum in 1938 was also due to this genus and other minute unidentified forms. No marked spring diatom population developed in that year. The spring population of *Oscillatoria* declined in May, and large numbers of *Dinobryon* appeared later in the month, as in the preceding year. Data available for the spring of 1939 indicate two blue-green maxima in the spring composed of *Oscillatoria* and, at least during May, *Aphanizomenon*. Two *Dinobryon* maxima occurred, one in March and April and the other in June; both appeared to be associated with a previous fall in the population of other algae. As in 1938, few diatoms were present. Among the minor species present in Linsley Pond, the most interesting is *Asterionella formosa*, though it has never been a dominant member of the phytoplankton. Its occurrence in the spring of 1937 has already been noted. A similar rise and fall of the species occurred in the early winter of the same year and again another maximum in May 1938. The bearing of these variations on the problem of the seasonal occurrence of diatoms is discussed later.

The Shropshire Meres. It is finally of interest to recall the numerous examples of the "breaking of the waters" or "breaking of the meres" recorded by Phillips (1884). These records refer to the production of water blooms in shallow lakes which during the middle nineteenth century were doubtless much less subject to human influence than are some of the lakes known to produce massive water blooms today. Phillips notes *Oscillatoria* sp. *Gomphosphaeria lacustris, Anabaena circinalis,* another species of the same genus referred to as *Dolichospermum ralfsii* and also less commonly *Aphanizomenon holsaticum.* *A. circinalis* was evidently the most important bloom-forming alga. Phillips noted that although the bloom was usually a summer phenomenon, winter breaking was possible in Newton Mere, in which a great quantity of *A. circinalis* and *G. lacustris* appeared in February 1882, and in Ellesmere, in which *Gloeotrichia echinulata* produced the summer and *A. circinalis* the winter breaking. Phillips collected several early records of water blooms, from Greville (1823–1828), Drummond (1838), Dickie (1860), Cohn (1878), and Hughes (1880). Several of these records refer to small bodies of water in mountainous regions not likely to have suffered from human interference. Phillips' account is so much more explicit and is based on so much more extensive study than the earlier works that he evidently deserves a more honorable place in the history of limnology than he is usually accorded.

INTERPRETATION OF THE OBSERVED SEASONAL CHANGES

The environmental factors that have been considered by various students to underlie the phenomena set out in the preceding pages may be considered in the following categories.

Partially independent physical factors
    Temperature
    Light
    Turbulence (page 287)
Interdependent biochemical factors
    Inorganic nutrients
    Accessory organic materials, vitamins, etc.
    Antibiotics
Biological factors
    Parasitism in the broad sense
    Predation
    Competition

This classification is not impeccable logically. Competition does not operate *in vacuo* but must always be competition for something. Moreover, the direction of competition, the nature of the winner in any particular struggle is so frequently determined by environmental factors that what seems to be the direct exclusion of a species by some unfavorable physicochemical factor may often turn out to be the result of the operation of this factor in determining the direction of competition; if the biological association is altered, the physical factor will appear to operate very differently. Throughout the entire discussion of the other factors, the possible role of competition must always be kept in mind, though it will be convenient to discuss some specific aspects of the phenomenon at the end of this chapter.

Considered in a most general way, the three most important environmental variables, temperature, light, and the concentration of inorganic nutrients, can evidently have suboptimal, optimal, and supraoptimal values. The optima, as will appear later, are not by any means absolute, and a more complicated consideration of their nature will be necessary. Considered in the simplest terms, however, when such variables have suboptimal values, an initially small population will increase less rapidly than when the values are optimal. An initially large population, however, may remain suspended in the water for a considerable time without reproduction in markedly suboptimal conditions. When the environmental variables are supraoptimal, active inhibition, injury, and death intervene. The nature of the action of the environmental factors on either side of the optima therefore tends to be asymmetrical. This is a point of particular importance in interpreting the autumnal phytoplankton when at least the values of the physical factors, illumination and temperature are falling. The fact that an already large population may remain static or even go on increasing slowly during a period of declining temperature is not evidence that the spectacular rise which produced the population at the warmest time of year was not favored by the high temperatures then prevailing. The times of maximum rate of increase are in fact of more interest than the times of maximum population.

Although only cell division can increase a population of unicellular organisms in a closed volume, the mechanisms of decrease are obviously diverse. When one of the environmental variables is supraoptimal, active inhibition or injury may produce a decline in the population. During periods of general suboptimal conditions or at times of low turbulence mere sedimentation unbalanced by reproduction may be sufficient to reduce a population, and death by parasitism or predation is also always occurring, though with varying intensity. Whatever may be the cause of loss of individuals from the population, it is essential to remember that without the loss no succession would take place. All available nutrients would be locked up in a maximum stationary population which would remain invariant from season to season. Though the Protista are potentially immortal, an immense mortality is obviously needed to produce the type of cycle commonly observed.

TEMPERATURE

Experimental studies of temperature. In view of the obvious importance of temperature, it is curious how little experimental work has been done on temperature optima for the development of populations of phytoplanktonic algae.

Rodhe (1948) alone has investigated the matter in the light of modern physiological concepts. He has stressed the fact that the concept of an optimum temperature or optimum illumination is not necessarily absolute, since a considerable increase in growth may be produced momentarily by an increase in temperature or illumination to levels that also produce a slow irreversible injury to the living system under investigation. We can therefore distinguish between momentary optima under any given set of conditions and the lower limit at which irreversible inhibition by light or temperature sets in. Whatever may happen at higher temperatures or higher illuminations over short periods of time, the lower limit of irreversible inhibition will always appear to be optimal if all environmental factors are held constant and the time of observation is sufficiently long. The apparent optimum will therefore fall during any set of experiments, approaching asymptotically the *limiting optimum* (*Grenzoptimum* or limit optimum of European authors) below which irreversible inhibition does not occur. These ideas were applied to whole organisms by Stalfelt (1939a,b) in his study of the lichen *Usnea dasypoga* but will be familiar to anyone acquainted with the behavior of enzymes. Since the rate of enzymatic inactivation nearly always increases with rising temperature above a certain limiting temperature, it is evident that temperatures may exist at which the rate of catalysis over a short initial period is greater than the mean rate over a long period, by the end of which a considerable amount of inactivation has occurred. Over very long periods of time the enzyme may produce more change in a substrate (present in excess) at a lower temperature at which there is no inactivation than at a higher temperature at which the enzyme preparation is initially more active but is rapidly losing activity.

Rodhe found just this type of effect in cultures of *Melosira islandica helvetica,* reared at 5, 10, and 20° C. The culture at 20° C. first produced the greatest growth, but development was somewhat abnormal and the diatom soon died. The culture at 10° C. was after one week rather better developed than that at 5° C. but later declined somewhat, so that at the end of the experiment the greatest population had been produced by the culture at the lowest temperature. It is evident that in this case the limiting optimum is near 5° C.

As far as can be determined, the limiting optima for the species studied by Rodhe had the values given in Table 23: If, as seems often to be the case, the species with high limiting optima hardly

TABLE 23. *Limiting thermal optima for various planktonic algae*

Melosira islandica helvetica	about 5° C.
Synura uvella	about 5° C.
Asterionella formosa	10 to 20° C., probably about 15° C.
Fragilaria crotonensis	apparently about 15° C.
Ankistrodesmus falcatus	about 25° C.
Scenedesmus, Chlorella, Pediastrum, Coelastrum	20 to 25° C.

grow at all at low temperatures and if, as there always must be, there are processes such as sedimentation and predation which continually remove part of the population, these results would suggest that a seasonal cycle in the phytoplankton might be formally explicable in terms of temperature alone. There would be, as there is in Erken, from which lake Rodhe's algae were obtained, a winter population of *Melosira islandica helvetica,* a summer population of green algae, and perhaps vernal and autumnal maxima of *Asterionella* and *Fragilaria.* Such an interpretation would be very much in line with the conclusions of Wesenberg-Lund (1904), based on field experience in Denmark. It is to be noted, however, that the concept of the limiting optimum, as the optimum for the slow, steady development of large stable populations, does not exclude occasional transient maxima at higher temperatures. It is quite possible to imagine a species restricted early in the year by some factor other than temperature. As the temperature rises, if the previous nonthermal restriction also be removed, a temporary maximum might occur well above the limiting optimum just as it did in Rodhe's experiments with *Melosira islandica helvetica.* This possibility suggests, in fact, that the interaction between thermal and photic or chemical control of the phytoplankton may produce complicated patterns that could hardly be analyzed without prolonged experimental study.

The only other detailed experimental study on the effect of temperature on a plankton alga is Lund's (1949a,b) study of *Asterionella formosa.* Lund found that with adequate enrichment of nutrients the logarithmic phase of the growth rate of *A. formosa* in lake water is about one division per day at 10° C. and two at 20° C. The apparent increase in the lake is never so high as this.

Studies in nature on the role of temperature in regulating phytoplankton periodicity. Though both the temperature of the water and the light available for photosynthesis are greater in summer than in winter, the temperature, depending to some extent on the integral of the radiation delivered to the lake surface, rises later in the spring than does the illumination. Moreover, at almost any time of year there may be periods at which there are variations in illumination during the course of a few days, with little or no change in temperature. Theoretically, it should be possible, with enough data, to separate by statistical methods the effects of temperature and light on the variation of the phytoplankton. This has so far been attempted only by Riley (1939b, 1940) in a study of photosynthetic rates, or rates of production of organic carbon. Since the rate of photosynthesis is related in a rather complex way to the total standing crop of phytoplankton,

Riley's results will best be considered when the problem of productivity is discussed in Volume III. Apart from statistical analysis, the difference in the illumination at equal autumnal and vernal temperatures should make some separation of the effects of light and temperature possible by inspection, for if a species underwent a great increase at a given temperature in both spring and autumn the fact that the temperatures were identical would be almost equivalent to indicating that the light available differed at the two periods of increase. Wesenberg-Lund (1904) stressed the point of view that temperature is the most important single factor regulating the fresh-water plankton. His most interesting case relates to *Fragilaria crotonensis*. This diatom was found by him to be an abundant component of the late spring and of the autumn plankton, reaching its greatest development between 13 and 16° C. In most years, along with the other diatoms of the spring plankton, it is largely replaced in summer by blue-green algae. However, in 1902, a year with an exceptionally cold summer, almost none of the lakes reached temperatures in excess of 16° C., instead of 21 to 23° C. as is usual in normal years in Denmark. In 1902 *Fragilaria crotonensis* continued as an abundant alga in these lakes until the investigation was terminated at the end of July, strongly suggesting that the summer temperature of the water is responsible for the persistence or disappearance of the diatom in different years.

More recently Ruttner (1937b) and Findenegg (1943b) have published lists of extreme temperature ranges of occurrence, temperature ranges of marked development, and mean temperatures for the maxima of a number of species found in the lakes of Austria. Findenegg investigated a group of Carinthian lakes and Ruttner investigated principally the lakes of the Salzkammergut. The extreme temperature range for the occurrence of a species is not significant, for at very high or very low temperatures the organism may merely be persisting in a nonreproductive state in a highly unsuitable environment. Ruttner's temperature data are largely based on vertical series, and his range of temperatures for well-marked development on any occasion is the range within which the population density is at least 10 per cent of the maximum density. Findenegg evidently uses a more subjective criterion, but his results are certainly comparable with those of Ruttner. Stankovič (1960) has published comparable data for Lake Ohrid. The mean temperatures of maxima and the ranges for well-marked development for species common to the lists of at least two of these authors are given in Table 24. Wesenberg-Lund's data relate to an entirely different kind of lake district and so mainly refer to species not encountered by Ruttner and Findenegg; these data

TABLE 24.   *Temperature ranges of common species of phytoplanktonic algae*

	Mean Temperature of Maximum			Approximate Optimal Temperature Range		
	Findenegg	Ruttner	Stankovic	Findenegg	Ruttner	Wesenberg-Lund
Oscillatoria rubescens	5.8° C	8.1° C	—	5–8° C	5.8–10.4° C	4–10° C
Stephanodiscus astraea	7.1	5.2	6.5	6–12	—	4–6 (rarely to 15)
Synedra acus delicatissima	7.2	5.2	—	5–10	4.8–7.6	1–10
Asterionella formosa var. *hypolimnetica*	—	5.9	8.0	—	5.1–8.0	—
A. formosa (Ossiacher See)	7.8	—	—	5–8	—	—
Dinobryon divergens	7.8	12.1	10.5	5–8	8.5–13.2	—
A. formosa var. *epilimnetica*	—	12.2	—	—	6.7–14.2	—
A. formosa (Millstätter See)	9.2	—	—	7–10	—	—
Gymnodinium helveticum	9.3	9.3	—	6–10	5.2–12.2	—
Gomphosphaeria lacustris	10.0	13.8	14.0	12–18	7.7–14.1	—
Fragilaria crotonensis	10.2	13.8	—	8–14	7.1–14.1	13–16
Peridinium willei	10.5	10.8	—	10–15	8.2–12.5	—
Rhodomonas lacustris	10.6	10.7	—	8–12	5.1–13.2	—
Cyclotella bodanica	11.8	9.5	—	10–15	6.3–12.3	—
Synedra acus angustissima	11.9	(13.7)	—	10–14	—	—
Cyclotella comensis	12.5	9.5	—	12–15	6.4–12.8	—
Dinobryon sociale	13.2	12.9	12.5	10–15	9.8–14.4	—
Cyclotella comta	13.5	11.7	—	8–15	(4.6)–14.7	—
Chroococcus minutus	13.6	13.6	—	12–18	9.3–14.1	—
Ceratium hirundinella	14.4	12.5	16.5–23.0	12–19	8.9–13.7	Warmest period of year
C. cornutum	14.5	13.8	—	12–19	8 upward	—
Staurastrum cingulum	14.6	11.0	—	10–15	6.5–13.2	—
Gloeococcus schroeteri	15.5	13.2	—	circa 15	9.5–13.8	—
Anabaena flos-aquae	17.2	14.5	—	16–21	11.6–14.5	16–18

are not summarized in any very systematic way, but a comparison is possible for a few species.

Apart from *Cyclotella comensis,* and *Dinobryon divergens* the incidence of which is probably not determined thermally within the range of 5 to 21° under consideration, there are no striking inconsistencies in the table. The ranges of *Anabaena flos-aquae* given by Findenegg and Ruttner do not overlap, but this is probably only because none of Ruttner's lakes really reached a temperature high enough for the maximal development of this blue-green alga. Most of the other common plankton species of Myxophyceae do not occur in either of the Austrian groups of lakes. Wesenberg-Lund indicates that for optimal development in Denmark *Anacystis cyanea, Anabaena spiroides, A. macrospora,* and *Gloeotrichia echinulata* all require temperatures between 18 and 22° C. to form water blooms, whereas *Gomphosphaeria lacustris* appears only in the late summer and early autumn when the temperature has fallen to 17 or 15° C.; the last species does not occur in the spring and its incidence must be regulated by some factor in addition to temperature. Wesenberg-Lund's observed values for the temperature at the time of increase of the summer bloom-forming algae are probably in agreement with or possibly somewhat below those of many other limnologists.

It would be reasonable to conclude that the first few species in Table 24, namely, *Oscillatoria rubescens, Stephanodiscus astraea, Synedra acus delicatissima,* and the cold stenotherm race of *Asterionella formosa,* are cold-water organisms and that their incidence, though not wholly explained by low or at least moderate temperatures, is in part determined by them. Even in the case of *Oscillatoria rubescens* there are complications (pages 441–442), whereas *Stephanodiscus astrea,* which is a late spring form in Mendota and some English localities, can exhibit maxima in Denmark from May to July (Jørgensen 1957). To the cold-water list can be added *Melosira islandica helvetica* as in Lake Ladoga (Petrova 1959) and possibly some additional species of blue-green algae such as *Oscillatoria prolifica* (Riley 1940), *Anacystes thermatis,* and the so-called winter *Aphanizomenon* spp. (Wesenberg-Lund 1904) of somewhat uncertain taxonomic status.

There is little doubt that high temperatures are involved in the incidence of summer Myxophyceae and some evidence that *Anabaena flos-aquae* develops massive populations at slightly lower temperatures than several other species of *Anabaena* and *Anacystis.*

In the intermediate range of temperatures a number of species, including several very important diatoms, would appear to be mesothermal, but much caution is needed in interpreting the data. It is important to note that only a minority of these mesothermal species are characteristic of both the spring and autumn periods of intermediate temperature. Among the nineteen species that Findenegg considers as having the upper limit of the optimal temperature range between 12 and 15° C. only four (*Tabellaria flocculosa, Fragilaria crotonensis, Cyclotella glomerata* and *Cryptomonas erosa*), certainly, and two others (*Melosira granulata* and *Uroglena volvox*), possibly, have both vernal and autumnal maxima. Among the remaining eleven species nine are late summer and autumn forms, two are spring and early summer species, and one appears to occur at appropriate depths throughout the whole period from June to September.

It may be noted that *Asterionella, Fragilaria,* and *Tabellaria* all tended to have spring and autumn maxima in Lake Mendota at the time of Birge and Juday's (1922) main study. Chandler found that *Fragilaria* behaved in the same way in Lake Erie, though here the pennate *Asterionella* and *Tabellaria* showed their greatest development only in spring, whereas the centric genera *Cyclotella, Stephanodiscus,* and *Melosira* tended to be autumnal. Findenegg, whose views are further elaborated later, has never considered temperature to be the only factor involved. Nevertheless, for the species that do show both vernal and autumnal maxima it would be reasonable to conclude that temperature is a most important factor in regulating the seasonal incidence, as Wesenberg-Lund, in particular, has urged. Further study, however, will show that other important factors must often be operating in such cases and that part, at least, of the argument that might be derived from existence of species with well-defined vernal and autumnal maxima cannot be substantiated.

Thermal races or ecotypes. Ruttner (1937a) has concluded from his prolonged experience of the lakes of the Austrian Alps that two races of *Asterionella formosa* exist in that region. One race has an optimum temperature for development between 5.1 and 8.0° C., probably in the vicinity of 5.9° C.; the other race is much less stenothermal and can form considerable populations between 6.7 and 14.2° C., with an optimum at about 12.2° C. The cold stenothermal race, termed by Ruttner var. *hypolimnetica,* is always small; the extreme range of variation of the length of the cells is 36 to 54 μ. This race exhibits no secular cycle of size reduction. The eurythermal race, termed var. *epilimnetica* by Ruttner but presumably the nominate race, is usually larger than the cold stenotherm race but apparently exhibits cycles of decrescence and supposed auxospore formation (see page 932), so that the length of the cells can vary from 39 to 90 μ. Ruttner found var. *hypolimnetica* in the Lunzer Untersee, Grundlsee, Toplitzsee, Vordere Lahngangsee, Leopoldsteiner See, Altausseer See, and probably in the Hallstätter See; in all of these lakes the diatom is restricted to the deeper water during the period of stratification. The eurythermal var. *epilimnetica* occurred in the upper layers of the Wolfgangsee and Krottensee; it is possible that the Traunsee contains both forms. Findenegg's populations from the Ossiacher See and Millstätter See may be referrable to var. *hypolimnetica* and var. *epilimnetica,* respectively, though the distinction between these two populations seems less great than between those studied by Ruttner. Stankovič (1960) believes that the form in Lake Ohrid is referrable to *hypolimnetica.*

Allen (1920) found two forms, referred by him to *Asterionella gracillima,* in the San Joachim River and associated canals near Stockton, California. These two forms were usually sharply separable but differed only in size, the cells of the larger being about half as long again as those of the smaller form. With the exception of the late summer period, July to October, both forms could occur throughout the year, but during the cooler months the larger form was sometimes absent and was never so common as the smaller form. The major maximum of the small form occurred in January and February between 4 and 10° C., but after a minimum in the spring both forms increased to considerable concurrent maxima in June at 20 to 26° C. Both forms, after being absent in the late summer, may produce well-marked pulses in the autumn, in November and December, at 9 to 12° C. The two forms evidently are adapted to temperature ranges somewhat different from those of Ruttner's two forms of *A. formosa;* indeed, it seems probable that the smaller Californian form is eurythermal and the larger form, warm stenothermal. Allen's two forms, however, clearly indicate the same sort of general phenomenon as Ruttner's. Gardiner (1941), moreover, suspected the presence of classes of two sizes comparable to Ruttner's thermal ecotypes in the Windermere population of *Asterionella,* though the material from this lake studied by Lund (1949a) obviously belongs exclusively to the eurythermal form.

Apart from these cases, Ruttner has also concluded that *Synedra acus delicatissima* and *S. acus angustissima* are differentiated in a comparable way in their thermal requirements, the first named being the cold stenothermal form. Findenegg's results are in accord with Ruttner's conclusion. The morphological differences are somewhat greater than those of the two forms of *Asterionella.*

Criticism of the theory of thermal control. Although many eminent investigators, notably Wesenberg-Lund and Kofoid in the early part of the twentieth century and Ruttner and Findenegg more recently, recognize the importance of temperature changes controlling the cycle, a contrary view has been expressed by certain other students of equal experience and ability, among whom Pearsall, the main proponent of the theory of chemical control, is the most important. Pearsall (1923) has emphasized the extreme variability of the temperature relations of particular species studied in a sufficiently wide range of localities. Thus *Melosira granulata* has been recorded by W. and G. S. West (1912b) as producing maxima at temperatures between 1.7 and 4.9° C. in Windermere and by West (1909) between 18.5 and 23.3° C. in the Yan Yean Reservoir in Australia. Other workers have assigned slightly higher temperature ranges to the maxima of the same species. This case involves certain special features to be discussed later (page 461). Pearsall believed that the same variability is exhibited by *Asterionella* and *Tabellaria,* and many less extreme cases are known from the literature.

It has already been pointed out that *Fragilaria crotonensis* was regarded by Wesenberg-Lund as a definitely mesothermal spring and autumn species not flourishing above 16° C. and present throughout the aestival months only during abnormally cool summers. Kofoid (1908) also regarded the species in the Illinois River as having an optimum at about 15.5° C. and as tending to disappear after the middle of May. In both Mendota and Lake Erie *Fragilaria* seems to be mesothermal. Jaag (1938) found the species in the Rhine at Schaffhausen dominant from June to September at temperatures of 14 to 20° C. Similar temperatures are indicated by Berardi and Tonolli (1953) for Lago Maggiore and for the Baldeggersee by Bachofen (1960), the maxima occurring in late summer and in June, respectively. Bachofen's data for the maximum in the Hallwilersee in July indicate that it occurred at slightly higher temperatures, around 22° C. Apstein (1896) found the species developing summer maxima in June and July at about 20° C. in some of the German lakes, whereas Steinecke (1923) considered *F. crotonensis* to be associated characteristically with blue-green algae in August and so possibly at somewhat higher temperatures. In Linsley Pond in 1937 a great increase in the species occurred during July and early August at temperatures between 25 and 29° C.

Though *F. crotonensis,* a widespread, abundant, and easily recognized plankter, probably provides the best example of apparent control by temperature in a manner that differs greatly from lake to lake, it is probable that other almost equally dramatic, cases exist. Vollenweider (1950) found that *Oscillatoria rubescens,* which, we have seen, ordinarily behaves as a cold stenotherm, can in fact be cultured at 20° C., though probably not at higher temperatures. In some cases, as in the Baldeggersee, this temperature appears actually to be the upper limit for populations in nature (Bachofen 1960), though more usually the limit seems to be much lower, well below 10° C.

Variations of the kind discussed can be explained in at least four different ways.

Firstly, the populations occasionally found at what may seem to be abnormally high temperatures may be due to the occurrence of transitory maxima above the limiting optimum temperature. Rodhe's (1948) experimental results suggest that this might sometimes be true in *Fragilaria crotonensis* at temperatures well above 20° C., though the population in Linsley Pond persisted for several weeks.

Secondly, variation in apparent tolerance may be due to genetically determined ecotypes of the kind that Ruttner believes to occur in *Asterionella formosa.*

Thirdly, the effects of temperature may be indirect, acting in different ways on different biological communities in regulating competition. One or two cases suggestive of such processes are presented in a later section.

Fourthly, the apparent dependence on temperature may be illusionary, the real independent variables being light or chemical parameters correlated with temperature. This appears to be the position, based on

vast experience, of Lund (1964). Such an explanation has been suggested by Vollenweider for *Fragilaria crotonensis* and *Oscillatoria
rubescens*. He supposes that the former species can be inhibited
by excess light at the height of summer, though Bachofen believes
that his own observations are not in accord with this suggestion. In
O. rubescens Vollenweider points out that the lakes in which large epilimnetic populations are present at moderate temperatures in summer
are small and productive. He suggests that if the nitrate requirement
of the species increases with increasing light intensity, as White (1937)
found for *Lemna,* the alga would flourish only in nitrate-rich localities
during the season of maximal illumination. The hypothesis is interesting but requires further study. In other species, which appear in
some lakes to be controlled by temperature, evidence exists that the
incidence is actually determined chemically. This is apparent particularly for *Dinobryon,* though here it is quite probable from Kozarov's
elegant observations in Lake Ohrid that different species of *Dinobryon*
may occur at different temperatures or light intensities under appropriate chemical conditions.

ILLUMINATION AS A FACTOR IN SEASONAL SUCCESSION

Although Calkins (1893) realized that the inception of the spring
diatom pulse could not be due to increasing temperature, for it often
occurred before the temperature had begun to increase, and although
Whipple (1894, 1895) came to a like conclusion, the earlier authors
were in general unable or unwilling to separate the effects of illumination from that of other seasonally variable factors. In his paper of
1896 Whipple emphasized the importance of adequate illumination
for diatom growth but regarded the chemical and mechanical effects
of full vernal circulation as the most important determinants of the
spring diatom maximum. A little later, and in full knowledge of the
American work, Zacharias (1899b) concluded that the main factor in
producing the great increase in the population of *Asterionella,
Fragilaria,* and *Melosira* in the Grosser Plöner See in April at a temperature of about 4 or 5° C. can only be the increase in available light.
It is unlikely that anyone would question this conclusion today, for
much more comparable evidence is now available. Chandler (1944),
for instance, observed that in 1941 the great vernal increase in phytoplankton in western Lake Erie occurred under an ice cover during
February and March. The water temperature remained at this time
nearly constant, but the plankton, mainly *Asterionella,* increased forty-
fold. The illumination, exceptionally great in 1941, delivered to the
lake was 3364 cal. per cm.[2] in January, 5849 cal. per cm.[2] in February,

and 10,201 cal. per cm.[2] in March, a threefold increase during the period when the lake was icebound.   At Wray Castle (Lund 1949a) the illumination may increase sixfold during the period from the minimum in December to the time of the vernal diatom maximum.   Lund indeed found that a good population of *Asterionella* can be developed in winter water from Windermere merely by exposing the water to a bright light.   He has also shown experimentally, as is indicated later, that although the potential division rate of the diatom in the lake can be influenced by temperature and illumination the illumination is likely to be of far greater importance than the temperature during the early spring.   It is reasonable to expect, as is indeed confirmed in practice (page 447), that shallow lakes will exhibit an increase due to increasing illumination earlier than deep lakes, for in the latter, at vernal circulation, the individual cells of the phytoplankton will spend a disproportionate time in deep unilluminated water.

In spite of the rather obvious significance of increasing illumination in the spring, it is important to realize that such an increase can be effective only when other factors, notably the chemical composition of the medium, are also favorable.   The result is doubtless often so striking because the period of low illumination and low temperature during the winter has permitted the accumulation of a store of nutrients in the water.   In general, we find high nitrate and high phosphate concentrations under the ice whenever the lake basin is fertile enough to permit their accumulation.   As the illumination increases in February and March, this accumulated store of nutrients supports the developing spring phytoplankton pulse.   It must be borne in mind, however, that large populations of certain photosynthetic organisms can appear and make some use of this store of accumulating nitrate and phosphate at the low light intensities occurring beneath the ice at the end of December or beginning of January.   The problem of these peculiar winter pulses which conceivably may be determined by the presence of certain accessory organic substances or vitamins in the water, is considered in a later section (see page 473).

A particularly interesting example of the effect of illumination is provided by Chandler's (1940, 1942a,b, 1944) admirable work on Lake Erie.   In the western part of the lake the turbidity of the water is variable.   Since there is little or no thermal stratification, strong winds at any time of year except when there is an ice cover can increase the turbidity by stirring the lake to the bottom; river discharge at the time of the melting of snow in the basin and after heavy rain also adds to the turbidity, which is thus greater after wet stormy weather than in dry calm seasons. Under the ice the suspended material producing the turbidity gradually settles out, so that the maximum transparency of the water is generally observed in March, though the nature of the ice cover will influence the amount of light that enters the water when the lake is frozen. The well-marked vernal diatom pulse normally begins to develop under the ice, but the degree of its development depends not only on how long reproduction occurs in the more pellucid waters of the frozen lake but also on what happens when the ice melts. In 1939 there was an enormous increase in turbidity and decrease in transparency just as the ice broke. The vernal diatom population, which had begun to develop under the ice, declined catastrophically and the total phytoplankton remained fairly low throughout the summer until September. In 1940 there was a far less well-marked period of turbidity when the ice broke, and except for a small check just at the end of March, corresponding to this slight increase in turbidity, the spring pulse developed in the open lake to a well-marked double maximum in late April and the first half of May. In 1941, a dry year with a uniformly lower turbidity during the spring and summer than had ever been observed, an enormous spring pulse consisting of *Asterionella* started developing very early under the ice, which was thinner than in preceding years. When the ice broke, there was no stormy period of enhanced turbidity and the phytoplankton continued to develop to an unprecedented maximum in April. The rapid decline in the population in May was uncorrelated with any change in turbidity and presumably was due in this year not to a temporary decline in illumination as in 1939 but to an exhaustion of some nutrient. Chandler indicates, however, that he could find no evidence of limitation of the phytoplankton by depletion of nitrogen and phosphorus. It is probable that in Lake Erie the decline in the autumn bloom is often due to increasing turbidity as the daily illumination declines, but it is quite clear that the marked summer minimum, largely responsible for the definite double-peaked seasonal development of phytoplankton in this lake, must be determined by factors other than illumination. Grazing by zooplankton or some chemical limitation is an obvious possibility. There is some evidence that variation in the qualitative composition of the autumnal pulse is determined by turbidity and so by illumination. Blue-green (largely *Aphanizomenon* and *Oscillatoria*) and green (largely *Dictyosphaerium*) algae appear to be favored by low turbidity and therefore by high illumination, whereas species of *Cyclotella* and *Stephanodiscus* seem to be among the forms most tolerant of high turbidities. The autumnal pulse over a temperature range of 23 to 10° C. would appear to be more aestival in less turbid seasons and more truly autumnal in more turbid seasons. No detailed treatment of the phenomenon is possible in the absence of specific determinations, for several of the more important organisms involved belong to genera containing species of the most diverse ecological requirements. Chandler was satisfied, from microscopic study, that the phenomena he observed was due solely to the effect of suspended matter in reducing underwater illumination and not to any deleterious mechanical effects of the suspended inorganic material on the plankton cells.

Experimental studies of light and temperature acting together. Lund (1949a) suspended subcultures of *Asterionella formosa* in enriched lake water at depths of 0.5, 5, and 7 m. from a buoy moored in Windermere and in this way has studied throughout a year the possible division rate of the diatom at the temperature and illuminations characterizing these three depths. Lund expressed his results graphically, but it is possible to read off the observed values from his histograms with sufficient accuracy to compute correlation coefficients. This has been done for the whole year and for the period between February 1 to April 30 when the diatom is usually entering its vernal period of intense population growth. The following simple correlation coefficients were obtained:

Division rate at 0.5 m.—surface temperature, whole year	.919
Division rate at 0.5 m.—surface illumination, whole year	.729
Division rate at 5.0 m.—surface illumination, whole year	.884
Division rate at 7.0 m.—surface illumination, whole year	.917
Mean division rate at 0.5 to 7.0 m.—surface temperature, whole year	.924
Division rate at 0.5 m.—surface temperature, February to April	.782
Division rate at 0.5 m.—surface illumination, February to April	.940

In making this investigation it is reasonable to assume that the illumination at any depth is closely proportional to the illumination at the surface but that because of variations in the development of the epilimnion the temperature in the deeper cultures will not be so closely related to the surface temperature. For this reason the variation in the division rate in the cultures set at 5 and 7 m. can be considered only in relation to illumination, but at the surface, where the temperature data can be properly used, partial correlations of division rate on temperature-excluding illumination, and on illumination-excluding temperature can be computed.

Division rate at 0.5 m.—temperature, whole year	.891
Division rate at 0.5 m.—illumination, whole year	.515
Division rate at 0.5 m.—temperature, February to April	.222
Division rate at 0.5 m.—illumination, February to April	.941

It is evident from these figures that for a single species of eurythermal alga at the surface of the lake temperature is more important than light in determining the potential rate throughout a whole year. However, during the early spring, the period in which *Asterionella* normally undergoes its great increase, the effect of light is more important than that of temperature. Moreover, when the mean division rate at increasing depths is considered, the effect of illumination at the surface becomes progressively more important, and the simple correlation coefficient on light is in consequence much greater for the mean division rate than for the division rate at the surface. Because of the effect of turbulent mixing in the epilimnion, it is probable that the mean division rate in the illuminated zone is more interesting a figure than the division rate at the surface. These results may be compared with those of Riley (1939b), who found that in Linsley Pond the photosynthetic production of oxygen in water samples containing the natural association of plankton was significantly correlated at the surface with the magnitude of the phytoplankton population, as measured by the chlorophyll content and with the temperature but not with the illumination; for the lake as a whole there was a significant correlation with temperature and illumination but not with the phytoplankton.

The pattern of seasonal succession produced by interaction of light and temperature. Although the interaction of light and temperature as determining factors is implicit in the discussion of the individual species in Ruttner's (1930) account of the Lunzer Untersee, the formal interpretation of the annual cycle of the phytoplankton in terms of this interaction has been developed primarily by Findenegg (1943b).

The asymmetry of the seasons may be easily realized if we compare the thermal conditions at the time of the vernal and autumnal equinoxes when, with a clear sky, the radiation is equal. In view of the differences between the times of the maxima and minima in the annual cycles of radiation and of temperature, four extreme periods can in fact be recognized; namely

> Winter, at its height in late January and early February
> > low radiation, low temperature
> Spring, at its height in late April and early May
> > high radiation, low temperature
> Summer, at its height in late July and early August
> > high radiation, high temperature
> Autumn, at its height in late October and early November
> > low radiation, high temperature

Findenegg points out that if, in the autecology of any group of planktonic organisms, the light and temperature requirements vary independently it is possible to obtain four kinds of organism, each characteristic of a season as just defined.

	Low Light Species	High Light Species
Low temperature species	Winter	Spring
High temperature species	Autumn	Summer

In a stratified lake, the Wörthersee (Fig. 109) being paradigmatic, winter conditions of low light and low temperature will be preserved throughout the summer in the upper part of the hypolimnion, and, provided that this region does not develop special chemical characters such as low oxygen content (a proviso that holds for large deep unproductive lakes), winter species should be able to persist perennially in this region. Any species that flourishes at the surface in winter and persists in an actively reproducing form in the upper hypolimnion throughout the summer is likely to be a winter form in Findenegg's sense, really requiring low light and low temperature, for it is reasonably certain that the species will be exposed to variation in chemical factors throughout the year without its having much effect on the seasonal cycle.

It is evident from Findenegg's data on *Oscillatoria rubescens* in the Keutschacher See, the Längsee, and the Wörthersee (the last named under study in four different years) that this species does behave as a winter or cold-water shade plant. It usually starts to decrease at the surface in April; it may undergo a great increase in the hypolimnion at a temperature of 6 to 10° C. during summer stratification and becomes uniformly distributed throughout the lake in winter. The same kind of behavior is recorded by Thomas (1949). It is highly probable that the even more widespread *Oscillatoria prolifica* will have similar photic and thermal requirements. Neither Findenegg's nor Thomas' account of the development and distribution of populations of *O. rubescens* suggests that it is sufficient simply to characterize the species as oligophotic and oligothermal. The metalimnion or upper hypolimnion in summer, and not the whole lake in winter, often appears to provide the optimal environment for the species. In a few localities such as the Rotsee (Bachmann 1910; Adam and Birrer 1943) it can be quite epilimnetic during the summer or form massive populations in the upper metalimnion (Bachofer 1960) at temperatures of 15 to 20° C. This also seems to be true in Lake Washington (Edmondson, personal communications). Either the winter temperatures around 4° C. are rather too low for rapid increase, even of this winter species, or some other factor such as illumination or the liberation of nutrients (from dead, sedimenting, epilimnetic plankton or from the sediments in contact with the layers of the hypolimnion in which the alga is growing) must be involved in determining the metalimnetic and hypolimnetic development of *O. rubescens*. Vollenweider's (1950) belief that the species requires more nitrate at moderate than at low temperatures has already been mentioned.

In contrast to such supposedly oligophotic cold-water forms, the spring phytoplankton is believed by Findenegg to consist of oligothermic, polyphotic species or cold-water sun plants, which should be unable to persist in the cold waters of the hypolimnion throughout the summer because of the low light intensity of the region. Findenegg regards *Dinobryon divergens* and *Uroglena volvox,* which show just such seasonal variation in vertical distribution in the Wörthersee, as good examples of spring forms. It will appear, however, at least in *D. divergens,* that powerful chemical influences are also operating to determine its seasonal occurrence. Usually *Asterionella* behaves superficially like a cold stenotherm polyphotic species, but, as will be apparent later, its decline in the late spring is usually due to silica deficiency; what prevents its later development in warm water in which SiO_2 has regenerated is quite obscure.

The summer warm-water polyphotic species occupy the fully developed epilimnion with maxima in June, July, and August, though in some cases limited populations are found throughout the year with little sign of reproduction. Findenegg regards *Cyclotella comensis* var. *melosiroides,* apparently the same organism as the *C. melosiroides* of Vetter (1937) and of Grim (1939), as a perennial summer form, but it is evident that if the interaction of light and temperature really determines the maximum of this species the race studied by Vetter is quite different from that observed by Findenegg and by Grim. A better example of a summer species is provided by *Ceratium hirundinella,* which throughout almost the whole of Europe is a conspicuously seasonal species overwintering as cysts and which always appears to have a pronounced summer maximum. Even in this species Vetter's observations (page 424) in the Schleinsee are a warning against easy acceptance of a false simplicity.

Findenegg regards *Dactylococcopsis smithii* as a late summer form, transitional to the autumnal group of warm-water shade species, of which *Lyngbya limnetica* is characteristic. From Ruttner's (1930) observations it is reasonable to suppose that *Staurastrum cingulum* in the Lunzer Untersee has temperature and light requirements not unlike those of *D. smithii* in the lakes studied by Findenegg. The perennial *Gomphosphaeria lacustris,* which produces great populations in the

Wörthersee and in some other Austrian lakes in September and in October and November in the Millstätter See and Lago Maggiore, is also regarded as a polythermic oligophotic species; if Findenegg's category is valid, *G. lacustris* is probably its best known example. Findenegg, however, thinks that the late autumnal maximum in the Millstätter See implies that the population of this lake belongs to an ecotype different from that present in the other Austrian lakes.

THE THEORY OF CHEMICAL DETERMINATION

Though the belief that seasonal variations in the chemical composition of lake water are in part responsible for the periodicity of the phytoplankton is at least as old as Whipple's studies (1894, 1895; Whipple and Jackson 1899) and was advanced in the special case of the blue-green algae by Apstein (1896), the chief protagonist of the idea has been Pearsall (1921, 1923, 1930, 1932). Pearsall certainly developed his concepts partly as the result of Whipple's work and partly, no doubt, in recognition of the importance of chemical factors in explaining the geographical distribution of the phytoplankton (West and West 1909, 1912b; Pearsall 1921a).

Pearsall (1923) considered chiefly the seasonal cycle of diatoms, attributing the development of maxima to the temporary presence of a high concentration of silicate, which, he supposed, was to be attributed to flood water entering the lake. Thus the melting of snow and the consequent influx of water which had had an opportunity to pick up large amounts of suspended and dissolved matter in the drainage basin of the lake was considered to be the chief cause of vernal diatom maxima. If, in his first paper on phytoplankton periodicity, Pearsall did not sufficiently consider either the occurrence of the phenomenon in organisms other than diatoms or the importance of internal chemical changes in the lake, he amply rectified these deficiencies in his later contributions (1930, 1932) in which he produced a complete and detailed theory of the periodicity of all the commoner autotrophic organisms in net plankton of the English lakes. His conclusions were as follows:

(1) The high nutrient level in the spring is the primary cause of the vernal diatom maximum. Individual species have different requirements and so succeed each other as the waxing population reduces the available nutrient supply. In particular, *Asterionella formosa*, which appears before *Tabellaria fenestrata*, was believed to require a higher nutrient level than the latter species.

(2) *Dinobryon divergens*, which often replaces the diatoms at the end of the spring bloom, appears when calcium and silicate fall and

when the ratio $N.NO_3:P.PO_4$ rises, as is usual in the English lakes in spring. Pearsall also concluded that the species could not develop in hard-water lakes unless the silicate content fell below 0.5 mg. per liter.

(3) Green algae occur mainly in the summer when the nutrient content is low. Low calcium concentration and a low $N.NO_3:P.PO_4$ ratio tend to favor desmids.

(4) Myxophyceae occur in quantity only in those lakes in which there is an appreciable amount of nitrogenous organic matter in solution. Pearsall measured this organic matter as albuminoid ammonia, that is to say, the ammonia liberated by permanganate oxidation. He concluded that at least 0.035 mg. per liter were needed to ensure a large development of blue-green algae. The blue-green algae, however, are clearly able to develop immense populations when the inorganic nutrients are at a very low level indeed. *Melosira granulata* is characteristic of lakes that produce important blue-green maxima, but the main development of this diatom need not coincide with that of the Myxophyceae. Pearsall believed that the seasonal incidence of large quantities of blue-green algae was due to the previous decay of either diatom plankton or littoral plants which liberated large amounts of nitrogenous organic matter into the water. A comparable explanation of late summer blooms of *Anacystis* was advanced as early as 1896 by Apstein.

Although none of Pearsall's four major propositions can be accepted unreservedly in the form in which he stated them, they are based on a wealth of observation and can constitute an excellent basis for further study. Their detailed evaluation can be undertaken best by considering separately the data available for the spring diatoms, for *Dinobryon,* and for the blue-green algae. Little critical work has yet appeared on the desmids, and Pearsall's data, apart from showing that they are summer forms, are largely used by him to explain differences from lake to lake (see page 331) rather than in the temporal sequence.

Diatom periodicity and the vernal maximum. There is little doubt that the true spring maximum, in March or April, whether composed of diatoms or of other forms, is, as we have seen, to be attributed largely to an increase in illumination at a time when the concentration of nutrients is high. The concentrations normally observed, namely 3 to 10 mg. m.$^{-3}$ phosphate phosphorus and up to 100 mg. m.$^{-3}$ nitrate nitrogen, are, as the experiments of Chu (1943) and Rodhe (1948) have shown, suboptimal for plankton diatoms cultured in purely synthetic media but are more than adequate, at least for *Asterionella,* growing in lake water. In the northern part of the temperate zone the illumination at midwinter is clearly not adequate for the maximal rate of growth of most planktonic algae, particularly if ice with a covering of snow is present on the lake. We should expect any species capable of dividing at 3 to 7° and of using nutrients in the concentrations just mentioned as generally occurring, to increase rapidly in the spring as soon as the ice cover disappeared, if not before. This does indeed often happen, as we have seen in the preceding section. Just such an increase at high nutrient levels underlay Whipple's theory of the spring diatom maximum, though he was concerned only with nitrate rather than with phosphate and silicate and believed that vernal circulation was necessary to bring spores or other resting stages from the mud into the free water before they could reproduce. In the simple slow rise to a midsummer maximum observed by Findenegg in his study of the phytoplankton of the Carinthian lakes much of the initial increase occurs at quite low temperatures. It is indeed probable that such a regular rise in a large, deep, and usually cool lake to a maximum in late June or July is comparable to the sudden and sharply defined spring maxima in smaller and more productive lakes rather than to the irregular summer pulses observed, for instance, in Mendota. The interaction of light and nutrient concentration admirably explains all such simple vernal increases, whether sudden or slow. It does not, however, explain why in some lakes, such as the Schleinsee, the vernal maximum may be absent, nor why in others, such as Linsley Pond, it may be so much less well developed than other pulses less easily explained on chemical grounds.

It is also reasonably certain that in some cases depletion of nutrients must slow and finally stop the growth of the population of a dominant species and at the same time of any rarer species requiring a concentration as high as or higher than the dominant. Moreover, we might suppose that species with a somewhat lower nutrient requirement might still be able to increase and by lowering the nutrient level still more would further inhibit the original dominant and might finally come to replace it. But in so doing they would have reduced the nutrient level nearer and nearer to their own limit of effective nutrition, though again a still more parsimonious form might be able to increase and so become a third in a series of dominants. This is evidently the sort of process that Pearsall believes takes place when *Tabellaria* replaces *Asterionella* and also, though there are other factors involved according to Pearsall, when *Dinobryon* succeeds the diatoms. It is necessary, however, when constructing a theory of plankton periodicity along these lines, to explain why the later, more parsimonious species is unable to develop in the presence of its spendthrift predecessor and must wait until the latter has exhausted its heritage of nitrogen and phosphorus to a point at which it can no longer thrive. This problem was not considered at all by Pearsall. It can be solved in the special case of *Dinobryon,* which, as Rodhe has found, is actually inhibited by levels of phosphate that are hardly adequate for the diatoms that precede it. This solution, however, is rather improbable for the majority of species of phytoplankton. It is possible that the specific production by the dominant species of antibiotics active against competitors may ultimately prove to play an important part. As indicated later, Jørgensen (1957) has suggested that the rather sharp alternation in occurrence of *Stephanodiscus hantzschii* in the plankton and of various epiphytic diatoms in the littoral of Furesø and Lyngby Sø can be explained by inhibition of the epiphytes by *Stephanodiscus.* It is quite likely, moreover, that no valid general explanation will ever be forthcoming, for the autecology of every species is so different from that of every other that no useful unitary thesis could ever cover every case. That idiosyncrasies of a remarkable sort may occur is abundantly demonstrated by present knowledge of the occurrence of *Asterionella.*

The problem of Asterionella. The periodicity of *Asterionella* has probably been studied more intensively than that of any other phytoplanktonic alga. Some confusion exists in the literature between the two supposed species *A. formosa* and *A. gracillima,* the distinctness of which has not been accepted by all workers. Ruttner has found, as has already been stated, that in the lakes of the Austrian Alps two physiological races or ecotypes of *A. formosa* exist, one a eurythermal and one a cold stenothermal form. Other observations have been made that appear to support this or some comparable distinction, but in most of the cases to be discussed it would seem that investigations have been made primarily on the eurythermal *A. formosa epilimnetica.*

The first work of importance was by Whipple (1892, 1895; Whipple and Jackson 1899), by whom the large spring maximum was attributed, without doubt correctly, to the vernal increase in light at a time when the water is well supplied with nutrients. Whipple probably overestimated the role of nitrate and of aeration and also attributed a special role to the mechanical effects of circulation in distributing supposed resting stages from the bottom throughout the lake. No evidence (Lund 1949a,b) has ever been obtained of the existence of such stages.

Among other early workers, Wesenberg-Lund (1908) made interesting contributions to the study of the seasonal cycle of *Asterionella,* but, beyond noting that he believed this diatom to be one of the few members of its group in which the seasonal incidence is not primarily regulated by temperature, his observations are best discussed after certain more modern work has been considered.

Among these more recent studies, those conducted in the English Lake District mainly by Lund (1949a,b, 1950a,b; Lund, Mackereth, and Mortimer 1963) are the most important. Lund's simplest and most prevalent type of cycle is conveniently described before any other more complex sort of behavior is stressed. It consists of a rapid increase in the population during February, March, and April in Esthwaite and a month or two later in the North and South Basins of Windermere to a pronounced spring maximum, after which there is a catastrophic decline. The earlier rise in the population in Esthwaite than in the larger, deeper Windermere is reasonably to be attributed to the greater mean illumination incident on a cell when circulating freely in the spring water of a small rather than a large lake. In all cases showing this simple pattern the maximum is reached and the decline initiated when the soluble silicate has been reduced by the diatoms to a concentration of 400 to 500 mg. SiO_2 m.$^{-3}$ (Fig. 113). In all it is reasonable to suppose that increasing illumination at a time when the water is rich in silicate and other nutrients initiated the increase in population in the lake as it did in the experimental bottles containing *Asterionella* cultures, whereas decreasing silicate terminated the growth period. During the later part of the process, when the temperature of the water is rising rapidly in late April and May, it is reasonable to suppose that temperature is also operating to increase the rate of production of new cells. There is no evidence of limitation by the exhaustion of any nutrient other than silicate. Though natural populations in the English Lake District appear to contain 0.28 to 0.81γ phosphorus per 10^6 cells, experiments have shown that division is possible in lake water containing less than 1 mg. P m.$^{-3}$, and that it continues until the cells contain only 0.06γ P per 10^6. Since the greatest population recorded is 1.2×10^7 cells per liter, the minimum amount of phosphorus that would have been required to produce such a population would be about 0.7γ. Actually, such a population is likely to have contained rather more phosphorus than this, but it evidently would not need to do so. It is therefore improbable that phosphorus would ever act as a limiting factor in the lakes that have been studied; Windermere in fact appears to contain about 1 mg. m.$^{-3}$ even at the time of spring *Asterionella* maximum. The nitrate nitrogen, the only important inorganic fraction of the combined nitrogen in the English Lakes, may fall at the time of the *Asterionella* maximum, but not infrequently it shows no sign of decrease. The diatom, moreover, can start increasing, as at the time of the autumnal period of

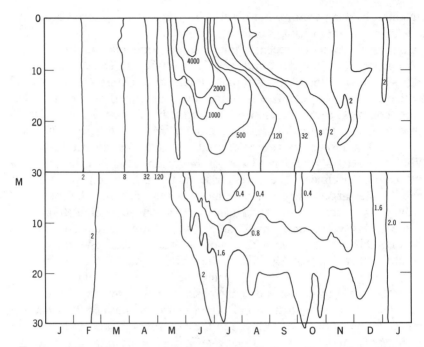

FIGURE 113. Upper panel, seasonal distribution (live cells, ml^{-1}) of *Asterionella formosa* in the top 30 m. of Windermere in 1942. Lower panel, soluble silicate (mg. SiO_2 · per liter) in the same. Note that the enormous bloom in late May and early June depletes the silica to 0.4 mg. S_1O_2 per liter and that relatively low concentrations are maintained till the winter. (Lund, Mackereth, and Mortimer.)

development in Esthwaite, at a concentration of 100 mg. $N \cdot NO_3$ m.$^{-3}$. This is a low value for the English Lakes, but it corresponds in concentration to the total nitrogen in 1.2×10^7 cells per liter, which number of cells would contain about 100γ of nitrogen. In nearly every case in the late spring the nitrate content is greater than 100 mg. m.$^{-3}$ when the decline in the population of *Asterionella* begins.

Lund found that *Asterionella* could develop between pH 6.5 and 9.0. The calcium requirements also appeared to be more than met by even the softest water of the English Lakes. No information regarding the iron and manganese requirements of the alga exists; it is probable that the waters inhabited by *Asterionella* contain suspended ferric hydroxide which may be an adequate source of Fe for the diatom. All the available evidence points to low illumination in winter and low silica at the time of the maximum as the only factors limiting *Asterionella* in Esthwaite and Windermere in the spring and early summer of most years. In order that this refreshingly simple situation may prevail, however, certain other conditions evidently have to be met, and their nature remains for the most part totally unknown.

A rather complicated sequence (Fig. 119) of events in Esthwaite in 1949 can be explained by competition for light with *Oscillatoria agardhii* var. *isothrix,* the occurrence of floods which wash the upper epilimnetic population away, and the incidence of parasitism (see page 474). The occurrence of a marked minimum in the population of the north basin of Windermere at the end of May 1948 is much less easily elucidated. This minimum occurred during fine dry weather, and growth was resumed when the lake level started to rise after rains. There was a less striking but perhaps comparable interruption of the development of *Asterionella* in 1949, and even more marked depressions of the population seem to have occurred in the less adequately studied seasons of 1938 and 1939. Lund thinks that the type of depression observed in 1948 may be connected with a reduced flow in the influents to the lake, but no suggestion of the nature of the material delivered to the lake and permitting *Asterionella* to develop has been made.

In a number of lakes in the English Lake District in which *Asterionella* occurs no large population ever develops. Lund notes that in 1949 this was true in Hawes Water, Derwent Water, Crummock Water, Grasmere, Rydal Water, Elter Water, and Lowes Water.

In Linsley Pond the same sort of situation was encountered in 1937–1938 when the phytoplankton was under continuous intensive investigation. The maxima in the *Asterionella* populations produced

TABLE 25. *Conditions for increase of Asterionella in Linsley Pond*

Spring Maxima	Temper- ature	$N \cdot NO_3$	$P \cdot PO_4$	SiO_2[a]	Maximal Observed Number of Cells ml.$^{-1}$
Increase from April 30 to May 17, 1937	11.1–17.5	n.d.–22	1–1	4200–2100	212
Increase from May 11 to May 18, 1938	16.5–15.5	23–20	0.5–1	6900–7000	113
Autumn Maximum					
Increase from December 20, 1937 to January 2, 1938	n.d.– 3.0	81–122	3–4	n.d.	147

[a] Rough order of magnitude figures, as it is now known that the method employed does not give unambiguous results. (Volume I, page 789.)

were of the order of one tenth of those normally produced in Windermere; the physicochemical conditions at the time of the increment and of the maximum indicate no limitation by exhaustion of silicate or any of the other nutrients that were determined.

Lund has not given any detailed account of the *Asterionella* maxima observed in the autumn. It would appear, however, that silicate begins to accumulate in excess of the minimal amount of 400 mg. m.$^{-3}$ shortly after the decline of the spring population, but because of the operation of some unknown factor no increase in *Asterionella* occurs in Esthwaite until the autumnal circulation period. Storey (1943) in studies of *Asterionella* in culture had concluded that, for a time after the maximum, division could not take place even with adequate nutrients in solution but had no explanation of the phenomenon. In Windermere the increase in silicate starts later than in Esthwaite, and the diatom begins to increase later than in Esthwaite but before full circulation occurs in Windermere. When Windermere is fully circulating, not enough light reaches the average cell to permit an autumnal maximum as large as that in Esthwaite (Lund 1949b).

A remarkable development of an autumnal maximum of *Asterionella* may be mentioned, namely, Tucker's (1957) observations on Douglas Lake. In this lake *Asterionella formosa* was, at least during the period between July 1950 and October 1951, primarily an autumnal organism, which reached its maximum population of about 687 cells ml.$^{-1}$ in the epilimnion in November 1950. The soluble phosphorus before the development of the maximum was about 5γ per liter, a value higher than that observed on other occasions and due undoubtedly to the mixing of particularly phosphorus-rich hypolimnetic water into the euphotic zone during autumnal circulation. A rise in silicate might be confidently expected at the same time. Later, smaller maxima of the diatom are less well correlated with nutrient content, but as far as the observations go they indicate the production of a maximum under chemical conditions not unlike those ordinarily found in spring, though light intensity was decreasing.

Extensive observations on variation in the size of populations of *A. formosa* have also been made by Gardiner (1941), who has described the changes in the diatom populations of a small English reservoir, Barn Elms No. 8, belonging to the Metropolitan Water Board in London. During the period of the investigation the reservoir was closed and received and lost water only by precipitation and evaporation. Observations were made on silicate and phosphate, but no nitrogen or temperature data have been published. In common with that of many of the other reservoirs supplying London, the water has a

remarkably high phosphorus content.   Barn Elms No. 8 does not strat-
ify thermally, so that none of the effects described can be attributed
to a restricted vernal circulation period.

In 1938 there was a single spring maximum of *Asterionella formosa*
in the first week in March, followed by a rapid decline, during which
Stephanodiscus astraea increased to a maximum in early April and
declined even faster than it had risen.   Phosphate and silicate were
present initially in high concentration, about 200 mg. $P \cdot PO_4$ and
10,000 mg. SiO_2 m.$^{-3}$.   Both substances declined as the populations,
first of *Asterionella*, then of *Stephanodiscus*, waxed and waned.   The
main decline in the population of *Asterionella* took place as the phosphate decreased from 85 to 60 mg. m.$^{-3}$ and the silicate from about
6500 to 5000 mg. m.$^{-3}$.   It is certain from experience both at Barn
Elms in other seasons, notably in May 1939, and in many other waters
that these concentrations are not limiting to *Asterionella formosa*.   The
Stephanodiscus maximum apparently occurred at a phosphate concentration[3] of the order of 5 mg. m.$^{-3}$ and a silicate concentration of the
order of 700 mg. m.$^{-3}$.   Gardiner thinks that these concentrations are
not limiting.   He rather tentatively suggests that some other substance,
first taken up by *Asterionella*, is liberated as the *Asterionella* dies
and so becomes available for *Stephanodiscus*.

In 1939 the diatom population rose early in January, the first peak
being due mainly to *Asterionella formosa* accompanied by a small
maximum of *Stephanodiscus astraea*.   The *Asterionella* population declined with great rapidity and was succeeded by *Fragilaria crotonensis*,
which species produced two maxima in February, whereas *S. astraea*
increased a little in February and then disappeared.   *Cyclotella* sp.
(probably *comta*) was also present from February to May as a subdominant.   During March the decline in the great *Fragilaria* population, which had consisted of about 7000 cells ml.$^{-1}$ on February 28,
led to a minimum in the total diatom population in the first half
of April.   Then *Asterionella* started to increase once again, rising to
a huge maximum of about 9000 cells ml.$^{-1}$ on May 2.   An immediate
decline was temporarily arrested in the second half of May, but after
the beginning of June all diatoms became scarce and remained so
throughout the summer.   In *Asterionella*, the first decline from the
January maximum occurred at a silicate content of about 6500 mg. m.$^{-3}$
and a phosphate phosphorus content of 220 to 250 mg. m.$^{-3}$.   In view

---

[3] There seems to be discrepancy between the data as presented in Gardiner's
figure and his Table 1, but it is certain that the phosphorus content was of the
order of a few milligrams per cubic meter at the end of March.

of the other observations that have been made on *Asterionella*, the replacement of this diatom in February by *Fragilaria* cannot have been due to the succession of a species requiring a high phosphate or silicate by one tolerating a much lower concentration. The main decline of *Fragilaria* occurred when the silicate concentration was about 2700 mg. m.$^{-3}$ and the phosphate, about 150 to 170 mg. m.$^{-3}$, as the silicate fell to about 200 mg. m.$^{-3}$; this again can hardly have been limiting. The final increase in *Asterionella*, however, is quite likely to have been terminated by silicate deficiency. The phosphate phosphorus fell suddenly to 10 mg. m.$^{-3}$ at the time of the final maximum but started to rise again. No increase in silicate occurred until June, when the diatom phase of the phytoplankton was evidently over for the season, but, since the initial fall from the May maximum was arrested for a week or two about the time when the phosphate began to increase, it is possible that silicate was also entering the water from the mud but was being used as fast as it was produced. Even if the final decline of the *Asterionella* population in 1939 was comparable to the normal sequence in the Lake District, as seems likely, the decline in 1938 and the first decline in 1939 seem much more like the exceptional depressions that sometimes occur in the Lake District populations. Gardiner also studied two other reservoirs in the spring of 1939. Both were being used for water storage and so the flow through them was considerable. In both there were two *Asterionella* maxima, the first maximum in each causing little change in the silicate or phosphate content, the silicate being about 3000 and the phosphate more than 40 mg. m.$^{-3}$. The second maximum in Island Barn Reservoir, however, caused the phosphate to decline to 1 mg. m.$^{-3}$ and the silicate to 1700 mg. m.$^{-3}$. This phosphate value is extraordinarily low for the Metropolitan Water Board catchment area, but neither it nor the concurrent silicate concentration is likely to have been limiting. Gardiner, without explaining the marked rises and falls in the populations of individual species of diatoms during the early spring, when nutrient depletion is most improbable, stresses the great rapidity of such changes and the curious fact that in several cases the decrease after the maximum proceeds even more rapidly than did the very great increase that produced the maximal population.

All the evidence assembled, and some less complete observations, such as those of Liepolt (1958) on the phytoplankton of the Zellersee, Salzburg, suggest that the existence of large populations of *Asterionella* depend on adequate illumination at times of high silicate concentration and adequate supplies of other nutrients. These are necessary but certainly not sufficient conditions, but it is not clear what other factor or factors are operating. As has already been indicated, in Esthwaite water, parasitism by the fungus *Rhizophidum planktonicum* may sometimes cause declines in the *Asterionella* population; this matter is discussed later in somewhat greater detail. Lund, however, is certain that neither parasitism nor cropping by zooplankton (Lund 1949b) was responsible for decreases in the north basin of Windermere at the time of his investigations. The anomalous minima in the London reservoirs and in Windermere in 1948 conceivably might be explained by sudden reductions in turbulence that would permit sedimentation of the diatoms at an abnormal rate; but this hypothesis is improbable, being in poor accord with the flotation of *Asterionella* under ice discussed subsequently, and does not explain the many lakes, apparently chemically suitable, that develop only small populations. It would seem far more likely, therefore, that unknown chemical or biological factors are involved in regulating the occurrence of the diatom. That the periodicity may involve less obvious factors than are apparent at first finds some support from other considerations. In the first place, it appears that *Asterionella* in Windermere (Pennington 1943) and Linsley Pond (Patrick 1943) has become an important constituent of the planktonic flora in relatively recent years and in both basins since the formation of sediments that are believed to be influenced by human settlement of a fairly intensive kind in the last two or three centuries (cf. also Hasler 1947). This fact strongly suggests that although *Asterionella* is apparently a true autotroph in culture and is in some regions (Rawson 1956) apparently characteristic of very unproductive lakes some accessory substance derived from land drainage or sewage may promote the development of large populations in nature. There is evidence, moreover, that natural waters may contain something that can be removed by filtration or sterilization and that this removal may impair the growth of *Asterionella* inoculated into the water (Lund 1959b) in spite of the supposed autoauxotrophic nature of the diatom. Rodhe's (1948) observation, which is in essence confirmed by Lund's work (1949a,b; cf. Chu 1943, 1945; Freshwater Biological Association 1946; and Volume I, page 734), that the phosphate requirements of *Asterionella* are far lower in lake water than in culture medium suggests that whatever is needed in lake water to permit colonization by the diatom acts at least in part by enabling the diatom to use its phosphate supply with reasonable efficiency. It is by no means unlikely that variations in the amount of the phosphorus-sparing factor, perhaps dependent on varying degrees of cultural eutrophication of the lake basin, may be responsible for the anomalous behavior of the diatom in many lakes.

Finally, it is necessary to return to Wesenberg-Lund's (1908) interesting observations on *Asterionella* in the Danish lakes in the winter of 1901–1902. Wesenberg-Lund observed that in three out of four lakes that acquired a transient ice cover during December *Asterionella* appeared in numbers when this ice melted during a mild spell a short time later. In the deep lakes that acquired no ice until late in January no such midwinter maximum in *Asterionella* occurred. However, when the ice that formed in January melted in the spring, enormous maxima of *Asterionella* developed in these deep lakes. Wesenberg-Lund therefore concluded that in some way the formation and melting of ice was involved in regulating the periodicity of this diatom. He seems to have concluded also that the ice cover acted indirectly in controlling competition with *Melosira crenulata,* normally the dominant winter diatom in Denmark. He considered *M. crenulata* as less well adapted to flotation than *Asterionella.* When the ice cover prevents disturbance of the water by the wind, *M. crenulata* will sediment out of the untroubled waters more rapidly than *Asterionella.* On the return of favorable conditions, when the ice melts, *Asterionella* will be found in undisputed possession of the water. That such an explanation is reasonable will appear when the peculiar nature of the cycles of the planktonic species of *Melosira* are considered.

Other diatoms. Jørgensen (1957), studying Furesø (Fig. 118) and Lyngby Sø near Copenhagen, found large spring pulses of *Stephanodiscus hantzschii.* The pattern of occurrence of this diatom in Furesø was in fact much like that of *Asterionella* in Windermere, though limitation appeared to occur at the rather lower SiO_2 contents of 30 to 35 mg. m.$^{-3}$. No large autumnal population developed in spite of rising silica and phosphorus from the late summer onward. *Asterionella* and *Stephanodiscus astraea* had small maxima in May, and large populations of epiphytic diatoms occurred in the summer. In Lyngby Sø an expansion and contraction of epiphytic diatoms preceded the spring maximum of *S. hantzchii,* which occurred later in April, rather than February or March, in Furesø. Later, in Lyngby Sø, the plankton was denser and epiphytes less important; in this lake *S. hantzschii* exhibited a marked autumnal bloom. Jørgensen (1957) suspects that antiobiotic competition between epiphytes and plankton may explain some of the details of the seasonal cycles and of the differences between the lakes.

Apart from Pearsall's observations that *Tabellaria* normally succeeds *Asterionella* in Windermere, Bassenthwaite, Ullswater, and Esthwaite Water and the observations given by Lund on the succession in the last-named lake (see page 447), it is unfortunate that little detailed information on the chemical circumstances surrounding the replacement of one diatom by another is available. Gardiner's views on the mechanism of such replacement have already been given. For the present it can be said only that Pearsall may be correct in believing that *Tabellaria* is more parsimonious than *Asterionella*. The field data certainly suggest such a conclusion. Any apparent objections to be derived from experiments with purely synthetic media are probably irrelevant, nor does Vollenweider's (1950) comparison between the two genera really bear on the question.

The problem of *Dinobryon divergens.* Pearsall's (1932) conclusions with regard to *Dinobryon* are certainly not correct in all details, but his observations now lend themselves to a reasonable explanation. There is no evidence from regions other than the English Lakes that calcium content is involved in the seasonal cycle or that in hard-water lakes the concentration of silicate limits this alga. On both points Hutchinson (1944) found definite evidence to the contrary (Fig. 114). If the alkalinity of the water of Linsley Pond, due mainly to calcium, changes at the time that a *Dinobryon* population is increasing, it becomes greater rather than less. The silicate content in 1937, when *Synedra* was abundant, did fall just before the *Dinobryon* maximum, as Pearsall indicated, though not to the low levels that he recorded. In 1938, however, when only filamentous Mysophyceae and almost no diatoms were present in the spring plankton, no great fall in silicate occurred, and at a silicate concentration said to have been 7000 mg. m.$^{-3}$ and certainly very much above Pearsall's critical limit, *Dinobryon* appeared in numbers considerably in excess of those at the maximum of the preceding year. Hutchinson concluded with Pearsall that *Dinobryon* tended in general to occur after the nutrients present in the spring had been largely removed by the major species of the vernal maximum and that in some but not all cases this nutrient depletion corresponded to a rise in the $N.NO_3 : P.PO_4$ ratio. This would be expected if the ratio were initially greater in the medium than in the algae, as seems often to be the case in inland waters. Hutchinson was not able to suggest why the *Dinobryon* did not develop until the major spring species had reduced the nutrient concentration and were themselves about to disappear. This point is now admirably explained by Rodhe's (1948) finding that 5 mg. m.$^{-3}$ $P \cdot PO_4$ inhibit *Dinobryon divergens* in culture, for although it is possible that different races in different lakes may have slightly different requirements, nearly all the available data suggest that *D. divergens* does not appear until

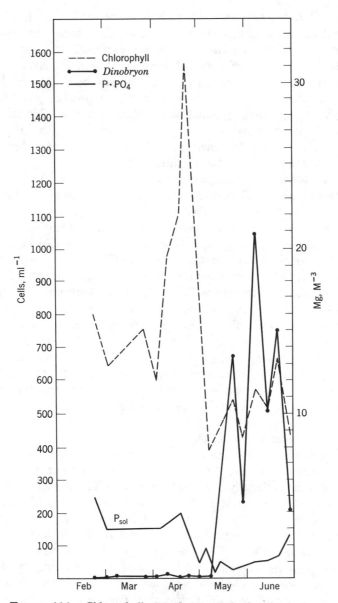

FIGURE 114. Chlorophyll, *Dinobryon,* and phosphate phosphorus in Linsley Pond, spring 1938. Note the rise of *Dinobryon* as the spring algal maximum depletes the phosphate. (Hutchinson, modified.)

the phosphate phosphorus is less than some critical concentration between 1 and 5 mg. m.$^{-3}$. *Uroglena americana* probably behaves in a similar manner.

The remarkable succession of species of *Dinobryon* (Fig. 115) observed by Kozarov (in Stankovič 1960) in Lake Ohrid, in which *D. divergens* is followed by *D. sociale* in the late spring and early summer and in turn by *D. bavaricum* later in the year, suggests competitive replacement to Stankovič, the direction of competition varying with environmental factors such as temperature and illumination.   Observations of this kind in other localities would be most interesting.

The problem of *myxophycean maxima* **and water blooms in summer.**   Pearsall's conclusions about the blue-green algae raise an entirely different and in some ways even more difficult set of problems than those already encountered.   There is no doubt that the group as a whole, with a few rare if interesting exceptions, is characteristic of waters that are highly productive.   Blue-green algae occur in quantity in the lakes of the cultivated plains rather than in those of virgin and rugged terrain.   In many cases (Hasler 1947), and Lake Mendota is evidently a most conspicuous example, lakes which once produced small water blooms, if any at all, now produce excessive quantities of *Anacystis, Anabaena, Aphanizomenon,* and other blue-green algae, apparently as the result of the leaching of agricultural fertilizer and

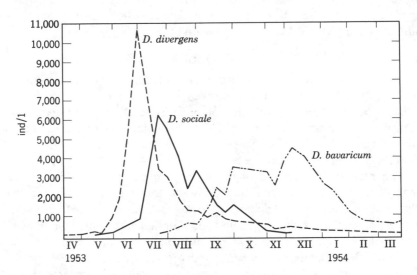

FIGURE 115.   Seasonal succession of the three species of *Dinobryon* in Lake Ohrid 1953–1954. (Stankovič, after Kozarov.)

of the inflow of sewage and other nutrient-rich contaminants. These facts are in themselves probably enough to account for the rather high organic content of lakes with abundant myxophycean plankton. There is no unequivocal evidence from outside the English Lake District to support Pearsall's (1932) contention that a rise in nitrogenous organic matter causes the rise in blue-green algae. The rather meager data on total organic nitrogen in Linsley Pond are definitely not in accord with Pearsall's hypothesis. In this particular case, in which the dominant organism is an *Anabaena* that almost certainly fixes nitrogen, it is much more likely that the rise of blue-green algae causes the rise in the organic nitrogen than the reverse.

Pearsall is certainly correct in his contention that the rise in the summer Myxophyceae tends to occur most rapidly at a time when the inorganic nutrients are present in minimal amounts. In view of the nature of the localities to which the blue-green blooms are restricted, this conclusion, though evidently valid, is highly paradoxical.

The only reasonable explanation of the phenomena of the water bloom would seem to be that in water of the right composition, usually fairly hard but with a reasonable sodium content, blue-green algae take up inorganic nutrients at very low levels and at high temperatures can grow rapidly, outcompeting other forms as a rather large amount of phosphate, and in some cases nitrate or ammonia, diffuses at a high temperature from the organic mud in contact with the water of the epilimnion. Evidence that such large nutrient input to the water can occur, even when the stationary concentration is very low, has already been given in Volume I (page 741).

PERENNATION AND THE PROBLEM OF *MELOSIRA*

The seasonal cycle of *Melosira*, as elucidated in particular by Lund (1954, 1955), is apparently controlled by quite different factors from those operating in any of the other adequately studied examples, though, as indicated earlier (p. 287) the regulation of succession of different sized species by changes in turbulence is likely to prove important in other less special cases.

The planktonic members of the genus can be divided into two ecological groups, namely one appearing at low temperatures and low light intensities and including *M. italica subarctica, M. islandica helvetica,* and *M. baikalensis,* and another appearing at high temperatures and high light intensities and including *M. granulata, M. ambigua, M. nyassae, M. agassizii,* and probably other tropical species. The taxonomy, as is usually the case in planktonic genera of any size, has been in a state of confusion, but Lund's investigations have clarified the matter considerably.

In the English Lake District *M. italica subarctica* is the only plank-tonic member of the genus.   It is a winter form, often appearing with great suddenness in September and forming considerable popula-tions, mainly in the more productive lakes.   If a lake freezes over, the population is likely to fall, only to rise again in the early spring when the ice cover breaks up.   More than fifty years ago, as already noted, Wesenberg-Lund (1904, 1908) recognized that the high autumnal populations of *Melosira* disappeared under the ice, although *Asterionella* might flourish in such circumstances; he attributed this fact to the greater sinking speeds of the members of the genus *Melosira.* In the English lakes a marked decline in the populations of *M. italica subarctica* occurs before thermal stratification is well established, and the organism is absent from the plankton during the summer months.

Lund finds, as Nipkow (1950) had found a few years earlier for *M. islandica helvetica,* that filaments which have sunk to the bottom remain alive in the mud.   In laboratory experiments *M. italica sub-arctica* can survive well in anaerobic mud from eighteen months and some filaments may live three years.   In the layered mud of the Lake of Zurich *M. islandica helvetica* apparently can survive in nature as long as five years.   *M. granulata,* from Frederiksborg Castle Lake, was found to survive in a resting condition in sulfide-rich mud in the laboratory for at least ten months and a similar resting condition was recorded by Lund in *M. ambigua.*   In Esthwaite Water live fila-ments of *M. italica subarctica* were found 5 cm. below the mud surface, under 15 m. of water.   In this lake the sediment is deposited at a rate of about 3 mm. year^{-1}, so that unless they have been buried by *Tubifex* or other benthic organisms, as is quite likely, the oldest living filaments in the mud might be sixteen years old.

The autumnal rise in the population of *Melosira* in the English lakes is sometimes so rapid that it cannot be explained by division of an undetectably low population in the water.   Such gains take place during the autumnal circulation, and Lund concludes that they are the result of the resuspension of filaments that had fallen to the bottom in the spring and had survived the summer in a resting condition at the surface of the mud.   Actual counts indicate that a rise in the *Melosira* population in the water is correlated with a fall in the resting population at the bottom.   Moreover, the new population in the water at first consists of filaments that contain condensed spherical chroma-tophores typical of resting populations (Fig. 116).

Under the ice cover the *Melosira* filaments tend to sink in the rela-tively nonturbulent water, as Wesenberg-Lund had found; when the ice breaks they are resuspended.   During the late spring the population declines as thermal stratification develops; it would inevitably sink out

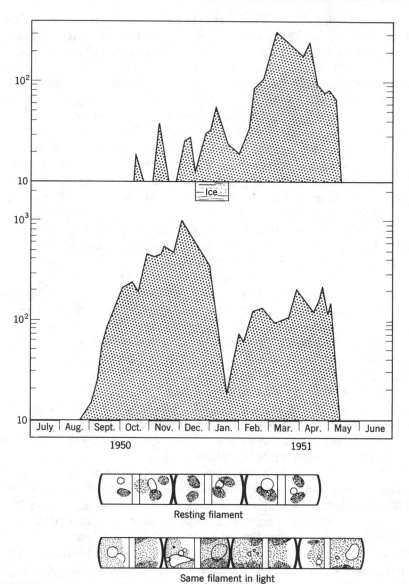

FIGURE 116. Upper panel, *Melosira italica subarctica,* variations in number of cells per milliliter, Windermere (South Basin), 1950–1951; middle panel, the same, Esthwaite Water, 1950–1951, showing decline when the lake is frozen; lower panel, resting filament from sediment at 13 m. in Blelham Tarn, and same filament after eight days in light. (Lund.)

of the deeper undisturbed layers. The decline in the epilimnion may be, but is not certainly (Lund 1955), due to the injurious effects on high light intensity on *M. italica subarctica.* Lund's preliminary experiments indicate that this diatom has a higher sinking speed than *Asterionella formosa* or *Cyclotella praetermissa.*

Most of the available data, summarized by Lund, suggests that the incidence of *M. islandica helvetica,* which is common in many productive lakes in western Europe and which ordinarily has a seasonal cycle comparable to that of *M. italica subarctica,* is determined in the same way. A prolonged period of ice which breaks up late may under the more continental conditions of parts of Europe suppress the early spring maturation.

In Lake Baikal *Melosira baikalensis* is by far the most important member of the phytoplankton, though it is sometimes accompanied by *Cyclotella striata* var. *magna* and other species in small numbers (Meier 1927); *M. baikalensis* shows temporal variation comparable to that of *M. italica subarctica.* The maximum abundance of the alga in the plankton is in the spring and early summer, when the surface waters of this immense lake are still very cold. When the superficial temperature rises to about 7° C., the *Melosira* sinks out of the surface layers. In the later summer strong winds, which cause the upwelling of cold water upwind, may at the same time bring up considerable quantities of the alga (Skabitschewsky 1929; see also Lund 1954). *Melosira baikalensis* has thick frustules and therefore probably a high sinking speed, though during the rapid multiplication of the alga in the spring thin-walled cells may be formed, so that the complete filament will have thick-walled cells only at its two ends. It is probable, as Lund points out, that the greater part of the population is slowly but continuously lost to the depths of the lake in which circulation is imperfect (see Volume I, page 460). The restoration of the population presumably must depend on filaments that have sunk to the bottom at moderate depths of not more than 400 to 600 m. According to Skabitschewsky, auxospore formation takes place under the ice.

The periodicity of the warm-water species is less well understood. In Europe the main populations of *M. granulata* and *M. ambigua* are usually but not invariably recorded in the warmer half of the year. In Florida Lake, an artificial lake not much more than 6 m. deep near Johannesburg, Transvaal, Schuurman (1932) found *M. ambigua* to be perennial but maxima in the southern winter and spring. The lake might be slightly stratified in summer, and the irregular appearance of a large population at the surface in winter at temperatures of 7.5 to 16° C. perhaps suggests resuspension and sedimentation. Lund (1954) found *M. ambigua* and *M. agassizii* to be most abundant under isothermal conditions in the Buvula Channel of Lake Victoria, a marginal part of the lake that stratifies; he also concludes that *M. nyassensis* and other species in Lake Nyasa are most abundant from June to November when steady trade winds produce a more or less freely circulating mixolimnion of maximal thickness. In this lake, as in Lake Baikal, much of the population must sink into the depths and be lost. As far as can be told at present, the hypothesis that the seasonal cycle of warm-water, high-light species depends on sedimentation and suspension as the thermal structure of the lake changes would probably explain otherwise rather puzzling features in the appearance of these organisms. The phenomena presented by *Melosira* can be regarded as constituting a special case of seasonal succession dependent on variations of sinking speeds as turbulence changes, which type of succession though little studied has already been mentioned as of some potential importance (see page 287).

ANTIBIOSIS, OR THE PRODUCTION OF SPECIFIC SUBSTANCES INHIBITING COMPETITORS, AS A POSSIBLE FACTOR IN SEASONAL PERIODICITY

Akehurst (1931), in a curious paper, developed a detailed theory of chemical interaction between the different species of the phytoplankton, involving both stimulation and inhibition.

He considers the unicellular algae as forming two metabolic groups: a *starch group,* storing a carbohydrate and containing Chlorophyceae (starch), Euglenineae (paramylum), Cryptophyceae (starch or allied polysaccharides), and Myxophyceae (sugar and glycogen), and an *oil group,* storing mainly fat and including Bacillariophyceae (oil and volutin), Chrysophyceae (oil and leucosin), Dinophyceae (oil and starch), Heterokontae (oil), and Chloromonadineae (oil). Akehurst believes that each member of the phytoplankton produces an autotoxin which limits the growth of its own population, a process that now would be termed autoantibiosis. This autotoxin is supposedly always stimulatory to organisms of the group (starch or oil) opposite to that producing the autotoxin; it may be stimulatory or inhibitory to members of different classes of the same group and may be stimulating to members of the same class as the producing organism. Intraclass heteroantibiosis is apparently not postulated by Akehurst. He examines a number of records of phytoplankton periodicity in various ponds in England and in one of the reservoirs supplying the water on Jersey in the Channel Islands and is able to interpret all the observations according to these hypotheses.

In the Jersey reservoir, in 1923, *Asterionella* had spring and autumn maxima. The spring maximum was followed by a development of *Synedra* in March. From April until the height of summer the dominant organism was *Coelastrum,* which disappeared in August when *Cosmarium* started to increase. *Dinobryon* appeared at the end of August and reached its maximum in September, during which time there was a slow rise in *Asterionella* to an autumnal maximum in November. Events in later years followed a similar pattern, somewhat disturbed by copper sulfate treatment. Dinophyceae, though often present, never were able to increase to any great extent. This successional history is interpreted in the following way. *Asterionella* inhibits itself when it reaches its maximum but stimulates the developing population of *Synedra.* Both diatoms, being members of the oil group, produce a substance that stimulates the members of the starch group—in this case *Coelastrum.* This alga in turn limits itself, but in so doing produces a stimulant to *Cosmarium.* Both *Coelastrum* and *Cosmarium,* members of the starch group, prepare the way for *Dinobryon* and *Asterionella* of the oil group. However, *Dinobryon* is itself inhibited by *Asterionella,* and as *Asterionella* develops *Dinobryon* declines. In a later year, when *Asterionella* was artificially destroyed by copper sulfate, *Dinobryon* developed in great numbers. Akehurst realized that *Dinobryon* usually occurs after the decline of some other abundant species. Since *Asterionella* and the Dinophyceae are believed to compete for the same nutrients, the presence of *Asterionella* is supposed to prevent the development of any large dinophycean population, even though other experience suggests that the substance produced by *Asterionella* stimulates the Dinophyceae.

It is reasonably certain that Akehurst's detailed conclusions are largely erroneous, even though they can be made to explain almost everything that is known about the variation of the phytoplankton. It is nevertheless quite likely that a residue of truth may underlie some of his hypotheses and represent a curiously intuitive understanding of certain of the events in lake or pond. As such, his contribution is of considerable historic interest.

Autoantibiosis in algal cultures has been known since the work of Harder (1917) on *Nostoc punctiforme.* Much attention has been given to the process in *Chlorella* (Pratt 1940, 1942, 1943; Pratt and Fong 1940; Pratt, Daniels, Eiler, Gunnison, Kummler, Oneto, Strait, Spoehr, Hardin, Milner, Smith, and Strain 1944), in cultures of which in some strains a substance termed *chlorellin* accumulates in the medium and inhibits photosynthesis by the alga. Later work (Spoehr, Smith, Strain, Milner, and Hardin 1949) indicates that inhibitory photo-oxidizable unsaturated fatty acids can be extracted from *Chlorella*.

A large number of experiments mainly with green and blue-green algae reported by Lefèvre and his co-workers (Lefèvre and Nisbet 1948; Lefèvre and Jakob 1949; Lefèvre, Nisbet, and Jakob 1949; Lefèvre, Jakob, and Nisbet 1950, 1951, 1952) are discussed in greater detail later. In diatoms Levring (1945) found autoantibiosis in *Skeletonema costatum* and von Denffer (1948), in *Nitzschia palea*. In the last-named work assimilation was not inhibited, and the effect of the autoantibiotic appeared to be due to the blocking of mitosis, also suggested in some of the heteroautibiotic experiments summarized in Table 2. In several studies, notably that of Lefèvre, Jakob, and Nesbit (1952) on *Scenedesmus quadricauda,* mere dilution (Fig. 117) of a culture that has ceased to divide with distilled water will stimulate division by reducing the concentration of the antibiotic.

Heteroantibiotic effects were first studied on a considerable scale by Lefèvre and his co-workers, whose results are reported in the papers just listed. Other work by Rice (1954), Jørgensen (1956), Talling (1957), Proctor (1957a,b), and Hartman (1960) has added considerably to an understanding of the phenomena observed. Hartman's

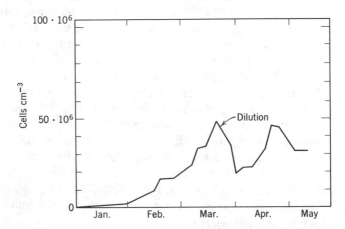

FIGURE 117. Effect of dilution with an equal volume of distilled water on a culture of *Scenedesmus quadricauda* that had stopped growing and was beginning to decline. Dilution halved the concentration of cells at the end of March, but by the end of April the previous maximal concentration was restored, followed again by a decline. (Lefèvre, Jakob and Nisbet.)

(1960) review may be consulted for various aspects of the matter not of direct limnological interest.

In their main paper Lefèvre, Jakob, and Nisbet (1952) used a large number of different algae, but only in *Scenedesmus quadricauda* and *Pandorina morum* were their cultures not only unialgal but bacteriologically sterile. Though they give evidence that bacterial action was not involved, it is obviously better to place more reliance on observations from which bacteria were excluded.

It is apparent from their experiments with media in which either of these two algae was grown and which they designate as containing "scenedesmine" and "pandorinine," respectively, that different and even closely allied species can react differently to the same heteroantibiotic and that the two antibiotic solutions have very different properties. This will be apparent in Table 26.

The active materials from both types of solution can be absorbed on charcoal, leaving solutions largely without antibiotic activity, though in a few cases some reduction of division rate is recorded.

After boiling, the "pandorinine" solution became stimulatory to nearly all the algae examined, including *P. boryanum* on which the effect was marked, but it largely retained its inhibitory action on *P. clathratum* var. *punctulatum*. Boiling the "scenedesmine" solution reduced its toxicity to all forms except the two species of *Cosmarium* studied, but in no case[4] was the growth normal, let alone better than in the control. It is reasonably certain that at least the "pandorinine" solution contained more than one active substance and that such substances are different from the active material in the "scenedesmine" solution.

There was considerable variation in the activity of various preparations; "pandorinine" preparations from subcultures of varying ages can be inactive toward *Pediastrum clathratum* but highly toxic to *Cosmarium lundellii,* active in producing morphological abnormality in *P. clathratum* but not fully toxic to *C. lundellii,* or, in very old cultures (four to five months), stimulatory to both test algae. Other comparable examples are given.

Several later studies have been made with diatoms. Rice (1954) found that mutual antibiosis occurred in mixed cultures of *Chlorella vulgaris* and *Nitzschia frustulum* but that a greater excess of *Chlorella* was needed to produce a given inhibition of growth rate of *N. frustulum* than of *N. frustulum* to produce a given inhibition of growth rate of *C. vulgaris.*

---

[4] *Achnanthes microcephala* not used.

TABLE 26. *Effects of media from cultures of* Scendesmus quadricauda *and* Pandorina morum *on other algae*

Test Organism	Scenedesmus quadricauda medium ("scenedesmine")	Pandorina morum medium ("pandorinine")
Myxophyceae *Phormidium uncinatum*	Immediate inhibition; cells become clear yellow	Good growth for ten days; then rapid degeneration
Chlorophyceae *Scenedesmus oahuensis*	At first a few divisions, then complete inhibition; the few daughter cells are yellow, with granular vacuolizsed contents	Several divisons, daughter cells at first, full of brown inclusions which disappear as the cells become vacuolized, hypertrophy, and die
Pediastrum boryanum	Division for a short time; then total inhibition, the daughter cells being unable to divide, hypertrophy.	Initial slight inhibition, division normal by 19th day
P. clathratum var. *punctulatum*	A single division that produces abnormal looking cells which, however, lived until the twentieth day.	Some multiplication but apparently no recovery by 19th day
Mesotaenia caldariorum	Fairly good multiplication for some days but inhibition by the ninth day	Excellent growth
Cosmarium lundellii and *C. ochthodes*	Some divisions, but ultimate inhibition with loss of inclusions, but not death.	Become black and opaque with inclusions, a few divisions, death
Bacillariophyceae *Achnanthes microcephala*	Somewhat slower reproduction, otherwise normal	Production of small forms, some half initial size

Jørgensen (1956) has studied the important planktonic diatom *Asterionella formosa* in mixed culture with *Nitzschia palea*. Both can produce antibiotics against one another, but at least sometimes, though not always, in old culture a stimulatory effect of *Asterionella* on *N. palea* was observed, and it is possible that *Asterionella* on occasion may produce a substance promoting its own growth, a type of phenomenon that has given rise to a large, controversial, and largely forgotten literature in protozoology.

A further experimental advance has been made by Proctor. Proctor (1957a) studied the growth of five different algae cultured together in pairs. His results are set out in Table 27, in which the relative growth of different pairs are compared with the growth in unialgal culture, taken as 100.

Anacystis marina (sub *nidulans*) inhibits every organism with which it was grown; *Haematococcus pluvialis* is inhibited by every other organism in the table. Mutual inhibition occurred in three of the ten combinations; in all other cases there was a drastic reduction, and in three cases, elimination of one alga after five days.

Proctor (1957b) studied, in particular, the pair *Haematococcus pluvialis-Chlamydomonas reinhardi* in which the first species was eliminated in the preliminary experiments within five days. The inhibition took place much less rapidly at low light intensities than at high. When ammonium nitrate is used as a nitrogen source there is evidence that *H. pluvialis* takes up nitrate preferentially, and so raises the pH, whereas *C. reinhardi* takes up ammonium ions preferentially, and so lowers the pH. In the original medium the nitrogen source was solely nitrate; when substituted by urea, elimination of the *Haematococcus* took twelve days under conditions in which, in the presence of nitrate, the process required three to five days. Although the conditions of most rapid elimination correspond to a rising pH in the culture, in unialgal culture *H. pluvialis* can tolerate a pH of 9.5, which is not

TABLE 27.   *Growth of algae listed on left in presence of species listed at top*

| Organism | Grown in the Presence of | | | | |
	Anacystis marina	*Chlorella vulgaris*	*Haematococcus pluvialis*	*Chlamydomonas reinhardi*	*Scenedesmus quadricauda*
Anacystis marina	100	100	100	100	100
Chlorella vulgaris	5	100	100	75	100
Haematococcus pluvialis	0	20	100	0	29
Chlamydomonas reinhardi	0	25	100	100	87
Scenedesmus quadricauda	10	25	38	12	100

reached until after it has been eliminated in mixed culture with nitrate at high light intensity.

Cell-free extracts of young *C. reinhardi* culture, slightly stimulatory to *H. pluvialis,* probably condition the medium by removing toxic heavy metals. Cell-free medium at a *p*H of more than 8.5 obtained from cultures of *C. reinhardi* more than four days old, however, was toxic, though hardly so at *p*H 7.5. The toxic substance is always much more abundant in cells than in the medium. It is thermostable, but can be lost in boiling, being volatile in steam, and has in fact the properties of a fatty acid, a class of substances in which many algae are rich, especially under poor conditions. Proctor concludes that the antibiotic acid is present primarily in cells from which it is liberated on death and then dissolved in alkaline water. Cultures on urea do not rapidly become alkaline enough to permit solution; cultures at low light intensity grow slowly and very few cells die. Proctor therefore suspects that at least one large class of algal heteroantibiotics, though they may be highly effective in controlling succession are produced only in old crowded populations in which many cells are dying. The effects can be imitated to some extent by long-chain fatty acids, palmitic being the most toxic of the saturated and linoleic of the unsaturated acids studied. The effects of these acids, however, do not provide a perfect model, for, though of the five species, *Haematococcus* is much the most sensitive alga to such compounds, as it is also to the *C. reinhardi* substance, *Anacystis marina,* very resistant in the paired culture experiments, is much more sensitive to straight-chain fatty acids than is *Scenedesmus quadricauda.* The last-named species was in fact the most resistant to the known acids but more susceptible in mixed culture than any species save *H. pluvialis.* In spite of this difficulty, Proctor's work is highly suggestive, and it is quite likely, as he believes, that many of his predecessors were employing cultures as sources of heteroantibiotics in which many dead cells were present.

Hartman (1960) has tabulated all of the cases of chemical coactions in laboratory experiments between two species of algae reported in the literature, adding some of his own. Though some uncertainty in interpretation and classification is inevitable, the over-all distribution given in Table 28 provides an idea of the commonness of inhibition or stimulation of one algal species by another.

It is evident that most algae produce inhibitors of other algae and that stimulation is rarer than either inhibition or no coaction but can occur in a reasonable minority of cases. Clearly so much bias has been present in planning experiments that nothing can be said about the prevalence of intraclass as against interclass antibiosis.

TABLE 28. *Incidence of inhibition or stimulation of one alga by another.*

Source of Material	Inhibitory to		No Effect	Stimulatory
Chlorophyta	Chlorophyta[a]	38	$9\frac{1}{2}$	$3\frac{1}{2}$
	Euglenophyta	1	1	—
	Bacillariaceae	5	1	—
	Myxophyceae	6	1	1
Bacillariaceae	Chlorophyta	5	—	2
	Euglenophyta	—	3	—
	Bacillariaceae	3	—	2
	Myxophyceae	—	—	—
Myxophyceae	Chlorophyta	10	2	1
	Euglenophyta	—	—	0
	Bacillariaceae	—	—	1
	Myxophyceae[b]	$16\frac{1}{2}$	$6\frac{1}{2}$	—
		$84\frac{1}{2}$	24	$10\frac{1}{2}$

[a] Includes one case given as "little affected," which is shared between the inhibitory and the no effect columns, and one case in which inhibition or stimulation depended on the conditions of culture, shared between the inhibitory and the stimulatory columns.

[b] Includes three cases of "little affected" shared between the inhibitory and no effect columns.

Few experiments have been conducted with two species of eulimno-planktonic algae. In the only good case, that of Talling's (1957) study of *Asterionella formosa* and *Fragilaria crotonensis,* no inhibition was observed.

Natural waters are often strikingly inhibitory when collected from localities rich in algae. Lefèvre, Jakob, and Nisbet (1952) made a large number of trials of the effect of water from ponds in Sologne and from a lake-like system of canals in the Parc de Rambouillet. Water collected at the time of large algal blooms proved to be inhibitory to a number of species used as test organisms.

In the localities in which large populations of blue-green algae developed (*Anabaena spiroides, Anacystis cyanea, Oscillatoria planctonica, Aphanizomenon gracile*) the filtrate from the water containing the algal bloom was always algistatic to all species studied. From a pond dominated by *Ceratium hirundinella* the water killed *Pediastrum boryanum* and *Cosmarium lundellii,* inhibited division of *Cosmarium obtusatum, Nitzschia palea,* and *Phormidium uncinatum,* was weakly inhibitory to *Scenedesmus quadricauda* and *Chlorella pyrenoidosa,* but stimulated *Pediastrum clathratum* var. *punctulatum.* The same investigators, who in July 1950 studied L'étang de la Bergerie, a pond at Brinon-sur-Sauldre (Sologne), which contained a highly diversified desmid and protococcal flora, found that the water had no effect on the culture of a variety of test organisms; but in October, when there was much decomposition of higher plants, the water was inhibitory to *Cosmarium lundellii, C. obtusatum,* and *Phormidium uncinatum,* markedly stimulatory to both species of *Pediastrum* studied, and had no effect on other test organisms. In one case, involving a considerable development of *Anabaena spiroides* var. *crassa* in the Lac d'Équizon (Cher), a body of water held by a dam and 30 m. deep at the point of study, water samples were studied at a series of depths from 0 to 20 m. No details are given, but the algistatic properties of the water are said to have decreased with depth as might be expected.

It is reasonably certain from this work that massive growths of algae do inhibit the development of many other species, and some field evidence, such as the production of a rich flora with twenty-six taxa of Euglenophyta and ten of other algae, after the decomposition of an almost unialgal *Anabaena* bloom suggests that the disappearance of a bloom may permit development in a few weeks of a great diversity of organisms. In the case just quoted the previous existence of the bloom doubtless prepared the way for an extraordinary assemblage of alloauxotrophic and perhaps partly heterotrophic forms. Even if as Proctor supposes, the antibiotic effects were due mainly to dead cells in the massive populations, they could have marked ecological effects in regulating succession. However, it is quite likely that these effects would be felt mainly in small and very eutrophic bodies of water. Lund (1964) seems unwilling to go even as far as that. Unfortunately, some of the cases recorded from the field, such as the relatively low level of *Fragilaria* until *Asterionella* started to decline in Barn Elms No. 8, which would be tempting to explain by antibiosis, relate to organisms for which the laboratory evidence suggests no such process.

The most interesting field evidence is probably that of Jørgensen (1957), who believes that a marked antibiotic interaction between epiphytic and planktonic diatoms may occur. In Furesø there is a spring pulse of *Stephanodiscus hantzschii* (Fig. 118) which reduces the silica content of the lake water. When this pulse declines, a great development of epiphytic diatoms on *Phragmites* stems occurs. The epiphytes have in their substrate a source of silica not available to the planktonic forms. Jørgensen suspects that the sudden rise of the epiphytic flora as the planktonic diatoms decrease is due to the latter having inhibited the former when soluble silica was high. In Lyngby Sø, in which the planktonic diatoms are more abundant than in Furesø, the epiphytic flora is poorer, again perhaps because of antibiotic inhibition.

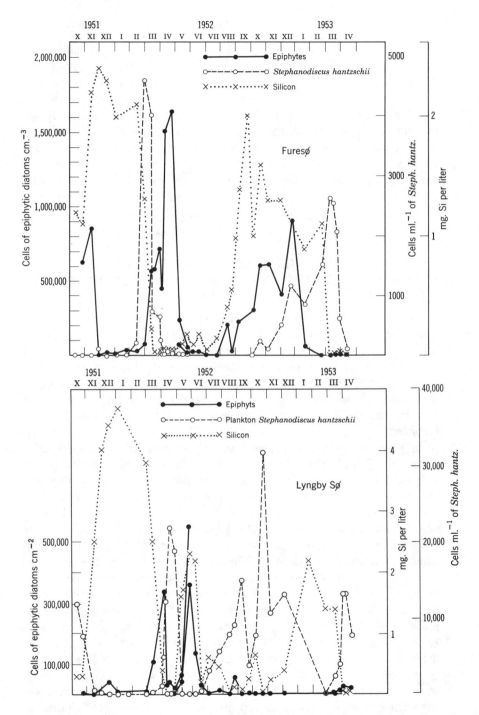

FIGURE 118. Seasonal incidence of *Stephanodiscus hantzschii* in the plankton, of epiphytic diatoms, and of soluble silicate in Furesøano Lyngby Sø. (Jørgensen.)

Another interesting case, though involving only very small bodies of fresh water, has been studied by Proctor (1957a). In nature *Haematococcus pluvialis* is largely confined to such minute habitats as the water that collects in the saucers of flower pots in greenhouses. Proctor concludes that in basins large enough to be permanently wet the species is rapidly removed by competitive antibiosis but that in small temporary waters the capacity of the alga to withstand desiccation permits its continued survival through an indefinite number of rapid cycles of filling and drying, with wet phases too short to permit antibiotic inhibition of the *Haematococcus* and with drying complete enough to destroy any competitor that may be accidently introduced.

ACCESSORY ORGANIC COMPOUNDS OR VITAMINS AS POSSIBLE DETERMINANTS OF SEASONAL SUCCESSION

Akehurst's scheme implied stimulation as well as inhibition, and the data obtained, notably by Lefèvre, Jakob, and Nisbet (1952), Jørgensen (1956), and one or two other investigators, suggest that though the coaction between algal species is usually inhibitory stimulation occurs in an appreciable minority of cases. In the rather miscellaneous list of organisms studied so far, it appears from Hartman's tabulation that in roughly 9 per cent of the cases studied some stimulation was recorded.

In view of the frequent need for an external source of vitamins in the nutrition of many algae, notably cobalamines, thiamin, and biotin, it is natural to suppose that marked stimulation is caused by liberation of one of these compounds from living or perhaps more often dead cells of one alga in concentrations stimulatory to another. It must be realized, however, that in the laboratory, although nearly all the Chrysophyceae, Dinophyceae, and Euglenophyta that have been cultivated are alloauxotrophic, the cultivated species include few eulimnoplankters. If we may infer that alloauxotrophy is a valid taxonomic character of these groups, then it is likely that stimulatory effects by liberation of vitamins not only permit multispecific associations, as already suggested, but also play a part in succession. Until more is known about the nutrition of specific members of the lake plankton, these concepts remain hypothetical, though probable. The a priori unlikeliness of alloauxotrophic forms occurring among the chrysophyceaen and dinoflagellate plankton of large, deep, unproductive lakes is to some extent reduced by the growing evidence of the importance of at least cobalamines in the blue waters of the open oceans.

One rather special situation deserves attention, though mainly to suggest a possible line of future research. There is evidence that an organic material needed by the marine diatom *Ditylum brightwelli* is present in the English Channel only in the winter; the diatom was able to flourish there only at that season (Harvey 1939). Since Hutchinson and Setlow (1946) found evidence that niacin and perhaps biotin were more concentrated in the surface waters of Linsley Pond under the ice than at any other time of year, it is worthwhile to consider whether any fresh-water organisms might behave like *Ditylum,* requiring some vitamin produced more abundantly or most probably decomposed less rapidly at low than at high temperatures. There is no direct or experimental evidence in favor of the existence of such forms, but the curious massive increases of certain species of unicellular algae in small lakes or ponds under ice, which produce on occasion quite extraordinary maxima, suggest that, as in the sea, winter species may be determined by some special chemical conditions that are likely to involve the accumulation of specific organic compounds. In view of the discovery (Hutner and Provasoli 1951) that *Chlamydomonas chlamydogama* requires not only B_{12} but also histidine and may be stimulated by aspartic acid, the occurrence of green flagellates under circumstances suggesting that organic accessory substances are involved is not surprising.

The following cases are recorded by Burkholder (1931) from the Cayuga Basin, New York:

Chicago Pond: maximum depth, 3 m.; water, brown; pH, 6.5; O_2, 0.2 mg. per liter. *Glenodinium neglectum* (December to March maximum) and *Carteria globosa* (January maximum).

Dryden Lake: maximum depth, 5 m.; water, yellow-green; pH, 7.9; O_2, 8.9 mg. per liter. *Peridinum inconspicuum* (January maximum).

Lowery Pond: maximum depth, 20 m.; water, slightly brown; pH, 8.2; O_2, 7.5 mg. per liter. *Chlamydomonas angulosa* (January maximum) and *Carteria globosa* (February and April maximum).

Phillips Pond: maximum depth, 10 m.; water, slightly brown; pH, 8.0; O_2, 11 mg. per liter. *Carteria globosa* (January maximum) and *Chlamydomonas angulosa* (February maximum).

The chemical data refer to water under ice in February and indicate the wide range of conditions in which *Carteria globosa* can form enormous populations. Moreover, Burkholder found that in another locality, Pout Pond, this species was always present, with a maximum in September at a temperature of 16.5° C. and a minimum in February under the ice. Pout Pond is peculiar in that it is set in a large swamp with a great accumulation of vegetable detritus and bears considerable amounts of *Lemna* and *Wolffia*. Because of the probability of high concentrations of organic extractives and products of bacterial metabolism in such a locality, it is tempting to suggest that the high winter concentrations of *Carteria, Chlamydomonas,* and certain Peridinians may be determined by the presence of accessory organic nutrient substances which can reach adequate concentrations only at low temperatures or when, as in Pout Pond, much organic matter is decomposing.

PARASITISM AS A CAUSE OF SEASONAL SUCCESSION

The phytoplanktonic algae are parasitized by fungi of the orders Chytriales, Lagenidiales, and Saprolegniales. In recent years critical attention has been given to the possible effects of such parasitism on the economy of the lakes, particularly by Canter (1949; Canter and Lund, 1948, 1951). The populations of *Asterionella formosa* in the English Lake District and elsewhere are usually infected to a slight extent by the chytrid *Rhizophidium planktonicum* (*A, B* of Fig. 119). The infection rate is often low and the presence of the parasite is then of no great importance in regulating the size of the population. In some eutrophic lakes (Lund 1950a), particularly in the late summer or autumn, epidemics may occur which tend to destroy the autumnal population of the diatom (*D* of Fig. 119). Thus in Esthwaite in October 1947 the incidence of parasitism rose from a fraction of one per cent to 40 per cent and the population of *Asterionella* then fell from 121 to 37 cells ml.$^{-1}$ Parasitism declined, but in the late autumn the diatom maintained a steady population of about 80 cells ml.$^{-1}$ Marked destruction never occurs in the less eutrophic North Basin of Windermere but can happen in the South Basin. In the small more productive lakes, Esthwaite Water and Blelham Tarn, spring epidemics have also been noted. In all cases there is a tendency for the number of *Asterionella* cells per colony to fall at the time of an epidemic. The same or some comparable species apparently attacks *Oscillatoria* (Canter 1949). Lund records that from January to May, 1949, *Oscillatoria agardhii* var. *isothrix* was common enough in Esthwaite to reduce the illumination and consequently the population of *Asterionella*. Moreover, in late February and March the *Asterionella* was badly parasitized, and the phytoplankton was also apparently washed out by floods in April. During May the *Oscillatoria* itself became badly

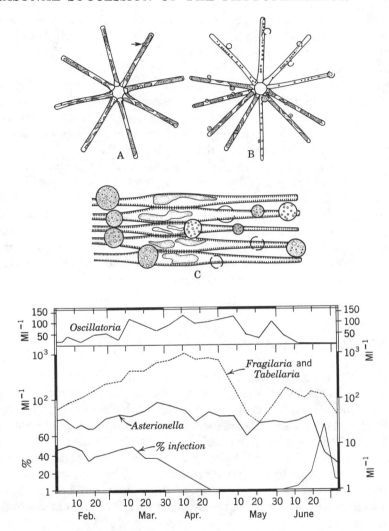

FIGURE 119. *A. Asterionella formosa,* healthy colony, but with a single zoospore of *Rhizophidium planktonicum* (arrow). *B.* A heavily parasitized colony with disorganized cells and sporangia, either alive or dehisced. *C. Rhizophidium fragilariae* on *Fragilaria crotonensis. D.* Upper line, variation in number of *Oscillatoria agardhii* var. *isothrix* filaments per milliliter in Esthwaite, spring 1950; second line, sum of *Fragilaria crotonensis* and *Tabellaria fenestrata* var. *asterionelloides;* third line, *Asterionella formosa,* all cells per milliliter; fourth line, percentage infection of *Asterionella* with *R. planktonicum.* (Canter and Lund.)

parasitized and disappeared; *Asterionella* suddenly increased, and a new *Rhizophidium* epidemic, severer than that in March, broke out. The diatom population suddenly declined just about the time that the silicate content was approaching the lower limit for *Asterionella*, but evidently enough nutrients were present for the sudden development of a great population of *Fragilaria crotonensis* and *Tabellaria fenestrata* var. *asterionelloides*. In this case *Rhizophidium* seems to have played a striking if complex role in determining the seasonal succession, but it must not be forgotten that apparently this can happen only in nutrient-rich lakes. *Fragilaria crotonensis* can also be controlled by parasitic infections, (*C.* of Fig. 119) and in severe epidemics 70 per cent of the cells may be infected, from forty to thirteen. (Freshwater Biological Association 1949b). Canter also believes that desmids and other green algae may be limited in their numbers by comparable parasites, and Vorstman (in Canter 1949) notes that the populations of *Oocystis lacustris* and *O. crassa* in the Zuidersee are to some extent regulated by a fungus, probably *Olpidium entophytum*.

GRAZING AS A DETERMINANT OF SEASONAL SUCCESSION

The increase in temperature during the summer in temperate regions ordinarily results in an increase in activity and reproduction of planktonic animals. As a result, we should expect a greater utilization of food in the summer; if a seasonal increase in primary photosynthetic productivity did not keep up with the demands placed on the phytoplankton, they would tend to be removed by the zooplankton, which in turn would begin to suffer from starvation. The situation that would develop has some of the properties of the well-known Lotka-Volterra prey-predator cycle, but it is essentially a forced oscillation controlled by the seasons.

In most lakes the form of the interaction, particularly if a number of species with different requirements and feeding mechanisms are present, is likely to make isolation of the grazing effect difficult. Riley (1940), who studied the interrelation of the zooplankton and phytoplankton of Linsley Pond statistically, obtained no significant results. It is often tempting to conclude that particular temporal features of the phytoplankton curve which can be correlated inversely with contemporary changes in the zooplankton imply grazing, but often other equally striking changes apparently do not. In Lake Ohrid the period dividing the main summer pulse of the phytoplankton into two principal maxima does correspond to a large maximum in rotifers; unfortunately, Kozarov's (in Stankovič 1960) phytoplankton data overlap those of Serafimova-Hadžišče (in Stankovič 1960) for the zooplankton only for a period of a few months. Certain cases are known (page 706) in which variation of the zooplankton is apparently produced by previous variations in phytoplankton but without any clear evidence of the inverse coaction.

The best cases perhaps are provided by the simplified associations found in the saline lakes of the Grand Coulee (Anderson, Comita and Engstrom-Heg 1955; Anderson 1958). In Lake Lenore (Volume I, pages 111–113), an alkaline lake of salinity 14 per mille, striking spring and late summer phytoplankton maxima (upper part of Fig. 120) are due to a small species of *Amphora* and to *Chaetoceras elmorei,* respectively; no other species contributed quantitatively to the phytoplankton. The minimum intervening between the two major phytoplankton pulses is characterized by the rapid development of a large population of *Leptodiaptomus sicilis,* which declined as the *Chaetoceras* pulse developed. During the latter part of the phytoplankton minimum *Moina hutchinsoni* also was present in moderate numbers. It exhibited a subsidiary maximum, accompanied by the rotifer *Hexarthra fennica,* as the *Chaetoceras* declined, so that there was possibly a small secondary total maximum in zooplankton biomass after the *Chaetoceras* pulse. A fair supply of phosphate existed in solution at the time of the absolute phytoplankton minimum, which can hardly have been due to nutrient deficiency. The most reasonably interpretation of the data is control of the phytoplankton by zooplankton grazing. Anderson (1958), however, expresses caution about accepting this interpretation because in bottles suspended in the lake with and without zooplankton the expected greater increase in the absence of grazing organisms did not take place.

In Soap Lake, a meromictic and even more concentrated (salinity 35 per mille) lake in the same region, only a winter and early spring phytoplankton pulse (lower part of Fig. 120) was recorded; because of the alkalinity of the water qualitative plankton samples disintegrated. As this pulse declined, the rotifer *Brachionus plicatilis* rose and fell, and throughout the summer the only important zooplankter was *Moina hutchinsoni.* Experiments showed that filtering off the animals led to an increase in chlorophyll-bearing organisms, as would be expected. Because the *M. hutchinsoni* is much larger than *B. plicatilis,* it is reasonable to attribute the grazing effect, if genuine, to the crustacean rather than to *B. plicatilis.* Anderson considers it possible that phytoplankton productivity was as great or greater during

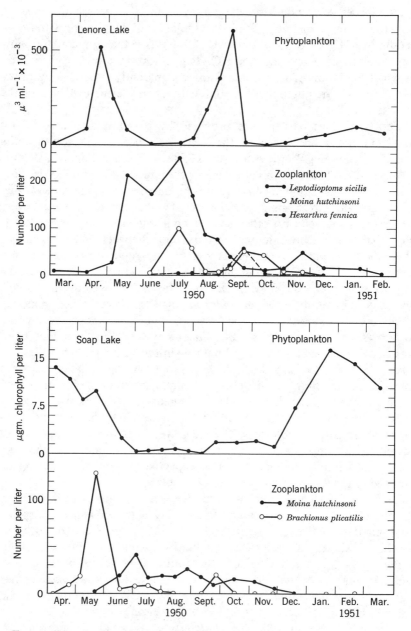

FIGURE 120. Seasonal incidence of phytoplankton and principle species of zooplankton in Lenore Lake and in Soap Lake, Grand Coulee, Washington. (Anderson, Comita and Engstrom-Heg.)

the warm periods with low phytoplankton and high zooplankton concentrations than during the cooler periods when the phytoplankton were much more abundant. The possibility (Hrbáček, Dvořakova, Kořínek, and Procházkóva 1961) that stocking with fish can modify the phytoplankton may imply differences in grazing by associations of small species when many fish are present and by large species when the fish fauna is small. Spodniewska (1962) however, found no constant effects of this kind.

COMPETITIVE RELATIONSHIPS DETERMINED BY ENVIRONMENTAL VARIABLES

Though competition between species is continually tending toward exclusive occupancy by the most efficient species present at a particular time (cf. page 355), it is reasonable to suppose that in the plankton community the environmental factors controlled by the march of seasons, temperature, illumination, nutrient content, and the various secondary and interdependent factors already discussed are continually altering the values of the competition coefficients. That such control occurs in nature has been clear since the work of Beauchamp and Ullyott (1932) on the factors determining the incidence of certain species of flatworms in streams. Experimental demonstration of the same principle was given by Gause, Nastukova, and Alpatov (1934), working on the environmental control of competition between *Paramecium caudatum* and *P. aurelia*. A population in a variable environment may now be moving toward a condition in which one species is the dominant and sole heir to the biotope and now toward a condition in which a quite different species potentially will inherit the trophogenic waters of the lake. What, in fact, we term the seasonal succession of the phytoplankton is therefore really a series of false starts in ever-changing directions toward a momentarily defined unispecific specificity which is never achieved because the environment immediately changes and alters the direction of competition pressure.

According to the hypothesis put forward in Chapter 22, that true equilibrium assemblages are largely lacking in the phytoplankton, we should expect rather irregular seasonal changes determined primarily by the effects of the environment on competition between whatever species happened to be present. The simultaneous co-occurrence of several species of a given genus at the same time in one lake and of several species of another genus in another lake may in fact be compared to the practice of holding cattle races in parts of Africa and horse races in parts of Europe and North America. A racing ox would be out of place at Belmont Park or Ascot, for custom decides the type of animal that may compete; yet, once a field of slightly different individual organisms has been lined up, there is a high probability that one individual, running not uncertainly (as St. Paul reminds us), will receive the prize. The condition in a lake, however, is always starting time on a race course in which it is never finally decided whether horses or racing oxen, carrier pigeons or human runners are to provide the sport. The competitors of a given genus or other higher taxon are from time to time lined up, and sometimes the race begins, but, as it might be in the works of Lewis Carroll, the event is always called off before it is completed and something entirely different is arranged in its place.

Apparent competitive relationships in *Staurastrum*. Within a given genus a rather more significant analysis can be made. The planktonic desmids of the genus *Staurastrum* are, in the Lunzer Untersee, as elsewhere, summer species, clearly thermophil at least by the standards of a relatively cool lake in an alpine landscape. At least six species increase together during the summer (Ruttner 1930); it is almost certain, therefore, that competitive equilibrium is never established. It is highly probable that each one of the six species has a slightly different set of ecological requirements; this can be demonstrated for the two commonest taxa, *S. cingulum* and *S. lütkemülleri*. If equilibrium were established, it is reasonable to suppose that, although the conditions obtained at the height of summer are, in general terms, favorable for *Staurastrum,* only one species would survive, provided, of course, that members of the genus persisted in the diminishing community as equilibrium is approached.

In the Lunzer Untersee *Staurastrum cingulum* clearly begins to increase in numbers at a slightly lower temperature than *S. lütkemülleri* (Fig. 121). In the very hot summer of 1911, when the surface temperature of the lake was higher than 15° C. continuously from the beginning of July to the middle of September, *S. lütkemülleri,* which started to increase a little later than *S. cingulum,* nevertheless outstripped that species at the beginning of September; at that time the rate of increase of *S. lütkemülleri* was almost maximal but that of *S. cingulum* was definitely lower than it had been in the preceding months. In the subsequent year the summer was cool, particularly during August. Temperatures higher than 15° C. did not occur continuously in that month and were quite unknown in September. *S. cingulum* rose in August almost to the level it had achieved in the preceding year, though the increase was rather irregular; *S.*

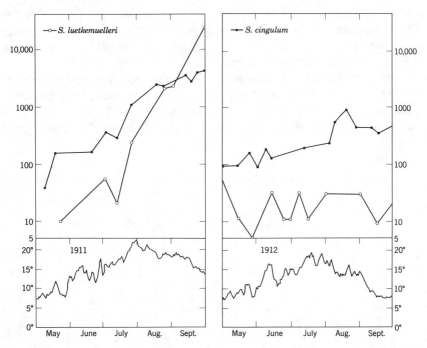

FIGURE 121. Seasonal incidence of *Staurastrum luetkemuelleri* and *S. cingulum* in the Lunzer Untersee in the very hot year 1911, in which the growth rate of the former species overtook that of the latter as the temperature rose above 20° C., and in the normal year 1912, in which *cingulum* increased slightly but *luetkemuelleri* maintained a rather constant low population. (Data from Ruttner.)

lütkemülleri remained scarce throughout the entire period. It is reasonable to suppose that if the temperature could remain permanently above 15° C. *S. lütkemülleri* would displace *S. cingulum,* whereas, if the temperature of the lake were maintained permanently between 12 and 15° C., *S. cingulum* would displace *S. lütkemülleri.* In nature, however, before the competition has proceeded very far, the temperature falls in the lake and doubtless other conditions alter; both species decline and remain of little importance throughout the winter. Though slight continuous variations in the populations show that factors other than temperature, and direction of competition controlled by temperature, certainly are operating, the evidence from other years definitely supports the interpretation just presented. *S. lütkemülleri* was an important constituent of the plankton not only in 1911 but also in 1909, in which year the species increased at the end of August and in

September surpassed *S. cingulum.* It is therefore particularly interesting to note that in the four-year period 1909 to 1912, 1909 and
1911 were the two years in which surface temperatures higher than
15° C. were recorded in September.

Nitrogen fixation and competition between *Anabaena* **and** *Fragilaria.*
Another case, involving what is probably the initial stage of competition
between two members of the phytoplankton community of widely different taxonomic position, is provided (Fig. 122) by the behavior of

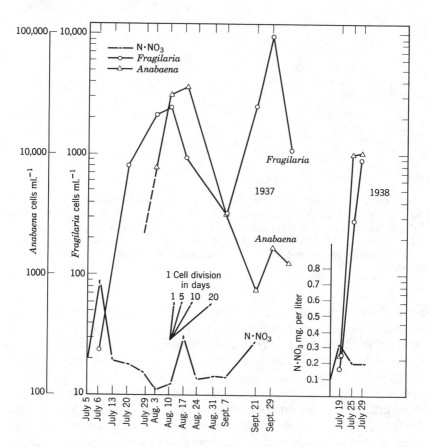

FIGURE 122. Rise and fall of populations of *Anabaena circinalis* and of
Fragilaria crotonensis in Linsley Pond in the summers of 1937 and 1938. In
the first year *Fragilaria* rose before *Anabaena* at a time when the nitrate
nitrogen was falling from slightly over 0.7 mg. per liter; in 1938 with an initially
much lower nitrate concentration the presumed nitrogen fixer *Anabaena* started
to rise before *Fragilaria.*

Anabaena circinalis and *Fragilaria crotonensis* in Linsley Pond in 1937 and 1938 (Hutchinson 1944). Unfortunately, in 1938 it was impossible to prevent the addition of copper sulfate to the lake early in August, so that the full potential history of the populations remained unrealized. Both species evidently can develop at low nutrient concentrations. *Anabaena circinalis* is almost certainly capable of fixing molecular nitrogen, whereas *Fragilaria crotonensis* requires an external source of inorganic combined nitrogen. At the height of summer, when the surface waters reach their maximum temperature and the soluble phosphate content is of the order of 1 mg. m.$^{-3}$, both species, which seem to have similar requirements, can begin to develop huge populations. In 1937 *Fragilaria* started to increase before *Anabaena* and no longer exhibited its maximum growth rate when the *Anabaena* was increasing at its greatest rate. In 1938 the opposite occurred, *Anabaena* having reached a maximum before *Fragilaria* stopped increasing. Among the physicochemical factors that might be responsible for this difference the nitrate content is likely to be the most interesting in view of the probable differences in nitrogen requirements of the two species. It is perhaps not surprising, though highly satisfactory, to find that the nitrate nitrogen content at the beginning of the process of increase was much higher when *Fragilaria* dominated than when *Anabaena* dominated in the succeeding year. Later in 1937, as *Fragilaria* reduced the nitrate, *Anabaena* increased to a maximum. It seems, however, that as *Anabaena* floats to the surface, drifts to the margins, dies, and decomposes more nitrate is liberated into the lake to produce a second *Fragilaria* maximum.

Neither of these examples really demonstrates competition, but both suggest that when antibiosis is not involved the competitive phenomena in the phytoplankton resemble the early stages in one of the artificial systems used by Gause and his associates. The behavior of the mixed *Staurastrum* populations is distinctly reminiscent of that of such experiments, that of *Anabaena* and *Fragilaria* rather less so, because *Fragilaria* reduces the nitrate and alters the environment, whereas *Anabaena* when dying probably increases the nitrate and again alters the environment. In *Staurastrum* competition, insofar as it occurs, is under external control; relatively simple and predictable changes may be expected. The *Fragilaria-Anabaena* system alters the environment itself and does so reversibly and rapidly in comparison with the time taken for the development and decline of populations. Such a system is therefore likely to provide more complex and less predictable changes. In any case, relatively gentle competition, never proceeding, at least in temperate regions, to a unispecific equilibrium, is presumably responsible for a good deal of the pattern actually observed in nature and also, no doubt, for a great deal of the difficulty in analyzing that pattern.

Species versus higher taxa in seasonal succession. One interesting aspect of succession involving competition remains to be noted. In the study of the ecology of terrestrial animals much discussion has been devoted to the rules that regulate the coexistence or mutual exclusion of closely or remotely related species. Although far too little attention has been given to the matter, it would seem likely that closely allied species tend to be mutually exclusive but that species of somewhat larger groups, perhaps of the size of the genera of botanists or of the more conservative zoologists,[5] tend to be adapted to particular ranges of ecological variables and may therefore occur together with a frequency greater than chance expectation (Williams 1947). Though it is not yet possible to make a precise analysis of this matter in the case of the phytoplankton, a cursory qualitative examination of the problem is not without interest.

Loose generalizations have frequently been made that diatoms dominate in the cooler seasons, whereas green and blue-green algae, with certain notable exceptions, tend to be summer plants. The general co-occurrence of the members of major groups is often sufficiently striking to enable investigators to lump together their species as diatoms, blue-green algae, green algae, autotrophic flagellates (often classified with the holozoic rhizopods and ciliates, merely as "protozoa") and still have something to show for their labors. For reasons that no doubt are more profound than obvious, though some may be deduced from the special physiological requirements of particular classes, it is evident that a certain degree of taxonomic relationship implies a significant degree of probability that related species will have ecological requirements within a delimited range. There are exceptions, of course, such as the cold-water species of *Oscillatoria* among the usually thermophil Myxophyceae, which provide a rare exception to prove the significance though not the validity of a rule, for these exceptions show that cold-stenotherm Myxophyceae can exist and so emphasize the problem that the blue-green algae are in general less successful as frigophil organisms than, for example, are diatoms. The occurrence of competition, or synergism, between members of groups of related species which have comparable ecological requirements and tend to form associations of the kind described in Chapter 22 requires study. A priori we might expect, from the working of competitive exclusion, successional phenomena in which these plankton types were virtually

---

[5] Zoologists more than botanists appear to be prone to split genera.

absent and in which each association consisted of species of widely different groups of an ecology as diverse as possible. Instead of a diatom plankton in spring and a blue-green plankton in late summer, with several subdominants of the same class as the dominant, we might expect a spring plankton of one species of blue-green, one species of diatom, one of *Staurastrum,* and so on, replaced by a summer plankton of different but equivalent and related species. The natural situation is certainly intermediate between these extremes, but the balance between like taxonomy implying like ecological requirements on the one hand and competitive exclusion of allied species on the other clearly provides a genuine problem that requires study.

SUMMARY

In large, deep, relatively unproductive lakes in temperate regions the total quantity of phytoplankton usually rises in the spring and early summer to a maximum in July and then decreases. In some lakes and in some years the summer maximum may be prolonged; with a lesser or greater indication of diacmy. The subalpine lakes of Austria illustrate these conditions. Lake Constance also exhibited a winter minimum and two hardly separable maxima in July and October in the years in which it was most intensely studied. In these lakes the major components of the phytoplankton are diatoms of the genus *Cyclotella.* In very soft-water lakes desmids perhaps behave similarly. An extension of the process that generates the diacmy would tend to produce late spring and autumn maxima with a summer minimum. This condition, well-known in the sea, is apparently developed in the Great Laurentian lakes and perhaps in the larger, more productive lakes of the English Lake District. Vernal and autumnal maxima in diatom populations are known in various other lakes, but in the few cases in which the nannoplankton has been adequately studied the spring and autumn maxima in diatoms retained in plankton nets are found to be only two events of a much more complicated sequence involving a considerable number of species of diverse systematic position. Comparison of the net and nannoplankton of Lake Mendota in 1916–1917 provides a striking example.

In smaller lakes there is generally a complicated succession of maxima and minima throughout the year; thus in the Schleinsee in Bavaria the total phytoplankton is minimal in March, June, late August, and November. Of the intervening maxima, that in the spring is the least developed. When a myxophycean water bloom is produced, the maximum phytoplankton population is usually in late July, August, or early September; in such lakes the vernal maximum is often inconsiderable relative to that in summer, but there may be striking subsidiary autumnal maxima.

The factors proposed to account for the seasonal variation of the phytoplankton as a whole and of its particular component species may be grouped roughly as (1) *partly independent physical factors*, that is, temperature, illumination, and turbulence; (2) *interdependent chemical factors*, for example, inorganic nutrients, accessory organic compounds (vitamins), and antibiotics; and (3) *biological factors*, or parasitism, predation, and competition. This classification is not logically impeccable; antibiotics may be the mechanism of competitive coactions and competition must always be for some requirement, usually one of the chemical factors or light; it is, however, a convenient way of grouping the various factors postulated to explain seasonal succession.

Light, temperature, and the concentration of inorganic nutrients may have suboptimal, optimal, and supraoptimal values. Under suboptimal conditions a mere retardation of division is probable; under supraoptimal conditions a definite injury is likely. A fall in temperature or illumination thus may leave a large population of a species characteristic of high temperature or illumination in possession of the water for a long time, whereas a rise in temperature or illumination may destroy a species adapted to cool and less intensely illuminated conditions. There is evidence, moreover, that an increase in division rate may continue, even though irreversible injury is slowly occurring, so that in experiments a rapid increase in a large population followed by a decline may occur at a high temperature, whereas a much slower rise which will produce a permanent population can occur at a low temperature. The highest temperature producing the most rapid increase not followed by injury and decline may be called the *limiting optimum*. The same term may be applied to illumination. It is known that *Melosira islandica helvetica* has a limiting optimum temperature of about 5° C., though it can make rapid initial growth at 20°, whereas many green algae have limiting optima at 20° C. or higher. In Lake Erken in Sweden *M. islandica helvetica* is a predominantly winter form, green algae occur in summer, and certain diatoms with intermediate limiting optima have intermediate spring and autumn maxima. A good many examples of supposed temperature optima in nature have been deduced from field observations. Such optima usually refer to maximal populations rather than to the more critical maximum division rates. They suggest, though not unequivocally, that a few species are definitely cold-water forms (e.g., *Oscillatoria rubescens*); others (*Anabaena flos-aquae* and other Myxophyceae,

Ceratium hirundinella usually, and *Staurastrum* spp.) are apparently warm-water species. It is probable, however, in some planktonic algae, notably *Asterionella formosa* and possibly *Fragilaria crotonensis,* that thermal ecotypes are adapted to different temperature ranges. In these species no regular behavior from lake to lake can be expected.

A thermally determined cycle would probably imply identical mesothermal species in spring and autumn, except insofar as temperatures rising to and above an optimum have a different effect from temperatures falling to and below an optimum. Some suggestion of maxima in both spring and autumn is in fact provided by certain diatoms, notably *Fragilaria crotonensis* in Denmark, but in general only a minority of the supposed mesothermal species behave in this way. Temperature is certainly important in controlling both the total quantity and the specific composition of the planktonic flora, but it cannot be the only important variable.

There can be no doubt that rising light intensity, hardly involving any initial change in temperature, is adequate to produce the spring maximum in the phytoplankton of any lake in the temperate regions in which the dissolved nutrients are adequate. This must be the ordinary mechanism underlying such maxima wherever they occur. In *Asterionella formosa* in Windermere there is experimental evidence that although temperature is important in controlling division rate when a whole annual cycle is considered during the rise of the population in the late winter and spring (February to April) increasing illumination is far more significant.

Since the thermal maxima and minima follow those in illumination by a month or two, it is reasonable to define

> Winter (January–February) as cold with relatively low light
> Spring (April–May) as cold with relatively high light
> Summer (July–August) as warm with relatively high light
> Autumn (October–November) as warm with relatively low light

A true winter species might be expected to persist in the feebly illuminated cool water of the upper part of the hypolimnion of a stratified lake. *Oscillatoria rubescens* does this, and therefore appears as a cold-stenotherm shade plant. In striking contrast to such an alga the summer Myxophyceae which develop blooms are probably warm-water polyphotic species. This very appealing classification runs into difficulties and though it is possible that good examples of all of the categories will be found it is probably too simplified to provide a complete explanation of what happens in nature.

The theory of the role of turbulence has already been explained in Chapter 20. In general, turbulence will decrease during the summer months, for even the surface layers will acquire some stability by day. A species with a high sinking speed but capable of forming an equilibrium population in the turbulent waters during the early stages of stratification might well be unable to persist as the turbulence declined. This phenomenon, which probably underlies some seasonal succession, needs thorough study in lakes. So far, only the special case of *Melosira* appears to involve this kind of seasonal cycle.

The role of inorganic nutrients is well established in some cases. In certain of the English lakes the rise in the population of *Asterionella* in the early spring is certainly determined by increasing light at relatively high nutrient levels; the end of the increase is set by the exhaustion of silicate when the concentration falls to about 500 mg. SiO_2 m.$^{-3}$ It is uncertain to what extent actual replacement of one species by another can be caused by falling nutrients, except in the case of *Dinobryon divergens* and perhaps *Uroglena americana,* which usually increase at times of decrease of other species as nutrients become limiting. *Dinobryon* not only can utilize very low concentrations of phosphate but appears actually to be inhibited by concentrations optimal for other species.

A majority of algae appear to produce substances that inhibit the growth of other algae. It is probable, however, that such substances are significant only in regulating the specific composition of the phytoplankton when they are being produced from very dense populations. It is not unlikely that some of these substances are liberated only by dead or dying cells. They appear, at least in part, to be fatty acids or their soaps. They have different effects on different, even quite closely allied organisms and are present in the water at the time of heavy blooms of blue-green algae, *Ceratium,* etc. They almost certainly limit the diversity of the phytoplankton in small productive lakes, but at present their general role in seasonal succession is uncertain.

The role of accessory nutrients or vitamins in lake water is little explored, though the known requirements of many algae and the small but definite evidence of seasonal variation in vitamin concentration suggest that such substances may prove important. Indirect evidence suggests that some winter plankton may be determined by vitamin concentration.

The removal of part of the phytoplankton, possibly a specifically well-defined part, by the feeding of zooplankton is an obvious possible cause of phytoplankton pulses, but it seems that in most cases both the plant or producer and animal or consumer parts of the community are so diversified and complicated that it is hard to see what is happening. In some saline or alkaline lakes with an abundant but specifically poorly diversified biota clear evidence of the effect of grazing by Cladocera or Copepoda can be made out. It is quite likely that the low concentration of phytoplankton often observed in the early summer may be produced in part in this way.

Parasitism, particularly by chytrid fungi, clearly can reduce the population of a dominant; it is then replaced by another species which sometimes suffers an epidemic itself. The available evidence suggests that this is most likely to happen in eutrophic lakes.

All types of environmental control are likely to affect competition, so that what often looks like the direct influence of an external variable on a species turns out to operate only in the observed way if a competitor species is present. This is likely to be a main cause of the confusing and unsystematic aspects of seasonal succession in the lake plankton. At present, however, there are almost no well-documented cases. It seems likely that temperature determines the outcome of competition among *Staurastrum* in the Lunzer Untersee. It is possible also that the direction and result of competition between *Fragilaria crotonensis,* which does not fix molecular nitrogen, and *Anabaena circinalis,* which almost certainly does, is determined by competition going in favor of *Anabaena* when the nitrogen content of the water is initially low but in favor of *Fragilaria* when it is initially high.

CHAPTER 24

The Nature and Biology

of the Zooplankton

T he important groups represented in the zooplankton of lakes belong to the free-living nonphotosynthetic Protista, to the Rotifera, and to the Crustacea. In addition to members of these groups, which form by far the greater proportion of both species and individuals, a few coelenterates, flatworms, gastrotrichs, mites, and larval insects at times may be found in the plankton. The very restricted number of planktonic larvae of soft-bodied invertebrates in fresh waters contrasts conspicuously with the situation in the sea, as has already been emphasized. The few fresh-water larvae, mainly of parasitic trematode flatworms, and one or two cases of limnoplanktonic fish eggs will be discussed after the animals, planktonic as adults, have been considered.

The various members of the zooplankton have, in spite of some convergent similarities, different types of life history. Since the differences are likely to involve different relationships in the timing of life cycles to the progression of the seasons, it has seemed convenient to treat the seasonal cycles of the various zooplanktonic groups separately in this chapter. The cognate problem of the trophic relationships of the zooplankton to the phytoplankton is discussed briefly, after the systematic presentation, in a largely qualitative way; the more quantitative aspects of the matter will be discussed in a chapter on trophic dynamics after the benthos and nekton have been considered in Volume III. The interesting questions of vertical migration and horizontal distribution and of seasonal form change or cyclomorphosis have been given separate chapters to themselves.

PLANKTONIC ANIMAL PROTISTA

The various groups of animal protists may occur in the plankton of lakes, though it is probable that the free-living zooflagellates are of little importance in the association, and none of the other groups has the significance that ordinarily may be accorded to the metazoan zooplankters.

Lobosa. Most of the records of naked forms in the plankton are unsatisfactory and some no doubt refer to amoeboid phases of otherwise flagellated and not necessarily holozoic forms such as *Chrysamoeba radians.* There is a possibility, however, that a group of small amoebae, with long pseudopodia, characteristically living suspended in the water, may be regularly meroplanktonic and as such make a small but definite contribution to the fresh-water plankton of shallow lakes. *Astramoeba radiosa,* a widely distributed species once found living limnetically (Francé 1897) in Kis Balaton, a small lake at the western end of Lake Balaton in Hungary, and *Amoeba* (sens. lat.) *gorgonia* found by Kutkuhn (1958) in North Twin Lake, Calhoun County, Iowa, are organisms of this kind.

Of far greater importance are several species of the thecate genus *Difflugia,* first recorded as planktonic by Asper and Heuscher (1887), whose *D. urceolata helvetica,* however, is not certainly recognizable, though it is sometimes supposed (Gauthier-Lièvre and Thomas 1958) to be identical with the later species *D. hydrostatica.*

The only adequately investigated species is *D. limnetica* (Fig. 123), studied by Schönborn (1962) in several lakes, both oligotrophic and eutrophic, near Rheinsberg. *D. limnetica* appeared in the plankton of these lakes in June and increased to a maximum in September. The subsequent decline in abundance is due to the sinking of the population, partly as the result of an increase in density as fat globules are metabolized. The majority of the sinking specimens apparently die, but some form cysts within their tests and others remain active in the littoral benthos throughout the winter. Excystment occurs in March, and during the spring the cells multiply as benthic organisms, covering their tests with quartz grains. These individuals lack a collar around the three-lobed pseudostome. In June the benthic specimens become planktonic, their density being reduced not only by the accumulation of fat droplets but also by the production of gas bubbles that lie between the cytoplasm and the test (see pages 253–255). The individuals formed by division in the plankton cover their tests with *Cyclotella* frustules and develop a collar. Late in the season such collared individuals use quartz grains, apparently carried into the surface layers as the epilimnion cools and is mixed into the deeper water. It is such collared but quartz-covered individuals that sink and within some of which cysts are formed. Since *D. limnetica* has often been recorded from quite minute bodies of water, Schönborn concludes that the cycle that he describes is not obligatory.

It is not unlikely that the earlier described but still inadequately known, though probably distinct, *D. hydrostatica* (*A* of Fig. 124) originally (Zacharias, 1897) from the Grosser Plöner See and differing in the six to eight lobes of the pseudostome, has a similar life history, *D. hydrostatica* f. *lithophila* being the benthic form. Both species are recorded from Denmark (Wesenberg-Lund 1904). Outside Europe *D. hydrostatica* is recorded from Lake Tjigombong, west Java, with a maximum population of 420 specimens per liter at 3 m. (Ruttner, 1952a). The planktonic populations from South Africa which have been referred to *D. schuurmanae* may well be conspecific with *limnetica*. *D. schuurmanae* is a perennial and sometimes an important element (Schuurman 1932, sub. *D. oviformis*) in the plankton of Florida Lake,

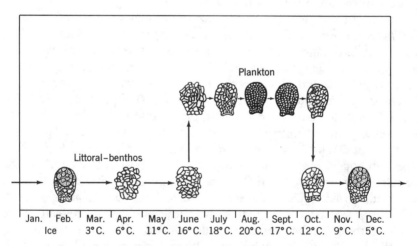

FIGURE 123. Seasonal cycle of *Difflugia limnetica;* encysted in littoral benthos in winter, active in littoral benthos in spring, ascending into free water, in which a trilobate collar is developed and tests of *Cyclotella* (August and September) are used instead of quartz grains as a covering, descending in October, and encysting late in November. (Modified from Schönborn.)

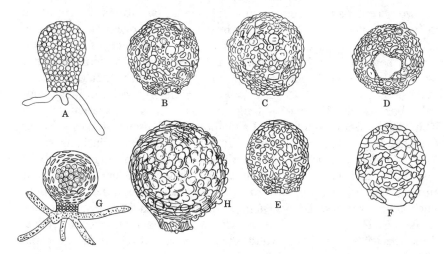

FIGURE 124. Planktonic species of *Difflugia*. *A. D. hydrostatica* (× 275) Grösser
Plöner See; *B, C, D. D. Limnetica* (× 158) L. Bocksjön, Sweden; *E. D. pelagiea*
(× 158) from the same; *F.* from L. Leken. *G. D. cyclotellina* (× 225), Lago
Maggiore. *H.* The same (× c. 400) to show asymmetrical collar. (Zacharias,
Pejler, Garbini, Grospietsch.)

a shallow artificial lake near Johannesburg. It is commonest in the
southern spring during September and October at temperatures of 15
to 19° C. The Florida Lake population[1] has grayish white tests, apparently with minute sand grains, and a three-lobed pseudostome. The
absence of *Cyclotella* frustules in the lake make such objects unavailable
as a covering.

---

[1] Specimens from Florida Lake, Transvaal, collected by Miss Schuurman, have
been re-examined and are deposited in the Peabody Museum, Yale University.
The specific name was introduced as *schuurmani* by van Oye (1931), first appearing as a *nomen nudum* and referring to material from Weltevreden West
Pan, Lake Chrissie district, Transvaal, from Leeuwkraalkuil, Heidelberg district,
Transvaal, and from Matsume Lake, S. Chopiland, Portuguese E. Africa. In the
later description of the species (van Oye 1932) the Florida Lake locality however, is mentioned first. There is thus some doubt as to the exact type
locality and its nature; none of the specimens recorded in the first paper came
from collections as genuinely planktonic as those from Florida Lake. Later
nonplanktonic specimens from central Africa are recorded by Gauthier-Lièvre
and Thomas (1958), who state, however, that the test is yellowish-brown, thus
negating the only known character permitting separation from *limnetica*. Whatever the nature of *D. schuurmanae,* spelling of the name with a feminine
genitive ending is mandatory under Article 31 of the International Code of
Zoological nomenclature.

Two recent papers seem to indicate one or two forms closely allied to but separated from *D. limnetica* in the plankton of lakes. Štěpánek and Jiří (1958) have given details of the occurrence of two forms which they refer to *D. gramen* and *D. gramen* f. *achlora* in ponds or small lakes in Czechoslovakia. *D. gramen* has zoochlorellae, a hyaline test, and is 71 to 92 μ long; *D. gramen* f. *achlora* lacks zoochlorellae, is but 53 to 65 μ long, and has a less transparent grayish white test. Both populations have three-lobed pseudostomes. Pejler (1962a) has recorded two forms from Sweden. One which he refers to *D. limnetica* is larger, has a three-lobed pseudostome (*B, C, D,* Fig. 125) and lacks zoochlorellae; the cement substance of its test varies from yellowish-brown to a more frequent colorless condition. The second species, *D. pelagica,* ordinarily has a four-lobed pseudostome (*E, F,* Fig. 125), is somewhat smaller, and has zoochlorellae. Both species are probably valid since they can co-occur in the same plankton collection.

In North America what may well be *D. limnetica* has been recorded, on the basis of C. H. Edmondson's (1918) key, as *D. lobostoma,* to which *D. limnetica* would probably run, from North Twin Lake, Calhoun County, Iowa, by Kutkuhn (1958) in his careful study of that shallow lake, largely based on more or less marginal samples.

Although *D. limnetica, D. schuurmanae,* the supposed *D. lobostoma* from North American lakes, and perhaps *D. gramen* f. *achlora* may well be conspecific, the *Cyclotella*-covered *D. cyclotellina* (*G, H,* Fig. 125), known so far (Garbini 1898; Grospietsch 1957) only from Lago Maggiore, appears to differ strikingly in its more globose shape and unlobed, but at times asymmetrical, pseudostome.

Kofoid (1896) recorded not only *D. lobostoma* from West Twin Lake, Michigan, but also, as *D. globulosa,* presumably on the basis of Leidy's (1879) figures, what is probably a quite different species from the upper 10 m. of Lake Michigan and from smaller lakes near Chalevoix. According to Cash and Hopkinson (1909), Kofoid's conception of *D. globulosa* corresponds to theirs of *D. globulus.* The American eulimnoplanktonic species of *Difflugia* evidently need even more study than the European. Another species that is clearly eulimnoplanktonic (Kikuchi 1930a) is *D. biwae* from Lake Biwa, Japan, but the original account of this organism has not been traced.

Several authors, notably Zacharias (1899a) and Kutkuhn (1958), have reported *D. corona, D. pyriformis,* and odd specimens of other species in the plankton of shallow waters.

The peculiar horned *Nebela bipes* (Fig. 125) is recorded from the plankton of Loch Ness, Loch Shiel, Inverness, and Sraheens Lough,

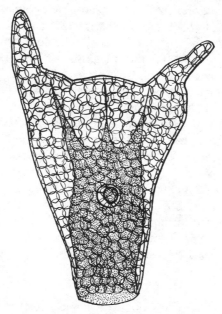

FIGURE 125. *Nebela bipes* (× 225), a facultative planktonic species.

Achill Island (Wailes and Penard 1911), as well as in a few small bodies of water in which it is evidently not planktonic. It is probable, as Deflandre (1936) suggests, that *N. triangulata* is the benthic form of the species. The status of *N. kizakiensis,* the original description of which has not been traced, reported (Kikuchi 1930a) as present in great numbers in the epilimnia of Lakes Kizaki and Aoki in Japan, requires further clarification.

Odd specimens of *Arcella* are often encountered in plankton samples. Davis (1962) has noted at times as many as 280 specimens per liter of an undetermined species of the genus in the plankton of the Bass Island region of Lake Erie. The production of a gas vacuole (pages 253–255) in *Arcella* obviously would facilitate a meroplanktonic habit. Zacharias (1899a) noted *A. vulgaris, A. discoides,* and *A. dentata* as found on occasion in the plankton of shallow ponds. *Cochliopodium bilimbosum* has also been recorded as a common spring form in the plankton of North Twin Lake, Calhoun County, Iowa, by Kutkuhn (1958). Most of these records and others of various testaceous Lobosa probably represent fortuitous pseudoplanktonic occurrences.

Filosa and Granuloreticulosa.   At most meroplanktonic, there are old records (Apstein 1896; cf. also Zacharias 1899a) of *Diplophrys archeri,* an anomalous member of the Granuloreticulosa, and of *Cyphoderia ampulla* of the Filosa, in the plankton of the Dobersdorfer See; Kutkuhn (1958) has recorded two species of *Pseudodifflugia,* one being *P. gracilis,* a genus also belonging to the Filosa, in the plankton of North Twin Lake, Calhoun County, Iowa.

Heliozoa.   Though the unstalked species look beautifully adapted to pelagic life, the group is usually not important in the plankton and reaches its greatest development in small pools.   *Actinophrys sol,* however, was reported miles from shore in the plankton of Lake Michigan by Kofoid (1896) and was also found by Apstein (1896) in the Dobersdorfer See.   Both this species and *Actinosphaerium eichorni* are recorded in the plankton of North Twin Lake by Kutkuhn, but *A. eichorni* seems definitely less planktonic elsewhere.   The only heliozoan that appears to be eulimnoplanktonic is *Raphidiocystis lemani*

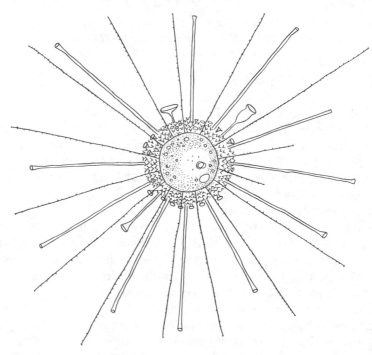

FIGURE 126.   *Rhaphidiocystis lemani* ($\times$ 750), a eulimnoplanktonic helizoan.   (Penard.)

(Fig. 126), recorded from the Lake of Geneva, Lago Maggiore, Lago di Garda, the Grosser Plöner See, and perhaps from other localities under various names (Penard 1904). Other allied species seem to occur casually, as in North Twin Lake. Recently Mason, Goldman, and Hobbie (1963) found unspecified heliozoa to be the main animal plankton in the fantastic meromictic Lake Vanda, South Victoria Land, Antarctica.

The pseudoheliozoan *Dimorpha mutans* has been recorded by Kutkuhn (1958) as fairly common in the early spring plankton of North Twin Lake, Calhoun County, Iowa.

Planktonic Ciliophora. Analogy with the oceanic plankton suggests that the ciliates might also play an important role in lakes. It is possible that for technical reasons they have been largely overlooked, but it is also not unlikely that their place is taken to a considerable extent in the limnoplankton by rotifers, which are much less important in the sea than in fresh waters. Some species of ciliates, however, appear to be truly limnoplanktonic animals of wide distribution and occasionally of some quantitative importance. Many of the rarer species, moreover, are biologically interesting and often very beautiful.

Eulimnoplanktonic ciliates. Among the Holotricha, the euplanktonic *Amphileptus trachelioides* feeds on algae and rotifers (Huber and Nipkov 1927). In Europe it is usually a spring form; Awerintzew (1908) finds that it encysts above 22° C., and according to Gajewskaja (1933) in Lake Baikal it is one of the extreme cold stenotherm species. Nauwerck (1963) finds *Trachelius ovum* (*C* of Fig. 127) to be a significant member of the plankton of Lake Erken in Sweden in the late winter and spring. *Askenasia elegans* (cf. *A* of Fig. 127) appears in Erken as a summer form; it shows a striking maximum in June at a time of minimal algal population and maximum detritus and two lower maxima in August and September, the intervening minimum coinciding with a maximum of *Ceratium hirundinella;* the September maximum corresponds to a second detritus maximum.

Paradileptus elephantinus (*B* of Fig. 127), which in Europe is mainly heleoplanktonic (Kahl 1935), occurs in the upper oxygenated layers of several lakes in Java (Ruttner 1952a), though not abundantly. This occurrence may be compared with several other cases (see pages 341, 506, 695) of tropical limnoplanktonic organisms with less planktonic temperate representatives.

Teuthophrys trisulca (*D* of Fig. 127), a most curious but widely distributed ciliate, first obtained from Lake Schiessrothried in the Vosges Mountains, has three proboscis-like extensions of the cell body

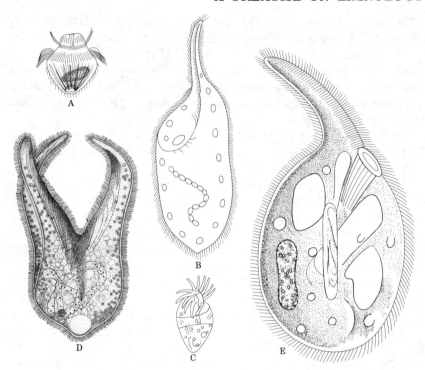

FIGURE 127. Eulimnoplanktonic ciliates. *A. Askenasia faurei*, recoraea in plankton in France and perhaps the species determined as *A. elegans* in the summer plankton of Lake Erken. *B. Paradileptus elephantinus*, heleoplanktonic in Europe but a eulimnoplankter in Java. *C. Trachelius ovum*, a cold-water species in Lake Erken. *D. Teuthophrys trisulca*, heleoplanktonic and sometimes limnoplanktonic, feeding on rotifers. *E. Strombidium viride*, recorded from alpine lakes of Europe (all × 300). (*A, B, C, E* after Kahl, D Wenrich.)

surrounding the cytostome, which give it an almost medusoid appearance (Chatton and de Beauchamp 1923). In most of its occurrences it is heleoplanktonic; it feeds on rotifers (Wenrich 1929).

One suctorian, *Staurophrya elegans,* is a definite free-living eulimnoplanktonic animal of wide distribution.

In the Spirotricha the heterotrich *Bursaria truncatella* is a late winter and spring form in Erken (Nauwerck 1963), and there are several eulimnoplanktonic members of the Oligotrichida. *Strombidium* has three species in Erken, none exactly determined (Nauwerck 1963): one a winter cold stenotherm is known also in Lappland; the second is an early summer form, which declines from a large maximum at the beginning of June as *Askenasia* increases; the third species occurs in late summer, and, as with *Askenasia,* has a secondary minimum at the time of the *Ceratium* maximum. These observations suggest interesting competitive relationships. *Strombidium viride* (*E* of Fig. 127), which may be one of the Erken species, occurs according to Ruttner (1952a) in the alpine lakes of Europe and has a closely allied species in Tségombong, Java. *Halteria grandinella,* in Erken, is an autumn and winter species of some importance. The genus *Strobilidium* also occurs as a planktonic organism in Europe and Indonesia.

There are also fresh-water members of the predominantly marine order Tintinnida. The most important lacustrine members of the group are referrable to *Tintinnopsis* and *Codonella*[2] in the Codanellidae and to *Titinnidium* (*A* of Fig. 128) in the Tintinnidiidae, the first-named family being largely fresh-water, the second, mainly marine. The marine *Coxliella* apparently occurs in Lake Baikal (Gajewskaja 1933). These organisms are case-bearing animals, often with foreign bodies attached to the cases. There is much variation in the group, perhaps partly determined by environmental factors (*B, C, D,* Fig. 128), and the limits of the various taxa appear somewhat uncertain. Davis (1954) finds *Tintinnidium* sp. in Lake Erie to be a spring form, and *Codonella cratera,* predominantly a summer species. Nauwerck (1963) likewise found *Tintinnidium fluviatile* in Erken to have a spring maximum, whereas *Codonella cratera* (sub. *Tintinnopis lacustris*) occurred from June to midwinter.

In addition to the more characteristic ciliates mentioned, a number of eurytopic forms such as *Didinium nasutum, Euplotes harpa,* and several species of *Stentor* may occur in the plankton, even in that of the open water of Lake Baikal (Gajewskaja 1933).

Heleoplanktonic ciliates. Species of a number of other holotrich genera are regarded by Fauré-Fremiet (1924) as planktonic but in his study of these organisms he gives neither ecological details nor in some cases specific localities. Among the lower gymnostome Holotricha such species are recorded in *Acropisthium, Actinobolus, Holophrya, Monodinium, Phascolodon, Prorodon,* and *Spathidium.* Nearly all the species considered by Fauré-Fremiet are doubtless heleoplanktonic, living in the open water of small ponds. In the holotrich order Hymenostomida the same may well be true of species of *Dichilum, Lembadion, Tetrahymena* (sub. *Leucophrys*), and *Urocentrum.*

---

[2] *Codonella cratera* is often called *Tintinnopsis lacustris* by European investigators. Both genera are maintained by Corliss (1961), but it is possible that only one species is really involved.

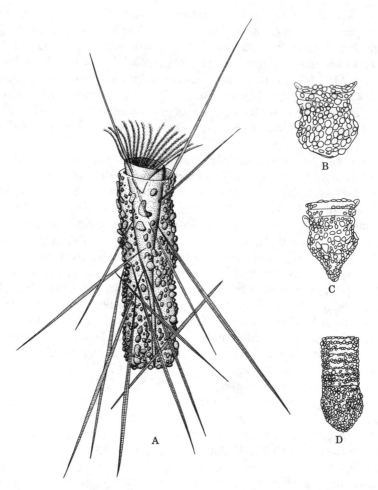

FIGURE 128. Fresh-water Tintinnida. *A. Tintinnidium fluviatile* f. *cylindrica* (× c. 500), Lake Baikal, with *Synedra* affixed to the shell. *B. Tintinnopsis lacustris* (× 300), Torneträsk, covered with *Cyclotella*. *C.* L. Sommen, Sweden, with pointed apex. *D.* Funbosjön, long annulated tube, characteristic of certain lakes on the loamy plains of central Sweden.

Some of these supposed planktonic ciliates, such as *Acropisthium mutabile* (sub *Dinophrya lieberkühni*), are regarded by Fauré-Fremiet as adapted to their mode of life by the rapidity of their movements; others, such as *Disematostoma* (sub *Leucophrys*) *tetraedrica*, contain enormous oil droplets and yet others, such as *Holophrya gargamellae*, which Kahl (1930) thinks is the *Prorodon morula* recorded as eulimno-planktonic in Lake Baikal by Gajewskaja (1933), have vacuolar cyto-plasm that is claimed to be hardly denser than the surrounding water.

Planktonic and epiplanktonic Peritrichida. Among the Holotricha Peritrichida are a number of epiplanktonic species in *Epistylis* and *Vorticella* and a few free-living members of these genera, of *Ophry-dium, Zoothamnium,* and *Telotrochidium* and of the peculiar family Astylozoidae, which swim unattached in the open water. It is reason-able to suppose that such free-living forms are derived from sessile ancestors.

Ophrydium naumanni appears to be a characteristic eulimnoplank-tonic species in the oligotrophic lakes of Sweden (Pejler 1962a), the allied *O. versatile* being a eurytopic, littoral, and, in part, sessile species. The most remarkable eulimnoplanktonic peritrich is doubtless *Gonzeela coloniaris,* in which a large number of cells are set in a gelatinous sphere as much as 2 mm. across. This colonial ciliate, referrable to the family Scyphidiidae, is an important member of the plankton of Lake Tumba in the Congo basin (Kufferath 1953; Marlier 1958).

Epistylis fluitans, at least in part of its life history, apparently attaches to the surface film by means of a slight expansion at the base of its stalk; if the animal is submerged, the margins of the expansion embrace an air bubble. A free-swimming stage is also known. It would seem from Fauré-Fremiet's (1924) account that this meropleus-tonic species is characteristic only of small and shallow localities.

Gajewskaja finds little preference in substrate selection in those peri-trich ciliates of Lake Baikal that are attached to Gammaridae and other arthropods. However, there often seems to be striking substrate specificity among the epiplanktonic peritrichs both in this immense lake (Fig. 129) and elsewhere. Stiller (1940) noted several cases of such preferences in the Grosser Plöner See among the epiplanktonic members of *Epistylis* and *Vorticella.* In some of these cases ecological factors other than substrate preference could have been operating. *Epistylis irregularis* was common in September as an epiplankter on an unidentified planktonic cyclopid, and at the same time *E. rotans* occurred as a free-swimming member of the plankton, *E. lacustris* was common on a benthic littoral cyclopid, and *E. plicatilis* was rare on benthic green algae. In the genus *Vorticella, V. dimorpha* was anchored in the jelly of the colonial planktonic rotifer *Conochilus uni-cornis; V. hyalina* without zoochlorellae and *V. chlorellata* with such symbionts occurred in a similar way in planktonic *Gloeotrichia* colonies, whereas *V. anabaenae* appeared in the plankton on a species of *Anabaena.*

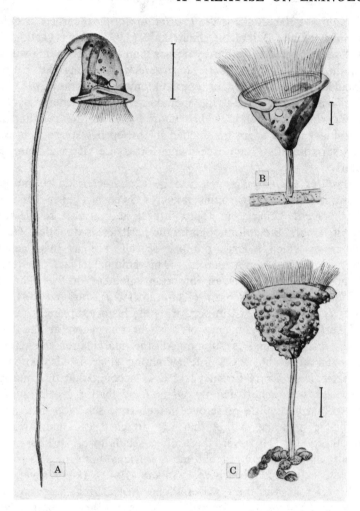

FIGURE 129. Pelagic species of *Vorticella*. *A. V. convallaria*
var. *natans*, free-swimming *B. V. pelagica* on *Melosira* and
occasionally other diatoms in the open water of Lake Baikal.
C. V. monilata var. lockwoodi on *Anabaena* in shallow enclosed
bays of Lake Baikal and recorded from other regions; line
represents natural length × 100. (Gajewskaia.)

The ciliate fauna of Lake Baikal. A special fauna appears to exist in the plankton of Lake Baikal (Fig. 130). The ciliates of this immense and remarkable lake have been the subject of a magnificent monograph by Gajewskaja (1933), some of whose observations have already been noted. In her study, which included marginal waters of all sorts, one hundred and ninety-one species were recorded. The really peculiar species belong mainly to cold-water pelagic fauna which disappears from the surface layers in the summer when the temperature rises above 8° C. The most abundant of these cold stenotherm ciliates are *Liliomorpha viridis, Longitricha flava, Marituja pelagica, Mucophrya pelagica, Spathidiosus bursa,* and *Sulcigera comosa,* all but one members of monotypic and not closely allied holotrich genera, along with

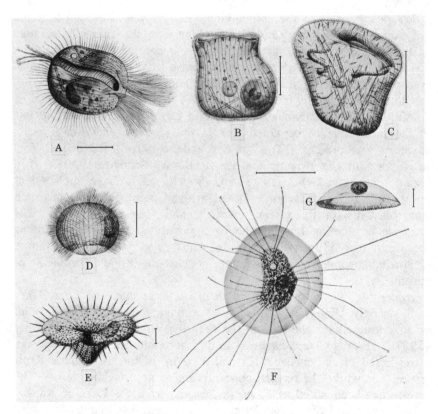

FIGURE 130. Endemic ciliates from the plankton of Lake Baikal. *A. Sulcigera comoso. B. Spathidiosus bursa. C. Marituja pelagica. D. Longitricha flava. E. Liliimorpha viridus. F. Mucophrya pelagica. G.* Cyst of same. Lines indicate natural length × 100. (Gajewskaia.)

Tintinnidium fluviatile f. *cylindrica* and rather less commonly *Ophryo-glena pelagica, Sphaerophrya melosirae, Vorticella pelagica, Amphi-leptus tracheloides,* and a few other widespread species. The summer fauna at the surface lacks strikingly peculiar forms, though most of them probably persist in the upper part of the cold, deep water between 100 and 300 m. This, however, is not true of *Liliomorpha viridis,* which apparently depends entirely on zoochlorellae and does not feed phagotrophically; it is found in the surface layers under ice in late winter and in early spring but its subsequent fate is not recorded. The other species feed mainly on phytoplanktonic algae. Although twenty-four species of ciliates are known as epibionts of *Gammaridae* below 500 m., *Marituja pelagica,* which extends at least to 700 m. is the only free-swimming ciliate to reach such depths. *Mucophrya pelagica,* a medusoid-like suctorian, extends downward to 500 m., *Sulcigera comosa* and the epiplanktonic *Sphaerophrya melosirae* found only on *Melosira* spp., to 400 m. Otherwise, the rich ciliate fauna of the lake, both planktonic and benthic, is confined to the top 300 m. Cysts are formed by *Ophroglena pelagica,* in which the encysted animal is surrounded by a gelatinous sheath, and by *Mucophrya pelagica,* in which an apparently more resistant cap-shaped structure is formed.

Although the peculiar cold stenotherm forms are doubtless endemic to the lake, it is to be noted that in the summer plankton *Bursella spumosa* was present. This species had been recorded once before in a pond in the Botanic Garden at Bonn (Schmidt 1920). If it had not been for Schmidt's record, *Bursella* might have appeared as an endemic Baikalian genus. Other rare and peculiar species recorded from only two or three very remote localities are also known. The presence of floating but, in one case, possibly resistant cysts among the peculiar cold stenotherm fauna must also be borne in mind; Gajewskaja comes to no definite conclusion on the problem of endemicity.

Anaerobic ciliates in the plankton of the hypolimnion. Some lakes with strongly clinograde oxygen curves have a special oligoaerophil or anaerobic ciliate plankton in the deep hypolimnetic water. Ruttner (1937b) records *Coleps hirtus* as an oligoaerophil species, which occurs in restricted layer between 21.5 and 22.0 m. in the meromictic Krot-tensee, just where the oxygen concentration is approaching zero. The species has been noted as *C. hirtus lacustris* from the Lake of Annecy (Fauré-Fremiet 1924), and Ruttner (1952a) found that it flourished in some Javanese lakes at oxygen concentrations above those to which it was restricted in the Krottensee. In the Krottensee, *Loxocephalus* sp. had a distribution like that of *Coleps hirtus,* whereas *Metopus fuscus* and *Caenomorpha medusula* occurred at deeper levels under quite anaerobic conditions. In Java Ruttner (1952) found that *Uronema* cf. *marinum, Prorodon* sp., and *Metopus contortus* flourished in water containing but 0.1 mg. O_2 per liter, as in Lake Pasir, while *Discomorphella pectinata* lived at low oxygen concentrations in the hypolimnion of Ranu Lamongan.

Juday (1919) has recorded an anaerobic ciliate living in the hypolimnion of Lake Mendota, and I have observed a large ciliate, which could not be further studied, in the hypolimnion of Lake Manasbal in Kashmir. A remarkable benthic ciliate fauna living on anaerobic mud rich in H_2S was discovered by Liebmann (1938) in the Bleiloch Dam in Germany, the deep layers of which are contaminated by sulfite wastes from a pulp mill. Symbiotic bacteria apparently occur in these ciliates.

COELENTERATA

The only true fresh-water planktonic forms are the Limnomedusae of the genera *Limnocnida* and *Craspedacusta* (Fig. 9). As has already been indicated (see page 42), there is more than a suggestion that *Craspedacusta* is adapted to ecological conditions in which the benthic hydroid lives in running water, the planktonic medusa, in lakes; an alternation of such conditions takes place in some of the river systems of China from which *Craspedacusta* appears to have been spread by man. In its capacity to bud medusae from the manubrium and so multiply without the inevitable intervention of the sessile stage, *Limnocnida tanganjicae* is evidently the most adapted to a limnoplanktonic existence of all these organisms.

Davis (1955) has reviewed a number of qualitative records of the food of *Craspedacusta sowerbyi* which indicate that it feeds largely on rotifers and planktonic Crustacea, the smaller prey being taken by the younger medusae. In specimens from Crystal Lake, Ravenna, Ohio, he found on an average 6.5 specimens of *Daphnia,* 3.7 specimens of *Diaphanosoma,* and 2.6 colonies of *Conochilus unicornis,* other food being of little importance. Since digestion is complete in an hour or so, this represents less than the daily food intake. The population that he observed, however, probably did not greatly affect the other zooplankton by predation: it is uncertain to what extent the medusae are eaten by fish.

PLANKTONIC TURBELLARIA

The genus *Mesostoma* of the Neorhabdocoelida may at times enter the plankton and occasionally is of some importance. It appears to

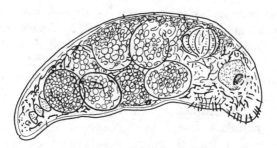

FIGURE 131. *Mesostoma inversum,* an apparently euplanktonic species of turbellarian from
Lake Bulera, with a concave pigmented ventral
surface. (de Beauchamp.)

be more planktonic in tropical than in temperate waters, though Kikuchi
(1930a) has noted an undetermined species in plankton in Japan.

Mesostoma productum, frequently a free-swimming species in
Europe, though only occasionally encountered in lakes, in Indonesia
is a minor but not inconspicuous component of the zooplankton. In
Lake Ranau in Sumatra, in which it achieved its greatest population,
it occurred by day at 2 to 40 m., with a maximum of five individuals
per liter at 5 m.; at night the population tended to rise (Ruttner
1952a).

A more remarkable case is provided by *M. inversum* from Lake
Bulera in Ruanda (de Beauchamp 1954), which occurred in great
abundance in a rich crustacean plankton. The animal is peculiar in
that its morphologically dorsal size is pale and concave, whereas its
ventral side is convex and pigmented (Fig. 131). It presumably takes
its position with the ventral side up when moving through the water.
A new genus may be required for this curious animal.

ROTIFERA

The rotifers are the most important soft-bodied invertebrates in the
fresh-water plankton.

Life history. Except for the small and exclusively bisexual marine
parasitic or epibiotic subclass, the Seisonaceae, the Rotifera are mainly
and characteristically inhabitants of inland waters. As such, they have
developed a type of life history that finds its closest analogue in that
of the Cladocera, another group of characteristically fresh-water animals. The life histories of these two groups, which are clearly adapted
to the same general range of ecological conditions in much the same

way, differ considerably, however in their underlying cytological and developmental mechanisms.

The typical life history of the subclass Monogononta, which includes all the planktonic species, begins with the hatching from a resting *mictic* egg of a parthenogenetic or *amictic* female (*A* of Fig. 132). This female produces a series of *subitaneous*[3] eggs that develop without fertilization and produce further parthenogenetic females. When a considerable population has been built up and certain ecological conditions, the nature of which has given rise to much discussion, have apparently been fulfilled, male-producing or mictic females may appear in the population. These seem to be morphologically indistinguishable from the ordinary amictic females. Ahlstrom (1940) has supposed that in some species the mictic and amictic females do differ in structure. This opinion was based partly on figures of *Brachionus leydigii* given by Hauer (1937), who, however, does not imply that the very small differences are significant in this respect, and partly on observations on *B. angularis* in which the variations are by no means properly understood (see page 906). More recently Parise (1961) found that *Filinia terminalis* in Lake Nemi had longer anterior appendages when carrying mictic than amictic eggs, but he was uncertain whether one or two genetically diverse populations were involved (cf. Hutchinson 1964).

The males, which hatch from the characteristically small unfertilized mictic eggs (*B* of Fig. 132), are always reduced in size and complexity (*C* of Fig. 132) relative to the females and may be very simplified indeed; they fertilize some of the mictic females, which then produce large mictic resting eggs that undergo a prolonged diapause. Such eggs are usually different in appearance (*D* of Fig. 132) from the ordinarily subitaneous eggs produced by parthenogenesis and tend to have thickened and often ornamented external walls. Meanwhile, at least in some cases, low temperature or drought may have produced unfavorable circumstances, which the resting egg permits the species to survive. Hatching of the resting eggs, when conditions become suitable, seems to be brought about by poorly understood chemical changes in the environment (Paresi *fide* Edmondson personal communication). This life cycle is obviously advantageous in small or temporary ponds in which many species of rotifers live. In a few cases the first-generation females hatching from mictic eggs differ from later amictic females. This is most noticeable in species of *Filinia* and *Polyarthra* in which such females lack appendages; *Anarthra* was

---

[3] This appears to be the correct form of a rather uncomfortable word.

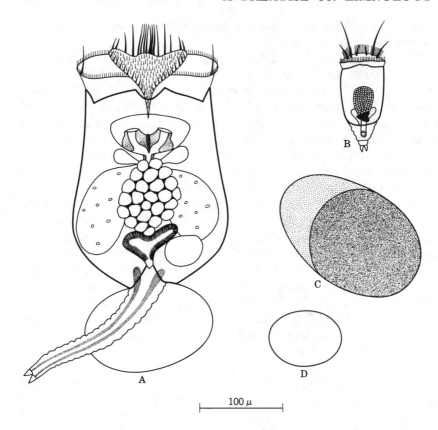

B

C

D

A

|——— 100 μ ———|

FIGURE 132. *Brachionus calyciflorus.* *A.* Amictic female carrying normal parthenogenetic subitaneous egg. *B.* Male. *C.* Resting mictic egg. *D.* Small male producing egg. (Gilbert.)

founded on specimens of *Polyarthra* hatched from resting eggs (Nipkov 1952). The planktonic species, which are far less numerous than the littoral benthic, though their populations are probably vastly greater in many cases, are less dependent on the production of resting eggs and sometimes appear to be *acyclic,* without the sexual phase. When the sexual phase intervenes, usually at the time of the maximum population density, once a year, the species is said to be *monocyclic,* as has been indicated in an earlier chapter (see page 242). When sexual eggs are produced twice a year, the species is *dicyclic.* In some species males and resting eggs are produced several times a year, often at irregular intervals; such species are termed *polycyclic.* It is probable that Wesenberg-Lund (1923, 1930), to whom so much of our knowledge of these matters is due, has somewhat overemphasized the regularity with which particular species exhibit sexual periods.

Eggs are ordinarily, though by no means always, produced one at a time. The large size of the egg relative to the female that produces it (*Frontispiece*) is to be noticed, as it is likely for this reason that maternal influences of a biochemical kind may have a greater effect on development in the rotifers than is usual among other groups of animals. Such an influence is discussed in a later paragraph concerned with aging.

In most planktonic rotifers (*Keratella, Kellicottia, Brachionus, Polyarthra, Filinia, Hexarthra, Pompholyx, Conochilus, Conochiloides,* and a few species of *Synchaeta*) the amictic female carries the subitaneous egg; in most species of *Synchaeta,* and in *Ploesoma* and *Gastropus,* the egg is laid free in the water (cf. Edmondson 1960). *Asplanchna* is viviparous. Some of the benthic littoral species, such as members of the genus *Euchlanis,* attach their eggs to water plants. In the largely heleoplanktonic species of *Trichocerca,* with the exception of *T. cylindrica* which carries its eggs, the eggs are laid on other organisms, *T. porcellus* attaching them to *Melosira* filaments and *T. stylata* (Wesenberg-Lund 1904, sub *Rattulus bicornis*) to the dorsal side of *Brachionus angularis.*

Nature and determination of the males. The rotiferan life history must involve special cytological phenomena. All workers (Shull 1921; Storch 1924; Tauson 1927; Whitney 1929; see also Remane 1929–1933) agree that the males are derived from haploid eggs, whereas the parthenogenetically produced daughters of amictic females are derived from diploid eggs which have given off but one polar body. The resting egg is merely a haploid egg that has been fertilized; it then undergoes special cortical development to form a resistant shell. Tauson concluded that in *Asplanchna intermedia* the male becomes diploid during embryogenesis and that reduction, without a maturational division, takes place during spermatogenesis to give haploid sperm. Whitney, on the contrary, found the somatic cells of male embryos of *A. sieboldii* (sub *amphora*) to be haploid. Normal sperm are produced in such animals without reduction, but in the same animal some immobile and functionless sperm, which may be hemiploid, are produced by a process involving an attempt at reduction. The proportion of the two types of sperm varies widely from individual to individual. At least insofar as the chromosomal constitution of somatic cells of the male is concerned, the published evidence seems equally good for the conclusions of both Tauson and Whitney; it therefore appears that within a single genus the male can be somatically haploid or diploid.

The widespread presence of abortive and presumably hemiploid sperm in the rotifers suggests, however, that at least the germ line of the male is normally haploid. The subject clearly requires renewed study on an extended scale.

According to Tannreuther (1920), there are two kinds of fertilized eggs in *Asplanchna sieboldii* (sub *ebbesborni*). Both produce two polar bodies, but they differ in the thickness of the eggshell; the thin-shelled eggs hatch immediately as if they were subitaneous eggs. The opposite situation has been observed by Ruttner-Kolisko (1946) who found that in *Keratella hiemalis* pseudosexual resting eggs are produced parthenogenetically at the height of the development of the population in the Lunzer Obersee in July.

The production of mictic females, and so males and resting eggs, correlated in at least some cases with the regular seasonal developments of a large population (Wesenberg-Lund 1930), is almost certainly under some sort of direct environmental control. Different workers, studying a variety of easily cultured species, have come to different conclusions regarding the nature of the controlling environmental factors. The stimulus has been variously identified as starvation (Nussbaum 1897), optimal feeding (Whitney 1916b, 1917, 1919), optimal feeding followed by brief starvation (Mitchell 1913a,b), changes in the nature of the food (Whitney 1914a,b, 1916a,b; Hodgkinson 1918; Luntz 1926; Watzka 1928), changes in culture medium, specifically with respect to accumulating excretory products (Shull 1910, 1911), pH or bicarbonate and oxygen content (Tauson 1925), or general vitality in dense population (Buchner 1936, 1941a,b). None of these factors is generally operable. Shull (1911), working with *Epiphanes senta,* found that both ammonia and urea tend to inhibit the production of mictic females, which is the opposite of what would be expected from experiments with *Brachionus* in which crowding produced mictic females. The emphasis that has been placed by various workers on the change rather than on the absolute amount or kind of food available recalls Slobodkin's (1954) similar observations on ephippial production in *Daphnia obtusa.*

Tauson (1925) found that eggs producing mictic females in *Asplanchna* are smaller than those producing the amictic. The difference is apparently due to the amount of yolk incorporated into the egg; this Tauson thought depended on the reaction of the water.

Only in *Asplanchna sieboldii* and *Brachionus calyciflorus* is there agreement between different workers; in these two species the immediate environmental stimulus to mictic female production is clearly different.

In *A. sieboldii* both Tannreuther (1920) and Gilbert (*personal communication*) found that transfer from colorless to vchlorophyll-bearing food produces a marked epidemic of mictic female production.

In *Brachionus calyciflorus* there can be no doubt that crowding (Buchner 1936, 1941a,b), or rather the ratio of amictic females to volume of medium (Gilbert 1963), is the significant factor. Buchner believed that dense populations developed at times when the species exhibited greatest vitality and that the production of males followed from that condition. Gilbert found that the effect is not social; a single amictic female reared in a small volume of water produces as many mictic daughters as would a female in a more populous culture with the same ratio of animals to volume as in the small single female culture. Gilbert concluded that the effect must be chemical and relatively specific. Crowding with *Paramecium* has no effect, but there was some hint that *B. angularis* in sufficient numbers can affect *B. calyciflorus*. It is evident from Carlin's (1943) detailed studies in the field that male production is limited to the species showing a maximum population at any given time and does not occur in other species present in small numbers (Fig. 141). This argues for considerable species specificity. Though the field observations, and the experiments of Buchner and Gilbert, clearly indicate that the development of dense populations is likely to be a determinant of mictic female production in many species and actually is so in *B. calyciflorus*, it is evident that in some members of the genus *Brachionus* another mechanism exists. Pourriot (1957c) found that in nominotypical *B. ureolaris*, but not in f. *sericus* nor in *B. calciflorus*, change of food from *Chlorella pyrenoidosa* to *Scenedesmus quadricauda, Selenastrum minutum,* or *Colacium arbuscula* was highly effective in producing mictic females. Change to several other algae, including *Scenedesmus falcatus*, however, was quite ineffective.

Anabiosis. The type of life history just outlined is characteristic of many planktonic and all benthic Monogononta, including most of the genera of rotifers; the rotifers of the subclass Bdelloida, none of which is planktonic, have lost all trace of sexual process and never produce resting eggs. This would appear paradoxical, since these animals live in the most uncertain aquatic habitats, wet moss, minute pools of rainwater, and the like, were it not for the fact that they have developed the most extraordinary capacity to persist as adults in a dried, anabiotic condition. The desiccated anabiotic bdelloid rotifers which are, in fact, among the most resistant of living Metazoa, survive temperatures little above absolute zero and little below boiling water; it may be doubted, however, to what extent the word living may be properly applied to organisms that seem rather to be reversibly dead.

Fecundity. In *Keratella valga* Kolisko (1938) found an average of 0.40 egg produced per individual per day in culture at 17° C. Edmondson's (1960) observations in Bare Lake, discussed in a later section (pages 516–517), suggest lower rates of 0.077 to 0.267 egg per day, the highest figure occurring at the warmest time of year at a temperature of 18.2° C. *Kellicottia longispina* in the same lake exhibited lower reproductive rates of 0.032 to 0.227 egg per day. Laboratory experiments have indicated that some littoral benthic species are much more fecund; *Epiphanes senta* produces as many as 6.22 eggs per day.

The newly hatched rotifer is ordinarily large and grows to maturity very rapidly. Reproduction in *K. valga* begins after the fourth day.

Expectation of life. In all cases it may be reasonably supposed that the life span is short compared to the length of the potentially favorable season for the growth and reproduction of the animal, so that at least several generations may be expected each year.

In the species which have been most widely cultivated in the laboratory (cf. Edmondson 1945) and which are not eulimnoplanktic the survivorship curve is often found to be of a markedly rectangular kind. *Lecane inermis* (Miller 1931; Edmondson 1945), in fact, yields the most striking example of such a curve, in which mortality is at first low but suddenly increases as all animals approach the end of a physiologically determined life span (Fig. 133). In this particular case the mean duration of life falls from ten days at 18° C. to four or five days at 29 to 35° C. (Finesinger 1926). Many rotifers probably have similar life spans; in *Brachionus calyciflorus, Epiphanes brachionus, E. senta, Proales decipiens,* and *P. sordida,* raised under various conditions, the mean life span lay between 5.6 and 11.3 days. Some species, however, live much longer. *Keratella valga* has an average life span of twenty-two days at 17° C. in laboratory culture (Kolisko 1938), whereas life spans of thirty-five and forty-two days are known for the bdelloid *Rotaria rotatoria* and the sessile *Cupelopagis vorax,* respectively. Life spans in nature are hard to determine, though it is possible in sessile species in some cases. Edmondson found values up to eleven days in *Floscularia conifera.* Solitary individuals live less long than individuals that form colonies. Kolisko's mean for *K. valga* may be taken as a typical figure for planktonic rotifers in default of better data, as it falls in the middle of the range of known longevities. In a perennial species, living twice as long in winter as in summer,

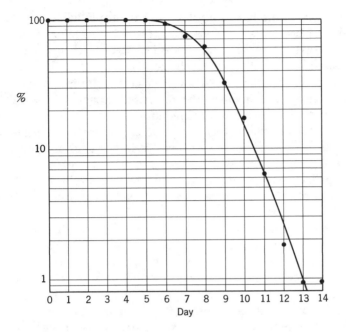

FIGURE 133. Survivorship curve for amictic females of the rotifer *Lecane inermis* in culture. (Data of Miller 1931.)

a mean life span of eleven days in the warmer half of the year would imply about twenty-five generations per annum.

Aging in rotifers. A particularly interesting aspect of the rotatorian life cycle is the process of aging, studied by Lansing (1942a,b, 1947, 1948), whose results have been applied to an interpretation of certain ecological properties of the group by Edmondson (1948). Lansing has found, as others had suspected, that in parthenogenetic reproduction the expectation of life of an individual from an egg produced early in its mother's reproductive career is greater than that of an individual born late. The effect, moreover, is cumulative as far as the individuals produced from old mothers are concerned, for their late offspring have a still lower expectation. Lansing investigated this phenomenon by breeding a series of orthoclones or clones consisting always of the nth daughter of the nth daughter in any line. In *Philodina citrina* if $n \geq 6$ the orthoclone dies out; when n equaled 6, the line lasted for seventeen generations, and when n equaled 16 it lasted for three. The orthoclone in which the maximum number for n permits indefinite survival, in this case 5, is termed the isoclone. In practice, nearly all orthoclones in the laboratory tend to produce individuals with either decreasing (geriaclones) or increasing (pediaclones) life spans. Lansing found that a spectacular increase in longevity occurred in animals treated periodically with sodium citrate solution, which he believes removes calcium as the well-known calcium citrate complex. The process of aging thus involves the reversible accumulation of calcium. It seems probable that this accumulation affects the egg as much as the somatic cells of the mother; since the egg is very large in proportion to the size of the adult animal, eggs that come from aged mothers might well produce adults that start life with an excess of calcium and are, in fact, physiologically aged before they are born. The possible bearing of some of these facts on cyclomorphosis in rotifers is made apparent later.

Population dynamics in nature. Numerous investigators have recently been concerned with attempting to analyze variations in zooplankton populations in terms of birth and death rates. The fundamental data required in the simplest case involves an estimate of reproductive rate, of some parameter relating to rate of development, and an estimate of the observed change to be analyzed. In view of the very short prereproductive life of a rotifer once it has hatched, Edmondson (1960), who has studied the matter in several planktonic rotifers, has used as the fundamental data (cf. Elster 1954) the number of eggs per female (l_r) which for rotifers carrying a single egg will be the proportion of ovigenous females and the time (r_e) taken for an egg to develop. For any population the rate of egg laying b_e can be taken as

$$(1) \qquad\qquad b_e = \frac{l_r}{r_e}$$

If we take the growth (positive or negative) of the population over a short finite interval to be given by

$$(2) \qquad\qquad N_t = N_0 e^t$$

the effective rate of increase

$$(3) \qquad\qquad \mathrm{r} = \frac{\ln N_t - \ln N_0}{t}$$

This quantity is essentially the difference between natality **b** and mortality **d** over the time interval t. If b_e is estimated at the beginning of an interval of unit time,

$$(4) \qquad\qquad \mathrm{b} = \ln (b_e + 1)$$

When b_e is small, b and b_e are almost identical and b_e almost an estimate of b. The treatment must be regarded as an approximation appropriate only to populations not changing greatly in numbers; in certain cases it apparently leads to erroneous results (Edmondson, personal communication).

In order to determine b_e from egg counts, the time of development r_e of the egg must be known. This is temperature-dependent and must be determined from laboratory experiments. Edmondson's data appear to indicate that for several planktonic lake species the time of development varies from about three to ten days at 10° C. to about one day at 27° C. (Fig. 134). For *Hexarthra fennica* from a small tank the rate of development at high temperatures, and its temperature-dependence, appeared to be much greater. It is not unlikely that

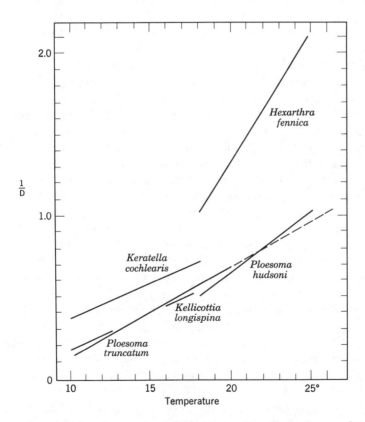

FIGURE 134. Mean rate of development (per day) of eggs of various rotifers related to temperature. (Edmondson.)

opportunistic species appearing at the height of summer would exhibit marked temperature-dependence (cf. pages 593–594).

Edmondson (1960) has analyzed the dynamics of *Keratella cochlearis, Kellicottia longispina,* and *Ploesoma truncatum* in Bare Lake, Alaska. In general, the temporal dynamic pattern is not dissimilar in the three species studied. The most extensive data refer to *K. cochlearis* in 1951, which is used as an example of the analysis. The numbers of individuals per unit volume were counted at intervals and the proportion of eggs per adult female was calculated. Knowing the temperature and the rate of egg development (Fig. 135) at that temperature (from experiments done at Pallanza), we can determine the rate of egg laying (b_e) and the birthrate coefficient (**b**). The effective rate of increase per unit time (**r**) can be determined from the counts by (3); since **b** is known, the death rate (**d**) can be obtained. There are wide variations in **d,** including one impossible negative value; Edmondson gives reasons for supposing that these irregularities are largely statistical artifacts and that a smoothed death rate (**d'**) is preferable.

The large initial rise in the population is the result of a relatively low birthrate exceeding a very low death rate. In the middle of July, when fertilization of the lake greatly increased the phytoplankton, the birthrate began to increase, but the death rate increased also; the smoothed value varied with the birthrate, though usually with a slightly greater value. The net result is that the population shows a marked if irregular decline. Edmondson's data for the rate of egg production b_e in both 1950 and 1951, temperature and Secchi disk transparency, as a rough inverse measure of the standing crop of phytoplankton, yield partial correlation coefficients for b_e on temperature-excluding transparency of .795 and on transparency-excluding temperature of —.660. It is to be noted that although the smoothed death rate (Fig. 135) is clearly highly correlated with the birthrate the death rate cannot be regarded as dependent on adult density; the highest death rates as well as the highest birthrates occur when the population had declined well below its maximum. A high, variable, and possibly density-independent mortality of young animals seems to be involved.

Edmondson (1964) in a study of the number of eggs carried by females in qualitative plankton samples from Windermere has also been able to examine the effects of temperature and variations in the population of food organisms on the reproduction of *Keratella cochlearis, Kellicottia longispina,* and *Polyarthra vulgaris.* The reproductive rates are in all cases temperature-dependent, as would be expected; the effect is most marked for *K. cochlearis.* The variation in the

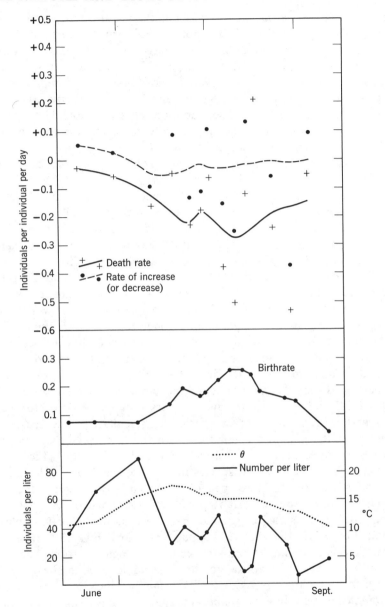

FIGURE 135. Population dynamics of *Keratella cochlearis* in Bare Lake, Alaska. Lowest panel, temperature and number of individuals per liter; middle panel, birth rate from egg number and temperature; upper panel, crude death rate and rate of increase and smoothed lines giving the most probable values. (Edmondson redrawn.)

abundance of *Chrysochromulina* was reflected in the reproductive rate of *Keratella* and *Kellicottia,* but not of *Polyarthra,* which, however, was sensitive to the variation of the larger *Cryptomonas* on which *Polyarthra* is known to feed (see page 527). There is a significant negative correlation between *Chlorella* and the rate of reproduction of *Kellicottia* which perhaps suggests antibiosis. *Eudorina* may also have an inhibitory effect. The variation of some abundant species of algae, within the size ranges apparently preferred for food by the rotifers studied, sometimes seemed to have no effect.

Principal planktonic rotifers and their ecology. The Monogononta, to which all the planktonic rotifers belong, is divided by Remane (1929–1933) into the Ploima, the Flosculariaceae, and the Collothecaceae. The last-named group consists entirely of sessile forms, but the other two groups have many planktonic members. In the Ploima a great many benthic forms belong to the Notommatidae, or to related families, which have given rise to four families of planktonic forms. The Gastropodidae, including *Gastropus, Ascomorpha,* and *Chromogaster,* are mainly small rotifers without any obvious adaptations to planktonic life. *Ascomorpha minima* (*A* of Fig. 136), which has a length of 50 μ, is the smallest known metazoan. *Trichocerca* (Trichocercidae), with one toe often elongate and sometimes a fairly long spine at the anterior end of the body, is elongate fusiform, fairly asymmetrical, and swims in a well-defined spiral. *Trichocerca cylindrica,* (*B* of Fig. 136), *T. capucina,* and, in some subtropical waters, *T. chattoni* (*C* of Fig. 136) are more or less limnoplanktonic (see page 548) and a great many other species are facultative and accidental members of the plankton community. In the family Synchaetidae the two genera *Synchaeta* (*E, F,* Fig. 136) and *Polyarthra* (Fig. 146) contain a number of difficult species in the lake plankton. The family Asplanchnidae contains three genera, of which *Asplanchna* is the best known and most abundant; it is predaceous and one of the few viviparous rotifers; it is also of great interest because of the polymorphism exhibited by some of its species (see pages 863–869, Figs 230, 231).

Another line within the Ploima leads from benthic forms, perhaps not unlike the Notommatidae, to the Brachionidae, a family of great importance in the plankton. This family includes the genus *Brachionus,* which has a number of species, nearly all found only in slightly if not extremely alkaline waters and, at least in large ponds and small lakes, often of great quantitative importance, though less eulimnoplanktic than some of the other genera. *Keratella* is another member of the family which is, if anything, more widespread, being capable of entering an even greater variety of habitats than *Brachionus*

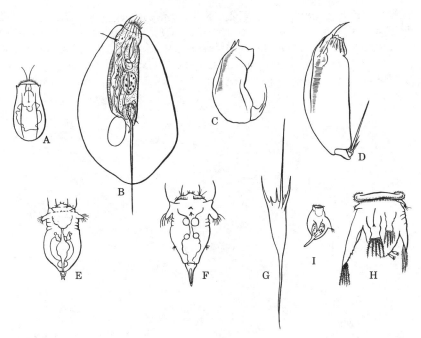

FIGURE 136. Diversity of form in planktonic rotifers. *A. Ascomorpha minima* (× 300), the smallest known metazoan, about 50 μ long. (von Hofsten, 1923). *B. Trichocerca cylindrica* (× c. 110), in gelatinous sheath and carrying egg, which is unusual in the genus. (Lauterborn, 1908.) *C. T. chattoni* (× c. 110), a warm-water planktonic species. (Hauer, 1938.) *D. T. porcellus* (× c. 110), facultative plankter for a short time in summer. (Jennings, 1903.) *E. Synchaeta oblonga* (× 75) and *F. S. stylata* (× 75), two members of a genus of numerous closely allied species. (Edmondson after Rousselet.) *G. Kellicottia longispina* (× c. 60), a very important plankter in the temperate northern hemisphere. (Weber, 1898.) *H. Hexarthra mira* (× 130), female, (Weber, 1898). *I.* Male. (Wesenberg-Lund, 1923.)

and containing in *Keratella cochlearis* one of the commonest members of the lake plankton in temperate regions. *Notholca, Kellicottia,* and *Anuraeopsis* are other planktonic genera of the same family. *Kellicottia longispina* (*G* of Fig. 136) and *Notholca striata* appear to be the dominant plankton rotifers of Lake Baikal (Jaschnov 1922). Allied families, notably the Euchlanidae, may contribute species to the heleoplankton and are of interest because of the studies of the biology of rotifers to which they have lent themselves.

The Flosculariaceae consist of four families, one of which, the Flosculariidae, contains exclusively sessile forms; another, the Conochilidae,

includes two genera, both planktonic; one genus, *Conochiloides* (*E,
F,* Fig. 137), is solitary and the other *Conochilus* (*C, D,* Fig. 137)
is colonial, but both in general form are close to the Flosculariidae.
The Hexarthridae contain the single genus *Hexarthra* (= *Pedalia*) (*H,
I,* Fig. 136), once supposed to show affinity with the arthropods:
eight species occur, all more or less planktonic, some specifically in

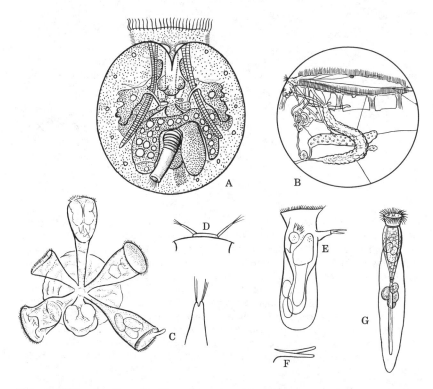

FIGURE 137. *A. Testudinella patina* (× 215), heleoplanktonic and littoral-
benthic but tolerant of very wide variation in other ecological conditions.
(Edmondson.) *B. Trochosphaera solstitialis* (× 60), a rare trochophore-like,
heleoplanktonic rotifer. (Edmondson after Rousselet.) *C. Conochilus unicornis* (× 50), limnoplanktonic with antenna fused as indicated. *D. C. hippocrepis,* paired antennae of this larger heleoplanktonic species forming
larger colonies, though introgressive hybridization is known in some small
lakes in Scandinavia. *E. Conochiloides dossuarius* (× c. 70), an autumnal
species of a noncolonial ally of the preceding, very closely allied to *C. natans,*
a cold-water species with, *F.* unfused ventral antennae. (*A-F,* Edmondson,
modified.) *G. Collotheca pelagica* (× c. 110), one of the more planktonic
of the few free-swimming members of the order Collothecaceae. (Rousselet,
1893.)

alkaline lakes. The fourth family, the Testudinellidae, although allied to the other two in the form of the trophi and ciliation of the corona, contains several planktonic genera of extraordinarily diverse facies. One of these genera, namely, *Trochosphaera* (*B* of Fig. 137), has in the past been adduced as evidence of the close affinity of the rotifers to the polychaet worms, but it is clearly very specialized and certainly does not bear on the problem of the taxonomic position of the group. In the same family the genus *Filinia* (including *Tetramastix* as a subgenus) possesses long spinous processes and sometimes a rather fusiform body (Fig. 139), so that it may be regarded as showing a parallel body form to that of the planktonic species of *Trichocerca*. The genera *Testudinella* (*A* of Fig. 137), *Horaella,* and *Pompholyx* include small round forms, whereas the rare *Voronkowia* is not unlike a tubicolous Flosculariid which has become free-swimming.

Distribution and ecology. Most of the planktonic rotifers appear to have a potentially world-wide distribution, but they vary greatly in their ecological tolerances and requirements. According to Edmondson (1944), who has provided the most recent discussion of the zoogeography of the group, *Testudinella patina* has an exceptionally wide range of tolerances and is to be regarded as truly cosmopolitan. More often some limitation to a particular, if broadly defined, class of habitats is to be observed. Several species (Fig. 138) of *Brachionus* are common in the plankton of hard-water ponds in both temperate and tropical regions, but other species, such as *B. falcatus* and *B. mirabilis,* are more abundant in warm than in cold water; if they occur in temperate regions, as *B. falcatus* does, they are unimportant summer forms. No species of the genus was recorded by Ahlstrom (1940) in water more acid than pH 6.6 and in nearly every locality in which its members are conspicuous the water is alkaline; some species, however, are able to tolerate much more alkaline waters than others, *B. satanicus* and *B. pterodinoides* are noteworthy, if rather rare, alkaline water species, and *B. plicatilis* is a very common and widely distributed inhabitant of saline waters. It is reasonably certain that within the genus some forms are more likely than others to be found in the open water of lakes, but in general most species probably prefer localities in which a solid substratum is available for attachment by the foot. A few cases of truly geographical limitation may occur in the genus; *Brachionus havanaensis* is found only in American waters, *B. diversicornis* and *B. forficula,* only in the Old World.

A similar picture is presented by *Keratella* (Figs. 232–236, 238, 239), but this genus, in which the foot is lacking, is far more likely to occur in the open water of deep lakes than is *Brachionus.*

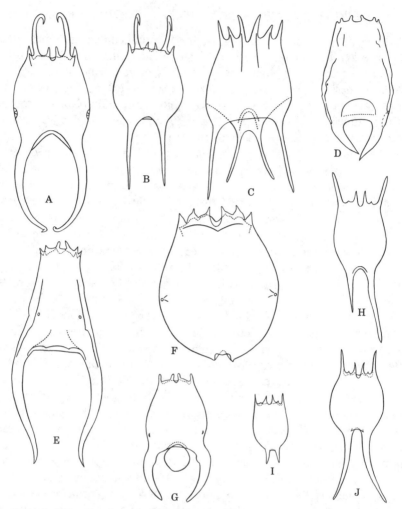

Figure 138. Species of the genus *Brachionus* with striking ecologically or geographically limited distributions. *A. Brachionus falcatus,* a warm-water species, exuberant form from Brazil. *B.* The same, more typical form from Panama. *C. B. mirabilis,* another warm-water species, Rio de Janiero, Brazil. *D. B. satanicus* only in alkaline and saline waters from a few New World localities, reduced form from Encadenadas, Argentina. *E.* The same, exuberant winter form from Devil's Lake, North Dakota. *F. B. plicatilis,* a widespread species in saline waters, Tso Nyak, western frontier of Tibet. *G. B. forficula,* apparently limited to the Old World from eastern Europe and central Africa, east to China and Japan, specimen figured from Taiwan. *H. B. havanaensis,* apparently found only in New World, possibly the allopatric representative of *forficula,* Illinois River at Havana, Illinois. *I.* The same, reduced form from New Orleans. *J.* The same, symmetrical but exuberant form from Arcadia, Florida (all roughly × 150). (Ahlstrom.)

K. cochlearis and *K. quadrata* are widespread temperate forms; *K. valga,* perhaps, occurs more in ponds than these species, *K. tropica* is tropicopolitan, and *K. hiemalis* is a cold-water form. *K. serrulata* and *K. taurocephala* inhabit somewhat acid waters. *K. irregularis* is probably the Old World representative of *K. earlinae,* both occurring in plankton. *K. reducta* is known only in Africa south of the Zambesi, but it is an important element in the aquatic fauna at least of the Transvaal and so distinct that it can scarcely have been overlooked elsewhere.

Among the members of other genera, *Filinia (Tetramastix) opoliensis* is probably mainly a warm-water, tropicopolitan species. In *Filinia* (s. str.) the common and closely allied *F. (F.) limnetica* and *F. (F.) longiseta* (Fig. 139) differ ecologically: *limnetica* is eulimno-planktonic, whereas *longiseta* is found in ponds and both occur mainly in summer. Though they seem to be distinct, a little introgression, perhaps mainly in fluviatile habitats, appears to occur in some regions. Differing mainly from *longiseta* in the terminal insertion of the posterior appendage, *F. (F.) terminalis* is a seemingly cold-water species, recorded only below 20° C. and usually in much colder water, in the plankton of lakes at all depths in winter but only in the hypolimnion in summer, except in localities at extreme altitudes in which the water is perennially cold. A fourth close ally, *F. (F.) pejleri,* previously confused (Hutchinson 1964a) with *F. (F.) terminalis,* is, in striking contrast, evidently a thermophil species, known only from warm temperate and subtropical localities in South Africa, India, and the southern United States.

In the other major genera of the planktonic rotifers, such as *Notholca, Hexarthra,* and *Synchaeta,* the same sort of ecological allopatry with a quite limited degree of geographical restriction is to be expected. In *Hexarthra, H. mira* is widely distributed in ordinary waters, *H. bulgarica* is a montane Old World species, and several other members of the genus are distributed in saline or alkaline lakes.

Most planktonic genera therefore appear to have geographically widespread species of varying tolerances, limited by broad thermal or chemical conditions, and sometimes also a few species with more restricted geographical ranges.

Nature of chemical limitation. The full meaning of the chemical limitations so characteristically exhibited by many rotifers is, unfortunately, quite inadequately understood. Several possible significant variables, pH, CO_2, Ca, and HCO_3' concentrations and the total mineral content, salinity or osmotically active concentration, may vary in a correlative way, so that when limitation by pH has been reported,

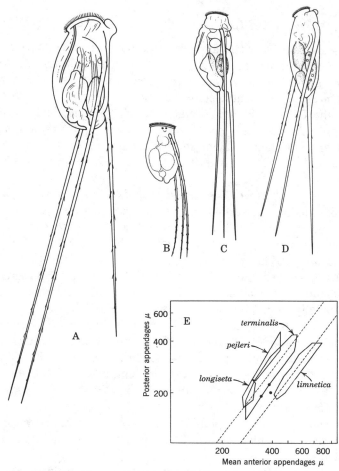

FIGURE 139. Four very closely allied species of *Filinia* (s. str.) with characteristic ecological requirements. *A. F. (F.) limnetica,* a limnoplanktonic species common in summer in temperate regions and in some warmer regions doubtless perennial (× 70). *B. F. (F.) longiseta,* a pond species in temperate regions. *C. F. (F.) terminalis* from cold water, occurring in winter, in the hypolimnion of stratified lakes and at great altitudes in summer (× 70). *D. F. (F.) pejleri* from warm temperate or subtropical localities (× 70). *E.* Plot on a double logarithmic grid of the length of the posterior appendage against the mean length of the anterior appendages, showing envelopes enclosing points for the four species and a few points (black circles), all from rivers or ponds in central Europe, suggesting introgression of *F. (F.) limnetica* into *F. (F.) longiseta* populations; broken lines have a slope of 1.33, indicating the value of the exponent in the heterauxetic relationship for the structures in question in both *limnetica, terminalis,* and perhaps also *pejleri.* (Edmondson and Hutchinson, Voigt, Carlin, Hutchinson, all somewhat modified.)

as is usually the case, the actual limiting factor may be one of the other variables. There seems to be no good information available for the planktonic species. Edmondson's (1944) field studies of the sessile rotifers of Wisconsin strongly suggest that there may be more than one factor operating to produce an apparent limitation to a given *p*H range.

Collotheca tenuilobata is found in water of *p*H 4.1 to 7.5; it is apparently sensitive to moderate or high HCO_3' content, for it never occurs above 17 mg. per liter HCO_3'. *Floscularia pedunculata,* recorded in the same study from *p*H 4.1 to 6.8, however, occurred at a bicarbonate concentration of 141.5 mg. per liter. These two species, though occupying series of waters with approximately the same *p*H range, nevertheless seem to be limited in different ways, a low total concentration or low concentration of HCO_3' or Ca^{++} being necessary for the former but not for *F. pedunculata.* In more alkaline water, from *p*H 6.2 to 8.8, *Collotheca algicola* did not occur above a bicarbonate concentration of 57 mg. per liter, whereas *Lacinularia flosculosa* over about the same *p*H range apparently required 40 mg. per liter HCO_3' and tolerated 318 mg. per liter. *Beauchampia crucigera* provided an example of a species seemingly limited to rather low bicarbonate (1.8 to 59 mg.$^{-1}$) and moderate *p*H round neutrality (5.8 to 7.6). Where low calcium is required, we may perhaps suspect that we are involved with species very sensitive to the aging process.

How far the same kinds of phenomena occur among planktonic species is still unknown, but it is not unreasonable to suppose that the various ecological groups of rotifers will all provide examples of this kind of limitation and will show a complex type of chemical determination analogous to that exhibited by the sessile species.

Though certain species appear limited to alkaline water, there seems to be no clearly defined acid-water plankton rotifer fauna comparable to the large benthic acidophilous assemblages described, for instance, by Myers (1931). Pejler (1957c) has in fact cast some doubt on the importance within rather wide limits of either *p*H or total ionic concentration in determining the distribution of planktonic rotifers, considering that the apparent correlations are due primarily to certain rotifers being characteristic of water of moderate and others of high primary productivity: variation in the latter being correlated with mineral content and *p*H. *Chromogaster ovalis, Asplanchna herricki, Synchaeta grandis,* and *Ploesoma hudsoni,* all transcursonial species in Myers (1931) terminology and all but the first occurring from *p*H at least as low as 5.7 and as high as 8.1, are regarded by Pejler and to some extent by other European workers whose observations he discusses in detail as characteristic of relatively unproductive waters, whereas most species of *Brachionus, Keratella quadrata, K. cochlearis tecta,* many species of *Trichocerca,* and *Pompholyx sulcata* are considered rather generally as species of eutrophic waters. There are, however, obvious difficulties. The information collected by Ahlstrom (1940) strongly suggests that *Brachionus* is limited almost entirely to alkaline water, and it is hard to avoid the conclusion that this very general chemical determination is primary and causal. Yet *B. angularis* is known (Strøm 1944) in high mountain lakes of low conductivity and with negligible phytoplankton, an exception that does not permit any choice between chemical and trophic determination. *K. quadrata* and *P. sulcata* have similarly been found on occasion in manifestly oligotrophic waters. The species of *Trichocerca,* like many of those of *Brachionus,* are found mainly in productive ponds in temperate lakes but may occur during summer myxophycean blooms. Pejler thinks some of the difficulties of the problem are due to species consisting of several ecogenotypes adapted to different sets of environmental variables. The whole matter clearly needs much further study.

Differentiation of planktonic from littoral fauna. Differences in behavior probably keep certain species in the littoral zone even though they are capable of swimming in the free water between stems or leaves of aquatic plants and other objects on which they live. Little work, however, has been done on this matter in the rotifers, but it is under investigation by Pennak (personal communication). Pejler notes that *Notholca* spp., including *N. foliacea,* and *Ploesoma lenticulare* are often more littoral than *Keratella* spp., *Brachionus* spp., and *Ploesoma hudsoni.* In a sample jar the littoral benthic species seem to collect near the bottom, but the planktonic remain swimming; a like behavioral difference in nature would lead to niche separation. More refined work on the taxes involved is clearly needed.

Food of planktonic rotifers. The food of rotifers has been investigated by several students of the group, both on the basis of observation in nature (Dieffenbach and Sachse 1911; Naumann 1923; Myers 1941) and on that of the food required or chosen by animals in laboratory culture (de Beauchamp 1938; Pourriot 1957b). Pejler (1957b) has discussed this work from an ecological point of view, and has added some observations of his own.

In general, the free-swimming species of the plankton feed either by sedimenting particles as the result of the action of the coronal cilia or they are raptorial. Most of the sedimenters (Fig. 140) have a malleate (Brachionidae, Proalidae) or malleoramate (Flosculariaceae) mastax.   The raptorial forms have obviously developed along more than one line.   Among forms with virgate trophi many species can suck in whole small planktonic animals; others can puncture the cell or body walls and suck in the contents.   Among planktonic species the incudate trophi of the Asplanchnidae can be extended to seize food; the same is true of the great variety of forcipate trophi in the Dicranophoridae which, however, are not euplanktonic.

Among the sedimenting forms *Filinia longiseta* appears (Naumann 1923) to eat smaller food, well below 10 μ in diameter, than most other species do.   Pejler suspects that the requirements of *F. terminalis* are comparable.   *Conochilus unicornis* is supposed to eat food less than 10 μ in diameter, including, according to Naumann, humic detritus, though its nutritive value, unless it is carrying bacteria, may be doubted. Naumann considers the species nonselective, but de Beauchamp noted that it refused *Chlorella*.   *Kellicottia longispina*, the various species of *Keratella* and the planktonic species of *Collotheca* (*C. libera, C. mutabilis*) also apparently eat particles smaller than 10 to 12 μ in diameter.   Pourriot (1963) finds that *Keratella* and *Kellicottia* can capture *Cryptomonas* 16×48 μ in size by seizing the flagellum with the mastax; this leads to rupture of the cell wall and ingestion of the particulate part of its contents.   Such a mode of feeding is apparently not very efficient, and on the whole ingestion of small whole organisms is of primary significance to these genera.   Dieffenbach and Sachse (1911) found that *Polyarthra* spp. (sub. *P. "platyptera"* and *P. euryptera*) feed mainly on *Cryptomonas*, whereas de Beauchamp concluded that *Chlorella* was not eaten.   These results on *Kellicottia, Keratella,* and *Polyarthra* are in line with Edmondson's (1964) findings on the effects of variation in food supply on reproduction in nature in these genera.   *Keratella cochlearis, Notholca* cf. *squamula,* and *Anuraeopsis* spp. have been reared by de Beauchamp on *Chlorella*, as has *K. heimalis* by Ruttner-Kolisko, (1940), but at least de Beauchamp found that an indefinite number of generations could not be maintained on this food, though it appeared complete for *Brachionus calyciflorus* raised in the same laboratory and by Pourriot (1957b).   Both *K. cochlearis* (Luntz 1926) and *Brachionus calyciflorus* can be raised on *Polytoma*.   Pourriot (1957b) used a variety of algae in culturing *Brachionus calyciflorus, B. u. urceolaris,* and *B. urceolaris sericus*. The last two taxa are not recognized as distinct by Ahlstrom (1940), but Pourriot found that *sericus* was not only more striated (though this character became less pronounced in culture than in nature) but also less likely to remain attached to the culture vessel and in general to be a more free-swimming form than *urceolaris* s. str.   Large species of algae (*Pandorina*), or those departing much from a simple spherical or elliptical form, particularly if they tend to aggregate (*Ankistrodesmus,* notably *A. falcatus* var. *acicularis,* and *Selenastrum*), are less suitable than small simply shaped algae under 20 μ in maximal diameter. *Chlorella pyrenoidosa* was the best food for all three taxa.   The free-swimming *calyciflorus* and *sericus* did less well on *Selenastrum minutum* than did *B. urceolaris* s. str., and, according to the text, better on the actively swimming stages of *Haematococcus pluvialis,* though the difference reported in the table for *urceolaris* and *sericus* on *Haematococcus* is negligible.   *Scenedesmus quadricauda,* which sank to the bottom, was second best for *B. calyciflorus* and the third best (after *S. minutum*) for *B. urceolaris,* but quite bad for *sericus.*   Though the behavioral difference between *urceolaris* and *sericus* probably makes *urceolaris* better able to take sedimented and *sericus* suspended algae, as Pourriot indicates, the data presented, in which the amount of agitation of the cultures is not indicated, do not bring out the difference clearly.   The great difference between the active *calyciflorus* and the sedentary *urceolaris,* on the one hand, and *sericus,* which behaves like the first, on the other, when the food is *Scenedesmus quadricauda,* is interesting but unexplained.   Observations of this sort, even though analysis has not gone far, are of great importance in emphasizing the possibility of niche specificity of a fairly subtle kind in the genus. It is interesting that *Scenesdesmus* can be a good food for *B. calyciflorus,* though it is a mediocre diet for *Daphnia* (see page 556). *Euchlanis dilatata,* another brachionid, feeds on diatoms and blue-green algae (de Beauchamp 1909; Carlin 1943; von Bülow 1954) but not on *Chlorella.*

Erman (1962) studied the feeding of *Brachionus calyciflorus* by an ingenious technique in which algal suspensions could flow slowly through a tube in which rotifers were maintained by an electric barrier. The rotifer was found to clear about the same volume, usually from 2 to 5.6 mm.³ individual^{-1} hr.$^{-1}$, provided the density of the *Lagerheimia ciliata* used as a food organism did not exceed 100 mg. per liter.   Although in different runs under different conditions the clearance rate varied considerably, and could reach 13.6 mm.³ individual^{-1} hr.$^{-1}$, depending on previous feeding and probably other conditions, in any given short run fair constancy was obtained and up to 100 mg. per liter of *L. ciliata* food intake was proportional to food concentration. At the maximum feeding rate animals took up 180 per cent of their wet weight as wet food or 360 per cent of their dry weight as dry food in a day.   The maximal food intake may greatly exceed the calofic equivalent of the daily respiration; in one case quoted the rotifer required 26×10^{-5} cal. day^{-1} but consumed 500×10^{-5} cal. day^{-1}, an almost twentyfold excess. Nothing is said about egg production. Above 100 mg. per liter of *L. ciliata,* which corresponds to a rich bloom in a natural water, the clearance rate, and perhaps also the rate of food intake, fell. The species of food makes a considerable difference. The greatest food intake of 550 per cent dry weight per day was given by a particular strain (No. 3) of *Scenesdesmus acuminatus. S. acuminatus* No. 2, *Cyclotella meneghiniana, Nitzschia communis,* and *Ankiskrodesmus angustus* all could support more than 400 per cent uptake of dry matter per day. *S. acuminatus* No. 1 and *Lageheimia ciliata* No. 1 were separated faster than the others, but containing less dry matter, they contributed less organic matter to the rotifer in unit time. *Chlorella terricola* no. 1, *Ankistrodesmus falcatus* and *S. quadricauda* were used less readily, as were other strains of *Chlorella, Oocystis lacustris* and *Coelastrum sphaericum. Gomphosphaeria* (sub *Dictyosphaerium ehrenbergianum*) sp. was the worst food, the colonies being too large and only odd detached cells being edible. Size, however, is by no means the only factor. Two strains of *L. ciliata* of mean dimension $6.8 \times 3.9 \ \mu$ and $6.0 \times 3.9 \ \mu$ gave clearance rates respectively corresponding to daily food uptake of 360 per cent and 250 per cent dry matter. Comparable differences were exhibited by different strains of *Chlorella terricola.* When mixed food was given, there was evidence of some inhibition by one strain of *Chlorella vulgaris.* Otherwise, the rate of removal of *L. ciliata* remains more or less unchanged if the other alga present was a preferred and rapidly removed species, such as *Scenesdesmus acuminatus* or *Cyclotella meninghiniana,* or might even be increased by the presence of nontoxic but unsuitable food such as *Oocystis lacustris* and *Coelastrum sphaericum.* Erman's results are often presented too briefly to make clear exactly what was happening, but it is evident that *B. caliciflorus* can exercise some powers of selection.

Among species with a virgate mastax *Ascomorpha saltans* (Wesenberg-Lund 1930; Myers 1941; Carlin 1943; Pejler 1957b) sucks out dinoflagellates, and *Chromogaster* eats the same food and *Chlorella.* The species of *Synchaeta* feed on both protists and other rotifers; *S. pectinata* eat *Mallomonas* (Rezvoj 1926), various protists (Rousselet 1902), and *Keratella* (de Beauchamp 1909), and *S. grandis* feeds mainly on rotifers such as *Polyarthra* (Rousselet 1902). The allied genus *Ploesoma* feeds similarly, *P. hudsoni* eating *Ceratium* and other rotifers, including its congener *P. lenticulare,* which itself eats *Synchaeta stylata* and *Chromogaster ovalis.*

The incudate mastax of *Asplanchna,* which includes the largest plank-tonic rotifers, is used to capture a variety of food, notably algae, *Keratella, Ascomorpha,* and other rotifers and even small Crustacea.

Pejler (1957b) points out that in many cases coexisting species of closely allied rotifers exhibit size differences which probably result in their feeding on slightly different food. The examples that he gives are

> *K. hiemalis, K. quadrata*
> *Notholca labis,* large and small forms
> *N. labis, N. acuminata*
> *N. labis, N. squamula* when they co-occur
> *Polyarthra remata, P. vulgaris, P. major* (Fig. 146)
> *Conochilus unicornis, C. hippocrepis*

It is also to be noted that within some genera, notably the more raptorial *Synchaeta, Asplanchna,* and *Trichocerca* but also in the pre-dominantly ecologically allopatric species of *Hexarthra* which is pre-sumably a sedimeter, the detailed form of the mastax (Fig. 140) differs in the different species. Though the planktonic genera in any locality have fewer sympatric species differing in this way than many benthic littoral genera such as *Notommata* or *Proales,* it is reasonable to suspect that such diversity in the trophi as is exhibited intergenerically is likely to be correlated with different food habits.

Although much probably remains to be learned about the role of the feeding of rotifers in the ecology of the fresh-water plankton, it is evident that the different species of the class can be herbivores and primary or even secondary carnivores and that even among the more microphagous forms there can be some selectivity, in part perhaps based on size but also on behavior. The planktonic rotifers as a whole may be expected to occupy fairly discrete niches in the plankton community as Pejler (1957b) has emphasized, and the paradoxical situation that has been discussed in relation to the phytoplankton is not likely to develop.

Seasonal succession. Prolonged systematic observations have been made by a number of investigators, notably Ahlstrom (1933), Beach (1960), Carlin (1943), Davis (1954), Kofoid (1903, 1908), Ruttner (1930), Schreyer (1920), and Wesenberg-Lund (1904, 1930) in vari-ous lakes and rivers of different character in the temperate regions of the Northern Hemisphere. Among these studies that of Carlin is of particular importance, since it is not only thorough but also based on far more refined taxonomic concepts than were available to his predecessors. Carlin's work inevitably provides the main factual basis

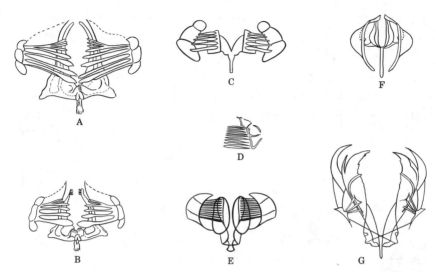

FIGURE 140. Main types of mastax found in the families of rotifers that contain planktonic species. *A–E* with a malleate mastax in the Brachionidae and a malleo-ramate in the Flosculariaceae, mainly feed by sedimenting; *F* and *G* with virgate or incudate mastax, respectively, are more selective and raptorial. *A. Euchlanis (Dapidia) calpidia*, littoral benthic. *B. E. (E.) dilatata*, littoral benthic and also a facultative plankter, the two species showing marked and specific variation within a genus. *C. Hexarthra intermedia*, the member of the genus with the fewest (5), and *D., H. jenkinae* (uncus only) with the most (9 to 10) teeth on the malleo-ramate mastax. *E. Filinia terminalis* with still finer more numerous teeth. *F. Synchaeta pectinata*, virgate mastax found in this genus and also in *Polyarthra*, another very important member of the plankton, as well as in numerous benthic littoral members of the Ploima. *G. Asplanchna herricki* incudate mastax of a large predaceous species. Other rarer types are known, but, except in the few free-swimming Collothecidae, they are found only in benthic species.

of the following discussion and a point of departure in the consideration of other studies.

Carlin's observations were made in the Motala River which flows eastward from Lake Vättern and in several lakes through which this river or its tributaries pass. The most important quantitative data came from the river at Fiskeby, just below Lake Glan; enough confirmatory data suggest that the seasonal succession at Fiskeby reflects the events in the upper layers of the lake. The water of the Motala is medium hard, containing 16.9 mg. Ca per liter, usually at a pH of 7.2 to 7.3 which rises to 8.0 to 8.5 in the late summer. The phytoplankton consisted largely of *Asterionella* and *Cyclotella* in the spring and early

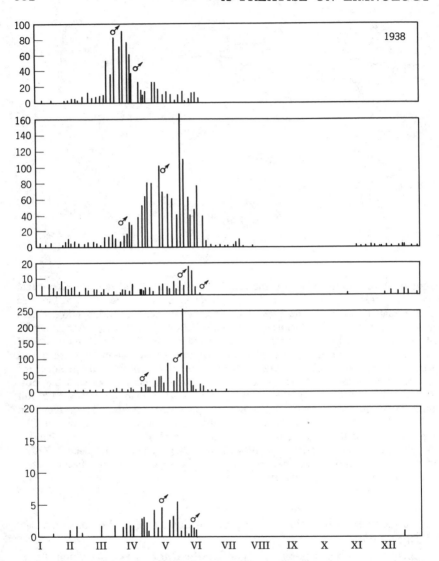

FIGURE 141. Seasonal incidence, as individuals per liter, of winter and early spring rotifers in the Motala River draining Lake Glan; ♂ signs indicate beginning and end of male production. Note that the production of males by the large population of *N. caudata* in 1938 does not affect male production in *F. terminalis*. (Redrawn from Carlin.)

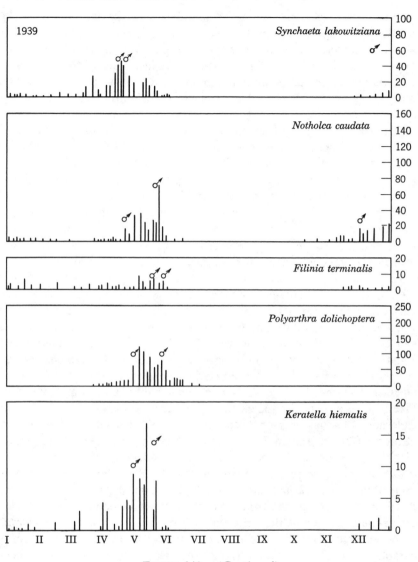

FIGURE 141. (*Continued*)

summer, of *Oscillatoria* in great abundance in the late summer, the maximum being in August, followed by *Lyngbya* and more irregularly by *Aphanizomenon,* and of a large population of *Melosira* in the autumn, maximal some time between August and October. There is a minimum of all phytoplankton in winter, as would be expected, and, as elsewhere, there often seems to be a subsidiary minimum in June.

The rotifer fauna of these waters is divided by Carlin into perennial and seasonal species, the seasonal being further divided into summer and winter forms. The perennial forms ordinarily have their maxima in the early summer, so that it seems logical to consider them between winter and seasonal late summer species.

Winter and early spring species. The winter and early spring species behave as if they were cold stenotherms (Fig. 141). The most striking is certainly *Synchaeta lakowitziana,* a little-known rotifer, previously recorded from cold water in Germany (Lucks 1930) and known also from hypolimnetic waters in central Sweden and Lapland (Pejler 1957b,c). *S. lakowitziana* appears in the Motala system as the temperature falls from 10 to 5° C. in the early winter and disappears when a temperature of 15° C. is reached in spring. The maximum can be at any temperature between 0 and 10° C.; males are produced about the time of the greatest population. *Keratella hiemalis,* which is known elsewhere as a hypolimnetic cold stenotherm in the Lunzer Obersee (Ruttner-Kolisko 1946), as an inhabitant of some mountain lakes, such as Mystic Lake, Banff, Alberta, Canada, though not in Colorado (Pennak personal communication), and under ice in the Ororotse Tso in Indian Tibet, one of the highest lakes from which plankton has been obtained (Ahlstrom 1943, Figs. 5, 7, 8, sub *K. quadrata* var *brevispina*), occurs in the Motala at approximately the same time as *S. lakowitziana,* though the detailed variations in abundance of the two species from year to year are not correlated. Males are produced at about the height of the maximum. *Notholca caudata,* a species discovered by Carlin in his investigation, also shows a comparable seasonal occurrence, though its temperature range is perhaps a little greater than those of the two preceding species. In some years there is a tendency for the species to develop two maxima, one at about 0° C. in winter, the other between 10 and 15° C. in the spring. Moreover, the production of males seems not to occur in the later part of the spring maximum, even though the population may then have its maximum density. Carlin thinks that this is due to rising temperatures which inhibit the effects of high population density in the later part of the spring maximum but admits that it is just possible that there are two races, one a more pronounced cold stenotherm with a sexual period, the other less restricted to cold water and perennially parthenogenetic, though vanishingly rare in summer. Several other winter species of *Notholca* co-occur with *N. caudata* (Fig. 142), among which *N. cinetura,* a large, possibly polyploid derivative of *N. caudata,* is an early form occurring between December and the beginning of May, with males at some time between the beginning of February and the end of April. *N. cornuta,* though rarer, shows a comparable seasonal distribution from January to April. *N. squamula* occurs from late October to late May, with a giant, possibly polyploid f. *frigida* appearing sporadically in January, February, and March. *N. foliacea,* a little less hivernal, occurs between December and early June. Only *N. limnetica,* recorded in November and December and from mid-March to late July, can occur during the warmer months of the year. Other workers generally have found *Notholca* to be a cold-water genus, and three species are important members of the zooplankton of Lake Baikal. *Filinia terminalis* has a temperature range comparable to that of *Notholca caudata* but with no clear tendency to more than one maximum. Males appear late in the periods of greatest abundance. In 1938 there is clear evidence that male production occurred earlier in *S. lakowitziana* and in *N. caudata*

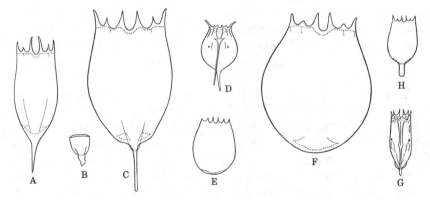

FIGURE 142. Species of the genus *Notholca* from the Motala River draining Lake Glan. *A. N. caudata* ♀, an important winter and spring species. *B.* The same ♂. *C. N. cinetura,* December to May, possibly a polyploid derivative of *N. caudata.* *D. N. cornuta,* January to April. *E. N. squamula* October to May. *F. N. squamula* f. *frigida,* perhaps a giant polypoid. *G. N. foliacea,* December to June. *H. N. limnetica,* November to December, and March to July, the only species surviving into the summer ($\times$ 100). (Carlin.)

than in *F. terminalis.* Elsewhere in temperate regions the latter is clearly a cold winter and early spring form (Słonimski 1926) which also occurs in the hypolimnia (Fig. 143) of stratified lakes (Ruttner *fide* Carlin), where it is doubtless often reported as *F. longiseta,* as in Lake Ohrid (Stanković 1960). Edmondson and Hutchinson (1934) record what is certainly *terminalis* from Togom Tso in Indian Tibet, a lake still frozen in July at an altitude of 5334 m.; they also note what is apparently the same species from the hypolimnion of Lake Manasbal and from Wular Lake, an unstratified lake in Kashmir, in which in April at 15 to 19° C. it is presumably at the upper limit of its temperature range.

Polyarthra dolichoptera has a cycle roughly comparable to that of the preceding species, but it remains uncommon from December to April, during which period *aptera* forms hatched from resting eggs may be present. It then increases rapidly and disappears as temperatures of 15 to 18° C. are reached. Male production is usually centered around the time of the maximum population, but in 1938, when an exceptionally dense population was recorded, males appeared in the early part of its development, though at densities comparable to the maxima of other years. The only other winter form in the Motala system is *Conochiloides natans,* which was ordinarily found very sparingly between the beginning of December and the end of May. It was not uncommon in April and May 1940, a year in which *Notholca* spp. and *Synchaeta lakowitziana* were much rarer than they had been before at this time. Edmondson (personal communication) noted *C. natans* in Linsley Pond from January to May in striking contrast to its congener *C. dossuarius,* which occurred with falling temperatures in September to November.

Perennial species with late spring and early summer maxima. A number of widely distributed species were found by Carlin to be perennially present in his localities but to increase to a striking early summer maximum (Fig. 144). A secondary maximum often occurs in July, and it is quite probable that it is related to the rise in phytoplankton in the spring and early summer, followed by a temporary decrease in June perhaps due in part to grazing by the numerous zooplankton nourished by the first spring bloom. Some suggestion of a Lotka-Volterra prey-predator cycle perhaps is implied by such sequences, but the cycle is probably partly forced by environmental changes and not a free oscillation.

Keratella cochlearis (sub *stipitata*), which is probably the commonest planktonic rotifer in temperate regions, developed a maximum in June in the Motala system; in most years there were indications of a sec-

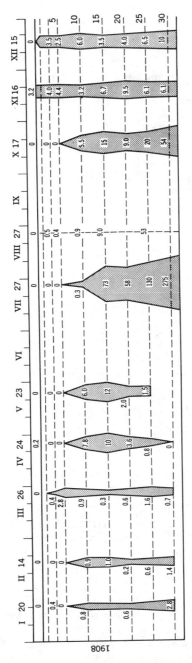

Figure 143. Restriction of *Filinia terminalis* to the hypolimnion of the Lunzer Untersee in summer, where however, a large population may develop. Numbers refer to individuals per liter. (Ruttner.)

537

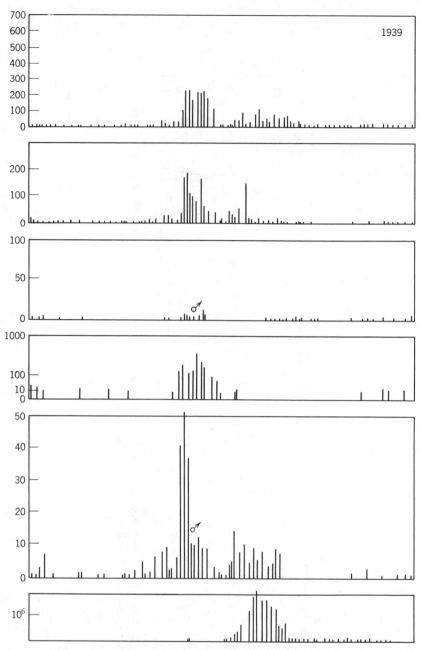

FIGURE 144. Seasonal incidence as numbers per liter of *Keratella cochlearis,*
*Kellicottia longispina, Asplanchna priodonta, Conochilus unicornis, Keratella
quadrata,* and *Oscillatoria* spp. in the Motala River draining Lake Glan in 1939
and 1940. Note the secondary abundance of the first four species and particularly the raptorial *A. priodonta* and the microphagous sedimenter *C. unicornis*
after the decline of *Oscillatoria* in 1940 but not in 1939. (Redraw from Carlin.)

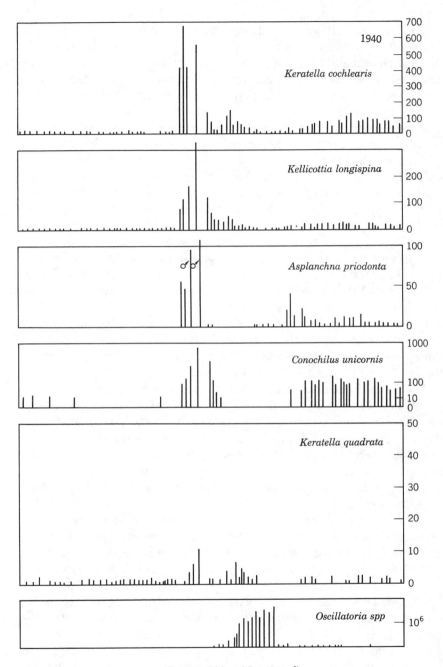

FIGURE 144.   (Continued)

ondary maximum in late July or August and a temporary increase in late autumn. It was rare during the winter but never disappeared. Male production was never recorded. In 1940, when the water remained abnormally cool throughout May, the rise to the spring maximum was somewhat delayed and took place with great abruptness; later in this year a larger autumnal population developed than ever before.

The general cycle exhibited in the Motala is comparable to what is known elsewhere, though in many cases more than one independent interbreeding or purely parthenogenetic population must exist (see page 879). Beach (1960) records a similar cycle in the lakes of the Ocqueoc River System, notably Lake Nettie. In this region *K. cochlearis* f. *robusta,* a fairly well-marked taxon of uncertain status, occurs with the typical *K. cochlearis* and varies more or less concomitantly in abundance, though it is possible that *robusta* inhabits slightly deeper, cooler layers of the lake than does *cochlearis* (s. str). *K. earlinae,* also present, varies in abundance in a comparable but rather less regular way. In marked contrast to these occurrences in small lakes, Davis (1954) found that *K. cochlearis* in Lake Erie, though essentially perennial, has a feeble secondary maximum in May and is really abundant only during August, September, and October. The maximal abundance in early August was clearly correlated with intense reproductive activity during the preceding few weeks. The species was essentially nonreproductive from October to May and in June.

Kellicottia longispina in the Motala system has a cycle of seasonal abundance comparable to that of *K. cochlearis.* As in *cochlearis,* the main maximum was late and sudden in 1940, and in that year the autumnal population was larger than before. Moreover, no males were observed. *Conochilus unicornis* follows a similar pattern but has small populations after the decline to the minimum; as in the preceding species, the autumnal population in 1940 was greater than in earlier years.

Asplanchna priodonta, which, as a large raptorial form, feeds quite differently from the preceding species, has, however, a comparable seasonal cycle; its first maximum in most years is so nearly synchronous with those of *K. cochlearis* and *K. longispina* that it is unlikely that a prey-predator relationship is involved in determining the incidence of the predator species. The maxima of *A. priodonta* appear to be somewhat sharper than those of the other perennial species; male production is at about the time of the densest populations. After the main maximum the species may disappear almost completely for a few weeks in July. The behavior in 1940 is comparable to that of the previously discussed perennial species.

Polyarthra vulgaris, a species confused by most investigators of seasonal succession with other members of the genus having narrow appendages under the name *P. trigla,* appears usually, but not always, to have its maximum later than that of the previously discussed perennial species and in 1940 had so marked an autumnal maximum that it actually exceeded that of late June. Whatever the size of the summer maximum, male production appears to be a strictly autumnal phenomenon, occurring between late September and early November.

Keratella quadrata, a less abundant species than the preceding, in some years exhibits a persistent sparse population but usually has a maximum in late May or June. Only in May 1939 was a quite considerable population developed and only at this time was male production noted. As in the other perennial species, the maximum in 1940 was a little delayed, but *K. quadrata* did not have a marked autumn increase in that year.

The rather striking similarity between the behavior of all perennial species save *K. quadrata* in producing greater autumnal maxima in 1940 than in any other years is interesting, particularly since species of diverse food habits are involved. Carlin observed that *Oscillatoria* declined a little earlier in 1940 than in other years, almost disappearing with great suddenness in a few days between August 29 and September 2. At the same time *Cryptomonas* increased, though his diagrams suggest no more so than at the same time the year before. Carlin was unable to provide a clear explanation of the annual differences in the behavior of the perennial species; at least we may conclude that the phenomena he describes must express underlying events which if more fully elucidated would be of great interest.

Seasonal summer species. *Gastropus stylifer* is present only from late May to August or very rarely in September. It usually has a maximum in June or July; the large population in 1940 developed a little later than that in 1939. *Ascomorpha ecaudis* has a similar seasonal cycle but may occur as early as March or April. This species usually has its maximum in June. In some years *A.* cf. *minima* and more often *A. saltans* occurred somewhat later, the latter species having well-marked maxima in August followed by *C. ovalis,* commonest in September.

Ploesoma hudsoni, a not particularly abundant species, occurred from April to mid-October, usually with a maximum in June or July. It appeared less frequently in the earlier than in the later years of the study.

Pompholyx sulcata is an important summer form generally maximal in the Motala system in August.

The species of *Synchaeta* are of considerable interest (Fig. 145). *S. oblonga* can have several maxima any time between the beginning of May and the first half of October. These maxima can be very pronounced, with almost complete disappearance of the species at the minima. The less common *S. pectinata* behaved in a comparable way, and at least in some years the two species seem to have had some simultaneous maxima. The same is true of *S. kitina,* but in this case the great abundance of the species was in July 1938 at the time of a very small *S. oblonga* maximum; in 1939, though its maximum coincided with one of the great *oblonga* maxima, *kitina* was not an abundant species. *Synchaeta stylata* was almost entirely confined to the late summer months, usually commonest in August. It is difficult to decide to what extent these species replace one another.

The genus *Trichocerca,* which is ordinarily largely heleoplanktonic, is an important element in the Motala system only in the late summer (Fig. 145). Both *T. porcellus* and *T. rousseleti,* which tend to vary in numbers together, may have two maxima in July and September or a single one in August, whereas *T. birostris* is commonest in August, sometimes replacing and sometimes co-occurring with the two first mentioned species of the genus.

Carlin points out that *Trichocerca* is usually a genus of eutrophic ponds, and Wesenberg-Lund (1930) concluded that *T. porcellus* (sub. *Diurella tenuior*) is planktonic only during a few weeks in the summer. In the Motala system the planktonic occurrence of the genus appears to correspond to the bloom of *Oscillatoria* when the phytoplankton is more nearly comparable to that of small productive lakes. Comparable behavior is exhibited by *T. capucina,* usually maximal with male production in September, and by *T. pusilla* mainly in August, but these species are rarer than the other three.

Euchlanis dilatata, like *T. porcellus,* appears to be planktonic only for a short period, usually in September. Male production occurs but only after the middle of August, so that in 1937, when the maximum occurred in July, males did not appear till well after the main population had declined. Among other summer species *Filinia limnetica* may be mentioned as occurring mainly in August and September, in striking contrast to its cold stenotherm ally *F. terminalis.* The various species of *Hexarthra,* other than the montane *H. bulgarica,* are in general probably warm-water species with summer maxima.

Autumnal species of Polyarthra. The genus *Polyarthra* is of great interest in (Fig. 146) that there seems to be fairly clear vicarious behavior when the species are compared; the temporal separation is not perfect, but, as Pejler (1957b) points out, size differences are also involved and it is reasonably certain that the different species would be found to occupy reasonably well-separated niches. *P. dolichoptera* has already been mentioned as a spring cold-water form, occurring over a temperature range of 0 to 20° C., with its maximum usually at a temperature between 5 and 15° C. *P. vulgaris,* a perennial species, usually has a late spring or early summer maximum in a temperature range below 15 to about 20° C. but with the possibility of an autumnal maximum between 5 and 10° C. Carlin thinks factors other than temperature are involved. The other three species in the lake are commonest in the late summer or autumn. *P. major* can have a single maximum in July, August, or September or marked double maxima in the first and last of these months, as in 1939. *P. euryptera,* which is less common, occurs at about the same season, whereas *P. remata,* appearing in June, slowly develops a large population which is maximal some time in September. At least in some years, the declining temperatures during the period of maximum abundance cover the same range as the rising temperatures during the maxima of *P. dolichoptera.* Both *P. major* and *P. remata* usually produce males at the time of maximum population density, though in 1940 no males of *P. major* occurred and in 1935 male production in *P. remata* was observed only once in November after the population had become sparse.

Bērziņś (1958), in his beautiful study (Fig. 201) of the distribution of zooplankton in Skärshultsjön (see page 798), found that in August *P. remata, P. vulgaris,* and *P. euryptera* were epilimnetic, the second and third having almost identical horizontal distributions. These three species differ in size, the lengths recorded in the literature being 80 to 120 μ, 100 to 145 μ, and 160 to 210 μ, so that a good separation is probable. In the same profile Bērziņś found *P. major* (136 to 180 μ) to be metalimnetic and *P. longiremis* (160 to 220 μ), hypolimnetic. At least in this case, size differences and the temperature ranges provided by thermal stratification clearly produce niches for a taxocene of five species.

Succession in a warm monomictic lake in temperate South Africa. Schuurman (1932) has published data on the rotifers of Florida Lake, Johannesburg, which are of considerable interest when compared with the results of studies in Europe and North America. The lake is a small artificial body of water with a maximum depth of 5 to 6 m. The surface temperature varies from 7.5° C. in June to 22.5° C. in January, though for nearly the whole year it is higher than 10° C. The lake therefore differs from the dimictic lakes discussed in the preceding paragraphs mainly in its high winter temperatures. As in other lakes, rotifers were commoner in summer than in winter. The

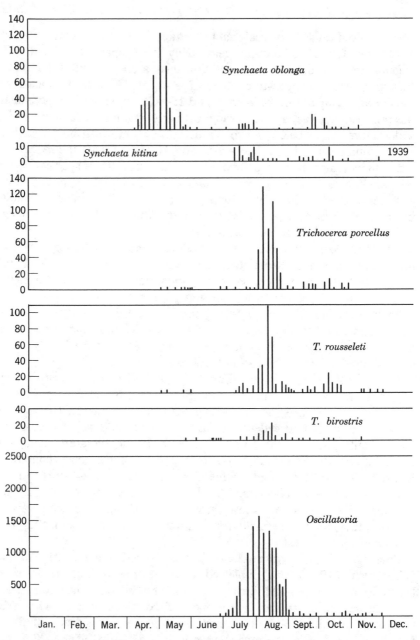

FIGURE 145. Seasonal incidence in the Motala River of two species of *Synchaeta*, *S. oblonga* and *S. kitina*, showing comparable variations in 1939, and of three species of *Trichocerca*, *T. porcellus*, *T. rousseleti* and *T. birostris*, also varying concomitantly and appearing maximally about the time of the *Oscillatoria* bloom. (Carlin.)

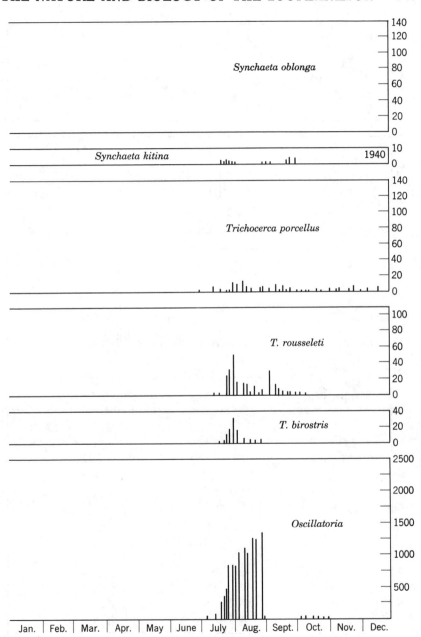

FIGURE 145.   (Continued)

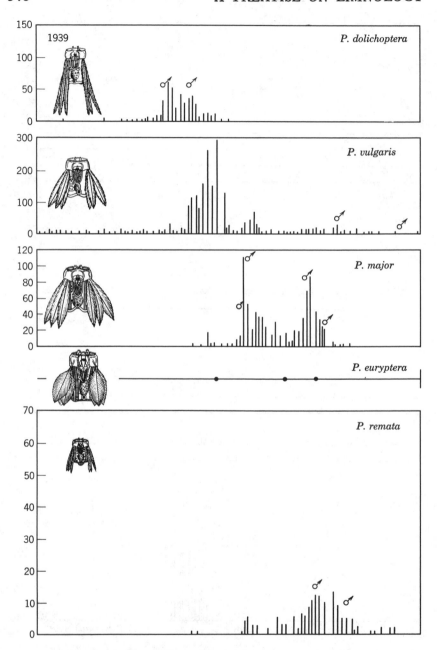

FIGURE 146. Largely asynchronous incidence of the five species of *Polyarthra* occurring in the Motala Rivers. (Redrawn from Carlin.)

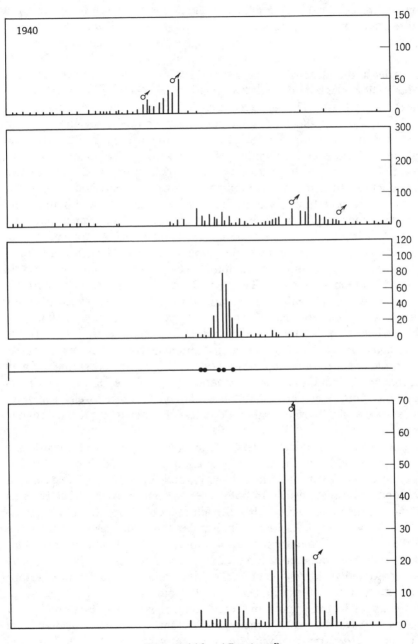

FIGURE 146.   (*Continued*)

only form confined to the colder months was an *Anuraeopsis* referred to *A. fissa,* which was always rare. *Keratella cochlearis, K. tropica,*[4] *Synchaeta pectinata,* and *Brachionus calyciflorus* were perennial, though the last-named has never been common. Both species of *Keratella* were least abundant when the strictly aestival species were commonest. The major maxima of the two species of *Keratella* appear not to be synchronous. The *K. cochlearis* population increased rapidly in October, whereas the *K. tropica* population showed irregular variation but was best developed rather earlier in the winter, between the beginning of June and beginning of September. The temperatures at this time were somewhat variable (12 to 16° C.); it would be a mistake to conclude that the tropicopolitan *K. tropica* flourished in Florida Lake at temperatures systematically lower than those favoring the strictly temperate *K. cochlearis.* Factors other than temperatures are doubtless involved. *Brachionus angularis* exhibited a large maximum in late August 1928; it was not quite perennial, being absent for a short time in the southern autumn. The other common species are clearly summer forms. They include *Hexarthra mira, Filinia (Tetramastix) opolienis, Brachionus falcatus,* and *Trichocerca chattoni,* all of which may be expected to be warm-water species on the basis of earlier records, though the last-named species is little known. *Conochiloides dossuarius* with a maximum in November occurs rarely from October to June; it is thus best developed in the spring, though probably at about the same temperatures as those of autumn in its dimictic North Temperate habitats. *Asplanchna brightwelli* is a summer species in Florida Lake; doubtless its maxima are determined mainly by food.

Succession in tropical waters. The only long series of records are Apstein's (1904) for the Lake of Colombo, Ceylon and Green's (1960) for the River Sokoto, Northern Nigeria, and for a pool isolated, except at flood, from that river. In both cases the major external determinant of faunal succession was clearly the alternation of wet and dry seasons.

In the vicinity of Colombo, November and May are normally the wettest, July and January the driest, months. The temperature is relatively constant between 26 and 28.3° C. The poorest rotifer fauna was recorded in May, but after the May rains a large fauna develops in June. It is difficult to learn much from Apstein's account about the changes in the populations of individual species. *Hexarthra mira* seems to have its greatest maximum in January and a subsidiary one in June. *Brachionus calyciflorus* apparently had maxima in June and

---

[4] Sub *K. quadrata valga;* specimens re-examined and found to be *K. tropica.*

September. Specimens lacking posterior spines were found mainly in the middle of the June maximum; earlier in May and in September the spined f. *amphiceros* was somewhat more abundant, though there seems to be no correlation with the irregular variation in the population of *Asplanchna brightwelli* (cf. page 902). From the fact that individuals of f. *amphiceros* did not carry small male-producing eggs in mid-June, when one sixth of the unspined population had such eggs while at the maximum of the spined form in September some were carrying male eggs, suggests that the spineless (*B. pala willeyi* of Apstein) and spined (*B. amphiceros borgerti* of Apstein) populations were independent and behaving as separate species, as Apstein implied by his taxonomic treatment. *Brachionus rubens* was enormously abundant in early July, *B. falcatus,* common in August, and *B. forficula* present only in May, September, and January, having a large maximum in the last month. *Keratella tropica,* absent in early May, rose to a large population in late June and then declined irregularly.

Apstein suggests that the washing in of food may play a part in the observed variation, the large populations developing in June and January shortly after the rains. It is also fairly clear that each of the sympatric species of *Brachionus* are reacting differently to the variations in external conditions.

Green's locality, which is fluviatile rather than lacustrine, varies in temperature more than does the Lake of Colombo; there is a cool spell in December, January, and February. During January the water temperature fell to 16° C., but from May to November it was higher than 25° C. The river level is maximal in October, after the principal rains, and the rotifer fauna at that time appears to be largely washed away. The common planktonic species, *Brachionus calyciflorus, B. falcatus, Keratella tropica, Asplanchna brightwelli,* and *Filinia (T) opolienis,* which clearly form an assemblage likely to occur commonly in Africa, if not in other tropical countries, all follow the same sort of cycle, disappearing almost entirely during the months from July to January or later. During the flood period adventitious specimens of benthic species, washed down from the backwaters of the river, constituted nearly all of the rotifer fauna; *Monostyla bulla, Lecane luna,* and *L. leontina* are the commonest. The marked cyclomorphosis of *K. tropica* in this locality is discussed in a later section (see pages 894–895).

Vertical distribution. As already indicated, the most striking feature of the vertical distribution of planktonic rotifers in thermally stratified dimictic lakes is the occurrence of supposedly cold-water forms such as *Polyarthra dolichoptera, Filinia terminalis,* and *Keratella hiemalis* in the hypolimnia of such lakes in summer. Pejler (1957b) finds that in subalpine tarns in Swedish Lapland, *Kellicottia longispina, Anuraeopsis fissa, Synchaeta truncata, Polyarthra dolichoptera,* and *Conochiloides natans* are able to live under the ice in oxygen-deficient water. All but *K. longispina* behave as cold stenotherm species in Swedish Lapland, though elsewhere *A. fissa* is usually a summer species. Pejler concludes that seemingly cold stenotherm rotifers are less likely to be sensitive to low oxygen concentrations than are more eurytherm or thermophilic species: they would therefore be better adapted to hypolimnetic life in the summer. Beadle (1963), however, has recently found in central Africa warm-water species such as *Keratella tropica* and *Horaella brehmi* living under anaerobic conditions deep in the mononolimnion of Lake Nkugute. Pejler thinks that niche separation between *Polyarthra vulgaris* and *P. dolichoptera* is in part mediated by differences in toleration of low oxygen concentration, particularly since *P. vulgaris* is more positively phototactic toward moderate light than are most planktonic rotifers. If both species co-occurred in a stratified lake, *P. vulgaris* would tend to stay in the upper illuminated waters at least by day and, at moderate temperatures, might well exclude *P. dolichoptera* by competition; *P. dolichoptera* would occur in deep cold and perhaps somewhat deoxygenated waters, from which *P. vulgaris* would tend to move upward by day and from the deeper layer of which this species might be excluded by low oxygen concentration. Comparable phenomena possibly occur in *Keratella hiemalis* and such epilimnetic members of the genus with which it may be associated. Pejler's (1961) most recent observations cast some doubt on the simple determination of all such distributions by temperature alone, for in Ösbysjön, a shallow unstratified lake, both *Filinia terminalis* and *Polyarthra dolichoptera* may sometimes appear in considerable numbers, usually near the bottom, at temperatures as high as 19° C. Pejler suggests that factors other than temperature may be involved and that the cold stenothermy of these species is apparent rather than real. The matter clearly requires further study; analogous phenomena probably occur in other planktonic groups.

It is to be noted that in almost every well-described case much greater populations of apparently cold stenotherm species are found under the thermocline in summer than throughout the lake in winter. This presumably reflects the increase in the rain of particulate food, such as faeces, and dead or inadequately buoyant organisms from the euphotic layers, as they became warmer. However much their trophic details may differ, *Oscillatoria rubescens, Keratella hiemalis,* and, as we shall see later, certain species of *Daphnia,* all exhibit a comparable type of vertical distribution.

POSSIBLE PLANKTONIC GASTROTRICHA

The gastrotrichs of the family Neogosseidae and, to a somewhat lesser extent, those of the family Dasydytidae are free-swimming, though the Dasydytidae swim mainly just above the bottom (Remane 1936). According to de Beauchamp (1933), *Neogossea antennigera* and *N. fasciculata* may occur in the plankton of ponds in various parts of France, living in the open water with rotifers of the genera *Anuraeopsis* and *Keratella,* though *N. antennigera* at least may also be found on the bottom. The animals, which often occur in great numbers in summer, produce parthenogenetic resting eggs in September, for male structures are unknown in the Chaetonotoidea. *N. antennigera* has also been found in the plankton of a tank in Nagpur in India (Visvesvara 1964).

It is possible, moreover, that *Kijanebalola dubia* (Fig. 147), a neogosseid from Lake Kijanebalola, a shallow body of water in Uganda, west of Lake Victoria, is rather more limnoplanktonic, though doubtless similar in its autecology, than the two species of Neogosseidae just

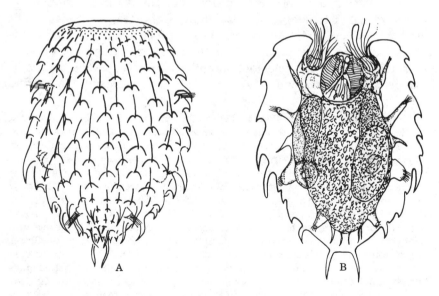

FIGURE 147.   *A.* Dorsal exterior view of the supposedly planktonic gastrotrich *Kijanebalola dubia* from central Africa.   *B.* Optical section of internal organs from ventral side.   (de Beauchamp.)

mentioned. *K. dubia* so far is known only from two specimens found in a plankton sample taken in the center of the lake (de Beauchamp 1932).[5]

CRUSTACEA

The great majority of the planktonic Metazoa, both in the sea and in fresh waters, belong to the Crustacea; in the fresh-water environment the planktonic Crustacea are represented mainly by species of the Cladocera in the subclass Branchiopoda and of the Copepoda in the subclass Maxillopoda. A few fresh-water Malacostraca and one or two Ostracoda may also be regarded as nektoplanktonic. All species of Crustacea considered here are more nektonic than any species of phytoplankton and are also less likely to be at the mercy of turbulent water movements than the planktonic rotifers.

In the following account it has proved convenient to begin with certain aspects of the nutritive physiology of the smaller Crustacea, since the existing information is most interesting if approached in a comparative manner and is also relevant to much of what comes later. The individual groups containing representatives in the fresh-water plankton are then considered systematically, emphasis being placed on general natural history, life history, and seasonal occurrence.

Nutritional physiology of planktonic Crustacea. The important work on the availability of various algae as food for the smaller Crustacea has been done primarily on three rather dissimilar groups of organisms, namely the halobiont anostracan *Artemia,* several species of the genera *Daphnia* and *Moina* in the Cladocera, and the marine

---

[5] Both de Beauchamp and Remane believe that an animal described and figured by Gosse (1886) as a rotifer, under the name of *Eretmia cubeutes,* is congeneric with *Kijanebalola dubia;* Remane (1936, page 203) in fact uses the combination *K[ijanebalola] cubeutes.* Murray (1906) found dead specimens of what he took to be Gosse's animal in Loch Ness and Lock Huna. In most of these specimens the "trophi of a rotifer, in the same definite position" were observed. Murray, in a footnote, says that Rousselet had concluded *E. cubeutes* to represent a rhizopod test inhabited by a rotifer. Murray, however, felt that the supposed test did not conform to that of any rhizopod known to him. De Beauchamp thinks that the supposed rotatorian characters of *E. cubeutes* given by Gosse, largely following Hood, are due to faulty observation. This is by no means self-evident, but even if *E. cubeutes* is not a rotifer its resemblance in external characters to *Kijanebalola dubia* is by no means so great that inclusion of the two species in the same genus is inevitable. Stranger beings than planktonic gastrotichs (cf., however, Arnold and Baker 1960; Baker 1960; Baker and Westwood 1960) have been supposed to inhabit Loch Ness, but Murray's observations certainly cannot be used as an argument that gastrotrichs do in fact occur in the plankton of the lake (cf. Hutchinson 1965).

littoral harpacticoids *Tigriopus californicus* and *T. japonicus*. Only the Cladocera studied belong to a family containing normal freshwater plankters, but even then the species used in experiments have been pond forms. Because this work gains much of its significance when the results are viewed comparatively, it seems best to consider the experiments on all three groups together. It is necessary, however, to discuss some aspects of the trophic natural history of several individual groups later in this chapter, but these aspects involve the morphology and behavior of food capture as much as the adequacy of the food captured to support the life of the feeding animals.

Tolerance of starvation. In *Daphnia magna* von Dehn (1930) finds that most specimens, in some experiments as high as 80 per cent, cannot survive more than about twelve hours when in bacteriologically filtered pond water. No attempt was made to sterilize the animals, and the results is attributed solely to the removal of nutritive particles. Addition of kieselgühr had no effect on survival; starch has only a very moderate effect, so that the lack of passage of materials, which would include the liquid medium, through the gut, is not involved. Only about 1 per cent of the unfed animals survived the whole experimental period of thirty-six hours. Controls, fed on dispersed fibrin or on ground *Anacharis,* all survived for the duration of the experiment. Earlier experiments, such as those of Wolff (1909) on *Simocephalus,* which gave contrary results, are criticized by von Dehn as technically imperfect (cf. also Kerb 1911). Gellis and Clarke, however (1935), obtained evidence that *D. magna* can grow and reproduce by using as food organic the particles that pass a Berkefeld filter and so are smaller than ordinary bacteria.

In contrast to von Dehn's results, Stuart, MacPherson, and Cooper (1931) claim to have kept *Moina macrocopa* for four days unfed in distilled water, a result that is surprising quite apart from the problem of starvation. Baylor and Smith (personal communication) have had comparable success with *Daphnia pulex.* The entire subject is probably of considerable ecological importance and should be reinvestigated.

Axenic culture. Two cases are known in which Crustacea have been raised in the laboratory on axenic but complicated diets. Triellard (1924) raised *Daphnia magna,* sterilized by repeated washings followed by rapid passage through H_2O_2 solution, in a 1 per cent solution of glucose to which red blood corpuscles of rabbits were added. Ten successive parthenogenetic generations were obtained. The red blood corpuscles, however, are no more easily analyzed nutritionally than the cells of an algal population, though the alleged bacteriological sterility of the culture is interesting.

Provasoli and Shiraishi (1959) found that Utah brine shrimps, referred to the amphigonic American race of *A. salina,* could be grown aseptically in a complex nonliving medium made up in sea water and containing trypticase (a commercial protein hydrolysate), liver infusion, hydrolyzed desoxyribose and ribose nucleic acids, serum, sucrose, cholesterol, glutathione, a mixture of B-vitamins, the so-called paramecium factor from yeast, and rice starch particles.   The starch particles apparently function to stimulate filter feeding and to increase transport through the gut, the nutrient medium going along with the solid particles.   Without such particles no development beyond the fifth instar occurred, although the controls with starch particles became adult. Thiamin and folic acid were proved to be essential vitamins, required in amounts in excess of, or in a form different from, the amount supplied by trypticase, serum, or liver.   Biotin, pyridoxine, nicotinic acid, and choline were needed to produce optimum development. Glutathione could be replaced by cysteine.   The required materials in horse or ox serum were heat stable.   Fatty acids, particularly as sodium salts, were markedly toxic.

Monoxenic culture.   Gibor (1956) found that *Stephanoptera gracilis, Dunaliella viridis,* and, to a lesser extent, *D. salina* and *Platymonas* sp. were much better foods for *Artemia* than *Stichococcus* sp., on which almost no growth was obtained.   All of these algae occur in the salt ponds from which the animals were obtained.   *Stichococcus,* as well as the other species, is ingested and experiments with mixed cultures indicate that it does not actively inhibit *Artemia.*   There is evidence, however, that *Stichococcus* is not well digested and can be recovered alive from the faecal pellets of the *Artemia.*   When *Artemia* is placed in a mixture of *Stichococcus and Dunaliella,* the result of its continuous filtering is to remove most of the *Dunaliella,* whereas the genus *Stichococcus* flourishes.   In controls without animals both species persist.   Gibor concludes that the rather usual dominance of *Stichococcus* in the more dilute salt ponds may be due to *Artemia* eating both species but mainly digesting *Dunaliella.*

Provasoli, Shiraishi, and Lance (1959) have extended this work by using nineteen different algal strains.   In this investigation a brine-living *Dunaliella* from La Jolla proved inadequate, whereas a marine strain permitted complete development.   Comparable differences occurred with two strains of *Platymonas.*   As before, *Stephanoptera* proved adequate, as did *Brachiomonas pulsifera,* but *Nannochloris oculata, Stichococcus fragilis,* and *Pyramidomonas inconstans* were inadequate foods.   Thus in the green algae five species or races were inadequate food and four were adequate.

In the Chrysophyceae three out of four and in the Cryptophyceae all three species, including the important plankter *Rhodomonas lens* were adequate foods. Neither *Eutreptia* sp. in the Eugleninae nor *Gyrodinium cohnii* nor *Peridinium* in the Dinophyceae permitted full development. In comparable experiments with the marine littoral harpacticoid *Tigriopus japonicus* four algal foods were adequate and seven inadequate for both Crustacea, whereas six were adequate for *Artemia* only and two for *Tigriopus* only. As they stand, these figures do not differ significantly from a chance distribution if it is assumed that it is equally likely that any strain is or is not adequate for *Artemia* or for *Tigriopus,* adequacy for one genus being independent of adequacy for the other. Further work will probably invalidate this assumption; it is quite likely that some groups will as a whole be better (Chrysophyceae, Cryptophyceae) and some worse (Dinophyceae) as crustacean food. It is clear, however, that the various crustacea differ in their reactions to the same food, and that very small taxonomic differences in the food may determine whether a species is adequate or inadequate for a given crustacean.

Although Gibor's work, as well as the investigations to be reported in a later section, on unialgal but not bacteriologically monoxenic cultures as sources of food for cladocera clearly indicates that digestibility, in part determined by the thickness of the cell wall of the food is of paramount importance, there must also be other factors. In *Tigriopus* it is possible to perform experiments involving many generations. In these, there is evidence that single species of food may not permit permanent cultures, whereas mixtures of two species do. Thus a particular strain of *Platymonas* was found (Provasoli, Shiraishi, and Lance 1959) to permit only two generations of *Tigriopus californicus* in monoxenic culture but indefinite reproduction when bacteria were present. The same *Platymonas* also proved to be an incomplete food for *T. japonicus* but only after eight generations.

Monochrysis lutheri, one of the species inadequate as food for *Artemia,* permitted indefinite (i.e., at least eighteen generations) survival of *Tigriopus japonicus.* However, *Isochrysis galbana* allowed but eight generations to survive and *Rhodomonas lens,* only six. When mixed, *I. galbana* and *R. lens* constituted a complete food on which twenty-six generations were reared. A mixture of B-vitamins, or glutathione, can correct the deficiency resulting from a diet of monoxenic *Isochrysis* (Shiraishi and Provasoli 1959). When animals are reared on such a supplemented diet, the effect lasts three generations on transfer back to unsupplemented *Isochrysis.* In the last generation females die more easily than males; all the adults in nonreproducing deficient cultures may belong to the latter sex. The biochemical meaning of the replacement of vitamins by glutathione is not clear, but it is also not certain which of the vitamins are involved.

The monoxenic cultures studied as food for Cladocera have been both bacterial and algal. Stuart, McPherson, and Cooper (1931) found that *Moina macrocopa* can live for six generations on *Aerobacter aerogenes* and *Escherichia* sp. (sub *B. coli mutabole*). *Bacillus subtilis* was an inadequate or perhaps toxic food; colored strains of *Chromobacterium violaceum* caused the death of the *Moina* but colorless strains can be eaten. One or two other chromogenic bacteria appeared unsuitable, but an unidentified *Flavobacterium* could be used as food. Pacaud (1939) has maintained sterile *Daphnia magna* and *D. pulex* on monoxenic cultures of both *Chlamydomonas agloeformis* and *Gonium pectorale*.

Unialgal cultures. The use of unialgal cultures as food for unsterilized Crustacea naturally throws much less light on the trophic requirements of the animals than the experiments just reported. The unialgal bacterially contaminated foods, however, are of interest in showing what algae can be eaten, to some extent despite their individual completeness as diets.

The main differences here are of digestibility, which on the whole seems to depend on the thickness and other properties of the algal cell wall.

The most extensive series of studies is that of Lefèvre (1942), though this work suffers somewhat from that author's distrust of quantitative presentation.

Among the lower green algae studied *Gonium pectorale, Pandorina morum, Chlamydomonas snowiae, Selenastrum bibraianum, Tetrallantos lagerheimii, Ankistrodesmus falcatus* var. *acicularis, Quadrigula* sp. *Chlorella vulgaris, C. pyrenoidosa,* and *Scenedesmus spinosa* permitted good development of both *D. magna* and *D. pulex* cultures fed on the algae. *Scenedesmus quadricauda* and *S. oahuensis,* however, were mediocre foods, and *Dimorphococcus lunatus* was almost or entirely inadequate. Lefèvre attributes the differences primarily to the thickness of the cell wall in these species relative to the green algae that are suitable as food.

Among three species of desmids tried *Staurastrum teliferum* and *Cosmarium impressulum* were adequate though less satisfactory than most of the green algae listed in the preceding paragraph. *Cosmarium tetraophthalmum* proved completely inadequate; like adult *Dimorphococcus lunatus* it was completely undigested, which Lefèvre again attributes to its thick cell wall.

Phacus pyrum proved to be adequate, but none of three species of *Euglena* was satisfactory; this is apparently partly caused by their size and shape, which make it difficult for the *Daphnia* to catch (*E. americana, E. mutabilis*) or to pass them from the prebuccal space (*E. deses*) into the alimentary tract. *E. americana* is also not easily digested. Field evidence that the shape and size of other algae, notably filamentous Myxophyceae, render them unsuitable as food for certain Crustacea, is presented later in this chapter (see pages 613, 616).

In most or in all cases in which digestion occurs (von Dehn 1930) the process involves passage of enzymes into and of products of digestion out of the algal cell, the wall of which tends to remain intact. It is therefore not surprising that the thicker-walled algae are not digested; what is remarkable perhaps is that any digestion occurs at all.

Experiments to determine whether the undigested algae such as *Dimorphococcus lunatus* and *Cosmarium tetraophthalmum* were really viable after passage through *Daphnia* were not undertaken. Though the aggregation of any living cells into faecal pellets would probably reduce their chances of survival, it is quite likely that, as in *Stichococcus* ingested by *Artemia,* some species of algae might possess a selective advantage in being able to pass intact through the gut of a *Daphnia*. This advantage, however, would be set against the obvious metabolic disadvantages of an overthick cell wall. Lefèvre concluded from the fact that *Staurastrum teliferum* is a fairly good diet for *Daphnia* that the spines of desmids provide little protection to their possessors, though other workers (e.g., Wagler 1925) have suggested that other spinous algae, such as *Rhaphidium* and *Scenedesmus,* might prove injurious to zooplankton attempting to eat them.

The fresh-water Branchiopoda. We turn now to a systematic survey of the planktonic crustacea of inland waters. The first group, the Branchiopoda, is best divided (page 118) into two living orders, the Anostraca and the Phyllopoda, the latter comprising the three suborders Notostraca, Conchostraca, and Cladocera. Ecologically, the life histories of the Anostraca, Notostraca, and Conchostraca (Fig. 148) are comparable; most members of all three groups live in waters that are either subject to periodic drying or extensive freezing or, as in the habitats of *Artemia,* are excessively saline. There can be little doubt that being on the whole large and rather slow-moving organisms few members of the three groups can live perennially in association with fishes (Bond 1934); *Lepidurus arcticus,* which may occur in permanent water in the mountains of Scandinavia and may be an important food for Salmonidae (Sømme 1934), appears to be the only significant exception. The general adaptation of most mem-

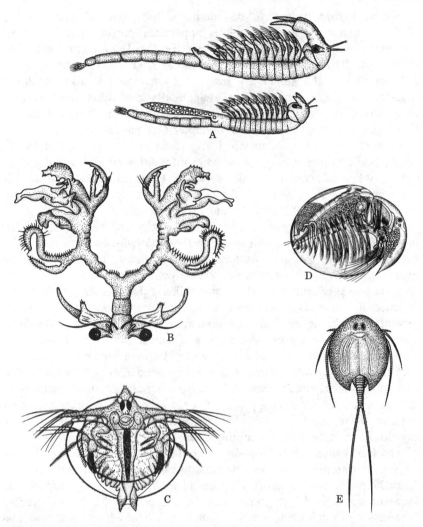

FIGURE 148.   Anostraca (*A, B*), Conchostraca (*C, D*), and Notostraca (*E*).
A. Branchinecta paludosa, ♂ (*above*), ♀ (*below*), with ventral surfaces
upward as in life.   *B.* Copulatory antennae of great complexity of ♂ of
Dendrocephalus denticornis.   *C.* Heliophora.   *D.* Adult of *Lynceus brachy-
urus.*   *E. Triops cancriformis.*   (Barnard, Calman, Sars, Weltner, Wesenberg-
Lund, redrawn and slightly modified.)

bers of all three groups clearly permits them to inhabit localities in which, because of drying, complete freezing to the bottom, or extreme salinity, neither fishes nor perhaps some other predators (Lundblad 1920; Lowndes 1933) are able to form permanent populations.

The Anostraca particularly are most characteristic of the inland waters of savannah, steppe, and semi-arid landscapes, occurring commonly in western North America, central Asia, the drier parts of Africa, and Australia. Such regions usually support more species of anostracans than comparable wetter areas. In New England four species (including an old occurrence of *Artemia* near New Haven, Connecticut) are recorded, in Texas, six (Mattox 1959); in France eight species or varieties are known (Mathias 1937), in Rumania, eleven (Botnariuc and Orghidan 1953); in the humid southwestern part of South Africa near Cape Town only two are noted, whereas in the arid terrain of Southwest Africa fifteen species are reported (Barnard 1929). A rather remarkable arctic fauna containing species of the genus *Polyartemia, Artemiopsis,* and of some less specialized and widely distributed genera such as *Branchinecta* also exists.

The distribution of the existing members of the Notostraca, of which there appear to be but two living genera and about seventeen species and subspecies[6] is not unlike that of the Anostraca. The Conchostraca are apparently widely distributed in both semi-arid and humid regions; in the latter they appear mainly in temporary rainfilled depressions often of small size, and at least in temperate eastern North America the described species are extremely local and sporadic in their occurrence.

Few members of the three groups can be regarded as lacustrine and fewer still as planktonic. The Anostraca live and feed in the free water, but usually in relatively small pools. They may occasionally occur in astatic pans or playa lakes of considerable area but very small depth. Thus Hutchinson, Pickford, and Schuurman (1932) record *Streptocephalus proboscideus* from Avenue Pan, Benoni Transvaal (Volume I, Plate 7), which had a diameter of the order of 0.5 km. and a depth, when studied, of about 60 cm.; they also note *Branchinella ornata* along with the conchostracan *Eocyzicus obliquus* in Eliazar Pan near Potchesfstroom, of like area but only 10 to 15 cm. deep when visited. Comparable occurrences are doubtless to be noted in Australia. *Artemia* can be regarded as a limnoplankter in the large but shallow saline lakes, such as the Great Lake, that it often

---

[6] Longhurst (1955) and Linder (1959) do not agree on the status of several forms.

inhabits. The Notostraca are essentially benthic. *Lepidurus arcticus,* as already noted, may live in permanent large bodies of water in the Norwegian mountains, whereas *Triops australiensis* in Australia and perhaps *Lepidurus lynchi* in North America (Linder 1952) can occur in shallow astatic lakes. A single record (Bowkiewicz 1923) of *Triops* sp. from Lake Baikal is due presumably to inadequate or incorrect labeling. The Conchostraca occur in large bodies of water only when the latter are very shallow and seasonal, as in the case of *Eocyzicus obliquus* just mentioned.

In spite of their lack of direct importance to the student of lake biology, some discussion of the life histories of these animals provides a useful background for an understanding of the biology of the Cladocera, a group of immense significance to the biological limnologist.

Life history. Nearly all the Anostraca and Conchostraca are bisexual. Parthenogenesis is certainly found in some populations of *Artemia salina* (*s. lat*), in which species complicated raciation involving polyploidy occurs (the most recent paper is Goldschmidt 1953, from which earlier literature can be explored, e.g., Artom 1931, Barigozzi 1946). It has been supposed that *Branchipus stagnalis* and *Streptocephalus torvicornis* may also occasionally reproduce parthenogenetically (Abonyi 1911), though Mathias (1937) is skeptical. Among the Conchostraca, males have never been found in *Limnadia lenticularis,* in which species Lereboullet (1866) concluded reproduction must be by parthenogenesis. The recently described and very local *Caenestheriella gynecia* from Ohio also lacks males (Mattox 1950). In the Notostraca males are almost but not quite unknown in *Lepidurus arcticus* (Longhurst 1955). In *L. apus* in Europe there seem to be no males in the northern part of the range of the species, but proceeding southward their incidence increases irregularly. Most European collections of *Triops cancriformis* lack males even as far south as Pavia, but in tropical latitutes males are common. A comparable increase southward in the proportion of males also seems to occur in the species inhabiting North America. It is to be noted that populations of *T. cancriformis* deficient in males exist to the south of the bisexual populations of *L. apus.* The latter species apparently always lacks males in rice paddies; Longhurst suggests that postglacial colonization northward and postagricultural colonization of new artificial habitats are less easily performed by bisexual populations. As Bernard (1891, 1896) pointed out long ago and as Longhurst (1954, 1955) has shown critically in more recent years, the populations lacking males actually consist of apparent females that are really self-fertilizing hermaphrodites; parthenogenesis seems not to occur in the Notostraca.

Mating (Mathias 1937) in the Notostraca appears to involve ventral apposition of the two individuals. In the Anostraca the antenna of the male is greatly, indeed sometimes fantastically (*B* of Fig. 148), modified to form a clasping organ by which the male, after preliminary pose below the female (Moore and Ogren 1962), seizes her just in front of the ovisac. In the Conchostraca the margin of the carapace of the female is held by the modified anterior pair of legs of the male. It is interesting that in the extinct Devonian branchiopod *Lepidocaris,* which may be referred to a special order the Lipostraca, the maxilla was apparently modified as a clasping organ, whereas in the Cladocera it is the antennule that has become differentiated in this way.

The great specific diversity of form in the male antennae of the Anostraca provides the best taxonomic characters within a genus. This might seem remarkable in organisms usually found in unispecific taxocenes, but Dexter (1953) has recorded a sufficient number of cases of two or more species living together to suggest that structural isolating mechanisms may have a survival value.

Eggs are ordinarily retained by the female in special structures for some days after laying, though, except in the ovoviviparous *Cyclestheria,* as in the Cladocera, the biological meaning of this retention is far from clear. In the Anostraca the eggs are discharged into an ovisac on the ventral side of the first two postpedigerous segments; in the Notostraca part of the eleventh postcephalic appendage of the female is modified to retain them; in the Conchostraca they are held in the cavity of the carapace above the trunk of the animal. In the Anostraca and Notostraca the eggs are scattered several days after entering the ovisacs. In the Conchostraca they may not be dispersed until the death or more usually the molt of the female. In the widely distributed subtropicopolitan genus *Cyclestheria* the egg hatch in the brood pouch much as in the Cladocera. According to Barnard (1929), though *Cyclestheria* may occur in perennial ponds in southern Africa, it is also apparently an inhabitant of basins only seasonally filled with water. How it survives in these localities is unknown.

Mathias (1937) records in various Anostraca around a hundred eggs in the ovisac and indicates that clutches of this size may be produced three or four times in the life of a female. Actually, the greatest recorded egg production is by a female of *Streptocephalus seali* kept by Gaudin (1960) which laid 488 eggs in forty-six days. In *Eubranchipus vernalis* Avery (1940) noted up to ninety-five eggs produced by a female. Gaudin thinks this figure is minimal, having found as many as 300 eggs present at one time in an ovisac of *S.*

seali. The Conchostraca, according to Mathias, produce as many eggs as the Anostraca, or even more, the Notostraca perhaps rather fewer.

In *Artemia* there appear to be two kinds of eggs, one essentially subitaneous or ovoviviparous, with an egg membrane but no shell, hatched within the ovisac, the other resting, with a tough shell, and liberated from the ovisac. As in *Branchipus,* the *Artemia* eggshell is three-layered, with gas-filled lacunae in the middle layer. Such eggs may float. The proportion of subitaneous to resting eggs varies seasonally in *Artemia,* but there seems to be no consistent environmental control of the cycle. In some localities, such as Cagliari, ovoviviparity occurs in winter and spring and resting eggs are produced in the summer; near Marseille the reverse is observed.

Though it has long been apparent that eggs of most species of the three groups under discussion can undergo a prolonged period of diapause, usually in a dry but sometimes in a wet or frozen condition, the biology of diapause and hatching are still not adequately understood. The idea frequently held, and as a formal concept due to Wolf (1908), that the species of Anostraca and of the larger Phyllopoda can be grouped into those occurring in cold countries producing eggs that must be frozen and those occurring in warm countries producing eggs that must be dried before they will hatch, is certainly an erroneous oversimplification.

It is quite likely that the fundamental change needed to produce hatching is always a lowering of osmotic pressure at the egg surface. This, in effect, may happen indirectly when a dried egg is wetted or a frozen egg is placed in a liquid medium. It may also happen more directly when the water in which an egg is lying is diluted by distilled or, in nature, rain water.

In *Chirocephalus diaphanus,* the most widespread Anostracan in western Europe, the eggshell is two-layered and splits during hatching to reveal a nauplius covered by the egg membrane. There is a considerable delay of twenty-four to forty-eight hours before the membrane ruptures and eclosion is complete (Hall 1953). In pondwater at 15° C. permanently wet eggs begin the hatching process by splitting the eggshell in nine to fourteen days, with a mode at twelve days after liberation from the ovisac; complete eclosion occurs about two days later. Drying the eggs for seven to fourteen days produces a delay of a day or two in the time taken for splitting to occur after the eggs are returned to water. Longer periods of drying, more than forty-two days, apparently decrease the time needed for the split to occur. In all the samples of dried eggs a certain proportion, as high as 40 per cent, showed a marked delay in proceeding to the next stage of egg membrane rupture and liberation of the nauplius. In these cases the egg appeared to continue developing normally; the nauplius may have molted in the membrane. Addition of distilled water caused a proportion of these delayed eggs to hatch; a further addition some days later produced another batch of nauplii.

In *Eubranchipus vernalis,* the common species in eastern North America, undried unfrozen eggs can hatch apparently after many months, and so much more slowly than do those of *C. diaphanus* (Avery 1939; Castle 1938). Moreover Dexter (1946) could find no evidence of any truly subitaneous eggs in this species. It is conceivable that cases in which diapause is broken in permanently wet eggs kept for months are due to unrecorded dilution of the medium. Either freezing or drying and then return to liquid water can terminate diapause. Eggs frozen or dried after three months in water hatch in about twelve days after return to a liquid medium (Weaver 1943). By no means all of the eggs in any group react to the stimulus of drying; it is probable that a second drying may cause a further batch of nauplii to appear. The absence of *E. vernalis* in shallow Canadian ponds which freeze to the bottom is apparently due to the timing of the life history. The early stages are ordinarily passed during winter under ice; if no liquid water were present in winter, the young hatched in the autumn would perish (Ferguson 1939). The time taken to hatch after wetting is similar in *Chirocephalus* and *Eubranchipus,* but in some species it is clearly much shorter. Moore (1955) found *Streptocephalus seali* to have hatched in nature, at a temperature of about 11°C., in less than sixty hours after the filling of a temporary pond by heavy rain.

Mathias (1937), faced by the same kind of irregularity that has been partly analyzed in *Chirocephalus* and *Eubranchipus,* concluded, probably wrongly, that *Branchipus* laid two kinds of eggs, subitaneous and resting, but that they are not distinguishable, though possibly laid by different females. He tended to generalize his results to all the larger Branchiopoda, but this is clearly incorrect. It is evident that some anostracan eggs such as those of *Chirocephalus diaphanus* hatch spontaneously, whereas in other species, for example *Eubranchipus vernalis,* diapause always occurs and can be broken by freezing or drying, or both, if not by merely diluting the medium. Even when eggs hatch spontaneously they can resist desiccation before hatching. In all cases after diapause there is probably great variability in the time taken to hatch. Whatever the genesis of this variability and, in some species, of the delay of about twelve days before hatching occurs, it is quite likely that these characters are adaptive, for one of the great dangers faced by an animal inhabiting temporary waters is that a very small rain may flood the bottom of the basin with a thin layer of water which could cause hatching or other kinds of emergence from resting stages, only to generate a catastrophe as the water evaporated in a day or two. Such false beginnings of a rainy season might be destructive, but could clearly be mitigated without prejudicing a species by introducing competitively dangerous delays if a great inherent variability existed in the emergence rate from the resting stages. Evidence, not only from the Anostraca, as has just been indicated, but from the Cladocera also, suggests that such variability exists; whether it is based on genetic polymorphism or, as is more probable, on nongenetic factors, is, of course, quite unknown.

In the Notostraca the older accounts are confusing. Brauer (1877) concluded that *Lepidurus apus* eggs die if dried. The species appears to be less confined to seasonal waters than *Triops cancriformis* (Lundblad 1920), but it can occur in seasonal ponds (Linder 1952). At least the eggs of the Australasian *L. apus viridis* have hatched after drying in the laboratory (Fox 1949; Longhurst 1955). *L. arcticus,* which is presumably the species most characteristic of perennial waters, produces eggs that can be hatched after drying (Longhurst 1955). In the species of *Triops,* including *cancriformis,* which have been well studied, some eggs can hatch immediately without drying. It is probable that the main ecological factors determining the occurrence of the different species of Notostraca in any continental area is the temperature achieved by the waters suitable for these animals. *L. arcticus* clearly develops at low temperatures, *L. apus* perhaps best in slightly warmer water, whereas *T. cancriformis* needs fairly warm water for its development and is a summer form in central Europe.

In Australia *Lepidurus apus viridis* occurs in the southwestern coastal region where there is winter rain, *T. australiensis* in the arid interior where rain is sparse. In North Africa *Triops granarius* apparently spreads into wetter regions than *T. cancriformis* (Longhurst 1955). In all cases a relationship to be incidence of rainfall and prevalent temperatures is probably involved in determining distributions.

Longhurst (1955) found that the addition of 30 per cent distilled water to the pond water in which the eggs of *T. cancriformis* were laid stimulated hatching in twelve to fourteen days but that the resulting larvae died if not placed in full-strength pond water. Alternate dilution by rain and concentration by evaporation might well provide the conditions for successful hatching in nature.

There are fewer observations on the breeding biology of the Conchostraca, but it is not unlikely that a number of species can pass through successive generations without drying and without diapause. This seems to be true *Eulimnadia stoningtonensis* (Berry 1926), which lives in the humid landscape of New England, and of *Cyzicus cycladoides* in France (Gravier and Mathias 1930), but it is apparently not true of the undetermined species of conchostracan from the Middle East, studied embryologically by Cannon (1924). In *C. cycladoides* at least the egg is known to be resistant to drying. In *Eulimnadia agassizii* the appearance of several broods in very shallow temporary waters which dried after a few days is attributed by Zinn and Dexter (1962) to successive hatching of parts of an old population of eggs. The species can achieve reproductive maturity in as little as four days. The species best studied is *Caenestheriella gynecia,* in which only females are known (Mattox 1950; Mattox and Velardo 1950). This animal occurred in temporary pools near Oxford, Ohio. Nauplii appeared on April 17, 1949, at a temperature of 17° C. Ovigenous adults first appeared on May 31, 1949, at the same temperature and remained present for at least a month, during the whole of which time nauplii were also present. Experiments showed the seasonal sequence to be interpretable in terms of temperature. Wet eggs could not hatch below 13° C., and eggs that had been dried, below 17° C.; at such temperatures hatching was slow, taking respectively, twenty-one and eighteen days. The most rapid development occurred between 24 to 28°, at which undried eggs hatched in thirty-two and previously dried eggs in thirty-five hours.

At higher temperatures hatching was somewhat delayed; no eggs hatched at temperatures in excess of 38° C.; at this temperature it apparently took 3.8 days for undried and 7.7 days for previously dried eggs to hatch. Though there are some obscurities in the account, it appears that drying always delays the hatching time. Nauplii hatching at supraoptimal temperatures are somewhat reduced in size. The eggs can be stored dry for at least five years and in hermetically sealed and presumably anaerobic water for eight years. Eggs stored in the latter way, when placed in fresh tap water hatched in six days at 21° C. It is possible that admission of oxygen was involved in breaking diapause, but an osmotic stimulus is also likely.

It is probable that the species of Conchostraca which inhabit relatively humid regions generally have life histories determined mainly by temperature. Since, however, these species, at least those in eastern North America, appear to be excessively local and to inhabit localities that are very shallow, of small area, and easily destroyed, they must be fugitive species whose persistence in nature clearly depends on the viability of dried eggs. The species of warmer and more arid countries doubtless normally appear after seasonal rains and in some cases drying may be necessary. In view of what is known about the hatching of Anostraca and Notostraca, we may suspect, however, that osmotic breaking of diapause is likely to be possible even when a pool has merely shrunk and not dried up. The main difference between the species of humid and arid regions, if there is such a difference, is most likely to lie in the facultative nature of the diapause in those species such as *C. gynecia* that inhabit humid areas.

According to Heath (1924), the first instar anostracan larva is a typical nauplius, but indications of several postmandibular segments are apparent in the second instar, and the rudiments of five or six postcephalic limbs are distinguishable in the third. In the Notostraca the carapace begins to form in the third instar, but at this stage the Conchostracan larva is a rather peculiar organism (*D* of Fig. 148) with a flat incipient carapace, large labrum, and rudimentary antennule termed the *heliophora* by Botnariuc and Orghidan (1953).

Artemia is sexually mature in the twelfth and *Branchinecta occidentalis* in the fifteenth instars. In *Artemia* at least one and in *B. occidentalis* certainly two molts occur after the instar in which maturity is attained. The achievement of sexual maturity in *Artemia* occurred in the laboratory in twenty-one days. Many generations per year are clearly possible in nature. Comparable rates of development probably characterize most Anostraca living under uniform favorable conditions at a temperature higher than 20° C. (cf. Mathias 1937). In the Anostraca having as is usual life histories regulated by both temperature and desiccation, the generation time is likely to be longer than in *Artemia*. This has already been indicated for *Eubranchipus*.

In *Branchinecta occidentalis* at Stanford, California, Heath found rapid growth from November 29, in the pond in which the animals lived, until December 24, when some animals were in the fourteenth or preadult instar, at temperatures of 8.5 to 10.3° C. Little further development occurred till April, when mating was noted; in May only a few seventeenth instar postreproductive individuals occurred at 23.3° C. It is suggested that the maturation of the gonads occurs only above 22° C. and that until this temperature is achieved the final molt cannot occur. In this species at Stanford the prereproductive period evidently lasts about four months, the reproductive period one to two months.

In *Polyartemia forcipata,* an arctic form, Ekman (1904) concluded that the total life span in the mountains of Sweden was two and a half to three months.

In the Conchostraca the scattered records collected by Mathias seem to indicate that reproduction often begins before the animal is more than half its adult size, so that the proportion of prereproductive life is often shorter than in the larger Anostraca. This trend is, of course, accentuated in the still smaller cladocera. At least in *Cyzicus cycladoides* the total life span in the laboratory may be four months.

Cladocera: subdivision and occurrence in the plankton. The Cladocera are best (Brooks 1959) divided into two suborders, the Haplopoda and the Eucladocera. The Haplopoda contain the single species *Leptodora kindtii*, the longest known cladoceran, which reaches a length up to 18 mm. (*A, B* of Fig. 149). It is an elongate predaceous plankter, with six prehensile cylindrical legs borne on a distinct thorax, behind which is a three-segmented abdomen. The carapace is reduced to a small brood pouch borne dorsally on the thorax. The young hatch from the sexual eggs as nauplii, which molt after two hours to become (Sebestyén 1949) metanauplii. Though the life history is typically cladoceran and the reduced carapace is obviously a secondary development from a more normal structure enclosing the animal, the retention of the partly segmented abdomen and the nauplius larva (Fig. 150) indicate that *Leptodora* has no close affinities among the other Cladocera, though the comparable specialization of the antennule of the male both in *Leptodora* and in the other members of the group may argue against such a diphyletic origin.

The Eucladocera are divided into three groups which are given the status of superfamilies by Brooks and termed by him the Sidoidea (= Ctenopoda), Daphnoidea[7] (= Anomopoda), and Polyphemoidea (= Onychopoda).

Sidoidea (*C, D, E,* Fig. 149). The Sidoidea appear to be primitive filter feeders, though like the Conchostraca they were doubtless originally not limnoplanktonic. They possess six phyllopod appendages, with little differentiation between the various pairs. Their appendages are used in such a way that the anterior ones are moving backward as the posterior ones move foward. Water is thus drawn in between the anterior appendages while it is being expelled between the posterior. As the water is drawn in, it is strained through setae on the endites of the limbs; as the water is expelled, the food is loosed from the setae of the more distal endites and brought to the region of the gnathobase, the setae of which constitute the definitive filter. Combs on the wall of the food groove between the appendages appear to scrape off the food collected by the gnathobase filters, and it is moved

---

[7] Brooks (*personal communication*) indicates that since a superfamily name is involved it should follow that of the oldest genus and family included; Daphnoidea therefore at his request replaces Chydoroidea.

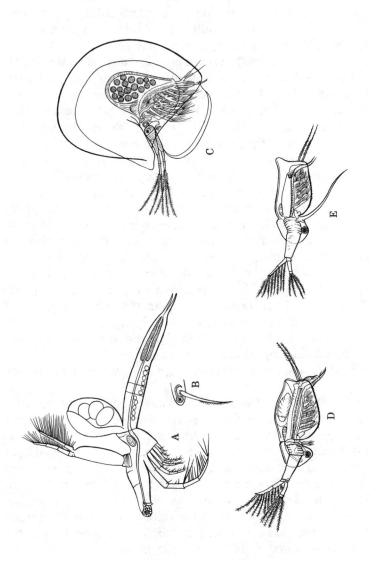

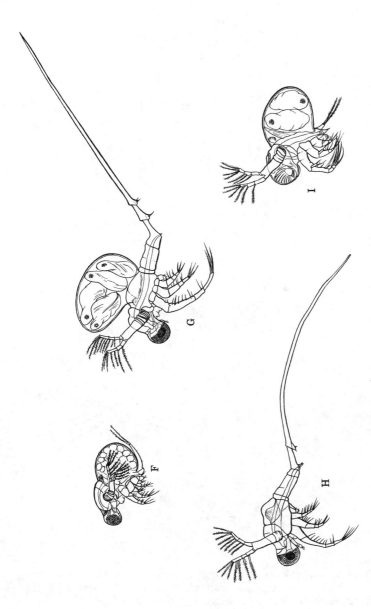

FIGURE 149. Haplopoda, Sidoidea, and Polyphemoidea. *A. Leptodora kindtii* (× 6). *B.* Head of ♂ of same. *C. Holopedium gibberum* (× c. 11). *D. Limnosida frontosa* ♀. *E.* The same ♂ (× 15). *F. Polyphemus pediculus* (× c. 14). *G. Byhotrephes longimanus* f. *brevimanus* ♀. *H.* The same, ♂ (× c. 13). *I. Podon intermedius* (× c. 26), all planktonic, *A, C, F,* holarctic or more widespread. *D, G,* palaearctic. *J.* A marine form occasionally found in inland localities. (Lilljeborg, slightly modified.)

569

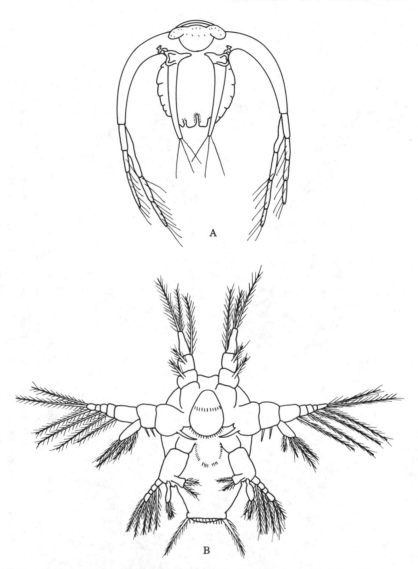

FIGURE 150. *A*. Branchiopodan nauplius of *Leptodora kindtii*. (Warren, modified, to show only external features, yolk granules omitted.) *B*. Copepodan nauplius of *Cyclops strenuus*. (Dukina, modified.)

forward by discontinuous spurts of water at the end of the backstrokes. The whole process has been elucidated for *Diaphanosoma* by Cannon (1933) on the basis of photographic studies by Storch (1929) and of his own examination of the anatomy of this genus; the original contribution must be consulted for the rather complex details of the process.

The majority of the Sidoidea, although filter feeders, are not limnoplanktonic. *Sida* has large cervical glands and can attach itself by means of these organs with its dorsal side toward a suitable substrate, filtering water in a stationary position *Latona* is a benthic littoral form that apparently feeds on the bottom. The euplanktonic species are *Sida crystallina* var. *limnetica,* possibly deserving specific rank, in the large subalpine lakes of Switzerland and Italy; *Diaphanosoma leuchtenbergianum,* a widespread holarctic species very close to and often not regarded as distinct from the more benthic *D. brachyurum;* a number of species of *Diaphanosoma* of restricted range in the lakes of the tropical Old World (Brehm 1933; Harding 1942) and the two species of the curious genus *Holopedium* (*C* of Fig. 149), in which a gelatinous capsule is developed on the carapace, *H. gibberum* occurring in calcium-deficient waters in the temperate Holarctic and *H. amazonicum* in the warm parts of the New World from Brazil northward into the United States.

Daphnoidea. The four families Chydoridae, Macrothricidae, Bosmidae, and Daphniidae have much more differentiated appendages than the Sidoidea; the group was probably originally benthic in shallow water, as the Chydoridae and Macrothricidae are today. Certain species of these families are likely to become planktonic (Fig. 163) when blue-green algal blooms develop, as Birge (1898) noted for *Chydorus sphaericus* in Lake Mendota. The phenomenon may be comparable to the occurrences of the rotifer *Trichocerca* (see page 542) under the circumstances and, as Dr. Clyde Goulden (personal communication) suggests, conceivably depends on the optical properties of a turbid but deep lake imitating those of shallow water.

In the other families filter feeding in free water has been redeveloped, and a number of species have left the small ponds and weed beds to become limnoplanktonic. The details of the filter-feeding mechanisms of *Daphnia* have been worked out by Cannon (1933), whose paper may be consulted by the interested reader.

In *Moina* which Goulden regards as the type of a separate family, including also *Moinodaphnia,* most species are heleoplanktonic and in fact generally occur in small temporary pools. *Moina micrura* (= *M. dubia, M. propinqua*) a quite small species is however often

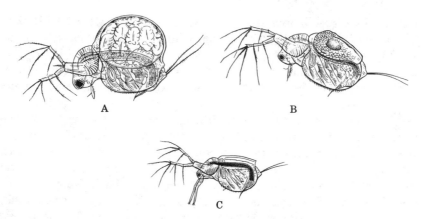

FIGURE 151. *Moina micrura*. *A*. Parthenogenetic female. *B*. Ephippial female. *C*. Male. Gracemere Lagoon, Queensland (× 25). (Sars sub *M. propinqua*.)

limnoplanktonic, particularly in tropical and warm temperate latitudes (Fig. 151). Other members of the genus are truly planktonic in saline or rather alkaline lakes in which there would be less competition from other cladocera and less risk of predation by fish.

In the Daphniidae the members of *Ceriodaphnia* are pond forms. *Scapholeberis* is associated with the surface film in relatively quiet bodies of water, whereas *Simocephalus* is benthic, often filtering like *Sida* when attached to a solid surface by its cervical gland.

The genus *Daphnia*[8] consists of about thirty species (Brooks 1957a), of which nine belong to *Ctenodaphnia* and the others to *Daphnia* s. str. Most species of *Ctenodaphnia* are large and probably suffer the same disadvantages as the Conchostraca in the presence of predators. They tend to inhabit ponds, often of a temporary nature and sometimes somewhat saline or alkaline. *D. (C.) tibetana*, a large dark species, occurs in the closed lakes of central Asia, in which fish are absent (Hutchinson 1937a). *D. (C.) lumholtzi*, from various

---

[8] Wherever possible, the species of *Daphnia* will be designated by their correct names, mainly following Brooks (1957a), who, as far as the North American species are concerned, has brought order out of unbelievable chaos. When a correct designation is not possible, the name used by the author of the work under discussion, followed by (sens lat.) will be employed. The taxonomy of *Moina* is at the time of writing still in great confusion, shortly to be resolved by the appearance of a monographic revision by Dr. Clyde Goulden whose advice has been available in preparing this chapter. Great difficulties still attend the classification of the species of *Bosmina*.

African localities, including Lake Victoria, is perhaps the most typically euplanktonic species of the subgenus.

In *Daphnia* s. str. the taxonomy is critically known only for North America (Brooks 1957a). Some species, such as *D. laevis* of the southern United States, apparently are strictly pond forms, and others, such as the widespread *D. galeata mendotae, D. catawba, D. retrocurva,* and *D. dubia* in eastern North America, are characteristically lacustrine, whereas *D. thorata* is endemic to the lakes, some very large, of the Pacific Northwest. A number of species, however, can occur in waters of a variety of depths. *D. pulex* is usually a pond form but is known in the plankton of some lakes. *D. schødleri* is likewise a well-known pond species in western North America, but it is also a conspicuous member of the plankton of Lake Mendota and occurs in inshore swarms in Great Slave Lake. The factors influencing these distributions are quite obscure, though in the case of *D. schødleri* in a very large, transparent body of water it is clear that some feature is present in the littoral that is significant to the animal and absent from the depths. Comparable cases in other planktonic groups are discussed in a later paragraph (see pages 799–804).

There is little doubt that a low temperature determines the distribution of *D. longiremis* and its European ally *D. cristata*. *D. longiremis* at the southern end of its range is a hypolimnetic form occurring in lakes large and infertile enough not to develop a very marked clinograde oxygen curve, though in the Arctic it can live in ponds. *D. middendorffiana* (Fig. 152), a large and dark pond form, is also found mainly in cold water, but it is seemingly not a species of large lakes and in its southern outposts occurs most often in mountain tarns. *D. ambigua*, which, if it were known only in the lakes of temperate eastern North America, would be recognized immediately as a cold stenotherm species found in summer in the hypolimnion, is also found in the same region in shallow ponds at quite high temperatures. It is evident that a great deal remains to be learned about the autecology of the species of *Daphnia*.

In Europe it is probable that much the same situation exists as in North America, but the taxonomy, though more advanced than in North America as late as 1957, is now seen in the light of Brooks's study to be inadequate. It is greatly to be hoped, in view of the importance of the genus in both theoretical and applied limnology, that the work of Hrbáček (1959b) will be extended to produce a satisfactory monographic revision for the temperate Old World. The main limnetic forms are clearly *D. hyalina, D. g. galeata,* and *D. cucullata;* the first is very close to *D. thorata,* the second, conspecific

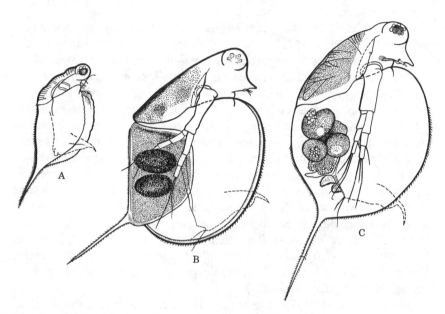

FIGURE 152. *Daphina middendorffana.* *A.* Parthenogenetic female, Manitoba.
B. Ephippial, probably pseudosexual, female, Alaska. *C.* Male, Grand Coulee,
Washington (× 30). (Brooks, somewhat diagrammatized.)

with *D. g. mendotae* of North America, and the third may be the
ecological equivalent of *D. retrocurva.*   The status of many populations
referred to *D. longispina* is quite obscure and the problem of limnetic
D. pulex, which may be complicated in both Old and New Worlds
by the existence of another species, *D. pulicaria,* not recognized by
Brooks, remains to be considered.   Outside the holarctic the taxonomy
of *Daphnia,* apart from comments by Brooks, remains essentially as
left by Wagler (1936).

   In the Bosminidae both genera, *Bosminopsis* and *Bosmina,* are essentially planktonic. *Bosminopsis,* containing a single species, *B. deitersi,*
tends to be warm temperate to tropical. *Bosmina* appears to consist
of at least two superspecies, *B. coregoni* and *B. longirostris,* which differ
in a number of minute but significant characters (Goulden and Frey
1963).   Within these superspecies the detailed taxonomy is still very
confused. A number of peculiar cyclomorphotic populations, found
mainly in the countries surrounding and east of the Baltic, are referrable
or allied to *B. coregoni* and are discussed in Chapter 26.   As will appear

in the discussion of biostratonomy in Volume III, the replacement of *B. coregoni* by *B. longirostris* is most characteristic of lakes that are passing from oligotrophy to eutrophy; contemporary field evidence in fact indicates that *B. longirostris* is found, when it leaves the littoral, in more fertile waters than *B. coregoni,* but the underlying physiological differences are unknown.

Polyphemoidea. The third superfamily of Eucladocera includes nine predaceous genera. As in *Leptodora,* the carapace is reduced and the limbs are free and able to grasp particulate objects. Unlike *Leptodora,* the Polyphemoidea have only four pairs of thoracic legs; the abdomen, of course, is unsegmented and reduced, and development is direct, so that no close affinity need be postulated between the two groups. *Podon* and *Evadne* are marine; *Evadne* is important in the Caspian basin, in which a number of Ponto-Caspian endemic genera also occur (Fig. 46). Several species of *Bythotrephes* (Fig. 149) are characteristic of the plankton of the lakes of the northern Palaearctic, whereas *Polyphemus* is widely distributed, mainly in ponds. The whole group might well be Sarmatian or Ponto-Caspian in origin.

Cladocera: life history. In general, the Cladocera have a life history which in its superficial characters resembles that of the Rotifera Monogononta. Presumably, in both groups the life history was evolved in the same kind of habitat—shallow fresh waters which dried or otherwise became uninhabitable by active individuals during part of the year.

Development of the parthenogenetic female. Parthenogenetic females may be hatched from resting eggs or produced by an earlier generation by parthenogenesis. Typically, this process continues for a number of generations until sexual reproduction intervenes. The sexual processes and certain exceptional life histories are discussed in detail in a later section.

In mature *Daphnia* the parthenogenetic egg is laid about half an hour after molting (Green 1956). The oviducts are very narrow, and the egg is squeezed out as a elongate structure into the brood pouch between the trunk and the carapace. Here the newly laid egg starts to swell and become rounded. The subsequent development (Table 29) in *Daphnia magna* is divided by Green into eight embryonic stages. With this system of stages, he established the remarkable fact that a collection of females of *D. magna* from a given pond contains a greater proportion of some stages and a lesser proportion of others than would be expected from random ovulation followed by development according to the timetable given in Table 29. The time of brooding, while the young are still in the pouch, depends on the length

TABLE 29.   *Embryonic stages D. magna (Green 1956)*

Stage		Duration in Hours $\sim15°(13–18°)$ C.	$22°$ C.
1	Opaque becoming translucent, with transparent edges	4–6	3
2	Markedly granular transparent edges, fat cells forming	26	17
3	Embryo apparent, head not defined, egg membrane cast off	18	10
4	Embryo head defined, with short antenna	18	10
5	Embryo with two very small pink eyes and longer antenna	7	4
6	Embryo with two distinct red eyes	4	2–3
7	Embryo with two large black eyes	13	8
8	Embryo with one large black eye	26	15
	Total embryonic period	~117	~70

of the instar, which increases progressively with age. A population of large old animals might therefore be carrying a disproportionate number of mature young. Thus at 25° the embryonic period (Obreskove and Fraser 1940) is forty-six hours, but Anderson and Jenkins (1942) found that such embryos might be retained for periods of forty-six to sixty-four hours. This retention, however, does not explain a disproportionate number of Stage 1 embryos. It would seem that eggs are laid more or less synchronously by the females of a pond, presumably in response to external variables, or that adults congregate in particular regions at different stages of each instar. These possibilities clearly require investigation; Hall (1962) found an essentially random temporal distribution of reproductive stages in *D. galeata mendotae* in Baseline Lake, Michigan.

The liberation of the young apparently occurs ordinarily a little before molting.

Green (1956) found that during development 16 to 25 per cent of the dry matter of the egg is lost in *Daphnia magna*. In this animal, as in most Cladocera, there is no mechanism by which the developing embryo is given additional food, but in the Polyphemoidea a nutritive fluid is secreted so that the brood pouch becomes a functional uterus, and in *Moina* blood plasma appears to be filtered into the chamber (Weismann 1877a).

The number of preadult or prereproductive instars is variable. Rammner (1926) observed but two in *Bosmina* and Agar (1930), occasionally two, but ordinarily three, in *Simocephalus gibbosus*. The same author records four as normal in *Daphnia carinata,* as in *D. laevis*[9] (Wood, Ingle, and Banta 1939) and in *D. galeata mendotae* (Hall 1962). In *Daphnia magna* there is considerable variation, four to eight preadult instars having been recorded (Anderson 1932; Anderson, Lumer, and Zupancic 1937; Anderson and Jenkins 1942).

Green (1956) found a tendency in females that are large in the first instar to mature earlier, in his case in the fifth rather than the sixth. This condition may, in part, be genetic, for a small Camargue strain always had five preadult instars, whereas larger British animals had four or five. It can also be determined environmentally in various ways. Green (1954) found that the young produced in the third brood had a maximal modal length and concluded that they were likely to include the minimal number of late breeding individuals. Anderson found that competition with ciliates or *Aeolosoma* reduced the growth increment per instar; reproduction occurred when the animal was 2.49 to 2.60 mm. long, irrespective of the instar, which might exceptionally be the ninth. Variation is known in all adequately studied species of *Daphnia* and would probably be found in *D. carinata* and *D. laevis* also. In general, when interspecific comparison is made, the correlation of large neonates with early breeding breaks down some, but not all, smaller species maturing at an earlier molt than large species.

In *Ceriodaphnia reticulata* Schubert (1929) found one population in a moorland pool near Seeon, Upper Bavaria, which had six instars and began reproduction in the fourth, whereas in another pool the population had five instars and usually began to reproduce in the third.

The rate of growth both absolute and per instar under given conditions increases to a maximum and then decreases. Anderson concluded that reproductive maturity ordinarily corresponded to the inflection point of the growth curve. In general, this is approximately true, but Green (1956) found that in many cases the maximum growth

---

[9] Brooks (1957a) has shown convincingly that the animals figured by Banta and his associates as *D. longispina* are *D. laevis.* There are good zoogeographic reasons for thinking that whether they were collected near Cold Spring Harbor or in Florida the only species that would have been regarded as comparable with the figured specimens would in fact be conspecific with them. It therefore seems legitimate to refer to *D. laevis* all the lines regarded by Banta as belonging to *D. longispina.* Banta's *D. pulex* unhappily cannot be identified. He claimed to have had ten morphologically recognizable clones of the species but evidently did little cross breeding. From Brooks's study it is reasonable to suppose that several species were involved, so that when the work from Banta's laboratory on these animals is discussed it is referred to *D. pulex* (sens. lat.).

TABLE 30. *Number of preadult molts and lengths of first instar and mature ♀*
in six species of Daphnia

	Length Mature ♀ ♀	Number of Preadult Molts	Mean Size at First Instar
D. (C.) *magna*	2.3–6.0 mm.	4	0.99 mm. (modal)
		5 (Britain)	0.88
		5 (Camargue)	0.77
D. (C.) *thomsoni*[a]	2.1–4.8	4	0.78 (modal)
		5	0.72
		6	0.70
D. (D.) *pulex*	1.5–3.5	4	0.73 (modal)
		5	0.53
D. (D.) *obtusa*	1.5–3.0	3	0.70
		4	0.62 (modal)
		5	0.57
D. (D.) *curvirostris*	1.3–2.2	4	0.58 (modal)
		5	0.53
D. (D.) *ambigua*	0.9–2.0	3	0.49 (modal)
		4	0.48

[a] Not necessarily the same species as recorded from S. Africa.

was not in the last prereproductive or adolescent but in the preadolescent instar, and, very rarely, even earlier.

The total number of instars in the life of a Cladoceran varies from four in some wild populations of *Bosmina longirostris* and the peculiar *B. thersites* (Rammner 1926), in which maximum longevity may not be achieved, to well over twenty in various species of *Daphnia* in culture, whereas in *Scapholeberis mucronata* Rammner found as many as nineteen. There is often a little negative growth during the instars of extreme old age.

In their experiments on longevity, MacArthur and Baillie (1929) found life spans for parthenogenetic females and for males of *Daphnia magna* fed on bacterial food at four temperatures, as set out in Table 31.

TABLE 31. *Longevity in days for* Daphnia magna

	Mean Longevity		Maximum Longevity	
	♀	♂	♀	♂
8°	108.41 ± 3.93	107.94 ± 2.81	202	179
10°	85.93 ± 3.33	89.67 ± 2.07	150	150
18°	44.73 ± 0.47	38.62 ± 0.22	99	92
28°	29.24 ± 0.44	21.93 ± 0.29	57	46

It will be noted that the male life span is almost the same as that of the female at low temperatures but markedly less at high. Initially, the survival of males is better than females at all temperatures, and produces a more rectangular survivorship curve for the male sex. The males, though somewhat reduced in size, are not ephemeral creatures, as may well be supposed in the rotifers.

MacArthur and Baillie realized that in their cultures crowding may not have permitted the development of maximal longevity. Anderson and Jenkins (1942), working with isolated individual females, in fact found a mean life span at 25° C. of forty days, which is rather more than would be expected from MacArthur and Baillie's observations. Anderson and Jenkins found a slight statistically insignificant increase in life span with increasing age, in instars, of maturity. The average number of instars was about seventeen; individuals primaparous in their fifth, sixth, and seventh instars had, respectively, an average of 12.5, 12.2, and 12.3 adult instars; the difference between the three kinds of individuals thus appears to be due to intercalation of preadult instars rather than to a reduction of the number in adult life. Such differences do not appear to be genetic but their origin is obscure.

Some interesting effects of variation in food intake were obtained by Ingle, Wood, and Banta (1937) on *Daphnia laevis.* Two standard media were made up, one about one thirty-sixth as rich in manure culture food as the other.

Though we have seen that complete starvation (page 553) can be rapidly fatal to *Daphnia,* animals fed on the poor medium may live considerably longer than those on the undiluted medium, though their total reproductive capacity is much reduced. When an animal is maintained for part of its expected life on the poor medium and then transferred to the undiluted medium, its life expectancy is enhanced over either well-fed or poorly fed individuals, and the longer it has been poorly fed the greater the effect. The reproductive capacity of such animals is a little below normal except when they have been on the poor food for a relatively short, mainly prereproductive, period. The enhanced mean life span then compensates for a slight decrease in mean number of offspring per brood, which also remains a little higher late, though less high early, in life. These effects are indicated in Table 32.

Growth was much retarded by the low food supply and greatly accelerated when the dilute medium was replaced by the undiluted medium, so that most of the growth normally to be expected may be made up in later life. At a given food level growth rate is not determined strictly by size, and still less by age, but by the nutritive history of

TABLE 32. *Effects of partial starvation of longevity and reproduction of*
Daphnia laevis *(data of Ingle, Wood and Banta 1937).*

	Number of ♀♀	Longevity in Days	Reproductive Life in Days	Number of Broods	Young per Brood	Young per ♀	Young per ♀ per Day of Reproductive Period
Undiluted medium	45	30.30	24.16	11.54	23.11	269.4	11.30
Diluted medium till 6th instar	44	37.08	29.65	15.2	18.95	293.7	9.82
Diluted medium till 9th instar	44	39.52	32.13	14.4	17.46	253.8	7.96
Diluted medium till 12th instar	48	41.03	33.80	16.4	15.90	263.4	7.71
Diluted medium till 15th instar	54	44.68	36.35	17.4	12.91	228.9	6.18
Diluted medium till 18th instar	41	48.93	40.30	17.1	9.30	163.0	3.95
Diluted medium throughout life	42	38.62	32.10	13.1	5.76	74.5	2.40

the individual. It is probable, however, as Slobodkin has insisted, that the age at which a certain size is achieved could be used as a measure of history on the one hand and growth and reproductive potential on the other, so that the state of a population could be expressed in tensor form in which each individual would be entered according to its size and age (Slobodkin 1953b).

Hall (1962) found no effect on the longevity of *D. galeata mendotae* in culture when the food was varied over a 64-fold range, but the quantities of algae used are given on an arbitrary scale and it is possible that his lowest ration was much more nutritious than the diluted medium of Ingle, Wood, and Banta.

Delay, by partial starvation, of growth, reproduction, and death is not unknown in other animals but can clearly, though perhaps not always, occur in a rather extreme form in *Daphnia*. In view of the opportunistic nature of population growth in the Cladocera and of the rapid variations likely to occur in their phytoplanktonic food, it is reasonable to suppose that this particular kind of adjustment to periods of reduced food supply would have great adaptive value in the group.

There is possibly a direct effect of light on growth and reproduction of some Cladocera because both Warren (1900), working with *D. magna,* and Parker (1959), with *Simocephalus vetulus,* had difficulty raising cultures in darkness.

Natality. The number of eggs in a clutch is variable, depending primarily on the age of the mother, nutrition, and genetic constitution, as made apparent in the examples in subsequent paragraphs.

De Kerhervé (1927), who seems to have studied the most prolific *Daphnia magna* or indeed the most prolific cladoceran on record, found nineteen eggs in the first mature instar and 105 in the seventh reproductive instar, after which the clutch size fell somewhat irregularly to twelve in the nineteenth reproductive (probably twenty-third actual) instar. The total egg production of this individual was 1072. No other worker (Green 1954) has recorded such great fecundity. Anderson and Jenkins (1942) found a much smaller maximum at the fifth instar and a more irregular decline. This general pattern of a rise to and fall from a maximum in fecundity in the first half of life, shown by de Kerhervé's extreme case, is apparently typical of most Cladocera.

Under optimal conditions, *D. laevis* (Wood, Ingle, and Banta 1939) has an average of fifteen eggs in the first clutch, thirty-three to thirty-six in the fifth and sixth clutches. The number then descends so that very old animals may have but four to six eggs in the eighteenth and nineteenth clutches. The maximum production may be a little later in *D. magna* maturing after six preadult instars than after four (Anderson and Jenkins 1942), and it is possible that when there are eight such instars the maximum is not exhibited.

Further examples are given in Slobodkin's (1954) work. Green (1954) found the same effect, but it was irregular and dependent on crowding, the congregation of animals, and settling of food in a way that makes his results hard to interpret.

Food supply influences clutch size in a direct way in all Cladocera that have been studied. In *Moina macrocopa* Brown (Banta and Brown 1939) showed that the influence of food was exerted before the third quarter of any instar under consideration, which at 21° C. means about 16.5 hours before egg laying.

Hall (1962) has made an important contribution to the analysis of the effects of food as well as temperature on the rate of reproduction of *D. galeata mendotae*. The average length of an instar during reproductive life is very dependent on temperature, being 48 hours at 25° C, 67.5 hours at 20° C, and 192 hours at 11° C, independent of food supply within the variation used. The number of instars in the average individual's life appears to be higher at 20 than at 25° and a little higher than at 11° C.

The number of eggs per brood and the amount of growth per instar are said to be unaffected by temperature within the limits studied but are positively correlated with the food supply. Since the rate of reproduction will depend on the rate of brood production and the size of the brood, it will be dependent on both food and temperature.

TABLE 33. *Rate of increase per individual per day in* Daphnia galeata
mendotae *at different temperatures and food levels*

	11°	20°	25°
Minimal food	0.07 ± 0.005	0.23 ± 0.002	0.36 ± 0.015
4 times minimal	0.10 ± 0.003	0.30 ± 0.002	0.46 ± 0.013
64 times minimal	0.12 ± 0.006	0.33 ± 0.016	0.51 ± 0.006

The results of Hall's study expressed in terms of the rate of increase
per average individual per day at his three temperatures and three
food levels, are given in Table 33. It is evident that rather large
variations in food supply are needed to produce increases in reproductive rate equivalent to those due to relatively small increases in
temperature.

Intraspecific genetic control of vital statistical properties. There are
likely to be recognizable genetic differences determining variations in
longevity and fecundity in various strains of the same species of
Cladocera in nature as there certainly are in the laboratory experiments
on *D. laevis* by Ordway (Banta and Wood 1939). Summarized in
Table 34 all the data given are said to be derived from clones whose

Table 34. *Mean values of certain vital statistics in different clones of*
Daphnia laevis *(data of Ordway in Banta and Wood 1939)*

	Number of ♀ ♀	Longevity, (days)	Age at Beginning of Reproduction (days)	Brood Interval (days)	Brood Size, all Broods	Young per ♀ per Day over Reproductive Period
Series I						
HN_2	38	37.47	7.00	2.66	12.9	4.89
HN_3	24	31.00	7.54	3.32	6.7	2.24
G_1	35	43.59	7.89	2.74	12.5	4.39
G_2	30	46.00	7.07	2.58	21.7	7.89
Series II						
G_1	9	—	7.25	2.69	13.40	5.01
G_2	21	—	7.11	2.51	24.18	9.46
G_6	13	—	9.13	2.40	13.43	5.70
G_{22}	21	—	7.15	2.40	25.93	10.72
G_{23}	27	—	7.33	2.51	22.62	8.96
G_{24}	24	—	7.20	2.40	21.25	8.71
G_{25}	18	—	7.46	2.60	16.62	6.21

fundatrices were hatched from sexual eggs produced by parents related to each other.   Clones designated by the same letters (HN, G) were derived from crossing parents from the same two clones.   Temperature and other culture conditions were not accurately controlled but were maintained as uniform as possible by keeping all culture vessels close together.   Within each series the results certainly are comparable; the similar behavior of clones G_1 and G_2 in the two series suggests that interserial comparison is also legitimate.

Since lines from the same crosses yield very different results, the great interest in these data is that they show that ordinarily parthenogenetic generations must consist of populations of quite heterozygous individuals, which carry a variety of genes controlling the quantitative details of the life history.   Obviously great opportunities for natural selection to operate successfully in different ways in different environments exist in the Cladocera.

Density dependence.   Pratt (1943), in his very important studies of population dynamics, found that at 25° C. the longevity of *D. magna,* fed on *Chlorella pyrenoidosa,* showed a marked density dependence (Fig. 153) with an optimum density of five animals per 50 cm.[3] when the mean longevity was about thirty days, a figure rather less than would be expected from the work already discussed.   At 18° the longevity is much greater, just over sixty days at optimal density, and the optimal density itself is much higher, namely, seventy-five individuals in 50 cm.,[3] and much less clearly expressed.   It is probable that the optimal density effects at the two temperatures have different explanations.   At 25° C. bacteria cloud the water slightly when only a single animal is present, and evidence was obtained that bacterial slime might seriously inconvenience the animals.   At higher densities this did not occur.   At 18° C. the much less marked optimum at a high density might be due to crowding, which causes partial starvation, and so increases longevity, or to lowering metabolism in other ways, though this alternative is unlikely, for it seems from Slobodkin's work (1954) that at the kind of densities studied no direct density-dependent effect is likely to be manifest.

Competitive phenomena in Cladocera.   Frank (1952, 1957) has studied the competitive relations between certain species of Cladocera in laboratory culture.   In his first experiments he used *Daphnia pulicaria* (later, but perhaps incorrectly regarded as *D. pulex* by Brooks) and *Simocephalus vetulus,* both animals being derived from the same locality near Chicago.   Both are pond species, but *S. vetulus* is more littoral than any species of *Daphnia* and spends much of its time fixed to solid substrates, in which position it can continue feeding.

In spite of this behavioral and consequent ecological difference between the species, when they are cultured in small volumes of water of the order of 165 ml., *D. pulicaria* invariably displaces *S. vetulus*. The natality of both species is strongly density-dependent in either unispecific or mixed cultures of the kind used by Frank, presumably primarily because of the greater demand placed by dense populations on the food supply. At high densities of the order of ten to twenty animals per millilitre the depression of natality is much greater for *S. vetulus* than for *D. pulicaria*. In mixed culture the density effect, at least at high densities, depended to some extent on the proportions of the species present. When 90 per cent of the population had a density of six animals of either species per millilitre, the natality of *D. pulicaria* was much less than in cultures of the same total density consisting of equal number of both species. A qualitatively reciprocal effect is shown by the natality of *S. vetulus,* but here it is necessary to reduce the proportion of that species to 1 per cent of the total before the effect is significant; with an absolute density of six animals per millilitre and this low proportion of *S. vetulus,* its natality is about five times what it would be if 90 to 100 per cent of the individuals belonged to that species. Moreover, the mean longevity of *D. pulicaria* rises slightly at this total density as the proportion of the species drops from 50 to 10 per cent, whereas in *S. vetlulus* the longevity is minimal in the 50 per cent mixture. This means that when *S. vetulus* becomes rare its length of life increases, and as it becomes very rare its reproductive rate does also. The causes of these phenomena are not entirely clear, but it is evident that rare specimens of *S. vetulus* in Frank's cultures had an advantage not possessed by the somewhat more abundantly represented individuals which were undergoing elimination. Frank thinks that the final extermination of *S. vetulus* in any culture was largely a chance matter and that in much larger populations, in larger volumes of water, some specimens might have persisted indefinitely. Since they would have spent much of their time attached to the walls of the culture vessel, they would have been littoral benthos, far removed from most of the free-swimming *D. pulicaria,* so imitating the situation in nature.

Frank's (1957) second study (Fig. 153) concerns *D. pulicaria* and *D. magna,* both species being derived from small ponds. The strains used came from near Chicago and near Pasadena, respectively. The identity of Frank's *D. pulicaria* with the original American or with the European populations to which that name has been given is not quite certain. It is probable, however, that the two species used by Frank do not regularly co-occur in nature.

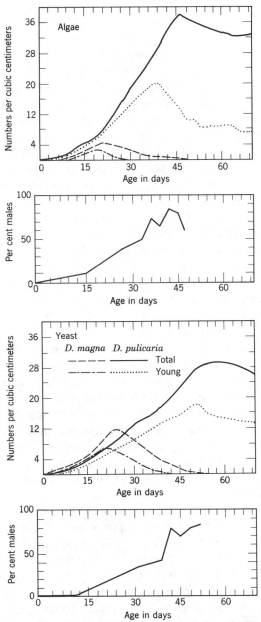

FIGURE 153. Competitive exclusion of *Daphnia magna* by *D. pulicaria,* in small volumes of water, showing the effect of the nature of the food on the initial stages of competition and the progressive increase in the number of males as the population of *D. magna* declined. (Frank, modified.)

Frank studied both species in single and in mixed cultures fed on two different foods, namely *Chlamydomonas moewusii,* and suspended baker's yeast. When fed on *Chlamydomonas* unispecific populations of *D. pulicaria* grow faster than those of *D. magna,* and it is therefore reasonable to find that in mixed culture in volumes of about 50 ml. *D. pulicaria* regularly and simply displaces *D. magna* in such food. When unispecific cultures fed on yeast are compared, *D. pulicaria* increases much more slowly than *D. magna.* Initially, this advantage is also observed in mixed cultures, but soon *D. pulicaria* begins increasing faster than *D. magna* and after about forty-five days *D. magna* is displaced in small cultures, whichever food is used. The explanation of these rather paradoxical findings appears to be that at first, when the *Daphnia* population is low, the yeast added as food is but slowly removed and meanwhile is respiring; in the cultures fed on *Chlamydomonas* such respiration is at least partly compensated by photosynthesis. Initially, therefore, yeast-fed cultures may suffer from low oxygen concentrations, to which adverse condition *D. magna* is less sensitive than *D. pulicaria.* So at the beginning of the experiment *D. magna* has a competitive advantage. Later, as larger populations keep either kind of food at a low concentration, oxygen deficiency does not develop and the inherently greater rate of increase of *D. pulicaria* permits that species to dominate. The actual mechanism of displacement of *D. magna,* however, is apparently not a direct effect of crowding on nutrition but rather on sex determination, so that late in the history of a culture a disproportionate number of specimens are males and the species fails as an active part of the assemblage through a lack of parthenogenetic females. Some ephippia, of course, will have been produced, so that *D. magna* is not really extinct in the culture but merely inactive.

Under quite different conditions, in mass cultures of 8 liters of water, fed weekly on a mixture of *Chlamydomonas* and yeast and maintained up to volume with distilled water, *D. pulicaria* gradually disappeared and only *D. magna* remained active. It is not unreasonable to suppose that in these cultures there was some deficiency of oxygen. However, if such a culture were allowed to dry up completely and then refilled, many more *pulicaria* than *magna* would appear from ephippia, the superior natality of the former species apparently being expressed also under these rather extreme conditions. It is evident from these experiments that in nature either *pulicaria* or *magna* might appear but that it would be extremely difficult to see merely from a study of two ponds, one with one species, the other with the other, what it was that determined the species present. In addition to this warning against oversimplification, Frank's important experiments also indicate how resting stages, doubtless developed in fresh waters primarily to avoid physically unfavorable periods, may also be of the greatest importance in tiding a species over biologically unfavorable circumstances involving the adverse effects of competition. This is likely to be generally important in the biology of the plankton (cf. Hutchinson 1964b).

In two series of experiments Parker (1960, 1961) has studied competition between Cladocera and Copepoda in laboratory cultures, using *Simocephalus vetulus* with *Megacyclops viridis* in one series and *Daphnia pulex* with *Eucyclops agilis*. All animals were fed on *Chlamydomonas moewusii*, which, as Parker points out in his second paper, is not a natural food for *M. viridis*. In both experiments the cladocerans suppressed the copepods without any apparent effect of the latter on the former.

Oscillating populations. Anyone who has kept Cladocera in the laboratory will know how frequently a new culture will flourish exceedingly and then without any apparent change in treatment decline, sometimes recovering and declining several times over. The periods of low population have sometimes been described as depressions (Berg 1934). There is also some old work on *Simocephalus vetulus* by Papanicolau (1910) in which the egg number decreased steadily from the first (mean 17.5) to the twentieth generation (mean 1.5) as the intervals between molts increased. Similar results were obtained with *Moina brachiata* (sub *M. rectirostris* var. *lilljeborgii*). The meaning of these experiments is far from clear, but they would not now be interpreted as indicating some inevitable degenerative change in genetic constitution.

Two important studies, namely, those of Pratt (1943) on *Daphnia magna* and of Slobodkin (1954) on *D. obtusa*, have thrown a great deal of light on the phenomenon of oscillating populations and depressions, which may well prove to be as important in nature as in the laboratory, but there still appears to be a good deal to learn. Frank's (1952, 1957) studies of single species controls in his competition experiments give results very like those of Pratt and Slobodkin, as he himself emphasizes.

Pratt (Fig. 154) found that *D. magna* cultivated in 50 cc. of pond water at 25° C., under conditions in which he believed the food, namely, *Chlorella pyrenoidosa,* not to be limiting, exhibited striking oscillation in a period of about forty days. The greatest population observed at a maximum was 126 animals, or 2.52 cm^{-3}; the smallest was zero, corresponding to spontaneous extinction at a minimum. As has been indicated, in artificial populations maintained at constant size the death rate

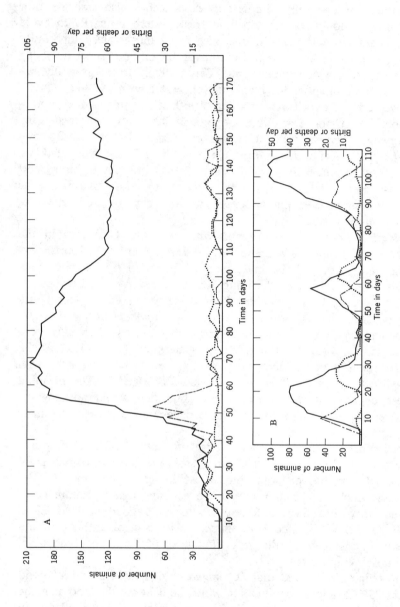

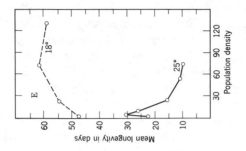

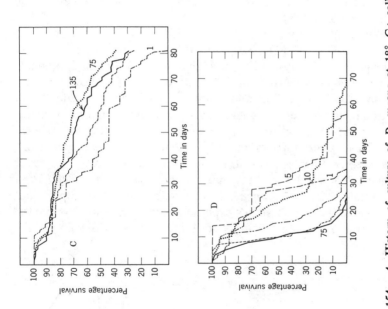

FIGURE 154. *A.* History of culture of *D. magna* at 18° C.; solid line total population, dotted line birth rate (smoothed by three-day periods); dotted and dashed line death rate (similarly smoothed). *B.* The same for 25° C. *C.* Survivorship curves at different constant densities (number of animals in 50 cc.) at 18° C. *D.* The same at 25° C. *E.* Mean longevity as a function of density. (Pratt, modified.)

was found to be strongly density-dependent, with maximum longevity at five animals in 50 cm.[3], while egg production was strongly and negatively density dependent.   Analysis of the populations at 25° C. indicate a maximum number of births at about the inflection point of the population as it rises to a maximum and a maximum number of deaths at about the inflection point of the descending curve.   Pratt supposes that in the varying populations the effect of population density, whatever its ultimate nature, on natality and mortality involves a lag in its operation so that the rates of birth and death at any time are appropriate to the density at a previous time.

At 18° the population rose to a maximum and then declined only a little.   The number of births rose and fell during the decline but the deaths showed no clear periodicity.   This is to be expected from the peculiar experimental fact that mortality at 18° is little density-dependent and, as far as it is so dependent, has a minimum value at a high optimum density.

The longest population studies (Fig. 155) in the laboratory are those of Slobodkin (1954), who carried cultures of *Daphnia obtusa* for as long as 363 days.   The animals were fed standard amounts of *Chlamydomonas moewusii* on alternate days and removed all the food in about ten hours.   The total number of individuals at equilibrium and the total biomass even under more or less nonequilibrium conditions are linearly dependent on the amount of food supplied at each feeding.   This effect appears to hold, up to ten animals per cm.[3], which is greater than is ever likely to occur in nature.   No direct density-dependent inhibition thus appears to occur.   Since *D. obtusa* is much smaller than *D. magna,* it is possible that the upper limit of density available in Slobodkin's experiments was, in terms of biomass, lower than the maximum biomass in Pratt's study.   In cultures the

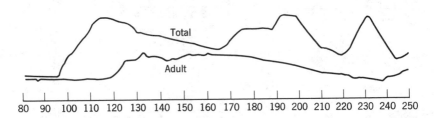

FIGURE 155. History of a culture of *Daphnia obtusa,* in a volume of 50 ml, which had reached an equilibrium population consisting almost entirely of adults during the first 80 days at 20 to 23° C, and was then transferred to a constant temperature at 14° C.   Note oscillations primarily involving preadult individuals.   (Slobodkin.)

population rises to a peak due to the birth of young animals and then declines. The number of larger animals remains essentially constant during the decline, many young animals dying, a few replacing old dying animals. Reproduction falls to a minimum and at about 18 to 20° C. an equilibrium population occurs. When such a population is transferred to 14° C. an oscillation is set up which is damped far more slowly than at the higher temperature, if at all. It is clear from Slobodkin's work that the initial increment depends on the food level that permits rapid reproduction in the founding females introduced into the culture. The newly produced young then put such a demand on the constant food supply that reproduction is almost inhibited. As the young grow, their food intake increases, and the starvation conditions become more intense until the young produced at the peak, gradually replacing their few old parents and older siblings, themselves become old and start dying. This permits a little reproduction to begin and leads at the higher temperature to an equilibrium population. At the lower temperature, in which slower growth and lower maintenance metabolism probably increase the efficiency of food utilization in reproduction, the production of new individuals clearly becomes too great for establishment of equilibrium and oscillations persist.

The simplest theoretical model which generates oscillations of the kind observed by Pratt at 25° C. in *D. magna* and by Slobodkin at 14° C. in *D. obtusa* is derived from the Verhulst-Pearl logistic[10] by inserting a time lag (τ) into the operation of the feedback term, giving.

$$(5) \qquad \frac{dN_t}{dt} = bN_t \left\{ \frac{K - N_{(t-\tau)}}{K} \right\}$$

The equation has been investigated by Cunningham (1954; Wangersky and Cunningham 1957), who finds that in spite of its intractability it can be shown by approximate methods, and by the use of the analogue computer, that when $b\tau$ is small, less than about 0.7, the population grows monatonically toward an asymptote, somewhat faster but in qualitatively the same way as in cases in which the unmodified logistic applies. With $0.7 < b\tau < 1.8$, the population initially develops to a value greater than K and then oscillates, but with decreasing amplitudes. With $b\tau > 1.8$ the oscillations constitute an undamped limit cycle.

---

[10] Smith (1963) in an important paper that appeared as this volume was being prepared for press has modified the logistic to include a term for maintenance metabolism. This treatment is specifically designed for equilibrium *Daphnia* populations.

In Pratt's (1943) experiments at 25° C., since the rate of production of young by isolated mothers was 2.4 per day per mother, the lag period would have to be at least three quarters of a day to maintain the oscillation. This would seem to be reasonable.

The use of (5) implies that the death rate may be considered as a negative birth rate, density-dependent in the same way as the actual positive birth rate. This may well be approximately true at 25° C., except at very low densities, but it is obviously untrue at 18° C., at which temperature oscillations did not occur in *D. magna.*

Slobodkin's (1953b) ingenious matrix notation, though it implies some rather artificial concepts, emphasizes the possibility of analyzing oscillations without the necessity of specifying the exact form of the function relating population size to time.

Comparative aspects of the effects of temperature. In general, as is apparent in several examples already given, the effect of increase in temperature over the tolerated range is to increase the rate of development, which in the Cladocera largely means the rate at which molts succeed each other, and to decrease not merely the instar length but also the length of life. Hall's work suggests a maximum number of instars at 20° C. in *D. galeata mendotae.*

Apart from this type of result, which is obviously of great ecological importance, certain observations have been made which throw some light on the effect of temperature differences on the life histories of different species of Cladocera. These observations are also applicable to the interpretation of seasonal succession and geographical distribution in nature.

Some species, namely *Moina macrocopa* and *Pseudosida bidentata,* pass into a chill coma when brought suddenly from room temperature to 0° C. (Brown 1929a). The race of *M. macrocopa* used by Brown was in fact inactivated by slow cooling to 10° C., and that of *P. bidentata* when slowly taken to 2.5 to 3.0°. In the same experiments *D. magna* and *Simocephalus vetulus,* though sluggish at 0°, continued active at 1° C. for an hour. The upper limit of life as indicated by one-minute thermal death point determinations is, for both *M. macrocopa* and *P. bidentata,* 48° C; for *S. vetulus* it is 43° C. and for *D. magna,* 41° C. *P. bidentata* is a south temperate and tropicopolitan form, *M. macrocopa* a summer form in small bodies of water in temperate regions; the other two species are more or less temperate perennial animals. It is evident that some warm-water species tend to have not merely higher upper limits but higher lower limits of viability. Brown also found that although *M. macrocopa* and *P. bidentata* could be bred up to 35° C. the other species survived at

30° C. but not at 35°.   However, for *D. magna,* as for an uncertainly determined species probably erroneously referred to *D. pulex,* there was a clear optimum at 25° C.   Brown studied the developmental rates of seven species, not all of which can be clearly identified.   The variation in rate of development of the identifiable species is shown in Fig. 156.   On the whole, the slopes of the lines vary in such a way that southern or summer species with a high thermal death point exhibit greater slopes than less southern perennial species with lower thermal death points (Brown 1929b).

Given a moderate, constant supply of food, it is obvious that a species like *M. macrocopa* would have an enormous selective advantage, other things being equal, over the other species at 30° or even 25°,

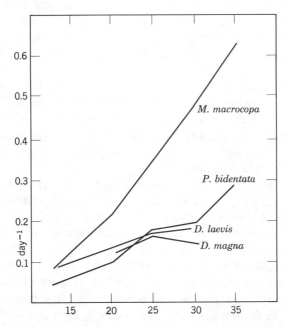

FIGURE 156.   Reciprocal of time of development up to beginning of first adult instar in *Moina macrocopa, Pseudosida bidentata, Daphnia laevis,* and *D. magna.*   Only the first two species, which are aestival and southern, respectively, can live at 35°.   Note the much greater response to temperature by *M. macrocopa* than by the other three. Several other species of uncertain identity also studied by Brown would fall in the general ranges of *Pseudosida* and *Daphnia.*   (Brown, modified.)

but not at 13° C., at which temperature it matures just a little more slowly that *D. laevis;* at 30° C. *M. macrocopa* develops two and a half times as fast as *D. laevis.* This kind of physiological interspecific difference may well underlie the behavior of some species as monacmic summer forms, whereas others are perennial and less regularly variable in numbers. A somewhat similar situation, as we have already seen, probably exists in rotifers.

The remarkable study of Banta and Wood (1928, 1939) on the spontaneous appearance of a line of *D. laevis* with greatly raised temperature requirements and tolerances must be mentioned. This line (Bg) was obtained from a sexual egg produced by inbreeding within a clone (XI) of normal thermal characteristics. The normal clone flourished best at 18 to 20° and could be reared only with difficulty at 27 to 29° C.; at first the thermal clone Bg was best reared at 27 to 29° C. and did very poorly below 20° C. After culture at 27° for about 110 parthenogenetic generations Bg became eurythermal and was then able to survive at 16 to 20° C., which was initially impossible.

Sexual reproduction in Cladocera. After a variable number of parthenogenetic generations, males appear. The males are ordinarily smaller than the females and are most reduced in the Daphniidae, though they are morphologically complete Cladocera and so comparable only in the most general terms with the minute and often greatly simplified males of the rotifers.

At the time that males appear, or a little later, females start producing resting eggs which are fertilized. These are ordinarily darker and have thicker shells than the subitaneous eggs. In the Haplopoda, Sidoidea, and Polyphemoidea they are passed into the rather soft-walled brood pouch and are apparently liberated on molting or at the death of the female, much as are the eggs of Conchostraca. In many of the Daphnoidea special modification (Fig. 151) of part of the carapace adjacent to the brood pouch may provide an extra covering for the resting egg. In *Bosmina,* except apparently in the peculiar *B. thersites* (Lilljeborg 1900), the posterodorsal wall of the carapace is thickened before resting eggs are produced and the eggs are retained in the molted exuvium. A comparable, less extensive thickening occurs in *Chydorus sphaericus* (Lilljeborg 1900), and in other chydorids fairly well defined ephippia occur. In the Daphniidae the posterodorsal carapace wall becomes modified to form an especially thick, dark envelop in which the resting eggs lie and which becomes detached from the rest of the exuvium to form the ephippium. At least in some species, such as *D. retrocurva,* the cellular wall of the ephippium forms a float.

The ephippial eggs of the Daphniidae have been the subject of some research in order to elucidate the conditions that lead to hatching. A sufficiently low osmotic pressure is probably involved, as in the Anostraca. Wood and Banta (1933, 1937) find that in *D. laevis* the addition of new medium or, in some cases, aeration can cause hatching. Short periods of drying may also be effective, but neither drying nor freezing is necessary to hatch the resting eggs of this species, of *D. pulex* (sens. lat.), or of *Moina macrocopa*. In *D. pulex,* and probably in *D. ambigua* (Pancella and Stross 1963), illumination appears to promote hatching, though apparently there is a latent period of length varying in different populations before which light is ineffective. Sodium hypochlorite treatment seems to reduce this latent period. In *D. obtusa,* as Slobodkin has shown me, ephippial eggs may undergo a direct and immediate development, just like subitaneous eggs; almost immediate hatching also appears to characterize a majority, but not all, of the ephippial eggs of *M. macrocopa* (Wood 1932).

Determination of male production. The similarity of the cladoceran life history to that of the Rotifera is obvious and, as will be seen, extends to the occasional production of pseudosexual resting eggs as well as occasional ephippial eggs that hatch immediately without undergoing diapause. It must be realized, however, that the cytological bases for the life history are quite different in the two cases. All Cladocera, parthenogenetic or sexual, male or female, seem to be functionally diploid, though it is possible that tetraploid races occur, as of *D. pulex* in Emilia (Bacci, Cognetti, and Vaccari 1961). The ordinary parthenogenetic egg appears to exhibit a single equational maturation division which produces one polar body; the normal sexual ephippial egg undergoes ordinary reduction with two polar bodies (Mortimer 1936; Banta and Wood 1939). Bacci, Cognetti, and Vaccari (1961), however, conclude that at least in their tetraploid *D. pulex* the maturation division is preceded by a pairing and separating of chromosomes within the nuclear membrane, as in the endomeiosis of certain aphids.

In striking contrast to the rotifers, the production of males has no genetic connection with the incidence of fertilizable females. Ordinarily, if the ephippial female is not fertilized, her resting eggs are resorbed; she may become parthenogenetic and possibly may again produce an ephippium late in life (Pacaud 1939).

There has been much controversy over the causes of the appearance of males and of ephippial eggs. Weismann (1876, 1879) believed that sexual reproduction had of necessity to intervene after a more or less fixed number of parthenogenetic generations. This idea of the internal determination of an inevitable cycle was widely adopted and for long influenced investigators of sexuality and seasonal form change in fresh-water animals (see page 887), but most modern workers (see particularly Wesenberg-Lund 1926; Banta and Brown 1939) have concluded, undoubtedly correctly, that environmental factors are primarily involved. It is certain that different species and races of Cladocera differ greatly in the ease with which they can produce males. Banta and Brown report that their stocks of *Moina affinis* and *Daphnia magna* produced males less easily than do the other species of *Moina, Daphnia,* and *Simocephalus* they investigated. Banta and Wood (1939) found it possible to extract clones from descendants of ephippial eggs of a population of *D. laevis* which differed greatly in their capacity to produce males. In many limnetic populations no males or ephippial eggs have been observed in nature. Berg (1931), however, studying in the laboratory, several such acyclic races of *D. cucullata* from lakes in Denmark, was able to obtain both males and ephippial females, but they are clearly less easily produced in such races than in pond forms. In *D. lacustris* from the Lake of Lucerne, he observed males in culture but no ephippia; in general, it is probable that these races are more resistant to the influences producing ephippia than to those producing males. The reverse, however, has been noted by Brooks (1957a) in *Daphnia thorata* from western North America.

The stimuli involved in male production and in the formation of ephippial eggs are apparently quite different, though they normally succeed one another in nature. This independence was apparently first shown by von Scharfenberg (1914), who found that if detritus from the bottom of Cladocera cultures were reused as food for *D. magna* ephippia but not males would be produced, though the reverse appeared to be true in *D. pulex.* The detailed meaning of these experiments is obscure.

Banta and Brown (1929a, 1939) found that male production occurred only in cultures that are fairly well fed; starvation and repletion are both inhibitory. Certain conditions of temperature must be fulfilled, but they are complex and not fully understood. In *Moina macrocopa* male production is greatest at 14 to 17° C. and at 30° C., with one well-defined minimum between 24 and 27° C. and another at low temperatures. Macomber (in Banta 1939) has observed that in an undetermined cold stenotherm species of *Daphnia* from Long Island indefinite parthenogenesis is possible at 3 to 7° C., and Banta, working with the same strain, found epidemic sexual reproduction and cessation of parthenogenesis at 20 to 24° C. Within the fairly wide limits of nutrition and temperature that permit male production, the well-established cause of the process is high population density, as Grosvenor and Smith (1913) had earlier found in *Simocephalus*. The effect of crowding operates to produce males for a limited time about four hours before passage of the egg into the brood pouch (Banta and Brown 1929c), which is two and a half to three and a half hours before the maturation division in *Moina macrocopa*. Banta and Brown (1929a,b, 1939) found evidence that the crowding operated through the concentration of a nonspecific excretory product, which is not carbon dioxide, ammonia, uric acid, or urea; it is not present in the urine of terrestrial vertebrates but is produced by fresh-water invertebrates other than the Cladocera. They found that aeration or even bubbling nitrogen through the cultures inhibited male production but were unable to recover the supposed volatile substance. They concluded that the substance acted by depressing metabolism at the critical period, but there is no independent evidence of this action.

Berg (1931) on the whole confirms Banta and Brown's work on the effect of crowding; a few minor discrepancies are unlikely to be resolved until the nature of the crowding effect is more fully understood.

It is however to be noted that in some populations, as in the Schleinsee (Kuntze 1938), male production occurs well after the peak in numbers of animals has been passed. By analogy with what seems to happen in ephippial production (Stross and Hill *in press*), it is possible that there is a photoperiodic factor involved in male production in addition to the effect of crowding. This might well explain the striking autumnal incidence of male production in certain lakes.

In *Simocephalus exspinosus* and in *D. laevis* clones producing sex intergrades are apparently not rare (Banta 1939). These strains are generally prone to produce numerous males. In *S. exspinosus* the sex intergrades can be either males with female characters or females with male characters. Ontogenetic change from male to female has been recorded. In *D. laevis* the intergrades are nearly always female. Hermaphrodites are known, but partial sex reversal, from hermaphrodite to male, has been recorded only once.

Determination of the production of resting eggs. Ephippial production takes place, according to Banta and Brown, when the food supply is low. Since, at persistent minimal nutrient levels, Slobodkin (1954) found that in *D. obtusa* a minority of females produced a single parthenogenetic egg, whereas most specimens were not reproducing at all, a constant low nutrient level cannot be the stimulus for ephippial production in this species. Moreover, low parthenogenetic egg production of the kind observed by Slobodkin in *D. obtusa* clearly occurs in nature at times of low food intake, as Brooks (1946) found for

D. retrocurva. It is therefore probable, as Slobodkin concluded from his incidental observations on ephippial egg production, that the stimulus for such production is a rapid decrease in available food rather than a low constant nutrient level. In nature large crowded populations that produce males would also reduce the food supply rapidly, causing sexual eggs to develop shortly after male production and so achieving adequate synchrony in the two prerequisites for sexual reproduction, though in a way quite different from what happens in rotifers.

In *Daphnia* sexual eggs can be produced experimentally at any time during the fertile life of a female, but in *Moina macrocopa,* unless the first eggs are sexual, only parthenogenetic eggs are laid (Banta and Brown 1939). It seems probable (Banta and Wood 1939) that the two ovaries can respond to the conditions of ephippial egg production rather differently when only one egg is produced in each ephippium. The period of the decline in food supply needed to stimulate sexual egg production is evidently very short, and if it is too intense and too prolonged the developing eggs may be resorbed. Stross and Hill (*in press*) find that the production of ephippial eggs by limnetic *D. pulex* is much greater under appropriate conditions of high density when the culture alternately is illuminated for twelve hours and kept in darkness for twelve hours, as in September at the equinox, than when a long day alternates with a short night. In a sixteen-hour period of illumination and an eight-hour period of darkness ephippia were not produced in the population under study. This, however, cannot be quite general or *D. middendorffiana* would be unable to produce ephippia in the Arctic.

Degeneration of parthenogenetic eggs (Brooks 1946) can occur, though Hall (1962) concludes that the causes are not obvious; starvation in fact may not be involved. In *Daphnia,* and in the *macrocopa* group of *Moina,* two sexual eggs are produced simultaneously so that the ephippium carries a pair of these eggs. In *Ceriodaphnia, Scapholeberis,* and *Simocephalus* and in the *brachiata* group of *Moina* there is but one egg in each ephippium. In *Acantholeberis curvirostris* there are usually two but occasionally (Fryer 1953) three, whereas *Eurycercus lamellatus* may have as many as thirteen. With the possible exception of two eggs in some *Camptocercus* ephippia, the other Chydoridae and Bosminidae produce but one sexual egg at a time (Scourfield 1901, 1902). At least in *Holopedium,* if not in other Sidoidea, the number of resting eggs, not being limited by ephippium formation, may exceed the maximum clutch of subitaneous eggs (Freidenfelt 1920).

Pseudosexual eggs. In a certain number of populations of *Daphnia*, mostly from the Far North, the ephippial eggs are pseudosexual (Banta 1926) or unfertilized parthenogenetic resting eggs. It is probable that all of these populations are referrable to *Daphnia middendorffiana* (Fig. 152). The available records refer to Spitzbergen (Olofsson 1918), Greenland (Poulsen 1940a,b), Alaska (Edmondson 1955), and to a locality on Long Island (Banta 1926).

The majority of the specimens in arctic localities appear to be *ex ephippio* females which produce further parthenogenetic ephippial eggs. A minority, at least in Alaska, may lay ordinarily parthenogenetic eggs which give rise to a small second generation of females. Males never occurred in Imikpuk Lake, studied in detail by Edmondson (1955), but either males or predominantly male intersexes seem to occur in crowded populations in tundra pools in the same general area. A more southern population from eastern Washington apparently produced males normally. Banta's population, which Schrader found to produce only diploid eggs, was apparently entirely thelytokous. It is most unfortunate that the taxonomic status of this, the most southern and apparently only obligate thelytokous but ephippial-producing, population is uncertain. The life histories of the other *D. middendorffiana* populations, real or putative, are obviously in accord with the environmental changes in the localities in which they occur. A clone ascribed to *D. pulex,* which produces pseudosexual eggs, has apparently been made available commercially (Pancella and Stross 1963).

Feeding and filtration in the Cladocera. As has just been indicated, although the Haplopoda and the Polyphemoidea are raptorial, the Sidoidea and Daphnoidea are filter feeders. However, the development of antennal swimming has permitted feeding and locomotion to become largely independent functions, unlike what is found in the Anostraca.

A number of measurements have been made on the rate at which Cladocera remove food from the water, either by direct observation of the decrease in number of suspended food particles or by determining the increase in radioactivity of animals feeding on particles labeled with a radioisotope such as C^{14}. The second, more modern technique, used by Nauwerck (1959) and by Monakov and Sorokin (1960) determines the rate of clearance of food during a shorter period that is necessary in direct counting and so in principle permits more accurate determination of the relation of food uptake to concentration of food particles. In all this work what is measured is the effective or apparent filtration rate, for there seems very generally to be a mechanism for rejection of excess food as well as a reduction in filtering speed in concentrated food suspensions.

In *Daphnia magna* Ryther (1954) found that the feeding rate per milligram of *Daphnia* fell with the size of the animal in such a way that in animals more than 2.5 mm long, or of dry weight more than 0.09 mg., there was little variation in intake, though perhaps a vague maximum occurs at a weight of about 0.115 mg. or length of 2.75 mm. Animals of about this size filtered about 75 ml. day^{-1} in Ryther's experiments, provided the *Chlorella* suspension contained less than 2.10^5 cells per liter. In more concentrated suspensions the rate fell off considerably. Ryther concluded that this was due to a substance produced by the algal food, the effect being independent of whether the *Daphnia* were starved or fed bacteria and detritus before the experiment. He found the inhibition greater when old *Chlorella* cultures were used but also noted some such effects with *Navicula* and *Scenedesmus*.

Sushtchenia (1958) found that *Daphnia magna* of unspecified size filtered from media containing 4 to 23.10^5 *Chlorella* cells ml.$^{-1}$ at a rate of 4.56 to 8.15 ml. day^{-1}; the slower rate may involve size and other differences in the experiments when compared with those of Ryther, but the divergence seems very large. There was almost no systematic fall in filtration rate with concentration (Fig. 157).

McMahon and Rigler (1963; Rigler 1961), found a decline in feeding rate which they measured in terms of the rate of movement of the thoracic appendages and of the mandibles, the rate of swallowing of food boluses and the rate of rejection of food, when the food concentration was about 2.10^5 *Chlorella* cells ml.$^{-1}$ (Fig. 158). When the food concentration is increased above this critical value, the rate of food intake initially rises, but after fifteen to twenty minutes falls to the old value, set no doubt by the rate of passage through the gut. This process involves two different types of behavior on the part of the organism. The rate of beat of the thoracic appendages falls, and food is rejected by being kicked from the food groove between these appendages by the postabdomen. Since starved animals put into a high food concentration continue to filter at a maximum rate for a much longer time than animals previously fed in water containing less than the critical concentration of food particles, it is probable that the rate of movement of the thoracic appendages is determined by the tension on the gut wall and not by exteroreceptive or chemical stimuli from the particles in the water.

McMahon and Rigler were unable to confirm a specific depression due to *Chlorella*. They feel that Ryther's animals fed on bacteria and detritus were not in the same physiological state as those fed *Chlorella* and that therefore Ryther's argument regarding the external determination of the fall in feeding rate is not valid. On this point

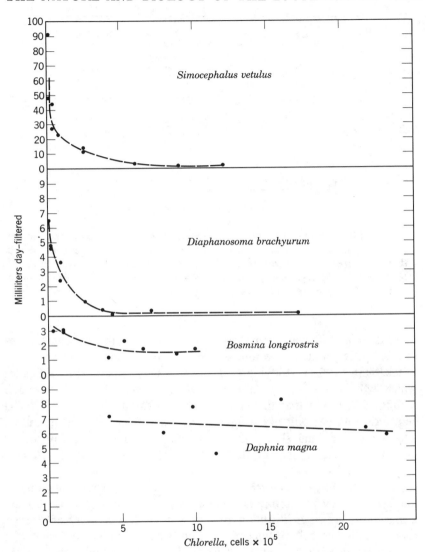

FIGURE 157. Rate of filtration in suspensions of varying concentration of *Chlorella* for four species of Cladocera, from the data of Sushtchenia. In these experiments the approximately hyperbolic relationship implies a fairly constant high food intake in *Simocephalus vetulus* and constant low intake in *Diaphanosoma brachyurum*, whereas the other two species, under the conditions of the experiments, appear to take in more food as the concentration increases.

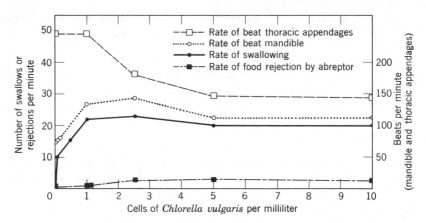

FIGURE 158. Trophic physiology of *Daphnia magna* as dependent on the concentration of *Chlorella vulgaris* used as food. The rate of beat of the thoracic appendages decreases as the food increases from 1.10^5 to 2.10^5 cells per ml.; the rejection of excess food accumulating increases in the same range. (Redrawn from McMahon and Rigler.)

McMahon and Rigler seem less convincing than throughout the rest of their paper, and for the moment the differences are unresolved.

Richman (1958), studying *D. pulex* in a very important investigation which will be considered at length when productivity is discussed in Volume III, found that this species behaved essentially like *D. magna*. Filtration rate per unit dry weight decreased with the size of the animal and per unit animal slowly increased, reaching 5 ml. per day for his largest specimens. It was independent of the quantity of *Chlamydomonas* used as food.

Unlike *D. magna* and *D. pulex*, *D. longispina* from reservoirs in the Volga valley was found by Monakov and Sorokin (1960) to show a marked decrease in filtration rate with increasing number of *Chlorococcum* cells per unit volume of medium, as given in Table 35.

The increase in food concentration is about tenfold, whereas the increase in rate of ingestion, though quite clear, is only threefold.

It is evident from scattered observations of other workers that further complications may be expected in the genus *Daphnia*. Nauwerck (1959) concluded that *Daphnia hyalina* fed more actively in twilight than in daylight. Slobodkin (personal communication) noticed that in unfed *D. obtusa* the gut contents were retained, defecation occurring only after more food was added.

In the other Cladocera studied (Sushtchenia 1958), namely, *Dia-*

phanosoma brachyurum, Simocephalus, and *Bosmina longirostris* (Fig. 157) the filtration rate at first falls rapidly with increasing food concentration and then more slowly. At any given food level filtration is most rapid in *Simocephalus* and least rapid in *Diaphanosoma.* When the actual food uptake is considered, it is almost constant and low in *Diaphanosoma* and, except at very low food concentrations, almost constant and high for *Simocephalus.* In *Bosmina* the rate of food intake rises slowly with the concentration over the range studied but doubtless would become asymptotic to some intermediate value at very high concentrations of food. The marked drop in filtration rate with increasing concentration of *Chlorella* in the experiments with *Simocephalus* and *Diaphanosoma* suggest that observed in *D. magna* in senescent *Chlorella* cultures in Ryther's work.

It is probable that genuine differences occur in the behavior of Cladocera toward varying food concentrations, but whether they reflect different sensitivities to inhibitory substances or different thresholds of reaction by tension receptors in the gut can be elucidated only by further study.

There is little evidence of sufficient differences in the size of the apertures in the filtering apparatus to produce marked differences in the size of food eaten. Coker and Hayes (1940), measuring the filtering setae in *Diaphanosoma brachyurum* and *Daphnia pulex* (s. lat.), find that in both animals they are set at about 1μ intervals and conclude that food 1μ across and upward could be eaten. The largest food they observed taken by other species was *Sphaerocystis* of diameter 52μ.

Pacaud (1939) finds that the time taken to refill the gut by filter feeding varies considerably in the different species that he studied. Using suspensions of carmine or Indian ink, he gives data from which

TABLE 35. *Filtration rate and food intake of* Daphnia longispina

Cells *Chlorococcum* (ml^{-1})		Filtration Rate (ml. indiv.$^{-1}$ day^{-1})	Daily Food Intake (as % body weight)
Number	Mass γ		
8.4×10^3	1.2	14.0	66
25.0	3.1	7.0	100
59.0	7.4	4.1	149
92.0	11.2	2.88	197

TABLE 36. *Clearance rate of gut in various Cladocera*

	Carmine	Indian Ink
Moina brachiata	(3 experiments) 12.3 min.	(1 experiment) 9.8 min.
Scapholeberis mucronata	(1 experiment) 13.3	
Daphnia pulex	(3 experiments) 16.2	(3 experiments) 10.5
Simocephalus vetulus	(3 experiments) 19.6	(3 experiments) 18.2
Sida crystallina	(2 experiments) 25.7	

the mean rates in Table 36 can be roughly determined for a temperature of 18 to 20° C.

In spite of irregularities, it is quite clear that *Moina brachiata* can fill its alimentary tract much faster than *Simocephalus vetulus, Sida crystallina,* or *Eurycercus lamellatus;* the results from the last named are not reported quantitatively. The slow feeding of *Simocephalus* in the experiments seems to contrast with the results of Sushtchenia; the two sets of observations are, however, not really comparable and so not necessarily inconsistent. No information about the sizes of the animals is given, though at least the *E. lamellatus* are likely to have been smaller than the specimens of some other species employed, Pacaud believes that in water containing large numbers of bacteria, and in which epibionts may form an abundant growth on the carapace of the Cladocera present, so impeding filtration, rapid removal of the food mass from the prebuccal space into the gut is advantageous, When the animal is clean and the seston present is algal or inorganic, the abreptor of the postabdomen is able to knock out excessive amounts of filtered material if it starts accumulating at the bases of the appendages or piles up in front of the mouth. If the filtering action is impeded by epibionts or if the surface of the filtration mechanism is modified by bacterial secretions, Pacaud believes that the rapid filtering species would be at an advantage. He considers *Eurycercus lamellatus* and *Simocephalus vetulus* species of clear water, whereas *Moina branchiata* and *Daphnia pulex* can live in water containing much bacterial seston.

Pacaud (1939) has made observations on the gut contents of the dominant Cladocera of small ponds, comparing the species present at any one time and also giving some data on temporal changes. *Leydigia leydigii* and *Macrothrix laticornis,* small chydorids living on the surface of organic detritus at the bottom, appear to thrive on decomposing vegetable matter; *Chydorus sphaericus* and the species of the genera (*Ceriodaphnia, Daphnia, Simocephalus, Moina*) of Daphniidae encountered all seem to do much better on a diet of small delicate flagellates than on detritus and bacteria.   There is a considerable variation, as in experimental work, in the digestibility of the algae present.   *Trachelomonas, Lepocinclis, Phacus,* and *Glenodinium* are poorly digested by *Daphnia* and *Moina,* whereas *Chlamydomonas, Mallomonas, Cryptomonas,* or small members of *Euglena* are easily digested.   If the algal population changes from one group to the other, there is correlative change in the fecundity of the Cladocera. Thus in a pond (La Mare Ronde, Greffiers, near Rambouillet) heavily fertilized by domestic animals, when the phytoplankton contained abundant *Chlamydomonas, Euglena,* or *Cryptomonas,* the brood pouches of *Moina brachiata* contained 12 to 30 eggs.   When the more edible genera were rare and *Trachelomonas* dominant the egg number fell to 7 to 9.   Similarly, in another pond (Mare de l'Eglise) in the same village, when *Lepocinclis texta* was the dominant flagellate, *Daphnia pulex* was observed to be carrying 0 to 2 eggs; when *Euglena* and *Chlamydomonas* appeared, the egg number rose to 6 to 10 per brood.

There was some evidence that *Simocephalus,* in which the exoskeleton is heavier than in *Daphnia* or *Moina* and in which the mandible is probably able to exert more pressure on food particles, can use the less digestible genera, notably *Glenodinium.*   Pacaud also suspected that the detritus-feeding benthic forms probably cannot filter particles as small as those used by the Cladocera filtering in the free water, though no direct evidence of this was given.

Although Pacaud's work does suggest some food specificity between the habitual detritus feeders, such as *Leydigia* and *Macrothrix,* the more planktonic forms, such as *Moina* and *Daphnia,* which thrive best on minute flagellates, and the rather well-skeletonized and partly sedentary *Simocephalus,* there is no hint of any dietary difference between *Daphnia pulex* and *Moina brachiata* when co-occurring. Insofar as niche specificity is developed in these two animals, it may well be temporal, dependent on temperature.   Very subtle differences in nutritional requirements can occur; Hrbáčková-Esslová (1963) has found that *D. pulex* (sens. str.) may die in filtered lake water containing no particles larger than 60 μ across but that *D. hyalina* and *D. pulicaria* can survive in the same medium, though they grow more slowly and reproduce later than in well-fed control cultures.   The fecundity of primiparous females was reduced to about one fifth of the controls in *D. hyalina* and to about one third of the controls in *D. pulicaria,* which species evidently is more resistant to partial starvation than the other two. Since *D. pulex* and *D. pulicaria* are so closely allied that their distinctness has not been recognized by most investigators, the marked difference in nutritional physiology between the species is interesting.

Chemical limitation of planktonic Cladocera. The Cladocera as a whole are a fresh-water group; a few marine (*Penilia, Podon, Evadne*) or thalassohaline brackish (*Bosmina maritima*) members have already been mentioned, as have also the peculiar Ponto-Caspian members of the Polyphemoidea. Within the large majority of species inhabiting inland waters few clear cases of chemical limitation have been recognized; it is possible that some shallow-water benthic bog species will prove in part limited to soft and acid water.

Holopedium gibberum is usually and no doubt rightly regarded as characteristically living in the plankton of soft-water lakes (Thienemann 1926; Yoshimura 1933). It is ordinarily found when the calcium content is less than 10 mg. Ca per liter, at *p*H 6.5 to 7.2. Strøm (in Hamilton 1958) notes that it occurs regularly in Steinsfjord with a $CaCO_3$ content up to 54 mg. per liter, or 13.5 mg. Ca per liter. It has been recorded over a *p*H range of 5.3 to 8.1 and in its occurrence in Lake Michigan (Wells 1960) must have been taken in water containing about 32 mg. Ca per liter. Many lakes within the calcium range (1.4 to 8.2 mg. per liter) of the species in Japan were found by Yoshimura not to contain *Holopedium*. Pennak (in Hamilton 1958, discussion) notes the same phenomenon in North America and concludes (personal communication) that something other than calcium is involved. The full determination of occurrence is clearly not understood.

Within the genus *Daphnia* there is evidently considerable variation in tolerance to high ionic concentration. In general, the members of *Ctenodaphnia* appear to be more able to survive in somewhat mineralized water than are the ordinary planktonic members of *Daphnia* (*s. str.*), though in the latter the supposed endemic *D. balchashensis* in Lake Balkash can live in saline water (Manuylova 1948). In *Ctenodaphnia D.* (*C.*) *similis* has been important in Devil's Lake, North Dakota, and the alkaline lakes of Saskatchewan (Brooks 1957a) and *D.* (*C.*) *dolichocephala,* in quite saline water in North Africa (Gauthier 1928). *D.* (*C.*) *gibba* occurs in moderately alkaline and saline closed lakes in the Transvaal but probably usually at lower concentrations than tolerated by *D.* (*C.*) *similis* and *D.* (*C.*) *dolichocephala.*

Hutchinson (1932), by adding small amounts of magnesium as chloride to ordinary pond water of known magnesium content, found the upper limit of tolerance to magnesium in persistent culture to vary within the Daphniidae as follows:

0.03 g. per liter or 0.0025N *Daphnia (D.) ambigua, D. (C.) thompsoni*
0.06 g. per liter or 0.005N   *D.(D.) laevis, Ceriodaphnia reticulata*
0.12 g. per liter or 0.01N   *D.(D.) pulex*
0.18 g. per liter or 0.015N   *Moina macrocopa*
0.24 g. per liter or 0.02N   *D.(C.) magna*

If the organism referred to *D. (C.) thompsoni* is really a low-crested laboratory variant of *D. (C.) gibba*, which is very likely, its low magnesium tolerance is most curious. This work, undertaken to test the now obviously invalid conclusion that Lake Tanganyika lacks Cladocera because of its magnesium content (0.0392 g. per liter), at least shows considerable variability in tolerance among related species in a single family.

In the genus *Moina* at least three species appear to be adapted to quite saline water and two of them may be limnoplanktonic. *M. mongolica* is limnoplanktonic (Rylov 1935 sub *microphthalma*, Goulden *in press*) in the Sea of Aral and also occurs in the saline temporary waters of the sebkhas or small interior drainage basins of Tunis and Algeria (Gurney 1909 sub *M. salinarum*), where it may accompany *Daphnia dolichocephala* (Gauther 1928). *M. hutchinsoni,* recorded originally from the plankton of Big Soda Lake and Winnemucca Lake, Nevada (Brehm 1937a), is now known to be a widespread species in saline alkaline lakes in the northwestern United States and adjacent parts of Canada (Moore 1952; Anderson, Comita, and Engstrom-Heg 1955). *M. eugeniae* appears to occupy saline environments in Argentina (Olivier 1954).

The eulimnoplanktonic species of Cladocera appear to require a higher oxygen concentration than the pond forms, though it is possible that the species occurring in the hypolimnia of lakes, which, has been indicated, are also often heleoplanktonic, are more tolerant of low oxygen than are the species of the epilimnion. Herbert (1954) found that *Leptodora kindtii* and *Bythotrephes longimanus* from Lago Maggiore required more than 2.0 and 2.4 mg. per liter, respectively, as 50 per cent of her animals died in four hours at 18° C at these concentrations. *Daphnia hyalina,* from a reservoir in the south of England, required more than 1.1 mg. per liter, but a number of pond species (cf. also Pacaud 1939), including *D. pulex, D. obtusa,* and *D. thomsoni* (from New Zealand ephippia), tolerated much less oxygen, half dying in four hours at concentrations between 0.3 and 0.6 mg. per liter. In the pond species of *Daphnia* hemoglobin develops under conditions of low oxygen concentration (Fox 1948; Fox, Gilchrist and Phear 1951; Fox and Phear 1953), but it seems to remain reduced at the limiting concentrations just recorded.

Seasonal succession of Cladocera. The general pattern of seasonal succession in the lakes of temperate regions is comparable to that already described in the Rotifera. There are essentially three main types of life history. In the *aestival species,* individuals appear sporadically in the late spring and multiply to produce a large population, usually during the time when the water is warmest or a little after that time. Males then appear and resting eggs are produced, from which the population is recruited in the following spring. Such species thus are ordinarily monocyclic and monacmic, though sometimes two maxima separated by a well-defined minimum may be observed. In the ordinary *perennial species* overwintering is done as adults rather than as resting eggs, so that the population in extreme cases is acyclic; the numbers increase greatly in spring and decline in the autumn and only a small part of the adult population appears to overwinter; these species may be monacmic, diacmic, or irregularly polyacmic. In the perennial but apparently *cold-water species,* in default of experimental studies they cannot be legitimately called stenothermal, there is usually an increase in the population in the upper waters in early spring; the animals then retire to the metalimnion or upper hypolimnion during the summer months and may here reach their maximum abundance. The maximum population may occur in the deep water during summer rather than in spring or winter. These species ordinarily appear to be acyclic.

Species that are aestival in some lakes may be perennial in others, for example, *Daphnia cucullata* in certain Danish lakes compared with the same species in the Austrian alps. Moreover, both types of behavior, production of resting eggs and overwintering of adult females, may occur in a single population as in the same species in some Danish localities. The categories employed therefore are not to be considered as more than convenient divisions of a rather complicated continuum of behavior.

Aestival species other than Daphnoidea. It is convenient first to consider the limited number of planktonic Cladocera not included in the Daphnoidea. Nearly all appear to be aestival forms that pass the winter as resting eggs. This is true of the North Palaearctic *Limnosida frontosa,* which Lilljeborg (1900) found to be typically aestival and monocyclic in Lake Mälar, as well as of the species living more generally at lower latitudes. *Diaphanosoma brachyurum,* and the closely allied *D. leuchtenbergianum,* often not separated or regarded merely as a more planktonic form of the former, is a most characteristic summer species throughout the temperate Holarctic. In Lake Lucerne Burckhardt (1900a) writes of the incidence of "dieser exquisit periodischen Planktoncladocere"; nearly all subsequent work substantiates his expression, whether in the clear deep waters of central Europe or in the very eutrophic Lake Mendota or Pymatuning Reservoir (Apstein 1896; Birge 1898; Marsh 1897; Wesenberg-Lund 1904; Kofoid 1908; Berg and Nygaard 1929; Vanini 1933; Pelloni 1936; Sheffer and Robinson 1939; Findenegg 1953; Pirocchi 1947a; and Borecky 1956). There is usually but a single striking summer maximum, a little after the warmest period, during which time the growth of the population may be supposed to be most rapid. The experience of Marsh, that the species appeared earlier in the shallow Lake Winnebago than in the much deeper Green Lake, is probably due to concomitant differences in temperature. Vetter (1937) noted that in the Schleinsee (Fig. 159) at the time of the maximum the phytoplankton had fallen to a low level and most specimens of *Diaphanosoma* had empty guts. This suggests that the ensuing decline of the population was the result of starvation. The maximum is usually a regular monacmic peak, as Birge found, though in some cases, as in Pirocchi's observations on Lago Maggiore, Wells's (1960) on Lake Michigan, and most notably the study of Patalas (1954) in Lake Charzykowo in Poland, a well-marked minimum separates two summer maxima, in the last-named lake in July and September.

Among the other Sidoidea, both *Sida crystallina* and *Holopedium gibberum* have been sporadically recorded as aestival forms by several of the afore-mentioned authors; *H. gibberum* has been the subject of a special study by Freidenfelt (1920). The species appears in the Örensee in South Sweden in April at temperatures of about 3.5° C. Only·young specimens are present until the second half of May, but these young increase greatly in abundance, so that the maximum number of individuals was present at the end of April; thereafter there was a decline in numbers so that in May only one third or one half of the individuals present at the maximum of juveniles appear to have survived. The initial large number of individuals is attributed to the hatching of an immense number of resting eggs; in *Holopedium* there may be more resting eggs in a clutch than there are subitaneous eggs. The great mortality of the young is attributed to predation by young fish; the young are said at this time to have rather small gelatinous capsules. The population in 1910 rose to a maximum of adults at the end of June and then declined so that no specimens were present in late September. In the preceding year, however, there was appar-

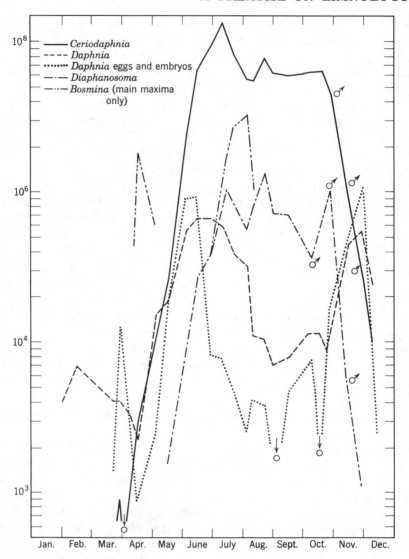

FIGURE 159. Seasonal distribution of Cladocera in the Schleinsee, as numbers per square meter of lake surface. Note the decline in reproductive
capacity of *Daphnia* as the characteristically aestival *Diaphanosoma* and
Ceriodaphnia develop, the suggestion of an alternation between *Bosmina*
and the other Cladocera, and the production of males by *Ceriodaphnia*
and *Diaphanosoma* late in the season, after the maximal population
densities. (Data of Kuntze.)

ently a secondary autumnal maximum in October. The egg number in 1909 increased throughout the season from a mean of 3.45 per egg-bearing female in June to 6.89 per egg-bearing female in September, but the resting eggs were still more numerous, namely a mean of 8.43 per egg-bearing female in October. It is unfortunate that the data for eggs per female and total population overlap for only a few months so that is is not possible to reconstruct the details of the population dynamics. The enormous production of young from resting eggs, which clearly hatch in deep water, appears to be unique in the Cladocera.

In the Polyphemoidea *Polyphemus pediculus* is usually a littoral species often occurring in swarms, but it may be limnoplanktonic even in a lake as large as Lake Michigan (Wells 1960); it appears to be characteristically aestival, as do the various members of the much more typically limnoplanktonic genus *Bythotrephes* in Europe (Wesenberg-Lund 1904; Patalas 1954).

Leptodora kindtii, the sole representative of the Haplopoda, is characteristically aestival and usually has its maximum in summer (Wesenberg-Lund 1904, Findenegg 1953), although sometimes, as in the Lake Winnipeg region (Bajkov 1935), it occurs in autumn. Being a large predaceous animal, it may play a significant role in regulating the seasonal occurrence of other species (Hall 1962).

Seasonal succession in the genus Daphnia. The genus *Daphnia* provides examples of all types of seasonal cycle, but in spite of the large number of records the confused state of the taxonomy of the numerous species involved makes many of the older accounts valueless.

In Birge's (1898) work on Lake Mendota three species of *Daphnia* are recorded which are now (Brooks 1957a) known to be referrable to *D. retrocurva, D. galeata mendotae,* and *D. schødleri.* This study forms a convenient point of departure (Fig. 160).

D. retrocurva in Lake Mendota is a typical aestival species, appearing at the end of June, reaching a maximum in August, and then fluctuating, with a second maximum in October when ephippial eggs are produced. The ephippia float and presumably hatch in the littoral in the subsequent spring. After ephippial production the females die off slowly, the last disappearing in January. Wells (1960), however, found no ephippia in Lake Michigan, though he supposed the species to survive the winter as resting eggs, and Brooks (1965) now believes overwintering as adults characterizes the cyclomorphotic species of temperate lakes.

Brooks (1946), in his study of the cyclomorphosis of *D. retrocurva* in Bantam Lake, Connecticut, was able to get a good idea of the

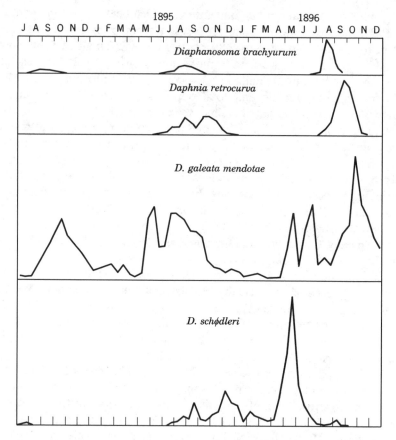

FIGURE 160. Seasonal incidence of *Diaphanosoma brachyurum,*
Daphnia retrocurva, D. galeata mendotae, and *D. schødleri* in Lake
Mendota 1894–1896. (Birge.)

succession of generations from the changes in head form and concluded
that if the young appearing in April are regarded as the first generation
of the year the young occurring in the middle of August represent
on an average the seventh generation. Since the species certainly persists, reproducing in the plankton after Brooks's study terminated in
August, it is reasonable to suppose that in this lake the species has
eight to ten generations per year.

Though not closely allied to *D. cucullata,* it is probable that *D.
retrocurva* is the ecological equivalent in North America of that well-
known European species. *D. cucullata,* like *D. retrocurva,* usually
tends to have a striking summer maximum, and in some localities

it is probably truly aestival, though Wesenberg-Lund (1904) suspected that in Denmark, where abundant ephippia may be produced and the species often seems to disappear in winter, a few females may survive to contribute to the population in the ensuing spring. This is also Berg's (1931) experience in the same region. In Carinthia (Findenegg 1953), where the species occurs mainly in deep lakes far less productive, at least per unit volume, than those of the Baltic countries, *D. cucullata* appears to be perennial and acyclic, ephippia being unknown, but it is evidently much commoner in summer. In Poland, two forms of the species occur in different lakes, one commonly referred to *f. kahlsbergensis,* though perhaps the typical *D. c. cucullata* with a very tall, straight helmet, the other, *D. c. procurva,* a local subspecies with a forwardly directed helmet, both appear in small numbers in May and disappear in November and December (Patalas 1954, 1956). There is considerable irregular variation in numbers, so that the population may tend to exhibit a diacmic cycle, as in Lake Charzykowo, in which a minimum is observed in this and other Cladocera during August, apparently because of unfavorable conditions at the time of the maximum myxophycean water bloom. Wiktor (1961), also in Poland, found a well-defined diacmic cycle with a maximum in June or early July and again in August and September. Wagler (1912) noted that in the small shallow basins that supported the species in Germany the populations also tended to a diacmic cycle, but with little sexual reproduction at the spring maximum. In the Schleinsee (Vetter 1937), in which the cycle is aestival and monacmic, the maximum is at the end of June, and *D. cucullata* with other planktonic Crustacea tend to decline as the myxophycean bloom approaches its maximum. Judging from some of Findenegg's (1953) figures, it is likely that in the Carinthian lakes, much less eutrophic than Bantam Lake, Connecticut, *D. cucullata* may have five or six generations per year.

Daphnia galeata mendotae behaved in Lake Mendota as a perennial species with a more or less diacmic cycle. A marked increase in the population to a maximum in late spring was followed by irregular oscillations which culminated in a second maximum in autumn. Birge concluded that above 18° C. the reproductive capacity of the species was impaired; in view of Hall's (1962) recent work, this seems unlikely. The species appears to behave in a comparable way in other localities, notably Pymatuning Reservoir (Borecky 1956), Leavenworth Lake, Kansas (Tash and Armitage 1960), and Base Line Lake, Michigan (Hall 1962). In none of these localities is the summer minimum likely to be due to competition with other species of *Daphnia,* though in Base Line Lake the minimum may be prolonged by the presence later in the summer of *D. retrocurva*. Wells (1960) found no evidence of the spring maximum in Lake Michigan, but it is possible that his sampling was not frequent enough to reveal all the changes in the population of this species; it was most abundant in late summer and autumn, for ephippia were produced in November.

As already indicated (pages 581–582), Hall (1962) has made an elaborate study of the population dynamics of *D. g. mendotae* in laboratory cultures and has applied his results to an analysis of the variation of the population of the species in Base Line Lake (Fig. 161). The initial spring maximum, culminating in late May and early June, is preceded by a maximum in brood production. At this time there is very little mortality and the rate of increase almost equals the birth rate. After the spring maximum a rise in mortality is followed by a second increase in birth rate, which, however, fails to produce a second maximum of breeding individuals. The loss of young stages, which greatly reduces the rate of increase, is apparently due to the appearance of the predatory *Leptodora* in the plankton at the height of summer. During the late summer some effect of competition with *D. retrocurva,* which greatly outnumbers *D. galeata mendotae* at its maximum in September, may also be involved in the depression of the natality of the latter species at this time. The small surviving population of *D. galeata mendotae* exhibits a third maximum in brood production in the autumn and thus leads to a second maximum in November and December. Reproduction then stops till March, as a low death rate slowly reduces the population during the winter.

Tappa (1964), in a study of Aziscoos Lake in Maine, in which there are six species of *Daphnia,* the greatest taxocene of this genus known anywhere, finds *D. g. mendotae* to be relatively less common at the height of summer than *D. catawba,* which is monacmic and aestival. The two species clearly occupy the same water and eat the same food. Competition must proceed very slowly, probably being in favor of *catawba* at high and *D. g. mendotae* at somewhat lower temperatures.

The equivalent European subspecies *D. g. galeata* was found by Wesenberg-Lund (1904) to be diacmic, with maxima in early summer and autumn much as its American representative has in Lake Mendota. In Lake Mälar, however, the species is regarded by Ekman (1907) along with *D. cucullata* and *D. cristata* as a summer form. Farther north, in Sweden, Axelson (1961a,b) found that *D. g. galeata* in Ransaren, where the temperature never exceeds 17° C. and is usually lower even at the height of summer, can exhibit a sharp monacmic

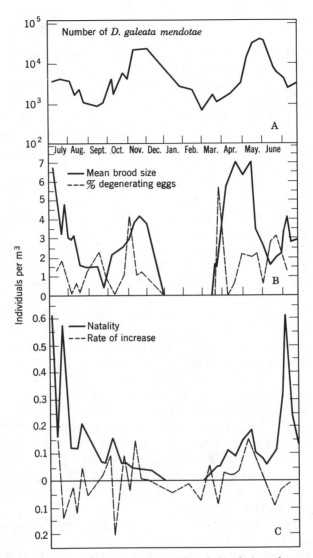

FIGURE 161. Population dynamics of *Daphnia galeata mendotae* in Base Line Lake, Michigan. *A*. Total population density. *B*. Mean number of eggs per female (solid line) and percentage of degenerating eggs (broken line). *C*. Natality (**b**) computed from number of eggs carried and from experimental studies on the dependence of rate of development on temperature (solid line) and observed rate of increase or decrease (**r**); note that high values of **b** imply high values of **r** only in the spring and autumn and that during the height of summer a very high natality can occur when the population is declining, apparently as a result of predation by *Leptodora*. (Redrawn from Hall.)

maximum at the beginning of July, though in some years the species is scarce throughout the summer.

Among the various species that have been referred to *D. longispina* in Europe both monacmic and diacmic cycles are known. In the Lunzer Untersee, in which temperatures rarely exceed 20° C., Ruttner (1930) found *D. longispina* (s. lat.) to have a simple maximum at the height of summer, followed by male production and the appearance of ephippia in the autumn.

Pirocchi (1947a) found a diacmic cycle in a species probably referrable to *D. hyalina* in Lago Maggiore.

In the lakes of Poland (Patalas 1954, 1956), *D. hyalina* may be either aestival or just perennial. In Lake Charzykowo it appears to be diacmic at most stations, with maxima in June and in September or October. These maxima, separated by heavy development of a myxophycean water bloom, may by synchronous with those of *D. cucullata;* as is usual with the less helmeted species of the genus, *D. hyalina* tends to occur in somewhat deeper water, though there is much overlapping of the vertical ranges of the two species. Wiktor (1961), also working in Poland, found a weak diacmic aestival cycle with the maxima in late spring and in late summer.

In Denmark Berg (1931) found *D. longispina* (s. str.) to be perennial in some localities, aestival in others, and more often diacmic than monacmic. It is, in fact, almost as often dicyclic as monocyclic; when there is a single sexual period, it is usually autumnal but can be vernal. In Ösbysjön, Djursholm, a very shallow, thermally unstratified lake, Pejler (1961) finds an irregular diacmic cycle in *D. longispina* (s. str.), with major maxima in June and October. The species however, is not perennial, for it disappears when the water below the ice becomes depleted of oxygen.

Wesenberg-Lund (Ostenfeld and Wesenberg-Lund 1906) found that in Thingvallavatn in southwest Iceland, an extreme example of an almost monomictic lake in an oceanic climate, which neither freezes nor reaches a summer temperature above 11° C., *D. longispina* (s. lat.) is dicyclic and probably diacmic. Ephippia were produced in July and August and then again in late September. The population between the two sexual periods was very reduced, and the exact relationship of the individuals present at the second sexual period to those constituting the July and August population is not quite certain. In a second Icelandic lake, Myvatn in northern Iceland, a more typically subarctic dimictic lake which is frozen from the end of October to the end of May, a comparable form of *D. longispina* apparently produced but two generations in the open season, one *ex ephippio* and parthenogenetic, the second, sexual. Not dissimilar cycles seem to have been encountered by Ekman (1904) in the species of *Daphnia* found in the mountains of Swedish Lapland.

The climax in the reduction in generation number with increasingly arctic conditions is achieved by *Daphnia middendorffiana* in arctic Alaska and probably in its other far northern, but not in its more southern montane, stations (for ♂ see Brooks 1957a). In Imikpuk Lake, Point Barrow, Edmondson (1955) found *ex ephippio* juveniles as the ice broke up in July. A few of these gave birth to normal parthenogenetic young but most of individuals produced parthenoge-netic or pseudosexual ephippia, so that there was for most of the popula-tion but one generation per year.

The third species reported by Birge in Lake Mendota, (sub. *D. pulicaria*) is now known to be referrable to *D. schødleri,* a species little known in Europe but of considerable importance and rather vari-able ecology in North America. In Lake Mendota it appears to behave as a cold-water form, increasing in spring to a marked maximum and then retiring to the metalimnion for the summer. *D. ambigua,* a well-known hypolimnetic form in many eastern North American lakes, clearly behaves similarly (Tappa 1964). Hall (1962) found a com-parable seasonal cycle in *D. pulex*[11] in Base Line Lake, and in Europe Findenegg (1943a, 1948) suspects the same sort of behavior supposedly of this species in the Millstätter See. It is most peculiar that this group of apparently cold-water forms belong to species that are pre-eminently inhabitants of ponds and whose general natural history gives no hint of their being cold stenotherm species; possibly tolerance of low oxygen concentration is involved.

A far more characteristic cold-water Cladoceran, and in fact the most striking apparently cold stenotherm plankter in the group, is *D. longiremis.* In the far north it is a pond form; on the southern boundary of its distribution in North America it appears to behave like the animals just discussed, for it maintains populations throughout the summer only in deep, cold water. It is acyclic and perennial almost everywhere, though the male has been described from a single specimen taken in Shuswap Lake, British Columbia (Brooks 1957a). Its detailed natural history has been worked out primarily by Freidenfelt (1913) in the Örensee in southern Sweden, where *D. longiremis* is accompanied by its close ally *D. cristata* (Fig. 162). The two species

---

[11] Some of the lacustrine populations referred to *D. pulex* in both Eurasia and North America are probably referrable to *D. pulicaria,* as understood by Hrbáček (1959b).

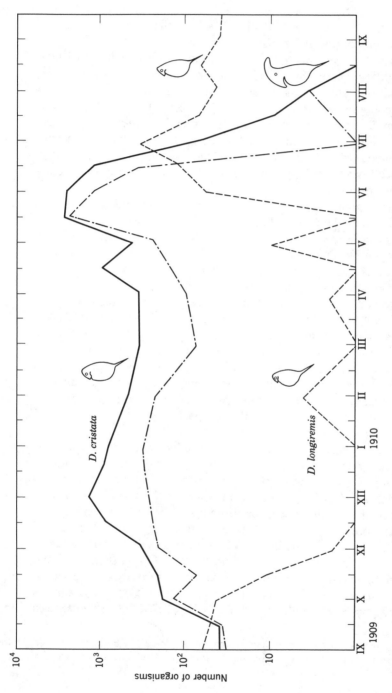

FIGURE 162. Seasonal distribution in the Orensee of *Daphnia cristata* (solid line) throughout water column and (dot and long dashed line) in the 0 to 10 m. epilimnetic layer and of *D. longiremis* (short dashed line), with figures showing the form of each species in winter and summer adjacent to the curves of the populations. Note that *cristata* (upper left and lower right-hand figures) is rarest most helmeted and, though epilimnetic, is most abundant at the coldest time of the year, whereas *longiremis* (lower left and upper right-hand figures), here not cyclomorphotic, is commonest in the hypolimnion but in summer. The numbers refer to an arbitrary undetermined constant volume dependent on the net used. (Data of Freidenfelt.)

618

differ from all other members of their genus by the length of their antennae and are separated primarily by the reduction of the basal seta on the three jointed ramus (endopodite) of the antenna in *D. longiremis,* contrasted with its complete absence in *D. cristata. D. longiremis* is almost certainly a cold stenotherm species. In the Örensee it is evidently perennial and acyclic, as almost everywhere else in its range, but becomes very rare in winter. It develops its maximum population in July at a temperature of about 7.3° C. at depths in excess of 25 m. In the same lake *D. cristata,* also a northern species but Palaearctic rather than Holarctic, is perennial, though with some sexual reproduction, strikingly diacmic, eurythermal at least up to 18° C., and highly cyclomorphotic. It is generally if somewhat irregularly distributed throughout the water column during winter and has a maximum in December at about 2.4° C.; in summer there is a comparable maximum in June, mainly in the epilimnon at temperatures between 10.8 and 18.0°C. There is a little sexual reproduction between October and January. The two species, though morphologically so similar, are clearly well separated physiologically and ecologically.

Seasonal periodicity in other Daphniidae. C. pulchella in the genus *Ceriodaphnia* is a widely distributed summer form in small lakes (e.g., Findenegg 1953; Pejler 1961), while *C. quadrangula* f. *hamata* is recorded as having an immense maximum in July and August in the Danish lake Mossø (Wesenberg-Lund 1904). The genus in general is more characteristic of ponds than of lakes. *Moina* is still more an inhabitant of ponds, but when limnetic, as in the saline lakes of northwestern United States and adjacent parts of Canada, shows a striking summer (Figs. 120, 121) periodicity (Anderson, Comita, and Engstrom-Heg 1955).

Seasonal periodicity in Bosmina. Insofar as adequate data exist, the species of *Bosmina* though clearly varying greatly in numbers, show maxima and minima less easily correlated with the seasonal cycle of temperature than the cycles of most of the species already discussed. Wesenberg-Lund (1904) found both *Bosmina coregoni* (s. lat.) and *B. longirostris* in most of the lakes that he studied. Both seemed to have maxima in the spring, *B. coregoni* persisting in the open water in somewhat reduced numbers and then producing a late autumnal maximum in November and sometimes persisting in numbers till January. Under the ice the species declined. Pejler (1961) noted comparable behavior by *B. longirostris,* but Wesenberg-Lund, and later Berg and Nygaard (1929), usually found the species to be purely littoral in late summer, reappearing in the open water in autumn but hardly producing an autumnal maximum.   Probably in accord, Patalas (1956) found a marked spring maximum in Lake Zamkowe.   Wiktor (1961) noted a diacmic cycle in *Bosmina coregoni* in Szezecin, Poland, with maxima in the spring and autumn.   In the Lunzer Untersee Ruttner (1930) found only moderate changes in abundance of *B. coregoni*, though in 1909 but not in later years there were well-marked maxima in February–March, June–July, and October–November. Ruttner found a rather feeble tendency toward increased egg production in spring and fairly clear minima in the average number of eggs per brood, which fell almost to zero in July and December.   At the time of the summer minimum *D. longispina* (s. lat.) was still breeding, though the broods were much smaller than in spring.

In contrast to these findings, Patalas (1954) found both the commoner *B. coregoni crassicornis* and the less common *B. (coregoni) gibbera* to be monacmic in Lake Charzykowo, though the maxima in August and September were only noticeable in the more eutrophic northern basin of the lake.   Possibly an aestival monacmic cycle may be characteristic of the more cyclomorphotic members of the genus.

Seasonal occurrence of other planktonic species of Cladocera.   The most important species so far not mentioned is *Chydorus sphaericus,* which seems to be perennial in the littoral but appears in quantity in the open water only during summer, usually in July or August (Apstein 1896; Birge 1898; Wesenberg-Lund 1904; Berg and Nygaard 1929; Patalas 1954).   Though, as Wesenberg-Lund points out, *Bosmina longirostris* in some localities can behave in a comparable way, living perennially in the littoral and entering the plankton only for a short period, Berg and Nygaard's detailed study (Fig. 163) shows the two great maxima in the Frederiksborg Slotsø to be well separated in time; that of *B. longirostris* occurred in May at the time of the spring phytoplankton maximum, largely due to *Scenedesmus armatus* var. *chodatii,* the other green algae, and diatoms, whereas *C. sphaericus* had its great maximum in the second half of August, as the late summer phytoplankton maximum, largely of *Anacystis,* was reaching its peak.   Comparable increases of *C. sphaericus* as myxophycean blooms began to develop are noted by Apstein (1896) and by Birge (1898).   Dr. Clyde Goulden (personal communication) has found comparable behavior in *Alona* in Indiana, and *Eurycercus lamellatus* perhaps behaves in this way in Lake Winnebago in some years (Marsh 1897).   Still further work on the littoral and pond species is needed.

General aspects of cladoceran periodicity.   It must first be noted that the number of generations of perennial species in temperate lakes

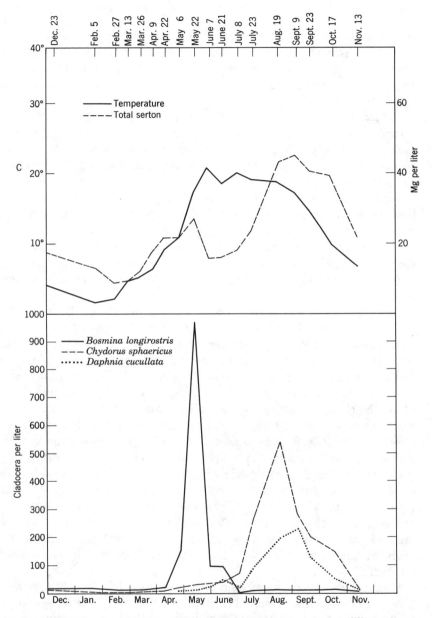

FIGURE 163. Seasonal incidence of temperature, total seston (milligrams per liter) and number of Cladocera of the more important species in the Frederiksborg Slotsø. Note the coincidence of the *Bosmina longirostris* peak with the spring seston maximum, due largely to green algae and diatoms, and that of the *Chydorus sphaericus* peak during the *Anacystis* bloom. *Daphnia cucullata* is also most abundant during the late summer bloom and so after the time when the helmet is likely to be best developed. (Berg and Nygaard.)

621

is likely to be of the order of magnitude of ten per year, falling to one or two in the arctic. These numbers may be somewhat less than occur in rotifers, but they are certainly more than would be expected in nearly all copepods under comparable ecological conditions.

Although the only feature common to almost all the annual cycles discussed is the occurrence of low populations in the late winter, usually under ice, it is possible at least to suggest some unifying hypotheses which may be the bases of future investigations of all the temporal changes that have been observed.

Hall's study of *D. g. mendotae* (1962) makes clear what might have been suspected from much earlier work; that given a moderate, constant supply of food at a moderately low temperature an increment in the rate of development of a population can be caused by increasing the temperature which increases the rate of molting and brood production or by increasing the food supply which increases the number of eggs per brood.

Starting with a low population at the end of winter, both the rise in phytoplankton in the spring and the usually somewhat later rise in temperature will be conducive to an increase in the cladoceran population, whether it is derived from a few overwintering adults, as in the ordinary perennial and the cold-water species, or from the hatching of resting eggs in the aestival species. The nature of the timing of the hatching of resting eggs is still obscure. The most reasonable mechanism, dilution of the littoral water by melt water, could act only when initially floating ephippia collect in the littoral zone in lakes that lie in landscapes in which much snow and ice melt in the spring. That melting in the lake basin can have impressive physical results in diluting the surface water under ice has already been indicated (Volume I, p. 456). In *Holopedium gibberum* in the Örensee, for example, it is clear, as Freidenfelt (1920) points out, that hatching must occur at considerable depths, if not all over the bottom; this might be the effect of spring circulation, but there is still no evidence for such a process.

However the spring population is started, its subsequent history may follow several paths, given the kinds of cycle that we have noted. In the simplest cases, those of monacmic aestival or perennial cycles, the population merely builds up to a peak and then declines. The decline may be due to starvation, as is probable in *Diaphanosoma* in the Schleinsee, or, in some other cases in which large numbers of males and ephippia are produced, to the change in reproductive pattern from one involving immediate hatching to one involving a delay.

Only in *Holopedium* does the large population in the autumn appear to imply, after diapause, a large population of young in the spring. In almost all cases there is likely to be competition between Cladocera and other filter-feeding or sedimenting animals. It is quite possible that in the strictly aestival species the phenomenon observed in *Moina* by Brown (1929b,c), of a great dependence of growth rate on temperature, may be operating, so that as the temperature rises these species are at a competitive advantage over the less sensitive species as long as food holds out. This may well underlie the sudden increases in *Diaphanosoma* so often recorded; the genus in general seems to flourish best at high temperatures and has a number of tropicopolitan species.

When a diacmic or polyacmic cycle is observed, it may be due to several causes which may act concomitantly. The usual fall in egg number after the spring, observed in every case for which there are data except that of *Holopedium gibberum* in the Örensee, indicates a lowered food supply as one probable factor in population control. Even this rather obvious aspect of the matter may have its complexities because Wiktor (1961) found that in *Daphnia hyalina* and *D. cucullata* the summer minimum in egg production could be partly explained by reproduction occurring at a smaller size than in the spring or later summer, whereas in the autumn low egg number apparently occurred in individuals as large as those producing many eggs in the spring. Evidently high temperature and low food are interacting to produce the summer minimum. Increasing temperature may increase the efficiency of utilization of this food in reproduction by increasing the rate of production of small broods, though this is not inevitable. In many cases either a decline in phytoplankton after the spring bloom or the development of massive blooms of species unsuitable as food may be a primary cause of the minimum between the peaks of a diacmic cycle. In the best analyzed case, that of *Daphnia galeata mendotae* in Base Line lake, predation by *Leptodora* appears to be the most reasonable explanation of the summer minimum. It is also reasonable to suppose that variation in predation by fishes, perhaps incident in the seasonal production of fry, could have a similar effect. Wiktor (1961) found that except at the height of summer her population of *D. cucullata* always and that of *D. hyalina* often consisted of a majority of immature individuals, suggesting very high mortality, and she supposed that variation in death rate was much more important than variation in birth rate in producing the observed changes in population. In *B. coregoni*, however, the excess of immature specimens was apparent only in the spring.

Competition with a more thermally sensitive species is likely to provide another mechanism for producing the summer minimum in diacmic populations; there is indeed a suggestion that *D. retrocurva* holds back the later development of *D. g. mendotae* until the autumn in Base Line lake.   Tappa's work may indicate that the same sort of thing is occurring very slowly in *D. catawba* and *D. g. mendotae*.   Inhibition of reproduction by high temperatures certainly can occur in the Cladocera and is under genetic control as made clear by the work of Banta and Wood on *D. laevis* already discussed (see page 594). Birge (1898) thought that this was happening in *D. g. mendotae* in Lake Mendota but it is now apparent that his explanation, if in any way correct, was too simple.   The only cases in which an inhibition appears reasonably likely is in some of the cold-water species, notably *D. longiremis*.   Freidenfelt's (1913) work on this animal provides a remarkable example of a summer maximum in cold water which can be reasonably explained only by an increase of food, either developing *in situ* or, more likely, falling from better illuminated layers.   The analogy with *Oscillatoria rubescens,* though its nutritive physiology is so different, and with rotifers such as *Filinia terminalis, Keratella hiemalis* and *Polyarthra dolichoptera* is obvious, but it should act also as a warning against too ready an acceptance even of *D. longiremis* as a true cold stenotherm species without experimental evidence.

The occurrence of *Chydorus sphaericus* as a summer plankter when water blooms develop is interesting.   Again rotiferan analogies exist, notably in *Trichocerca*.   It would be interesting to know whether the movement of such species from the littoral is in any way dependent on an increase in the back-scattering of light from the open water in which much seston is suspended, for it seems likely that part of the separation of species into littoral and open-water forms depends on the intensity of reflected upward illumination by day in their habitats (see page 804).

PLANKTONIC COPEPODA

The free-living fresh-water Copepoda are conveniently grouped in three orders, the Harpacticoida, Cyclopoida, and Calanoida.   The first has no representatives in the plankton; it is not unimportant in the benthos, and some benthic species have life histories in which a diapause comparable to that of the Cyclopoida intervenes.   It is convenient to mention this phenomenon in a section devoted primarily to the diapause of the Cyclopoida.   The Cyclopoida are mainly littoral benthic and pond forms, but the less numerous planktonic species are of immense importance and a considerable discussion of the natural history of the whole group is needed to provide a proper perspective for their consideration.   The Calanoida are all in some sense at least meroplanktonic and nearly all species truly and permanently belong to the plankton, though some inhabit very small ponds.

For the general classification of the group Gurney's monograph of the British forms may be consulted.   It has appeared desirable, however in this book to depart from Gurney's conservative position regarding generic subdivision and the recognition of species sometimes called critical and difficult.   It is reasonably certain, from Price's (1958) study of *Acanthocyclops vernalis* and its allies in North America, in which six noninterbreeding species almost or entirely devoid of any distinguishing morphological characters appear to exist, that in certain cases nature has split species far more than any taxonomist.

Though ecologists are, as amateur taxonomists, almost always lumpers, for the discussion of ecological problems the more finely drawn professional taxonomy of the splitter frequently gives greater clarity and insight.   This is particularly true of a book such as this which attempts to present the limnology of the earth as impartially as possible without undue preference for particular continents.   To use the single genus *Diaptomus* for almost all the Diaptomidae would obscure the fact that most of the forms occurring in Europe are not related to those in North America and that the fauna of South America again is quite different.   Subgeneric rather than generic distinction can, of course, be employed, as by Light and by Wilson in the Diaptomidae and Gurney in the Cyclopidae.   However, it increases the typographic complexity of the page merely to emphasize that the groups in question are not very different morphologically.   Having stated that the differences are small at the beginning of the discussion, it will be more convenient throughout the rest of the section of the copepods to follow Kiefer and other European workers and treat nearly all the recognizable species groups as genera.

Cyclopoida: Classification and Biology.   The free-living freshwater Cyclopoida are referred to a single family, the Cyclopidae, which may be divided provisionally (Kiefer 1929) into three subfamilies,[12] the Halicyclopinae, Eucyclopinae, and Cyclopinae.   The first of these subfamilies contains the genus *Halicyclops,* widely distributed in brackish waters along the coasts of both the Old and New Worlds and

---

[12] Gurney (1933), from a consideration of naupliar characters, regards Kiefer's subfamilies as artificial, concluding in particular that *Macrocyclops* is not allied to the other Eucyclopinae.   It is clear, however, that most of the genera of the Eucyclopinae and all those of the Cyclopinae are allied.

recorded from inland saline lakes in North Africa, from a mineralized hot spring in Java and very occasionally from supposedly fresh water (Marsh 1913).

The other two subfamilies are primarily fresh-water, though a few of their members may occur in mineralized lakes. Each includes a number of closely allied genera (*Macrocyclops, Tropocyclops, Eucyclops, Paracyclops,* and *Ectocyclops* in the Eucyclopinae and *Cyclops, Megacyclops, Acanthocyclops, Diacyclops, Microcyclops, Bryocyclops, Orthocyclops, Mesocyclops, Thermocyclops* and a few little known genera in the Cyclopinae), the species of which are for the most part benthic and littoral. Nearly all of these genera have an immensely wide distribution, and a few species are nearly cosmopolitan, being absent only from very unfavorable regions. The most widely distributed species appear to be *Macrocyclops albidus, Eucyclops agilis* (= *serrulatus*), *Microcyclops varicans,* and *Mesocyclops leuckarti.* A rather definite assemblage of Holarctic forms may enter the northern parts of the Ethiopian and Oriental regions but are for the most part confined to temperate North America, Asia, and Europe. The genus *Cyclops* (s. str.) and some of its component species such as *C. scutifer* (Scandinavia and Poland through Russia into northern North America, Novaya Zemblaya) have such a distribution. *Microcyclops rubellus,* if distinct from *M. varicans,* is another characteristic Holarctic species, and the list is likely to grow as the North American fauna becomes more critically understood. There are certainly some North American endemics such as *Acanthocyclops carolinianus, A. exilis, A venustoidea,* and *Diacyclops nearcticus.* There are also a number of species of restricted distribution in the Old World. *Diacyclops alticola* and *Cyclops ladakanus* from great altitudes in southwestern central Asia may be mentioned as examples, in addition to several species referred to in the discussion of the planktonic Cyclopidae. In many of the more important cases the taxonomy has been elucidated so recently that the distributions are still inadequately known.

A considerable number of species of *Tropocyclops, Microcyclops, Mesocyclops,* and *Thermocyclops* are of tropical distribution. Some, such as *Microcyclops linjanticus,* known from southern tropical Africa eastward through India and Indonesia into Formosa, have immense ranges, whereas others, such as the numerous species of *Microcyclops* endemic to Tanganyika, *Mesocyclops tobae* from Lake Toba in Sumatra and *Thermocyclops wolterecki* from Lake Lanao in the Philippine Islands, are perhaps confined to single lakes. In the genus *Thermocyclops* Kiefer (1938a) finds considerable local differentiation; in the Ethiopian region there are twelve species, all but one confined to Africa, whereas in Indonesia there are six species, of which four are confined to the southeast Oriental region. Only *T. decipiens* is common to the two areas. There also appear to be a number of tropical American species in the genus (Coker 1943).

Food and feeding habits of the Cyclopoida. Though the group has been familiar to the amateur naturalist no less than to the professional biologist for more than two centuries, information regarding their food has been extremely inadequate until recently. The earlier work (Naumann 1923; Lowndes 1928; Roy 1932), which in part is misleading, has been reviewed by Fryer (1957a,b) to whom most of our modern knowledge is due. Casual observations in the century and a quarter since Jurine (1820) noted that some species were carnivorous has given evidence of feeding on both animal and plant material; the selective nature of feeding by various species, however, was not really clear. A few species, however, had been known to attack young urodeles and fishes vastly larger than themselves, an interesting observation in view of the probable derivation of most if not all of the parasitic copepods from the Cyclopoida.

All the free-living Cyclopoida must be regarded as raptorial (Fig. 164), whether the food seized is animal or plant; no filtration mechanisms are developed in the group. Fryer's (1957b) study of gut contents led him to the following conclusions as to food eaten, the animals being arranged roughly in descending order of size:

Macrocyclops albidus (a benthic littoral and pond form). Carnivore, feeding largely on other cyclopoid and calanoid copepods and various Cladocera, with dipterous larvae and other arthropods, rotifers, oligochaets, and probably flat-worms; may take a little algal food.

Macrocyclops fuscus (a benthic littoral and pond form). Carnivore, probably rather more specialized than *M. albidus,* taking predominantly chydorid Cladocera, but not others, nor eating calanoids; dipteran larvae unimportant; algal food as in preceding.

Megacyclops viridis (ordinarily a benthic littoral and pond form). Carnivore, feeding on both groups of copepods, and nonchydorid Cladocera, but not chydorids; the species appears to have a preference for large food such as dipteran larvae and oligochaetes; possibly more algal food ingested than in two preceding species.

Acanthocyclops vernalis. Markedly carnivorous but little studied.

Acanthocyclops venustus. Carnivorous in general, possibly also somewhat herbivorous but little studied.

Diacyclops bisetosus. Herbivorous, feeding on indeterminate algae.

Cyclops strenuus strenuus (ordinarily a pond form but sometimes in lakes). Carnivorous, little studied but food primarily crustacean.

C. abyssorum (planktonic). Carnivorous, mainly on calanoid copepods.

Mesocyclops leuckarti (planktonic). Mainly carnivorous, on cyclopoid and other copepods, various other Crustacea (cf. Naumann 1923) including *Diaphanosoma,* and rotifers, but also large diatoms probably actively seized and ingested.

Eucyclops spp. All appear to be mainly herbivorous. At least in *E. macruroides* algae less than 20 μ long are eaten for the most part only if they form clumps. The European species examined by Fryer, namely *E. agilis, E. macruroides,* and *E. macrurus,* all of which may occur as in the littoral of Esthwaite Water, feed mainly on filamentous green algae and diatoms; *E. macrurus* possibly takes more protozoa and rotifers than the other two. In central Africa *E. gibsoni* was found eating mainly filamentous green algae and some rotifers, *E. dubius,* a Nyasa endemic, mainly diatoms.

Microcyclops spp. Mainly herbivorous, though *M. nyasae* also eats rotifers.

In general it appears that the carnivorous species are larger than the herbivorous. Among the larger carnivorous species, in the best studied case, namely *Macrocyclops albidus, M. fuscus,* and *Megacyclops viridis,* all of which can co-occur, there seems to be some specific food preferences. *M. fuscus* avoids calanoid copepods or nonchydorid Cladocera but eats a disproportionate amount of chydorids when compared with *M. albidus; M. viridis* appears to favor larger food than the other two species. Among the herbivorous species Fryer could find no specific food preferences in *Eucyclops agilis, E. macruroides,* and *E. macrurus* when they co-occurred. He suggests that the carnivorous habit is primitive, as seems probable, and that the development of algal feeding provided so great an increase in potential nutriment that co-occurring herbivorous species are not limiting each other by competition for food. Speciation appears to have been more intense in the herbivorous group of genera.

If Kiefer's classification of the Cyclopidae is adopted, the herbivorous mode of feeding has probably evolved in both Eucyclopinae and Cyclopinae; if *Macrocyclops* is removed from the former subfamily, this conclusion is less sharp but still not unlikely. It is quite clear that the large carnivores *Macrocyclops fuscus* and *M. albidus,* on the one hand, and *Megacyclops viridis,* on the other, are not very closely allied.

It is interesting to note that although diatoms, even when eaten whole, are digested, green algae such as *Spirogyra, Scenedesmus,*

Micractinium, Cosmarium, and *Tetraspora* appear to pass intact through the gut if not torn by the mouth parts. Comparable phenomena have been observed among the Cladocera. It is most curious, however, that when the extruded parts of *Spirogyra* cells are eaten by *Eucyclops,* as is frequently the case, the chloroplasts pass out undigested, though presumably the rest of the cell contents are of value to the copepod, since feeding on filamentous algae is common in *Eucyclops.* *E. macruroides* can squeeze out the contents of a broken *Hydrodictyon* cell, and both it and *E. macrurus* have been observed to puncture such cells with their maxillules and suck out the contents.

The only two planktonic species studied by Fryer, members of the two most important planktonic genera, are mainly carnivores. Further work is clearly needed, particularly since occasionally a cyclopid copepod of one of these genera may be almost the only zooplankter present in some lakes. This is true, for instance, of the open surface water of Panggong Tso, in which *C. ladakanus* was the sole animal form collected, with *Daphnia tibetana* only in the littoral (Hutchinson 1937a), though it also occurred in deeper open water. It is just possible that if these cases prove valid the nauplii feed on ultraplankton (mu-flagellates, etc.), whereas the more mature stages are largely cannibals. Work on the food of the earliest stages is clearly desirable. Loeffler (personal communication) has found a case of this sort in the Anostraca.

In feeding, the benthic cyclopids apparently swim over their substratum of mud, stones, or vegetation by a series of discontinuous jerks or hops and encounter their food in a random manner. On contact their behavior changes as attempts are made to seize the prey. If the food is lost, a series of circular searching movements may be undertaken, but apparently they are not very effective (Fig. 164). In view of the close affinity of the benthic and planktonic cyclopids, it is probable that the latter also encounter their food as the result of random swimming movements.

Planktonic Cyclopidae. Although the Cyclopidae are among the most important members of the zooplankton, only a small proportion of the species of the family are planktonic. In this the family is reminiscent of the Cladocera or Rotatoria. It is remarkable how easily the transition between the benthic or littoral and the planktonic mode of life apparently can be made in the group. The most extraordinary case is perhaps that of *Megacyclops viridis,* recorded as planktonic in Lake Van by Kiefer (1957), who remarks that thoughout his long experieice of the group in no other locality is the species a regular

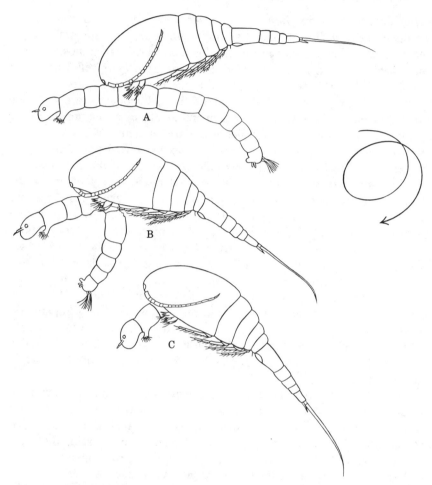

FIGURE 164. Semidiagrammatic drawings of a carnivorous cyclopoid feeding on a chironomid larva and a typical pattern of searching movement when a copepod has lost its prey. (Fryer.)

member of the plankton. This case may be comparable to that of the nektoplanktonic *Gammarus,* to be discussed in a later section (see page 696).

A common planktonic cyclopid of the Great Lakes and some of the lakes of Wisconsin is referred by American workers to *Tropocyclops prasinus.* It is quite possibly somewhat different from the typical *T. prasinus* of Europe, a pond species which, at least in Britain (Gurney 1933), does not occur in lake plankton; it is certain, however, that

the American eulimnoplanktonic species is closely allied to the European.   Similarly, *Diacyclops bicuspidatus thomasi,* another form common in the American lake plankton, is very close not only to *D.
b. bicuspidatus* found in shallow pools and ditches in Europe but also
to another American form, *D. b. navus,* which lives mainly in subterranean waters. (Yeatman 1944).   *Cyclops strenuus,* even in its nominotypical form (Roen 1957; Elgmork 1959), appears to be both a
member of the fauna of temporary ponds and of small but quite permanent lakes, though minute statistical differences may be discernable
between pond and lake populations in the same region.   Remarkable
ecological separation of closely allied planktonic, littoral, and benthic
species may occur in lakes.

The majority of the planktonic members of the Cyclopidae belong
to *Cyclops* (Fig. 165) and *Mesocyclops* (Fig. 166).   In the other

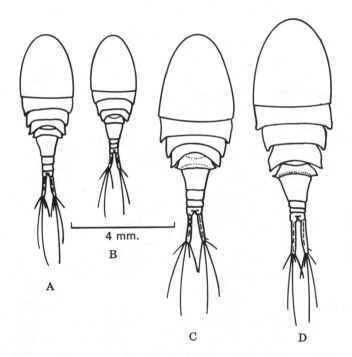

FIGURE 165.   Members of the *Cyclops strenuus* group from
Lake Constance.   *A. C. s. praealpinus* planktonic in main
basin.   *B. C. s. landei* in the more eutrophic Untersee.   *C.
C. bohater* rare and of uncertain ecology.   *D. C. abyssorum
bodanus,* sublittoral and profundal benthos.   (Kiefer.)

genera of the Cyclopinae and in the Eucyclopinae the planktonic habit is sporadic; indeed the only important eucyclopine in the zooplankton is the North American *Tropocyclops* already mentioned. In the cooler waters of the North Temperate Zone a perplexing number of closely allied species of *Cyclops* are important members of the eulimnoplankton. The best studies of these organisms are Koźmiński's (1936), Kiefer's (1939a), and particularly Dussart's (1958) treatments of the European forms, but even in Europe much remains to be done. It is not possible, for instance, to equate all the forms described by Gurney (1933) and by Roy (1932) with those discussed by Koźmiński. The various species differ mainly in minute details of body form, shape of the posterior thoracic segments, and spination. None of these characters appears to be clearly adaptive to particular habitats and must be accompanied by physiological differences of greater significance.

In Scandinavia *Cyclops lacustris* occurs in large oligotrophic lakes occupying the epilimnion and having a maximum population in late summer. In the same region *Cyclops abyssorum* occurs. In Tyrifjord, where Strøm (1932) found both species, *abyssorum* occurred in the deep water near the bottom, and this seems in general to have been the experience of Scandinavian collectors. Koźmiński, however, found *C. abyssorum* in Poland in more or less eutrophic and slightly dystrophic lakes, localities evidently differing greatly from some of the Norwegian stations for the species. Outside Scandinavia and Poland the distribution of the species is problematical; Gurney thought *abyssorum* was the principal *Cyclops* of the lakes of Scotland and the English Lake District; Koźmiński is doubtful of his determination, but it is acceptable to Fryer (1954). *C. a. abyssorum* is rare in France; it is apparently found in Lake Constance as *C. a. bodanus* and further subspecies occur eastward in Carinthia (*carinthica*) and in the Altai Mts. (*gracilipes*). A third limnetic species in the North of Europe, *C. scutifer,* has a wide distribution from Norway and Poland eastward through the subarctic to North America eastward and southward at least to New York State and northern Connecticut. In Poland there appear to be two forms, f. *scutifer* (s. str.) found mainly in unproductive waters and f. *wigrensis* in moderately productive or eutrophic waters; the two differ not only in size, *wigrensis* being larger, but in the relative length of the dorsally set seta at the end of the furca, which is shorter in *wigrensis*. *Cyclops kolensis* has in the Wigry region in which Koźmiński worked a wider distribution than any of the foregoing forms. It occurs in all types of lake except the extremely dystrophic or the extremely eutrophic in which the winter fauna may be limited by oxygen deficiency under the ice. Three other species were encountered by Koźmiński in the lakes of the Wigry region. *C. bohater* is probably a bottom form found in moderately eutrophic lakes, both in Poland and in some French localities (Dussart 1958). *C. vicinus* appears in Europe in three forms: *C. v. vicinus* is widespread in the plankton of eutrophic lakes and spreads farther south than any of the species hitherto mentioned but does not occur around Lake Wigry. *C. v. kikuchii* appears to be littoral and has colonies in a few East German, Polish, and Japanese lakes. The last limnetic form of the region is *C. strenuus landei,* which is characteristic of extreme dystrophic forest lakes. Also in the same area are *C. s. strenuus* and *C. furcifer,* which, as elsewhere, are found mainly or, in the case of *furcifer,* exclusively in small temporary waters.

In Lake Constance Kiefer (1939a, 1954) has found three ecologically separated forms (Fig. 165). One, *C. strenuus praealpinus,* is planktonic in limited numbers at all depths and seasons. A second taxon, *C. abyssorum bodanus,* is benthic, and the third is a littoral form that may be referred (Dussart 1958) to *C. strenuus landei,* though it obviously occurs in a quite different habitat to that favored by Koźmiński's specimens. The detailed distribution of *Cyclops* in the other parts of Europe is less well known. The most important additional forms to be considered are *C. tatricus,* which apparently replaces the other planktonic species in the lakes of the Tatra Mountains, the Alps, and the Pyrenees, *C. s. vranae* and *C. a. maiorus* which may be the main members of the genus in the subalpine lakes of Italy. In Poland *C. scutifer* appears to breed in summer; *C. vicinus,* in winter or spring, though Gurney finds it a summer species in England; *C. abyssorum* is markedly diacmic in the Wigry lakes with maxima in February-March and July-August; *C. kolensis* breeds from December to March, and the species of temporary waters perforce are regulated by the filling of their habitats which dry in summer. Koźmiński states *C. tatricus* is anacmic. It is probable that most species of *Cyclops* are not tolerant of very high temperatures, but this may be a false impression given by their disappearance in summer during diapause. Resistance to desiccation, reaction to temperature directly or in terms of the annual thermal cycle of the lake, possibly reaction to light in determining vertical position in the water, reaction to the quantity if not the nature of the food, and perhaps some special reactions to chemical composition are all likely to be involved in determining the ecological distribution of the species of *Cyclops* in Europe, but an immense amount of additional work would be needed before the interaction of these factors could be understood. The general impression given by the multiplicity of forms that have been described is rather like that given by the systematics of the white fishes of the subfamily Coregoninae.   This similarity extends to the occurrence of ecologically separated species within a lake such as Lake Constance.

Farther east several geographically restricted species of the genus occur in the lake plankton, such as *C. ochridanus* from Lake Ohrid and *C. ladakanus* from the high altitude lakes of the western part of the Tibetan plateau.   In the northern part of Palaearctic Asia certain European species are, as has been indicated, certainly present, but their distribution and ecology is little known.   The same unfortunately is also true of the North American species in spite of Yeatman's (1944, 1959) admirable revision.

Farther south in the southern part of the Holarctic and in the tropical parts of both New and Old Worlds the genus is replaced in the plankton by *Mesocyclops* and *Thermocyclops.*

The genus *Mesocyclops* (Fig. 166) and the closely related, or indeed doubtfully distinct, genus *Thermocyclops* also offer instructive cases of the ecological specialization of very closely allied species.   These copepods tend to replace the species of *Cyclops* as the chief planktonic Cyclopoida in subtropical and tropical regions and often in the summer in temperate latitudes.   The most important Old World species is *M. leuckarti,* which is represented in most, though not all, North American localities (Coker 1943) by *M. edax* (Forbes).   In certain of the subalpine lakes of central Europe (Constance, Neuchâtel, Bieler) and in Eğrirdir-Göl in the Cilician mountains (Mann 1940) *M. leuckarti* is replaced in the plankton by an extremely closely allied species *M. bodanicola* which differs in its slightly smaller size, slenderer form, and longer, narrower furca.   In addition one of its terminal spines on the fourth pleopod is more or less seta-like.   In Lake Constance, in which *M. bodanicola* occurs in the plankton, it may also be found accompanied by *M. leuckarti* in the more productive Untersee and in the littoral benthos; in the Bielersee or Lac de Bienne the separation between the two forms is perhaps sharper (Thiébaud 1931).   In the open water of these lakes competition evidently favors the smaller, slenderer, local species.

Over a very large part of its range, however, *M. leuckarti* seems to be a well-established member of the eulimnoplankton.   In some samples from the littoral of tropical lakes, such as Toba, *M. l. aequitorialis* indeed appears to replace the typical form (Kiefer 1933).   Kiefer's observations of the associations involving the species in Indonesia (Heberer and Kiefer 1932; Kiefer 1933) are of considerable ecological interest.   *M. leuckarti* was observed in the plankton of

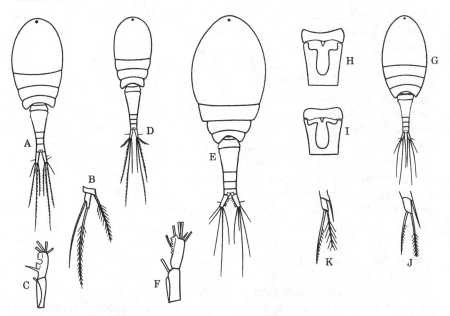

FIGURE 166. Ecologically important species of *Mesocyclops* and *Thermocyclops*
A. M. leuckarti ♀ (× 50). *B.* Fifth foot of same to show medial insertion
of inner seta on the apical podomere, characterizing *Mesocyclops*. *C.* Terminal
segment of attennule of same. *D. M. bodanicola* ♀ (× 50), the more limnoplanktonic ally of *M. leuckarti,* when they co-occur in the large subalpine
lakes of Europe. *E. M. edax* ♀ (× 50), the usual species in North American
lakes. *F.* Apex of antennule of same. *G. Thermocyclops hyalinus* ♀ (× 50).
H. Genital segment of same. *I. T. decipiens* genital segment, showing the much
narrower anterior diverticula of the spermatheca, one of the main differences
in these very closely allied but ecologically largely separate species. *J. T.
hyalinus* ♀ fifth leg, with subapical inner seta, characteristic of *Thermocyclops*.
K. T. decipiens, the same with a relatively shorter outer seta on the apical
podomore. (Kiefer, Yeatman, Gurney, and Coker.)

eleven lakes studied by the German Limnological Expedition to
Sumatra, Java, and Bali; in ten lakes there were other species of *Mesocyclops* and *Thermocyclops* present with it. If collections made in
localities other than the open water of lakes are considered, there
is more evidence of *M. leuckarti* occurring by itself, but in the lake
plankton it usually exists with certain other very closely allied species.
The commonest associates are one of two species of *Thermocyclops*,
T. hyalinus, or *T. decipiens*. Neither of these two species differs much
in size from *M. leuckarti,* and although they are placed by Kiefer
in a different genus the generic character is minute. Presumably some

difference in vertical distribution (cf. Worthington and Ricardo 1936), food, or behavior is sufficient to permit the association to exist. In contrast to this there is clear evidence that *T. hyalinus* and *T. decipiens* nearly always exclude each other. In the whole of southeastern Asia Kiefer (1933, 1938a; Heberer and Kiefer 1932) has recorded *T. decipiens* in eighteen localities and *T. hyalinus* in thirty-five; in only one case did the species co-occur. Although the size ranges of *T. hyalinus* (♀ 0.56 to 0.98 mm. long) and of *T. decipiens* (♀ 0.76 to 0.92 mm.) as a whole overlap completely, in the one locality, Lake Beira in Ceylon, in which they co-occur *T. decipiens* is slightly larger (♀ 0.82 to 0.86 mm. long) than *T. hyalinus* (♀ 0.72 to 0.80 mm. long), suggesting character displacement. Apparently there is a tendency for *T. decipiens* rather than *T. hyalinus* to occur in artificial waters, though no other obvious ecological differences in habitat can be recognized from the available data. Hutchinson (1951) has suggested that perhaps *T. decipiens* is a fugitive species, just as Elton (1929) supposed *Eurytemora velox* to be. On this hypothesis *T. decipiens* would be better adapted to dispersal than *hyalinus* but would be unable to compete successfully when the two species entered the same water, except perhaps when the potentially competing populations already differed in size.

In addition to *T. hyalinus*, *M. leuckarti* is associated in Lake Toba with the endemic *M. tobae*. The *M. leuckarti* population is here composed of large individuals (♀ 1.12 mm. long) whereas *M. tobae* is small (♀ 0.7 to 0.8 mm. long). Similarly, in Lake Lanao, Mindanao, Philippine Islands, *T. hyalinus* is accompanied by a smaller endemic species *T. wolterecki* (♀ 0.56 to 0.60 mm.) long. In both lakes, therefore, the two endemics are smaller than their sympatric allies.

It is interesting to note that Gurney (1933) records *M. leuckarti* and *T. hyalinus* with *Cyclops vicinus* in the plankton of the Norfolk Broads. Here, as elsewhere, *T. hyalinus* appears to be little smaller than *M. leuckarti,* though no measurements of populations from the same localities are given. It is quite certain, however, that *M. leuckarti* is much smaller (♀ 0.96 to 1.13 mm. long) than *C. vicinus* (♀ 1.49 to 1.80 mm. long) with which it co-occurs. Another interesting pair of species differing in size and occurring together may be mentioned here, though the animals in question are probably littoral or benthic. They are members of the genus *Eucyclops, E. parvicornis* (Harding 1942) and *E. spatharum,* apparently endemic to Lake Young in Northern Rhodesia. Both are closely allied to *E. agilis* (= *serrulatus* auct.), from which they differ in their relatively shorter antennules. *E. spatharum* also has rather spathulate spines on the pleopods which may be correlated with some special behavior. Apart from these and other minor characters of taxonomic importance, there is a marked size difference, the adult ♀ of *E. parvicornis* being 1.02 to 1.06 mm. long and that of *E. spatharum* 0.64 to 0.71 mm. long. Lake Young lies at an altitude of 1400 m. in latitude 1°10′ S. and longitude 31°50′ E. It is about 6.4 km. long, has a maximum depth of 8 m., and drains into the Chambezi River. The two endemic species of *Eucyclops* occurred together in some samples, and the lake also yielded the two widespread African species *E. gibsoni* and *E. euacanthus*.

Longevity. Walter (1922) found that *Megacyclops viridis* in the laboratory might live eight to nine months. She suggested ten to fourteen months as likely maximum physiological life spans for the larger cyclopoids and four to six months for the smaller. Elgmork's careful study of *C. s. strenuus* (1959), discussed in a later paragraph, indicates under different ecological conditions life spans of one to two months if a spring generation develops directly to the adult instar and at least twelve months if diapause intervenes.

Reproduction. All cyclopoids are bisexual. Elgmork concludes that the male *C. s. strenuus* lives for a shorter time than the female and so appears rarer. Eggs are always carried in two egg sacs attached to either side of the genital segment.

In *C. s. strenuus* egg number varies from fifteen to seventy-two. The number exhibits a weak inverse correlation with temperature. A given population may tend to have a higher mean egg number than that of an adjacent pond; this is apparently not correlated with the density of the population. Females that reach maturity from diapausing copepodite larvae in the spring are larger and carry more eggs than their daughters if the latter form a second spring generation. The largest number of eggs, however, is carried by the autumnal population produced from the early hatching of diapausing copepodites. Food is said not to be a likely determinant, but the incidence of epizoic organisms may be.

A detailed account of the variation of egg number and egg diameter in *Mesocyclops leuckarti* and *T. oithonoides* in certain Polish lakes has been given by Czeczuga (1960). The two species, though closely allied and both spring and summer forms, appear to have slightly different niches (Patalas 1954) in Lake Charzykowo, the larger *M. leuckarti* occurring more commonly in the northern eutrophic part and the smaller *T. oithonoides* more commonly in the southern, less eutrophic part; *M. leuckarti* appears also to be almost entirely epilimnetic, whereas a fair portion of the *T. oithonoides* population lived

TABLE 37.   *Egg number and egg diameter in* M. leukarti *and* T. oithonoides

		Spring		Summer	
		Egg Number	Egg Diameter (μ)	Egg Number	Egg Diameter (μ)
Białe	*T. oithonoides*	14 (12–16)	83.6 (83.0–91.3)	6.4 (2–8)	74.2 (66.4–83.0)
	M. leuckarti	28 (27–30)	91.3 (78.0–96.2)	22 (20–24)	82.0 (78.0–88.0)
Rajgrodzkie	*T. oithonoides*	15.5 (14–16)	81.3 (74.7–83.0)	8.5 (8–10)	78.8 (74.7–83.0)
	M. leuckarti	83.7 (28–88)	78.1 (72.0–88.0)	15 (14–20)	78.0 (72.0–88.0)
Dręstwo	*T. oithonoides*	15.1 (12–16)	83.5 (83.0–91.3)	13.3 (12–16)	74.2 (66.4–83.0)
	M. leuckarti	28.2 (28–30)	78.0 (72.0–88.0)	21 (16–26)	80.0 (72.0–84.0)
Krywe	*T. oithonoides*	15.1 (14–16)	76.4 (74.7–83.0)	14.6 (14–16)	78.0 (74.7–83.0)
	M. leuckarti	30 (28–32)	74.7 (70.0–78.0)	18 (12–20)	78.0 (72.0–82.0)

in or below the metalimnion.   Unfortunately, it is not possible to check these differences in Czeczuga's data, which are presented in Table 37, for four lakes in increasing order of eutrophy.

A tendency to reduction in egg number in summer was shown in all lakes by *M. leuckarti,* but it is rather irregular; in *T. oithonoides* a marked reduction occurs only in the two less productive lakes.   The egg diameter in both species falls from spring to summer in the most oligotrophic lake and in *T. oithonoides* a little in the two mesotrophic lakes.   Only in the most eutrophic lake is there a slight rise in egg diameter in both species in summer.   The marked difference in behavior of egg number in the most productive locality is particularly interesting, but no clear interpretation is available.

Elgmork (1959) finds in *C. s. strenuus* natality from 0.03 to 5.66 eggs per female per day.   There is some evidence of egg mortality, and it is almost certain that there is a high naupliar mortality.

Margalef (1953, 1955) concludes that the number of eggs is proportional to the cube of the ratio of body length without terminal setae to the naploid chromosome number; when body length is in millimeters, the constant of proportionality is 15.

Stages.   In the Cyclopoida there are now known to be six naupliar stages (Dukina 1956; Elgmork, personal communication), usually numbered 1 to 6, and five copepodite stages numbered I to V,[13] the adult being VI.

Dispersal and resting stages.   Because most of the genera and a number of the species of fresh-water Cyclopidae are far more widely

---

[13] German workers usually use CY. 1 to CY. 5.

distributed than the equivalent taxa of the fresh-water Calanoida, it might be supposed that the species of the family would have very efficient dispersal mechanisms. Actually, not very much is known about them in most species.

Henry (1924) reared a species referred to *Eucyclops agilis* (sub. *Leptocyclops*) from dried mud in Australia; it may not be correctly determined by modern standards; *Megacyclops viridis, Diacyclops crassicaudus, Microcyclops rubellus, M. minutus,* and *M. inopinatus* have also appeared in mud culture or recently dried pools (Claus 1894; Sars 1927; Gurney 1933). In the last-named South African species there is presumptive evidence that the resting stage was copepodite V. Roen (1957) believed that *M. viridis* could aestivate in drying mud in an early copepodite stage or even as a metanauplius, though the evidence is indirect.

Roy (1932) found that in *Cyclops furcifer* both eggs and last copepodite or adult stages could undergo desiccation, provided they were dried in a protective layer of mud. There is, however, clearly some special circumstance required to permit the eggs to hatch after drying, for the experiment was successful in only two out of twelve cases. It is probable that the embryo must be at a certain critical stage in development if it is to survive. About one third of the subadult and adult individuals that were dried survived; all were females, and it seems likely that the male cannot recover after drying. No cyst is produced. Roy found that *Diacyclops bisetosus* behaves in a similar way; some viscid secretion is produced around the drying adult, but there is no true cyst wall. In *Diacyclops bicuspidatus* subadult stages can apparently appear in temporary pools after rain; *Eucyclops agilus* did not stand desiccation in Roy's experiments.

Although these cases indicate possible ways in which cyclopids in a dried state may be transported accidentally by wind or animals, they do not throw much light on the widespread distribution of the group. It may be noted moreover that in *Macrocyclops albidus, Eucyclops agilis,* and in *Microcyclops varicans,* three of the most widely spread species, there is no certain evidence of a resting stage than can be dried. Such evidence as the observations afford suggests that in the Cyclopidae stages other than the egg are often involved in diapause, desiccation, and possible transport.

The role of diapause in the life history. Of far greater limnological importance is the intercalation of a diapause in the life histories of the planktonic cyclopids living in permanent ponds and lakes. The phenomenon was discovered by Birge and Juday (1908) in *Diacyclops bicuspidatus thomasi* in Lake Mendota. This animal encysts in the mud of Lake Mendota from June to September and has subsequently been discovered behaving in the same way in Douglas Lake, Michigan, and Crystal Lake, Hennipin County, Minnesota (Moore 1939; Cole 1953a,b, 1955). It appears to remain in the mud in a cocoon-like cyst during the stage copepodite IV. A similar type of behavior is known for *Cyclops strenuus strenuus* in Norway, where Elgmork (1959) has demonstrated aestivation (Fig. 167) in the mud of ponds and of the small Lake Bergstjern, also in copepodite IV but without a cyst. Several benthic harpacticoids of the genus *Canthocamptus* have been found to have such a summer resting stage.

In contrast, a winter resting stage is known in *Mesocyclops leuckarti* (Fryer and Smyley 1954) in Esthwaite Water and Windermere (copepodite V, with a very few IV), *T. oithonoides* (Elgmork 1958) in Bergstjern (copepodite V), and in *Cyclops bicolor* (Fryer and Smyley

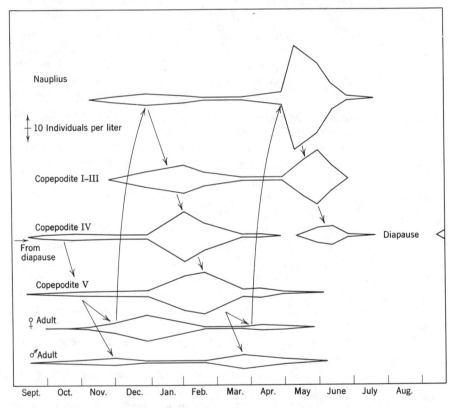

FIGURE 167. Annual cycle of *Cyclops s. strenuus* in Bergstjern 1952–1953, redrawn in somewhat diagrammatic form from the figures and data of Elgmork, showing the incidence of diapause.

1954) in Scale Tarn (copepodite III and IV), a pond in the English Lake District. The first of these species certainly and the second probably lack cysts, but in *Cyclops bicolor* a cyst is formed.

The phenomenon of diapause in these animals, which has a profound effect on their seasonal cycles, has been studied most satisfactorily by Elgmork (1959), in *Cyclops s. strenuus* in several ponds and in the small, thermally stratified, but very shallow (Z_m = 5m.) Lake Bergstjern,[14] Ringerike, Norway. The maximum population of adults is in April when reproduction proceeds rapidly. The fourth copepodite stage is reached by most individuals in June. Though in some years, apparently those of considerable addition of flood water and low temperatures, the first spring generation in part matures and produces a second spring generation, in most years development stops in the population of the first spring generation at the fourth copepodite instar. When a second generation is produced, development proceeds till the fourth copepodite instar but no further. During the summer, notably in late June, the species disappears from the lake. This is due to a descent of the copepodite IV population into the mud, in which such individuals undergo diapause. During autumn there is an irregular emergence from diapause, sometimes involving almost all of the population and sometimes none of it. Maturation and some reproduction takes place, but there is heavy mortality of the nauplii. The development of a large autumnal generation is promoted by flood water. In localities that have little water under the ice, and that water depleted in oxygen, the only method of survival over winter is in diapause. Emergence of diapausing copepodites can therefore be the main source of the spring population. The considerable variability observed demonstrates that diapause is facultative. It also indicates that one to three generations per year can be expected, in some cases individuals of the third generation living with survivors of the first. The ease with which diapause is broken increases with the length of previous diapause. In nature it is probably terminated in part by rising spring temperatures. There is no clear indication that either temperature or oxygen determine whether and when a fourth instar copepodite should seek the bottom and enter diapause. Elgmork thinks that the animals are driven downward by the bright light of summer and that this may be the major stimulus involved. The diapausing copepodites

---

[14] Dussart who examined the material agrees that the pond populations are *C. s. strenuus,* whereas the little lake contained a transitional population to *C. s. landei.* Elgmork considers that the subspecific name *C. s. strenuus* may justifiably be applied to all his populations, though admitting a tendency toward *landei* in the lake.

occur mainly in the top 10 cm. of mud but may be found in small numbers down to 27.5 to 30.0 cm.; they apparently survive anaerobically under conditions of complete oxygen depletion and H_2S formation. The maximum number of copepodites may be found in the mud in the deepest water in some years but at lesser depths in others.

On the whole, the seasonal cycle of *C. s. strenuus* observed by Elgmork is likely to be found in other parts of the range of the species in temperatue latitudes.

Smyly (1962) finds that, as in *C. strenuus,* the diapausing copepodite V of *M. leuckarti* is revived by raising the temperature but that the amount of warming needed falls progressively throughout the winter. Agitation also breaks diapause, again progressively more easily as the season advances. The population in Esthwaite is apparently more easily revived than that in Windermere. Increasing illumination has no effect in breaking the winter diapause in this species.

Seasonal cycles in other cyclopoid populations. The type of life cycle described by Elgmork probably explains the seasonal incidence of *C. strenuus* in the temperate Northern Hemisphere, and it is not unlikely that several other closely allied species of *Cyclops* (s. str.) behave in the same way. It must be noted, however, that exceptional populations occur even in *C. s. strenuus,* which in the Frederiksborg Slotsø (Berg and Nygaard 1929) showed winter, spring, and summer maxima in February, May, and July-to-August. The summer minimum corresponds to an early summer minimum in phytoplankton and to a maximum in *Eudiaptomus graciloides,* though competition between the cyclopoid and diaptomid is unlikely. Wesenberg-Lund (1904) found in Esrom Sø and Viborg Sø autumnal and vernal maxima but a summer minimum which may well correspond to diapause.

In the small mountain lakes of Swedish Lapland Ekman (1904) found that *Cyclops scutifer* appeared in the spring as nauplii which matured in four weeks. A second generation was then produced. In large lakes the progression of stages was much less regular.

In the lakes of Jämtland, in north central Sweden, Lindström (1952, 1958) found *C. scutifer* normally present in winter both as nauplii and as late copepodite stages. In May the nauplii are replaced by young copepodites, the late copepodites by adults, which bear eggs in June. In August the young copepodites of the spring have become adults and present with them are the young from the eggs laid in June. The population thus appears to consist of two heterochronic fractions which develop side by side but with a considerable difference in their relation to the seasons of the year. Not all the lakes of the region appear to contain both type of population, and Lindström believes that the dualism, as he terms it, is not so sharp that it would permit genetic divergence. A few of the early summer adults might be delayed in breeding to overlap the breeding of the August adults. The relationship of these events to the winter diapause recently discovered in the species (Elgmork 1962) should be interesting. In the extreme arctic conditions of Lake Hazen, Ellesmere Land, *C. scutifer* is biennial (McLaren 1961); even more surprising is Elgmork's (personal communication) finding of a triennial cycle in some Norwegian lakes.

In general in the North Temperate Zone it is usual to find species of *Cyclops* present in the cooler seasons, whereas species of *Mesocyclops* are more abundant in summer. Thus in the Libiszow lakes of east Poland, Kowalczyk (1957), who has paid special attention to the occurrence of a number of species of copepods, notes among the truly planktonic species

Cyclops kolensis	Mature egg-bearing ♀ ♀	Dec.–April, adults absent July–Aug.
C. vicinus vicinus	Mature egg-bearing ♀ ♀	Oct., Dec., adults absent May–Sept. Feb., Mar.
Mesocyclops leuckarti	Mature egg-bearing ♀ ♀	Mar.–Oct. adults absent Dec.–Feb.
Thermocyclops oithonoides	Mature egg-bearing ♀ ♀	May–Oct. adults absent Nov.–Mar.
Thermocyclops crassus	Mature egg-bearing ♀ ♀	June–Sept. adults absent Oct.–May

All five species occur, though somewhat sparingly, in Lake Białe. However, *C. kolensis,* which, as a very close ally of *C. strenuus,* might be expected to have a summer diapause stage, may exhibit summer maxima in other parts of Poland, perhaps comparable to the exceptional summer maxima in *C. strenuus.* The *leuckarte* and *T. oithonoides* of *Mesocyclops* are known to have a winter diapause; elsewhere, just east of the Urals, *T. crassus* passes the winter as eggs and exhibits no diapause after hatching (Ulomskii 1960).

In North America there is possibly a seasonal alternation between *Diacyclops bicuspidatus thomasi,* which has a summer diapause, and *Mesocyclops edax,* which appears to leave the sediment in the early summer and becomes common in the plankton (Cole 1955, 1961).

Beadle (1963) found that in Lake Nkugute, in western Uganda, *Thermocyclops schuurmanae* occurred in all stages in the mixolimnion, but in the deep water of the 40-m. oxygen-free monimolimnion nauplii and cyclopoid stages were present, though adults quickly died under such conditions. This survival of the young stages may be comparable to survival in the mud but clearly does not involve diapause.

A beautiful alternation (Fig. 168) in life history between a *Cyclops* of the *strenuus* group now referrable to *C. abyssorum maiorus* and *M. leuckarti* has been described in Lago Maggiore by Ravera (1954), the former breeding in September or October throughout the winter, with its peak naupliar population in December and copepodite population in April and May, the latter from April to October, with the largest number of immature stages in August. It is not easy to fit any diapause stage into Ravera's data, and it seems quite likely that the alternation between the two seasonal types of cyclopid is independent of, though in many cases facilitated by, the presence of such stages.

In warm temperate and tropical latitudes the tendency is perhaps toward a perennial multivoltine life history. Schuurman (1932) found an irregular variation in numbers, with breeding throughout the year, in *Thermocyclops macracanthus* in Florida lake, Transvaal. Apstein

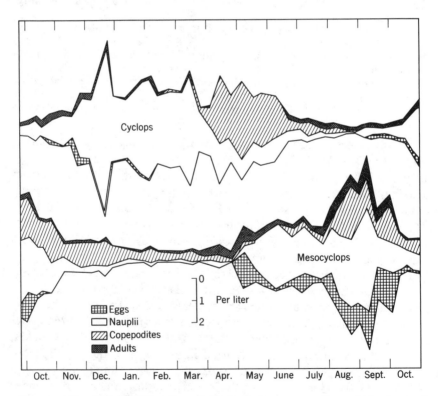

FIGURE 168. Annual alternation of *Cyclops abyssorum maiorus* (sub *strenuus*) and *Mesocyclops leuckarti* in Lago Maggiore 1951–1952. (Ravera modified.)

(1907), however, noted *M. "leuckarti"* in the Lake of Colombo most abundantly in January in the dry season.

Calanoida. The entire suborder lives in an essentially nektoplanktonic manner in the free water, though this water may be that of a tiny temporal pond, a lake, or the ocean. In the species of very small bodies of water temporary association with the solid substrate is possible, but it is achieved in a way quite different from what occurs in Cyclopoid and Harpacticoid copepods (see page 678). All the fresh-water families whose origin has been discussed at length in Chapter 18 contribute to the zooplankton of lakes. No taxonomic review is needed in the present chapter, since all the necessary detail has already been given (pages 122–129).

Life history. As with other copepods the animal is hatched as a nauplius; subsequent larval stages are classified as naupliar and copepodite. There are six naupliar stages ordinarily designated by Arabic numbers. The first two, often referred to as orthonaupliar (ON 1 and 2), possess antennule, antenna, and mandible. The maxillule is added in naupliar instar 3, the maxilla in instar 5, the maxilliped and the rudiments of the first two pleopods in instar 6; these four later naupliar instars are frequently termed metanaupliar (MN 1–4). At the next molt, giving copepodite instar I, there is usually a distinct change in form, the animal becoming elongate and obviously copepodan; Gurney (1931) noted in *Eudiaptomus vulgaris*[15] that the maximum relative increase in length (Table 38), by a factor of 1.4, occurs at this molt. Subsequent stages are usually denoted in publications in English by Roman numerals,[16] the adult normally corresponding to copepodite VI.

The absolute amount of growth achieved during the development from nauplius to adult is known in marine copepods to be inversely related to temperature and directly related to food supply. The importance of these two main controlling factors varies with the species and locality considered. A full discussion has been given in an excellent paper by G. B. Deevey (1960). Comparable phenomena will be observed in fresh-water species.

It has been supposed by Gurney (1931) that in some species individuals of the ordinary adult instar VI may molt to give a second adult instar VII. In both sexes of *Arctodiaptomus wierzejskii* from Tunis and in the females of the same species from the Shetland Islands he noted two discontinuous size groups. He gives a table of the mean

---

[15] Perhaps indistinguishable from a variety of *Eudiaptomus coeruleus.*

[16] German workers use CY. 1–CY. 5 for the copepodite stages prior to the adult.

TABLE 38. *Mean length and the growth factor or ratio of length of any instar to that of previous instar for* Eudiaptomus vulgaris *from Norfolk, England, and for the later stages of* Arctodiaptomus wierzejskii *from Tunisia (Gurney 1931)*

	E. vulgaris				A. wierzejskii			
	Mean Length		Growth Factor		Mean Length		Growth Factor	
N1 (ON1)	0.175 mm.				—		—	
N2 (ON2)	0.205		1.17		—		—	
N3 (MN1)	0.230		1.09		—		—	
N4 (MN2)	0.280		1.25		—		—	
N5 (MN3)	0.322		1.15		—		—	
N6 (MN4)	0.370		1.15		—		—	
C I	0.52		1.40		—		—	
C II	0.69		1.33		0.760		—	
C III	0.89		1.29		0.818		1.07	
C IV	♂1.06	♀1.15	1.19	1.29	♂0.960	♀1.090	1.3	
C V	1.32	1.44	1.24	1.25	1.160	1.303	1.20	1.17
Ad (VI)	1.55	1.72	1.17	1.19	1.338	1.590	1.15	1.20
? Ad (VII)	—	—	—	—	1.635	2.150	1.20	1.30

lengths of these size groups and of the copepodite stages that were available to him; the measurements for the population from Tunis are given in Table 38. *A. laticeps* in the Orkney Islands seem to exhibit the same phenomenon. Data suggesting a comparable phenomenon in a few marine species are recorded by Gurney.

It must be noted, however, that Comita (1956) has observed in a population of *Limnocalanus johanseni* in Imikpuk Lake in arctic Alaska a weak but significant bimodality in size in the adult instar. This he attributes to a sudden fall in phytoplankton probably due to overgrazing earlier in the season. He supposes that the larger individuals had got ahead by starting development when a rich source of food was available. He found that the first specimens entering any of the later copepodite stages were larger than those that entered that stage later, suggesting that a differential increase in size acquired early in life as a result of good nutrition was retained in later instars. Though the dimorphism is not very striking, it indicates that growth involving a second adult instar is not the only way that two size classes of adults can be produced.

A dimorphism involving not merely length but also the development of the apical process on the twenty-third joint of the geniculate antennule and the size of the spine on its fifteenth segment is recorded in some South American species (Wright 1938), notably *Argyrodiaptomus azevedoi, A. furcatus,* and *Notodiaptomus nordestinus.* At present there are not adequate data to decide whether the "long" form of Wright is really an adult in stage VII, a dimorphic form of the normal adult instar VI, or a sibling species. Kiefer (1939b) considers that the last-named explanation is correct in the case of *Neodiaptomus physalipus* and *N. diaphorus* from the Nilghiri Hills of southern India, which differ (Fig. 169) in a way comparable to the supposedly dimorphic forms of the South American diaptomids. Some seasonal variation in both size and in the details of the geniculate antennule apparently occurs (see pages 860–861) in certain Diaptomidae (Hartmann 1917) and the significance of the variations just discussed is clearly still problematic.

Calanoids being in general harder to rear in captivity than cyclopoids, there are few data on longevity; Walter (1922) found that *E. vulgaris* matured in one and a half to two months and lived for ten to thirteen months, whereas Dietrich (1915) found that in the same species and in *A. wierzejskii* about a month is needed for maturation. The variations are doubtless due to food and temperature differences.

Reproduction. The male in all fresh-water members of the group

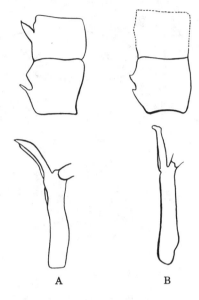

FIGURE 169. *A. Neodiaptomus physalipus,* fourteenth and fifteenth podomeres (*above*) and antepenultimate podomere (*below*) of geniculate antenna of ♂. *B.* The same for *diaphorus,* a a slightly less exuberant form or sibling species, illustrating a phenomenon found in various scattered species of Diaptomidae but neither adequately figured nor well understood. (Kiefer, fifteenth podomere of *diaphorus* described as without spine but only its lower margin figured.)

except *Senecella* (and the dubious *Calanus finmarchicus telezkensis*) has the right antennule geniculate or modified as a clasping organ. The fifth pleopods of the male are also highly and asymmetrically modified, and in *Epischura* the male abdomen bears on right side, on the fifth and, in most species, other segments, projections which supposedly can be used against other segments in clasping the female. In the Diaptomidae, as described by Wolf (1905) for *Eudiaptomus gracilis* (*A, B,* Fig. 170) the male is oriented in the opposite direction to that of the female in copulation, at first poised above her abdomen,

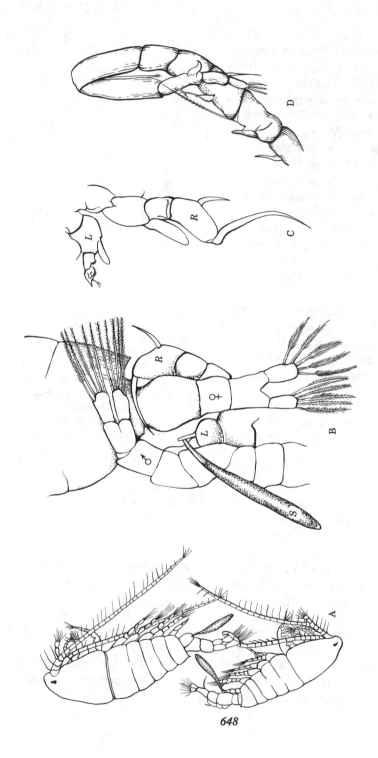

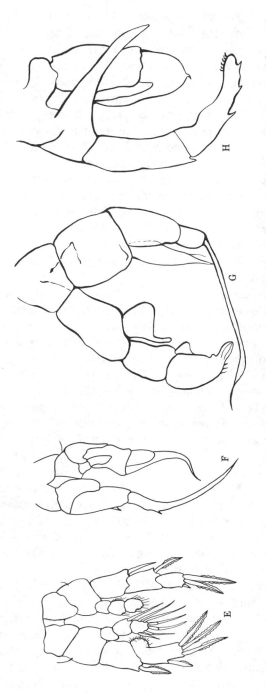

FIGURE 170. Asymmetrical structures and mating in fresh-water calanoid copepods. *A. Eudiaptomus gracilis*, ♂ seizing apex of urosome of ♀ by the geniculation of his right antennule. *B.* ♂ has seized genital segment of ♀ (both sexes marked on the penultimate urosome segments ♂ and ♀, respectively), with the right (*R*) fifth leg and is about to affix a spermatophore (*S*) with the left fifth leg (*L*). *C.* Fifth legs of ♂, the same joints marked as in *B*; *D.* Apex of right antennule of ♂ to show geniculation. (*A, B, D*, Wolf redrawn and slight modified, C. Gurney.) *E.* Slight asymmetrical fifth legs of *Osphranticum labronectum* (Centropagidae). *F.* Modified, fairly symmetrical fifth legs of ♂ *Boeckella orientalis* (Centropagidae). *G.* The same *Senecella calanoides* (Pseudocalanidae). *H.* Very asymmetrical fifth legs in ♂ of *Epischura nordenskiölda* (M. S. Wilson, Kiefer, Juday, and C. B. Wilson.)

which is clasped by the geniculate antennule. The male then swings round, seizing the base of the female abdomen by the highly modified right pleopod while affixing a spermatophore at her genital opening with the left pleopod. There is a strong tendency to minor asymmetries in the group, but very little indication of *situs inversus.* In *Pleuromamma gracile,* a marine species, the left antennule is geniculate; in *P. abdominale,* as in other calanoids, the right. Gurney (1931) follows older ideas that this can be interpreted in terms of the probable geniculation of both antennules in an ancestral form. It is more probable that *P. gracile* represents an evolutionary *situs inversus.* The last cephalothoracic segment and the genital segment of the female are often slightly asymmetrical. In the Pseudodiaptomidae, in which there are clearly two primitive egg sacs, that on the right may be almost or completely suppressed, whereas Dietrich (1915) in *Eudiaptomus vulgaris* and Cole (1961) in *Mastigodiaptomus albuquerquensis* find the right furcal seta in some of the naupliar stages to be longer and stouter than that on the left.

In most fresh-water calanoids the eggs are carried in a single egg sac (*A* of Fig. 171). This is true of all the Diaptomidae, of *Osphranticum* (Marsh 1933), *Boeckella* (Fairbridge 1945a), *Calamoecia* (Jolly 1955), *Gladioferens* (Henry 1919), and probably all the other fresh-water genera of Centropagidae, except *Limnocalanus* and possibly *Sinocalanus,* and of *Eurytemora* among the Temoridae. In the brackish water Pseudodiaptomidae, as has just been noted, there are basically two egg sacs as in *P. forbesi* of the Yangtse River region (Burckhardt 1913), but, in several species, for example, *P. coronatus* and *P. pelagicus,* the right sac is reduced and in *P. culebrensis* completely atrophied (Marsh 1933). Wilson (1959) says most species of *Epischura* lay eggs free in the water and the same is true of *Heterocope.* In the latter genus *H. caspia* and *H. soldatovi* (Rylov 1935) carry eggs stuck to the genital segment, but not in a sac. *H. saliens* produces masses of eggs (Matschek 1909) embedded in a sticky jelly, which may become temporarily attached to the setae of the furca. In *H. borealis* the eggs are dropped singly or in groups of two or three, but not more than six eggs are extruded one after another.

The eggs that are dropped by those calanoids that do not form egg sacs appear to be resting eggs undergoing diapause. In the Diaptomidae, however, resting eggs may also be found that are carried for a time, and in some species of temporary waters the egg sacs themselves may be thickened. As indicated in a later section, certain Diaptomidae and apparently also *Eurytemora velox* produce resting eggs and subitaneous eggs, both, however, by normal sexual processes.

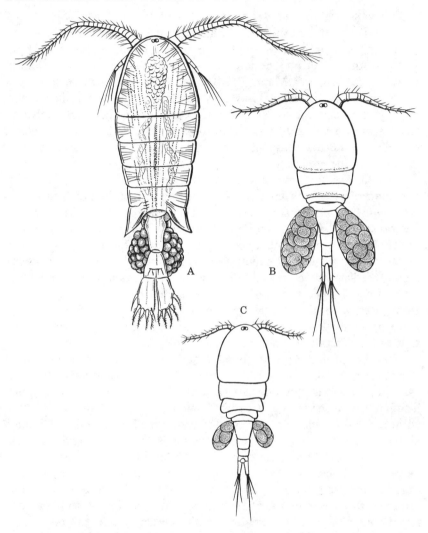

FIGURE 171. Ovigerous females of fresh-water copepods (× c. 65). *A. Paradiaptomus lamellatus,* widespread in South Africa, ovigerous female bearing the typical single egg sac of the Diaptomidae (Sars). *B. Microcyclops varicans bitaenia* and *C. M. nyasae,* congeneric littoral species apparently differing in number of eggs and so size of the paired cyclopoid egg sacs. (Fryer.)

The variation in clutch size in the fresh-water Calanoida. When, as in most of the fresh-water Calanoida, the female carries the eggs in an egg sac, it is easy to observe the variation in the size of the clutch within a given species. This variation may on occasion be quite extraordinary. Mann (1940) noted in samples from Turkish populations of *Arctodiaptomus bacillifer* mean values for the number of eggs per egg sac ranging from six to sixty; the total individual range was two to sixty-four. Unfortunately, his data are not extensive enough to indicate seasonal trends in a single locality or to suggest the cause of the variation.

Schacht (1897), one of the first workers to comment on the phenomenon of variable clutch size in copepods, which he noted in *Leptodiaptomus siciloides,* supposed that, since the species was originally described from Lake Tulare in California as carrying four eggs but in the Illinois River in summer carried up to eighteen eggs, high temperatures and good food supply were responsible for the large clutches. About the same time Birge (1898) noted in *Skistodiaptomus oregonensis* in Lake Mendota a seasonal trend in the number of eggs carried, from twenty to thirty per egg sac in the spring to nine to fifteen per egg sac in summer. It would therefore appear that a direct correlation with temperature is not inevitably involved.

Wesenberg-Lund (1904) found that *Eudiaptomus gracilis,* in most of the Danish lakes in which it occurred, carried twenty-five to thirty eggs in the spring but only six to eight in the summer and autumn. Though there was some irregularity in the second half of the year, with a possible hint of a second or autumnal maximum, the main and very striking feature of the variation is the rise to and fall from the well-marked spring maximum. Rzóska (1925) found a similar variation in Poland where *E. gracilis* carries twenty-two to twenty-eight eggs in the spring but only four to six in July. Within a given region, however, there is some variation from lake to lake, for Wesenberg-Lund found that although the variation in the clutch size in the population of Furesø[17] followed the same seasonal pattern as in the other Danish lakes containing the species the absolute number of eggs per clutch was smaller, falling from fourteen to twenty at the time of the spring maximum to four or five at the time of the minimum.

Czeczuga (1960) found a clear indication of a relationship between

---

[17] The Furesø population is, by an obvious misprint, referred to *E. graciloides* in that part of the English text dealing with egg number; reference to the earlier paragraphs of the English section or to the extended Danish text make clear that *E. gracilis* is meant.

TABLE 39. *Relationship between egg number and environmental variables in* Eudiaptomus *in Poland*

Lake	z_m.	Secchi Disk Trans- parency	Chlorophyll (August)	Mean Clutch Size, All Seasons	
				E. graci- lis	*E. graci- loides*
Białe	34.0	4.5 m.	5–14 mg. m.^{2}	8.2	4.8
Rajgrodskie	51.0	3.5 m.	5–36 mg. m.^{2}	9.3	6.8
Dręstwo	25.0	3.0 m.	15–39 mg. m.^{2}	12.5	7.5

egg number and trophic status of the habitat, which type of relationship is suggested by the work of most other investigators on other species. His results are summarized in Table 39.

Roen (1955, 1957) found that in *Eudiaptomus vulgaris,* living in very eutrophic ponds in Denmark, the number of eggs produced fell from around thirty in June to less than ten in October; he writes as if this represents primarily a decrease in clutch size occurring progressively through the life of the individual female. The same species in acid (*p*H 5.2 to 7.0) humic pools in a *Sphagnum* bog never produced, during the same period, more than eleven eggs per sac, the mean slowly declining from nine to six.

Von Klein (1938), in his admirable study of the planktonic Crustacea of the Schleinsee, found that *E. gracilis* carried a mean clutch of 11.0 to 11.4 eggs in April. This sank to a minimum of 3.0 per eggsac on August 2 and then rose to an autumnal maximum of 8.8 on November 5. Throughout the winter five or six eggs were carried. The variation follows that in abundance and reproductive capacity of *Daphnia cucullata* in the same lake. As already indicated, the production of eggs by *Daphnia* is certainly regulated by the food supply, and it is reasonable to conclude that the same is true of *diaptomids,* even though von Klein does point out that the populations of *Diaphanosoma* and *Geriodaphnia,* which like *Daphnia* are filter feeders, rise and fall at the very time that the reproductive capacities of both *D. cucullata* and *Eu. gracilis* decline and then increase.

Gurney (1931) considers that fewer eggs are carried by the females of *E. gracilis* in large lakes than by those inhabiting small ponds; these remarks may in part be influenced by Burckhardt's statement (1900a) that in the Swiss lakes two, three, or most frequently, four eggs are carried. It is not impossible, however, that there is also confusion between very close sibling species.

Brehm and Zederbauer (1902) recorded as many as seven eggs in the egg sacs of *E. gracilis* in the lakes of the Austrian Alps in summer but usually only four in winter. The winter population (Brehm and Zederbauer 1906) was found to have a much reduced spinous process on the twenty-third segment of the geniculate antenna. Haempel (1918) also noted this apparent cyclomorphosis, which is comparable to a number of cases described by Hartmann (1917). Gurney (1931) failed to find such cyclomorphosis in English populations, but a reduction of the antennal process is found in a small slender form described as *E. pusillus* and present in Loch Ness, in which lake typical *gracilis* also occurs. It is not impossible that two species are involved, one always with a produced twenty-third joint to the right ♂ antennule, slightly larger and more fertile, the other either cyclomorphotic or with a reduced process, smaller and less fertile. The endopodite of the fifth right leg of the ♂ is also proportionately slightly narrower in *pusillus*. The case deserves much closer study, for neither the slight cyclomorphotic phenomena in diaptomids nor the possibility of genetic variation in egg number is at all well understood.

Elster (1954) has studied the egg production (Fig. 172) of the population of *Eudiaptomus gracilis* in Lake Constance (Obersee). The species is perennial, and females with eggs can be found at any time of year. Ordinarily, adults are least abundant in the autumn and most abundant in the late winter and early spring (January to March), but there is often a subsidiary maximum about midsummer. The number of eggs carried at a time is variable, from one to fifteen, the modal number being six. In the much more eutrophic Untersee, higher numbers, up to thirty, are recorded, and it is at least reasonable to suppose that in part the number is determined by nutrition. The maximum number of eggs per female occurs in May and June (mean seven to ten in different years); an irregular fall occurs in the late summer, with a minimum when the population is at its lowest in October to December, suggesting perhaps lowered fertility in aging populations. In January the mean number increases again to five or six, with a secondary minimum in March to April of three or four. Part of the summer minimum may be due to the tendency of the animals to live in deep water deficient in food at this season. In April 1935, when the spring circulation was especially prolonged, a very low egg number was recorded, the modal number of eggs per sac being one.

Czeczuga's data (1960), given in Table 39, indicate that though *E. graciloides* is in any given circumstances less fecund than *E. gracilis*

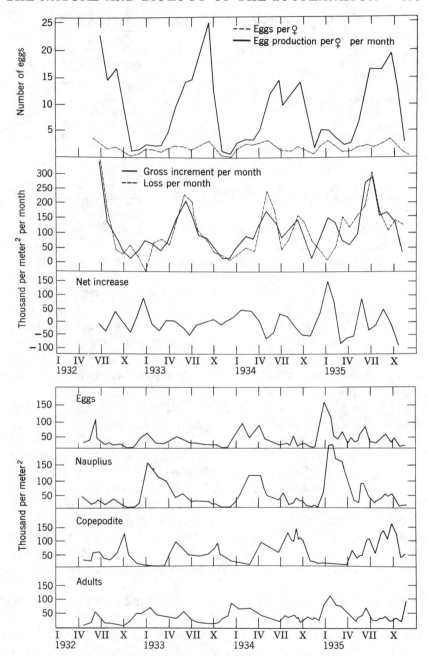

FIGURE 172. Population dynamics of *Eudiaptomus gracilis* as computed by Elster for Lake Constance 1932–1935. The eggs carried and egg production refer to the average ♀ and not the average egg-bearing ♀.

TABLE 40. *Breeding seasons of* Leptodiaptomus *in Lake Erie*

	Time of Maximum Clutch	Time of Maximum Number of Ovigerous ♀ ♀
minutus	March	April
ashlandi	May	April
sicilis	May	January

the two species react to the variations in the environment in essentially the same way.

Wesenberg-Lund (1904) noted a variation in *Eudiaptomus graciliodes* comparable to that exhibited by *E. gracilis,* save that in the case of the *graciliodes* so little reproduction occurred after the late spring that individuals carrying any eggs in the summer were very rare. In Esromsø, at the beginning of the breeding season in April, at a temperature of 3° C. the females carried seven to nine eggs, in the first week in May at 8° C., ten to twelve eggs, at the end of May at 13° C., four to five eggs. The very small number of females reproducing later in the year generally carried four, five, or six eggs. In other lakes comparable variations were noted; occasionally as many as sixteen eggs may be carried.

Davis (1961) studied the variation of egg number in five sympatric diaptomids in Lake Erie. In *Leptodiaptomus minutus* the mean egg number per ovigerous female fell from 7.33 in March to 2.00 in August. The maximum reproduction occurred in April and May at intermediate egg numbers. In *L. ashlandi,* which also started breeding in March, the mean number rose from 9.67 in the beginning of the breeding season to 26.4 in May and then declined to 8.0 in August. In *L. sicilis,* a winter breeding species, the number increased as the number of ovigerous females decreased, from fifteen in January to twenty-four in May; a little reproduction occurred in October with a mean of three eggs per female. All three species seem to be reacting somewhat differently.

The other two species are predominantly summer rather than spring breeders. In *Leptodiaptomus siciloides* the maximal clutch size is at the height of summer, the mean number being fifteen to seventeen in June and July in different parts of the lake.[18] There is a slight

---

[18] A single ovigerous female with twenty-eight eggs is recorded in May.

minimum of eleven eggs in August, when most ovigerous females are present, and another maximum in September of 19.5 eggs, after which clutch size and number of ovigerous females decline.

Skistodiaptomus oregonensis exhibits a comparable cycle, with a minor minimum in mean egg number (9.26) when most ovigerous females are present. The July maximum is higher (18.48) than that in September (12.85). It must be pointed out, however, that there is much variation in a given month throughout the lake.

Insofar as comparisons can be made, the egg numbers of *S. oregonensis, L. siciloides,* and perhaps *L. ashlandi* were lower in the more eutrophic western end of the lake than in the eastern less eutrophic part of the basin, but this effect seems to have been reversed in August, at least for the first two species.

A statistical demonstration of the effect of variation in food supply has been given by Comita and Anderson (1959), who investigated the clutch size in *Leptodiaptomus ashlandi* in Lake Washington. They found that the mean number of eggs carried by the ovigerous females was significantly correlated ($r = 0.598$, $P \simeq 0.01$) with the chlorophyll content of the water two weeks earlier. Obviously, factors other than nutrition are also involved. Comita and Anderson also found evidence that females taken in the hypolimnion carried more eggs than those from the epilimnion. They suggest that these more fecund females may be weighed down or otherwise impeded in swimming, by their large burden of eggs and so sink into the hypolimnion. If this is correct, an excessive production of eggs might result in selection against high fecundity.

In the Centropagidae of the Southern Hemisphere seasonal variation in egg number is reported by Bayly (1962a) for *Boeckella propinqua,* which had a maximum number of eggs in the autumn (March to May) in Lake Aroarotamahine, Mayor Island, Bay of Plenty, in part apparently correlated with low temperatures. Bayly found a slight decrease in egg number from 4.8 to 3.5 with depth at the height of summer. The animal is epilimnetic, and the distribution of the number of eggs per female corresponds roughly to the abundance of females at any level and so presumably to the general favorableness of the environment. Bayly quotes unpublished work by Byars which shows marked seasonal variation in *B. triarticulata.* Variation of egg number in tropical Calanoida appears to have been studied only in *Phyllodiaptomus annae* in the Lake of Colombo, Ceylon, by Apstein (1907). Here, when reproduction begins in June, the females carry about thirty eggs, but this number sinks progressively as the population rises to a maximum and falls again, till in September, when little reproduction is occurring and the species is becoming scarce, the egg sacs contain only seven to nine eggs.

Eichhorn (1959) has given some data on the variation of clutch size in monocyclic species which produce resting eggs at certain seasons.

In *Mixodiaptomus laciniatus* in the Titisee the number of subitaneous eggs carried rose from five to six at the beginning of the breeding season to ten to eleven in April or May and then fell to a minimum of five. Resting eggs were produced in May and June, but the number carried was always less than the subitaneous eggs, the mean value varying from 2.7 to 3.5.

In *Acanthodiaptomus denticornis* the highest mean number carried, namely 35.5 in the Titisee and 26.8 in the less productive Feldsee, occurred early in the breeding season in June or early July. The number fell to a moderate minimum of ten to eleven at about the time that resting eggs appeared in August but then rose to a maximum of about eighteen at the height of resting-egg production. It is clear that, unlike *M. laciniatus*, *A. denticornis* may produce clutches of resting eggs as large as those laid late in the season of subitaneous egg production.

Matschek's (1909) observation that *Heterocope saliens* in the Titisee lays ten to twenty-six eggs in a large gelatinous mass, but the same species in the less productive Feldsee lays ten to twelve eggs in a smaller mass, suggests that if the total egg production could be ascertained in the species that drop their eggs egg number would be found to depend on environmental factors just as in the species with egg sacs.

There is very little evidence of the number of clutches, or egg sacs full of eggs, that a female can produce. Fairbridge (1945a) thinks that in *Boeckella* only one clutch is normal, though in *B. opaqua*, in which the egg sac ordinarily contains five to nine eggs, he obtained evidence that on rare occasions there may be more than one clutch.

Ravera and Tonolli (1956) indicate that in the mountain lakes of the Italian Alps, in which *Arctodiaptomus bacillifer* and *Acanthodiaptomus denticornis* both usually produce only resting eggs, sporadically in the warmer lakes the same female may lay first a clutch of subitaneous and then a clutch of resting eggs, though very rarely is there evidence that the subitaneous eggs contribute a second generation of adults to the population.

Populations in which one or two eggs only are carried at any one time must obviously produce several such clutches if they are to survive, and it is probable that the production of but a single clutch of four or five eggs would imply a mortality, if the population is to be maintained, far lower than is likely to occur in nature.

Elster (1954) implies the probability of a number of successive clutches and indicates the importance of determining whether the primaparous female lays a small clutch and whether there is a decline in fecundity in old age; both phenomena are known in the Cladocera.

Clutch size and egg size. Wesenberg-Lund (1904) claims that in the three copepods that he studied not only does clutch size decrease but egg diameter increases as spring gives place to summer. He gives measurements only for *E. graciloides,* whose spring eggs have a diameter of about $120\,\mu$ and rare summer eggs of 160 to $200\,\mu$. If a spherical form is assumed, the volume of a spring egg must be about $9 \times 10^5\ \mu^3$ and of a summer egg $21\text{--}42 \times 10^5\ \mu^3$. Thus a summer clutch of four eggs might contain as much egg substance as a spring clutch of nine to eighteen. More recently the same type of phenomenon has been studied in several Polish lakes by Czeczuga (1959, 1960), whose results for *Eudiaptomus gracilis* and *E. graciloides* (1960) are tabulated in Table 41. The phenomenon is clearly real but is less dramatic in the Polish localities than it would appear to be in certain Danish lakes. Comita (1964) has found comparable phenomena in *Leptodiaptomus siciloides,* the egg volume increased irregularly at least twofold from the end of May to the end of June and then declining, whereas the number of eggs declined four- to fivefold. Czeczuga gives some data for *Eurytemora lacustris* which does not exhibit the phenomenon.

It is evident that the effect varies from lake to lake and that this variation is not identical in the two species. In *E. graciloides* in Drestwo the total mean egg volume per female is almost as great (87.5%) in the summer as in spring, and a similar relation (86.0%) is observed for the same quantities in Rajgrodzkie. In the other cases the compensation of egg number by egg size is less great.

It may be suggested that the variation in clutch size throughout the breeding season is not necessarily merely a passive effect of low food intake and that part of the variation may prove to have some adaptive meaning. Wesenberg-Lund suggests that the large summer eggs may hatch at a naupliar stage later than the first. The observations of Birge, Langford, and, most notably, von Klein, Elster, and Eichhorn indicate that the considerable total number of eggs produced by the large though rather infertile summer populations of a diaptomid copepod may add far fewer individuals to the population than the much more modest total egg production by the small fertile but increasing spring population. Kuntze (1938) indeed estimated that only

TABLE 41. *Egg number and egg volume in two species of* Eudiaptomus *in three Polish lakes*

	Mean No. Eggs Carried	Mean Mass ♀ (μg.)	Egg Mass (%) ♀	Mean and Range, Egg Diameter (μ)	Mean Egg Volume Assuming Spherical Form (mm.^{3} 10^{-3})	Egg Volume Per ♀
			SPRING			
Eudiaptomus gracilis						
Białe	15.5	61	18.0	107.9 (99.6 −110.2)	0.657	10.2
Rajgrodzkie	17.3	69	18.0	106.2 (99.6 −116.2)	0.627	10.9
Dręstwo	19.8	59	24.0	109.1 (83.0 −117.9)	0.681	13.1
Eudiaptomus graciloides						
Białe	6.4	40	12.5	116.2 (99.6 −132.8)	0.825	5.3
Rajgrodzkie	9.5	45	17.7	112.9 (99.6 −124.5)	0.749	7.1
Dręstwo	11.0	64	12.5	107.9 (99.6 −116.2)	0.654	7.2
			SUMMER			
Eudiaptomus gracilis						
Białe	6.4	64	8.0	116.2 (99.6 −132.8)	0.825	5.3
Rajgrodzkie	7.5	51	13.7	118.3 (105.4 −122.2)	0.865	6.5
Dręstwo	9.7	57	15.8	120 (116.2 −132.8)	0.902	8.8
Eudiaptomus graciloides						
Białe	4.5	36	8.0	122.8 (108.0 −132.8)	0.971	4.4
Rajgrodzkie	6.8	52	13.7	119.5 (116.2 −132.8)	0.895	6.1
Dręstwo	6.0	41	14.6	126.2 (108.0 −141.1)	1.052	6.3

TABLE 41. *Egg number and egg volume in two species of* Eudiaptomus *in three Polish lakes (Continued)*

	Mean No. Eggs Carried	Mean Mass ♀ (μg.)	Egg Mass (%) ♀	Mean and Range, Egg Diameter (μ)	Mean Egg Volume Assuming Spherical Form (mm.³ 10⁻³)	Egg Volume Per ♀
			AUTUMN			
Eudiaptomus gracilis						
Białe	2.6	43	5.0	113.6 (99.6 −116.2)	0.765	2.0
Rajgrodzkie	3.0	55	5.5	116.2 (99.6 −120.2)	0.825	2.5
Dręstwo	8.0	59	10.2	112.0 (99.6 −132.8)	0.735	5.9
Eudiaptomus graciloides						
Białe	3.7	35	9.0	114.5 (108.0 −132.8)	0.786	2.9
Rajgrodzkie	4.0	58	5.2	116.2 (110.0 −126.0)	0.825	3.3
Dręstwo	5.5	58	8.6	116.2 (108.0 −124.5)	0.825	4.5

about one fifth of the copepodite larvae of *Eu. gracilis* present in the Schleinsee at the end of May become adult. In the spring, when the population is rapidly expanding into a cool food-laden environment, the maximum production of nauplii is doubtless advantageous. In the summer when a low food supply, large population, and high metabolic rate must conspire against the newly hatched nauplius, anything that promotes the expectation of life of the nauplius is likely to be advantageous, and the presence of an increased amount of food in the egg may well be more important than a large egg production per individual female. This explanation of the variation of egg size inversely with egg number and with food supply supplies a loose temporal analogy to geographical variations noted in the char *Salmo alpinus* by Svärdson (1949) and Määr (1949) (also cf. Brooks 1950b).

Clutch size and body size. Ravera and Tonolli (1956) have made a remarkable study of the clutch size of thirty-two populations of

Arctodiaptomus bacillifer and fourteen of *Acanthodiaptomus denticornis* from forty-five lakes in the Italian and adjacent Swiss Alps; only one lake contained both species. In general, the habitats for *A. bacillifer* lay at a greater altitude (1971 to 2890 m.) than those of *A. denticornis* (562 to 2131 m.). The mean clutch size, based here clearly on resting eggs, varies from 1.54 ± 0.14 to 23.83 ± 0.69 in different lakes for *A. bacillifer* and from 5.90 ± 0.23 to 52.43 ± 0.60 for *A. denticornis*. The standard errors indicate that in any particular lake the variation is not great. In both species the clutch size is significantly correlated with the size of the female, as indicated by the length of the cephalothorax. Comparable observations have been made for *Oithona* and *Pseudocalanus* in the sea (Marshall 1949). There is also, in *A. bacillifer,* a significant rank correlation between number of eggs and the ratio of drainage basin to lake area, which gives a rough estimate of the rate of replacement of water in the lake. This is interpreted to mean that one selective force acting on the population is loss of copepods from the outlet, which will depend on the rate of flow through the lake. A balancing selective force operating in favor of small females when the rate of replacement of water is small must be postulated. This is curious because there is apparently a tendency for the lakes with least rapid replacement to be the most productive.

Bayly (1962a) found a comparable relationship between body length and egg number in populations of *Boeckella propinqua* in the North Island of New Zealand. On Mayor Island, Bay of Plenty, the mean body length and egg number were higher in the less productive humic Lake Te Paritu than in the more productive Lake Aroarotamahine. Bayly suspects that rate of replacement is involved in explaining his observations but, in view of the large seasonal variations in egg number, thinks that whatever factors are involved operate directly and not as agents of natural selection producing local races.

The only other study specifically concerned with egg number and size is that of Davis (1961) on the subitaneous eggs of five species of Diaptomidae in Lake Erie. The results refer to separate monthly collections when the species present were breeding abundantly. In the two collections (April and May) of *Leptodiaptomus ashlandi,* in the July and September but not in the August, October, and November collections of *Skistodiaptomus oregonensis,* in the January and March but not in the April collections of *L. sicilis,* and in the August and September but not the June and July collections of *L. siciloides* there is a clear positive correlation between egg number and the length of the ovigerous female. In the other cases mentioned there is no significant correlation, but in *Leptodiaptomus minutus* in both months (April and May) studied a negative correlation was observed. The meaning of these variations is far from clear.

Resting eggs. In most Diaptomidae inhabiting permanent waters, such as the common *Eudiaptomus gracilis* of Europe, only one kind of egg is produced, and it is invariably carried by the female in an egg sac. In localities in which this species (or the possibly distinct *E. pusillus*) breeds in winter the eggs apparently may be carried by the female for a period of several months. They appear, however, to differ in no way from the eggs laid in summer, save that developing at a very low temperature, or existing below some critical temperature, they develop slowly or, for a time, not at all. In other eulimnoplanktonic species, such as *Limnocalanus macrurus* or *Heterocope borealis*, the eggs are not carried by the female, as has been indicated, but are dropped to the bottom. In this case also hatching does not take place until spring, but since the eggs have spent some time in the mud, independent of the parent, they are commonly and legitimately termed resting eggs. The use of this term, as it is applied to *Limnocalanus*, does not necessarily imply any greater resistance to desiccation or other unfavorable conditions than is possessed by the subitaneous eggs of *E. gracilis*.

In several other genera (see page 650), none belonging to the Diaptomidae, there is a comparable production of eggs which are not carried in egg sacs and so fall to the bottom. This occurs, for instance, in most species of *Epischura* and *Heterocope* (Temoridae) and in *Senecella* (Pseudocalanidae).

In a number of Diaptomidae, and also apparently in *Eurytemora velox* in Sweden (Ekman 1907), two kinds of eggs may be produced, though both are carried initially in egg sacs. Thus Häcker (1902) found that *Acanthodiaptomus denticornis* in the Titisee produced subitaneous eggs during the first half of the breeding season, at the height of summer, but later laid a morphologically distinct type of resting egg with a two-layered shell. Comparable observations were made by Wolf (1903, 1905) on some populations of *Eudiaptomus vulgaris* and by Häcker (1902; Eichhorn 1959) on *Mixodiaptomus laciniatus*. Ekman (1907) recorded comparable behavior by *E. graciloides* in L. Mälar. In general, *Arctodiaptomus laticeps, A. denticornis,* and *M. laciniatus* in the northern parts of their ranges, in which they are often pond species, produce only resting eggs; in the southern parts, in which they tend to be limnoplanktonic, they produce only subitaneous eggs. In intermediate latitudes at least *A. denticornis* and *M. laciniatus* can produce both sorts of eggs in the same lake: subitaneous eggs early in the year, resting eggs later. In the Feldsee Eichhorn finds two clearly distinct generations, the adults that appear in the early summer being derived from resting eggs and producing subitaneous eggs. The generation hatching from the subitaneous eggs produce resting eggs, mainly in August and September. It is possible that adults derived from resting eggs maturing late in the spring may contribute to resting-egg production. In 1953 in the Titisee the entire crop of immature specimens from subitaneous eggs seems to have perished, but enough resting eggs were laid at the end of the summer breeding season to ensure a new generation. When, as in *A. denticornis,* resting eggs are always produced in the north, late in the breeding season in intermediate latitudes, and never in the south, it is natural to suppose temperature controls the type of egg. There is no direct evidence for this, however, and Eichhorn's observations on the production of subitaneous eggs by *M. laciniatus* in April and May, followed by resting eggs in May and June, show that control by low temperature cannot be general.

In a great many cases resistant resting eggs are produced by species inhabiting temporary waters, and here they may be produced when the habitat is drying up, which often corresponds to a time of rising temperature. In some cases not only is the eggshell thickened but the egg sac also, as in *Hemidiaptomus amblyodon* (Rylov 1935). As will be apparent in the discussion of the species of the temporary waters of North Africa, some produce only resting eggs; others lay both types of egg. It is reasonable, following Tollinger (1911), to suppose that the reduction in the size of the habitat is in some way the stimulus to resting-egg production in these cases. Tollinger suggests that increasing salt concentration, as the level of the saline waters inhabited by this species falls, is the effective stimulus for resting-egg production in *Arctodiaptomus salinus.* It is also not inconceivable that increasing concentration of the more dilute electrolytes in the waters inhabited by other species such as *E. numidicus, Mixodiaptomus incrassatus,* and *Metadiaptomus chevreuxi* acts as a stimulus in a lake manner. Not merely is experimental work required to elucidate these matters but further study of more complex situations may be necessary. In Europe Wolf (1903, 1905) concluded that *Diaptomus castor,* a species characteristic of temporary pools filling in winter and drying in late spring or summer, laid only resting eggs. Gurney (1931) found some colonies in Britain which produced such eggs, carried initially in very tough egg sacs, but he also observed other colonies in which only subitaneous thin-shelled eggs were produced. Such active eggs hatched in about four days in an aquarium. Gurney was convinced by a continuous study of the localities from which only females with subitaneous eggs were known that no resting eggs were ever produced by the populations inhabiting them.   However, the water dried up in such localities, and when they refilled nauplii appeared just as if resting eggs had been produced.   A somewhat similar paradox seems to be provided by the data of Fairbridge (1945a) on *Boeckella opaqua,* a Western Australian species known only from temporary pools in exfoliated granite.   In captivity nothing but subitaneous eggs were produced.   These eggs apparently did not hatch until long after they had been dropped from the egg sac and the females had died; nevertheless they could not withstand desiccation, though in nature the fate of any eggs so dropped would seem to be drying and so death.

Apart from the examples already discussed, or to be discussed later in specific ecological contexts, there is a certain amount of information about the mere existence of resting eggs in the species of fresh-water Calanoida that have been hatched from eggs included in dried mud, collected in many parts of the world.   Thus Sars (1901) obtained the South American *Argyrodiaptomus furcatus,* "*Diaptomus*" *coronatus,* and "*D.*" *conifer* from São Paulo by this method and he likewise reared *Tropodiaptomus australis* and *Eodiaptomus lumholtzi* from mud forwarded to him from Australia.   In his cultures of *A. furcatus* he observed the production of both active and later resting eggs.   Among the species of the genus *Boeckella, B. triarticulata* from New Zealand (Sars 1894), *B. minuta* (Sars 1896; Henry 1924), *B. fluvialis* (Henry 1924), *B. coronaria* (Henry 1924) from Australia, and *B. bergi* from the Argentine (Sars 1901) have been reared from mud.   It is reasonably certain moreover that *B. oblonga* must have some resistant resting stage, and the same is presumably also true of *Calamoecia subattenuata* if not of other members of the genus.   Evidently both the Centropagidae of the temperate parts of Australia and South America and the Diaptomidae of the rest of the world make extensive use of the production of resistant resting eggs as a method of colonizing temporary waters.   Whether a similar adaptation occurs in any of the other families of fresh-water Calanoida is apparently still uncertain.   Very little information appears to exist relating to the conditions under which the resting eggs hatch.   In the larger permanent waters such as Lake Constance or the Titisee it is obvious that neither drying nor freezing are likely to be necessary to ensure eclosion; Brehm (1927) considers that at least in *Hemidiaptomus (Gigantodiaptomus) amblyodon,* a large species characteristic of temporary waters in eastern Europe and western Asia, drying is necessary before hatching can occur.   He suggests that this may also be true of some other species.

Elster (1954) found that short exposure to room temperature followed by cooling to 4 to 5° C. produced 100 per cent hatching of the eggs of *Heterocope borealis* from the bottom of Lake Constance. Since the main period of emergence of nauplii is in April, the meaning of the temperature shock found to be effective is problematic. In *Aglaodiaptomus stagnalis,* a univoltine species living in temporary ponds near Chicago, the eggs hatch in March and the new generation of adults start reproducing in April. The eggs go into diapause before embryogenesis can occur and are then resistant to drying. Development is rapid when the temperatures decline in the autumn and a second but facultative diapause intervenes in which the egg is resistant to cold but apparently not to freezing. Hatching is stimulated by a decline in oxygen concentration, probably due to bacterial respiration in the mud (Brewer 1964).

Temperature and the rate of egg development. The rate of egg development is, of course, highly temperature-dependent. Laboratory studies by Elster (1954) and by Eichhorn (1959) have produced important data on the length of the egg stage in *Eudiaptomus gracilis, Acanthodiaptomus denticornis,* and *Mixodiaptomus laciniatus;* these data are summarized in Fig. 173, following Edmondson, Comita, and Anderson (1962). In the first-named species changes in pH and cation content of the kind likely to occur over the range of habitats of the species had no effect on the rate of embryonic development. Eichhorn in the second and third species found that naupliar instar 1 lasted respectively 4.5 and 5.5 times as long as the egg stage at a given temperature, so that the temperature dependence remains the same in passing from the egg to the first naupliar instar. At this first naupliar stage much yolk is still present and the animal appears independent of an external food supply. The duration of instar 2 after feeding had begun depended not merely on temperature but on other environmental factors. Taking the mean number of eggs per adult female and the mean length of embryonic life *in ovo,* it is possible to obtain by simple division an approximate measure of the rate of reproduction of the population, equivalent to b_e, used by Edmondson in his later studies of comparable phenomena in rotifers. In default of a value for the mean expectation of life of an adult female one cannot estimate the mean number of clutches produced, but for the study of the development of new populations the resulting figure is of great value. As will appear in a later paragraph, the variation in egg production could not be predicted from the static data on the average number of eggs carried without information on reproductive rate. Although egg number per female clearly depends partly on food,

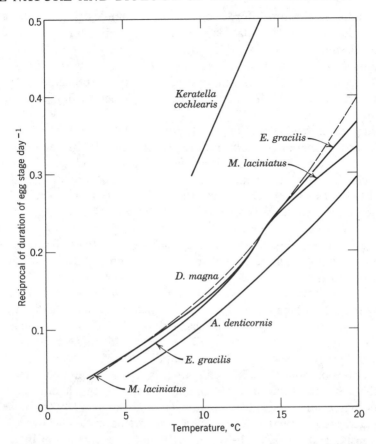

FIGURE 173. Relation of egg development rate (as reciprocal of duration of egg stage in days) to temperature in the diapto-mids *Acanthodiaptomus denticornis, Mixodiaptomus laciniatus, Eudiaptomus gracilis,* compared with approximate values for *Daphnia magna* and with the more rapidly developing smaller rotifer *Keratella cochlearis.* (Slightly diagrammatized from Edmondson, Comita, and Anderson's treatment of the available data, largely that of Elster and Eichhorn.)

the rate of reproduction as reflected in actual egg production is very temperature-dependent, and at seasons when few females carry eggs, and then not many, the reproductive rate may often be higher than under cooler conditions when most females carry many slowly maturing eggs.

Types of seasonal cycle. The breeding cycle of the calanoid cope-pods of lakes has been studied by a number of workers, whose results are summarized below in order of increasing complexity. Though most of the life cycles are relatively long, considerably longer than those of planktonic Cladocera and often somewhat longer than those of the cyclopoid copepods, there are striking differences between the species, and Ravera (1954) has emphasized (cf. Fig. 168) the role that these differences play in permitting several species to inhabit the same basin.

Only in a few cases, in which the absolute duration of the egg stage as a function of temperature is known, or can reasonably be inferred, is it possible to arrive at any real understanding. Such cases obtained by using the data of the previous section are discussed later in detail. Meanwhile it is important to realize that when eggs are carried by a very large proportion of the females of a perennial species in winter while in summer few females may be ovigerous and with smaller clutches, the more rapid development in the warmer months may imply a higher reproductive rate in summer than in winter, in spite of contrary superficial appearances.

In the simple case of *Limnocalanus macrurus,* studied in Ekoln in Sweden by Ekman (1907), breeding occurs throughout the winter. In this genus the eggs are not carried by the female and must fall to the bottom and hatch there. Eggs were laid only when the water temperature was below 7° C.; the nauplii hatched in March and April about the time that the adult generation is dying. The new generation reaches the adult instar in May and June, but egg laying does not occur again till about five months later. Marsh (1897) in a less extended study found evidence of a comparable cycle in Green Lake. Wisconsin, and Gurney (1931) thought that the population in Ennerdale Water behaved in the same way. Langford (1938), however, concluded that in the hypolimnion of Lake Nipissing the species reproduces throughout the summer. All authors agree that *L. macrurus* is a cold stenotherm and in particular that reproduction occurs only at low temperatures. It is to be noted, however, that Elgmork (1964) found that the species, when carried into quite small ponds, might survive several weeks at 14 to 20° C. In the allied but rather distinct species *L. johanseni,* living in the arctic Lake Imikpuk, in which temperatures higher than 7° C. are noted only in the first two weeks of August, Comita (1956) found that eggs hatched under the ice early in July. Adults began to appear early in August and by August 25 only mature specimens were present. No females bore spermatophores until the middle of September. Eggs must be laid in the succeeding months and the adults die off during the early winter.

The genus *Heterocope* provides another univoltine life cycle, most recently studied for *H. borealis* in Lake Constance by Elster (1954) and by Eichhorn (1959) and Bossone and Tonolli (1954) for *H. saliens* in Bavaria and the Italian Alps.   In Lake Constance the eggs of *H. borealis* hatch in the early spring, beginning in late February or March.   The maximum population of immature stages is at the end of April, with adults appearing in the second half of May and persisting in small numbers to the middle of winter.   The mortality of the early stages may be very great; its variation appears not to have any great influence on the final adult population.   In the Titisee Eichhorn found that *H. saliens* develops in six to seven weeks, beginning in early May.   The rate of development appears to be rather greater than for *H. borealis* in Lake Constance, but this development is taking place in somewhat warmer water.   The mortality of early stages seems to be much less than for *H. borealis* in Lake Constance.

Several species of European diaptomids have a univoltine cycle in the north, or at great altitudes, where they lay only resting eggs but, as has been indicated, they produce only subitaneous eggs in their southern lowland habitats.   Ekman (1904), in his classical paper on the crustacean plankton of the northern mountains of Sweden, believed that *M. laciniatus, A. denticornis,* and *E. graciloides* completed their life cycles in Swedish Lapland in two months or somewhat more, whereas at least the first and third usually had a ten-month period between hatching and reproduction further south.   The length of the life cycle of *E. graciloides* in the south is perhaps more variable than Ekman thought, and it is not quite certain from his evidence that these species really develop in the far north quite so fast as he believed. There may be genetic differences or unanalyzed environmental factors which permit a more rapid life cycle when the open season is very short, but the depths of Tornträsk, for instance, are not (Rodhe 1955) so metabolically uneventful in winter as would be expected.   No unequivocal judgment is possible until more work is published.   It is clear that on going south the different species react in different ways.

Mixodiaptomus laciniatus, which in its most northern colonies reproduces at the end of summer by resting eggs, remains essentially univoltine.   In the Titisee in the Black Forest (Eichhorn 1959) adults do not appear until September.   There they overwinter, producing subitaneous eggs in the spring and resting eggs at the end of the breeding season in May or June.   The role of the resting eggs in this lake appears to be unimportant, and there is no evidence of the breaking up of the population into two groups, those descended from subitaneous and those from resting eggs. First instar nauplii are present from late March to the end of May or early June. The species appears at all stages to extend into deep cold water.

In the Lake of Lucerne *M. laciniatus* is hypolimnetic (Burckhardt 1900b) during summer stratification and does not begin to breed until the autumnal circulation raises the temperature at 50 m. to 7° C. In Lago Maggiore (Ravera 1954) the cycle is somewhat similar, the species breeding from December to April in the upper 50 m., which are freely circulating for part of the time. Adults are absent at the height of summer, the main aestival form being copepodite V. When the adults appear, they tend to occupy initially cold water below 50 m.

Lindström (1952, 1958) has studied the annual cycle of *Arctodiaptomus laticeps* in the lakes of the province of Jämtland in the upland parts of north-central Sweden. In the best studied lake, Oltsjön, the cycle was evidently comparable to that of *M. laciniatus* or *A. denticornis* in the northern part of their range, nauplii derived from resting eggs hatching in the spring and maturing into adults from which resting eggs were again derived in the late summer of autumn. In at least two lakes, Ånn and Holtön there was evidence that part of the population overwintered as adults. In Ånn naupliar and copepodite stages present in August were probably descendents of such adults, whereas the adults present in August were, as in Ottsjön, derived from resting eggs. In Hottön there was evidence of some overwintering but not of a late summer population of immature forms. It is possible that variations in vertical distribution may, by varying the temperature of development, blur the distinctions between the two populations. Lindström regards the case for dualism in the population of *A. laticeps* as less well established than in *Cyclops scutifer* (see page 642), in which it may be due to varying diapause length, and does not consider these populations as genetically isolated. A somewhat comparable case (see page 675) which involves the centropagid *Boeckella propinqua* has been described quite independently by Bayly (1962a).

A true bivoltine life history with regular monocyclic alternation of a resting-egg-producing generation with one producing subitaneous eggs is known in *Acanthodiaptomus denticornis* and has been described notably by Eichhorn (1959) for the population of the Feldsee in the Black Forest. Adult females lay subitaneous eggs in June and July, leading to a generation that matures in August and September and lays resting eggs mainly in October. Such eggs ultimately fall to the bottom, hatch in spring, and produce the summer generation. The species also occurs in the lower-lying, more productive Titisee along with *Mixodiaptomus laciniatus*.   In 1953, the year of Eichhorn's main study, the subitaneous eggs of the summer generation led to the production of only a few individuals of copepodite stages; all were transparent blue comparable to unfed specimens kept in vials.   It is not unlikely that at least part of the autumnal population that produced resting eggs had itself been derived from such eggs.   A small number of CV individuals certainly persisted from June till October in decreasing numbers, whereas adults were also present throughout this whole period.

Comita and Anderson (1959) found that *Leptodiaptomus ashlandi* produces eggs in Lake Washington, Washington, over a long period from November to the following August but with a pronounced maximum in production in January to March.   A maximum population of Copepodite III appears in May, of copepodite IV in June, and of Copepodite V in September.   The immature stages must extend over a period as long as seven months.   Adults are probably also long-lived, up to five or six months, and occur throughout the year though commonest in November to January.   The mean number of eggs per sac carried by the female was significantly correlated, as has already been indicated, with the chlorophyll concentration of about two weeks earlier. The proportion of egg-bearing females, however, is not necessarily maximal at times of high chlorophyll, and the maximum number of eggs present seems to have been at a low chlorophyll concentration in the winter.   The nauplii hatching from these eggs may perhaps have encountered optimal nutritive conditions in the early spring.

The prolonged type of breeding cycle in *L. ashlandi* easily leads to cases in which low egg production occurs over most of the year but with more than one maximum in the process.

In Europe *Eudiaptomus gracilis* may breed throughout the season, but there are often periods in which the species is particularly abundant or the production of eggs and nauplii particularly intense.   In Ekoln (Ekman 1907) and the Lake of Lucerne (Burckhardt 1900b) there appears to be a single breeding season in the spring.   In the Danish lakes, in which Wesenberg-Lund found the species, breeding occurs throughout the year, whereas in some of the south German localities studied by Wolf (1905) there were two breeding seasons, a less intense one from the beginning of December to the middle of April and a more intense one from the middle of May to the end of June.   Tollinger (1911), Haempel (1918), and Gurney (1931) summarize a good deal of the less detailed information that suggests that in many localities *E. gracilis* is more or less bivoltine, or even multivoltine, with two or more maxima in reproductive activity occurring at somewhat different temperatures. A very clear case is provided by the later work on the population of the species in the Schleinsee (Kuntze 1938; von Klein 1938). Here egg production reaches a spring maximum in early April. Von Klein found that the eggs hatched in the laboratory in six days at 10° C. and in four days at 17° C. About seven days later than the spring peak in egg production there is a maximum in the nauplius population and seventeen days later a maximum in copepodite stages, followed by a maximum in adults at the end of May. The whole period from egg laying to the last molt took about fifty-five days. The large adult population in May produced an immense number of eggs; although the number carried per egg sac is smaller than in the spring, the large number of females ensures that the total egg production at this time is much greater than in April. There is no subsequent obvious maximum in nauplii, but later there is an increase in copepodite stages and a maximum in adults about fifty days after the summer maximum in egg production. Von Klein seems certain that resting eggs, which are not known in the species, were not produced at the time of the summer maximum, so that a very great mortality must be assumed for the eggs or for the nauplii hatched from them. Kuntze (1938), working on the same lake in 1935, the year after von Klein concluded his work, gives evidence of maximal naupliar production in April after a moderate maximum in the egg production in March. In June there was another maximum in egg production followed by a relatively small maximum in nauplii; the process was then repeated again in August and September and apparently in December. It is probable from Kunze's data that the survival of the spring eggs was better than those laid at any other time of year, just as seems to have been the case in preceding years. Haempel (1918) clearly implies that in the Hallstätter See the development of the spring brood follows the main period of diatom production.

A detailed study of the population of *Eudiaptomus gracilis* in Lake Constance has been made by Elster (1954). Since the temperature dependence of the rate of egg development is known, it is possible (see page 514) to compute the natality from egg counts made at intervals, in this case, ten days, in a series of depth zones for which temperatures are known. If the population at all stages subsequent to hatching is also known at each time interval, the losses of all stages between one time interval and the next give an estimate of mortality. The relevant data derived from Elster's study are presented graphically in Fig. 173. The mortality follows in general the same temporal pattern as natality and is clearly the result of a very great loss of young nauplii. The subsequent population in the lake clearly depends on a fine balance between reproductive rate and death rate, which varies with environmental conditions. At least part of the juvenile mortality can be catastrophic and density-independent; since the nauplii remain in the surface waters throughout the diel cycle, on exceptional occasions when the lake is stratified and the surface water is being blown down stream almost the entire naupliar population (97 per cent in September 1934) may be carried down the Rhine. In general the balance between natality and mortality is most conspicuously and persistently positive in winter and very early spring, from December till late February or into March. Most of the adults that survive are likely to have been born at this time, though by far the greater number of nauplii are hatched later in the season.

Ekman's (1904) observations of the breeding of the allied *Eudiaptomus graciloides* (Lillj.) in Lapland seem to indicate two reproductive periods, but in Denmark, where Wesenberg-Lund (1904) found *E. gracilis* breeding throughout the year, *E. graciloides* is markedly univoltine; nearly all the females breed in spring, from March to May, though a few, perhaps not more than one in a thousand, were observed to carry egg sacs at other times of year. In striking contrast, in Lake Erken Nauwerck (1963) concluded that the species is multivoltine with generation times of 35 to 40 days during the summer; most of the life cycle is spent in prereproductive stages.

Ravera (1954) found *Eudiaptomus vulgaris* to be perennially present in almost all stages in every sample taken from the top 50 m. of Lago Maggiore throughout the year. Particularly large naupliar populations occurred in late February, April, and August. Although it is not easy to disentangle the generations, Ravera concluded that the corresponding adult populations were present in May and June, August and September, and December and January. It is probable from his data that the population maturing between April and August was the least successful of the three (see also Tonolli 1961 and page 860).

Birge (1898) found that *Skistodiaptomus oregonensis* did not reproduce in Lake Mendota during the winter. Young, presumably copepodite, stages appear in May and in some years at least there is evidence of another brood later in the season. In Lake Erie the breeding season lasts throughout the summer and autumn from May into November (Davis 1961). It is clear, however, from Birge's account that although reproduction continues during the summer the later eggs may add little to the population. Birge suspected that a marked fall in temperature below 18 to 20° C. at the end of September put a stop to the reproductive activity of the species, though it is quite obvious that the activity responsible for the main rise in spring must occur at much lower temperatures, below 15° C.   Langford (1938), who investigated this species more recently in Lake Nipissing and who supplies more detail about the younger stages than did Birge, found that reproduction began in May and that the nauplii then produced were responsible for a maximum in the adult population in late June and early July. Nauplii are again produced in numbers in July, and these nauplii are responsible for an increase in the adult population in August.   Although Langford was not able to separate the early stages of the several copepods present, it is evident that the chance of an egg produced by *S. oregonensis* in the summer becoming an adult is far less than that of an egg laid in May.   The seasonal biology of *S. oregonensis* in Lake Nipissing is thus very similar to that of *E. gracilis* in the Schleinsee.   Langford notes that, although most of the adults present in September appear to overwinter, a few may give rise to nauplii that presumably mature slowly throughout the winter.   In the same lake Langford found that *Leptodiaptomus minutus* exhibited a somewhat similar cycle, but in this species the maximal population develops more rapidly in the spring, the peak being in June.   Evidence of a second generation swelling the population late in August was obtained in 1934, but the spring generation obviously survived much better than did that produced in summer.   In 1935, in which year *S. oregonensis* was more abundant than in the preceding year, *L. minutus* was present in smaller numbers and showed little fluctuation, throughout the summer.   In Lake Erie *L. minutus* starts breeding in March, much earlier than *S. oregonensis,* and no eggs are found after August (Davis 1961).

The only important observations on a diaptomid population in a perennial tropical water still appear to be those presented by Apstein (1907) on *Phyllodiaptomus annae* in the Lake of Colombo, Ceylon. Here the species, which is absent early in the year, first becomes common in June and reaches its maximum in July after the monsoon rains have given place to the driest season of the year.   In the relatively small June population one in two or three females are carrying egg sacs, but as the population increases and then decreases the proportion of reproducing females declines.

In the artificial south-temperate Florida Lake in the Transvaal Schuurman found *Metadiaptomus colonialis* to be perennial with a rather irregular breeding season.   The species was not common, and its temporal variations cannot be analyzed from the available data.

The only significant observations on the seasonal cycle of one of the centropagids of the Southern Hemisphere are due to Bayly (1962a), who has reported an interesting situation in *Boeckella propinqua* in Lake Aroarotamahine on Mayor Island off the North Island of New Zealand in the Bay of Plenty.   Lake Aroarotamahine is a warm mono-mictic lake, circulating in May at a temperature of 14.3° C. and having surface temperatures as high as 27.9° C. in January.   The lake is often colored by a bloom of *Anabaena* and is clearly eutrophic. *Boeckella propinqua* appears to breed at all seasons in the lake.   During most of the year the body length of the copepod varies inversely with temperature, but in November there appears to be a distinct bimo-dality in body length, with the smaller individuals becoming relatively more abundant in the deeper cooler water.   A far smaller proportion of the females of the smaller class of adults carried eggs than of the larger class.   Bayly supposes that the larger class represents the main population reproducing in a continuous multivoltine manner, whereas the smaller class represents recruitment from resting eggs, not, however, morphologically distinguishable, all hatching at about the same time and producing a group of adults which at the time of sampling were largely still too young to breed.   The explanation is admittedly hypo-thetical, but it is interesting because it shows how a rather confusing set of relationships in a series of vertical samples might develop fairly simply.

Chemical ecology of the Calanoida.   Except for the restriction of a few species to very concentrated waters, little is known about the chemical ecology of the fresh-water calanoids.   Most of the commoner limnoplanktonic species of Diaptomidae seem to have fairly wide toler-ances of soft and moderately hard water.   In western North America *Leptodiaptomus ashlandi* appears as a species characteristic of dilute unproductive lakes (Whittaker and Fairbanks 1958; see also Moore 1952), but it also occurs in Lake Erie.   In this genus *L. sicilis, L. tenuicaudatus* and *L. siciloides* are recorded (Moore 1952; Whittaker and Fairbanks 1958) in waters containing more than 12 per cent salinity, but at least the first and last of these also co-occur with *L. ashlandi,* as, for instance, in Lake Erie.   It is quite likely that real physiological differences in tolerance occur between these species, but it is also evident that in a given region much of the apparent restriction to waters of a given chemical composition is actually deter-mined by other characters of lakes which in a particular landscape will not exhibit all the properties of the set of inhabitable basins within the geographical range of the species concerned.   In the Columbia Basin of southeastern Washington the sharp separation of *L. ashlandi* from four lakes and *L. sicilis* from thirteen lakes, none being the same, even though the salinity ranges of 170 to 305 ppm. for *L. ashlandi* and 210 to 12,100 ppm. for *sicilis* overlap, is clearly due to the existence of rather well defined groups of lakes resembling each other in ways other than salinity.

In the western Palaearctic the eurytopic but apparently fugitive *A. wierzejskii,* the northern and often alpine *A. bacillifer,* and the very large *Hemidiaptomus amblyodon* appear to extend into alkaline and saline water, notably in Austria (Loeffler 1959) and Hungary (Ponyi 1956), but at least the first two can certainly also occur in very dilute waters. Of particular interest, however, is the existence of two species that are obligate inhabitants of different kinds of concentrated water, *Arctodiaptomus salinus* occurring primarily in chloride waters and *spinosus,* in carbonate waters. Löffler (1957, 1959, 1961, personal communication) finds that the last-named may occur in a concentration of more than tenth-normal soda, and although it can survive in both moderate sodium chloride and sodium bicarbonate solutions, it is capable of breeding only when an appreciable amount of the anion present is carbonate and bicarbonate rather than chloride. The species is distributed from Austria eastward through Armenia into Japan. Parker and Hazelwood find high positive correlations between the incidence of *Aglaodiaptomus leptopus* and the concentrations of chromium (0 to 13 mg. m.$^{-3}$ and vanadium (0 to 11 mg. m.$^{-3}$) in Kepple Lake, Washington, whereas there are high negative correlations with iron and strontium. Further studies are need to interpret these curious results.

Swimming and feeding movements. As in nearly all species of all three fresh-water orders of Copepoda, jerky swimming movements involving almost all the appendages can be observed. Bennett (1927), who has considered the movements of the fresh-water centropagid *Boeckella triarticulata,* the commonest fresh-water calanoid in New Zealand, concluded that the most rapid darting movement involved mainly a violent flexion of the antennules. This impression, however, would be obtained by anyone watching any calanoid without special equipment and may well be erroneous in part; a forward flexure of the metasome is equally likely to be involved.

A second type of swimming movement, a gentle gliding through the water, is very well developed in the Calanoida. This second type of movement is produced by the rotary action of the antennae, mandibular palps, maxillules, and perhaps maxillipeds. It has been studied by various authors, notably by Storch and Pfisterer (1925), by Cannon (1928), and most completely by Lowndes (1935). The last-named author states that a comparable type of locomotion is exhibited by the cyclopoid *Tropocyclops prasinus.* There are variations in the tendency to perform the first or jerky type of movement and in the position adopted during the second type of swimming, even when species of the same genus of Diaptomidae are compared.   Thus Wright (1938) notes that *Argyrodiaptomus azevedoi* glides on its back in contrast to many other members of the family and that unlike all other South American species of both *Argyrodiaptomus* and the related genera he studied it never exhibits the jerky type of movement spontaneously, but only when disturbed.   Fairbridge (1945b) also found that two fresh-water western Australian Centropagids of the genus *Calamoecia, C. attenuata* and *C. subattenuata,* exhibited far more violent jerky movements when disturbed than did the species of the allied genus *Boeckella* inhabiting the same region.

The main organ of propulsion during the gliding type of movement is certainly the antenna, and the action of this organ has been studied most carefully in *Eurytemora velox* by Lowndes (1935).   In this species the animal is likely to hold onto filamentous algae and similar bodies with its antennule and while in this position continue to produce the current that accompanied gliding locomotion when the animal is free to move.   Presumably, when anchored, the main purpose of the movement is to permit filter feeding.   As Lowndes indicates in his illustrations, the antennae undergo a rotatory motion; the exopodite and endopodite both act as inclined blades of a propellor during parts of the motion and so set up a pair of vortices on either side of the animal.   These vortices are reinforced and modified by the action of the mandibular palps, the maxillules, and perhaps the maxillipeds.

In *Boeckella triarticulata* Bennett (1927) noted that if the antennary movement, which normally occurred at a rate of ten to seventeen times per second, ceased, the animal might move back because of the action of the maxillipeds.   If the trophi stop moving and only the antennae operate, the animal moves forward and upward more rapidly than if the cephalic appendages were all working.   Bennett found that although *Boeckella* is usually poised in a semierect position with the antennules extended, it can also swim with either the dorsal or the ventral surface downward, and in the latter position it may feed on material on the bottom.

The vortical currents set up by the antennae carry particles suspended in the water to the maxillae, which all authors agree are the principle organs of filtration in the Calanoida.   Lowndes found that in *Eurytemora velox* the rotatory beat of the antennae occurred at a rate of about forty-five per second.   Most authors have supposed that under these circumstances filtration is likely to be entirely automatic, though it has been widely realized that variation in the size of the available food may impose a certain degree of selective feeding. Lowndes concludes, however, that the trophic behavior of *Eurytemora* is considerably more selective than has ordinarily been supposed; active predation may occur in *E. velox,* for fragments of crustacean limb as well as diatom valves may be found in the feces. Annemarie Burckhardt (1935), moreover, has made an extensive study of the feeding habits of the brackish water *E. affinis.* When this animal is given pollen, *Lycopodium* spores, or plant detritus in suspension, this material is filtered out of the water and taken into the alimentary tract, but in nature there is evidence of some selectivity in feeding. The chief food of the species in the mouth of the Elbe consists of diatoms, *Coscinodiscus, Melosira,* and *Cyclotella* being the most important genera eaten. No organisms larger than *Coscinodiscus* are ingested; this diatom has a relative mean abundance of 16 per cent in the gut contents against 13 per cent in the plankton. *Cyclotella* is enriched considerably more in the food, of which it comprises 31 per cent, as against 16 per cent in the plankton. *Melosira,* concentrated from 18 to 26 per cent, occupies an intermediate place. Green and blue-green algae are greatly under-represented in the food. This is partly because of the large size of *Dictyosphaerium* colonies, which are an important part of the summer nondiatomaceous plankton. There is evidence, however, of a more specific selection, apparent when the intakes of males and of females are compared. Males not only eat less than females but also take a disproportionately large amount of the smaller forms such as *Cyclotella* and tend to reject *Coscinodiscus.* A simple size difference in the predators is not involved, for the sexes do not differ significantly in size in *E. affinis.* There may be some anatomical explanation of the exclusion of *Coscinodiscus,* but Burckhardt thinks this unlikely inasmuch as the exclusion is imperfect. It is interesting to note that both in experiments with pollen suspensions and in the study of collections made in nature evidence was obtained that not only do females eat more than males but reproductive females with eggs or spermatophores or both take in more than nonreproductive females.

Lowndes (1935) found that *Diaptomus castor,* a species of small temporary pools, produced a food current in much the same sort of way as *Eurytemora velox.* Like *E. velox, D. castor* can engage in filter feeding while grasping solid bodies such as filamentous algae. The rate of antennary beat, namely ten to twenty-seven strokes per second when the animal is anchored and about twenty-two strokes per second when gliding, is definitely less than exhibited by *E. velox.* Feeding is not entirely unselective, for starch grains added to the water are avoided and large particles are discarded by a separation of the maxillae which allows them to be carried away by the food current. *D. castor* can also feed by scraping the surface of *Lemna* and other solid bodies. The feces consist mainly of detritus with a few larger objects.

Eudiaptomus gracilis was also studied by Lowndes. This species, though it can occur in small bodies of water, is frequently eulimnoplanktonic and was never observed by Lowndes to anchor itself. The maxillipeds do not appear to be held in quite the same position as do those of the preceding species nor to be used as prehensile organs, as they may be in *D. castor*. It is possible, however, that they play a somewhat greater part in producing the food current in *E. gracilis*. The rate of antennary beat is high; twenty-nine to forty-four strokes per second, in spite of which the animal moves only about 0.75 mm. sec.$^{-1}$ Lowndes thinks that the very rapid vibration of the trophi may shake the smaller particles off the maxillae and so prevent them from being taken up. He found that the species was highly selective in its feeding; when living in a pond containing vast quantities of *Kirchneriella*, the copepod was nevertheless feeding solely on a species of benthic desmid.

Schöder (1961) finds that in *Mixodiaptomis laciniatus, Eudiaptomus vulgaris,* and *E. gracilis* the frequency falls with increasing size of the organism. In the first named organism the frequency rises from 32.5 per second at 9° to 61 at 23°, and then falls, death occurring at a little above 25° C. He also thinks that the low sound produced may be used in echolocation; the wavelength of such sound would, however, be very long.

The most impressive example of selective feeding in the Calanoida is described by Fryer (1954), who compared the nutrition of *Eudiaptomus gracilis* and *Arctodiaptomus laticeps* living together in Windermere during February, March, and April. The larger species *A. laticeps,* which is 1.54 to 1.65 mm. long and proportionately more robust than *E. gracilis,* was feeding at this time almost exclusively on *Melosira italica,* whereas *E. gracilis* of length 1.14 to 1.23 mm. was consuming mainly tiny spherical green algae and plant detritus with very few *Melosira, Cyclotella,* and naviculoid diatoms and on one occasion apparently a rotifer. It is clear that the food taken in by the two species is quite different, and, equally significantly, neither was using any appreciable quantity of *Asterionella,* which was plentiful at the time.

Finally the members of the genus *Heterocope,* although they may filter out coarse particles selectively, as Naumann (1923) found for *H. appendiculata,* are to a large extent predaceous (Elster 1936; Burckhardt 1944). Bossone and Tonolli (1954) indeed attribute mutilations of the appendages of *Arctodiaptomus bacillifer* to attacks by the much larger *Heterocope saliens*.

It is quite evident that within the Calanoida as a whole, or even within a single family such as the Diaptomidae, there are likely to be small differences in feeding habits from species to species, dependent on size or in some cases on structural differences, particularly in the maxillipeds, and also on the rates of movement of the antennae, the posture of the trophi, and other purely physiological or behavioral characters. These differences, though so little investigated, are just well enough known to suggest that two species of Calanoida, though occupying the same volume of water, are by no means necessarily living in the same niche.

Size differentiation and the association of species of Calanoida. It has been observed by several investigators that when two species of calanoid copepods occur together in a single lake they are often markedly different in size. This was emphasized by Hutchinson, Pickford, and Schuurman (1932) with reference to the co-occurrence of *Lovenula excellens* of length 3.0 to 3.5 mm. and *Metadiaptomus transvaalensis* of length about 1.9 mm. in the alkaline perennial pans of the transvaal and of *Lovenula falcifera* of length 3.5 to 5.0 mm. and its usual associates *M. transvaalensis, M. meridianus,* or *M. colonialis* of length 1.0 to 2.1 mm. in the less mineralized temporary pans of the same region (Fig. 174). In some of the more complex associations of the pluvial zone of North Africa a very marked size differentiation is also observable, *Hemidiaptomus ingens* being about 5.2 mm. long, *Diaptomus cyaneus,* about 3.2 mm., and both *Eudiaptomus numidicus* and *Mixodiaptomus lilljeborgi,* about 1.5 mm. long. The large species of *Hemidiaptomus* evidently often have smaller associates. Mann (1940) noted *H. brehmi* of length 4.5 to 5.0 mm. living with *Arctodiaptomus belgrati* of length 0.98 to 1.24 mm. in a pool in Turkey; a third species, *Arctodiaptomus byzantinus,* was also present, but its size is not recorded. Mann similarly records *H. kummerlöwei* of length 3.5 to 4.45 mm. with *A. bacillifer* of length 0.87 to 1.19 mm.

A comparable case in North American temporary waters was noted by Hutchinson (1937b) for[19] *Hesperodiaptomus nevadensis* (3.5 mm.) and *Leptodiaptomus sicilis* (1.1 to 1.5 mm.) in Big Washoe Lake, Nevada, a playa lake specifically comparable to the Transvaal pans. The same association is recorded by Comita (Cole 1961) in saline

---

[19] Incorrectly recorded as *H. franciscanus* and *L. tenuicaudatus*. Mrs. M. S. Wilson has most kindly identified the original material.

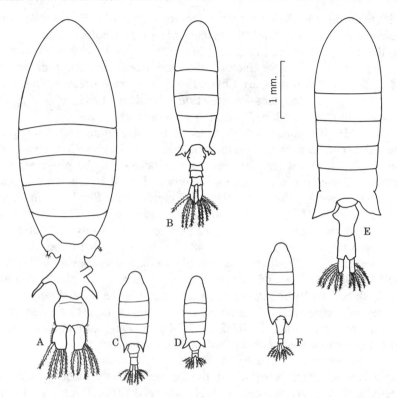

FIGURE 174.   *A–D*, diaptomid taxocene from a temporary pond in the pluvial zone of North Africa (Mare No. 1, Forêt de la Réghaia, Algiers).   *A. Hemidiaptomus (Gigantodiaptomus) ingens. B. Diaptomus cyaneus. C. Mixodiaptomus lilljeborgi D. Eudiaptomus numidicus. E–F*, diaptomid taxocene characterizing the more dilute pans of the Transvaal.   *E. Lovenula falcifera; F. Metadiaptomus transvaalensis.*   (Gurney, Gauthier, and Kiefer; original *A* is somewhat conjectural, no adequate figure of the anterior end being available.)

water.   Other American associations in temporary waters involve *Hesperodiaptomus wardi* (1.24 to 1.6 mm.), *H. novemdecimus* (3.8 to 4.0 mm.) from Montana (Wilson 1953), and *Onychodiaptomus sanguineus* (1.0 to 2.1 mm.), with both *Aglaodiaptomus conipedatus* (1.3 mm.) and *A. stagnalis* (3.0 to 4.0 mm.) from Louisiana (Wilson and Moore 1953), *A. clavipoides* (2.0 to 2.3 mm.) with *Leptodiaptomus moorei* (1.15 mm.) from the same state (Wilson 1955), and *Hesperodiaptomus shoshone* (2.59 to 3.33 mm.) with *Leptodiaptomus nudus* (1.1 mm.) from Arizona (Cole 1961).

Hutchinson (1937b) suspected that this type of differentiation was mainly characteristic of astatic waters; this appears not to be so, though all the really dramatic cases in which the larger species differs from the smaller by a linear factor of 2 are from temporary or somewhat mineralized waters, as Cole (1961) has emphasized. This may be attributed to the lack of selection (Hrbáček 1958, 1960, 1962; Brooks and Dodson 1965, Dodson *in preparation*).

Carl (1940) concluded that moderate size differences were of general occurrence in the permanent waters of British Columbia. In this region the most frequent calanoid association to be observed was between *Epischura nevadensis* of length 1.8 to 2.1 mm. and *Leptodiaptomus tyrelli,* which is 1.2 to 1.3 mm. long. Similarly, in Kora Göl in Turkey Mann (1940) found *Arctodiaptomus (Rhabdodiaptomus) sensibilis* which is 1.76 to 2.58 mm. long, living with *A.(A.) pectinicornis* of length 1.14 to 1.48 mm. Cole (1961) has recently reviewed much of the literature, notably the important papers of Wilson (1953, 1954, 1955) and has added some cases of his own. Among the fresh-water Centropagidae of West Australia the very closely allied species of *Calamoecia, C. attenuata* of length 1.48 to 1.65 mm. and *C. subattenuata,* which is 0.95 to 1.28 mm. long, can co-occur and probably provide a good case of the same phenomenon.

In one or two cases there is a hint of the occurrence of character displacement. This is exhibited by the very closely allied species of *Arctodiaptomus, A. laticeps,* and *A. wierzejskii* (*C, D* of Fig. 175). In general there is no consistent size difference between the two species, but on the rare occasions on which they co-occur there is a small difference. In Loch Moracha, North Uist, the mean lengths of the populations of the two species were (Gurney 1931)

	♂	♀
A. wierzejskii	1.4 mm.	1.53 mm.
A. laticeps	1.3	1.43

Cole suggests that comparable character displacement may also occur in *Mastigodiaptomus albuquerquensis* and *Leptodiaptomus novamexicanus.*

It will be apparent from Fryer's (1954) work, discussed in the preceding section, that a factor of 1.35 in length, possibly assisted by some other differences, is enough to give nearly nonoverlapping food niches in the Diaptomidae. In many of the cases already discussed comparable differences exist. It is also worth noting that a size difference of this magnitude often separates sympatric vertebrates of the same genus (Hutchinson 1959a), an empirical observation for which there is apparently some theoretical justification (E. O. Wilson

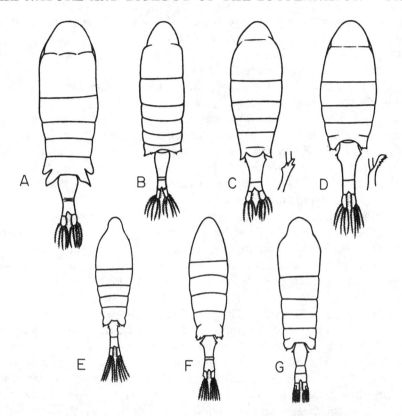

FIGURE 175. Diaptomine copepods of ecological interest. *A. Mixo-diaptomus laciniatus,* with a more or less boreo-alpine distribution and of great importance in the plankton of some subalpine lakes. *B. Eudiaptomus gracilis,* probably the most important species in the plankton of Palaearctic lakes. *C. Arctodiaptomus laticeps,* with apex of podomere 23 of right antennule of ♂. *D. A. wierzejskii,* the same. These two species differ significantly only in the slightly more anterior position of the greatest width in *laticeps* and in the minute characters of the antennules. They are drawn at approximately the relative sizes found in some Scottish localities, *laticeps* being proportionately larger in Windermere when occurring with *E. gracilis. E. Leptodiaptomus minutus,* ordinarily the smallest planktonic species in temperate North America and often co-occurring with other species, including *F. L. siciloides. G. Skistodiaptomus pallidus,* a seemingly slightly rheophil species. All females, except in detail of *C* and *D,* appendages and egg sacs omitted ($\times$ 50). (Gurney and Wilson, somewhat modified.)

personal communication). It is also of interest to note that passage through two, or in some cases one, instar would be sufficient in many species (cf. page 646) to produce the same sort of size difference that separates *E. gracilis* and *A. laticeps*.

Other types of niche specificity in sympatric calanoids. It is clear from the preceding section that size differences in sympatric calanoids can be correlated with differences in food, at least when the larger form is about 35 per cent greater than the smaller, as is true of *A. laticeps* and *E. gracilis*. It is also evident that in many diaptomid associations size differences of this sort are common: In spite of the frequency of this kind of niche specificity, it is evident that it is not the only kind found among the calanoids.

Several striking cases of the sympatric occurrence of species of essentially the same size are given by Cole (1961). Of particular interest are the species of Diaptomidae, all about 1.1 mm. long, that have occurred in Reelfoot Lake. The lake (Volume I, p. 12), which was formed by subsidence as the result of an earthquake in 1812, is very shallow (Hoff 1943), in most places under 3 m. deep, and tree trunks still stand in the water. Eddy (1930), who studied the lake in 1928–1929, found that the commonest diaptomid *L. siciloides* occurred in all samples. *L. sicilis* was less common and appeared only in November. The other data on these species, notably from Lake Erie, suggest that a slight size difference and a more marked difference in breeding season may be involved. Later Hoff (1944) found that only *S. mississipensis* and *S. pallidus* were present in Reelfoot Lake, both most abundant in shallow weedy water; *S. pallidus* was confined to the upper part of the lake, which appears from his map to be narrower and so more fluviatile than the lower basins. Yeatman (1956) makes the interesting observation that *S. pallidus* seems to be more rheophil than other American diaptomids and that in Woods Reservoir on the Elk River, Tennessee, in which there was presumptive evidence of competition with *S. reighardi,* another species of the same size, *S. pallidus* appeared to persist only along the old channel, where the current was greater than in the more peripheral parts of the lake.

In Lake Erie eight sympatric Calanoida have been reported; the five common Diaptomidae have been studied by Davis (1961) and probably represent the most complex taxocene of this family so far investigated (see Table 42).

The first three species and the last two in Table 42 are clearly separated by markedly different breeding cycles, a difference that Ravera has stressed as of general importance in reducing competition. Among the spring species there are slight size differences. The early

TABLE 42. *Breeding periods and size of Diaptomidae in Lake Erie*

Species	Extent of Breeding Seasons	Maximal Breeding	Mean Length of Cephalothorax in Lake Erie
Leptodiaptomus sicilis	January–May; October	January–March	1.113 mm.
L. minutus	March–August	April–May	0.675
L. ashlandi	March–August; October	April	0.800
L. siciloides	May–December	July–September	0.899
Skistodiaptomus oregonensis	May; July–November	August–November	0.965

breeding and evidently very eurytopic *L. sicilis* is clearly differentiated from *L. ashlandi* in average size, being 138 per cent the length of the latter, and *L. ashlandi* may also be differentiated, though less strikingly, from the much rarer *L. minutus*. The two summer breeding species, however, are feebly differentiated from each other in size or in breeding season. In the parts of Lake Erie studied by Davis there is very little opportunity for vertical differences in temperature to provide niche separation so that the nature of the ecological differences, particularly those separating *L. siciloides* and *S. oregonensis,* are quite obscure.

In the more usual taxocenes of two or three species, particularly in stratified lakes, it is often fairly easy to see at least roughly how niche specificity is achieved. Thus in the Titisee *Mixodiaptomus laciniatus* co-occurs with a small population of *Acanthodiaptomus denticornis.* Though at identical temperatures the egg and first naupliar instar of *M. laciniatus* take less time than they do in *A. denticornis.* *M. laciniatus* grows more slowly in the lake as it is distributed in colder, deeper water. A comparable case is probably provided by the univoltine *M. laciniatus* and multivoltine *E. gracilis* in Lago Maggiore (Ravera 1954). Moreover, at least in Ravera's study, it seems that the single breeding period of *M. laciniatus* begins a little earlier than the winter and early spring breeding of *E. gracilis* and that all stages will be larger at any given time from January to April.

A fascinating case has been described by Bossone and Tonolli (1954). In the Italian and adjacent parts of the Swiss Alps two diaptomid copepods occur: *Arctodiaptomus bacillifer* and *Acanthodiaptomus denticornis.* *A. bacillifer* tends to be found at higher altitudes than are usual for *A. denticornis*. Though there is an appreciable range of altitudinal overlap of 1971 to 2131 m., co-occurrence of the two species is excessively rare, only one case being known in the great series of Alpine lakes studied by Tonolli, and four others in other parts of the Alps, recorded in the literature.

In the Lago di Monscera, studied by Bossone and Tonolli, *A. bacillifer* hatches in May and only adults are present toward the end of June (Fig. 176). Meanwhile, in this lake *Heterocope saliens,* a much larger and mainly predaceous calanoid, also appears and matures. When the maximum population of *H. saliens* is present in July, many mutilated specimens of *A. bacillifer* are observed. This species dies out early in August in Lago di Monscera, probably because of predation by *Heterocope,* though in most other alpine lakes *A. bacillifer* persists as adults much longer. Nauplii of *A. denticornis* appear in June and adults in late July just before *A. bacillifer* disappears. Both *A. denticornis* and *H. saliens* persist at least till October. Bossone and Tonolli suggest that the presence of *H. saliens* curtails the life span of *A. bacillifer* and this prevents competition between that species and the developing *A. denticornis*. In three of the five localities containing both *A. bacillifer* and *A. denticornis, H. saliens* is known to occur, which in view of its rarity (Tonolli and Tonolli 1951) in such lakes can hardly be accidental. Since *A. denticornis* probably develops best at a slightly higher temperature than *A. bacillifer,* unusual variations in temperature regime due in part to the lowering of the water level in summer may also be involved. Bossone and Tonolli suggest that since, in spite of its short life history, *A. bacillifer* appears to persist in Lago di Monscera, selection has produced a more rapidly maturing and reproducing race in that lake. Attempts to establish the taxocene by introducing numerous specimens of the other two species into a neighboring lake rich in *A. bacillifer,* failed completely.

In concluding this section it would be well to note that in Fryers (1954) fundamental observations on *E. gracilis* and *A. laticeps* neither species eat *Asterionella* in spite of its abundance. This, along with a few other observations, notably those of Lowndes on *E. gracilis,* already given, suggests that some sort of truly behavioral selection of food is possible. This should be considered in all cases in which two species of the same size are living together.

Some aspects of the ecological zoogeography of the fresh-water Calanoida. It is a matter of considerable interest that the calanoid copepods of inland waters, particularly within the dominant family of the Diaptomidae, appear to exhibit far greater regional endemicity than do any other group of planktonic organisms. This is dramatically

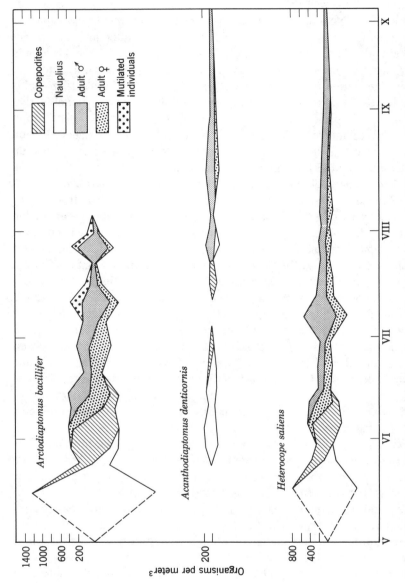

FIGURE 176. Seasonal incidence of various stages of *Arctodiaptomus bacillifer, Acanthodiaptomus denticornis,* and *Heterocope saliens* in the Lago di Monscera. (Bossone and Tonolli.)

687

illustrated by the fact that in the latest lists of North American copepods (Wilson and Yeatman 1959) approximately half the Cyclopoida are known either in essentially identical form or as allopatric subspecies in the Old World, whereas only about 10 per cent of the Calanoida show such a distribution. It is possible that part of the difference lies in the better taxonomic characters presented by male diaptomids than by either sex of the Cyclopoida, but it seems indisputable that a number of species of the latter order have a far wider distribution than any calanoids. Not only is the fauna of the Palaearctic and Nearctic largely distinct, but even greater differences are apparent when the Oriental and Ethiopian regions are compared with the Palaearctic. The development of the Paradiaptominae, with the genera *Paradiaptomus, Lovenula,* and *Metadiaptomus,* in Africa is striking, though one or two species of the subfamily do occur in the Mediterranean region and in Asia. Moreover, within Africa probably no species is common to the tropical region and the area south of the Limpopo. This greater specific endemicity in the Diaptomidae is rather puzzling. Though many eulimnoplanktonic species do not produce resting eggs, the numerous diaptomids of astatic waters produce eggs that can be dried. Superficially, at least, the species that appear to produce eggs most easily transported accidentally are no less restricted than those of eulimnoplanktonic habit. It is indeed possible that the former group will be found to contain the greater number of restricted endemic species. In view of the interest of the problem of diaptomid distribution, some of the cases derived from a study of the diaptomids of astatic waters may be considered before proceeding to other problems of ecological zoogeography.

The Diaptomidae of astatic waters. In western Europe *Diaptomus castor,* already noted as a facultative producer of resting eggs, is the main species in temporary waters which dry up in summer. Further east it is replaced by *D. mirus,* found in European Russia, east to Siberia and south to the eastern Balkans. *Mixodiaptomus tatricus* inhabits temporary waters in the eastern Alps and other mountains of eastern Europe; in the plains of Hungary and Poland, eastward into central Asia, the large *Hemidiaptomus (Gigantodiaptomus) amblyodon,* often accompanied by smaller species, is found in astatic basins. In the New World *Onychodiaptomus sanguineus,* occurring in spring in temporary pools, is the ecological equivalent of *D. castor;* in the far west several species have already been noted (see page 680) as forming taxocenes comparable to those to be described from Africa.

The most careful investigation of the Diaptomidae of astatic waters is certainly that conducted by Gauthier (1928, 1931, 1933a, 1933b) in Palaearctic North Africa.   In the pluvial zone of Algeria and Tunis, in which the rainfall exceeds 500 mm. per year, the temporary ponds contain (*A-D,* Fig. 174) *Eudiaptomus numidicus, Diaptomus cyaneus, Mixodiaptomus lilljeborgi,* and *Hemidiaptomus (Gigantodiaptomus) ingens.*   In the substeppic zone, which has rainfall between 500 and 300 mm. per annum, any of these species may be found on occasion; *D. cyaneus* and *E. numidicus* may co-occur with *Metadiaptomus chevreuxi* typical of the steppic zone, but the characteristic and abundant species of the substeppic zone are *Arctodiaptomus wierzejskii* and *Mixodiaptomus incrassatus.*   In the steppic zone only *M. chevreuxi* is found, and in the few pools that occur in the true desertic zone it is replaced by *Lovenula (Neolovenula) alluaudi,* a species with fairly wide distribution in the dry parts of North Africa but which also occurs in the Iberian Peninsula, the Balkans, Hungary, and Galicia.

Among the North African species already discussed it appears that *E. numidicus, Mixod. incrassatus,* and *Metad. chevreuxi,* typical species of the pluvial substeppic and steppic zones, can produce both kinds of eggs, whereas *H. ingens* and perhaps *D. cyaneus* produce only resting eggs.   It is at first paradoxical that the last-named species should both belong to the pluvial zone, but the paradox is resolved when it is realized that *H. ingens* is one of the largest known freshwater copepods and *D. cyaneus* is more than 3 mm. long.   Evidently even in the pluvial zone the period during which water occurs in the temporary ponds is not long enough to permit such large diaptomids to complete more than one life cycle.

Outside the Palaearctic region the most extensive observations on the ecology of the Diaptomidae of a semiarid region are those made in the Transvaal by Hutchinson, Pickford, and Schuurman (1932; see also Kiefer 1934).   Here the degree of permanence and length of time that a basin contains water is doubtless as important as elsewhere, but there are also certain chemical variables in the environment that are hard to disentangle from the microclimatic factors.

Two large species of the characteristically Ethiopian genus *Lovenula* (s. str.) are found in the southern Transvaal and are associated with various other species, mainly of *Paradiaptomus* and *Metadiaptomus.* The Calanoid fauna as a whole thus consists largely of members of the Paradiaptominae rather than of the Diaptominae.   *Lovenula excellens* is found mainly in the perennial if astatic lakes that occur in the shallow deflation basins or pans in the eastern Transvaal, though it was once found in a comparable body of water in a pan that is known to have been quite dry some time before.   All of the localities are slightly saline and alkaline and all contain a great deal of suspended or colloidal inorganic as well as organic matter. They lack all trace of phanerogamic vegetation and cyclopoid copepods are never found in them. In every case *Lovenula excellens* was accompanied by *Metadiaptomus transvaalensis.*

In the same general region *Lovenula falcifera* occurs exclusively in temporary waters, some of which are turbid and slightly alkaline but others support a good phanerogamic flora and then were often inhabited by cylopoid copepods. When *L. falcifera* occurred in turbid and frequently vegetationless habitats, it was accompanied by *Metadiaptomus transvaalensis* (*E, F,* Fig. 176), but in the other localities, in which there was fairly clear open water with aquatic plants, *M. colonialis, M. meridianus,* and *Paradiaptomus lamellatus* replaced *M. transvaalensis.* The ecological factors determining the distribution of these species are by no means obvious. It is known that *P. lamellatus* produces resting eggs, Sars (1927) having reared the species from dried mud; from the history of the localities in which they occur it is reasonably certain that this is true of *Lovenula falcifera* and of the species of *Metadiaptomus* with which it is associated. It seems probable that *Metadiaptomus transvaalensis* is characteristic of turbid water with little or no higher vegetation and the other two species of *Metadiaptomus,* of less turbid water with water weeds. *Metadiaptomus colonialis* is a common species in the artificial lakes of the Transvaal and *M. meridianus* has also been found in one such locality. These species therefore evidently can live in the open water of permanent lakes. Another species, *Thermodiaptomus syngenes,* also occurs in this type of habitat in the Transvaal. In the vicinity of Cape Town, the only other well-studied part of South Africa, an entirely different fauna occurs; *M. meridianus* appears to be the only calanoid common to the Transvaal and to the Cape Province, in the southwestern part of which *Lovenula simplex, Metadiaptomus capensis,* and *M. purcelli* are the characteristic species. Because a number of South African species can obviously withstand desiccation and so would be susceptible to various kinds of passive transport overland, it is obvious that whatever the natural mechanism of transport of dried eggs may be there are considerable obstacles to the process. In view of the curious distributions recorded in other less well known parts of the world, the conclusion that although many Diaptomidae potentially can be distributed as dried eggs considerable if unelucidated difficulties are in the way of this dispersal would appear to be generally applicable.

Littoral and eustatic pond species of Calanoida. Apart from the obvious significance of permanence of the habitat in the ecology of the Calanoida, it must not be forgotten that there is evidence of differentiation between pond and lake species and between the species of the eulimnoplankton contrasted with those of the littoral of the lake. Marsh (1893) long ago indicated that among the common species in Wisconsin, *Onychodiaptomys sanguineus* and *Aglaodiaptomus leptopus* are species of stagnant pools, *Skistodiaptomus pallidus* is supposedly a littoral species, and the others, namely *Leptodiaptomus sicilis, L. minutus, Leptodiaptomus ashlandi,* and *Skistodiaptomus oregonensis,* together with *Epischura lacustris* and *Limnocalanus macrurus,* are eulimnoplanktonic.

In the western Palaearctic several species inhabiting temporary pools have already been mentioned. *E. vulgaris* (s. str.) is mainly a pond form, but either subspecies or closely allied species become important members of the plankton of large lakes in the southern part of Europe. *Arctodiaptomus bacillifer* is widespread in small permanent bodies of water, though it can also occur in large lakes; with *Acanthodiaptomus denticornis* it is particularly a species of small, high, mountain lakes in the Alps. *E. gracilis* is generally a lake form, though of wide distribution both ecologically and geographically. *E. graciloides* is a eulimnetic form most frequent in eutrophic lakes of the plains. Wesenberg-Lund's (1904) data tend to show that *E. gracilis* and *E. graciloides* exclude each other in Denmark, even though they have been recorded as both living in a number of European lakes.

A difference in the habitats of widespread species is not unusual in different parts of their range. *Mixodiaptomus laciniatus* is found only in large deep lakes such as Maggiore in the southern part of its range, but in the north it is a pond species. This is presumably explicable by *M. laciniatus* being a cold-water form. *A. wierzejskii,* which inhabits temporary pools in the substeppic zone of North Africa and lakes in the highlands of Scotland and the Shetland Islands, provides a more curious and difficult case.

In Indian Tibet (Kiefer 1939b), where *Arctodiaptomus stewartianus* is the ordinary calanoid in large bodies of water, *A. parvispineus* was found in small ponds, which, though they do not dry, presumably freeze solid in winter. In the Nilgiri Hills in South India, where the ordinary species in ponds and artificial lakes are *Neodiaptomus diaphorus* and *N. physalipus,* a tiny, richly vegetated pool contained a hitherto undescribed species, *Tropodiaptomus euchaetus.* Comparable cases of special species in small weedy pools could no doubt be multiplied almost indefinitely.

Special littoral species in large lakes are less well known, but Brehm (1937b) has given good examples from South China, notably L. Ssujfun, littoral *E. pachypoditus,* pelagic *Neodiaptomus handeli;*

Ningyuen, littoral, *E. mariadvigae,* pelagic *Neodiaptomus handeli;* and Talifu, littoral, *E. mariadvigae,* pelagic, *Tropodiaptomus episcopus.*

In all of these cases of littoral limnetic species, and in those of small but permanent weedy ponds in which the production of resting eggs is likely to be of little importance, it may be suggested that the capacity to grasp solid objects such as algal filaments, which capacity *D. castor* and *E. velox* are known to possess, may be a critical factor in determining the local distribution of species.   As indicated in a later section on the horizontal distribution of the zooplankton, it is not unlikely that photostatic and photokinetic reactions to differences in illumination in the littoral compared with the open water may also determine the distribution of planktonic Crustacea.

Altitudinal and latitudinal zonation in the Calanoida.   In both Europe and North America a number of species of Diaptomidae are distributed in a way that suggests that temperature plays an important part in regulating their occurrence.   It can hardly be doubted that this is true in a general way; for example, when the members of the Holarctic genera as a whole are compared with those of the numerous genera characteristic of the tropics.   It is also reasonable to suppose that the arctic *Diaptomus glacialis* the Boreo-British *Arctodiaptomus laticeps,* and the strikingly Boreo-alpine *Mixodiaptomus laciniatus* are actually more or less cold stenotherm organisms: the same conclusion would be reasonable about *A. bacillifer* in both its alpine (Pirocchi 1944) and North American (Marsh 1929) stations, though not necessarily so of all populations in intervening parts of the range of the species.

When, however, certain rather obvious cases of what looks superficially like thermally controlled distributions are analyzed, curious discrepancies are likely to be encountered.   Thus in the restricted genus *Diaptomus* in western Europe and North Africa there appears to be a neat zonation with *D. glacialis* in the extreme north which reaches its southern limit in Iceland; *D. castor* in the western part of transalpine Europe and *D. mirus* occupy comparable latitudes further west, and in the Mediterranean countries several species are known, *D. castaneti* from the Pyrenees, *D. kenitraensis* from Morocco, and *D. cyaneus* from Algeria, Tunis, Corsica, and the Italian Maritime Alps.   On Pirocchi's (1947b) map, as has been indicated, the zonation appears obvious, but reference to her text indicates that the only localities for *D. cyaneus,* the most southerly ranging species, on the mainland of Europe, are two alpine lakes, the upper and lower Laghi di Pierafica at altitudes of 2355 and 2336 m., respectively.

Comparable situations are encountered among the species of western

North America studied by Carl (1940). One of the most striking features of the series of distribution maps given by that author is the restriction of *Skistodiaptomus oregonensis* to the extreme southern part of the region he studied. The species is widespread in North America. Within the limited area of British Columbia, namely Vancouver Island and the adjacent mainland, in which it occurs, it has a considerable altitudinal range from sea level to rather over 1000 m., which is unexpected if its northward extension is really limited by low temperatures. Carl also discusses the data relating to *Leptodiaptomus ashlandi*, *Aglaodiaptomus leptopus*, *Leptodiaptomus tyrelli*, *L. sicilis*, *L. nudus*, and *Hesperodiaptomus shoshone*. Though he still considers temperature to be the main factor in determining the distribution of these species, in every case some feature that would be paradoxical according to this interpretation emerges when all the information, which is often meager enough, is examined.

Insular distributions in western Europe. A series of most curious distributions, inexplicable solely in terms of the physical factors of the present environments, is provided by the Calanoida of the inland waters of the islands of the western coast of Europe, a region that has probably been as well studied as any other area of equal size in the world.

In Iceland (Poulsen 1939) two species are present, namely, *D. glacialis* and *Leptodiaptomus minutus*. The former is of Old World origin; the latter is widespread in northern North America and occurs in Greenland, in which country *D. castor* is also known (Haberbosch 1920).

In the Faeroe Islands (Poulsen 1937) the only fresh-water calanoid recorded is a single specimen of *Eurytemora velox,* though a number of lakes have been studied.

In the Shetland Islands (Gurney 1931) *Diaptomus castor* has been recorded from a single small pool, but the common species is *Arctodiaptomus wierzejskii,* known from twenty-seven lochs in the group.

In the Orkney Islands (Gurney 1931) *A. wierzejskii* is known from two lochs on Rousay. These lochs have more rocky shores than the several lochs on Pomona or Mainland, in which only *A. laticeps* has been taken.

In the Outer Hebrides the distribution is most peculiar. Lewis, the largest and northernmost is known to have *A. wierzejskii* in two lochs, but the common species appear to be *Eudiaptomus gracilis* and *Mixodiaptomus laciniatus*. On North Uist *A. wierzejskii* occurred alone in eight lochs, most of which lacked the cladoceran *Holopedium*. *A. laticeps* was present without *A. wierzejskii* in nine lochs and the two species appeared together in three; all localities for *A. laticeps* supported *Holopedium* populations. *Eurytemora velox* has been reported from the island. On Barra, the smallest and southernmost of the four principal islands of the chain, only *A. wierzejskii* has been recorded.

In the Inner Hebrides only *M. laciniatus* is recorded by Gurney (1931) from the Isle of Skye. *A. wierzejskii* and *E. gracilis* occurred on Mull and on Jura the same two species were present, the former in one loch, the latter in sixteen.

On the mainland of Scotland *A. wierzejskii*, *A. laticeps,* and *M. laciniatus* are known, but all are northern, hardly occurring, if at all, in the lowlands. Together with *E. gracilis,* they have been recorded once together, but more usually *M. laciniatus* with or without *E. gracilis* is found either by itself or with *A. wierzejskii* or *A. laticeps.* In southern England the calanoid fauna of permanent waters consists of *E. gracilis, E. vulgaris,* and *Eurytemora velox,* whereas *D. castor* occurs in the temporary pools. The calanoid copepods of Ireland require further study; it is not unlikely that *E. gracilis* and *A. laticeps* are the most important species in the Irish lakes.

The ecological interpretation of these distributions would appear impossible unless historic, biotic, and competitive factors were considered, and the nature of such factors is almost completely unknown. Part of the irregularity may be the result of chance historic events; this is particularly likely in the Hebridean islands. It is reasonable to suppose, as has already been indicated, that the Boreo-British *A. laticeps* and the Boreo-alpine *M. laciniatus* are cold-water forms. The finding of *M. laciniatus* in the smaller upland lakes of the Great Glen region, however, does not prepare us for the fact that *M. laciniatus* is one of the most important planktonic Crustacea in Lago Maggiore. This is probably correlated with the added facts that *A. laticeps* does not occur in the Alps, whereas *A. bacillifer* and *Acanthodiaptomus denticornis,* the commonest species in the small high-altitude lakes of the Alps, are absent from Scotland.

Of *A. wierzejskii* it can be said only that it seems to be adapted to every contingency save perhaps that of meeting its fellow Diaptomidae. It is, as has already been indicated, a characteristic species of the temporary waters of the substeppic zone of North Africa. Between that region and the Shetlands it is found sporadically in Europe, reaching north to the Kola Peninsular, east to Mongolia, and southwest to the Azores. There is almost nothing in its distribution or ecology on the Eurasiatic continent to suggest that it would be the dominant member of the zooplankton of the lochs of the Shetlands. It is just possible, however that because of good means of dispersal and the freedom from competition that small islands afford the species is more insular than would be expected from chance alone.

A. laticeps and *A. wierzejskii* differ only in minute details. The region of the maximum width of the cephalothorax is somewhat farther forward in *A. laticeps,* as its name implies. In the female of *A. laticeps* there are two setae on segment 13 of the antennule, whereas in *A. wierzejskii* there is only one. In the male the apex of the twenty-third segment of the right or geniculate antennule is multidenticulate, which is not the case in *A. laticeps.* These minute antennular characters are the safest way of separating the species; the fifth pleopods of the males provide no characters, and though the species tend to differ in size when they co-occur, the dimensions give no criteria for separating allopatric populations. It is reasonable to suppose that there must be physiological or autecological differences between the two species, but it is impossible to guess what they may be. The more constant association of *A. laticeps* with *Holopedium* in North Uist suggests that this species may be more calcifuge than *A. wierzejskii,* but the distribution in the Orkneys implies that the latter species occurs in even more oligotrophic lakes than the former, and we would rather suspect such lakes to contain the minimal amounts of calcium present in the waters of the region. Gurney thinks that in these insular and highland localities the life cycles of the species run parallel. The occurrence of *A. wierzejskii* in the Shetlands and the Azores suggests considerable powers of dispersal and persistence in the absence of competition, but it cannot be a completely fugitive species. Its absence from the Faeroe Islands is curious. The co-occurrences with *A. laticeps,* which would seem a priori particularly unlikely, may involve size differences of significance, but *A. wierzejskii* also can co-occur with *M. laciniatus* and *E. gracilis.* For the present the species is primarily a warning against over-facile explanations in ecological zoogeography.

Other Planktonic Crustacea. Two species of Ostracoda of the genus *Cypria,* one, *C. javana pelagica,* from the Lake of Colombo (Apstein 1907) and various lakes in Indonesia (Klie 1933), the other, *C. petenensis* (Fig. 177), from Lago de Petén, Guatemala (Brehm 1939; Ferguson, Hutchinson, and Goulden 1964), seem to be limnoplanktonic, but they are little known either systematically[20] or ecologically.

---

[20] The Asiatic species referred by Apstein to *C. purpurascens* is clearly not the species described under that name by Brady (1886). Klie regards it as a subspecies of *Cypria javana,* to which he gave the name *pelagica.* Since *pelagica* and *javana* are more or less sympatric, it is more likely that *pelagica* is a full species, or alternatively an ecophenotype, rather a subspecies.

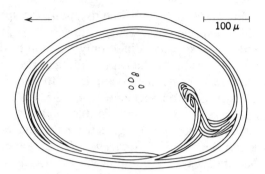

FIGURE 177. *Cypria petensis* ♂, one of the few planktonic ostracods in inland waters, Lake Peten, Guatemala. (Ferguson, Hutchinson and Goulden.)

Several benthic malacostraca, notably *Mysis relicta,* may appear as free-swimming nectoplanktonic animals at night; these species will be considered when the benthos is discussed. *Caridina nilotica* lives planktonically during its immature stages in Lake Victoria (Worthington 1931).

Some amphipoda are able to live as nektoplankton. This is strikingly true of *Gammarus (Rivulogammarus) pulex* (s. lat.) in some of the lakes at the western end of the Tibetan plateau (Ueno 1934b; Hutchinson 1937a). In closed lakes that lacked all species of fish this amphipod is evidently abundant in the free water, feeding on smaller Crustacea. In specimens from Khyargar Tso the gut contained fragments of the characteristically dark exoskeleton of *Daphnia tibetana,* another crustacean that seems unable to survive in the presence of fish. In the lakes in which *G. pulex* occurs pelagically it may settle in great numbers at intermediate depths on the anchor rope while a station is being made. It is reasonable to suppose that the absence of fish permits the extension of the ecological niche of the amphipod to include the conditions presented by the free water. The pelagic specimens appear to have a longer dactylus on the periopods and to differ slightly in other ways, in part shared by *G. pulex* from small bodies of water in the region, from typical members of the species from Europe. The name *stoliczkae* is available as a subspecific designation if nomenclatorial recognition is desirable.

A far more specialized case is provided by the Baikalian *Macrohectopus branickii,* a very large amphipod, up to 33 mm. long, transparent and slender, exhibiting striking diurnal migration in the top few hundred meters of the open water of the lake (Zachvatkin 1932; see page 735). *Macrohectopus* is the main food of the endemic fishes of the genus *Comephorus* in the lake (Werestschagin and Sidorytschev 1929).

MISCELLANEOUS PLANKTONIC ARTHROPODA

Planktonic Hydracarina. There are a number of records of water-mites occurring in plankton collections. Since certain species appear to recur, these occurrences are probably not entirely accidental; a few species are evidently more likely to stray into the open water than are the majority of members of the group. None, however, can be regarded as more than pseudoplanktonic.

Apart from casual records of *Limnesia maculata, Neumannia* spp., and *Heutfeldia rectipes,* collected together by Viets (1924), the water-mites found in the plankton belong to the genera *Piona* and *Unionicola*.

Viets (1924) found *Piona rotunda* to be one of the commonest and most eurytopic mites in the lakes of North Germany. It occurred in the plankton of the Grosser Plöner See on several occasions between the beginning of April and the end of September. Nymphs were commoner than adults in the planktonic habitat. It is possible that there is a tendency for the mite to move into the upper layers of water during the early part of the night. In Russian Carelia, where Sokolow (1930) found *P. rotunda* to be rarer and *P. variabilis* apparently to be commoner than in the lakes of Holstein studied by Viets, it is the latter rather than the former species that can appear on occasion in the plankton of Lake Sandal and Lake Ssonozero. Viets also notes odd planktonic records of *Piona brehmi, P. carnea,* and *P. nodata*.

The genus *Unionicola,* and particularly *U. crassipes* (A of Fig. 29), is often supposed to be particularly likely to turn up in the plankton, though Viets suspects some of the records are based on misidentifications. Sokolow records *U. crassipes* as a very rare and sporadic form in the plankton of Lake Sandal, and the same species was found in the plankton of the Grosser Plöner See by Koenicke (1896). These records are certainly correct, and it is most curious that Viets found *Piona rotunda* rather than *U. crassipes* in the plankton of the last-named lake.

Planktonic Insect Larvae. The insects have contributed extremely little to the plankton. The only good cases of limnoplanktonic insects are in fact the larvae of the various species of the genus *Chaoborus*. As far as hydrostatic mechanisms are concerned, they are by far the most perfectly adapted of any planktonic organism, but in lakes, paradoxically, they spend at least the day on the bottom or in some cases enter soft sediments and only become planktonic by night. In ponds they may hug the bottom less than in lakes. Because of their mode of life they are more frequently encountered in bottom samples than in collections of plankton and have been almost exclusively treated by investigators of the benthos. For this reason further treatment of these remarkable animals is postponed to the appropriate place in Volume III. A few littoral larvae, notably very young stages of the Trichoptera, are also occasionally recorded in planktons.

PLANKTONIC EGGS AND LARVAE

As indicated in an earlier chapter (see page 188), one of the greatest differences between marine and fresh-water plankton is the great number of meroplanktonic eggs and larvae of benthic and nektonic animals found in the marine and their rarity in the fresh-water association. A few cases of fresh-water planktonic eggs and larvae, growing as adults into sedentary, parasitic, or, in the case of fishes, large nektonic animals have been described.

Larvae of Free-living Invertebrates. The fresh-water phylactolae-matous Polyzoa, in which the life history ordinarily involves production of resting statoblasts or internal buds as the main means of reproduction, may produce sexually ciliated free-swimming larval colonies. The life history of these animals, though of a generalized fresh-water type involving resistant and subitaneous reproductive products, is thus the reverse of that of the rotifers and Cladocera in which the resting stages are produced sexually. When the larvae of a polyzoan are produced in the lake littoral, they must be, for a short time, at least, marginal members of the plankton. Their incidence is apparently little known; examples from Hastings (1929) and Brien (1953) are given in Fig. 178.

Of far greater importance, in those lakes in which the species occurs, are the veliger larvae (Fig. 179; cf. also Fig. 154) of *Dreissena polymorpha*. This animal, which seems to have entered western Europe during the Pleistocene and to have existed there in late prehistoric times, became extinct sometime within the historic era and was reintroduced from the Ponto-Caspian basins during the nineteenth century. The animal is now spread over northwestern Europe largely in rivers, north to Lake Mälar. The larva is a common constituent of the plankton of many lakes, including the Grosser Plöner See and others in North Germany during summer (Apstein 1896; Rylov 1935). In Lake Ohrid it is perennially present but may form a large maximum in the late summer (Serafimova-Hadžišce in Stanković 1960). This

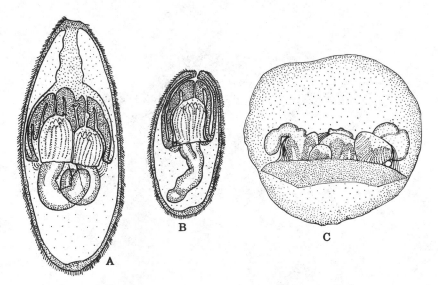

FIGURE 178. Free swimming larvae or larval colonies of phylactolaematous Polyzoa. *A. Plumatella fungosa. B. Fredericella sultana.* (Brien.) *C. Lopho-podella capensis.* (Hastings.)

case seems to provide the best limnetic analogy to the vast number of larvae of soft-bodied invertebrates ordinarily observed in marine plankton. It is possible, however, that they tend to have a longer pelagic life than the frequently fluviatile veliger of *Dreissena*. A transitory veliger stage also exists in the life history of at least some freshwater species of *Corbicula* (Sinclair 1963), though the brackish members of the genus have much more definitely planktonic larvae.

The highly modified larvae of the Unionaceae appear to be parasitic almost entirely on fishes, though one or two cases of infestation of amphibia or of direct development are known. The glochidium larvae of these animals sink rapidly and have no locomotor or flotation organs (Schierholz 1888; Latter 1891; Lefevre and Curtis 1912). The host fishes in general appear to frequent the bottom in the vicinity of mussel beds and must become infected without the glochidium being transported any significant distance. In special cases the coloration and spawning movements of the female may attract fish or the host may regularly feed on mussels (Coker, Shira, Clark, and Howard 1921). The significance of causal records of glochidia of *Anodonta* and *Unio* in the limnoplankton is therefore not clear, but such occurrences may reasonably be supposed accidental. The extraordinary tentaculate larva

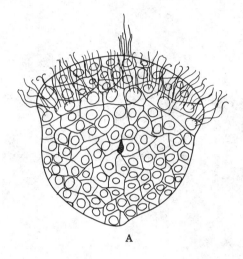

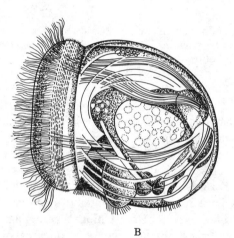

FIGURE 179. *Dreissena polymorpha.*
A. Ventral view of trochophore larva;
B. Lateral view of young veliger larva.
(Meisenheimer, cf. also Fig. 15.)

of *Mutela bourguignati* is also apparently not free-floating (Fryer 1961).

Cercariae of Trematodes. The cercariae of digenetic trematodes lead a brief free-living existence; although nearly all species have tails and can swim, the great majority can be regarded at most as transient members of the littoral planktobenthos, occurring for a short while in the immediate vicinity of the molluscan hosts from which they emerged. In a few families, in which the cercaria encysts on fish, quite elaborate adaptations to pelagic life have been developed. Although the literature of the trematodes is immense, such adaptation to planktonic existence has been specifically considered by few investigators; Wesenberg-Lund (1934), having worked on the group, is naturally the most important of the students of this aspect of the Trematoda. In the following account the systematics follow Dawes (1946).

Bucephalus, and presumably other fresh-water genera of the Bucephalidae, pass their sporocyst stage in fresh-water bivalves of the superfamily Unionaceae. The cercaria emerging from such a host is provided with two very extensible processes which may represent the furca of a more normal furcocercaria. The processes are attenuated and extended upward and then curved forward and contracted as they lift the body of the cercaria a little. The animal can thus maintain an almost constant position in the medium by a process (Fig. 180) that Wesenberg-Lund picturesquely but most aptly describes as "uninterrupted treading water." If within about twenty-four hours the larva comes in contact with a small fish, in Denmark usually *Scardinius erythrophthalmus*, it encysts, most often under the scales, and attains its adult form when this host is eaten by a larger carnivorous fish, in whose alimentary tract the mature *Bucephalus* lives. Wesenberg-Lund regards *Bucephalus* as having the most truly planktonic of all cercariae.

A number of planktonic furcocercariae, doubtless all referrable to the Diplostomidae, have been studied by Cort and Brooks (1928) and by Wesenberg-Lund (1934). They seem to be more or less pelagic, occasionally swimming to maintain their position, but most of the time floating quietly suspended in the water in strikingly different and characteristic poses (Fig. 182). They probably survive a day or two if they do not come in contact with an appropriate fish or, in some species, a tadpole. When they invade such a secondary host, they usually migrate to the lens of the eye. The impaired vision resulting from this site of infestation probably makes the fish more easily eaten by the final avian or in some cases reptilian host. *Diplostomum spathaceum*, of which the cercaria (cercaria C of Szidat 1924; Wesenberg-Lund 1934) holds the body and base of the tail bent at an acute angle to the main part of the tail, finally becomes a parasite of gulls and other water birds. A very closely allied *Cercaria flexicauda* from Douglas Lake (Cort and Brooks 1928), Michigan, adopts a similar pose. Two other quite different poses taken by other species not yet identified are also illustrated in Fig. 181. Cort and Brooks,

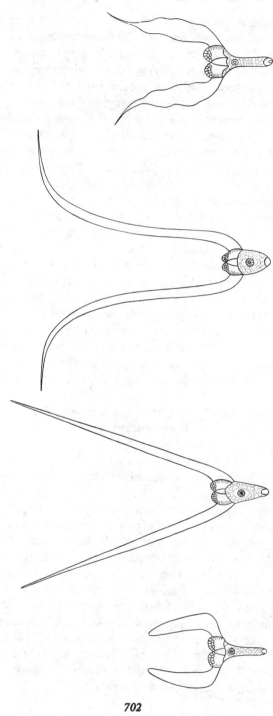

FIGURE 180. Four stages in the movements of the cercaria larva of *Bucephalus polymorphus* (× c. 25). (Wesenburg-Lund.)

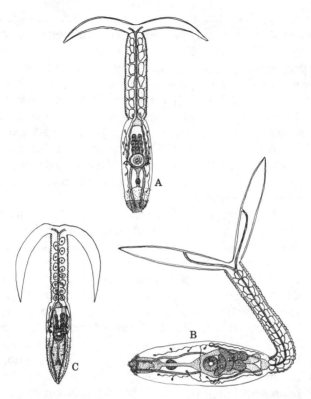

FIGURE 181. Cercariae of diplostomid trematodes, showing characteristic positions in the water of these limnoplanktonic larvae. *A.* "Furcocercaria helvetica xxxi." *B. Diplostomum spathaceum.* *C.* "Furcocerca 1 Petersen." ($\times$ c. 50). (Wesenberg-Lund.)

in discussing their *Cercaria douglasi,* which is probably the larva of a member of the allied family Strigeidae, indicate that the cercariae of the latter group, in which encystment is in a mollusk rather than a fish, tend to be less pelagic than the diplostomid cercariae.

The cercaria of *Sanguinicola,* a genus living as an adult in the blood stream of fishes, is known as *Cercaria cristata* because of the curious crest on the dorsal side of the body. The cercaria lives pelagically, usually taking the form of an incomplete ring; the animal is twisted at the base of the tail so that although the plane of the ring is vertical the crest lies horizontally on one side, increasing the form resistance of the animal.

The larvae of the Schistosomatidae are less strikingly adapted to a planktonic life than are the specialized cercariae which have just been considered. They are, of course, of enormous medical importance because of the genus *Schistosoma,* which contains three species infecting man, mainly in tropical and subtropical countries. The bird-infesting species of other genera not infrequently can infect human beings temporarily. When their cercariae are abundant in the littoral regions of lakes, they may produce the dermatitis commonly known as swimmer's itch (Cort 1928a,b; see also Wesenberg-Lund 1934 and Brackett 1941 for summary of older accounts). In Europe the main species involved appears to be *Trichobilharzia ocellata,* but Cort (1928b) thinks that a fair proportion of the known avian schistosomes can probably cause the condition.

The allied bird parasite, *Bilharziella polonica,* has a more or less neustonic cercaria which secretes an organic slime at the surface film. Ducks passing will entrap this slime and from it the cercaria can easily penetrate the skin of the bird.

In addition to the various genera just discussed, each of which has some species of limnological interest, Wesenberg-Lund (1934) describes finding in the plankton of Lake Furesø the peculiar *Cercaria splendens,* in which the tail ends in a pair of sucker-like lobes. This larva probably belongs to the genus *Azygia,* parasitic in the alimentary tract of fishes, but does not seem to be regularly planktonic. Odd specimens of other cercariae swept up by currents as they emerge from their molluscan hosts are almost certain to occur in the plankton from time to time.

Planktonic Fish Eggs. Graham (1929) found eggs and a few larvae of various sizes, which he thought to form a continuous set of stages of one species, in the plankton of Lake Victoria in central Africa. The largest larval fish was certainly a member of the genus *Eugraulicypris.* Worthington (1931) obtained similar eggs and concludes that they are almost certainly those of *E. argenteus,* a pelagic fish very common in the lake. It is reasonable to suppose that the other pelagic species in the genus, *E. stellae* in Lake Rudolf, *E. minutus* in Tanganyika, and *E. sardella* in Lake Nyasa, will prove to have comparable life histories.

An interesting planktonic fish egg has been described by Davis (1959). It is produced by the common bottom-living sciaenid fish, the sheepshead or fresh-water drum, *Aplodinotus grunniens,* in Lake Erie. The earliest stages of the life history of the fish are poorly understood, but it has been known for some time that the eggs when laid in captivity float to the surface. Davis obtained ten eggs (Fig. 182) on various occasions in July by towing a half-submerged plankton

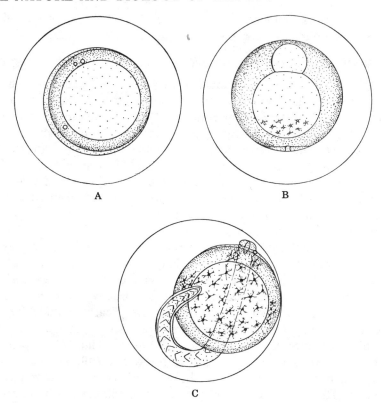

FIGURE 182. Planktonic eggs of *Aplodinotus grunniens* (× c. 20). *A*. Early stage, with large oil droplet, hatching 22 hours later. *B*. An egg about three hours later than *A*, differentiated head region visible just below the larger of the two oil drops. *C*. Egg just before hatching. (Davis.)

net at the surface of Put-in-Bay, western Lake Erie, when an onshore wind was blowing. These eggs matched exactly the eggs produced by a female *A. grunniens* spawning in captivity. He was able to rear one egg to hatching and estimates the over-all duration of the egg stage at 22° C. as twenty-five to thirty hours, which is extremely short for a temperate fresh-water teleost. The newly hatched larva, 4.2 mm. long, is, like the egg, much less dense than water and floats to the top at an angle of 45°, ventral side uppermost, whenever it relaxes its swimming movements, being buoyed up by the oil droplet in the yolk which lies just before the anus.

Davis reviews a number of other cases of floating eggs of fresh-water fishes but no others appear to be really planktonic.

INTERRELATIONS OF THE DIFFERENT GROUPS OF ZOOPLANKTON

Ideally it should now be possible, by using the detailed information available from the various groups of zooplanktonic organisms, to construct a model of seasonal succession. However, too little information exists to produce any convincing interpretation. It is reasonably clear how the elements in such a design would have to be classified and how in a most general way their synthesis would be achieved.

Initially, it is also clear that there are three major sets of categories, those involving the timing of the life history, those involving the nature of the food, and those involving other sorts of other specific relationships, mainly predation and competition.

Temporal relationships. In the first set we have a rough progression beginning with the rotifers, which have a life span perhaps of the order of ten to thirty days, dependent on the temperature, rapid egg development, and rapid production of eggs, usually, however, in small clutches of one or two. Such animals are clearly multivoltine and tend to build up large opportunistic populations. In the one study in which the rates of the production of different kinds of zooplankton have been estimated, namely, that of Nauwerck (1963), the rotifers, though numerically abundant, are in terms of biomass relatively unimportant. This suggests that in general they will be much influenced by their food supply, though we would expect that when crustacean competitors are present the control of the food would be due primarily to them. Dieffenbach (1911) found that in ponds near Munich the total planktonic rotifer population followed that of the nannoplankton very closely, with a lag of from five to ten days. Sládeček (1958) has found a similar relationship in a pond near Prague. In both cases short pulses of Cladocera developed but did not seem to have a significant relationship to the phytoplankton. It is quite likely that in such cases there is a good deal of selective feeding. It seems not impossible that the production of a very few eggs at a time, usually one or at most two, would permit the rotifers to react in a more linear way to their food supply, avoiding the type of oscillation due to time lag that can develop in the Cladocera.

The Cladocera probably tend to be longer-lived as well as larger than the rotifers, but like them are multivoltine except at extremely low temperatures. Their prereproductive period is short and eggs are produced quite fast under optimal conditions. The large potential size of the brood determined by the food supply, but not instantaneously, clearly predisposes the group to oscillation due to time lag effects, though the additional opportunities for exploiting a rapidly increasing food supply, for example, during the spring increase in phytoplankton, may compensate in the long run for the dangers of oscillation.   Like the rotifer, the Cladoceras can produce sexual resting eggs.

The calanoid copepods ordinarily have still longer life histories. Their reproductive activity is clearly under marked environmental control, as in the Cladocera, but the longer times involved are likely to damp oscillations that might arise.   There is often evidence of immense juvenile, usually naupliar, mortality.   Though Wesenberg-Lund's (1904) account of *Eudiaptomus graciloides* in Denmark suggests a largely univoltine cycle, Nauwerck (1963) found that the species in Lake Erken had generations living thirty-five or forty days, much as in Cladocera, though with a relatively longer prereproductive period.   The cyclopoid copepods, as far as their life histories are elucidated, apparently provide examples of even longer life cycles in extreme conditions, even becoming biennial or triennial and complicated in many cases by diapause.

Pejler (1926b) in his study of Ösbysjön has emphasized that life history plays a major part in the development of the zooplankton in the spring as temperatures increase and food becomes more abundant, the rotifers rising to a maximum in early June, the Cladocera in early July, and the Copepoda not till August.

Trophic relationships. The second set of categories tends to parallel the first.   The majority of planktonic rotifers are sedimentary suspension feeders, and there are differences in the sizes of particles taken: *Synchaeta* and *Polyarthra* eat living algae, generally larger than the fine detrital particles eaten by *Filinia* and *Conochilus*.   Pejler believes that the tendency for *Synchaeta* and *Polyarthra* to appear early, in the spring or beginning of summer, can be explained by the abundance of their living food, whereas the detritus feeders are largely aestival, autumnal, or, in some cases, hypolimnetic species.   Most cladocera also feed on fine suspended matter, though by filtration rather than by active sedimentation.   The Calanoida, however, appear to be *coarse* rather than *fine suspension feeders*, though in particular cases such as *Eudiaptomus* there may be some uncertainty.   The scheme (Straškraba 1963) is completed by the raptorial species, mainly Cyclopoida but also a few rotifers such as *Asplanchna*, Cladocera such as *Leptodora* and the Polyphemoidea, and *Heterocope* in the Calanoida.

In general, therefore, we seem to find in the zooplankton a continuum from animals with short multivoltine life histories feeding on very small food and developing large transitory opportunistic populations to animals with much larger, occasionally even biennial or triennial life histories, which may feed on larger food and have, at least as adults, fairly persistent stable populations. This ordering is, of course, not evolutionary; there may in fact be evidence of change in both possible directions in the evolution of cyclopoid copepods.

Heinrich (1962) has considered the types of life cycle found in marine Copepoda, mainly belonging to the Calanoida, and recognizes even within this group two extreme types, one largely independent of the season of food production and the other regulated by it. A third intermediate type can be recognized. This scheme, which was the original basis for the discussion of the last few paragraphs, could probably be adapted ultimately to fresh water, but in lakes we have a rather different group of animals, the rotifers taking the place of the invertebrate larvae that would occur in the sea. A more comparative study of planktonic life histories is clearly needed in limnology.

Among the external factors determining seasonal incidence, the influence of temperature is significant, not merely because of differences in thermal tolerance, which are clearly involved in the occurrence of the cold stenotherm species such as *Limnocalanus,* but more importantly because of differences in reaction to temperature increases which must give many summer forms an enormous competitive advantage.

Predation and size. Biotic factors involving competition and predation are obviously also important. Enough cases have been given, notably among the various congeneric sympatric planktonic rotifers, to suggest the importance of the timing of life cycles in mitigating the effects of competition, whereas a few cases of cycles greatly influence by predation, as by *Heterocope* (see page 686) and *Leptodora* (see page 614), are known. As in other aspects of the seasonal succession of the zooplankton, the picture here is fragmentary. It is reasonable, however, to suppose that we can now just see enough to know in what directions to investigate.

One particularly significant aspect of predation relates to its effect on the size, and, so indirectly, on other properties of the organisms composing the zooplankton. Although it has been pointed out that large diatoms are probably less easily eaten than are small, it appears that in the zooplankton avoidance of predators is often achieved by a reduction in size rather than an increase. Hrbáček (1958, 1960, 1962; Hrbáček, Dvorakova, Kořiner, and Procházkóva 1961) working on backwaters of the Elbe in Czeckoslovakia has pointed out that the presence of fish tends to reduce or eliminate the populations of large Cladocera such as *Daphnia pulicaria* and *D. longispina,* both more than 1 mm. long when adult, whereas the dominant species in very well stocked ponds are the small *D. cucullata* 0.68 to 0.76 mm. and *Bosmina longirostris* 0.33 to 0.38 mm. long. A possible hint of the same process is observed in the Copepoda, where *Cyclops vicinus* is confined to localities with small fish populations and the species of *Thermocyclops* are more uniformly distributed. Rotifers in general seem to increase with increasing fish stock; Spodniewska (1963), however, did not observe such an effect. Comparison of various localities indicate a dependence of both fish stock and zooplankton on the total organic and ammonia nitrogen in water and nannoseton. An increase in fish stock in a given pond tends to produce no significant quantitative change in zooplankton but merely an increase in small and a decrease in large species. Hrbáček and Hrbačekova-Esslová (1960) also obtained evidence from culture experiments that the populations of *Daphnia* of various species, ranging from the large *D. pulicaria* to the small but helmeted *D. cucullata,* tended to occur in such a way that a fast-growing population reaching a large size lived in poorly stocked localities, whereas where fish populations were large the *Daphnia* tended to be slow-growing and dwarf. The presence of large fish populations clearly exerts a selection in favor of small size. Brooks and Dodson (*in press*) have further provided dramatic evidence of a striking change in zooplankton caused by the introduction of alewives (*Pomolobus aestivalis*) into a lake in New England. This affected both copepods and cladocerans, the large species being replaced by the small.

SUMMARY

The main groups contributing to the zooplankton of lakes are the free-living nonphotosynthetic Protista, the Rotatoria, and several subclasses of Crustacea; there are in addition a few coelenterates, flatworms, gastrotrichs, mites, and larval insects that may be planktonic to varying degrees but are rarely taken with ordinary sampling methods. The very restricted number of larvae of soft-bodied invertebrates in fresh waters contrasts markedly with the ocean; most of the planktonic larvae found transitorily in fresh waters belong to parasitic trematode flatworms.

A few amoebae may occur in the lake plankton, but by far the most important planktonic lobose Protista belong to the genus *Difflugia*. Several species are regularly meroplanktonic, probably living in the littoral benthos in the winter and invading the open water in the summer when they are buoyed up by gas vacuoles and fat droplets. A few Heliozoa also may occur in the plankton, *Raphidiocystis lemani* being the most characteristic.

The Ciliophora are sometimes quite common but probably less important than in the ocean. *Amphileptus tracheloides, Trachelius ovum, Bursaria truncatella,* and some species of *Strombidium* are evidently cold stenotherm species. *Askenasia elegans* and other species of *Strombidium* are summer forms. There are a few important freshwater species of the predominantly marine planktonic order, the Tintinnida, *Tintinnidium fluviatile* being a spring species, *Codonella cratera,* a summer plankter. A number of Peritrichida of the genera *Epistylis* and *Vorticella* are epiplanktonic, often with rather specific preferences for the substrate organism, and some members of these and other allied genera may be free-swimming. There is an extraordinary cold-water and apparently endemic ciliate fauna found in the plankton of lake Baikal, the most important species being members of isolated monotypic genera (*Liliomorpha, Longitricha, Marituja,* etc.). In some lakes a rather well developed microaerophil or anaerobic taxocene has been recorded from the deeper parts of the hypolimnion, at the top of the monimolimnion in the meromictic Krottensee; in that lake *Coleps hirtus* appears to be microaerophil, whereas *Metopus fuscus* and *Caenomorpha medusula* inhabit supposedly quite anaerobic levels.

The coelenterate plankton consists of the medusae of *Limnocnida* and *Craspedacusta,* the latter having achieved a very wide distribution during the last half century.

A few plankton flatworms are known in the genus *Mesostoma; M. inversum,* which probably swims ventral side uppermost, is important in the plankton of Lake Bulera in Ruanda.

The rotifers share with the Cladocera and Copepoda the major part in the animal limnoplankton. The planktonic species all belong to the Monogononta. In this subclass a series of parthenogenetic or amictic generations occurs until a large population is built up. Usually at the time of maximum population density mictic females lay small male-producing parthenogenetic eggs or, if fertilized, large thick-shelled resting eggs. The determination of mictic female production and the subsequent appearance of males and of resting eggs is environmentally determined but not always caused by the same external stimulus. In *Brachionus calyciflorus* the stimulus is certainly due to crowding, probably mediated by the accumulation of a substance, produced by a female, which can determine that this female's daughter will be mictic. In *Asplanchna sieboldii* it seems equally certain that mictic female production is determined by the inclusion of chlorophyll-bearing organisms in the food. The male germ line is almost certainly haploid, but there is some disagreement as to whether this is true of the somatic tissues. The males are always much reduced in morphology and sometimes excessively so.

Planktonic rotifers produce from one egg every thirty days to at least almost one egg every other day, depending on the species, temperature, and food supply. Some littoral species are apparently much more fecund. There is a suggestion that the reproduction of some species, notably *Kellicottia longispina,* may be inhibited by the presence of much *Chlorella* and perhaps *Eudorina* in the plankton.

The planktonic rotifers belong to the orders Ploima and Flosculariaceae of the subclass Monogononta. The primitive members of the latter group are littoral benthic animals which have given rise to five families containing planktonic genera (Gastropodidae, Trichocercidae, Synchaetidae, Asplanchnidae, and the not too closely related Brachionidae) in the Ploima and three families (Hexarthridae, Testudinellidae, and Conochilidae) in the Flosculariaceae. The remaining family of the Flosculariaceae is sessile, and both *Conochilus* and *Conochiloides* are undoubtedly derived from such attached forms.

The majority of species of planktonic as of other ecological groups of rotifers appear to be potentially cosmopolitan, but a few species such as *Brachionus havanaensis* in the New World, *B. diversicornis* and *B. forficula* in the Old, *Keratella reducta* in South Africa, and perhaps some species of the same genus in America, probably exhibit real geographical limitations not dependent exclusively on ecology. A few species such as *Testudinella patina* are immensely eurytopic, but more usually some ecological limitations can be observed. *Brachionus falcatus, Keratella tropica, Filinia (Tetramastix) opoliensis,* and *F. (F.) pejleri* seem to be warm-water species, whereas in the last two genera *K. hiemalis* and *Filinia (F.) terminalis* are cold-water animals. Comparable examples could be found in other genera. *Brachionus* is almost limited to water on the alkaline side of neutrality, with some species such as *B. plicatilis* being marked halobiont. Species of *Hexarthra* are also often found in alkaline waters; this is not necessarily true of the common *H. mira* or the presumably cold stenotherm alpine *H. bulgarica,* though in fact this species is probably chemically eurytopic.

The nature of the chemical limitations acting on planktonic rotifers is not well investigated. By analogy with sessile species, it seems likely that a hard water may favor or exclude different species for different chemical reasons. There seems to be a far less marked tendency for planktonic than for benthic littoral rotifers to develop special acid-water species. Some investigators believe most of the apparent chemical limitation of the planktonic rotifers is really a limitation imposed by varying food levels.

There may be behavioral differences between the littoral and pelagic free-swimming species in a lake, but little definite information is available on the matter.

The planktonic rotifers in general feed either by sedimenting fine particles as a result of the beating of the coronal cilia or are raptorial. In the former group the Brachionidae (e.g. *Brachionus, Keratella, Kellicottia, Notholca*) with malleate and the planktonic Flosculariaceae (e.g., *Filinia, Hexarthra, Conochilus*) with malleoramate trophi are the most important, whereas in the raptorial group *Ascomorpha* and *Synchaeta* with virgate and *Asplanchna* with incudate trophi are particularly characteristic. This division, however, is not absolute, for *Kellicottia* and *Keratella* can use the mastax to catch the flagellum of *Cryptomonas,* which then disintegrates, and the particulate matter liberated is eaten by the rotifers. There is a good deal of size variation in the food particles taken by the sedimenting species. *Filinia, Conochilus, Kellicottia, Keratella,* and the free-swimming species of *Collotheca* eat mainly food of diameter under 12 μ. *Polyarthra* eats mainly *Cryptomonas,* which may be as much as 50 μ long. There may be much finer adjustments to specific food sizes than this in the trophic variables of the niche, for there seems to be a good deal of size differentiation of allied and frequently sympatric species. Moreover, there is clear experimental evidence that certain foods are more easily or readily taken than others. The shape of algal cells, as well as their size, is evidently involved, but chemical stimuli probably also play a part. There is some evidence that at least some species of *Chlorella* are not eaten or may even be inhibitory to some rotifers, but others can live more or less permanently on algae of the genus. Even if the subject is still inadequately explored, it seems likely that an ordinary phytoplankton association could provide niches for a number of species of rotifers.

The rotifers of the lake plankton show a complex series of seasonal changes. In temperate regions there is a small group of apparently cold-water species, notably *Synchaeta lakowitziana, Keratella hiemalis, Filinia terminalis,* and several less well known species of *Notholca* which appear in the lake in autumn or winter and usually reach their maximum population, at least in the upper layers of the lake, in early spring, though in stratified lakes they may persist abundantly in the hypolimnion. With these species are found a number of perennial limnoplanktonic rotifers, which, however, are ordinarily far more abundant in summer than in winter. Such species often show a large maximum in late spring or early summer, a minimum in June, and a secondary maximum in July, probably related to the cycle of available food. *Keratella cochlearis*, perhaps the commonest planktonic rotifer in temperate regions, provides an excellent example of these perennial species, as do *Kellicottia longispina* or *Conochilus unicornis*. The raptorial *Asplanchna priodonta* may rise and fall with these microphagous species, the synchrony sometimes being clear enough to preclude a prey-predator relationship determining the variation.

A number of special species also develop in summer. *Gastropus stylifer, Ascomorpha* spp., *Ploesoma hudsoni, Pompholyx sulcata,* and *Hexartha* spp. may be mentioned. In *Synchaeta* several species often exhibit two or more very sharp maxima during the summer. There is also a group of rotifers, including *Trichocerca porcellus* and other members of its genus, and *Euchlanis dilatata,* which, though more or less perennial or at least exhibiting little seasonal restriction in the littoral, appear in the plankton only in the late summer, generally associated with blue-green algal blooms. A comparable phenomenon is known in the Cladocera.

In the genus *Polyarthra* and perhaps less strikingly in *Synchaeta* in which the polyacmic maxima obscure the seasonal cycle there is clearly a succession of species throughout the year which must vary in environmental tolerances and probably in food as well as in size.

In the warm temperate regions little is known of seasonal succession: in South Africa there are perennial species such as *K. cochlearis* where it occurs, *K. tropica, Brachionus calyciflorus,* and *Synchaeta pectinata,* and summer species such as *Hexarthra mira, Filinia (Tetramastix) opoliensis, Brachionus falcatus,* and *Trichocerca chattoni* all probably also warm-water species in Europe. Washing out of fauna and washing in of food as the result of seasonal rains may be of paramount importance in the rather atypical bodies of water which have been adequately studied in tropical regions.

Among the Gastrotricha are a few heleoplanktonic species of *Neogossea,* and the little-known allied central African *Kijanebalola dubia* is probably an inhabitant of the plankton of shallow lakes.

The planktonic Crustacea belong for the most part to the Cladocera among the Branchiopoda and to the Copepoda among the Maxillopoda. There are also a very few tropical planktonic species of *Cypria* in the Ostracoda. In the Malacostraca some of the benthic Peracarida such as *Mysis relicta* may be described as meronektoplanktonic; *Caridina nilotica,* when immature lives as a planktonic animal in Lake Victoria, *Gammarus pulex stoliczkae* occurs as a nektoplanktonic animal in lakes without fish in southwest central Asia, and the very large *Macrohectopus branickii* is important in Lake Baikal.

Like the rotifers, the planktonic Crustacea are either raptorial or microphagous, feeding, however, by filtration rather than sedimentation. The Cladocera, except for the raptorial *Leptodora* and the Polyphemoidea, are filter feeders, as are the pseudoplanktonic Anostraca and Conchostraca. The Calanoida are essentially filter feeders, though *Heterocope* seizes individual particles of food, and most Cyclopoida are raptorial. More is known of the physiology of feeding in Crustacea than in rotifers, but this knowledge comes mainly from studies of the ecologically aberrant anostracan *Artemia,* the littoral marine harpacticoid copepod *Tigriopus,* and the more heleoplanktonic species of *Daphnia* and *Moina,* hardly a representative sample of the lake plankton. There is clear evidence that some algal cells are not complete foods for *Artemia* and *Tigriopus;* four kinds of algae were adequate and seven inadequate in pure culture for both Crustacea, whereas six were adequate for *Artemia* and two for *Tigriopus* alone. Sometimes mixing two inadequate types produces an adequate food. Thiamin and folic acid are certainly required by *Artemia,* and biotin, pyrodoxine, nicotinic acid, and choline were required to produce an optimal yield on an elaborate axenic diet. Some Crustacea, notably *Daphnia magna,* seem to starve to death in a day or two if unfed; it is claimed, however, that colloidal organic matter can be used as food.

The Branchiopoda, though systematically best divided into the Anostraca, on the one hand, and the Phyllopoda, including the Notostraca, Conchostraca, and Cladocera, on the other, are ecologically separable into the larger forms (Anostraca, Notostraca, and Conchostraca) found for the most part in astatic waters and the Cladocera found in both astatic and eustatic waters, including the open water of lakes. Some Anostraca and Conchostraca can be regarded as nektoplanktonic in the shallow pans or playa lakes of semiarid regions. Nearly all lay eggs which can be dried and probably hatch on wetting as the result of the setting up of an osmotic gradient. The Cladocera have retained this type of egg as the sexual resting egg, but most of their reproduction is parthenogenetic and leads to the production of subitaneous eggs that pass into a brood pouch between the dorsal side of the animal and its carapace, in which they develop and hatch. Eggs pass into the brood pouch just after a molt; after developing and hatching, the young are liberated at the end of the instar.

There are two (*Bosmina*) to eight (some *Daphnia magna*) preadult instars; four is the usual number. The total number of instars varies from four in some wild *Bosmina* populations to well over twenty in

Daphnia in culture. The length of life and the length of the average instar are negatively correlated with temperature but apparently not proportionately so; at least in *D. galeata mendotae* there is a suggestion of more instars at 20° C. than at 25 or 11° C. In some cases a reduction of food intake increases longevity, though fecundity is much reduced. The longest-lived *Daphnia laevis* are those partly starved for a large part of their lives and then placed on a highly nutritious medium. The most fecund over a whole lifetime had been partly starved for a short time. It is uncertain how general these effects may be.

The total number of young produced by a single parthenogenetic female *D. magna* can be more than a thousand, but for most Cladocera numbers of the order of a few hundred under favorable circumstances are probable. The size of the clutch at any instar or brood varies greatly with food intake and a little with age, being maximal in middle life. The rate of production of broods is directly temperature-dependent. Both food and temperature therefore influence the reproductive rate of parthenogenetic females. In *Daphnia laevis* there is considerable genetic control of longevity and fecundity and considerable heterozygosity with respect to the genes involved in the populations from which the lines studied were obtained.

The regulation of fecundity by food intake is not instantaneous, and this may well underlie the phenomenon of oscillatory populations so often observed in cultures of Cladocera. At any given time the brood size may be appropriate, not to the actual trophic potential, but to that of some hours earlier, during which time the population has increased. This leads to the production of a population of more than equilibrium size, even if a constant rate of supply of food is maintained. The population exceeding the food supply will become starved, reproduction will fall, and adults will die without being replaced. When the population falls below the expected equilibrium size, it will after an appropriate lag period start increasing again. Theory suggests that when the product of the instantaneous birth rate b and the time lag τ is small, the oscillation is not observed; when it is large, the oscillation persists as a limit cycle, whereas at intermediate values the oscillation appears but is damped. The limits for the last-named situation can be ascertained empirically from models on an analogue computer and appear to be $0.7 < b\tau < 1.8$.

Sexual reproduction occurs when males appear and at least in most of the Daphnoidea when special fertilizable females with somewhat modified carapaces also occur. The modified part of the carapace, when it occurs, encloses a resting egg and is known as an ephippium.

In the Haplopoda, Sidoidea, and Polyphemoidea the resting egg is carried until the molt or death of the female in the unmodified brood pouch.

Males appear to be produced by crowding, as in *Brachionus calyciflorus* among the rotifers, but in spite of much work the efficient stimulus is still unknown. Ephippium production has no genetic relationship to male production. It seems likely to be due to a fairly rapid reduction of food supply during the life of the female. Maintenance of parthenogenetic animals at a low constant food level merely inhibits reproduction. Since rapid reduction in food supply is likely to occur as the population increases, males and ephippial females will develop about the same time. The probability of these processes apparently also depends on day length and differs from species to species, and quite likely from clone to clone, but there is clearly no internal genetic determination of the rhythm.

Some species, such as *Moina macrocopa* and *Pseudosida bidentata,* are clearly better adapted physiologically to high temperatures and other species to lower temperatures. In general, as far as studies on thermal tolerance go, the species surviving high laboratory temperatures are south temperate, tropicopolitan, or transitory summer forms. The warm-water species such as *M. macrocopa* tend to show a steeper curve relating rate of development to temperature than, for example, *D. laevis.* The great temperature dependence in *Moina* would permit the opportunistic development of a large population in a pond at the height of summer. Within a single species genetic differences in thermal tolerance are known. Chemical limitations certainly exist. *Holopedium* is ordinarily confined to soft waters, though there are difficulties in setting limits to its appearance. In the genus *Daphnia* the large species of *Ctenodaphnia* seem more halophil than the majority of the members of *Daphnia* (s. str.), though *D. balchashensis* is moderately halophil. There are several planktonic species of *Moina* found mainly in alkaline or saline lakes; the common fresh-water species of the genus are less planktonic. The pond species appear to be less sensitive to oxygen lack than are the limnoplanktonic, and haemoglobin may be formed under conditions of oxygen deprivation in pond-living species of *Daphnia.*

In the temperate Northern Hemisphere *Diaphanosoma* is ordinarily a monacmic summer form, as is *Holopedium gibberum, Limnosida frontosa, Polyphemus pediculus* mainly in the littoral, the various species of *Bythotrephes,* and *Leptodora kindtii.* Though the highly helmeted species *D. cucullata* in Europe and *D. retrocurva* in North America are ordinarily monacmic and often produce ephippia in numbers, they apparently tend to overwinter as large autumnal females with fairly low helmets, which can start reproduction rapidly in spring.

Daphnia galeata mendotae is often diacmic in North America and the same seems true of *D. g. galeata* in various European lakes. In Base Line Lake, Michigan, it is known that the vernal and autumnal maxima are due to relatively high birth rates at the time of low death rates; a third maximum in production of young does not lead to a marked summer maximum, filling in between the vernal and autumnal maxima, because of the very high mortality, in part probably due to the rise of the predatory *Leptodora,* observed in summer. It is also probable that the autumnal peak is delayed in this species by competition with *D. retrocurva* at the time of its single late summer maximum. Elsewhere *D. catawba* may produce a like effect.

A few cold-water species which appear to be perennial develop quite large populations in the hypolimna of lakes in summer. *Daphnia pulex* may do so, as may *D. schødleri* and *D. ambigua* in North America. These species are inhabitants also of ponds, and *D. pulex* at least has a greater tolerance of lower oxygen concentration than the eulimnoplanktonic species studied. It is possible that these cold-water species are characteristically hypolimnetic because of this tolerance. The most characteristic cold-water species, however, is *D. longiremis,* found in small and shallow waters in the north but in temperate latitudes a definitely hypolimnetic lake species in summer. The closely allied but solely Old World *D. cristata* may develop an epilimnetic population in summer in the same lake inhabited by *D. longiremis* in the deeper water. The hypolimnetic population of the latter in summer may be considerably greater than the more evenly distributed winter populations. As with other cold stenotherm organisms in lakes, it is probable that optimal conditions are provided by the cold water of the hypolimnion during seasons when much food material is falling from above.

A few littoral Cladocera tend to become planktonic for a restricted part of the season; *Chydorus sphaericus* provides an example of a species primarily planktonic at times of blue-green algal blooms in much the same way as are some species of *Trichocerca.*

The Copepoda are at least as important as the Cladocera and often more so. The Cyclopoida are mainly benthic as far as number of species is concerned, but the transition from a benthic life to living in the plankton is evidently very easy in the group. The majority of the planktonic species belong to *Cyclops, Mesocyclops, Thermocyclops,* and *Microcyclops* in the Cyclopinae, though in this subfamily *Diacyclops bicuspidatus thomasi* is a planktonic New World subspecies

of a widely spread benthic-littoral and pond species; in Lake Van
the normally littoral-benthic *Megacyclops viridis* is recorded as a planktonic animal. These examples along with the closely related species
of the *Cyclops strenuus* group from habitats of various sizes emphasize
how the transition from benthic to planktonic life can be taken by
the Cyclopinae with very little if any morphological modification. In
the Eucyclopinae nearly all species are benthic-littoral or pond forms,
but *Tropocyclops prasinus* appears to be planktonic in North America.
The free-living Cyclopoida are all more or less raptorial; the larger
more primitive genera such as *Macrocyclops* in the Eucyclopinae (if
indeed it belongs in the subfamily) and *Megacyclops, Cyclops,* and
Mesocyclops in the Cyclopinae seize and eat animal prey, whereas
Eucyclops in the Eucyclopinae and *Diacyclops* and *Microcyclops* in
the Cyclopinae feed mainly on algal cells. In the genus *Cyclops*
numerous species are related to *C. strenuus,* itself typically an inhabitant
of small lakes and ponds. These species of *Cyclops* occur under different but often rather obscurely characterized ecological conditions; in
many large lakes there is a good deal of replacement of one very
closely allied species by another in different associations. Thus in
Lake Constance *Cyclops strenuus praealpinus* is planktonic, occurring
in limited numbers at all depths and all seasons. A second taxon
C. a bodanus occurs on the bottom, and a third near *C. s. landei*
is littoral. In the genus *Mesocyclops* a comparable distinction between
the planktonic *M. bodanicola* and the less planktonic *M. leuckarti*
has been noticed in Lake Constance and elsewhere in Switzerland,
though in many lakes the last-named species is a typical member of
the eulimnoplankton.

When two or more species of the same genus co-occur in the same
part of the lake there may be size differences suggestive of character
displacement. There is, moreover, a temporal separation of *Cyclops*
and *Mesocyclops; Cyclops* usually reproduces during the colder times
of year and undergoes diapause as a copepodite larva in the mud
during the summer as does also *Diacyclops bicuspidatus thomasi,* whereas *Mesocyclops* may show a winter diapause. In some species at
least diapausing copepodites are resistant to drying and may constitute
dispersal stages. In *Cyclops scutifer* in Sweden coexistence of diapausing and nondiapausing individuals may split the population into two
partly independent sections.

Little is known about longevity in copepods; the littoral-benthic and
pond species *Megacyclops viridis* can live for nine months in the
laboratory and possibly longer in nature. The smaller copepods prob-

ably live one to six months without diapause, but if diapause intervenes life in some cases may last two or three years.

Reproduction is always bisexual and eggs are carried by the female. There are great variations in egg number, but they are less clearly associated with environmental conditions than in the Cladocera and in the Copepoda Calanoida.

The Calanoida are essentially planktonic animals, though some live in very small bodies of water and may come to rest by grasping algal filaments with an antennule. The life history consists of six naupliar (or strictly two orthonaupliar and four metanaupliar) instars (N1–6) and five copepodite instars (CI–V), the sixth (CVI) being the adult. There is some uncertainty whether a second adult instar may not occur in some species. Slight variation in size and in the structure of a process on the twenty-third joint of the right or geniculate antenna of the ♂ and of some other details has been explained variously as due to the dimorphism of stage CVI, a molt to CVII, or the existence of sibling species. The laboratory observation of life history has been difficult, but the fresh-water Calanoida, probably fairly long-lived, take perhaps three to four weeks to achieve maturity. The adult reproductive period can be quite short, only a week or two in summer, at least in *Eudiaptomus graciloides,* but it is undoubtedly much longer in colder seasons. All Calanoida are bisexual; two kinds of eggs, subitaneous and resting, may be found in some species, but both have been fertilized. The eggs are usually carried in a single egg sac but are laid loose in *Limnocalanus* and perhaps *Sinocalanus* in the Centropagidae and in most species of *Heterocope* and *Epischura,* but not of *Eurytemora,* in the Temoridae. The eggs dropped directly to the bottom seem to be resting eggs that undergo diapause; resting eggs, at first carried in egg sacs and then dropped to the bottom, are common in the Diaptomidae. Little is known of the total fecundity or number of clutches in the Calanoida, but the clutch size varies greatly. In limnoplanktonic species it is nearly always maximal in spring, falling to a minimum at sometime in the summer and sometimes with a secondary autumnal maximum. There is also evidence that the clutch size usually has a greater average value in eutrophic than in oligotrophic lakes in the strict sense of these terms, though occasional cases of the reverse being true are known. In some species the reduction in number of eggs in summer is partly compensated by an increase in their size. It is suggested that in spring under conditions of high food supply selection favors the largest possible brood but that later, when stratification is well established and the epilimnion somewhat nutrient-deficient and poor in phytoplankton, selection will favor the strongest and most mature nauplius possible.   Species which lay both resting and subitaneous eggs often produce the former at high latitudes and the latter at low, whereas in intermediate regions subitaneous eggs may be laid during the height of the summer and resting eggs later in the year.   Two distinct generations can be, but are not invariably, involved; sometimes the early production of subitaneous eggs occurs at a lower temperature than the later resting eggs, so that low temperatures cannot be a general efficient cause of resting-egg production. In some species that inhabit astatic and somewhat mineralized waters increasing salt concentration may well be involved.   Little is known of the mechanism by which egg diapause is terminated, but at least in the eggs of *Heterocope borealis* from Lake Constance rapid warming followed by cooling was completely effective experimentally, though the process is not clearly related to anything likely to happen in nature. Perhaps oxygen deficiency is also involved.   Few data relating to fecundity of individual females exist, but from the number of eggs carried by the average female and the length of the egg stage determined in the laboratory a reproductive rate can be computed.   This is dependent, as in the Cladocera, on the food that determines the clutch size and on the temperature that determines the rate of egg production in such a way that a small proportion of females carrying a few eggs for a short time can still be producing more nauplii than many fertile females carrying large clutches for long periods at winter temperatures.

Univoltine life cycles in which eggs are hatched in the spring occur in *Limnocalanus* and *Heterocope* and in the Diaptomidae that produce only resting eggs at high latitudes.   Some species such as *Mixodiaptomus laciniatus* remain essentially univoltine farther south.   In a few cases (*Arctodiaptomus laticeps* in the uplands of central Sweden, *Boeckella propinqua* in New Zealand) part of the adult population may overwinter and late in the summer produce young stages contemporary with adults derived from the overwintering of resting eggs. The late-summer young may mature to produce another overwintering generation, but no case yet appears known in which the dualism is so clear that two sympatric but heterochronous genetically isolated populations have occurred in the same lake.

In central Europe *Acanthodiaptomus denticornis* has a bivoltine cycle, in which the generations alternate between production of subitaneous and resting eggs.   Most species in the temperate region, however, appear to have prolonged reproductive periods with several more or less indistinguishable generations per year.   Naupliar survival may often be better in the spring than at the height of summer, and rapid reproduction at the warmest time of year may make little contribution to the population. In Lake Constance the annual cycle of *Eudiaptomus gracilis* is determined by a delicate balance between natality and mortality. Much of the mortality of nauplii can be catastrophic and density independent; as with other copepod nauplii those of *E. gracilis* occur in the surface waters, and if a wind is blowing toward the outlet they may be carried out of the lake down the Rhine. The balance of births and deaths is favorable to the species mainly in winter, from December to March, when most of the adults that survive will have been born, even though far more nauplii are hatched later. In a few cases, as in *E. vulgaris* in Lago Maggiore, a fairly definite multivoltine, in this case actually trivoltine, cycle, with adult maxima in May-June, August-September, and December-January, can be distinguished.

Most species of Diaptomidae seem fairly eurytopic chemically. A few species are characteristic of mineralized water; particularly interesting cases are provided by *Arctodiaptomus salinus* and *A. spinosus* in eastern Europe, *A. salinus* being characteristic of chloride and *A. spinosus,* of carbonate waters, a specialization which, though it might be expected, has rarely been established.

Swimming and feeding are primarily the result of movements of the second antennae, which produce vortices carrying food particles past the setae of the maxillae. As in the rotifers, such feeding is more selective than would at first be supposed. In *Eudiaptomus gracilis* specimens living in a pond containing vast numbers of *Kirchneriella* have been found to reject this alga in favor of much less abundant desmids. When living with *Arctodiaptomus laticeps,* *E. gracilis* took mainly small green algae, whereas *A. laticeps* fed largely on *Melosira,* both species avoiding *Asterionella.* When compared with *Daphnia longispina, E. gracilis* was far less effective in the filtration of bacteria, though when all else is excluded it seems able to live on minute detritus particles. Sympatric species of Calanoida frequently show a size difference probably corresponding to a food difference comparable to that between *E. gracilis* and *A. laticeps,* which is about one third larger. The size difference seems maximal in astatic waters, such as playa lakes, the zooplankton of which is characteristically composed of a species of *Daphnia* and two Diaptomidae, one much larger than the other. Comparable phenomena of a less striking kind, however, are to be noted in less extreme types of lake.

Apart from size differences and concomitant differences in food, differences in seasonal incidence of behavior relative to depth distribution, migration, and time of feeding and in a few cases probably in reaction to current may provide considerable niche specificity.   In the most complex taxocene recorded in the group, the eight sympatric calanoids, of which five are Diaptomidae, known in Lake Erie, the latter family has three spring breeding species slightly but significantly separable in mean size and two late summer breeding species, the larger only 7 per cent longer than the smaller and so probably not ecologically separated in this way.

The Calanoida present curious zoogeograhic features.   In spite of the rather large number of species that can produce resting eggs in astatic localities and so stages that might be supposedly dispersed, the species of Calanoida are much more localized than are those of the Cyclopoida, which seem to have less effective dispersal mechanisms. About half the Cyclopoida but only one tenth of the Calanoida of North America are represented by identical or allopatric subspecies in the Old World.   In Africa, in which the Paradiaptomidae are well developed in astatic waters, no species of Calanoida is common to tropical regions and the area south of the Limpopo.   The species of astatic waters such as playa-lakes are clearly distributed largely according to the way their life histories fit into the seasonal patterns of rainfall.   The specialization of some species as pond and others as lake forms and, within the latter, the existence of littoral and pelagic species in a group that is essentially free-swimming, must imply behavior differences which have not been adequately elucidated.   In any region of considerable altitudinal difference there are often upland species in high cold lakes, presumably determined in their distribution by temperature, but when an analysis of any area is at all detailed the obviousness of temperature control becomes less apparent.   In some cases as in the islands of western Europe, it is difficult to make much sense of the distributions however they are studied.   Here *Arctodiaptomus wierzejskii,* a characteristic species of the substeppic zone of Algeria, with an annual rainfall of 300 to 500 mm. is also the main species in the lakes of the Shetland Islands; it is absent from the Faeroes but occurs in two lochs out of many on the Orkneys and some in the Hebrides.   On the mainland of Britain it becomes scarce and absent on going south.   It occurs sporadically from the Kola Peninsula and Mongolia to North Africa and would seem to be adapted to every contingency save meeting diaptomid competitors.

Apart from the Crustacea, a few insects are meroplanktonic, notably the larvae of *Chaoborus,* which are members of the benthic community by day when they are usually caught.   A few caddisfly larvae when very young may be significantly dispersed by currents as meroplanktonic animals. Among the watermites records of certain species of *Piona* and *Unionicola* in plankton tows are frequent enough to suggest that the planktonic occurrence of these animals is at times not accidental.

The only mollusk to have a planktonic larva in fresh waters is *Dreissena polymorpha*. The littoral phylactolaematous Polyzoa produce free-swimming larvae or larval colonies which may occasionally get into the plankton. The cercariae of certain Trematodes mainly of the families Bucephalidae, Diplostomidae, and Schistosomatidae are meroplanktonic, and even though their sojourn in the plankton may be no more than twenty-four hours they appear very specialized as well as specifically differentiated in posture and swimming movements for life in free water.

A very few fish eggs, namely those of species of *Engraulicypris* in the Great Lakes of central Africa and of *Aplodinotus grunniens* in North America are typically planktonic.

In very simple cases in which, because of chemical restrictions or the diversity of the flora and fauna, the relationships between the few species present are particularly clear the rise of a zooplankton population following and grazing on a phytoplankton population which is thus greatly reduced, can be followed. More often specific diversity in both plants and animals obscure these relationships. There is clearly some competition among quite distantly related species. This must occur between the Cladocera, if not other Crustacea, and the Rotifera, which overlap considerably in the size range of available food particles even though the animals usually differ in size by a factor of 10. In both Cladocera and rotifers the life histories are similar in rapid parthenogenetic reproduction and in a relatively short prereproductive or juvenile period in the life span. Cladocera are doubtless ordinarily longer-lived than rotifers but probably not dramatically so. The perennial multivoltine populations of diaptomid copepods such as *Eudiaptomus graciloides* in Erken apparently have generations of thirty-five to forty days, as in the Cladocera, but a much greater proportion of their life histories is spent in the immature prereproductive stages. As such they would seem a priori to be less efficient organisms than the Cladocera; the latter moreover presumably have an equal or possibly greater range of filterable food. *Eudiaptomus graciloides* in Denmark seems to have a largely univoltine life history. In several other species, as we have seen, there are indications of very prolonged life histories within the family. It is therefore just possible that in some ways the diaptomid life cycle is more adjustable to varying and in part unfavorable conditions than is the cladoceran.

The cyclopoids probably have still longer life histories and are no doubt often univoltine, though this frequently involves diapause. Their feeding habits, however, are quite different from those of the Rotifera, Cladocera, and Calanoida, all of which groups at some stages of their life histories are no doubt eaten by the raptorial zooplanktonic species of *Cyclops* and *Mesocyclops*.

It is clear that fish exercise a considerable control on the nature of zooplankton associations. In general where fish are abundant large species are at a disadvantage in the zooplankton. This effect can occur at all levels leading to increase in rotifers at the expense of Crustacea, or to the substitution of one species of *Daphnia* such as *cucullata* for the larger *D. pulicaria.*

CHAPTER 25

The Vertical Migration
and Horizontal Distribution
of the Zooplankton

As already implied in Chapter 24, the various species of zooplankton are often confined to relatively restricted zones and so exhibit an uneven distribution in a vertical direction. This is, of course, most strikingly observed in those lakes in which a full summer thermal stratification is developed. In such lakes the zooplankton of the hypolimnion may be quite different from that inhabiting the epilimnion. Whatever differences of this sort may exist, there is frequently a considerable variation in the distribution as night succeeds day. The illumination is obviously the main variable to be considered, though there has been disagreement as to how it may operate; nor is there any reason to suppose that it always does so in the same way. Cases also are known in which the vertical distribution by day or by night varies systematically with the seasons and particularly with the development of thermal stratification in summer. It therefore may be supposed that thermal and chemical gradients as well as differences in illumination determine the position of the zooplankton. In view of the probability of multiple causation, it is simplest to begin with the problem of diurnal differences in vertical distribution, in which we can be reasonably certain that only changes in light, directly or indirectly underlie the phenomenon, before proceeding to the study of the limited amount of data on the seasonal changes due to a number of interacting primary causes.

OBSERVATIONAL DATA

The daily movement of the zooplankton was first recorded in lakes by Weismann (1877b) in Lake Constance and perhaps by Forel (1877) in Geneva. Other early work was done by Pavesi (1882) on the Italian lakes, Francé (1894) on Lake Balaton, Blanc (1898) on the Lake of Geneva, Steuer (1901) on ponds in the old bed of the Danube near Vienna, and by Lozeron (1902) on Lake Zurich. These contributions established the phenomenon as very widespread. Franz (1912), long after these and other investigators had laid the foundation of our knowledge of the migration both in fresh water and the sea, concluded that the phenomenon was an artifact because the organisms in the upper layers of the water saw and avoided the collecting gear when those layers are illuminated. Although it is certain that such behavior can introduce errors into the quantitative study of vertical distribution by day, it is equally certain that Franz's hypothesis is totally inadequate to explain the various more complicated types of nocturnal movement that are now known to occur. No serious modern worker doubts the reality of vertical diurnal migration.

In general diurnal migration is best observed in deep unproductive transparent lakes. Lozeron (1902) noted this when comparing his observations on Lake Zurich, in which the amplitude of the migration is about 13 m., with results on other large Swiss lakes, which are more transparent and in which the zooplankton often moves 20, 30, (Weismann 1877b), or even 50 m. per day, as Burckhardt (1900b, 1910) and later Worthington (1931) found in the Lake of Lucerne. In the much less transparent lakes of Wisconsin, Juday (1904) seldom found amplitudes in excess of 5 m., whereas in Lake Mendota, Birge (1895) found little evidence of the phenomenon at all. The distribution of *Diaphanosoma* in the various Japanese lakes studied by Kikuchi (1930a, 1930b) very clearly shows the dependence of the depth of the maximum population by day on transparency[1] (Fig. 183).

The Migration of Copepoda in the Lake of Lucerne. It is evident that the extent to which migration takes place varies not only from lake to lake but from species to species within a given lake. Sometimes the two sexes or the young and old individuals of a given species may differ greatly in their diurnal movements. Before these special aspects of the problem are considered, it may be well to study a

---

[1] One rather irregular set of observations on Lake Biwa is omitted as the numbers of *Diaphanosoma* encountered are obviously too small to provide any valid information regarding the form of the distribution.

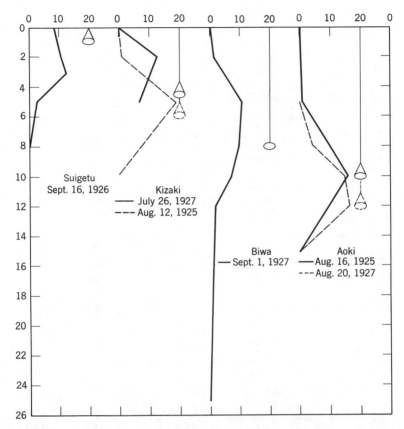

FIGURE 183. Vertical distribution, at 12 to 15 hours, in fine weather, of *Diaphanosoma brachyurum* in four Japanese Lakes, with indications of the Secchi disk transparency at the times of collection. (Data of Kikuchi.)

typical case of vertical migration in a deep transparent lake. No data are better for this purpose than those of Worthington (1931) on the copepodan plankton of the Lake of Lucerne (Fig. 184).

Two species of Calanoida, *Eudiaptomus gracilis* and *Mixodiaptomus laciniatus,* and three species of Cyclopoida, *Cyclops strenuus* (*s. lat.*), *C. abyssorum,* and *Mesocyclops leuckarti,* are present in this lake.[2] The simplest type of movement is exhibited by *Cyclops strenuus* (s. lat.), of which species the specimens present at the time of Worthing-

---

[2] The exact determination of all the Cyclopidae of the Lake of Lucerne probably requires further study.

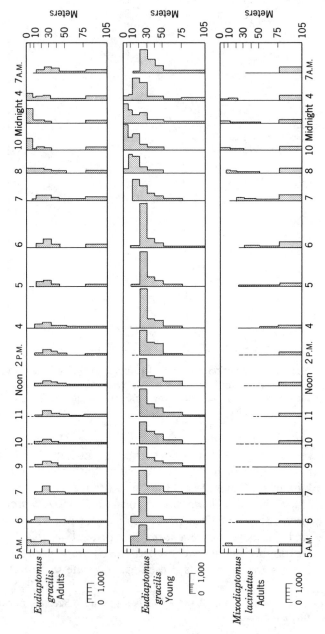

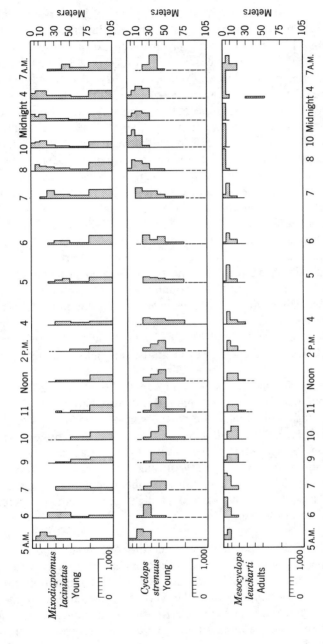

FIGURE 184. Vertical distribution throughout a twenty-four hour period of the planktonic copepods of Lake Lucern (*Eudiaptomus gracilis*, *Mixodiaptomus laciniatus*, *Cyclops strenuus ? praealpinus* and *Mesocyclops leuckarti*) Sept. 10–11, 1928. (Worthington.)

729

ton's study (September 10–11, 1928) appear to have been almost, is not entirely, in copepodite stages. The mode in the population at midday from 10 A.M. to 2 P.M. lay between 40 and 50 m.; during the afternoon there was a little scattering downward so that between 4 and 5 P.M. the population was more regularly distributed between 20 and 75 m. A definite upward movement started between 6 and 7 P.M. and by 10 P.M. almost the entire population was present in the top 30 m., with a mode between 5 and 10 m. After midnight the concentration was more evenly distributed between 5 and 30 m.; a clear descent started after 5 A.M. and continued strongly until 7 A.M.

Mesocyclops leuckarti occupied the region above *C. strenuus* (s. lat.); by day few specimens occurred below 20 m., and from 2 P.M. on through the afternoon the mode lay between 5 and 10 m. The rise in the population started rather later in the evening, between 7 and 8; from 8 P.M. until 4 A.M. the population occupied the top 5 m. and the descent to the daytime level was somewhat irregular. Only at midnight do significant parts of the two cyclopoid populations coincide and even then the mode in population of *C. strenuus* (s. lat.) is a little lower than that of *M. leuckarti*. Worthington states that the third species of Cyclopoid, referred to *C. abyssorum,* occurred below 50 m. by day and could rise to 20 m. by night but was never present in sufficient numbers to provide statistically adequate data.

The movements of the two Diaptomidae are more complex. *Eudiaptomus gracilis* in both adult and copepodite stages had a strong mode in population between 20 and 30 m. throughout the day, and after 6 P.M. most individuals rose to the surface, the mode being between 0 and 5 m. between 10 P.M. and midnight. A minor part of the population, particularly of the adults, broke away and descended to below 75 m. in the evening of Worthington's study. *M. laciniatus* remained below 30 m. by day, most of the population lying below 75 m. Part migrated to the surface at night but an equal or greater part exhibited no movement.

These movements in the Lake of Lucerne are of particular interest in showing how the three cyclopids on the one hand and the two diaptomids, on the other, can inhabit different layers of the same lake so that the species of a family do not come into competition to any great extent. It is obvious that the vertical distributions imply the existence of regions into which one species, but not another, goes and therefore that the conditions for stable co-occurrence of the groups of closely allied species exist. The data obtained by Worthington on the migration of the Cladocera in the Lake of Lucerne are discussed in the appropriate places in the succeeding paragraphs.

Organisms exhibiting diurnal movement. Although the reactions of phytoflagellates to light are well known, few cases have been studied in open water of any depth. *Volvox aureus* in North Germany (Utermöhl 1924) and *Eudorina elegans* in Lake Ranau, Sumatra (Ruttner 1943), are maximal at the surface at noon and later descend into deeper water. Similarly, *Gonyostomum semen,* a chloromonad living in Cedar Pond, Woods Hole, Mass. (Cowles and Brambel 1936) (Fig. 185), moves upward in the early morning when the maximum density is in the top meter of the pond. In the afternoon, often as early as 1 P.M. a descent begins, which is attributed to the metabolic results of photosynthesis by Cowles and Brambel. This case in which CO_2 as well as light gradients may be operating as causal agents is probably very different from those involving purely animal holotrophic plankters. Thienemann (1919) found *Ceratium hirundinella* moving down by night in the Ulmener Maar, but because, in this case, several rotifers and Crustacea (see page 740) behaved similarly, it is probable, as he suggests, that convectional streaming was partly involved. This is probably true also in the next cases to be described.

Among the truly animal protists of the plankton regular diurnal movement is hardly recorded. Kikuchi (1930a) noted that *Difflugia biwae* in Lake Biwa and, to a lesser extent, *Nebela kizakiensis in* Lake Kizaki and Lake Aoki tended to decrease at the extreme surface by day, moving up again at night. Whatever the cause of the decline above 2 m. by day, it is possible that the surface increase at night is a purely passive one, depending on the cooling and consequent loss of stability at the extreme surface in the evening, leading to more complete epilimnetic mixing at night. The possible presence of gas-

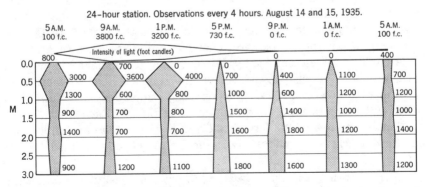

FIGURE 185. Upward movement of *Gonyostomum semen* early in day and descent in afternoon. (Cowles and Brambel.)

vacuoles in such rhizopods might, however, permit a more elaborate type of movement.

Marked diurnal migration has sometimes been recorded among the rotifers, though the group as a whole would appear to exhibit the phenomenon far less strikingly than the Crustacea. In the Grosser Plöner See Ruttner (1905) found that *Conochilus*[3] rose to the surface at night. Thienemann (1919) observed some downward movement of rotifers, notably *Hexarthra mira,* at nightfall in the Ulmener Maar, and Ruttner (1943) recorded movement away from the surface on the part of *Polyarthra* in the Lunzer Untersee and of *Hexarthra intermedia* in Lake Toba but not in Lake Ranu in Sumatra. Kikuchi (1930a) noted upward nocturnal movement of *Ploesoma truncatum* in Lake Aoki but not in other lakes, in which if there had been any movement, the species scattered downward a little at night. *Ploesoma hudsoni* living in the deeper water of Lake Aoki exhibited no migration. There can be no doubt about the reality of the migration of *P. truncatum* in Lake Aoki; *Keratella cochlearis* in the same lake also showed a well-marked upward nocturnal movement, but since this involved only the top two meters it may have been due to a retreat from the surface by day and subsequent restoration of a random distribution by night as the stability of the extreme surface layers declined. In addition to these cases Ruttner (1937a) has noted that a cold stenotherm *Synchaeta*[4] in the lakes of the Austrian Alps exhibits vertical diurnal movement.

Among the Crustacea it is probable that every genus of truly planktonic Cladocera contains members performing vertical diurnal movements and the same is doubtless true of all the more widespread genera of planktonic copepods. Moreover, the larvae of the prawn *Caridina nilotica* were found by Worthington (1931) to exhibit a moderate but perfectly definite upward nocturnal movement from the bottom into the middle water of Lake Victoria during the night. In the sea many decapods of course are known to perform such migrations. *Mysis relicta,* largely benthic by day, becomes nektoplanktonic by night (Dakin and Latarche 1913; Juday and Birge 1927; Southern and Gardiner 1926a,b, 1932). The pelagic amphipod *Macrohectopus branickii* shows marked migrations in Lake Baikal (Zachvatkin 1932).

Among the insects the larvae of *Chaoborus* (= *Corethra*) frequently perform marked migrations, living on or in the bottom mud by day

---

[3] The species is given as *C. volvox,* i.e., *C. hippocrepis,* but *C. unicornis* would seem to be more likely.

[4] Referred to *pectinata* but not improbably *lakowitziana.*

and moving upward to feed on the planktonic Crustacea at night. In Lake Victoria at least part of the population ascended to the surface at night from the bottom at a depth of more than 50 m.; most other lakes in which the genus occurs are less deep than Lake Victoria, so that the migration of these larvae is generally less remarkable. Some migration may be exhibited by fish; the most striking cases are probably provided by the endemic *Comephorus* and species of Cottocomephoridae in Lake Baikal (Zachvatkin 1932).

The types of vertical migration. The types of movement observed are numerous and depend not only on the species and locality under consideration but also on season, age, and sex. Moreover, the various possible kinds of behavior cannot, of course, be sharply delimited, so that any formal classification such as that of Kikuchi (1930b) is likely to be somewhat artificial.

The majority of species tend to rise at night from deeper water into the more superficial layers of the lake. If a single maximum is observed at the surface of the lake sometime between sunset and sunrise, the migration is often spoken of as *nocturnal*. If two maxima occur in the shallower layers, one maximum associated with the decline in illumination at about the time that the sun is setting and the other about dawn, the term *twilight* migration is sometimes used. In a few cases in which the animal moves downward during the night and upward to the surface during the morning, the term *reversed* migration emphasises the exceptional nature of the occurrence; diurnal migration as the opposite of nocturnal is not a satisfactory term because of the ambiguities associated with the words day and diurnal.

As will appear in later discussions of individual species, there are often minor elaborations of the simple up and down movements implied in the preceding paragraph. Cushing (1951) and, less formally, Siebeck (1960) regard as the paradigm a kind of twilight migration in which there is in the very early morning a slight upward movement as the sky begins to lighten, a marked descent after sunrise to a level at which the population remains for a few hours, an upward movement beginning in the afternoon and continuing for a varying time in the evening, and some downward scattering during part of the night. At least three different kinds of nocturnal migration as well as twilight migration have been described; all four types can be regarded as modifications of the paradigm, but it must be remembered that in many cases the upward morning twilight movement and the downward midnight movement are feeble and might easily be missed. Although there are certainly great differences among species from locality to locality, and even among individuals of different sexes or ages within a species, the formally separable types of nocturnal migration may in part be due to inadequate observation.

In the supposedly simplest type the organisms start moving upward before or shortly after sunset, reaching the upper layers of the lake some time before midnight (*A* of Fig. 186). Here they remain for several hours, only to descend again as the sky begins to lighten in the early morning. The young specimens of *Daphnia longispina* studied by Southern and Gardiner in Lough Derg (1926b, 1932) conform to this pattern, as in broad detail do nearly all the migratory forms of the Lake of Lucerne studied by Worthington (1931).

In many cases, such as that of the young *D. longispina* in Lough Derg and those of the young *D. longispina* among the Cladocera and both *Mixodiaptomus laciniatus* and *Cyclops strenuus* (s. lat.) among the copepods of Lake Lucerne, it appears that the upward movement continues well after dark, as recorded subjectively by the observer. Ullyott (1939), however, points out that at the end of August a slight decrease in illumination after 9 P.M. can be recorded at the surface of Windermere and that intensities of the order of 1.43 to 14.33×10^{-6} cal. cm.$^{-2}$ (1 to 10 erg. cm.$^{-2}$) can produce orientation movements in *Volvox* and so may be physiologically significant to other organisms also. Later in the night there is often what Cushing (1951) calls midnight sinking, but because it usually starts an hour or so before midnight it is better termed *nocturnal sinking*. Russell (1927) emphasized this process as the development of a random distribution in the dark, and Cushing points out that in the sea it is a more conspicuous feature of the movement of large than of small plankters, which suggests passive sinking. In fairly pure nocturnal migration it is exemplified by Worthington's data for *D. longispina* and young *Bosmina coregoni* in the Lake of Lucerne. The main descent began about an hour before the rays of the rising sun first struck the water.

A second possible type of nocturnal migration differs from the first in that the upward movement continues throughout the night, the surface layer containing its maximum population before dawn or, in most cases in temperate latitudes in summer, about 4 A.M. This type of movement was observed by Blanc (1893) in Lake Constance, by Fordyce (1900) for *Leydigia fimbriata* in a pond 3 m. deep in Nebraska, and by Southern and Gardiner (1926b, 1932) for *Eudiaptomus gracilis* (*B* of Fig. 186) and most strikingly for *Eurytemora velox* in Lough Derg. In these cases the upward ascent appears to continue through the time of minimum illumination. Though Southern and Gardiner's observations made on a very dark cloudy night are convincing, it is conceivable that other cases represent evening ascent, a very little nocturnal sinking, and a morning twilight movement upward, the three phases being inadequately separated.

The third type of nocturnal migration involves an accentuation not of upward movement in darkness but rather of the nocturnal sinking which, starting early, continues throughout the night. Kikuchi (1930a) gives a most striking example (C of Fig. 186) from the population of Cyclops strenuus (s. lat.) in Lake Noziri. Here the depth of the maximum population, which is obviously not always properly sampled, seems to fall regularly from 0 m. to 18 m. during the period from 7 P.M. to 7 A.M. Zachvatchin (1932) found that the pelagic amphipod Macrohectopus branickii in Lake Baikal in August appeared in the top 25 m. between 7 and 10 P.M. The adults then disappeared by 1 A.M., and the juveniles became very scarce. The latter but not the former reappeared again about 3 A.M.

Twilight migration was recorded by Ruttner (1905) for both Daphnia cucullata and Bosmina coregoni in the Plöner See in July, but later in August there was little migration of the first species and the morning but not the evening maximum of the second species was suppressed. During the period in which both species of Cladocera showed marked twilight migration, Leptodora kindtii, Conochilus, and the copepoda of the lake exhibited a marked nocturnal maximum at the surface. Ullyott's (1939) data for Cyclops strenuus (s. lat.) in Windermere indicate that in April there is a slight movement away from the surface at midnight and a return very early in the morning, so that although his observations suggest a fairly typical nocturnal migration, there is a slight tendency toward the twilight type of movement as in Cushing's paradigm. The same pattern is indicated by Siebeck (1960) in the Lunzer Untersee for Daphnia longispina, for the females of Cyclops tatricus, which can also exhibit striking twilight migration (D, E, of Fig. 186) in the Schluchsee (Schröder 1959), and perhaps for the copepodites and females of Eudiaptomus gracilis. The males of E. gracilis exhibited a striking concentration at the surface at 6 P.M. and then became more evenly distributed; they exhibited less upward twilight movement in the morning than the females (Fig. 187). A striking concentration of males is recorded at the surface in the Schluchsee by Schröder (1959) but it took place at ten minutes after eleven, more than three hours after sunset, though on a moonlight night. The difference between the sexes in respect to the temporary surface concentration is as great as in Siebeck's case.

A very instructive case has been described by Kikuchi (1930a,b). He found that on August 6 and 7, 1929, in lake Kizaki, Bosminopsis dietersi appeared three times at the surface. The first appearance

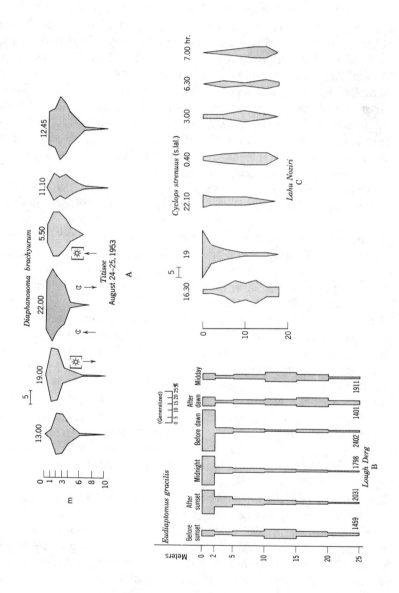

736

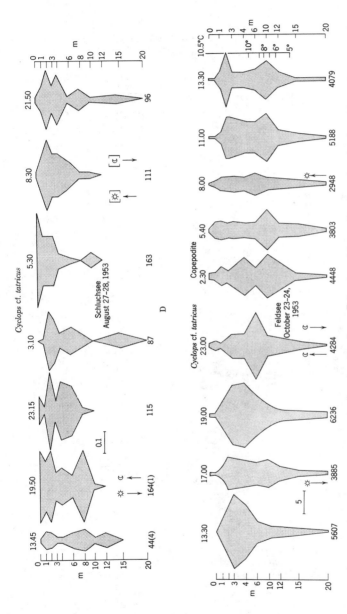

FIGURE 186. *A*. Vertical movement of *Diaphanosoma brachyurum* in the Titisee, showing the simplest kind of nocturnal movement with a maximum at the surface about midnight, subsequent slow descent, and very little hint of downward nocturnal scattering or of twilight migration (Schröder). *B*. *Eudiaptomus gracilis*, whole population in Lough Derg, showing continued upward movement throughout the night (Southern and Gardiner). *C*. *Cyclops strenuus* s. lat. Lake Noziri, rising to surface in evening and descending throughout the night (Kikuchi). *D*. Typical twilight migration with surface maxima in morning and evening exhibited by *Cyclops* cf. *tatricus* in the Schluchsee, Black Forest, Bavaria. *E*. The same species in another but neighboring lake (Feldsee), showing feeble twilight upward movement in the evening followed by pronounced nocturnal sinking and a rise in the morning, suggesting an intermediate condition between *D* and the reversed migration of, for instance, Fig. 188. Scale in *A*, *C*, *D*, and *E* indicated as numbers per liter, in *B* as percentage of total. Sun- and moonrise indicated by appropriate symbols in *A*, *D*, and *E*.

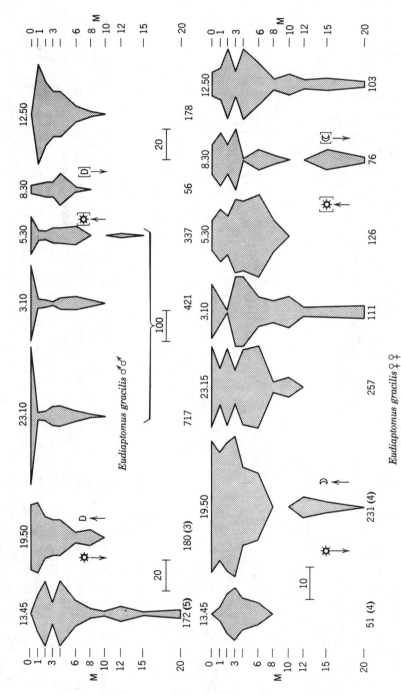

Eudiaptomus gracilis ♂♂

Eudiaptomus gracilis ♀♀

738

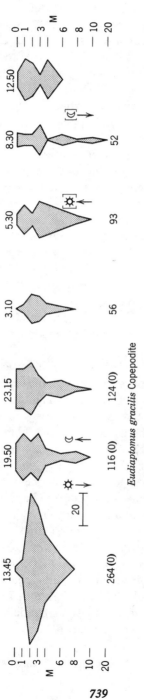

Eudiaptomus gracilis Copepodite

FIGURE 187. Vertical migration of ♂, ♀ and copepodite stages of *Eudiaptomus gracilis* in the Schluchsee, August 27–28, 1953. The males show a striking nocturnal migration with little descent during the night; the females and perhaps the copepodites exhibit a slow irregular descent after 20 hr. with suggestions of upward twilight movement in the morning. (Schröder, rearranged.)

was about 5:30 P.M. as the sun was sinking behind the mountains
to the west of the lake; the population then moved away from the
surface so that at astronomical sunset at 6:30 P.M. almost no specimens
were present.   As the light failed rapidly, the species reappeared at
the surface of the lake, only to decline again after a transitory maximum
at 7:30 P.M.   Throughout the night the numbers remained low, but
at 4:30 P.M. half an hour before sunrise, which is not obscured by
mountains, many individuals again appeared in the surface plankton
Copepod nauplii behaved in the same sort of way, though they rose
in the late afternoon earlier than *B. dietersi*.   It is evident that the
rate of change of illumination in either direction constituted the effective
stimulus to upward movement.   In the same lake, however, *Cyclops
strenuus* (s. lat.) showed two surface maxima at 6:30 P.M. and 4 A.M.;
Polyphemus pediculus also exhibited a very small maximum at
5:30 P.M. and a much greater one at 5:30 A.M.   *Acanthodiaptomus
pacificus yamanacensis*[5] and *Diaphanosoma brachyurum* exhibited typical if slightly irregular nocturnal patterns; the maximum of *A. p.
yamanacensis* at the surface came between 8:30 and 10:30 P.M. and
of *D. brachyurum* between midnight and 3 A.M.   These observations
emphasize that even if there is one fundamental pattern of behavior,
its modifications under the same physical conditions in different species
can be very great.

Very rarely planktonic animals have been observed to exhibit reverse
migration as if they were phototrophic plants, rising during the day
and sinking at night.   The limited movement downward from about
2 to about 4 m., between afternoon and early morning, recorded by
Thienemann (1919) for the rotifers of the Ulmener Maar on the
night of August 9–10, 1913, and the rather more striking descent
of *Daphnia longispina* and *Eudiaptomus graciloides* at the same
time seem to imply reversed migration, though they could in part
be the result of convectional streaming.   The sharpness of all the
maxima, whether in the afternoon or early morning, however, suggests
that this is not the whole story.   Ruttner's (1943) record of *Hexarthra
intermedia* moving downward in the evening in Lake Toba, *Ceriodaphnia dubia, Mesocyclops tobae,* and *Tropodiaptomus doriai* all moving
up, and *Conochiloides dossuarius* showing little change provides a better
case; *H. intermedia* apparently did not migrate in Lake Ranau.

Tropodiaptomus banforanus, (Fig. 188) studied in Lake Rudolf in
East Africa by Worthington and Ricardo (1936), provides an excellent

---

[5] Referred to *A. pacificus* but in later papers (Kikuchi 1937, 1938) regarded as
A. pacificus var. *yamanacensis*.

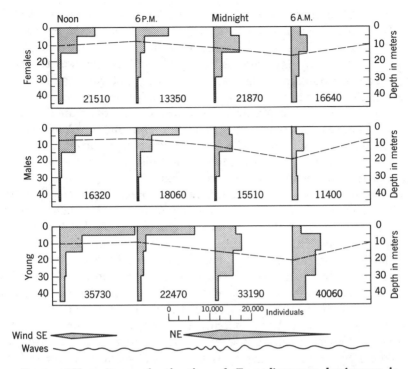

FIGURE 188. Reversed migration of *Tropodiaptomus banforanus* in Lake Rudolf. Dotted line inverted position of average individual. (Worthington and Ricardo.)

case of reversed migration. Adults of both sexes, as well as copepodite stages, were observed to be strongly concentrated in the top 5 m. between noon and 6 P.M. and to descent into deeper water, with the maximal population between 5 and 15 m. during the night.

Bayly (1962a) has recorded reversed migration by the males but not by the females of *Boeckella propinqua* in Lake Aroarotamahine, Mayor Island, New Zealand, at the height of summer. In the same lake the phenomenon was also exhibited by *Daphnia* (*Ctenodaphnia*) *carinata,* both when the lake was strongly stratified in January and when it was isothermal in May. There is evidence, however (Jolly 1952; Bayly 1962a), that elsewhere in New Zealand this species performs an ordinary nocturnal migration, as it does in Lake Ranau in Sumatra (Ruttner 1943). Worthington (1931) found that the allied *D.* (*C.*) *lumholtzi* showed some reversed migration in Lake Victoria. Bayly (1963) suggests that the phenomenon may be characteristic of alkaline waters, of *p*H 8.2 or more.

Reversed migration is recorded by Maloney and Tressler (1942) for *Diacyclops bicuspidatus*[6] in Caroga Lake, New York, on August 7 and 8, 1934. The main population of both adults and copepodite stages lay at 3 m. by day, whereas by night there was a marked tendency to random distribution. This case is probably not comparable to the others. Pennak (1944) found reversed migration of *Holopedium gibberum,* as well as of *Keratella cochlearis* and *K. quadrata,* in Silver Lake, Colorado. At the same time *Bosmina* and *Cyclops* showed typical nocturnal migration with some descent at midnight. Weather conditions precluded a full study. It is unlikely that the *p*H effect postulated by Bayly was operating in this case.

A curious case has been recorded by Southern and Gardiner (1926b, 1932) in their studies of the crustacean plankton of Lough Derg in Ireland. In this lake the immature specimens of *Daphnia longispina* migrate normally, rising to the surface after sunset and remaining there till shortly before dawn. The adult population, however, behaves in an entirely different manner. The population is concentrated in the top few meters of the lake at midday and appears to descend in the early evening (Fig. 189). On some nights the nocturnal population would seem to be rather randomly distributed throughout the lake and on the others the descent seems to go on from sunset to dawn.

Kikuchi has recorded that in Lake Hiruga, which has received its water and fauna from the ocean, *Acartia clausi* and *Oithona nana* rise to the surface from the early afternoon onward and start to descend shortly after sunset. This case seems intermediate between reversed and twilight migration. Reverse migration has occasionally been reported in the open sea.

Kikuchi (1937) has recorded a slight upward movement of *Bosminopsis dietersi* and *Polyphemus pediculus* at midday in Lake Kizaki, movements apparently superimposed on twilight migration and probably attributable to light adaptation. This case is comparable to but less conspicuous than the double cycle recorded by Schröder (1959) in the migration of *Daphnia longispina* and *Bosmina coregoni* in the Titisee early in August, in which the maxima lay at the surface at 8 P.M.; a downward movement observed until 4:30 A.M. was followed by a movement upward of part of the population which terminated in a midday maximum at 1 m. depth in the case of *D. longispina* and at the surface in that of *B. coregoni.* The rather large differences in absolute numbers at the different sampling times suggest that lateral variations in numbers may have contributed some irregularity in this

---

[6] Doubtless the American planktonic race *D. bicuspidatus thomasi.*

case, which therefore is not quite so diagrammatically clear as could be wished.

Variation with age. Differences in the behavior of immature and adult animals have been freely recorded; some examples have already been given. In the Lake of Lucerne (Fig. 190) Worthington found that adults of *Daphnia longispina* live by day at somewhat higher levels than the young. Since a considerable proportion of all ages reaches the surface layer at night, the daily excursion performed by the young animals is considerably longer than that performed by the older individuals. The average immature *D. longispina* in this lake travels from 60 m. to 5 m. during the early part of the night, whereas the average adult moves only from 40 m. to 10 m.

An even more striking variation occurred in the *Bosmina coregoni* population in the same lake. The adult members of the population remain stationary, with the modal distribution between 10 and 20 m. by both day and night. The young, however, undergo extensive migration, the average individual descending to 55 m. by day and ascending to 7 m. at night. Worthington found much less variation in the behavior of the two copopods studied both as adults and copepodite stages than was exhibited by the Cladocera.

A quite different picture is presented by Lough Derg. Here the adult specimens of *D. longispina,* as has already been indicated, exhibit reverse migration, whereas the young concentrate at the surface during the night, the proportion at the surface changing little from midnight to dawn. In the same lake there is little difference in the migration of the copepodite and adult *Eudiaptomus gracilis* which rise throughout the night. The young specimens of *Cyclops strenuus* (s. lat.) concentrate at the surface by night but seem to be randomly distributed by day.

In some diaptomids, notably *Eudiaptomus gracilis,* studied by Elster (1954) in Lake Constance, the nauplius appears to be confined to the surface waters, in this case the top 5 m., which may lead, with winds blowing the surface water toward the effluent, to considerable surface losses of nauplii.

In *Acanthodiaptomus pacificus yamanacensis* in Lake Kizaki the nauplius performs twilight migration, exhibited in the numerous samples taken at the surface and also demonstrated by the more infrequent vertical series, which indicate that the copepodite and adult stages undergo typical nocturnal migration.

Langford (1938) found that in *Skistodiaptomus oregonensis* in Lake Nipissing the large copepodite population was mainly hypolimnetic and stationary, and the smaller adult population in the epilimnion frequently but not invariably exhibited some nocturnal migration.

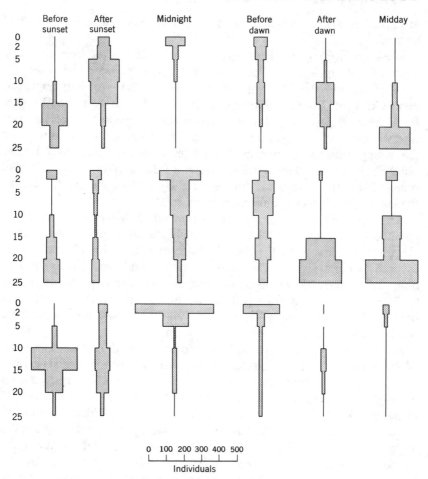

FIGURE 189. Left-hand panels, vertical distribution of young *Daphnia longispina* on three consecutive days in the plankton of Lough Derg, showing fairly typical nocturnal migration; right-hand panels, the same for large adult specimens, showing a striking tendency to reversed migration. (Southern and Gardiner.)

Variation with sex. Southern and Gardiner (1926b, 1932) noted that in *Cyclops strenuus* (s. lat.) in Lough Derg the males remained at the bottom and showed very little of the migration characteristic of the copepodite stages and the adult females. Kikuchi found no difference in the behavior of the sexes in Lake Noziri, in which an animal referred to *C. strenuus* had a marked maximum at the surface in the early evening.

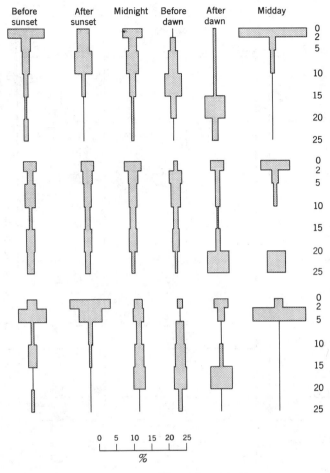

FIGURE 189. (*Continued*)

In the Lunzer Untersee Siebeck (1960) found that females of *Cyclops tatricus* show a moderate twilight migration, being most abundant at the surface at 4 A.M. and 6 P.M. and having a fairly well defined maximum at 4 to 6 m. between 8 A.M. and noon (Fig. 191). The males are almost nonmigratory; the maximal population occurs perennially between 6 and 10 m., but a few more casual specimens occur by night at the surface than by day.

A striking accumulation of males but not of females or juveniles of *Eudiaptomus gracilis* at the surfaces of some lakes for a short time in the evening or at night has already been noted. It is not

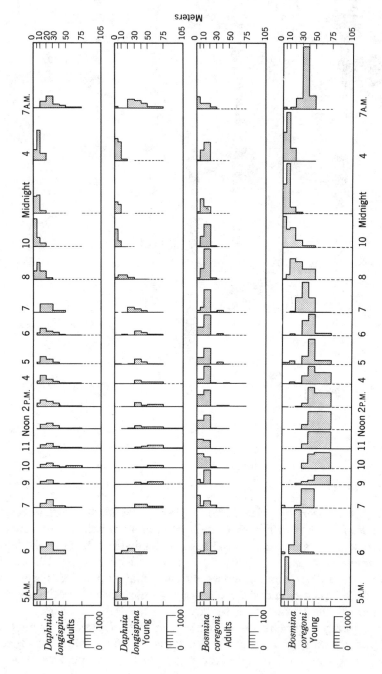

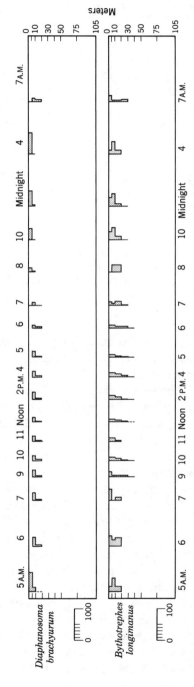

FIGURE 190. Vertical movement of certain Cladocera in Lake of Lucerne showing greater amplitude of movement by young than by adult specimens, which in the case of *Bosmina coregoni* are essentially nonmigratory. (Worthington.)

747

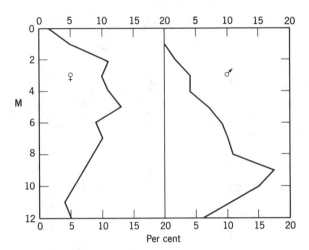

FIGURE 191. Position if the two sexes of *Cyclops tatricus* in the Lunzer Untersee at time of greatest descent, expressed as a percentage of the total (1197 ♂ ♂, 2856 ♀ ♀) catch. (Siebeck, redrawn.)

unlikely that whenever a close enough study of a copepod is made some significant difference in the vertical migratory behavior of the two sexes is likely to be disclosed.

A most complicated pattern is recorded for *Leptodiaptomus minutus* in Lake Nipissing by Langford. In general, the population is concentrated at about 3 m. by day. As the surface illumination fell from about 0.25 gm.-cal. cm.$^{-2}$ min.$^{-1}$ (10,000 foot-candles) to about one fifth that amount, a large proportion of the males broke away from the population and descended to about 8 m., leaving a great excess of females nearer the surface, toward which they apparently were moving. Between 6 and 8 P.M. there was a marked difference in the distribution of the sexes, the females being above the males. During the next few hours of the night there appeared to be an upward movement of males and a downward movement of females, the sexes changing places, so that from 10:30 P.M. to 4:00 A.M. the number of males at the surface greatly exceeded the number of females. In the early morning the males tend to sink and the females to rise; by 10 A.M. both sexes showed a maximum at 5 m. On the second day of the investigation there was a tendency for the males to begin their downward descent earlier in the afternoon than on the first day.

Special behavior of part of the population. On some occasions relatively large parts of the main population, not differing in sex or age, have been observed to break away from the main population and to behave in an idiosyncratic manner. A striking example (Fig. 192) is given by Worthington (1931) in his study of Lake Victoria in central Africa. Worthington found in a two-day study of the lake that *Daphnia longispina* (s. lat.), *Thermodiaptomus gabeloides* and, to a lesser extent, some of the other Crustacea showed a marked nocturnal migration, with a maximum in the top 16.5 m. at 9 P.M. Between this time and midnight, particularly on the first night of the investigation, a considerable part of the population broke away from the surface and descend below 50 m., only to rise again by 3 A.M. In *D. longispina,* which showed the phenomenon most strikingly, about half the adults behaved in this way on the night of September 22, 1927; the other half of the population remained in the upper zone throughout the night.

Other observers have also noted the sporadic breaking away of parts of the population to form groups that behave in a peculiar manner. On one day of observation Southern and Gardiner (1926b, 1932) found that part of the adult *D. longispina* population in Lough Derg descended to the bottom at midday, instead of performing the reverse migration characteristic of the rest of the population on that day and the whole of the adult population on the other days of their study.

Relationship of migration to underwater illumination and other environmental variables. Some information is available regarding the relationship of the movement to the diurnal changes in light intensity in the water. Ullyott found that in April the modal depth of the population of *Cyclops strenuus* (s. lat.) in Windermere followed approximately the isophot, for light of wavelength 4400 to 5200 A, 2.3×10^{-4} cal. cm.$^{-2}$ sec.$^{-1}$ (9600 ergs cm.$^{-2}$ sec.$^{-1}$), whereas in June the modal depth lay somewhat below the isophot 7.3×10^{-6} cal. cm.$^{-2}$ sec.$^{-1}$ (305 ergs cm.$^{-2}$ sec.$^{-1}$) for the same wavelengths. The great difference in sensitivity is not explained. The distribution over a diurnal cycle is better correlated with the short wave than with the total illumination.

In more recent work Siebeck (1960) has found less clear evidence of a regular descent and ascent with the isophots. In the Lunzer Untersee the distribution of the plankton of the top 4 m. descended significantly between 7:30 and 9:30 A.M., but not certainly later in the morning. When movement started upward again from the depths, which had been rather stably occupied during the later hours of the morning, the illumination was decreasing less rapidly than it was increasing during the descent between 7:30 and 9:30 A.M. The upward

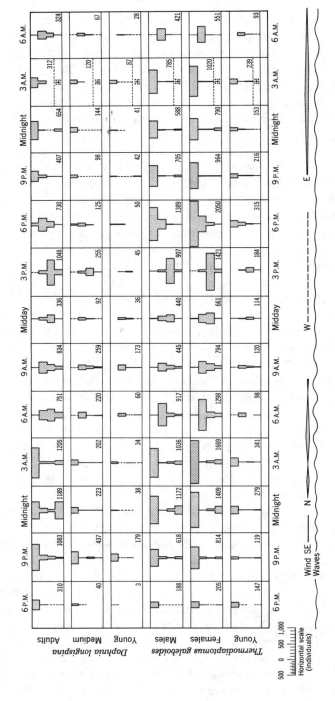

FIGURE 192. Vertical movement of *Daphnia longispina* (s. lat.) and *Thermodiaptomus gabeloides* in Lake Victoria September 22–24, 1927, showing part of the population breaking away to undergo nocturnal descent and then returning to the major part remaining at the surface. (Worthington.)

750

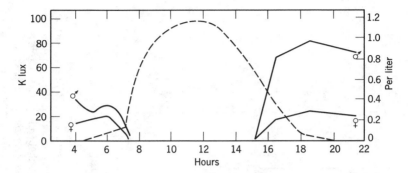

FIGURE 193. Diurnal variation of illumination at the surface of the Lunzer Untersee, September 6, 1958 (broken line), and variation in surface population of ♂ and ♀ *Eudiaptomus gracilis*, showing the departure of the dark adapted animals in the morning after a slight initial twilight rise and a return in the evening after some light adaptation at a time when the surface light intensity is greater than at the time of morning descent. (Siebeck, modified.)

movement in the afternoon may therefore take a species such as *Eudiaptomus gracilis* back into the upper layers at a time when the light intensity there is two or three times the value that the animal experienced when it left these layers earlier in the day (Fig. 193).

Less rigorously quantitative studies, however, suggest a fairly definite range of values for the intensity at the level of the maximum population of a given species at the time when this population is at its deepest and relatively stationary during the late morning. There is some fairly definite species specificity involved, for *Eudiaptomus gracilis* lies above *Daphnia longispina* or *Bosmina c. longispina,* whatever the absolute depths. On the whole the illumination measured in the region of the maximum, when the latter lies deep with a high summer sun, is for a given species about the same as that when it lies less deep with a lower autumnal sun, was clearly shown for *E. gracilis* and *D. longispina* on June 16 and October 1, 1958. The relative differences in the blue and red illumination between the two dates depend, of course, on changes in the optical properties of the lake water. When, as on October 8, 1958, there is a cloud cover, which implies a greater proportion of scattered light, the maxima not only lie higher but are less definite (Fig. 194).

Effect of the moon. Both Waldvogel (1900) who studied the Lützelsee and Lozeron (1902), the Lake of Zurich, found that the zooplankton tends to become concentrated in a thin superficial layer

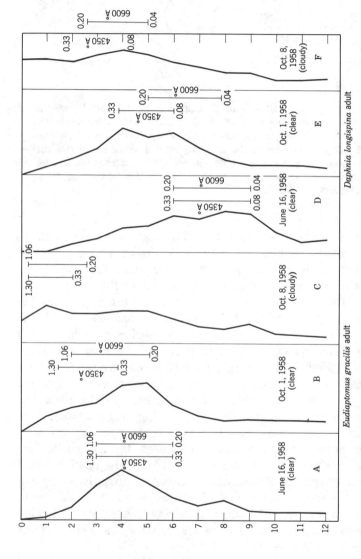

FIGURE 194. *A*. Distribution of *Eudiaptomus gracilis* in the top 12 m. of the Lunzer Untersee (as percentage of total catch) at time of deepest descent (10 to 12.30 hr.). The vertical lines embrace the region containing half the population; the figures at the top and bottom give the light intensity in millecalories cm.⁻² min.⁻¹ for a brand 50Å wide centered about the blue (4350 Å) and red (6600 Å) part of the spectrum. *B*. Distribution of the same October 1, vertical lines now indicating the same light intensity ranges as in *A*. *C*. The same October 8, under cloudy sky. *D.-E.-F*. The same for *Daphnia longispina*, parthenogenetic adult females. (Redrawn from Siebeck.)

when the moon is shining. Lozeron noted that *Leptodora kindtii Bythotrephes longimanus, Daphnia hyalina, Bosmina coregoni, Eudiaptomus gracilis,* and *Cyclops strenuus* (s. lat.) all behave in this way, collecting in the top 30 cm. when the moon is shining and scattering downward when her face is obscured by a cloud. The upward movement in moonlight is, of course, analogous to the ascent just before dawn in twilight migration.

Nonoptical effects of weather. France (1894) noted in Lake Balaton that the population tended to avoid the surface at night during a storm. Whether this was a passive result of great turbulent mixing or because of an active reaction to the enhanced turbulence of the water, requires investigation, as indeed does the reality of the phenomenon itself.

Seasonal variation. Apart from the cases to be discussed, in which the depletion of the oxygen in the lower hypolimnion influences the vertical distribution of zooplankton, particularly cold stenothermal forms confined to that region, there is evidence that seasonal change, presumably operating through temperature, can influence the vertical migration pattern. Unfortunately, few investigations have studied vertical movement repeatedly throughout a season. The most convincing case is reported by Langford (1938) for *Epischura lacustris* in Lake Nipissing (Fig. 195). In late June 1935 this copepod exhibited no vertical movement, for the population extended from the surface to a maximum at 1 to 3 m. and then slowly declined to zero at 30 to 35 m. Almost all of the population was living at temperatures of 16° or 18° to 12° C. In early July the distribution was similar but had a hint of a little reversed migration. In July and August, when a warm epilimnion at 20 to 24° occupied

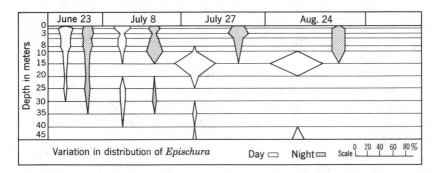

FIGURE 195. Seasonal variation in the vertical migration of *Epischura lacustris* in Lake Nipissing. (Langford.)

the top 15 to 20 m. of the lake, a marked migration developed; the species left the top 10 m. by day and congregated at the bottom of the epilimnion, with a maximum at 15 m., to rise at night, when no specimens were left at the 15 m. level.

Behavior under ice and snow. Ruttner (1909), working at Lunz, made an important study of the distribution and movement of the zooplankton under varying thicknesses of ice and snow. Two rather extensive series of observations of the distribution of the zooplankton down to 32 m. are given for midday, January 20, 1909, under about 20 cm. of ice, and for 3 P.M. February 13, 1907, under about 25 cm. of ice covered with 9 cm. of snow. Three species of Crustacea and one of Rotatoria occurred with sufficient frequency in both samples to justify comparison of the behavior under conditions of fair illumination and of fairly persistant darkness.

Under ice the adults *Eudiaptomus gracilis* were largely concentrated in a zone from 1 to 5 m. and were almost absent in the surface sample. Under snow the distribution was far more uniform, and although the greatest number of specimens were found at 1 m. the numbers present at the different levels vary little and may well represent a random distribution. The nauplii of *Eudiaptomus,* present in significant numbers only on February 15, 1907, show a less uniform distribution than the adults and tend to be concentrated between 3 and 10 m. On certain other occasions, when the distribution of *E. gracilis* in the top 3 m. was studied under snow-covered ice, there appeared to be a slight concentration of the adults immediately under the ice.

Cyclops strenuus (s. lat.) was found irregularly on both occasions down to 20 m. and much more abundantly at 32 m. No nauplii occurred in the top 3 m. on either date.

Bosmina c. longispina was absent at the surface and less abundant at 20 and 32 m. than in the intermediate water on both occasions, but the population under the snow appeared to lie at a somewhat higher level than under uncovered ice.

Polyarthra sp. (sub *platyptera*) showed a marked maximum at 1 m. under uncovered ice and at the surface under snow-covered ice.

None of the few other species present under snow-covered ice shows any evidence of the rather uniform vertical distribution exhibited by *E. gracilis.*

Ruttner found that *E. gracilis* underwent ordinary nocturnal migration in the ice-covered lake. On February 14, 1908, a sunny day without snow, the maximal population density for this species lay at 5 m. at 11 A.M., but by 9 P.M. the maximal population had risen to 1 m.

Though at 9 P.M. the copepod must have experienced darkness for several hours at least as great as under 9 cm. of snow over 25 cm. of ice, there was no trace of a random distribution; the population evidently moved systematically upward in the evening. It is most unfortunate that no information exists to indicate how rapidly the random or uniform type of distribution is established.

When snow is cleared from a limited area of frozen lake surface to produce a window, the zooplankton tends to move downward in the region of the lake illuminated by it. In the evening *Eudiaptomus gracilis* moved up again to give a distribution under the window at 7 P.M. essentially like the day distribution before the snow had been cleared off. The upward evening movement of the rotifers that had descended was much less marked. *Bosmina c. longispina,* however, became greatly concentrated at the surface below the cleared area.

It is evident from these observations that neither high summer temperatures nor a marked temperature gradient are needed to produce diurnal migration. There is some evidence of positive phototaxis to weak light which presumably concentrates *E. gracilis* just under the ice when there is moderate snow cover and brings great numbers of *Bosmina* up under an artificial clearing as the light fails in the evening. The evidence for a tendency to random distribution in enduring darkness is fairly striking in the adult *E. gracilis* but not in those of the other organisms studied.

Velocity of migratory movement. Worthington (1931), whose data for the Crustacea of the Lake of Lucerne are doubtless the best for a really long vertical path in fresh waters, calculated the mean speed of ascent and descent of these animals during their migratory excursions. His results recalculated as velocities rather than as times taken to move 1 m. are given in Table 43.

The swimming rates of a number of marine organisms were recorded by Hardy and Bainbridge (1954) in their plankton wheel. The most rapid ascent was performed by *Meganyctiphanes norvegica* which could maintain an average upward movement of 2.6 cm. sec.$^{-1}$ for a period of an hour and could do 4.8 cm. sec.$^{-1}$ over short periods of 2 minutes duration. *Centropages* sp., which is nearer in size to the animals of Table 43, could maintain 0.85 cm. sec.$^{-1}$ for an hour upward or 1.5 cm. sec.$^{-1}$ during short spurts of 2 minutes duration. Other copepods and larval Crustacea all swam up more slowly. Kikuchi (1938) found *Acanthodiaptomus pacificus yamanacensis* capable of upward speeds of 1.33 cm. sec.$^{-1}$ in the laboratory. Cushing (1955), however, noted in his experiments with a vertically submersible tube 3.5 m. long that both a diaptomid, apparently *E. gracilis,* and a speci-

TABLE 43. *Velocities of vertical movement of migrating plankton in the Lake of Lucerne*

Organism	Length mm.	Mean Depth Range m.	Velocity of Descent cm. sec.$^{-1}$	Velocity of Ascent	
				Before Dusk cm. sec.$^{-1}$	After Dusk cm. sec.$^{-1}$
Daphnia longispina adult	c. 2.2	10–40	0.33	0.067	0.27
Daphnia longispina juv.	1.0–1.5	5–60	0.54	0.096	0.39
Bosmina coregoni adult	c. 1.1	15–15	no movement		
Bosmina coregoni juv.	0.6–0.9	7–55	0.37	0.064	0.22
Diaphanosoma brachyurum	0.6–1.4	3–10	0.17	0.00	0.14
Eudiaptomus gracilis adult	1.1–1.2	13–30	0.14	0.028	0.11
Eudiaptomus gracilis juv.	0.7–1.1	20–37	0.083	0.028	0.14
Mixodiaptomus laciniatus adult	1.5–1.8	25–75(?)	0.36	0.12(?)	0.21
Mixodiaptomus lacinatus juv.	1.1–1.6	12–46	0.38	0.060	0.21
Cyclops strenuus juv.	1.1–1.6	12–46	0.24	0.028	0.30
Mesocyclops leuckarti adult	1.0	3–15	0.083	0.016	0.14

men of *Daphnia,* apparently *D. hyalina,* could "climb the full length of the long tube in 1 min., i.e., 3.5 m./min." This speed of 5.8 cm. sec.$^{-1}$ appears out of line with any others recorded. For *E. gracilis* in Windermere, Ullyott (1939) found a rate of ascent before dusk of 0.07 cm. sec.$^{-1}$ in nature.

Hardy and Bainbridge give a number of determinations of downward movement. They were convinced that in general these represented true downward swimming. They were not consistently greater than the rates of ascent for the same animals.

The rates of descent calculated by Worthington and given in the table are not very different from the velocities of sinking of *Daphnia* given by Eyden (1923), Bowkiewicz (1929), Brooks and Hutchinson (1950), and Hantschmann (1961). Brooks and Hutchinson give sinking speeds for *Daphnia galeata mendotae* and *D. dubia,* 1 to 2 mm. long, of the order of 0.1 to 0.3 cm. sec.⁻¹, and Hantschmann, for *D. schødleri,* over about the same size range, speeds of descent of 0.06 to 0.45 cm. sec.⁻¹. The speed of descent given by Worthington for the sinking of the adult *Daphnia,* just over 2 mm. long without the tail spine, could therefore be due to passive sinking. If this is granted for the adults, it is certain that the descent of the immature specimens must involve active downward swimming, for such specimens would fall passively less rapidly than the mature stages because they are smaller (see page 279). If it is supposed that the young are sinking passively and that the adults are trying to maintain their position in the water in spite of an incomplete inhibition of movement, the velocity of the descent of the young must imply a much greater excess density than that of the animals studied by Brooks and Hutchinson or by Hanschmann. The discrepancy is accentuated by the fact that in nature the animals are falling through a turbulent medium; it is indeed likely that the turbulence of the epilimnion implies active descent by *Daphnia* of all ages and species. In *Bosmina* the adults must work to maintain their position by day as by night, but here again the rate of descent of the small (< 1 mm.) immature specimens would seem to imply active downward swimming.

In contrast to the behavior of the two Cladocera, of which both immature and adult stages were studied by Worthington, the behavior of *Eudiaptomus gracilis* is by no means inconsistent with the hypothesis that the descent is passive. The adults of *Mixodiaptomus laciniatus,* however, which are about the same length as *D. longispina* and fall at about the same speed, descend slightly less rapidly than the immature specimens, so that here it is almost certain that the young are swimming downward or the adults are attempting feebly to swim upward during the period of descent. It is to be noted that although the range of the migration of *M. laciniatus* is much greater than that of *E. gracilis* the general relation of the movements of both species to the onset of night and of dawn is essentially the same and gives no hint of great qualitative differences in the determination of the behavior.

LABORATORY STUDIES OF VERTICAL MOVEMENT

The earliest workers interpreted the movement solely in adaptive terms. Forel (1877) considered that the animals descended by day to avoid onshore winds, whereas Weismann (1877b), with greater perspicacity, believed the crustacean eye to be adapted to low light intensities and that in consequence the migrating plankton were seeking those regions in which the illumination was most suitable for their vision. This view has often been revived and in somewhat less explicit terms is essentially that later held by Russell (1927). Other investigators have restated the concept of the optimal light intensity in more mechanistic terms, but, as is so often the case, a Weismannian concept has been the basis of a whole series of ideas in more modern biology, though the original author of the idea might not approve of its modern form. In the translation of these early views into more mechanistic concepts the possible adaptive significance of the migratory behavior has been largely forgotten. Such a change in emphasis is in accord with the general development of biological science since the third quarter of the nineteenth century, but it would be foolish for the modern worker to disregard the interest of both causal and adaptive interpretation. The various causal hypotheses will be considered first and an attempt will be made to evaluate them. It will then be possible to consider briefly the adaptive meaning, if any, of the movement.

Early work on reactions to light. Since the time of Trembley (1744), who observed that the *polypes* living in a vessel of pond water were attracted by the light of the sun, it has been realized that reactions to light played a part in the lives of small aquatic organisms. Because the initial laboratory reaction is often a positive phototaxis,[7] which provides no clear explanation of the ordinary kinds of vertical migration, and because such phototaxes may undergo curious reversals, a simple explanation of the migration is by no means always forthcoming from the experimental data.

The first relatively complete theory was due to Groom and Loeb (1890), who observed that the newly hatched nauplii of *Balanus perforatus,* though initially positively phototactic, rapidly become negatively phototactic in the bright light, the rapidity of the reversal depending on the intensity of the illumination. A positive response to weak light is, however, retained. Intense vertical illumination should therefore cause a descent of the animals by day, but in the evening the population will ultimately find itself in a layer in which the illumination is so weak that it will evoke a positive response. As the light fails, such a positive response will be given at progressively higher levels and the animals will finally arrive at the surface. Whenever an adequate amount of light is present, the animals will occupy a zone that is sufficiently illuminated to prevent the upward response but not to permit the negative response. From a strictly physiological point of view there is a great difference in the concept of an optimal illumination expressed solely in terms of intensity, as in Weissmann's

---

[7] The terminology adopted by Fraenkel and Gunn (1940) is employed.

theory that the Crustacea sought a level appropriate to their visual requirements, and the concept of a neutral zone in which neither positive nor negative phototaxis occurs, for these are primarily responses to the direction of illumination though the nature of the response may be controlled by intensity. The concept of optimum illumination necessarily implies the idea of adaptation; the neutral zone of balanced phototactic response may be the mechanism by which this adaptation is achieved.

During the last decade of the nineteenth century and the first decade of the twentieth a good deal of work was done on the problem whether the reaction of the smaller crustacea was to illumination as such or solely to the direction of the incident light. Both views had been expressed by botanists working with motile spores and flagellates. Davenport and Cannon (1897) showed conclusively that *Daphnia,* when positively phototactic, could react to direction independent of intensity, and Yerkes (1899), obtained some evidence that *Simocephalus,* when reacting positively, could response to intensity as well as to direction. This particular problem, however, is of little importance in relation to the behavior of the zooplankton in a lake, for in nature the intensity is reasonably certain to decline along the path of the main component of the illumination.

More serious consideration must be given, to the allied problems of the significance of the diffuseness of light and of the possible difference between the reaction of an animal to two beams of the same intensity and divergence, one horizontal, the other vertical. This difference was emphasized particularly by Bauer (1908, 1909), who found that the benthic marine mysids might react markedly to horizontal illumination but not at all to a vertical beam. Bauer therefore concluded that the phototactic theory of vertical migration was erroneous. There is no doubt that the earlier workers were too prone to assume, as Loeb (1908) did on the basis of experiments with unidentified copepods and cladocerans, that if a planktonic organism reacts in a particular way to a horizontal beam it will react in the same way to a vertical beam. In spite of such criticism, it appears that the original theory of Groom and Loeb is in essence the ancestor of the most satisfactory modern theories of vertical migration.

Effect of external factors on the phototactic sign. Loeb (1893, 1904, 1906, 1908) soon found that factors other than a change in the intensity of the incident light could change the sign of the phototactic response. In some copepods shaking was found to produce a transient positive phototaxis. Towle (1900) observed that the mechanical disturbance of sucking *Cypridopsis* into a pipette could produce a like effect.   Yerkes (1900) noted the opposite action when *Daphnia* were disturbed by pipetting, when positively phototactic animals became negative.

In general, Loeb and most subsequent workers have found that raising the temperature accentuates negative and reverses positive phototaxis, whereas lowering the temperature has the reverse effect. Adding carbon dioxide to the medium also produces a marked reversal of negative phototaxis, this often being the easiest way to obtain strongly positive specimens of *Daphnia*.   Loeb therefore concluded that in nature the warming of the surface waters and the photosynthetic removal of CO_2 by day would, like increasing illumination, accentuate negative responses to light, whereas in the evening falling temperature and increasing CO_2 content, like the falling illumination, would promote a positive response.   Unfortunately, it is most unlikely that the thermal and chemical changes are ever great enough in the open water of a lake or the ocean to act effectively in this way.   The phenomena observed in Loeb's experiments with temperature, however, may well underlie seasonal variations in migration, such as those of *Epischura lacustris* in Lake Nipissing.

Negative photokinesis in a gravitational field.   Ewald (1912) found that the nauplii of *Balanus* though positively phototactic to a horizontal or vertical beam, could be made to sink when illuminated obliquely.   This action was apparently due to an inhibition of movement by the oblique illumination.   He also interpreted certain earlier (1910) experiments on other Crustacea in terms of the same sort of inhibition. Schallek (1942, 1943) made the important discovery that the marine copepod *Acartia tonsa,* though positively phototactic to direct vertical illumination, sinks passively when, after a period of darkness, it is illuminated by diffuse light.   By introducing a photometer, Schallek was able to measure the intensity of the illumination from different directions, set up by an oblique beam refracted and reflected from the walls of the experimental vessel.   He found that the degree of diffuseness necessary to inhibit the swimming movements of *Acartia* was no greater than that always existing by day in nature.   The response to a vertical, more or less parallel, beam of light in the laboratory may therefore be totally irrelevant to the behavior of the organism in nature.   It is reasonably certain that the oblique illumination employed by Ewald in his experiments with *Balanus* nauplii produced a diffuse illumination comparable to that employed by Schallek.   It is convenient to refer to the particular type of negative photokinesis produced in planktonic Crustacea by diffuse light as the Ewald-Schallek inhibition.

Spontaneous reversal of phototaxis. Ewald (1910) found a spontaneous rhythmical movement in *Daphnia, Ceriodaphnia, Leptodora,* and probably other Cladocera, in a light gradient, so that the animals, though perhaps reacting negatively to the light for longer periods than they react positively, undergo continuous reversal of phototactic sign and make repeated excursions to the top of tube illuminated from above or toward the illuminated end of a tube in a horizontal beam. The movement does not take place in a red light to which the animals were supposed to be insensitive. It can be accentuated by certain chemical conditions, notably low oxygen content. Other observers, notably Clarke (1930, 1932) and particularly Smith and Baylor (1953) have observed similar behavior.

Effect of illumination on geotaxis. A good many observers have claimed that planktonic animals tend to be negatively geotactic in darkness and positively geotactic in the light. It is evident that great care must be exercised in distinguishing such behavior from negative photokinesis or the Ewald-Schallek inhibition of movement in the light. In some cases it seems reasonably certain that true active geotaxes really occur. Fox (1925) obtained evidence of this in *Paramecium,* Esterly (1907) observed that *Macrocyclops albidus* is negatively geotactic in the dark and positively geotactic in light; though it reacts negatively to a strong horizontal beam, it moves downward toward the source of light if the light is directed upward through the transparent floor of the vessel. Comparable observations made by many investigators of the Cladocera are best described in connection with the interaction of phototactic and so-called geotactic responses in *Daphnia,* considered in the next few paragraphs. Harper (1907) found that larvae referred to *Corethra plumicornis*[8] developed a marked positive geotropism when illuminated from either above or below.

Laboratory studies of phototaxis and geotaxis in *Daphnia.* Because of the ease with which some of the species of *Daphnia* can be cultivated in the laboratory, much work has been done on their reactions to light, gravity, and other stimuli. It is particularly convenient to introduce the theories of vertical migration based on the interaction of such reactions in terms of the experiments on this genus. It is important, however, to remember that the forms most studied, such as *D. magna* and pond-living races referred to *D. pulex,* are doubtless among the least interesting members of the genus from the standpoint

---

[8] This is a synonym of *Chaoborus crystallinus;* it is not clear, however, that this species occurs in North America, and Harper may well have worked on some allied member of the genus.

of the study of migration in lakes, though it would be absurd to dismiss all of this work as irrelevent to the problem.

Reactions to light. Von Frisch and Kupelwieser (1913) and later Heberday (1949) found *Daphnia* to be red-blind and to react to much lower intensities of blue than of green light. Smith and Baylor likewise found clear evidence of responses at 5700 but not at 6800 A. It is not unlikely that the crustacean eye in general is far more sensitive to the short wave part than to the rest of the visible spectrum and that the light penetrating to the deepest parts of clear unstained lakes is likely to include the most effective wavelengths. As von Frisch and Kupelwieser (1913) noted, it is possible for animals to be negatively phototactic to blue and positively phototactic to longer wavelengths. Herberdey and Kupka (1942) concluded that at high light intensities three types of retinal cell are maximally sensitive to ultraviolet, blue-violet, and yellow, respectively. Scheffer, Robert, and Médioni (1958) conclude that the eye as a whole is most sensitive to 5200 to 5400 A, and nonocular receptors to 4000 to 4200 A. A specific negative phototaxis exhibited by *Daphnia* to ultraviolet of wavelength less than 3341 A (Moore 1912) is, as Clarke (1930) points out, unlikely to be of any importance in the interpretation of the behavior of the animal in nature.

Further physiological details may be found in the excellent summary of crustacean vision by Waterman (1961).

If the direction of the light incident on the eye of a *Daphnia* is changed, the position of the eye is altered to minimize the change in the angle of incidence on the eye (Radl 1903; Ewald 1910). The change in the position of the eye is followed by a change in the plane of beat of the antenna, so that the position of the body of the animal is adjusted in such a way that the eye maintains not only its original orientation relative to the incident beam but also its original position relative to the body. The effect of these changes normally is to maintain the animal oriented with its back to the light, the dorsal light reaction of von Buddenbrock (1915). Von Frisch and Kupelwieser (1913) found that increasing a vertical light source had an effect on posture so that the animal adjusted by swimming downward to about the previous intensity but decreasing the intensity produced the opposite change with upward movement.

If *Daphnia* or *Bythotrephes* are illuminated from both above and below, they show circus movements, first swimming away from the nearer light and turning round and then, with their backs to the initially farther light, from which they likewise move away, swimming round to the original position (Ewald 1910). Ewald also found that sometimes sudden changes in phototaxis could involve a rapid but quite transitory change in the orientation of the body relative to the light, but this was never observed by Clarke.

Unlike the first investigators of the genus, but in agreement with Ewald and with Dice (1914), Clarke (1930, 1932) found that both *Daphnia magna* and *D. pulex* are normally negatively phototactic to a parallel horizontal beam. He noted, however, that certain individuals appeared to behave spontaneously and persistently in a contrary manner, being primarily positively phototactic. Examination of the swimming position showed no difference in the orientation of the body when the negatively and positively phototactic animals were compared; both groups have their backs to the light, and the difference in the direction of movement apparently depends on the plane of beat of the antennae. Clarke (1932) later found that very young animals are almost always strongly positively phototactic, and as they mature the sign of the response changes in an irregular manner until the mature negative reaction is established. A sudden fall in temperature will cause a negative animal to become positive in a horizontal beam but will not affect a positive animal; a sudden rise in temperature will cause a positive animal to become negative. This is in accord with the considerable body of older work, some already mentioned, done mainly on marine species.

Even under quite constant conditions, striking reversals of phototactic sign, lasting for varying periods, were recorded by Clarke and are evidently identical with the rhythmic reversals observed by Ewald. Clarke felt, however, that in nearly every case it is possible in a sufficiently long experiment to designate the animal under observation as primarily positive or negative, with a more or less periodic reversal superimposed on this primary tendency.

Provided that the change was made rapidly enough, reduction in light intensity was found by Clarke to reverse negative phototaxis. Under ordinary conditions, the rate of change in intensity of illumination along a parallel horizontal beam in water is not great enough for a *Daphnia* swimming down the light gradient to experience this type of reversal.

Nature of the supposed geotaxis. When *Daphnia* is swimming, the beat of the second antennae ordinarily produces a movement that is upward and forward. At the same time, there is a tendency for the animal to sink passively. If the antennary beats are feeble and the rate of beating slow, the over-all direction of the animal will be obliquely downward; if they are strong and fast, it will be obliquely upward. Under steady-state conditions, when the animal is moving horizontally, it appears from the work of Grosser, Baylor, and Smith (1953) and of Hantschmann (1961) that the rate of the antennary beat is determined by the sinking speed, so that a feed-back system is achieved to permit steady horizontal swimming. The antennae of large animals under given conditions beat faster than those of small, but the relationship between rate of beat and sinking speed is not linear (Fig. 196) and possibly not quite monotonic, there being a hint of a minimum in the small size range and of a plateau in the large range. The efficiency of the beat as well as the rate clearly must be involved.

It is hard to immobilize active *Daphnia* in the free water without causing injury, but Hantschmann concluded that when suspended on minute hooks inserted into the brood pouch *Daphnia schødleri* either performed violent and irregular escape movements or remained quiescent. The downward fall between beats apparently is necessary to maintain the ordinary steady locomotor movement.

Grosser, Baylor, and Smith, and Hantschmann, found that animals placed in glucose solutions of sufficient strength exhibited swimming movements which, in relation to the normal, were inverted, as if reflected in the water surface. Between strokes the animal rose, and the beat brought the inverted animal down again so that a horizontal course could be maintained. These observations preclude the response to gravity in normal swimming being due to a statocyst. Both Grosser, Baylor, and Smith and Hantschmann also believe that the

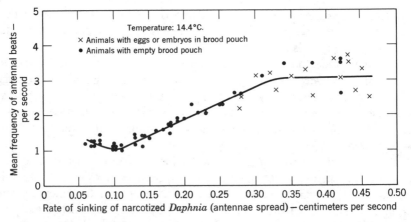

FIGURE 196. Relationship of mean frequency of antennal beat required to keep animal at a given horizontal position in the water and the rate of passive sinking in *Daphnia schødleri* of increasing size, at 14.4° C. (Hantschmann.)

receptor organ is on the antennae and that it is stimulated by the change of the position of the antenna during the free fall between strokes or presumably by any upward water current. Hantschmann suspected that a pair of very short basal bristles were involved. She also noted behavior on the part of an undetermined ostracod that suggested a comparable regulatory system which may therefore be general in the smaller Crustacea.

Interrelationship of responses to gravity and light. By its nature the type of response just discussed implies that any kinetic effect added to the regulated antennary beat will cause a movement in the opposite sense from passive sinking, or, in experimental situations, rising.

Dice (1914) studied the effect of light on the responses to gravity of *Daphnia pulex* (s. lat.) from Californian ponds. He concluded that the sinking induced by increasing illumination is a true positive geotaxis in which, when the effect is sufficiently marked, the head of the animal is actually pointed downward. Strong light from any direction could produce a downward movement which immediately gave place to an upward movement when the light was diminished. Dice believed that this change in geotactic sign with changing illumination is the most important type of behavior underlying vertical migration. This effect is clearly different from the passive sinking just described but possibly comparable to the behavior observed by Schröder (1959) when *D. longispina* was illuminated from below.

Clarke (1930) observed that a reduction in light intensity appeared to produce a negative geotaxis, again provided that the reduction proceeded fast enough. This response was transitory, particularly in *D. magna*, and Clarke seemed worried lest his experiments imply that in nature, as appeared to be the case in the laboratory, the animal, under the influence of a primary positive response to gravity and a negative phototaxis, would continue to sink indefinitely but never with a sufficient velocity to reverse the signs of these directed movements. It is reasonably certain that Dice's *D. pulex* (s. lat.) showed a much more stable positive response to gravity in light than did the *D. pulex* used in England and, a fortiori, the *D. magna* studied by Clarke.

The red and blue dances of Smith and Baylor. Smith and Baylor (1953) found that all the Cladocera belonging to the genera *Sida, Daphnia, Simocephalus, Ceriodaphnia, Moina, Bosmina, Kurzia, Chydorus,* and *Leptodora* react qualitatively in the same way to a change in wavelength. If the illumination is from above and is more or less monochromatic, when the wavelength is changed from less than 5000 to more than 5000 A, the animals start swimming upward into the light. This is in accord with observations of von Frisch and Kupelwieser (1913). When the reverse change is made, there is a downward movement after a period of brief lateral scattering. Though short-wave light acts as if it appeared brighter than long-wave of the same energy, this is not the full explanation, for going from yellow to white evokes a short-wave response, from blue to white, a long-wave response (Brown in Smith and Baylor 1953). The descent, when light of short-wave length is added horizontally to animals in a vertical beam, is due both to inhibition of swimming movements and a tendency toward active downswimming. In *D. magna* the greatest response occurs in passing from 4500 to 3500 A. *Moina affinis* is extremely sensitive to blue light, being almost completely inhibited by the intensities used. *Ceriodaphnia reticulata* responded more slowly than the other ten species studied.

When the animals are illuminated from above by short-wave light, the upswimming starts when a horizontal beam of wavelength 4700 A is added, but a maximal effect is not produced till 5700 A is reached. At 6800 A the effects disappear. No evidence for more than a two-color pattern of response was obtained.

When populations are exposed to a constant light source, they show one or the other of two types of behavior, dependent on the wavelength. Under long wavelengths they move backward and forward in the direction of the beam, under short wavelengths, at right angles to the beam. Smith and Baylor observed these contrasting responses in *Daphnia, Ceriodaphnia, Moina,* and *Bosmina* and term them the red dance and blue dance, respectively, though yellow dance for the backward and forward pattern would have been better. The dances appear to be discrete responses; under white light at any moment an individual is engaged in either a red or a blue dance, changing spontaneously from one to the other. The proportion of time spent in either activity is very sensitive to the relative energies of short and long components. With vertical light, dimming causes up swimming, whatever the wavelength. Brightening a blue source caused marked downward movement, but increasing the intensity of a yellow light produces very little descent. Any increase in illumination was found to cause agitation, but in steady illumination the movements were stronger at low intensities. After fifteen or twenty minutes the whole population lies higher in the tank in dim than in bright light, though the type of dance remains determined by wavelength. After long periods almost no difference in position due to intensity is observed. Since the animal has no way of measuring its depth, but can only be sensitive to the streaming produced by changes in depth, this result is not unexpected. The red dance to and from the source can be manipulated by slow changes of intensity, a falling intensity increasing the positive phototactic movement, a rising intensity the negative movement. By rhythmical changes in the light from a vertical source, the population can be made to move up and down synchronously. The blue dance completely disappears in a few seconds, when *D. magna* is cooled below 10° C.; no adaptation to this change occurs over a period of a month.

When *D. magna* is illuminated from above with a white or blue light that is slowly increased in intensity, the population moves down slowly, usually exhibiting a descending or almost horizontal blue dance which must involve a very slight reduction of the beat of the antenna, controlled by the animal falling through the water. If the bottom of the vessel is cooled, the animals entering the cold layer begin to move up and down in a red dance and no further downward movement takes place. A few individuals may move up out of the cooled region but immediately start a descending blue dance and return to the cold water. If a horizontal light is projected into the lower part of the cold water, the red dance is oriented to this horizontal beam.

There is evidence that as the intensity falls the proportion of short-wave energy required to produce a particular type of response in mixed illumination also falls.

The to and fro movements that earlier workers observed and interpreted as reversals of phototrophic sign are clearly the same as the dances of Smith and Baylor. It is evident that striking qualitative differences in response to horizontal and vertical beams played no part in their experimental results with Cladocera. This, of course, does not exclude such differences in other Crustacea such as the Mysidacea. The paralysis induced in *Moina* by blue light is clearly comparable to the Ewald-Schallek inhibition; that this response should vary from species to species is interesting but perhaps not unexpected. The effect of diffuse light of various wavelengths requires further study.

Experimental models simulating migration in Daphnia. Harris and Wolfe (1955) studied the movements of *Daphnia magna* in water, the transparency of which was decreased by the addition of Indian ink. By this method it is possible, in a small tank 60 cm. high, to produce a gradient in light intensity in which the intensity at the bottom of the tank is 10^{-3} to 10^{-5} that of the surface value, so imitating over a very short vertical range the differences observed over great depth ranges in nature (Fig. 197).

In total darkness the activity of the animals was much inhibited and they remained at the bottom. If the tank was then illuminated by a slowly increasing parallel beam directed vertically downward, the population started to move up to the surface and accumulated there

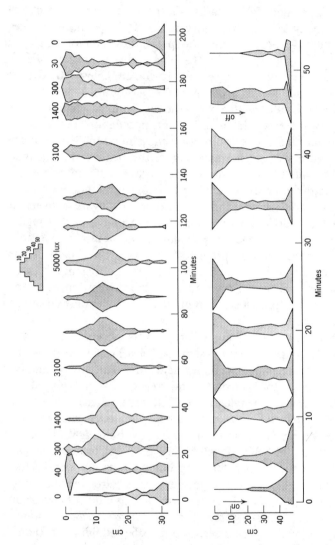

FIGURE 197. Upper panel, experimental migration of intact *D. magna* in artificially turbid water, starting from a condition comparable to nocturnal sinking followed by a twilight rise as the light increases slightly above the vessel and a typical downward descent in bright light with reverses as the light decreases, producing upward twilight migration followed by nocturnal sinking. Lower panel, photokinetic effect of feeble horizontal illumination turned on at the beginning and off 45 minutes later. (Harris and Wolfe.)

when the surface intensity was about 40 lux. Further increase in illumination caused an active descent, in which animals occupied a zone in the tank between the isophots for about 300 and 100 lux. The illumination was allowed to reach a maximum and then to decline. During the decline the animals rose to the surface and then descended as the light became more feeble and was reduced to zero. This pattern of movement could be reproduced with the total experimental cycle as short as two hours or as long as thirteen hours. The experimental migration induced is in fact a twilight migration with marked descent at the time of absolute darkness.

The zone of maximal population when the light is on, lying between 100 and 300 lux, had its center at 15 cm. in one of the experiments; the intensity at this depth in the tank might correspond to 25 m. in Windermere or 100 to 200 m. in the open ocean. The ratio of the maximum to minimum intensities within which 90 per cent of the population lay is of the order of 30:1 and reproduces the considerable scatter often found in nature.

Harris and Wolfe were able to show that the initial "dawn" rise in their experiments, which occurs at very low intensities of 0.8 to 40 lux at the surface, is a purely photokinetic effect, acting on animals that are set in the water with the antennae spread and the body vertical in a position characteristic of animals in very feeble light. The photokinesis is easily inhibited by increasing light intensities, but the susceptibility to inhibition depends directly on the time of exposure to darkness, so that adaptation occurs. Below about 40 lux the animal swims upward, becomes inhibited, sinks for a short distance, becomes photokinetically sensitive, is then inhibited at a higher intensity, and finally reaches the surface, at which at the appropriate intensity a balance between photokinesis and inhibition takes place. The effect occurs whether illumination comes from above, below, or laterally. It occurs (Fig. 198) in animals from which the eye has been removed (Harris and Mason 1956), and by analogy with the earlier work of Schulz (1928) it is probably independent of the ocellus and is due to some general kind of photosensitivity. The photokinesis is also independent of wavelength, for it occurs in red photographic light. The resulting vertical up and down movement is clearly what Smith and Baylor (1953) had independently described as the red dance.

The general photokinetic response, alternating with inhibition, can produce in eyeless *Daphnia magna* an extreme type of twilight migration with a descent to the bottom at very low light intensities, some adaptation to about 2.5 lux while the light is maximal, and an evening twilight migration upward with the 2.5 lux isophot. The sensitivity of these

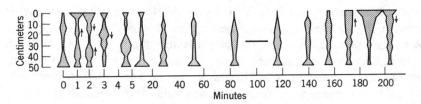

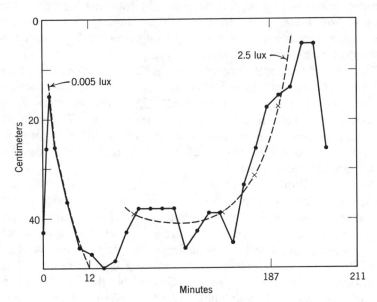

FIGURE 198. Upper panel, overhead illumination. Distribution of a population of fifty eyeless *Daphnia magna,* illuminated by a light whose intensity was varied through a 5-hr. cycle of intensity changes. The abscissa varies in scale in order to show the more rapid movements at the beginning of the cycle. The horizontal block in the center indicates the scale of 10 animals. In certain ambiguous distributions the direction of movement (shown by unrecorded observations) is indicated by arrows. Lower panel, movement of the median of the eyeless population of the upper panel. The full line indicates the position of this median. The dotted line on the left indicates the position of the zone of intensity of illumination 0.005 lux; that on the right the position of the zone of intensity of illumination equal to 2.5 lux. The time scale along the horizontal axis is expanded fourfold from 0 to 12 minutes and from 187 to 210 minutes in order to exhibit the more rapid movements during these two periods. (Harris and Mason.)

eyeless individuals appears to be much greater than that of normal specimens.

At higher light intensities above about 100 lux a quite different type of response, which is entirely absent in the eyeless forms, intervenes in normal *D. magna.* It is well known that if a *Daphnia* is adapted to a parallel beam at moderate intensities, a change in the direction of the beam will cause the eye to move so that its relation to the light vector remains unchanged. The animal then reacts to maintain its body position unchanged in relation to the eye. Ordinarily the adapted animal at such intensities has its long axis normal to the beam and with its dorsal surface directed toward it. This type of orientation is often called the dorsal light reaction (von Buddenbrock 1915). Von Frisch and Kupelwieser (1913) concluded that for such an adapted animal an increase in illumination is equivalent to an increase in the angle of light from the zenith. The animal is tilted down and will swim down until the illumination is about that to which it had been exposed. This type of orientation response evidently produces the relatively stationary distribution of the normal *Daphnia* at an intermediate depth under conditions in which the eyeless forms would be at a much lower intensity. Harris and Wolfe, moreover find that within the ranges of illumination in which the *Daphnia* is thus oriented, and in which any sudden change in illumination leads to a change in depth, the change is followed by a striking adaptation to the new intensity. If the light is increased from, say, 50 to 300 lux at the level of the center of gravity of the population, the population will descend so that the center of gravity will lie about 50 lux and then after a few minutes rise asymptotically to almost the original level, now at 200 lux. The reverse process occurs but with an even more rapid adaptive descent. These adaptational changes presumably imply that passing clouds and other transitory changes in illumination have little effect on the over-all depth distribution in nature inhibition.

The horizontal movements of the oriented *Daphnia* clearly depend on the eye and are not influenced by red light. They correspond in fact to the blue dance of Smith and Baylor.

Harris and Wolfe found that in *D. obtusa,* the only species other than *D. magna* they studied, the threshold for oriented movement was much higher than in *D. magna.* Neither species lives in deep water, but in two deep-water species comparable difference could correspond to an extremely pronounced, as compared with a moderately pronounced, twilight type of migration.

Migration in the laboratory with artificially manipulated natural daylight. Schröder (1959) made experiments in which plankton was placed in plexiglass tubes 100 cm. long and 10 cm. in diameter, which when placed vertically could be illuminated from the sides by diffuse light and could be set horizontally and illuminated at one end. During the day the illumination at the entering surface was somewhat more than 100 lux; during the night it was less than 0.03 lux (Fig. 199).

With *D. longispina* in the tube illuminated from above, the animals tended to be fairly high but not quite at the surface during the day, as in the Titisee and Schluchsee, from which they were presumably derived. During the night they became randomly distributed, rising and descending in the early morning and then rising again. The pattern in the experimental vessel, though variable, appears to parallel the rather peculiar behavior of the *Daphnia* observed in nature. When the tube was illuminated from below, the animals swam down actively to the bottom by day, tending to turn over, with the antennae beating. This was also observed in *Bosmina coregoni* and *Bythotrephes longimanus*. As the light declined, an upward movement was followed by sinking during the middle of the night and another ascent before dawn. The general pattern, in fact, was one of twilight migration. With diffuse lateral light, a striking twilight migration was achieved, but it was not possible to ascertain whether the downward movements were active or passive.

When the tube was on its side, the animals were randomly distributed during the day, doubtless performing a horizontal or blue dance. As the light was reduced, they became concentrated in a definitely positively phototactic state at the illuminated end. As the illumination increased in the morning, the animals scattered.

A form of dorsal light reaction, which leads to circus movements when illumination is from below, certainly exists in fresh-water Diaptomidae, and there is also evidence in other calanoids of the Ewald-Schallek inhibition. In Schröder's experiments with *E. gracilis* in decreasing and increasing diffuse lateral light no typical twilight migration occurred; in fact the migration that took place tended to be reversed. It is therefore not impossible that in the calanoid copepods the more phototactic aspects of the migration mechanism dominate and photokinensis is of less importance.

Effects of environmental conditions on Daphnia. Smith and Baylor have restudied the effects of environmental conditions on the movement of *D. magna*. They find that when an animal reared at 15° C. is brought to a low temperature, around 0 to 5° C., it becomes strongly positively phototactic, but raising the temperature to 30° C. induces marked apparent negative geotaxis, which presumably means more

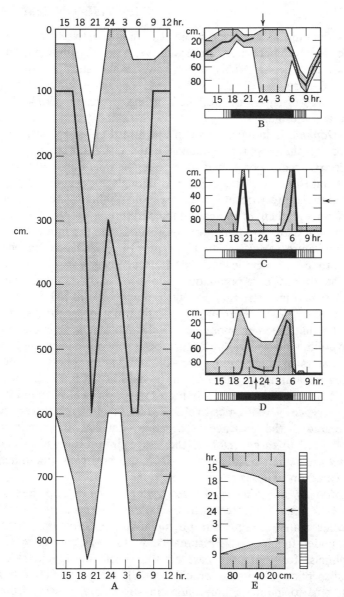

FIGURE 199. *A*. Vertical position of the upper and lower limits and level of the maximum density of the population of *Daphnia longispina* in the Titisee, Aug. 1–2, 1953. *B*. The same for *D. longispina* in tube, illuminated from above with natural daylight, the small panel below indicating full daylight (>100 lux), twilight (100–0.03 lux), and darkness (<0.03 lux). *C*. The same with diffuse lateral illumination giving photokinetic twilight migration. *D*. The same with illumination from below giving a comparable but less marked twilight migration. *E*. Horizontal position with tube set horizontally and weakly illuminated from one end with animals accumulating at the illuminated end as the light fell and remaining there in darkness. (Schröder, modified.)

rapid swimming at whatever speed it is falling between antennal beats. At the low temperatures employed very strong illumination causes paralysis and death. This effect can be abolished by the presence of $10^{-7}M$. atropine in the water. At ordinary temperatures a like effect of strong light is produced in the presence of $10^{-7}M$. acetyl choline and eserine.

If a *Daphnia* is induced to swim downward either by negative phototaxis or by the apparent geotaxis induced by horizontal blue light, entry into a region of slowly decreasing temperatures accelerates the descent until the water becomes cold enough to reverse the phototaxis. A comparable effect occurs in an ascending animal which is stimulated as it enters a gradient of increasing temperature.

In the dark an apparent positive geotaxis can be induced by raising the temperature to 25 or 30° C. The pH dependence may be strong enough to be of significance, but it is of such a kind that its effect might be to divide a population into two groups. *D. magna* raised at pH 8.0 becomes strongly positively phototactic at pH 8.5, and negatively so at pH 7.0. Part of a population moving at random in a light gradient would become more negatively phototactic if it entered more acid hypolimnetic waters, and more positively so if it entered more alkaline waters in the region of maximum photosynthesis. There is a tendency for animals to move up in the dark at pH 9.0 and down at pH 7.0. Some experiments by Smith (1954), in which animals reared in the presence of redox-positive vital dyes or of catechol were positive, those reared in the presence of negative vital dyes or cystein negative were believed to throw light on the less definite observation that when *Daphnia* are reared on algae they are more positively phototactic than when reared on bacteria.

Relation of the blue dance to feeding. Smith and Baylor point out that in a tank covered with lucite trays, some of which are filled with clear water, some with phytoplankton suspension, the animals collect under the latter. Presumably, in the spectral range visible to the animals, for the most part on the short wavelength side of the main absorption maximum of chlorophyll, the phytoplankton remove enough blue to induce a blue dance in any animal passing from under phytoplankton into a region under clear water. This increased horizontal movement will tend to cause animals to collect under the shaded areas which in nature would be the more nutritious. Hunger can under any wavelength cause a blue dance that will lead to horizontal exploration.

Experimental studies on euplanktonic Crustacea other than *Daphnia*. The most important work on fresh-water species is still that of Kikuchi

(1938), who studied one calanoid copepod, namely, *Acanthodiaptomus pacificus yamanacensis,* and five species of Cladocera other than *Daphnia.* All the material was obtained from Lake Kizaki. Kikuchi examined the photic responses in both horizontal and vertical tubes and also the responses to gravity in horizontal tubes in diffuse light and in darkness. Unfortunately, not all combinations of stimulus field could be applied to all species.

Holopedium gibberum is generally positively phototactic in the horizontal tube, but at high light intensities an increasing proportion, apparently more than 10 per cent of the animals, become indifferent or negative. In a vertical tube with a directed beam from above most animals move upward with a velocity of 18 to 32 cm. min.$^{-1}$ Increase in illumination, however, can cause a temporary downward movement. In a vertical tube in diffuse light there is a marked tendency for the animals to accumulate at the bottom, which tendency entirely disappears in darkness. Such behavior in the gravitational field is doubtless due to the Ewald-Schallek.

Increase of temperature in darkness evoked an apparent positive geotaxis. *Leptodora kindtii* and *Diaphanosoma brachyurum* appeared to resemble *Holopedium* in their reactions to both light and gravity. *Polyphemus pediculus* was found to be negatively phototactic in a strong horizontal beam and positive in a weak beam. Comparable behavior was observed in the vertical tube. Unlike *Holopedium,* the *Polyphemus* became more positively geotactic in darkness than in diffuse light. *Bosminopsis dietersi,* indifferent to moderate constant illumination, becomes temporarily positively phototactic when the intensity is decreased and temporarily negatively phototactic when the intensity is increased. Comparable responses were observed in the vertical tube in which a reduction in intensity could produce an upward movement of 12 to 30 cm. min.$^{-1}$ at 27.5° C. No geotactic reactions appear to be exhibited either in darkness or diffuse light.

Acanthodiaptomus pacificus yamanacensis is always positively phototactic in both the horizontal and vertical tubes at constant temperature. In the vertical tubes the upward movement is 50 to 80 cm. min.$^{-1}$. Raising the temperature, however, retards this movement, at least temporarily.

In darkness the animals appear almost neutral to gravity and are distributed fairly evenly in a vertical tube with little tendency to accumulate at the bottom. Diffuse illumination causes an immediate marked positive apparent geotaxis, which, however, tends to pass off in a few hours, though in diffuse light there is always a greater proportion of specimens at the bottom of the tube than there is in darkness.

When an equilibrium distribution is established in diffuse light, either cooling from 22.4 to 16.8° C. or warming from 14.3 to 18.5° C. can cause the reappearance of the apparent geotaxis.

In the considerable body of work on marine species, most notably *Calanus finmarchicus,* the most important paper is that of Hardy and Bainbridge (1954). In general, these authors conclude that the most significant factors are a photokinesis or phototaxis leading to upward movement at low light intensities and a strong active downward swimming at high light intensities. They conclude that the predominantly horizontal movement seen for instance in the blue dance of *Daphnia* is probably characteristic of animals with well-developed compound eyes such as the Cladocera and Malacostraca, in contrast to the Copepoda. It is now clear that it would also be legitimate to inquire how far reflexes, based on free fall and streaming between strokes of the antennae are represented in animals with a more continuous type of locomotion involving the metachronal rhythm of several appendages. It is not unlikely that there is more than one type of downward movement, though it is important to remember that since the movement at least in the epilimnion will always occur in a turbulent medium with a greatly increased apparent viscosity, the occurrence of sinking speeds in nature comparable to those of passive free fall in the laboratory probably implies that in the field the descent is an active process, as Hardy and Bainbridge believe.

It is interesting to note that in their experiments the calanoid *Centropages* sp. tended to exhibit reversed migration, as it may in nature.

Rather complicated effects have been recorded by Schröder (1959) when diaptomid copepods were used in his plexiglass tubes, but it is not clear whether some of the irregularities observed were due to variations in species, *Mixodiaptomus laciniatus, Eudiaptomus gracilis,* and *Acanthodiaptomus denticornis* all having been employed, though in the variable control runs illuminated from above there is no indication of the species used. In these runs the characteristic pattern seemed to be a kind of reversed twilight migration. When illuminated from below, *E. gracilis* apparently behaved more or less like *D. longispina.* Lateral diffuse illumination in two experiments produced a diffuse distribution, in two others a striking down and upward reversed twilight migration at dusk, and in one case a similar excursion at dawn. Lateral illumination with the tubes horizontally set tended to bring all populations to the illuminated end, but there were considerable irregularities and perhaps some indication of specific differences, since *M. laciniatus,* which is a deeper-living form in nature, seemed more positively phototactic than *E. gracilis,* whereas *A. denticornis* behaved essentially like *Daphnia longispina.* These experiments certainly point the way to a more extended series of studies using refinements of Schröder's basic method. The demonstration of twilight migration under conditions in which only photokinesis and a gravitational field are involved is interesting, though not unexpected in the light of the work of Harris and Mason (1956). The active downward movement when the illumination is from below can be paralleled by observations of the older investigators but is not easily explained in terms of the observations of Grosser, Baylor, and Smith (1953) and Hantschmann (1961). It is possible that all movement described by Schröder of the more or less inverted animals stimulated from below is of the nature of escape movement, observed by Hantschmann in animals in an isopycnal sugar solution.

Experimental evidence of adaptation to particular levels. Cushing (1955) found that the behavior of planktonic Crustacea transferred from the free water of the lake to submerged vertical tubes depended within a given species on the level of origin of the individuals. Specimens of *Eudiaptomus gracilis* obtained in Windermere from the top 30 m. tended to move upward in tubes set in the top 20 m., and specimens from below 30 m. tended to move down as if they were seeking their previous level.

Possible effects of pressure. Hardy and Bainbridge (1951) have shown that mixed decapod larvae, mainly the zoeae and megalopae of *Portunus* and *Carcinus,* develop a negative geotropism in the dark if the pressure is raised from a little over 1 atm. to the equivalent of 5 m. of water. Further increase in pressure gave little or no increase in the rate of ascent; the curves for the number of individuals in the top half of the experimental tube plotted against time after raising the pressure showed no significant differences when the rise was equivalent to 10 and to 20 m. of water.

Knight-Jones and Qasim (1955) found that a number of marine animals, including Hydromedusae, polychaets, and various Crustacea, became more active and swam upward when the pressure was increased, but when a decrease took place the animals became less active and sank passively. In the fish *Blennius* and the ctenophore *Pleurobrachia* there was also an active swimming downward when the pressure was decreased; both animals are buoyant, though in the case of *Pleurobrachia* this must be due to ionic or other chemical properties, since there is no gas-containing structure in this animal. No response was given by *Calanus, Tomopteris,* or *Sagitta,* although the first two are pelagic members of groups, other species of which show the effect.

The sensitivity of some of these animals is considerable, *Blennius* responding to changes of 5 millebars and decapod larvae to changes of 10 millebars, or 5 and 10 cm. of water, respectively.

Digby (1961), who found that *Palaemonetes varians* living in salt water is sensitive to pressure changes, has produced an ingenious theory, in which he supposes that the animal responds to changes in thickness of a very thin layer of hydrogen produced electrolytically at the body surface. Surface-active materials, producing in water a positive complex and a negative small ion, may abolish the response, but if a negative complex is introduced into the medium there may be an enhancement of the effect. Crowding may abolish and washing with bicarbonate and then acid solutions can restore the response. Such treatment followed by the addition of a little salt can produce the response in *Daphnia,* which Digby says when newly caught showed only a trace of the reactions. Hantschmann found no pressure sensitivity in *Daphnia schødleri,* though she was interested only in a very limited range corresponding to increases of up to 20 millebars. It seems probable that only plankton in marine or at least somewhat saline waters are pressure sensitive.

Enright (1963), who found that the littoral marine amphipod *Synchelidium* is sensitive to a pressure change of 5 millebars, when at ordinary pressures, concludes from compressibility determinations that this change would correspond to a decrease in thickness of the hypothetical gas film of less than 2 Å, a decrease not likely to affect receptor organs very easily. This conclusion makes Digby's attractive theory seem improbable.

Effects of polarized light. Baylor and Smith (1953) have observed that Cladocera, watermites, and trichopterous larvae are markedly sensitive to the plane of polarization of polarized light. They suppose that this sensitivity may have some directional effect, particularly near dawn and sunset, at which the greatest polarization is exhibited by the skylight from the zenith.

In general, *Daphnia* tends to orient itself at about 90 degrees to the plane of the electric vector, but there is evidence of some individual asymmetry in *D. schødleri* (Waterman 1960) and of further preferred orientations in the plane of this vector, ordinarily spoken of as the plane of polarization, and at 45 degrees either way to this plane. These four basic directions appear in a great variety of arthropods (Jander and Waterman 1960). In spite of the inherent fascination of these discoveries, which may play a role in the behavior of small Crustacea near reflecting surfaces, the responses to polarized light are not likely to be of great importance in regulating vertical movement.

Effect of feeding on density. Eyden's (1923) observations that the specific gravity of *Daphnia* is raised by feeding suggests that if organisms fed vigorously in an upper layer rich in algae they would have progressively greater difficulty in remaining high in the water and might start slowly to sink. Fox and Mitchell (1953) failed to confirm Eyden's results. However, when an animal has been feeding on cells containing material of high density, such as the silica frustules of diatoms, and has rapidly digested and metabolized compensating sheath materials or fat, it is conceivable that the phenomenon allegedly noted by Eyden may play a small part in regulating vertical movement.

The effect of feeding on the responses of *Chaoborus* larvae. Berg (1937) found that in darkness unfed larvae of *C. flavicans* performed continued vertical movements up and down in his aquarium. In light they buried themselves in the mud placed at the bottom of the tank. This response was entirely independent of the previous nocturno-diurnal rhythm experienced by the animals and could be easily elicited when the larvae were kept in the dark by day and illuminated by night. The movement, however, was abolished by providing an abundant supply of food; when many specimens of *Cyclops* sp. or of *Diaptomus castor* were provided as prey, the *Chaoborus* larvae remained in the water of the aquarium regardless of the illumination. It is not certain that the well-fed larvae in a deep body of water would show no vertical movement, but it is evident that some of the details of the behavior of the insect in nature are to be explained in this way. Berg noted that larvae from Sorte Dam, near Hillerød exhibited no migration in the laboratory, though they appeared to belong to *C. flavicans*. It is uncertain whether these larvae were well fed or were racially distinct from those from Lake Esrom used in the experiments.

INTERPRETATION OF THE OBSERVATIONS MADE IN NATURE

When the body of experimental data just reviewed is considered in relation to events in nature, it would seem that enough elements of behavior have been isolated in the laboratory to provide a formal model of ordinary types of migration, though there is still a great deal to be learned about the nature of the differences between groups of animals migrating in various ways. The formal models that are satisfactory are constructed mainly in terms of photokinensis, phototaxis, and passive sinking in a gravitational field. Pressure changes are probably not important, and perhaps, as we have seen cannot be important in fresh water. How far feeding and its effects play a significant role in modifying the cycle obviously requires further work.

Purely phototactic and photokinetic models. Given a slight excess density and a swimming mechanism that involves discontinuous acceleration moving the animal upward, with sinking between strokes, a model of twilight migration is achieved by Harris and Mason's eyeless *Daphnia,* which are positively photokinetic in weak light and inhibited in strong light and in total darkness. When the eye is present, the same general pattern of behavior is observed but with far greater refinement in adjustment. The downward movement then becomes an active negative phototaxis, which, because of the dorsal light reaction or blue dance and of adaptation, tends to cause the animals to remain at a rather constant depth throughout the period of illumination. The observed differences in the threshold of this negative phototaxis between *D. magna* and *D. obtusa* provides a model of the different behavior of planktonic species which descend to different depths during the day. What this model does not explain is the relative feebleness or complete absence of nocturnal sinking in many cases. This, however, is probably not a real difficulty because there seems to be evidence that behavior in total darkness differs from species to species. At least in *D. schødleri,* the animal can keep a horizontal course in the dark by responding to the water movement relative to its body as it falls between strokes. In this case all that is needed to produce a typical nocturnal migration is a positive photokinesis to dim light that will bring the animals to the surface at dusk, where they would remain through the night, and a negative phototaxis to bright light to submerge the animals by day. This model depends for its details almost entirely on experiments on *Daphnia.* To be general, the model would require that other migrating Crustacea have a physiology essentially similar to that of the Cladocera.

It will be observed that the old theory of Groom and Loeb differs from this model only in that these early workers did not distinguish between photokinesis and phototaxis; given a swimming mechanism like that of *Daphnia,* a photokinesis will inevitably lead to upward movement, in nature toward the light. The model is also more or less equivalent to the optimal light intensity theories in that the simplest way of achieving aggregation at a given light intensity is by a positive reaction to low light intensities and a negative one to high. This is all that optimum light intensity can mean in default of information as to the animal's subjective preferenda, if any, and the effects of maintenance of plankton animals for long periods of time at different intensities.

Models involving geotaxis. Models involving geotaxis have been popular in the past and were particularly favored by Dice (1914) and Worthington (1931). In view of what we know about swimming movements and the sensory equipment of the plankton, it is probably necessary to distinguish between passive and active reactions to gravity. By a passive reaction we mean the free fall that takes place in any organism having excess density over that of the medium, when swimming movements are insufficient to maintain it afloat. An example is provided by the Ewald-Schallek inhibition in diffuse light; it is also probable that passive gravitational sinking occurs in total darkness in some species of *Daphnia,* as in *D. magna* in the experiments of Harris and Wolfe (1955), but not in others, as in *D. schødleri* studied by Hantschmann (1961). The same variability is implied in Kikuchi's experiments with other Cladocera. When a passive reaction to gravity occurs, it may exist with a sensory mechanism which is stimulated by water currents and which leads to a response modifying the movement of the animal. Such a response, however, will be given when the animal is moving upward in a dense medium, just as if it were moving down in a less dense medium. An active response to gravity, or true geotaxis, implies as a sensory mechanism the working of a statocyst, in which a dense object is oriented by gravity and stimulates appropriate sensory cells as the animal's position relative to the gravitational field is changed. Since the statolith is largely isolated from the external medium, placing the animal in a dense solution will not alter its orientation. It is evident that the Cladocera can react only passively. The various free-swimming Malacostraca are in general likely to be able to respond directly and actively to gravity, but in the fresh-water plankton these animals will be represented only by the meronektoplanktonic mysids and immature *Caridina* and perhaps other Caridea. It does not seem that geotropism in any strict sense of the term is likely to play an important part in the migration of the fresh-water plankton, though the process may be much more important in the sea, in which there is probably a greater proportion of nektoplanktonic Malacostraca.

When a negative geotactic type of response seems to be involved, as in some of Southern and Gardiner's and of Worthington's observations of the continuation of upward movement in the darkest times of the night, the most reasonable explanation would seem to be that we are dealing with animals oriented and activated by the water currents produced by sinking and in which the swimming movements in the dark are strong enough to produce an upward movement either because of an endogenous swimming rhythm or more likely because the response to the water currents produced by a free fall is adjusted to produce a small vertical upward movement rather than a horizontal path.

Regarding the downward descent, there are three possibilities. First, there might be passive sinking, presumably of the Ewald-Schallek type, which evidently can occur in some marine copepods, as Schallek has stressed. Second, there is certainly a negative phototaxis, as emphasized in the preceding paragraph, that causes animals to swim downward, as Harris and his co-workers and Hardy and Bainbridge have found respectively, in *Daphnia* and in marine crustacea of various groups. Third, in a very few groups of no importance in most lakes there may be a true positive response to gravity mediated by a statocyst. The main problem in the fresh-water plankton is clearly the relative importance of active downward negative phototactic swimming and passive sinking which results from negative photokinetic inhibition. It seems likely that there will be considerable variation from species to species and from stage to stage in the importance of these two processes, though, as has already been indicated, it is not unlikely from what is known of sinking speeds of narcotized animals in the laboratory that a turbulent epilimnion may require active down swimming if migration is to occur over the distances ordinarily recorded in transparent lakes.

Role of photic adaptation. As will already be apparent from Siebeck's observations in the field and Harris and Wolfe's experiments, the history of the organism over the full cycle of light and darkness has to be considered, as some adaptation clearly occurs. The extent to which it is present, however, is presumably variable from species to species; for instance, in Ullyott's observations on *Cyclops strenuus,* following on any particular day a particular narrow range of isophots and Siebeck's observation suggesting that the adapted animal may rise into progressively brighter regions in the late afternoon. A striking effect of physiological adaptation is clearly indicated by Kikuchi's (1930 a,b) observations on *Bosminopsis dietersi* which rose to the surface three times during the twenty-four hour cycle as the light underwent two discontinuous decreases in intensity and one increase at the surface of Lake Kizaki.

Vertical migration in hypolimnetic crustacea. Wherever it occurs in stratified lakes, *Limnocalanus macrurus* is a hypolimnetic form. Juday (1904) found that in Green Lake, Wisconsin, the species did not occur above 50 m. by day but migrated to the top of the hypolimnion at 15 m., but not higher, by night.

Several forms of hypolimnetic *Daphnia,* once referred to *D. longispina* but actually for the most part belonging to other species as well as more legitimate representatives of *D. pulex* (see page 617), have been recorded as hypolimnetic. The diurnal migration of one such form in Lake Aoki has been studied by Kikuchi, who found the main concentration of the species to be at 15 m., where the temperature was 10.6° C. by day, and to rise by night to 10 m., where the temperature was 15.5° C. At no time were any specimens found in the epilimnion, from 0 to 5 m., throughout which layer the temperature was about 25° C. The *Daphnia* of Lake Kizaki (Kikuchi 1938) behaves similarly. It is evident that in this and comparable cases upward migration stops when a certain temperature is reached.

In his experiments with the Lake Kizaki form Kikuchi (1938) found that transient negative phototaxis is induced either by an increase in light intensity or less effectively by any change in water temperature. Of greater importance is the fact that at least in darkness any marked change in temperature produces a downward movement which persists one or two hours. On of Kikuchi's experiments, conducted in a vertical tube 30 cm. long and showing the distribution of hypolimnetic *Daphnia* from Lake Kizaki, is recorded in part in Table 44. The effect certainly occurs with a falling temperature in diffuse light; an experiment with rising temperature unfortunately seems not to have been done under these conditions.

If, as is probable, the production of a downward movement when the temperature is changed is independent of the illumination, it is easy to see that no individuals could ever enter the epilimnion, for as soon as they attempted to do so they would become apparently positively geotactic and start descending. The probable distribution would be a scattered population of adapted animals ascending in the hypolimnion and of markedly geotactic animals either descending or accumulated near the bottom.

TABLE 44. *Distribution of hypolimnetic* Daphnia *from Lake Kizaki in darkness before and after temperature is raised (three specimens caught at surface film were omitted after heating)*

Time	12.35 P.M.	12.55	1.20		1.40	1.50	2.15	2.40	3.10
Temperature	14.° C.	13.8	14.0	(temperature raised)	19.2	24.4	25.2	25.5	25.3
0–6 cm.	12	7	8		0	0	0	1	3
6–12	2	3	3		0	0	0	1	2
12–18	0	2	5		0	0	0	0	2
18–24	0	4	3		0	0	0	3	4
24–30	11	18	15		31	31	31	26	20

Kikuchi obtained evidence of the induction of a supposed positive geotaxis by temperature changes in certain other Crustacea. This occurred in *Acanthodiaptomus pacificus yamanacensis* in diffuse light but not in darkness, both heating and cooling being effective. Unlike the Lake Kizaki *Daphnia, A. p. yamanacensis* is a normal nocturnal migrant, and Kikuchi concludes that this is because positive geotaxis is not induced by a change of temperature in darkness and that the species can move freely up the temperature gradient at night.

Holopedium gibberum was found by Kikuchi to show a marked positive reaction to gravity, or more probably the Schallek-Ewald inhibition in diffuse light, but at constant temperature to be indifferent to gravity in the dark. In a stratified lake a hypolimnetic population of *Holopedium* would not be able to reach the surface by night, for it would develop an apparent positive geotaxis when it crossed the thermocline. Kikuchi notes that this sometimes happens in Lake Kizaki, though on other occasions the population is purely epilimnetic.

It is reasonable to suppose that the induction of apparent positive geotaxes by increasing temperature will prove to be a fairly general way in which hypolimnetic species are prevented from entering the epilimnion. The nature of the apparent geotaxis, which in the essentially nonturbulent hypolimnion may merely involve passive sinking, remains to be elucidated.

Limitation of movement to the epilimnion. Cases are known of Crustacea limited to the epilimnion. Langford's (1938) data on *Leptodiaptomus minutus* indicate very few specimens in Lake Nipissing below 20 m. and the bulk of the population above 10 m. It seems reasonably certain here that the thermocline is operating as a more or less effective barrier to downward movement. Certain cases among species of the genus *Daphnia* are discussed in Chapter 26 (see pages 835–842). Smith and Baylor's (1953) experiment on the abolition of the descending blue dance by low temperature provides a reasonable explanation. Similar distributions can doubtless be found in the rotifers, such as *Chromogaster ovalis* in the Lunzer Untersee or the *Keratella*, determined as *cochlearis* but probably referrable to *K. earlinae* Ahlstrom, found by Juday (1904) to be characteristic of the epilimnia of the lakes he studied in Wisconsin. Comparable behavior is presumably involved in the separation of *Cyclops* and *Mesocyclops* in waters such as the Lake of Lucerne.

It is clear, however, that in other cases a vertical separation even of closely allied species may be effected by light alone. An extreme example may be provided by *Diaphanosoma sarsi,* which seems to occur at or below 10 m. by day and at or below 5 m. at night in Lake Toba, in which the whole of the top 10 m. is also inhabited by the nonmigratory *D. modigliani*. The temperature gradient was not more than 0.15° C. in 10 m. at the time of investigation (Ruttner 1943).

Seasonal movement of hypolimnetic Crustacea. A characteristic type of movement throughout the period of summer stagnation is exhibited by populations of hypolimnetic Crustacea in lakes in which a more or less clinograde oxygen curve develops. Langford found that the copepodite stages of *Skistodiaptomus oregonensis,* which are, unlike the adults, confined to the hypolimnion of Lake Nipissing, moved upward during July and August, so that the zone of maximum density which was near the bottom at 45 m. in July lay during late August at 35 m. Subjection of *Cyclops* and *Daphnia* in a gradient tank to low O_2 or high CO_2 at the bottom produced an upward movement and provided a model of the behavior observed in the field. Langford concluded that a local fall in O_2 content to 5 mg. per liter and a rise of CO_2 to 9 mg. per liter concurrently constitute an adequate stimulus to produce the movement in the lake (Fig. 200).

The final result of these movements is sometimes seen in the very localized zonal distribution of cold stenotherm plankton just below the thermocline. The distribution of the helmeted form of *Daphnia longiremis* in Muskellunge Lake, Wisconsin, (see page 218) provides a good example. An even more striking case has been presented by Strøm (1946). In Steinsfjord, near Oslo, a well-stratified lake 22 m. deep develops a well-defined clinograde oxygen curve during summer stagnation. *Limnocalanus macrurus* inhabits the lake; during the height of stagnation it is confined to the zone below 10 m., in which the temperature is less than 14° C., and above 17 m., where the oxygen content is more than 5.7 mg. per liter. The main population was found to be sharply localized at 13 m., where the temperature was about 12° C. and the oxygen, 6.4 mg. per liter. Strøm characterizes this striking distribution as providing a superb example of an ecological niche. The localization of phytoplankton to particular stable layers is, of course, frequently noticed, and in some cases the presence of abundant food may accentuate the sharpness of the limitation observed in the distribution of animals.

In very extreme cases of metalimnetic oxygen minima the zooplankton populations of the epilimnion and hypolimnion may for a time be completely separated from each other (Patalas 1963).

Possible adaptive significance. It is necessary first to inquire into the probable metabolic economics of the process of vertical migration, as Worthington (1931, page 433) has remarked on the "astonishing expense of energy" that seemed to him to be involved. A rough

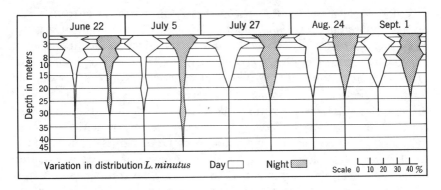

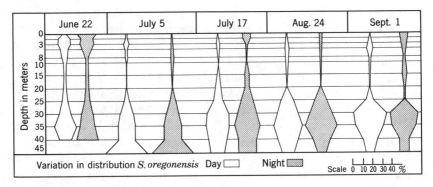

FIGURE 200. Separation of *Leptodiaptomus minutus* and *Skistodiaptomus oregonensis* to form, respectively, epilimnetic and hypolimnetic populations in Lake Nipissing. (Langford.) The large deep-water population of *S. oregonensis* is composed primarily of copepodites, the adults entering the epilimnion.

calculation shows that this expense is in reality not particularly serious. For an organism of excess density 0.015, a force of $15 \times 10^{-6} \times 981$ or 14.7×10^{-3} dyne must be overcome per milligram of organism in moving it upward. For a migration path of 50 m. or 5000 cm. the energy needed will be 73.5 ergs or 1.76×10^{-6} g.-cal. Since 1 mM. of glucose in solution yields on oxidation 699 g.-cal., the minimum amount of glucose that would have to be oxidized to produce this energy would be 4.5×10^{-7} mg. Since we may reasonably assume that an organism of mass 1 mg. and of excess density 0.015 would contain at least 1.2 per cent or 12×10^{-3} mg. of organic matter, it is evident that, even assuming an efficiency as low as 1 per cent, less than one two-hundredth of the organic matter of the body would be oxidized in performing the ascent. This would not appear to be a very serious drain on the biochemical economy of a migrating organism eating its body weight of food each day (see for instance Table 35). A relatively small advantage conferred by vertical migration would in fact make that process an economically significant one, favored by natural selection. Actually, as Riley (personal communication) has pointed out, in a species with purely passive descent it would not cost more to migrate than to remain perennially at a given high level.

Harris (1953) considered the whole process merely as a by-product of adaptation to maintaining a subsurface position, but, in view of their great variability, this cannot be the whole explanation of the observed migrations. In a very general way, perhaps, the phenomenon may be seen as a special case of the tendency of animals toward nocturnalism, a tendency presumably reflecting prey-predator relationships in the water as well as on land. Weismann no doubt was right in a sense in believing that the crustacean eye is adapted to low light intensities; this, however, is probably true of most eyes, for diurnal animals are less abundant than are nocturnal, but this optic sensitivity is most unlikely to be the complete story.

The recent theory put forward by Wynne-Edwards (1962) that aggregation of plankton near the surface permits social interactions to reduce the reproductive rate as resources are exhausted, is so out of line with what is known about the physiology of reproduction in copepods and Cladocera that it can hardly be taken seriously, even though the book in which the idea is put forward raises important and difficult issues in the many pages devoted to the vertebrates.

Apart from the hypothesis of nocturnalism, or of the safety conferred on small and edible animals by living in the dark, the recent treatment of the adaptive significance of vertical migration proposed by McLaren (1963) must be taken very seriously. McLaren's idea is that it is more efficient to feed at high temperatures and to grow at low temperatures. Moore and Corwin (1956) had earlier expressed the view that for some reason a rhythmical change in temperature might be beneficial. Making reasonable assumptions and considering the data are available for the temperature dependence of physiological processes in the Crustacea, McLaren concludes that as soon as a fairly marked thermal stratification develops vertical migration is metabolically advantageous. At temperatures around $20°$ a difference of as little as 1 or even $0.5°$ between the surface temperature and that of the layers visited by day may be enough to confer some advantage over an animal remaining perennially in a warm phytoplankton-rich epilimnion. This explanation of the migration is most attractive to the limnologist for whom variations in pressure are small and to whom the supposed advantages of upward and downward movements in avoiding unfavorable and entering favorable water masses are less important than to the oceanographer (Hardy and Gunther 1935; Bainbridge 1961; David 1961). It is possible, however, that in the ocean animals undergoing very extensive migration may have their metabolism raised by increasing pressure as fast as it is lowered by decreasing temperature during a prolonged descent (Napora 1964). To me the most likely explanation of the migration process is that it was initially an expression of photic sensitivity which has acquired additional significance because of protective nocturnalism, niche diversification, and the mechanisms examined by McLaren.

HORIZONTAL DISTRIBUTION

In the early days of research on lake plankton there was a good deal of discussion whether the zooplankton was regularly distributed in any given layer throughout a lake or whether it tended to collect in plankton swarms. In large or very irregular lakes, such as Lough Derg (Southern and Gardiner 1926a), a considerable degree of heterogeneity may be expected and indeed actually occurs. In such lakes more or less independent water masses, which will have different relations with the littoral and the bottom and with influents, and indeed which may have somewhat different temperatures and chemical properties, naturally are likely to differ somewhat in their plankton.

Until recently very little was known about how far these phenomena occurred over comparatively small distances, though Moberg (1918) had recorded considerable diversity in various transects across Devils Lake, North Dakota. The development of modern statistical methods has now made the problem an accessible, though laborious one, but it cannot be said that the phenomena that have been discovered are adequately understood. None of the fresh-water studies yet made appears to be so extensive as those of Barnes and Marshall (1951) in the seas of the southwest coast of Scotland.

Superdispersion and infradispersion. The most convenient way of treating irregularities in the horizontal distribution of the plankton was first used by Ricker (1937) following statistical procedures developed by earlier statisticians, notably Fisher (1934).

The Poisson distribution and the coefficient of dispersion. If in any horizontal layer the plankton organisms are randomly dispersed and not too abundant in any series of samples taken in that layer, the probability of a sample containing 0, 1, 2, 3, . . . organisms is given approximately by a Poisson series,

$$e^{-m}, \qquad e^{-m}m, \qquad e^{-m}\frac{m^2}{1.2}, \qquad e^{-m}\frac{m^3}{1.2.3}, \cdots,$$

where m depends on the concentration of the organisms and the size of the sample. In this type of distribution the variance is equal to the mean, so that

(1)
$$\bar{N} = \sigma^2 = \sum_0^n \frac{(\bar{N} - N)^2}{(n - 1)},$$

and for n samples

(2)
$$\chi^2 = \sum \frac{(\bar{N} - N)^2}{\bar{N}} = \frac{(n - 1)\sigma^2}{\bar{N}}.$$

The ratio $\sigma^2/\bar{N}$ is termed the *coefficient of dispersion* of the population under study. When this quantity is significantly less than unity, the individuals counted in the samples were spaced more regularly than at random; such a population is said to be *infradispersed*. When the coefficient of dispersion is significantly greater than unity, the individuals of the population were spaced less regularly than at random in the sense that they were aggregated in some and rarified in other regions of the area or volume sampled; a population of this sort is said to be *superdispersed*. The significance of the departure of the coefficient from unity can easily be assessed from the related value of χ^2. To compare the significance of differences in variance in two series, the most convenient statistic is $\frac{1}{2}(\ln \sigma_1{}^2 - \ln \sigma_2{}^2)$, tabulated by Fisher (1934, Table 6).

Double Poisson and negative binomial distribution. In general, a superdispersed distribution may be regarded as one of a class of *contagious distributions* in which the presence of one individual in a sample increases the probability that other individuals will be present in the same sample. The number of samples containing no individuals will be greater and that of samples containing one individual less than in a random distribution, and there will be a corresponding increase in samples containing some number of individuals greater than one. Two main kinds of superdispersed distributions have been considered in the study of the plankton.

In Neyman's Type A distribution and in Thomas' double Poisson distribution it is supposed that the population consists of clumps of individuals. In Thomas' formulation, which is the most convenient in plankton studies, the number of individuals in a clump is randomly distributed about a mean value of $(1 + c)$, the clumps themselves being randomly distributed, and the mean number of clumps per sample is n_c. The probability $P(\hat{x})$ of a sample containing $\hat{x} = 0, 1, 2, 3, \ldots$ individuals is given by

$$(3) \qquad P(\hat{x}) = \sum_{r=1}^{\hat{x}} \frac{m^r e^{-m}}{r!} \left[e^{-r b'} \frac{b'^{(\hat{x}-r)} r^{(\hat{x}-r)}}{(\hat{x}-r)!} \right].$$

The individual terms in the series are thus the sums of the products of the successive terms of two Poisson series.

The second type of distribution, which has been widely used in ecology (Bliss and Fisher 1953), is the negative binomial distribution. The basic model is a mixture of Poisson distributions, the means of which follow the simplest possible frequency distribution, the Eulerian or Pearson type IV. The form has long been recognized but first became important in the study of accidents in factories by Greenwood and Yule (1920). If each worker could have 0, 1, 2, . . . independent accidents in a specified time interval, the distribution of the occurrence of accidents over an individual life history would give a Poisson distribution, but since the liability to accidents is found to vary the temporal distribution of accidents in the whole population is based on the sum of Poisson distributions with varying means.

The probability $P(\hat{x})$ of a sample containing $\hat{x} = 0, 1, 2, 3, \ldots$ is now given by

$$(4) \qquad P(\hat{x}) = \frac{(k + \hat{x} - 1)!}{k!(k-1)!} \left(\frac{\bar{N}_x k^k}{(k + \bar{N})^k_x} \right).$$

The population is thus characterized by $\bar{N}$ as a measure of the average density per sample and by k, which gives a measure of dispersion. The details of the method of fitting may be found in Bliss and Fisher's paper. In this distribution the variance is given by

$$(5) \qquad \sigma^2 = \bar{N} + \frac{1}{k} \bar{N}^2$$

or the coefficient of dispersion by

$$(6) \qquad \frac{\sigma^2}{\bar{N}} = 1 + \frac{\bar{N}}{k}.$$

When the mean is small, the distribution appears nearly random; when it is large, superdispersions become increasingly apparent. Barnes and

Marshall, though they regard the Thomas double Poisson distribution as giving the best fit for their data, point out that very generally their results can be expressed by a coefficient of dispersion of the form of (6), where $1/k = 0.12$. Further developments may be found in Cassie (1962) who favors a Poisson log-normal distribution which can also be used in environmental and correlation analyses.

A third distribution has recently been suggested by Taylor (1961) on purely empirical grounds; in this

$$(7) \qquad\qquad \sigma^2 = a'N^{k'}.$$

The constant a' is apparently dependent on the size of the sample, whereas k' is a statistic giving information regarding the dispersion. In a single application to the distribution of fresh-water plankton, based on the observations of Littleford, Newcombe, and Shepherd (1940), Taylor obtains a value of the exponent k' of 1.57 irrespective of the sample size.

The statistical treatment of departures from random distribution may be regarded as having two aims. The less ambitious aim is the derivation of some statistic or statistics which permit easy and precise characterization of the distribution observed in any given case and easy comparison between different cases. The writer is unconvinced that anything more elaborate than the coefficient of dispersion is justified by the existing development of the subject. Though the coefficient gives no information regarding the nature of the departure from the random other than its amount and sense, its estimation involves no theoretical constructs that later may prove incorrect.

The second and more ambitious aim is an attempt to construct models of various kinds and then to see what theoretical distributions implied by these models do in fact fit the observations made in nature. In a few cases in which both distributions have been considered it has been found that the negative binomial may fit the data somewhat better than the double Poisson distribution; this is true of the *Microcalanus* populations sampled by Barnes and Marshall and considered by Bliss and Fisher and also of the attached eggs of *Skistodiaptomus oregonensis* studied by Comita and Comita (1957), though in the latter case we should have thought that the conditions for the double Poisson distribution would have been neatly exemplified.

A much more hopeful method of analysis is probably provided by the multivariate techniques recently introduced by Cassie (1961, 1962, 1963) so far applied mainly, but not exclusively (see p. 795) to marine plankton.

Observed horizontal distributions. The first worker to attempt a statistical examination of the problem was Naber (1933), who used a statistical technique less appropriate than the ones just outlined. His results, however, are easily treated by the more appropriate modern method. Naber took six 2-liter samples at a central station in the Plöner basin of the Grosser Plöner See, and eight other samples within the 10-m. isobath at distances not more than 800 m. apart. The results in Table 45 are typical of his findings.

In no case is there any clear evidence of superdispersion; the *Daphnia* are slightly infradispersed unlike the *Keratella;* but even in the case of the latter animal, the departure from randomness is quite insignificant. Naber was undoubtedly correct in concluding that when he studied the locality, on October 28, 1931, the plankton was more or less randomly distributed.

Ricker (1937) was primarily interested not in horizontal distribution as an aspect of the biology of the zooplankton but in assessing the value of a few samples in estimating the total zooplankton population of a lake. He gave some results for twenty vertical tows over a depth range of 40 m. to the surface at a single station in Cultus Lake, British Columbia (see Table 46). This case is of historic importance as the first clear demonstration of the existence of the various types of distribution, but in view of the great vertical range of the tows its interpretation is uncertain. In general, Ricker concluded that the superdispersion did not increase when distant samples rather than those taken at a single station were used, but in at least one case the Cyclopidae were random at the distant stations rather than infradispersed.

Langford (1938) made a number of studies on the horizontal distribution of various planktonic Crustacea in Lake Nipissing, Ontario.

TABLE 45. *Random horizontal distribution of Keratella and Daphnia in the Grosser Plöner See*

		$\bar{N}$	σ^2	$\dfrac{\sigma^2}{\bar{N}}$	P
Single station	*Keratella* (mainly *cochlearis*)	130.6	146.2	1.12	$\sim$0.4
Eight separate stations	*Keratella* (mainly *cochlearis*)	142.0	182.3	1.28	$\sim$0.2
Single station	*Daphnia* spp.	2.8	1.21	0.43	$\sim$0.8
Eight separate stations	*Daphnia* spp.	2.6	1.44	0.55	$\sim$0.8

TABLE 46. *Types of horizontal distribution in Cultus Lake*

	$\bar{N}$	σ^2	$\sigma^2/\bar{N}$	χ^2	P	
Epischura	5.05	5.95	1.18	22.4	∼0.25	random
Bosmina	1.15	1.51	1.30	24.9	∼0.15	random
Notholca	13.6	18.1	1.33	25.3	∼0.15	random
Daphnia	71.4	144.0	2.66	38.3	<0.01	superdispersed
Cyclopidae adults	122.2	54.3	0.154	8.4	∼0.02	infradispersed
Copepod nauplii	58.2	34.3	0.324	11.2	∼0.1	random or slightly infradispersed

Langford studied the variance in ten samples at a depth of 8 m. taken at a single station with a 10-liter plankton trap, and here described as *close,* and of ten samples taken at the same depth at different points all within 1200 m. of each other, and here described as *distant.* The plankters in the close samples were presumably living within a few meters of each other at the time of capture, those in the distant samples, within a few hundred meters or more. The results are given in Table 47.

The various stages of the Diaptomidae, at least as far as the adults are concerned apparently mainly *Leptodiaptomus minutus,* are clearly superdispersed at the single station and still more so at the distant stations. The difference between the close and distant collections, however, is not so great as might appear, since the values in the last column would have to be at least 0.8494 for the difference to be significant to the 1% level. There is also a suggestion of superdispersion in the Cyclopidae, whereas the two Cladocera are clearly randomly distributed.

Langford also studied several vertical series; unfortunately, the dates when the collections were made are not given in his paper. Though there is strong superdispersion in a number of cases, there are great and unsystematic differences in the behavior of the different organisms studied for Diaptomidae and *Daphnia* at a single station in 1934 and at a line of ten stations a little over 50 m. apart. At the single station the close samples indicate that superdispersion is significant throughout, though maximal at 5 m., in the adult diaptomid population. There is superdispersion in the top 5 m., with random or very slightly superdispersed distribution at 8 and 10 m. in the *Daphnia* population, and moderate superdispersion ($\sigma^2/N = 3.8\text{–}7.2$) rising to a high value of 18.5 at 10 m. for the Cyclopidae. In the longitudinal

TABLE 47. *Distribution of plankton in the 8-m. layer in Lake Nipissing*

	Close				Distant				
	$\bar{N}_1$	σ_1^2	$\dfrac{\sigma_1^2}{\bar{N}_1}$	P	$\bar{N}_2$	σ_2^2	$\dfrac{\sigma_2^2}{\bar{N}_2}$	P	$\frac{1}{2}(\ln\sigma_1^2 - \ln\sigma_2^2)$
Diaptomidae adults (mainly *L. minutus*)	168.5	703	4.16	<0.01	103.7	3033	28.1	<0.01	0.73
Diaptomidae metanauplii	5.5	18	3.28	<0.01	10.9	76	6.97	<0.01	0.72
Diaptomidae nauplii	510.5	3485	6.82	<0.01	405.9	12500	30.8	<0.01	0.64
Epischura lacustris	6.0	7.6	1.27	~0.3	4.6	15.1	3.29	<0.01	0.34
Cyclopidae	27.3	52	1.81	0.05–0.10	26.5	154	5.8	<0.01	0.54
Daphnia	33.7	33.1	0.99	~0.5	38.8	55	1.42	~0.2	0.25
Bosmina	4.0	5.0	1.25	~0.25	3.9	5.6	1.44	~0.2	0.06

series the *Daphnia* are almost random at the surface, though becoming somewhat superdispersed at 3 m.; the Diaptomidae superdispersed throughout are most markedly so in the deeper layers.

Tonolli (1949b) made a number of horizontal tows over 50 m. courses at a number of depths in Lago Maggiore, three tows being made on each parallel course spaced about 5 m. from the next. Considering all depths and species, ninety-three horizontal samples are available. Infradispersion significant at the 1 per cent level occurred twice; in view of the number of samples studied, we should expect it by chance at this level perhaps once, so that the phenomenon need not be regarded as particularly importance.

Superdispersion at a very significant level, however, occurred in a number of cases. *Eudiaptomus vulgaris, Daphnia longispina* (s. lat.), and *Asplanchna priodonta* were clearly more superdispersed in the epilimnetic layers down to 10 m. than in the deeper water. *Cyclops abyssorum maiorus* (sub *strenuus*) exhibited very irregular variations, different in the two sexes and in the juvenile stages.

In a later paper Tonolli (1958) analyzed a series of horizontal tows at a depth of 11 m., near the expected daytime maximum for the commonest species, over long courses divided into twenty sections; the sections, sampled by drawing off plankton by a tube from a Clarke Bumpus type sampler, would each be 65 m. long. In nearly every case striking superdispersion was found which corresponded to variations in population density over distances of the order of 100 m. It is evident from the plots of the counts that part of the horizontal variation is common to more than one species, and when these species belong to the same trophic level it is reasonably attributed to unelucidated hydrographic factors. Part of the variation, however, is highly idiosyncratic and is perhaps due in part to biological factors. The two large predatory Cladocera, *Bythotrephes longimanus* and *Leptodora kindtii,* taken together are significantly correlated with the total number of other planktonic crustacea on which they feed.

Cassie (1961), using an ingenious correlation technique, has found that in samples taken in the top 50 m. at twelve stations throughout the length of Lago Maggiore, there is, on any given day, a strong correlation between populations of the larval stages of copepods, both *Eudiaptomus vulgaris* and *Cyclops,* and the total population of rotifers. The correlation of adult copepods and rotifers is much less apparent. Cassie concludes that the observed strong correlation is due to both rotifers and copepod larvae having comparable ecological requirements. This particular way of approach may be of considerable interest in relation to competitive relationships.

The most obvious mechanisms for producing the kind of superdispersion described would be based on the Langmuir spirals (Volume I, page 279) that develop in an epilimnetic wind-driven current. If water is brought up from a depth of several meters in the linear divergences of such a current pattern and the plankters then tended to descend as the result of negative phototaxis to bright light (see pages 758–759) in the intervening convergences, a very superdispersed distribution could result, which could differ from species to species, depending on fine differences in phototactic reactions and possibly in reaction to the water movements.

Comita and Comita (1957) examined the distribution of all stages of *Skistodiaptomus oregonensis* in Severson Lake, Minnesota, collected on July 27, 1955, by fifty-one oblique tows with a modified Clarke-Bumpus sampler from near the bottom at not more than 5 m. to near the surface. The spacing of the individual samples is not indicated. Each sample represented the filtration of 56.8 liters. It should be borne in mind that it is not very likely that the copepods would be uniformly distributed vertically, so that, as in Langford's study, a rather complicated distribution may have been sampled by the mode of collection.

Comita and Comita found that ovigerous females were randomly distributed within their samples but that their attached eggs were not, as would be expected. The attached eggs in fact tended to follow the negative binomial distribution. Unattached eggs were also superdispersed, apparently even more than the attached eggs. It is supposed that the eggs collected after their release still showed much of the original distribution of randomly distributed clutches of various sizes. None of the individual later stages shows any indications of superdispersion, and in N1, N5, N6, CI, CII (very markedly), CIII, CIV, and CV are infradispersed well below the 1 per cent level of significance. The total adult population is randomly distributed. Both the CV and adult populations are infradispersed when the individual sexes are considered, the males being more so than the females. If all immature stages other than eggs are considered together, there is evidence of marked superdispersion. Adding the adults decreases this superdispersion; adding the adults and eggs, increases it.

Comita and Comita conclude that for the individual stages the usual pattern is one of infradispersion caused by competition for food; they do not go further in elaborating this conclusion. The randomness of the total adult population, the two sexes being infradispersed, they attribute to the disruption by sexual attraction of the spaced pattern produced by food competition. They conclude that a large number of samples must be considered to demonstrate departures from randomness. When five sets of three randomly selected samples were examined, 87 per cent of the dispersion coefficients calculated indicated a random effect, but from all fifty-one samples only 33 per cent of the coefficients were in the random range. Both infradispersion and superdispersion occurred in greater proportions in the full set. They also concluded with Barnes and Marshall (1951) that low absolute-population numbers correspond to low values of the coefficient of dispersion. In a series of five samples from Mary Lake in the same region the dispersion increased enormously but irregularly with the sample size, perhaps reaching an asymptotic value of the order of 1100 in the largest volumes studied.

Hydrodynamic origins of patchiness. In an earlier chapter it was pointed out (see page 293) that two rather different mechanisms are likely to produce patchiness in plankton. Ragotzkie and Bryson (1953) conclude that since an animal may keep its level phototactically, although the water entering a convergence must move vertically downward, convergences will act as plankton concentrators. They observed great patchiness in the distribution of *Daphnia* in the top meter of Lake Mendota and concluded that in general the concentrations corresponded to areas of convergence. Ragotzkie and Bryson adapted the equations, ordinarily used for considering nonconservative properties in oceanography, and concluded that if adequate discretion is used it is possible to predict accumulations from the data on water movements.

Although the same kind of behavior is to be expected in the convergences of Langmuir spirals as in the larger convergences studied by Ragotzkie and Bryson, it is also possible that in the spiral structure of the wind drift slow passive descent may lead to accumulation of particles in divergences (Fig. 83); floating particles will, of course, appear in convergences giving rise to visible wind rows of blue-green algae or any other objects less dense than water (Fig. 84). McNaught and Hasler (1961) obtained striking evidence of the distribution of *Daphnia* in relation to wind rows in Lake Mendota, where the White Bass, *Roccus chrysops,* swims along the wind row, feeding largely on *Daphnia* that have become entangled in the surface film. There is evidence, however, of more than twice as many *Daphnia* in the water under the areas of high numbers of such airlocked specimens in the surface film, so that the animals are reacting to the pattern of water movements. This may involve Ragotzkie and Bryson's mechanism, but there is also (Baylor 1964) a possibility of zooplankton being concentrated in the convergences by the slight evaporative lowering of temperature in such regions.

Types of horizontal distribution. Bērziņš (1958) has studied the distribution of the zooplankton in a cross section of Skärshultsjön in the Aneboda region of south Sweden. The lake, which is well known limnologically, is an elongate basin running north and south, the deepest part being at the upper narrow northern end, where it is about 13 m. deep. In this part of the lake the western shore is very steep; on the shallower eastern shore are *Nuphar, Potamogeton, Myriophyllum,* and various emergent plants. The lake contains humic or chthoniotrophic water. The distribution of the plankton rotifers and Crustacea was studied on August 7, 1950, in a section across the deepest part. At this time there was a freely circulating epilimnion about 3 m. thick with a temperature of 19.0 to 19.2; below 7 m. the temperature of the hypolimnion lay at 8 to 10° C. At the time of study the isotherms were nearly horizontal. Most of the epilimnetic rotifers tended to occur in two regions of concentration on either side of the long axis of the lake, as indicated in the diagrams (Fig. 201) for *Polyarthra remata* and even more clearly *P. vulgaris* and *P. euryptera. Kellicottia longispina* and *Collotheca mutabilis* showed comparable patterns. It will be seen that in some species the western and in other species the eastern populations were better developed. In *Keratella cochlearis* the western population was large; the more eastern was displaced centrally and somewhat deeper in the metalimnion. Among the epilimnetic rotifers only *Collotheca pelagica* showed its main population in the middle of the lake, to some extent avoiding the main concentrations of its congener *C. mutabilis.* Bērziņš found that in the top 2 m. the mean length of *Kellicottia longispina* is 570 to 580 μ, but with increasing depth the mean length increased to 620 to 630 μ in the deepest region at about 5 m., where the species occurred (*F* of Fig. 202). This suggests a partly passive distribution, though the animals can clearly avoid the deeper cold water. Although the total count for all live rotifers shows the two regions of concentration, the pattern of distribution of the identifiable dead rotifers, largely *K. cochlearis, K. longispina,* and *Polyarthra* spp., shows a single deeper and more central zone of concentration. If the distribution of the living specimens were purely passive, this difference would not occur. Bērziņš thinks that reaction to turbulent movement of the water is involved in explaining the rotifer distributions, but a full elucidation is not yet available. Clear cases of fairly regular metalimnetic (*Polyarthra major*) and hypolimnetic (*P. longiremis, Keratella hiemalis,* and a *Filinia* referred to *longiseta* but perhaps *terminalis*) species were observed.

The distribution (*A–D*, Fig. 202) of the larger and more powerful crustacean members of the zooplankton appeared to be quite different. Most species tended to be distributed at moderate depths, with a marked concentration on the eastern side of the lake. The extreme vegetated littoral was avoided by the more characteristically planktonic species such as *Eudiaptomus gracilis, Daphnia cristata cederströmi,* and *Limnosida frontosa. Polyphemus pediculus* was strikingly littoral, as is usually the case, (Axelson 1961a; Lindström 1952). The two species of *Bosmina, B. longirostris similis* largely littoral, and *B. coregoni longispina* largely pelagic, form an interesting case of two closely allied species replacing each other. Farther north in Scandinavia *B. coregoni* occurs alone and may occupy both the littoral and pelagic habitats (Lindström 1957; Axelson 1961a). The concentration of the majority of species in the deeper water of Skärshultsjön, on the eastern side, is attributed at least in part to lower light intensity at any given temperature in this region of the lake because of the presence of a good deal of suspended organic detritus of littoral origin.

Avoidance of the littoral by eulimnoplankton. It has often been observed that near the margin of a lake the true limnoplankton is poorly developed. Burckhardt (1910) records in Lago Lugano an almost complete absence of planktonic Cladocera near the shore in the eastern arm, a considerable reduction in the number of copepods and also of some rotifers, notably *Kellicottia longispina, Keratella cochlearis,* and *Filinia* sp., whereas *Polyarthra* sp. and *Asplanchna priodonta* became more abundant.

The same phenomenon was noticed in the Silsersee, for which quotients were computed, one of littoral avoidance based on the ratio of open water (51 m.) nocturnal populations at 2.5, 5, and 10 m. to the populations at these depths along the shore, the other of vertical migration, based on the ratios of the populations at the selected levels at the 51-m. station at night to that by day. The figures are very irregular and hardly justify Burckhardt's belief that species showing strongest vertical movement are most likely to avoid the littoral. Burckhardt supposes that any plankter descending in the morning in the littoral will, when it reaches the bottom, continue its descent obliquely down the slope until it reaches an optimal level of illumination. At night, however, the organism will ascend vertically; even if some random redistribution takes place, the process should deplete the population in the free water near the bank. Lindström (1957), however, was not able to detect this type of migration in *Daphnia hyalina* in a locality in which it might have been expected.

Lindström (1952), who has investigated the phenomenon in some

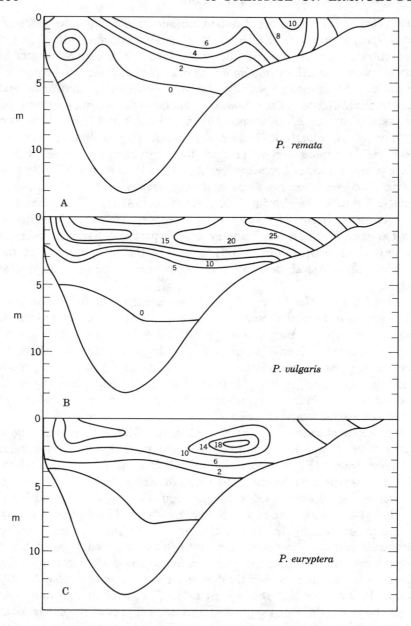

FIGURE 201. Distribution contours (specimens per liter, based on 10-liter samples) of five species of *Polyarthra* and of temperature in a profile across the deepest part of Skarshultsjön, Aug. 7, 1950. *A.-B.-C.* The small-, medium-, and large-sized species, respectively, *P. remata, P. vulgaris,* and *P. euryptera,* the last two having almost identical distributions. *D.* The metalimnetic *P. major.* *E.* The hypolimnetic *P. langiremis,* living in a region containing at all levels less than 2.2 mg. O_2 per liter. *E.* Temperature. (Redrawn from Bērzinš.)

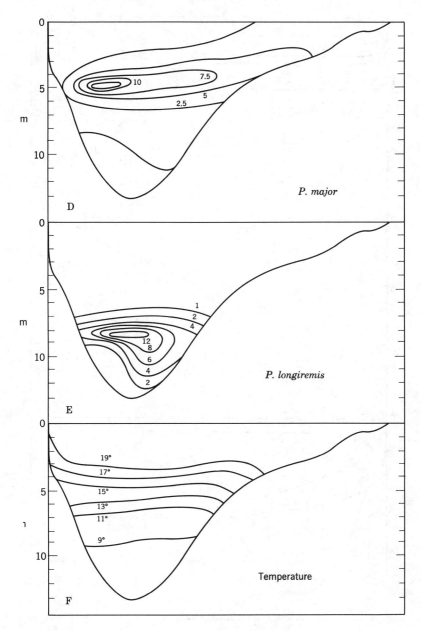

FIGURE 201. (*Continued*)

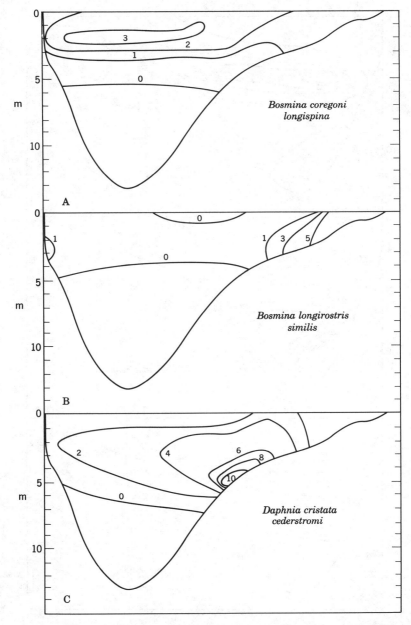

FIGURE 202. Distribution of various zooplanktonic organisms in Skärshultsjön, Aug. 7, 1950.   Scale and temperature as in Fig. 201. A. The
predominantly pelagic *Bosmina coregoni longispina*. B. The much more
littoral *B. longirostris similis*. C. *Daphnia cristata cederströmi*, exempli-
fying a distribution shown by most of the planktonic crustacea. D. The
extreme littoral distribution of *Polyphemus pediculus*. F. The distribution
of *Kellicottia longispina*, typical of the epilimnetic rotifers and in striking
contrast to C. F. Distribution of mean length in *Kellicottia longispina*.

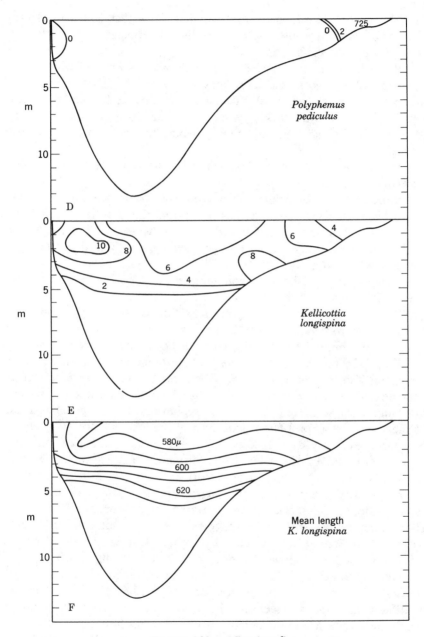

FIGURE 202. (*Continued*)

of the lakes of Jänstland, central Sweden, finds that adults of *Arctodiaptomus laticeps* show signs of it, though apparently the effect is not statistically significant. All stages of *Cyclops scutifer* may be greatly reduced in littoral shallow water, but on occasion exceptional dense populations near the shore may be observed. The phenomenon was not clearly shown by the Cladocera in this study, but in a later paper (1957) it was strikingly indicated for *Daphnia hyalina* living at a depth of about 2 m., but mainly where the water was more than 3 m. deep.

Pennak (personal communication) notes no cases of littoral avoidance in Colorado.

Ruttner (1914), by confining trout and char fry in live boxes through which lake water can circulate, has found that the quantity of plankton in the littoral zone of the Lunzer Untersee is so small that the fry cannot be successfully raised, though they grew well farther from the shore or when fed artificially.

A special case of littoral avoidance is provided by those cases in which planktonic organisms avoid littoral vegetation. Hasler and Jones (1949) conclude that an antagonistic action of such vegetation may be exerted on rotifers but not on Crustacea.

A peculiar but quite instructive case has been reported by Hutchinson (1937a). In the large, closed, very oligotrophic and somewhat saline Panggong Tso, on the borders of Tibet and Ladak, *Daphnia tibetana* appeared to be concentrated in the littoral region in which the water, no doubt because of reflection from the bottom, looked green instead of the vivid blue (Forel-Ule II) of most of the lake. Since *D. tibetana* is a large and very dark species, it is easily seen from the boat. At the western end of the lake, some distance from shore, there was a patch of greenish water over a former islet, which in 1932 was submerged under about 5 m. of water; here *D. tibetana* occurred, as in the littoral. The cladoceran appeared to be commoner in 5 to 10 m. than at the surface in the deep water. Insofar as this case could be studied, it would seem to suggest that in the deep water the animal is forced down by day, presumably by light, whereas in the shallow water a balanced response to incoming light and light diffusely reflected from the bottom leads *D. tibetana* to take up a position nearer the surface than in deep water. The whole subject clearly needs much more detailed study than it has hitherto received.

Avoidance of outlets. There is some evidence that the zooplankton tends to avoid the outlets of lakes. Woltereck (1908a) noted the phenomenon in the outlet of the Lunzer Untersee; the water leaving the lake contained less plankton, particularly zooplankton, than the open water of the lake. The Crustacea showed the effect more markedly than the smaller animals. André (1926) noted a similar phenomenon in the Rhone, leaving the Lake of Geneva. It is uncertain, however, whether these observations indicate a phenomenon comparable to the reduction of eulimnoplankton in the littoral or whether a rheotactic phenomenon is involved. Chandler (1939) found that on an average the total plankton in the water of the outlet of Base Line Lake, Michigan, was about 55 per cent of that in the lake. The composition, in terms of the percentage representation of the major groups of organisms, hardly differed from that of the lake plankton and showed no indication of a deficiency of the Crustacea. In a later study Hall (1962) seems to have found no avoidance of the outlet by the *Daphnia galeata mendotae* population of Base Line Lake.

Apparently quite critical evidence of the phenomenon is provided by the observations of Brook and Woodward (1956), who found in Loch Kinardochy clear evidence, from the ratio of the number of organisms in surface samples taken near the point of exit of the outlet to the number in the water at the outlet, that copepods and Cladocera do actually avoid being washed out of the lake. The observed ratios were

Copepod nauplii	2.5
Copepodite stages of Diaptomidae	4.3
Eudiaptomus gracilis	9.2
Copepodite stages of Cyclopidae	4.0
Cyclops strenuus	11.9
Bosmina coregoni obtusirostris	14.1
Daphnia hyalina	13.8

Capacity to avoid the current seems to develop ontogenetically in the copepods. Senescent individuals, however, are said to loose some of the capacity. *D. hyalina* in a small nearby lochan and its effluent did not show the avoidance effect. In experiments the Cladocera were observed to swim steadily against a current, whereas copepoda were found to react primarily to an acceleration in water movement.

Chandler found that the plankters that left the lakes he studied were usually rapidly filtered out of the effluent rivers by macroscopic plants, though some were caught on other sorts of solid bodies. In rivers lacking much rooted vegetation the loss of plankton is much slower and is probably due largely to its mechanical destruction in riffles and rapids (Reif 1939; Hartman and Himes 1961). It seems possible that the unselective reduction observed in the effluent of Base

Line Lake at the time of Chandler's work was caused by factors other than the responses of the organisms to illumination or current. It is perhaps possible that filtration by littoral plants may be of some importance in reducing the limnoplankton in the marginal parts of lakes, particularly near the effluent, in which unidirectional water movements would be marked.

SUMMARY

Most of the Crustacea, some of the rotifers, and a few other members of the fresh-water plankton perform considerable vertical movement in the course of the twenty-four hour period. This movement is, in general, more accentuated, the more transparent the lake, and is certainly controlled by the diurnal light cycle. A given species is frequently found to behave differently in different lakes, at different stages and at different times of year or according to sex differences. When two allied species occur in the same lake, they often occupy different strata and behave differently. Photosynthetic mobile forms (*Volvox, Eudorina, Gonyostomum*) rise in the water in the morning and descend during the afternoon or evening. The typical migratory behavior of animal plankton is to move upward at night (nocturnal migration) or at dawn and sunset (twilight migration). Rare cases in which the animal behaves like the autotrophic plankton (reversed migration) have been recorded. Nocturnal migration itself is of several kinds. Often there is a slight tendency for animals which have reached the surface about 10 P.M. to scatter downward, possibly tending toward a random distribution. It has been supposed that the paradigmatic case would be an upward movement in the late afternoon and evening, a nocturnal scattering downward, an upward ascent just before dawn, and then a descent with a prolonged sojourn in deep water during the day. Cases of upward movement, however, have been recorded as continuing throughout the night until the sky begins to lighten. Occasionally parts of the main population break off and behave individually.

A great deal of experimental work has been done to elucidate the vertical migration of plankton in lakes, much of it on species living in ponds and exhibiting little migration in nature. In the simplest cases there appears to be a positive photokinesis in very dim light which will cause the upward movement of any animal ordinarily oriented relative to the earth's gravitational field either directly or by means of currents set up as it sinks. At greater light intensities a negative photokinesis or inhibition of movement, or a negative phototaxis involving a change of posture relative to the light and swimming away from the source, intervenes. These alternate modes of reaction will give ordinary nocturnal migration if the animal maintains its position during complete darkness but some form of twilight migration if it sinks in the dark.

Very few fresh-water animals are likely to respond directly to gravity, since few possess statocysts. In *Daphnia* the apparent response to gravity is actually a response to the relative movement between water and animal as the latter sinks passively between beats of the second antennae. No fresh-water animals are clearly recorded as being sensitive to pressure, as is common in the sea; it is possible that the mechanism of pressure sensitivity requires a saline external medium.

In addition to the positive photokinesis at low intensities, apparently dependent on a general photosensitivity of the nervous system and not on the eye, and the inhibition at higher intensities, the forms that have been studied all exhibit more complex reactions. The most important is the dorsal light reaction, by which any animal illuminated from above will alter its posture if the direction of the light is changed so that its relation to the illumination is maintained. Moreover, if the light is altered in intensity, the animal swims down when the intensity is increased or up when it is decreased, moving obliquely, as if the angle had been changed, until it has regained the position at which the intensity is restored to its previous value. In the Cladocera, at least, there is a marked difference in response to short- and long-wave light. Shortening the wavelength causes movement to and fro in a horizontal plane and at right angles to the direction of the incident beam, the so-called blue dance. Increasing the long-wave component causes movement to and fro in the line of incidence, whatever its angle. All the reactions are sensitive to environmental conditions.

A very good model of migration can be made by putting *Daphnia* in water which is made artificially less transparent with India ink. A cylical change in illumination produces all the main features observed in nature, apparently in terms of the superposition of the dorsal light reaction due to the eye, on the extreme twilight migration undertaken by eyeless animals.

Within the pattern of response characteristic of the Cladocera as a whole there are great differences in the strength of individual reactions in different species. Thus *Holopedium gibberum* in diffuse light descends but in darkness becomes randomly distributed, whereas *Polyphemus pediculus* becomes more apparently geotactic in the dark, and *Bosminopsis dietersi* appears to be indifferent to the gravitational field in either diffuse light or darkness. It is reasonably certain that the differences observed in nature are related to this kind of difference, but so far there are not enough data to build up a consistent model of the specific differences. In some special cases, in which the population in nature is confined to a particular layer in the lake, often in the upper hypolimnion, a marked reversal of behavior on entering a region of rapidly changing temperature must occur and is in general in accordance with laboratory findings.

There has been uncertainty regarding the relative importance of passive sinking and downward swimming in nature. Although the descent speeds are of the order of magnitude of those of freely falling, narcotized animals in the laboratory, these animals are falling in an almost nonturbulent medium. At least in the epilimnion the descent in nature is in a turbulent medium of far greater virtual viscosity than obtains in a laboratory vessel. The descent rates observed in nature therefore probably imply active downward movement. Detailed observations under the best experimental conditions support this conclusion. In some cases the passive component may be increased by a rise in density after feeding on dense organisms such as diatoms.

Changes in the migratory pattern throughout the season can in some cases be shown to depend on the decline of the O_2 and increase in the CO_2 content of the deep water.

The process of vertical migration involves a relatively small expenditure of energy. It is most reasonably regarded as a special case of the nocturnalism which gives some degree of protection to small animals of the presence of predators. In addition to this possible function of the process, there is reason to believe that at least at high temperatures descent from the trophogenic layer even into a very slightly cooler, deeper layer after feeding provides the most economical metabolic conditions for growth and reproduction.

In very large lakes, particularly of complicated form, in which there is a possibility of existence of largely independent water masses, the zooplankton populations would not be expected to be uniform horizontally. There is, however, apparently also much horizontal diversity within quite small bodies of water. The most practical measure of this diversity can be developed from the fact that in a series of samples of a randomly distributed species we should expect the numbers of samples containing 0, 1, 2, 3, . . . specimens to follow a Poisson distribution e^{-m}, $e^{-m} \cdot m$, $e^{-m}m^2/1.2$, . . . , where m is a number dependent on the sample size and average concentration of the species. In such a distribution the variance of the samples is equal to their mean. If the variance significantly exceeds the mean, the specimens are said to be superdispersed, forming clumps separated by spaces lacking individuals; if the variance is less than the mean, the specimens are more regularly spaced than would be expected on the hypothesis of randomness and are said to be infradispersed. Of the various distributions which might be expected in a superdispersed population, the Thomas double Poisson distribution in which the size of the clumps varies randomly around a mean greater than unity, although their distribution in the water is random, has been regarded by some planktologists as reasonable. The negative binomial distribution, in which the population is made up of classes of individuals, each with a different Poisson distribution, as in the array of accidents in time exhibited by a human population of individuals of different degrees of accident proneness, has also been regarded as a possible model. Superdispersion certainly exists, though many populations of plankton organisms seem to be horizontally random. When attempts are made to fit one or another of the more elaborate models, so far little of biological significance emerges. The observed superdispersion may be due to hydrographic factors; rheotactic movements in a Langmuir spiral could well give plankton streaks, which would appear as clumps if sampled in transverse section. It is possible that at least in very stable metalimnetic or hypolimnetic layers the reproductive activity of plankters might lead to a little superdispersion. Some nonhydrographic explanation is probably needed for the 8-m. layer of Lake Nipissing in which, over short distances, all stages of the Diaptomidae, but not *Epischura* or the Cladocera, were superdispersed. Infradispersion is much less frequently reported; it has been noticed, however, and has been explained as due to competition. Such competition would involve a sort of territoriality, which seems unlikely in copepods, but of enormous interest if it were established.

A special kind of horizontal distribution is provided by those cases in which plankton animals appear to avoid the shore. The mechanisms involved are uncertain. It has been suggested that in downward migration over a sloping bottom the descending animal might continue moving downward, hugging the bottom, and so reach deep water in a more central spot than it had inhabited previously. Subsequent upward movement would lack any such constraint and the population would tend to collect in the pelagic region. It is possible that the littoral also possesses certain optical properties due to the reflection of light from the bottom that would cause some species to aggregate and others to avoid the region. In some cases planktonic organisms appear to avoid the outlets of lakes.

CHAPTER 26

Cyclomorphosis

The term cyclomorphosis was introduced by Lauterborn (1904) to designate the seasonal polymorphism that is frequently observed in planktonic organisms. The phenomenon has been noted in the dinoflagellates, rotifers, Cladocera, to some extent in the Lobosa (see pages 491–492), and to a much less striking degree in the copepods. In its extreme form it appears to be exhibited primarily by organisms that reproduce during most of the year by asexual or parthenogenetic methods, so giving rise to genetically homogeneous clones. The degree to which cyclomorphosis is developed within different populations of the same species is variable, and although the seasonal incidence of the change is clearly determined by environmental factors, partly known, partly still unknown, there may also be inherited diversity in the capacities of different races of a species to react to these environmental factors. There is a rather general belief that the cyclomorphotic changes must have some adaptive significance, but difficulties have arisen in the adaptive interpretation of nearly every case and no agreement on the matter yet appears possible. The study of cyclomorphosis is evidently a part of limnology that can make interesting contributions to other branches of biology, notably morphogenesis and evolutionary genetics. For this reason and because, outside the recent admirable works of Brooks (1947, 1957a) on *Daphnia,* no adequate review of the subject exists, it is treated at considerable length in this book, but it must be admitted that the treatment, though lengthy, may well appear fragmentary and disappointing to any embryologists and geneticists who may chance to consult it.

In presenting an account of this puzzling though fascinating subject, it is convenient first to call attention to a method of analysis of particular value in the study of morphological change and then to proceed to a discussion of the organisms exhibiting cyclomorphosis, beginning with the best studied cases.

HETERAUXESIS IN THE PRODUCTION
OF SEASONAL FORM CHANGE

It is commonly found that when a marked change in shape occurs in a growing organism the size of the part mainly responsible for the change in shape (**y**) can be related to some standard dimension (**x**) by the equation

$$(1) \qquad y = bx^k.$$

The particular kind of relative growth implied by this relationship has been the subject of many studies in the last three decades, the most important being those of Huxley (1932) and of Needham (1942). A rather confusing terminology has grown up during this period. At first the process was termed heterogony, later allometry; neither term is free from objection, and at present *heterauxesis* (Needham and Lerner 1940) appears to be generally acceptable as a name for the process described by the equation. When **k** is less than unity, so that the part is relatively smaller in large than in small stages of growth, the term *bradyauxesis* may be used; when **k** is greater than unity, the corresponding term is *tachyauxesis*. For the special case in which $k = 1$ and in which no change in shape occurs, the term *isauxesis* is convenient.

The simplest method of determining this kind of relative growth is to convert to logarithms

$$(2) \qquad \log y = \log b + k \log x$$

If a series of measurements of the part (**y**) is plotted against those of the standard dimension (**x**) on a double logarithmic grid, the points will fall on or close to a straight line if there is heterauxetic growth; the slope of the line will give the value of **k**. It is often useful to apply this process to populations, each individual representing either a growth stage or an adult that has stopped growing at a particular value of x and so exemplifies the appropriate value of y. The concept of heterauxesis has been particularly valuable in the analysis of cyclomorphosis in *Daphnia* (Brooks 1946); as will be shown later, it can be applied to certain aspects of cyclomorphosis in another cladoceran, namely, *Bosmina coregoni,* and with even greater success to certain cyclomorphotic changes in the rotifers of the genus *Keratella* (Margalef 1947).

Other methods of graphical analysis of the local and seasonal variation of planktonic organism have been employed from time to time (e.g., Woltereck 1924; Rammner 1926), but in my opinion the study of heterauxesis is the only one of these methods that provides any real insight into cyclomorphosis.

CYCLOMORPHOSIS IN CLADOCERA

The cyclomorphosis of species of the genus *Daphnia.* Daday (1885–1888) was apparently the first to appreciate the great variability expressed in certain planktonic populations of *Daphnia,* but the temporal aspect of this variation was not clearly indicated until the slightly later work of Zacharias (1894). A considerable number of investigators observed the phenomenon in the next ten years. The full details of this early work may be explored by means of the bibliographies compiled by Wesenberg-Lund (1900, 1904, 1908).

The phenomenon in the genus *Daphnia* consists in the main of a seasonal variation in the structure of the head. In some species of *Daphnia* the antennary muscles are inserted almost entirely along the anterodorsal margin of the head, but in many forms there is a thin ridge or carina, formed from the adpressed body walls of the left and right sides, lying dorsal and anterior to the insertion of the muscles. This crest can be produced forward as a helmet of varying size. Cyclomorphosis in *Daphnia* consists mainly in the seasonal variation of the crest or helmet. There is often, however, a slight change in size; the carapace length, which is the most convenient standard measure, decreases between the spring and summer and then increases again. Small changes in the shape of the rostrum and adjacent posteroventral parts of the head have also been noted. The posterior spine is often variable, but its variation is little understood and may have nothing to do with that of the head. There is frequently a marked tendency for the number of parthenogenetic eggs to be much greater in the large spring forms than in the smaller summer forms with exaggerated helmets. Wesenberg-Lund thinks that the number of eggs laid in an individual lifetime may perhaps be as great in the summer generations as in the spring because of a greater number of egg-producing instars.   He does hint that sometimes summer females with long helmets and narrow bodies may be sterile, though the evidence is certainly not conclusive.

The extreme diversity of the shape of the helmet, when different species are compared, must be emphasized.   In some it is an evenly rounded structure and in others produced forward as a triangular or pointed process; in yet others it is retrocurved or more rarely procurved and is sometimes better developed dorsally than anteriorly.   Even within a given species the extreme form of the helmet may differ somewhat in different geographic races.

Cyclomorphosis in holarctic species of Daphnia (s. str.).   Striking development of a helmet occurs in a number of species of holarctic *Daphnia* (s. str.) which have been studied mainly in the Baltic region (Fig. 203), the central European massif, and the eastern half of North America (Fig. 204).   *Daphnia hyalina* and *D. galeata,* two of the species involved in Europe, have in recent years usually been regarded as at most only subspecifically distinct from *D. longispina.*   Brooks (1957a), following Richard (1896), separates *hyalina* and *galeata* as distinct eulimnoplanktonic species from the less limnetic *D. longispina* (s. str.).   Both species may be helmeted, though the helmet is more pointed apically in *D. galeata.*   There is often more than one form present in a lake, one being helmeted, the other round-headed, as in the Millstätter See (Haempel 1924), the Chiemsee in Bavaria (Berger 1934), and Lago Lugano in Switzerland (Berger 1934).   A good deal of additional taxonomic work is needed before all the numerous described forms can be assigned to reasonably valid species.   The commonest European form with a well-developed helmet appears to be *galeata,* but the rounded though elongate head of f. *ceresiana* from Lago Lugano indicates *D. hyalina.*   In North America neither *longispina* nor *hyalina* occur, though the relatively noncyclomorphotic *D. thorata* from Montana, Washington, and British Columbia is very near *hyalina. D. galeata,* however, with somewhat broader if still pointed head and fewer anal teeth, occurs as *D. g. mendotae,* an important cyclomorphotic form in North America east of the Rocky Mountains from Alaska to Central America.

Daphnia cucullata, the most conspicuously cyclomorphotic of the European species, is a Palaearctic form, the eastern limits of its distribution being somewhat uncertain.   It does not occur naturally in Italy, though it was introduced in Lake Nemi by Woltereck (1928, Volterra 1924, Volterra D'Ancona 1927, 1931, 1938, D'Ancona 1942).   It is known, however, in at least one Balkan locality (Wagler 1923). *D. cucullata* is presumably an offshoot of the same stock which gave

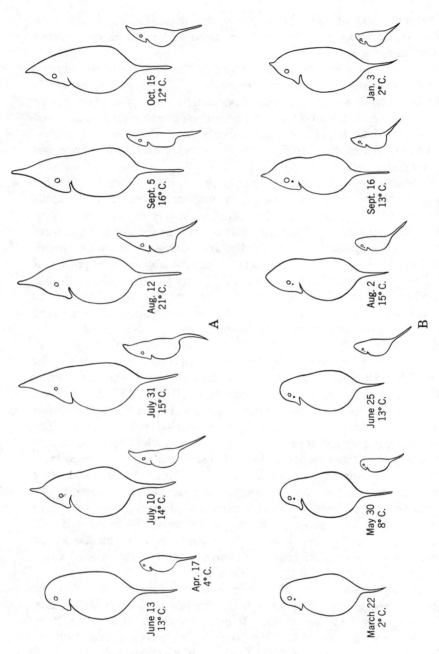

FIGURE 203. *A.* Cyclomorphosis of *Daphnia cucullata*, Esrom Sö. *B. D.* Same of *galeata*, Haldsö, Denmark (Wesenberg-Lund). The small individuals are first instar juveniles drawn to the same scale as the adults.

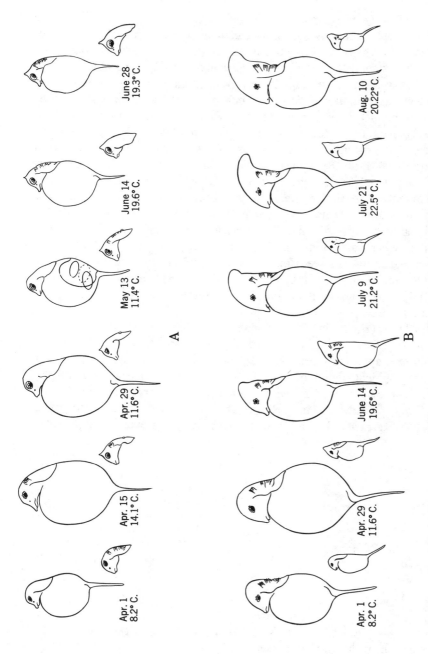

A

B

FIGURE 204. *A*. Cyclomorphosis of *D. ambigua*, Bantam Lake, Conn. *B*. Same of *D. retrocurva*, Bantam Lake, Conn. (Brooks.) The small heads or individuals at the right and below the adults are first instar juveniles.

rise to *D. longispina,* from which species it differs primarily in lacking an eye spot. There is a round-headed noncyclomorphotic ecotype of *D. cucullata* which inhabits ponds, but, unlike the forms of *D. longispina* living in ponds, *D. cucullata* f. *hermanni* is a dwarf rather than a large heavy form (*A* of Fig. 206). Woltereck (1928) claims to have produced a comparable form by culturing *D. cucullata* with a pronounced helmet in a cement tank. Wagler considers that in general large lakes contain large forms of *D. cucullata* with high helmets. A reanalysis of his data by Brooks (1946) shows that apart from the occurrence of the round-headed dwarf f. *hermanni* in shallow or small bodies of water, Wagler's great array of facts do not support his contention in this matter. If, however, a homogeneous series of lakes in a single region is compared, it seems quite possible that a relationship of this sort may be apparent. Findenegg (1943a) quotes a rather clear case from the Carinthian lakes (Fig. 205). Here the Lanser See, Rauschelesee, Keutschacher See, Ossiacher See, and Wörther See, a ranging in depth from 11 to 84 m. and in area from 0.03 to 19.4 km.², contain populations of increasing size and relative head length.

The least (*B,C,* Fig. 206) markedly (f. *berolinensis* and f. *vitrea*) and most (*E* of Fig. 206) markedly (f. *cucullata*) cyclomorphotic populations apparently are scattered more or less at random throughout the geographical range of the species, and the same is true of the much less common form with a well-developed retrocurved (*F.* of Fig. 206) helmet (f. *incerta*). A form with a markedly procurved (*G,H* of Fig. 206) helmet (*D. c. procurva*), however, is apparently

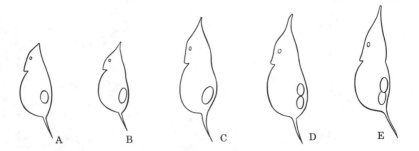

FIGURE 205. *Daphnia cucullata,* extreme form taken at height of summer, primparous parthenogenetic female. *A.* Lanser See, 11 m. deep, 0.03 km.² area. *B.* Rauschelesee, 11 m. deep, 0.2 km.² area. *C.* Keutschacher See, 15 m. deep, 1.4 km.². *D.* Ossiacher See, 46 m. deep, 10.6 km.² area. *E.* Wörther See, 84 m. deep, 19.4 km.² area. (Findenegg.)

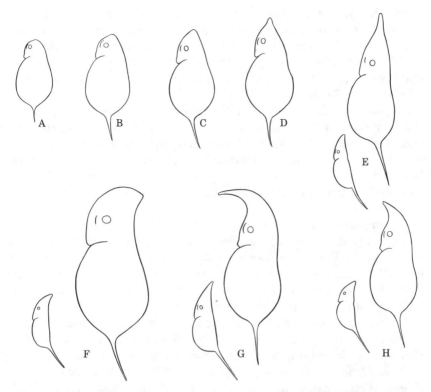

FIGURE 206. Form of *Daphnia cucullata* as developed in various populations at the height of summer. *A.* Wroxham Broad, Sept. 29, 1898, sixth instar adult, f. *hermani.* *B.* Borsdorf, near Leipzig, Saxony, July 17, 1911, sixth instar adult, f. *berolinensis.* *C.* Rohrbach, near Leipzig, millpond, June 21, 1902, sixth instar adult, f. *vitrea.* *D.* Otterwisch, near Leipzig, millpond, Sept. 2, 1918, sixth instar adult, f. *apicata.* *E.* Zarnowitzer See, July 23, 1906, first instar juvenile and sixth instar adult f. *cucullata* (= *kahlsbergensis* auctt.). *F.* Wannsee, July 20, 1909, first instar juvenile and adult f. *incerta.* *G.* Lake Witoschtno, Poland, Aug. 27, 1909, first instar juvenile and sixth instar adult, extreme form of *D. c. procurva.* *G.* Lake Plensno, Poland, first instar juvenile and sixth instar adult, less extreme population of *D. c procurva.* (Magnification × 50–80, length in μ given within drawing of each specimen.) (Wagler.)

a geographically defined subspecies, found only in the lakes of the middle Brahe valley in Poland.

Daphnia cristata, another species that seems to be confined to the Palaearctic and is particularly common in northern Europe, exhibits strong cyclomorphosis (Fig. 162). The extreme summer form (f. *cederstromi*) has a well-developed retrocurved helmet. The closely allied *D. longiremis* shows much less temporal variation in Europe than does *D. cristata,* though both may occur in the same lake (Freidenfelt 1913). *D. longiremis* is widespread in northern North America, usually as a cold-water hypolimnetic form, and in two instances a retrocurved helmet has been recorded in American *D. longiremis* (Woltereck 1932; Brooks 1957a).

D. pulex in Europe is of little importance in the lake plankton and is very slightly cyclomorphotic. There is an indication, however, from the work of Hsi-Ming (1942) that slight changes in head form are to be observed when populations of this relatively stable and un-helmeted form are followed throughout the year. Berg (1931) has also given interesting data, to be discussed later, on the variation of the tail spine.

D. retrocurva, though not very closely allied to *D. cucullata,* appears to represent it ecologically in North America; the extreme form best developed in Wisconsin has a much more retrocurved helmet than is usual in *cucullata.* As in that species, a persistent round-headed form, *D. retrocurva* f. *breviceps,* is found in some localities within the range of the cyclomorphotic form.

D. dubia, another North American species, has a rather limited distribution in the northeastern United States and adjacent parts of Canada. It develops a fairly tall, somewhat retrocurved helmet.

D. ambigua, in which only a spine-like helmet is developed, is probably a species native to North America, though described originally from Kew in England. Under the names *longispina* and *galeata* it has played an important part in the experimental elucidation of cyclomorphosis.

It is a matter of considerable interest that although permanently helmeted species of *Ctenodaphnia* occur in Africa the most strikingly cyclomorphotic forms in the subgenus *Daphnia* (s. str.) appear mainly in localities in which there is a great temperature change between winter and summer. *D. retrocurva* is distributed across north temperate North America from Puget Sound to New England (Brooks 1957a). The helmet is highest in Wisconsin, though quite variable in the Middle West, where f. *breviceps* may also occur. In the most western localities, under an oceanic climate, the helmet is much less striking than in Wisconsin or New England. In the southern United States the species is replaced by *D. parvula,* very like *D. retrocurva* f. *breviceps.* Similarly, *D. dubia,* is replaced from Long Island south, by the allied attenuate but relatively unhelmeted *laevis.* The most southern species in North America to show obvious cyclomorphosis is *D. ambigua,* in which only a sharp spike is produced, and *D. galeata mendotae* with its broad, rounded helmet, both are much less impressive than the summer forms of *D. retrocurva* or *D. dubia*. The rarity of *D. cucullata* in Europe south of the Alps is probably an example of the same phenomenon. If the helmet were primarily an adaptation for life in warm water this particular type of distribution would not be expected.

Findenegg (1943a) alone seems to have considered quite critically the events throughout an entire annual cycle, though Brooks's (1946) work is even more detailed for the spring and summer seasons. Findenegg found that *D. cucullata* is acyclic in the Wörther See. The overwintering generation is born in November and December at a time when the epilimnetic temperature is falling from 16 to 4° C. In most years the mean temperature at this time appears to be about 8 to 10° C. (Findenegg 1938); since reproduction stops at 6° C., most of the neonatae born after November 1 are likely to be produced at temperatures above rather than below this mean value. Such animals have low pointed heads which are retained as they grow slowly throughout the winter. They start reproducing in April as soon as the temperature reaches 5° C. and produce round-headed progeny which retain their round heads throughout life. The round-headed generation reproduces in May and June, when the epilimnetic temperature is rising from 10 to 20° C.; they produce helmeted progeny which, however, are not so extreme as the later generations, born when the temperature of the epilimnion is more than 20° in July and August. There is no evidence of an increase in the proportionate head length within the life span of a single individual early in the summer, growth being, as is indicated clearly by Findenegg's figures, isauxetic until June. Later in the summer it appears that the helmet does grow more rapidly than the carapace, as Wesenberg-Lund considered to happen quite generally; this aspect of Findenegg's observations is apparent only from his figures but fits in perfectly with what Brooks has discovered in American species (see page 829). As indicated later, in the discussion of relative growth rates, there is some uncertainty whether the height of the helmet relative to the carapace decreases slightly during ontogeny at low winter temperatures, but Findenegg's careful observations and figures give no evidence of a dramatic decline in the relative height of the head during ontogeny when the temperature is falling, as Zacharias (1903) believed to take place.

Seasonal variation in tail spine. The seasonal variation of the spine in *D. pulex* in a small circular pond, Hylderupgaard Dam, near Lyngea, in northern Zealand, has been considered by Berg (1931). The neonate are always provided with a long spine, 200 to 320 μ long or 22.7 to 43.2 per cent of the length of the rest of the animal. The proportions of the spine at birth appear to vary irregularly and without regard for season. During the summer the spine may increase in length so that in July primaparae of length 1570 μ can have a spine of length 390 μ and old females of length 2620 μ, a spine of length 470 μ. During the winter and early spring, however, the spine length actually decreases during ontogeny. At this season primaparae of length 1610 μ have a spine of only 60 μ, or about one third the length that they presumably had as neonatae, and the very large old females, which may be as long as 3000 μ, lack spines entirely.

Olofsson (1918) came to an entirely different conclusion regarding spine length in *D. pulex* in Spitzbergen; he believed that the length in the neonata varied directly with the temperature at which development took place and that the spine retrogressed in the course of post-natal ontogeny only if the temperature were low. The field evidence on which these conclusions are based evidently leaves much to be desired.

Minor variations of form in Daphnia. In addition to the major cyclomorphotic changes in size, helmet, and tail spine, a few minor types of variation have been recorded. In some otherwise rather stable populations of *D. longispina* a marked excavation of the ventral contour of the head between the rostrum and the vertex may develop (Fig. 207). Wagler (1912) concluded from the occurrence of this form in old poorly fed cultures that it is the result of malnutrition, and the contour in question has been referred to as the hunger curve. Berg (1931) found that in Egekjaergaard Dam, near Sorø, a pronounced hunger curve developed in the population of *D. longispina* during the autumn at a time when sexual reproduction was becoming increasingly frequent, which probably indicates a decline in the quantity of available food. Comparable phenomena have been noted by Uéno (1934a) in Japanese populations of *D. longispina* and by Volterra (1924) in a population of Lake Nemi when the density of the zooplankton was very great. Banta's (1939) "dish-faced" *Daphnia* probably show a comparable condition, but he has also studied genetically determined excavated heads of very similar shape.

A curious elongate form (*A* of Fig. 208) of *D. laevis* with a marked helmet, recorded occasionally in numbers in nature and as single individuals in culture, has been studied by Banta (sub *D. longispina*). In parthenogenesis the offspring may be slightly attenuate but rapidly become normal in form. No attenuation is shown in later generations, and no clear idea of the origin of the form can be obtained from the available data. Brooks (1957a) points out that immature females

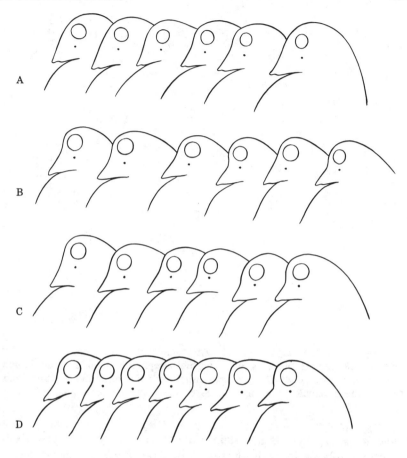

FIGURE 207. *Daphnia longispina,* Egekjaergaard Dam, near Sorø, Denmark, profiles of heads of populations. *A.* Parthenogenetic females, July 22, 1925. *B.* Parthenogenetic females, Sept. 8, 1925. *C.* Ephippial females from the same collection as *B.* *D.* Ephippial females, Nov. 3, 1925. The hunger curve appears to have increased as a greater proportion of the population became ephippial. (Berg.)

of this species are often somewhat helmeted but tend to become less so as they grow up, even in entirely natural environments.

Manuylova (1948) has concluded that in Lake Balkasch the helmet of the taxonomically problematic *D. balchashensis* (Fig. 209) is progressively lowered as the salinity of its habitat increases.

Helmeted and other exuberant forms of tropical and subtropical species of the subgenus Ctenodaphnia. In addition to the relatively detailed field and laboratory studies of members of the subgenus

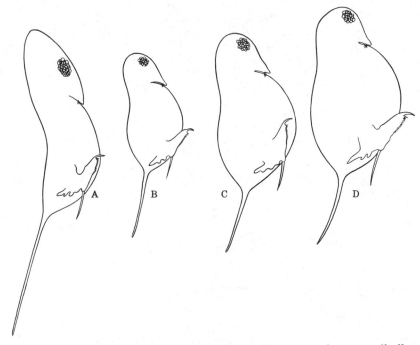

FIGURE 208. *Daphnia laevis.* *A.* Attenuate form occurring sporadically among 428 normal offspring of a normal exephippio female in an inbred laboratory strain. *B.-C.-D.* Specimens from successive generations reared from this attenuate female, showing absence of the attenuate condition. (Banta.)

Daphnia (s.s.), there is a certain amount of evidence (Wagler 1936) of cyclomorphosis in various species of *Ctenodaphnia*, particularly in the tropics and subtropical regions of the Old World. Among these species *D. lumholtzi* Sars, which may be a pelagic derivative of *D. (C.) magna,* is known to produce a helmet both in the Volga Delta region and in Australia, localities at the opposite ends of its area of distribution. The fully helmeted form has a very sharp pointed projection on the head, inevitably reminiscent of a narwhal, if not a unicorn (*B* of Fig. 209). Sars obtained such animals from mud cultures; later ephippial females without helmets were produced (*C* of Fig. 210). *Daphnia (C.) barbata* is in form not unlike some of the less extreme races of *D. cucullata;* it is exclusively African and appears to be more or less helmeted (*A* of Fig. 210) in all the collections that have been made. Biologically *D. barbata* probably differs from *D. cucullata* in producing a marked helmet in quite shallow weedy swamps, such as Brakpan, Transvaal, from which locality Wagler figures the most extreme of six illustrated populations of the species. In many localities in South Africa, especially in the perennial, slightly alkaline, turbid, shallow saucer-shaped pans of the Transvaal (Hutchinson, Pickford, and Schuurman 1932), *D. (C.) gibba,* in which a marked dorsal crista (Fig. 211) is developed, is very flat animal with a wide oval profile. In some collections there is evidence of considerable variation in the development of this crest even in adults; this is noticeable in the late autumnal series, May 27, 1928, figure by Wagler from Eliazar Pan, Potchefstrom, Transvaal. A specimen reared from mud sent from this locality has been figured by Hutchinson (1933) as *D. thomsoni.* It is reasonable to suppose that this form represents the most reduced phase of *D. gibba* developing in nonturbulent water. *D. (C.) cephalata* (*E* of Fig. 211) is a still more extraordinary form from India, Ceylon, Flores, Australia, Tasmania, and New Zealand. Nothing seems to be known about the swimming positions of these remarkable animals. It is not clear why a dorsal crest is of any

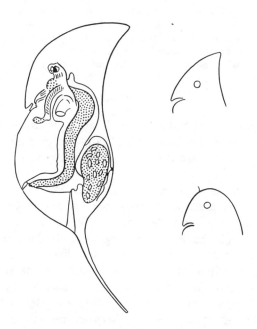

FIGURE 209. *Daphnia balchashensis* (cf. *incerta* of *F* of Fig. 206), whole animal with marked retrocurved helmet from dilute part of the lake and heads of two progressively reduced forms from water of increasing concentration. (Manuylova.)

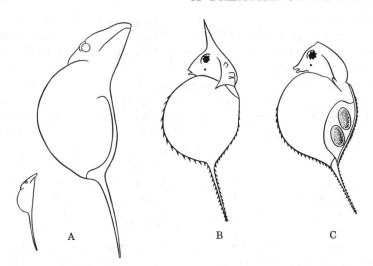

FIGURE 210. Helmeted species of the subgenus *Ctenodaphnia* apparently not exhibiting typical cyclomorphosis. *A. Daphnia* (*C.*) *barbata* juvenile and adult from a late winter (Aug. 30, 1928) collection from the Vereiniging Dam on the Vaal River, S. Africa. (Wagler.)   *B.-D.* (*C.*) *lumholtzi,* parthenogenetic female raised by Sars from mud from Gracemere Lagoon, west of Rockhampton, N. Queensland, Australia. *C.* Ephippial female of a later generation, without helmet.   (Sars.)

special significance in these shallow and rather astatic localities.   The growth series of *D.* (*C.*) *gibba* from Blaauwater Pan given by Wagler indicates great relative growth in a dorsal-ventral direction during ontogeny.

The efficient causes of cyclomorphosis in Daphnia.   The history of the theoretical interpretation of cyclomorphosis is somewhat confused by a failure, particularly on the part of Wesenberg-Lund, who was the most notable of the early investigators, to distinguish between efficient and final causes.   Wesenberg-Lund first saw in the decline in density of the medium and consequent supposed increase in sinking speed of the plankton in summer the final cause of cyclomorphosis as an adaptation to increase form resistance.   He later accepted Ostwald's (1902) idea that the fall in viscosity rather than density with increasing temperature is of primary significance.   In 1904 Ostwald published the results of some rather crude experiments with *Daphnia cucullata,* which showed quite definitely that the embryos in the brood pouch of a female in cold water developed far less striking helmets than those developing in warm water.   Ostwald had great difficulty in keeping his animals alive, and it is reasonably certain that none was adequately fed, a fact that actually should have given support to his conclusion. In 1901, however, Krogh discussed the problem in a lecture summarized by Wesenberg-Lund (1908) and suggested, without supplying evidence, that the level of nutrition was the important determinant of cyclomorphosis. The problem of the efficient cause of the growth of the helmet thus settled down into

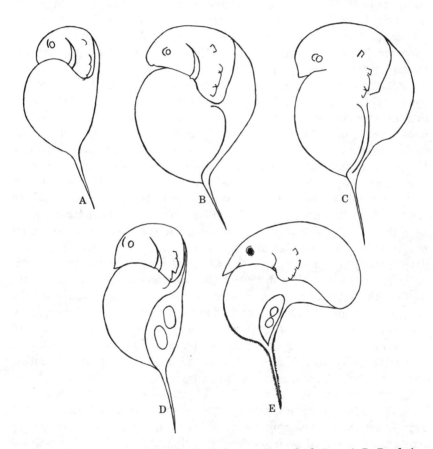

FIGURE 211. Extreme forms in the subgenus *Ctenodaphnia*. *A.-B. Daphnia (Ctenodaphnia) gibba,* Eliazar Pan near Potchefstrom, Transvaal, May 27, 1928, adults, showing minimum and maximum development of the helmet. *C.-D. (C.) gibba,* Blaauwater Pans, Lake Chrissie district, Transvaal, Feb. 27, 1928. *D.-D. (C.) gibba* moderately helmeted ephippial female, Magdalenas-meer, Lake Chrissie district, Transvaal. *E. D. (C.) cephalata,* Australia. (A.-D., Wagler, E., Sars.)

the issue of temperature versus nutrition, and Wesenberg-Lund, however he may have felt originally about the nature of the stimulus provided by warm, less dense, less viscous as opposed to cold, more dense, more viscous water, became the protagonist of the idea that temperature was of paramount importance. Several German authors, most notably Woltereck, developed the view that the major factor operating was increased nutrition in the summer. Initially, the only experimental work, namely that of Ostwald, seemed to give results unequivocally in favor of the view that temperature was of primary importance, but for a quarter of a century after Ostwald's work nearly all experimental publications came from Woltereck's laboratory (Woltereck 1908b, 1909, 1911, 1913, 1921) and they put forward entirely different ideas on the subject.

According to Woltereck, the main environmental variable controlling cyclomorphosis in *Daphnia cucullata* is the food supply. The limitation of cyclomorphosis in *D. g. mendotae* by starvation has indeed been confirmed (Jacobs 1961). Woltereck, however, went much further than the simple conclusion that food supply could be involved. He believed that increased or reduced temperatures operate to control head form only indirectly by increasing or reducing the food intake. The effect of an optimal food supply, moreover, was supposed by Woltereck to be cumulative. The individuals that hatch from resting eggs, the so-called *ex ephippio* generation, are never helmeted, and about six generations of optimal feeding were needed to produce the highest development of the helmet in the so-called *vertex* generation. Subsequently, there is a decline in helmet size, even under optimal conditions, so that the later *clivus* generations, though more helmeted than the *ex ephippio,* are definitely less helmeted than the *vertex* generation. Woltereck claims that if during the process of the development of the helmet in the early generations the nutrition of the developing eggs is radically disturbed, as by amputating an antenna of the mother and so producing disharmonious swimming movements, the young hatch with very reduced helmets, and two generations may be needed to bring the helmet back to the condition of that of the operated animal. This effect of food supply on the size of the helmet in succeeding generations is termed *preinduction* by Woltereck. Certain observations suggest that effects of variation in the food supply can produce changes in the offspring in the Cladocera. In *Simocephalus* Agar (1913) showed that a peculiar gaping of the valves can be produced by feeding the mother a particular, most unfortunately unidentified, alga. Rather similar induction of physiological characters by changes in the quantity of the food has been reported by Bradshaw (1949). Woltereck's pre-induction therefore should not be dismissed as an impossible case of the inheritance of an acquired character. Indeed, the only difficulty in the way of accepting this and all of Woltereck's other conclusions is the fact that whenever attempts have been made by other workers (Berg 1936; Coker and Addlestone 1938; Coker 1939; Brooks 1946) to confirm Woltereck's experimental results on cyclomorphosis in *Daphnia* the results have almost always (cf. Jacobs 1961) been found to be incorrect. None of Woltereck's experiments was reported in detail with full information as to the technique, number of animals, density of food, and actual measurements of specified specimens. At best, a single typical individual of each of the more significant genera-tions was figured. It is impossible therefore to gain any clear idea of what really happened in his cultures. The only policy that can be adopted with regard either to his experimental work or to the studies of behavior on which his theory of the adaptive value of cyclomorphosis is based in part is to pass them over with a relatively brief mention, without forgetting Woltereck's great services to limnology in other directions, and to proceed to the more recent work of other investigators who have provided more objective information about their experimental procedure and results.

The first unequivocal evidence of the correctness of the view that temperature as such was at least one of the efficient causes of the development of the helmet was derived from experiments on *D. cucullata* by Ostwald (1904; see also Zacharias 1903). Subsequent to Woltereck's rejection of temperature as an efficient cause, almost no further experiments except his own were conducted until the investi-gations of Coker and Addlestone (1938) and Coker (1939). Berg, it is true, had made the rather surprising discovery that a small helmet can develop on the head of an *ex ephippio* specimen of *D. cucullata* during its lifetime, though the round-headed *ex ephippio* neonata can never develop so high a helmet as can her helmeted neonate daughter. These experiments, though they run counter to one of Woltereck's conclusions, do not throw light on the factors required to produce a helmet, for both temperature and food were maintained at a high level in Berg's experiments. The first direct confirmation of Ostwald's discovery was produced by Coker and Addlestone (1938) working on *D. ambigua*. Coker and Addlestone found that when specimens of this species, brought in during the early summer from a shallow lake at Durham, North Carolina, produced young at temperatures below 11° C. the neonate were invariably round-headed, but when the temperature was above 15° C. the neonatae were provided with a small spike-like helmet.   Between 11 and 15° C. a rather variable population, some with and some without small spikes, was produced.   It was possible to show that the period during which the embryo is sensitive to the effects of temperature is the middle third of embryonic life.

When Coker and Addlestone reared their spike-headed neonatae, even at normal summer temperatures, it was found that they lost all trace of the helmet after a few molts and that the reproductive adults were round-headed.   Examination of the population in the lake indicated, however, that this retrogression did not happen in nature, in which the whole population remained helmeted.   Retrogression in the laboratory was also reported by Banta in the case of *Daphnia retrocurva* from Lake Mendota; an adult female was actually observed in the process of becoming less helmeted when brought into culture.   Coker and Addlestone concluded that the size of the vessel, the difference in fact between a shell vial and a lake, was probably involved in regulating this retrogression.   They believed that in large vessels in the laboratory there was some indication of a better retention of neonate head form than in the small vials used in most of their experiments. The more complete elucidation of retrogression is considered later in relation to Brooks' work, but it is desirable to point out here that Woltereck's supposed inevitable decline in the development of the helmets from the vertex to the clivus form is probably based on culture experiments in which specimens with fully developed helmets had been brought into the laboratory.

Analysis of relative growth in D. ambigua *and* D. retrocurva. During the last twenty years Brooks and his associates have been engaged in a most penetrating study of cyclomorphosis in *Daphnia*. Brooks (1946) studied, in his first contribution, the cyclomorphosis of *D. ambigua* and of *D. retrocurva* both in field studies and laboratory culture, making extensive use in his analysis of the concept of heterauxetic growth.

The Connecticut populations of *D. ambigua* which Brooks examined were found to behave essentially as described by Coker and Addleston, except that the critical temperature for the appearance of the spine on the apex of the head was very low, about 7° C. rather than 11 to 15° C.   This is in accord with the development of a small head spike in populations of this species in the hypolimnia of several Connecticut lakes at about 8° C.   As in Coker and Addlestone's experiments, the head spike was always lost as growth occurred in the laboratory, even though it was retained by later instars in nature.   In spite of some difficulties in measurement, Brooks was able to obtain evidence that the growth of the head relative to the length of the carapace in nature was markedly bradyauxetic (k = 0.64) in April and May and somewhat less so in June (k = 0.80).

Daphnia retrocurva in the same lake proved to be particularly amenable to relative growth studies and could be cultured with a fair degree of success in the laboratory. Brooks obtained his most important results (Fig. 212) with this species.

1. The neonate head length is a function of temperature. Laboratory experiments and observations on the populations in the lake agree in indicating a continuous and roughly linear relationship between the temperature and the ratio of head length to carapace length, over the range of 3 and 22° C.

2. In the lake during the spring and early summer, when the temperature lay between 8 and 16° C., the growth in length of the head is isauxetic with respect to the length of the carapace and the animal maintains essentially the same shape that it had on hatching.

3. In the lake, during the height of summer, from June 28 to August 10, when the water temperatures lay above 19° C., the growth of the head became markedly tachyauxetic. In the middle of July values of k as high as 1.30 to 1.44 were observed. This implies that at the height of summer the head grows relative to the carapace throughout the ontogeny of the animal.

4. In the laboratory, no values of k approaching unity were observed, growth being invariably bradyauxetic. In cultures raised at various temperatures from 6 to 24° C. the value of the exponent varied from 0.52 to 0.74, apparently exhibiting a minimum value at about 10 to 12° C.

It would appear from these very clear-cut results that at least three factors must be considered in any analysis of the cyclomorphotic change of head shape in *D. retrocurva*. The first is obviously the temperature during embryonic life, which determines the shape of the head in the neonata. The second is some factor operating in the lake but absent in the laboratory which permits isauxetic or tachyauxetic growth in the former environment, whereas in the latter, under the conditions employed, the head always grew bradyauxetically. The third factor apparently induces tachyauxesis in the lake at the height of summer (see page 802). It is reasonably certain from the work already presented on *D. ambigua* that three factors of the same sort operate on this species also, though the increase in nature of the value of k, which for this species is always less than unity, takes place in this species about a couple of weeks before k rises from unity to about

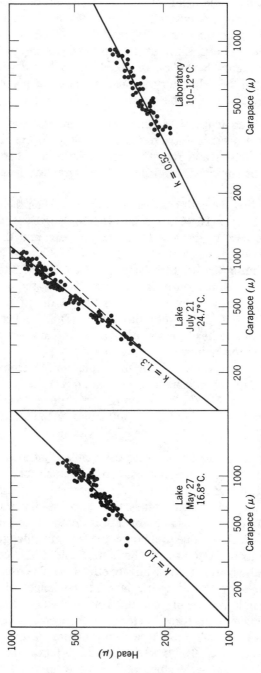

FIGURE 212. Variation in heterauxetic relationship of helmet to carapace length in *D. retrocurva* under different conditions. *A*. In lake population, in late spring, relative growth is essentially isauxetic. *B*. In lake at the height of summer, relative growth is somewhat tachyauxetic. *C*. In still laboratory culture, relative growth is markedly bradyauxetic. (Redrawn from Brooks.)

1.4 in *D. retrocurva*. A rather elaborate indirect estimation of the variations in the food supply available to the *Daphnia* in Bantam Lake strongly suggested that the variations in availability and intake of food in the lake had no effect whatever on the cyclomorphosis of *D. retrocurva*. This and some other aspects of the relative growth have been explored in greater detail in *D. (D.) galeata mendotae* in the later work of Brooks (1947) and of Jacobs (1961).

Comparison with D. cucullata. With regard to other species, and particularly *D. cucullata*, unfortunately the evidence of the detailed pattern of relative growth is scanty. Wesenberg-Lund thought that the helmet increased in length relative to the carapace during ontogeny at the height of summer, but the evidence of his measurements is far less convincing than would appear at first sight. Zacharias (1903) considered that in the autumn a bradyauxetic reduction in helmet size occurred during ontogeny; this is evidently quite possible, as indicated by Jacobs (1961), when temperature is falling. The best figures purporting to show the actual ontogenetic changes in *D. cucullata* are certainly those of Findenegg (1943a). Measurements of these figures (Fig. 213) clearly indicate that the length of the head of both the round-headed generation developing in April and May and of the moderately helmeted generation developing in June and July must increase more or less isauxetically, at least after the first instar, with respect to the carapace length, whereas in the generation developing the highest helmet during August and September the head grows tachyauxetically, the value of **k** being about 1.5. The general pattern of growth is thus clearly comparable to that of *D. retrocurva* in Bantam Lake, in spite of the fact that the *D. cucullata* population in the Wörther See even in the period between the end of July and beginning of September lives by day largely at temperatures below 10° C. and enters the warm only epilimnetic water only at night. There is a possibility that Findenegg's drawings of the development of the low-helmeted form, which is hatched in November and December and grows over the winter period, indicates a very slight bradyauxesis.

The effects of temperature and nutrition on D. galeata mendotae. Jacobs (1961) has examined the behavior in the laboratory of three populations, two being clonal, of *D. galeata mendotae* from Bantam Lake, cultivated for a long period in the laboratory. In this animal in nature the carapace length of the neonate is maximal in winter, whereas the head length and ratio of head to carapace follow the temperature curve. The value of **k** varies from 1.42 at the end of June to 0.78 in February. This value depends on estimates involving individuals hatched on various dates; it is much less well related

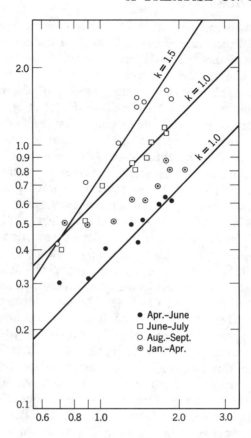

FIGURE 213. *D. cucullata,* Wörther See, growth of helmet relative to body length in spring and summer generations, showing increase in **k** in the late summer. (Data from figures of Findenegg.)

to temperature than the measurements made on neonatae hatched about the time that the temperature was taken.

Jacobs found that in the laboratory one of his clones showed a relatively constant value of **k** at both 14 (**k** = 0.76) and 24° C. (**k** = 0.83); the relative growth curves plotted on a double logarithmic are almost parallel straight lines. In his other clone the value of **k** was significantly greater at 24° C. (**k** = 0.87) than at 14° C. (**k** = 0.69), and the relative growth curves are strongly divergent.

If an animal is reared for its first few instars at a low temperature, and then transferred to a higher temperature, the relative growth of the head after the transfer is greatly increased (Fig. 214) so that the form of the animal at any given later carapace length tends gradually to approach what it would have been if reared throughout its life at the higher temperature. In this process the value of **k** appears considerably increased. In one case transfer at the fifth instar from 14 to 24° C. caused animals initially growing with **k** at approximately the standard laboratory value of **k** = 0.76 to grow with **k** = 1.22. The reciprocal effect in going from a warm to a cold environment also occurs. Under conditions of extreme starvation, the food supply being 1 to 2 per cent of that ordinarily used, at 24° C. the effect on head length was almost identical to a transfer to 14° C. without reducing food. However, the molting rate was much more reduced by temperature than by starvation, whereas egg production was not reduced at 14° but was completely stopped by starvation. Although Jacobs believes that the absolute specific growth rate probably determines the relative growth, the dependence of molting, growth of the

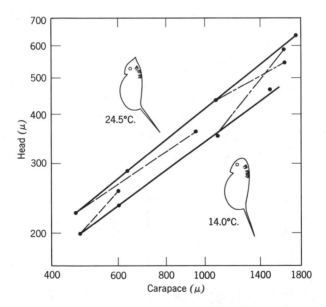

FIGURE 214. *Daphnia galeata mendotae;* newly hatched young developed at 14.0 and 24.5° C.; standard heterauxetic relationships between head and carapace length, given by upper solid line for 24.5 ° C. and by lower solid line for 14.0° C.; broken lines indicate relative growth on transfer from one temperature to the other. (Jacobs.)

head and egg production all seem somewhat differently related to any hypothetical metabolic activity that may be postulated; in his most recent work he concludes that other factors, notably turbulence may override the effects of absolute growth rate (personal communication).

The effects of turbulence. Coker and Addlestone (1938) first demonstrated critically that life in the laboratory is very different for a *Daphnia* from life in nature; some of Woltereck's results can be seen in retrospect to imply something similar. Brooks observed the same effect in *D. g. mendotae* and discovered that if in the laboratory two cultures were kept at the same temperature and illumination and with the same food supply and oxygen concentration the animals growing in one culture vessel in which the water was at rest showed a bradyauxetic growth of the head, whereas in the other vessel in which the water was kept in a turbulent state by means of the rotation of a slightly excentric stirring rod the growth of the head was almost isauxetic (Fig. 215). It would therefore seem that the second factor

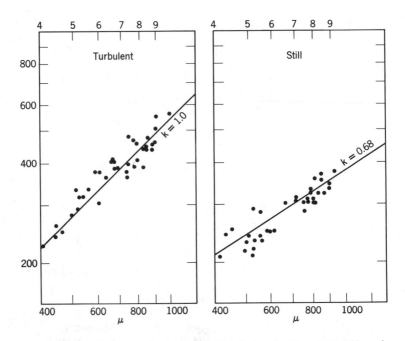

FIGURE 215. *Daphnia galeata mendotae;* heterauxetic relationship of helmet to carapace length in turbulent (k = 1.0) and still (k = 0.68) water in laboratory. (Brooks.)

involved in the cyclomorphosis of *Daphnia,* for the lack of which retrogression occurs in the laboratory, is the mechanical agitation produced by turbulent water. Though it would appear to be the experience of marine aquarists that some degree of agitation is necessary, over and above that needed for aeration, for the maintenance of the health of marine planktonic organisms, and though it is quite likely that the difficulties of keeping some lake plankters in culture could be traced to the same cause, the discovery that turbulence can have an effect on morphology is rather surprising. Unpublished work by Frank and by Yeatman (personal communications) suggests that a morphogenetic effect of turbulence may prove to occur in other Cladocera and in copepods. The nearest comparable case would appear to be the effects of mechanical stimulation due to crowding in the production of the migratory phase of locusts. Later field work supports the conclusion that turbulence is second only to temperature in promoting cyclomorphosis. Brooks (1957b) described a case in which *D. longiremis* developed in helmet only in a year when its habitat was freely circulating, whereas Hrbáček (1959a) found in some cases that the helmet of *D. cucullata* increased during the late summer as the temperature of its epilimnetic habitat fell and the depth of the initially very thin epilimnion increased. It is reasonably certain that turbulence was increasing as the epilimnion increased in depth and the helmets of the *Daphnia* became higher.

Jacobs (1962) found in experiments that turbulence produced by a spinning rod was effective at 14° C. but apparently not at 24° C. The effect is apparent only in illuminated cultures. There was evidence of an effect on the embryos presumably mediated through the mother. The turbulent movement of the water increased antennal beat rate by 40 per cent, probably because the animal was attempting continually to adjust to disturbances in its relation to light and to sinking in a gravitational field. Possibly some effect of over-all activity on feeding rate is ultimately involved in the morphogenetic action of turbulence on *Daphnia.*

Stratification of unhelmeted and helmeted Daphniae. It has long been known, at least from the time of the observations of Juday (1904), that when helmeted and round-headed *Daphnia* occur in the same lake, the helmeted ones, almost without exception, have their maxima of occurrence at lesser depths than the unhelmeted. This was confirmed by Ter Poghossian (1928) and by Aurich (in Woltereck 1930) for various Bavarian lakes. Table 48 lists Aurich's findings in the Klostersee, July 2.

TABLE 48.   *Stratification of* Daphnia *in the Klostersee*

Depth (m.)	Temperature (°C.)	*longispina** (per liter)	*cucullata* (per liter)
0.3	23.5	0	0
1	23.2	0.9	0.5
3	23.0	2.1	0.9
5	20.8	8.0	**1.9**
7	15.0	12.9	0.9
10	12.0	**53.3**	0
13	9.9	0.3	0

* See Footnote 1.

A comparable distribution is recorded (Findenegg 1943a) by day in the Wörther See on September 1, 1937 (Fig. 216).   Here the maximum at midday for *D. cucullata* lay at about 12 m., for *longispina*,[1] which is a strictly hypolimnetic form in this lake, at about 14 m. By night a very large part of the *cucullata* population has ascended into the top few meters, whereas the *longispina* maximum had risen only to about 12 m.   In Victoria Nyanza Worthington (1931) found that the markedly helmeted *Daphnia lumholtzi* in general remains in the upper waters of the lake but that *D. longispina*[1] undergoes very pronounced vertical migration, nearly all the individuals being in the region below 50 m. at 3 P.M. and in the region above 17 m. at midnight. Worthington therefore suspected that helmeted individuals migrated less than the round-headed, but this is not borne out by Findenegg's results in the Wörther See.

In North America, where Juday made his original discovery of the phenomenon, interpretation of the data has been made difficult until recently by the complexity of the taxonomy.

The usual situation during the summer months in the lakes of northern Wisconsin, which have so far been more intensively studied than those of other regions in America, is the occurrence either of *Daphnia longiremis* or a round-headed form of *D. pulex* in the hypolimnion and of one or more helmeted forms in the thermocline or in the epilimnion (Fig. 217).   With regard to these helmeted forms, it is probably commonest to have *D. retrocurva* and *D. galeata mendotae* both present; to Woltereck (1932), the former, lacking an ocellus and usually being smaller, represented *D. cucullata* of Europe,

---

[1] The species is probably actually *hyalina,* at least in the Klostersee.

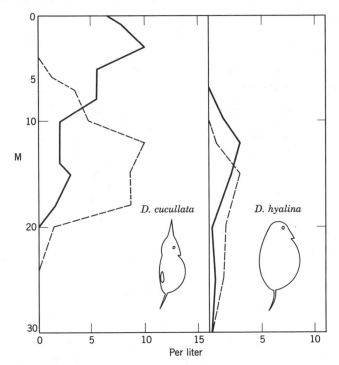

FIGURE 216. Vertical distribution of *Daphnia cucullata* and of *D. hyalina* at 11.00 hr. (broken lines) and 23.00 hr. (solid lines) on Sept. 1, 1937, in the Wörther See. (Redrawn from Findenegg.)

the latter, with an ocellus and usually larger, *D. longispina.* There is one case known in Wisconsin, namely Muskellunge Lake, in which a form of *D. longiremis* develops quite a definite retrocurved helmet. Woltereck stresses the fact that in this lake the stratum of water available for the cold stenotherm, polyoxibiont *D. longiremis* is very narrow indeed. Muskellunge Lake is inhabited by three quite distinct species of *Daphnia,* all of which are cyclomorphotic. *D. galeata mendotae,* with a tall helmet, inhabits the epilimnion, and *D. retrocurva* and *D. longiremis,* the thermocline region, the last-named species having the least extended vertical range of the three (Fig. 218). A similar distribution is exhibited by the same three species in Trout Lake, but there *D. longiremis* is round-headed and occurs in great numbers at the very bottom of the lake. There is some information regarding the diurnal migration (Fig. 219) of these three species in Trout Lake

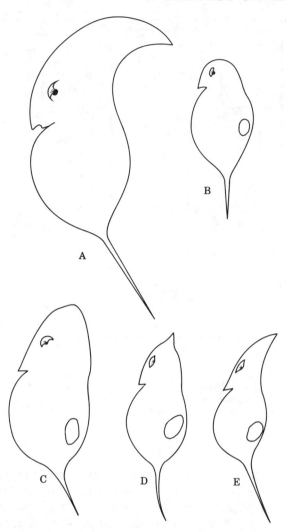

FIGURE 217. The more important planktonic species of *Daphnia* inhabiting the lakes of Wisconsin (all × 25). *A. D. retrocurva* f. *retrocurva,* extreme helmeted race from Lake Mendota, from the original material of Forbes. *B. D. retrocurva* f. *breviceps,* Fence Lake. *C. D. galeata mendotae,* Day Lake. *D. D. galeata mendotae* (presumed), Crawling Stone Lake. *E. D. dubia,* Day Lake. (Woltereck, but cf. Brooks 1957a.)

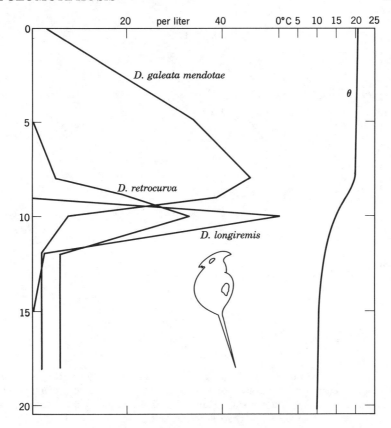

FIGURE 218. Muskellunge Lake, Vilas County, Wisconsin. Vertical distribution of three species of *Daphnia* with a figure of the rare helmeted form of *D. longiremis*. (Data of Woltereck.)

D. g. mendotae has its maximum at 5 m. by day and at 3 m. at 9 at night, a migration that corresponds to a shift in the center of gravity of the population of not more than 2 or 3 m. *D. retrocurva* has its maximum at 10 m. and the center of gravity of its population at about 12 m. by day, both moving up to about 5 m. by night. The majority of the individuals of *D. retrocurva* therefore move about twice as far in their upward evening migration than most specimens of *D. g. mendotae*. The population of *D. longiremis*, which is very dense just over the bottom, is almost stationary.

A superficially different but fundamentally similar distribution was apparently provided by Weber Lake on August 28, 1931. This lake

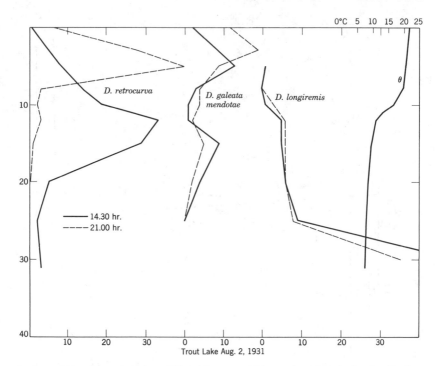

FIGURE 219. Trout Lake, Vilas County, Wisconsin. Vertical distribution during the early afternoon and the evening of three species of *Daphnia* and of temperature in the afternoon. Here the hypolimnetic nonmigratory *D. longiremis* is not helmeted. (Data of Woltereck.)

was transparent and had a good supply of oxygen at the bottom. The deep-water form reaching its maximum at 13 m. and living entirely in the hypolimnion is here referred by Woltereck to a round-headed form of *D. pulex* (s. str.). The upper hypolimnion contains *D. retro-curva* f. *breviceps,* ranging from 8 to 13 m., with a maximum at 10 m. *D. galeata mendotae* occurs in the epilimnion, and in fact extends throughout the depth of the lake from 1 to 13 m., with a marked maximum between 5 and 8 m. The distribution is therefore comparable to that in Muskellunge and Trout lakes, save that *D. longiremis* is replaced by *D. pulex* and *D. retrocurva* is not helmeted. The significance of the latter fact will be considered further (see page 846). The distribution in Fence Lake (Fig. 220) differs from the pattern found in Trout and Muskellunge lakes mainly in the presence of four co-occurring forms; *D. g. mendotae* is abundant in the eplimnion (maximum 3 m.) and *D. retrocurva breviceps* lives intercalated between *D.*

retrocurva f. *retrocurva* (maximum 8 m.) and *D. longiremis* (maximum 20 m.). Not all the lakes containing *D. g. mendotae, D. retrocurva,* and *D. longiremis* show the species stratified from above in that order, though *D. longiremis* is always hypolimnetic. Presque Isle Lake has the maximum of *D. retrocurva* at 3 m. and of *D. g. mendotae* at 5 m. In Crawling Stone Lake, which has five different populations, *D. pulex* (s. str.), *D. retrocurva* f. *breviceps, D. longiremis,* and the adult specimens probably referable to *D. galeata mendotae* (D of

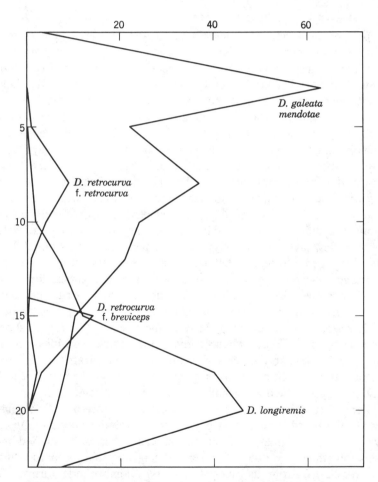

FIGURE 220. Fence Lake, Vilas County, Wisconsin. Distribution by day of *D. galeata mendota, D. rectrocurva* f. *retrocurva, D. retrocurva* f. *breviceps,* and *D. Iongiremis.* (Data of Woltereck.)

Fig. 217) all have their diurnal maxima between 13 and 14.5 m.   The
D. retrocurva f. *retrocurva* population has its maximum at 8 m. and
the immature group of *D. g. mendotae* at 5 m. (Fig. 221).   Finally,
in Day Lake a population of uncertain affinity (referred by Woltereck
to *D. l. longispina*) occurs on the bottom, the maximum moving from
14 to 11 m. between 4 and 9 P.M.   Above this level *D. retrocurva*
f. *breviceps* and *D. g. mendotae* co-occur in the thermocline with
maxima at 9 to 10 m., neither species migrating much, whereas *D.
dubia* occupies the epilimnion, its maximum moving upward from 7
to 5 m. between 5 and 9 P.M.

In New England Brooks (1957a) has found taxocenes fully as
complex as those recorded in Wisconsin and has been able to follow
them over the entire summer.   At least in the southern part of the
area, the typical inhabitant of the deeper part of the hypolimnion
is not *D. longiremis* but *D. ambigua,* in which a short spike-like helmet
is developed except at the lowest temperatures.   The most interesting
conclusions were reached in Bantam Lake, which is little stratified
thermally (Brooks *in press*).   Here *D. retrocurva* nearly always has
a maximum at some depth between 2 and 5 m.; it is always almost
or quite absent at the surface and usually much rarer at the bottom
than in the intermediate layers.   The position of the maximum may
vary with the brightness of the day and may differ when adults and
young are considered separately, but Brooks found absolutely no indication of a difference in behavior in the late spring when the helmet
was very low and at the height of summer when the animal had a
fully developed retrocurve helmet.   On some days, when a form of
D. galeata mendotae was present in the lake, the population of this
form lay above that of *retrocurva,* but on other occasions the *D.
g. mendotae* were very evenly distributed or occupied the water below
the *D. retrocurva* maximum.   Early in the season *D. ambigua* showed
little stratification but when some temperature stratification had
developed in the lake the species was usually concentrated in the cooler
bottom water, which, however, is much warmer than the water occupied
by *D. ambigua* in the deep stratified lakes of Connecticut.

Adaptive significance of cyclomorphosis in Cladocera.   Two general
views of the possible adaptive significance of cyclomorphosis in *Daphnia*
have been widely held, and since that genus has been studied more
extensively than any other, the general tendency has been to transfer
ideas developed from its study to other cyclomorphotic forms.   The
two views are related to, though logically independent of, the two
theories of the efficient cause of the process.   It is important to separate
the ideas relating to efficient and final causes here as in the rest of

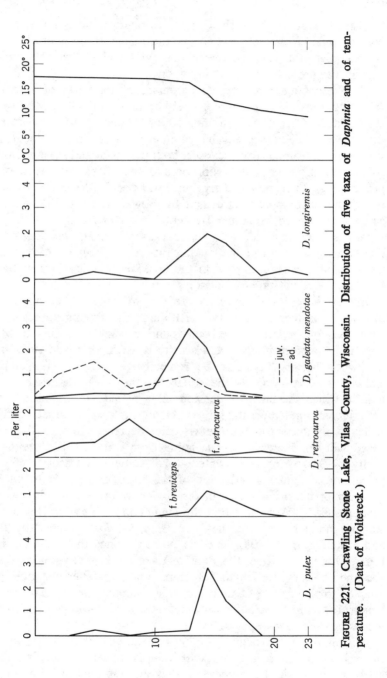

FIGURE 221. Crawling Stone Lake, Vilas County, Wisconsin. Distribution of five taxa of *Daphnia* and of temperature. (Data of Woltereck.)

biology; though we have seen that Wesenberg-Lund is far more nearly correct than Woltereck, as far as the efficient cause of the process is concerned, this does not mean that Woltereck's ideas about adaptive significance are necessarily wholly wrong.

To Wesenberg-Lund the development of the crest of *Daphnia* is, following Ostwald's terminology, either in its original meaning or as redefined in Chapter 19, a method of increasing form resistance as the viscosity decreases in the summer, just as many other spinous processes and similar excrescences, both perennial and seasonal, in other planktonic organisms, are supposed to be.   This view is reasonable *prima facie,* but unfortunately has little positive evidence to support it.   Some experiments performed by Bowkiewicz (1929), and added as an appendix to his paper by Woltereck indicate that *D. longispina* sank passively at a rate of about one-third and *D. cucullata* of about one-fifth to one-sixth of the sinking rate of *D. magna.*   Mature narcotized females were used, but in the absence of indications of the sizes the data are rather uninformative.

In certain cases there is no reason to believe that any increase in form resistance is required.   Although the sinking speeds of small specimens of *Daphnia* have been found to vary according to Stokes's law, and therefore imply that viscosity is as important to such animals as it can be, it must also be remembered that this law implies a sinking speed varying as the square of the linear diameter.   In any case in which the linear dimensions of a planktonic organism decrease with increasing temperature by a factor at least as great as that by which the square root of the viscosity decreases, the passive sinking speed will not increase.   Thus going from water at 4° C. with a viscosity of 0.0157 poise to water at 25° C. with a viscosity of 0.0089 poise would require a reduction of the linear dimensions to only 75.3 per cent of the original size in the colder water to maintain a constant sinking speed.   Assuming that the *Daphnia retrocurva* in Bantam Lake sink in the same way as the *D. galeata mendotae* studied by Brooks and Hutchinson (1950), which is probable since the latter are if anything a little larger than the Bantam Lake *D. retrocurva,* a large spring specimen of carapace length 1.34 mm. is neither better nor more poorly adapted to water at 4° C. than the largest summer specimen of carapace length 1.00 mm. would be to water at 25° C. if no change in shape whatever were to take place.

It is moreover particularly to be remembered that the phenomenon of cyclomorphosis is almost exclusively epilimnetic, whereas it has already been shown (Chapter 19) that the apparent sinking speeds in the turbulent epilimnion are likely to be far less than in the relatively nonturbulent hypolimnion. It is in fact only in that region that the apparent or ecological sinking speed derived from field studies will approach the true passive sinking speed of a narcotized animal in a quiescent column of water in the laboratory. These considerations naturally induce some degree of scepticism of the position held so tenaciously by Wesenberg-Lund.

Woltereck took an entirely different view of the matter. He believed that the major function of the helmet is to set the animal, by changing the position of the center of gravity in relation to the insertion of the antennae, in such a position that the direction of movement as the animal swims is horizontal rather than vertical. Jacobs (1964) finds that the modal position of D. *galeata mendotae* to be 80° from the vertical for helmeted and 70° from the vertical for short-headed individuals under like conditions of illumination. Woltereck's diagrammatic drawings implied a greater difference in tilt and consequent swimming path. The significance of this horizontal movement was found by Woltereck to lie in the belief, which is undoubtedly largely correct, that the nannoplankton has a marked zonal distribution in depth and that therefore in any lake the best pastures for *Daphnia* are necessarily of limited thickness. Moreover, by night the surface film and by day the cold and anaerobic hypolimnion in eutrophic lakes in summer constitute dangers that can easily be avoided by horizontal swimming. Woltereck's ideas of the adaptive significance of cyclomorphosis fit neatly with those that he held about the efficient cause of the process, for if a rich supply of food induces the helmet and the helmet causes the animal to swim horizontally it will stay in the strata that are best supplied with food.

Woltereck regarded the whole of the evidence relating to vertical distribution as favorable to his theory but took a special pleasure in cases such as that of D. *longiremis* in Muskellunge Lake, in which a species that in North America rarely develops a helmet becomes helmeted when it is forced to live in an extremely narrow stratum between the overhot water of the epilimnion and the underoxygenated water of the hypolimnion. Nevertheless, however appealing the distributions in Muskellunge Lake and some other localities may be, it is very difficult to accept Woltereck's ideas *in toto;* it is never clear to what extent the different swimming positions really represent observed and not deductive postures.

Although it is quite clear that when a lake contains a number of different forms of *Daphnia* each tends to have its maximum development in a different depth zone, the conclusion that the cyclomorphotic forms undergo less migration than the noncyclomorphotic, though supplied by the evidence in certain cases, such as that of the populations of *D. longispina* and of *D. lumholtzi* in Victoria Nyanza, is definitely contradicted in others. Thus in Day Lake the epilimnetic and extremely cyclomorphotic *D. dubia* migrates, according to Woltereck himself, more than the other species in the lake, whereas the moderately cyclomorphotic *D. galeata mendotae* occupies the same strata as the round-headed *D. retrocurva* f. *breviceps* and neither migrates appreciably. In Trout Lake *D. retrocurva* is evidently highly migratory, although it is possible that Woltereck did not regard the *retrocurva* helmet as acting in the same way as the straight helmet. In the Wörthersee *D. cucullata* migrates far more than the hypolimnetic form of *D. longispina* which lives below it. Brooks, moreover, has found that there is almost no difference in the vertical distribution of *D. retrocurva* in Bantam Lake in the spring when the animal is round-headed and in the height of summer when it is helmeted. Brooks (1957b) indeed suspects that in some very transparent lakes, such as Weber Lake, the hypolimnetic population f. *breviceps* represents a population forced down by light into the essentially nonturbulent hypolimnion, in which region the full development of the helmet, insofar as it depends on the turbulence of the medium, cannot be achieved.

A third possibility has been suggested briefly by Jacobs (1961) and by Einsele (personal communication), namely that the high helmeted form is less easily eaten by the very young stages of plankton feeding fish, which as fry pick out individual prey. This hypothesis has some support from the geographical distribution of the extreme cyclomorphotic species of *Daphnia* (s. str.) as well as of *Bosmina*, which live in areas in which they are more likely to meet young *Coregonus* than they would farther south. There is, of course, no correlation from lake to lake, since many lakes containing such *Daphnia* must lack plankton feeding fish. A regional possibility of natural selection of this sort, however, does exist, and the possibility of its occurrence should be seriously considered.

Somewhat influenced by this point of view, Brooks (1965), in a paper published after this chapter was completed, has suggested a new and most ingenious theory of the adaptive significance of cyclomorphosis in *Daphnia*. He believes that the typical life history of a cyclomorphotic species in a dimictic lake with a winter ice cover involves a period of winter motile diapause on the part of immediately preadult instars. As soon as the ice breaks and the water becomes turbulent and rich in phytoplankton, these animals enter the first mature instar and start reproducing. A strong selective premium would be put on rapid assimilation and rapid reproduction at this time. Later, as the water becomes warm, the rate of predation, negligible in winter, increases. Under these conditions there will be a strong selection in favor of inconspicuousness, which is greater the smaller the body of the *Daphnia*. Brooks supposes that the development of the helmet, a thin plate that is not so visible as the body, provides a compromise by which relative inconspicuousness can be achieved without necessitating a radical modification of the assimilative process.

The cyclomorphosis of *Scapholeberis mucronata.* The genus *Scapholeberis* contains several species of small Cladocera which feed on neuston, passing the surface film between the valves of the carapace which have straight ventral margins. *S. mucronata* is found not infrequently with an anterior spine pointing forward from the head, the general appearance of the spiked form being reminiscent of *D. lumholtzi,* in which the helmet takes the form of a sharp narrow spine. The form of *S. mucronata,* with a very marked spine (*C* of Fig. 222), is commonly referred to f. *cornuta,* the form without any trace of the spine to f. *fronte-laevi* (*A* of Fig. 222), the typical form of the species having a small spine (*B* of Fig. 222).

Lilljeborg (1900) noted that f. *cornuta* often occurred in May in localities in Sweden which yielded adults with reduced horns or without horns later in the summer; in such cases, however, the neonatae were horned at all seasons. He believed that *cornuta* was characteristic of the larger bodies of water he had studied. Gruber (1923) in Bavaria came to a like conclusion with respect to the correlation of habitat size and horn; he regarded *cornuta* as a lake form and *fronte-laevi* as characteristic of moorland pools, the typical intermediate form being an inhabitant of ponds. Unfortunately, Rammner (1927) found a population fully as horned as Gruber's lake form living in the smaller of two moorland pools he examined, the larger pool having f. *fronte-laevi.* It is evident that any correlation between the size of the environment and the length of the spine is no more exact than the comparable correlations supposed to exist in the genus *Daphnia.*

Gruber (1913) found that at all times between May and October the length of the horn decreases absolutely with the increase in length of the animal in nature, though it would seem likely from culture experiments that the earliest spring forms, supposedly *ex ephippio,* have both the absolute and relative maximum horn length, at least in some cases, in an early but not in the first instar. The length of the mucro appears always to increase throughout life. At least from May to August there is a marked decrease in the horn length and a slight increase in mucro length in nature. The data for September and October are rather imperfect but may indicate a slight increase in horn length at the end of the season.

Experimental rearing of lines derived from three females captured at the end of March indicate that except perhaps in the first brood produced by such females the horn is always somewhat shorter than in the corresponding instar of the mother. Gruber believed that there was a progressive decline in length not merely from generation to generation but from brood to brood within a given generation.

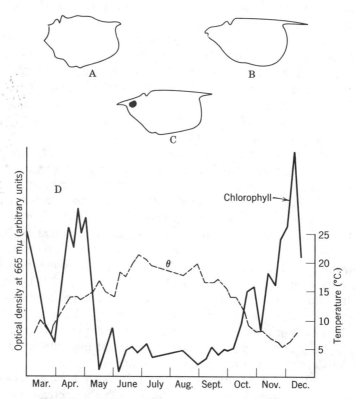

FIGURE 222. *A. Scapholeberis mucronata* f. *fronte-laeve. B.* Typical form *mucronata C.* f. *cornuta* (Gruber) *D.* Temperature (θ) and chlorophyll in 1961 in Hampton Court Long Water. *E.* Body length from base of head spine to base of mucro in *S. mucronata* in this locality in 1961. *F.* Absolute and percentage length of head spine, believed to vary as chlorophyll. *G.* The same from mucro, believed to vary as temperature. (Green.)

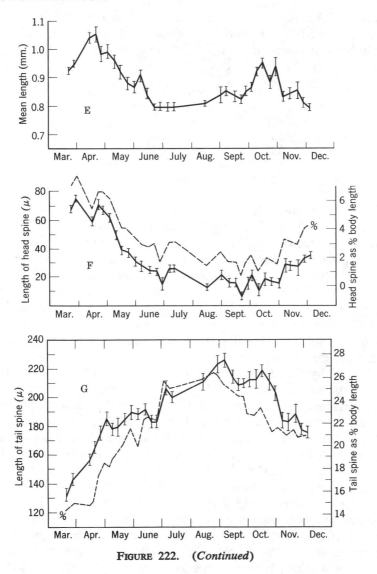

FIGURE 222.   (*Continued*)

Measurements of four individuals or broods,[2] namely the fundatrix of line B, members of her first brood termed B1, of the first brood of B1 termed B1², of the first brood of B1² termed B1³, and of

---

[2] It is not clear whether the measurements given are means or of single, supposedly representative specimens.

a brood B3.2.8 (or the eighth brood of an individual from the second brood hatched from an individual of the third brood of B), are given in Gruber's arbitrary units, which are not the same for the carapace length and the other two dimensions.

	Carapace	Horn	Mucro
B	51	24	38
B1	46	18	40
B1^2	47.5	13	40
B1^3	43	15	39
B3.2.8	48.5	12	52

Experiments conducted at different temperatures, namely at room temperature and at 22 to 26° C., indicated that the animals reared at the higher temperature were at any instar smaller than those reared at the lower temperature; the proportions of the horn and mucro to the body, however, were unaltered. Starvation was found to have no effect on the initial length of the horn, but after the third molt starved animals of any given size had much shorter horns and mucrones than the controls. Since in nature the horn and mucro do not vary together in the same direction, it seemed unlikely that the food supply determines the cyclomorphosis.

A study by Green (1963) indicates that in nature the length of the tail spine or mucro (*G* of Fig. 222) is primarily a direct function of temperature, which reaches its maximal absolute and relative value in August when the total body length is minimal or nearly so (*E* of Fig. 222). The head spine, however, tends to vary in the opposite way, being longest in the spring, falling to a minimum in September, and increasing somewhat in the autumn (*E* of Fig. 222). Green concludes that the variation of the head spine follows that of the food supply. Green found that *S. mucronata* from Fondo Toce, Lago Maggiore, at temperatures above those at which his English material was collected, nevertheless had shorter mucrones. He supposes that at temperatures of 25° C. or more the tail spine begins to retrogress. It is quite possible, however, that local genetic differences would also be involved. In view of the negative correlation between temperature and chlorophyll in Hampton Court Long Water, (*D* of Fig. 222) where Green worked, the isolation of the two supposed causal factors in determining cyclomorphosis may prove too simple an interpretation.

The cyclomorphosis of species of the genus *Bosmina.* The genus *Bosmina* is commonly regarded as containing two widely distributed taxa here regarded as super- or coenospecies, *longirostris* and *coregoni;*

at least one species, *fatalis* from China, of more restricted distribution (Rühe 1912; Rylov 1935; Austin 1942; Burckhardt 1943) is often recognized. Other species apparently exist in America (Goulden unpublished). The genus is probably cosmopolitan, and *B. longirostris* is certainly very widely distributed. *B. coregoni* is a very composite coenospecies; some of its component species or subspecies are highly cyclomorphotic and reach their greatest development in the countries south of the Baltic; the less variable *B. coregoni longispina* occurs in northern Europe, in the Alpine Massif and in North America, but it is to some extent replaced by *B. c. coregoni* in a broad belt running from Britain across the plains of northern central Europe and Russia into Siberia. Very little is known about the distribution of these or of other possibly valid species of *Bosmina* in Africa, Oriental Asia, South America, and Australasia. In general, *B. longirostris* appears to be characteristic of the smaller and more productive lakes of any region and *B. c. longispina,* of the larger and less productive, though this rule is not of absolute validity. Findenegg (1943a) finds that in some localities *B. longirostris* is epilimnetic, *B. c. longispina,* hypolimnetic. Austin has shown how the latter species has been replaced by the former in the developmental history of Linsley Pond. In other localities *B. longirostris* may be littoral, *B. c. longispina* pelagic.

Cyclomorphosis in B. longirostris *and* B. c. longispina. In *Bosmina longirostris* the cyclomorphosis appears to be moderate and to involve a decrease in the length of the antennule and mucro during the summer (Stingelin 1897; Wesenberg-Lund 1908; Huber 1906; Schmidt 1925; Vanini 1933). The total length of the summer form is apparently often less than that of the winter form, but the reduction in the length of the antennule is not merely the result of tachyauxesis combined with this over-all seasonal change in size, for Rammner's (1926) data on a population from a pond near Munich clearly show almost no growth in the antennule during the last two instars of the summer form. There is a general tendency for such summer forms to have the antennule strongly curved and hooklike (f. *cornuta*) as well as reduced in length.

B. c. longispina[3] undergoes little cyclomorphosis, and what is observed is comparable to that in *B. longirostris,* the antennule being shorter in summer than in winter.

---

[3] Supposedly, *longispina* and *coregoni* are connected by forms designated *seligoi* and *reflexa,* but their status requires investigation. There are apparently American forms, *hagmanni* and *tubicen,* in the same coenospecies; their interrelations are under study by Goulden.

Cyclomorphosis in B. coregoni *and related forms.* A number of well-marked races which have been referred to this species exist (Fig. 223). The work of Lieder (1953) indicates that at least some of them behave toward each other almost if not quite as full species. Pending more complete investigations, they are best treated as subspecies of *B. coregoni,* though they are not allopatric representative forms but rather ecologically isolated and often sympatric. The problem of their introgressive hybridisation will be considered in Volume III.

In *B. c. longicornis* a mucro is present, but the form change involves only an immense increase in the length of the antennule in summer. In *B. c. berolinensis* the lengths of the antennule and the mucro increase. In *B. c. coregoni,* in which the mucro is absent, only the antennule increases in length; in *B. c. lilljeborgi,* which has a small mucro, the height of the animal increases somewhat in summer. In *B. c. gibbera,* which lacks a mucro, the increase in height is very marked and the animal appears humped; the antennule is also markedly elongate. In *B. c. thersites* the hump is still more developed and retrocurved. In these humped forms the column cells forming the dorsal wall of the brood pouch are greatly elongate. When the antennule varies from race to race, the number of incisures along its edge, representing cell boundaries, may provide convenient meristic characters in racial analysis.

The most striking races of *B. coregoni,* notably *berolinensis, gibbera* and *thersites,* appear to be limited to the Baltic countries and to European Russia, but the moderately cyclomorphotic *lilljeborgi* apparently occurs, with the relatively noncyclomorphotic *coregoni* (s.s.), throughout the entire territory between Ireland and Kossogol. According to Lieder (1950, 1953), *thersites* is characteristic of shallow warm eutrophic lakes in northern Germany, whereas *coregoni* and *berolinensis* are found in deeper eutrophic lakes which do not reach the high summer temperatures of the *thersites* lakes.

The only detailed series which has been published of measurements of *B. coregoni* taken at different seasons are those given by Wesenberg-Lund for several populations of *B. c. coregoni* and for one or two populations of quite extreme *B. c. gibbera.* Double logarithmic plots of the total length and total height have been made from the published measurements of some of these populations. In most cases the data are inadequate to permit an unequivocal judgment of the course of the growth in height relative in length, but it would seem that in the slightly cyclomorphotic population of Furesø there is strong tachyauxesis, the height increasing in both winter and summer as the square of the length. The antennae are relatively longer in summer but no regular

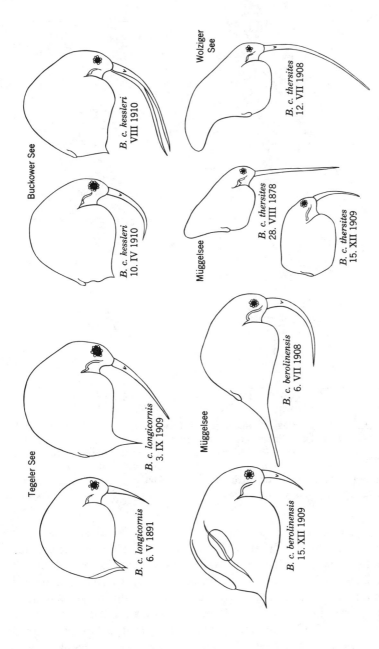

FIGURE 223. Cyclomorphosis in the superspecies *Bosmina coregoni* in lakes of Baltic Germany, showing summer elongation of the antenna (*kessleri*), the mucro and antenna (*berolinensis* and to some extent *longicornis*), and of the humped back of the animal directed somewhat backward (*thersites*); See also Fig. 224. (Rühe.)

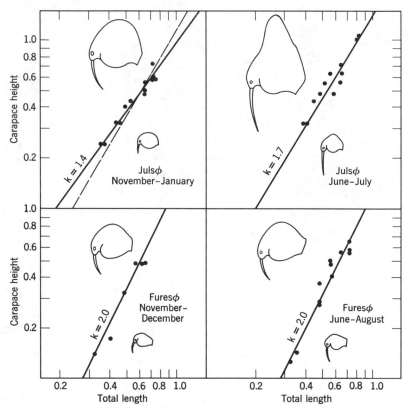

FIGURE 224. Heterauxesis in *Bosmina*. (*Above*) *B. coregoni gibbera* in Julsø, showing a small increase in the dorsal curvature at birth and a definite increase in tachyauxesis during the summer, producing an exuberant form. (*Below*) *B. c. coregoni*, markedly tachyauxetic in both winter and summer but with no seasonal variation of importance either in birth form or relative growth. (Data from Wesenberg-Lund.)

relationship between antennal and total length can be derived from the data. In Julsø, in which an extreme form of *gibbera* occurs, the tachyauxesis is paradoxically less well marked than in Furesø, but it seems to show a slight seasonal variation, **k** having a value of about 1.4 in winter and 1.7 in summer. The ratio of height to length is initially so much greater in the summer Julsø population that in spite of the higher value of **k** the Furesø specimens never get large enough to equal those in Julsø in their relative height. Though rather definite results (Fig. 224) can be obtained from Wesenberg-Lund's data for these two lakes, it is important not to generalize prematurely on the exact role of heterauxesis in the cyclomorphosis of *Bosmina coregoni,* for some of the other measurements that Wesenberg-Lund gives, for populations from other lakes, appear to indicate that a variety of different relations may ultimately be discovered. The numerical data given by Lieder (1950) for the cyclomorphosis of *B. c. thersites* in the Gr. Müggelsee are unfortunately all in terms of mean relative coordinates of certain dimensions averaged over the entire population at a given date. His illustrations, however, strongly suggest slight ($k = 1.3$) heterauxetic growth of the recurved hump during the height of summer and, as far as they go, perhaps hint at stronger heterauxesis very early in the season.

Seasonal size variation and minor cyclomorphotic changes in other Cladocera. A small amount of cyclomorphosis is known in certain populations of species of the genus *Ceriodaphnia.* Hartmann (1915) records one such population of *C. reticulata* in small weedy pools near Graz in which there was a marked reduction in size in summer, with a striking increase in the relative height of the head and of the diameter of the eye, the minimum size and the maximum relative head dimensions being exhibited on 31 July, with very symmetrical changes before and after that date. The small large-eyed, high-headed forms are referred to f. *kurzii* (Fig. 225). Another population from the Teichhüttenteich near Graz behaved quite differently, increasing in size throughout the summer from June to October and showing a slight relative increase in the length of the fornices during this period.

In the Teichhüttenteich two populations of *Ceriodaphnia pulchella* were apparently present. One of these populations was littoral, the other pelagic. Both showed a decrease in size throughout the period from May to July; the planktonic form declined in size still further and disappeared after September 8, whereas the littoral form became

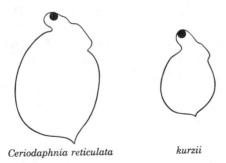

Ceriodaphnia reticulata   *kurzii*

FIGURE 225. *Ceriodaphnia reticulata* and its apparently summer form *kurzii.* (Hartmann.)

larger between July and September.   Stingelin (1897) had previously noted a minimal size in summer and comments on the fact that the posterior spine or rudimentary mucro is better developed in the small summer specimens.   A very similar variation in size accompanied by less change in shape is recorded by Huber (1906).

Hartmann (1915) has studied two populations of *Chydorus sphaericus* in which there is a marked seasonal variation in size, the animals reaching a length in winter some 30 per cent greater than that achieved in summer.   Irregular fluctuations occur in the ratio of height to length, but it is by no means certain that they have statistical significance. A small amount of additional data given by Hartmann in the same paper indicates that *Alonella nana, A. exigua,* and *Alona rectangula* also reach smaller sizes in the summer than in the late autumn.   A similar variation in size was noted in *Acroperus harpae;* in this species, moreover, there appear to be slight variations in the height of the head, which is relatively greater in the summer, with the distance of the ocellus from the vertex and from the apex of the rostrum likewise proportionally greater at this season in relation to the other dimensions. Lilljeborg (1900) figures a number of specimens varying considerably in outline and supposed to be autumnal forms, but it is hard to appreciate the essential differences between them and the single vernal form he illustrated.   According to Hartmann in a pool near Graz in November, the parthenogenetic females had a normal head contour, whereas ephippial females present in the same population had the anterioi keel-like margin of the head much reduced, as in f. *frigida.*

SEASONAL VARIATION IN THE COPEPODA

The seasonal variation exhibited by the pelagic copepods of inland waters is far less striking than that of the Cladocera discussed in the foregoing pages.   It is perhaps reasonable to correlate the extreme cyclomorphosis of some of the pelagic Cladocera and Rotatoria with the prevalent parthenogenesis of these groups, as Hartmann long ago (1917) suggested, but the meaning of the correlation remains unknown. There is, however, a well-defined change in size with the seasons in successive broods of a few species of both calanoid and cyclopoid copepods and a small amount of evidence indicating seasonal changes in form or proportions as well.   Though such changes have been studied a little better in the ocean (see Seymour Sewell 1948, pp. 367–386) than in lakes, some of the limnological material is by no means lacking in interest.

The first record of a seasonal variation in size in a fresh-water copepod is apparently provided by Lilljeborg's (1901) observations on the larger spring and smaller summer forms of a species which he referred to *Cyclops strenuus;* in view of the taxonomic confusion then surrounding this species and its allies, Lilljeborg's observations, though confirmed by later workers, such as Hartmann (1917) and Rzóska (1927), have, of themselves, little evidential value.

A population certainly referrable (Koźmiński 1936) to *C. s. strenuus,* living in a pond in the Zoological Gardens at Poznan, Poland, was studied by Rzóska (1927). This population showed great variability in size, the largest specimens occurring in April and May when the water temperature appears to have been between 15 and 20° C., whereas the smallest occurred in June and August during which months temperatures in excess of 25° C. probably occurred. A similar reduction in size is reported by Koźmiński in the allied *C. vicinus;* both species become rarer as well as smaller in summer. Rzóska referred to an optimum temperature for the production of large specimens; Koźmiński gives no water data, so that it is not possible to ascertain whether *C. vicinus* behaves in the same way as *C. strenuus* in this respect.

In an allied form from the Lunzer Untersee, referred by Koźmiński (1936) to *C. tatricus,* Rzóska found a different seasonal cycle, the smallest specimens occurring in February and the largest in June and October. Since the surface temperatures of the Lunzer Untersee were never in excess of 19° C. during the period studied and exceeded 16° C. only between June and October, Rzoska concluded that the two species he had studied probably reacted to temperature in the same way but that the differences in the temperature cycles in his two localities were expressed by differences in the season at which the largest and smallest specimens occurred. Heuscher (1917) appears to have observed a cycle comparable to that in the Lunzersee in a population referred to *C. strenuus,* and possibly really *C. tatricus* in the Lake of Zurich. Rzóska gives measurements of the furcal setae of both species studied and of the cephalothorax and abdomen separately for *C. strenuus,* but these measurements give no indications of significant seasonal changes in proportions.

Coker (1933) found that the size of *Acanthocyclops vernalis* varies inversely with the temperature at which the animals are reared in a most striking way (Fig. 226). Comparable but less remarkable evidence of the same sort of variation was obtained for *Eucyclops agilis* and also for *Megacyclops viridis.* Starvation reduced the rate of development but had far less effect on the final adult size than the temperature; indeed the only certain final effect of starvation was a slight and not certainly significant reduction of the length of *M. viridis,* an effect not observed in experiments on the other two species.

Seymour Sewell (1935) has made a careful study of the variation

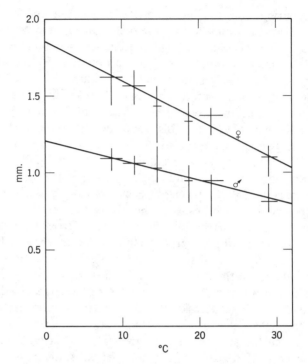

FIGURE 226. Length of *Acanthocyclops vernalis* as a function of temperature. The oblique lines give the temperature range and the vertical lines, the size range for each experimental group. (Coker.)

of the mean total length of *Heliodiaptomus viduus, H. contortus, Meso-cyclops leuckarti,* and also *Thermocyclops rylovi* in an artificial tank in the grounds of the Indian Museum at Calcutta; a series of physiochemical observations were made at the same time by Pruthi (1933). All four species exhibit their greatest length in December and January when the water is at its coolest (Fig. 227). All species exhibit a minimum length in March, after which the Diaptomidae increase in size very slightly and remain fairly small until the water starts to cool in November. The cyclopoids, particularly *T. rylovi,* show a marked secondary maximum in length in May and a minimum in October. These remarks refer primarily to the females, but the cycle exhibited by the males, insofar as it was observed, follows almost the same pattern. It is difficult to avoid the conclusion that part of the size variation is inversely related to temperature, but other factors must be involved at least in the variation of the cyclopoids. Seymour Sewell

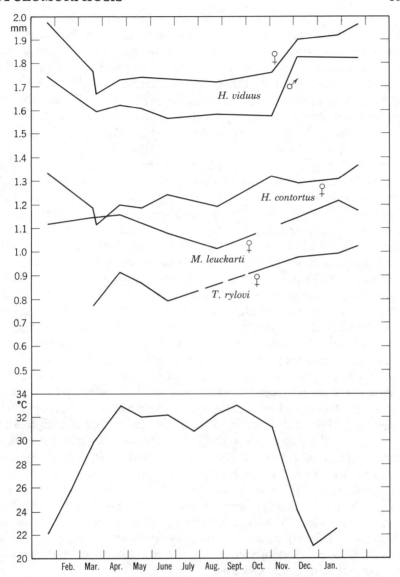

FIGURE 227. Seasonal variation in length of *Heliodiaptomus viduus* ♀ and ♂ *H. contortus* ♀ *Mesocyclops leuckarti,* and *T. rylovi* in an artificial tank in the grounds of the Indian Museum, Calcutta; the temperature variations are indicated in the lower panel. The males of *H. contortus* are not common enough to plot but follow below the females as in *H. viduus*. (Seymour Sewell, Pruthi.)

suggests that the fall in *p*H from March to June indicates some trophic change in the pond that may be reflected in the secondary maximum in size in June, but the nature of such a change cannot be elucidated from the available data. Tonolli (1949a) has observed a slight falling off in various dimensions throughout the summer and autumn in an alpine population of *Arctodiaptomus bacillifer;* the size reduction did not affect all characters equally so that November adults were proportionately narrower than those taken in July. The effect is apparent in the Copepodite I stage.

In considering possible seasonal dimorphism in size in Diaptomidae, it must be remembered that Gurney (1929, 1931) has given suggestive evidence that in some species, notably in the closely allied *Arctodiaptomus laticeps,* and *A. wierzejskii,* the animal not unfrequently may molt once after achieving its adult form; variations in the proportions of individuals in the first and second adult instars might therefore give a spurious appearance of a seasonal dimorphism in size.

Tonolli (1961) more recently has found that in *Eudiaptomus vulgaris* populations of Lago Maggiore not only have the total lengths of the antennule and of the anterior part (head and first thoracic segment) of the cephalothorax maximal values in the winter and spring, but there is a shift in proportions, so that in March the cephalothoracic measurement is about 33.5 per cent of the antennule, while in July it is about 37 per cent. It is possible by a study of the distribution of measurements to ascertain which animals have recently been added to the population and so to elucidate some aspects of the population dynamics. Using this method, Tonolli concludes that a complete replacement of the population takes 88 to 136 days in winter and early spring (November to April) and 58 to 76 days in late spring and summer (April to July). It is also possible to calculate from longevity data and such information on eggs carried that about half the larvae produced, namely about 6 out of 11 eggs laid per liter per year, reach stage CV., an astonishingly high proportion for an invertebrate. This utilization of minor cyclomorphotic variation in the analysis of population dynamics suggests remarkable developments in the future study of the zooplankton.

Brehm and Zederbauer (1906) first observed differences in the structure of the antepenultimate joint of the geniculate antenna of *Eudiaptomus gracilis* from the Hallstätter See in Austria. The form present in summer has the distal end of the hyaline lamella produced outward as a spine that is much reduced in the winter form. Brehm (1927) later indicated that this change is found only in those populations of *E. gracilis* that inhabit a few lakes in Austria. As already indicated (see page 654), it is possible that the supposed cyclo-morphosis results from a confusion of two closely allied species. Hart-mann (1917), however, observed in *Eudiaptomus zachariasi,* in *Acanthodiaptomus denticornis,* and in *Mixodiaptomus tatricus* com-parable but less striking changes in the development of the lamella and in these species and in *Diaptomus castor* very minute differences between the fifth feet of the summer and winter females.

A curious seasonal variation has been noted in the population of *Eudiaptomus vulgaris* in Lago di Trasimeno by Baldi (1935). In April 92 per cent of the males had processes developed on the fourteenth, fifteenth, and sixteenth segments of the geniculate antennae (Fig. 228).

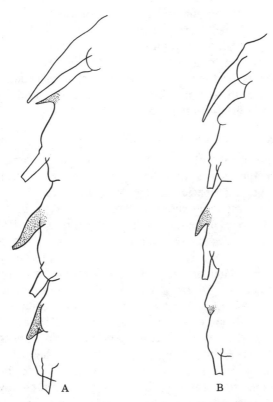

FIGURE 228. Margin of joints 14–16 of the geniculate antenna of the ♂ of *Eudiaptomus vulgaris* from Lake Trasimene. *A.* Extreme development of processes (April). *B.* Re-duced spination (July.)

As the population reached its maximum in May, the proportion of males maturing with complete armature declined to 60 per cent, rising again to 72 per cent as the species declined in June. The meaning of this variation is not understood.

Hartmann (1917) also believed that in *Cyclops strenuus* the form of the fifth foot varies with the temperature, but the figures indicating this variation in natural populations and in laboratory cultures are far from convincing.

Coker (1934) found that the proportions of the caudal furca of *Acanthocyclops vernalis* were markedly influenced by temperature, the small forms reared at 28 to 30° C. having proportionally much wider furcae than the large forms reared at 7 to 10° C. More than half the individuals reared at the lower temperature had a spine in the place of a seta on the outer border of the terminal segment of the endopodite of the fourth pleopod, but it was never observed in the culture at the high temperature. A slight tendency to a broadening of the pleopods and to a reduction in the number of spines on the terminal segments of their exopodites was also suspected to occur in the high temperature cultures but the number of specimens available was insufficient to establish these changes.

CYCLOMORPHOSIS IN THE ROTIFERA

The phenomena presented by the Rotatoria are seemingly even more complex than the events that have been described in the seasonal cycles of the various species of *Daphnia* and *Bosmina*. Part of the complexity is undoubtedly more apparent than real; in *Daphnia* it is usually possible to pick out a genetically continuous series of forms within a single species by comparing the young in the brood pouch of a form early in the series with the adults present in the same locality somewhat later in the season. There is no doubt that the extreme specimens of *D. cucullata* and of *D. retrocurva* are actually the descendants of round-headed forms, because even when working only with preserved material it is possible to find round-headed adults carrying embryos with moderate helmets and in later samples moderately helmeted adults carrying embryos with extreme helmets. Except in the case of the genus *Asplanchna,* it is seldom possible, in the rotifers to be considered, to ascertain from ordinary samples anything about the relevant details of morphology of the embryo from the eggs laid or carried by the amictic female. In consequence of this, there is still some uncertainty whether a given form is a member of a cyclo-morphotic series of a well-known species or is an independent species of limited seasonal occurrence. Difficulties of this sort can be resolved in theory by culturing the forms in question; this in practice is unfortunately often extremely difficult.   Moreover, it is quite certain that, as in *Daphnia,* so in certain genera of rotifers, events do not run parallel in culture and in nature.   However, if the doubtful cases such as the supposed cyclomorphosis in the genera *Filinia* be excluded and the possibility of the coexistence of several races of a species in a single locality be borne in mind in certain other cases in which supposed cyclomorphotic forms may turn out to have no genetic connection, even so the unquestioned cyclomorphotic phenomena in the group are sufficiently diverse and sufficiently interesting.

Size and form-change in the genus *Asplanchna*

In the Montiggler Lakes of the Italian Tyrol Huber (1906) noted *A. priodonta* to be smaller in summer than in winter, the range of variation being about 470 to 730 μ.   This variation in size was not accompanied by any significant form change.   Such a simple reduction of body size with increasing temperature, which Huber also records in *Pompholyx sulcata* in the same lakes and which may be compared with the decreases in size with increasing temperature discussed in the Copepoda and some Cladocera, probably underlies part of the cyclomorphotic phenomena exhibited by *Keratella* and perhaps by other genera of rotifers.   A few observations by Schreyer (1920), however, may indicate that the reduction in size observed by Huber in *A. priodonta* is not a simple effect of temperature.   In the Moosseedorfsee near Bern, Schreyer observed two maxima in the length of *A. priodonta,* one in May, the other in September (Fig. 229).   The incidence of these maxima corresponded to the times of maximal abundance of the species.   The long animals present at these times are said to be mictic and to produce a new series of parthenogenetic generations which increase in length until the time of a new maximum with a sexual period.   The few measurements of breadth that are given indicate that the phenomenon described is quite different from that discussed in the next paragraph.   The animals observed in April and May clearly have a slightly more elongate form that those of the later generations, but the minimum ratio of length to width does not correspond to the minimum length, and the September specimens are actually proportionately less elongate than the specimens taken in June when the absolute length is at a minimum.   It is evident that much more study of the relatively small changes in size and shape of this species in central Europe would be worthwhile.

In the populations inhabiting certain lakes in Denmark, namely Haldsø, Farumsø, and Thorssø near Silkeborg and occasionally in

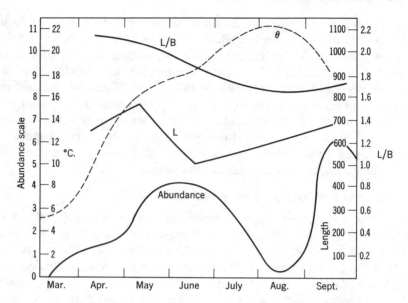

Figure 229. Abundance (relative scale), length (L) in μ, ratio of length to breadth (L/B) of *Asplanchna priodonta*, and temperature in °C., Moosseedorfsee, Bern, Switzerland. The tabular and graphical indications of abundance in the original paper seem not to agree, and the presentation in this figure is somewhat schematic. (Schreyer.)

Furesø, Wesenberg-Lund (1900, 1908, 1930) has observed a quite different and very remarkable type of cyclomorphosis (Fig. 230). The same kind of change has been recorded by Voigt (1904) in the Grosser Plöner See. In these lakes during May the population of *A. priodonta* consists of normal animals about 1.5 times as long as wide. At the end of May such animals produce embryos which are two and a half to three times as long as wide, and during June long animals of like proportion are present as adults. Throughout succeeding weeks the embryos are always proportionately more elongate than the adults carrying them, until in July very elongate vermiform and somewhat curved individuals, about five times as long as wide, make their appearance in the plankton. These extreme forms are usually sterile, and the population tends to decline markedly as they die off in August. Once, in Farumsø, Wesenberg-Lund observed very elongate amictic females containing small round embryos, so that it seems possible that the whole cycle can be completed parthenogenetically. In Haldsø, however, from which lake abundant material was available, the elongate form apparently always died out in late summer, leaving no descendants; the new population of the next season must in

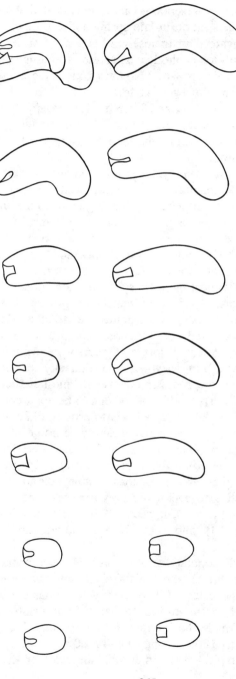

FIGURE 230. *Asplanchna priodonta*, Haldsø, Denmark; least and most elongate specimens taken on a series of dates in the spring and summer of 1902. (Wesenberg-Lund.)

| May 3 | May 30 | June 11 | June 17 | June 25 | July 3 | July 9 |
| 7° | 8° | 12° | 12° | 13° | 14° | 15°C |

this case be derived from resting eggs produced by mictic females present earlier in the summer. Wesenberg-Lund concluded that not only did the females of each generation in late spring and early summer produce daughters more elongate than themselves but that in such daughters in the period of major cyclomorphotic change the ratio of length to breadth actually increased during growth.   No very convincing evidence of this ontogenetic elongation, however, is forthcoming.   In view of the remarkable changes to be described later in *A. sieboldi,* it is interesting that Wesenberg-Lund specifically notes that the elongate forms have but four flame cells and trophi that are indistinguishable from those of their normal ancestors.

This case is the only one yet reported among the rotifers which shows any great resemblance to the cyclomorphosis of *Daphnia.*   In both animals the elongation occurs with rising temperature; it is initiated in the embryo and is accompanied by a decrease in fertility.   In *Daphnia,* however, the lower number of embryos per brood in the summer is quite likely due to a failing food supply and is presumably not confined to cyclomorphotic forms, whereas in *Asplanchna* it would seem likely that many of the most elongate specimens are actually intrinsically sterile.   If this intrinsic sterility ultimately be proved it will be very hard to attribute any adaptive significance to the elongation in *Asplanchna priodonta.*   Indeed, Wesenberg-Lund tends to believe that the extreme forms represent a sort of hypertely or evolutionary overshooting of the mark.   In view of the difficulty of devising any mechanism to account for processes of this sort, most students of evolution now consider such hypertelic phenomena to be apparent rather than real.   The present case, however, is likely to provide difficulties in whatever way it is viewed; a renewed study is clearly desirable.

Asplanchna sieboldi.   This species, unlike the preceding, but like most of its congeners, is primarily an inhabitant of ponds rather than of lakes.   It has been the subject of considerable taxonomic confusion; in the literature relating to cyclomorphosis it appears both as *amphora* and as *sieboldi.*   In the present account these two species, following Wesenberg-Lund (1930) and other recent authors, are regarded as identical.

In many but not all populations of *A. sieboldi,* two forms are known (Fig. 231).   One of these, the typical or saccate form, according to Powers (1912), is normally 500 to 1200 μ long, has 20 to 40 flame cells, and is the only form known to hatch from resting eggs.   The other well-known form is the humped f. *ebbesborni,* which, according to Powers, is 1000 to 1800 μ long, has 40–60 flame-cells, and differs strikingly from the typical form in having the sides of the body pro-

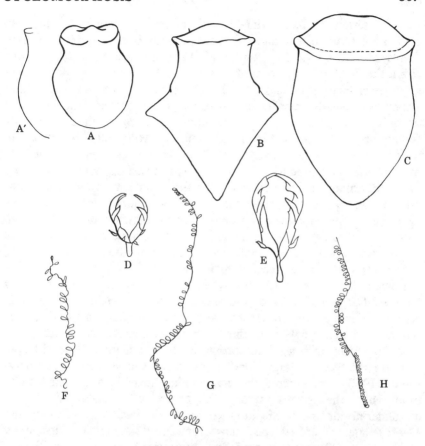

FIGURE 231. Polymorphism of *Asplanchna sieboldi*. A saccate form (Gravelly Run), probably somewhat contracted and not quite typical but apparently the only specimen figured from a population also producing the other forms. *A'*. A more typical profile Montigny-Beauchamp, Seine et Oise, France. (Both somewhat diagrammatized from de Beauchamp.) *B*. Humped form *ebbesborni* outline. (Powers.) *C*. Campanulate form, outline (Powers). *D*. Trophi of humped form, comparable to those of saccate but rather larger. *E*. Trophi of campanulate form, large and showing some heterauxetic form change. (Powers.) *F*. Moderate number of loosely spaced flame cells on nephridium of saccate form. (Montigny-Beauchamp.) *G*. Large number of moderately space flame cells in humped form (Gravelly Run.) *H*. Very closely set flame cells in French (Réville) campanulate specimens. (de Beauchamp.)

duced laterally to form wing-shaped processes which in lateral view form a posterodorsal hump. It would appear from the culture experiments of Lange (1911), Powers (1912), and Mitchell (1913a) and from the field observations of Wesenberg-Lund that, as populations grow from saccate ancestors hatched from resting eggs, sooner or later *ebbesborni* usually appears and in many cases becomes the dominant or only form present. A few populations, such as that studied by Wesenburg-Lund in Torkeri Pond south of Hillerrød, are known in which no form other than the typical saccate is ever present. In most of the Danish ponds in which the species occurs Wesenberg-Lund found that humped forms appear when the water temperatures are highest. However, he concluded that not only summer temperatures but also dense populations were needed for this production of f. *ebbesborni*. In certain localities the species was common in some summers and rare in others; during the summers when the populations were small all specimens were typical saccate individuals.

Powers (1912) and Mitchell (1913a; Mitchell and Powers 1914) have studied populations of the species from the vicinity of Lincoln, Nebraska, which are trimorphic and which produce as well as the saccate and humped form a third very large campanulate form, 1800 to 2500 μ long, with a broad corona, 80 to 115 flame cells, and trophi that are relatively enlarged and of characteristic shape. Wesenberg-Lund (1923) encountered the companulate form in a Danish locality from which the species rapidly disappeared after discovery, so that no material for thorough study could be obtained. De Beauchamp (1951) later described specimens from North America collected by Myers, and Gilbert (personal communication) has also found the form.

According to Mitchell, production of the humped form is due primarily to the nature of the food. When isolation cultures were fed on *Paramecium*, in nearly all lines only saccate specimens were produced. When *Euglena* or *Oxytricha* were substituted for the *Paramecium*, humped individuals soon made their appearance and continued as the only form as long as these kinds of food were given. It appears, indeed, that on returning a line fed on *Euglena* or *Oxytricha* to a diet of *Paramecium* the change back to the saccate form does not take place immediately, but after several generations, and that when the change is taking place in either direction intermediate forms occur. The question of the possibility of the effect of feeding on one generation being carried by some sort of cytoplasmic inheritance through several subsequent generations requires further study, as Mitchell's data are not presented in such a way that the details of the process are apparent.

The campanulate form is produced, according to Powers, primarily by cannibalism in mass culture or in dense natural populations. It would seem possible, however, that a low level of cannibalistic feeding is responsible for the production of humped forms in mass culture. The consumption of young *Moina* also apparently produces almost identical campanulate forms (Powers 1912), though Mitchell produced humped forms with this food. No agency other than crowding, which may act by permitting cannibalism, and the quality of the food had any effect on the morphology of *A. sieboldi;* temperature changes and quantitative variation of the food supply were quite ineffective as morphodynamic agents. It is reasonably certain, however, that genetic differences in the ease with whch the form changes can occur exist within the species. Mitchell found that some lines never produced *ebbesborni* in mass culture; such lines were generally composed of unenergetic animals that tended to leave a good deal of food unconsumed in the culture vessels. In their last contribution Mitchell and Powers (1914) claimed that the resting eggs derived from campanulate mictic females produced humped lines, whereas the resting eggs of humped females, derived from the same parthenogenetic line as the campanulate mictic females, produced, under identical conditions of culture, only saccate lines. The number of lines studied was far too few to substantiate the very improbable claim that the production of inherited humpedness is due to the prior induction of campanulateness, but it is reasonably certain from the experiments that lines can be isolated with a genetic tendency to humpness under conditions in which other lines produce only the typical saccate form. In view of the rarity or at least the very localized known distribution of the campanulate form, it would seem not unlikely that it can be produced only by certain races.

De Beauchamp (1928) has suggested that the production of the humps in *A. sieboldi* involves a change in this soft-bodied form that is comparable to the production of posterolateral spines in the loricate genus *Brachionus,* to be discussed at length below.

The cyclomorphosis of the species of the genus *Keratella*. The genus *Keratella* (= *Anuraea* auctt.) contains a large number of species, some of which exhibit, or appear to exhibit, striking seasonal form change. There has been a great deal of confusion in the taxonomy of these animals, not all of which is fully resolved, though the work of Carlin (1943), Ahlstrom (1943), Ruttner-Kolisko (1949), Gillard (1948), and Bērziņš (1955) has brought considerable order out of chaos. As will be made apparent in the present discussion, it is quite possible that several of the species to be considered are still composite. The recent important contribution of Sudzuki (1964) appeared too late to be used but must be consulted carefully by the serious student of the subject.

For the purposes of this book, it is convenient to consider the supposed cyclomorphotic species in three groups.

K. cochlearis and other species (*irregularis, earlinea, taurocephala,* and *americana*) with a single posterior spine. *K. tecta* is probably best regarded as a separate species of this group distinguished from *cochlearis* in lacking the spine altogether but otherwise very close to the species with a single posterior spine. At least one other sibling species (*hispida* or *micracantha*), with a short spine, may be involved.

K. quadrata, K. hiemalis, and *K. testudo,* three species formerly confused, but clearly distinct, with the body wide at the posterior end and with two posterior spines (rarely absent).

K. valga, K. tropica, K. procurva and *K. lenzi* with the body narrowed at the posterior end and usually with two asymmetrical posterior spines (absent in *K. lenzi*).

Early work on Keratella cochlearis *and its allies.* The best studied though far from fully understood case of cyclomorphosis in the rotifers is probably provided by *K. cochlearis,* which in many parts of the world must be the commonest fresh-water metazoan. *K. cochlearis* is a small well-armored form with six anterior and one posterior spine; its general appearance, characteristic dorsal sculpture, and range of variation are indicated in Fig. 232. The cyclomorphosis of *K. cochlearis* was studied in great detail by Lauterborn (1898, 1901, 1904) in a series of papers that initiated the serious study of seasonal form change in the rotifers; most subsequent work has confirmed and extended Lauterborn's observations, though much modification of his interpretation has been necessary because of changing concepts of the taxonomy of *Keratella.*

Lauterborn's work supposedly allows the distinction of three types of habitat, in which the seasonal variation of the species proceeds differently. Lauterborn himself, however, was primarily concerned only with the second and third type of habitat, discussed later.

In large lakes, such as Lake Constance, a relatively small amount of change occurs, the absolute length of the lorica and both absolute and relative lengths of the posterior spine being less during the summer than during the winter months. Vialli (1924), who studied the species in Lago di Como, points out that in large lakes the maximum over-all length and spine length occur usually in March or April, the minimal lengths in August.

In natural ponds in which there is considerable free water surface Lauterborn noted a much more complicated series of events than was indicated by his rather restricted material from large lakes. In general, in winter the species is long-spined, most specimens being referrable to f. *macracantha* Lauterborn. During May and June the mean absolute size, as measured by the lorica length, falls markedly, and, in addition to this simple reduction in size, true cyclomorphosis or change in shape also occurs. According to Lauterborn, three possible types of cyclomorphosis can take place in populations of *K. cochlearis;* indeed, all may supposedly be exhibited simultaneously and independently by the population of a single pond. The first type of change is an accentuation of that which was noted in Lake Constance and involves a decrease in lorica length accompanied by a marked reduction in the length of the posterior spine, from f. *macracantha* through typical f. *cochlearis* and the forms termed f. *micracantha* and f. *tuberculata* by Lauterborn to *tecta* in which there is no spine at all. The second type of cyclomorphosis is the production of minute spinelets all over the surface of the lorica, giving f. *hispida,* the supposed intermediate form being called f. *pustulata.* The third type of change reported by Lauterborn is the production of irregular plaques in the dorsal sculpture, such animals being referrable to f. *irregularis,* which can undergo reduction of the posterior spine to give the spineless f. *ecaudata.* Lauterborn found that in the Altrhein near Neuhofen his best studied locality, *micracantha* and *tuberculata* appeared early in June with a few *tecta* late in the month; early in July *tecta* constituted 20 per cent of the total population or 30 per cent of the *macracantha-tecta* (or *cochlearis* s.s.) series. No *tecta* occurred after the first half of October. In the *hispida* series the intermediate *pustulata* appeared in the first half of June and true *hispida* at the end of the month. Supposed transitions between *irregularis* and *cochlearis* occurred in spring and winter, but the fully developed *irregularis* was found only from May to November. The events in Lauterborn's other large natural ponds followed essentially the same course. Schreyer, moreover (1920), has described almost the same progression in the Moosseedorfsee near Bern, a small plankton-rich lake, though he claims that in other comparable lakes that he studied *macracantha* occurred throughout the year, the largest individuals being taken in December under ice and in July at 25° C. It is most unfortunate that no detailed measurements were presented. Apart from Schreyer's observations, Gallagher (1957) finds long-spined forms to occur in summer in a pond in Philadelphia. He claims that contrary to the results of most of his predecessors spine length varies directly with temperature and accordingly erects an elaborate theory of temperature-dependent enzyme systems controlling cyclomorphosis. This theory obviously will not explain events in other

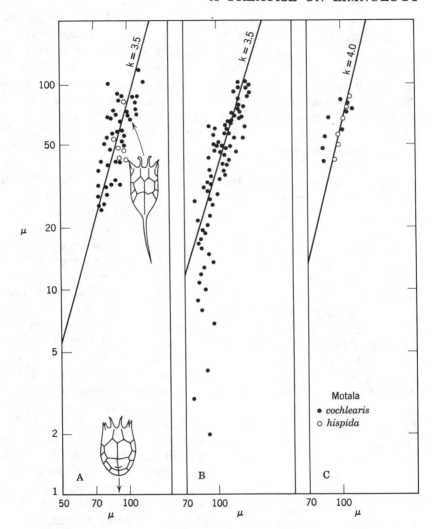

FIGURE 232. Variation of *Keratella cochlearis* and its allies or variants *tecta* and *hispida*. *A*. Relationship of posterior spine length (ordinate) to carapace length (abscissa) for all specimens recorded by Ahlstrom as *cochlearis*, except those referred to *tecta* without a tail spine and so not appearing on the logarithmic plot, and to *faluta* which appears not to belong with the other members of the species: (*above*) f. *macracantha* (*below*), *tecta*. *B*. Similar plot for all Lauterborn's specimens of the *cochlearis* series, giving a comparable slope to that of *A* when the lorica is more than 100 μ long but with a number of very short spined forms not following the kind of tachyauxetic relationship usual in the genus. *C*. Similar plot for f. *cochlearis* (s. str.) and the finely spinulose *hispida* in the Motala River. (After Carlin.) *D*. Similar plot for three assemblages from central Swedish lakes studied by

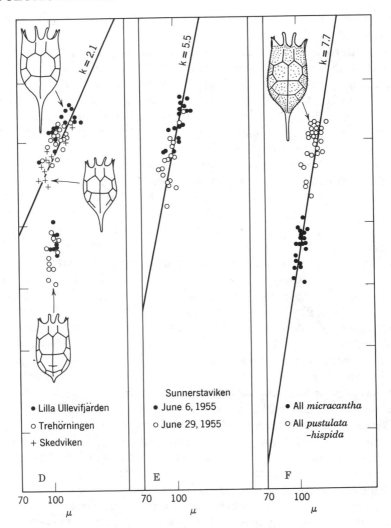

Pejler. All three lakes have a population of fairly large specimens with moderate spines (f. *cochlearis*), which exhibit tachyauxesis relative to the lorica; in two lakes there are populations of smaller very short spined animals clearly not related to the f. *cochlearis* populations. *E*. A population apparently spanning the gap between *cochlearis* and *micracantha*, exhibiting marked tachyauxesis with a high value of **k** and showing a marked fall in spine length with a slight decrease in body length as summer progresses. *F*. All true f. *micracantha* and all hispid (*pustulata* and *hispida*) forms in Pejler's data, suggesting that *micracantha* represents a small and nonhispid form of *hispida*.

populations of the species. He does not give lorica lengths and so his data are unanalyzable.

In a third type of locality studied by Lauterborn, namely a group of recent artificial excavations which tended to become filled with higher vegetation progressively invading the free-water surface, Lauterborn found another form, *K. cochlearis* f. *robusta*. This is a large form distinguished from *cochlearis* and *macracantha* by its much deeper body. Through somewhat variable in the length of the lorica and of the posterior spine, f. *robusta* undergoes no regular decrease in size in summer and so no concomitant regular cyclomorphotic reduction of the relative length of the posterior spine.   Voigt (1904) confirmed Lauterborn's observations to the extent of finding a comparable form in a carp pond near Plön, and Wesenberg-Lund says he knows of noncyclomorphotic population of *K. cochlearis* in ponds in Denmark. Schreyer records f. *robusta* as inhabiting the Geistsee near Bern. This locality is a moorland pond 10 m. deep with marginal sphagnum.   Here the largest spined forms are said to occur in December and June, the smallest at the end of August and in September, so that the population is not really non-cyclomorphotic.   Ahlstrom (1943) described and figured material referred to f. *robusta* mainly from large lakes and states that the form normally co-occurs with *cochlearis* (s.s.). The specimens that Ahlstrom figures appear to be definitely less gibbous than Lauterborn's examples, and it is conceivable that Ahlstrom did not have the true f. *robusta* before him.   Pejler (1957a) has recorded and figured *robusta* from Lapland but is doubtful of its status.

Taxonomic problems. Before considering the observations of seasonal variation recorded by other investigators, it is necessary to discuss the work of Ahlstrom (1943) and subsequent students of the taxonomic problems involved.

Ahlstrom concludes that *irregularis* should be given specific rank, the supposed intermediates between this form and *cochlearis* (s. str.) being illusionary. *Keratella irregularis* is apparently confined to the Old World but is represented by the closely allied *K. earlinae* in North America (*B* of Fig. 233).   Both are evidently warm-water species which doubtless usually show a reduction in the absolute length of the lorica and in the relative length of the spine at the time of maximum temperatures, just as *K. cochlearis* (s. str.) does.   It is evident that the *irregularis* series of Lauterborn must be considered separately and not as part of the cyclomorphotic variation of *K. cochlearis* (s. str.).

The status of the forms referred to *pustulata* and *hispida* is obscure. Ahlstrom treats *hispida* as a variety of *cochlearis,* concluding, presumably partly on the authority of Lauterborn, that the two probably intergrade but admitting that he had seen no true intermediates. Schreyer merely says that *pustulata* and *hispida* intergrade. Carlin (1943) thinks two separate forms are involved, though the hispid or pustulate characters exhibited by one of these forms are apparent only in summer. A comparable, though perhaps not identical, conclusion seems to emerge from Pejler's work, discussed later in greater detail. As with *cochlearis, irregularis,* and *earlinae, hispida* (or *pustulata*) has a relatively shorter posterior spine when the lorica length is short.

A third taxonomic problem is raised by *tecta,* the form without a posterior spine but otherwise identical to *cochlearis.* Kofoid (1908) found that in the Illinois River, at Havana, Illinois, forms with long posterior spines occurred from October to May. Early in April short-spined forms began to appear, and such animals constituted three quarters of the population at the time of maximum abundance on May 10. Later, during a second maximum in June, about half the population was entirely spineless, but such *tecta* froms were also present on almost every date throughout the year, except during August and early September, when no representatives of the *K. cochlearis* group were recorded.

Thus *tecta* in some localities, such as the Altrhein, Neuhofen, or Moosseedorfsee or in the localities near Munich studied by Buchner and Mulzer (1961), appears solely as a summer form and in others, such as the Illinois River, it is perennial or almost so. Wesenberg-Lund (1930) states that although *tecta* is a common form in the ponds of Denmark and often co-occurs with *cochlearis,* in many local-ities there is no evidence of the one form having been derived from the other. On only two occasions did he observe any evidence of transition between them. When they are present in the same pond, one form may produce a large maximum in one year, the other in the succeeding year. If both produce maxima in the same year, *tecta* is usually dominant in the summer and *cochlearis* in the spring or autumn. Pourriot (1957a), studying populations near Paris, concluded that *cochlearis, robusta,* and *tecta* are three often coexistent forms; he does not give them specific status, though their separateness seems to be implied by his brief discussion.

In America Ahlstrom (1943) reports that in large lakes such as Lake Michigan, Lake Washington, and central Lake Erie long-spined forms occurred throughout the year. In Terwillinger's pond, an em-bayment of Lake Erie, the posterior spines showed more reduction, having a minimum length in July. A few *tecta,* however, appeared in Lake Erie in the spring, but they did not appear to have anything to do with the cyclomorphosis of the main population. In a pond on an island in the lake *tecta* occurred throughout the summer, outnumbering the form with a short posterior spine, presumably *micracantha,* by about one hundred to one. This pond is supposed to receive water from the lake in stormy weather, and it or similar ponds may well be the source of *tecta* observed in the lake in the spring.

Pejler (1957a) found *tecta* present in central Swedish lakes, along with one or more forms with a posterior spine. His work indicates that in such lakes *tecta* behaves as an independent taxon. Pejler is unwilling to recognize it as a species on the grounds that it differs from the other forms only in lacking a posterior spine which could have been lost repeatedly. He feels that if *tecta* were admitted as a species it might prove to be a polyphyletic.

In general, these rather fragmentary observations suggest that although at times animals referred to *tecta* may be the extreme results of cyclomorphotic reduction most populations of *tecta* behave as if they were a separate species. These facts are expressed by Wesenberg-Lund by his treatment of *tecta* as a subspecies, but his use of that term is idiosyncratic and not in accord with accepted taxonomic procedure. It may be equally reasonable to conclude that *tecta* is a composite taxonomic category, some specimens referred to it perhaps being recent descendants of *cochlearis,* whereas most belong to a distinct if still morphologically inadequately characterized species.

The problem of *K. tecta,* however, does not exhaust the taxonomic difficulties which have to be faced in studying cyclomorphosis in *K. cochlearis.* The observations of Vialli (1924) on the species in Lago di Como, and more recently those of Parise (1958) on the population present in Lake Nemi before the lake level was lowered, suggest that the very modest amount of supposed seasonal variation, which involves a shift from *macracantha* to *cochlearis* (s. str.) during the month of May, is perhaps the replacement of one population by another genetically distinct population rather than a true cyclomorphotic change within a single line. In Como the distribution curve for the total length of the *cochlearis* present in August and September, is symmetrical and unimodal, with the mode at 159 μ. The curve for the predominantly *macracantha* population present in April and May has its main mode at 207 μ, but a subsidiary mode corresponds to the August *cochlearis* population, suggesting that the ancestors of the summer population were already present as short individuals in the spring. The histograms for the total lengths of the Nemi populations suggest a similar replacement.

Heterauxesis in the K. cochlearis *group.* More than 20 years ago Dr. W. T. Edmondson (personal communication) suggested that part of the cyclomorphosis of *K. cochlearis* and its allies could be simply explained by assuming that the primary change was a slight reduction in the size of the animal with increasing temperature, in line with Huber's (1906) observations on *Asplanchna priodonta* and *Pompholyx sulcata,* and that the changes in proportion in the spine depended on a tachyauxetic relationship form of the length of the spine to the length of the lorica. If the exponent **k** departs sufficiently from unity, pronounced changes in the proportions of the adult will result merely from slight changes in adult size and, because of the very large size of the animal at hatching, considerable uniformity is likely to be exhibited by any population at a given time. This hypothesis is, of course, merely a particular quantification of the old observations of Lauterborn and especially of Hüber, that the small specimens of *K. cochlearis* have relatively shorter spines than the larger. Dr. Edmondson concluded that a population of *cochlearis* or some allied species in Lake Mendota actually behaved in this way, the spine length increasing tachauxetically with increasing body length. Ahlstrom's discovery that several species had been confused as *cochlearis* rendered the identification of some of the Mendota specimens uncertain, though it now seems likely that they were mostly *earlinae*. For this reason the measurements made by Dr. Edmondson were never published. Margaleff (1947) first published measurements to show the heterauxetic growth of the spines of *K. quadrata,* while Green (1960) and Pejler (1962c) have emphasized the dependence of the relative lengths of the spines on absolute body size but without using a formal heterauxetic analysis.

In Figs. 232 and 233 logarithmic plots of the spine length against the lorica length of various collections of *cochlearis* and its allies are given. The diagram designated *A* in Fig. 232 refers to all specimens measured and recorded by Ahlstrom (1943) as *cochlearis,* except those referred to f. *tecta,* which obviously cannot appear on a logarithmic plot, and to f. *faluta,* which do not seem to belong with the other forms. The series thus includes *macracantha,* Ahlstrom's *robusta* (which may not be Lauterborn's), typical *cochlearis, micracantha,* and *hispida.* In view of the uncertain status of the last-named form, it has been indicated by open circles. The whole collection of points falls fairly close to a straight line corresponding to **k** = 3.5.

In the same figure all measurments for the *cochlearis* series (s. str.) that is, *macracantha, cochlearis, micracantha,* and *tuberculata,* given by Lauterborn (1901) are plotted. In the upper part of the range the points fall fairly close to a line of slope 3.5, as in the series measured by Ahlstrom, but when the spine length is less than 20 μ and the lorica length less than 100 μ the relationship appears to break down.

If Vialli's (1924) mean measurements for various samples from the Lago di Como are plotted in the same way, the very modest relative variation in spine length corresponds moderately well to a heterauxetic relationship with $k = 2.5$. If the individual measurements, which are given only for April and September are considered, those for the spring follow a similar pattern of variation, but the smaller form present in September, like Lauterborn's *micracantha* seems not to conform to the heterauxetic relationship. In both series of data the smaller forms may well be genetically different from the larger.

In C of Fig. 232 the points representing the mean monthly dimensions of *K. cochlearis* in the Motala are given. The mean data for these specimens are presented by Carlin (1943) in graphical form and the actual measurements are given as a series of histograms. The summer population is composed of members of what Carlin calls the *tecta*[4] series, that is to say the *cochlearis* (s.s.) series and the *hispida* series. True *tecta* or even *tuberculata* appear to be absent. It is evident from Carlin's presentation that *hispida*, which first appears at the end of May, undergoes a cyclomorphotic reduction in spine length throughout the summer but that the corresponding reduction in body length is less great than in *cochlearis* (s.s.). The data for the two forms together would fall within the envelop defined by Ahlstrom's measurements in A of Fig. 232, but it is clear from Carlin's discussion and illustrations and from the mode of treatment here adopted that *hispida* is behaving somewhat differently from *cochlearis*. The points for *cochlearis* would appear to fall somewhat irregularly on either side of a line with a slope somewhat less than 3, whereas the *hispida* points fit with considerable precision along the line corresponding to $k = 4$. It is thus reasonably certain that the Motala population contains two genetically different races or species. Carlin thinks that one, dominant in winter, is larger and produces *hispida* in summer and that the other is smaller and commoner in summer. On the basis of Carlin's data it might be equally reasonable to suppose that one race is true *cochlearis* and is present throughout the year, whereas the other, *hispida,* is present only in summer, both showing cyclomorphosis with different heterauxetic relationships between spine and body

---

[4] This unfortunate designation for the main cyclomorphotic series of *K. cochlearis* should be abandoned.

length. A similar interpretation would fit the data of Robert (1925) for *K. cochlearis* (s. str.) in Lac de Neuchâtel, in which large forms (120 to 129 μ, spine 75 to 905 μ) occur in winter and somewhat smaller forms (110 to 118 μ, spine 66 to 73 μ), in summer. The measurements of these animals fit a heterauxetic relationship in which **k** is just under 3.5, whereas the few records of *hispida*, occurring only in summer, though somewhat irregular, indicate a higher value of **k**.

Sibling species and the problem of hispida. Pejler's (1957a, 1962c) recent study adds considerably to our knowledge of these animals. In central Sweden (*D* of Fig. 232) three forms, separated on the basis of the posterior spine as *tecta, micracantha,* and *cochlearis* (s. str.) can occur together as discrete populations. In Lilla Ullevifjärden such co-occurrence was observed on September 26, 1955, and in Trehörningen on August 22, 1955. Trehörningen had been visited on July 7, when *cochlearis* alone was common, *micracantha* very rare, and *tecta* absent The difference in the populations observed on the two dates could be explained as a cyclomorphotic reduction in the posterior spine. It is more likely, however, to be due to the seasonal variations in the relative abundance of three genetically independent stocks, for within the *cochlearis* (s. str.) population in Trehörningen there is no evidence of a systematic change in lorica length and in relative spine length when the two dates are compared.[5] Other lakes studied by Pejler have a population of *tecta* accompanied by one rather than two populations of animals with a single posterior spine. In Skedviken such a population appears to represent the *cochlearis* of Lilla Ullevifjärden and Trehörningen, but the average length is somewhat less and the relative spine length is correspondingly reduced. In another lake Sunnersta-viken (*E* of Fig. 232) the spined population appears to be intermediate between the *cochlearis* and *micracantha* populations of Lilla Ullevifjärden and Trehörningen; in this locality there is clear evidence of a decrease in absolute size and of relative spine length during the month of June. In general, in this region, *tecta* appears primarily in lakes of low transparency and high productivity (Pejler 1962c).

In addition to the forms just described, there are, in most lakes, animals referable to *pustulata* or more rarely to the more extreme *hispida*. No very small specimens are recorded, but when all the measurements of individuals referred by Pejler to *pustulata* or *hispida* are plotted on a double logarithmic grind (*F* of Fig. 223) it appears

---

[5] This reasoning assumes that cyclomorphosis does not proceed by discontinuous steps; if it does, much of the argument of these paragraphs would need modification, but no investigator seems to have taken this as a serious possibility.

that the nonpustulate populations referred to *micracantha* are likely to represent, not small specimens of *cochlearis* (s. str.), but rather of the *pustulata-hispida* series. This, of course, may be an accidental and not a general relationship. Carlin suspected that a winter nonhispid form of *hispida,* distinguished from *cochlearis* only by its large size, occurred in the Motala system. Pejler thinks that the spine arises more sharply from the lorica in *pustulata* than in true *cochlearis* of the same general shape; there is nothing, however, in his figures of *micracantha* that precludes their having this character. It is evident that although much more work is needed there is a considerable possibility that *hispida,* with which *pustulata* may presumably be placed, constitutes an independent species in which the posterior spine varies heterauxetically with lorica length, the value of **k** being rather greater than in the associated populations of *cochlearis* (s. str.). This hypothesis would imply that small summer specimens of *K. hispida* f. *pustulata* are not pustulate and in fact constitute the *micracantha* populations of Pejler.[6]

The intermediate populations, such as that in Sunnerstaviken, although not *pustulata,* may represent introgression of the supposed *micracantha-hispida* stock into *cochlearis* (s. str.)

Conclusions on the problem of cyclomorphosis in the K. cochlearis *group.* In a summary of the phenomena presented by *K. cochlearis* it would appear likely that insofar as true cyclomorphosis occurs in the species it primarily involves a heterauxetic variation of spine length with lorica length, the latter being usually, though not always, minimal in summer. The evidence for a complete loss of the posterior spine in cyclomorphosis is inadequate, for most populations without the spine, referrable to *tecta,* appear to behave as a species separate from the spined form. The very short-spined forms appear to vary in a non-heterauxetic manner in some populations, and it is probable that at least in some cases two taxa are involved here also, short-spined *micracantha* and *tuberculata* replacing an independent population of long-spined *cochlearis* and *macracantha* rather than being descended from them. The production of pustulate or hispid forms may represent a cyclomorphotic process, presumably occurring in a sibling species *K. hispida* which occurs with *K. cochlearis* (s. str.) If cyclomorphosis of this sort really occurs, it seems probable that the hispid or pustulate condition is limited to large specimens in summer. In *K. hispida,*

---

[6] The arrangement of Pejler's (1962c) figures perhaps suggest that he was aware of this possibility. The figure that he gives of a pustulate *irregularis* obviously has no bearing on the problem.

whatever its true nature, there is a heterauxetic relationship between spine and lorica, the value of **k** being greater than in *K. cochlearis* (s. str.). There is no evidence that the other forms referred to *K. cochlearis* are in any way involved in cyclomorphotic changes.

The other species of *Keratella* with a single posterior spine exhibit heterauxetic relationship between spine length and lorica length, though the seasonal implications of the variation are unstudied. In *A* of Fig. 233 the measurements for *K. taurocephala* given by Ahlstrom (1943) are plotted. These measurements appear to indicate a quite definite heterauxetic relationship with a fantastic value of **k** of about 23.5. In *B* of Fig. 233 Ahlstrom's measurements for *K. earlinae* and *K. irregularis,* which are probably American and European allopatric species, are seen to fall nicely along the line corresponding to **k** = 6. As already indicated, *K. earlinae* is almost certainly cyclomorphotic in Lake Mendota, slight reduction in lorica size at the height of summer being correlated with a marked reduction in spine length. The last diagram, (*C* of Fig. 233) shows the relationship in the predominantly southern North and Central American species *K. americana,* in which the points follow closely the line representing **k** = 4.5.

There is some evidence, moreover, that a slight cyclomorphotic variation in spine length in one of these species can be independent of the heterauxetic mechanism just suggested. Ordway (unpublished MS essay) in a study of *K. earlinae* in Bantam Lake, Connecticut, found that the lorica length fell progressively from the beginning of April to the middle of July, whereas the length of the posterior spine rose to a maximum in the middle of June and thereafter declined. The maximum relative spine length is therefore on June 14, when the modal spine length is 95 per cent of the body length; the minimum relative spine lengths are on April 1, May 13, and July 20, when the modal relative spine length is 70 per cent of the body length. Ordway's points fall for the most part in the middle third of the envelope defining Ahlstrom measurements in *A* of Fig. 233 but show no trace of a heterauxetic relationship.

Keratella quadrata *and related forms.* Confusion between *K. quadrata, K. testudo,* and the more recently described short-spined cold stenotherm *K. hiemalis* probably vitiates a good deal of the earlier work on the cyclomorphosis of *K. quadrata,* including the observational part of Krätzschmar's paper (1908) in which seasonal form change in this species was first seriously treated. It is therefore best to present a little information on races that are nearly invariant in the field and on the simplest and most certain cases of cyclomorphosis in *K. quadrata* first, irrespective of the historic order of appearance of the work.

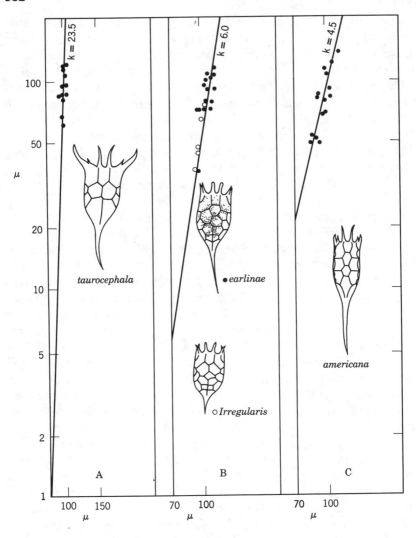

FIGURE 233. Heterauxetic relationship between lorica length and length of posterior spine in *Keratella taurocephala* from acid waters in North America, in *K. irregularis* and *K. earlinae* allopatric species in Europe and North America, respectively, and in *K. americana*. (Data from Ahlstrom.)

In the Lago di Como, *K. quadrata,*[7] though rather uncommon, exists as an apparently invariant population (Vialli 1924). An almost invariant population is also present in the Lunzer Obersee during the autumn and winter, declining in spring and almost absent in summer. In the same lake both *K. hiemalis* and *K. testudo* occur, the latter being present throughout almost all of the year. Confusion of the three species, which occur in different proportions at different depths as well as at different seasons, gives a spurious appearance of seasonal form change. The nearly invariant *quadrata,* however, undergoes a striking reduction of spine length as soon as it is cultured in the laboratory (Ruttner-Kolisko 1949).

The simplest situation involving appreciable cyclomorphosis is provided by the population of *K. quadrata* studied by Carlin (1943) in the Motala system. This population may be referred to f. *frenzeli,* which is a large long-spined form, usually with relatively poorly marked dorsal sculpture, occurring for the most part in the plankton of lakes. The main cyclomorphotic change consists of a slight reduction in the dimensions of the lorica accompanied by a marked heterauxetic decline in the length of the posterior spines during the summer; the value of k for the tachyauxetic growth of the spines relative to the length of the lorica is about 3 (Fig. 234). It is evident that this case provides a most striking example of the heterauxetic type of cyclomorphosis. Voigt (1904) observed what is almost certainly an identical type of variation.

Hartmann (1920) has described the cyclomorphotic changes in a form of *K. quadrata* found in a number of ponds near Graz; the extreme long-spined forms are evidently similar to those studied by Carlin and termed by him *frenzeli.* They have, however, more divergent posterior spines, at least under some conditions, than the long-spined form in the Motala and are referred by Hartmann to f. *divergens.*[8] When a pond had been completely dry during the preceding autumn, the population appearing in the spring, and certainly derived from resting eggs, was long-spined. During the course of the summer there was a gradual but slight reduction in lorica length and a considerable decline in spine length. This reduction continued smoothly, independ-

---

[7] In view of the importance of accurate taxonomy, it is desirable to record that specimens isolated from a sample preserved in the Istituto Italiano di idrobiologia at Pallanza, and made available through the kindness of Dr. Livia Pirocchi Tonolli, proved that on May 27, 1921, one of Vialli's dates, the species present was certainly *K. quadrata.*

[8] Carlin compares Hartmann's form with his f. *dispersa,* which, however, he says is not cyclomorphotic.

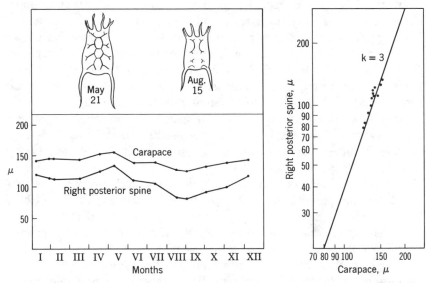

FIGURE 234. *Keratella quadrata.* (*Upper left panel*) Large late spring form referrable to f. *frenzeli* and small summer form with relatively shorter posterior spines. (*Lower left panel*) Seasonal variation of mean carapace and right posterior spine length. (*Right panel*) Length of right posterior spine relative to carapace, showing tachyauxesis with $k \simeq 3$. (Redrawn and plotted from Carlin.)

ent of the temperature or other seasonal changes in the pond, until in winter the population was all short-spined; in a few cases even single spineless specimens were found. In subsequent years, when the pond had not been dry, the spring forms were short-spined. An increase in spine length until June, followed by a decline from summer into the late autumn, then occurred (Fig. 235). Hartmann supposed, on the basis of experiments by Krätzschmar and others, that the resting eggs always produce long-spined forms, that there is an inevitable internal tendency for such forms to produce descendents, which in each succeeding generation are smaller and have shorter spines, until sexual reproduction restores the vigor of the strain, but that this inevitable degeneration can be modified, either reversed or accelerated, by environmental factors.

Changes involving elongation of the posterior spines followed by reduction have been recorded by subsequent investigators as well as by Hartmann in the later years of his study. This was noted by Klausener (1908) in the population of the Pascumin Pool at 2000 m. in the Swiss Alps. Ruttner-Kolisko (1949) has further described such

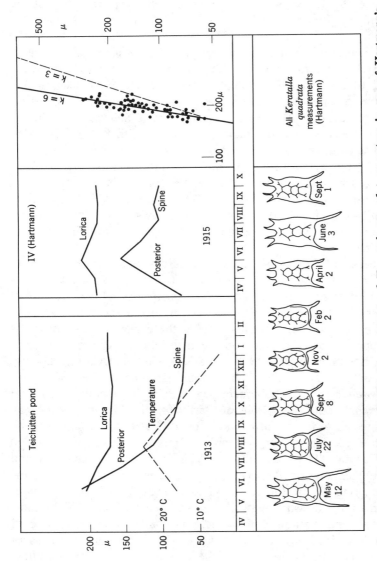

Figure 235. (*Left upper panels*) Measurements of *K. quadrata* and temperature in one of Hartmann's localities. (*Lower panel*) Typical individuals from this population. (*Right panel*) Length of posterior spine relative to carapace for all the specimens recorded by Hartmann, showing a more extreme tachyauxesis (k = 6) than occurred in the Motala River (k = 3).

a sequence in the Seebachlacke, a small but deep (z_m 4m.) pond near Kienberg, Lower Austria. Here in 1944 *K. quadrata* appeared in February, reached a maximum abundance in May and June, and declined in the late summer, disappearing after a slight recrudescence and a sexual period in October. Absolute size and relative spine length increased to the time of maximal abundance and then decreased to a minimum in September. The variation is not great, and the graphical mode of presentation of the data makes further analysis difficult, but the case appears to involve simple heterauxesis with $k \simeq 3$. It seems certain that in these cases the long-spined forms are derived, not from resting eggs, but by an increase in spine length through several parthenogenetic generations, for the intermediate stages between the spring form with short spines and the early summer form with longer spines are present on an intermediate date. Hartmann's conclusions therefore appear to rest more on an erroneous theoretical viewpoint than on observed facts.

A sequence of forms varying from medium-spined in the spring to long-spined at the height of summer is also recognized by Carlin as characteristic of some ponds races of *K. quadrata*. He considers that such races should be considered as constituting a series *quadrata* (s. str.) distinguished from *frenzeli* in which the cyclomorphosis proceeds in the opposite direction, the short-spined form being aestival. It is admitted, however, that the two forms look alike and Carlin's nomenclature would involve calling Hartmann's populations *frenzeli* in one year and *quadrata* in a subsequent year. It is reasonably certain that all Hartmann's data, like those of Carlin for the Motala system, can be explained in terms of variations in over-all size, which may depend on nutritive conditions, coupled with a marked heterauxetic variation in spine length. The main difference between Ruttner-Kolisko's and Carlin's populations on the one hand and those of Hartmann on the other seems to be in the much greater value of k, namely about 6, implied by the measurements of the last-named worker. Since most shallow pond populations appear more cyclomorphotic than those of deep ponds and lakes, the high value of k in the smaller bodies of water may itself be determined by the environment or may constitute a genetic adaptation of unelucidated significance to life in very shallow water.

Wesenberg-Lund (1930), working with the populations of the Danish lakes and ponds, finds in general that the long-spined forms occur in winter, whereas reduced and variable forms which produce resting eggs at the maxima occur in the warmer seasons. In some localities, however, notably the Frederiksborg Castle Lake, he observed the sudden production of long-spined forms in the early summer and later a slow reduction in size and spine length, which apparently proceeded independently of the environmental temperature in much the same way that Hartmann had reported.

Carlin distinguished a f. *dispersa* with divergent posterior spines, said to be stouter than in *divergens,* and the dorsal sculpture of the lorica is also described as being more marked than in that form. This is supposed to be a noncyclomorphotic pond form analogous to *K. cochlearis* f. *robusta.*

An entirely different concept of the cyclomorphosis of *K. quadrata* was derived from the study of the species in Spitsbergen by Olofsson (1918). He was able to visit most of his localities but once or twice and attempted to complete his picture by trying to recognize young and old animals by their reproductive states. He reached the conclusion that resting eggs hatching at a low temperature produced spineless or shortspined specimens, whereas a little later, as the water warmed up, the resting eggs which had remained unhatched now produced longer-spined individuals. Later there appeared to be a reduction of the spine length in the descendents of these long-spined individuals. The data given by Olofsson can certainly be interpreted in this way, but they are so meager that for the present it is hardly necessary to conclude that temperature changes have any effect on the morphology of the individuals hatching from resting eggs.

Amrén (1964a) has indeed recently confirmed the fact that cyclomorphosis of *K. quadrata* in Spitsbergen involves increase in spine length, but in his study this evidently occurred over four or five generations.

Experimental work on *K. quadrata* has been done mainly by Krätzschmar (1913), by Dieffenbach (1911) and by Kolisko (1938, Ruttner-Kolisko 1946). The general result of the extensive experiments of Krätzschmar and of Kolisko is that when long-spined specimens are brought into culture the length of the spines rapidly diminished, not merely with succeeding generations, but in succeeding offspring of the same amictic female brought in from nature. As Kolisko points out, this process may take place in the laboratory during a period in which the natural population remaining in the field persists in an unaltered form. In fact, in the Lunzer Obersee, from which she obtained her material for culture, there is almost no cyclomorphosis. Krätzschmar, under the influence of Weismannian ideas current at the time of his work, supposed that he was witnessing an inevitable reduction series, the most perfect, large, long-spined specimens being produced from fertilized resting eggs and the degenerative decrease in size and relative spine length proceeding until mictic females, males, and fertilized resting eggs are again produced. Whatever the interpretation, there is no doubt about the reality of the phenomena observed by Krätzschmar, since Kolisko has confirmed the earlier observations in all important details. It should be observed, however, that the production of spineless forms in Kolisko's cultures did not often happen and that when one-spined forms were produced they could give rise to descendents with two very short spines. Beyond a certain degree of reduction the variation in fact is random, and although sexual resting eggs are produced only when this degree of reduction has been achieved, the most reduced, entirely spineless females or those with one short spine are no more likely to be mictic than the specimens with two equal but very short spines. Krätzschmar tried a number of different environmental modifications in his cultures, varying temperature, nutritive conditions, and osmotic pressure without being able to influence the reduction process in any important way. Dieffenbach (1911) criticized Krätzschmar's conclusions, claiming that they were the result of the use of unsuitable food. He found that the type of reduction observed by Krätzschmar occurred only when pure *Chlamydomonas* was used to feed his cultures of *K. quadrata;* with a mixed diet of natural nannoplankton, no reduction is said to have taken place. Krätzschmar (1913) repeated his experiments, using a form of *K. quadrata* similar to that previously cultured but feeding it with natural nannoplankton; he obtained reduction exactly as in his earlier experiments. In view of Dieffenbach's somewhat unsatisfactorily reported experiments, Krätzschmar concluded that two races were involved, one in which the reduction was inevitable and the other in which it could be controlled by external factors. These were supposed to be lake and pond races and to lay smooth resting eggs and resting eggs covered with minute spines, respectively. It is clear from the later accounts of Hartmann and of Carlin that the form with the spiny resting eggs is likely to have been *K. testudo,* in which, according to Ruttner-Kolisko, laboratory reduction does not take place.

Dieffenbach states definitely that the animals he cultured, which were certainly not *K. testudo,* did not in culture produce resting eggs. It is therefore unlikely that they were *K. hiemalis,* which Ruttner-Kolisko (1949) found underwent no laboratory reduction but which could not be carried for more than two or three generations because of the production of pseudosexual resting eggs. It is perhaps unprofitable to pay too much attention to Dieffenbach's work, for it is reported so inadequately with such poorly labeled figures that it is by no means certain what really happened in his experiments, though it does seem clear that he bred several generations of some rotifer allied to *K. quadrata* without reduction occurring.

In Krätzschmar's (1913) second paper measurements are given of an amictic female and some of her descendents. These measurements are plotted on a double logarithmic grid in Fig. 236 from which it is apparent that a very marked reduction occurred not only from generation to generation but between successive offspring. This reduction involves the lorica length as well as the posterior spine length, which appears to decline heterauxetically. Indeed, if the very short-spined second daughter (III 22) of the second daughter (II 2) of the original female (I) is excepted, all the points for the individuals measured lie very close to the line $k = 7.5$. Rapid reduction in spine length has been recorded in nature; Ahlstrom (1943) noted individuals with spines equal in length to the lorica apparently producing offspring in which the spine was but one sixth the lorica in length.

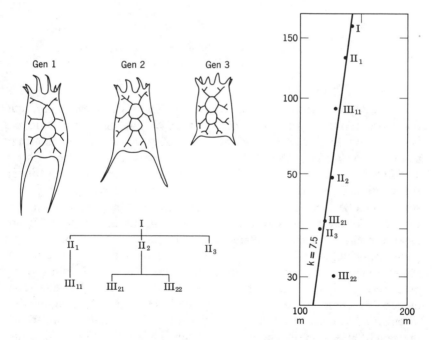

FIGURE 236. *Keratella quadrata,* typical specimens of successive generations in laboratory culture, with pedigree and double logarithmic plot of individuals in one clone, showing progressive heterauxetic spine reduction with both increasing generation and increasing parity. (Modified and computed from Krätzschmar.)

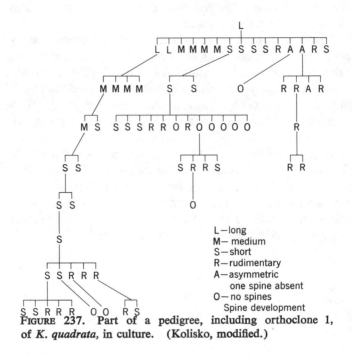

FIGURE 237. Part of a pedigree, including orthoclone 1, of *K. quadrata,* in culture. (Kolisko, modified.)

Considering particularly the later experiment of Kräzschmar, and the more modern works of Kolisko, part of one of whose pedigrees is produced in Fig. 237, we are inevitably reminded of Lansing's studies on aging in rotifers (1942a, b, 1947, 1948; Edmondson 1948). Yet Lansing's scheme will not cover the entire phenomenon. Although the continual decline in spine length in the successive offspring of a single amictic female is precisely the kind of thing that would be expected from Lansing's work, in his experiments the youngest orthoclone remains in a juvenile condition. The line on the left of Kolisko's pedigree, which should correspond to such an orthoclone, nevertheless shows marked reduction in spine length, no long-spined young being produced after the first laboratory-born generation. Lansing concluded, as indicated earlier (see page 514) that the accumulation of calcium was at least one of the biochemical mechanisms underlying the remarkable aging phenomena in rotifers. It is therefore interesting, as Edmondson (1948) points out, to note that Behrens (1933) concluded that the shorter-spined forms in the vicinity of Plön characterized water containing more calcium than that in which the longer-spined *K. quadrata* occurred. The evidence presented appears inadequate to support this conclusion, but it is so suggestive of the mechanism discovered by Lansing that further investigation is obviously most desirable. The possibility that a specific chemical stimulus, comparable to that produced by *Asplanchna* and acting on *Brachionus calyciflorus* (see page 901), is involved is a less attractive hypothesis in view of the almost complete lack of variation in the wild population in the Lunzer Obersee from which Ruttner-Kolisko obtained the lines that undergo such rapid reduction in the laboratory. She obviously doubts that her results and those of her predecessors have anything very much to do with natural cyclomorphosis, but Wesenberg-Lund, never one to stress the significance of experimental work unduly, seems to regard a hypothesis of inevitable degeneration as the best explanation of his observations that *K. quadrata*[9] in the Danish lakes may undergo a gradual reduction.

In summarizing the phenomena in *K. quadrata,* it appears that a heterauxetic variation in spine length underlies the major variation in nature; maxima in relative spine length may occur in winter, late spring, or early summer, depending on the incidence of optimal conditions for the production of large individuals. The amount of cyclomorphosis can vary greatly, **k** being about 3 in some populations in fluviatile lakes and deep ponds, and about 6 in Hartmann's shallow ponds which dried up in some seasons. The laboratory reduction, which can occur rapidly in some races, almost noncyclomorphotic in nature, may well have little bearing on the natural cycle but implies the action of some factor in nature that in not acting in the laboratory or vice versa. At least in the best studied cases this factor must be more constantly present in nature than the peculiar substance produced by *Asplanchna* which regulates spine length in *Brachionus calyciflorus*.

In contrast to this situation, in the cold stenotherm *K. hiemalis* and in the perennial *K. testudo* in the Lunzer Obersee, Ruttner-Kolisko found neither clear cyclomorphosis in nature nor reduction in the laboratory. In *K. hiemalis* the production of pseudo-sexual eggs makes prolonged culture impossible, but it seems quite clear that this species differs from *K. quadrata* with which it co-occurs in not exhibiting declining spine length under cultural conditions.

In Swedish Lapland there seems to be a relationship between the length of the posterior spine of *K. hiemalis* and the altitude or maximum surface summer temperature, irrespective of the temperature at the time of collection. A single observation by Pejler (1957a) indicates what may be true cyclomorphosis, the spines being longer in

---

[9] These observations may be vitiated by the occurrence of *K. hiemalis.*

October than in April in a population in Katterjaure. Unfortunately, very few specimens were available for measurement. The temperature was under 4° C. on both occasions but had been 15.7° C. at an intermediate date. None of the variation, either seasonal or altitudinal, is related to lorica length, which shows little and unsystematic variation. Heterauxesis is clearly not involved. In two tarns individuals lacking one or both posterior spines were encountered. Specimens lacking one spine (f. *valgoides*) are also recorded by Edmondson and Hutchinson (1934) from high altitude lakes in Ladak or Indian Tibet and by Amrén (1964b) from Spitsbergen (Fig. 238).

K. valga *and related species.* The *valga* group, as currently understood, Bērziņš 1955), consists of four species: *K. valga, K. tropica, K. procurva* (Fig. 239), and *K. lenzi.* The first-named is temperate Palaearctic, the other three mainly tropicopolitan, entering temperate regions in the Southern Hemisphere. *Keratella lenzi* lacks posterior spines, and two other forms, *brehmi* and *aspina*, also lacking posterior spines, have been described as reduction stages of other species of the group. According to Bērziņš, *aspina* is the spineless form of *tropica.* Bērziņš believes that the form described by Klausener (1908) as *K. curvicornis* var. *brehmi* is properly ascribed to *K. curvicornis,* which is not involved in the *valga* group; Klausener in spite of his reference of the new variety to another species clearly believed it to be genetically anticedent to *valga.* It is evident that the existence of a true *valga* without posterior spines requires further substantiation.

Klausener observed cyclomorphosis of *K. valga* in the Stätzerhorn Blutsee, a circular pond, 25 m. across and 50 cm. deep at an altitude of 2200 m. in the Swiss Alps. The locality is frozen and largely under snow from November to June. During the summer it has a variable

FIGURE 238.   *Keratella hiemalis* (× 160) from Spitsbergen.   (Amrén.)

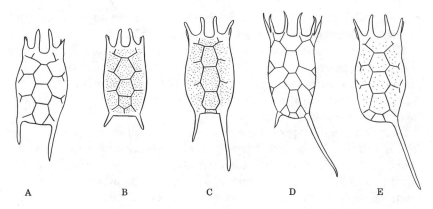

A   B   C   D   E

FIGURE 239. *Keratella valga*-group, to show specific character of the polygonal pattern on the posterior margin of the lorica and the degree of development of the left posterior spine in *K. tropica*. A. *Keratella valga*, Aneboda, Sweden (× 160). B. *K. procurva* (× 160), Sohawa, Pakistan. C. *K. tropica*, Ootacamund, Nilgiri Hills, southern India. D. *K. tropica* f. *asymmetrica*, Sohawa. E. *K. tropica*, completely monospinous f. *monstrosa*, Wular Lake, Kashmir.

temperature that may reach 27° C. by day. On July 12 at 13° C. only a few *brehmi* occurred; five days later at 22° C. a large number of *monospina* with single posterior spines had made their appearance and a week later at 18° C. both *monospina* and *valga* were present without *brehmi*. This condition continued throughout the summer, the two-spined symmetrical *valga* becoming progressively commoner. Resting eggs carried by *monospina* and *valga* appeared at the beginning of September. At the end of October *brehmi* reappeared, all three forms being present on October 24 at 7° C. As already mentioned, Klausener believed *brehmi* to be a form of *K. curvicornis,* but he also believed it to be the ancestor of the other forms he encountered. The exact extent of the cyclomorphosis depends on the status accorded to *brehmi,* but there is obviously a partial replacement of *monospina* by *valga* during the summer season.

Hartmann (1920) has given an account of the cyclomorphosis of *K. valga,* similarly involving partial replacement of *monospina* by *valga,* in a small pool lacking higher vegetation near Graz. In April at a temperature of 7.5° C. *monospina* accompanied by some *valga* occurred. In May and in September and therefore supposedly throughout the summer there was a strong tendency toward production of fairly long and relatively symmetrical *valga* forms which gave place to

asymmetrical *valga* forms in the winter.   Hartmann regards this cycle
of changes as probably determined environmentally.

Plotting the various measurements for the lorica and for the right
and the left posterior spine which have been given by Klausener, Hartmann, and Ahlstrom on double logarithmic paper provides no indication of the control of the length of either posterior spine as a heterauxetic
function of lorica length in *K. valga*.

Luzzatti (1935) evidently found both *K. quadrata* and *K. valga*
in Lago di Trasimeno, but her failure to distinguish between them prevents analysis of her data.   It is evident that *valga* was dominant from
April to October.   The measurements for these months suggest a maximum relative length for the posterior spines in June, a summer maximum comparable to that found in ponds.   The evidence presented,
however, is quite inadequate to establish such a conclusion.

Ruttner-Kolisko (1949) has cultured a typical race of *valga* from
an alpine meadow pool near Lunz.   The cultured line showed a reduction in length and in relative length of the right spine, the left spine
disappearing altogether, in the first two generations.   All later generations were monospinous with no further reduction.

The common tropical and subtropical species *K. tropica* is interesting
because of the markedly asymmetrical posterior spines, the right spine
often being very long, the left short or absent.   Not enough data
exist to indicate anything certain about cyclomorphosis, but it is clear
that the exuberant structure of *tropica* must be correlated with peculiari-
ties in its swimming behavior.

Green (1960) has made an important study of the cyclomorphosis
of a population of *K. tropica* in the Sokoto River in tropical West
Africa.   As is characteristic of the species, the posterior spines are
asymmetrically developed, the right always being longer than the left,
which may be absent (f. *monstrosa*); Green's population did not contain
specimens exhibiting the most exuberant development of the right spine
as recorded elsewhere.   He found that in any collection the right spine
length was correlated with carapace length.   In a new plot of Green's
figures k appears to be about 1.6; the value of $\hat{b}$ varies in the different
collections but it is not possible to learn if this were systematic.

On the whole the animals appearing as the floods subside in
December have short or no left spines, and as the season progresses
the length of the left spine increases and then apparently decreases.
In the years for which temperature records are given there seems to
be little obvious correlation between temperature and spine length,
though the cycle seems not unlike that described by Hartmann; the
progression of ecological change seems to have been very different.

In 1956 the spine length was maximal in March when observation began and fell steadily till May, with virtually no change in temperature; in 1954 there was a marked reduction in the number of f. *monstrosa* and an increase in spine length at the time when temperatures were probably rising; in 1957 the very feeble change seemed to produce a maximum length of the left spine when the temperatures were low.

The cyclomorphosis of species of the genus *Brachionus*. The phenomena in the genus *Brachionus* are probably partly comparable to those elucidated for the various species of *Keratella*. More culture experiments, however, have been performed, although in view of the number of species involved, there is a dearth of long well-illustrated series of field observations, those of Sachse (1911) and of Wesenberg-Lund (1930) being by far the best. The emphasis on this genus in the various publications therefore appears to be somewhat different from that of those on *Keratella*.

In a population of *Brachionus falcatus* in Steindorf Pond I, near Graz, Hartmann (1920) found larger specimens with longer anteromedian and posterior spines in July than in September. In another pond, Steindorf Pond II, which was much overgrown, particularly with *Marsilea,* the smaller form with shorter spines occurred in July, though the temperatures in the two ponds are said not to have differed. In a third pond specimens found in September seem to have been intermediate in form between the July and September populations in Steindorf Pond I. Each of the three populations differed in minute details of spination and of the surface ornamentation of the lorica. As far as the lengths of the posterior spines go, the three sets of measurements given for mean populations in the two Steindorf ponds strongly suggest a heterauxetic relationship with carapace length. It is most unfortunate that no temperatures are available, but the temperature variations recorded graphically for other ponds near Graz, as well as general climatological considerations, would suggest that the large long-spined form occurring in July was living in warmer water in Steindorf Pond I than the succeeding September population. This case, though probably comparable to that of most of the species of *Keratella* in the heterauxetic mechanism of cyclomorphosis, thus appears to differ in its temperature relationships. Green (1960) in tropical West Africa found a heterauxetic type of relationship between spine length and carapace length in his collections but no systematic seasonal variations.

Brachionus calyciflorus (= *pala* auctt.) is a species that varies considerably in the relative lengths of the anterior and most strikingly of the posterolateral spines; animals in which the latter are well developed tend moreover to have the lips of the pedal aperture of the lorica produced backward as a pair of posterior processes. The posterolateral spines are absent in the typical form of the species, but when they are present the animal is referrable to f. *amphiceros* (Fig. 240). Specimens lacking posterolateral spines but with the anteromedian spines greatly elongated are referrable to f. *dorcas;* f. *spinosus* is the corresponding *amphiceros* form.

Kofoid (1908) found that the species had a major maximum in the Illinois River in April and May and a minor maximum in August and September, but it was entirely absent from his collections during the second half of June. The specimens present in March are without posterolateral spines, most being referrable to f. *dorcas.* Throughout April the percentage of f. *amphiceros* steadily increased, till at the maximum nearly all specimens had posterolateral spines. No specimens of f. *dorcas* were present during the first half of May, but on May 24 both this form and f. *spinosus* occurred. During the late summer maximum *amphiceros* was again dominant, but neither *dorcas* nor *spinosus* appeared after the peak in numbers had been passed. A few *dorcas* reappeared in December, and Kofoid believed it to be a cold-water form, persisting until the spring maximum.

Sachse (1911) similarly found that near Leipzig *B. calyciflorus* is dicyclic, with sexual periods at the maxima. The few specimens present in winter were all *calyciflorus* (s.s.); as the number of individuals increased in April, more and more short-spined *amphiceros* were observed, and as this form replaced the type its development became more extreme. By the end of May, at a temperature of

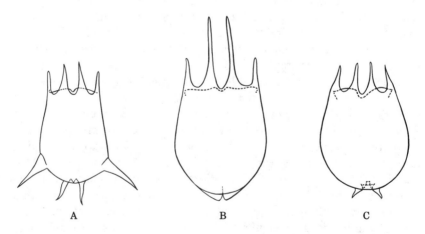

FIGURE 240. *Brachionus calyciflorus. A.* f. *amphiceros. B.* f. *dorcas, C.* f. *calyciflorus.* (Ahlstrom.)

19.4° C. only *amphiceros,* mostly with very long spines, were present. Asymmetrical development of the posterolateral spines was occasionally observed. During mid-June f. *dorcas* and f. *spinosus* were common, the latter form representing the most exuberant development of the species. From the end of June until September the species was rare and only *calyciflorus* occurred. During the autumnal maximum a short-spined form of *amphiceros* was present, but at this season neither extreme *amphiceros* nor *dorcas* and *spinosus* occurred. Sachse concluded that *amphiceros* could appear at any temperature and was associated with the abundant food supply which he believed determined the incidence of the population maxima.

Wesenberg-Lund (1936) records considerable variation in the appearance of *amphiceros* in the Danish populations of the species he has observed. In some localities it never occurs; in other ponds both *amphiceros* and *calyciflorus* could be found together. The species is said to be generally diacmic in Denmark and to be absent during the warmer months. In diacmic populations the extreme form of *amphiceros* with large posterolateral spines and a greatly developed foot sheath does not appear. In some localities, however, monacmic populations, reaching their maximum development at the height of summer, are found, and these populations produce the extreme type of *amphiceros* at the time of the maximum population. The history of one such population, in Bistrup Pond near Lake Furesø, is presented in Fig. 241. The species was first observed on July 22, 1929, at the height of a maximum, when long-spined *amphiceros* were almost the only form present. On August 2 a few short-spined *amphiceros* were observed, and by August 22 the proportion of these individuals had increased. Early in September *calyciflorus* (s.s.) appeared, and when the species was observed for the last time of the season on September 22 only such small spineless specimens were taken. During the winter the species was absent, but a few small *calyciflorus* (s.s.) reappeared on May 2, 1930. The population steadily increased throughout May, mictic females appearing on May 15 and consitituting the bulk of the population during the second half of the month. Throughout this period there was an increase in size, but all specimens still lacked posterolateral spines. On June 7 the water was milky, so abundant were the rotifers. On this date the population was again largely amictic and a few short-spined *amphiceros* had made their appearance. The species remained excessively common throughout June; on June 16 perhaps *amphiceros* was slightly commoner than it had been two weeks before and by June 24 nearly all specimens were of this form. They carried amictic, male, and resting eggs.

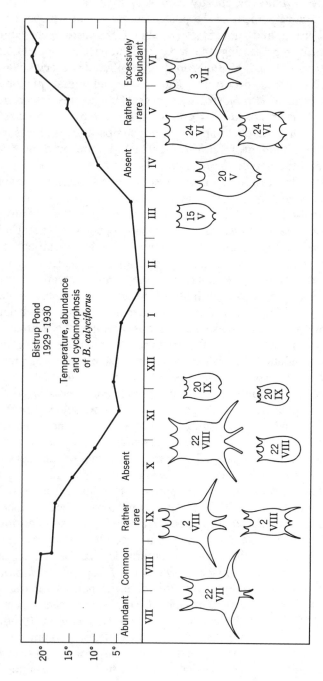

FIGURE 241. Temperature of water and seasonal incidence and form variation of *Brachionus calyciflorus* in Bistrup Pond. (Wesenberg-Lund.)

When the locality was last examined on July 3, very long-spined specimens constituted the whole population and many carried resting eggs. This type of annual cycle differs from that recorded by Sachse mainly in the absence of the autumnal maximum.

Laboratory investigations have been made by Whitney (1916c), Sachse (1911), Buchner (1936), de Beauchamp (1937, 1938, 1952a,b), Schneider (1937), and Buchner, Mulzer, and Rauh (1957). Both Sachse and de Beauchamp find that long-spined *amphiceros* produce offspring with reduced spines, the reduction being successively greater in the succeeding offspring of a wild-caught female transferred to the laboratory, as well as in later generations. With reference to the orthoclone of first-born offspring, reduction in the size of the animal and in relative spine length proceeds over three or four generations. At this stage a culture can be obtained in which *amphiceros* with short spines and the typical spineless *calyciflorus* are present. According to de Beauchamp, the population in such a culture, derived from an isolated specimen of long-spined *amphiceros,* is indeed identical with a population produced by culturing the descendents of an isolated *calyciflorus;* at least in the lines that he studied *calyciflorus* always produces a certain proportion of short-spined *amphiceros* appearing in an apparently irregular way among its descendents. A like irregularity in the production of spined and spineless forms in culture was noted by Buchner (1936). The nature of the food supply appears to influence the rate of reduction of the spines in successive generations derived from wild long-spined *amphiceros;* three subcultures, fed on natural nannoplankton, nannoplankton mixed with *Chlorella,* and pure *Chlorella,* had modal spine lengths of 110, 110, and 70 μ, respectively. Very occasionally long-spined individuals may occur without apparent cause in cultures otherwise producing only short-spined and spineless individuals. One such sporadic long-spined specimen, isolated by de Beauchamp, produced nothing but spineless descendents. Schneider (1937) believed that there was a greater tendency for such long-spined forms to occur among the descendants of *amphiceros* than among those of *calyciflorus* (s.s.). This is demonstrated in three experiments in which, however, the absolute numbers of long-spined forms are variable from series to series. It is not clear from Schneider's brief publication how the series differed in age; the experiment should be repeated using orthoclone technique. Both Schneider and Buchner, Mulzer, and Rauh found that a greater proportion of long-spined individuals appeared in cultures maintained at a low temperature than in those maintained at a high. There is obviously some genetic difference in sensitivity to low temperature, since only three of the five lines

TABLE 49. *Effect of starvation on spination of* B. calyciflorus

Descendents of	amphiceros (%)	Long-spined amphiceros (%)
Before starvation		
f. *calyciflorus*	65	3
f. *amphiceros*	66	7
After eight days starvation		
f. *calyciflorus*	74	22
f. *amphiceros*	82	25

used by Buchner, Mulzer, and Rauh exhibited environmentally controlled effects. In two series of experiments in which Schneider attempted to determine the effect of starvation, a fairly convincing result was obtained because starvation unexpectedly increased the number of long-spined *amphiceros* in lines in which relatively few such well-developed specimens had occurred in preceding generations. Buchner, Mulzer, and Rauh obtained a comparable effect in the lines that also were sensitive to temperature. As the effect of culture is to reduce the spination, any increase in spination is likely to be due to the environmental changes to which the line is subjected. One of Schneider's experiments gave the results in Table 49.

A second experiment, though giving different proportions, showed the same sort of differences between them. In spite of Schneider's finding that the reduction of body size incident on starvation is accompanied by a tendency toward long posterolateral spines and of a rather vague statement by Sachse (1911) indicating a like inverse correlation between the size of the animal and the length of the posterolateral spines, the measurements given by Ahlstrom (1940) of animals derived from a great variety of sources indicate, that provided spines are present they are likely to be longer when the animal is large than when it is small. Ahlstrom's measurements of total length presumably include the anterior spines and do not give a satisfactory measure of the size of the lorica; if the lengths of the posterolateral spines are plotted against width, the correlation is not striking but it is certainly not negative.

Whitney (1916c) believed that he had produced *amphiceros* from *calyciflorus* by treating cultures with sodium silicate, but in view of the general instability of the lines studied by other workers and of the fact that de Beauchamp was unable to confirm Whitney's findings, they evidently need not be considered very seriously.

In his most recent contributions de Beauchamp (1952a,b) has gone far toward explaining the most striking phenomena observed in *Brachionus calyciflorus* in an unexpected way. He noted that specimens of certain clones which had hatched from eggs laid in cultures of *Asplanchna brightwelli* developed long, often very long, spines. This observation appears to have been made first when the *Brachionus* was being used for food for the carnivorous *Asplanchna,* but it was repeated in a critical way in mixed cultures from which the *Asplanchna* was thinned out to prevent too much predation. An experiment in such a culture is reported in Table 50; the separation between the various long and short categories is based on the spines, in the long categories being longer than the distance between their bases and those of the nuchal (anterodorsal) spines, whereas in the short categories they are shorter than this difference. All de Beauchamp's long-spined animals clearly have much longer posterolateral spines than any of those in the experiments of Buchner, Mulzer, and Rauh. It is probable that the variation observed by the last-named workers is not the same type of phenomenon as that exhibited in the presence of *Asplanchna* (see Table 50).

Further studies indicate that a spine-stimulating substance accumulates in the water in which *Asplanchna brightwelli* lives. This material is nonvolatile and can be dried at a low temperature; it is, however, fairly thermolabile, for it resists boiling for a few seconds but not for a few minutes.

Not all races of *B. calyciflorus* respond equally to the substance produced by *A. brightwelli.* From the Bois de Boulogne and the gardens of the Luxembourg lines started in March give only a feeble reaction, whereas lines started in summer give a strong reaction, which

TABLE 50. *Effect of Asplanchna on spination of* A. calyciflorus

	Unarmed	Very Short	Short	Fairly Short	Fairly Long	Long	Very Long
Original B. *calyciflorus* clone	131	20	7	0	0	0	0
Part cultured alone as control	72	19	9	2	0	0	0
Part cultured with *Asplanchna*	1	1	1	4	33	16	11

in some cases leads to the production of f. *spinosus* rather than f. *amphiceros*. The apparent cyclomorphotic production of long-spined forms in these localities therefore represents a replacement of a less reactive winter by a more reactive summer race, which in the presence of *A. brightwelli* gives rise to the long-spined summer forms. The winter race seems to have a lower thermal death point than the summer.

The production of the long spines of the fully developed f. *amphiceros* is clearly adaptive, for *A. brightwelli* has great difficulty in feeding on such animals, though it is able to feed on the fairly short-spined type of animal produced as a maximum reaction by the winter races.

The spine-stimulating substance is apparently present in the males and resting eggs of *A. brightwelli*. It is also produced, though in smaller quantities than in *A. brightwelli,* by the parthenogenetic females of *A. girodi* and *A. priodonta,* though these two species are said not to be able to eat large rotifers. Another carnivorous pelagic rotifer, *Eosphora naias,* gave no positive indication of producing the substance.

Gilbert (personal communication) has confirmed and greatly extended de Beauchamps observations. He suspects the active substance involved to be a protein. He believes that the various records of long-spined specimens are in general concordant with a determination by the presence of *Asplanchna*.

Brachionus quadridentatus (= *bakeri* auctt.), though extremely variable (Fig. 242), does not appear to exhibit a well-marked and regular cyclomorphotic progression of forms with the seasons. Jennings (1900) noted short-spined forms in landlocked pools cut off from the Great Lakes and long-spined forms in swamps on some of the islands in the lakes. He suspected that the shortness of the spines in the landlocked pools was due to the relatively high electrolyte concentration of their waters, an idea which may be of interest in view of the possible relation of spine length to calcium in *Keratella quadrata*.

Kofoid (1908) records a number of forms from the Illinois River; none is common except between the beginning of June and the end of September. Unfortunately, Kofoid clearly did not distinguish *B. falcatus* from the typical form of *quadridentatus* with long posterior spines and it is by no means certain that all the other forms he records are to be referred without question to *B. quadridentatus*. Kofoid's data for the seasonal incidence are therefore less valuable than are those for *B. calyciflorus*. They seem to show that f. *cluniorbicularis* with a rounded posterior margin is the earliest form to appear, though

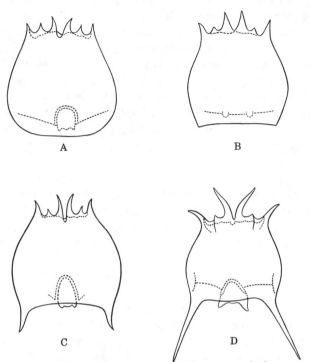

FIGURE 242. Variation of *Brachionus quadridentatus.*
A. f. *cluniorbicularis. B.* f. *rhenanus. C.* f. *brevispina.*
D. f. *quadridentatus.* (Ahlstrom, Lunt.)

it persists sporadically throughout the summer. *F. rhenanus,* in which the posterolateral spines are absent but the posterior margin of the lorica is obtusely angulate laterally, appears in Kofoid's statistics as a late summer form (August 9 to September 13). A very wide form without posterolateral spines, referred to f. *obesus* Barrois and Daday, occurred rarely and sporadically and was supposed by Kofoid to be produced by rapid reproduction. The status of the forms with posterolateral spines in Kofoid's collection is so doubtful that nothing significant can be said about them except that they were certainly rarer than the spineless forms early in the season and possibly as common or commoner than the spineless forms late in the season.

According to Sachse (1911), *cluniorbicularis* appears in the spring, and the specimens with the longest posterolateral spines are mainly autumnal. The cyclomorphosis, however, is said to be less regular than in *Brachionus calyciflorus*. Sachse indicates that there is an inverse relationship between body length and spine length, but he gives no measurements and the figures that might illustrate the relationship are not very convincing. The measurements of specimens from many widely spread localities given by Ahlstrom (1940) appear to indicate exactly the opposite relationship, for, provided posterior spines are present, they tend to be longer in the larger individuals; the data so far as they go strongly suggest heterauxesis. Ahlstrom thinks that when both long- and short-spined forms occur in a single locality there is usually continuous variation between f. *cluniorbicularis* and f. *brevispina* but a discontinuity between the latter and the typical f. *quadridentatus*. The meager reports of experimental studies of the species seem to indicate that different lines behave quite differently in culture and perhaps confirm the suspicion that arises from observations in the field that what passes for cyclomorphosis may partly be the replacement of one form of a genetically polymorphic coenospecies by another.

Luntz (1928) found f. *rhenanus* at Berlin-Dahlem in the spring and the typical long-spined form in the same locality in summer. It appears, however, that he was able to culture f. *rhenanus* but not the typical form. The appearance of spined forms in summer may therefore have been due to the replacement of a spineless by a spined race with different ecological requirements and not to true cyclomorphosis.

Sachse (1911) found that in culture only specimens with moderate or short spines occurred, an observation apparently confirmed by de Beauchamp (1927), but Luntz found f. *rhenanus* to be stable in culture; variation in the temperature at which the culture was kept was quite without morphological influence, though the density of the animal was altered, as previously (pages 278–279) indicated. Buchner (1936), using material from the same pond that furnished Luntz's material, found great variability in culture, in which specimens both without spines and with long spines were produced.

Because the optimal culture conditions and the food requirements of Buchner's animals differed greatly from those found best by Luntz, it is reasonably certain that the two workers had very different animals, albeit from the same locality. Later, Buchner, Mulzer, and Rauh (1957) studied five lines of the species which did not respond to environmental change in temperature and food supply. One line produced no spines, and in another from the same pond the spines varied from fairly short to long. It is evident that there is considerable genetic diversity as well as unidentified environmental determination involved in producing the variability of this species.

In *Brachionus bidentatus* (= *furculatus* auctt.) no significant studies of seasonal variation in the field appear to have been made, but de Beauchamp (1924a,b, 1927) has used this species, which at first he regarded as a form of the preceding one, for a series of experimental studies. He has reared several lines which give an irregular distribution of spineless or short-spined forms in his pedigrees. Individuals derived from resting eggs are always spineless, and the spineless form appears to be somewhat more abundant in all cultures. Different lines, however, give very different proportions of spined forms. One series of mass cultures derived from three resting eggs yielded 39 per cent spined forms, whereas the line from which the eggs were obtained, when continued under the same conditions parthenogenetically, gave only 1 per cent spined forms. Transfer from old to new *Polytoma* culture tends to produce spined forms, and an even surer method of producing the transition is to transfer a line from *Polytoma* to *Chlamydomonas* as food. If such a line is kept on *Chlamydomonas* indefinitely, the number of spined forms decreases but appears to remain above the proportion that would occur in a *Polytoma*-fed culture. Transfer back to *Polytoma* does not produce an increase in spined forms, but de Beauchamp does not record whether it produces the decrease that might be expected. It is evident that both genetic and environmental factors must be operating in this as in the preceding case, though what the environmental factors may be is not at all clear, save that they are mainly effective as the results of abrupt changes in the quality of the food. It must be borne in mind in this case, as in so many others, that the most exuberant forms occurring in nature have yet to be produced or maintained in culture. Further work may well indicate a situation like that in *B. calyciflorus*.

In *Brachionus angularis* Sachse (1911) records animals with an unsculptured lorica referrable to f. *bidens* (*C* of Fig. 243) as being present in October near Leipzig. During the autumn and winter, as the water temperature fell from 10.6 to 2.6° C. the size of the members of this population increased, reaching a maximum in January or February. There appears to be a slight irregular decrease in size in the spring, though the specimens present on May 11 as the population was increasing toward a maximum, at 12.5° C. were larger than the specimens present in the previous October. After the May maximum a very small form (*D* of Fig. 243), with ridges running dorsoventrally at the lateral angles of the lorica, appears in June and remained until the end of August. In the next month or two small specimens of f. *bidens* reappeared. A cycle comparable to that described by Sachse was reported from the Volga by Meissner (1901 *fide* Sachse). It

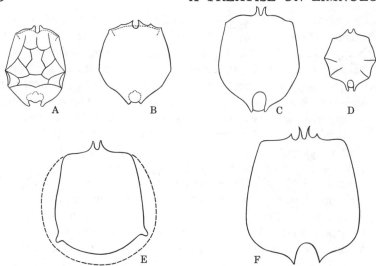

FIGURE. 243. *Brachionus angularis*. *A*. Swan Creek, northwestern Ohio, mictic ♀. *B*. The same, amictic. *C*. f. *bidens* winter form, Leipzig. *D*. Small summer form, near f. *aestivus*, Leipzig. *E*. Frederiksborg Castle Lake, August 7, with gelatinous sheath. *F*. The same locality, November 28, winter form without sheath (× 120). (Ahlstrom, Sachse, Wesenberg-Lund.)

is probable, moreover, that Sachse's small summer form is close to or identical with f. *aestivus* Skorikov, which is known not only from south Russia in summer but also from India (Ahlstrom 1940).

Ahlstrom concluded that at Swan Creek, Ohio, there was a definite dimorphism in the females, amictic females being larger and relatively unsculptured (*B* of Fig. 243), mictic, smaller and well-sculptured (*A* of Fig. 243). It may be pointed out, however, that in Sachse's population at the maximum in May the specimens of f. *bidens* were evidently mictic and carried resting eggs. If some of these specimens had persisted into June, when the smaller summer form had already appeared as a population of amictic females, exactly the opposite of Ahlstrom's finding would be produced. Unfortunately, Ahlstrom gives no indication of the time of year when his material was collected nor of the number of times that the locality was visited. It is also to be noted in all these researches that the evidence of the production of the summer form from the unsculptured winter form is very meager.

A quite different cycle was reported by Wesenberg-Lund (1930) in the Fredericksborg Castle Lake and in some but not all of the other Danish ponds in which the species was observed. In these local-ities *B. angularis* is perennial but is likely to have a spring maximum with mictic females at about 10 to 12° C. and another smaller maximum at a higher temperature in August. The late summer maximum is composed of animals (*E* of Fig. 243) which are coated with a gelatinous sheath through which the lateral antennae set on raised papillae project. This form disappears in the early part of September and is replaced later by the winter form, without papillae or other sculpture, which is probably referrable to *bidens* as interpreted by Sachse. The summer papillate form is evidently a member of Ahlstrom's variation series *β*, the most extreme member of which he characterizes as pseudo-dolobratus (this seems not to be introduced as a formal name), which is recorded from Florida and is doubtless a warm-water form. Wesenberg-Lund found that both his winter form f. *bidens* and his summer form with the gelatinous sheath could be either mictic or amictic. Voigt (1904) evidently observed a comparable cycle which involved the production of a gelatinous sheath in the Heiden-See, Edeberg-See, and Kleiner Madebröcker-See near Plön. These sheath-bearing sum-mer forms were much smaller than the spring forms and of yellowish color.

In *Brachionus caudatus* the cyclomorphosis has been studied by Green (1960) in the Sokoto River. *B. caudatus,* which some authors have placed under *B. angularis,* shows considerable variation in the length of a pair of posterior spines arising on either side of the foot opening and so are not homologous with the posterolaterals of *B. calyciflorus* or *quadridentatus.* The over-all general picture is one of an irregular seasonal increase in lorica length and of spine length from December to June. The variation on any given day is considerable (Fig. 244). The larger animals seem to exhibit heterauxesis with a value of **k** around 2.85, being greater late in the season than earlier. However, there appears to be a lower limit to this relationship, below which the incipient spine grows very fast relative to the lorica. In the rather small animals present in December this period of rapid growth of the spine is perhaps followed by a much less marked, possibly brady-auxetic growth.

Adaptive significance of cyclomorphosis in rotifers. The cyclomor-phosis and comparable form changes in the rotifers seem to be of four general kinds.

1. The elongation of *Asplanchna priodonta* in Denmark and north Germany is comparable in its seasonal incidence and in its over-all nature to the production of an extreme type of helmet, as in *Daphnia*

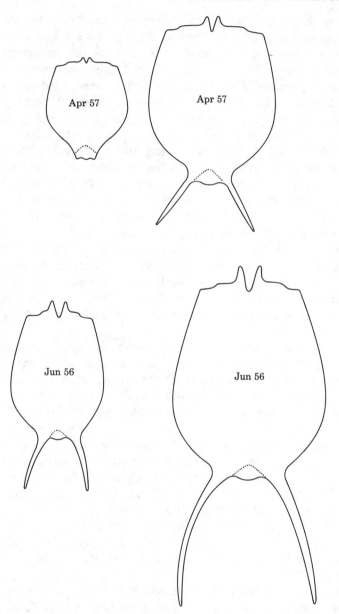

FIGURE 244. *Brachionus caudatus,* Fesafari, River Sokoto, extreme range of form in April and in June (× 200).

cucullata in the same waters. The elongate *A. priodonta,* however, are almost always sterile, and the adaptive meaning of the change is obscure. It is possible that Gallagher's (1957) colony of *Keratella quadrata* also behaved in a similar way to the Cladocera; no other investigator has found this species with longer spines in summer than in winter, though this sort of change perhaps occurs in *K. valga* and *Brachionus caudatus.*

2. The production of form changes in *A. sieboldi* by changes in food. It is just possible that the change is an adaptation to the food producing it, but there is no evidence of this.

3. The production of lateral spines in *Brachionus calyciflorus,* which appears to be due to the presence of *Asplanchna* and may reasonably be regarded as a defensive adaptation against predation by the latter.

4. Reduction in size generally at high temperatures with a consequent tachyauxetic reduction in spine length, as in *Keratella.* The production of a fine pubescent-like spination may also occur in summer in the genus as in the form referred to *hispida.* Carlin (1943) and Pejler (1962c) point out that the heterauxetic control of the spine, which is proportionately longer when the water is cold and its viscosity high, rules out the significance of direct adaptation to decreasing viscosity in these forms. Carlin thinks that reduced molecular viscosity, other things being equal, leads to increased turbulence or eddy viscosity. This attractive hypothesis requires empirical study, as the lower wind-drift velocities but greater stabilities in summer are likely to be involved in complicated ways. Pejler's discussion, which is somewhat muddled mechanically, makes the interesting suggestion that when the animal becomes larger projections that increase its form resistance must be increased to a greater degree. In the Stokes law range, appropriate for a rotifer, if the viscosity did not change, increasing the linear dimensions from d_1 to d_2 increases sinking speed by $(d_2/d_1)^2$. If this coincided with an increase in viscosity say from 0.0089 poise in summer at 25° to 0.0157 poise in winter at 4° C. the allowable increment in length would be by a factor of 1.3. In most species of *Keratella,* in which the value of **k** for posterior spine growth relative to lorica length is rather more than 2, an increase in length by a factor of 1.3 would imply a doubling of the spine length even though no such change in shape appears to be necessary. Actually, it is not obvious without experiments exactly how the increase in spine length would affect sinking speed. Pejler is certainly right in rejecting the cycle of change in viscosity as being reflected in the seasonal form changes in *Keratella,* but his own explanation, though it is, in the terms in which it has been restated, by no means impossible logically, seems

not to be fully in accordance with the meager available quantitative data. More information applying to genetically continuous populations over a full cycle from the coldest to the warmest time of year is clearly needed. As already indicated (see page 270), there is no reason to suppose that the peculiar condition of *hispida* would be effective in reducing sinking speed.

CYCLOMORPHOSIS IN THE PHYTOPLANKTON

Cyclomorphosis in the limnoplanktonic species of Ceratium. Two or three of the few fresh-water species of the genus *Ceratium* are pre-eminent among the rather limited number of phytoplanktonic organisms showing cyclomorphosis. The genus is distinguished from allied dino-flagellates in bearing an anteriorly directed apical horn and usually two or three posterior horns. Among the posterior horns the longest is called the antapical; and the right posterior horn, following the convention that the sulcus is ventral, is termed the postequatorial, and the left posterior horn, which is frequently absent, is often simply the fourth horn. According to Krause (1911), the variation in horn length on three-horned specimens of *C. hirundinella* is such that the length of the apical horn is equal to three times the difference between the antapical and postequatorial horns. Though there is probably a widespread tendency for the apical horn to be long when the difference between the posterior horns is considerable, Krause's relationship does not hold for all populations, for it breaks down notably when the mean measurements of three-horned specimens studied by List (1914a) from localities near Darmstadt are examined.

Certain species of *Ceratium* are widely distributed in the limnoplankton. In the temperate regions of the Northern Hemisphere they appear mainly in summer and in autumn produce asexual resting stages or cysts, the forms of which are fairly characteristic of the various species. It would appear from the work of Huber and Nipkov (1922) that at least three distinguishable taxa can co-occur abundantly in a single lake, but in most cases the occurrence of but one or two is doubtless more usual. The taxonomic differentiation of the valid species depends not merely on the shape of the cell and the disposition of the horns but also on the details of the arrangement of the cellulose plates which form the armor or exoskeleton of the cell. Entz (1927a) has figured such details for some of the cyclomorphotic fresh-water taxa which have been confused in the past and has clearly shown *C. hirundinella* and *C. furcoides* to be distinct. There is, however, still considerable uncertainty regarding how many species should be recognized.

Cyclomorphosis (Fig. 245) in *Ceratium* involves changes (1) in

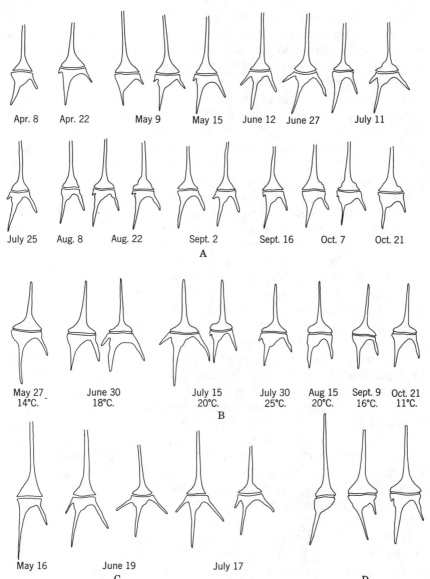

Apr. 8   Apr. 22   May 9   May 15   June 12   June 27   July 11

July 25   Aug. 8   Aug. 22   Sept. 2   Sept. 16   Oct. 7   Oct. 21

A

May 27   June 30   July 15   July 30   Aug 15   Sept. 9   Oct. 21
14°C.   18°C.   20°C.   25°C.   20°C.   16°C.   11°C.

B

May 16   June 19   July 17

C   D

FIGURE 245. Cyclomorphosis in *Ceratium hirundinella* and its relatives. *A*. In Kirchbergteich, Darmstadt, Germany, showing addition of fourth horn early in summer and minimum width in August. (List.) *B*. A comparable but less extensive series from Furesø, Denmark. (Wesenberg-Lund.) *C*. A narrow *furcoides* form from the Pfaffenloch, Erfelden, Germany. (List.) *D*. Two-spined form from the Federsee, Württemberg, Germany, in early spring, with a third and traces of a fourth spine developing later in the season. (Ammann.)

the over-all size of the cell, best measured in terms of its width, to
avoid complications due to partly independent variation in horn length,
(2) in the lengths of the various horns, (3) in the presence or absence
of the fourth horn and sometimes of the postequatorial horn, (4)
in the degree of divergence of the posterior horns, and probably (5)
in the development of sculpture on the thickened plates of the cell
wall. The evidence relating to the last type of variation is unsatisfactory, and all that can be said is that scattered observations, notably
those of List, suggest that the reticulate sculpture is often best developed
when the horns are of maximal length in the late spring or early
summer. The cyclomorphosis of populations referred to *C. hirundinella* was investigated by a number of workers at the end of the
nineteenth and beginning of the twentieth centuries. The results of
these investigations have been admirably summarized by Wesenberg-
Lund (1908). In general a fairly consistent picture emerges from
this work to which the most important contributions were those of
Entz (1904) and of Wesenberg-Lund himself. Unfortunately, both
authors and several others probably included more than one species
under the name *C. hirundinella,* and part of the phenomena that
they describe may be due to the replacement of one species by another
rather than to a true transformation within a genetically homogeneous line.

The various possible types of cyclomorphosis in *Ceratium hirundinella* are well indicated in the detailed study made by List (1914a)
of populations in ponds near Darmstadt and Erfelden, a little to the
west of Darmstadt. The populations of *Ceratium* in three ponds near
Darmstadt were found to exhibit very varied degrees of cyclomorphosis.
In the Waltersteich, an artificial pond dating from 1686, surrounded
with woodland and covered with *Lemna* during the late summer, there
was very little change in form, save that three horned specimens did
not occur except in April. No marked and consistent variation occurred in cell breadth or in the lengths of any of the horns in either
the normal summer of 1909 or in the abnormally hot summer of
1911.

In the Kirchbergteich, a more exposed pond, a well-marked series
of changes occurred. The breadth of the cell is maximal in May
or early June and again in October, with a definite intervening minimum in August. No indications of the proportions of three- and four-
horned specimens on most dates are given, but it is evident that three-
horned individuals were common early and late in the season and
rare or absent in June when none was obtained for measurement.

The angle of divergence of the antapical and postequatorial horns varies in a systematic way with the temperature, increasing to a maximum in July; a similar, less regular variation is exhibited by the angle between the fourth and antapical horns. The lengths of all the horns increase to a maximum in May or June before the water temperature is maximal and thereafter decline without the development of any secondary maxima of significance. List believed that this decline was more marked in the very hot year 1911 than in the normal year 1909, which may be so in the cases of the apical and fourth horns, though some reduction of the lengths in 1911 below those of 1909 was already apparent at the end of May before the abnormal nature of 1911 had become evident in the water temperatures. The very high temperatures of 1911 seem to have had no effect on the breadth or on the length of the antapical horn; both quantities varied similarly in the two years in spite of the marked temperature differences.

In a third locality near Darmstadt, the Kranichstein Castle Pond, the general trend in 1909 was similar to that in the Kirchbergteich, but the maximal length of the fourth horn was achieved in August, much later than that of the other horns. In the very hot summer of 1911 a marked minimum in the length of the apical and a rather less marked minimum in the length of the antapical horns was observable at the end of July and the beginning of August, followed by definite secondary maxima.

Two of the five localities near Erfelden were studied by List on only three occasions and add little of interest to the investigation. A third locality contained several forms, certainly including both *C. furcoides* and *C. hirundinella;* the variation in this pond is hard to disentangle. Two other rather deep open ponds, namely the Vorderes and Hinteres Bruderloch, contained reasonably uniform populations. The population of *Ceratium* in the Vorderes Bruderloch in April consisted of large forms, the width and the length of the apical horns decreased throughout the summer but in August and September began to increase a little to produce a low secondary maxima in October. The length of the fourth horn increased to a maximum in July and August and slowly decreased in the autumn. Apart from early incidence of the primary maxima and the occurrence of secondary maximum in the length of the apical horn, this pattern of variation is reminiscent of the conditions in the Kirchbergteich. In the neighboring but smaller Hinteres Bruderloch the water temperatures varied throughout the year in almost exactly the same way as in the Vorderes Bruderloch. The *Ceratium* population of this pond, however, differed markedly in having a distinct secondary maximum in the lengths of apical and antapical horns and in the breadth in August, followed in each case by a decline in the dimensions in November with slight indications of an increase in width and in the length of the apical horn in December. The postequatorial and fourth horns increase and decrease in length with temperature, as does the angle of divergence, just as in the other cases considered by List.

In the later paper List (1914b) records the result of an experiment in which the Ludwigsteich at Darmstadt, which previously lacked *Ceratium,* was seeded with material from the Vorderes Bruderloch. The experiment was started on May 24, 1913. During the summer of this year the cyclomorphosis in the Vorderes Bruderloch consisted primarily in a marked elongation in the postequatorial and fourth horns, which reached maximal lengths in August. There is evidence of a spring maximum in the length of the apical horn but after a small decrease in length it increases again to a secondary maximum in August; the changes involved, however, are very small. The antapical horn also decreased at first but reached its absolute maximum in August. The width declined a little in June and increased again irregularly in late summer. The changes in general are less marked than in the earlier year discussed in List's first paper. In the new population established in the Ludwigsteich the general trend appears to differ little from that of the parent population, save that the lengths of all three posterior spines are consistently shorter. This change does not affect the apical horn or the breadth. At the same time a group of small individuals appears and remains present with the main population until the end of July. It is to be noted that the temperature in the Ludwigsteich is consistently lower than in the Vorderes Bruderloch.

In most European localities studied by other workers a general sequence of events comparable to that observed by List in the Kirchbergteich has been observed, though the date of the spring maximum in breadth and horn length is variable and the extent of secondary autumnal maxima in these dimensions also differs from lake to lake. Comparable temporal variation was noted by Jean C. Smith (personal communication) in the population of Bantam Lake, Connecticut.

In Furesø, a lake in which Wesenberg-Lund (1908) considered but a single form to be present, the species appeared as a fairly wide type lacking a fourth horn in April. The population was very small until the end of June, when enough *Ceratium* appeared to provide a satisfactory indication of the statistical aspects of cyclomorphosis. There is a steady fall in modal width throughout July; by September there is some indication that this quantity is increasing. Early in the season many specimens, presumably derived from cysts, with very long apical horns, are present, but after the middle of July they become rare and during September the modal length of this horn is about equal to the minimum length at the end of June. Considering only mean values, the variation in the width and the decline in length of both the apical and antapical horns would be strictly comparable to the changes observed in the Kirchbergteich by List; however, when Wesenberg-Lund's distribution curves are examined, the decrease in the lengths of the apical and antapical horns suggest a discontinuous change which would not have been apparent if single mean values alone had been considered. It is therefore not quite impossible that some replacement of one form by a genetically diverse form of different origin underlies the apparent cyclomorphosis in apical and antapical horn lengths. The variation in the fourth horn, however, is a truly cyclomorphotic process. This horn is indicated only as a very slight protruberance in May and early June, but by the end of the latter month it is well developed, most notably in the long-spined forms, which there is every reason to suppose are directly descended from the spring population. On the fifteenth of July about one quarter of the specimens present lack fourth horns, and thereafter the number of four-horned individuals decreases so that by September 7, 59 per cent of the population are three-horned and in all the four-horned specimens the fourth horn is short. It is evident that even if the late summer and autumn population with short apical and antapical horns is of different origin and genetic constitution from the long-horned spring form the incidence of the fourth horn in both populations is greatest when the temperature is highest.

In all the Danish lakes in which *Ceratium hirundinella* occurs it appears that the specimens with the longest horns, and the greatest total length, occur in the late spring and early summer. A decline in horn length always takes place in July and there are no secondary maxima in these dimensions, though the breadth may increase slightly at the end of the season.

In some Danish lakes two quite distinct forms occur, though Wesenberg-Lund was uncertain of their status. In Haldsø a predominantly summer form with a very divergent and well-marked fourth horn accompanied the ordinary type of *C. hirundinella*. This second form, which Langhans (1925) tentatively regarded as a separate species, is clearly related to *scoticum* and *piburgense*. In marked contrast to this summer form, the second and predominantly aestival *Ceratium* in Sorøsø and Tjustrupsø is always three-horned and slender with subparallel posterior horns. It is referred to *C. furcoides* by Langhans. In the two Danish lakes in which this species and *C. hirundinella* occurs both appear to have their maximum horn length late in May or early in June and at least in Sorøsø both undergo a slight reduction in width between the end of May and the beginning of August. Such mixed populations of *C. furcoides* and *C. hirundinella* are evidently not uncommon in European lakes, occurring, for example, in the Grosser Plöner See and in Lake Balaton. *C. furcoides,* it is to be noted, is by no means always three-horned. Entz (1927) mentions a population from Tata in Hungary in which four-horned individuals formed 76 per cent of the population in April but disappeared entirely in June, reappearing in small numbers in July.

A very important study of the cyclomorphosis of *Ceratium hirundinella* was made by Pearsall (1929) on material collected by Tattersall and Coward in Rostherne Mere in Cheshire, England, in 1912 and 1913. In this locality the species is perennial, though there is considerable production of cysts in the autumn and the small winter population is probably replaced in spring by a population derived from the hatching of such cysts. In April and May there is a very rapid increase in the means dimensions of the species. Examination of the distribution of lengths in the population shows that it is due at least in part to a large form replacing a small form. Pearsall interprets this as the replacement of the overwintered population by individuals that have hatched from cysts. By the beinning of May in 1913 only one mode in the distribution of lengths was present, but since the mean length continued to increase until the beginning of June there is clearly a real increase in length and breadth in the spring population. About the time that the maximum length is reached a few specimens with short fourth spines appear. During June there is a marked decline in length and a rather less rapid decline in breadth. The low winter values for the length are already apparent in July but the breadth goes on declining until August. Meanwhile the proportion of specimens with four horns increases to a maximum of more than 80 per cent and then declines again through August and September. The individuals that do not encyst at this time continue as a small three-horned population throughout the winter.

A graphical presentation of the seasonal variation in total length of *C. hirundinella* inhabiting a large pond at Lágymányos, Hungary, has been given by Entz (1931). The graph is obviously much smoothed but clearly indicates that the population, which is perennial, has a maximum length in April, certainly corresponding to the spring maximum in the length of the whole cell and of the apical and antapical horns in nearly all the other populations that have been studied. After reaching a value of about 260 μ in late April the mean length declines slowly to a winter value of about 170 μ. The rise in length in spring is evidently very rapid and may well represent the emergence from cysts of a new population, as in Rostherne Mere.

Earlier work by Entz (1904) on the *Ceratium* populations of lake Balaton is confused by the failure to separate the two taxa present. Large specimens of *furcoides* appear in the spring and disappear in June. Individuals purporting to have the apical end of one species and the antapical of the other are figured, but they are not very convincing. The late summer population consists of true *C. hirundinella* in which the proportion of four-spined individuals increases from 16 per cent in July and August to 94 per cent in September, falling again to 22 per cent in October.

An interesting variant of the general cyclomorphotic sequence in fresh-water species of *Ceratium* is exhibited by a population, apparently referred to *C. hirundinella* by Entz (1927), which occurs in the Federsee in Württenberg and was studied by Ammann (1922). Here the earliest spring forms are normally two-horned, only the apical and antapical horns being developed. Later in the season there are three- and even four-horned specimens which form cysts in the autumn. Ammann indicates that in the Federsee, as well as in three Bavarian lakes in which the two-horned form evidently does not occur, the over-all variation in size found by other workers in other localities also takes place, the largest specimens occurring in May and the smallest at the end of summer as the species is disappearing from the plankton. Huber-Pestalozzi (1927) has noted that the two-horned form occurs sporadically in nature and has been found in some numbers in cultures hatched from cysts at low temperatures by Huber and Nipkov (1923).

The only experimental study bearing on the production of cyclomorphosis in *Ceratium* is that of Huber and Nipkov (Huber-Pestalozzi 1950). They reared individuals at a variety of temperatures from cysts obtained from the mud of Lake Zurich. The early stages of the cell hatching from a cyst is unarmored, and the heavy cell walls and horns are not developed for some hours after the so-called *Gymnodinium*-stage has begun its free life in the water. Huber and Nipkov, working apparently with *C. hirundinella* f. *austriacum,* found that the proportion of individuals developing two, three, and four horns differ markedly according to the temperature at which the cultures are reared.

It is evident that the proportion of four-horned specimens is greatest at 23 to 26°, which is the temperature for most rapid development. At temperatures below 9° C. all specimens develop abnormally and

TABLE 51. *Effect of temperature on number of horns in* Ceratium hirundinella

	10–12° C.	13–14° C.	15° C.	16–18° C.	23–26° C.	28–30° C.
Two-horned (%)	14	6	0	0	0	10
Three-horned (%)	86	72	61	62	35	67
Fourth horn present but rudimentary (%)	0	19	35	34	35	9
Fourth horn well developed (%)	0	3	4	4	30	14

three-horned forms are rare; no development takes place above 35° C. and even at 28 to 30° C. most specimens are somewhat abnormal. The divergence of the horns was most marked in specimens developing at 23 to 26° C.; the total length of the cells which probably implies a maximal length for the apical and antapical spines was greatest in specimens reared at 15–18° C.

Huber and Nipkov also did a considerable number of experiments on the effect of colored lights on the development and form of *C. hirundinella,* but these experiments mainly show that a reduction in intensity in some part of the spectrum may reduce the length of the cell and the incidence of the fourth horn. A number of experiments on the effect of various dissolved substances indicate that various abnormalities can be produced by the addition of potassium nitrate, sodium chloride, and other solutes to the cultures of *C. hirundinella* but they throw little light on variation in nature.

It is reasonable to conclude from these experiments, in conjunction with most of the field data presented, that the incidence of the fourth horn and the summer increase in the divergence of the horns depends on the high temperature at this season. It is obvious, however, that different populations respond differently to a given increase in temperature. In some cases no fourth horn is added to any specimens; in others nearly all of the population is four-horned in summer. Moreover, there are at least two populations known, namely that of *C. hirundinella* from Lake Balaton and that of *C. furcoides* from Tata, in which the maximum in the proportion of four-horned specimens is autumnal and vernal, respectively, and certainly not associated with the warmest time of year.

It would be reasonable to regard the rather general decline in width, probably the best criterion of cell size, in early summer, when this decrease is followed by an increase in the autumn, as inversely related to temperature. The division rate greatly increases in the late spring and decreases again in the autumn, while at the same time it is not unlikely that anabolic processes, dependent on combined nitrogen, phosphorus, or other nutrients, would proceed more slowly at the height of summer. The very marked maximum in total length in the Lágymányos population occurs just before the division rate rises to its summer maximum, and a general explanation based on division rate and anabolic growth in relation to temperature and the nutrient cycle would be tempting in other cases. In Huber and Nipkov's experiments, however, there is some indication of an optimal temperature for the growth of the excysted cell, regardless of division. There are, moreover, well-studied cases in which there is no indication of a second autumnal maximum in width or length, for the temperature falls through the same range that coincided with the maximum in cell size in the spring. Though the possibility of some control by light as well as by the nutrient cycle cannot be overlooked, especially in a photosynthetic form, it seems not impossible that the lowness or in some cases the total absence of the second maximum in size in autumn involves internal change comparable to that observed in the variation of size and of spine length in rotifers of the genus *Keratella* in culture. This is indeed a more modern and noncommittal, if more cumbersome, way of expressing what was formerly described as degeneration, though when Entz (1904) used the term in this context he indicated that he believed environmental changes to control the process.

Pearsall alone has attempted to investigate the meaning of the size changes in *Ceratium* in terms of the nutrition and physiology of the cell. Pearsall believes that the two antagonistic processes involved in determining the apparent optimum for cell size in the spring are the rate of growth of the protoplast and the rate of hardening of the cell wall rather than any division rate controlled by the nucleus. The rate of protoplasmic growth he regards as varying directly with the temperature. The rate of hardening of the cell wall is regarded as dependent on a supply of the raw materials from which the wall, mainly of cellulose, is formed. If no starch is being stored in the cell it implies that there may be a deficiency of carbohydrate that could be used to construct a wall. Under these conditions, the wall remains soft and expansile for a long time. The cell therefore grows large, and the higher the temperature the faster it will grow, whereas hardening may be progressively inhibited. At the end of June starch appears in the cells and at the end of July, fat that can be stained with Sudan III. Under these conditions the cell wall may be expected to harden rapidly and division must occur at a progressively earlier, smaller stage than in the spring. At the same time there is evidence of a change in the composition of the cell wall. In the spring specimens the various cellulose tests are given easily; in the late summer a reaction is obtained with some difficulty unless the cell has first been treated with HCl. At the same time the wall looks thicker, swells less easily and is found to give a progressively increasing test for fatty acids. Pearsall believes that the small late summer forms are not only the result of excessive photosynthesis leading to early cessation of growth and early division but that the microchemical evidence substantiates the progressive impregnation of the cell wall with a calcium soap or salt of some fatty acid which contributes to the premature hardening process. This process, which becomes more and more extreme, is believed to end in some lines in the production of a rigid wall before any appreciable growth has occurred; in such a case a cyst has been formed. Pearsall also suspects that the more swollen outline of the spring form is due to a higher water content, possibly implying a lower acidity in the cell at that season. It is highly probable from the data obtained in other localities that at the time of maximal protoplasmic synthesis combined nitrogen would be more available than later in the season when most of the products of photosynthesis seem to form carbohydrate or fat. Pearsall's whole approach is most ingenious and interesting and is also somewhat reminiscent of the ideas that have been put forward in relation to senescence and calcium in the rotifers. It receives support moreover from the fact that in Ennerdale Water and Crummock Water in the English Lake District, in which the calcium content is very low, the *Ceratium* tend to be larger than in Windermere and Esthwaite in which the calcium content is higher. Pearsall also notes that there is great variability in the extent to which the tests of *Ceratium* remain intact after preservation and suspects that the well-preserved specimens come from the more calcareous lakes. In spite of this supporting evidence, it is obvious that a great deal more work must be done before the validity of the theory can be judged.

Adaptive significance of cyclomorphosis on Ceratium. To Wesenberg-Lund the development of the fourth horn is primarily a device to reduce the passive sinking speed, though he admitted that its presence, particularly when the postequatorial horn was strongly divergent, would slow the rate of active movement through the water. He also implied that the production of long apical and antapical horns acted in the same way as flotation organs, in spite of the fact that these outgrowths tend to decrease in most localities at the time when the temperatures are reaching their maximal values and so the viscosities their minima.

Krause (1911) regarded the outgrowth of the horns as one method by which the fall in viscosity is compensated, the other methods being outgrowth of pseudopodia, formation of gelatinous sheaths, chain formation, and exuviation. None of these other processes appears to have any widespread occurrence. List considered that the growth of the fourth horn, the divergence of the posterior horns, the decrease in size, and an increase of the scale-like sculptural roughness of the surface all contributed to the reduction of sinking speed in summer. The data on the production of marked sculpture are inadequate, but otherwise these contentions are reasonable. The only published data on passive sinking speeds unfortunately relate to fixed material. Fritz (1935) found that such material in two forms, having presumably about the same cell volumes, but one narrow long-horned from the Schleinsee and the other slightly wider and short-horned from Lake Constance, fell at markedly different rates through a column of water. The Schleinsee specimens (width 36 μ, apical horn 124 μ, antapical 104 μ) had a passive rate of sinking 64.5 per cent of that of the Lake Constance specimens (width 47 μ, apical horn 58 μ, antapical 43 μ). Caution must be used in applying such results from sinking of fixed cells, particularly in the present case for Pearsall has claimed that *Ceratium* may under certain conditions contain fat droplets which would lower its density. Such fat droplets are likely to be variable in their incidence, and under some conditions of fixation might disappear post-mortem. A priori, however, Fritz's results look reasonable.

It is therefore very probable that some advantage in the reduction of the sinking speed is conferred on the cell by the elongation of the main horns as well as by the presence of the fourth horn. It is also very probable that the development of the fourth horn and the increase in the angle between the horns in the summer tend to decrease the rotation of the *Ceratium* as it swims. Huber-Pestalozzi (1938) influenced by Wolterecks' views on *Daphnia,* believes that a reduction in the vertical component of motion is as important to the functioning of the three divergent posterior horns in summer as is the decrease in passive sinking speed. It is clear, however, that here as in other cases considered no final opinion is as yet possible.

Cyclomorphosis and other temporal changes in the Bacillariophyceae. The problem of the existence and meaning of seasonal changes in size and shape among diatoms is complicated not merely by the probability of the replacement of one related form by another adapted to slightly different ecological conditions, as has been observed to produce spurious cases of cyclomorphosis in *Ceratium* and in the rotifers, but also by a peculiar phenomenon dependent on the structure and mode of growth of the bacillariophycean cell.

The cycle of decrescence, auxopore formation, and restitution of size in planktonic diatoms. The exoskeleton of the diatom cell is formed essentially of an opal box with an almost identical lid (Fig. 246). This lid, or epitheca, differs from the box or hypotheca only in being large enough to fit over it. The internal diameter of the epitheca is thus slightly larger than that of the hypotheca. The bottom of the box and the top of the lid are termed the valves and the line joining the middle of the valves is the pervalvar axis. The sides of the box form the connecting bands or girdle bands, the two together being the girdle. On dissolution of the cell the girdle bands frequently become detached from the valves and must be regarded as structurally distinct from them. During the normal growth of the cell the greatest and often the only increase is in the direction of the pervalvar axis, the girdle increasing in width but the valves remaining at the same

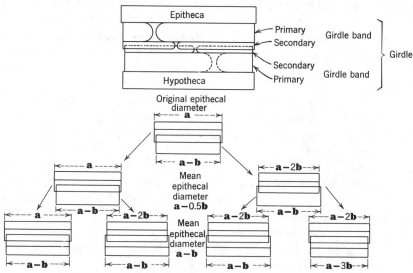

FIGURE 246. (*Above*) Highly schematized diagram, based on Palmer and Keeley's (1900) study of *Surirella*, of the epitheca, hypotheca, and girdle bands of a diatom, showing in this genus, the cleat-like structures of the interrupted secondary girdle fitting a gap in the primary girdle, which may permit expansion. (*Below*) Diagram of the supposed mechanism of reduction in mean size and increase in variance in populations of successive generations of a diatom according to Pfitzer's binomial scheme.

TABLE 52. *Pfitzer's binomial distribution of sizes*

Genera-ation						Diameters							
	a	a–b	a–2b	a–3b	a–4b	a–5b	a–6b	a–7b	a–8b	a–9b	a–10b	a–11b	a–12b
0	1												
1	1	1											
2	1	2	1										
3	1	3	3	1									
4	1	4	6	4	1								
. . . .													
12	1	12	66	220	495	792	**924**	792	495	220	66	12	1

diameter. At division the epitheca of the mother cell becomes the epitheca of one daughter, the hypotheca of the mother cell, the epitheca of the other daughter. New hypothecae are then formed to fit inside these epithecae. If there is no provision for the growth of the girdle of the daughter whose epitheca is the maternal hypotheca, it is clear that the diameter of this daughter will be slightly less than that of its sister. It is evident that this process would lead in every generation to the production of individuals that are smaller than any previously present in the line.

If the growth of the population were unrestricted and all daughter cells matured and divided equally fast, the descendants of a single cell of epitheca diameter **a** and hypothecal diameter **a–b** will have the distribution by size classes given in Table 52, the numbers of individuals in each class being proportional to the coefficients of the binomial expansion $(x + y)^n$, where n is the generation number. This was first pointed out by Pfitzer (1869). It is extremely doubtful that it has ever been shown that a population of noncolonial diatoms is actually built up in this way, but Hustedt (1929) agrees with Pfitzer that at least limited sections of the chains of certain species of *Eunotia* and *Achnanthes* do show the arrangement implied by the so-called binomial law. In one of the few other attempts to check the validity of the law on a colonical species, namely Müller's (1884) investigation of *Melosira arenaria,* it was found that the smaller of the two daughter cells divided, not at the same time that its sister divided, but when the larger daughter of this sister divided. The result of this lag, which implies that the smaller cell takes twice as long to mature and divide as the larger, is to produce a distribution of the kind shown in Table 52.

The distribution in the nth generation in any column is obtained by adding to the value for the $(n-1)$th generation in that column

TABLE 53.   *Müller's distribution for* Melosira arenaria

Generation	a	a–b	a–2b	Diameters a–3b	a–4b	a–5b	a–6b
0	1						
1	1	1					
2	1	2					
3	1	3	1				
4	1	4	3				
. . . .							
12	1	12	55	120	126	56	7

the value for the $(n-2)$th generation in the column immediately to the left. Histograms showing the change in the percentage composition by classes **b** units apart, proceeding from the original population (0), in the third, sixth, ninth, and twelfth generations are given in Fig. 247 for both the Pfitzer binomial type of distribution and for that observed in Müller. It has usually been supposed that when the size is reduced to a particular critical limit auxospores are formed by the cell contents extending themselves from the frustrule. Such auxospores, which are formed asexually in the Centrales and pseudorhaphic Pennales, including most planktonic species, are without silica skeletons and can grow to form cells of maximum size for the species under consideration.

Actually, it is becoming increasingly evident that although reduction in size often occurs in lines of diatoms reproducing asexually it is by no means inevitable. Geitler (1932) found no evidence of the phenomenon in *Eunotia pectinalis minor* in culture; Ruttner (1937a), as is indicated below, records a population of *Asterionella* which has persisted in nature for years without evidence of size reduction.

Working with various species of *Nitzschia,* (Fig. 248) Wiedling (1948) has discovered a quite astonishing variety of behavior within a single genus. In *Nitzschia kützingiana exilis* (*A* of Fig. 248) and in *N. communis* typical decline in mean size followed by rapid restitution occurred in some but not all of the lines studied. Auxospores were apparently not actually observed, but the general history of one culture of each species strongly suggests that auxospore production took place at a definite limited period in the history of the line when the size had declined to a certain critical value. In one line of *N. subtilis paleacea* and in one of *N. palea debilis* (*B* of Fig. 248) no change in dimensions whatever occurred in culture. In other lines of these four species and of other species studied declines in mean

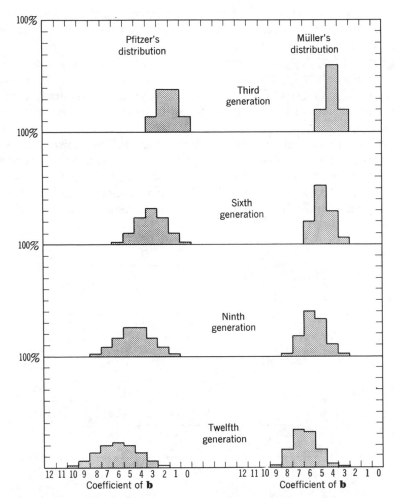

FIGURE 247. Distribution in size classes in terms of the original epithecal diameter (a) less multiples of the difference (b) between the original diameters of epitheca and hypotheca, expressed as percentages of total population in the third, sixth, ninth and twelfth generations of an original fundatrix, according to Pfitzer's binomial scheme (*left*) and Müller's scheme (*right*). Note the increasing variance apparently not indicated in natural populations as the mean size and modal size class decline and the reduced skewness in the later generations exhibiting the Müller distribution.

size at a decreasing and at an increasing rate are recorded. Sometimes the rate of decrease changed quite abruptly in the course of the experiment. In one line of *N. kützingiana exilis* (*C* of Fig. 248) the mean length declined regularly for about six months and then remained constant for about two and a half years. In a line of typical *N. kützingiana* (*D* of Fig. 248) after two years of essentially constant size the dimensions began to decline with increasing speed in the third and fourth year.

It is evident that in many diatoms there are mechanisms by which the size can be regulated during periods of ordinary cell division but that these mechanisms come into operation rather irregularly and sporadically. Geitler concluded that his strain of *Eunotia pectinalis minor* had girdle bands that diverged slightly in section so that by sufficient

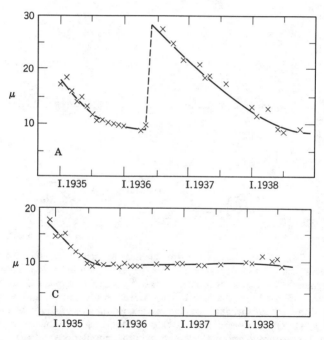

FIGURE 248. Various reduction patterns in mean length of *Nitzschia* spp. *A. N. kützingiana exilis*, typical size reduction followed by restitution. *B. N. palea debilis*, exhibiting essentially no change. *C. N. kützingiana exilis*, reduction followed by long period of constant size. *D. N. kützingiana*, constant size, followed by increasingly rapid reduction and death of clone. (Wiedling.)

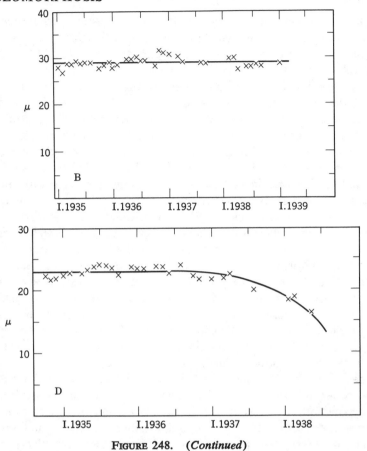

Figure 248. (*Continued*)

growth in the pervalvar axis the diameter of the girdle could be increased to any value determined by considerations other than the geometry of the frustrules. Moreover it was pointed long ago out by
Palmer and Keeley (1900) that isolated girdle bands usually did not
consist of uninterrupted endless rings of opal, but gave evidence of
a break which would permit expansion. Miquel (1894) had pointed
out that there is often a very considerable change in the proportions
of the major to the minor axis of pennate diatoms undergoing reduction
of such a kind that the minor axis is much less reduced than would
be expected. This has been confirmed by Geitler (1932). Palmer
and Keeley stressed the fact that such a form change implies some
sort of readjustment even when size reduction is occurring. The more
recent results of Geitler, Ruttner, and Wiedling indicate that readjust-

ments, not necessarily always of the same nature, permit complete reproduction of the maternal diameter in both daughter cells in a considerable number of cases. It is most unfortunate that so little information other than the arithmetic mean is available on the statistics of size distribution in cultures; the experimental work in the laboratory unlike the observations of limnologists in the field is definitely deficient in this respect. It is noteworthy that whenever auxospore production is believed to occur the culture goes back very quickly to the original size, which seems to imply either that a number of quite large cells can form auxospores in culture or that the distribution of sizes near the minimum shows in culture, as it certainly does in nature, far less spread than would be expected even from the Müller, let alone the binomial law. Bachmann and Wesenberg-Lund concluded from events in nature and Geitler from laboratory cultures that though a critical upper limit of size for auxospore formation appears to exist from laboratory cultures reduction to this size is a necessary rather than a sufficient condition for the production of auxospores. Some cells give rise to lines that continue to decrease in size, become too small for auxospore production, and finally produce degenerate forms that die.

It is rather unfortunate that in a number of cases of apparent restitution of the original size both in culture and in nature no evidence of auxospore formation is forthcoming. It is certain that qualitatively the geometrically determined reduction originally postulated by Pfitzer must occur in many species and that it is reversed occasionally by the production of auxospores; it is equally certain that in other species it does not occur and that the quantitative aspects of the process, except in a few chain-forming species, are known inductively hardly at all. In spite of this unsatisfactory situation, the material to be presented in the next few paragraphs demonstrates that among some of the planktonic diatoms the phenomena of size reduction when carefully studied over a number of years can prove to be very impressive indeed. Such careful studies have been made by Lozeron (1902), Bachmann (1904), and Wesenberg-Lund (1908) on plankton samples and most spectacularly by Nipkow (1927) on the diatom frustules incorporated in the easily dated recent varved sediments of the Lake of Zurich and the Baldeggersee. For the sake of completeness reference may also be made to Wimpenny's (1936, 1946) studies on comparable phenomena in the marine diatom *Rhizosolenia.*

Melosira granulata. During the rather limited periods of observation, in no case as much as two years, Wesenberg-Lund did not find any systemic temporal variation in size in three populations of this species.

M. crenulata. Wesenberg-Lund studied this species in three lakes in Denmark. In Furesø the populations showed a feebly marked cycle in size reduction with a rather definite tendency for large specimens, which draw out the distribution curve toward high values, to appear in May and October. The general movement of the mode is so feeble that it is hard to estimate its speed. A comparable sequence occurs in Haldsø (*C* of Fig. 249). There is clearly a marked falling off in diameter just after the major period of restitution—in October 1900 in Furesø and in the first half of 1901 in Haldsø. The population in Julsø presents a different picture. Here the restitution and decline in size seem to have constituted an annual cycle with a maximum modal diameter of 17 μ which rapidly declines during the spring and summer to a minimum modal value of 9 μ.

M. islandica helvetica. This species was studied by Nipkow (1927) in the varved sediment of Lake of Zurich. The species appeared in 1901 as a very narrow form 6 μ wide; in subsequent years the modal width declined to 4 μ. In the spring of 1905 auxospores were formed, and a mode at 16 μ appeared, representing cells derived from such auxospores. The descendents of these very wide colonies fell rapidly in modal diameter to 12 μ so that for a short time a trimodal distribution was apparent, the lower mode representing the old population, the middle mode representing the main new population derived from individuals developed from auxospores, and the top mode representing sporadic type large individual. The main mode of the new population declined from 12 μ to 6 μ in fourteen years, thus at a rate of about 0.4 μ per year. In subsequent years, until the end of the study in 1924, the mode remained at 6 μ. All through this long period of declining or stationary mode there is evidence of the formation of large individuals in small numbers at rather sporadic intervals.

Cyclotella bodanica lemanica. This diatom was investigated in the Lake of Lucerne by Bachmann (1904), who appears to have found an annual cycle. The maximum in numbers occurs in the autumn, at which time the modal diameter is low, betweeen 30 and 40 μ. Auxospore production occurs during the winter, and the distribution curve of size becomes very irregular. The small spring maximum consists mainly of large individuals derived from auxospores. During the summer the distribution curve becomes more regular and the mode shifts gradually to the left.

Cyclotella comta tenuis. This diatom was studied in the varved sediments of the Baldeggersee by Nipkow (1927). In 1903 the modal diameter is 5 μ, but there is evidence that marked restitution in size had recently occurred, the distribution curve being much skewed. The

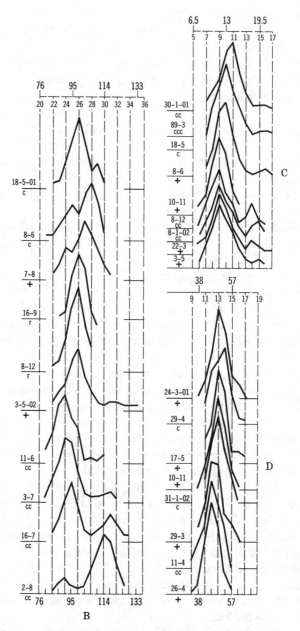

FIGURE 249. *A. Fragilaria crotonensis,* length frequency of populations in Furesø; at least in later part of the history two populations seem present,

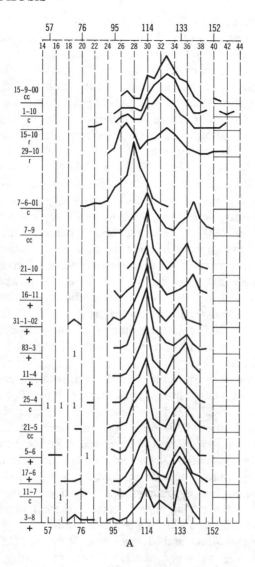

A

the smaller relatively invariant, the larger declining in size. *B*. The same in Esromsø, showing a rather irregular decline with striking restoration of large size late in the process. *C*. *Melosira crenulata*, diameter frequency showing decline in size in Haldsø. *D*. *Stephanodiscus astraea* diameter frequency showing decline in size in Furesø. (Wesenberg-Lund.)

main mode declined to 2 μ in the autumn of 1905, although many large specimens were present in the population of the preceding spring. In the spring of 1906 the curve is multimodal, and a new period of decline of the main mode from 5 to 2 μ sets in. The latter modal diameter is reached in the autumn of 1908 and in the next year the main mode is displaced to 4 μ. It appears therefore that in spite of the presence of many large specimens at various times the greater part of the population goes through a three-year cycle of slow decrescence and sudden restitution.

Stephanodiscus hantzschii. Nipkow (1927) found in the varved sediments of the Lake of Zurich evidence of the variation of this species during the period from 1900 to 1924. From 1900 to 1907 the size distribution was unimodal and declined slowly from 8 to 6 μ. In the spring of 1907 auxospores first appeared, and a new mode at 12 to 14 μ was present in subsequent years. Until 1915 the two modes persist but vary in relative magnitude in an irregular manner. The higher mode drifts to the left and is at 10 μ when it was last recognizable in 1915. Meanwhile the mode at 6 μ persisted and can be followed in almost all the curves to the end of the period. In 1917, 1919, and 1920 this lower mode was absent, but in 1918 and 1921 it alone was present. In all the later years some large specimens were always present.

Nipkow thinks that the main periods of restitution resulted in large representatives of young lines being dominant in 1908, 1914, 1917, 1919, 1920, and 1922. It is evident that the species behaved differently before and after 1907, but it is not easy to see what really happened between 1918 and 1921, in which period the mode moved to the right in two successive years and then fell in a year from 14 to 6 μ.

In the much shorter periods studied by Wesenberg-Lund (1908) *S. astraea* showed less systematic size variation in the three Danish lakes in which large populations occurred, though at least in Furesø (*D* of Fig. 249) there is a hint of a reduction process.

Asterionella gracillima and *A. formosa.* Lozeron (1902) found in the Ober-Zürichsee a rather steady movement of the modal length of *Asterionella gracillima,* from 73 μ in September 1896 to 63 μ in June 1901, a reduction of 2.1 μ per annum. In the main or lower Lake of Zurich the picture was much more complex. In 1896 there were two distinct modes, at 66 and 94 μ. By the end of 1897 the upper of these two modes had almost disappeared, though a few large individuals continued to occur sporadically. A new mode at 49.5 μ appeared in the spring of 1899, and thereafter the individuals represented by this mode are usually though by no means always dominant.

Meanwhile both the mode originally at 66 μ in 1896 and the new mode at 49.5 μ in 1899 drifted slowly to the left so that at the end of the investigation in the autumn of 1901 they were at 46 and 59 μ, respectively. A feebly marked mode at 73 μ appeared in September 1901.

There can be little doubt that three entirely different populations with somewhat different seasonal cycles are present in the lake. The possible genetic connection between the large forms (population I) present in 1896 and those of 1901 is too uncertain to justify discussion. There is little doubt, however, about the history of the population (II) having its mode at 66 μ in 1896; it can be traced throughout the whole series of observations, but when the new small population (III) appears in 1899 there is evidence of alternation in the relative abundance of the individuals belonging to II and III, though no regular seasonal replacement of one by the other. Lozeron also obtained evidence that on some occasions the relative proportions of populations II and III varied vertically in the lake. The mean rate of reduction in the length of II appears to be 1.3 μ per annum and of III, 1.7 μ per annum, but in view of the fact that the micrometer employed was divided only to units of length of 3.3 μ and that the period of observation was short it is doubtful if these two estimates differ significantly from each other or from that for the upper lake. Lozeron noted a few small irregularities, most notably a slight movement in 1898 of the mode of population II to the left and then back again to the right. No periods of auxospore formation were noted, nor on the basis of the interpretation of the threefold nature of the *Asterionella* populations in this lake would any period of auxospore formation be expected while all three populations were decreasing in size.

The subsequent history of the *Asterionella* populations of the Lake of Zurich can be ascertained from Nipkov's study of the biometry of diatoms present in the annual layers of sediment. Unfortunately, *Asterionella* was not common in the sediments representing the years studied by Lozeron. Nipkow's diagram shows the small population (III) apparently persisting until 1907. In 1906 a very considerable mode appeared at about 58 μ and two years later one at about 70 μ. The former persisted for many years; the latter disappeared at once. In 1909–1910 very large individuals again appeared and apparently underwent a rapid decline in length so that except in 1916 and 1917, when the mode was slightly displaced to the right, the main population from 1912 to 1920 had a modal length just under 60 μ. In 1922 and 1923 larger specimens again appeared. Nipkov interprets all the increases in size as due to auxospore formation and the addition of the new small mode, as in 1899, as due to unfavorable conditions. No independent evidence of auxospore formation is available and it appears much more likely that most, though not all, of the history observed is explicable in terms of the changing proportions of several populations, any one of which can give rise to auxospores when sufficiently reduced. Apart perhaps from slight differences in their environmental requirements, there is no need to suppose that these populations are genetically very different but merely that each has reached a different stage in reduction at any given time. On this theory the small cells of 1899 probably represent a new introduction of the species. It is possible, however, that the small form arose by mutation, and Nipkow actually has figured such a mutation in size in a colony of *Asterionella* which he observed in the plankton of the lake. It is probable that the persistent mode at about 58 μ observed in 1906 and for many subsequent years represents an invariant population derived from II by a process comparable to that observed by Wiedling in *Nitzschia kützingiana exilis*.

Wesenberg-Lund (1908) has given a good deal of information about the size reduction in *Asterionella gracillima* in the Danish lakes (Fig. 250). The most impressive series of data relate to Furesø; the mode of the very steep, regular distribution curve of the population in this lake migrated slowly from 72 to 57 μ in the period from August 1900 to October 1905. During 1906 and most of 1907 there was little changes in the position of the mode, so that at that time the population was behaving very much like population II in the Lake of Zurich when it reached a comparable modal length. In August 1907 a new mode appeared at 72 μ, whereas the old mode at 57 μ disappeared. The new mode was reasonably attributed to the mass formation of auxospores by most of the population. The very smallest specimens, however, were, according to Wesenberg-Lund, too small to form auxospores, and therefore there is a small mode at 49 μ the history of which is not further considered.

In the other Danish lakes the rate of decrescence is greater than in Furesø, so that although the periods of observation are all less than two years long marked changes occurred in most of the populations. In Haldsø the mean rate of decrescence was about 4.6 μ per annum, in Sorøsø about 6.2 μ per annum, and in Esromsø about 7.5 μ per annum. Only in Tjustrupsø was there no clear evidence of a regular decline in size, but here the whole sequence of events is obscure, though some changes certainly took place. In both Esromsø and Sorøsø (Fig. 250) very marked periods of restitution in size, pre-

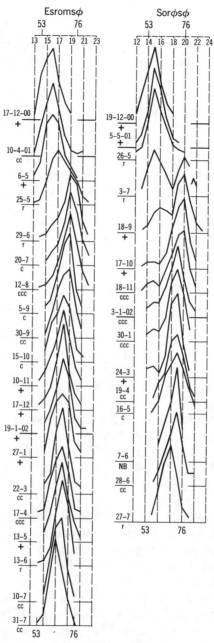

FIGURE 250. *Asterionella gracillima* length frequency histograms, showing replacement of a small population by a larger one and the slow decline of the latter over a period of two years. *A.* In Esromsø. (Wesenberg-Lund.)

935

sumably accompanied by auxospore formation occurred. In Esromsø the mode jumped from 60 to 72 μ during April and May 1900 and in Sorøsø from 57 to 76 μ during May 1901.

Le Roux (1908), working on the Lake of Annecy concluded that a longer form of *A. gracillima* (about 90 μ) occurred in summer and a shorter wider form (about 68 μ), in winter. The evidence of such a seasonal change, however, is inadequate.

A somewhat different picture is presented by the *Asterionella* of the lakes of the Austrian Alps, referred by Ruttner (1937a) to *A. formosa*. Here two forms apparently occur, one a more or less eury-thermal form termed by Ruttner var. *epilimnetica,* the other a cold stenothermal form designated var. *hypolimnetica. Epilimnetica* is fairly large (63 to 90 μ) and evidently undergoes a cycle of reduction and restitution in size, though the data are inadequate to permit a full description of the process. The cold-water hypolimnetic form, on the other hand, is apparently invariant; in the Lünzer Untersee the mean length varied from 39.8 to 38.1 μ between 1906 and 1914 and was still almost the same, namely 38.9 μ, in 1935. Ruttner compares the behavior of this race with *Eunotia pectinalis minor* studied by Geitler, but comparison with several species of *Nitzschia* would now also be appropriate. It is possible that Virieux (1916) encountered an *Asterionella* in the Lac de Saint Point in the Jura which did not show diminution, but his study extended for only three years and he is obviously uncertain of the significance of the observations.

Tabellaria fenestrata. The first notice of a reduction in size in this species is given by Lozeron (1902), who found the modal length to have been 53 μ in 1897 and 46 μ in 1901, corresponding to a diminution of 1.7 μ per annum. Wesenberg-Lund concludes that in Furesø the species is littoral benthic throughout most of the year and that the large transitory populations that may occur in the plankton are derived from the littoral stock and are not self-perpetuating. He found, in 1900, predominantly small forms of modal length 57 μ. Late in the year a small mode appeared at 68 μ, and the subsequent history of the population is primarily the replacement of the population represented by the 57-μ mode by that represented by the 68-μ mode, and the gradual decline in length of this new population so that at the end of 1905 the mode was at 53 μ. The rate of this decrement is therefore about 3 μ per year. It is to be noted, however, that half way through the process, when the main mode lay at 62 μ, a small new mode appeared at 83 μ, even though the reduction of the main population had by no means proceeded to the limit. It is unfortunate that this small mode at 83 μ appeared just before a period in which the lake had not been studied for about two years, but it is reasonable to assume that the unimodal curves of 1905 are related to the main mode at 62 μ and not to this subsidiary mode.

A longer and more complete record is given by Nipkow (1927) who found that *Tabellaria* in the Lake of Zurich behaved more regularly than any of the other species he studied. Marked restitution in size occurred in 1897, 1903, 1909–1911, 1915–1916, 1921, or approximately every fifth or sixth year. In the most regular periods the decrescence occurred at a rate of about 3 to 5 μ per year. In the Baldeggersee a rather less regular cycle which Nipkow considers to be slightly shorter was also recorded in the frustrules buried in the varved sediments. In the most regular period, from 1908 to 1912, the rate of deminution was about 3 μ per year. It is of interest to note that in the case of the sediments of the Lake of Zurich there is clear evidence that the restitution in size of the cells is likely to be followed by a considerable increase in the size of the population.

Fragilaria crotonensis. The temporal variation in length was first studied in the Lake of Zurich by Schröter and Vogler (1901), whose data (Fig. 251) are confirmed and extended by Nipkow's studies of the species in the varved sediments of the lake. Considering all the material at their disposal, Schröter and Vogler came to the conclusion that four discontinuous size classes could be recognized. These they designated *curta* of modal length 57 to 60 μ, *media* of modal length about 78 μ, *subprolongata* of modal length about 104 μ, and *prolongata* of modal length 126 to 129 μ. All but the last were supposed to occur in the Lake of Zurich. At the beginning of their investigation in 1896 the distribution of lengths in the lake was rather irregular, at least *curta* and *subprolongata* being present. In 1898 all three of the forms inhabiting the lake are supposed to have been found together, but after that year only *subprolongata* seems to have occurred. This form, however, which is clearly dominant in Nipkow's sedimentary layers laid down at the beginning of the twentieth century, declined in size slowly until in 1907 it had passed over into *curta*. This decrescence occurred at a rate of about 5 μ per year. In 1908 and 1909 an irregularly distributed array of larger forms was found and by 1910 *subprolongata* was again the only form present. At the next cycle the decline was more rapid, and the minimum size was reached in 1913 with restitution of size in 1914. In later years the behavior of the species became irregular and cannot be analyzed. It is quite evident that the three supposed forms recorded by Schröter and Vogler

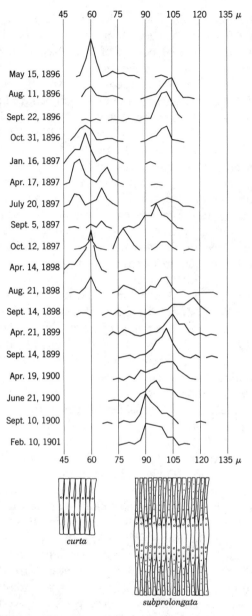

FIGURE 251. Temporal variation of *Fragilaria crotonensis* in the Lake of Zurich from 1896 to 1901, with figures of the small (f. *curta*) and large (f. *subprolongata*) forms composing the population, respectively, dying out and coming in during 1898. (Schöröter and Vogler.)

in the Lake of Zurich can be merely stages in reduction cycles. In the sediments of the Baldeggersee, also studied by Nipkow, the same type of cycle, with but a single irregularity in the rhythm of slow decrescence and sudden reconstitution, is even more clearly apparent. In this lake the reconstitution periods were in 1908, 1912, and 1920, and during the regular declines of the first two periods the rate of decrescence was about 8 to 10 μ per year; in the third period the rate was initially high but appears to have fallen off after two or three years.

Wesenberg-Lund's experience with *Fragilaria* was in general comparable (*A, B,* Fig. 249), but in Furesø after 1901 it is obvious that two genuinely distinct populations were present. One was undergoing reduction in size from 140 to 133 μ over a period of about a year, whereas the mode of the other remained quite stationary over the same period at 112 μ.

Brutschy (1922) in November 1913 found in the Hallwyler See a small mode at 83 μ which disappeared in the course of 1914 and a large mode at 103 μ which moved regularly to the left so that in January 1915 it lay at 90 μ. The rate of decrescence was thus about 12 μ per year.

Synedra a. delicatissima and *S. longissima acicularis*. With their intermediate forms, these diatoms show a fairly well marked reduction cycle extending three to seven years in the Lake of Zurich and a less sharply defined cycle in the Baldeggersee.

General remarks on size reduction in planktonic diatoms in nature. Considering the available information as a whole, the behavior of the planktonic diatoms is much as would be expected from laboratory experience. Cases of very regular decrement are followed by sudden restitution, as is beautifully indicated in the *Tabellaria* populations of the Lake of Zurich and by some of the *Asterionella* populations in the Danish lakes. Cases also occur, the most notable being Ruttner's *Asterionella formosa hypolimnetica,* in which no displacement of the modal length whatever occurs over a long period of time. When reduction does occur, it may lead to a stationary mode at a certain size little above the absolute minimum for the species, as in the population of *Asterionella gracillima* in the Lake of Zurich (population II) and in Furesø and in the culture of *Nitzschia kützingiana exilis* studied by Wiedling. It is indeed not unlikely that many species will show rather prolonged periods with a stationary mode before auxospore formation occurs. It is probable that most or all of the irregularities encountered in nature would find parallels in laboratory experience, but since so many sources of error are likely to be present in studies

of large natural populations there is no justification for insisting on
the parallelism in the minor variations in the rate of reduction that
may be observed.

Both Bachmann and Wesenberg-Lund agree that the smallest individuals in a population are often too small to produce auxospores and
that their degenerate decendents inevitably perish; this has been confirmed in the laboratory by Geitler.

It is evident that there are certain aspects of auxospore formation
and restitution that are not adequately understood.   Much of the evidence obtained in the field is in accordance with the hypothesis that
a certain amount of auxospore formation can occur at any stage in
the reduction cycle, though not necessarily at any time of year.   In
this way it is easy to explain the sporadic occurrence of large forms
at times when the main population has certainly not reached the critical
limit for auxospore formation.   Since both the binomial scheme and
Müller's scheme of size reduction provide for the appearance of small
numbers of specimens of any designated size long before the mode
reaches the size in question, there is nothing very surprising in this.
What is remarkable is the fact that what Nipkow calls total rejuvenation,
the rapid movement to the right of the whole distribution curve, occurs
so often.   This is also evident in the mean figures given, for the cultures
that show a size restitution, by Wiedling, who, however, gives no adequate idea of the distribution of sizes within the rejuvenated culture.
It is very hard to avoid the conclusion that either environmental factors
or some control by a diffusible substance produced by the cells themselves is involved.   It is probable that in some species auxospore formation cannot occur at all seasons.   Bachmann concluded this to be
the case in *Cyclotella bodanica lemanica,* in which the phenomenon
occurs only in winter, as Wesenberg-Lund also found with *Melosira
crenulata.*   Ostenfeld (1906 *fide* Wesenberg-Lund 1908) found in the
Faeroes that *M. islandica* likewise formed auxospores in winter and
M. italica in summer.   Such a seasonal determination coupled with
a division rate of the right magnitude would produce the kind of
seasonal size cycle observed in *C. bodanica lemanica* in the Lake
of Lucerne or of *Melosira crenulata* in Julsø.   It would, however,
seem much more usual for the population to decline and then stay
stable at the critical modal size for a greater or lesser time, which
in *Asterionella gracillima* in Furesø was at least two years; it then
allegedly undergoes a pandemic auxospore production which leads to
total or almost total rejuvenation.   The most reasonable explanation
is that some minor but relatively rare environmental change is needed

to initiate restitution processes and that either the whole population is so unstable that nearly all except the smallest degenerating cells partake in the process or that some cells, having started to produce auxospores, stimulate others chemically to do so, setting up a sort of chain reaction. The postulated auxospores are rarely observed.

As is readily seen when Figs. 249–251 and 247 are compared, in the best natural cases there is no evidence at all of the increase in variance that should result from the geometry of size reduction, even if Müller's reduction scheme is accepted. Since any reduction scheme must involve some increase in variance, there is clearly a compensatory selection against both the largest and smallest specimens. The largest will at any time have a greater sinking speed and lower nutrient uptake per unit volume than the other cells and there will perhaps be a little selection against such cells on this account. The small cells will tend to take part in the continuous small-scale auxospore production, and under some circumstances this alone might produce an equilibrium population. It is evident from the observations made by Nipkow on the relation of the total number of frustules in the sediments and the time of size restitution that large populations tend to follow restitution. Wesenberg-Lund came to a like conclusion that the large cells derived from auxopores were really rejuvenated and divided more rapidly than the small members of old lines. This hypothesis would perhaps accord with the progressive decrease in the rate of reduction as the line gets smaller, though one can hardly suppose that the mode can be held at a low value for several years merely by the cells not dividing. Such observations in fact solve no problems, but they do emphasize that problems exist and that in spite of the patient work of the first explorers of this interesting field there is still a great deal to be learned.

Cyclomorphosis in colony form and size in diatoms. The cells of *Tabellaria fenestrata* form colonies. When the daughter cells separate so that sometimes one end, sometimes the other, remains in contact, irregular chains are built up, whereas if the daughter cells of successive divisions always adhere at the same end partial or complete stars result. The adherence of the cells is connected with the secretion of jelly or mucilage, but the way in which regular stars, or in some cases spirals, as opposed to irregular chains, are constructed does not seem adequately investigated. The number of cells forming a colony presumably bears some relation to the amount of jelly produced, for when there is little mucilage the colony would seem likely to be less stable than when much is present. In *T. fenestrata* there is often evidence of a seasonal change in the proportions of stars and chains and in the number of specimens in the colonies. It has sometimes been supposed that the stars belong to a special genetically determined var. *asterionelloides* Grun. This seems to be unlikely, for Lozeron (1902) has shown that in the Lake of Zurich the population present in May 1901 gave precisely the same length frequency distribution for stars as for chains, with a mode at 46 μ, whereas the mode for the whole population in 1897 had been 53 μ. It is extremely unlikely that the star and chain curves of 1901 would superimpose so exactly if they were derived from two independent populations.

Schröter (1896) concluded that var. *asterionelloides* is a summer form in the Lake of Zurich, occurring between June and September, whereas during the rest of the year the colonies consisted mainly of chains. Lozeron (1902), however, in the same lake found that in 1901 on all but one occasion not more than 10 per cent of the colonies were var. *asterionelloides*. On the one occasion in May, at the time of the spring maximum of the species, on which 90 per cent of the colonies belonged to the star-shaped variety the temperature had risen to only 8° C., and as it rose again later chains replaced the stars.

In Furesø Wesenberg-Lund (1908) found that the species was a benthic littoral form during most of the year and that before it appeared in the plankton of the open water it could be obtained as free-floating chains in the littoral above the benthic colonies. In May chains and stars are equally common, but in June and July the free water contains almost only stars, which Wesenberg-Lund supposed to perish without leaving descendents.

More recently the matter has been investigated again in Switzerland. Quartier (1948), who finds that in the Lac de Morat, in which *Tabellaria fenestrata* forms an appreciable maximum in April, star-shaped colonies predominate at all seasons. In the spring, from March to May, the modal number of individuals in these colonies, however, is only three so that the distinction from chains is a rather artificial one. At this time a few real chains with five or six cells are also present. In June a new mode at eight cells per star develops, and throughout the rest of the summer and autumn there is a strong, if rather irregular, development of this bimodality.

A similar progression occurs in the Lac de Bienne (Fig. 252), in which the modal type of colony in March and April is three-celled and designated as a star but is accompanied by many chains containing as many as five or six cells. In May, at the time of an immense maximum, there is rather feeble evidence of bimodality. In June the mode is an eight-pointed star, followed in the later summer by bimodal

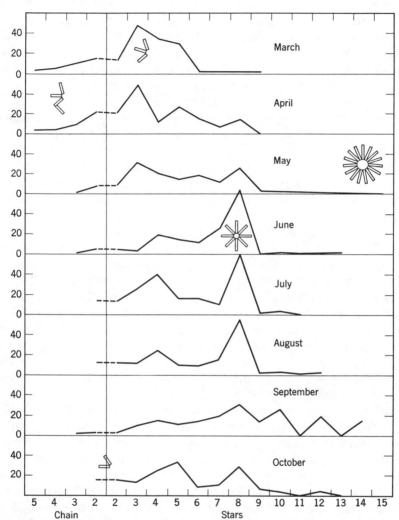

FIGURE 252. Frequency of number of cells in colonies of *Tabellaria fenestrata* in the plankton of the Lac de Bienne. Irregular chains are plotted to the left of the dividing line, incomplete and complete stars to the right. Note the high incidence of chains and incomplete three-celled stars in the spring and autumn, the high incidence of complete stars, with a modal number of eight cells in the summer. Very large stars may be characteristic of intermediate months (May, September). Note that a two-celled chain cannot be distinguished from a two-celled star. (Quartier, modified.)

distributions with the modes at four and eight cells. In the Lac de Neuchâtel a less developed population exhibited a less regular cyclomorphosis along the same lines as in the other lakes. It is probable that Schröter's sequence was not very different from that described by Quartier. In *Asterionella* Schröder (1899) noted in the River Oder colonies of two to six cells in March and of six to ten or more in June or July. Marsson (1901) in an artificial lake near Berlin observed only eight-rayed stars in October. During the winter smaller colonies occurred, in some cases consisting of only three cells, which might form chains in December and January. After May no colonies appear to have contained fewer than four cells. Wesenberg-Lund (1908) likewise concluded that the star-shaped colonies had fewer rays in spring than later in the year. He noted chains of three-rayed stars in April and May in Julsø, Mossø, and Skanderborgsø, whereas in Tjustrupsø and the Frederiksborg Castle Lake the May population consisted mainly of single-four-rayed stars. Many-rayed spiral colonies occurred chiefly after the great spring diatom maximum but sometimes persisted until later in the year. In Esromsø in August the abundant *Asterionella* were mainly eight-rayed, whereas in late September the stars of most colonies had sixteen to thirty rays. From July throughout the late summer and autumn, and sometimes as late as February, the usual form in Denmark, as in Germany, is evidently an eight-rayed star. In some lakes, such as Haldsø, the eight-rayed star appears to be more stable than in others in which the many-rayed spiral forms are not uncommon in summer.

In contrast to this account of Wesenberg-Lund, LeRoux (1908), working on the *Asterionella* population of the Lake of Annecy, regarded the many-celled spiral colonies as mainly winter forms. Virieux (1916) in his study of the Jura lakes concluded that the size of the colony in *Asterionella* is not directly determined by seasonally variable physiochemical factors but depends on the rate at which the organism reproduces, a rate that certainly varies with the seasons, but in an irregular way, in the present genus. Because the sheath substance of at least some diatoms (Lewin 1955; see Volume I, p. 798) is produced under conditions of nutrient deficiency, the form of the colony is likely to depend in part on the ratio of production of new cells to that of sheath; the hypothesis of Virieux, as far as it goes, appears reasonable. It is evident that in this form the hypothesis resembles that put forward to explain cyclomorphosis in *Ceratium*. In Windermere however the most recent work (Lund, Mackereth and Mortimer 1963) indicates maximal colonies at the time of greatest expansion of the population (Fig. 253).

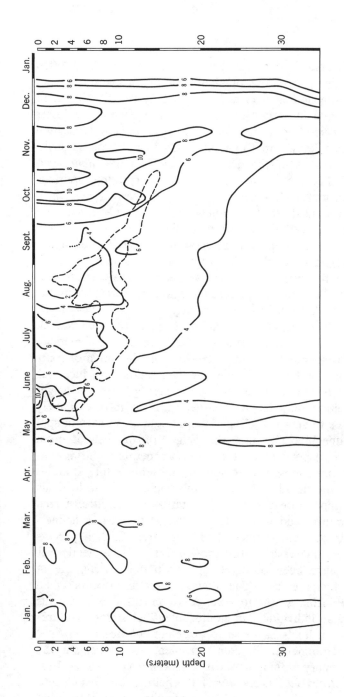

FIGURE 253. Variation of mean number of individuals in the star shaped colonies of *Asterionella formosa* in the top-shaped colonies of *Asterionella formosa* in the top 35 m. of Windermere in 1947. Note that the maximum number of cells per colony is at the surface when the species is most abundant (cf. Fig. 113), falling to very low values when the species is limited by nutrient deficiency in the epilimnion in August but increasing again to a maximum in October, though there is no major increase in the total population. Dashed line embraces region of stability in excess of $2 \times 10^{-4} \text{gr.cm.}^{-3} \text{m.}^{-1}$, metalimnion.

945

SUMMARY

The term cyclomorphosis refers to the seasonal polymorphism often exhibited by successive generations of planktonic organisms throughout a year. The phenomenon is very well marked in the Cladocera of the genus *Daphnia* and in certain species of *Bosmina* and to a lesser degree in *Scapholeberis* and a few other genera. There is a little seasonal form change known in Copepods. Cyclomorphosis is strikingly exhibited by rotifers of the genera *Asplanchna, Keratella,* and *Brachionus,* though here some supposed cases are certainly spurious because of the confusion of allied species of differing seasonal incidence. Striking cases are provided by the dinoflagellate genus *Ceratium,* and some peculiar temporal variations known in diatoms are conveniently discussed in this chapter, though for the most part they are not analogous to the cyclomorphosis of the other organisms considered.

In the Cladocera the genus *Daphnia* has been extensively studied. In temperate lakes the ordinary pattern of seasonal variation in the species showing the phenomenon most clearly (e.g., *D. hyalina, D. g. galeata, D. cucullata,* and *D. cristata* in Europe; *D. ambigua, D. retrocurva, D. galeata mendotae* and *D. dubia* in North America) is the gradual development of an extension of the anterior part of the head to form a helmet during the generations produced as the water increases in temperature during the spring. At the height of summer this helmet may be very large; in form it is extremely variable from species to species. Experimental studies on several species have shown unequivocally that the height of the helmet on a newborn specimen is dependent on the water temperature during embryogenesis; in the cases most clearly studied the temperature stimulus is effective during the middle third of embryonic life. Individuals reared at a low temperature for several instars and then transferred to a higher temperature tend to develop in their later instars helmets appropriate to their new environmental temperature. Such a process involves a regulatory increase in the growth rate of the head over that which would have been observed at any period during life at a uniformly high temperature. The reverse situation occurs when *Daphnia* are moved from warm to cold environments. Experiments have also shown that individuals in nonturbulent laboratory cultures never grow such tall helmets as those in turbulent cultures or in their natural turbulent environments. In some turbulent environments in the laboratory the antennary swimming beat is accelerated. In at least some species (*D. retrocurva, D. cucullata*) the helmet grows isoauxetically with carapace length in nature at moderate temperatures but tachyauxetically at high temperatures. During postembryonic growth any influence, notably temperature or food supply which changes the specific growth rate of the carapace, has a relatively greater effect on that of the head and so alters the relative growth.

Helmeted *Daphnia* are generally more epilimnetic than unhelmeted species living in the same lake; this may be merely because the turbulent epilimnion provides optimal conditions for the development of the helmet. One case is known in which the normally unhelmeted hypolimnetic *D. longiremis* developed a helmet when the lake in which it occurred became, during a single cool summer, a freely circulating thermally third class lake.

The adaptive significance of cyclomorphosis, if any, has been widely debated. It has been supposed to be an adaptation to lower viscosity in the summer by increasing form resistance. It is very uncertain, however, how far the increase in form resistance is important to an actively swimming species in the turbulent epilimnion; it would make active movements both up and down more costly even if downward passive movements were slowed. Moreover, the almost universal decline in absolute size of planktonic organisms between the spring algal maximum and the height of summer, when egg production, and so almost certainly nutrition, is usually reduced, in itself constitutes a mechanism of reducing sinking speed within the Stokes law range that would often be commensurate with the viscosity change. The hypothesis that cyclomorphosis enforces a more horizontal direction in swimming has a certain plausibility, but at present it is supported largely by unconfirmed assertions of little value about the behavior of *Daphnia*.

The extreme variability of the form of the helmet must be taken into account in any hypothesis of adaptive value; pointed, rounded, procurved, retrocurved, and very broad helmets occur in different species but can hardly be equivalent hydromechanically. A further difficulty is that in *D. ambigua* the lowest temperature at which the neonata develops its full helmet is apparently lower in New England than in North Carolina, though the change of viscosity with temperature must be the same in both localities. It is quite possible that the helmet is developed as an adaptation against predation by the very young fry of certain fishes or reduces visibility and so predation by larger predators.

As well as variation in head form, there is often seasonal variation in the length of the posterior spine. Very few observations are available; in one population of *D. pulex* the spine length at birth showed no systematic seasonal variation. During winter the spine actually decreased in length during ontogeny, whereas in summer it increased.

In the peculiar semineustonic genus *Scapholeberis,* which belongs to the Daphniidae, a small spine on the head is often developed and would reasonably be regarded as homologous with the spine-like helmet of *Daphnia ambigua* and some forms of *D. g. galeata.* In *Scapholeberis,* however, the horn, if present, ordinarily appears to decrease in length throughout the summer. It is always relatively smaller in large than in small animals in the same population and apparently always undergoes striking reduction in culture, just as do the posterior spines of some rotifers. The mucro, homologous with the posterior spine in *Daphnia,* behaves quite differently, increasing in length through the summer. The meaning of these changes is quite obscure.

In *Bosmina* there is a great variety of seasonal changes. In *B. longirostris* the little cyclomorphosis that does occur is limited to a reduction in length of the antennule and mucro in the summer. The summer forms are often small, but at least in the case of the antennule the effect is not simply bradyauxesis because the antennule almost stops growing before the rest of the animal approaches its full size. There is also a tendency for the antennule to become hook-like.

In *B. c. longispina* there is again very little cyclomorphosis; as in the preceding form, the antennule appears to be relatively short in summer.

In *B. c. coregoni,* and in allied forms particularly developed in the Baltic region, extraordinarily diverse types of cyclomorphosis appear. The antennule may be enormously elongated, whereas the mucro tends to disappear (*coregoni*), or both antennule and mucro may be very long (*berolinensis*) or the animal becomes humped, the hump being at right angles to the long axis (*gibbera*) or retrocurved (*thersites*). The shape of the brood pouch of the humped forms is apparently determined by immense elongation of the cells in its dorsal wall. Evidence of both seasonal variation in the shape of the neonata, and of tachyauxesis, is forthcoming, but the problems presented by this genus for the most part still require elucidation.

Slight seasonal differences in size, the small forms occurring in summer, have been noted in *Ceriodaphnia, Chydorus,* and *Alonella,* and in some cases small form changes have been noted but are little understood.

The phenomena observed in the Copepoda are likewise largely a decrease in size usually with rising temperature and sometimes affecting different parts of the body rather differently. A few cases of form change in the geniculate antenna of male Diaptomidae are known.

In the Rotifera cyclomorphosis has frequently been reported; it is, however, much harder to study in the field than in the Cladocera because preserved material, except in one case, never gives any indication of the genetic continuity of two different seasonal forms. In consequence, it is extremely easy to confuse the varying seasonal incidence of closely related species with seasonal form change within a species.

A. priodonta in the genus *Asplanchna* exhibits slight seasonal variation in size in ponds in central Europe, the largest specimens occurring either in winter or at times of maximum abundance in spring and autumn. In some of the lakes of Denmark and the adjacent parts of Germany the same species exhibits a striking type of cyclomorphosis, which alone of the rotatorian cases bears any clear resemblance to what occurs in *Daphnia*. In May the animal is about 1.5 times as long as it is wide; during the end of that month embryos appear in those individuals that are two or three times as long as wide. During succeeding weeks the embryos are always relatively longer than their mothers, so that in July vermiform, slightly curved individuals five times as long as they are wide constitute the population. Such individuals are nearly always sterile; on the very rare occasions when they have been found carry embryos these are short and round. The elongate forms die off in August and the species does not reappear till the succeeding spring. Whether the spring specimens are descended from very rare short offspring of elongate parthenogenetic females or whether resting eggs have been produced before the latter occur is not known.

In *A. sieboldi,* a second species of *Asplanchna,* a quite different kind of form change, dependent primarily on nutrition, is recorded. The typical form is a sac-shaped animal 500 to 1200 μ long with 20 to 40 flame cells; the so-called humped form, or f. *ebbesborni,* is rather longer and has 40 to 60 flame cells and two large wing-shaped projections from the body wall on the sides of the body; a third or campanulate form, 1800 to 2500 μ long with a very broad corona and 80 to 115 flame cells, is recorded from North America, France, and Denmark. The saccate forms are ordinarily hatched from resting eggs. Fed a diet exclusively of *Paramecium,* they remain saccate, and transferred to *Euglena* or *Oxytricha* they become humped; several parthenogenetic generations are said to be required to complete the change. Mild cannibalism in mass culture fed on *Paramecium* may have the same effect, and extreme cannibalism, or a diet of young *Moina,* produces a campanulate form. The resting eggs of campanulate animals are said to hatch as humped animals.

Very extensive work on cyclomorphosis has been performed on members of the genus *Keratella*. The most significant changes in the much-studied *Keratella cochlearis* (s. str.) and probably in the related *K. earlinae* involve a slight reduction in the lorica length, usually in summer, coincident with a marked heterauxetic decrease in the length of the median posterior spine. In at least one locality, Lake Nemi in central Italy, there is some evidence that the smaller shorter-spined form is racially distinct from and not genetically continuous with the larger winter form. Total loss of the posterior spine to produce the form referrable to *tecta* is by no means certainly established in any genetically continuous population. Other supposed cyclomorphotic changes in the species are probably due to confusion of other species, notably *K. irregularis*, with *K. cochlearis*. The forms known as *pustulata* and *hispida* with minute excrescences on the lorica may develop cyclomorphotically from nonpustulate individuals present in winter, and the available data suggest that possibly this occurs only in the larger specimens of the race or species involved.

In *Keratella quadrata* some populations which have been referred to f. *divergens* and f. *frenzeli* appear to show a marked heterauxetic dependence of the two posterior spines on lorica length; the latter is usually minimal in the late summer but may undergo a progressive increase in the spring and early summer before the size reduction sets in. In culture there is a very striking tendency for the posterior spines to decrease in length, not merely from generation to generation, but in the successive offspring of an isolated female. Finally, very short-spined individuals or specimens with only one or even no spine appear. Mictic females appear in such cultures, but there is no correlation between their incidence and spine length. It is possible that this phenomenon is related to the peculiar processes of senescence in Rotifera. It is probably a laboratory effect analogous to the bradyauxesis recorded in the helmet development of *Daphnia* in nonturbulent water, though there is no evidence that turbulence is involved here. The regular occurrence of noncyclomorphotic populations which undergo immediate reduction in culture argues against a biologically produced substance, such as that from *Asplanchna* which affects *Brachionus calyciflorus*, being present in nature, but does not exclude such a possibility.

In *Keratella hiemalis*, a population which lives in nature with *K. quadrates* may be cultured without any reduction in the short spines which characterize the species, even though the sympatric *quadrate* shows striking laboratory reduction. It is possible, though not by any means certain, that *K. hiemalis* shows slight cyclomorphosis in nature in some localities.

In *K. valga*, populations in which a single (right) posterior spine is present early in the spring, although two spined forms occur later, are recorded in ponds in Europe. In culture there is a very rapid reduction in the right spine and total loss of the left. The related tropicopolitan *K. tropica,* which may have a very long right spine, also occurs in a variety of forms differing in the presence and degree of development of the left spine. In the less extreme populations that have been studied the right posterior spine varies a little with the length of the body, whereas the left seems to show striking but rather irregular increase and decrease independent of the absolute size or behavior of the right spine.

B. falcatus in the genus *Brachionus* probably exhibits a kind of cyclomorphosis like that in *Keratella cochlearis* and *K. quadrata* f. *frenzeli,* in which the anteromedian and posterior spines are proportionately shorter in the small individuals which occur primarily at the height of summer.

In *B. calyciflorus* the presence of posterior spines, and the largely independent length of the anteromedian spines, is variable, the posterior spines at least apparently being best developed when the species is or has recently been at its maximum abundance. A clear correlation with temperature appears to be excluded by the data. The form (*dorcas*) without posterior spines but with long anteromedians has been identified as a winter form. In laboratory culture there is a strong tendency to spine reduction and loss in lines started from a female with large posterior spines. In such cultures individuals with short spines are mixed with spineless ones, irrespective of the nature of the fundatrix of the population. In some cases changes in the qualitative nature of the food produces a slight change in modal spine length. Occasional long-spined forms may occur very sporadically in any culture, for there is apparently adequate evidence that their incidence can, rather unexpectedly, be increased by starvation or low temperature. The only certain factor involved in spine production in culture, however, is the presence of *Asplanchna brightwelli,* which appears to produce a diffusible nonvolatile, fairly thermolabile material and stimulates the production of long spines in the rotifer. The reaction is adaptive, the carnivorous *Asplanchna* finding the spined forms hard to eat. The very irregular incidence of spined forms in nature is doubtless explained by the concurrent incidence of *Asplanchna.* There is also much racial difference in the reaction to *Asplanchna;* reactive strains may replace less reactive ones in summer.

Several other species of *Brachionus* exhibit cyclomorphosis, but the data are less complete and even more confusing than in other rotifers. In the field observations evidence of genetic continuity is nearly always lacking, and few of the more extreme cyclomorphotic forms ever appear in culture.

In *Brachionus bidentatus* there is evidence that different lines produce different proportions of spined and spineless forms and that a change of food from *Polytoma* to *Chlamydomonas* can increase the incidence of spined forms; though the incidence of spined specimens on a *Chlamydomonas* diet is always greater than on *Polytoma,* the proportion is greatest just after the shift in diet. The partial control of posterior spination by diet is reminiscent of the situation in the production of the humped form *ebbesborni* of *Asplanchna sieboldi.* The lateral expansions of *ebbesborni* may in fact be morphologically comparable to the posterior spines of *Brachionus.*

In general in the Rotifera cyclomorphosis appears to involve (a) the addition of posterolateral spines or processes, partly under the influence of food, partly of other organisms, but in most cases uninvestigated; (b) the heterauxetic growth of spines so that small individuals, often characteristic of summer populations, have relatively reduced spines; (c) the occasional development of exuberant summer forms as in the Baltic *Asplanchna priodonta.*

Usually the first two types of cyclomorphosis are more marked in ponds than in lakes. The addition of posterior spines is known to reduce sinking speed in *B. quadridentatus,* but its spineless specimens reared at high temperatures have a lower excess density than those reared at low temperatures, which provides just as effective a mechanism of reduction in sinking speed. The heterauxetic dependence of spine length on lorica length cannot provide a mechanism to compensate low summer viscosity, for the relative spine length is usually minimal at times of maximum temperature. The small over-all size at such times, however, may well compensate for the viscosity change. The development of hairy pustules on *Keratella* (*cochlearis* f.) *hispida* probably involves too fine a structure to influence resistance in falling or swimming. The one case comparable to *Daphnia,* namely *Asplanchna priodonta,* is complicated by the fact that the extreme individuals are usually sterile and so adaptation is questionable. Only in *Asplanchna*-produced spines on *Brachionus calyciflorus,* which makes predation by the *Asplanchna* more difficult, is any adaptive meaning apparent.

A marked cyclomorphosis of a quite complicated type appears in the fresh-water species of *Ceratium.* As in the Rotifera there is often great difficulty in ascertaining how many distinct taxa are present in an apparently cyclomorphotic population. The characters showing variation are the numbers of posterior horns, the width of the cell, the length of the whole cell determined mainly by the length of the horns, the degree of divergence of the posterior horns, and the ornamentation of the cell wall. The resting cyst hatches to give an unarmored Gymnodinium stage, and the number of posterior horns developed as this stage becomes armored has been shown experimentally to be a function of temperature. In nature, although this temperature dependence can explain the variation in horn number, particularly the incidence of the fourth or left posterior horn, the absolute number, which may in some localities be two in the spring but in most is three, seems also to be under some genetic control. The absolute size of the cell, best measured as width, is nearly always maximal in late spring or early summer; there may be a later maximum in the autumn. The maximum length, determined mainly by the apical and antapical horns, is also maximal in the late spring, but there is not necessarily a secondary maximum in length corresponding to the autumnal increase in width. The divergence of the posterior horns is usually greatest in summer, and the ornamentation of the cell wall also increases after the length and width have begun to decrease. It has been suggested that these changes reflect a change from rapid protoplasmic synthesis and cell division in spring to rapid cell-wall formation in late summer. In the summer individuals the cell-wall is supposed to form too fast to permit the protoplasmic contents to reach a maximum size. In view of the usual fall in available nitrogen during the spring and early summer, a change from predominantly protein synthesis to predominantly carbohydrate synthesis would be reasonable, but increased uptake of calcium may also be involved. There is some experimental evidence suggesting that the development of the fourth horn may increase form resistance.

In diatoms two very different types of temporal variation have been recorded. One of these depends on the fact that when a diatom cell divides the hypotheca (or box) of the parent becomes the epitheca (or lid) of one daughter. One daughter is thus the same size as its mother cell, the other slightly smaller. This process should lead to a mean decrease in size and an increase in variance as time goes on. When a certain limiting size is reached, auxospore formation, by producing a stage that has no exoskeleton, may permit restitution of the original size. The problem is complicated by the fact that the girdle bands which fit over each other are not always so rigid that the hypothecal daughter is inevitably smaller than the epithecal. Some diatoms therefore do not show the reduction, and others which do exhibit it may do so in a most irregular way. Moreover, there are species that apparently undergo reduction and restitution in which auxospores are unknown, and the great increase in variance that should occur early in the process seems to be unknown in nature and unstudied in the laboratory.   In some species of *Melosira, Cyclotella, Stephano-discus, Asterionella, Tabellaria,* and *Synedra* a decline in mean size followed by restitution has been recorded in a number of European lakes, either in plankton samples taken over a series of years or in lakes in which annually varied sediments are being laid down.   In some cases very closely related forms, as in the genus *Asterionella,* appear to behave quite differently, one form being invariant, the other undergoing reduction.   *Melosira crenulata* in at least one Danish lake seems to go through the cycle annually, simulating true cyclomorphosis.

A second type of temporal change is recorded in colonial diatoms in which there is often a tendency to produce chains or star-shaped colonies of different sizes at different seasons.   This is most notable in the genera *Asterionella* and *Tabellaria.*   It is probable that the larger colonies are produced when the rate of formation of sheath substance relative to new cells is maximal.   There is some evidence that this is likely to occur at times of low nutrient concentration, which may explain why large stars of *Asterionella,* containing eight or more cells are in many localities found predominantly in the summer, small stars of three or four cells in the spring.   Some analogy to the supposed mechanism of cyclomorphosis in *Ceratium* is here apparent.

A	area of a lake
A	coefficient of turbulence, used also as an approximation to coefficient of eddy viscosity and diffusivity (cf. vol. I)
B_1, B_2	defined as

$$B_1 = \int_0^\infty \frac{1}{(a^2 + \xi)(b^2 + \xi)(c^2 + \xi)} \cdot d\xi$$

$$B_2 = a^2 \int_0^\infty \frac{1}{(a^2 + \xi)\sqrt{(a^2 + \xi)(b^2 + \xi)(c^2 + \xi)}} \cdot d\xi$$

in the theory of a slowly sinking ellipsoid

B	biotope space, the ordinary three dimensional space occupied by organisms
$B_1, B_2 \cdot \cdot \cdot$	biotope space or ordinary three dimensional space occupied by species $S_1, S_2 \cdot \cdot \cdot$
C_b	concentration of blood
C_m	concentration of medium
C_u	concentration of urine
D_N	information per individual
D_V	information per unit volume
D	information (Shannon-Weiner) approximated as

$$D = - \sum_{i=1}^{n} p_i \log_2 p_i$$

where p_i is taken to be N_i/N_s.

955

E	constant of integration in the theory of vertical distribution of plankton
F_c, F_d, F_p, F_s	absorption rate functions for cylinder, disk, plate and sphere respectively, in terms of d/d_0
F	force of gravity (acting on body immersed in water)
K	constant, as in equations for resistance and velocity of sinking bodies
K	molecular diffusivity
K, K_1, K_2, K_m $\cdots$	saturation populations in general or of S_1, S_2, $\cdots$ S_m
[L]	dimension of length
[M]	dimension of mass
N, N_1, $N_2 \cdots N_m \cdots$	populations, in general or of species S_1, $S_2 \cdots$ $S_m \cdots$
N_i	populations of S_i species
$N_{i \neq m}$	populations of all S_i species except S_m
$N_m(t)$	population of S_m at time t
N_n	where $n = 1$, 2, $3 \cdots$ population sample containing n specimens
N_0	initial population
N_s	total number of specimens in an assemblage of m species

$$N_s = \sum_{i=1}^{m} N_i$$

N_t	population at time t
$\bar{N}$	mean number of individuals in a series of samples
N	hyperspace in which niche is constructed as a hypervolume whose angles are defined by X' X'', $X''' \cdots X_1'$, X_2'', $X_1''' \cdots$, X_2', X_2'', $X_1''' \cdots$ etc.
N_1, $N_2 \cdots$	niches of species S_1, $S_2 \cdots$
$P(\hat{x})$	probability of a sample containing $\hat{x}$ specimens
P	probability that a given distribution or correlation is due to chance
R	gas constant
R_e	Reynolds number

$$R_e = dv\rho\mu^{-1}$$

R	resistance or drag on a body in a fluid
R_a	resistance of a nonspherical body
S, S_1, $S_2 \cdots S_m$	species whose populations are N, N_1, $N_2 \cdots N_m$
S_M	number of species in modal octave in the Preston theory
S_s	total number of species in an assemblage

S_t	total number of species in the log normal universe of the Preston theory
S_x	number of species in an octave in the Preston theory
S	species when involved in formal niche theory
$S_1, S_2 \cdots$	species whose niches in the hyperspace N are N_1, $N_2 \cdots$
[T]	dimension of time
V_s	volume of water containing a population of planktonic organism
W_0	minimum work performed by an organism in maintaining osmotic equilibrium under specified conditions
W_{oe}	part of W_0 done by excretory organs
W_{os}	part of W_0 done at body surface
$X', X'' \cdots$	variables defining the coordinates of points in the multidimensional niche space N.
$X_1', X_2', X_1'', X_2'' \cdots$	upper and lower limits of values of $X', X'' \cdots$ defining a niche.
a, b, c, d,	numbers; by itself a is a small area, b is an unrestricted birth rate; a, b, and c used together semidiameters of ellipsoids
â	relative increment of the radius of a planktonic organism due to secretion of a gelatinous sheath a', a coefficient in an expression for contageous distribution of plankton.
a	in theory of vertical distribution of plankton defined as $a = v/2A$
a, a − b	used together, epithecal and hypothecal diameters of diatoms
$b_1, b_2 \cdots b_m, b_i$	unrestricted birth rates or Malthusian parameters of populations $N_1, N_2 \cdots N_m$ or the set of populations N_i.
b′	constant in Thomas distribution
b̂	constant giving size (y) of part of an organism of linear dimension x when $x = 1$.
b_e	rate of egg laying
b	natality in treatment of zooplankton; on page 268 number of beads in a chain
b_1, b_2, b_3	in theory of vertical distribution of plankton, defined as

$$b_1 = \sqrt{a^2 - \hat{w}_1/A}$$
$$b_2 = \sqrt{a^2 - \hat{w}_2/A}$$
$$b_3 = -ib_1 \quad \text{or} \quad -\sqrt{-1}\,(b_1).$$

c   number in excess of unity of individuals in a clump

c   constant in an expression for d_a

d   linear dimension; if a sphere, the diameter

d_1, d_2 similar linear dimensions of objects of the same shape but different sizes

d' critical diameter

d_a linear dimension in Allen's formulation of sinking speeds at moderate values of R_e, defined as $d_a = d - cd'$

d_c diameter of a cylinder of l_c and volume equal to a sphere of diameter d_s

d_c' diameter of a cylinder of length l_c' and volume equal to a disc sinking at a velocity $\hat{v}_d$

d_0 in theory of nutrient uptake by a sinking cell, defined as

$$d_0 = \sqrt[3]{g(72\mu k)^{-1}\rho(\rho' - \rho)^{-1}}$$

or for water at 20° C, $0.24 \times 10^{-2}(\rho' - \rho)^{-1/3}$.

d   death rate

d' computed or smoothed death rate

f   function

f coefficient of sliding friction

g acceleration due to gravity

i   set of numbers

i the root of -1

k, k', k_b constants in expressions for resistance to and velocity of sinking of bodies immersed in a fluid; note that k is such that for any appropriate d, kd^3 is the volume of the object.

k measure of dispersion in negative binomial distribution

k' measure of dispersion in plankton distribution defined by $\sigma^2 = a'N^{k'}$

k   relative growth or heterauxetic growth exponent.

l_r number of eggs per female

$m, n,$ numbers

n_c mean number of clumps of plankton per sample

n   concentration of nutrient in medium

n' concentration of nutrient in cell

p, p, r, s,   numbers, used as exponents in dimensional analysis

p_s mean permeability of the surface of an organism

p_e probability of extinction

p_i probability of occurrence of the ith species

$\hat{p}$ phytoplankton concentration

$\hat{p}_1, \hat{p}_2$ phytoplankton concentrations in the euphotic and dysophotic zones respectively.

r	effective rate of increase
r	rank order
r_i	rates of increase for i species
r_m, $r_m(t)$	relative rate of increase in general, or at time t
r	radius
r_a	nominal radius defined as the radius of a sphere sinking at the same speed as a specified non-spherical body of the same density
r_e	time for egg to develop
r_s	radius of a sphere
t	time
$\hat{v}$, $\hat{v}_s$, $\hat{v}_a$	terminal velocity of a body sinking (or rising) in a fluid, of a sphere or of other body.
$\hat{v}_{min}$, $\hat{v}_{max}$	minimum and maximum values of v when a body varies in density
w	velocity of vertical current
$\hat{w}$	net rate of increase of a phytoplankton population
$\hat{w}_1$	net rate of increase in euphotic zone $(0 - z_l)$
$\hat{w}_2$	net rate of decrease in dysphotic zone $(z > z_l)$
$\hat{w}'$	computed rate of increase
x	number, slightly less than unity, dependent on the size of sample, in the Fisher, Corbet, and Williams distribution.
$\hat{x}$	number of specimens in a sample
x, y	horizontal distances; where x is used alone it can be any variable on the abscissa
x_M	value of x at the mode in the Preston theory
x, y	dimensions of organisms and their parts
z	depth, measured downwards
z_l	depth of euphotic zone
z_m	maximum depth
δ	in the Preston theory, defined by

$$\delta = (\sqrt{2}, \sigma^2)^{-1}$$

Ψ	stream function
Ψ_1	function giving trajectory of a particle
Ψ_0	value of Ψ equal and opposite to θ
α'	index of diversity
$\hat{\alpha}$	competition coefficient
$\hat{\beta}$	competition coefficient
$\hat{\gamma}$	predation coefficient
$\hat{\delta}$	efficiency coefficient in predation
θ, θ_1, θ_2...	temperatures
θ_z	temperature at depth z
θ_K	absolute temperature
μ	viscosity

ξ	arbitrary variable of integration
ρ	density of medium
ρ'	density of organism or other body
ρ''	density of gelatinous sheath
$\rho_{min}' \rho_{max}'$	minimum and maximum densities of an organism undergoing change
$\sigma, \sigma_1, \sigma_2$	standard deviations
τ	time lag
τ_n	time taken by a cell containing $n'gr \cdot gr^{-1}$ of a nutrient to absorb an equal amount from medium containing $n \cdot gr \cdot gr^{-1}$ of the nutrient and so be able to divide
τ_0	defined as $\tau_0 = (\mu^2/g^2K)^{1/3}$
ϕ_r	coefficient of form resistance $= \vartheta_s/\vartheta_a$
ϕ_a	coefficient of form resistance of a disc falling normal to flat surface
ϕ_b	coefficient of form resistance of a disc falling parallel to flat surface

Bibliography

and Index

of Authors

Let us now praise famous men
Ecclesiasticus 44.1.

When an alternative title in English, French or German is given in the summary of a paper in a Slavonic language or is listed in the part of the periodical in which the paper appears, it has usually proved convenient to use that title. The abbreviations used for the titles of periodicals are those of the *World List of Scientific Periodicals* . . . 4th ed. Butterworths, London 1965.

The asterisk (*) indicates works of exceptional limnological importance, either on account of the discoveries reported or because of extensive lists of references.

Aaronson, S., and Baker, H., 1959. A comparative biochemical study of two species of *Ochromonas*. *Protozool*, 6:282–284.† **348**

Abonyi, A., 1911. Ueber die Entwicklung der Phyllopodeneier. *Allat. Kozl.*, 10:171–176 (not seen: ref. Mathias 1937). **560**

Absolon, K., and Hrabe, S., 1930. Über einen neuen Süsswasser-Polychaeten aus den Höhlengewässern der Herzogowina. *Zool. Anz.*, 88:249–264. **73**

Adam, F., and Birrer, A., 1943. Biologisch-chemische Studie am Baldeggersee. *Mitt. naturf. Ges. Luzern,* 14:21–98. **441**

Agar, W. E., 1913. Transmission of environmental effects from parent to offspring in *Simocephalus vetulus*. *Phil. Trans. R. Soc.*, 203B:319–350. **826**

Agar, W. E., 1930. A statistical study of regeneration in two species of Crustacea. *Br. J. exp. Biol.*, 7:329–369. **576**

† Numbers in bold type are references to pages of this volume.

Ahlstrom, E. H., 1933. A quantitative study of Rotatoria in Terwilliger's Pond, Put-in-Bay, Ohio. *Bull. Ohio Biol. Surv.,* 30 (vol. 6 no. 1), 36 pp. **530**

Ahlstrom, E. H., 1940. A revision of the rotatorian genera *Brachionus* and *Platyias* with descriptions of one new species and two new varieties. *Bull. Am. Mus. Nat. Hist.,* 77:143–184. **507, 521, 522, 526, 527, 900, 903, 904, 906**

Ahlstrom, E. H., 1943. A revision of the rotatorian genus *Keratella* with descriptions of three new species and five new varieties. *Bull. Amer. Mus. Nat. Hist.,* 80:411–457. **534, 869, 872, 874, 875, 877, 878, 881, 882, 889, 894**

Akehurst, S. C., 1931. Observations on pond life, with special reference to the possible causation of swarming of phytoplankton. *Jl R. micros. Soc.* (ser. 3), 51:237–265. **462, 463, 472**

Aldrich, F. A., 1962. Results of the Catherwood Foundation Peruvian Amazon Expedition. The distribution of *Acetes paraguayensis* Hansen (Crustacea; Decapoda). *Notul. Nat.,* 351, 7 pp. **111**

Allee, W. C., and Schmidt, K. P., 1951. *Ecological Animal Geography.* Second edition of an authorized edition, rewritten and revised based on *Tiergeographie auf oekologischer Grundlage* by Richard Hesse. New York, John Wiley & Sons, xiii, 715 pp. **239**

Allen, J. F., 1854. *Victoria regia or the great water lily of America.* Boston, printed and pub. for the author by Dutton and Wentworth, 16 pp., 6 pl.

Allen, H. S., 1900. The motion of a sphere in a viscous fluid. *Phil. Mag.* (ser. 5), 50:323–338; 519–534. **257, 258, 271, 272**

Allen, W. E., 1920. A quantitative and statistical study of the plankton of the San Joaquin River and its tributaries in and near Stockton, California in 1913. *Univ. Cal. Publs. Zool.,* 22:1–292. **433, 434**

Allman, G. J., 1856. *A monograph of the fresh-water Polyzoa, including all the known species both British and foreign.* London, Ray Society. viii, 119 pp. **36**

Ammann, H., 1922. Zum Formenkreis von Ceratium hirundinella O.F.M. und Anuraea cochlearis. Vorläufige Mitteilung. *Arch. Hydrobiol.,* 13:92–96. **911, 917**

Amren, H., 1964a. Temporal variation of the rotifer *Keratella quadrata* (Müll.) in some ponds on Spitsbergen. *Zool. Bidr. Upps.,* 36:161–191. **887**

Amren, H., 1964b. Ecological and taxonomical studies on zooplankton from Spitsbergen. *Zool. Bidr. Upps.,* 36:209–276. **892**

An-der-Lan, H., 1939. Zur Rhabdocoelenfauna der Ochridasees (Balkan). *Sber. Akad. Wiss. Wien,* Math-nat. Kl., Abt. I., 148:195–254. **44**

Anderson, B. G., 1932. The number of pre-adult instars, growth, relative growth, and variation in Daphnia magna. *Biol. Bull. mar. Biol. Lab., Woods Hole,* 63:81–98. **577**

Anderson, B. G., and Jenkins, J. C., 1942. A time study of events in the life span of Daphnia magna. *Biol. Bull. mar. Biol. Lab., Woods Hole,* 83:260–272. **576, 577, 579, 581**

Anderson, B. G., Lumer H., and Zapancic, L. J., 1937. Growth and variability in Daphnia pulex. *Biol Bull. mar. Biol. Lab., Woods Hole,* 73:444–463. **577**

Anderson, G. C., 1958. Seasonal characteristics of two saline lakes in Washington. *Limnol. Oceanogr.,* 3:51–68. **477**

Anderson, G. C., Comita, G. W., and Engstrom-Heg, V., 1955. A note on the phytoplankton-zooplankton relationship in two lakes in Washington. *Ecology,* 36:757–759. **477, 478, 607, 619**

André, S., 1926. Sur le plancton du Rhône. [*Schweiz.*] *Z. Hydrol.,* 3:259–266. **805**

Annandale, N., 1911. Note on a rhizocephalous crustacean from fresh water and on some specimens of the order from Indian seas. *Rec. Indian Mus.,* 6:1–4. **130**

Annandale, N., 1912. Preliminary description of a freshwater medusa from the Bombay Presidency. *Rec. Indian Mus.,* 7:253–256. **42**

Annandale, N., 1922. The marine element in the fauna of the Ganges. *Bijdr. Dierk.* (Feest.-Numb. M. Weber): 143–154. **87**

Annandale, N., 1923. Animal life of the Ganges. *J. Bombay Nat. Hist. Soc.,* 29:633–642. **195**

Apstein, C., 1896. *Das Süsswasserplankton. Methode und Resultate der quantitativen Untersuchung.* Kiel, vi + 200 pp. **400, 435, 496, 609, 620, 698**

Apstein, C., 1907. Das Plankton im Colombo-See auf Ceylon. *Zool. Jb.* (Abt. Syst.), 25:201–244. **443, 444, 548, 549, 644, 645, 657, 674**

Arber, A., 1920. *Water Plants. A Study of Aquatic Angiosperms.* Cambridge, Univ. Press, xvi + 463 pp. **32n, 34, 36**

Arnold, H. D., 1911. Limitations imposed by slip and inertia terms upon Stokes law for the motion of spheres through liquids. *Phil. Mag.* ser. 6, 22:755–775. **258, 270**

Arnold, R., and Baker, P. F., 1960. The mystery of Loch Ness. *The Scotsman,* Monday, September 12, 1960. **552n.**

Arnon, D. I., 1958. The role of micronutrients in plant nutrition with special reference to photosynthesis and nitrogen assimilation. In C. A. Lamb, O. G. Bentley, and J. M. Beattie, *Trace Elements.* New York, Academic Press, pp. 1–32 **317, 322**

Arnon, D. I., and Wessel, G., 1953. Vanadium as an essential element for green plants. *Nature, Lond.,* 172:1039–1040. **322**

Artom, C., 1931. L'origine e l'evoluzione della partenogenesi attraverso i differenti biotipi di una specie collecttiva (*Artemia salina* L.), con speciale riferimento al biotipo diploide partenogenetico di Sète. *Memorie R. Acc. Ital.* Classe Scienze, 2:1–57. **560**

Asper, G., and Heuscher, J., 1887. Zur Naturgeschichte der Alpenseen. *Jber. St Gall. naturw. Ges.,* 1885/6 [27]:145–187. **491**

Austin, T. S., 1942. Appendix 1. The fossil species of Bosmina. In E. S. Deevey Studies on Connecticut Lake Sediments. III. The Biostratonomy of Linsley Pond. *Am. J. Sci.,* 240:325–331. **851**

Avery, J. L., 1939. Effect of drying on the viability of fairy shrimp eggs. *Tr. Am. micros. Soc.,* 58:356. **563**

Avery, J. L., 1940. Studies on the egg-laying habits of the fairy shrimp. *J. Wash. Acad. Sci.,* 30:31–33. **561**

Awerintzew, S., 1908. Beiträge zur Kenntnis der Süsswasserprotozoen. *Annls. Biol. lacustre,* 2:163–170. **497**

Ax, P., 1954. *Thalassochaetus palpifoliaceus* nov. gen. nov. spec. (Archiannelia, Nerillidae), ein mariner Verwandter von *Troglochaetus beranecki* Delachaux. *Zool. Anz.,* 133:64–75. **67**

Ax, P., 1963. Relations and phylogeny of the Turbellaria. In *The Lower Metazoa. Comparative Biology and Phylogeny,* ed. E. O. Dougherty et al. pp. 191–224. **43, 44, 46**

Axelson, J., 1961a. Zooplankton and impoundment of two lakes in northern Sweden (Ransaren and Kultsjön). *Rep. Inst. Freshwat. Res. Drottningholm,* 42:84–168. **614, 799**

Axelson, J., 1961b. On the dimorphism in *Cyclops scutifer* (Sars) and the cyclomorphosis in *Daphnia galeata* (Sars). *Rep. Inst. Freshwat. Res. Drottningholm,* 42:169–182. **614**

Bacci, G., Coletti, G., and Vaccari, A. M., 1961. Endomeiosis and sex determination in *Daphnia pulex. Experientia,* 17:505–506. **595**

Bachmann, H., 1904. *Cyclotella bodanica* var. *lemanica* O. Müller im Vierwaldstättersee und ihre Auxosporenbildung. *Jb. wiss. Bot.,* 39:106–133 **928, 929, 940**

Bachmann, H., 1910. Eine Wasserblüte von Oscillatoria rubescens DC in Rotsee. *Verh. schweiz. naturf. Ges.,* 93:254–255. **441**

Bachmann, R. W., and Odum, E. P., 1960. Uptake of Zn^{65} and primary productivity in marine benthic algae. *Limnol. Oceanog.,* 5:349–355. **313**

Bachofen, R., 1960. Stoffhaushalt und Sedimentation im Baldegger- und Hallwilersee. Univ. Zürich, Inaug. Dissert., 118 pp. **435, 441**

Bainbridge, R., 1961. Migrations. In *The Physiology of Crustacea,* ed. T. H. Waterman. New York and London, Academic Press, II:431–463. **788**

Bajkov, A. D., 1935. The plankton of Lake Winnipeg drainage system. *Int. Revue ges. Hydrobiol. Hydrogr.,* 31:239–272. **611**

Baker, P. F., 1960. Objects seen in Loch Ness. *The Scotsman,* Tuesday, September 13, 1960. **552n.**

Baker, P. F., and Westwood, M., 1960. Under-water detective work. *The Scotsman,* Wednesday, September 14, 1960. **552n.**

Baldi, E., 1935. Sul problema delle forme locali di *Eudiaptomus vulgaris* Schm. nel Lago di Garda e in altri laghi italiani. *Mem. Mus. Stor. nat. Venetia trident.,* 3:247–283. **861**

Baldwin, E., 1948. An Introduction to Comparative Biochemistry. Cambridge, Univ. Press, xii + 164 pp. **15**

Balss, H., 1957. Decapoda. H. G. Bronns *Klassen und Ordnungen des Tierreichs,* 5 Band, 1 Abt., 7 Buch, 12 Lief.:1505–1672. **111n.**

Banner, A. H., 1953. On a new genus and species of mysid from southern Louisiana (Crustacea, Malacostraca). *Tulane Stud. Zool.,* 1:3–8. **95n., 98**

Banta, A. M., 1926. A thelytokous race of Cladocera in which pseudo-sexual reproduction occurs. *Z. indukt. Abstamm. u. VererbLehre,* 40:28–41. **599**

* Banta, A. M., 1939. Studies on the Physiology, Genetics, and Evolution of Some Cladocera. *Carnegie Inst. Washington, Pub.* 513, Paper No. 39, Dept. Genetics, x + 285 pp. **596, 597, 820, 822, 828**

Banta, A. M., and Brown, L. A., 1929a. Control of sex in Cladocera. I. Crowding the mothers as a means of controlling male production. *Physiol. Zoöl.,* 2:80–92. **596, 597**

Banta, A. M., and Brown, L. A., 1929b. Control of sex in Cladocera. II. The unstable nature of the excretory products involved in male production. *Physiol. Zoöl.,* 2:93–98. **597**

Banta, A. M., and Brown, L. A., 1929c. Control of sex in Cladocera. III. Localization of the critical period for control of sex. *Proc. natn. Acad. Sci. U.S.A.,* 15:71–81. **597**

Banta, A. M., and Brown, L. A., 1939. Control of male and sexual-egg production. In Banta 1939, pp. 106–130. **581, 598**

Banta, A. M., and Wood, T. R., 1928. Genetic evidence that the Cladocera male is diploid. *Science, N.Y.,* 67:18–19. **594**

Banta, A. M., and Wood, T. R., 1939. General studies in sexual reproduction. In Banta 1939, pp. 131–181. **582, 594, 595, 598, 624**

Barbour, T., and Ramsden, C. T., 1919. The herpetology of Cuba. *Mem. Mus. Comp. Zool. Harv.,* 47:69–213. **141**

Barghoorn, E. S., and Tyler, S. A., 1965. Microorganisms from the Gunflint Chert. *Science, N.Y.* 147:563–577. **4**

Barigozzi, C., 1946. Ueber die geographische Verbreitung der mutanten von Artemia salina Leach. *Arch. Julius Klaus-Stift. VererbForsch.,* 21:479–482. **560**

Barker, D., 1959. The distribution and systematic position of the Thermosbaenacea. *Hydrobiologia,* 13:209–235. **94**

Barnard, J. L., 1958. Index to the families, genera, and species of the gammanidean Amphipoda (Crustacea). *Occ. Pap., Allan Hancock Fdn,* no. 19., 145 pp. **107n., 108n.**

Barnard, J. L., 1959. The number of species of gammaridean Amphipoda (Crustacea). *Bull. Sth. Calif. Acad. Sci.,* 58:16. **107**

Barnard, K. H., 1914. Description of a new species of *Phreatoicus* (Isopoda) from South Africa. *Ann. S. Afr. Mus.,* 10:231–240. **106**

Barnard, K. H., 1927. A study of the freshwater isopodan and amphipodan Crustacea of South Africa. *Trans. R. Soc. S. Afr.,* 14:139–215. **106**

Barnard, K. H., 1929. Contributions to the crustacean fauna of South Africa. No. 10. A revision of the South African Branchiopoda (Phyllopoda). *Ann. S. Afr. Mus.,* 29:181–272. **558, 561**

Barnard, K. H., 1940. Contributions to the crustacean fauna of South Africa, XII. Further additions to the Tanaidacea, Isopoda and Amphipoda, together with keys for the identification of hitherto recorded marine and freshwater species. *Ann. S. Afr. Mus.,* 32:381–543. **109**

Barnes, H., and Marshall S., 1951. On the variability of replicate plankton samples and some applications of 'contagious' series to the statistical distribution of catches over restricted periods. *J. mar. biol. Ass. U. K.,* 30:233–263. **788, 791, 797**

Barrell, J., 1916. Influence of Silurian-Devonian climates on the rise of air-breathing vertebrates. *Bull. geol. Soc. Amer.,* 27:387–436. **134**

Bateson, W., 1889. On some variations of *Cardium edule* apparently correlated to the conditions of life. *Phil. Trans. R. Soc.,* 180B:297–330. **59n.**

Bauer, V., 1908. Ueber die reflektorische Regulierung der Schwimmbewegungen bei den Mysiden. *Z. allg. Physiol.,* 8:343–369. **759**

Bauer, V., 1909. Vertikalwanderung des Planktons und Phototaxis. *Biol. Zbl.,* 29:77–82. **759**

Baumberger, J. P., and Olmstead, J. M. D., 1928. Changes in the osmotic pressure and water content of crabs during the molt cycle. *Physiol. Zoöl.,* 1:531–544. **172**

Bayer, F. M., and Fehlmann, H. A., 1960. The discovery of a freshwater opisthobranchiate mollusc *Acochlidium amboinense* Strubell in the Palau Islands. *Proc. biol. Soc. Wash.,* 73:183–194. **66**

Baylor, E. R., and Smith, F. E., 1953. The orientation of Cladocera to polarized light. *Am. Nat.,* 87:97–101. **778**

Bayly, I. A. E., 1961. A revision of the inland water genus *Calamoecia* (Copepoda:Calanoida). *Aust. J. mar. Freshwat. Res.*, 12:54–91. **126**

Bayly, I. A. E., 1962a. Ecological studies on New Zealand lacustrine zooplankton with special reference to *Boeckella propinqua* Sars (Copepoda:Calanoida). *Aust. J. mar. Freshwat. Res.*, 13:143–197. **657, 662, 670, 674, 741**

Bayly, I. A. E., 1962b. Additions to the inland water genus *Calamoecia* (Copepoda:Calanoida). *Aust. J. mar. Freshwat. Res.*, 13:252–264. **129**

Bayly, I. A. E., 1963. Reversed diurnal vertical migration of planktonic Crustacea in inland waters of low hydrogen ion concentrations. *Nature, Lond.*, 200:704–705. **741**

Bazikalova, A. R., 1945. Les amphipodes du Baikal (In Russian, French summ.). *Trudȳ baĭkal'. limnol. Sta.*, 11:1–440. **108**

Beach, N. W., 1960. A study of the planktonic rotifers of the Ocqueoc River system, Presque Isle County, Michigan. *Ecol. Monogr.*, 30:339–357. **530, 540**

Beadle, L. C., 1931. The effect of salinity changes on the water content and respiration of marine invertebrates. *J. exp. Biol.*, 8:211–27. **159**

Beadle, L. C., 1937. Adaptation to changes in salinity in the polychaetes. I. Control of body volume and of body fluid concentration in *Nereis diversicolor. J. exp. Biol.*, 14:56–70. **159**

* Beadle, L. C., 1943. Osmotic regulation and the faunas of inland waters. *Biol. Rev.*, 18:172–183. **151, 220, 221**

* Beadle, L. C., 1957. Comparative physiology: osmotic and ionic regulation in aquatic animals. *A. Rev. Physiol.*, 19:329–358. **151, 168**

Beadle, L. C., 1963. Anaerobic life in a tropical crater lake. *Nature, Lond.*, 200:1223–1224. **550, 643**

Beadle, L. C., and Cragg, J. B., 1940. The intertidal zone of two streams and the occurrence of *Gammarus* spp. on South Rona and Raasey (Inner Hebrides). *J. Anim. Ecol.*, 9:289–295. **154, 200, 201, 202**

Beadle, L. C., and Shaw, J., 1950. The retention of salt and the regulation of the non-protein nitrogen fraction in the blood of the aquatic larva *Sialis lutaria. J. exp. Biol.*, 27:96–109. **183**

Beauchamp, P. de, 1909. Recherches sur les rotifères: les formations tégumentaires et l'appareil digestif. *Arch. Zool. exp. gén.*, (ser. 4ᵉ) 10:1–410. **528, 529**

Beauchamp, P. de, 1924a. Sur l'apparition de la variation dans les conditions expérimentales chez les rotifères du genre *Brachionus. C. r. hebd. Séanc. Acad. Sci., Paris*, 179:1207–1209. **905**

Beauchamp, P. de, 1924b. Sur la transmission de la variation chez les rotifères du genre *Brachionus. C. r. hebd. Séanc. Acad. Sci., Paris*, 179:1290–1291. **905**

Beauchamp, P. de., 1927. A propos des formes réduites de *Brachionus Bakeri* Müller et *Br. furculatus* Thorpe. *Bull. Soc. zool. Fr.*, 52:61–67. **904, 905**

* Beauchamp, P. de, 1928. Coup d'oeil sur les recherches récentes relatives aux rotifères et sur les méthodes qui leur sont applicables. *Bull. biol. Fr. Belg.*, 62:51–125. **869**

Beauchamp, P. de, 1929. Triclades terricoles, triclades paludicoles, némertien. *Treubia*, 10:405–430. **47**

Beauchamp, P. de, 1932. Scientific results of the Cambridge Expedition to the East African lakes, 1930–31. 6. Rotifères et Gastrotriches. *J. Linn. Soc. (Zool.)*, 38:231–248. **551, 552**

Beaucamp, P. de, 1933. Sur la morphologie et l'ethologie des *Neogossea* (Gastrotriches). *Bull. Soc. zool. Fr.,* 58:331–342. **551**

Beauchamp, P. de, 1937. Quelques remarques sur la variation de *Brachionus pala* Ehrenberg. *C. r. hebd. Séanc. Soc. Biol.,* 125:448–450. **899**

Beauchamp, P. de, 1938. Les cultures de rotifères sur chlorelles Premiers résultats en milieu septique. *Trav. Sta. zool. Wimereux,* 13:27–38. **526, 899**

Beauchamp, P. de, 1951. Sur la variabilité spécifique dans le genre *Asplanchna* (Rotifères). *Bull. biol. Fr. Belg.,* 85:137–175. **867, 868**

Beauchamp, P. de, 1952a. Un facteur de la variabilité chez les rotifères du genre *Brachionus. C. r. hebd. Séanc. Acad. Sci., Paris,* 234:573–575. **899, 901**

Beauchamp, P. de, 1952b. Variation chez les rotifères du genre *Brachionus. C. r. hebd. Séanc. Acad. Sci., Paris,* 235:1355–1356. **899, 901**

Beauchamp, P. de, 1954. Un rhadocoele (sic) pélagique dans un lac du Ruanda *Mesostoma inversum* n. sp. *Revue Zool. Bot. afr.,* 50:157–164. **506**

Beauchamp, R. S. A., and Ullyott, P., 1932. Competitive relationships between certain species of fresh-water triclads. *J. Ecol.,* 20:200–208. **479**

Behning, A., 1924. Studien über die Malakostraken des Wolgabassins. I. Systematische Übersicht der im Wolgabassin bis jetzt aufgefunderen Malacostraca. II. Über den Ursprung und die Verbreitung der Malakostraken Fauna der Wolga. *Int. Revue ges. Hydrobiol. Hydrogr.,* 12:228–247. **216**

Behning, A., 1925. Studien über die Malakostraken des Wolgabassins. III. Über einige morphologische Merkmale und über die Variation der selben bie den Malakostraken der Wolga. *Int. Revue ges. Hydrobiol. Hydrogr.,* 13:46–77. **218**

Behning, A. L., 1928a. Das Leben der Wolga. *Binnengewässer* 5, 162 pp. **237**

Behning, A. L., 1928b. Ueber das Plankton des Tschalkar-Sees (in Russian with German summ.). *Russk. gidrobiol. Zh.,* 7:219–228. **219**

Behrens, H., 1933. Rotatorienfauna ostholsteinischer Tümpel. *Arch. Hydrobiol.,* 25:237–260. **890**

Beklemischew, W., 1914. Über einiger acöle Turbellarien des Kaspischen Meeres. *Zool. Anz.,* 45:1–7. **214**

Beliaev, G. M., and Birstein, J. A., 1944. A comparison between the osmoregulatory ability in Volga River and Caspian Amphipods. *Dokl. Akad. Nauk. USSR,* 45:304–306. **175**

Bellairs, A. d'A., and Underwood, G., 1951. The origin of snakes. *Biol. Rev.,* 26:193–237. **141**

Bennett, E. W., 1927. Biological notes on the copepod *Boeckella triarticulata. Trans. Proc. N. Z. Inst.,* 58:114–124. **676, 677**

Benoit, R. J., 1957. Preliminary observations on cobalt and vitamin B_{12} in freshwater. *Limnol. Oceanogr.,* 2:233–240. **317**

Berardi, G., and Tonolli, V., 1953. Clorofilla, fitoplancton e vicende meteorologiche (Lago Maggiore). *Memorie Ist. ital. Idrobiol.,* 7:165–187. **404, 411, 412, 435**

Berg, K., 1931. Studies on the genus Daphnia O. F. Müller with especial reference to the mode of reproduction. *Vidensk. Meddr. dansk Naturh. Foren.,* 92:1–222. **596, 597, 613, 616, 818, 819, 820**

Berg, K., 1934. Cyclic reproduction, sex determination and depression in the Cladocera. *Biol. Rev.,* 9:139–174. **587**

Berg, K., 1936. Reproduction and depression in the Cladocera, illustrated by the weight of the animals. *Arch. Hydrobiol.,* 30:438–464. **827**

Berg, K., 1937. Contributions to the Biology of *Corethra* Meigen (*Chaoborus* Lichtenstein). *Biol. Meddr.*, 13, Nc. 11, 101 pp. **779**

* Berg, K., and Nygaard, G., 1929. Studies on the plankton in the lake of Frederiksborg Castle. *Mém. de l'Acad. Roy. des Sci. et des Lett. de Danemark, Sec. des Sci.*, ser. 9 1:223–316. **609, 619, 620, 621, 642**

Berg, L. S., 1913. A review of the clupeoid fishes of the Caspian Sea, with remarks on the herring-like fishes of the Russian Empire. *Ann. Mag. nat. Hist.*, ser. 8 11:472–480. **215**

Berg, L. S., 1916. *Les poissons des eaux douces de la Russie*. Moscow, xxvii + 563 pp. **214**

Berg, L. S., 1933. *Les poissons des eaux douces de l'U.R.S.S. et des pays limitrophes*. 3-e édition, revue et augmentée Partie I (in Russian) Leningrad 1932, 543 pp., Partie II, 1933, pp. 547–899, figs. 475–762, maps. Also publ. as *Freshwater Fishes of the U.S.S.R. and Adjacent Countries* 1948 [1962], Vol. I 4th ed. Publ. for Nat. Sci. Fdr., Wash., D.C. by the Israel Program for Scientific Translations, Jerusalem, 1962, 504 pp. **178, 214, 215**

Berg, L. S., [1940] 1947. *Classification of fishes, both recent and fossil*. [lithoprint of Russian text from *Trav. Inst. Zool. Acad. Sci. URSS.*, 5:87–345 and English trans. 346–517], Ann Arbor, Mich., J. Edwards. **17n., 137**

Berg, L. S., and Popov, A., 1932. A review of the forms of *Myoxocephalus quadricornis* (L.). *Dokl. Akad. Nauk SSSR*, 1932:152–160. **179, 180**

Berger, K., 1934. Die Art *Daphnia longispina*. *Int. Revue ges. Hydrobiol. Hydrogr.*, 30:306–370. **813**

Bergeron, J. A., 1956. A histophysiological study of osmotic regulation in the euryhaline teleost *Fundulus heteroclitus*. *Diss. Abstr.*, 16:2192–2193. **170, 171**

Bernard, H. M., 1891. Hermaphroditismus bei Phyllopoden. *Jena Z. Naturw.*, 25:337–338. **560**

Bernard. H. M., 1896. Hermaphroditism among the Apodidae. *Ann. Mag. nat. Hist.*, ser. 6, 17:296–309. **560**

Berrill, N. J., 1955. *The origin of vertebrates*. Oxford, Clarendon Press, viii + 257 pp. **134**

Berry, E. W., 1925. The environment of the early vertebrates. *Am. Nat.*, 59:354–362. **134**

Berry, E. M., 1926. Description and notes on the life history of a new species of *Eulimnadia*. *Am. J. Sci.*, ser. 5 11:429–433. **565**

Bērziņš, B., 1951. Contribution to the knowledge of the marine rotatoria of Norway. *Univ. Bergen Árb.* Naturv. R., 6:1–11. **48**

Bērziņš, B., 1955. Taxonomic und Verbreitung von *Keratella valga* und verwandten Formen. *Ark. Zool.* ser. 2, 8:549–559. **869, 892**

* Bērziņš, B., 1958. Ein planktologisches Querprofil. *Rep. Inst. Freshwat. Res. Drottningholm*, 39:5-22. **543, 798, 800**

Bessey, E. A., 1950. *Morphology and taxonomy of fungi*. Philadelphia and Toronto, Blakiston Co., ixx 791 pp. **8**

Bigelow, H. B., and Schroder, W. C., 1948. Sharks. In *Fishes of the Western North Atlantic*. Part I. Lancelets, Cyclostomes, Sharks., Sears Found. Mar. Res., Yale University, pp. 59–546. **136**

Bigelow, N. K., 1928. The ecological distribution of microscopic organisms in Lake Nipigon. *Univ. Toronto Stud. biol. Ser.*, 31:57–74. **238**

Birge, E. A., 1895. Plankton studies on Lake Mendota. I. The vertical distribution of the pelagic Crustacea during July, 1894. *Trans. Wis. Acad. Sci. Arts Lett.,* 10:421–484. **726**

* Birge, E. A., 1898. Plankton studies on Lake Mendota. II. The Crustacea of the plankton from July 1894, to December 1896. *Trans. Wis. Acad. Sci. Arts. Lett.,* 11:274–451. **571, 611, 612, 619, 624, 652, 659, 673**

Birge, E. A., and Juday, C., 1908. A summer resting stage in the development of *Cyclops bicuspidatus* Claus. *Trans. Wis. Acad. Sci. Arts Lett.,* 16:1–9. **639**

* Birge, E. A., and Juday, C., 1922. The inland lakes of Wisconsin. The plankton. I. Its quantity and chemical composition. *Bull. Wis. Geol. Nat. Hist. Surv.,* 64 (Sci. ser. 13), 222 pp. **402, 405, 406, 420, 421, 422, 432**

Black, V. S., 1948. Changes in density, weight, chloride, and swimbladder gas in the killifish, Fundulus heteroclitus, in fresh water and sea water. *Biol. Bull. mar. biol. Lab. Woods Hole,* 95:83–93. **170**

Black, V. S., 1957. Excretion and osmoregulation. Ch. 4 in M. E. Brown *The Physiology of Fishes.* New York, Academic Press, pp. 163–205. **168**

Blanc, H., 1898. Le plankton nocturne du lac Léman. *Archs Sci. phys. nat.,* 6:182. **726, 734**

Blanchard, E., 1847. Recherches sur l'organisation des vers. *Annls Sci. nat.,* (ser. 3) 8:119–149. **47**

Bles, E. J., 1929. Arcella. A study in cell physiology. *Q. Jl micro. Sci.,* 72:527–648. **253, 254**

Bliss, C. I., and Fisher, R. A., 1953. Fitting the negative binomial distribution to biological data and note on the efficient fitting of the negative binomial. *Biometrics,* 9:176–200. **790, 791**

Boas, J. E. V., 1899. Kleinere carcinologische Mitteilungen. 2 Über den ungleichen Entwicklungsgang der Salzwasser-und Süsswasser form von *Palaeomonetes varians. Zool. Jb.* Abt. Syst., 4:793–805. **113**

Bock, W. J., 1958. A generic view of the plovers (Charadriinae, Aves). *Bull. Mus. comp. Zool. Harv.,* 118:27–97. **174**

du Bois-Reymond, R., 1914. Volumänderungen organischer Gewebe mit Berücksichtigung der "Schwebefauna." *Sber. Ges. naturf. Freunde Berl.,* 1914:373–378. **251**

du Bois-Reymond Marcus, E., 1948. An amazonian heteronemertine. *Bolm Fac. Filos. Ciénc. Univ. S. Paulo.*Zoologia, 13:93–109. **47**

Bond, R. M., 1934. Report on Phyllopod Crustacea (Anostraca, Notostraca and Conchostraca) including a revision of the Anostraca of the Indian Empire. *Mem. Conn. Acad. Arts Sci.,* 10:29–62. **118, 557**

Bond, R. M., 1935. Investigations of some Hispaniolan lakes. (Dr. R. M. Bond's Expedition.) I. Hydrology and hydrography. *Archiv. Hydrobiol.,* 28:137–161. **220**

Borcea, J., 1924. Faune survivante de type caspien dans les limans d'eau douce de Roumanie. *Annls scient. Univ. Jassy,* 13:207–232. **59**

Borecky, G. W., 1956. Population density of the limnetic Cladocera of Pymaluning Reservoir. *Ecology,* 37:719–727. **609, 613**

Borradaile, L. A., 1907. On the classification of the decapod crustaceans. *Ann. Mag. nat. Hist.,* (ser. 7) 19:457–480. **111n.**

Borutskii, E. V., 1952. Harpacticoida presnykh vod. *Fauna USSR,* No. 50 Crustacea, 3(4):1–424 [in Russian]. [Trans. as Freshwater Harpacticoida, tr.

A. Mercado; Jerusalem, Israel Program for Scientific Translations, XII:396 pp.] **129, 130**

Boschma, H., 1933. The Rhizocephala in the collection of the British Museum. *J. Linn. Soc.* (Zool.), 38:473–552. **130, 131, 189**

Boschma, H., 1934. Notes on the Rhizocephala in the collection of the British Museum. *Proc. Linn. Soc. London*, 146:24–25. **130, 189**

Bossone, A., and Tonolli, V., 1954. Il problema della convivenza di *Arctodiaptomus bacillifer* (Koelb), di *Acanthodiaptomus denticornis* (Wierz.) et di *Heterocope saliens* Lill. *Memorie Ist. ital Idrobiol.*, 8:81–94. **669, 680, 685, 686, 687**

Botnariuc, N., and Orghidan, T., 1953. Phyllopoda. *Fauna Repub. pop. rom. Crustacea*, Vol. IV, Fasc. 2:99 pp. **559, 566**

Bouillon, J., 1957a. Étude monographique du genre Limnocnida (Limnoméduse). *Annls Soc. r. zool. Belg.*, 87:253–500, **41, 42**

Boulloin, J., 1957b. *Limnocnida congoensis nouvelle espèce de Limnoméduse du bassin du Congo. Revue Zool. Bot. afr.*, 56:388–395. **42**

Boulenger, C. L., 1908. On Moerisia lyonsi, a new hydromedusan from Lake Qurun. *Q. Jl micros. Sci.*, 52:357–378. **39, 41**

Boulenger, G. A., 1895. On the type specimen of *Boulengerina stormsi. Proc. Zool. Soc. Lond.*, 1895, 865–866. **141**

Boulenger, G. A., 1909. *Catalogue of the Fresh-water fishes of Africa in the British Museum* (*Natural History*), I, xi, 373 pp. **138**

Bousfield, E. L., 1958. Fresh-water amphipod crustaceans of glaciated North America. *Can. Fld. Nat.*, 72 No. 2:55–113. **107, 109**

Bovee, E. C., and Jahn, T. L., 1965. Mechanisms of movement in Taxonomy of Sarcodina. II. The organization of subclasses and orders in relation to the classes Autotractea and Hydraulea. *Am. Midl. Nat.*, 73:293–298. **11n.**

Bouvier, E.-L., 1925. *Recherches sur la morphologie, les variations, la distribution geographique des crevettes de la famille Atyidés* (Encyclop. Entom 4). Paris, Chevalier, 370 pp. **113**

Bowerbank, J. S., 1874. *A monograph of the British Spongiadae*, London, Ray Society, III; xvii, 362 pp. **36**

Bowkiewicz, J., 1923. Biologische Beobachtungen über das Vorkommen von Apusiden in Siberien. *Int. Revue ges. Hydrobiol. Hydrogr.*, 11:317–321. **560**

Bowkiewicz, J., 1929. Schwebephase in der Bewegung der Cladoceren und Viskosität des Wassers. *Int. Revue ges. Hydrobiol. Hydrogr.*, 22:146–156. **279, 280, 756, 844**

Brackett, S., 1941. Schistosome dermatitis and its distribution. *Symposium on Hydrobiology.* Madison, Univ. Wis. Press, pp. 360–388. **704**

Bradshaw, A. S., 1949. A case of cytoplasmic transmission in Daphnia pulex De Geer. *Anat. Rec.*, 105:531–532. **827**

Brady, G. S., 1886. Notes on Entomostraca collected by Haly in Ceylon. *J. Linn. Soc.* (Zool.), 19:293-317. **695n.**

Brauer, F., 1877. Beiträge zur Kenntniss der Phyllopoden. *Sber. Akad. Wiss. Wien*, Abt. 1, 5:583–619. **564**

Brehm, V., 1927. 3. Ordung der Crustacea Entomostraca: Copepoda. W. Kükenthal, und T. Krumbach, *Handbuch der Zoologie*, 3 Bd., Erste Hälfte., pp. 435–496. **665, 860**

Brehm, V., 1933. Die Cladoceren der Deutschen Limnologischen Sunda-Expedition. *Arch. Hydrobiol. Suppl.*, 11:631–771. **571**

Brehm, V., 1937a. Zwei neue Moina-Formen aus Nevada, U.S.A. *Zool. Anz.,* 117:91–96. **607**

Brehm, V., 1937b. Die tiergeographischen Beziehungen der Diaptomiden des "Wallacea" Zwischengebietes. *Int. Revue ges Hydrobiol. Hydrogr.,* 34:287–293. **691**

Brehm, V., 1939. La Fauna microscópica del Lago Petén, Guatemala. *An. Esc. nac. Cienc. biol. Méx.,* 1:173–202. **695**

Brehm, V., and Zederbauer, E., 1902. Untersuchungen über das Plankton des Erlaufsees. *Verh. zool.-bot. Ges. Wien,* 52:388–402. **654**

Brehm, V., and Zederbauer, E., 1906. Beiträge zur Planktonuntersuchung alpiner Seen. *Verh. zool.-bot. Ges. Wien,* 56:19–32. **654, 860**

Brewer, R. H., 1964. The phenology of *Diaptomus stagnalis* (Copepoda: Calanoida): The development and the hatching of the egg stage. *Physiol. Zoöl.,* 37:1–20. **666**

Brien, P., 1952. Étude sur les phylactolémates. *Annls r. Soc. zool. Belg.,* 84:301–444. **698, 699**

Bristowe, W. S., 1930. Notes on the Biology of Spiders. II. Aquatic Spiders. *Ann. Mag. nat. Hist.,* (ser. 10) 6:343–347. **87**

Bristowe, W. S., 1931. Notes on the Biology of Spiders. IV. Further notes on aquatic spiders with a description of new species of Pseudoscorpion from Singapore. *Ann. Mag. nat. Hist.,* (ser. 10) 8:457–465. **87**

Broekema, M. M. M., 1942. Seasonal movements and the osmotic behaviour of the shrimp, *Crangon crangon. Arch. néerl. Zool.,* 6:1–100. **166, 173**

Brook, A. J., 1959a. The status of desmids in the plankton and the determination of phytoplankton quotients. *J. Ecol.,* 47:429–445. **307, 328, 330, 393**

Brook, A. J., 1959b. *Staurastrum paradoxum* Meyen and *S. gracile* Ralf in the British fresh-water plankton, and a revision of the *S. anatinum*-group of radiate desmids. *Trans. R. Soc. Edinb.,* 63:589–628. **307, 328, 329**

Brook, A. J., and Woodward, W. B., 1956. Some observation on the effects of water inflow and outflow on the plankton of small lakes. *J. Anim. Ecol.,* 25:22–35. **805**

Brooks, H. K., 1962. On the fossil Anaspidacea, with a revision of the classification of the Syncarida. *Crustaceana,* 4:229–242. **94, 97**

* Brooks, J. L., 1946. Cyclomorphosis in Daphnia. I. An Analysis of *D. retrocurva* and *D. galeata. Ecol. Monogr.,* 16:409–447. **597, 598, 611, 612, 812, 815, 816, 819, 827–830**

Brooks, J. L., 1947. Turbulence as an environmental determinant of relative growth in *Daphnia. Proc. natn. Acad. Sci. U.S.A.,* 33:141–148. **831, 834**

* Brooks, J. L., 1950a. Speciation in ancient lakes. *Q. Rev. Biol.,* 25:30–60, 131–176. **38, 197**

Brooks, J. L., 1950b. Recent advances in limnology. *Ecology,* 31: 659–660. **661**

* Brooks, J. L., 1957a. The systematics of North American Daphnia. *Mem. Conn. Acad. Arts Sci.,* 13:5–180. **572–574, 577, 583, 596, 606, 611, 617, 813, 818, 820, 838, 842**

Brooks, J. L., 1957b. The species problem in freshwater animals *in* The Species Problem. A Symposium presented at the Atlanta Meeting of the American Association for the Advancement of Science, Dec. 28–29, 1955, ed. Ernst Mayr. *Publs Am. Ass. Advmt Sci.,* 50:81–123. **835, 846**

Brooks, J. L., and Dodson, S. I., 1965. Predation, body size, and composition of plankton. *Science, N.Y.* 150:28–35. **682**

Brutschy, A., 1922.   Die Vegetation und das Zooplankton des Hallwiler Sees. Teil 2.   *Int. Revue ges. Hydrobiol. Hydrogr.*, 10:271–298. **939**

Buchner, H., 1936.   Experimentelle Untersuchungen über den Generationswechse der Rädertiere.   *Z. indukt. Abstamn.-u. VererbLehre*, 72:1–49.   **510, 511, 899, 904**

Buchner, H., 1941a.   Freilanduntersuchungen über den Generationswechsel der Rädertiere.   *Zool. Jb.* Abt. allg. Zool. Physiol., 60:253–278. **510, 511**

Buchner, H., 1941b.   Experimentelle Untersuchungen über den Generationswechsel der Rädertiere.   II. *Zool. Jb.* Abt. allg. Zool. Physiol., 60:279–344. **510, 511**

Buchner, H., and Mulzer, F., 1961.   Untersuchungen über die Variabilität der Rädertiere.   II. Der Ablauf der Variation im Freien.   *Z. Morph. Ökol. Tiere*, 50:330–374. **875**

Buchner, H., Mulzer, F., and Rauh, F., 1957.   Untersuchungen über die Variabilität der Rädertiere.   I. Problemstellung und vorläufige Mitteilung über die Ergebniss.   *Biol. Zbl.*, 76:289–315. **899, 900, 901, 904**

Buddenbrock, W. von, 1915.   Die Tropismentheorie von J. Loeb Ein Versuch ihrer Widerlegung.   *Biol. Zbl.*, 35:481–506. **762, 771**

Bülow, T. von, 1954.   Ernährungsbiologische Studien an Euchlaniden (*Rotatoria*).   *Zool. Anz.*, 153:126–134. **528**

Burckhardt, A., 1935.   Die Ernährungsgrundlagen der Copepodenschwärme der Niederelbe.   *Int. Revue ges. Hydrobiol. Hydrogr.*, 32:432–500. **678**

Burckhardt, G., 1900a.   Faunistische und systematische Studien über des Zooplankton der Grosseren Seen der Schweiz und ihrer Grenzgebiete.   *Revue suisse Zool.*, 7:353–713. **609, 653**

Burckhardt, G., 1900b.   Quantitative Studien über das Zooplankton des Vierwaldstäter See.   *Mitt. naturf. Ges. Luzern*, 3:129–437. **670, 671, 726**

Burckhardt, G., 1910.   Hypothesen und Beobachtungen über die Bedeutung der vertikalen Planktonwanderung.   *Int. Revue ges. Hydrobiol. Hydrogr.*, 3:156–172; 335–338. **726, 799**

Burckhardt, G., 1913.   Wissenschaftliche Ergebnisse einer Reise um die Erde von M. Pernod und C. Schröter.   III. Zooplancton aus ost- und süd-asiatischen Binnengewässern.   *Zool. Jb.* Abt. Syst. Ökol. Geogr., 34:341–472. **127, 128, 650**

Burckhardt, G., 1920.   Zum Worte Plankton.   *Schweiz. Z. Hydrol.*, 1:190–192. **236**

Burckhardt, G., 1943.   Esistono forme intermedie fra "le due specie di Bosmina."   *Schweiz. Z. Hydrol.*, 9:128–148. **851**

Burckhardt, G., 1944.   Verarmung des Planktons in kleinen Seen durch *Heterocope*.   *Schweiz. Z. Hydrol.*, 10:121–124. **679**

Burden, C. E., 1956.   The failure of hypophysectomized Fundulus heteroclitus to survive in fresh water.   *Biol. Bull. Mar. biol. Lab., Woods Hole*, 110:8–28. **170, 171, 172**

Burger, J. W., 1962.   Further studies on the function of the rectal gland in the spiny dog fish.   *Physiol. Zool.*, 35:205–217. **169**

Burger, J. W., and Hess, W. N., 1960.   Function of the rectal gland in the spiny dogfish.   *Science, N.Y.*, 131:670–671. **169**

* Burkholder, P., 1931.   Studies in the phytoplankton of the Cayuga Lake basin, New York.   *Bull. Buffalo Soc. nat. Sci.*, 15:21–181. **473, 474**

Buxton, P. A., 1926. The colonization of the sea by insects with an account of the habits of *Pontomyia*, the only known submarine insect. *Proc. zool. Soc. Lond.*, 1926(2):807–814. **86**

Cain, A. J., 1959. The post-Linnaean development of taxonomy. *Proc. Linn. Soc. Lond.* 170:234–244. **14**

Calman, W. T., 1909. *Crustacea*. In *A Treatise on Zoology*, ed. R. Lankester. London, Black, 346 pp. **97, 558**

Canabaeus, L., 1929. Über die Heterocysten und Gasvacuolen der Blaualgen und ihre Beziehung zueinander. *Pflanzenforschung* 13. **252, 253**

Cannon, H. G., 1924. On the development of an estherid crustacean. *Phil. Trans. R. Soc.*, 212B:395–430. **565**

Cannon, H. G., 1928. On ꞌthe feeding mechanism of the copepods, *Calanus finmarchicus* and *Diaptomus gracilis*. *Br. J. exp. Biol.*, 6:131–144. **676**

Cannon, H. G., 1933. On the feeding mechanism of the Branchiopoda. *Phil. Trans. R. Soc.*, 222B:267–352. **571**

Canter, H. M., 1949. The importance of fungal parasitism in limnology. *Verh. Int. Verein. theor. angew. Limnol.*, 10:107–108. **474, 476**

* Canter, H. M., and Lund, J. W. G., 1948. Studies on Plankton Parasites. I. Fluctuations in the numbers of *Asterionella formosa* Hass. in relation to fungal epidemics. *New Phytol.*, 47:238–261. **474, 475**

Canter, H. M., and Lund, J. W. G., 1951. Studies on plankton parasites. III. Examples of the interaction between parasitism and other factors determining the growth of diatoms. *Ann. Bot.*, 15:359–371. **474**

Canter, H. M., and Lund, J. W. G., 1953. Studies on plankton parasites. II. The parasitism of diatoms with special reference to lakes in the English Lake District. *Trans. Brit. mycol. Soc.*, 36:13–37. **475**

Capart, A., 1952. Crustacés décapodes brachyures. *Explor. Hydrobiol. du Lac Tanganika* (1946–1947), vol. III fasc., 3:41–67. **117**

Carăusu, S., 1943. Amphipodes de Roumanie I. Gammarides de type Caspien. *Monographia Institutul de Cerceyari Piscicole al Romaniei*, No. 1, 293 pp., 85 pl. **218**

Carl, G. C., 1940. The distribution of some Cladocera and free-living Copepoda in British Columbia. *Ecol. Monogr.*, 10:55–110. **682, 693**

* Carlin, B., 1943. Die Planktonrotatorien des Motalaström: zur Taxonomie und Ökologie der Planktonrotatorien. *Meddn. Lunds Univ. Limnol. Inst.* No. 5., 255 pp. **511, 522, 528–547, 869, 872, 875, 878, 880, 883, 884, 886, 887, 888, 909**

Caroli, E., 1937. *Stygiomysis hydruntina* n.g., n. sp., Misidaceo cavernicolo di Terra d'Ótranto, rappresentante di una nuova famiglia. Note preliminare. *Boll. Musei. Zool. Anat. comp. R. Univ.*, Torino, 8:219–227. **99**

Carpenter, K. E., 1928. *Life in Inland Waters, with especial reference to animals*. New York, Macmillan, xviii + 267 pp. **37n.**

Carter, G. S., and Beadle, L. C., 1930. The fauna of the swamps of the Paraguayan Chaco in relation to its environment. I. Physico-chemical nature of the environment. *J. Linn. Soc. Zool.*, 37:205–258. **136**

Castle, W. A., 1938. Hatching of the eggs of the "fairy shrimp." *Science, N.Y.*, 87:531. **563**

Carter, G. S., and Beadle, L. C., 1931. The fauna of the swamps of the Paraguayan Chaco in relation to its environment. II. Respiratory adaptations in the fishes. *J. Linn. Soc. Zool.*, 37:327–368. **136**

Cash, J., and Hopkinson, J., 1909. *The British freshwater Rhizopoda and Heliozoa.* London, Ray Society, II: xviii + 166 pp. **494**

Caspary, R., 1891. Nymphaeaceae in Engler A., and Prantl, K. *Die natürlichen Pflanzenfamilien,* III, Abt., 2:1–10. **34**

Cassie, R. M., 1961. The correlation coefficient as an index of ecological affinities in plankton populations. *Memorie Ist. ital. Idrobiol.,* 13:151–177. **791, 796**

Cassie, R. M., 1962. Frequency distribution models in the ecology of plankton and other organisms. *J. Anim. Ecol.,* 31:65–92. **791**

Cassie, R. M., 1963. Multivariate analysis in the interpretation of numerical plankton data. *New Zeal. J. Sci.,* 6:38–59. **791**

Chamberlin, T. C., 1900. On the habitat of the early vertebrates. *J. Geol.,* 8:400–412. **134**

Chandler, D. C., 1939. Plankton entering the Huron River from Portage and Base Line Lake, Michigan. *Trans. Am. microsc. Soc.,* 58:24–41. **805, 806**

* Chandler, D. C., 1940. Limnological studies of Western Lake Erie. I. Plankton and certain physical-chemical data of the Bass Islands region, from September 1938 to November 1939. *Ohio J. Sci.,* 40:291–336. **404, 437–439**

* Chandler, D. C., 1942a. Limnological studies of western Lake Erie. II. Light penetration and its relation to turbidity. *Ecology,* 23:41–52. **404, 432, 437–439**

* Chandler, D. C., 1942b. Limnological studies of western Lake Erie. III. Phytoplankton and physical-chemical data from November, 1939 to November, 1940. *Ohio J. Sci.,* 42:24–44. **402, 404, 432, 437–439**

* Chandler, D. C., 1944. Limnological studies of western Lake Erie. IV. Relation of limnological and climatic factors to the phytoplankton of 1941. *Trans. Amer. microsc. Soc.,* 63:203–236. **402, 404, 432, 436, 437**

Chappuis, P. A., and Delamare Deboutteville, C., 1954. Recherches sur les Crustacés souterrains. *Arch. Zool. exp. gén.,* 91:1–194. **192**

Chappuis, P. A., Delamare Deboutteville, C. and Paulian, R., 1956. Crustacés des eaux souterraines littorales d'une résurgence d'eau douce à la Réunion. *Mém. Inst. Scient. Madagascar* A(Biol.), II:51–78. **101, 192**

Chase, F. A., Mackin, J. G., Hubricht, L., Banner, A. H. and Hobbs, H. H., 1959. Malacostraca. In H. B. Ward and G. C. Whipple *Fresh-water Biology,* 2nd ed., ed. W. T. Edmondson. New York and London, John Wiley & Sons, 869–901. **174**

Chatton, E., 1952. *Traité de Zoologie,* ed. P. Grassé. Tome I. *Phylogénie. Protozoaires: Généralities. Flagellés,* Paris, Masson et cie. xii, 1071 pp. **25**

Chatton, E., and Beauchamp, P. de, 1923. *Teuthophrys trisulca* n.g.n. sp. Infusoire pélagique d'eau douce. *Arch. Zool. exp. gén.,* 61, Notes et revues: 123–129. **498**

Child, C. M., 1901. The habits and natural history of Stichostemma. *Am. Nat.,* 35:975–1006. **47**

Chodorowski, A., 1959. Ecological differentiation of turbellarians in Harsz-Lake. *Polskie Archwm Hydrobiol.,* 6:33–73. **231**

Chopra, B., and Tiwari, K. K., 1950. On a new genus of Phreatoicid Isopod from wells in Benares. *Rec. Indian Mus.,* 47:277–289. **106**

* Chu, S. P., 1942. The influence of the mineral composition of the medium on the growth of planktonic algae. I. *J. Ecol.,* 30:284–325. **320, 337**

Chu, S. P., 1943. The influence of the mineral composition of the medium on the growth of planktonic algae. II. The influence of the concentration of inorganic nitrogen and phosphate phosphorus. *J. Ecol.*, 31:109–148. **444, 453**

Chu, S. P., 1945. Phytoplankton. *Rep. Freshwat. biol. Ass.*, 13:20–23. **453**

Cicak, A., McLaughlin, J. J. A., and Wittenberg, J. B., 1963. Oxygen in the gas vacuole of the rhizopod protozoan, *Arcella*. *Nature, Lond.*, 199:983–985. **253**

Clark, E., 1936. The freshwater and land crayfishes of Australia. *Mem. natn. Mus., Melb.*, 10:5–58. **114**

Clarke, F. W., 1924. The data of geochemistry. 5th ed. *Bull. U. S. geol. Surv.*, 770, 841 pp. **213**

Clarke, G. L., 1930. Change of phototropic and geotropic signs in *Daphnia*. *J. exp. Biol.*, 7:109–131. **761, 762, 763, 765**

Clarke, G. L., 1932. Quantitative aspects of the change of phototropic sign in *Daphnia*. *J. exp. Biol.*, 9:180–211. **761, 763**

Claus, A., 1937. Vergleichend-physiologische Untersuchungen zur Ökologie der Wasserwanzen mit besonderer Berücksichtigung der Brackwasserwanze *Sigara lugubris*. *Zool. Jb.* Abt. allg. Zool. Physiol. Tiere, 58:365–432. **220**

Claus, C., 1894. Über die Wiederbelebung im Schlamme eingetrockneter Copepoden und Copepodeneier. *Arb. zool. Inst. Univ. Wien*, 11:1–12. **639**

Clements, F. E., and Shelford, V. E., 1939. *Bio-ecology.* New York, John Wiley and Sons; London, Chapman and Hall, vi + 425 pp. **230, 231**

Clench, W. J., 1959. Molluscs. In H. B. Ward and G. C. Whipple, *Freshwater Biology*, 2nd ed. Ed. W. T. Edmonson. New York, and London, John Wiley & Sons, pp. 1117–1160. **64, 68**

Cloud, P. E. Jr., 1960. Gas as a sedimentary and diagenetic agent. *Am. J. Sci.*, Bradley Vol. 258A:35–45. **14**

Cloudsley-Thompson, J. L., 1948. *Hydroschendyla submarina* (Grube) in Yorkshire: with an historical review of the marine Myriopoda. *Naturalist, Hull*, October–December, 1948, pp. 149–152. **79**

Coe, W. R., 1943. Biology of the Nemerteans of the Atlantic Coast of North America. *Trans. Conn. Acad. Arts Sci.*, 35:129–328. **46, 47**

Cohn, F., 1878. Rivularia fluitans ad. int. *Hedwigia*, 17:1–5. **425**

Coker, R. E., 1933. Influence of temperature on size in freshwater Copepods (Cyclops). *Int. Revue ges. Hydrobiol. Hydrogr.*, 29:406–436. **857, 858**

Coker, R. E., 1934. Influence of temperature on form of the freshwater Copepod, *Cyclops vernalis* Fischer. *Int. Revue ges. Hydrobiol. Hydrogr.*, 30:411–427. **862**

Coker, R. E., 1939. The problem of cyclomorphosis in Daphnia. *Q. Rev. Biol.*, 14:137–148. **827**

Coker, R. E., 1943. Mesocyclops edax (S. A. Forbes), M. leuckarti (Claus) and related species in America. *J. Elisha Mitchell scient. Soc.*, 59:181–200. **627, 635**

* Coker, R. E., and Addlestone, H. H., 1938. Influence of temperature on cyclomorphosis in Daphnia longispina. *J. Elisha Mitchell scient. Soc.*, 54:45–75. **827, 828, 834**

Coker, R. E., and Hayes, W. J. Jr., 1940. Biological observations in Mountain Lake, Virginia. *Ecology*, 21:192–198. **603**

Coker, R. E., Shira, A. F., Clark, H. W., and Howard, A. D., 1921. Natural history and propagation of freshwater mussels. *Bull. Bur. Fish. Wash.*, 37:75–182. **699**

Colditz, F. V., 1914. Beiträge zur Biologie des Mansfelder Sees mit besonderen Studien über das Zentrifugenplankton und seine Beziehungen zum Netzplankton der pelagischen Zone. *Z. wiss. Zool.*, 108:520–630. **422, 423**

Cole, G. A., 1953a. Notes on copepod encystment. *Ecology*, 34:208–211. **640**

Cole, G. A., 1953b. Notes on the vertical distribution of organisms in the profundal sediments of Douglas Lake, Michigan. *Am. Midl. Nat.*, 49:252–256. **640**

Cole, G. A., 1955. An ecological study of the microbenthic fauna of two Minnesota lakes. *Am. Midl. Nat.*, 53:213–230. **640, 643**

* Cole, G. A., 1961. Some calanoid copepods from Arizona with notes on congeneric occurrences of *Diaptomus* species. *Limnol. Oceanogr.*, 6:432–442. **643, 650, 680–682, 684**

Comita, G., 1956. A study of a calanoid copepod population in an Arctic lake. *Ecology*, 37:576–591. **646, 668**

Comita, G., 1964. The energy budget of *Diaptomus siciloides* Lilljeborg. *Verh. int. Verein theor. angew. Limnol.*, 15:646–653. **659**

Comita, G. W., and Anderson, G. C., 1959. The seasonal development of a population of *Diaptomus ashlandi* Marsh, and related phytoplankton cycles in Lake Washington. *Limnol. Oceanogr.*, 4:37–52. **657, 671**

Comita, G. W., and J. J. Comita, 1957. The internal distribution patterns of a calanoid copepod population, and a description of a modified Clarke-Bumpus plankton sampler. *Limnol. Oceanogr.*, 2:321–332. **791, 796**

Cooke, A. H., 1895. Molluscs. In *Cambridge Natural History,* ed. S. F. Harmer, and E. A. Shipley. Molluscs, Brachiopods (recent) Brachiopods (fossil). London and New York, Macmillan, pp. 1–459. **55**

Copeland, Herbert Faulkner, 1956. *The classification of lower organisms.* Palo Alto, Calif., Pacific Books, ix + 302 pp. **2**

Corliss, J. O., 1956. On the evolution and systematics of ciliated protozoa. *Syst. Zool.*, 5:68–91, 121–140. **11**

Corliss, J. O., 1961. *The Ciliated Protozoa: Characterization, Classification, and Guide to the Literature.* New York, Oxford, London, and Paris, Pergamon Press, 310 pp. **11, 499n**

Corrêa, D. D., 1948. A polychaete from the Amazon-Region. *Bolm Fac. Filos. Ciénc. Univ. S. Paulo. Zoologia*, 13:245–252. **70**

Cort, W. W., 1928a. Schistosome dermatitis in the U. S. (Michigan). *J. Am. med. Ass.*, 90:1027–1029. **704**

Cort, W. W., 1928b. Further observations on Schistosome dermatitis in the U. S. (Michigan). *Science, N.Y.*, 68:388. **704**

Cort, W. W., and S. T. Brooks, 1928. Studies on the Holostome cercariae from Douglas Lake, Michigan. *Trans. Am. microsc. Soc.*, 47:179–221. **701**

Cowles, R. P., and C. E. Brambel, 1936. A study of the environmental conditions in a bog pond with special reference to the diurnal vertical distribution of Gonyostomum semen. *Biol. Bull. mar. biol. Lab., Woods Hole*, 71:286–298. **731**

Credner, R., 1887. Die Reliktenseen. Eine physisch-geographische Monographie. I. Teil: Über die Beweise für den marinen Ursprung der als Reliktenseen bezeichneten Binnengewässer. *Petermanns Mitt.* Ergänzungsheft 19, No. 86, 110 pp. **211**

Croghan, P. C., 1958a. The survival of *Artemia salina* (L.) in various media. *J. exp. Biol.,* 35:213–218. **173**

Croghan, P. C., 1958b. The osmotic and ionic regulation of *Artemia salina* (L.). *J. exp. Biol.,* 35:219–233. **173**

Croghan, P. C., 1958c. The mechanism of osmotic regulation in *Artemia salina* (L.): The physiology of the branchiae. *J. exp. Biol.,* 35:234–242. **173**

Croghan, P. C., 1958d. The mechanism of osmotic regulation in *Artemia salina* (L.): The physiology of the gut. *J. exp. Biol.,* 35:243–249. **173**

Crosse, H., 1886. Description du nouveau genre *Quadrasia. J. Conch.,* 34:159–163. **65**

Crowell, K., 1961. The effects of reduced competition in birds. *Proc. natn. Acad. Sci. U.S.A.,* 47:240–243. **364**

Cullison, J. S., 1938. Dutchtown fauna of southeastern Missouri. *J. Paleont.,* 12:219–228. **133**

Cunningham, W. J., 1954. A nonlinear differential-difference equation of growth. *Proc. natn. Acad. Sci. U.S.A.,* 40:708–713. **591**

Cupp, E. E., 1943. Marine plankton diatoms of the west coast of North America. *Bull. Scripps Instn Oceanogr.,* 5:1–238. **300n**

Cushing, D. H., 1951. The vertical migration of plankton crustacea. *Biol. Rev.,* 26:158–192. **733–735, 755**

Cushing, D. H., 1955. Some experiments on the vertical migration of zoo-plankton. *J. Anim. Ecol.,* 24:137–166. **777**

Czeczuga, B., 1959. Oviposition in *Eudiaptomus gracilis* G. O. Sars and *E. graciloides* Lilljeborg (*Diaptomidae, Crustacea*) in relation to season and trophic level of lakes. *Bull. Acad. pol. Sci. Cl. II, Sér. Sci. biol.,* 7, 227–230. **637, 638, 659**

Czeczuga, B., 1960. Zmiany płodności niektórych przedstawicieli zoo-planktonu. I. *Crustacea* Jezior Rajgradzkich. *Polskie Archwm Hydrobiol.,* 7:61–89, (Eng. summ. 90–91) **637, 638, 652, 654, 659**

Daday, E. v., 1885. Beiträge zur Kenntniss der Plattensee-Faune. *Math. naturw. Ber. Ung.,* 3:179–184. **812**

Daday, E. v., 1888. *A Magyarországi Cladocerák Maganrajka. Crustacea Cladocera Faunae Hungaricae.* Budapest (Magyar text, Latin diagnoses vi, 128 pp.) **812**

Dahl, E., 1954. Some aspects of the ontogeny of *Mesamphisopus capensis* (Barnard) and the affinities of the Isopoda Phreatoicoidea. *K. fysiogr. Sällsk. Lund Förh.,* 24: No. 9, 1–6. **105**

Dahl, E., 1956. Some crustacean relationships. Bertil Hanström, Zoological papers in honour of his sixty-fifth birthday, Nov. 20, 1956, pp. 138–147. **93, 94, 119, 130**

Dahl, E., 1958. Fresh and brackish water amphipods from the Azores and Madeira (Report No. 1 Lund University Expedition in 1957 to the Azores and Madeira). *Bolm Mus. munic. Funchal,* 11, Art. 27:22 pp. **109**

Dahl, E., 1962. Main evolutionary lines among recent Crustacea (abstract). Conference on the Evolution of Crustacea reported upon by Isabella Gordon. *Crustaceana,* 4:163–164. **93, 94**

Dahl, F., 1894. Die Copepodenfauna des unteren Amazonas. *Ber. naturf. Ges. Freiburg. -i-B.,* 8:10–23. **126**

Dahl, F., 1908. Grundsätze und Grundbegriffe der biocönotischen Forschung. *Zool. Anz.,* 33:349–353. **227**

Daily, W. A., 1938. A quantitative study of the phytoplankton of Lake Michigan collected in the vicinity of Evanston, Illinois. *Butler Univ. bot. Stud.,* 4:65–83. **402, 404**

Dakin, W. J., and LaVarche, M., 1913. The plankton of Lough Neagh: a study of the seasonal changes in the plankton by quantitative methods. *Proc. R. Ir. Acad.,* 30:20–96. **732**

Dallavalle, J. M., 1948. *Micromeritics. The Technology of Fine Particles,* 2nd ed. New York and London, Pitman Publishing Corp., xxviii + 555 pp. **259, 260n, 269, 272**

Damann, K. E., 1941. Quantitative study of the phytoplankton of Lake Michigan at Evanston, Illinois. *Butler Univ. bot. Stud.,* 5:27–44. **404**

Damant, G. C. C., 1925. The adjustment of the buoyance of the larva of *Corethra plumicornis. J. Physiol., Lond.,* 59:345–356. **254, 255**

Damas, H., 1939. Sur la présence dans la Meuse belge de *Branchiura sowerbyi* (Beddard) *Craspedacusta sowerbyi* (Lankester) et *Urnatella gracilis* (Leidy). *Ann. Soc. r. Zool. Belg.,* 69:293–310. **52n**

Davenport, C. B., and Cannon, W. B., 1897. On the determination of the direction and rate of movement of organisms by light. *J. Physiol., Lond.,* 21:22–32. **759**

David, P. M., 1961. The influence of vertical migration on speciation in the oceanic plankton. *Systematic Zool.,* 10:10–16. **788**

Davis, C. C., 1954. A preliminary study of the plankton of the Cleveland Harbor area, Ohio. III. The zooplankton and general ecological considerations of plankton production. *Ohio J. Sci.,* 54:338–408. **499, 530, 540**

Davis, C. C., 1955. Notes on the food of *Craspedacusta sowerbii* in Crystal Lake, Ravenna, Ohio. *Ecology,* 36:364–365. **505**

Davis, C. C., 1957. *Cordylophora lacustris* Allman from Chagrin Harbor, Ohio. *Limnol. Oceanogr.,* 2:158–159. **39**

Davis, C. C., 1959. A planktonic fish egg from fresh water. *Limnol. Oceanogr.,* 4:352–355. **704, 705**

Davis, C. C., 1961. Breeding of Calanoid Copepods in Lake Erie. *Verh. int. Verein. theor. angew. Limnol.,* 14:933–942. **656, 662, 673, 674, 684, 685**

* Davis, C. C., 1962. The plankton of the Cleveland Harbor Area of Lake Erie in 1956–1957. *Ecol. Monogl.,* 32:209–247. **495**

Dawes, B., 1946. *The Trematoda with special reference to British and other European Forms.* Cambridge University Press, xvi + 644 pp. **701**

Dawydoff, G., 1937. Une métanémerte nouvelle, appartenant à une groupe purement marin, provenant du Grand Lac du Cambodge. *C. r. hebd. Séanc. Acad. Sci., Paris,* 204:804–806. **47**

* Deevey, G. B., 1960. Relative effects of temperature and food on seasonal variations in length of marine copepods in some eastern American and western European waters. *Bull. Bingham oceanogr. Coll.,* 17, No. 2:54–85. **645**

Deflandre, G., 1936. Etude monographique sur le genre *Nebela* Leidy (Rhizopoda-Testacea). *Annls Protist.,* 5:201–286. **495**

von Dehn, M., 1930. Untersuchungen über die Verdauung bei Daphnien. *Z. vergl. Physiol.,* 13:334–358. **553, 557**

Dejdar, E., 1934. Die Süsswassermeduse *Craspedacusta sowerbii* Lankester in monographisher Darstellung. *Z. Morph. Ökol. Tiere*, 28: 595–691. **42**

Delachaux, T., 1927. Un polychète d'eau douce cavernicole *Troglochaetus beranecki* nov. gen. nov. sp. *Bull. Soc. neuchât. Sci. nat.*, 45:1–7. **71**

Delamare-Debouteville, C., 1960. Biologie des eaux souterraines littorales et continentales. *Actual. scient. ind.*, 1280. 740 pp. **97, 98, 192**

von Denffer, D., 1948. Über einen Wachstumshemmstoff in alternden Diatomeenkulturen. *Biol. Zbl.*, 67:7–13. **464**

Denison, R. H., 1956. A review of the habitats of the earliest vertebrates. *Fieldiana: Geol.*, 11:359–457. **134**

Denton, E., 1960. The buoyancy of marine animals. *Scient. Amer.*, 203, No. 1:118–128. **247**

Derzhavin, A. (= Derjavin), 1912. *Caspionema pallasi*, eine Meduse des Kaspischen Meeres. *Zool. Anz.*, 39:390–396. **39**

Derjavin, A. N., 1924. Fresh water peracarida from the coast of the Black Sea of Caucasus. *Russk. gidrobiol. Zh.*, 3:113–128 (Russ. text.); (Engl. summ. 129). **196**

Derjavin, A. N., 1925. Materials of the Ponto-Azoph Carcinofauna (Mysidaceae, Cumacea, Amphipoda). *Russk. gidrobiol. Zh.*, 4:10–35 (Russian with Eng. res.). **217**

Derjavin, A. N., 1930. The freshwater Malacostraca of the Russian Far East. *Russk. gidrobiol. Zh.*, 9:1–8 (Russ. and Eng.). **100**

Derouet, L., 1952. Influence des variations de salinité du milieu extérieur sur des crustacés cavernicoles et épigés. 2. Etude des teneurs en chlore du milieu intérieur et des tissues. *C.r. hebd. Séanc. Acad. Sci., Paris*, 234:888–890. **193**

Dexter, R. W., 1946. Further studies on the life history and distribution of Eubranchipus vernalis (Verrill). *Ohio J. Sci.*, 46:31–44. **563**

Dexter, R. W., 1953. Studies on North American fairy shrimps with the description of two new species. *Am. Midl. Nat.*, 49:751–771. **561**

Dice, L. R., 1914. The factors determining the vertical movement of *Daphnia*. *J. Anim. Behav.*, 4:229–265. **763, 765, 781**

Dice, L. R., 1952. *Natural Communities.* Ann Arbor, Univ. Michigan Press, x + 547 pp. **230**

Dickie, G., 1860. *Botanists's Guide to the Countries of Aberdeen, Banff and Kincardine.* Aberdeen, A. Brown & Co., xxxii, 344 pp. **425**

Dieffenbach, H., 1911. See Dieffenbach and Sachse. **706, 887, 888**

Dieffenbach, H., and Sachse, R., 1911. Biologische Untersuchungen an Rädertieren in Teichgewässern. *Int. Revue ges. Hydrobiol. Hydrogr.*, Biol. Suppl. 3:1–93. (Dieffenbach: Studien an pelagische Rädertiere, 9–42; Sachse: Beiträge zur Biologie litoraler Rädertiere 43–87.) **526, 527**

Dietrich, W., 1915. Die Metamorphose der freilebenden Süsswassercopepoden. I. Die Nauplien und das erste Copepodidstadium. *Z. Wiss. Zool.*, 113:252–324. **647, 650**

Digby, P. S. B., 1961. Mechanism of sensitivity to hydrostatic pressure in the prawn, *Paleomonetes varians* Leach. *Nature, Lond.*, 191:366–368. **778**

Doudoroff, M., 1940. Experiments on the adaptation of Escherichia coli to sodium chloride. *J. gen. Physiol.*, 23:585–611. **20**

Dougherty, E. C., 1955. Comparative evolution and the origin of sexuality. *Syst. Zool.*, 4:145–169, 190. **2, 3**

Dougherty, E. C., and Allen, M. B., 1959. Speculations on the position of the crypotomonads in protistan phylogeny, 15th *Int. Cong. Zool.*, London, 184–186. **6**

Droop, M. R., 1958. Optimum relative and actual ionic concentrations for growth of some euryhaline algae. *Verh. int. Verein. theor. angew. Limnol.*, 13:722–730. **28**

Drouet, F., 1959. Myxophyceae. In H. B. Ward, and G. C. Whipple *Fresh-water Biology*, 2nd ed., ed. W. T. Edmondson. New York and London, John Wiley & Sons, pp. 95–114. **307, 313**

Drouet, F., and Daily, W. A., 1956. Revision of the coccoid Myxophyceae. *Butler Univ. bot. Stud.*, 12:1–218. **313, 394**

Drummond, J. L., 1838. On a new *Oscillatoria,* the coloring substance of Glaslough Lake, Ireland. *Ann. nat. Hist.*, 1:1–6. **425**

Dugdale, R., Dugdale V., Neess, J., and Goering, J., 1959. Nitrogen fixation in lakes. *Science, N.Y.*, 130:859–860. **318**

Dugdale, V. A., and Dugdale, R. C., 1962. Nitrogen metabolism in lakes. II. Role of nitrogen fixation in Sanctuary Lake, Pennsylvania, *Limnol. Oceanogr.*, 7:170–177. **318**

Dukina, V. V., 1956. Specific differences in Cyclop larvae. *Zool. Zk.*, 35:680–690 (Russ. text; Eng. sum. pp. 6–7 of summaries). **570, 638**

Dusi, H., 1932. In Lwoff A. *Recherches biochimiques sur la nutrition der Protozoaires.* Paris, Masson et Cie., 158 pp. **312**

Dussart, B., 1958. Remarques sur le genre Cyclops s. str. (Crust. cop.) *Hydrobiologia,* 10:263–292. **632, 633**

Duval, M., 1925. Recherches physicochimiques et physiologiques sur le milieu intérieur des animaux aquatiques. Modifications sous l'influence du milieu extérieur. *Annls Inst. Océanogr., Monaco,* Nouv. série, 2:32–407. **155, 169**

Dybowski, W., 1900. Beschreibung einer Hinterkiemer-Schnecke aus dem Baikalsee. (*Ancylodoris baicalensis m.*) *Nachrbl. dt. malakozool. Ges.*, 32: 143–152. **65**

Eddy, S., 1930. The plankton of Reelfoot Lake, Tennessee. *Trans. Am. microsc. Soc.*, 49:246–251. **684**

Edmonds, G., 1935. The relations between the internal fluid of marine invertebrates and the water of the environment, with special reference to Australian crustacea. *Proc. Linn. Soc. N.S.W.*, 60:233–47. **172**

Edmondson, C. H., 1918. Amoeboid Protozoa (Sarcodina). In H. B. Ward, and G. C. Whipple *Fresh-Water Biology*, New York and London, John Wiley & Sons. 1918, pp. 210–237. **494**

* Edmondson, W. T., 1944. Ecological studies of sessile Rotatoria. Part 1. Factors affecting distribution. *Ecol. Monogr.*, 14:31–66. **521, 525**

* Edmondson, W. T., 1945. Ecological studies of sessile Rotatoria. II. Dynamics of populations and social structures. *Ecol. Monogr.*, 15:141–172. **512**

Edmondson, W. T., 1948. Ecological applications of Lansing's Physiological Work on Longevity in *Rotatoria. Science, N. Y.,* 108:123–126. **513, 890**

Edmondson, W. T., 1955. The seasonal life history of *Daphnia* in an arctic lake. *Ecology,* 36:439–455. **599, 617**

Edmondson, W. T., ed., 1959. H. B. Ward and G. C. Whipple *Fresh-water Biology,* 2nd ed. New York and London, John Wiley & Sons, xx, 1248 pp. **307, 519, 520**

* Edmondson, W. T., 1960. Reproductive rates of rotifers in natural populations. *Memorie Ist. ital. Idrobiol.*, 12:21–77. **509, 512, 514, 515, 516, 517**

* Edmondson, W. T., 1964. Reproductive rate of planktonic rotifers as related to food and temperature in nature. *Ecol. Monogr.*, 35:61–111. **516, 527**

Edmondson, W. T., Comita G. W., and Anderson, G. C., 1962. Reproductive rate of copepods in nature and its relation to phytoplankton population. *Ecology*, 43:625–634. **666, 667**

Edmondson, W. T., and Hutchinson, G. E., 1934. Report on Rotatoria of Yale North India Expedition. New Haven, *Mem. Conn. Acad. Arts Sci.*, 10: 153–186. **524, 536, 892**

Edwards, F. W., 1926. On marine Chironomidae (Diptera); with descriptions of a new genus and four new species from Samoa. *Proc. zool. Soc. Lond,* 1926 (2):779–806. **86**

Ege, R., 1915. On the respiratory function of the air stores carried by some aquatic insects. *Z. allg. Physiol.*, 17:81–124. **185**

Eggleton, F. E., 1931. A limnological study of the profundal bottom fauna of certain fresh-water lakes. *Ecol. Monogr.*, 1:231–331. **240**

* Eichhorn, R., 1959. Zur Populationsdynamik der calanoiden Copepoden in Titisee und Feldsee. *Arch. Hydrobiol.*, Supp. 24:186–246. **658, 659, 663, 664, 666, 669, 670**

Einsele, W., and Grim, J., 1938. Über den Kieselsäuregehalt planktischer Diatomeen und dessen Bedeutung für einige Fragen ihrer Ökologie. *Z. Bot.*, 32:545–590. **247, 277**

* Einsele, W., and Vetter, H., 1938. Untersuchungen über die Entwicklung der physikalischen und chemischen Verhältnisse im Jahreszyklus in einem mässig eutropher See (Schleinsee bei Langenargen). *Int. Revue ges. Hydrobiol. Hydrogr.*, 36:285–324. **424**

* Ekman, S., 1904. Die Phyllopoden, Cladoceren und freilebenden Copepoden der nordeschwedischen Hochgebirge. *Zool. Jb., Abt. Syst.* 21:1–170. **566, 617, 642, 663, 668, 669, 671, 673**

Ekman, S., 1907. Über das Crustaceenplankton der Ekoln (Mälaren) und über verschiedene Kategonien von marinen Relikten in schwedischer Binnenseen. Zoologiska Studier, tillägnade Professor T. Tullberg, pp. 42–65. **125, 614, 663, 668**

* Ekman, S., 1913. Studien über die marinen Relikte der nordeuropäischen Binnengewässer II Die Variation der Kopfform bei *Limnocalanus grimaldii* (deGuerne) und L. *macrurus* G.O. Sars. *Int. Revue ges Hydrobiol. Hydrogr.,* 6:335–371, (German text) 371–372 (Eng. summary). **176**

Ekman, S., 1914. Artbildung bei der Copepodengattung *Limnocalanus* durch akkumulative Fernwirkung einer Milieuveränderung. *Z. indukt. Abstamm.-u., VererbLehre,* 11:39–104. **176**

Ekman, S., 1915. Die Bodenfauna des Vättern, qualitativ und quantitativ untersucht. *Int. Revue ges. Hydrobiol. Hydrogr.*, 7:146–204, 275–425. **240**

Ekman, S., 1917. Allgemeine Bemerkungen über die Tiefenfauna der Binnenseen. *Int. Revue ges. Hydrobiol. Hydrogr.*, 8:113–124. **241**

Elgmork, K., 1958. On the phenology of *Mesocyclops oithonoides* (G.O. Sars). *Verh. int. Verein. theor. angew. Limnol.*, 13:778–784. **640**

* Elgmork, K., 1959. Seasonal occurrence of Cyclops strenuus strenuus. *Folia limnol. scand.* 11, 196 pp. **631, 637, 638, 640, 641, 642**

Elgmork, K., 1962. A bottom resting stage in the planktonic freshwater copepod *Cyclops scutifer* Sars. *Oikos*, 13:306–310. **643**

Elgmork, K., 1964. Dynamics of zooplankton communities in some small inundated ponds. *Folia limnol. scand.*, 12, 83 pp. **668**

* Elster, H. J., 1936. Einige biologische Beobachtungen an *Heterocope borealis*. *Int. Revue ges. Hydrobiol. Hydrogr.*, 33:357–433. **679**

* Elster, H. J., 1954. Über die Populations dynamik von *Eudiaptomus gracilis* Sars und *Heterocope borealis* Fischer in Bodensee-Obersee. *Arch. Hydrobiol.*, Suppl. 20:546–614. **514, 654, 655, 659, 666, 672, 743**

Elton, C., 1927. *Animal ecology*. London, Sidgwick and Jackson, xxi 207 pp. **125**

Elton, C., 1929. The ecological relationships of certain freshwater copepods. *J. Ecol.*, 17:383–391. **125, 636**

Enderlein, G., 1908. Biologisch-faunistische Moor-und Dünen-Studien. *Ber. westpreuss. Bot. -zool. Ver.*, 30:54–238. **227**

Enright, J. T., 1963. Estimates of the compressibility of some marine crustaceans. *Limnol. Oceanogr.*, 8:382–387. **778**

Entz, G., 1904. Beiträge zur Kenntniss des Planktons des Balatonsees. *Resultate wiss. Erforsch. Balatonsees* II, Bd. 1. **912, 917, 919**

Entz, G., 1927. Beiträge zur Kenntnis der Peridineen II. resp. VII. Studien an Süsswasser-Ceratien. *Arch. Protistenk.*, 58:344–440. **910, 916, 917**

Entz, G., 1931. Analyze des Wachstums und der Teilung einer Population sowie eines Individuums des Protisten *Ceratium hirundinella* unter den natürlichen Verhältnissen. *Arch. Protistenk.*, 74:310–361. **916**

Erman, L. A., 1962. On the quantitative aspects of feeding and selection of food in the planktonic rotifer *Brachionus calyciflorus* Pallas *Zool. Zh.*, 41:34–47 (Russian text). **528**

Esaki, T., and China, W. E., 1927. A new Family of aquatic heteroptera. *Trans. ent. Soc. Lond*, 75:279–295. **85, 184n.**

Esterly, C. O., 1907. The reactions of *Cyclops* to gravity. *Am. J. Physiol.*, 18:47–57. **761**

Evitt, W. R., 1963. Occurrence of freshwater alga *Pediastrum* in Cretaceous marine sediments. *Am. J. Sci.*, 261:890–893. **19**

Ewald, W. F., 1910. Ueber Orientierung, Lokomotion und Lichtreaktion einiger Cladoceren und deren Bedeutung für die Theorie der Tropismen. *Biol. Zbl.*, 30:1–16, 49–63, 379–384, 385–399. **760–762**

Ewald, W. F., 1912. On artificial modification of light reaction and the influence of electrolytes on phototaxis. *J. exp. Zool.*, 13:592–612. **760**

Ewer, D. W., and Hattingh, I., 1952. Absorbtion of silver on the gills of a freshwater crab. *Nature, Lond.*, 169:460. **163**

Eyden, D., 1923. Specific gravity as a factor in the vertical distribution of plankton. *Proc. Camb. phil. Soc. biol., Sci.*, 1:49–55. **250, 279, 280, 756, 779**

Eyster, C., 1952. Necessity of boron for *Nostoc muscorum*. *Nature, Lond.*, 170:755. **317**

Fairbridge, W. S., 1945a. West Australian species of fresh-water calanoids. I. Three new species of *Boeckella* with an account of the developmental stages of B. *opaqua* n.sp. and a key to the genus. *J. Proc. R. Soc. West. Aust.*, 29:25–65. **650, 658, 665**

Fairbridge, W. S., 1945b. West Australian species of freshwater calanoids. II. Two new species of *Brunella* with an account of the developmental stages of B. *subattenuata* n.sp. *J. Proc. R. Soc. West. Aust.*, 29:67–89. **677**

Falkenhagen, H., 1931. Klassische Hydrodynamik. In *Handbuch der Experimental Physik.*, Bd. 4. 1 Teil. *Hydro-und Aero-dynamiks.* Strömungslehre und allgemeine Versuchstechnik, pp. 47–237. **258, 259, 261**

Fänge, R., and Fugelli, K., 1963. The rectal salt gland of elasmobranchs, and osmoregulation in chimaeroid fishes. *Sarsia*, 10:27–34. **169**

Fänge, R., Schmidt-Nielsen, K., and Osaki, H., 1958. The salt gland of the herring gull. *Biol. Bull. mar. biol. lab., Woods Hole*, 115:162–171. **174**

Fassett, N. C., 1940. *A manual of aquatic plants.* New York and London, McGraw-Hill, vii, 382 pp. **31**

Fauré-Fremiet, E., 1924. Contribution à la connaissance des infusoires planctoniques. *Bull. biol. Fr. Belg.*, Suppl. 6:1–171. **499, 501, 504**

Fauvel, P., 1932. Annelida Polychaeta of the Indian Museum, Calcutta. *Mem. Indian Mus.*, 12:1–262. **73**

Felföldy, L. J. M., 1960. Photosynthetic experiments with unicellular algae of different photosynthetic type. *Ann. Biol. Tihany*, 27:193–200. **309**

Feller, W., 1939. Die Grundlagen von Volterraschen Theorie des Kampfes ums Dasein in wahrscheinlichkeitstheoretischer Behandlung. *Acta Biotheoret.*, 5:11–40. **357**

Ferguson, E. Jr., Hutchinson, G. E., and Goulden, C. E., 1964. *Cypria petenensis*, a new name for the ostracod *Cypria pelagica*, Brehm 1932. *Postilla* 80, 4 pp. **695, 696**

Ferguson, M. S., 1939. Observations on Eubranchipus vernalis in southwestern Ontario and Eastern Illinois. *Am. Midl. Nat.*, 22:466–469. **563**

Fetcher, E. S. Jr., 1939. The water balance in marine mammals. *Q. Rev. Biol.*, 14:451–459. **175**

Fetcher, E. S. Jr., and Fetcher, G. W., 1942. Experiments on the osmotic regulation of dolphins. *J. Cell. Comp. Physiol.*, 19:123–130. **175**

Feuerborn, H. J., 1931. Ein Rhizocephale und zwei Polychaeten aus dem Süsswasser von Java und Sumatra. *Verh. int. Verein. theor. angew. Limnol.*, 5:618–660. **70, 71, 74, 130**

Findenegg, I., 1938. Sechs Jahre Temperaturlotungen in den Kärntner Seen. *Int. Revue ges. Hydrobiol. Hydrogr.*, 37:364–384. **819**

Findenegg, I., 1943a. Zur Kenntnis der planktischen Cladoceren Kärntens. *Carinthia II.*, 1943:47–67. **617, 816, 819, 831, 832, 836, 837, 851**

* Findenegg, I., 1943b. Untersuchungen über die Ökologie und die Produktionsverhältniss des Planktons in Kärntner Seengebiete. *Int. Revue ges. Hydrobiol. Hydrogr.*, 43:368–429. **385, 401, 402, 407–410, 430, 431, 434, 440–443**

Findenegg, I., 1948. Die Daphnia-Arten der Kärntner Gewässer und ihre Beziehung zur Grösse des Lebensraumes. *Öst. Zool. Z.*, 1:519–532. **617**

* Findenegg, I., 1953. Kärntner Seen naturkundlich betrachtet. *Carinthia II*, 15, Sonderheft. Klagenfurt, 101 pp. **609, 611, 613, 619**

Finesinger, J. E., 1926. Effect of certain chemical and physical agents on fecundity and length of life, and on their inheritance in a rotifer, Lecane (Distyla) inermis (Bryce). *J. exp. Zool.*, 44:63–94. **512**

Fisher, R. A., 1934. *Statistical methods for research workers*, 5th ed. London, Oliver and Boyd, xiii, 319 pp. **788, 789**

Fisher, R. A., Corbet, A. S., and Williams, C. B., 1943. The relation between the number of species and the number of individuals in a random sample of an animal population. *J. Anim. Ecol.*, 12:42–58. **360, 361, 369**

Florkin, M., 1949. *Biochemical Evolution.* New York, Academic Press, vi, 157 pp. **151, 195**

Foged, N., 1954. On the diatom flora of some Funen Lakes. *Folia limnol. Scand.,* 6, 75 pp. **368–372, 374**

Fogg, G. E., 1941. The gas-vacuoles of the Myxophyceae (Cyanophyceae). *Biol. Rev.,* 16:205–217. **251**

Fogg, G. E., 1964. *Algal cultures and phytoplankton ecology.* Univ. Wisconsin Press, xiii, 126 pp., 1965. **292, 307**

Fordyce, C., 1900. The Cladocera of Nebraska. *Trans. Am. microsc. Soc.,* 22:119–174. **734**

Forel, F. A., 1877. Faune profonde du Léman xxxii. Faune pélagique. *Bull. Soc. Vaud. Sci. nat.,* 14:210–223. **726, 734, 757**

Fox, H. M., 1920. Methods of studying the respiratory exchange in small aquatic organisms with particular reference to the use of flagellates as an indicator for oxygen consumption. *J. gen. Physiol.,* 3:565–573. **184**

Fox, H. M., 1925. The effect of light on the vertical movement of aquatic organisms. *Proc. Camb. Phil. Soc. biol. Sci.,* 1:219–224. **761**

Fox, H. M., 1948. The haemoglobin of *Daphnia. Proc. R. Soc.,* 135*B*: 195–212. **608**

Fox, H. M., 1949. On Apus: its rediscovery in Britain, nomenclature and habits. *Proc. zool. Soc. Lond.,* 119:693–702. **564**

Fox, H. M., Gilchrist, B. M., and Phear, E. A., 1951. Functions of haemoglobin in *Daphnia. Proc. R. Soc.,* 138B:514–528. **608**

Fox, H. M., and Mitchell, Y., 1953. Relation of the rate of antennal movement in *Daphnia* to the number of eggs carried in the brood pouch. *J. exp. Biol.,* 30: 238–242. **250, 279, 779**

Fox, H. M., and Phear, E. A., 1953. Factors influencing haemoglobin synthesis by *Daphnia. Proc. R. Soc.,* 141B:179–180. **608**

* Fraenkel, G., and Gunn, D. L., 1940. *The Orientation of animals; kineses, taxes and compass reactions.* Oxford, The Clarendon Press, vi + 353 pp. **758n**

Francé, R. H., 1894. Zur Biologie des Planktons. *Biol. Zbl.,* 14:33–38. **726**

Francé, R. H., 1897. Protozoen. *Resultate wiss. Erforsch. Balatonsees,* Bd 1, Tiel 1, Sect. 1:1–64. **491**

Frank, P. W., 1952. A laboratory study of intraspecific and interspecific competition in *Daphnia pulicaria* (Forbes) and *Simocephalus vetulus* O. F. Müller. *Physiol. Zoöl.,* 25:178–204. **583, 587**

* Frank, P. W., 1957. Coactions in laboratory populations of two species of *Daphnia. Ecology,* 38:510–519. **583–587**

Frankenberg, G. v., 1915. Die Schwimmblasen von Corethra. *Zool. Jb. Abt. allg. Zool. Physiol.,* 35:505–592. **254, 255**

Franz, V., 1912. Zur Fragen der vertikalen Wanderungen der Planktontiere. *Arch. Hydrobiol.* 7:493–499. **726**

Freidenfelt, T., 1913. Zur Biologie von Daphnia longiremis G. O. Sars und Daphnia cristata G. O. Sars. *Int. Revue ges. Hydrobiol. Hydrogr.,* 6:229–242. **617, 618, 624, 818**

Freidenfelt, T., 1920. Zur Kenntnis der Biologie von *Holopedium gibberum* Zadd. *Arch. Hydrobiol.,* 12:725–749. **598, 609, 622**

Frenzel, J., 1892. Untersuchungen über die mikroskopische Fauna Argentiniens. *Salinella salve* nov. gen., nov. spec. ein viel zelliges infusorienartiges Tier (Mesozoon). *Arch. f. Naturges,* 58:66–96. **14**

Freshwater Biological Association, 1946. Algae. In Director's Progress Report. *Rep. Freshwat. biol. Ass.,* 14:25–28. **453**

Freshwater Biological Association, 1949a. Algology. In Report of the Director. *Rep. Freshwat. Biol. Ass.* 17:18–21. **401**

Freshwater Biological Association, 1949b. Mycology. In Report of the Director. *Rep. Freshwat. Biol. Ass.* 17:21–22. **476**

Friederichs, K., 1927. Grundsätzliches über die Lebenseimheiten höherer Ordung und den ökologischen Einheitsfaktor. *Naturwissenschaften,* 15:153–157, 182–186. **227**

Frisch, K. V. von, und Kupelwieser, H., 1913. Ueber den Einfluss der Lichtfarbe auf die phototaktischen Reaktion niederer Krebse. *Biol. Zbl.,* 33:518–552. **762, 765, 766, 771**

Fritsch, F. E., 1935. *The structure and reproduction of the algae.* New York, Macmillan; Cambridge Univ. Press, I: xvii, 791 pp. **307, 320**

Fritsch, F. E., 1945. *The structure and reproduction of the algae.* Cambridge University Press, II: xiv, 939 pp. **6, 22, 251, 307**

Fritsche, H., 1916. Studien über die Schwankungen des osmotischen Druckes der Körperflüssigkeit bei *Daphnia magna. Int. Revue ges. Hydrobiol. Hydrogr.,* 8:22–80, 125–203. **162**

Fritz, F., 1935. Über die Sinkgeschwindigkeit einiger Phytoplanktonorganismen. *Int. Revue ges. Hydrobiol. Hydrogr.,* 32:424–431. **276, 287, 292, 921**

Fryer, G., 1953. Notes on certain freshwater crustaceans. *Naturalist, Hull,* 1953:101–109. **598**

Fryer, G., 1954. Contributions to our knowledge of the biology and systematics of the freshwater Copepoda. *Schweiz. Z. Hydrol.,* 16:64–77. **632, 679, 682, 686**

Fryer, G., 1957a. The feeding mechanism of some freshwater Cyclopoid Copepods. *Proc. zool. Soc. Lond.* 129 (1):1–25. **627, 630**

Fryer, G., 1957b. The food of some freshwater cyclopoid copepods and its ecological significance. *J. Anim. Ecol.,* 26:263–286. **627, 629**

Fryer, G., 1957c. Freeliving freshwater crustacea from Lake Nyasa and adjoining waters. Part 1. Copepoda. *Arch. Hydrobiol.,* 53:62–86. **651**

Fryer, G., 1959. Development in a mutelid lamellibranch. *Nature, Lond.,* 183:1342–1343. **57**

Fryer, G., 1961. The developmental history of *Mutela bourguignati* (Ancey) Bourguignat (Mollusca: Bivalvia). *Phil. Trans. R. Soc.,* 244B:259–298. **55, 700**

Fryer, G., and Smyley, W. J. P., 1954. Some remarks on the resting stages of some freshwater cyclopoid and harpacticoid copepods. *Ann. Mag. Nat. Hist.,* (Ser. 12) 7:65–72. **640, 641**

Fuhrmann, O., 1939. Sur *Craspedacusta sowerbyi* Lank. et un nouveau coelentéré d'eau douce *Calpasoma dactyloptera,* n.g.n. sp. *Revue Suisse Zool.,* 46: 363–368. **41, 43**

* Gajewskaja, N., 1933. Zur Oekologie, Morphologie und Systematik der Infusorien des Baikalsees. *Zoologica, Stuttg.,* 83, 298 pp. **497, 499, 501, 502, 503**

Gallagher, J. J., 1957. Cyclomorphosis in the rotifer Keratella cochlearis (Gosse). *Trans. Am. microsc. Soc.,* 76:197–203. **871, 909**

Gams, H., 1918. Prinzipienfragen der Vegetationsforschung. *Vjschr. Natur. Ges., Zürich,* 63:293–493. **234, 237, 238**

Ganapati, S. V., 1940. The ecology of a temple tank containing a permanent bloom of *Microcystis aeruginosa* (Kütz) Heufr. *J. Bombay. Nat. Hist. Soc.,* 42:65–77. **388**

Ganapati, S. V., 1960. *Proc. Symposium on Algology I.C.A.R.,* 204 (not seen: ref. in George 1962). **388**

Gans, R., 1911. Wie fallen Stäbe und Scheiben in einer reibenden Flüssigkeit. *Sitz. bayer. Akad. Wiss.,* Math.-phys. Kl., 1911: 191–203. **264**

Garbini, A., 1898. Due nuovi rhizopodi limnetici (*Difflugia cyclotellina-Heterophrys Pavesii*). *Zool. Anz.,* 21:667–670. **493, 494**

Gardner, A. C., 1941. Silicon and Phosphorus as factors limiting development of diatoms. *J. Soc. Chem. Ind., Lond., Trans.,* 60:73–78. **434, 450, 452, 455**

Garman, S., 1913. The Plagiostoma (Sharks, Skates, and Rays) *Mem. Mus. Comp. Zool.,* 36, 515 pp., 77 pl. (in separate volumes). **137**

Gaudin, F. A., 1960. Egg production of Streptocephalus seali Ryder, with notes on the distinctions between certain North American Streptocephalids. *SWest. Nat.,* 5e:61–65. **561**

Gause, G. F., 1934. *The Struggle for Existence.* Baltimore, Williams and Wilkins, 163 pp. **357**

Gause, G. F., 1935. Vérifications expérimentales de la théorie mathématique de la lutte pour la vie. *Actual. scient. ind.,* 277: 62 pp. **357**

Gause, G. F., Nastukova, O. K., and Alpatov, W. W., 1934. The influence of biologically conditioned media on the growth of a mixed population of *Paramecium caudatum* and *P. aurelia. J. Anim. Ecol.,* 3:222–230. **479**

Gause, G. F., and Witt, A. A., 1935. Behavior of mixed populations and the problem of natural selection. *Am. Nat.,* 69:596–609. **197, 231**

* Gauthier, H., 1928. *Recherches sur la faune des eaux continentales de l'Algérie et de la Tunisie.* Alger, Imp. Minerva, 419 pp., 3 pls., 6 maps. **606, 607, 681, 688**

Gauthier, H., 1931. Catalogue des Entomostracés récoltés par M. Seurat au Sahara central. *Bull Soc. Hist. nat. Afr. N.,* 22e:370–389. **688**

Gauthier, H., 1933a. Nouvelles recherches sur la faune des eaux continentales de l'Algérie et de la Tunisie. *Bull. Soc. Hist. nat. Afr. N.,* 24:63–68. **688**

Gauthier, H., 1933b. Faune aquatique du Sahara Central. Récoltes de M. Th. Monod dans l'Emmidir et l'Ahnet. *Bull. Soc. Hist. nat. Afr. N.,* 24:127–132. **688**

Gauthier-Lièvre, L., and Thomas, R., 1958. Les genres *Difflugia, Pentagonia, Maghrebia* et *Hoogenradia* (Rhizopodes testacés) en Afrique. *Arch. Protistenk.,* 103:241–370. **491, 493n.**

Gaw, H. Z., and Kung, L. H., 1939a. Freshwater medusae found in Kiating Szechuen, China. *Science, N.Y.,* 90:299. **42**

Gaw, H. Z., and Kung, L. H., 1939b. Studies on the freshwater medusae found in Kiating Szechuen. *Sci. rep. natn Wuhan Univ.,* Biol. Sci., 1:1–12. **42**

Geitler, L., 1932. Der Formwechsel der pennaten Diatomeen (Kieselalgen). *Arch. Protistenk.,* 78:1–226. **924**

Geitler, L., 1942. Zur Kenntnis der Bewohnes des Oberflächenhäutchens einheimischer Gewässer. *Biologia gen.,* 6:450–475. **238**

Geitler, L., and Ruttner, F., 1936. Die Cyanophyceen der Deutschen Limnologischen Sunda-Expedition, ihre Morphologie, Systematik und Ökologie. Dritter Teil. *Arch. Hydrobiol.,* Suppl 14:557–721. **22**

Gellis, S. S., and Clarke, G. L., 1935. Organic matter in dissolved and in colloidal form as food for *Daphnia magna*. *Physiol. Zoöl.*, 8:127–137. **553**

George, M. G., 1962. Occurrence of a permanent algal bloom in a fish tank at Dehli with special reference to factors responsible for its production. *Proc. Indian Acad. Sci.*, 56:354–362. **388**

Gerloff, G. C., Fitzgerald, G. P., and Skoog, F., 1950a. The isolation, purification, nad nutrient solution requirements of blue-green algae. *Symposium on the culturing of algae*, Charles F. Kettering Found. Dayton, Ohio, pp. 27–44. **317**

Gerloff, G. C., Fitzgerald, G. P., and Skoog, F., 1950b. The mineral nutrition of *Coccochloris peniocystis*. *Am. J. Bot.*, 37:835–840. **316, 317**

Gerloff, G. C., Fitzgerald, G. P., and Skoog, F., 1952. The mineral nutrition of *Microcystis aeruginosa*. *Am. J. Bot.*, 39:26–32. **316, 317**

Gerloff, G. C., and Skoog, F., 1957a. Availability of iron and manganese in southern Wisconsin lakes for the growth of *Microcystis aeruginosa*. *Ecology*, 38:551–556. **318**

Gerloff, G. C., and Skoog, F., 1957b. Nitrogen as a limiting factor for the growth of *Microcystis aeruginosa* in southern Wisconsin lakes. *Ecology*, 38:556–561. **318, 319**

Gessner, F., 1948. The vertical distribution of phytoplankton and the thermocline. *Ecology*, 29:386–389. **286, 288**

Giambiage De Calabrese, D., 1923. Una nueva especie de "Tanais." *Physis, B. Aires*, 6:248–253. **100**

Gibor, A., 1956. Some ecological relationships between phyto- and zooplankton. *Biol. Bull. mar. biol. Lab., Woods Hole*, 111:230–234. **554**

Gilbert, J. J., 1963. Mictic female production in the Rotifer *Brachionus calyciflorus*. *J. exp. Zool.*, 153:113–124. **508, 511**

Gillard, A., 1948. De *Brachionidae* (*Rotatoria*) van België ned Beschouwingen over de Taxonomie van de Familie. *Naturwet. tijdschr.*, 30:159–218. **869**

Glaessner, M. F., 1958. New fossils from the base of the Cambrian in South Australia. *Trans. R. Soc. S. Aust.*, 81:185–188. **15**

Glaessner, M. F., 1959. Precambrian Coelenterata from Australia, Africa and England. *Nature, Lond.*, 183:1472–1473. **15**

Goldman, C. R., 1960. Molybdenum as a factor limiting primary production in Castle Lake, California. *Science, N.Y.*, 132:1016–1017. **312**

Goldman, C. R., 1961. Primary productivity and limiting factors in Brooks Lake, Alaska. *Verh. int. Verein. theor. angew. Limnol.*, 14:120–124. **311**

Goldsmith, E., 1952. Fluctuation in chromosome number in *Artemia salina*. *J. Morph.*, 91:111–131. **560**

Goldstein, S., 1929. The steady flow of a viscous fluid past a fixed spherical obstacle at small Reynolds numbers. *Proc. Roy. Soc.*, 123A:225–235. **259**

Goodey, T., 1951. *Soil and freshwater nematodes*. A Monograph. London, Methuen; New York, John Wiley & Sons, xxvi + 390 pp. **49**

Goodge, W. R., 1957. Locomotion and other behavior of the Dipper. *Condor*, 61:4–17. **144**

Goodwin, T. W., 1964. The plastid pigments of flagellates. In *Biochemistry and Physiology of Protozoa*, 3:319–339. **7n.**

Gordon, I., 1957. On *Spelaeogriphus*, a new cavernicolous crustacean from South Africa. *Bull. Br. Mus. nat. Hist. Zool.*, 5:31–47. **95, 99**

Gordon, I., 1960. On a *Stygiomysis* from the West Indies, with a note on *Spelaeogriphus* (Crustacea, Peracarida). *Bull. Br. Mus. nat. Hist. Zool.,* 6:283–324. **99**

Gordon, M. S., 1962. Osmotic regulation in the green toad (*Bufo viridis*). *J. exp. Biol.,* 39:261–270. **140, 175**

Gordon, M. S., Schmidt-Nielsen, K., and Kelly, H. M., 1961. Osmotic regulation in the crab eating frog (*Rana cancrivora*). *J. exp. Biol.,* 38:659–678. **140**

Goryunova, S. V., 1955. Yavlenie khishchnichestva u sinezelenykh vodoroslei [Predatory behavior in blue-green algae]. *Mikrobiologiya,* 24:271–274. **316**

Gosse, P. H., 1886. In C. T. Hudson and P. H. Gosse, *The Rotifera or Wheel Animalcules, both British and foreign.* I: vi + 113 pp. II, 144 and Suppl. vi, 64. **552n**

Goulden, C. E., and Frey, D. G., 1963. The occurrence and significance of lateral head pores in the Genus *Bosmina* (Crustacea, Cladocera). *Int. Revue ges. Hydrobiol. Hydrogr.,* 48:513–522. **574**

Grafflin, A. L., 1937. The problem of adaptation to fresh and salt water in the teleosts viewed from the standpoint of the structure of the renal tubules. *J. cell. comp. Physiol.,* 9:469–476. **170**

Graham, M., 1929. *The Victoria Nyanza and its fisheries—a report on the fishing survey of Lake Victoria.* London, Crown Agents of the Colonies. 1927–28: 255 pp. **704**

Grassé, P., 1948. *Traité de Zoologie,* Vol. I–XVII. Pub. sous la direction de P. Grassé. Paris, Masson et Cie., **37n.**

Grassé, P., 1952. Généralités *in Traité de Zoologie,* ed. P. Grassé, Tome I, Fasc. 1. Paris, Masson et Cie., pp. 37–152, 154. **8, 10, 19**

Gravely, F. H., and Agharker, S. P., 1912. Notes on the habits and distribution of *Limnocnida indica* Annandale. *Rec. Indian Mus.,* 7, No. 3:399–403. **42**

Gravier, C., 1901. Sur trois nouveaux Polychètes d'eau douce de la Guyane francaise. *Bull. Soc. Hist. nat. Autun,* 14:353–371. **70**

Gravier, C., and Mathias, P., 1930. Sur la reproduction d'un crustacé phyllopode (*Cyzicus cycladiodes* (Joly)). Paris, *C. r. hebd. Séanc. Acad. Sci.,* 191:183–185. **565**

Green, J., 1954. Size and reproduction in *Daphnia magna* (Crustacea: Cladocera). *Proc. zool. Soc. Lond.,* 124:535–545. **577, 581**

Green, J., 1956. Growth, size and reproduction in *Daphnia* (Crustacea: Cladocera). *Proc. zool. Soc. Lond.,* 126:173–204. **575, 576, 577**

Green, J., 1960. Zooplankton of the River Sokoto. The Rotifera. *Proc. zool. Soc. Lond.,* 135:491–523. **548, 549, 877, 894, 895**

Green, J., 1963. Seasonal polymorphism in *Scapholeberis mucronata* (O. F. Müller) (Crustacea: Cladocera). *J. Anim. Ecol.,* 32:425–439. **848, 849, 850**

Greenwood, M., and Yule, G. U., 1920. An inquiry into the nature of frequency distributions representative of multiple happenings with particular reference to the occurrence of multiple attacks of disease or of repeated accidents. *Jl R. statist. Soc.,* 83:255–279. **790**

Greville, R. K., 1823–28. Scottish Cryptogamic Flora. (not seen: ref. 1884.) **425**

Griffiths, B. M., 1928. On desmid plankton. *New Phytol.,* 27:98–107. **330**

Griffiths, B. M., 1939.   Early references to waterbloom in British lakes.   *Proc. Linn. Soc. Lond.,* 151:12–19. **398**

* Grim, J., 1939.   Beobachtungen am Phytoplankton des Bodensees (Obersee) sowie deren rechnerische Auswertung. *Int. Revue ges. Hydrobiol. Hydrogr.,* 39:193–315. **247, 277, 295, 385, 401, 402, 410, 411, 442**

Grim, J., 1951.   Ein vergleich der Produktionsleistung des Bodensee-Untersees, des Obersees und des Schleinsees. *Abh. Fisch. Hilfswiss.,* Lief 4:787–841. **277, 285–288**

Grinnell, J., 1904.   The origin and distribution of the chestnut-backed Chickadee. *Auk,* 21:364–382. **357**

Grobben, C., 1880.   Die Antennendrüse der Crustaceen. *Arb. zool. Inst. Univ. Wien,* 3:93–110. **160**

Groom, T. T., and Loeb, J., 1890   Der Heliotropismus der Nauplein von *Balanus perforatus* und der periodischen Teifenwanderungen pelagischer Tiere. *Biol. Zbl.,* 10:160–177, 219–220. **758, 780**

Grospietsch, T., 1957.   Beitrag zur Rhizopodenfauna des Lago Maggiore. *Arch. Hydrobiol.,* 53:323–331. **493, 494**

Gross, F. and Raymont, J. E. G., 1942.   The specific gravity of *Calanus finmarchicus. Proc. R. Soc. Edinb.,* Sect. B., 61:288–296. **249**

Gross, F., and Zeuthen, E., 1948.   The buoyancy of plankton diatoms: a problem of cell physiology. *Proc. R. Soc.,* 135B., 382–389. **246, 277**

Grosser, B. I., Baylor, E. R., and Smith, F. E., 1953.   Analysis of geotactic responses in *Daphnia magna. Ecology,* 34:804–805. **764, 777**

Grosvenor, G. H., and Smith, G., 1913.   The life cycle of *Moina rectirostris. Q. Jl microsc., Sci.,* 58:511–522. **597**

Gruber, K., 1913.   Studien an *Scapholeberis mucronata* O.F.M.   I Beiträge zur Frage der Temporalvariation der Cladoceren und ihrer Beeinflussung durch das Experiment. *Z. indukt. Abstamm.-u. VererbLehre,* 9:301–342. **847, 848, 856**

Gruber, K., 1923.   Beobachtungen an Lokalrassen der Cladocera. *Int. Revue ges. Hydrobiol. Hydrogr.,* 41:345–408. **847**

Guillard, R. L., 1960.   A mutant of *Chlamydomonas moewusii* lacking contactile vacuoles. *J. Protozool.,* 7, 262–268. **27, 160**

Gunter, G., and Christmas, J. Y., 1959.   Corixid insects as part of the offshore fauna of the sea. *Ecology,* 40:724–725. **86**

Günther, R. T., 1893.   Preliminary account of the fresh-water medusa of Lake Tanganyika. *Ann. Mag. nat. Hist.,* (ser. 6) 11:269–275. **42**

Gurjanova, E., 1946.   Individual and growth-conditioned variation in Mesidotea entomon and its role in the evolution of the genus Mesidotea Rich. *Trudy zool. Inst., Leningr.,* 8:105–143 (Russian text); (Eng. summ. 143–144). **180, 181**

Gurney, R., 1909.   On the fresh-water Crustacea of Algeria and Tunisia. *Jl. R. micros. Soc.,* 1909:273–305. **607**

* Gurney, R., 1913.   The President's Address [Origin and conditions of existence of the fauna of freshwater]. *Trans. Norfolk Norwich Nat. Soc.,* 9:461–485. **189, 195**

Gurney, R., 1929.   Dimorphism and rate of growth in Copepoda. *Int. Revue ges. Hydrobiol. Hydrogr.,* 21:189–207. **860**

* Gurney, R., 1931.   *British Freshwater Copepoda.*   Calanoida, I: London, The Ray Society.   lii + 238 pp. **121, 122, 124n, 125, 625, 645, 646, 650, 653, 664, 668, 671, 681, 682, 693, 694, 860**

Gurney, R., 1932. *British Fresh-water Copepoda*. II. Harpacticoida, ix, 336 pp. London, The Ray Society. **129, 625**

*Gurney, R., 1933. *British Fresh-water Copepoda*. III. Cyclopoida, London, The Ray Society, xxix, 384 pp. **121, 625, 632, 633, 636, 639**

Gurney, A. B., and Parfin, S., 1959. Neuroptera in H. B. Ward, and G. C. Whipple, *Fresh-water Biology* 2nd ed., ed. by W. T. Edmondson. New York and London, John Wiley and Sons, pp. 973–980. **81**

Guseva, K. A., 1935. *Mikrobiologiya*, 4: (ref. Lastochkin 1945). **340**

Guseva, K. A., 1937. Deistvie mangantsa na razvitie vodoroslei (Influence of manganese on the development of algae). *Mikrobiologiya*, 6:292–307. **318**

Guseva, K. A., 1939. The blooming of the Uchinskii reservoir. *Byull. mosk. Obshch. Ispÿt. Prir. Biol.*, 48:30–32 (German abstract, 32). **318**

Haas, F., 1932, 1955. Mollusca Bivalva, Teil II. H. G. Bronns *Klassen und Ordnungen der Tierreichs*, 3 Abt. 3, teil ii: xii + 923 pp. **52, 53, 59**

Haberbosch, P., 1920. Die Süsswasser-Entomostracen Grönlands; eine faunistische, oecologische und tiergeographische Studie. *Z. Hydrol. Hydrogr. Hydrobiol.*, 1:136–184, 245–349. **693**

Häcker, V., 1902. Über die Fortpflanzung der limnetischen Copopoden des Titisees. *Ber. natur. Ges. Freiburg i.B.*, 12:1–33. **663**

Hadzi, J., 1953. A reconstruction of animal classification *Syst. Zool.*, 2:145–154. **14**

Haeckel, E., 1866. *Generelle Morphologie der Organismen*, xxxii, Berlin G. Reimer, 1: *Allgemeine Anatomie der Organismen*, xxxii, 574 pp. 2. Band: *Allgemeine Entwicklungsgeschichte der Organismen*, clx, 462 pp. **1**

Haeckel, E., 1891. Plankton-Studien. *Jenaische Zeitschr. f. Naturw.*, 25:232–336. **236**

Haeckel, E., 1894–6. *Systematische Phylogenie*, 3 vols. Berlin G. Reimer. 1. Theil: *Systematische Phylogenie der Protisten und Pflanzen*, xv, 400 pp., 1894. 2. Theil: *Systematische Phylogenie der Wirbellosenthiere*, xviii, 720 pp., 1896. 3. Theil: *Systematische Phylogenie der Wirbelthiere*, xx, 660 pp., 1895. **1**

Haempel, O., 1918. Zur Kenntnis einiger Alpenseen, mit besonderer Berück-sichtigung ihrer biologischen und Fischerei-Verhaltnisse. I. Der Hällstätter See. *Int. Revue ges. Hydrobiol. Hydrogr.*, 8:225–306. **654, 671, 672**

Haempel, O., 1924. Zur Kenntnis einiger Alpenseen. III. Der Millstättersee. *Arch. Hydrobiol*, 14:346–400. **813**

Hairston, N. G., 1959. Species abundance and community organization. *Ecology*, 40:404–416. **367, 368**

Haldane, J. B. S., 1924. A mathematical theory of natural and artificial selection. II. *Trans. Camb. phil. Soc.*, 23, p. 19–41. **357**

* Hall, D. J., 1962. An experimental approach to the dynamics of a natural population of *Daphnia galeata mendotae*. Ph. D. Thesis, Univ. Michigan 1962, *see also Ecology*, 45:94–112, 1964. **576, 577, 580–582, 592, 598, 611, 612–615, 617, 622, 805**

Hall, R. E., 1953. Observations on the hatching of eggs of *Chirocephalus diaphanus* Prévost. *Proc. zool. Soc. Lond.*, 123:95–109. **562**

Hamilton, J. D., 1958. On the biology of *Holopedium gibberum* Zaddach (Crustacea: Cladocera). *Verh. int. Verein. theor. angew. Limnol.*, 13:785–788. **248, 249, 606**

Hansgirg, A., 1886. Beiträge zur Kenntnis des Salzwasser-Algenflora Böhmens. *Öst. bot. Z.*, 36:331–336. **333**

Hanson, E. D., 1958. Origin of the Eumetazoa. *Syst. Zool.*, 7:16–47. **14**

Hantschmann, S. C., 1961. Active compensation for the pull of gravity by a planktonic cladoceran *Daphnia schødleri* Sars. Ph.D. Thesis, Yale University. **250, 271, 279, 281, 282, 756, 757, 764, 765, 777, 778, 781**

Harder, R., 1917. Ernährungsphysiologische Untersuchungen an Cyanophyceen, hauptsächlich dem endophytischen Nostoc punctiforme. *Z. Bot.*, 9:145–242. **463**

Hardin, G., 1960. The competitive exclusion principle. *Science, N.Y.*, 131:1292, 4297. **357**

Harding, J. P., 1942. Cladocera and Copepoda collected from East African lakes by Miss C. K. Ricardo, and Miss R. J. Owen. *Ann Mag. nat. Hist.*, (ser. 11) 9:174–191. **571, 636**

Hardisty, M. W., 1956. Same aspect of osmotic regulation in lampreys. *J. exp. Biol.* 33:431–447. **169**

Hardy, A. C., and Bainbridge, R., 1951. Effect of pressure on the behavior of Decapod larvae (Crustacea). *Nature, Lond.*, 167:354–355. **777**

Hardy, A. C., and Bainbridge, R., 1954. Experimental observations on the vertical migrations of plankton animals. *J. mar. biol. Ass. U.K.* 33:409–448. **755, 756, 776, 782**

Hardy, A. C., and Gunther, E. R., 1935. The plankton of the South Georgia whaling grounds and adjacent waters, 1926–1927. *'Discovery' Rep.*, 11:1–456. **788**

Harnisch, O., 1930. Daten zur Respirationsphysiologie Hämoglobinführender Chironomidenlarven. *Z. Vergl. Physiol.*, 11:285–309. **184**

Harnisch, O., 1937. Die Funktion der präanalen Oberflächenvergrösserungen (Tubuli) der Larve von Chironomus thummi bei sekundärer Oxybiose. *Z. Vergl. Physiol.*, 24:198–209. **184**

Harper, E. H., 1907. The Behaviour of the phantom larvae of *Corethra plumicornis*. *Journ. Comp. Neurol.*, 17:435–456. **761**

Harris, J. E., 1953. Physical factors involved in the vertical migration of plankton. *Q. Jl microsc. Sci.*, 94:537–550. **787**

Harris, J. E., and Mason P., 1956. Vertical migration in eyeless *Daphnia*. *Proc. R. Soc.*, 145B:280–290. **769, 770, 777, 780, 782**

Harris, J. E., and Wolfe, V. K., 1955. A laboratory study of vertical migration. *Proc. R. Soc.*, 144B:329–354. **767–769, 771, 781, 782**

Harrison, L., 1928. On the genus *Stratiodrilus* (*Archiannelida: Histriobdellidae*) with a description of a new species from Madagascar. *Rec. Aust. Mus.*, 16:116–121. **75, 76**

Hartman, Olga, 1938. Brackish and fresh-water Nereidae from the northeastern Pacific, with the description of a new species from Central California. *Univ. Cal. Publs Zool.*, 43:79–82. **67, 72**

Hartman, Olga, 1951. Fabricinae (Feather-duster Polychaetous Annelids) in the Pacific. *Pacif. Sci.*, 5:379–391. **69, 75**

Hartman, Olga, 1959a. Capitellidae and Nereidae (marine annelids) from the Gulf side of Florida, with a review of the freshwater Nereidae. *Bull. mar. Sci. Gulf. Caribb.*, 9:153–168. **70**

Hartman, Olga, 1959b.   Catalogue of the polychaetous annelids of the world. Part 1 *Occ. Pap. Allan Hancock Fdn*, 23, 353 pp. **70**

Hartman, Olga, 1959c.   Polychaeta.   In H. B. Ward, and G. C. Whipple *Fresh-water Biology*, 2nd ed., ed. W. T. Edmondson.   New York and London, John Wiley & Sons, pp. 538–541. **75**

Hartman, R. T., 1960.   Algae and metabolites of natural waters.   The Pymatuning symposia in ecology: The ecology of algae. *Spec. Publs Pymatuning Lab. Fld Biol.*, 2:38–55. **464, 465, 468, 469, 472**

Hartman, R. T., and Himes, C. L., 1961.   Phytoplankton from Pymatuning Reservoir in downstream areas of the Shenango River. *Ecology*, 42:180–183. **805**

Hartman, W. D., 1958.   Natural history of the marine sponges of Southern New England. *Bull. Peabody Mus. nat. Hist.*, 12, 155 pp. **38**

Hartmann, Otto, 1915.   Studien über die Cyclomorphose bei Cladoceren. *Arch. Hydrobiol.*, 10:436–519. **855, 856**

Hartmann, Otto, 1917.   Über die temporale Variation bei Copepoden (Cyclops, Diaptomus) und ihre Beziehung zu der bei Cladoceren. *Z. indukt. Abstamm.-u. VerebLehre*, 18:22–43. **647, 654, 856, 857, 862**

* Hartmann, Otto, 1920.   Studien über den Polymorphismus der Rotatorien mit besonderer Berücksichtigung von Anuraea aculeata. *Arch. Hydrobiol.*, 12:209–310. **883–886, 891, 893–895**

Harvey, E. N., 1952. *Bioluminescence*.   New York, Academic Press, xvi, 649 pp. **195**

Harvey, H. W., 1939.   Substances controlling the growth of a diatom. *J. mar. biol. Ass. U.K.*, 23:499–520. **473**

Hasler, A. D., 1947.   Eutrophication of lakes by domestic drainage. *Ecology*, 28:383–395. **453, 457**

Hasler, A. D., and Jones, E., 1949.   Demonstration of the antagonistic action of large aquatic plants on algae and rotifers. *Ecology*, 30:359–364. **804**

Hastings, A. B., 1929.   Phylactolaematous Polyzoa from the "Pans" of the Transvaal. *Ann. Mag. nat. Hist.*, (ser. 10) 13:129–137. **698, 699**

Hauer, J., 1937.   Zur Kenntnis der Rotatorienfauna des Eichener Sees. *Beitr. naturk. Forsch. Südwdtl*, 2:165–173. **507**

Hauer, J., 1938.   Die Rotatorien von Sumatra, Java und Bali. *Arch. Hydrobiol.*, Suppl. 15:296–384, 507–602. **519**

Heath, H., 1924.   The external development of certain Phyllopods. *J. Morph.*, 38:453–83. **566**

Heberdey, R. F., 1949.   Das Unterscheidungsvermägen von Daphnia für Helligkeiten farbiger Lichter. *Z. vergl. Physiol.*, 31:89–111. **762**

Heberdey, R. F., and Kupka, E., 1942.   Das Helligkeitsunterscheidungsvermögen von *Daphnia pulex. Z. vergl. Physiol.*, 29:541–582. **762**

Heberer, G., and Kiefer, F., 1932.   Zur Kenntnis der Copepodenfauna der Sunda-Inseln. *Arch. Naturgesch.* N.F., 1:225–274. **634, 636**

Hedgpeth, J. W., 1957.   Classification of Marine Environments.   In *Treatise on Marine Ecology & Paleoecology*, 1:17–27. **239**

Heinrich, A. K., 1962.   The life histories of plankton animals and seasonal cycles of plankton communities in the oceans. *Journ. du Conseil International pour l'explor. de la Mer*, 27:15–24. **708**

Hendey, N. J., 1937.   The Plankton Diatoms of the Southern Seas. *Discovery Rep.*, 16:153–364. **340**

Henry, M., 1919. On some Australian Freshwater Copepoda and Ostracoda. *J. Proc. R. Soc. N.S.W.*, 53:29–48. **650**

Henry, M., 1924. Notes on breeding Entomostraca from dried mud and their habits in aquaria. *Proc. Linn. Soc. N.S.W.*, 49:319–323. **639, 665**

Hensen, V., 1887. Ueber die Bestimmung des Planktons oder des in Meere treibenden Materials an Pflanzen und Thieren. *Ber. Kommn wiss. Unters. dt. Meere*, 5:1–109. **235**

Herbert, M. R., 1954. The tolerance of oxygen deficiency in the water by certain Cladocera. *Memorie 1st. ital. Iarobiol.*, 8:99–107. **607**

Hesse, R., 1924. *Tiergeographie auf ökologischer Grundlage.* Jena, Verl. v. Gustav Fischer, xii, 613 pp. **227**

Heuscher, H., 1917. Das Zooplankton des Zürichsees mit besonderer Berücksichtigung der Variabilität einiger Planktoncladoceren. *Arch. Hydrobiol,* 11:1–81, 153–240. **857**

Hind, W., 1896–1905. *A monograph of the British carboniferous Lamellibranchiata,* 2 vols. London, Palaeontographical Society. **55**

Hinton, M. A. C., 1936. Some interesting points in the anatomy of the fresh water dolphin *Lipotes* and its allies. *Proc. Linn. Soc. Lond.,* 148:183–185. **147**

Högbom, A. G., 1917. Ueber die arktischen Elemente in der aralokaspischen Fauna, ein tiergeographischer Problem. *Bull. geol. Instn Univ. Upsala,* 14:241–260. **211**

Hodgkinson, E. E., 1918. Some experiments on the rotifer *Hydatina. J. Genet.,* 7:187–192. **510**

Hof, T., and Frémy, P., 1933. On Myxophyceae living in strong brines. *Recl Trav. bot. néerl.,* 30:140–162. **22**

Hoff, C. C., 1943. The cladocera and ostracoda of Reelfoot Lake. *J. Tenn. Acad. Sci.,* 18:49–107. **684**

Hoff, C. C., 1944. The copepoda, amphipoda, isopoda, and decapoda (exclusive of the crayfishes) of Reelfoot Lake. *J. Tenn. Acad. Sci.,* 14:16–28. **684**

Hofsten, N. von, 1923. Rotatorien der nordschwedischen Hochgebirge. *Naturwiss. Unters. Sarekgebirges in Schwedisch-Lappland.* 4 (zool.), pt. 8, pp. 829–894. **519**

Höll, K., 1928. Oekologie der Peridineen. Studien über den Einfluss chemischer und physikalischer Faktoren auf die Verbreitung der Dinoflagellaten im Süsswasser. *Pflanzenforschung,* 11, vi, 105 pp. **343, 345–347, 387**

Holm-Hansen, O., Gerloff, G. C., and Skoog, F., 1954. Cobalt as an essential element for blue-green algae. *Physiologia Pl.,* 7:665–675. **317**

Holmquist, C., 1959. *Problems on Marine-glacial relicts on account of investigations on the genus Mysis.* Lund, Berlingska Boktryckeriet, 270 pp. **97, 206, 207, 212**

Holtedahl, O., 1924. Studier over israndterrassene syd for de store østlandske sjøer. *Skr. norske Vidensk.-Akad.,* I. Mat-naturvk. Kl., 1924, No. 14, 110 pp. **211**

Holthuis, L. B., 1952. A general revision of the Palaemonidae (Crustacea Decapoda Natantia) of the Americas. II. The Subfamily Palaemoninae. *Occ. Pap., Allan Hancock Fdn Publications,* 12, 396 pp. **114**

Holthuis, L. B., 1956. An enumeration of the Crustacea Decapoda inhabiting subterranean waters. *Vie Milieu* (= *Actual. scient. ind.* 1249), 7:43–76. **114, 193**

Home, Sir E., 1826.   On the production and formation of pearls.   *Phil. Trans. R. Soc.*, 116 pt. 3:338–341. **55**

Honigberg, B. M., and Balamuth, W., 1963.   Subphylum Sarcomastigophora nom. nov. to embrace the flagellate and amoeboid assemblages of protozoans (Abstr.).   *J. Protozool.*, 10 (Suppl.):27. **8**

Honigberg, B. M., Balamuth, W., Bovee, E. C., Corliss, J. O., Gojdics, M., Hall, R. P., Kudo, R. R., Levine, N. D., Loeblich, A. R., Jr., Weiser, J., Wenrich, D. H., 1964.   A revised classification of the Phylum Protozoa.   *J. Protozool.*, 11:7–20. **8**

Hooker, W. J., 1861.   *The British ferns; or, coloured figures and descriptions, with the needful analyses of the fructifications and venation of the ferns of Great Britain and Ireland, systematically arranged.*   London, L. Reeve, 130 pp, 60 col. pl. **31**

Hrbáček, J., 1958.   Typologie und Produktivität der teichartigen Gewässer. *Verh. int. Verein. theor. angew. Limnol.*, 13:394–399. **708**

Hrbáček, J., 1959a.   Circulation of water as a main factor influencing the development of helmets in *Daphnia cucullata* Sars.   *Hydrobiologia*, 13:170–185. **835**

Harbáček, J., 1959b.   Über die angebliche Variabilität von *Daphnia pulex* L. *Zool. Anz.*, 162:116–126. **573, 617n.**

Hrbáček, J., 1960.   Density of the fish population as a factor influencing the distribution and speciation of the species in the genus *Daphnia* xv[th] *Intern. Congr. Zool., London* (1958), Sect. x, No. 27. **708**

Hrbáček, J., Dvořaková, M. Kořínek, V., and Procházková, L., 1961. Demonstration of the effect of the fish stock on the species composition of zooplankton and the intensity of metabolism of the whole plankton association. *Verh. int. Verein. theor. angew. Limnol.*, 14:192–195. **708**

Hrbáček, T., and Hrbáčková-Esslová, M., 1960.   Fish stock as a protective agent in the occurrence of slow developing dwarf species and strains of the genus *Daphnia. Int. Revue ges. Hydrobiol. Hydrogr.*, 45:355–358. **708**

Hrbáčková-Esslová, M., 1963.   The development of three species of *Daphnia* in the surface water of the Slapy Reservoir.   *Int. Revue ges. Hydrobiol. Hydrogr.*, 48:325–333. **605**

Hsi-Ming, K., 1942.   Über die Cyclomorphose der Daphnien einiger Voralpen-seen.   *Int. Revue ges. Hydrobiol. Hydrogr.*, 41:345–408. **818**

Hubault, E., 1938.   *Sphaeromicola sphaeromidicola*, nov. sp, commensal de *Sphaeromides virei* Valle, en Istrie et considérations sur l'origine de diverses espèces cavernicoles périméditerraniennes.   *Archs Zool. expt. gén.*, 80:11–24. **104, 193**

Huber, G., 1906.   (See also Huber-Pestalozzi, G.).   Monographische Studien im Gebiete der Montigglerseen (Südtirol) mit besonderer Berücksichtigung ihrer Biologie.   *Arch. Hydrobiol.*, 7:1–81, 123–210. **851, 856, 863, 877**

Huber, G., and Nipkow, F., 1922.   Experimentelle Untersuchungen über die Entwicklung von Ceratium hirundinella O.F.M. *Z. Bot.*, 14:337–371. **910**

Huber-Pestalozzi, G., 1927.   Gedanken über Ceratium hirundinella.   *Arch. Hydrobiol.*, 18:117–128. **917**

Huber-Pestalozzi, G., 1938.   Das Phytoplankton des Süsswassers. Systematik und Biologie. Allgemeiner Teil. Blaualgen, Bakterien, Pilze.   *Binnengewässer*, 16, Teil 1, 342 pp. **272, 306**

Huber-Pestalozzi, G., 1941. Das Phytoplankton des Süsswassers: Systematik und Biologie. Chrysophyceen, Farblose Flagellaten, Heterokonten. *Binnengewässer*, 16, Teil 2., 1. Hälfte, 365 pp. **306, 339**

Huber-Pestalozzi, G., 1950. Das Phytoplankton des Süsswässers: Systematik und Biologie. Cryptophyceen, Chloromonadinen, Peridineen. *Binnengewässer*, 16, Teil 3., ix, 305 pp. **306**

Huber-Pestalozzi, G., 1955. Das Phytoplankton des Süsswassers: Systematik und Biologie. Euglenophyceen. *Binnengewässer*, 16, Teil 4, ix, 606 pp. **306, 334**

Huber-Pestalozzi, G., 1961. Das Phytoplankton des Süsswassers: Systematik und Biologie. Chlorophyceae (Grünalgen) Ordnung: Volvocales. *Binnengewässer*, 16, Teil 5., xii, 744 pp. **306, 321**

Huber-Pestalozzi, G., and Hustedt, F., 1942. Das Phytoplankton des Süsswassers: Systematik und Biologie. Diatomeen. *Binnengewäasser*, 16, Teil 2. 2 Hälfte x, 183 pp. **296, 298, 341, 342, 343**

Huber, G., and Nipkow, F., 1923. Experimentelle Untersuchungen über Entwicklung und Formbildung von Ceratium hirundinella O. F. Müller. *Flora Jena* (n.s.), 16 (whole 116):114–215. **917**

Huber, G., and Nipkow, F., 1927. Beobachtungen am Plankton des Zürichsees. *Vjschr. naturf. Ges. Zürich* 72:312–325. **497**

Hughes, T. McK., 1880. On the transport of fine mud and vegetable matter by conferva. *Proc. Camb. phil. Soc.*, 3:339–341. **425**

Hungerford, H. B., 1922. Oxyhaemoglobin present in backswimmer *Buenoa margaritacea* Bueno (Hemiptera). *Can. Ent.*, 1922:262–263. **256**

Hungerford, H. B., (with material by R. I. Sailer), 1948. The Corixidae of the western hemisphere. *Kans. Univ. Sci. Bull.*, 32:827 pp. **185**

Hustedt, F., 1929. Untersuchungen über den Bau der Diatomeen IX. Zur Morphologie und Zellteilungsfolge von *Eunotia didyma* Grun. *Ber. dt. bot. Ges.*, 47:59–169. **923**

Hustedt, F., 1945. Die Diatomeenflora norddeutscher Seen mit besonderer Berücksichtigung des holsteinischen Seengebiets. *Arch. Hydrobiol.*, 41:392–414. **387**

Hutchinson, G. E., 1930. Restudy of some Burgess shale fossils. *Proc. U.S. natn. Mus.*, 78, Art. 11, 24 pp. **119**

Hutchinson, G. E., 1931. On the occurrence of *Trichocorixa* Kirkaldy (Corixidae, Hemiptera-Heteroptera) in salt water and its zoo-geographical significance. *Am. Nat.*, 65:573–574. **86**

Hutchinson, G. E., 1933. Experimental studies in ecology. I. Magnesium tolerance of *Daphniidae* and its ecological significance. *Int. Revue ges. Hydrobiol. Hydrogr.*, 28:90–108. **606, 823**

Hutchinson, G. E., 1937a. Limnological studies in Indian Tibet. *Int. Revue ges. Hydrobiol. Hydrogr.*, 35:134–176. **386, 629, 696, 804**

Hutchinson, G. E., 1937b. A contribution to the limnology of arid regions primarily founded on observations made in the Lahontan Basin. *Trans. Conn. Acad. Arts. Sci.*, 33:47–132. **239, 680, 682**

Hutchinson, G. E., 1941. Limnological studies in Connecticut. IV. Mechanism of intermediary metabolism in stratified lakes. *Ecol. Monogr.*, 11:21–60. **311, 318**

Hutchinson, G. E., 1944. Limnological studies in Connecticut. VII. A critical examination of the supposed relationship between phytoplankton periodicity and chemical changes in lake waters. *Ecology,* 25:3–26. **402, 406, 424, 425, 455, 456, 482, 483**

Hutchinson, G. E., 1950. The biogeochemistry of vertebrate excretion. *Bull. Amer. Mus. nat. Hist.,* 96, xviii, 554 pp. **173**

Hutchinson, G. E., 1951. Copepodology for the ornithologist. *Ecology,* 32:571–577. **125, 636**

Hutchinson, G. E., 1957. Concluding remarks. *Cold Spr. Harb. Symp. Quant. Biol.,* 22:415–427. **229, 232**

Hutchinson, G. E., 1959a. Homage to Santa Rosalia *or* Why are there so many kinds of animals. *Am. Nat.,* 93:145–159. **682**

Hutchinson, G. E., 1959b. Il concetto moderno di nicchia ecologica. *Memorie Ist. ital. Idrobiol.,* 11:9–22. **229, 232**

Hutchinson, G. E., 1960. On evolutionary euryhalinity. *Am. J. Sci.,* 258-A Bradley vol.):98–103. **18, 182**

Hutchinson, G. E., 1961. The paradox of the plankton. *Am. Nat.,* 95:137–146. **349, 367, 374, 375**

Hutchinson, G. E., 1964a. On *Filinia terminalis* (Plate) and *F. pejleri* sp.n. (Rotatoria:family *Testulinellidae*). *Postilla,* 81, 8 pp. **507, 523, 524**

Hutchinson, G. E., 1964b. The lacustrine microcosm reconsidered. *Am. Scient.,* 52:334–341. **189, 376, 587**

Hutchinson, G. E., 1965. *Eretmia* Gosse 1886 (?Rotatoria) proposed suppression of this generic name under the plenary powers. *Bull. zool. Nomencl.,* 22: 60–62. **552n.**

Hutchinson, G. E., and Pickford, G. E., 1932. Limnological observations on Mountain Lake, Virginia. *Int. Revue ges. Hydrobiol. Hydrogr.,* 27:252–264. **330**

Hutchinson, G. E., Pickford, G. E., and Schuurman, J. F. M., 1932. A contribution to the hydrobiology of pans and other inland waters of South-Africa. *Arch. Hydrobiol.,* 24:1–154. **333n, 386, 388, 559, 680, 689, 823**

Hutchinson, G. E., and Setlow, J. K., 1946. Limnological studies in Connecticut. VIII. The niacin cycle in a small inland lake. *Ecology,* 27:13–23. **388, 473**

Hutchinson, J., 1959. *The families of flowering plants.* Vol. I. Dicotyledons, xi, 1–510. Vol. II. Monocotyledons, viii, 511–792. Oxford at the Clarendon Press. **32n.**

Hutner, S. H., Baker, H., Aaronson, S., Nathan, H. A., Rodriguez, E., Lockwood, S., Sanders, M., and Petersen R. A., 1957. Growing *Ochromonas malhamensis* above 35°C. *J. Protozool.,* 4:259–269. **348**

Hutner, S. H., and Provasoli, L., 1951. The phytoflagellates. In *Biochemistry and Physiology of Protozoa,* I: ed. A. Lwoff. New York, Academic Press, 27–128. **473**

Hutner, S. H., and Provasoli, L., 1955. Comparative biochemistry of flagellates. In S. H. Hutner, and A. Lwoff, *Biochemistry and Physiology of Protozoa,* New York, Academic Press, II:17–43. **348**

Hutner, S. H., Provasoli, L., and Filfus, J., 1953. Nutrition of some phagotrophic freshwater chrysomonads. *Ann. N.Y. Acad. Sci.,* 56:852–862. **348**

Huxley, J. S., 1932. *Problems of Relative Growth.* New York, L. Mac Veagh, 276 pp. **811**

Huxley, J. S., 1958. Evolutionary processes and taxonomy with special reference to grades. *Uppsala Univ. Årsskr.,* 1958:6, 21–39. **2**

Hyman, L. H., 1938. North American Rhabdocoela and Alloeocoela II, Rediscovery of *Hydrolimax grisea* Haldeman. *Am. Mus. Novit.*, No. 1004:19 pp. **45**

Hyman, L. H., 1940. *The Invertebrates: Protozoa through Ctenophora,* 1st ed. McGraw-Hill Book Co., I:xii + 726 pp. **14, 37n.**

Hyman, L. H., 1951a. *The Invertebrates: Platyhelminthes & Rhynchocoela. The acoelomate Bilateria,* McGraw-Hill Book Co., II:vii + 550 pp. **14, 37n., 43**

Hyman, L. H., 1951b. *The Invertebrates: Acanthocephala, Aschelminthes, and Entoprocta,* McGraw-Hill Book Co., vii + 572 pp. **14, 37n., 47, 49n.**

Hyman, L. H., 1955. *The Invertebrates. Echinodermata,* McGraw-Hill Book Co., IV:vi + 763 pp. **14, 37n.**

Hyman, L. H., 1959. *The Invertebrates: Smaller Coelomate Groups.* New York. McGraw-Hill Book Co. V:viii + 783 pp. **14, 37n., 47**

Hynes, H. B. N., 1954. The ecology of *Gammarus duebeni* Lilljeborg and its occurrence in freshwater in Western Britain. *J. Anim. Ecol.,* 23:38–84. **161, 200**

Ingle, L., Wood, T. R., and Banta, A. M., 1937. A study of longevity, growth, reproduction, and heart rate in Daphnia longispina as influenced by limitations in quantity of food. *Jour. exp. Zool.,* 76:325–352. **579, 580**

Irving, L., Fisher, K. C., and McIntosh, F. C., 1935. The water balance of a marine mammal, the seal. *J. cell. comp. Physiol.,* 6:387–391. **175**

Issel, R., 1901. Saggio sulla fauna termale italiana. Nota I. *Atti Accad. Sci. Torino,* 36:53–74; Nota II, 36:265–277. **191**

Itô, T., 1951. A new athecate hydroid, *Cordylophora japonica* n. sp. from Japan. *Mem. Ehime Univ.,* Sect 2, 212:81–86. **39**

Ivanov, A. V., 1955. Pogonophora. Translation by A. Petrunkevitch. *Syst. Zool.,* 4:170–178. **17n.**

Jaag, O., 1938. Die Kryptogamenflora des Rheinfalls und des Hochrheins von Stein bis Eglisau. *Mitt. naturf. Ges. Schaffhausen,* 14:1–158. **435**

* Jacobs, J., 1961. Cyclomorphosis in Daphnia galeata mendotae Birge, a case of environmentally controlled allometry. *Arch. Hydrobiol.,* 58:7–71. **826, 827, 831, 833, 846**

Jacobs, J., 1962. Light and turbulence as co-determinants of relative growth in cyclomorphic *Daphnia. Int. Revue ges. Hydrobiol. Hydrogr.,* 47:146–156. **835**

Jacobs, J., 1964. Hat der hohe Sommerhelm zyklomorpher Daphnien einen Anpassungswert. *Verh. Int. Verein. theor. angew. Limnol.,* 15:676–683. **845**

Jacobs, W., 1943. Das Problem des specifischen Gewichtes bei Wassertieren. *Arch. Hydrobiol.,* 39:432–457. **249**

Jander, R., and Waterman, T. H., 1960. Sensory discrimination between polarised light and light intensity patterns by arthropods. *J. cell. comp. Physiol.,* 56:137–160. **778**

Järnefelt, H., 1952. Plankton als Indikator der Trophiegruppen der Seen. *Suomal. Tiedeakat. Toim. (Annls Acad. Sci. Fenn.),* ser A. IV Biol., 18., 29 pp. **332, 380, 393, 394**

Järnefelt, H., 1953. Einige Randbemerkungen zur Seetypennomenklator *Schwiez. Z. Hydrol.,* 15:198–212. **380**

* Järnefelt, H., 1956. Zur Limnologie einiger Gewässer Finnlands. XVI. Mit besonderer Berücksichtigung des Planktons. *Suomal. eläin-ja kasirt. Sevr.*

van. eläin. Julk. (*Annls Zool. Soc. Vanamo*), 17, No. 7:201 pp. **375, 380, 382, 383, 385, 387, 393**

Järnefelt, H., 1958. On the typology of the northern lakes. *Verh. int. Verein. Theor. angew Limnol.,* 13:228–235. **380**

Jaschnov, W. A., 1922. Das Plankton der Baikalsees nach dem material der Expedition der Zoologischen Museums der Moskauer Universität im Jahre 1917. *Russk. gidrobiol. Zh.* 7:225–238 (Germ. summ. 238–241). **519**

Jaschnova, M. K., 1929. Ueber das Vorkommen von Limnocalanus macrurus im Delta der Drina *Russk. gidrobiol. Zh.,* 8:304–306 (German summ. 307–308). **127**

Jenkin, P. M., 1957. The filter feeding and food of Flamingoes (Phoenicopteri). *Phil. Trans. R. Soc.* 240B:401–493. **148**

Jennings, H. S., 1900–1903. Rotatoria of the United States. *Bull. U.S. Fish. Commn,* 19:67–104, 1902: 272–352. **519**

Jepps, M. W., 1947. Contribution to the study of the sponges. *Proc. R. Soc.,* 134B:408–417. **38**

Johnson, H. P., 1903. Fresh-water Nereids from the Pacific coast and Hawaii, with remarks on fresh-water Polychaeta in general. *Mark Anniversary Volume.* New York, Henry Holt & Co., pp. 205–223. **69, 70**

Johnson, M. W., 1939. *Pseudodiaptomus (Pseudodiaptallous) euryhalinus.* A new subgenus and species of copepoda with preliminary notes on its ecology. *Trans. Am. microsc. Soc.,* 58:349–355. **126**

Jolly, V. H., 1952. A preliminary study of the limnology of Lake Hayes. *Aust. J. mar. Freshwat. Res.,* 3:374–91. **741**

Jolly, V. H., 1955. A review of the genera Calamoecia and Brunella (freshwater copepoda). *Hydrobiologia,* 7:279–284. **129, 650**

Jones, L. L., 1941. Osmotic regulation in several crabs of the Pacific coast of North America. *J. cell. comp. Physiol.,* 18:79–92. **172**

Jørgensen, E. G., 1956. Growth inhibiting substances formed by algae. *Physiol. Pl.,* 9:712–726. **464, 466, 472**

Jørgensen, E. G., 1957. Diatom periodicity and silicon assimilation. *Dansk bot. Ark.,* 18(1) 54 pp. **432, 446, 454, 471, 472**

Juday, C., 1904. The diurnal movement of plankton crustacea. *Trans. Wis. Acad. Sci., Arts, Lett.,* 14:534–68. **726, 782, 784, 835, 836**

Juday, C., 1919. A freshwater anaërobic ciliate. *Biol. Bull. mar. biol. Lab., Woods Hole,* 36:92–95. **505**

Juday, C., and Birge, E. A., 1927. *Pontoporeia* and *Mysis* in Wisconsin lakes. *Ecology,* 8:445–452. **732**

Juday, C., Birge, E. A., and Meloche, V. W., 1939. Mineral content of the lake waters of northeastern Wisconsin. *Trans. Wis. Acad Sci., Arts, Lett.,* 31:223–276. **331**

Jurine, L., 1820. *Histoire des monocles, qui se trouvent aux environs de Genève.* Genève and Paris, 260 pp. (ref. fr. Dussart 1958). **627**

Kahl, A., 1930. *Urtiere oder Protozoa.* I. *Wimpertiere oder Ciliata (Infusora).* 1—Allgemeiner Teil und *Prostomata,* pp. 1–180. **501**

Kahl, A., 1931. 2—*Holotricha,* pp. 181–398. **498**

Kahl, A., 1932. 3—*Spirotricha,* pp. 399–650. **498**

Kahl, A., 1935. 4—*Peritricha und Chonotricha,* pp. 651–886. **497, 498**

Karaman, S., 1933a. Microcerberus stygius, der dritte Isopod aus dem Grundwasser von Skoplje, Jugoslavien. *Zool. Anz.*, 102:165–169. **102**

Karaman, S., 1933b. Ueber zwei neue Amphipoden Balcanella und Jugocrangonyx aus dem Grundwasser von Skoplje *Zool. Anz.*, 103:41–47. **102**

Karaman, S., 1940. Die unterirdischen Isopoden Südserbeins. *Glasnik Bull. Soc. Sc.*, Skoplje 22 (not seen: ref. Delamare Debouteville 1960). **102**

Karaman, S., 1953. Über die Jaera-Arten Jugoslaviens. *Acta adriat.*, 5, No 5, 20 pp. **101**

Karling, T. G., 1940. Zur Morphologie und Systematik der Alloeocoela Cumulata und Rhabdocoela Lecithophora (Turbellaria). *Acta zool. fenn.*, 26:260 pp. **43**

Kelecnikov, V. P., 1950. Akchagilskie i Apsheronskie Molliuski. *Paleontologiya SSSR* 10, Pt. III, Sect. 12:259 pp. **59**

Kellogg, R., 1928. The history of Whales—their adaptation to life in the water. *Q. Rev. Biol.*, 3:29–76. **146**

Kemp, S., 1917. Notes on the fauna of the Matlah River in the Gangetic Delta. *Rec. Indian Mus.*, 13:233–241. **195**

von Kennel, J., 1891. On a Freshwater Medusa. *Ann. Mag. nat. Hist.* (ser. 6) 8:259–263. **39**

Kepner, W. A., and Thomas, W. L., 1928. Histological features correlated with gas secretion in Hydra oligactis Pallas. *Biol. Bull. mar. biol. Lab.*, Woods Hole, 54:529–533. **256**

Kerb, H., 1911. Über den Nährwert der im Wassergelösten Stoffe. *Int. Revue ges. Hydrobiol. Hydrogr.*, 3:496–505. **553**

de Kerhervé, J. B., 1927. La descendance d'une Daphnie (D. magna) ou ses millions de germes en une saison. *Annls Biol. lacustre*, 15:61–73. **581**

Keys, A., and Willmer, E. N., 1932. "Chloride secreting cells" in the gills of fishes, with special reference to the common eel. *J. Physiol., Lond.*, 76:368–378. **17**

Kiefer, F., 1929. *Das Tierreich; Crustacea Copepoda.* II. *Cyclopoida Gnathostoma.* Berlin & Leipzig, Walter de Gruyter & Co. Lieferung, 53, xvi, 102 pp. **625**

Kiefer, F., 1933. Die freilebenden Copepoden der Binnengewässer von Insulinde. *Arch. Hydrobiol.*, Suppl. 12:519–621. **634–636**

Kiefer, F., 1934. Die freilebenden Copepoden Südafrikas. *Zool. Jb. Abt Syst.*, 65:99–192. **681, 689**

Kiefer, F., 1937. Eine kleine Copepodenausbeute aus der östlichen Mongolei. *Zool. Anz.*, 119:293–298. **128**

Kiefer, F., 1938a. Die von Wallacea-Expedition gesammelten Arten der Gattung *Thermocyclops* Kiefer. *Int. Revue ges. Hydrobiol. Hydrogr.*, 38:54–74. **626, 636**

Kiefer, F., 1938b. Bemerkungen zur Pseudodiaptomidenausbeute der Wallacea-Expedition. *Int. Revue ges. Hydrobiol. Hydrogr.*, 38:75–98. **126**

Kiefer, F., 1939a. Zur Kenntnis des Cyclops "strenuus" aus dem Bodensee. *Arch. Hydrobiol.*, 36:94–117. **632, 633**

Kiefer, F., 1939b. Scientific results of the Yale North India Expedition. Biol. Rep. No. 19, Crustacea, Copepoda. *Mem. Indian Mus.*, 13, Pt. II:83–203. **647, 691**

Kiefer, F., 1954. Zur Kenntnis der freilebenden Ruderfusskrebse der Bodensees. *Beitr. naturk. Forsch. Südw Dtl.* 13:86–92. **631, 633**

Kiefer, F., 1957. Freilebende Ruderfusskrebse (Crustacea Copepoda) aus einigen ostanatolischen Seen. *Zool. Anz.,* 159:25–33. **629**

* Kikuchi, K., 1930a. A comparison of the diurnal migration of plankton in eight Japanese lakes. *Mem. Coll. Sci. Kyoto Univ.,* ser. B., 5:27–46. **494, 495, 506, 726, 731, 732, 735, 737, 782, 783**

Kikuchi, K., 1930b. Diurnal migration of plankton crustacea. *Q. Rev. Biol.,* 5:189–206. **726, 733, 735, 782**

Kikuchi, K., 1937. Studies on the vertical distribution of the plankton crustacea I. A comparison of the vertical distribution of the plankton crustacea in six lakes of middle Japan in relation to the underwater illumination and the water temperature. *Rec. oceanogr. Wrks, Japan,* 9:63–85, **740n., 742**

Kikuchi, K., 1938. Studies on the vertical distribution of the plankton Crustacea II. The reversal of phototropic and geotropic sign in the reference to vertical movement. *Rec. oceanogr. Wrks, Japan,* 10:17–42. **740n., 755, 774, 775, 781, 783, 784**

King, C. E., 1964. Relative abundance of species and MacArthur's model. *Ecology,* 45:716–727. **367**

Kinne, O., 1954. Die Gammarus-Arten der Kieler Bucht. *Zool. Jbr. Ab. Syst.,* 82:405–424. **199**

Kinne, O., and H-W., Ratthauwe. 1952. Biologische Beobachtungen und Untersuchungen über die Blutkonzentration an *Heteropanope tridentatus* Maitland (Dekapoda). *Kieler Meeresforsch.,* 8:212–217. **166**

Kitching, J. A., 1938. Contractile Vacuoles. *Biol. Rev.,* 13:403–444. **28, 160**

Kitching, J. A., 1952. Contractile vacuoles. *Symp. Soc. exp. Biol.,* 6:145–165. **27**

Kjellesvig-Waering, E. N., 1958. The genera, species and subspecies of the family Eurypteridae, Burmeister, 1845. *J. Paleont.,* 32:1107–1148. **87**

Klausener, C., 1908. Die Blutseen der Hochalpen. *Int. Revue ges. Hydrobiol. Hydrogr.,* 1:359–424. **389, 884, 892, 894**

Klebahn, H., 1895. Gasvacuolen, ein Bestandtheil der Zellen der wasserblüthe-bildenden Phycochromacean. *Flora, Jena,* 80:241–282. **247, 252**

Klebahn, H., 1922. Neue Untersuchungen über die Gasvacuolen. *Jb. wiss. Bot.,* 61:535–589. **252**

von Klein, H., 1938. Limnologische Untersuchungen über das Crustaceen-plankton der Schleinsees und zweier Kleingewässer. *Int. Revue ges. Hydrobiol. Hydrogr.,* 37:176–233. **653, 659, 672**

Kleinenberg, S. E., 1958. The origin of the Cetacea. *Doklady Akad. Nauk SSSR,* 122:950 [trans. *Dokl. (Proc.) Acad. Sci. USSR* Biol. Sci. sect., 122:752–754]. **146**

Klie, W., 1933. Die Ostracoden der Deutschen Limnologischen Sunda-Expedition. *Arch. Hydrob.,* Suppl. 11:447–502. **695**

Klotter, H.-E., 1955. Die Algen in den Seen des südlichen Schwartzwaldes. II. Eine ökologisch-floristische Studie. *Arch. Hydrobiol.,* Suppl. 22:106–252. **347**

Klomp, H., 1961. The concepts "similar ecology" and "competition" in animal ecology. *Arch. néerl Zool.,* 14:90–102. **373**

Klugh, A. B., 1923. A common system of classification in plant and animal ecology. *Ecology,* 4:366–377. **226, 227**

Knaysi, G., 1951. The structure of the bacterial cell Ch. II (pp. 28–66) of C. H. Werkman, and P. W. Wilson *Bacterial Physiology.* New York, Academic Press, xiv, 707 pp. **21**

Knight-Jones, E. W., and Qasim, S. Z., 1955. Responses of some marine plankton animals to changes in hydrostatic pressure. *Nature, Lond.,* 175:941–942. **777**

Knipowitsch, N., 1922. Hydrobiologische Untersuchungen in Kaspischen Meere in den Jahren 1914–15 *Int. Revue ges. Hydrobiol. Hydrogr.,* 10:394–440, 561–602. **213n.**

Knudson, B., 1952. In report of the Director. *Rept. Freshwat. biol. Ass.,* 20:17. **285**

Koch, H., 1934. Essai d'interpretation de la soi-disant "reduction vitale" de sels d'argent par certains organes d'Arthropodes. *Ann. Soc. scient. Brux.,* (sèr. B) 54:346–61. **163**

Koch, H., 1938. The absorbtion of chloride ions by the anal papillae of diptera larvae. *J. exp. Biol.,* 16:152–60. **164**

Koch, H. J., and Evans, J., 1956. On the absorption of sodium from dilute solutions by the crab *Eriocheir sinensis* (M. Edw.). *Mededel. K. vlaam. Acad.,* Kl. Wet., 18, No. 7:15 pp. **162**

Koch, H. and Krogh, A., 1936. La fonction des papilles anales des larves de Dipteres. *Ann. Soc. scient., Brux.* (sèr. B) 56:459–61. **164**

Koenicke, F., 1896. Holsteinische Hydrachniden. *Forschber. biol. Stat. Plön,* 4:207–247. **697**

Kofoid, C. A., 1896. A report upon the protozoa observed in Lake Michigan and the Inland Lakes in the neighbourhood of Charlevoix, during the summer of 1894. *Bienn. Rep. Mich. St. Bd. Fish Commnrs.,* Fish. Comm. Bull. No. 6:76–84. **494, 496**

Kofoid, C. A., 1903. Plankton studies. IV. The plankton of the Illinois River, 1894–1899, with introductory notes upon the hydrography of the Illinois River and its basin. Part I. Quantitative investigations and general results. *Bull. Ill. State. Lab. nat. Hist.,* 4:95–629. **530**

* Kofoid, C. A., 1908. The plankton of the Illinois River, 1894–1899. Part II. Constituent organisms and their seasonal distribution. *Bull. Ill. State Lab. nat. Hist.,* 8:3–361. **434, 530, 609, 875, 896, 902**

Kohn, A. J., 1959. The ecology of *Conus* in Hawaii. *Ecol. Monogr.,* 29:47–90. **367**

Kohn, A. J., 1960. Ecological notes on *Conus* (Mollusca:Gastropoda) in the Trincomalee region of Ceylon. *Ann. Mag. nat. Hist.,* (ser. 13) 2:309—320. **367**

* Kolisko, A., 1938 [see also Ruttner-Kolisko, A.]. Beiträge zur Erforschung der Lebensgeschichte der Rädertiere auf Grund von Individualzuchten. *Arch. Hydrobiol.,* 33:165–207. **512, 887, 888, 890**

Kolkwitz, R., 1912. Plankton und Seston. *Ber. dt. bot. Ges.,* 30:334–346. **235**

Komai, T., 1963. A note on the phylogeny of the Ctenophora pp. 181–188. In *The Lower Metazoa. Comparative Biology and Phylogeny,* ed. E. C. Dougherty et al Univ. California Press. **15n.**

Komarovsky, B., 1959. The plankton of Lake Tiberias. *Bull. Res. Coun. Israel,* B8:65–96. **333**

Korinek, J., 1926. Ueber Süsswasserbakterien im Meere. *Zentbl. Bakt. Parasitkde* Abt. II., 66:500–505. **20**

Kowalczyk, C., 1957. Widłonogi (*Copepoda*) jezior Libiszowkich. *Annls Univ. Mariae Curie-Skłodowska,* 12C:57–101. **643**

Kozarov, G., 1957. Qualitative and quantitative study on the phytoplancton of Lake Ohrid in the course of two years study. Unpublished thesis in Yugoslav, *see* Stankovič 1960. **404, 411, 457**

* Kozhov, Mikhail, 1963. Lake Baikal and its life. *Monographiae Biologicae.* The Hague, Dr. W. Junk, Vol. XI. **45**

Koźmiński, Z., 1936. Morphometrische und ökologische Untersuchungen an Cyclopiden der *strenuus*-Gruppe *Int. Revue ges. Hydrobiol. Hydrogr.,* 33:161–231. **632, 857**

Kramp, P. L., 1950. Freshwater Medusae in China. *Proc. zool. Soc. Lond.,* 120:165–184. **41, 42**

Kratz, W. A., and Myers, J., 1955. Nutrition and growth of several blue-green algae. *Am. J. Bot.,* 42:282–287. **316, 317, 318**

Krätzschmar, H., 1908. Über den Polymorphismus von Anuraea aculeata Ehrbg. Variationsstatistische und experimentelle Untersuchung. *Int. Revue ges. Hydrobiol. Hydrogr.,* 1:623–675. **881, 884, 888**

Krätzschmar, H., 1931. Neue Untersuchungen über den Polymorphismus von Anuraea aculeata Ehrbg. *Int. Revue ges. Hydrobiol. Hydrogr.,* 6:44–49. **887–890**

Krause, F., 1911. Studien über die Formveränderung von Ceratium hirundinella O. F. Müll. als Anpassungserscheinung an die Schwebefähigkeit. *Int. Revue ges. Hyrobiol. Hydrogr.,* Biol. Suppl., 3 ser (No. 2) 1–32. **910, 921**

Krieger, W., 1933. Die Desmidiaceen der Deutschen limnologischen Sunda-expedition. *Arch. Hydrobiol.,* Suppl. 11:129–230. **330, 333**

Krieger, W., 1937–39. *Die Desmidiaceen.* Rabenhorst, L. Kryptogamen-Flora von Deutschland, Osterreich und der Schweiz. 13: Abt. 1, Teil 1, vi, 712 pp., Teil 2, 1–117 (incomplete). **307**

* Krogh, A., 1939. *Osmotic Regulation in Aquatic Animals.* Cambridge, England, vii + 242 pp. **151, 153, 161, 162, 164, 165, 171, 173, 190, 210, 247**

Kromhout, G. A., 1943. A comparison of the protonephridia of freshwater, brackish-water and marine specimens of Gyratrix hermaphroditus. *J. Morph.,* 72:167–177. **45**

Kufferath, H., 1953. *Gonzeella coloniaris* n. gen., n. spec. Cilie coloniaire peritriche du Congo belge. *Rev. Zool. Bot. afr.,* 48:30–34. **501**

Kükenthal, W., and Krumbach, T. editors (1923–1959). *Handbuch der Zoologie.* Berlin, Leipzig, W. de Gruyter 8 vols. **37n.**

Kunkel, W. B., 1948. Magnitude and character of errors produced by shape factors in Stokes' Law estimates of particle radius. *J. appl. Phys.,* 19:1056–1058. **261, 268**

Kuntze, H., 1938. Limnologische Untersuchungen über das Crustaceenplankton des Schleinsees und zweier Kleingewässer. *Int. Revue ges. Hydrobiol. Hydrogr.,* 37:164–233. **597, 610, 659, 672**

Kutkuhn, J. H., 1958. The plankton of North Twin Lake, with particular reference to the summer of 1955. *Iowa St. Coll. J. Sci.,* 32:419–450. **491, 494, 495, 496, 497**

Lack, D., 1942. Ecological features of the bird faunas of British small islands. *J. Anim. Ecol.,* 11:9–36. **375**

Lamb, H., 1932. *Hydrodynamics.* London, Cambridge Univ. Press, 738 pp. **261, 265**

Lange, A., 1911. Zur Kenntnis von Asplanchna sieboldii Leydig. *Zool. Anz.,* 38:433–441. **868**

* Langford, R. R., 1938. Diurnal and seasonal changes in the distribution of limnetic crustacea of Lake Nipissing, Ontario. *Univ. Toronto Stud. biol., ser.* No. 45 (Publs Ontario Fish. Lab. 56), 1–42. **659, 668, 674, 743, 748, 753, 784, 785, 786, 792, 793**

Langhans, V. H., 1925. Gemischte Populationen von *Ceratium hirundinella* (O.F.M.) Schrank und ihre Deutung. *Arch. Protistenk.*, 52:585–602. **915**

Lansing, A. I., 1942a. Increase of cortical calcium with age in the cells of the rotifer, Euchlanis dilatata, the planarian, Phagocata sp., and the toad, Bufo fowleri, as shown by the microincineration technique. *Biol. Bull. mar. bio. Lab., Woods Hole*, 82:392–400. **513, 890, 891**

Lansing, A. I., 1942b. Some effects of hydrogen ion concentration, total salt concentration, calcium and citrate on longevity and fecundity of the rotifer. *J. exp. Zool.*, 91:195–211. **513, 890, 891**

Lansing, A. I., 1947. Calcium and growth in aging and cancer. *Science, N.Y.*, 196:187–188. **513, 890, 891**

Lansing, A. I., 1948. Evidence for aging as a consequence of growth cessation. *Proc. natn. Acad. Sci. U.S.A.*, 34:304–310. **513, 890, 891**

Lastochkin, D., 1945. Achievements in Soviet hydrobiology of continental waters. *Ecology*, 26:320–331. **990**

Latter, O. H., 1891. Notes on Anodon and Unio. *Proc. zool. Soc. Lond.*, 1891:52–59. **699**

de Laubenfels, M. W., 1955. Porifera. In *Treatise on invertebrate paleontology*, ed. R. C. Moore. Lawrence, Kansas, Univ. of Kansas Press. Vol. E, pp. 21–122. **13**

Lauterborn, R., 1898. Vorläufige Mitteilung über den Variationskreis von Anuraea cochlearis Gosse. *Zool. Anz.*, 21:597–604. **870, 871, 874, 875**

Lauterborn, R., 1901. Der Formenkreis von Anuraea cochlearis. Ein Beitrag zur Kenntnis der Variabilität bei Rotatorens. I Teil: Morphologische Gliederung des Formenkreises. *Verh. naturh.-med. Ver. Heidelb.* n.f., 7:412–448. **870–872, 874, 875, 877, 878**

Lauterborn, R., 1904. II Teil. Die cyklische oder temporale Variation von Anuraea cochlearis. *Verh. naturh-med. Ver. Heidelb.* n.f., 7:529–621. **810, 811, 874, 875**

Lauterborn, R., 1908. Gallerthüllen bei loricaten Plancton-Rotatorien. *Zool. Anz.*, 33:580–584. **519**

Lauterborn, R., 1915. Die sapropelische Lebewelt. *Verh. naturh.-med. Ver. Heidelb.* n.f., 13:395 (not seen: ref. in Fogg 1941). **252**

Leech, H. B., and Sanderson, M. W., 1959. Coleoptera. In H. B. Ward, and G. C. Whipple, *Fresh-water Biology*, 2nd ed., ed. W. T. Edmondson. New York, John Wiley & Sons, pp. 981—1023. **85**

Lefèbre, —, 1908. Notice sur le *Penaeus brasiliensis*, crevette du Bas Dahomey (crevette du lac Ahémé). *Bull. Mus. Hist. nat., Paris*, 14:267–270. **111**

Lefevre, G., and Curtis, W. C., 1912. Studies on the reproduction and artificial propagation of fresh-water mussels. *Bull. Bur. Fish., Wash.*, 30:103–201. **55, 699**

Lefèvre, M., 1942. L'utilisation des algues d'eau douce par les Cladocères. *Bull. biol. Fr. Belg.*, 76:250–276. **556**

Lefèvre, M., and Jakob, H., 1949. Sur quelques propriétés des substances actives tirées des cultures d'algues d'eau douce. *C. r. hebd. Séanc. Acad. Sci., Paris*, 229:234–236. **464**

Lefèvre, M., Jakob, H., and Nisbet, M., 1950. Sur la sécrétion, par certaines cyanophytes, de substances algostatiques dans les collections d'eau naturelles. *C. r. hebd. Séanc. Acad. Sci., Paris,* 230:2226–2227. **464**

Lefèvre, M., Jakob, H., and Nisbet, M., 1951. Compatibilités et antagonismes entres algues d'eau douce dans les collections d'eau naturelles. *Verh. int. Verein. theor. angew. Limnol.,* 11:224–229. **464**

* Lefèvre, M., Jakob, H., and Nisbet, M., 1952. Auto- et heteroantagonisme chez les algues d'eau douce. *Annls St. Cent. Hydrobiol. appl.,* 4:5–197. **464, 465, 466, 469, 472**

Lefèvre, M., and Nisbet, M., 1948. Sur la sécrétion, par certaines espèces d'algues, de substances inhibitrices d'autres espèces d'algues. *C. r. hebd. Séanc. Acad. Sci., Paris,* 226:107–109. **464**

Lefèvre, M., Nisbet, M., and Jakob, H., 1949. Action des substances excrétees en culture, par certaines espèces d'algues, sur le métabolisme d'autres espèces d'algues. *Verh. int. Ver. theor. angew. Limnol.,* 10:259–264. **464**

Lehmann, W. M., 1955. *Vachonia rogeri* n.g. n.sp. ein Branchiopod aus dem underdevonischen Hunsrückschiefer. *Paleont. Z.,* 39:126–130. **91, 118**

Leidy, J., 1879. Freshwater Rhizopods of North America. Washington, *Rep. U.S. Geol. Surv.,* 12, xi, 324 pp. **494**

Lemmermann, E., Brunnthaler, J., and Pascher, A., 1915. *Chlorophyceae. II. Tetrasporales, Protococcales, Einzellige Gattungen unsicherer Stellung.* Süsswasser-Flora Deutschlands, Heft 5, iv, 250 pp. **306**

von Lengerken, H., 1929. Die Salzkäfer der Nord- und Ostseeküste mit Berücksichtigung der angrenzenden Meere sowie des Mittelmeeres, der Schwarzen und des Kaspischen Meeres. Eine ökologisch-biologisch-geographische Studie. *Z. wiss. Zool.,* 135:1–162. **86**

Lenz, F., 1928. Zur Terminologie der limnischen Zonation. *Arch. Hydrobiol.,* 19:748–757, **239, 240**

Lereboullet, A., 1866. Observations sur la génération et le développement de la limnadie de Hermann (*Limnadia Hermanni* Ad. Brogn.). *Annls Sci. nat.,* Zoologie (Ser. 5) 5:283–308. **560**

LeRoux, M., 1908. Recherches biologiques sur le lac d'Annecy. *Annls Biol. lacustre,* 2:220–387. **936, 944**

Leslie, P. H., and J. C. Gower, 1958. The properties of a stochastic model for two competing species. *Biometrika,* 45:316–330. **357**

Leston, D., 1956. Systematics of the marine-bug. *Nature, Lond.,* 178:427–428. **86**

Levanidov, V. J., 1945. The application of Gaievskaia's method to large number weighings and weighings of large water invertebrates. *Zool. Zh.,* 24:337–340 (Russ. Text); (Eng. summ. 340). **249**

Leventhal, E. A. (*in press*). Chrysophycean cysts. In Ianula: A study of the history of Lago di Monterosi. **339, 374, 386**

Levring, T., 1945. Some culture experiments with marine plankton diatoms. *Göteborgs K. Vetensk. o. VitterhSamh. Handl.,* 6 Föl; ser. B. 3, 12:3–18. **464**

Lewin, J., 1953. Heterotrophy in diatoms. *J. gen. Microbiol.,* 9:305–313. **343**

Lewin, J. C., 1955. The capsule of the diatom *Navicula pelliculosa*. *J. gen. Microbiol.,* 13:162–169. **944**

Lewin, J. C., 1958. The taxonomic position of *Phaeodactylum tricornutum*. *J. gen. Microbiol.,* 18:427–432. **25**

Liebmann, H., 1938. Biologie und Chemismus der Bleilochsperre. *Arch. Hydrobiol.*, 33:1–81. **505**

Lieder, U., 1950. Beiträge zur Kenntnis der Genus Bosmina I. *Bosmina coregoni thersites* Poppe in den Seen der Spree-Dahme-Havelgebietes. *Arch. Hydrobiol.*, 44:77–122. **852, 855**

Lieder, U., 1953. Beiträge zur Kenntnis der Genus *Bosmina* II Über Bastarde zwischen einigen Formtypen der *Coregoni*-Kreises. *Arch. Hydrobiol.*, 47:453–469. **852**

Liepolt, R., 1958. Zur limnologischen Erforschung des Zellersees in Salzburg. *Wasser und Abwasser*, 1958:1–84. **452**

* Lilljeborg, W., 1900. Cladocera Sueciae oder Beitrage zur Kenntniss der in Schweden lebenden Krebsthiere von der Ordnung der Branchiopoden und der Unterordnung der Cladoceren. *Nova Acta R. Soc. Scient. upsal.*, (ser. 3) 19:1–701. **594, 608, 847, 856**

Lilljeborg, W., 1901. Synopsis specierum huc usque in Suecia observatorum generis Cyclops. *K. svenska Vetensk-Akad. Handl.*, 35:1–118. **568, 569**

Linder, F., 1945. Affinities within the Branchiopoda, with notes on some dubious fossils. *Ark. Zool.*, 37A, No. 4., 28 pp. **119**

Linder, F., 1952. Contributions to the morphology and taxonomy of the Branchiopoda Notostraca, with special reference to the North American Species. *Proc. U.S. natn. Mus.*, 102:1–69. **560, 564**

Linder, F., 1959. Notostraca. In H. B. Ward, and G. C. Whipple *Fresh-water Biology*, 2nd ed., ed. W. T. Edmondson. New York and London, John Wiley & Sons, pp. 572–76. **559**

Lindholm, W., 1927. Kritischen Studien zur Molluskenfauna des Baikalsees. *Trudÿ-Kom. Izuch. Ozera Baïkala*, 2:139–186. **65**

Lindström, T., 1952. Sur l'écologie du zooplankton crustacé. *Rep. Inst. Freshwat. Res. Drottningholm*, 33:70–165. **642, 643, 670, 799**

Lindström, T., 1957. Sur les planctons crustaces de la zone littorale. *Rep. Inst. Freshwat. Res. Drottningholm*, 38:131–153. **799, 804**

Lindström, T., 1958. Observations sur les cycles annuels des planctons crustacés. *Rep. Inst. Freshwat. Res. Drottningholm* 39:99–145. **642, 643, 670**

List, T., 1914a. Über die Temporal und Lokalvariation von *Ceratium hirundinella* O. F. M. aus dem Plankton einiger Teiche un der Umgegend von Darmstadt und einiger Kolke des Altrheins bei Erfelden. *Arch. Hydrobiol.*, 9:81–126. **910, 911, 912, 914, 915**

List, T., 1914b. Hat der künstliche Wechsel der natürlichen Umgebung einen formverändernden Einfluss auf die Ausbildung der Hörner von Ceratium hirundinella O. F. Muller? I. Mitteilung. *Arch. Entw. Mech. Org.*, 39:375–383. **914**

Littleford, F. A., Newcombe, C. L., and Shepherd, B. B., 1940. An experimental study of certain quantitative plankton methods. *Ecology*, 21:309–322. **791**

Lloyd, F. E., 1928. The contractile vacuole. *Biol. Rev.*, 3:329–358. **28**

Lockwood, A. P. M., and Croghan, P. C., 1957. The chloride regulation of the brackish and fresh-water races of *Mesidotea entomon* (L). *J. exp. Biol.*, 34:253–258. **210**

Loeb, J., 1893. Ueber die künstliche Umwandlung positiv heliotropischer Tiere in negativ heliotropischer und umgekehrt. *Pflügers Arch. ges. Physiol.*, 54:81–107. **759**

Loeb, J., 1904. The control of heliotropic reactions in freshwater Crustacea by chemicals, especially CO₂. *Univ. Calif. Publs Physiol.,* 2:1–3. **759**

Loeb, J., 1906. Ueber die Erregung von positiven Heliotropismus durch Säure, insbesonder Kohlensäure, und negativen Heliotropismus durch ultraviolette Strahlen. *Pflügers Arch. ges. Physiol.,* 115:151–181. **759**

Loeb, J., 1908. Über Heliotropismus und die periodischen Tiefenbewegungen pelagischer Tiere. *Biol. Zbl.,* 28:732–736. **759**

Löffler, H., 1953. Limnologische Ergebnisse der österreichischen Iranexpedition 1949/50. *Naturw. Rdsch. Stuttg.,* 2:64–68. **26n.**

Löffler, H., 1955. Die Boeckelliden Perus, Ergebnis der Expedition Brundin und der Andenkundfahrt unter Prof. Dr. Kinzl 1953/54. *Sber. öst. Akad. Wiss.,* M.-N. Kl. (Abt I), 164:723–746. **128**

Löffler, H., 1957. Vergleichende limnologische Untersuchungen an den Gewässern des Seewinkels (Burgenland) I. *Verh. Zool.-bot. Ges. Wien,* 97:27–52. **676**

Löffler, H., 1959. Zur Limnologie, Entomostraken-und Rotatorien Fauna des Seewinkelgebietes (Burgenland, Österreich). Öster. Akad. Wiss. Sonderheft für den . . . XIV Internat. Limnologenkongress. *Sber. Öst. Akad. Wiss. M.-N. Kl.* (Abt. I), 168:315–362. **676**

Löffler, H., 1961. Beiträge zur Kenntnis der Iranischen Binnengewässer II. Regional-limnologische Studie mit besonderer Berücksichtigung der Crustaceenfauna. *Int. Revue ges. Hydrobiol. Hydrogr.,* 46:309–406. **676**

Lohmann, H., 1911. Über das Nannoplankton und die Zentrifugierung kleinster Wasserproben zur Gewinnung desselben in lebendem Zustande. *Int. Revue ges. Hydrobiol. Hydrogr.,* 4:1–38. **235, 401**

Lomakina, N. B., 1952. The origin of the glacial-relict amphipods and the question of a late-glacial connection between the White Sea and the Baltic. *Vchon. Zap. Karelo-Finsk. Univers. Biol. Nauki,* 4:110–127 (Russian text). **182**

Longhurst, A. R., 1954. Reproduction in Notostraca (Crustacea). *Nature, Lond.,* 173:781–782. **560**

Longhurst, A. R., 1955. A review of the Notostraca. *Bull. Br. Mus. Nat. Hist. Zoology* 3:1–57. **559, 560, 564**

Lönnberg, E., 1932. Some remarks on the relict forms of *Cottus quadricornis* L. in Swedish freshwater lakes. *Ark. Zool.,* 24A, No. 7, 23 pp. **78, 179, 180**

Lönnberg, E., 1933. Notes on some relict races of *Cottus quadricornis* living in lakes in Finland. *Ark. Zool.* 24A, No. 12, 16 pp. **180**

Lovén, S., 1862. Om några i Vettern och Venern funna Crustaceer. *Öfvers. K. svenska Vetenks-Akad. Förh.,* 18:285–314. **204**

Lowndes, A. G., 1928. The Result of Breeding Experiments and other Observations on Cyclops vernalis Fischer and Cyclops robustus G. O. Sars. *Int. Revue ges. Hydrobiol. Hydrogr.,* 21:171–188. **627**

Lowndes, A. G., 1929. The occurrence of *Eurytemora lacinulata* and *Diaptomus gracilis.* *J. Ecol.,* 17:380–382. **125**

Lowndes, A. G., 1930. Some fresh-water calanoids. Direct observation *v.* indirect deduction. *J. Ecol.,* 18:151–155. **125**

Lowndes, A. G., 1931. *Eurytemora thompsoni,* A. Willey: a new European record. *Nature. Lond.,* 128:967. **124**

Lowndes, A. G., 1933. The feeding mechanism of *Chirocephalus diaphanus* Prévost, the fairy shrimp. *Proc. zool. Soc. Lond.,* 1093–1118. **559**

Lowndes, A. G., 1935. The swimming and feeding of certain calanoid copepods. *Proc. zool. Soc. Lond.,* 1935. pp. 687–715. **125, 676, 677, 678, 679, 686**

Lowndes, A. G., 1937. Body orientation in crustacea. *Nature, Lond.,* 140: 241–242. **271**

Lowndes, A. G., 1938. The density of some living aquatic organisms. *Proc. Linn. Soc. Lond.,* 150:62–72. **248, 249, 271**

Lowndes, A. G., 1942. The displacement method of weighing living aquatic organisms. *J. mar. Biol. Ass. U.K.,* 25:555–574. **248, 249**

Lozeron, H., 1902. Sur la répartition verticale du plankton dans le lac de Zürich, de décembre 1900 à décembre 1901. *Vjschr. naturf, Ges. Zürich,* 47:115–198. **283, 726, 751, 928, 932, 933, 936, 942**

Lucas, W. J., 1930. *The Aquatic (naiad) stage of the British dragonflies (Paraneuroptera).* London Ray Society, xi, 132 pp. **81**

Lucks, R., 1930. Synchaeta lakowitziana n. sp., ein neues Rädertier. *Zool. Anz.,* 92:59–63. **534**

Ludwig, W., 1928. Der Betriebsstoffwechsel von *Paramecium caudatum* Ehrbg. zugleich ein Beitrag zur Frage nach der Funktion der kontraktilen Vakuolen. *Arch. Protistenkunde,* 62:12–40. **262**

* Lund, J. W. G., 1949a. Studies on *Asterionella* I The origin and nature of the cells producing seasonal maxima. *J. Ecol.,* 37:389–419. **429, 434, 437, 439, 440, 446, 447–450, 453**

Lund, J. W. G., 1949b. The dynamics of diatom outbursts, with special reference to *Asterionella Verh. int. Verein. Theor. angew. Limnol.,* 10:275–276. **429, 446, 447–450, 453**

* Lund, J. W. G., 1950a. Studies on *Asterionella formosa* Hass. II Nutrient depletion and the spring maximum. Part I observations on Windermere, Esthewaite Water and Blelham Tarn. *J. Ecol.* 38:1–14. **447–450**

* Lund, J. W. G., 1950b. Studies on *Asterionella formosa* Hass. II Nutrient Depletion and the spring maximum. Part II Discussion. *J. Ecol.,* 38:15–35. **447–450**

* Lund, J. W. G., 1954. The seasonal cycle of the plankton diatom *Melosira italica* (Ehr.) Kütz subsp. *subarctica* O. Müll. *J. Ecol.,* 42:151–179. **375, 458–462**

* Lund, J. W. G., 1955. Further observations on the seasonal cycle of *Melosira italica* (Ehr.) Kütz subsp. *subarctica* O. Müll. *J. Ecol.,* 43:90–102. **375, 458–462**

Lund, J. W. G, 1959a. Buoyancy in relation to the ecology of the freshwater phytoplankton. *Br. phycol. Bull.,* No. 7, 17 pp. **271, 277**

Lund, J. W. G., 1959b. Biological tests on the fertility of an English reservoir water (Stocks Reservoir, Bowland Forest). *J. Instn Wat. Engrs.,* 13:527–549. **453**

* Lund, J. W. G., 1954. Primary production and periodicity of phytoplankton. (Edgardo Baldi Memorial Lecture). *Verh. int. Verein. theor. angew. Limnol.,* 15:37–56. **358, 436, 471**

* Lund, J. W. G., Mackereth, F. J. H., and Mortimer, C. H., 1963. Changes in depth and time of certain chemical and physical conditions and of the standing crop of *Asterionella formosa* Hass. in the North Basin of Windermere in 1947. *Phil. Trans. R. Soc.* 246B:255–290. **447, 448, 944, 945**

Lundbeck, J., 1926. Die Bodentierwelt norddeutscher Seen. *Arch. Hydrobiol.,* Supp. 7:1–473. **240**

Lundblad, O., 1920. Vergleichende Studien über die Nahrungsaufnahme einiger schwedischen Phyllopoden, nebst synonymischen, morphologischen und biologischen Bermerkungen. *Ark. Zool.,* 13, No. 16, 114 pp. **559, 564**

Lundh, A., 1951. Studies on the vegetation and hydrochemistry of Scanian lakes. III. Distribution of macrophytes and some algal groups. *Bot. Notiser*, Supp. 3:138 pp. **387**

Luntz, A., 1926. Untersuchungen über den Generationswechsel der Rotatorien. I. Die Bedingungen des Generationswechsels. *Biol. Zbl.*, 46:233–256, 257–278. **510, 527**

Luntz, A., 1928. Über die Sinkgeschwindigkeit einiger Rädertiere. Zugleich ein Beitrag zur Theorie der Zyklomorphose. *Zool. J. Abt.* Allgem. Zool. u. Physiol., 44:451–482. **248, 249, 278, 282, 903, 904**

Luntz, A., 1929. Weitere Untersuchungen über die Sinkgeschwindigkeit von Süsswasser Organismen. *Zool. Jb.* Abt. Allgem. Zool. und Physiol. der Tiere. 46:465–482. **278, 282**

Luzzatti, E., 1935. La variabilità del genere *Keratella* nel Lago Trasimeno. *Riv Biol.*, 19:21–46. **894**

Lytle, C. F., 1960. A note on distribution patterns in *Craspedacusta*. *Trans. Am. microsc. Soc.*, 79:461–469. **42**

Määr, A., 1949. Fertility of Char (Salmo alpinus L.) in the Faxälven water system, Sweden. *Rep. Inst. Fresh-wat. Res., Drottingholm*, 29:57–70. **661**

MacArthur, J. W., and Baillie, W. H. T., 1929. Metabolic activity and duration of Life I. influence of temperature on longevity in Daphnia magna. *Jour. exp. Zool.*, 53:221–242. **578, 579**

MacArthur, R. H., 1955. Fluctuations of animal populations and a measure of community stability. *Ecology*, 36:533–536. **364**

MacArthur, R. H., 1957. On the relative abundance of bird species. *Proc. natn. Acad. Sci. U.S.A.*, 45:293–295. **366–368**

MacArthur, R. H., 1958. Population ecology of some warblers of northeastern coniferous forests. *Ecology*, 39:599–619. **366**

MacArthur, R. H., 1960. On the relative abundance of species. *Am. Nat.*, 94:25–36. **366–368**

McFarland, W. N., and Munz, F. W., 1958. A re-examination of the osmotic properties of the Pacific ha, fish, Polistotrema stouti. *Biol. Bull. mar. biol. Lab.*, Woods Hole, 114:348–356. **134**

McIlrath, W. J., and Skok, J., 1957. Influence of boron on the growth of *Chlorella*. *Pl. Physiol.*, suppl. 32:xxiii. **322**

McLaren, J. A., 1960. On the origin of the Caspian and Baikal seals and the paleoclimatological implication. *Am. J. Sci.*, 258:47–65. **144, 215**

McLaren, I. A., 1961. A biennial copepod from Lake Hazen, Ellesmere Island. *Nature, Lond.*, 189:774. **643**

McLaren, I. A., 1963. Effects of temperature on growth of zooplankton and the adaptive value of vertical migration. *J. Fish. Res. Bd., Canada*, 20:685–727. **787, 788**

McMahon, J. W., and Rigler, F. H., 1963. Mechanisms regulating the feeding rate of Daphnia magna Straus. *Can J. Zool.*, 41:321–332. **600, 602**

McMichael, D. F., and Hiscock, I. D., 1958. A monograph of the freshwater mussels (Mollusca:Pelecypoda) of the Australian region. *Aust. J. mar. Freshwat. Res.*, 9:372–508. **56**

McNaught, D. C., and Hasler, A. D., 1961. Surface schooling and feeding behavior in the White Bass *Roccus chrysops* (Rafinesque), in Lake Mendota. *Limnol. Oceanogr.*, 6:53–60. **797**

McNown, J. S., and Malaika, J., 1950. Effects of particle shape on settling velocity at low Reynolds numbers. *Trans. Am. geophys. Un.*, 31:74–82. **259, 260–262, 267, 268, 270**

Maggenti, A. R., 1963. Comparative morphology in nemic phylogeny. In *The Lower Metazoa Comparative Biology and Phylogeny*, ed. E. C. Dougherty Univ. California Press. pp. 273–282. **49**

Maloney, M. T., and Tressler, W. L., 1942. The diurnal migration of certain species of zooplankton in Caroga Lake, New York. *Trans. Am. microsc. Soc.*, 61:40–52. **742**

Maluf, N. S. R., 1940. The uptake of inorganic electrolytes by the crayfish. *J. Gen. Physiol.*, 24:151–167. **162**

Mann, A. K., 1940. Über prelagische Copepoden türkischer Seen. *Int. Revue ges. Hydrobiol. Hydrogr.*, 40:1–87. **634, 652, 680**

Manton, I., 1952. The fine structure of plant cilia. *Symp. Soc. exper. Biol.*, 6:306–319. **7**

Manton, I., 1959. Electron microscopical observations on a very small flagellate: the problem of *Chromulina pusilla* Butcher. *J. mar. biol. Ass. U.K.*, 38: 319–333. **5n.**

Manton, I., and Parke, M., 1960. Further observations on small green flagellates with special reference to possible relatives of *Chromulina pusilla* Butcher. *J. mar. biol. Ass. U.K.*, 39:275–298. **5n.**

Manton, S. M., 1930. Notes on the habits and feeding mechanisms of *Anaspides* and *Paranaspides* (Crustacea, Syncarida). *Proc. zool. Soc. Lond.*, 1930:791–800. **97**

Manton, S. M., 1964. Mandibular mechanisms and the evolution of arthropods. *Phil. Trans. R. Soc.*, 247B:1–183. **16n., 88**

Manuylova, E. F., 1948. K izucheniyu izmenchivosti cladocera soobshchenie I. Izmenchivost *Daphnia* u Oz. Balkhash. *Izr. Akad. Nauk SSSR.*, Seriya Biologicheskaya, 5:595–606. **606, 821, 823**

Margalef, R., 1947. Notas sobre algunos Rotiferos. *Publ. Inst. Biol. apl., Barcelona*, 4:135–148. **812, 877**

Margalef, R., 1953. Caracteres ligados a las magnitudes absolutas de los organismos y su significado sistemático y evolutivo. *Publnes Inst. Biol. apl., Barcelona*, 12:111–121. **638**

Margalef, R., 1955. Temperature and morphology in freshwater organisms *Verh. int. Verein. theor. angew. Limnol.*, 12:507–514. **638**

Margalef, R., 1956. Información y diversidad específica en las communidades de organismos. *Investigación pesq.*, 3:99–106. **362**

Margalef, R., 1957a. Nuevos aspectos del problema de la suspensión en los organismos planctónicos. *Investigación pesq.*, 7:105–116. **275**

* Margalef, R., 1957b. La Teoría de la Información en Ecología. *Mems. R. Acad. Cienc. Artes, Barcelona*, (3v.) 32, No. 13, 79 pp. **362, 363**

Margalef, R., 1961. Velocidad de sedimentación de organismos pasivos del fitoplancton. *Investigación pesq.*, 18:3–8. **287**

Marlier, G., 1958. Recherches hydrobiologiques au lac Tumba. *Hydrobiologia*, 10:352–385. **501**

Marsh, C. D., 1893. On the Cyclopidae and Calanidae of Central Wisconsin. *Trans. Wis. Acad. Sci. Arts Lett.*, 9:189–224. **691**

Marsh, C. D., 1897. On the limnetic Crustacea of Green Lake. *Trans. Wis. Acad. Sci. Arts Lett.*, 11:179–224. **609, 620, 668**

Marsh, C. D., 1913. Report on freshwater Copepoda from Panama, with descriptions of new species. *Smithson. misc. Collns,* 61, No. 3, 30 pp. **626**

Marsh, C. D., 1929. Distribution and key of the North American copepods of the genus *Diaptomus,* with the description of a new species. *Proc. U.S. natn Mus.,* 75, Art. 14:1–27. **692**

Marsh, C. D., 1933. Synopsis of the calanoid crustaceans, exclusive of the Diaptomidae, found in fresh and brackish waters, chiefly of North America. *Proc. U.S. natn Mus.,* 82, Art. 18:58 pp. **122, 124, 650**

Marshall, E. K., and Smith, H. W., 1930. The glomerular development of the vertebrate kidney in relation to habitat. *Biol. Bull. mar. Lab., Woods Hole,* 59:135–153. **134**

Marshall, S. M., 1949. On the biology of the small copepods in Loch Striven. *J. mar. biol. Ass. U.K.,* 28:45–122. **662**

Marsson, M., 1901. Zur Kenntnis der Planktonverhältnisse einiger Gewässer der Umgebung von Berlin. *ForschBer. biol. Stn. Plön.,* 8:86–119. **944**

Martens, E. von, 1857. Ueber einige Fische und Crustaceen der süssen Gewässer Italiens. *Arch. Naturgesch.,* 23:149–210. **190**

Martens, E. von, 1858. On the occurrence of marine animal forms in fresh water. *Ann. nat. Hist.,* (ser. 3) 1:50–63 (trans. von Martens 1857). **190**

Martinson, G. G., 1940. Contributions to the study of the Circumbaikalian fossil microfauna and spongiofauna. *Trudÿ baïkal'. limnol. Sta.,* 10:425–450. **144, 214**

Martinson, G. G., 1958. Origin of the Baikal Fauna in the Light of Paleontological Studies. *Dokl. Akad. Nauk. SSSR.,* 120:115–118. *See also Dokl. (Proc.) Acad. Sci. U.S.S.R.,* Biol. Sci. Sect., 120:499–502 (Amer. Inst. Biol. Sci. Trans. pp. 499–502). **109, 144**

Mason, D. T., Goldman, C. R., and Hobbie, J. E., 1963. Light penetration and biology of two Antarctic dry valley lakes (abstr.). *Bull. ecol. Soc. Am.,* 44:39. **497**

Mathias, P., 1937. Biologie des Crustacés phyllopodes. *Actual. scient. ind.,* 447, 106 pp. **559, 560, 561, 563, 566**

Mathiesen, O. A., 1953. Some investigations of the relict crustaceans in Norway with special reference to Pontoporeia affinis Lindström and Pallasea quadrispinosa G. O. Sars. *Nytt Mag. Zool.,* 1:49–86. **211**

Matschek, H., 1909. Zur Kenntnis der Eireifung und Eiablage bei Copepoden. *Zool. Anz.,* 34:42–54. **650, 658**

Mattox, N. T., 1950. Notes on the life history and description of a new species of conchostracan phyllopod *Caenestheriella gynecia. Trans. Am. microsc. Soc.,* 69:50–53. **565**

Mattox, N. T., 1959. Conchostraca. In H. B. Ward, and G. C. Whipple *Fresh-water Biology* 2nd ed., ed. W. T. Edmondson. New York and London, John Wiley & Sons, pp. 577–586. **559, 560, 565**

Mattox, N. T., and Velardo, J. T., 1950. Effect of temperature on the development of the eggs of a conchostracan phyllopod *Caenestheriella gynecia. Ecology,* 31:497–506. **565**

Medwedeva, N. B., 1927. Über den osmotischen Druck der Hämolymphe von Artemia salina. *Z. vergl. Physiol.,* 5:547–554. **220**

Meier, K. I., 1927. Über das Phytoplankton der Baikalsees. *Russk. gidrobiol. Zh.,* 6:128–136 (Russian text); (German summ. 137). **461**

Meisenheimer, J., 1901. Entwicklungsgeschichte von Dreissensia polymorpha. *Z. wiss. Zool.,* 69:1–137. **700**

Meissner, U., 1901. Das Flussplankton der Wolga bei Saratow (in Russian). *Comptes rendus des travaux des vacances 1901 de la Station biologique du Volga.* (*Soc. des Naturalistes à Saratow*), p. 1 (not seen: ref. Sachse 1911). **905**

Meixner, J., 1938. Turbellaria (Strudelwürmer) 1. (Allgemeiner Teil). Grimpe and Wagler *Die Tierwelt der Nord u. Ostsee.* Leipzig, Teil 4b: pp. 1–146. **45**

Menzies, R. J., 1954. A review of the systematics and ecology of the genus "Exosphaeroma" with the description of a new genus, a new species, and a new subspecies (Crustacea; Isopoda, Sphaeromidae). *Amer. Mus. Novit.,* 1683, 24 pp. **103, 104**

Messikommer, E., 1927. Biologische Studien im Torfmoor von Robenhausen unter besonderer Berücksichtigung der Algenvegetation. *Mitt. bot. Mus. Univ.,* Zürich, 122, vi, 171 pp. **378**

Meuche, A., 1939. Die Fauna im Algenbewuchs. *Arch. Hydrobiol.,* 34:349–520. **237**

Meyer, D. K., 1948. Physiological adjustments in chloride balance of the goldfish. *Science, N.Y.,* 108:305–307. **165**

Miall, L. C., 1895. *The natural history of aquatic insects.* London, Macmillan, xi, 395. **81, 85**

Michael, A. D., 1888. *British Oribatidae.* London, The Ray Society, II: xi + 337–657 pp. **88**

Michaelsen, W., 1928. *Clitellata = Gürtelwürmer. Dritte Klasse der Vermes Polymera (Annelida).* Kükenthal & Krumbach, Handb. Zool., Berlin, 2(8), No. 8:1–112. **16n, 78**

Miller, A. K., Cullison, J. S., and Youngquist, W., 1947. Lower Ordovician fish remains from Missouri. *Am. J. Sci.,* 245:31–34. **133**

Miller, H. M., 1931. Alternation of generations in the rotifer Lecane inermis Bryce. I. Life histories of the sexual and non-sexual generations. *Woods Hole, Biol. Bull. mar. biol. Lab.,* 60:345–381. **512, 513**

Miller, P. L., 1964. Possible function of haemoglobin in *Anisops. Nature, Lond.,* 201:1052. **256**

Millot, J., and Anthony, J., 1958. Crossoptérygiens actuels. Latimeria chalumnae. *Traité de Zool.* ogie Ed. P. P. Grasse, Masson et Cie pp. 13:2553–2597. **136**

Miquel, P., 1894. Du rétablissement de la taille et la rectification de la forme chez les Diatomées. *Le Diatomiste,* 2:61–69. **927**

Mitchell, C. W., 1913a. Experimentally induced transitions in the morphological characters of Asplanchna amphora Hudson, together with remarks on sexual reproduction. *J. exp. Zool.,* 15:91–130. **510, 868**

Mitchell, C. W., 1913b. Sex-determination in Asplanchna amphora. *J. exp. Zool.,* 15:225–255. **510**

Mitchell, C. W., and Powers, J. H., 1914. Transmission through the resting egg of experimentally induced characters in Asplanchna amphora. *J. exp. Zool.,* 16:347–396. **868, 869**

Mitchell, P., 1949. The osmotic barrier in bacteria. *Symp. Soc. gen. Microbiol.,* 1949:55–73. **21**

Miyawaki, M., 1951. Notes on the effect of low salinity on an actinian, *Diadumene luciae. J. Fac. Sci., Hokkaido Univ.,* (ser. IV) 10:123–126. **43**

Moberg, E. G., 1918. Variation in the horizontal distribution of plankton in Devils Lake, North Dakota. *Trans. Am. microsc. Soc.,* 37:239–267. **788**

Möbius, K., 1877.   *Die Auster und die Austernwirtschaft.*   Berlin, Verlag von Wiegandt, Hempel & Parey, v, 126 pp. **227**

Modell, H., 1942.   Das natürliche System der Najaden.   *Arch. Molluskenk.,* 74:161–191. **56**

Modell, H., 1949.   Das natürliche System der Najaden. 2.   *Arch. Molluskenk.,* 78:29–48. **56**

Moll, F., 1936.   Les animaux rongeurs de bois sur côtes de l'Indochine. *J. Conch., Paris,* 80:296–301. **62**

Monakov, A. V., and Sorokin, Yu. I., 1960.   An Experimental investigation of Daphnia nutrition using C^{14}.   *Dokl. Akad. Nauk SSSR,* 135:1516–1518.   *See also Dokl. (Proc.) Acad. Sci. U.S.S.R.,* Biol. Sci., Sect. 135:925–926. **599, 602**

Monod, Th., 1927.   Sur le Crustacé auquel le Cameroun doit son nom (*Callianassa turnerana* White).   *Bul. Mus. Hist. nat.,* Paris, 33:80–85. **114**

Monro, C. C. A., 1937.   On some Freshwater Polychaets from Uruguay. *Ann. Mag. nat. Hist.,* (ser 10) 20:241–250. **74**

Moore, A. R., 1912.   Concerning negative phototropism in *Daphnia pulex.* *J. exp. Zool.,* 13:573–575. **762**

Moore, G. M., 1939.   A limnological investigation of the microscopic benthic fauna of Douglas Lake, Michigan.   *Ecol. Monogr.,* 9:537–582. **236, 640**

Moore, H. B., and Corwin, E. G., 1956.   The effects of temperature, illumination and pressure on the vertical distribution of zooplankton.   *Bull. Mag. Sci. Gulf Caribb.,* 10:430–443. **787**

Moore, J. E., 1952.   The Entomostraca of southern Saskatchewan.   *Can. J. Zool.,* 30:410–450. **607, 615**

Moore, W. G., 1955.   The life history of the spiny-tailed fairy shrimp in Louisiana.   *Ecology,* 36:176–184. **563**

Moore, W. G., and Ogren, L. H., 1962.   Notes on the breeding behavior of Eubranchipus holmani (Ryder).   *Tulane Stud. Zool.,* 9:315–318. **560**

Mordukhai-Boltovskoi, Ph. D., 1962.   Pontokaspiiskie polifemidy.   *Byull. mosk. Obshch. Ispȳt. Prir., Biol.,* 67, No. 6:131–132. **119**

Mordukhai-Boltovskoi, Ph. D., 1964.   Caspian fauna beyond the Caspian sea. *Int. Revue ges. Hydrobiol. Hygrogr.,* 49:139–176. **121**

Morris, R., 1956.   The osmoregulatory ability of the lampern (*Lampetra fluviatilis* L.) in sea water during the course of its spawning migration. *J. exp. Biol.,* 33:235–248. **168**

Mortimer, C. H., 1936.   Experimentelle und cytologische Untersuchungen über den Generationswechsel der Cladoceren.   *Zool. Jb.,* Abt. allg. Zool., 56:323–388. **595**

Müller, O., 1884.   Die Zellhaut und des Gesetz der Zelltheilungsfolge von Melosira arenaria Moore.   *Jb. wiss. Bot.,* 14:232–290. **923–925**

Munk, W. H., and Anderson, E. R., 1948.   Notes on a theory of the thermocline.   *J. mar. Res.,* 7:276–295. **288**

Munk, W. H., and Riley, G. A., 1952.   Absorption of nutrients by aquatic plants.   *J. mar. Res.,* 11:215–240. **265, 294–301, 349**

Murray, J., 1906.   The Rotifera of the Scottish Lochs.   *Trans. R. Soc. Edinb.,* 45:151–191. **552n.**

Myers, F. J., 1931.   The distribution of Rotifera on Mount Desert Island. I.   *Am. Mus. Novit.,* 494:1–12. **525**

Myers, F. J., 1936.   Three new brackish water and one new marine species of Rotatoria.   *Trans. Am. microsc. Soc.,* 55:428–432. **48**

Myers, F. J., 1941. *Lecane curvicornis* var. *miamensis,* new variety of Rotatoria, with observations on the feeding habits of rotifers. *Notul. Nat.,* 75, 8 pp. **526**

Myers, J., 1951. Physiology of the algae. *A. Rev. Microbiol.,* 5:157–180. **310**

Naber, H., 1933. Die Schichtung des Zooplanktons in holsteinischen Seen und der Einfluss des Zooplanktons auf den Sauerstoffgehalt der bewohnten Schichten. *Arch. Hydrobiol.,* 25:81–132. **792**

Nagel, H., 1934. Die Aufgaben der Exkretionsorgane und der Kiemen bei der Osmoregulation von *Carcinus maenas. Z. vergl. Physiol.,* 21:468–491. **162**

VanName, W. G., 1936. The American land and fresh-water isopod Crustacea. *Bull. Am. Mus. nat. Hist.,* 71, vii, 535 pp. **101, 104**

Napora, T. A., 1964. Studies on the biology of the zooplankton of the Bermuda area. Yale Ph.D. thesis (submitted November 1963). **788**

Naumann, E., 1917a. Beiträge zur Kenntnis des Teichnannoplanktons II Über das Neuston des Süsswassers. *Biol. Zbl.,* 37:98–106. **234, 238**

Naumann, E., 1917b. Undersökningar öfver fytoplankton och under den pelagiska regionen forsiggående gyttje-och dybildningar inom vissa syo-och mellasvenska ubergsvatten. *K. svenska VetenskAkad. Handl.,* 56, No. 6:1–165 (Swedish text); (German summ. 124–157). **379**

Naumann, E., 1919. Några synpunkter angående planktons ökologi. Med särskild hänsyn till fytoplankton. *Svensk, bot. Tidskr.,* 13:129–158 (Swedish text); (German summ. 158–163). **379**

Naumann, E., 1923. Spezielle Untersuchungen über die Ernährungsbiologie des Tierischen Limnoplanktons II. Über den Nahrungserwerb und die natürliche Nahrung der Copepoden und der Rotiferen des Limnoplanktons. *Acta Univ. lund.* N.F., Avd. II (K. fysiogr. Sällsk. Lund., Handl. N.F.), 19, Nr. 6:17 pp. **380, 526, 527, 627, 628, 679**

Naumann, E., 1925. Die Gallertbildungen des pflanzlichen Limnoplanktons. Eine morphologisch-ökologische Übersicht. *Acta Univ. Lund.,* N.F. Avd. II., 21, Nr. 5:26 pp. **272, 273**

Naumann, E., 1928. Die eulimnische Zonation. Einige terminologische Bemerkungen. *Arch. Hydrobiol.,* 19:744–747. **239, 240**

Naumann, E., 1931. Limnologische Terminologie. E. Abderhalden: *Handbuch der biologischen Arbeitsmethoden.* Berlin and Wien, Urban & Schwarzenberg, Abt lx, Teil 8:776 pp. **235, 236, 380, 381**

Naumann, E., 1932. Grundzüge der regionalen Limnologie. *Binnengewässer,* 11, ix, 176 pp. **380**

Nauwerck, A., 1959. Zur Bestimmung der Filtrierrate limnischer Planktontiere. *Arch. Hydrobiol.,* Suppl. 25 (Falkau-Schriften 4):83–101. **599, 602**

* Nauwerck, A., 1963. Die Beziehungen zwischen Zooplankton und Phytoplankton in See Erken. *Symb. bot. upsal.,* 17 Nr. 5, 163 pp. **497, 498, 499, 673, 706, 707**

Needham, J., 1930. On the penetration of marine organisms into fresh water. *Biol. Zbl.,* 50:504–509. **190**

Needham, J., 1942. *Biochemistry and morphogenesis.* Cambridge Univ. Press., 785 pp. **811**

Needham, J., and Lerner, J. M., 1940. Terminology of relative growth-rates. *Nature, Lond.,* 146:618. **811**

Neese, J. C., Dugdale, R. C., Dugdale, V. A., and Goering, J. J., 1962. Nitrogen metabolism in lakes I Measurement of nitrogen fixation with N 15. *Limnol. Oceanogr.,* 7:163–169. **318**

Newcombe, C. L., and Slater, J. V., 1950. Environmental factors of Sodon Lake—a dichothermic lake in Southeastern Michigan. *Ecol. Monogr.,* 20:207–227. **315**

Newell, I. M., 1945. *Hydrozetes* Berlese (Acari, Oribatoidea): The occurrence of the genus in North America, and the phenomenon of levitation. *Trans. Conn. Acad. Arts Sci.,* 36:253–268. **88, 254**

Nicholls, A. G., 1944. Littoral copepoda from South Australia II Calanoida, Cyclopoida, Notodelphyoida, Monstrilloida and Caligoida. *Rec. S. Aust. Mus.* 8:1–62. **129**

Nicholls, G. E., 1943. The Phreatoicoidea Part I.—The Amphisopidae. *Pap. Proc. R. Soc. Tasm.,* 1942. 1–45. **105**

Nicholls, G. E., 1944. The Phreatoicoidea. Part II, The Phreatoicidae. *Pap. Proc. R. Soc. Tasm.,* 1943. 1–156. **105**

Nicol, J. A. C., 1960. The biology of marine animals. London, Sir Isaac Pitman & Sons Ltd., xi, 707 pp. **151**

* Nipkow, F., 1927. Über das Verhalten der Skelette planktischer Kieselalgen im geschichteten Tiefenschlamm des Zürich-und Baldeggersees. *Z. Hydrol. Hydrogr., Hydrobiol.,* 4:71–120. **928, 929, 932–934, 937, 939**

Nipkow, F., 1950. Ruheformen planktischer Kieselalgen im geschichteten Schlamm des Zürichsees. *Schweiz Z. Hydrol.* 12:263–270. **459**

Noodt, W., 1963. Anaspidacea (Crustacea, Syncarida) in der südlichen Neotropis. *Verh. dt. zool. ges.* (*Zool. Anz.* Suppl.), 26:568–578. **94**

Norojilov, N., 1957. Un nouvel ordre d'Arthropodes particuliers: Kazacharthra, du Lias des monts Ketmen. *Bull. Soc. geol. Fr.,* (sér 6) 17:171–185. **118**

Nussbaum, 1897. Die Entstehung des Geschlechtes bei Hydatina senta. *Arch. mikrosk. Anat. EntwMech.,* 49:227–308. **510**

* Nygaard, G., 1938. Hydrobiologische studien über dänische Teiche und Seen. 1. Teil: Chemische—physikalische Untersuchungen und Plankton—wägungen. *Arch. Hydrobiol.,* 32:523–692. **407**

* Nygaard, G., 1949. Hydrobiological studies of some Danish ponds and lakes II. [*K. danske Vidensk. Selsk.*] *Biol. Skr.,* 7, 293 pp. **391, 392**

Nygaard, G., 1955. On the productivity of five Danish waters. *Verh. int. Verein. theor. angew. Limnol.,* 12:123–133. **392, 393**

Nygaard, G., 1956. Ancient and recent flora of diatoms and Chrysophyceae in Lake Gribsø pp. 32–94. In K. Berg, and Ib. C. Petersen. Studies on the humic acid Lake Gribsø. *Folia Limnol. Scand.,* No. 9, 273 pp. **339, 374, 386**

Obreshkove, V., and Frazer, A. W., 1940. Growth and differentiation of *Daphnia magna* eggs in vitro. *Biol. Bull. mar. biol. Lab., Woods Hole,* 78:428–436. **576**

Odum, H. T., 1953. Factors controlling marine invasion into Florida fresh waters. *Bull. mar. Sci. Gulf Caribb.,* 3:134–156. **173**

Okuda, S., 1943. Occurrence of a freshwater polychaete in Central China (Reports on the Limnological Survey of Central China. IV). *J. Shanghai Sci. Inst.* n.s., 2:99–103. **73**

Olivier, S. R., 1954. Una nueva especie del género "Moina" (Crust. Cladocera). *Notas Mus. La Plata* (Universidad Nacional de Eva Peron Notas del Museo), 17 (No. 148): 81–86. **607**

Olofsson, O., 1918. Studien über die Süsswasserfauna Spitzbergens. *Zool. Bidr. Upps.,* 6:183–646. **599, 887**

Olson, E. C., 1952. The evolution of a Permian vertebrate chronofauna. *Evolution, Lancaster, Pa.,* 6:181–196. **196**

Ostenfeld, C. H., 1913. De danske farvandes Plankton i aurene 1898–1901. Phytoplankton og Protozoer *K. danske Vidensk. Selsk. Schr.,* 7. Raekke, Naturvidensk. og Mathem. Afd., 9, No. 2:113–478. **241**

Ostenfeld, C. H., and Wesenberg-Lund, C., 1906. A regular fortnightly exploration of the plankton of the two Icelandic lakes, Thingvallavatn and Myvatn. *Proc. R. Soc. Edind.,* 25, No. 113:1092–1167. **616**

Osterhout, W. V., 1933. Permeability in large plant cells and in models. *Ergebn. Physiol.,* 35:967–1021. **29**

Österlind, S., 1947. Growth of a planktonic green alga at various carbonic acid and hydrogen-ion concentrations. *Nature, Lond.,* 159:199–200. **309**

Österlind, S., 1948. Influence of low bicarbonate concentrations on the growth of a green alga. *Nature, Lond.,* 161:319–320. **309**

Ostwald, W., 1902. Zur Theorie des Planktons. *Biol. Zbl.,* 22:596–605, 609–638. **251, 260, 824**

Ostwald, W., 1904. Experimentelle Untersuchungen über den Saisonpolymorphismus bei Daphniden. *Arch. EntwMech. Org.,* 18:415–451. **824, 827**

Otto, J. P., 1934. Über den osmotischen Druck der Blutflüssigkeit von Heteropanope tridentata (Maitland). *Zool. Anz.,* 108:130–135. **166**

Otto, J. P., 1937. Über den Einfluss der Temperatur auf den osmotischen Wert der Blutflüssigkeit bei der Wollhandkrabbe (Eriocheir sinensis H. Milne-Edwards). *Zool. Anz.,* 119:98–105. **166**

Overbeck, A., 1876. Ueber stationare Flüssigkeitsbewegungen mit Berücksichtigung der inneren Reibung. *J. reine angew. Math.,* 81:62–80. **261, 264**

Owen, G., 1959. Observations on the Solenacea with reasons for excluding the family Glaucomyidae. *Phil. Trans. R. Soc.,* 242B:59–97. **61**

Oye, P. van, 1931. Rhizopoda from South Africa. *Revue Zool. Bot. afr.,* 21:54–73. **493n.**

Oye, P. van, 1932. Neue Rhizopoden aus Africa. *Zool. Anz.,* 99:323–328. **493n.**

* Pacaud, A., 1939. Contribution à l'Ecologie des Cladocères. *Bull. biol. Fr. Belg.,* Suppl. 25:260 pp. **556, 595, 603, 604, 605, 607**

Palmer, T. C., and Keeley, F. J., 1900. The structure of the diatom girdle. *Proc. Acad. nat. Sci. Philad.,* 1900:465–479. **922, 927**

Pancella, J. R., and Stross, R. G., 1963. Light induced hatching *Daphnia* resting eggs. *Chesapeake Sci.,* 4:135–140. **595, 599**

Panikkar, N. K., 1940. Influence of temperature on osmotic behaviour of some crustacea and its bearing on problems of animal distribution. *Nature, Lond.,* 146:366–367. **166**

Panikkar, N. K., 1941a. Osmotic behaviour of the fairy shrimp *Chirocephalus diaphanus* Prévost. *J. exp. Biol.,* 18:110–114. **163**

Panikkar, N. K., 1941b. Osmoregulation in some palaemonid prawns. *J. mar. biol. Ass. U.K.,* 25:317–359. **172, 173, 220**

Panknin, W., 1947. Zur Entwicklungsgeschichte der Algensoziologie und zum Problem der "echten" und "zugehörigen" Algengesellschaften. *Arch. Hydrobiol.,* 41:92–111. **376**

Pantin, C. F. A., 1931. The adaptation of *Gunda ulvae* to Salinity. *J. exp. Biol.*, 8:82–94. **157**

Papanicolau, G., 1910. Experimentelle Untersuchungen über die Fortpflanzungsverhältnisse der Daphniden (*Simocephalus vetulus* und *Moina rectirostris* var. *Lilljeborgii*). *Biol. Zbl.*, 30:689–692, 737–750, 752–774, 785–802; 31:81–85. **587**

Parise, A., 1958. Variabilità di *Keratella cochlearis* (Gosse) (Rotatoria) nel Lago di Nemi. *Mem. Acc. Patavina*, 70:11–26. **876**

Parise, A., 1961. Sur les genres Keratella, Synchaeta, Polyarthra et Filinia d'un lac italien. *Hydrobiologia*, 18:121–135. **507**

Parker, R. A., 1959. *Simocephalus* reproduction and illumination. *Ecology*, 40:514. **580**

Parker, R. A., 1960. Competition between *Simocephalus vetulus* and *Cyclops viridis*. *Limnol. Oceanogr.*, 5, No. 2:180–189. **587**

Parker, R. A., 1961. Competition between *Eucyclops agilis* and *Daphnia pulex*. *Limnol. Oceanogr.*, 6, No. 3:299–301. **587**

Parker, R. A., and Hazelwood, D. H. Some possible effects of trace elements on fresh-water microcrustacean population. *Limnol. Oceanogr.*, 7:344–347. **676**

Pascher, A., 1913. Flagellatae 2. Süsswasser-Flora Deutschlands. Heft 2. Jena, Gustav Fischer iv, 192 pp. **335**

Pascher, A., 1927. Volvocales = Phytomonadinae. Flagellatae 4. = Chlorophyceae I (mit dem allgemeinen Teile zu den Chlorophyceen). Süsswasser-Flora Deutschlands. Heft. 4. Jena, Gustav Fischer iv, 505 pp. **306**

Patalas, K., 1954. Skorupiaki planktonowe jako baza pokarmowa w gospodarce sielawowej na jeziorze Charzykowo. *Polskie Archwm Hydrobiol.*, 2:259–276. **609, 611, 613, 616, 619, 637**

Patalas, K., 1956. Sezonowe zmiany w zespole skorupiaków pelagicznych w jeziorze Zamhowym. *Polskie Archwm Hydrobiol.*, 3:203–251. **613, 616**

Patalas, K., 1963. Seasonal changes in pelagic crustacean plankton in six lakes of Wegozewa distinct. *Roczn. Naukro In.* (ser. B) 82:209–234 (Polish text, English summ.). **785**

* Patrick, R., 1943. The Diatoms of Linsley Pond, Connecticut. *Proc. Acad. nat. Sci. Philad.*, 95:53–110. **453**

Patrick, R., 1949. A proposed biological measure of stream conditions, based on a survey of the Conestoga Basin, Lancaster County, Pennsylvania. *Proc. Acad. nat. Sci. Philad.*, 101:277–341. **389**

Patrick, R., 1959. Bacillariophyceae. In H. B. Ward, and G. C. Whipple *Fresh-water Biology*, 2nd ed., ed. W. T. Edmondson. New York, John Wiley & Sons, pp. 171–189. **307, 340**

Patrick, R., Hohn, M. H., and Wallace, J. H., 1954. A new method for determining the pattern of the diatom flora. *Notul. Nat.* (Acad. Nat. Sci. Phil.), No. 259, 12 pp. **361**

Paulian, R., and Delamare-Deboutteville, C., 1956. Un cirolanide cavernicole à Madagascar. *Mem. Inst. scient. Madagascar* (sér. A) 11:85–88. **100**

Pavesi, P., 1882. Altra serie di ricerche e studi sulla fauna pelagica dei laghi italiani. *Att. Soc. Veneto-Trentina Sci. nat.*, 8:340–403. **726**

Pearce, E. J., 1945. The water-bug Aphelocheirus montandoni Horv. in North Wales. *Entomologist's mon. Mag.*, 81:139. **188**

* Pearsall, W. H., 1921. The development of vegetation in the English Lakes, considered in relation to the general evolution in glacial lakes and rock basins. *Proc. R. Soc.,* 92B:259–284. **330, 331, 381, 389, 418, 433**

Pearsall, W. H., 1923. A theory of diatom periodicity. *J. Ecol.,* 11:165–183. **434, 443**

Pearsall, W. H., 1929. Form variation in Ceratium hirundinella O. F. M. *Proc. Leeds. Phil. lit. Soc.,* 7:432–439. **916, 919, 920**

* Pearsall, W. H., 1930. Phytoplankton in the English Lakes I. The proportions in the water of some dissolved substances of biological importance. *J. Ecol.,* 18:306–320. **418, 443**

* Pearsall, W. H., 1932. Phytoplankton in the English Lakes II. The composition of the phytoplankton in relation to dissolved substances. *J. Ecol.,* 20:241–262. **317, 381, 401, 419, 420, 443–446, 455, 457, 458**

Pearse, A. S., 1950. *The Emigrations of animals from the Sea.* New York, Sherwood, Dryden, xii, 210 pp. **199**

* Pejler, B., 1957a. On variation and evolution in planktonic rotatoria. *Zool. Bidr., Upps.,* 32:1–66. **873, 875, 876, 879, 880, 891**

* Pejler, B., 1957b. Taxonomical and ecological studies on planktonic rotatoria from northern Swedish Lapland. *K. svensk Vetensk Akad. Handl.,* Fjärde Ser. Bd. 6, No. 5, 68pp. **526, 529, 530, 542, 550, 874**

Pejler, B., 1957c. Taxonomical and ecological studies on planktonic rotatoria from Central Sweden. *K. svensk Vetensk Akad. Handl.,* Fjärde Ser. Bd., 6 No., 7, 52pp. **525**

Pejler, B., 1961. The zooplankton of Ösbysjön, Djursholm. I. Seasonal and vertical distribution of species. *Oikos,* 12:225–248. **550, 619**

Pejler, B., 1962a. Notes on some limnoplanktic protozoans with descriptions of two new species. *Zool. Bidr. Upps.* 33:447–452. **493, 494, 501**

Pejler, B., 1962b. The zooplankton of Ösbysjön, Djursholm. II. Further ecological aspects. *Oikos,* 13:216–231. **707**

Pejler, B., 1962c. On the variation of the rotifer Keratella cochlearis (Gosse). *Zool. Bidr. Upps,* 35:1–17. **877, 879, 880, 909**

Pelloni, E., 1936. Contributo all'indagine idrochimica e idrobiologica del Verbano (Bacino di Locarno). *Trav. Inst. zool. Univ. neuchâtel.* (not seen: ref. in Pirocchi 1947a). **609**

Pelseneer, P., 1905. L'origine des animaux d'eau douce. *Bull. Acad. r. Belg. Cl. Sci.,* 1905:699–741. **191**

Penard, E., 1904. *Les héliozoaires d'eau douce.* Genève H. Kündig., 341 pp. **497**

Pennak, R. W., 1944. Diurnal movements of zooplankton organisms in some Colorado mountain lakes. *Ecology,* 25:387–403. **742**

* Pennak, R. W., 1946. The Dynamics of Fresh-water Plankton Populations. *Ecol. Monogr.,* 10:339–356. **405, 407**

* Pennak, R. W., 1949. Annual Limnological Cycles in Some Colorado Reservoir Lakes. *Ecol. Monogr.,* 19:233–267. **402, 405, 407**

Pennak, R. W., 1963. Ecological affinities and origins of free-living acelomate fresh-water invertebrates. In *The Lower Metazoa. Comparative Biology and Phylogeny,* ed. E. C. Dougherty et al. Univ. California Press, pp. 435–451. **52, 195, 198**

Pennington, W., 1943. Lake Sediments: The bottom deposits of the north basin of Windermere, with special reference to the diatom succession. *New Phytol.,* 42:1–27. **453**

Pérès, J., 1957.   Le problem de l'étagement des formations benthiques.   *Recl. Trav. Stn mar. Endoume.*, 21, Bull. 12:4–21. **239**

Peters, H., 1935.   Ueber den Einfluss des Saltzgehaltes in Aussenmedium auf den Bau und die Funktion der Exretionsorgane dekapoder Crustaceen (nach Untersuchungen an Potamobius fluviatilis und Homarus vulgaris).   *Z. Morph. Ökol. Tiere*, 30:355–81. **159**

Petrova, N. H., 1959.   K. kharakteristike fitoplanktona Yakimvarskogo zaliva Ladozhskogo ozera (Contribution to a characterization of the phytoplankton of Yakimvarsk Bay, Lake Ladoga). *Bot. Zhur.*, 44:1311–1314. **432**

Pettibone, M. H., 1953.   Freshwater polychaetous annelid, Manayunkia speciosa Leidy, from Lake Erie. *Biol. Bull.*, 105:149–153. **69, 72**

Pfitzer, E., 1869.   Über Bau und Zelltheilung der Diatomaceen.   *Sber. niederr-bein. Ges. Nat. u. Heil.* 26:86–89. **922, 925**

Philipose, M. T., 1960.   *Proc. Symposium on Algology* I.C.A.R., 1959:272 (not seen: ref. George 1962). **388**

Phillips, W., 1884.   The breaking of the Shropshire Meres.   *Trans. Shrops. archaeol. nat. Hist. Soc.*, 7:277–300. **425, 426**

Picard, J., 1951.   Contributions à l'étude des meduses de la famille des Moerisi-idae.   *Bull. Inst. océanogr. Monaco*, No. 994, 16 pp. **39, 41**

Picken, L. E. R., 1937.   The excretory mechanism in certain Mollusca.   *J. exp. Biol.*, 14:20–34. **155**

Pickford, G. E., and Atz, J. W., 1957.   *The physiology of the pituitary gland of fishes.*   New York, New York Zool. Soc., xxiii, 613 pp. **168, 169n.**

Pickford, G. E., and Phillips, J. G., 1959.   Prolactin, a factor in promoting survival of hypophysectomized killifish in freshwater.   *Science, N.Y.*, 130:454–455. **171, 172**

Pierantoni, U., 1908.   Protodrilus. *Fauna Flora Golf. Neapel. Monogr.*, 31:vii, 226 pp. **67**

Pintner, I. J. & Provasoli L, 1958.   Artificial cultivation of a red-pigmented marine blue-green alga, *Phormidium persicinum*. *J. gen. Microbiol.*, 18:190–97. **317**

Pintner, I. J., and Provasoli, L., 1959.   The nutrition of *Volvox globator* and *Volvox tertius*. *Proc. Int. bot. Congr.*, Abs. II:300–301. **322, 325**

Pirocchi, L., 1944.   Distribuzione nella Penisola Italiana di tre specie di Diap-tomidi (*Eud. vulgaris* Schmeil, *Arctod. bacillifer* Koelb., *Acanthod. denticornis* Wierz.).   *Atti R. Accad. Ital. Memorie*, 14:859–888. **692**

* Pirocchi, L., 1947a.   Struttura e vicenda delle biocenosi mesoplanctiche del Lago Maggiore.   *Mem. Ist. ital. Idrobiol.*, 3:57–119. **609, 616**

Pirocchi, L., 1947b.   Diaptomidi d'alta montagna. III. Il diaptomide di Pierafica (Alpi marittime).   *Mem. Ist. ital. Idrobiol.*, 3:469–476. **692**

Pirozhnikov, P. L., 1937.   A contribution to the study of the origin of the northern element in the fauna of the Caspian Sea.   *Dokl. Akad. Nauk SSSR.*, 15:521–524. **212**

Poisson, R., 1926.   *L'Anisops producta* Fieb. (Hemiptère, Notonectidae) Ob-servations sur son anatomie.   *Arch. Zool. exp. gén.*, 65:181–208. **256**

Ponyi, E., 1956.   Die *Diaptomus*-Arten der Natrongewässer auf der grossen Ungarischen Tiefebene. *Zool. Anz.*, 156:257–271. **676**

Popham, E. J., 1961.   *Some aspects of life in freshwater.*   Harvard Univ. Press, vii, 127 pp. **188**

Potts, W. T. W., 1954a. The rate of urine production of *Anodonta cygnea*. *J. exp. Biol.*, 31:614–617. **155**

* Potts, W. T. W., 1954b. The energetics of osmotic regulation in brackish- and fresh-water animals. *J. exp. Biol.*, 31:618–630. **151, 155, 156, 157, 161, 167, 168, 202**

Potts, W. T. W., 1958. The inorganic and amino acid composition of some lamellibranch muscles. *J. exp. Biol.*, 35:749–764. **155**

* Potts, W. T. W., and Parry, G., 1964. *Osmotic and ionic regulation in animals*. New York, Macmillan, xiii, 423 pp. **151**

Poulsen, E. M., 1937. Freshwater Crustacea in *The Zoology of the Faroes*. Copenhagen, Andr. Fred. Høst & Søn, II Pt 1 No. xxxi, 21 pp. **693**

Poulsen, E. M., 1939. Freshwater Crustacea. in *The Zoology of Iceland*. Copenhagen and Reykjavik, Ejnar Munksgaard, III, Pt. 35:1–50. **693**

Poulsen, E. M., 1940a. *The zoology of East Greenland*. Freshwater Entomostraca. *Meddr Grønland*, 121:1–73. **599**

Poulsen, E. M., 1940b. Biological remarks on Lepidurus articus Pallas, Daphnia pulex de Geer and Chydorus sphaericus O.F.M. in East Greenland. *Meddr Grønland*, 131, No. 1:1–50. **599**

Pourriot, R., 1957a. Contribution à la connaissance des rotifères et des cladocères de la région parisienne. *Hydrobiologia*, 9:38–49. **875**

Pourriot, R., 1957b. Sur la nutrition des rotifères à partir des algues d'eau douce. *Hydrobiologia*, 9:50–59. **526, 527**

Pourriot, R., 1957c. Influence de la nourriture sur l'apparition des femelles mictiques chez deux especes et une variété de Brachionus. *Hydrobiologia*, 9:60–65. **511**

Pourriot, R., 1963. Utilisation des algues brunes unicellulaires, pour l'élèvage des rotifères. *C. r. hebd. Séanc. Acad. Sci. Paris*, 256:1603–1605. **527**

Powers, J. H., 1912. A Case of Polymorphism in Asplanchna, simulating Mutation. *Am. Nat.*, 46:441–462, 521–552. **866, 867, 868, 869**

Pratt, D. B., and Waddell, G., 1959. Adaptation of marine bacteria to growth in media lacking sodium chloride. *Nature Lond.*, 183:1208–1209. **20**

* Pratt, D. M., 1943. Analysis of population development in Daphnia at different temperatures. *Biol. Bull. mar biol. Lab., Woods Hole*, 85:116–140. **583, 587, 588, 589, 591, 592**

Pratt, R., 1940. Influence of the size of the inoculum on the growth of Chlorella vulgaris in freshly prepared culture medium. *Am. J. Bot.*, 27:42–56. **463**

Pratt, R., 1942. Studies on Chlorella vulgaris. V. Some properties of the Growth-inhibitor formed by Chlorella cells. *Am. J. Bot.*, 29:142–148. **463**

Pratt, R., 1943. Studies on Chlorella vulgaris. VI. Retardation of photosynthesis by a growth-inhibiting substance from Chlorella vulgaris. *Am. J. Bot.*, 30:32–33. **463**

Pratt, R., Daniels, T. C., Eiler, J. J., Gunnison, J. B., Kummler, W. D., Oneto, J. F., Spoehr, H. A., Hardin, G. J., Milner, H. W., Smith, J. H. C., and Strain, H. H., 1944. Chlorellin, an antibacterial substance from Chlorella. *Science, N.Y.*, 99:351–352. **463**

Pratt, R., and Fong, J., 1940. Studies on Chlorella vulgaris. II. Further evidence that Chlorella cells form a growth-inhibiting substance. *Am. J. Bot.*, 27:431–436. **463**

Prescott, G. W., 1951. Algae of the Western Great Lakes Area. *Bull. Cranbrook Inst. Sci.,* 30: xiii. 946 pp. **307**

Prescott, G. W., and Magnotta, A., 1935. Notes on Michigan desmids, with descriptions of some species and varieties new to science. *Pap. Mich. Acad. Sci.,* 20:157–170. **331**

Prescott, G. W., and Scott, A. M., 1952. The algal flora of southeastern United States. V. Additions to our knowledge of the desmid genus Micrasterias 2. *Trans. Am. microsc. Soc.,* 71:229–252. **331**

Preston, F. W., 1948. The commonness, and rarity, of species. *Ecology,* 29:254–283. **361, 362, 371**

Preuss, G., 1951. Die Verwandtschaft der Anostraca und Phyllopoda. *Zool. Anz.,* 147:49–64. **93**

Price, J. L., 1958. Cryptic speciation in the vernalis group of Cyclopidae. *Can. J. Zool.,* 36:285–303. **625**

Pringsheim, E. G., 1926. Über das Ca-Bedürfnis einiger Algen. *Planta,* 2: 555–568. **332**

Pringsheim, E. G., 1949. The relationship between bacteria and Myxophyceae. *Bact. Rev.,* 13:47–98. **3**

Pringsheim, E. G., 1953. Die Stellung der grünen Bakterien im System der Organismen. *Arch. Mikrobiol.,* 19:353–364. **315**

Proctor, V. W., 1957a. Some controlling factors in the distribution of *Haematococcus pluvialis. Ecology,* 38:457–462. **464, 467, 468, 472**

Proctor, V. W., 1957b. Studies of algal antibiosis using *Haematococcus* and *Chlamydomonas. Limnol. Oceanogr.,* 2:125–139. **464, 467, 468**

* Provasoli, L., 1958. Nutrition and ecology of Protozoa and Algae. *A. Rev. Microbiol.,* 12:279–308. **310, 320, 323, 343, 345, 347, 348**

Provasoli, L., 1960. *See* Provasoli, and Pintner 1960. **310, 312, 322, 331**

Provasoli, L., and Gold, K., 1962. Nutrition of the American strain of Gyrodinium cohnii. *Arch. mikrobiol.,* 42:196–203. **345**

* Provasoli, L., McLaughlin, J. J. A., and Droop, M. R., 1957. The development of artificial media for marine algae. *Arch. Mikrobiol.,* 25:392–428. **344, 345**

* Provasoli, L., and Pintner, I. J., 1953. Ecological implications of *in vitro* nutritional requirements of algal flagellates. *Ann. N.Y. Acad. Sci.,* 56:839–851. **313, 340, 348**

Provasoli, L., and Shiraishi, K., 1959. Axenic cultivation of the brine shrimp Artemia salina. *Biol. Bull. mar. biol. Lab., Woods Hole,* 117:347–355. **554**

Provasoli, L., Shiraishi K., and Lance, J. R., 1959. Nutritional idiosyncrasies of Artemia and Tigriopus in monoxenic culture. *Ann. N.Y. Acad. Sci.,* 77:250–261. **554, 555**

Pruthi, H. S., 1933. Studies on the Bionomics of Fresh-Waters in India. I. Seasonal changes in the physical and chemical conditions of the water of the tank in the Indian Museum Compound. *Int. Revue ges. Hydrobiol. Hydrogr.,* 28:46–67. **858, 859**

Pulikovski, N., 1924. Metamorphosis of *Deuterophlebia* sp. (Diptera Deuterophlebiidae). *Trans. ent. Soc. Lond.,* 1924, 45–62. **81**

Purasjoki, K. J., 1958. Zur Biologie der Brackwasserkladocere Bosmina coregoni maritima (P. E. Müller) *Suomal. eläin-ja kasvit. Seur. van. elain. Julk. (Ann. Zool. Soc., "Vanamo"),* 19, No. 2, 117 pp. **119**

Quartier, A., 1948. Sur le comportement de Tabellaria fenestrata (Lyngb.) Ktz. dans les trois lacs sub-jurassiens. *Schweiz. Z. Hydro.,* 10:13–22. **942, 943**

Radl, E., 1903. Untersuchungen über den Phototropismus der Tiere. Leipzig, viii, 188 pp. **762**

Ragotzkie, R. A., and Bryson, R. A., 1953. Correlation of Currents with the Distribution of Adult *Daphnia* in Lake Mendota. *J. mar. Res.,* 12:157–172. **291, 293, 797**

Rammner, W., 1926. Formanalytische Untersuchungen an Bosminen. *Int. Revue ges. Hydrobiol. Hydrogr.,* 15:89–136, 145–203. **576, 578, 812, 851**

Rammner, W., 1927. Zur Lokalvariation von Scapholeberis mucronata und deren Abhängigkeit von der Gewässergrösse. *Zool. Anz.,* 72:218–224. **847**

Rand, A. L., 1952. Secondary Sexual Characters and Ecological Competition. *Fieldiana. Zool.,* 34, No. 6:65–70. **357**

Rankama, K., 1948. New evidence of the origin of pre-Cambrian carbon. *Bull. geol. Soc. Am.,* 59:389–416. **11**

Rao, H. S., 1931. The supposed resting stage of *Limnocnida indica* Annandale. *Nature, Lond.,* 127:971. **42**

Rao, H. S., 1933. Further observations on the fresh-water medusa, *Limnocnida indica* Annandale. *J. Bombay nat. Hist. Soc.,* 36:210–217. **42**

Rapoport, E. H., and Sanchez, L., 1963. On the epineuston or the superaquatic fauna. *Oikos,* 14:96–109. **238**

* Ravera, O., 1954. La struttura demografica dei Copepodi del Lago Maggiore. *Mem. Ist. ital. Idrobiol.,* 8:109–150. **644, 668, 670, 673, 684, 685**

Ravera, O., and Tonolli, V., 1956. Body size and number of eggs in diaptomids as related to water renewal in mountain lakes. *Limnol. Oceanogr.,* 1:118–122. **658, 661**

Rawson, D. S., 1953. The limnology of Amethyst Lake, a high alpine type near Jasper, Alberta. *Can. J. Zool.,* 31:193–210. **385**

Rawson, D. S., 1956. Algal indicators of trophic lake types. *Limnol. Oceanogr.,* 1:18–25. **385, 387, 393, 453**

Raymond, P. G., 1935. Leanchoilia and other Mid-Cambrian Arthropoda. *Bull. Mus. comp. Zool. Harv.,* 76:203–230. **119**

Raymond, P. E., 1946. The genera of fossil Conchostraca—an order of bivalved Crustacea. *Bull. Mus. comp. Zool. Harv.,* 96:215–307. **118**

Reif, C. B., 1939. The effect of stream conditions on lake plankton. *Trans. Am. microsc. Soc.,* 58:398–403. **805**

Reindl, 1912. (not seen: incomplete ref. in Huber-Pestalozzi 1938, p. 297). **389**

Remane, A., 1929–33. Rotatorien. H. G. Bronns *Klassen und Ordnungen der Tierreichs,* 4 Bd, II Abt, 1 Buch:1–576 (incomplete). **509, 518**

Remane, A., 1936. Gastrotricha und Kinorhyncha. H. G. Bronns *Klassen und Ordnungen der Tierreichs,* 4 Bd., II Abt., 1 Buch, Teil 2: vi, 385 pp. **551, 552n.**

Remane, A., 1963. The enterocelic origin of the coelom. In *The Lower Metazoa. Comparative Biology and Phylogeny,* ed. E. C. Dougherty et al. Univ. California Press, pp. 78–90. **46**

Remane, A., and Schlieper, C., 1958. Die Biologie des Brackwassers. *Binnengewässer*, 22; viii, 348 pp. **151**

Remington, C. L., 1955. The "Apterygota". In *A Century of Progress in the Natural Sciences*—1853–1953. California Academy of Natural Sciences, pp. 495–505. **16n.**

Rezvoj, P., 1926. Über den Nahrungserwerb bei Rotiferen. Arb. d. naturf. Ges. zu Leningrad, Bd. 56 (Russian, German summ.). *Trudȳ leningr. Obschch. Estest. (Trav. Soc. Nat. Léningr.)*, 56:73–89 (not seen, ref. Pejler 1957b). **529**

Rice, T. R., 1954. Biotic influences affecting population growth of planktonic algae. *Fishery Bull. Fish Wildl. Serv., U.S.*, 87:227–245. **464, 465**

Rich, F., 1931. Notes on *Arthrospira platensis*. *Rev. algol.*, 6:75–79. **388**

Rich, F., 1932. Contributions to our knowledge of the freshwater algae of Africa. *Trans. R. Soc. S. Afr.*, 20:11–188. **333**

Richard, J., 1896. Revision des Cladocères. Deuxième partie. *Annals Sci. nat.*, (sér. 8) 2:187–360. **813**

Richardson, H., 1897. Description of a new species of Sphaeroma. *Proc. biol. Soc. Wash.*, 11:105–107. **104**

Richman, S., 1958. The transformation of energy by *Daphnia pulex*. *Ecol. Monogr.*, 28:273–291. **602**

Ricker, K. E., 1959. The origin of two glacial relict crustaceans in North America, as related to Pleistocene glaciation. *Can. J. Zool.*, 37:871–893. **205**

Ricker, W. G., 1937. Statistical treatment of sampling processes useful in the enumeration of plankton organisms. *Arch. Hydrobiol.*, 31:68–84. **788, 792**

Riegel, J. A., 1959a. Some aspects of osmoregulation in two species of Sphaeromid isopod crustacea. *Biol. Bull. mar. biol. Lab., Woods Hole*, 116:272–284. **103, 104, 167**

Riegel, J. A., 1959b. A revision in the sphaeromid genus Gnorimosphaeroma Menzies (Crustacea: Isopoda) on the basis of morphological, physiological and ecological studies of two of its "subspecies." *Biol. Bull. mar. biol. Lab., Woods Hole*, 117:154–162. **103**

Rigler, F. H., 1961. The relation between concentration of food and feeding rate of Daphnia magna Straus. *Can. J. Zool.*, 39:857–868. **600**

Riley, G. A., 1939a. Limnological studies in Connecticut. *Ecol. Monogr.*, 9:53–94. **406**

Riley, G. A., 1939b. Correlations in Aquatic Ecology with an example of their application to problems of plankton productivity. *J. mar. Res.*, 2:56–73. **429**

Riley, G. A., 1940. Limnological Studies in Connecticut. III. The Plankton of Linsley Pond. *Ecol. Monogr.*, 10:279–306. **402, 406, 424, 425, 429, 476**

Riley, G. A., 1963. Marine Biology, I. *Proceedings of the First International Interdisciplinary Conference*, ed. G. A. Riley. Washington D.C., Amer. Inst. Biol. Sci., 286 pp. *(See* pp. 69–70). **357, 358**

* Riley, G. A., Stommel, H., and Bumpus, D. F., 1949. Quantitative Ecology of the Plankton of the Western North Atlantic. *Bull. Bingham oceanogr. Coll.*, 12, No. 3, 169 pp. **277, 283–285, 287, 288, 289**

Roach, A. W., and Silvey, J. K., 1959. The occurrence of marine Actinomycetes in Texas Gulf Coast substrates. *Am. Midl. Nat.*, 62:482–499. **21**

Robbins, W. J., Hervey, A., and Stebbins, M. E., 1950. Studies on Euglena and vitamin B_{12}. *Bull. Torrey bot. Club*, 77:423–41. **349**

Robbins, W. J., Hervey, A., Stebbins, M. E., 1951. Further observations on Euglena and B_{12}. *Bull. Torrey bot. Club.*, 78:363–375. **317**

Robert, H., 1925. Sur la variabilité de quelques espèces planktoniques du lac de Neufchâtel. *Ann. Biol. lacustre,* 14:5–38. **879**

Robertson, G. M., 1950. Some paleoecological speculations regarding the earliest vertebrates. *Proc. Iowa Acad. Sci.,* 57:491–497. **134**

Robertson, J. D., 1949. Ionic regulation in some marine invertebrates. *J. exp. Biol.,* 26:182–200. **152**

Robertson, J. D., 1954. The chemical composition of the blood of some aquatic chordates including members of the Tunicata, Cyclostomata and Osteichthyes. *J. exp. Biol.,* 31:424–442. **168, 169**

Robertson, J. D., 1957. The habitat of the early vertebrates. *Biol. Rev.,* 32:156–187. **132, 134, 195**

Robertson, R. N., 1960. Ion transport and respiration. *Biol. Rev.,* 35:231–264. **151, 152**

* Rodhe, W., 1948. Environmental requirements of freshwater plankton algae. Experimental studies in the ecology of phytoplankton. *Symb. bot. upsal.,* 10, No. 1, 1–149. **317, 340, 386, 427–429, 435, 444, 455, 457**

Rodhe, W., 1955. Can plankton production proceed during winter darkness in subartic lakes? *Verh. int. Verein. theor. angew. Limnol.,* 12:117–122. **347, 669**

Røen, U., 1955. On the number of eggs in some free-living freshwater copepods. *Verh. int. Verein. theor. angew. Limnol.,* 12:447–454. **653**

Røen, U., 1957. Contributions to the biology of some Danish free-living freshwater copepods. *Biol. Skr.,* 9:1–100. **631, 639, 653**

Rogick, M. D., 1959. Bryozoa. In W. B. Ward, and G. C. Whipple, *Freshwater Biology,* 2nd ed., ed. W. T. Edmondson. New York and London, John Wiley & Sons, pp. 495–507. **53**

Roll, H., 1939. Zur Terminologie des Periphytons. *Arch. Hydrobiol.,* 35:59–69. **237**

Romer, A. S., 1955. Fish origins–fresh or salt water? *Deep-sea Res.,* Suppl. to Vol. 3:261–80. **134**

Rosenberg, M., 1938. The Culture of algae and its applications. *Rep. Freshwat. biol. Ass.,* 6:43–46. **340**

Rousselet, C. F., 1893. On Floscularia pelagica *n.sp.* and notes on several other rotifers, *Jl R. microsc. Soc.,* 1893:444–449. **520**

Rousselet, C. F., 1902. The Genus *Synchaeta:* A Monographic Study, with Descriptions of Five New Species. *Jl R. microsc. Soc.,* pp. 269–290, 393–411. **519, 529**

Roy, Jean, 1932. Copépodes et Cladocères de l'Ouest de la France. Recherches biologiques et faunistiques sur le plancton d'eau douce des vallées du Loir et de la Sarthe. Thèse. Faculté des Sciences, Univ. Paris, No. 2207, sér. A., No. 1338:1–226. **627, 639**

Ruffo, S., 1951. *Ingolfiella Leleupi* n. sp. nuovo anfipodo troglobio del Congo Belga (Amphipoda-Ingolfiellidae). *Rev. Zool. Bot. afr.,* 44:189–209. **107**

Rühe, F. E., 1912. Bosmina coregoni im baltischen Seengebiete. *Zoologica, Stuttgart,* 63:1–141. **851**

Russell, F. S., 1927. The vertical distribution of plankton in the sea. *Biol. Rev.,* 12:213–262. **734**

Ruttner, F., 1905. Ueber das Verhalten des Oberflächenplankton zu verschieden Tageszeiten im Grösser Plönersee. *ForschBer. biol. Stn Plön,* 12:35–62. **732, 735**

Ruttner, F., 1909.   Über tägliche Tiefenwanderungen von Planktontieren unter dem Eise und ihre Abhängigkeit vom Lichte. *Int. Revue ges. Hydrobiol. Hydrogr.*, 2:397–423. **754**

Ruttner, F., 1914.   Uferflucht der Planktons und ihr Einfluss auf die Ernährung der Salmonidenbrut. *Int. Revue ges. Hydrobiol. Hydrogr.*, Biol. Suppl. 6:7 pp. **804**

* Ruttner, F., 1930.   Das Plankton des Lunzer Untersees seine Verteilung in Raum und Zeit während der Jahre 1908–1913. *Int. Revue ges. Hydrobiol. Hydrogr.*, 23:1–138, 161–287. **247, 411, 413, 414, 415, 416–418, 440, 442, 480–482, 530, 537, 616, 620**

Ruttner, F., 1937a.   Ökotypen mit verschiedener Vertikalverteilung an Plankton der Alpenseen. *Int. Revue ges. Hydrobiol. Hydrogr.*, 35:7–34. **433, 434, 446, 732, 924, 927, 936**

* Ruttner, F., 1937b.   Limnologische Studien an einigen Seen der Ostalpen. *Archiv. Hydrobiol.*, 32:167–319. **416, 417, 430, 431, 434, 504**

Ruttner, F., 1943.   Beobachtungen über die tägliche Vertikalwanderung des Planktons in tropischen Seen. *Arch. Hydrobiol.*, 40:474–492. **731, 740, 741, 785**

* Ruttner, F., 1940, 1952b, 1962 [1953, 1963]. *Grundriss der Limnologie.* 1. aufl. W. deGruyter Berlin, 167 pp., 1940. 2. aufl. W. deGruyter Berlin, 1952b. 3. aufl. 1962, 332 pp., 1962. trans. of 2nd. [1953] and 3rd [1963] eds. as *Fundamentals of Limnology* by D. G. Frey, and F. E. J. Fry, Univ. Toronto Press, xi + 242 pp. **239, 240, 272, 273, 283**

Ruttner, F., 1947.   Zur Frage der Karbonatassimilation der Wasserpflanzen. I, Teil: Die beiden Haupttypen der Kohlenstoff auf nahme. *Öst. bot. Z.*, 94: 265–294. **309**

Ruttner, F., 1948.   Zur Frage der Karbonatassimilation der Wasserpflanzen. II. Teil: Das Verhalten von *Elodea canadensis* und *Fontinalis antipyretica* in Losungen von Natrium-bzw. Kaliumkarbonat. *Öst. bot. Z.*, 95:208–238. **309**

* Ruttner, F., 1952a.   Planktonstudien der Deutschen Limnologischen Sundan-Expedition. *Arch. Hydrobiol.*, Suppl. 21:1–274. **375, 383, 384, 386, 497, 499, 504, 505**

Ruttner, F., 1959.   Einige Beobachtungen ueber das Phytoplankton norditalienischer Seen. *Mem. Ist. ital. Idrobiol.*, 11:73–111. **385**

Ruttner-Kolisko, A., 1946.   Über das Auftreten unbefructeter "Dauereier" bei *Anuraea aculeata* (*Keratella quadrata*). *Öst. zool. Z.*, 1:179–191. **510, 534**

* Ruttner-Kolisko, A., 1949.   Zum Formwechsel-und Artproblem von Anuraea aculeata (*Keratella quadrata*). *Hydrobiologia,* 1:425–468. **869, 883, 884, 886–888, 891, 894**

Ruttner-Kolisko, A., 1955.   *Rheomorpha neiswestnovae* und *Marinella flagellata*, zwei phylogenetische interessante Wurmtypen aus dem Süsswasserpsammon. *Öst. zool. Z.*, 6:55–69. **16n.**

* Rylov, W. M., 1935.   Das Zooplankton der Binnengewässer. *Binnengewässer*, 15:ix + 271 pp. **125, 126, 607, 650, 664, 698, 851**

Ryther, J. H., 1954.   Inhibitory effects of phytoplankton upon the feeding of *Daphnia magna* with reference to growth, reproduction, and survival. *Ecology* 35:522–533. **600, 603**

Ryther, J. H., 1956.   The measurement of primary production. *Limnol. Oceanog.*, 1:72–84. **24**

Ryther, J. H., and Guillard, R. R. L., 1959. Enrichment experiments as a means of studying nutrients limiting to phytoplankton production. *Deep-Sea Res.,* 6:65–69. **312**

Rzóska, J., 1927. Contribution à l'étude des Copépodes de la Grande Prologne. *Bull. Soc. Amis Sci. Lett. Poznan,* B1:34–43, 1 pl. **652**

Rzóska, J., 1927. Einige Beobachtungen über temporale Grössenvariation bei Copepoden und einige andere Fragen ihrer Biologie. *Int. Revue ges. Hydrobiol. Hydrogr.,* 17:99–114. **857**

Sachse, R., 1911, *see* Dieffenbach & Sachse. **895, 896, 900, 903, 904, 905, 906**

Sailer, R. I., 1948. *The Genus Trichocorixa (Corixidae, Hemiptera).* In Hungerford 1948, pp. 289–407. **86**

Sanders, H. L., 1963. The Cephalocardia. Functional morphology, larval development, comparative external anatomy. *Mem. Conn. Acad. Arts Sci.,* 15:80 pp. **91, 92**

Sars, G. O., 1894. Contributions to the knowledge of the fresh-water Entomostraca of New Zealand, as shown by artificial hatching from dried mud. *Skr. VidenskSelsk. Christiania,* I Math-naturvid. Kl 1894. No. 5:62 pp. **665**

Sars, G. O., 1894–5. Crustacea Caspia. Pt. I. Mysidae. *Izv. imp. Akad. Nauk. (Bull. Akad. imp. Sci.)* (4) 26:51–74, Pt II. Cumacea *ibid.* 297–338, Pt III. Amphipoda *ibid.* (5) 1:179–223, 343–378, *ibid.* (5) 3:275–314. **207, 208, 209, 217**

Sars, G. O., 1896. On fresh-water Entomostraca from the neighborhood of Sydney, partly raised from dried mud. *Arch. Math. Naturv.* 18; No. 3, 81 pp. **572, 665, 824**

Sars, G. O., 1901. Contributions to the knowledge of the freshwater Entomostraca of S. Am. Part II Copepoda-Ostracoda. *Archiv Math. Naturv.,* 24; No. 1, 52 pp. **665**

Sars, G. O., 1902. On the Polyphemidae of the Caspian Sea. *Ezheg. zool. Muz (Annu. Mus. zool. Acad. Sci. St. Ptsb.),* 6:31–54. **119, 120**

Sars, G. O., 1903. On the crustacean fauna of Central Asia. Part III. Copepoda and Ostracoda. *Ezheg. zool. Muz (Annu. Mus. zool. Acad. Sci. St. Ptsb.),* 8:195–232. **123**

Sars, G. O., 1914. *Daphnia carinata* King, and its remarkable varieties. *Ark. Math. Naturv,* 34, No. 1, 14 pp. **825**

Sars, G. O., 1927. The fresh-water Entomostraca of the Cape Province (Union of South Africa). Part III: Copepoda. *Ann. S. Afr. Mus.,* 25:85–149. **639, 656, 690**

Sassuchin, D. N., Kabanov, N. M., and Neiswestnova, K. S., 1927. Über die mikroskopische Pflanzen und Tierwelt der Sandfläche der Okaufers bei Nurom. *Russk. gidrobiol. Zh.,* (Russ. text 6:59–87); (Germ. summ. 81–83). **237**

Saunders, G. W., 1957. Interrelations of dissolved organic matter and phytoplankton. *Bot. Rev.,* 23:389–409. **316, 320**

Sauramo, M., 1939. The mode of the land upheaval in Fennoscandia during late quaternary time. *C. r. Soc. geol. Finl.* (in *Bull. Commn géol. Finl: Suom. geol. Sour. Julk.,* No. 13:26 pp. (not seen ref. Segerstråle 1957a). **179**

Schacht, F. W., 1897. The North American species of *Diaptomus. Bull. Ill. St. Lab. nat. Hist.,* 5:97–203. **652**

Schaffer, D., Robert, P., and Médioni, J., 1958. Réactions oculo-motrices de la Daphnie (Daphnia pulex DeGeer) en repose à des lumières monochromatiques d'égale énergie. Sensibilité visuelle et sensibilité dermatoptique. *C. Séanc. Soc. Biol.*, 152:1000–1003. **762**

Schallek, W., 1942. The vertical migration of the copepod Acartia tonsa under controlled illumination. *Biol. Bull. Mar. biol. Lab., Woods Hole,* 82:112–126. **760, 782**

Schallek, W., 1943. The reaction of certain crustacea to direct and to diffuse light. *Biol. Bull. Mar. biol. Lab., Woods Hole,* 84:98–105. **760, 782**

Scharfenberg, V. von, 1914. Weitere Untersuchungen an Cladoceren über die experimentelle Beeinflussung des Geschlechts und der Dauereibildung. *Int. Revue ges. Hydrobiol. Hydrogr.,* Biol. Suppl., 6, No. 4, 34 pp. **596**

Schellenberg, A., 1942. Flohkrebse oder Amphipoda. *Tierwelt Dtl.,* 40, 252 pp. **112**

Schellenberg, A., 1937. Kritische Bemerkungen zur Systematik der Süsswassergammariden. *Zool. Jb.* Abt. Syst. Ök Geogr. 69:469–516. **109**

Schelske, C. L., 1962. Iron, organic matter, and other factors limiting primary productivity in a marl lake. *Science, N.Y.,* 136:45–46. **312**

Schelske, C. L., Hooper, F. F., and Haertl, E. J., 1962. Responses of a marl lake to chelated iron and fertilizer. *Ecology,* 43:646–653. **312**

Scher, W. I., and Vogel, H. J., 1957. Occurrence of ornithine S-transaminase: a dichotomy. *Proc. Natn. Acad. Sci. U.S.A.,* 43:796–803. **3n.**

Schierholz, C., 1888. Ueber Entwicklung der Unioniden *Denkschr. Akad. Wiss. Wien.,* Math. Natl. Kl., 55:183–214. **699**

Schiller, J., 1925. Die planktonischen vegetation des adriatischen Meeres. *Arch. Protistenk.,* 53:59–123. **24**

Schlieper, C., 1929. Ueber die Einwirkung niederer Salzkonzentrationen auf marine Organismen. *Z. vergl. Physiol.,* 9:478–514. **153, 159**

Schmarda, L. K., 1859. *Neue wirbellose Thiere beobachtet und gesammelt auf einer Reise un die Erde 1853 bis 1857.* Erster Band. *Turbellarien, Rotatorien und Anneliden.* Erste Hälfte. Leipzig, Verlag von Wilhelm Englemann. 66 pp., xv pl. **47**

Schmidt, C., 1925. Die Phyllopoden und Copepoden des Obinger und Seeoner-Sees. *Arch. Hydrobiol.,* 15:179–208. **851**

Schmiedel, J., 1928. Experimentelle Untersuchungen über die Fallbewegung von Kugeln und Scheiben in reibenden Flüssigkeiten. *Phys. Z.,* 29:593–610. **259, 264**

Schmidt-Nielsen, K., and Fänge, R., 1958. Salt glands in marine reptiles. *Nature, Lond.,* 182:783–785. **174**

Schmidt-Nielsen, K., Jorgensen, C. B., and Osaki, H., 1957. Secretion of hypertonic solutions in marine birds. *Fedn Proc. Fedn Am. Socs exp. Biol.,* 16:113–114. **174**

Schmidt-Nielsen, K., Jorgensen, C. B., and Osaki, H., 1958. Extrarenal salt excretion in birds. *Am. J. Physiol.,* 193:101–107. **174**

Schmidt-Nielsen, K., and Sladen, W. J. L., 1958. Nasal salt secretion in the Humboldt penguin. *Nature, Lond.,* 181:1217–1218. **174**

Schmitt, W. L., 1942. The species of *Aegla,* endemic South American Freshwater Crustaceans. *Proc. U.S. natn. Mus.,* 91:431–520. **115**

Schneider, P., 1937. Sur la variabilité de *Brachionus pala* Ehrenberg dans les conditions expérimentales. *C. r. Séanc. Soc. biol.,* 125:450–452. **899**

Schönborn, W., 1962. Über Planktismus und Zyklomorphose bei Difflugia limnetica (Levander) Penard. *Limnologica*, 1:21–34. **491, 492**

Schrader, F., 1926. The cytology of pseudo-sexual eggs in a species of Daphnia. *Z. indukt. Abstamm.-u. VererbLehre*, 40:1–27. **599**

Schreyer, O., 1920. *Die Rotatorien der Ungebung von Bern.* Inaug. Dissert., Univ. Bern. (Leipzig: J. Kluskardt.), 93 pp. **530, 863, 864, 871**

Schröder, B., 1899. Das pflanzliche Plankton der Oder. *ForschBer. biol. Stn Plön*, V:29–66. **944**

* Schröder, R., 1959. Die Vertikalwanderungen der Crustaceenplanktons der Seen des südlichen Schwarzwaldes. *Arch. Hydrobiol.*, Suppl. 25:1–43. **735, 737–739, 742, 765, 771–773, 776, 777**

Schröder, R., 1960. Echoorientierung bei *Mixodiaptomus laciniatus*. *Naturwissenchaften*, 23:548–549. **679**

Schröder, R., 1961. Über die Schlagfrequenz der 2. Antennen und Mundgliedmassen bei calanoiden Copepoden. *Arch. Hydrobiol.* Suppl. 25:348–349. **679**

Schröter, C., 1896. Die Schwebeflora unserer Seen. Zürich, Zurcher and Furrer, 57 pp. **942**

* Schröter, C., and Kirchner, O., 1896. Die Vegetation der Bodensees. *Lindau, Bodensee Forsch.*, 9, Tl. I:1–122, Tl. II:1–86. **238**

Schröter, C., and Vogler, P., 1901. Variationstatistische Untersuchung über *Fragilaria crotonensis* (Edw.) Kitton im Plankton des Zürichsees in den Jahren 1896–1901. *Vjschr. naturf. Ges. Zürich*, 46:185–206. **937, 938**

Schubert, A., 1929. Über die (postembryonale) Formenturchlung bei zwei Lokalrassen von Ceriodaphnia reticulata Jurine. *Int. Revue ges. Hydrobiol. Hydrogr.*, 22:111–125. **577**

Schütt, F., 1892. Analytische Plankton-Studien. Kiel und Leipzig (not seen; ref. Lohmann, *Int. Revue ges. Hydrobiol. Hydrogr.*, 4:1–38). **235**

Schuurman, J. F. M., 1932. A seasonal study of the microflora and microfauna of Florida Lake, Johannesburg, Transvaal. *Trans. Roy. Soc. South Africa*, 20:333–386. **461, 492, 493, 543, 644, 674**

Schwabe, E., 1933. Ueber die Osmoregulation verscheidene Krebse (Malacostracen). *Z. vergl. Physiol.*, 19:522–554. **159, 161**

Scourfield, D. J., 1901. The Ephippium of *Bosmina*. *J. Queckett microsc. Club.*, (ser. 2) 8:51–56. **598**

Scourfield, D. J., 1902. The Ephippia of the Lynceid Entomostraca. *J. Queckett microsc Club.*, (ser. 2) 8:217–244. **598**

Scourfield, D. J., 1937. An anomalous fossil organism, possibly a new type of chordate, from the Upper Silurian of Lesmahagow, Lanarkshire—*Ainiktozoon loganense* gen. et sp. nov. *Proc. R. Soc.*, 121B:533–547. **133**

Sebestyén, O., 1949. On the life method of the larva of Leptodora kindtii (Focke) (Cladocera, Crustacea). *Hung. Acta biol.*, 1:71–81. **567**

Sebestyén, O., 1951. Epibiontok Balavtoni Diaphanosomari. *Archiva biol. hung.* (Inst. Biol. Tihany), 20:161–165 (Magyar text); (Russ. & Eng. summ. 165–166). **275**

Segerstråle, S., 1950. The amphipods of the coast of Finland—some facts and problems. *Commentat. Biol.*, 10, No. 14:28 pp. **199**

Segerstråle, S., 1951. The recent increase in salinity off the coasts of Finland and its influence upon the fauna. *J. Cons. perm. int. Explor. Mer.*, 17:103–110.

* Segerstråle, S. G., 1957a. On the Immigration of the Glacial Relicts of Northern Europe, with Remarks on their Prehistory. *Commentat. Biol.*, 16, No. 16, 117 pp. **176, 179, 203, 205, 207, 211, 212**

Segerstråle, S. G., 1957b. Baltic Sea. Chap. 24 in Treatise on Marine Ecology and Paleoecology, Vol. 1. Ecology, J. W. Hedgpeth. *Mem. geol. Soc. Am.*, 67:751–800. **203, 204**

Sernander, R., 1917. De Nordeuropeiska Hafvens Växtregioner. *Svensk. bot. Tidskr.*, 11:72–124. **239, 240**

Seurat, L. G., 1927. Sur la presence d'une Serpule *Merceriella enigmatica* Fauvel, dans une rivière de la Tunisie. *C. r. hebd. Séanc. Acad. Sci., Paris*, 185: 549–556. **75**

Seurat, L. G., 1933. Considérations sur la faune des Estuaires de la Tunisie Orientale et la pénétration de certaines formes animales dans la région des grands Chotts. *Arch. Zool. exp. gén.*, 75:269–79. **62**

Seymour Sewell, R. B., 1935. Studies on the bionomics of fresh-waters in India II. On the fauna of the tank in the Indian museum compound and the seasonal changes observed. *Int. Revue ges. Hydrobiol. Hydrogr.*, 31:203–238. **857–859**

Seymour Sewell, R. B., 1948. The free-swimming planktonic copepoda. Geographical distribution. *Sci. Rep. John Murray Exped.*, 1933–34, 8:317–592. **856**

Sexton, E. W., 1928. On the rearing and breeding of *Gammarus* in laboratory conditions. *J. mar. biol. Ass. U.K.*, 15:33–55. **175, 203**

Sexton, E. M., and Spooner, G. M., 1940. An account of *Marinogammarus* (Schellenberg) gen. nov. (Amphipoda) with a description of a new species, *M. pirloti*. *J. mar. biol. Ass. U.K.*, n.s., 24:633–682. **198**

Shapeero, W. L., 1961. Phylogeny of Priapulida. *Science, N.Y.*, 133:879–880. **15n.**

Shaw, J., 1959a. The absorption of sodium ions by the crayfish, *Astacus pallipes* Lereboullet. *J. exp. Biol.*, 36:126–144. **162, 165**

Shaw, J., 1959b. Solute and water balance in the muscle fibres of the east African fresh-water crab, *Potamon niloticus* (M. Edw.). *J. exp. Biol.*, 36:145–156. **155**

Shaw, J., 1959c. Salt and water balance in the east African fresh-water crab, *Potamon niloticus* (M. Edw.). *J. exp. Biol.*, 36:157–176. **155, 162–165, 167**

Sheffer, V. B., and Robinson, R. J., 1939. A limnological study of Lake Washington. *Ecol. Monog.*, 9:95–143. **609**

Shiraishi, K., and Provasoli, L., 1959. Growth factors as supplements to inadequale algal foods for *Tigriopus japonicus*. *Tohoku J. agric. Res.*, 10:89–96. **555**

Shull, A. F., 1910. Studies in the life cycle of Hydatina senta: I. Artificial control of the transition from the parthenogenetic to the sexual method of reproduction. *J. exp. Zool.*, 8:311–354. **510**

Shull, A. F., 1911. Studies in the life cycle of Hydatina senta: II. The role of temperature of the chemical composition of the medium, and of internal factors upon the ratio of parthenogenetic to sexual forms. *J. exp. Zool.*, 10:117–166. **510**

Shull, A. F., 1921. Chromosomes and the life cycle of Hydatina senta. *Biol. Bull. mar. biol. Lab., Woods Hole*, 41:55–61. **509**

Shulz, H., 1928. Über die Bedeutung des Lichtes im Leben niederer Krebse. *Z. vergleich. Physiol.*, 7:488–552. **769**

Siebeck, O., 1960. Untersuchungen über die Vertikalwanderung planktischer Crustaceen unter besonderer Berücksichtigung der Stralungsverhältnisse. *Int. Revue ges. Hydrobiol. Hydrogr.*, 45:381–454. **733, 734, 745, 748, 749, 751, 752, 782**

Siewing, R., 1958a. Anatomie und Histologie von *Thermosbaena mirabilis.* Ein Beitrag zur Phylogenie der Reihe Pancarida (Thermosbaenacea). *Abh. math.-naturw. Kl., Acad. Wiss., Mainz,* 1957:197–226. **94**

Siewing, R. 1958b. *Ingolfiella ruffoi* n. sp. eine neue Ingolfiellide aus dem Grundwasser der Peruanischen Küste. *Kieler Meeresforsch,* 14:97–102. **107**

Siewing, R., 1959. Syncarida. In Bronns *Klassen und Ordungen des Tierreichs.* 5th Bd., I Abt., 4 Buch, II Teil, 121 pp. **94**

Simroth, H., 1894. Über einige Aetherien aus den Kongofällen. *Abh. senckenb. naturforsch. Ges.,* 18:278–288. **55**

Sinclair, R. M., 1963. Effects of an introduced clam (Corbicula) on water quality in the Tennessee River Valley. Tenn. Stream Pollution Control Board, 12 pp. (mimeogr). **699**

Sioli, H., 1951. Alguns resultados e problemes da limnologia Amazonica. *Bolm téc. Inst. agron. N.,* 24:3–65. **70**

Skabitschewsky, A. P., 1929. Ueber die Biologie von Melosira baikalensis (K. Meyer) Wisl. *Russk gidrobiol. Zj.,* 8:93–113 (Russian text); (German summ. 113–114). **461**

Skellam, J. G., 1955. The mathematical approach to population dynamics. In *The Numbers of Man and Animals,* ed. J. G. Cragg and N. W. Pirie. Edinburgh, Scotland, Oliver & Boyd Ltd. for the Institute of Biology, pp. 31–45. **373**

Skuja, H., 1948. Taxonomie des Phytoplanktons einiger Seen in Uppland, Schweden. *Symb. bot. upsal.* 9 Nr. 3, 399 pp. **321, 335**

Sládeček, V., 1958. A note on the phytoplankton-zooplankton relationship. *Ecology,* 39:547–549. **706**

Slevin, J. R., 1934. The Templeton Crocker Expedition to Western Polynesian and Melanesian Islands, 1933. No. 15 Notes on the reptiles and amphibians, with the description of a new species of sea snake. *Proc. Calif. Acad. Sci.,* (4 ser.) 21:183–188. **142**

Slobodkin, L. B., 1953. An algebra of population growth. *Ecology,* 34:513–519. **580, 592**

* Slobodkin, L. B., 1954. Population dynamics in *Daphnia obtusa* Kutz. *Ecol. Monogr.,* 24:69–88. **510, 581, 583, 587, 590, 591, 597**

Słonimski, P., 1926. Sur la variation saisonnière chez Triarthra (Filinia) longiseta E. *C. r. Séanc. Soc. Biol.* 94:543–545. **36**

Smith, E. A., 1892. On the shells of the Victoria Nyanza or Lake Oukéréwé. *Ann. Mag. nat. Hist.,* (ser. 6) 10:121–128. **55**

Smith, F. E., 1954. Analysis of interaction of pH and redox in the diurnal migration of pelagic organisms. *Fed. Proc.* 13:141. **774**

* Smith, F. E., 1963. Population dynamics in *Daphnia magna* and a new model for population growth. *Ecology,* 44:651–663. **591n.**

* Smith, F. E., and Baylor, E. R., 1953. Color responses in the Cladocera and their ecological significance. *Am. Nat.,* 87:49–55. **761, 765–767, 769, 771, 772, 774, 784**

* Smith, G. M., 1920. Phytoplankton of the Inland Lakes of Wisconsin. Part I. Myxophyceae, Phaeophyceae, Heterokonteae and Chlorophyceae. Exclusive of the Desmidaceae. *Bull. Wis. geol. nat. Hist. Surv.,* 57, Part I, 243 pp. **306, 315, 322–324, 337, 338**

* Smith, G. M., 1924.   Phytoplankton of the Inland Lakes of Wisconsin.   Part II. Desmidiacea.   *Bull. Wis. geol. nat. Hist. Surv.,* 57, Part II, 227 pp. **307, 328, 331, 333**

* Smith, G. M., 1950.   *The Fresh-water Algae of the United States,* 2nd ed.   New York, McGraw-Hill Book Co., vii, 719 pp. **6, 10, 307**

Smith, H. W., 1932.   Water regulation and its evolution in the fishes.   *Q. Rev. Biol.,* 7:1–26. **134**

Smith, H. W., 1936.   The retention and physiological role of urea in the Elasmobranchii.   *Biol. Rev.,* 11:49–82. **135**

Smith, H. W., 1953.   *From fish to philosopher.*   Boston, Little, Brown & Co., xiii, 304 pp. **134**

Smith, J. B., Tatsumoko, M., and Hood, D. W., 1960.   Carbamino carboxylic acids in photosynthesis.   *Limnol. Oceanog.,* 5:425–431. **308–310**

Smith, M., 1926.   Monograph of the sea-snakes (Hydrophiidae).   London, British Museum, xvii, 130 pp. **142**

Smith, R. I., 1950.   Embryonic development in the viviparous nereid polychaet, *Neanthes lighti* Hartman.   *J. Morph.,* 87:417–466. **72**

Smith, R. I., 1953.   The distribution of the polychaete Neanthes lighti in the Salinas River estuary, California in relation to salinity, 1948–1952.   *Biol. Bull. mar. biol. Lab., Woods Hole,* 105:335–347. **72**

Smith, R. I., 1957.   A note on the tolerance of low salinities by nereid polychaets and its relation to temperature and reproductive habit.   *Annls biol., Copenh.,* (3 ser.) 33, (61st year) 93–107. **164**

Smith, R. I., 1958.   Reproductive patterns in nereid polychaets.   *Syst. Zool.,* 7:60–73. **72**

Smyly, W. J. P., 1962.   Laboratory experiments with stage V copepodids of the freshwater copepod, *Cyclops leuckarti* Claus, from Windermere and Esthwaite Water.   *Crustaceana,* 4:273–280. **642**

Soar, C. D., and Williamson, W., 1927.   *The British Hydracarina,* London, Ray Society; II:viii, 215 pp. **88**

Sokolow, I., 1930.   Die Hydracarinen von Russisch-Karelien.   *Zool. Jb.* Ab. Syst. Ök. Geogr., 59:139–232. **697**

Sollas, W. J., 1884.   On the origin of freshwater faunas: a study in evolution.   *Sci. Trans. R. Dubl. Soc.,* (ser. 2) 3:87–118. **188, 196**

Sømme, S., 1934.   Contribution to the biology of Norwegian fish food organisms.   II. *Lepiduruis arcticus* Pallas, 1793, syn. *L. glacialis* Kroger, 1847. *Avh. norske Vidensk.Akad. Oslo,* Math. Naturv. Kl., 1934, No. 6, 36 pp. **557**

Southern, R., and Gardiner, A. C., 1926a.   The seasonal distribution of the crustacea of the plankton in Lough Derg and the River Shannon.   *Scient. Invest. Minist. Fish. Irish free St.,* 1:1–170. **788**

Southern, R., and Gardiner, A. C., 1926b.   A preliminary account of some observations on the diurnal migration of the crustacea of the plankton of Lough Derg.   *Int. Revue ges. Hydrobiol. Hydrogr.,* 15:323–326. **732, 734, 737, 742, 744, 749, 781**

Southern, R., and Gardiner, A. C., 1932.   Reports from the Limnological Laboratory.   II. The diurnal migrations of the crustacea of the plankton in Lough Derg.   *Proc. R. Ir. Acad.,* 40:121–159. **732, 734, 737, 742, 744, 781**

Sowerby, A. deC., 1941.   The romance of the Chinese fresh-water jellyfish. *Hongkong Nat.,* 10:186–189 (not seen: ref Kramp 1950). **42**

Spandl, H., 1923. Zur Kenntnis der Süsswasser-Mikrofauna Vorderasiens Wiss. Erb. Exped. nach Armenien und Mesopotamien. *Ann. Naturh. Mus., Wien,* 36:124–149. **121**

Spadniewska, I., 1962. Phytoplankton development in ponds varying in the density of carp fry population. *Bull. Acad. pol. Sci., Cl. II,* Sér. Sci. Biol., 10: 305–309. **479, 709**

Spoehr, H. A., Smith, J. H. C., Strain, H. H., Milner, H. W., and Hardin, G. J., 1949. Fatty acid antibacterials from plants. *Publs Carnegie Instn,* 586, 67 pp. **464**

Spooner, G. M., 1947. The distribution of Gammarus species in estuaries, Part I. *J. mar. biol. Ass. U.K.,* n.s., 27:1–52. **199**

Spooner, G. M., 1959a. The occurrence of *Microcharon* in the Plymouth offshore bottom fauna, with description of a new species. *J. mar. biol. Ass. U.K.,* n.s., 38:57–63. **101, 108**

Spooner, G. M., 1959b. New members of the British marine bottom fauna. *Nature, Lond.,* 183:1695–1696. **107, 108**

Stålfelt, M. G., 1939a. Licht-und Temperaturhemmung in der Kohlensäureassimilation. *Planta,* 30:384–421. **428**

Stålfelt, M. G., 1939b. Licht-und Temperaturhemmung in der Kohlensäureassimilation. *Svensk. bot. Tidskr.,* 33:383–417. **428**

Stålberg, G., 1931. Eine Calanus-Form aus dem Telezker See in Altai. *Zool. Anz.,* 95:209–220. **122**

Stammer, H. J., 1935. Desmoscolex aquaedulcis n. sp. der erste Süsswasserbewöhnende Desmoscolecide aus einer slowenische Höhle Nemat. *Zool. Anz.,* 109:311–318. **51**

Stammer, H. J., 1936. Ein neuer Hohlenschizopode, Troglomysis vjetrenicensis n.g. n. sp. *Zool. Jb.* Abt. Syst. Ökol. Geogr., 68:53–104. **95n., 100**

Stanier, R. Y., 1959. Introduction to the Protista. In H. D. Ward, and G. C. Whipple *Fresh-water Biology,* 2nd ed. ed. W. T. Edmondson. Pp. 7–15. **3**

* Stankovič, S., 1960. The Balkan Lake Ohrid and its living world. *Monographiae biol.,* 9, 357 pp. **44, 393, 404, 411, 430, 431, 457, 477, 536, 698**

Steere, J. B., 1894. On the distribution of genera and species of non-migratory land-birds in Philippines. *Ibis,* 1894; pp. 411–420. **357**

Steemann Nielsen, E., 1946. Carbon sources in the photosynthesis of aquatic plants. *Nature, Lond.,* 158:594–596. **309**

Steemann Nielsen, E., 1947. Photosynthesis of aquatic plants with special reference to the carbon sources. *Dansk. Bot. Arkw.,* 12, No. 8, 71 pp. **309**

Steemann Nielsen, E., and Jensen, P. K., 1958. Concentration of carbon dioxide and rate of photosynthesis in *Chlorella pyrenoidosa. Physiol. Pl.,* 11:170–180. **309**

Steinecke, F., 1923. Das Phytoplankton masurischer Seentypen *Bot. Arch.* 3: 209–213. **435**

Stěpánek, M., and Jiří, J., 1958. *Difflugia gramen* Penard. *Difflugia gramen* var. *achlora* Penard and *Difflugia gramen* f. *globulosa* f.n. (Morphometrical and Statistical Study). *Hydrobiologia,* 10:138–156. **494**

Stephenson, J., 1930. *The Oligochaeta.* Oxford, Clarendon Press, xiv, 978 pp. **76, 77**

Stephensen, K., 1936. A tanaid (*Tanais stanfordi* Richardson) found in the Kurile Islands, with taxonomic remarks on the genus *Tanais* sensu lat. (*Tanais* Audouin et Milne-Edwards 1829, and *Anatanais* Nordenstam 1930). *Annotnes zool. jap.,* 15, No. 3:361–373. **100**

Steuer, A., 1901. Die Entomostrakenfauna der "Alten Donau" bei Wien. *Zool. Jb.*, Abt. Syst., 15:1–156. **726**

Stiasny-Wijnhoff, G., 1938. Das Genus *Prostoma* Duges, eine Gattung von Süsswasser-Nemertinen. *Archs néerl. Zool.* 3 Supp. 1938, pp. 219–230. **46**

Stiller, J., 1940. Beitrag zur Pertrichenfauna des Grossen Ploner Sees in Holsteins. *Arch. Hydrobiol.*, 36:263–285. **501**

Stingelin, Th., 1897. Ueber jahreszeitliche, individuelle und locale Variation bei Crustaceen, nebst Bemerkungen über die Fortpflanzung bei Daphniden und Lynceiden. *ForschBer. biol. Stn Plön*, 5:150–165. **851**

Stokes, G. G., 1851. On the effect of the internal friction of fluids on the motion of pendulums. *Trans. Camb. phil. Soc.*, 9, Pt. II (8)–(14). **258, 259**

Stommel, H., 1949. Trajectories of small bodies sinking slowly through convection cells. *J. mar. Res.*, 8:24–29. **289–291**

Storch, O., 1924. Die Eizellen der heterogonen Rädertiere. Nebst allgemeinen Grarterungen uber die Cytologie des Sexualuorganges und der Parthenogenese. *Zool. Jb.*, Abt. Anat., Ont., 45:309–401. **509**

Storch, O., 1929. Die Schwimmbewegung der Copepoden auf Grund von Mikro-Zeitlupenaufnakmen analysiert. *Verh. deuts. Zool. Ges.* 33 Jahresversammlung. *Zool. Anz.*, 1929, Suppl. 4:118–129. **571**

Storch, O., and Pfisterer, O., 1925. Der Fangapparat von *Diaptomus*. *Z. vergl. Physiol.*, 3:330–376. **676**

Storey, J. E., 1943. Algae. *11th Annual Report Freshwater Biol. Association of the British Empire*, pp. 16–18. **450**

Stømer, L., 1944. On the relationships and phylogeny of fossil and recent Arachnomorpha. *Skr. norske Vidensk-Akad.*, 1 Math.-naturv Kl., 1944. No. 5, 158 pp. **119**

Størmer, L., 1963. *Gigantoscorpio willsi*. A new scorpion from the lower carboniferous of Scotland and its associated preying microorganisms. *Skr. norkse Vidensk-Akad.*, I Mat.-Naturv. Kl. N.S., 1963, No. 8, 171 pp. **87**

Straelen, V. van, 1942. Apropos de la distribution des écrevisses, des homards et des crabes d'eau douce. *Bull. Mus. r. Hist. nat. Belg.*, 18, No. 56, 11 pp. **114, 115**

Straelen, V. van, 1943. *Gilsonicaris rhenanus* nov. gen., nov. sp., branchiopode anostrace de l'eodevonien du Hunsruck. *Bull. Mus. r. Hist. nat. Belg.*, 19, No. 56. 10 pp. **119**

Strain, H. H. 1958. Chloroplast pigments and chromatographic analysis. *Ann. Prestley Lectures*, 180 pp. **10**

Stråskraba, M., 1963. Share of the littoral region in the productivity of two fishponds in southern Bohemia. *Rozpr. čsl. Akad.*, Ved. Rada MPV, 73, No. 13, 63 pp. **707**

Strøm, K. M., 1926. Norwegian mountain algae. *Skr. norske Vidensk-Akad.*, I Mat-Naturv. Kl., 1926, No. 6, 263 pp. **330, 386**

Strøm, K. M., 1932. Tyrifjord, a limnological study. *Skr. norske Vidensk-Akad.*, I Math.-Naturv. Kl., 1932, No. 3, 84 pp. **632**

Strøm, K. M., 1944. High mountain limnology. Some observations on stagnant and running waters of the Rondane area. *Avh. norske Vidensk-Akad. Oslo*, Math.-naturv. Kl., 1944, No. 8, 24 pp. **526**

Strøm, K., 1946. The ecological niche. *Nature, Lond.*, 157:375. **785**

Stross, R. G., and Hill, J. C., (1965). Diapause induction in Daphnia requires two stimuli. *Science, N.Y.*, 150:1462–1464. **597, 598**

Stuart, C. A., McPherson, M., and Cooper, H. J., 1931. Studies on bacteriologically sterile *Moina macrocopa* and their food requirements. *Physiol. Zool.*, 4:87–100. **553, 556**

Stummer-Traunfels, R. R. von, 1902. Eine Süsswasser-Polyclade aus Borneo. *Zool. Anz.*, 26:159–161. **44**

* Sudzuki, M., 1964. New systematical approach to the Japanese planktonic Rotatoria. *Hydrobiologia*, 23:1–93. **869**

Sugawara, K., Naito, H., and Yanada, S., 1956. Geochemistry of Vanadium in Natural Waters. *J. Earth Sci.*, 4:44–61. **322**

Sushtchenia, L., 1958. Kolichestvennȳe dannȳe o fil'tratsionnom pitanii-planktonnykh rachkov. *Nauch. Dokl. vȳssh. Shk.*, Biol. Nauki, 1:241–260. **600, 601, 602**

Svärdson, G., 1949. Natural Selection and Egg Number in Fish. *Rep. Inst. Freshwat. Res. Drottingholm*, 29:115–122. **661**

Svärdson, G., 1957. The Coregonid Problem. VI. The Palearctic species and their intergrades. *Rep. Ins. Freshwat. Res. Drottingholm*, 38:267–356. **205**

Szidat, L., 1924. Beiträge zur Entwicklungsgeschichte der Holosterniden. II. *Zool. Anz.*, 61:249–266. **701**

Taliev, D., and Bazikalova, A., 1934. Preliminary results of a comparison of the Baikal and Caspian faunae by the method of precipitin reaction. *Dokl. Akad. Nauk. SSSR*, N.S. 1934 (Pt 2):512–515 (Russian text), (Eng. summ. 515–517). **215**

Talling, J. F., 1957. The growth of two plankton diatoms in mixed cultures. *Physiologia Pl.*, 10:215–223. **464, 469**

Tannreuther, G. W., 1920. The development of Asplanchna ebbesbornii (rotifer). *J. Morph.*, 33:389–437. **510, 511**

Tansley, A. G., 1935. The use and abuse of vegetational concepts and terms. *Ecology*, 16:284–307. **230**

* Tappa, D. W., 1964. The dynamics of the association of six limnetic species of Daphnia in Aziscoos Lake, Maine. Yale, Ph.D. thesis (now published *Ecol. Monogr.*, 35:395–423, 1965). **614, 617**

Tash, J. C., and Armitage, K. B., 1960. A seasonal survey of the vertical movements of some zooplankters in Leavenworth County State Lake, Kansas. *Kans. Univ. Sci. Bull.*, 41:657–690. **613**

Tattersall, W. M., 1951. A review of the Mysidacea in the United States National Museum. *Bull. U.S. natr. Mus.*, 201, 292 pp. **95n., 97, 99**

Tattersall, W. M., and Tattersall, O. S., 1951. *The British Mysidacea.* London, Ray Society: viii + 460 pp. **97, 213**

Tauson, A. O., 1925. Wirkung des Mediums auf das Geschlecht des Rotators Asplanehna intermedia Huds. *Int. Revue ges. Hydrobiol. Hydrogr.*, 13:130–170, 282–325. **510**

Tauson, A., 1927. Die Spermatogenese bei Asplanchna intermedia Huds *Z. Zellforsch. mikrosk. Anat.*, 4:652–681. **509**

Taylor, L. R., 1961. Aggregation, variance and the mean. *Nature, Lond.*, 189:732–735. **791**

Teiling, E., 1916. En Kaledonisk fytoplanktonformation. *Svensk. bot. Tidskr.*, 10:506–519. **381**

Teiling, E., 1947. Staurastrum planctonicum and St. pinque. A study of planktic evolution. *Svensk. bot. Tidskr.*, 41:218–234. **330, 417n.**

Teiling, E., 1948. Staurodesmus, genus novum. *Bot. Notiser,* 1948, Häfte 1, 49–83. **307, 328**

Teiling, E., 1950. Radiation of desmids, its origins and its consequences as regards taxonomy and nomenclature. *Bot. Notiser,* 1950, Häfte 2, 299–327. **325, 328**

Ter-Poghossian, A., 1928. Über die räumliche und zeitliche Verteilung von Daphnia longispina und cucullata sowie von Bosmina coregoni und longirostris im Klostersee bei Seeon. *Int. Revue ges. Hydrobiol. Hydrogr.,* 20:73–88. **835**

Thiebaud, M., 1931. Sur quelques copépodes des environs de Bienne. *Bull. Soc. nuechâtel Sci. nat.,* 55:11–34. **634**

Thiele, J., 1925–1926. Solenogastres. Mollusca. Echinodermata. In Kükenthal u. Krumbach *Handbuch der zoologie:* Band 5:256 pp. **52, 69**

Thienemann, A., 1919. Über die vertikale Schichtung der Planktons im Ulmener Maar und die Planktonproduktion der anderen Eifelmaare. *Verh. naturh. Ver. Preuss. Rheinl.,* 74:103–134. **731, 732, 740**

* Thienemann. A., 1925. Die Binnengewässer Mitteleuropas. *Binnengewässer* 1, 255 pp. **240, 380**

Thienemann, A., 1926. Holopedium gibberum in Holstein. *Z. Morph. Ökol. Tiere,* 5:755–776. **606**

Thienemann, A., 1939. Grundzüge einer allgemeinen Ökologie. *Arch. Hydrobiol.,* 35:267–285. **389**

Thienemann, A., 1950. Umbreitungsgeschichte der Süsswassertierwelt Europas Binnengewässer, 18: xvi, 809 pp. **52n.**

Thomas, E. A., 1951. Neuere hydrobiologische Forschungsergebnisse aus dem Gewässersystem Walensee-Linth-Zürich, *Wass. Energiewirt.* Nr. 10: Oktober 1951, Zürich, 1–12. **441**

Thomas, E. A., 1953. Zur Bekämpfung der See-Eutrophierung: Empirische und experimentelle Untersuchungen zur Kenntnis der Minimumstoffe in 46 Seen der Schweiz und angrenzender Gebiete. *Mbull. schweiz. Ver. Gas-v. Wass Fachm.,* 1953, No. 2–3, 15 pp. **311**

Thomas, E. A., 1955. Sedimentation in oligotrophen und eutrophen Seen als Ausdruck der Produktivität. *Verh. int. Verein. theor. angew. Limnol.,* 12:383–393. **311**

Thomasson, H., 1925. Methoden zur Untersuchung der Mikrophyten der limnischen Litoral-und Profundalzone. *Handbuch der biolo gischen Arbeitmethoden,* ed. E. Abdenhalden, Abt. ix. Teil 2., 1 Hälfte, Heft 4, pp. 681–712. **240**

Thomasson, K., 1955a. Studies on South American fresh-water plankton. I. Plankton from Tierra del Fuego and Valdivia. *Acta Horti gothoburg.,* 19:193–225. **385, 386, 393**

Thomasson, K., 1955b. A plankton sample from Lake Victoria. *Svensk. bot. Tidskr.,* 49:259–270. **343**

Thompson, R. H., 1959. Algae. In H. B. Ward, and G. C. Whipple *Freshwater Biology* 2nd ed., ed. W. T. Edmondson, New York, John Wiley & Sons, pp. 115–170. **307**

Thomson, G. M., 1879. Description of a new species of isopodous crustacean (Idotea). *Trans. Proc. N.Z. Inst.,* 11:250–251. **106**

Thorpe, W. H., 1950. Plastron respiration in aquatic insects. *Biol. Rev.,* 25:344–390. **79, 184, 187, 188**

Thorpe, W. H., and Crisp, D. J., 1947a. Studies on plastron respiration. I. The biology of *Aphelocheirus* (Hemiptera, Aphelocheiridae (Naucoridae) and the mechanism of plastron retention. *J. exp. Biol.,* 24:227–269. **188**

Thorpe, W. H., and Crisp, D. J., 1947b. Studies on plastron respiration. II. The respiratory efficiency of the plastron of *Aphelocheirus*. *J. exp. Biol.,* 24:270–303. **188**

Thorpe, W. H., and Crisp, D. J., 1947c. Studies on plastron respiration. III. The orientation responses of *Aphelocheirus* (Hemiptera, Aphelocheiridae (Naucoridae) in relation to plastron respiration: together with an account of specialized pressure receptors in aquatic insects. *J. exp. Biol.,* 24:310–328. **188**

Thorpe, W. H., and Crisp, D. J., 1949. Studies on plastron respiration. IV. Plastron respiration in the Coleoptera. *J. exp. Biol.,* 26:219–260. **188, 256**

* Thunmark, S., 1945. Zur Soziologie des Süsswasser-planktons. Eine methodologisch-ökologische Studie. *Fol. limnol. Scand.,* 3, 66 pp. **377–379, 389, 390**

Tiegs, O. W., and Manton, S. M., 1958. The evolution of the Arthropoda. *Biol. Rev.,* 33:255–337. **16n., 88, 93**

Tippo, O., 1942. A modern classification of the plant kingdom. *Chronica bot.,* 7, No. 5:203–206. **12**

Tiwari, K. K., 1958. Nichollsidae, a new family of Phreatoicoidea (Crustacea: Isopoda). *Rec. Indian Mus.,* 53:293–295. **106**

Tollinger, M., Sr. Annuziata, S. St. V., 1911. Die geographische Verbreitung der Diaptomiden. *Zool. Jb.* Abt. Syst., Geogr. Biol., 30:1–302. **127, 664, 671**

Tonnoir, A., 1922. Le cycle evolutif de *Dactylocladius commensalis* sp. nov. chironomide à larve commensale d'une larve de Blepharocéride (Diptera). *Ann. biol. lacustre.,* 11:279–290. **81**

* Tonolli, L., and Tonolli, V., 1951. Osservazioni sulla biologia ed ecologia di 170 popolamenti zooplanctonici di laghi italiani di alta quota. *Mem. Ist. ital. Idrobiol.,* 6:53–136. **686**

* Tonolli, V., 1949a. Ciclo biologico, isolamento e differentiamento stagionale in popolazioni naturali di un copepode abitatore di acque alpine (Artodiaptomus bacillifer Koelb.). *Mem. Ist. ital. Idrobiol.,* 5:97–144. **686, 860**

Tonolli, V., 1949b. Struttura spaziale del popolamento mesoplanctico. Eterogeneità delle densità dei popolamenti orizzontali e sue variazione in funzione della quota. *Mem. Ist. ital. Idrobiol.,* 5:191–208. **795**

Tonolli, V., 1958. Richerche sulla microstruttura di distribuzione dello zooplancton nel Lago Maggiore. *Mem. Ist. ital Idrobiol.,* 10:125–152. **795**

* Tonolli, V., 1961. Studio sulla dinamica del popolamento di un copepode (*Eudiaptomus vulgaris* Schmeid.). *Mem. Ist. ital. Idrobiol.,* 13:179–202. **860**

Towle, E. W., 1900. A study in the heliotropism of Cypridopsis. *Am. J. Physiol.,* 3:345–365. **759**

Tregouboff, G., 1953. Classe des Heliozoaires. In Pierre-P. Grassé *Traité de Zoologie,* Masson & Cie, Paris) Tome I, Fascicule II: 437–489. **26**

Treillard, M., 1924. Sur l'elevage en culture pure d'un crustacé cladocere: *Daphnia magna. C.r. hebd. Séanc. Acad. Sci., Paris,* 179:1090–1092. **553**

Trelease, W., 1889. The "working" of the Madison Lakes. *Trans. Wis Acad. Sci. Arts Lett.,* 7:121–129. **422**

Trembley, A., 1744. Mémoires pour servir à l'histoire d'un genre de polypes d'eau douce, à bras en forme de cornes. Leiden. J & H. Verbeck, xv. 324 pp. **758**

Tucker, A., 1957. The relation of phytoplankton periodicity to the nature of the physico-chemical environment with special reference to phosphorus. *Am. Midl. Nat.* 57:300–370. **450**

Tyler, S. A., and Barghoorn, E. S., 1954. Occurrence of structurally preserved plants in Pre-Cambrian rocks of the Canadian Schield. *Science, N.Y.,* 119:606–608. **4**

Uchida, T., 1955. Dispersal in Japan of the freshwater medusa, *Craspedacusta sowerbyi* Lankester, with remarks on *C. iseana* (Oka and Hara) *Annotnes zool. Jap.,* 282:114–120. **42**

Udvardy, M. F. D., 1959. Notes on the ecological concepts of habitat, biotope and niche. *Ecology,* 40:725–728. **357**

Uéno, M., 1934a. The freshwater Branchiopoda of Japan III. Genus *Daphnia* of Japan. 1. Seasonal succession, cyclomorphosis and reproduction. *Mem. Coll. Sci. Kyoto Univ.,* (Ser. B) 9:289–320. **820**

Uéno, M., 1934b. Yale North India Expedition. Report on Amphipod Crustacea of the genus *Gammarus. Mem. Conn. Acad. Arts Sci.,* 10:63–75. **111, 696**

Uéno, M., 1936. Crustacea malacostraca collected in the lakes of the island of Kunasiri. *Bull. biogeog. Soc. Japan,* 6:247–252. **100, 104**

Uéno, M., 1943. *Kamaka biwae,* a new amphipod of marine derivative found on Lake Biwa. *Bull. biogeogr. Soc. Japan,* 13, 130–143. **108**

Ullyott, P., 1939. Die täglichen Wanderungen der planktonischen Süsswasser Crustaceen. *Int. Revue ges. Hydrobiol. Hydrogr.,* 38:262–284. **734, 735, 749, 756, 782**

Ulomskii, S. N., 1960. Ecology of *Mesocyclops crassus* (S. Fisch.), 1853 (Crustacea, Copepoda). *Dokl. Akad. Nauk SSSR,* 134:453–456; *See also Dokl. (Proc.) Acad. Sci. U.S.S.R.,* Biol. Sci., Sect. 134:697–700. **643**

Utermöhl, H., 1924. Tiefenwanderungen bei *Volvox. Schrft. f. Süsswasser- und Meereskunde.* 9 (*sic;* not seen: ref. Ruttner 1943). **731**

Utermöhl, H., 1925. Limnologische Phytoplankton-studien Die Besiedlung ostholsteinischer Seen mit Schwebpflanzen. *Arch. Hydrobiol.,* Suppl., 5:1–524. **247**

Utterback, C. L., Phifer, L. D., and Robinson, R. J., 1942. Some chemical, physical and optical characteristics of Crater Lake. *Ecology,* 23:97–103. **315**

Van Niel, G. B., and Stanier, R. Y., 1959. Bacteria. In H. B. Ward, and G. H. Whipple Fresh-water Biology, 2nd ed., ed. W. T. Edmondson. New York and London, John Wiley & Sons, pp. 16–46. **315**

Vannini, E., 1933. Contributo alla conoscenza dei Cladoceri dell'Italia Centrale. Il Diaphanosoma e la Bosmina del Laghetto di Poggio ai Pini presso Siena. *Int. Revue ges Hydrobiol. Hydrogr.,* 29:360–405. **609, 851**

Verber, J. L., Bryson, R. A., and Suomi, V. E., 1949. Abstract of Part II Currents in Lake Mendota. *Report to the University of Wisconsin Lake Investigation Committee.* 4 pp. (mimeogr.) **291**

Vetter, H., 1937. Limnologische Untersuchungen über das Phytoplankton und seine Beziehungen zur Ernährung der Zooplanktons un Schleinsee bei Langenargen am Bodensee. *Intern. Revue ges. Hydrobiol. Hydrogr.,* 34:499–561. **402, 423, 424, 442, 609, 613**

Vialli, M., 1924. Richerche sui Rotiferi pelagici del Plancton lariano. *La limnologia del Lario.* **870, 876, 878, 883**

Viets, K., 1924. Die Hydracarinen der norddeutschen, besonders der holsteinischen Seen. *Arch. Hydrobiol.,* Suppl. 4:71–180. **697**

Villadolid, D. V., and Manacop, P. R., 1934.   The Philippine Phallostethidae, a description of a new species, and a report on the biology of *Gulaphallus mirabilis* Herre. *Philiph. J. Sci.*, 55:193–220. **138**

Vinogradov, M. Ye., 1957.   Ozera antarkticheskogo oazisa.   *Priroda*, 1957(10) 89–92. **316**

Virieux, J., 1916.   Recherches sur le plancton des lacs du Jura central.   *Ann. biol. lacustre*, 8:5–192. **936, 944**

Visvesvara, G., 1964.   On some Gastrotricha from India with description of two new species.   *Ann. Mag. nat. Hist.* (ser. 13) 6:435–443. **551**

Vlk, W., 1938.   Ueber den Bau der Geissel.   *Arch. Protistenk.*, 90:448–488. **9**

Vogel, H. J., 1959a.   Lysine biosynthesis in *Chlorella* and *Euglena*: phylogenetic significance.   *Biochim. biophys. Acta*, 34:282–283. **7**

Vogel, H. J., 1959b.   On biochemical evolution: lysine formation in higher plants.   *Proc. natn. Acad. Sci., U.S.A.*, 45:1717–1721. **7**

Vogel, H. J., 1964.   Distribution of lysine pathways among fungi: evolutionary implications.   *Am. Nat.*, 98:435–446. **8**

Voigt, M., 1901.   Ueber Gallerthaute als Mittel zur Erhohung der Schwebfahigkeit bei Planktondiatomeen.   *ForschBer. biol. Stn. Plön*, 8:120–124. **273**

Voigt, M., 1904.   Die Rotatorien und Gastrotrichen der Umgebung von Plön. *ForschBer. biol. Stn. Plön*, 11:1–180. **524, 874, 883**

* Vollenweider, R. A., 1950.   Oekologische Untersuchungen von planktischen Algen auf experimentelles Grundlage.   *Schweiz. Z. Hydrol.*, 12:194–262. **316, 318, 319, 435, 442, 455**

Volterra, L., 1924.   La variabilita delle dafnie pelagiche nei laghi di Albano e di Nemi.   *Daphnia cucullata. Atti Accad. naz. Lincei Rc.*, (Ser. 5) 33 No. 2: 131–136. **813, 820**

Volterra, V., 1926.   Variazioni e fluttuazioni del numero d'individui in specie animali conviventi.   *Atti Accad. naz. Lincei Memorie*, (Ser. 6) 2:31–113. **357**

Volterra D'Ancona, L., 1927.   Ulteriori osservazioni sulla Daphnia cucullata del Lago di Nemi.   *Int. Revue ges. Hydrobiol. Hydrogr.*, 18:261–295. **813**

Volterra D'Ancona, L., 1931.   Ancora della Daphnia cucullata di Frederiksborg, amb. nel Lago di Nemi.   *Int. Revue ges. Hydrobiol. Hydrogr.*, 25:347–354. **813**

Volterra D'Ancona, L., 1938.   Un nuovo periodo di richerche sulle dafnie di Nemi (1930–1935).   *Int. Revue ges. Hydrobiol. Hydrogr.*, 37:571–603. **813**

Wadell, H., 1934.   The coefficient of resistance as a function of Reynolds number for solids of various shapes.   *J. Franklin Inst.*, 217:459–490. **259**

Wagler, E., 1912.   Faunistische und biologische Studien an freischwimmenden Cladoceren Sachsens.   *Zoologica, Stuttg.*, 67:305–366. **613, 820**

Wagler, E., 1923.   Über die Systematik die Verbreitung und die Abhängigkeit der Daphnia cucullata von physikalischen und chemischen Einflussen des Milieus.   *Int. Revue ges. Hydrobiol. Hydrogr.*, 11:41–88, 262–316. **813, 816, 817**

Wagler, E., 1925.   Zucht von Krebsen und Würmern.   *Handbuch der Biologischen Arbeitsmethoden*, ed. E. Abdenhalden, Abt. ix, Teil 2, 1 Hälfte, Heft 2, pp. 319–358. **557**

Wagler, E., 1936.   Die Systematik und geographische Verbreitung des Genus *Daphnia* O. F. Müller mit besonderer Berücksichtigung der südafrikanischen Arten.   *Arch. Hydrobiol.*, 30:505–556. **822, 825**

Wailes, G. H., and Penard, E., 1911.   Clare Island Survey, Rhizopoda.   *Proc. R. I. Acad.*, 31, Pt. 65, 64 pp. **495**

Waldvogel, T., 1900. Der Lützelsee und das Lautikerreid, ein Beitrag zur Landeskunde. *Vjschr. naturf. Ges. Zürich,* 45:277–350. **751**

Walter, E., 1922. Über die Lebensdauer der freilebenden Süsswasser Cyclopiden und andere Fragen ihre Biologie. *Zool. Jb.* 44:375–420. **637, 647**

Wangersky, P. J., and Cunningham, W. J., 1957. Time lag in population models. *Cold Spring Harb. Symp. quant. Biol.,* 22:329–338. **591**

Waris, H. (Waren, H.), 1933. Über die Rolle des Calciums im Leben dei Zelle auf Grund von Versuchen an Micrasterias. *Planta,* 19:1–45. **332**

Warming, E., 1895. *Plantesamfund. Grundträk af den ökologiske Plantegeographi. Kjobenhavn* (German trans. 1896, English as Ocology of Plants, new ed. with assistance of M. Vahl, trans. P. Groom, and I. B. Balfour. Oxford, 1909). **234**

Warming, E., 1923. Økologiens Grundformer. Vakast til en systematisk Ordning. *K. danske Vidensk. Selsk. Skr.,* Naturw. Mat. Afd. 8., Raekke IV:119–187. **234, 237**

Warren, E., 1900. On the reaction of Daphnia magna (Straus) to certain changes in its environment. *Q. Jl. microsc. Sci.,* 43 (n.s.):199–224. **580**

Warren, E., 1901. A preliminary account of the development of the free swimming nauplius of Leptodora hyalina (Lillj.). *Proc. R. Soc.,* 68:210–218. **571**

Waterman, T. H., 1960. Interaction of polarised light and turbidity in the orientation of *Daphnia* and *Mysidium. Z. vergl. Physiol.,* 43:149–172. **778**

Waterman, T. H., 1961. Light sensitivity and vision. *The Physiology of Crustacea,* ed. T. H. Waterman, New York and London, Academic Press, II:1–64. **762**

Watzka, M., 1928. Die Rotatorienfauna der Cakowitzer Zucher-fabrikteich und Versuche über das Auftreten von Rotatorien-Männchen und über die Entwicklungszeit der Dauereier. *Int. Revue ges. Hydrobiol. Hydrogr.,* 19:430–451. **510**

Weaver, C. R., 1943. Observations on the life cycle of the fairy shrimp *Eubranchipus vernalis. Ecology,* 24:500–502. **563**

Weber, E. F., 1898. Faune rotatorienne du bassin de Léman. *Rev. suisse Zool.,* 5:263–785. **519**

Weismann, A., 1876. Zur Naturgeschichte der Daphniden. I: Ueber die Bildung von Wintereiern bei Leptodora hyalina. *Z. wiss. Zool.,* 27:51–112. **595**

Weismann, A., 1877a. Beiträge zur Naturgeschichte der Daphnoiden. III Abhängigheit der Embryonalenturcklung von Frichtwasser der Multes. *Z. wiss. Zool.,* 28:176–211. **756**

Weismann, A., 1877b. Das Tierleben im Bodensee. *Schr. Gesch. Bodensees Umgebung,* 7:1–31. **726, 757**

Weismann, A., 1879. Beiträge zur Naturgeschichte der Daphnoiden. VII. Die Entstehung der cyclischen Fortpflanzung bei den Daphnoiden. *Z. wiss. Zool.,* 33:111–270. **595**

*Welch, P. S., 1952. Limnology, 2nd ed. New York, McGraw-Hill Book Co., xi, 538 pp. **236, 238**

Wells, L., 1960. Seasonal abundance and vertical movement of planktonic crustacea in Lake Michigan. *Fishery Bull. Fish WildL. Serv. U.S.,* 60:343–369. **609, 611, 614**

Wenrich, D. H., 1929. Observations on some freshwater ciliates (Protozoa)

I. *Teuthophrys trisulea* Chatton and de Beauchamp and *Stokesia vernalis* n.g., n.sp. *Trans. Am. microsc. Soc.,* 48:221–241. **498**

Werestschagin, G. J., and Sidorytschev, I. P., 1929. The biology of Comephoridae. *Dokl. Akad. Nauk SSSR,* (A) 1929:126–130. **697**

Werntz, H., 1957. Osmotic adaptations in marine and freshwater species of *Gammarus.* Yale Thesis, 1957, 55 pp. (in large part published in *Biol. Bull. mar. biol. Lab., Woods Hole,* 124:225–239). **159, 161, 166, 201, 202**

Wesenberg-Lund, C., 1900. Von dem Abhängigkeitsverhältnis zwischen dem Bau der Planktonorganismen und dem spezifischen Gewicht des Süsswassers. *Biol. Zbl.,* 20:606–619, 644–656. **812, 824, 864**

* Wesenberg-Lund, C., 1904. *Plankton investigations of The Danish Lakes.* Special Part. Copenhagen, 223 pp. (English summ. 44 pp). **429, 431, 432, 459, 492, 509, 530, 609, 611, 613, 614, 619, 620, 642, 652, 656, 659, 673, 691, 707, 812**

* Wesenberg-Lund, C., 1908. *Plankton investigations of the Danish lakes.* General Part: *The Baltic freshwater plankton, its origin and variation.* Copenhagen (Gyldendalske Boghandel), 389 pp. **272, 446, 454, 459, 812, 814, 844, 851, 852, 854, 855, 864, 865, 911, 912, 914, 915, 920, 928–932, 934–936, 939, 940, 941, 942**

* Wesenberg-Lund, C., 1923. Contributions to the biology of the Rotifera I. The males of the Rotifera. *K. danske Vidensk. Selsk. Skr.,* Naturw. Math. Afd., (ser. 8) 4:189–345. **508, 509, 863**

* Wesenberg-Lund, C., 1926. Contributions to the biology and morphology of the genus Daphnia. *K. danske Vidensk. Selsk. Skr.,* Naturw. Math. Afd., (ser. 8) 11, No. 2:92–250. **596**

* Wesenberg-Lund, C., 1930. Contributions to the biology of the Rotifera. Part II. The periodicity and sexual periods. *K. danske Vidensk. Selsk. Skr.,* Naturw. Math. Afd., (ser. 9) 2(1), No. 1, 230 pp. **508, 529, 530, 542, 864, 866, 875, 886, 892, 895, 897, 898, 906**

* Wesenberg-Lund, C., 1934. Contributions to the development of the Trematoda Digenea. Part II. The biology of the fresh-water cercariae in Danish fresh-waters. *K. danske Vidensk. Selsk. Skr.,* Naturw. Math. Afd., (ser. 9) 5, No. 3, 223 pp. **700, 702, 703, 704**

* Wesenberg-Lund, C., 1939. *Biologie der Süsswassertiere: Wirbellose Tiere,* Ed. O. Storch. Wien, J. Springer, xi + 817 pp. **54, 54, 558**

West, G. S., 1909. The algae of the Yan Yean Reservoir, Victoria; a biological and ecological study. *J. Linn. Soc.* Botany, 39:1–88. **134**

West, G. S., and Fritsch, F. E., 1927. *A treatise on the British fresh-water algae.* New and revised edition. Cambridge, 534 pp. **307, 333**

West, W., and West, G. S., 1904. *A Monograph of the British Desmidiaceae,* London, Ray Society, 1:xxxvi, 224 pp. **307**

West, W., and West, G. S., 1905. *A Monograph of the British Desmidiaceae,* London, Ray Society, 2:x, 204 pp. **307**

West, W., and West, G. S., 1908. *A Monograph of the British Desmidiaceae,* London, Ray Society, 3:xv, 274 pp. **307**

West, W., and West, G. S., 1909. The British freshwater phytoplankton, with special reference to the Desmid-plankton and the distribution of British Desmids. *Proc. R. Soc.,* 81*B*:165–206. **330, 443**

West, W., and West, G. S., 1912a. *A Monograph of the British Desmidiaceae.* London, Ray Society, 4:xiv, 194 pp. **307**

West W., and West, G. S., 1912b. On the periodicity of the phytoplankton of some British lakes. *J. Linn. Soc.* Botany, 40:395–432. **401, 418, 419, 434**

West, W., West, G. S., and Carter, N., 1923. *A monograph of the British Desmidiaceae*, London, Ray Society, 5:xxi, 300 pp. **307, 333**

Westblad, E., 1948. Studien über Skandinavischen Turbellaria Acoela V. *Ark. Zool.,* 41A, No. 7, 82 pp. **43**

Westblad, E., 1955. Marine "Alloeocoels" (Turbellaria) From North Atlantic and Mediterranean coasts, I. *Ark. Zool.* (n.s.), 7:491–526. **45**

Westoll, T. S., 1943. The origin of the tetrapods. *Biol. Rev.,* 18:78–98. **136, 140n.**

Wheeler, B. M., 1950. Halogen metabolism of Drosophila gibberosa. I. Iodine metabolism studied by means of I^{131}. *J. exp. Zool.,* 115:83–107. **163**

Whipple, G. C., 1894. Some observations of the growth of Diatoms. *Technol. Q.,* 7:214–231. **436**

Whipple, G. C., 1895. Some observations of the temperature of surface waters; and the effect of temperature on the growth of microorganisms. *J. New Engl. Wat. Wks Ass.,* 9:202–222. **436, 443, 446**

Whipple, G. C., 1896. Some observations on the relation of light to the growth of diatoms. *J. New Engl. Wat. Wks Ass.,* 11:1–24. **282**

Whipple, G. C., and Jackson, D. D., 1899. Asterionella—its biology, its chemistry, and its effect on water supplies. *J. New Engl. Wat. Wks Ass.,* 14:1–25. **443, 446**

White, C. M., 1946. The drag of cylinders in fluids at slow speeds. *Proc. R. Soc.,* 186*A*:472–479. **265**

White, H. L., 1937. The interaction of factors in the growth of Lemna. XI. The interaction of nitrogen and light intensity in relation to growth and assimilation. *Ann. Bot.* (n.s.), 1:623–647. **319, 436**

White, P. R., 1954. *The cultivation of animal and plant cells.* New York, Ronald Press, xi, 239 pp. **349**

Whitney, D. D., 1914a. The production of males and females controlled by food conditions in Hydatina senta. *Science, N.Y.,* 39:832–833. **510**

Whitney, D. D., 1914b. The influence of food in controlling sex in Hydatina senta. *J. exp. Zool.,* 17:545–558. **510**

Whitney, D. D., 1916a. Parthenogenesis and sexual reproduction in rotifers. Experimental research upon *Brachionus pala. Am. Nat.,* 50:50–52. **510**

Whitney, D. D., 1916b. The control of sex by food in five species of rotifers. *J. exp. Zool.,* 20:263–296. **510**

Whitney, D. D., 1916c. The transformation of Brachionus pala into Brachionus amphiceros by sodium silicate. *Biol. Bull. mar. biol. Lab., Woods-Hole,* 31:113–120. **899, 900**

Whitney, D. D., 1917. The relative influence of food and oxygen in controlling sex in rotifers. *J. exp. Zool.,* 24:101–138. **510**

Whitney, D. D., 1919. The ineffectiveness of oxygen as a factor in causing male production in Hydatina senta. *J. exp. Zool.,* 28:469–492. **510**

Whitney, D. D., 1929. The Chromosome cycle in the Rotifer Asplanchna amphora. *J. Morph.,* 47:415–433. **509**

Whittaker, R. H., and Fairbanks, C. W., 1958. A study of plankton copepod communities in the Columbia Basin, southeastern Washington. *Ecology,* 39:46–65. **675**

Wiedling, S., 1948. Beiträge zur Kenntnis der vegetativen Vermehrung der Diatomeen. *Bot. Notiser,* 1948: pp. 322–354. **924, 926, 927, 939**

Wigglesworth, V. B., 1933a. The effect of salts on the anal gills of the mosquito larva. *J. exp. Biol.,* 10:1–15. **164**

Wigglesworth, V. B., 1933b. The function of the anal gills of the mosquito larva. *J. exp. Biol.,* 10:16–26. **164**

Wigglesworth, V. B., 1933c. The adaptation of mosquito larvae to salt water. *J. exp. Biol.,* 10:27–37. **164**

Wigglesworth, V. B., 1938. The absorption of fluid from the tracheal system of mosquito larvae at hatching and moulting. *J. exp. Biol.,* 15:248–254. **164**

Wigglesworth, V. B., 1939, 1950 (4th ed. revised). *The Principles of Insect Physiology.* London, Methuen, viii, 434 pp. **184, 254**

Wikgren, B. J., 1953. Osmotic regulation in some aquatic animals with special reference to the influence of temperature. *Acta zool. fenn,* 71, 102 pp. **162, 164, 165, 167**

Wiktor, K., 1961. The influence of habitat factors on variations in the populations of *Bosmina coregoni, Daphnia hyalina* and *Daphnia cucullata.* (*Ekol. Pol.,* (ser. A) 9:1–19 (Polish text, English summ.). **613, 616, 620, 623**

Wilhelmi, J., 1917. Plankton und Tripton. *Arch. Hydrobiol.,* 11:113–150. **235, 236**

Wilkie, D. R., 1953. The coefficient of expansion of muscle. *J. Physiol.,* 119:369–375. **251**

Willey, A., 1923. Ecology and the partition of biology. *Trans. R. Soc. Can.* Sect. V., (ser. 3) 17:1–9. **125**

Williams, C. B., 1947. The generic relations of species of small ecological communities. *J. Anim. Ecol.,* 16:11–18. **484**

Willmer, E. N., 1956. Factors which influence the acquisition of flagella by the amoeba, *Naegleria gruberi. J. exp. Biol.,* 33:583–603. **182**

Willmer, E. N., 1960. *Cytology and Evolution.* New York and London, Academic Press, x, 430 pp. **182**

Wilson, C. B., 1932. Copepods of the Woods Hole Region, Massachusetts. *Bull. U.S. natn. Mus.,* 158, 635 pp. **124**

Wilson, E. O., 1961. The nature of the taxon cycle in the Melanesian ant fauna. *Am. Nat.,* 95:169–193. **375**

Wilson, L. R., and Hoffmeister, W. S., 1953. Four new species of fossil *Pediastrum. Am. J. Sci.,* 251:753–760. **18, 19**

Wilson, M. S., 1953. New and inadequately known North American species of the copepod genus *Diaptomus. Smithson. misc. Collns,* 122, No. 2, 30 pp. **681, 682**

Wilson, M. S., 1954. A new species of *Diaptomus* from Louisiana and Texas with notes on the subgenus *Leptodiaptomus* (Copepoda, Calanoida). *Tulane Stud. Zool.,* 2:51–60. **682**

Wilson, M. S., 1955. A new Louisiana copepod related to *Diaptomus* (*Aglaodiaptomus*) *clavipes* Schacht (Copepoda, Calanoida). *Tulane Stud. Zool.,* 3:35–47. **681, 682**

Wilson, M. S., 1959. *See* Wilson and Yeatman, 1959. **122, 123, 650, 683**

Wilson, M. S., and Moore, W. G., 1953. New records of *Diaptomus sanguineus* and allied species from Louisiana, with the description of a new species (Crustacea:Copepoda). *Jour. Wash. Acad. Sci.,* 43:121–127. **681**

* Wilson, M. S., and Yeatman, H. C., 1959.   Free-living Copepoda (Calanoida by M. S. Wilson; Cyclopoida by H. C. Yeatman; Harpacticoida by M. S. Wilson and H. C. Yeatman). In H. B. Ward, and G. C. Whipple, *Fresh-water Biology*, 2nd ed, ed. W. T. Edmondson. New York and London, John Wiley & Sons. Pp. 735–861. **688**

Wimpenny, R. S., 1936.   The size of Diatoms. I. The diameter variation of *Rhizosolenia styliformis* Brightw. and *R. alata* Brightw. in particular and of pelagic marine diatoms in general. *J. mar. Biol. Ass. U.K.* n.s., 21:29–60. **928**

Wimpenny, R. S., 1946.   The Size of Diatoms. II. Further observations on *Rhizosolenia styliformis* (Brightwell). *J. mar. Biol. Ass. U.K.* n.s., 26:271–284. **928**

Wolf, E., 1903.   Dauereier und Ruhezustände bei Copepoden. *Zool. Anz.*, 27:98–108. **663, 664**

* Wolf, E., 1905.   Die Fortpflanzungsverhältnisse unserer einheimischen Copepoden. *Zool. Jb.* Ab. Syst., Geog. Biol., 22:101–280. **647, 648, 649, 664, 671**

Wolf, G., 1908.   Die geographisch Verbreitung der Phyllopoden, mit besonderer Berücksichtigung Deutschlands. *Verh. dt. zool., Ges.* 18:129–140. **561**

Wolff, M., 1909.   Ein einfacher Versuch zur Pütterschen Theorie von der Ernährung der Wasserbewohner. *Int. Revue ges. Hydrobiol. Hydrogr.*, 2:715–740. **553**

Woltereck, R., 1908a.   Plankton und Seenausfluss *Int. Revue ges. Hydrobiol. Hydrogr.*, 1:303–304. **804**

Woltereck, R., 1908b.   Über natürliche und künstliche varietätenbildung bei Daphniden. *Verh. dt. zool. Ges.*, 18:234–240. **826**

Woltereck, R., 1909.   Weitere exper. Unters. üb. Artveränderung, speciell über das Wesen quant. Artunterschiede bei Daphniden. *Verh. dt. zool. Ges.*, 19:110–173. **826**

Woltereck, R., 1911.   Beitrag zur Analyse der "Vererbungerworbener Eigenschaften": Transmutation und Präinduktion bei Daphnia. *Verh. dt. zool. Ges.*, 21:141–172. **826**

Woltereck, R., 1913.   Über Funktion, Herkunft u. Entstehungsursachen der sogen. Schwebefortsätze pelagischer Cladoceren. *Zoologica, Stuttgart*, 67:475–550. **826, 845**

Woltereck, R., 1921.   Variation u. Artbildung. Teil I. *Int. Revue ges. Hydrobiol. Hydrogr.*, 9:1–150. **826**

Woltereck, R., 1924.   Beiträge zur Variationsanalyse tierischer starrer Formen I. Die Methode der "Raster-Analyse" bei Crustaceen. *Int. Revue ges. Hydrobiol. Hydrogr.*, 12:13–16. **812**

Woltereck, R., 1928.   Über die Population Frederiksborger Schloss-see von *Daphnia cucullata* und einige daraus neuentstadene Erbrassen, besonders diejenige des Nemisees. *Int. Revue ges. Hydrobiol. Hydrogr.*, 19:172–203. **813, 816**

Woltereck, R., 1930.   Alte und neue Beobachtungen über die geogr. und die zonare Verteilung der helmlosen und helmtragenden Biotypen von Daphnia. *Int. Revue ges. Hydrobiol. Hydrogr.*, 24L:358–380. **835**

Woltereck, R., 1932.   Races, Associations and Stratification of Pelagic Daphnids in some lakes of Wisconsin and other Regions of the United States and Canada. *Trans. Wis. Acad. Sci. Arts. Lett*, 27:487–522. **818, 836, 838, 839, 840, 841, 843–845**

Wood, T. R., 1932.   Resting eggs that fail to rest. *Am. Nat.*, 66:277–281. **595**

Wood, T. R., and Banta, A. M., 1933. Observations on procuring and hatching sexual eggs of Daphnia longispina. *Int. Revue ges. Hydrobiol. Hydrogr.*, 20:437–454. **595**

Wood, T. R., and Banta, A. M., 1937. Hatchability of Daphnia and Moina sexual eggs without drying. *Int. Revue ges. Hydrobiol. Hydrogr.*, 35:229–242. **595**

Wood, T. R., Ingle, L., and Banta, A. M., 1939. Growth and Reproductive Characteristics of Daphnia longispina. In Banta 1939, pp. 182–200. **577, 581**

Worthington, E. B., 1931. Vertical movements of fresh-water macroplankton. *Int. Revue ges. Hydrobiol. Hydrogr.*, 25:394–436. **696, 704, 726, 727–730, 732, 734, 741, 743, 747, 750, 755, 781, 785, 836**

Worthington, E. B., and Ricardo, C. K., 1936. Scientific results of the Cambridge expedition to the East African lakes 1930-1, No. 17. The vertical distribution and movements of the plankton in Lakes Rudolf, Naivasha, Edward, and Bunyoni. *J. Linn. Soc.* Zoology, 40:33–69. **740, 741**

Wright, E. P., 1864. On a new genus of *Teredininae*. *Trans. Linn. Soc. Lond.*, 24:451–454. **62**

Wright, S., 1936. A revision of the South American species of Pseudodiaptomus. *Anais Acad. bras Cienc.*, 8:1–24. **126**

Wright, S., 1938. A review of the *Diaptomus bergi* group with descriptions of two new species. *Trans. Am. microsc. Soc.*, 57:297–315. **646, 647, 677**

Wundsch, H. H., 1912. Eine neue spezies der Genus *Corophium* Latr. aus oem Müggelsee bei Berlin. *Zool. Anz.*, 39:729–738. **218**

Wynne-Edwards, V. C., 1962. *Animal Dispersal in Relation to Social Behavior*. New York, Hafner Publishing Co., xi, 653 pp. **787**

Yeatman, H. C., 1944. American Cyclopoid Copepods of the viridis-vernalis group (including a description of Cyclops carolinianus, n.sp.). *Am. Midl. Nat.*, 32:1–90. **631, 634**

Yeatman, H. C., 1956. Plankton studies on Woods Reservoir, Tennessee. *J. Tenn. Acad. Sci.*, 31:32–53. **684**

Yeatman, H. C., 1959, *See* Wilson and Yeatman, 1959. **634, 635**

Yerkes, R. M., 1899. Reaction of Entomostraca to stimulation by light. *Am. J. Physiol.*, 3:157–182. **759**

Yerkes, R. M., 1900. Reaction of Entomostraca to stimulation by light. II. Reactions of *Daphnia* and *Cypris*. *Am. J. Physiol.*, 4:405–422. **760**

Yonge, C. M., 1949. On the structure and adaptations of the Tellinacea deposit-feeding Eulamellibranchia. *Phil. Trans. R. Soc.*, 234B, 29–76. **62**

Yoshimura, S., 1933. Calcium in solution in the lake waters of Japan. *Jap. J. Geol. Geogr.*, 10:33–60. **606**

Zacharias, O., 1894. Beobachtungen am Plankton des Grossen Plöner Sees. *ForschBer. biol. Stn Plön*, 2:91–137. **812**

Zacharias, O., 1895. Ueber die wechselnde Quantität des Plankton im Grossen Plöner See. *ForschBer. biol. Stn Plön*, 3:97–117. **282**

Zacharias, O., 1897. Neue Beiträge zur Kenntniss der Süsswasserplanktons. *ForschBer. biol. Stn Plön*, 5:1–9. **492, 493**

Zacharias, O., 1899a. Die Rhizopoden und Heliozoen des Süsswasserplanktons. *Zool. Anz.*, 22:49–53. **494–496**

Zacharias, O., 1899b. Ueber die Verschiedenheit der Zusammensetzung des Winter planktons in grossen und kleinen Seen. *ForschBer. biol. Stn Plön,* 7:64–74. **436**

Zacharias, O., 1899c. Ueber einige biologische Unterschiede zwischen Teichen und Seen. *Biol. Zbl.,* 19:313–319. **388**

Zacharias, O., 1903. Über die jahreszeitliche Variation von Hyalodaphnia kahlbergensis Schoedl. *ForschBer. biol. Stn Plön,* 10:293–295. **819, 827**

Zachvatkin, A., 1932. Contribution à la connaissance des migrations journalières verticales du zooplancton du lac Baical. *Trudȳ baïkal, Limnol. Sta,* (Russian). 2:55–106. **697, 732, 733, 735**

Zenkevich, L. A., 1957. Caspian and Aral Seas. Chap. 26 in Treatise on Marine Ecology and Paleoecology, Vol. 1. Ecology, ed. J. W. Hedgpeth. *Mem. Geol. Soc. Amer.,* 67:891–916. **213n.**

* Zenkevich, L. A., 1963. *Biology of the Seas of the U.S.S.R.* (trans. B. Botcharskaya). New York, Interscience Publishers, 955 pp. **213n.**

Zeuthen, E., 1939. On the hibernation of Spongilla lacustris (L). *Z. vergl. Physiol.,* 26:537–547. **38**

Zimmerman, W., 1928. Über Algenbestände aus der tiefenzone des Bodensees. Zur Ökologie und Soziologie der Tiefenpflanzen. *Z. Bot.,* 20:1–35. **28**

Zinn, D. J., and Dexter, R. W., 1962. Reappearance of Eulimnadia agassizii with notes on its biology and life history. *Science, N.Y.,* 137:676–677. **565**

ZoBell, C. E., 1946. *Marine microbiology.* Waltham, Mass., Chronica Botanica Co., 240 pp. **20, 21**

ZoBell, C. E., and Upham, H. C., 1944. A list of marine bacteria including descriptions of sixty new species. *Bull. Scripps Instn Oceanogr.,* 5:239–292. **20**

Index of Lakes

The symbol † indicates a lake of the geological past.

In the case of a large lake the bearings refer to the approximate center. When bearings are not given, the exact position of the lake is not available.

Index of Genera and Species of Organisms

(Italic paging refers to figures)

The authors of species are given according to botanical conventions for plants and monerans and according to zoological conventions for animals. There is currently no consistent practice, applicable to all lower organisms, for the treatment of infraspecific taxa; sometimes, when large uniform populations of what are commonly called varieties are being considered, it has proved convenient to use simple trinomials as if the taxa in question were equivalent to zoological subspecies, as indeed, from the standpoint of population dynamics, they usually are. When there is doubt about specific status, this is indicated in the index, either explicitly or by placing parentheses around the specific name in citing a subspecies.

General Index